# Microbial Diversity and Biotechnology in Food Security

R. N. Kharwar • R. S. Upadhyay
N. K. Dubey • Richa Raghuwanshi
Editors

# Microbial Diversity and Biotechnology in Food Security

Springer

*Editors*
R. N. Kharwar
Department of Botany
Banaras Hindu University Centre of Advanced study in Botany
Varanasi
India

N. K. Dubey
Department of Botany
Banaras Hindu University Centre of Advanced study in Botany
Varanasi
India

R. S. Upadhyay
Department of Botany
Banaras Hindu University Centre of Advanced study in Botany
Varanasi
India

Richa Raghuwanshi
Department of Botany,
Mahila Mahavidyalaya
Banaras Hindu University
Varanasi
India

ISBN 978-81-322-1800-5 ISBN 978-81-322-1801-2 (eBook)
DOI 10.1007/978-81-322-1801-2
Springer New Delhi Dordrecht Heidelberg London New York

Library of Congress Control Number: 2014937320



Printed on acid-free paper

Springer is part of Springer Science+Business Media (www.springer.com)

# Foreword

Microbiology is an inherently valuable and an immensely broad-based discipline that offers an intimate view of a world that is invisible to the unaided eye. Although the microbes themselves are very small, the microbial world is quite large, almost all-pervasive, and diverse, and the capabilities of its members are immense. Apart from ethical values, microbial diversity is of enormous economic significance to the mankind. Microbes have always been the invisible villains that have caused enormous economic losses by spoiling food, infecting domesticated plants and animals, and causing a variety of ailments to humans. But microbes, as it turns out, are also the champions that work incessantly to remove wastes, improve the nutritional status of soil, enable the generation of a variety of fermented foods, and some of them are used as food. More recently, humans have deliberately domesticated a variety of microbes that are used as workhorses to achieve varied but highly specialized and extremely valuable ends, including production of an increasingly large number of drugs and food supplements.

Therefore, microbiology/mycology/plant pathology has been an active area of both basic and applied research. Apart from the specialized field of industrial microbiology that is dominated by the private sector, research activities in several other areas including agriculture and environment remain primarily in the public domain. The knowledge generated by research gets disseminated through education and finally trickles down to the farmers and other practitioners in these fields. This knowledge is often used for resolving various issues related to the environment and food security at the grassroots level of policy makers.

The present book, entitled "Microbial Diversity and Biotechnology in Food Security," is a compilation of the contributions from leading workers in different areas of microbiology, and is the outcome of an International Conference—ICMPB-2012—organized by the Department of Botany, Banaras Hindu University, Varanasi. It is admirable that the book begins with the basic concept of the important fundamental themes and effortlessly moves to the advanced ideas of microbiology, including biotechnology relevant for a sustainable development and food security that is under challenge by the incessant growth of the human population worldwide.

The present book is organized into four sections—Endophytes and Mycorrhizae, Microbial Diversity and Plant Protection, Microbial Functions and Biotechnology, and Microbes and the Environment—which together contains 53 chapters contributed by acknowledged experts in the concerned field from various parts of the world. The chapters are designed to cover the general as well as specific aspects of microbes' potential, and their exploitation for plant protection, environmental conservation, and biotechnology applications.

The book chapters contain numerous beautiful and revealing illustrations, and cover both traditional as well as recent developments in the field of microbiology, mycology, and plant pathology. Throughout the book, the approach has been of a critical evaluation, which is an integral part of the approach of science professionals. I am confident that this excellent effort will serve the needs of both researchers as well as the students of this fascinating field of enquiry.

It is a matter of delight that this useful collection of highly informative contributions from scientists of repute in their own fields has been given a final shape in a relatively short period of time, for which the editors deserve our appreciation.

B. D. Singh
School of Biotechnology,
Banaras Hindu University,
Varanasi, India

## Contents

# About the Editors

**Prof. R.N. Kharwar** is currently serving as a Professor in the Department of Botany, Banaras Hindu University, Varanasi. More than 50 research articles, 3 reviews in international journals, and 6 book chapters are to his credit. He is a fellow of MSI and IPS and recipient of Shome Memorial Lecture Award, 2012. His core areas of research include fungal and actinobacterial endophytes diversity, ecology, bioactive molecules, antioxidants, and myconanotechnology. Other areas of interest include epigenetic modulations for cryptic and enhanced metabolites production from endophytes.

**Prof. R.S. Upadhyay** received his M.Sc. (1976) and Ph.D. (1980) degrees from Banaras Hindu University, Varanasi. His main focus on research has been on biological control of plant pathogens, programmed plant cell death in response to pathogens, bioremediation of toxic effluents, induced resistance in plants and their immunization, chitinase production, mycorrhizal technology for reclamation of wastelands, and molecular basis of plant-microbe interaction. He is recipient of five national awards in the area of science, one conferred by the Prime Minister of India. In addition, he has also worked as a visiting scientist to The Royal Society, London, Research Associate of NIH, USA and INSA- JSPS, Japan.

**Prof. N.K. Dubey** has significantly contributed to the important area of botanical pesticides. He has formulated several novel plant based preservatives that exhibit significant potency in control of biodeterioration of food from fungi, mycotoxins, and insects as well as from lipid peroxidation. He has filed 4 patents, published 170 research papers and review articles, 5 books and is a recipient of several awards including Prof M J Narshimhan award and Young Scientist award. He acted as a Chairperson, Session Coordinator and Key speaker in 9$^{th}$ ICPP 2008, held Aug 24-29, 2008, Torino, Italy.

**Dr. Richa Raghuwanshi** is an Assistant Professor at the Department of Botany, Banaras Hindu University, Varanasi, since 2005. Her research is focused on microbes as biofertilizers and biocontrol agents in cultivated crops and Indian medicinal plants. She has published 28 research papers, book chapters and also edited one book.

# Contributors

**Rashmi Aggarwal** Division of Plant Pathology, Indian Agricultural Research Institute, New Delhi, India

**Maroof Ahmed** School of Biotechnology, University of Jammu, Jammu, India

**Gautam Anand** Department of Biotechnology, D.D.U Gorakhpur University, Gorakhpur, Uttar Pradesh, India

**T. S. Avinash** Department of Studies in Microbiology, University of Mysore, Mysore, Karnataka, India

**João Lúcio Azevedo** Faculty of Agriculture (ESALQ), Department of Genetics, University of São Paulo, São Paulo, Brazil

State University of Maringá, Maringá, Brazil

**D. J. Bagyaraj** Center for Natural Biological Resources and Community Development (CNBRCD), Bangalore, India

**Sagar Banerjee** Division of Plant Pathology, Indian Agricultural Research Institute, New Delhi, India

**B. M. Bashyal** Division of Plant Pathology, Indian Agricultural Research Institute, New Delhi, India

**Divya Bhatia** Department of Biotechnology, University Institute of Engineering & Technology, Kurukshetra University, Kurukshetra, Haryana, India

**Tarun C. Bora** Biotechnology Division, North East Institute of Science and Technology (NEIST), Council of Scientific and Industrial Research (CSIR), Jorhat, Assam, India

**Bandana Bose** Seed Physiology Laboratory, Department of Plant Physiology, Institute of Agricultural Sciences, Banaras Hindu University, Varanasi, India

**Preeti M. Chaudhary** Biochemical Sciences Division, National Chemical Laboratory, Pune, India

**S. Chitra** Department of Biotechnology, School of Life Sciences, Bharathidasan University, Tiruchirappalli, Tamil Nadu, India

**Preeti Chutani** Department of Microbiology, Maharshi Dayanand University, Rohtak, Haryana, India

**Manab Deka** Biotechnology Division, Gauhati University, Guwahati, Assam, India

**Sunil Kumar Deshmukh** Department of Natural Products, Piramal Enterprises Limited, Mumbai, India

**Mukund V. Deshpande** Biochemical Sciences Division, National Chemical Laboratory, Pune, India

**Sunita R. Deshpande** Organic Chemistry Division, National Chemical Laboratory, Pune, India

**Mahesh Dhar** Department of Microbiology, University of Delhi, New Delhi, India

**Manoj K. Dhar** School of Biotechnology, University of Jammu, Jammu, India

**Sanjeev Kumar Diwakar** Department of Biotechnology, D. D. U. Gorakhpur University, Gorakhpur, U.P., India

**M. S. Dkhar** Microbial Ecology Laboratory, Department of Botany, North-Eastern Hill University, Shillong, Meghalaya, India

**Amit Kumar Dubey** Department of Biotechnology, D.D.U Gorakhpur University, Gorakhpur, Uttar Pradesh, India

**Manish Kumar Dubey** Laboratory of Mycopathology and Microbial Technology, Centre of Advanced Study in Botany, Banaras Hindu University, Varanasi, India

**Nawal Kishore Dubey** Laboratory of Herbal Pesticides, Centre of Advanced Study in Botany, Banaras Hindu University, Varanasi, India

**S. K. Dwivedi** Department of Environmental Science, Babasaheb Bhimrao Ambedkar (A Central) University, Lucknow, Uttar Pradesh, India

**Mohamed A. Gad** Biotechnology Division, North East Institute of Science and Technology (NEIST), Council of Scientific and Industrial Research (CSIR), Jorhat, Assam, India

**R. Garg** Department of Botany, Banaras Hindu University, Varanasi, Uttar Pradesh, India

**Arti Goel** Amity Institute of Microbial Biotechnology, Amity University, Noida, U.P., India

**K. Gokul Raj** Department of Molecular Microbiology, School of Biotechnology, Madurai Kamaraj University, Madurai, Tamil Nadu, India

**S. K. Gond** Department of Botany, Visva-Bharati University, Shantiniketan, India

**Ram Gopal** Department of Environmental Science, Babasaheb Bhimrao Ambedkar (A Central) University, Lucknow, Uttar Pradesh, India

**J. Goutam** Mycopathology and Microbial Technology Laboratory, Centre of Advanced Study in Botany, Banaras Hindu University, Varanasi, India

**R. P. Gupta** Department of Plant Pathology, N.D. University of Agriculture & Technology, Faizabad, U.P., India

**Sangeeta Gupta** Division of Plant Pathology, Indian Agricultural Research Institute, New Delhi, India

**Swapna Gupta** Department of Botany, DDU Gorakhpur University, Gorakhpur, UP, India

**S. Rais Haider** Department of Botany, Shia P. G. College, Lucknow, India

**Naglaa A. Ibrahim** Department of Biology, Alexandria University, Alexandria, Egypt

**Jinender Jain** Department of Microbiology, Maharshi Dayanand University, Rohtak, Haryana, India

**C. Jeyalakshmi** Department of Plant Pathology, Pandit Jawaharlal Nehru College of Agriculture & Research Institute, Karaikal, U.T. of Puducherry, India

**S. R. Joshi** Microbiology Laboratory, Department of Biotechnology & Bioinformatics, North-Eastern Hill University, Shillong, Meghalaya, India

**György J. Kövics** Institute of Plant Protection, University of Debrecen, Debrecen, Hungary

**Sanjana Kaul** School of Biotechnology, University of Jammu, Jammu, India

**B. D. Kaushik** Anand Engineering College Keetham, Agra, U.P., India

**H. Kayang** Microbial Ecology Laboratory, Department of Botany, North Eastern Hill University, Shillong, India

**Oliver Kayser** Department of Biochemical and Chemical Engineering, TU Dortmund, Dortmund, Germany

**S. H. A. Kazmi** Department of Botany, Shia P. G. College, Lucknow, India

**Akash Kedia** Laboratory of Herbal Pesticides, Centre of Advanced Study in Botany, Banaras Hindu University, Varanasi, India

**M. M. Abid Ali Khan** Department of Botany, Shia P. G. College, Lucknow, India

**R. N. Kharwar** Mycopathology and Microbial Technology Laboratory, Centre of Advanced Study in Botany, Banaras Hindu University, Varanasi, India

**Polashree Khaund** Microbiology Laboratory, Department of Biotechnology & Bioinformatics, North-Eastern Hill University, Shillong, Meghalaya, India

**D. K. Kishore** IARI, Shimla-4, Himachal Pradesh, India

**Emese Kiss** Department of Biotechnology, Plant Protection Institute, Hungarian Academy of Sciences, Budapest, Hungary

**Kuldip** Department of Botany, Banaras Hindu University, Varanasi, India

**A. Kumar** Department of Botany, Buddha PG College, Kushinagar, India

**Amit Kumar** Department of Microbiology, Maharshi Dayanand University, Rohtak, Haryana, India

**Jitendra Kumar** Mycopathology and Microbial Technology Laboratory, Centre of Advanced Study in Botany, Banaras Hindu University, Varanasi, India

Department of Science, S. S. University, Varanasi, U.P., India

**Nilima Kumari** Department of Bioscience and Biotechnology, Banasthali University, Banasthali, Rajasthan, India

**I. K. Kunwar** Mycology and Molecular Plant Pathology Laboratory, Dept. of Botany, Osmania University, Hyderabad, A.P., India

**Parijat Kusari** Department of Biochemical and Chemical Engineering, TU Dortmund, Dortmund, Germany

**Souvik Kusari** Institute of Environmental Research (INFU), Department of Chemistry and Chemical Biology, TU Dortmund, Dortmund, Germany

Department of Plant Sciences, University of Oxford, Oxford, United Kingdom

**Ruth Laldinthar** Microbial Ecology Laboratory, Department of Botany, North-Eastern Hill University, Shillong, Meghalaya, India

**M. Loganathan** Division of Crop Protection, Indian Institute of Vegetable Research, Varanasi, India

Indian Institute of Vegetable Research, Varanasi, Uttar Pradesh, India

**Anjisha R. Maharshi** Centre for Advanced Studies in Plant Biotechnology and Genetic Engineering, Department of Biosciences, Saurashtra University, Rajkot, Gujarat, India

**Sherif S. Mahmoud** Department of Biology, Alexandria University, Alexandria, Egypt

**S. Malathi** Centre for Advanced Studies in Botany, University of Madras, Chennai, India

**Deepak Kumar Malik** Department of Biotechnology, University Institute of Engineering & Technology, Kurukshetra University, Kurukshetra, Haryana, India

**Smriti Mall** Department of Botany, St. Andrew's College, Gorakhpur, UP, India

**C. Manoharachary** Mycology and Molecular Plant Pathology Laboratory, Dept. of Botany, Osmania University, Hyderabad, A.P., India

**Mukesh Meena** Laboratory of Mycopathology and Microbial Technology, Centre of Advanced Study in Botany, Banaras Hindu University, Varanasi, India

**Ashish Mishra** Mycopathology and Microbial Technology Laboratory, Centre of Advanced Study in Botany, Banaras Hindu University, Varanasi, India

**Prashant Kumar Mishra** Laboratory of Herbal Pesticides, Centre of Advanced Study in Botany, Banaras Hindu University, Varanasi, India

**Sarad Kumar Mishra** Department of Biotechnology, D. D. U. Gorakhpur University, Gorakhpur, U.P., India

**D. Mohana Priya** Centre for Advanced Studies in Botany, University of Madras, Chennai, India

**Sananda Mondal** Seed Physiology Laboratory, Department of Plant Physiology, Institute of Agricultural Sciences, Banaras Hindu University, Varanasi, India

**S. A. Musanna** Department of Botany, Shia P. G. College, Lucknow, India

**J. Muthumary** Centre for Advanced Studies in Botany, University of Madras, Chennai, Tamil Nadu, India

**D. N. Nagaraju** Mycology and Molecular Plant Pathology Laboratory, Dept. of Botany, Osmania University, Hyderabad, A.P., India

**Ekta Narang** Department of Microbiology, Maharshi Dayanand University, Rohtak, Haryana, India

**David B. Olufolaji** Department of Crop, Soil and Pest Management, The Federal University of Technology, Akure, Nigeria

**P. Palani** Centre for Advanced Studies in Botany, University of Madras, Chennai, India

**M. Pandi** Department of Molecular Microbiology, School of Biotechnology, Madurai Kamaraj University, Madurai, Tamil Nadu, India

**Urja Pandya** Department of Microbiology, University School of Sciences, Gujarat University, Ahmedabad, Gujarat, India

**Kamlesh Kumar Prajapati** Department of Genetics and Plant Breeding, Narendra Deva University of Agriculture and Technology, Faizabad, India

**Bhanu Prakash** Laboratory of Herbal Pesticides, Centre of Advanced Study in Botany, Banaras Hindu University, Varanasi, India

**J. Pratibha Singh** Rajiv Gandhi Cancer Institute & Research Center, New Delhi, India

**Preeti** Department of Bioscience and Biotechnology, Banasthali University, Banasthali, Rajasthan, India

**Richa Raghuwanshi** Department of Botany, Mahila Mahavidyalaya, Banaras Hindu University, Varanasi, India

**A. B. Rai** Division of Crop Protection, Indian Institute of Vegetable Research, Varanasi, India

**Ashwani K. Rai** Department of Botany, Banaras Hindu University, Varanasi, India

**Ravishankar V. Rai** Department of Studies in Microbiology, University of Mysore, Mysore, Karnataka, India

**P. Rajapriya** Department of Microbial Technology, School of Biological Sciences, Madurai Kamaraj University, Madurai, Tamil Nadu, India

**T. Ramanadane** Pandit Jawaharlal Nehru College of Agriculture and Research Institute, Karaikal, Union Territory of Puducherry, India

**Jyoti Rani** Department of Zoology, Mahila Mahavidyalaya, Banaras Hindu University, Varanasi, India

**G. P. Rao** Division of Plant Pathology, Indian Agriculture Research Institute, New Delhi, India

**Meenu Rathi** Department of Botany, University College, Kurukshetra University, Kurukshetra, Haryana, India

**K. Ravindran** Department of Biotechnology, School of Life Sciences, Bharathidasan University, Tiruchirappalli, Tamil Nadu, India

**C. Rettinassababady** Department of Plant Pathology, Pandit Jawaharlal Nehru College of Agriculture & Research Institute, Karaikal, U.T. of Puducherry, India

Pandit Jawaharlal Nehru College of Agriculture and Research Institute, Karaikal, Union Territory of Puducherry, India

**Arshad H. Rizvi** Department of Botany, Shia PG College, Lucknow, India

**M. Zahid Rizvi** Department of Botany, Shia P. G. College, Lucknow, India

**S. Arshad Hasan Rizvi** Department of Botany, Shia P. G. College, Lucknow, India

**B. K. Roy** Department of Botany, Banaras Hindu University, Varanasi, Uttar Pradesh, India

**R. Sagar** Department of Botany, Banaras Hindu University, Varanasi, India

**S. Saha** Indian Institute of Vegetable Research, Varanasi, Uttar Pradesh, India

**Sujoy Saha** Division of Crop Protection, Indian Institute of Vegetable Research, Varanasi, India

**Pradeepa V. Samaga** Department of Studies in Microbiology, University of Mysore, Mysore, Karnataka, India

**Sangeeta** Department of Environmental Science, Babasaheb Bhimrao Ambedkar (A Central) University, Lucknow, Uttar Pradesh, India

**Sapna** Department of Microbiology, Maharshi Dayanand University, Rohtak, Haryana, India

**Meenu Saraf** Department of Microbiology, University School of Sciences, Gujarat University, Ahmedabad, Gujarat, India

**Gauri Saxena** Department of Botany, University of Lucknow, Lucknow, India

**Asha Sharma** Department of Botany, MDU, Rohtak, Haryana, India

**Kanika Sharma** Department of Botany, Mohanlal Sukhadia University, Udaipur, Rajasthan, India

**K. K. Sharma** Department of Microbiology, Maharshi Dayanand University, Rohtak, Haryana, India

**S. K. Sharma** YSP UH&F, Kinnaur, Himachal Pradesh, India

**Sapna Sharma** Division of Plant Pathology, Indian Agricultural Research Institute, New Delhi, India

**Tanwi Sharma** School of Biotechnology, University of Jammu, Jammu, India

**Vijay K. Sharma** Mycopathology and Microbial Technology Laboratory, Centre of Advanced Study in Botany, Banaras Hindu University, Varanasi, India

**Vinay Sharma** Department of Bioscience and Biotechnology, Banasthali University, Banasthali, Rajasthan, India

**Kumar Shivam** Department of Biotechnology, D. D. U. Gorakhpur University, Gorakhpur, U.P., India

**Swati Shukla** Department of Biotechnology, D. D. U. Gorakhpur University, Gorakhpur, U.P., India

**Arpita Singh** Division of Crop Protection, Indian Institute of Vegetable Research, Varanasi, India

**Bijender Singh** Department of Microbiology, Maharshi Dayanand University, Rohtak, Haryana, India

**D. K. Singh** Mycopathology and Microbial Technology Laboratory, Centre of Advanced Study in Botany, Banaras Hindu University, Varanasi, India

**Deepti Singh** Department of Microbiology, Maharshi Dayanand University, Rohtak, Haryana, India

**Dharmendra K. Singh** Department of Botany, Banaras Hindu University, Varanasi, India

**Dhirendra K. Singh** Department of Genetics and Plant Breeding, Institute of Agricultural Sciences, Banaras Hindu University, Varanasi, Uttar Pradesh, India

**Karuna Singh** Department of Zoology, Mahila Mahavidyalaya, Banaras Hindu University, Varanasi, India

**Prakash Singh** Department of Genetics and Plant Breeding, Institute of Agricultural Sciences, Banaras Hindu University, Varanasi, Uttar Pradesh, India

**S. K. Singh** Department of Plant Pathology, N.D. University of Agriculture & Technology, Faizabad, U.P., India

**S. P. Singh** Department of Plant Pathology, N.D. University of Agriculture & Technology, Faizabad, U.P., India

**Sanjeev Singh** Department of Agricultural Botany, Udai Pratap Autonomous College, Varanasi, Uttar Pradesh, India

**Vivek Kumar Singh** Laboratory of Mycopathology and Microbial Technology, Centre of Advanced Study in Botany, Banaras Hindu University, Varanasi, India

**Shilpam Sinha** Department of Botany, Mahila Mahavidyalaya, Banaras Hindu University, Varanasi, India

**S. Sivaramakrishnan** Department of Biotechnology, School of Life Sciences, Bharathidasan University, Tiruchirappalli, Tamil Nadu, India

**L. S. Songachan** Microbial Ecology Laboratory, Department of Botany, North Eastern Hill University, Shillong, India

**Michael Spiteller** Institute of Environmental Research (INFU), Department of Chemistry and Chemical Biology, TU Dortmund, Dortmund, Germany

**R. Srinivasamurthy** Institute Technology Management Unit, Indian Agricultural Research Institute, New Delhi, India

**Haribashai Swer** Microbial Ecology Laboratory, Department of Botany, North-Eastern Hill University, Shillong, Meghalaya, India

**Gábor Tarcali** Institute of Plant Protection, University of Debrecen, Debrecen, Hungary

**Vrinda S. Thaker** Centre for Advanced Studies in Plant Biotechnology and Genetic Engineering, Department of Biosciences, Saurashtra University, Rajkot, Gujarat, India

**A. K. Tiwari** Central Lab, U P Council of Sugarcane Research, Shahjahnapur, UP, India

**Arti Tiwari** Laboratory of Mycopathology and Microbial Technology, Centre of Advanced Study in Botany, Banaras Hindu University, Varanasi, India

**Santosh G. Tupe** Biochemical Sciences Division, National Chemical Laboratory, Pune, India

**Rashmi Tyagi** Department of Applied Sciences & Humanities, ITM University, Gurgaon, Haryana, India

**R. S. Upadhyay** Laboratory of Mycopathology and Microbial Technology, Centre of Advanced Study in Botany, Banaras Hindu University, Varanasi, India

**P. P. Upadhyaya** Department of Botany, DDU Gorakhpur University, Gorakhpur, UP, India

**Shilpa A. Verekar** Department of Natural Products, Piramal Enterprises Limited, Mumbai, India

**Jitender K. Verma** IARI, Shimla-4, Himachal Pradesh, India

**O. P. Verma** Department of Genetics and Plant Breeding, Narendra Deva University of Agriculture and Technology, Faizabad, India

**Praveen C. Verma** National Botanical Research Institute (CSIR), Lucknow, India

**Punita Verma** Department of Botany, Banaras Hindu University, Varanasi, India

**R. K. Verma** Forest Pathology Division, Tropical Forest Research Institute, PO-RFRC, Jabalpur, MP, India

**S. K. Verma** Department of Botany, Visva-Bharati University, Shantiniketan, India

**Jugsharan S. Virdi** Department of Microbiology, University of Delhi, New Delhi, India

**A. Wilson** Department of Biotechnology, School of Life Sciences, Bharathidasan University, Tiruchirappalli, Tamil Nadu, India

**B. C. Yadav** Department of Plant Pathology, N.D. University of Agriculture & Technology, Faizabad, U.P., India

**Dinesh Yadav** Department of Biotechnology, D.D.U Gorakhpur University, Gorakhpur, Uttar Pradesh, India

**Sangeeta Yadav** Department of Biotechnology, D.D.U Gorakhpur University, Gorakhpur, Uttar Pradesh, India

**S. N. H. Zaidi** Department of Botany, Shia P. G. College, Lucknow, India

**Andleeb Zehra** Laboratory of Mycopathology and Microbial Technology, Centre of Advanced Study in Botany, Banaras Hindu University, Varanasi, India

# Part I
# Endophytes and Mycorrhiza

# Recent Advances in Research on *Cannabis sativa* L. Endophytes and Their Prospect for the Pharmaceutical Industry

1

Parijat Kusari, Michael Spiteller, Oliver Kayser and Souvik Kusari

**Abstract**

Endophytic microorganisms residing within plants are constantly communicating with them and the external environment by means of various interaction mechanisms and biochemical processes. Although plants have evolved their own defense mechanisms, sometimes they fail to defend themselves from the constant attack of devastating and emerging pathogens. Thus, research involving endophytes that aid the defense responses of their host plants can be useful in biocontrol and pest management strategies. Elucidating the chemistry of endophyte–plant interactions can provide new insights into the production of target and/or nontarget metabolites, thereby enabling a better understanding of the metabolic processes *in planta* and *ex planta*. In this chapter, we highlight the interactions of endophytes harbored in the medicinally important plant *Cannabis sativa* L. with the host plant as well as with the pathogens. The various endophyte–plant–pathogen defense–counter defense crosstalk would aid in exploring the biocontrol potential of endophytes in thwarting pathogens attacking the plants, and thus, effectively decrease the loss of such therapeutically relevant medicinal plants. Such interactions will further lead to the discovery of bioactive compounds, including the ones exclusive to the host plants. This chapter deals with the recent advances made

S. Kusari (✉) · M. Spiteller
Institute of Environmental Research (INFU),
Department of Chemistry and Chemical Biology,
TU Dortmund, Otto-Hahn-Str. 6,
44221 Dortmund, Germany

S. Kusari
Department of Plant Sciences, University of Oxford,
South Parks Road, OX1 3RB, Oxford, United Kingdom
(Visiting Researcher)

S. Kusari
e-mail: Souvik.Kusari@infu.tu-dortmund.de

P. Kusari · O. Kayser
Department of Biochemical and Chemical Engineering,
TU Dortmund, Emil-Figge-Str. 66, 44227 Dortmund,
Germany

R. N. Kharwar et al. (eds.), *Microbial Diversity and Biotechnology in Food Security*,
DOI 10.1007/978-81-322-1801-2_1, © Springer India 2014

in bioprospecting endophytes harbored in *C. sativa* L. with regard to their efficacies in thwarting phytopathogens. When endophytes are challenged with host-specific phytopathogens, they show an assortment of physical and chemical defense responses under different media conditions. This supports the concept of one strain many compounds (OSMAC) approach. Using cues from the current investigation, future research can maximize the possibility of a holistic understanding of endophyte–endophyte, endophyte–plant, and endophyte–pathogen relationships.

**Keywords**

Fungal endophytes · Bioprospecting · Biocontrol · Phytocannabinoids · Medicinal Plants

## 1.1 Introduction

*Cannabis* is an annual herbaceous plant genus of Cannabaceae family, mainly from Central Asia. *Cannabis* and *Humulus* are the only two recognized genera in the Cannabaceae family (Fernald 1950; Flores-Sanchez and Verpoorte 2008). In *Cannabis,* mainly one species is famously recognized, namely *Cannabis sativa* (Linnaeus 1753), however, three other species (*C. indica, C. ruderalis,* and *C. afghanica*) have also been described recently (McPartland et al. 2000). *Humulus lupulus* is the only species recognized in the genus *Humulus*. However, *C. sativa* L. (Fig. 1.1) is the most rigorously studied plant that has been in use all over the planet since ages either in the form of narcotic or medicinal preparations or as a source of food and fiber (Jiang et al. 2006; Wills 1998; Murray et al. 2007). It is also the most controversial plant in the human history with a strongly divided medical, research, and political community with respect to its use. *C. sativa* L. is commonly called "hemp," and it is said that "hemp has no enemies" (Dewey 1914). However, this misleading notion is far from the truth since this plant is beleaguered by a plethora of specific and generalist microbial pathogens (Kusari et al. 2013a). A couple of infrequent attempts have been made so far for the eradication of the fungal pathogens attacking this plant (Ungerlerder et al. 1982; Kurup et al. 1983; Levitz and Diamond 1991; Bush Doctor 1993). However, a holistic, cost-effective, and environmentally friendly means to eradicate the pathogen-mediated diseases in *Cannabis* is essential.

**Fig. 1.1** *Cannabis sativa* L. plants sampled from the Bedrocan BV Medicinal Cannabis (The Netherlands)

In this chapter, we address this issue in detail. We demonstrate with some recent examples, how it might be possible to efficiently utilize unique *C. sativa*-associated microorganisms (called "endophytes") to thwart the loss of these therapeutically significant plants and considerably reduce the expanse of vulnerabilities caused by phytopathogens.

## 1.2 Overview of Phytocannabinoids in *C. sativa* L.

The major secondary metabolites of *C. sativa* L. constitute cannabinoids, terpenoids, flavonoids, alkaloids, and lignans (Flores-Sanchez and Verpoorte 2008). Among them, cannabinoids are the ones most extensively studied. Cannabinoids are terpenophenolics found in the Indian hemp (*C. sativa* L.) constituting a class of chemical compounds that include phytocannabinoids (i.e., oxygen-containing $C_{21}$ aromatic hydrocarbon compounds found in *Cannabis* plant) and related chemical compounds which mimic the actions of phytocannabinoids or have a similar structure (e.g., endocannabinoids). Cannabinoids are known to occur naturally in significant measure in the plant. In general, all plant parts are known to contain cannabinoids (Flemming et al. 2007). However, these phytochemicals are more concentrated in a viscous resin that is produced in glandular trichomes. Table 1.1 summarizes the major cannabinoids and related precursors that have been isolated from *Cannabis* plants.

Although the plant is mainly regarded as a drug of abuse due to the high content of delta 9-tetrahydrocannabinol ($\Delta^9$-THC), the main psychoactive compound, cannabinoids are known to have important therapeutic effects (Williamson and Evans 2000; Baker et al. 2003; Grotenhermen 2002, Grotenhermen and Müller-Vahl 2012; Musty 2004; Flores-Sanchez and Verpoorte 2008) such as analgesic, antispasmodic, anti-tremor, anti-inflammatory (Gomes et al. 2008), antioxidant, antineoplastic (Carchman et al. 1976; Mojzisova and Mojzis 2008), neuroprotective (Ameri 1999), immunosuppressive, antinociceptive, antiepileptic, antidepressants, and appetite stimulant. From 450 secondary natural product constituents in total (including 20 flavonoids, 15 polyketides), more than 108 cannabinoids have been discovered so far (ElSohly and Slade 2005; Radwan et al. 2008; Ahmed et al. 2008; Hazekamp et al. 2004, 2005; Taura et al. 2007; Fischedick et al. 2010; and refer to Natural Product Database, Nov. 2012). Due to such therapeutic potential of cannabinoids and the plant extracts themselves, several *Cannabis*-based medicines have already made their way to the pharmaceutical industries. Some prominent examples include Marinol® (Solvay Pharmaceuticals, Belgium), Sativex (GW Pharmaceuticals, UK), and Nabilone (Cesamet®, Veleant Pharmaceuticals International, USA). Although $\Delta^9$-THC is considered to be one of the major psychoactive compounds (Taura et al. 1995; Sirikantaramas et al. 2005; Pertwee 2006), other cannabinoids such as cannabigerol, cannabidiol, cannabinol, olivetol, and cannabichromene prove to be therapeutically beneficial either alone or synergistically. These cannabinoids are also known to be effective against various pathogenic bacteria and fungi of clinical importance thereby signifying the antifungal and antibacterial potency of the compounds (Appendino et al. 2008; Elsohly et al. 1982; Pollastro et al. 2011). However, more studies are still required to confirm the potential benefits of whole plant extracts compared to that of pure cannabinoids (Williamson and Evans 2000; Wachtel et al. 2002; Russo and McPartland 2003; ElSohly et al. 2003). $\Delta^9$-THC and other cannabinoids are also subjected to directed biosynthesis, or in other words, induced in the medium by biotransformation of structurally related compounds using various fungal isolates or plant cell suspension cultures (Miyazawa et al. 1997; Kawamoto et al. 2008; Toniazzo et al. 2005; McClanahan and Robertson 1985; Hartsel et al. 1983; Tanaka et al. 1997; Saxena 2009; Flores-Sanchez et al. 2009; Happyana et al. 2013).

## 1.3 Necessity for Discovering Endophytes Harbored in *C. sativa* L. Conferring Plant Fitness Benefits

Plants have been bioprospected for therapeutic potential since ages. Plants are known to contain various bioactive molecules with relevant biological functions such as chemical defense of the plant (Chen and Schmidt 1924; Lopez-Lazaro et al. 2003; Li et al. 2001; Holler et al. 2012; Wink 2008). However, due to the continuous coevolution of the attack–defense, counter defense, and other forms of crosstalk between plants and interacting organisms (including microorganisms, herbivores, feeders, pests, etc.), plants alone are unable to defend themselves against parasites, pathogens, and predators (Kusari et al. 2013b). For example, despite the significant quantity of

**Table 1.1** Important natural cannabinoids and metabolic precursors found in *Cannabis* plants

| Name of compound | Molecular formula | Structure |
|---|---|---|
| Olivetol | $C_{11}H_{16}O_2$ | 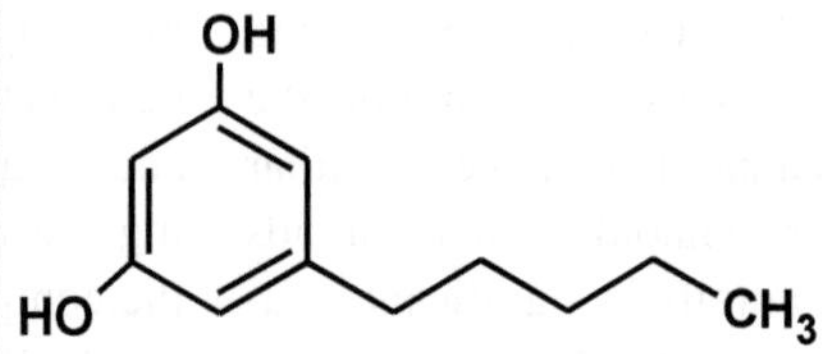 |
| Olivetolic acid | $C_{12}H_{16}O_4$ | 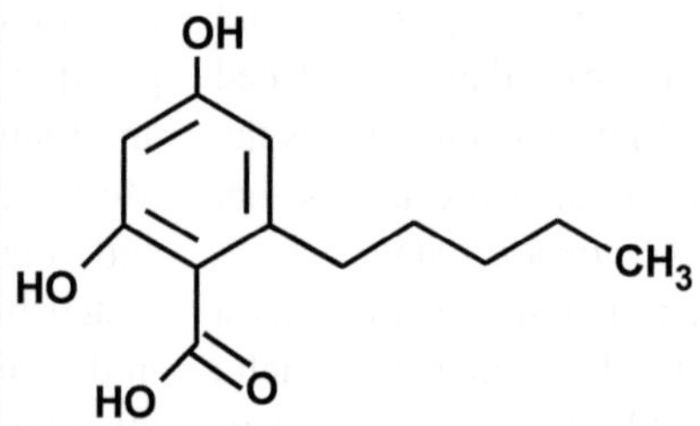 |
| Cannabigerol | $C_{21}H_{32}O_2$ | 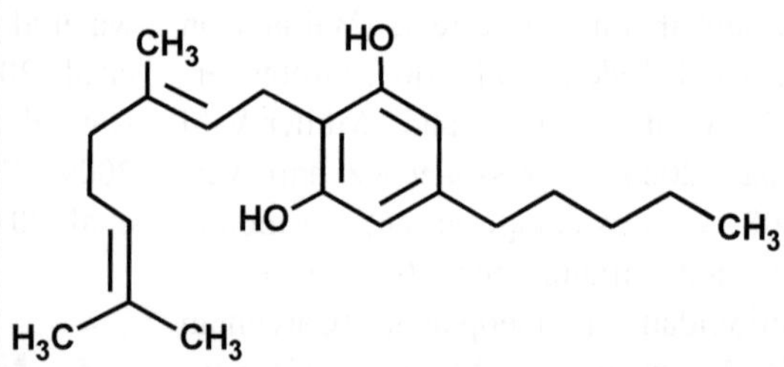 |
| Cannabigerolic acid | $C_{22}H_{32}O_4$ | 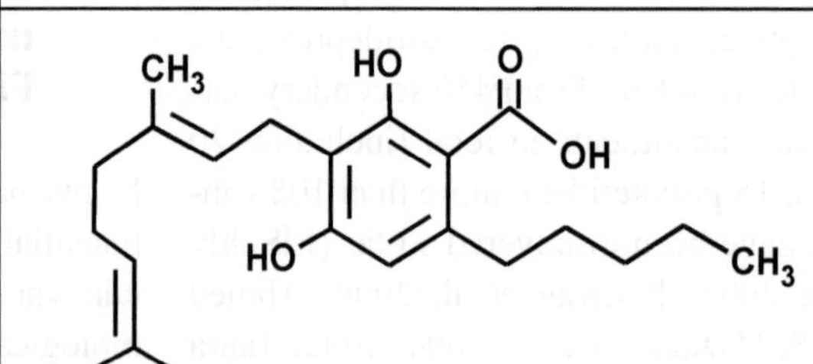 |
| Cannabichromene | $C_{21}H_{30}O_2$ | 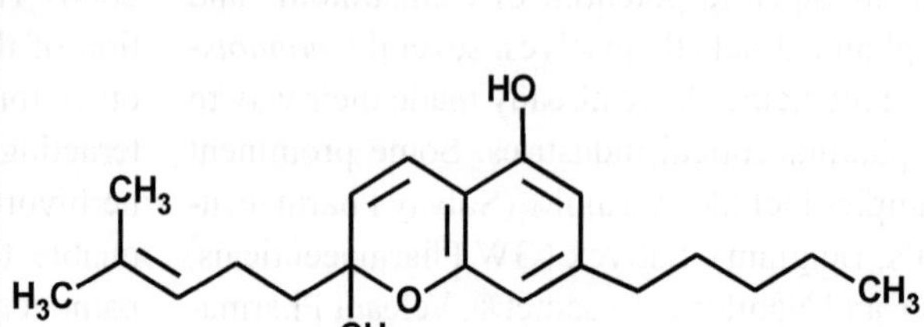 |

**Table 1.1** (continued)

| | | |
|---|---|---|
| Cannabichromenic acid | $C_{22}H_{30}O_4$ | 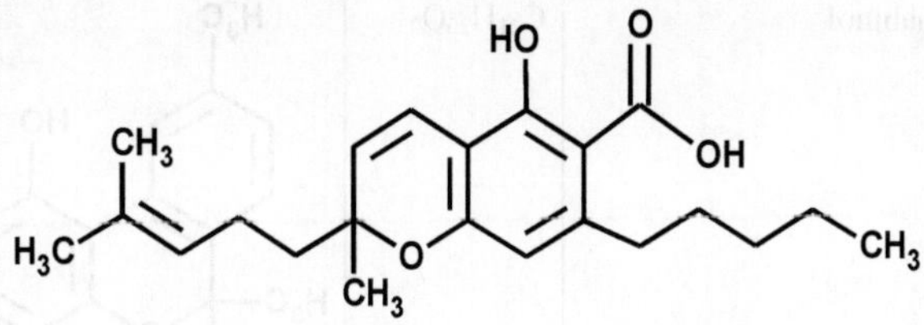 |
| Cannabidiol | $C_{21}H_{30}O_2$ | 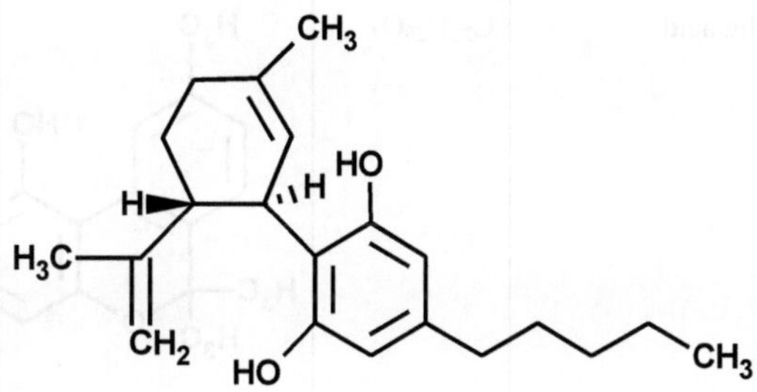 |
| Cannabidiolic acid | $C_{22}H_{30}O_4$ | 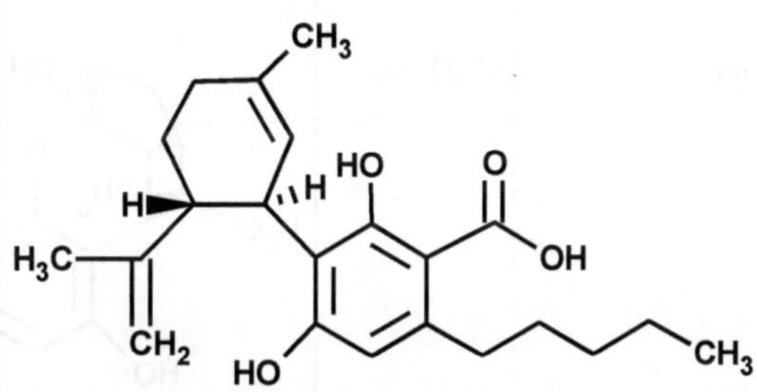 |
| $\Delta^9$-tetrahydrocannabinol | $C_{21}H_{30}O_2$ | 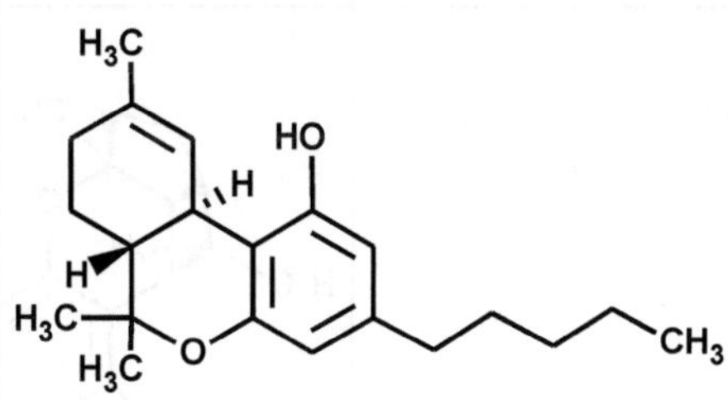 |
| $\Delta^9$-tetrahydrocannabinolic acid | $C_{22}H_{30}O_4$ | 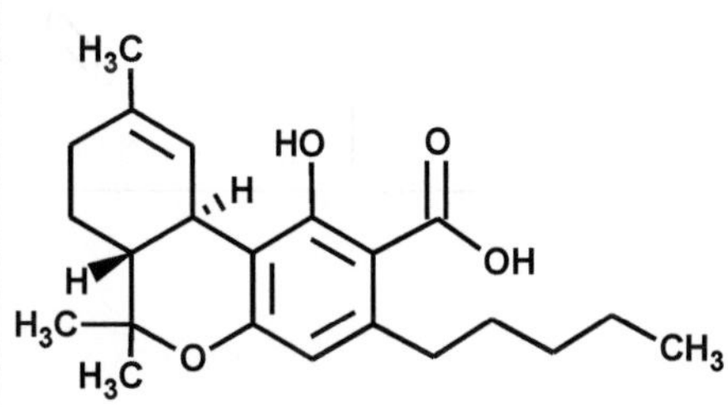 |

**Table 1.1** (continued)

| | | |
|---|---|---|
| Cannabinol | $C_{21}H_{26}O_2$ | 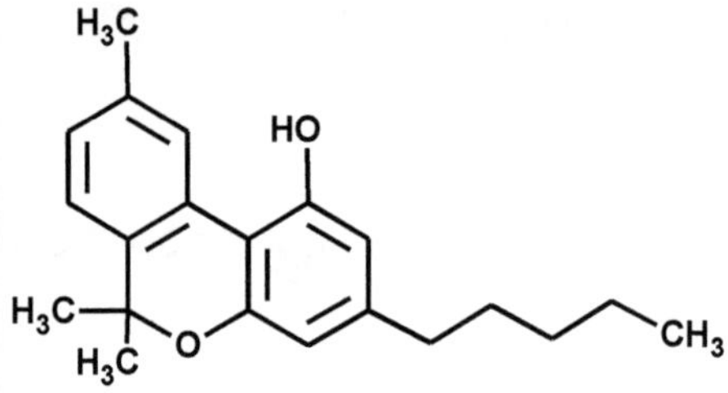 |
| Cannabinolic acid | $C_{22}H_{26}O_4$ | 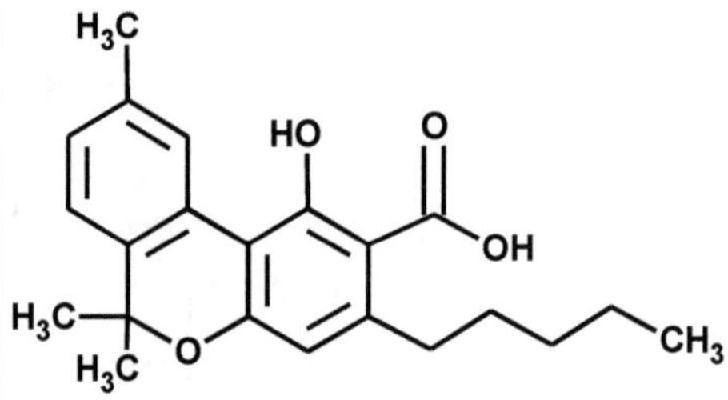 |
| Cannabicyclol | $C_{21}H_{30}O_2$ | 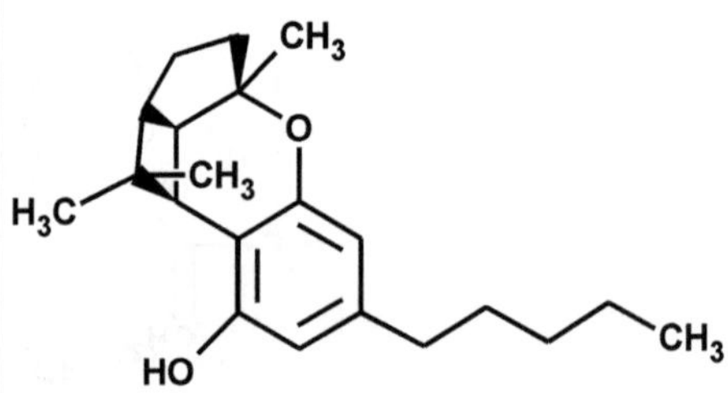 |
| Cannabicyclolic acid | $C_{22}H_{30}O_4$ | 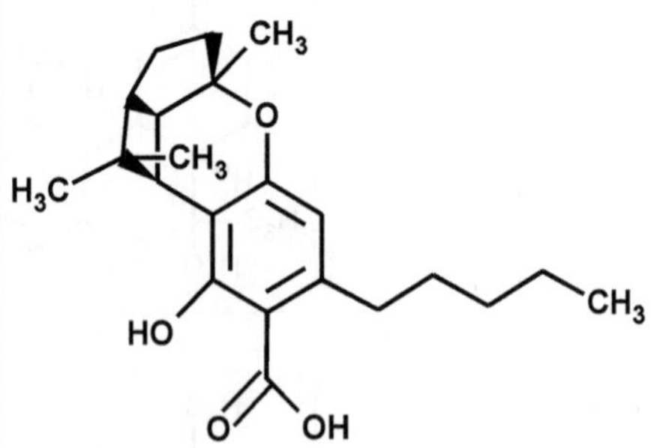 |

**Table 1.1** (continued)

| | | |
|---|---|---|
| Cannabielsoin | $C_{21}H_{30}O_3$ | 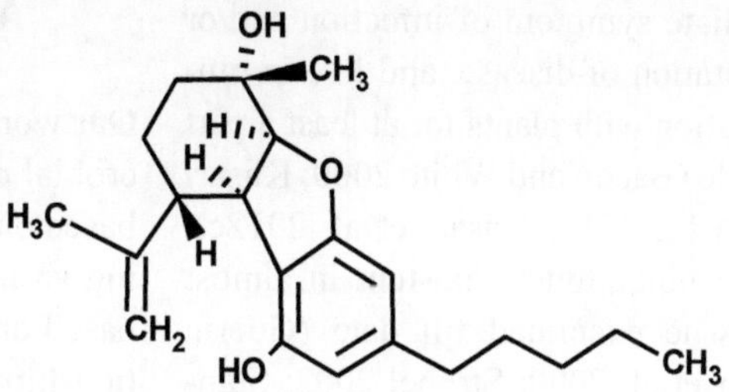 |
| Cannabitriol | $C_{21}H_{30}O_4$ | 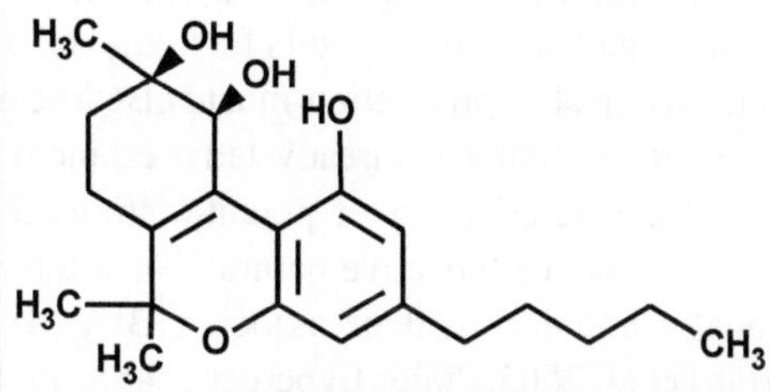 |
| Cannabichromanone | $C_{20}H_{28}O_4$ | 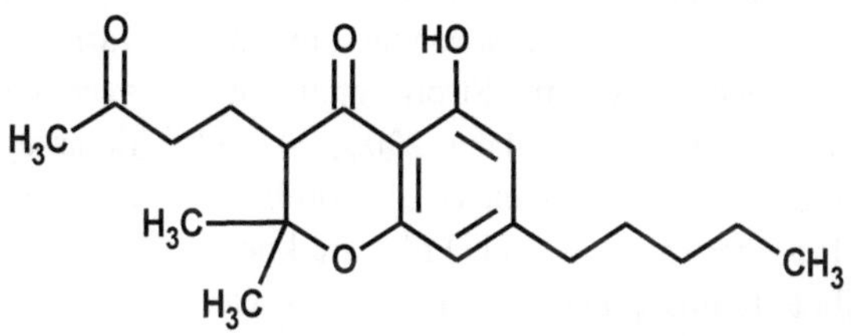 |

cannabinoids in the *C. sativa* L. plant, there are still reports of numerous phytopathogens attacking the different organs of the plant starting from seedling to even a mature plant (McPartland 1996). A plethora of bacteria and fungi are known to be responsible for the devastating infections caused to the plant (Hockey 1927; McPartland 1991, 1983, 1994, 1995). As a case in point, the two major phytopahogens, namely *Botrytis cinerea* and *Trichothecium roseum*, are potent greenhouse threats for the *Cannabis* cultivars and are known to cause localized to (potentially) epidemic disasters (Barloy and Pelhate 1962; Bush Doctor 1985). Although elimination attempts against many pathogens have been made so far (Ungerlerder et al. 1982; Kurup et al. 1983; Levitz and Diamond 1991; Bush Doctor 1993), for total eradication of causative agents and/or prevention of their pathogenicity to *Cannabis* plants, future investigation is required.

## 1.4 Strategies for Discovering Pharmaceutically Relevant Endophytes

In the last decade, discovery and intensive investigation of plant-associated microorganisms, termed endophytic microorganisms (or endophytes) have led to the possibility of exploring the potential benefits of these promising organisms in agriculture, medicinal, and pharmaceutical sectors. Endophytes can be defined, in a generalist manner, as a group of microorganisms that infect the internal tissues of plant without causing any immediate symptom of infection and/or visible manifestation of disease, and live in mutualistic association with plants for at least a part of their life cycle (Bacon and White 2000; Kusari and Spiteller 2011, 2012; Kusari et al. 2012c). Endophytes are ubiquitously existent in almost every plant tissue examined till date (Guerin 1898; Redecker et al. 2000; Strobel 2002; Staniek et al. 2008). With the increasing enormity of global health problems, and the incidence of drug-resistant microorganisms and new diseases, it has become clear that faster and effective pursuits for drug discovery and sustainable production must be made. This cumulative crisis has already led to the discovery and characterization of potent endophytes which can produce bioactive natural products, occasionally mimetic to their associated host plants (Puri et al. 2005, 2006; Eyberger et al. 2006; Kour et al. 2008; Kusari et al. 2008, 2009a, b, c, 2011, 2012b; Shweta et al. 2010). Endophytes are also known to produce a diverse range of biologically active secondary metabolites (Strobel and Daisy 2003; Strobel et al. 2004; Zhang et al. 2006; Gunatilaka 2006; Staniek et al. 2008; Suryanarayanana et al. 2009; Aly et al. 2010; Kharwar et al. 2011) that are known to contribute to host plant tolerance against various environmental stress herbivory, heat, salt, disease, and drought (Stone et al. 2000; Redman et al. 2002; Arnold et al. 2003; Rodriguez et al. 2004, 2008; Waller et al. 2005; Márquez et al. 2007; Rodriguez and Redman 2008; Porras-Alfaro and Bayman 2011). Even with such colossal amounts and breadth of successful discoveries of potentially beneficial endophytes, it has still not been possible to utilize them commercially for the "sustained production" of the desired pharmaceutically valuable compounds (Kusari et al. 2012c, 2013b). Therefore, understanding of the multitude of endophyte relationships with host plants needs more attention and investigation in various related aspects such as the endophyte–plant interactions, multispecies crosstalk, and links with herbivores and predators.

## 1.5 Endophytic Microorganisms Associated with *C. sativa* L.

Our work on the investigation of endophytic microbial community harbored in *C. sativa* L. was based on the recent advancements made in devising various strategies of discovering endophytes based on the rationale of their cost–benefit relationship with their hosts in order to exploit their potential beneficial efficacies. Since this plant is protected by national and international legislations and regulations, we sampled and imported the *C. sativa* L. plants from the legal farmer Bedrocan BV Medicinal Cannabis (The Netherlands) with the permission of the Federal Institute for Drugs and Medical Devices (Bundesinstitutfür Arzneimittel und Medizinprodukte, BfArM), Germany under the license number 458 49 89. Plant specimens have been deposited at the Bedrocan BV with voucher numbers (A1)05.41.050710. We then isolated a plethora of endophytes (Fig. 1.2) and subjected them to various culture conditions and parameters and even challenged them (dual-culture antagonistic assays of the fungal isolates) with two major phytopahogens of the *Cannabis* plant, namely *B. cinerea* and *T. roseum*, which are potent greenhouse threats for the cultivars and known to cause disasters at epidemic scales (Barloy and Pelhate 1962; Bush Doctor 1985). Our target was to evaluate the endophytes within the ecological and biochemical contexts, especially focusing on their biocontrol potential to thwart the host-specific phytopathogens. This led us towards the identification of potent endophytes that not only proved to be promising biocontrol agents against the specific phytopathogens, but also demonstrated qualities of being a natural reservoir of bioactive secondary metabolites (Kusari et al. 2013a). To the best of our knowledge, this work was the first to report the incidence, diversity analysis, and qualitative biocontrol potential of endophytic fungi harbored in *C. sativa* L. plants. Eleven different kinds of antagonistic interactions are observed when the endophytes were challenged with the phytopathogens in five different media, namely Sabouraud agar (SA), nutrient agar (NA), potato dextrose agar (PDA), malt extract

**Fig. 1.2** Representative plates showing emergence of endophytic fungal mycelia from surface-sterilized *Cannabis sativa* L. plant tissues on water agar media amended with antibiotic (streptomycin, 100 mg/L)

agar (MEA), and water agar (WA), respectively. This highlights the fact that endophytes are capable of producing different compounds under varying conditions which are otherwise "cryptic" metabolites. All the endophyte isolates showed antagonistic potency to some extent against either one or both of the phytopathogens in varying the media, but three of the isolates proved to exhibit prominent complete inhibition (Kusari et al. 2013a). Many endophytes started sporulating in NA, as expected, revealing their response to the unfavorable condition while countering the confronting pathogen. Interestingly, the same endophyte isolates showed various other interesting inhibition patterns such as formation of a clear halo (inhibition zone), release of exudates without even physical contact of mycelia, and change of mycelia color among others, which accompanied the inhibitions.

Plant–fungal associations are always accompanied by various physical and chemical interactions thereby establishing them either in a localized and/or systemic manner (Kusari et al. 2012c). The varying assortment of antagonisms demonstrated by the endophytes against the host phytopathogens indicates that their efficacies are either due to the production of secondary metabolites or the immediate intermediates in the biosynthetic pathway of those metabolites, triggered upon pathogen challenge. The interactions were in complete agreement with the well-known "one strain many compounds (OSMAC)" approach (Kusari et al. 2012c), thereby revealing that endophytes are capable of producing cryptic metabolites when elicited under certain selective interacting conditions apart from the normal metabolites produced under normal fermentation conditions. Our work not only reports endophytic fungi as potent biocontrol agents under suitable conditions but also provides a platform to compare the endophytes of the same plant from different wild populations and collection centers (if accessible) for global-scale diversity analysis and the production of successive bioactive secondary metabolites (target and/or nontarget) with strong therapeutic potential.

## 1.6 Outlook

The potential of inimitable fungal endophytes adept in biosynthesizing bioactive metabolites, occasionally those imitative to their host plants, has irrefutably been recognized. Endophytes can be accepted as new sources for gene- and drug discovery in medical sciences and will provide, by distinct genomic blueprints, new insights in gene assembly and expression control. Nonetheless, there is still no known breakthrough in the biotechnological production of these bioactive natural products using endophytes. It is imperious to expound the metabolome in endophytes correlating to their host plants on a case-by-case basis to comprehend how the biogenetic gene clusters are regulated and their expression is affected *in planta* and *ex planta* (i.e., by environmental changes and axenic culture conditions). Only a deeper understanding of the host–endophyte relationship at the molecular level might help to induce and optimize secondary metabolite production under laboratory conditions to yield desired metabolites in a sustained manner using endophytes. This can be achieved by challenging the endophytes by specific and nonspecific pathogens, especially those attacking their host plants, by devising suitable coculture and dual-culture setups (qualitative, followed by suitable quantitative experiments). The pathogens encountered can serve as an inducer that might trigger the production of defense secondary metabolites with prodrug-like properties. Once the production of a target or nontarget natural product with a desired biological activity has been achieved, techniques such as genome mining, metabolic engineering, and metagenomics could be utilized to influence the manipulation of secondary metabolite production by endophytic fungi or the plant itself by directed infection with beneficial endophytes (Kusari et al. 2012c, d). Such directed investigation with the scientific rationale of mimicking the natural plant–endophyte–pathogen interactions should be pursued to warrant a virtually incessant discovery and sustained supply of bioactive prodrugs against the current and emerging diseases.

**Acknowledgments** Research at the Institute of Environmental Research (INFU) of the Faculty of Chemistry and Chemical Biology is supported by the International Bureau (IB) of the German Federal Ministry of Education and Research (BMBF/DLR), Germany, the Ministry of Innovation, Science, Research and Technology of the State of North Rhine-Westphalia, Germany, the German Academic Exchange Service (DAAD; "Welcome to Africa" initiative), and the German Research Foundation (Deutsche Forschungsgemeinschaft, DFG). S. K. gratefully acknowledges M. S. for approving and authorizing, Gail M. Preston for hosting, and TU Dortmund for supporting his stay at the University of Oxford, UK, as a Visiting Researcher. Research at the Department of Biochemical and Chemical Engineering is supported by the Ministry of Innovation, Science and Research of the German Federal State North Rhine-Westphalia and the CLIB-Graduate Cluster Industrial Biotechnology (CLIB). We are thankful to Bedrocan BV for kindly providing us with the *Cannabis sativa* L. plants.

## References

Ahmed SA, Ross SA, Slade D, Radwan MM, Zulfiqar F, ElSohly MA (2008) Cannabinoid ester constituents from high-potency *Cannabis sativa*. J Nat Prod 71:536–542

Aly AH, Debbab A, Kjer J, Proksch P (2010) Fungal endophytes from higher plants: a prolific source of phytochemicals and other bioactive natural products. Fungal Divers 41:1–16

Ameri A (1999) The effects of cannabinoids on the brain. Prog Neurobiol 158:315–348

Appendino G, Gibbons S, Giana A, Pagani A, Grassi G, Stavri M, Smith E, Rahman MM (2008) Antibacterial cannabinoids from *Cannabis sativa*: a structure-activity study. J Nat Prod 71:1427–1430

Arnold AE, Mejia LC, Kyllo D, Rojas EI, Maynard Z, Robbins N (2003) Fungal endophytes limit pathogen damage in a tropical tree. Proc Natl Acad Sci USA 100:15649–15654

Bacon CW, White JF (2000) Microbial endophytes. Marcel Deker Inc, New York

Baker D, Pryce G, Giovannoni G, Thompson AJ (2003) The therapeutic potential of cannabis. Lancet Neurol 2:291–298

Barloy J, Pelhate J (1962) PremiËres observations phytopathologiques relatives aux cultures de chanvre en Anjou. Ann Epiphyties 13:117–149

Bush Doctor (1985) Damping off. Sinsemilla Tips 5:35–39

Bush Doctor, The (1993) How to preserve pot potency. High Times No 213:75, 77–78

Carchman RA, Harris LS, Munson AE (1976) The inhibition of DNA synthesis by cannabinoids. Cancer Res 36:95–100

Chen KK, Schmidt CF (1924) The action of ephedrine, the active principle of the Chinese drug ma huang. J Pharmacol Exp Ther 24:339–357

Dewey LH (1914) "Hemp". In: U.S.D.A. yearbook 1913 United States Department of Agriculture, Washington, DC, pp 283–347

Elsohly HN, Turner CE, Clark AM, Elsohly MA (1982) Synthesis and antimicrobial activities of certain cannabichromene and cannabigerol related compounds. J Pharm Sci 71:1319–1323

ElSohly MA, Slade D (2005) Chemical constituents of marijuana: the complex mixture of natural cannabinoids. Life Sci 78:539–548

ElSohly MA, Wachtel SR, de Wit H (2003) *Cannabis* versus THC: response to Russo and McPartland. Psychopharmacology (Berl) 165:433–434

Eyberger AL, Dondapati R, Porter JR (2006) Endophyte fungal isolates from *Podophyllum peltatum* produce podophyllotoxin. J Nat Prod 69:1121–1124

Fernald ML (1950) Gray's manual of botany, 4th ed. American Book Company, New York, p 556

Fischedick JT, Hazekamp A, Erkelens T, Choi YH, Verpoorte R (2010) Metabolic fingerprinting of *Cannabis sativa* L., cannabinoids and terpenoids for chemotaxonomic and drug standardization purposes Phytochemistry 71:2058–2073

Flemming T, Muntendam R, Steup C, Kayser O (2007) Chemistry and biological activity of tetrahydrocannabinol and its derivatives. In: Khan MTH (ed) Topics in heterocyclic chemistry, vol. 10. Springer, Berlin, pp 1–42

Flores-Sanchez IJ, Verpoorte R (2008) Secondary metabolism in *Cannabis*. Phytochem Rev 7:615–639

Flores-Sanchez IJ, Pec J, Fei J, Choi YH, Dusek J, Verpoorte R (2009) Elicitation studies in cell suspension cultures of *Cannabis sativa* L. J Biotechnol 143:157–168

Gomes A, Fernandes E, Lima JLFC, Mira L, Corvo ML (2008) Molecular mechanisms of anti-inflammatory activity mediated by flavonoids. Curr Med Chem 15:1586–1605

Grotenhermen F (2002) Review of therapeutic effects. In: Grothenhermen F, Russo E (eds) Cannabis and cannabinoids: pharmacology, toxicology and therapeutic potential. The Haworth Integrative Healing Press, New York, pp. 123–142

Grotenhermen F, Müller-Vahl K (2012) The therapeutic potential of *Cannabis* and cannabinoids. Medicine Dtsch Arztebl Int 109:495–501

Guerin P (1898) Sur la presence d'un champignon dansl'ivraie. J Botanique 12:230–238

Gunatilaka AAL (2006) Natural products from plant-associated microorganisms: distribution, structural diversity, bioactivity, and implications of their occurrence. J Nat Prod 69:509–526

Happyana N, Agnolet S, Muntendam R, Van Dam A, Schneider B, Kayser O (2013) Cannabinoid analysis of laser-microdissected trichomes of *Cannabis sativa* L. by LC-MS and cryogenic NMR. Phytochemistry. 87:51–59

Hartsel SC, Loh WH, Robertson LW (1983) Biotransformation of cannabidiol to cannabielsoin by suspension cultures of *Cannabis sativa* and *Saccharum officinarum*. Planta Med 48:17–19

Hazekamp A, Choi YH, Verpoorte R (2004) Quantitative analysis of cannabinoids from *Cannabis sativa* using $^{1}$H-NMR. Chem Pharm Bull 52:718–721

Hazekamp A, Giroud C, Peltenburg A, Verpoorte R (2005) Spectroscopic and chromatographic data of cannabinoids from *Cannabis sativa*. J Liq Chrom Rel Technol 28:2361–2382

Hockey JF (1927) Report of the Dominion field laboratory of plant pathology, Kentville Nova Scotia. Canada Department of Agriculture: 28–36

Holler JG, Sondergaard K, Slotved HC, Guzman A, Molgaard P (2012) Evaluation of the antibacterial activity of Chilean plants traditionally used for wound healing therapy against multidrug-resistant *Staphylococcus aureus*. Planta Med 78:200–205

Jiang HE, Li X, Zhao YX, Ferguson DK, Hueber F, Bera S, Wang YF, Zhao LC, Liu CJ, Li CS (2006) A new insight into *Cannabis sativa* (Cannabaceae) utilization from 2500-year-old Yanghai Tombs, Xinjiang, China. J Ethnopharmacol 108:414–422

Kawamoto M, Utsukihara T, Abe C, Sato M, Saito M, Koshimura M, Kato N, Horiuchi CA (2008) Biotransformation of (±)-2-methylcyclohexanone by fungi. Biotechnol Lett 30:1655–1660

Kharwar RN, Mishra A, Gond SK, Stierle D (2011) Anticancer compounds derived from fungal endophytes: their importance and future challenges. Nat Prod Rep 28:1208–1228

Kour A, Shawl AS, Rehman S, Qazi PH, Sudan P, Khajuria RK, Sultan P, Verma V (2008) Isolation and identification of an endophytic strain of *Fusarium oxysporum* producing podophyllotoxin from *Juniperus recurva*. World J Microbiol Biotechnol 24:1115–1121

Kurup VP, Resnick A, Kagen SL, Cohen SH, Fink JN (1983) Allergenic fungi and actinomycetes in smoking materials and their health implications. Mycopathologia 82:61–64

Kusari P, Kusari S, Spiteller M, Kayser O (2013a) Endophytic fungi harbored in *Cannabis sativa* L.: diversity and potential as biocontrol agents against host plant-specific phytopathogens. Fungal Divers. 60:137–151

Kusari S, Spiteller M (2011) Are we ready for industrial production of bioactive plant secondary metabolites utilizing endophytes? Nat Prod Rep 28:1203–1207

Kusari S, Spiteller M (2012) Metabolomics of endophytic fungi producing associated plant secondary metabolites: progress, challenges and opportunities. In: Metabolomics. Roessner U (ed) In Tech, ISBN 978–953-51–0046-1, pp 241–266

Kusari S, Lamshöft M, Zühlke S, Spiteller M (2008) An endophytic fungus from *Hypericum perforatum* that produces hypericin. J Nat Prod 71:159–162

Kusari S, Zuehlke S, Spiteller M (2009a) An endophytic fungi from *Camptotheca acuminata* that produces camptothecin and analogues. J Nat Prod 72:2–7

Kusari S, Lamshöft M, Spiteller M (2009b) *Aspergillus fumigatus* Fresenius, an endophytic fungus from *Juniperus communis* L. Horstmann as a novel source of the anticancer pro-drug deoxypodophyllotoxin. J Appl Microbiol 107:1019–1030

Kusari S, Zühlke S, Kosuth J, Cellarova E, Spiteller M (2009c) Light-independent metabolomics of endophytic

*Thielavia subthermophila* provides insight into microbial hypericin biosynthesis. J Nat Prod 72:1825–1835

Kusari S, Zühlke S, Spiteller M (2011) Effect of artificial reconstitution of the interaction between the plant *Camptotheca acuminata* and the fungal endophyte *Fusarium solani* on camptothecin biosynthesis. J Nat Prod 74:764–775

Kusari S, Verma VC, Lamshöft M, Spiteller M (2012b) An endophytic fungus from *Azadirachta indica* A. Juss. that produces azadirachtin. World J Microbiol Biotechnol 28:1287–1294

Kusari S, Hertweck C, Spiteller M (2012c) Chemical ecology of endophytic fungi: origins of secondary metabolites. Chem Biol 19:792–798

Kusari S, Pandey SP, Spiteller M (2013b) Untapped mutualistic paradigms linking host plant and endophytic fungal production of similar bioactive secondary metabolites. Phytochemistry. 91:81–87

Levitz SM, Diamond RD (1991) Aspergillosis and marijuana. Ann Int Med 115:578–579

Li SH, Zhang HJ, Yao P, Sun HD, Fong HHS (2001) Taxane diterpenoids from the bark of *Taxus yunnanensis*. Phytochemistry 58:369–374

Linnaeus C (1753) Species plantarum. T. I–II. Laurentius Salvius, Stockholm

Lopez-Lazaro M, de la Pena NP, Pastor N, Martin-Cordero C, Navarro E, Cortes F, Ayuso MJ, Toro MV (2003) Anti-tumour activity of *Digitalis purpurea* L. subsp. *heywoodii*. Planta Med 69:701–704

Márquez LM, Redman RS, Rodriguez RJ, Roossinck MJ (2007) A virus in a fungus in a plant: three-way symbiosis required for thermal tolerance. Science 315:513–515

McClanahan RH, Robertson LW (1985) Microbial transformation of olivetol by *Fusarium roseum*. J Nat Prod 48:660–663

McPartland JM (1983) Fungal pathogens of *Cannabis sativa* in Illinois. Phytopathology 72:797

McPartland JM (1991) Common names for diseases of *Cannabis sativa* L. Plant Dis 75:226–227

McPartland JM (1994) Microbiological contaminants of marijuana. J Int Hemp Assoc 1:41–44

McPartland JM (1995) *Cannabis* pathogens X: *Phoma*, *Ascochyta* and *Didymella* species. Mycologia 86:870–878

McPartland JM (1996) A review of *Cannabis* diseases. J Int Hemp Assoc 3:19–23

McPartland JM, Clarke RC, Watson DP (2000) Hemp diseases and pests: management and biological control. CABI Publishing, Wallingford

Miyazawa M, Nankai H, Kameoka H (1997) Enantioselective cyclization of (±)-lavandulol to (-)-(2S, 4S)-1, 5-epoxy-5-methyl-2-(1-methylethenyl)-4-hexanol by *Glomerellacingulata*. Nat Prod Lett 9:249–252

Mojzisova G, Mojzis J (2008) Flavonoids and their potential health benefits: relation to heart diseases and cancer. Recent Prog Med Plants 21:105–129

Murray RM, Morrison PD, Henquet C, Di Forti M (2007) *Cannabis*, the mind and society: the hash realities. Nat Rev Neurosci 8:885–895

Musty RE (2004) Natural cannabinoids: interactions and effects. In: Guy GW, Whittle BA, Robson PJ (eds) The medicinal uses of cannabis and cannabinoids. Pharmaceutical Press, London, pp. 165–204

Pertwee RG (2006) Cannabinoid pharmacology: the first 66 years. Br J Pharmacol 147:163–171

Pollastro F, Taglialatela-Scafati O, Allar M, Munoz E, Marzo VD, Petrocellis LD, Appendino G (2011) Bioactive prenylogous cannabinoid from fiber hemp (*Cannabis sativa*). J Nat Prod 74:2019–2022

Porras-Alfaro A, Bayman P (2011) Hidden fungi, emergent properties: endophytes and microbiomes. Annu Rev Phytopathol 49:291–315

Puri SC, Verma V, Amna T, Qazi GN, Spiteller M (2005) An endophytic fungus from *Nothapodytes foetida* that produces camptothecin. J Nat Prod 68:1717–1719

Puri SC, Nazir A, Chawla R, Arora R, Riyaz-ul-Hasan S, Amna T, Ahmed B, Verma V, Singh S, Sagar R, Sharma A, Kumar R, Sharma RK, Qazi GN (2006) The endophytic fungus *Trametes hirsuta* as a novel alternative source of podophyllotoxin and related aryl tetralin lignans. J Biotechnol 122:494–510

Radwan MM, Ross SA, Slade D, Ahmed SA, Zulfiqar F, ElSohly MA (2008) Isolation and characterization of new *Cannabis* constituents from a high potency variety. Planta Med 74:267–272

Redecker D, Kodner R, Graham LE (2000) Glomalean fungi from the Ordovician. Science 289:1920–1921

Redman RS, Sheehan KB, Stout RG, Rodriguez RJ, Henson JM (2002) Thermotolerance conferred to plant host and fungal endophyte during mutualistic symbiosis. Science 298:1581

Rodriguez R, Redman R (2008) More than 400 million years of evolution and some plants still can't make it on their own: plant stress tolerance via fungal symbiosis. J Exp Bot 59:1109–1114

Rodriguez RJ, Redman RS, Henson JM (2004) The role of fungal symbioses in the adaptation of plants to high stress environments. Mitig Adapt Strat Glob Change 9:261–272

Rodriguez RJ, Henson J, Van Volkenburgh E, Hoy M, Wright L, Beckwith F et al (2008) Stress tolerance in plants via habitat-adapted symbiosis. ISME J 2:404–416

Russo EB, McPartland JM (2003) Cannabis is more than simply delta(9)-tetrahydrocannabinol. Psychopharmacology (Berl) 165:431–432

Saxena S (2009) Fungal biotransformations of cannabinoids: potential for new effective drugs. Curr Opin Drug Discov Develop 12:305–312

Shweta S, Zühlke S, Ramesha BT, Priti V, Kumar PM, Ravikanth G, Spiteller M, Vasudeva R, Shaanker RU (2010) Endophytic fungal strains of *Fusarium solani*, from *Apodytes dimidiata* E. Mey. exArn (Icacinaceae) produce camptothecin, 10-hydroxycamptothecin and 9-methoxycamptothecin. Phytochemistry 71:117–122

Sirikantaramas S, Taura F, Tanaka Y, Ishikawa Y, Morimoto S, Shoyama Y (2005) Tetrahydrocannabinolic acid synthase, the enzyme controlling marijuana psychoactivity, is secreted into the storage cavity of the glandular trichomes. Plant Cell Physiol 46:1578–1582

Staniek A, Woerdenbag HJ, Kayser O (2008) Endophytes: exploiting biodiversity for the improvement of natural product-based drug discovery. J Plant Interact 3:75–93

Stone JK, Bacon CW, White JF (2000) An overview of endophytic microbes: endophytism. In: Bacon CW, White JF (ed) Microbial endophytes. Marcel Dekker Inc, New York, pp 3–30

Strobel GA (2002) Microbial gifts from rain forests. Can J Plant Pathol 24:14–20

Strobel GA, Daisy B (2003) Bioprospecting for microbial endophytes and their natural products. Microbial Mol Biol Rev 67:491–502

Strobel GA, Daisy B, Castillo U, Harper J (2004) Natural products from endophytic microorganisms. J Nat Prod 67:257–268

Suryanarayanana TS, Thirunavukkarasub N, Govindarajulub MB, Sassec F, Jansend R, Murali TS (2009) Fungal endophytes and bioprospecting. Fungal Biol Rev 23:9–19

Tanaka H, Takahashi R, Morimoto S, Shoyama YA (1997) New cannabinoid, Δ6-tetrahydrocannabinol 2Δ-O-β-d-glucopyranoside, biotransformed by plant tissue. J Nat Prod 60:168–170

Taura F, Morimoto S, Shoyama Y, Mechoulam R (1995) First direct evidence for the mechanism of 1-tetrahydrocannabinolic acid biosynthesis. J Am Chem Soc 117:9766–9767

Taura F, Sirikantaramas S, Shoyamaa Y, Shoyamaa Y, Morimotoa S (2007) Phytocannabinoids in *Cannabis sativa*: recent studies on biosynthetic enzymes. Chem Biodivers 4:1649–1663

Toniazzo G, de Oliveira D, Dariva C, Oestreicher EG, Antunes OA (2005) Biotransformation of (-)-β-pinene by *Aspergillusniger* ATCC 9642. Appl Biochem Biotechnol 121–124:837–844

Ungerlerder JT, Andrysiak T, Tashkin DP, Gale RP (1982) Contamination of marijuana cigarettes with pathogenic bacteria. Cancer Treatment Rep 66:589–590

Wachtel SR, ElSohly MA, Ross SA, Ambre J, de Wit H (2002) Comparison of the subjective effects of D9-tetrahydrocannabinol and marijuana in humans. Psychopharmacology (Berl) 161:331–339

Waller F, Achatz B, Baltruschat H, Fodor J, Becker K, Fischer M et al (2005) The endophytic fungus *Piriformospora indica* reprograms barley to salt-stress tolerance, disease resistance and higher yield. Proc Natl Acad Sci USA 102:13386–13391

Williamson EM, Evans FJ (2000) Cannabinoids in clinical practice. Drugs 60:1303–1314

Wills S (1998) *Cannabis* use and abuse by man: an historical perspective. In: Brown DT (ed) *Cannabis*: the genus *Cannabis*. Harwood Academic Publishers, Amsterdam, pp 1–27

Wink M (2008) Plant secondary metabolism: diversity, function and its evolution. Nat Prod Commun 3:1205–1216

Zhang HW, Song YC, Tan RX (2006) Biology and chemistry of endophytes. Nat Prod Rep 23:753–7711 Recent Advances in Research on *Cannabis sativa* L. Endophytes …

# Endophytic Fungi from Brazilian Tropical Hosts and Their Biotechnological Applications

2

João Lúcio Azevedo

**Abstract**

Endophytic microorganisms are defined in different ways, and a recent definition considers them as all of the microorganisms culturable or not that inhabit the inner parts of plant tissues and cause no harm to their hosts. They can be divided into two groups: those that do not generate external structures from the host (group I) and those which are able to develop external structures such as the nodules of $N_2$-fixing bacteria and mycorrhizal fungi (group II). Endophytic microorganisms such as fungi and bacteria play important roles in their plant hosts. The first studies with endophytes were conducted in temperate regions but have recently also been studied in plants from tropical countries. This chapter provides selected data obtained in Brazil mainly for endophytic fungi and focuses on their agricultural applications including the biological control of diseases and insect pests, and the promotion of plant growth. The biotechnological potential of the endophytic fungi isolated from not yet fully explored Brazilian environments such as the Amazon and Atlantic rain forests and mangrove forests is also discussed.

**Keywords**

Endophytic fungi · Plant growth hormone · Biological control · Brazilian forest · Mangroves

## 2.1 Introduction

Endophytic microorganisms are defined by Carroll (1986) as asymptomatic microorganisms living inside plants whereas Petrini (1991) defined them as microorganisms that inhabit plant's inner tissues at least for one period of their life cycle, without causing any apparent harm to the host. Other definitions were also proposed to describe endophytic microorganisms (Wilson 1995; Hallmann et al. 1997). Using molecular techniques, it has been shown that bacteria and fungi that are not culturable on standard media and under normal condition can also be found inside plants. Thus, a modified version of the previous definitions was proposed by Azevedo and

J. L. Azevedo (✉)
Faculty of Agriculture (ESALQ), Department of Genetics, University of São Paulo, São Paulo, Brazil
e-mail: jlazevedo@usp.br

State University of Maringá, Maringá, Brazil

R. N. Kharwar et al. (eds.), *Microbial Diversity and Biotechnology in Food Security*,
DOI 10.1007/978-81-322-1801-2_2, © Springer India 2014

Araujo (2007) that considers that endophytes are all microorganisms that are culturable or not, inhabit the interior of plant tissues, cause no harm to their hosts, and do not develop external structures. This definition was modified by Mendes and Azevedo (2007) dividing the endophytes in two groups, the first group (group I) does not generate external structures from the host and the second group (group II) develops external structures from the host plant and includes symbiotic nitrogen-fixing bacteria and mycorrhizal fungi.

Endophytic microorganisms have important roles in providing protection to the plant host by acting against predators and pathogens including cattle and insect pests (Azevedo et al. 2000). Endophytes may also increase the resistance of plants against biotic and abiotic stresses and produce plant growth hormones, antibiotics, enzymes, and many other compounds of biotechnological interest.

Endophytic microorganisms have been isolated from practically all plants studied to date. They were found in plants growing in different environmental settings including forests, mangrove swamps, pastures, agricultural fields, etc. The first studies with endophytes were conducted with host plants from temperate regions; however, more recent research was dedicated to the endophytic fungi and bacteria inhabiting the plants from tropical countries such as Brazil and India (Pereira et al. 1993; Rodrigues 1994; Suryanarayanan and Vijaykrishina 2001; Mishra et al. 2012); indeed, examples of endophytic species of fungi from tropical plant hosts were recently reviewed (Azevedo and Araujo 2007). Endophytic microorganisms have been used for the biological control of insect pests and plant diseases and production of vitamins, enzymes, antibiotics, and anticancer drugs. In Brazil, several laboratories are engaged in research for the isolation of and using fungi and bacteria for biotechnological, applications. In this chapter, selected data obtained in Brazil are presented, mainly with regard to fungal endophytes and their useful roles in agriculture.

## 2.2 Endophytes Versus Pathogenic Microorganisms Isolated from Cultivated Plants

The endophytes isolated in our laboratory (Faculty of Agriculture, ESALQ, University of São Paulo, Brazil) from citrus plants indicated that the isolates from the genus *Guinardia* were morphologically very similar to the pathogenic *Guignardia citricarpa* (anamorph *Phyllosticta citricarpa*). *G. citricarpa* causes citrus black spot disease and this pathogenic fungus is subjected to phytosanitary legislation in the European Union and the USA. The distinction between the pathogenic and endophytic *Guignardia* isolated from citrus was investigated by amplified fragment length polymorphism (AFLP) analysis. The results have also shown that the pathogenic and endophytic isolates from citrus were similar though with slight differences, regarding the conidia sheaths and colony color on oat meal agar. The molecular analysis allowed the classification of one pathogenic group of isolates as *G. citricarpa* and the endophytes as *Guignardia mangiferae* (*Phyllosticta capitalensis*). The endophyte *G. mangiferae* occurs in the European Union and the USA on many host species including citrus and does not cause symptoms of citrus black spot, justifying its exclusion from quarantine measures (Baayen et al. 2002). The molecular differences allowed the construction of DNA primers for use in a diagnostic kit to distinguish the pathogenic from the endophytic species of *Guignardia*. To avoid spreading of the pathogen, this kit was used to safely export the healthy citrus from Brazil to other countries. More recently, new and efficient primers for the diagnosis of citrus black spot were developed by Stringari et al. (2009). The authors cloned exclusive random amplified polymorphic DNA (RAPD) markers of *G. citricarpa* that were used to obtain "sequence-characterized amplified regions " (SCARS) that allowed the development of specific primers for the identification of pathogenic strains. In addition, Romao et al. (2011) showed that *G. citricarpa* produces great-

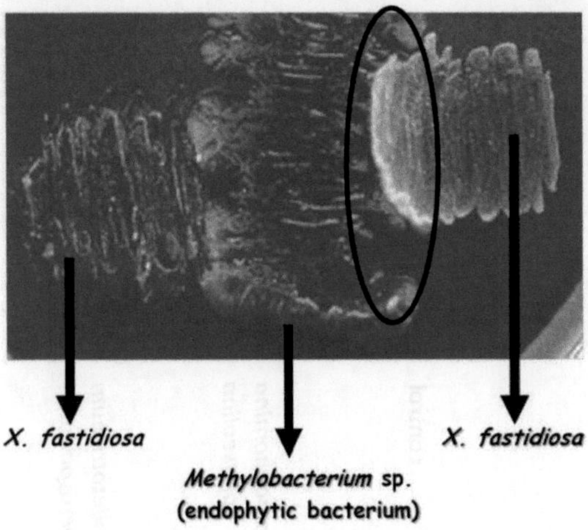


**Fig. 2.1** Increasing growth of *Xylella fastidiosa* in the presence of the endophyte *Methylobacterium*. (Lacava 2000)

er amounts of certain enzymes such as amylases, endoglucanases, and pectinases compared to *G. mangiferae,* suggesting that these enzymes could be the key in the development of citrus black spot, mainly pectin lyases, which makes the pathogenic strains more effective for pectin degradation. Although not involving fungi, we also studied bacterial endophytes from healthy plants and plants with symptoms of citrus variegated chlorosis (CVC), a disease caused by the bacterium *Xylella fastidiosa*. We observed a relationship between the symptoms of CVC and the frequency of isolation of species from the genus *Methylobacterium*, which were frequently isolated from symptomatic plants. In contrast, *Curtobacterium flaccumfaciens* was more frequently isolated from asymptomatic plants (Araujo et al. 2001). This and other findings (Araujo et al. 2002) permitted us to conclude that *X. fastidiosa* could in fact be an endophyte that with the assistance of certain *Methylobacterium* isolates, changes its state from endophytic to pathogenic and returns to the endophytic state via interference of *Curtobacterium* endophytic isolates (Fig. 2.1).

## 2.3 Endophytes and Plant Growth Hormone Production

Endophytic microorganisms affect plant growth directly or indirectly and can provide the hosts with compounds that are produced by the fungi for facilitating the uptake of nutrients from the environment. Endophytes can also act by decreasing or preventing the deleterious effect of pathogens. Varma et al. (1999) demonstrated that the fungus *Piriformospora indica* increases the growth of various hosts suggesting that it may be useful for the promotion of plant growth.

Experiments conducted at our university have shown that two varieties of soybean (*Glycine max*) are colonized by several genera of endophytic fungi (Mendes et al. 2001; Mendes and Azevedo 2007). The endophytic fungi were isolated from the leaf, stem, and root tissues and some of the endophytic fungi were able to increase the growth of plantlets (Fig. 2.2). Similar results were obtained by Pimentel et al. (2006) and by Kuklinsky–Sobral et al. (2004) using soybean seeds treated with endophytic bacteria and fungi.

Some endophytes isolated from *Eucalyptus* were also able to promote the growth of seedlings thereby preventing diseases in the early stages of plant development (Procopio 2004). The inoculation of endophytes in *Eucalyptus* is being successfully employed, thus promoting better growth of the plantlets. Indeed, such endophytes have been used by cellulose and paper companies in Brazil to increase the viability and growth of *Eucalyptus* plantlets.

## 2.4 Endophytes and Biological Control of Plant Pathogens, Insects, and Ticks

Endophytic microorganisms colonize an ecological niche similar to that of phytopathogens which might favor endophytes as candidates for use as biocontrol agents. Several studies demonstrate the ability of endophytes to control pests and

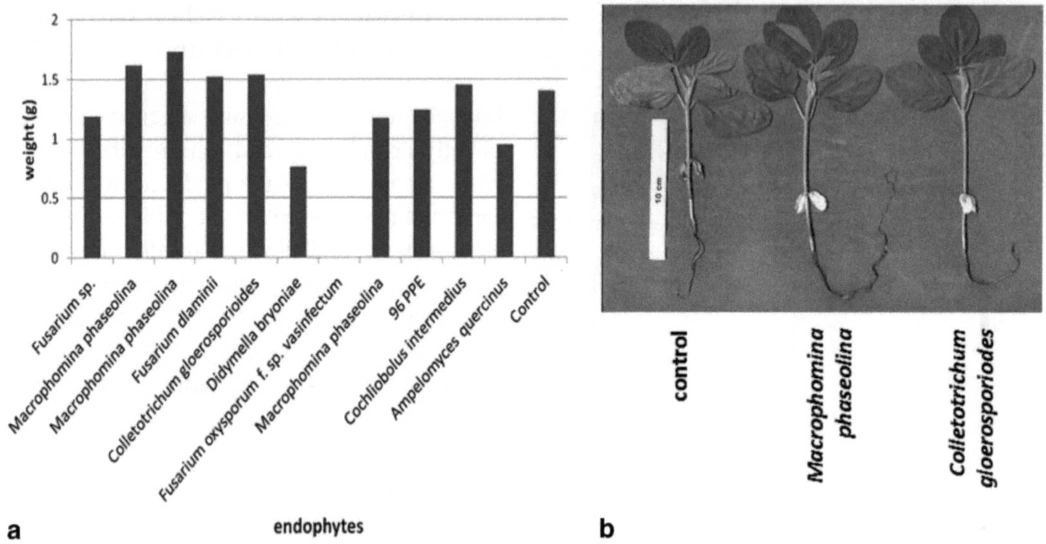


**Fig. 2.2** **a** Wet weight (in g) using several endophytic fungi from soybean. **b** Photo showing growth promotion of soybean (*Glycine max*) plantlets that were treated with two endophytic fungi. (Mendes and Azevedo 2007)

diseases vectors (Carroll 1986; Azevedo et al. 2000). In Brazil, the *Basidiomycete Moniliophtora perniciosa,* the causal agent of witches' broom disease of cacao (*Theobroma cacao*) is one of the main limiting factors for cacao production and is considered the most important pathogen of this crop (Griffith and Hedger 1994). The endophytic fungal communities of infected and healthy Brazilian *T. cacao* plants were isolated and evaluated both in vitro and in vivo by their ability to inhibit *M. perniciosa*. Among these isolates, some were identified as potential antagonists and the fungus *Gliocladium catenulatum,* reduced the incidence of the disease in cacao seedlings to 70 % (Rubini et al. 2005). Among the isolated fungi from cacao, *M. perniciosa* was found colonizing healthy parenchymatic tissues showing for the first time that this fungus may also behave as an endophyte (Lana et al. 2011).

Another important disease that occurs in Brazil is the leaf anthracnose of guarana (*Paullinia cupana*) caused by *Colletotrichum gloeosporioides* and related species of this genus. This crop is an important Amazon plant used in the production of soft drinks and several medicinal products that are used in Brazil and exported to other countries mainly Europe and the USA. This crop is cultivated by small farmers in the Maués region of the Amazon, and the disease is causing severe economic and social losses. In collaboration with a research group from the Federal University of Amazonas, we isolated fungi and bacteria from the host plant. The initial results (data not yet published) indicate that some endophytic fungi and bacteria have the potential to control the pathogen.

It is known that some entomopathogenic fungi behave as endophytes. Bing and Lewis (1991, 1992) isolated *Beauveria bassiana* from maize (*Zea mays*) and the fungus was used to control the European corn borer (*Ostrinia nubilalis*). In Brazil, some laboratories isolated endophytic fungi that are known as insect and nematode controllers; these isolates were obtained from plant hosts including, among others, sugar cane (*Saccharum* sp.), maize, and soybean (Pimentel 2001; Pimentel et al. 2006; Stuart et al. 2010). Some of the *Beauveria* strains isolated from maize were used against the insect pest *Spodoptera frugiperda* and the results showed that these endophytes are as good or even better biocontrol agents than the commercial entomopathogenic strains used in Brazil. These endophytic *Beauveria,* belonging to the *Beauveria bassiana* species and *Beauveria amorpha*, are also able to control the bovine tick *Rhipicephalus microplus*, an ectoparasite that causes substantial economic losses due to the reduced productivity caused by anemia, toxicity, and the transmission of various diseases to their hosts. The *Beauveria* strains were effective in laboratory bioassays and under field test conditions. (Campos et al. 2010) It was also shown that both chitin and tick cuticle, induced fungal chitinase production. A scanning electron microscopy (SEM) analysis of the endophytic *Beauveria* infecting *R. microplus* showed appressorium formation during the penetration on cattle tick's cuticle (Campos et al. 2005).

## 2.5 Endophytic Fungi Isolated from Not Yet Fully Explored Environments Such as Brazilian Forests and Mangroves

Several laboratories in Brazil have been searching for endophytes in plants of not yet fully explored environments such as mangrove plants and plants from the Amazon and Atlantic rain forests. Recently some endophytic fungi producing apparently new antimicrobials were isolated from mangroves (Sebastianes et al. 2012). Additionally, some endophytes from petrol-contaminated mangroves were found to be able to reduce oil contamination. Endophytes from host plants growing in the Brazilian Amazon region and Atlantic forest were also isolated (Cassa-Barbosa 2001; Costa-Neto 2002; Souza et al. 2004), and more recently not yet published results demonstrate the biotechnological potential of endophytes from unexplored Brazilian plant hosts and they may result in new valuable products for agricultural, medical, and other applications.

## 2.6 Final Considerations

Brazil is one of the few countries in the world that still retains a large animal, plant, and microbial diversity. Approximately 20% from 300,000 plant species in our planet are found in Brazil. Endophytes inhabiting these hosts are poorly studied and, as source of important compounds of biotechnological value, remain to be discovered. A major problem is the rapid reduction of forests and mangroves in tropical areas of Brazil, a situation that could result in the extinction of many fungi and other endophytic microorganisms with the loss of potentially important products for use in agricultural, pharmaceutical, environmental, and other fields of interest.

## References

Araujo WL, Maccheroni W Jr, Aguilar-Vildoso CI, Barroso PAV, Saridakis HO, Azevedo JL (2001) Variability and interactions between endophytic bacteria and fungi isolated from leaf tissues of citrus rootstocks. Can J Microbiol 47:229–236

Araujo WL, Marcon J, Maccheroni W Jr, van Elsas JD, van Vuurde JWL, Azevedo JL (2002) Diversity of endophytic bacterial populations and their interactions with *Xylella fastidiosa* in Citrus plants. Appl Environ Microbiol 10:4906–4914

Azevedo JL, Araujo WL (2007) Diversity and applications of endophytic fungi isolated from tropical plants. In: Ganguli BN, Deshmukh SK (eds) Fungi multifaceted microbes pp. 189–207 Anamaya, New Delhi

Azevedo JL, Maccheroni W Jr, Pereira, JO, Araujo WL (2000) Endophytic microorganisms: a review on insect control and recent advances on tropical plants. Elect J Biotechnol 3:40–65

Baayen RP, Bonants PJM, Verkley G, Carroll GC, Van Der Aa HA, De Weerdt M, Van Brouwershaven IR, Schutte GC, Maccheroni Jr W, Glienke-Blanco C, Azevedo JL (2002) Nonpathogenic isolates of the Citrus Black Spot fungus, *Guignardia citricarpa* identified as a cosmopolitan endophyte of woody plants. *G mangiferae* (*Phyllosticta capitalensis*) Phytopathol 92:464–477

Bing LA, Lewis LC (1991) Suppression of *Ostrinia nubilalis* (Hubner) (Lepidoptera: Pyralidae) by endophytic *Beauveria bassiana* (Balsamo) Vuillemin. Environ Entomol 20:1207–1211

Bing, LA, Lewis LC (1992) Temporal relantionships between *Zea mays, Ostrinia nubilalis* (Hubner) (Lep: Pyralidae) and endophytic *Beauveria bassiana* (Balsamo) Vuillemin. Entomophaga 37:525–536

Campos RA, Arruda W, Boldo JT, Silva MV, Barros NM, Azevedo JL, Schrank A, Vainstein MH (2005) *Boophilus microplus* infection by *Beauveria amorpha* and *Beauveria bassiana*: SEM analysis and regulation of subtilisin-like proteases and chitinases. Current Microbiol 50:257–261

Campos RA, Boldo JT, Pimentel IC, Dalfovo V, Araujo WL, Azevedo JL, Vainstein MH, Barros NM (2010) Endophytic and entomopathogenic strains of *Beauveria* sp to control the bovine tick *Rhipicephalus* (*Boophilus*) *microplus*. Genet Mol Res 9:1421–1430

Carroll G (1986) Fungal associates of woody plants as insect antagonists in leaves and stems. In: Barbosa P, Krischik VA, Jones, CG (eds) Microbial mediation of plant herbivore interactions. Wiley, New York, pp 253–271

Cassa-Barbosa LA (2001) Enzimas de interesse biotecnologico produzidas por fungos endofiticos de *Copaiba multijuga*. Dissertation, Universidade de Brasilia, Brazil, 110 p

Costa-Neto PQ (2002) Isolamento e identificação de fungos endofiticos da pupunha (*Bactris gasipae*) e caracterização por marcadores moleculares. Dissertation, Federal University of São Carlos, Brazil 86 p

Griffith GW, Hedger JN (1994) The breeding biology of biotypes of the witches' broom pathogen of cacao, *Crinipellis perniciosa*. Heredity 72:278–289

Hallmann J, Quadt-Hallman A, Mahaffee WF, Kloepper JW (1997) Bacterial endophytes in agriculture crops. Can J Microbiol 43:895–914

Kuklinsky-Sobral J, Araujo, WL, Mendes R, Geraldi IO, Pizzirani-Kleiner AA, Azevedo JL (2004) Isolation and characterization of soybean-associated bacteria and their potential for plant growth promotion. Environ Microbiol 6:1244–1251

Lacava PT (2000) Isolamento e caracterização genetica por RAPD e resistencia a antibioticos em *Xylella fastidiosa*. Dissertation, University of Sao Paulo Brazil, 108 p

Lana TG, Azevedo JL, Pomella AWV, Monteiro RTR, Silva CB, Araujo WL (2011) Endophytic and pathogenic isolates of the cacao fungal pathogen *Moniliophthora perniciosa* (Tricholomataceae) are indistinguishable based on genetic and physiological analysis. Genet Mol Res 10:326–334

Mendes R, Kuklinsky-Sobral J, Geraldi IO, AraujoWL, Azevedo JL, Pizzirani-Kleiner AA (2001) Monitoring soybean endophytic fungal community associated with glyphosate. In: XXIV Annual Congress of Microbiology. Brazilian Society of Microbiology, Foz do Iguassu, p 242

Mendes R, Azevedo JL (2007) Valor biotecnológico de fungos endofíticos isolados de plantas de interesse economico. In: Costa-Maia L, Malosso E, Yano-Melo AM (eds) Micologia:avanços no conhecimento. Brazilian Society Microbiology Publ., Recife, pp 129–140

Mishra A, Gond SK, Kumar A, Sharma VK, Verma SK, Kharwar RN, Sieber TN (2012) Season and tissue type affect fungal endophyte communities of the Indian medicinal plant *Tinospora cordifolia* more strongly than geographic location and their antimicrobial potential. Microb Ecol 64:388–398

Pereira JO, Azevedo JL, Petrini O (1993) Endophytic fungi of *Stylosanthes*: a first report. Mycologia 85:362–364

Petrini O (1991) Fungal endophytes of tree leaves In: Andrews J, Hirano SS (eds) Microbial ecology of leaves. Springer, New York, pp 179–197

Pimentel IC (2001) Fungos endofiticos de milho (*Zea mays* L.) e de soja (*Glycine max* L.) Merril e seu potencial valor biotecnologico no controle de pragas agrícolas. PhD Thesis, Universidade Federal do Paraná, Curitiba, Brazil 154 p

Pimentel IC, Glienke-Blanco C, Gabardo J, Stuart RM, Azevedo JL (2006) Identification and colonization of endophytic fungi from soybean (*Glycine max* L.) Merril under different environmental conditions. Brazilian Arch Biol Technol 49:705–711

Procopio REL (2004) Diversidade bacteriana endofítica de *Eucaliptus* spp e avaliação do seu potencial biotecnologico. PhD Thesis, University of São Paulo, Brazil 101 p

Rodrigues KF (1994) The foliar fungal endophytes of the Amazonian palm *Euterpe oleraceae*. Mycologia 86:376–385

Romao AS, Sposito MB, Andreote FD, Azevedo JL, Araujo WL (2011) Enzymatic differences between the endophyte *Guignardia mangiferae* (Botryosphaeriaceae) and the citrus pathogen *G. citricarpa*. Genet Mol Res 10:243–252

Rubini MR, Silva-Ribeiro R, Pomella AWV., Maki C, Araújo WL, Santos DR, Azevedo JL (2005) Diversity of endophytic fungal community of cacao (*Theobroma cacao*) L. and biological control of *Crinipellis perniciosa* causal agent of Witches' broom disease Int J Biol Sci 1:24–33

Sebastianes FLS, Cabelo N, El Aouade N, Valente AMMP, Lacava PT, Azevedo JL, Pizzirani-Kleiner AA, Cortes D (2012) 3-Hydroxypropionic acid as an antibacterial agent from endophytic fungi *Diaporthe phaseolorum*. Curr Microbiol 65:622–632

Souza AQL, Souza ADL, Astolfi-Filho S, Belém-Pinheiro L, Sarkis MM, Pereira JO (2004) Atividade antimicrobiana dse fungos endofíticos isolados de plantas tóxicas da Amazônia, *Paulicorea longiflora* and *Strychnus cogens*. Acta Amazônica 34:185–195

Stringari D, Glienke C, Christo D, Maccheroni W Jr, Azevedo JL (2009) High molecular diversity of the fungus *Guignardia citricarpa* and *Guignardia mangiferae* and new primers for the diagnosis of the Citrus Black Spot. Brazilian Arch Biol Technol 52:1063–1073

Stuart RM, Romao AS, Pizzirani-Kleiner AA, Azevedo JL, Araujo WL (2010) Culturable endophytic filamentous fungi from leaves of transgenic imidazolinone-tolerant sugarcane and its non-transgenic isolines. Arch Microbiol 192:307–313

Suryanarayanan TS, Vijaykrishina D (2001) Fungal endophytes of aerial roots of *Ficus benghalensis*. Fungal Divers 8:155–161

Varma A, Verma S, Sudha, Sahay N, Butehorn B, Franken P (1999) *Piriformospora indica*, a cultivable plant-growth-promoting root endophyte. Appl Environ Microbiol 65:2741–2744

Wilson D (1995) Endophyte: the evolution of a term and clarification of its use and definitions. Oikos 73:274–276

# 3

# Diversity and Biopotential of Endophytic Fungal Flora Isolated from Eight Medicinal Plants of Uttar Pradesh, India

R. N. Kharwar, Ashish Mishra, Vijay K. Sharma, S. K. Gond, S. K. Verma, A. Kumar, Jitendra Kumar, D. K. Singh and J. Goutam

**Abstract**

Endophytic fungi are hidden diversity mines of microbes that reside in the healthy and symptomless interior of plant tissues without causing any harmful effects. This chapter focuses on fungal endophytic diversity of eight medicinal plants of Uttar Pradesh, India with their biopotential ability. Total of 4,002 (38.38 % CF) endophytic isolates were recovered from 10,425 segments representing 131 endophytic fungal species belonging to different fungal classes. Out of 4,002 isolates, hyphomycetes were more pronounced with 71.43 % recovery followed by coelomycetes 16.61 %, ascomycetes 6.59 %, mycelia sterilia or unidentified 5.32 % and least from zygomycetes 0.020 %. Among total endophytic fungal species isolated, *Cladosporium cladosporioides* (3.39 % CF) was found to be the most dominated taxa followed by *Alternaria alternata* (2.35 % CF), *Curvularia lunata* (2.13 % CF), *Aspergillus niger* (1.95 % CF), *Chaetomium globosum* (1.85 % CF), *Nigrospora oryzae* (1.57 % CF) and *Phoma glomerata* (1.09 % CF). From a total of 131 endophytic species, 101 were tested for their antimicrobial and antioxidant activity. Out of 47 active species, 29.78 % displayed antibacterial activity, 27.65 % showed antifungal activity, 38.29 % exhibited antibacterial and antifungal activity both while only 4.25 % displayed antimalarial as well as antioxidant activity. Twenty-one endophytic fungal species were tested for extracellular production of amylase, xylanase and phosphate solubilization where 76.19 % found to produce amylase, 23.80 % for xylanase and 14.28 % exhibited phosphate-solubilization activity.

R. N. Kharwar (✉) · A. Mishra · V. K. Sharma · J. Kumar · D. K. Singh · J. Goutam
Mycopathology and Microbial Technology Laboratory, Centre of Advanced Study in Botany, Banaras Hindu University, Varanasi, India
e-mail: rnkharwar@gmail.com

S. K. Gond · S. K. Verma
Department of Botany, Visva-Bharati University, Shantiniketan 731235, India

A. Kumar
Department of Botany, Buddha PG College, Kushinagar, India

R. N. Kharwar et al. (eds.), *Microbial Diversity and Biotechnology in Food Security*,
DOI 10.1007/978-81-322-1801-2_3, 

**Keywords**

Antimicrobial · Diversity · Enzyme activity · Fungal endophytes · Medicinal plants

## 3.1 Introduction

Microorganisms are an important component of the environment, they affect their surroundings in various ways and forms, one of them are endophytes. The term endophyte was first introduced by de Bary in 1866 for all those microbes that reside inside the living healthy tissues. Many workers define endophytes in various ways, but the definition given by Bacon and White (2000) was perhaps most acceptable as 'microbes that colonize living, internal tissues of plants without causing any immediate and overt negative symptoms'. This is a topographical term and includes bacteria, fungi, actinomycetes and algae, which spend their whole or a period of life cycle either in symplast or apoplast region of healthy plant tissues without producing any disease or clinical symptoms. On the basis of their nature, endophytes may be categorized in three groups: (1) pathogens of another host that are nonpathogenic in their endophytic relationship, (2) nonpathogenic microbes, (3) pathogens that have been rendered nonpathogenic but are still capable of colonization by selection methods or genetic alteration (Backman and Sikora 2008). Among all the endophytic microbes, fungi are the most studied group so far. Endophytic fungi play an important role in plant/host community health by providing resistance from herbivores (Brem and Leuchtmann 2001), pathogenic fungi, bacteria, viruses, insects, nematodes (Gond et al. 2010), illness (Clay 1990), reduced seed production (Rice et al. 1990), temperature and salinity (Redman et al . 2002) and also against drought and minerals (Malinowski et al. 1997), heavy metal (Li et al. 2012). Endophytic fungi are also able to produce a considerable number of useful enzymes and this ability can make enzymes cost effective because approximately 60 % of the currently used industrial enzymes are of fungal origin (Østergaard and Olsen 2010). Interestingly, Suryanarayanan and his colleagues observed the number of foliar fungal endophytes associated with trees of forests in the Western Ghats mountain (in Southern India) produced a range of extracellular enzymes including amlyases, cellulases, chitinases, chitosanases, laccases, lipases, pectinases and proteases (Suryanarayanan et al. 2012).

An irrational and irregular use of antibiotics makes pathogen more resistant and it is a serious impediment for microbiologists providing the required demand of antibiotics. To cope with this problem, there is ultimate necessity for an alternative and novel source of effective drugs without destroying biodiversity. In such respect, endophytic fungi became an effective solution because one can isolate the compound of plant/host origin without destroying the plant population. After the discovery of taxol (billion dollar drug) from the endophytic fungus *Taxomyces andreanae* (Stierle et al. 1993), it proved itself as a novel source of taxol production without loss of the *Taxus* plant. After this discovery, the endophytic research came to light and microbes have been considered as a novel and alternative source for new biologically active compounds and/or compounds of host origin such as taxol (Stierle et al. 1993), vincristin (Tung et al. 2002), camptothecin (Shweta et al. 2010), piperin (Verma et al. 2011), azadirachtin (Kusari et al. 2012), etc. Today, credits go to endophytic microbes for producing a number of new and effective bioactive natural compounds that can be used in agriculture, medicine and industry. In addition, more than 100 anticancer compounds have been (57 % novel and 43 % known) isolated only from endophytic fungi (Kharwar et al. 2011). In this chapter, we have focused mainly on the diversity of endophytic fungi of eight medicinal (*Azadirachta indica, Agele marmelos, Catharanthus roseus, Eucalyptus citriodora, Nyctanthes arbor-tristis, Adenocalymma alliaceum, Tinospora cordifolia, Cinnamomum camphora*) plants of Uttar Pradesh, India, with their antimicrobial potential.

## 3.2 Transmission of Endophytic Fungi

Transmission describes the spreading of microbes within and among host population. Endophytic fungi have two transmission modes, vertical and horizontal. Vertical transmission occurs when fungi travel from host to their offspring via host tissues such as host seeds and vegetative propagules. Systemically infected endophytic fungi have vertical transmission mode that differs from horizontal transmission where fungus travels by its sexual or asexual spores.

## 3.3 Ecology and Biodiversity of Endophytic Fungi

Endophytic fungi are important, hidden, highly diverse, less exploited and highly potential component of the environment. Almost all plant species studied to date for endophytic diversity were found to act as a reservoir for potential of microbes to be used to resolve the problems of mankind. The endophytes were observed in all green biota ranging from algae (Yang et al. 2006), bryophytes (Chambers et al. 1999), pteridophytes (Schmid and Oberwinkler 1995), gymnosperms (Huang and Wang 2011) and to angiosperms (Mishra et al. 2012a), including underground root to all aerial parts of host (Kharwar et al. 2008). Endophytic fungi isolated from water-stressed deserts (Bashyal et al. 2005), cold-stressed arctic (Fisher et al. 1995), Antarctic ocean (Rosa et al. 2009; Wang et al. 2006), geothermal soils (Redman et al. 2002), highly diverse rain forests (Strobel 2002), dry deciduous and coastal forests (Suryanarayanan et al. 2003) and mangrove swamps (Maria et al. 2005). Fungal endophytes were isolated from either all or specific organs of selected hosts showing the impacts of environmental variables on their colonization frequency (CF), diversity and antimicrobial activity (Hyde and Soytong 2008; Mishra et al. 2012a; Verma et al. 2011; Verma et al. 2013). Our earth harbours almost 300,000 higher plants species, and each species represents either one or plethora of endophytic community which is well proved by the previous reports of higher plants fungal endophytes (Strobel 2002). Out of these plants that exist on the earth, only a few dozen, have been studied related to their endophytic biology, and every plant studied has an endophytic community. Including fungal endophytes, the ratio of fungal to plant species will reach up to 33:1 from 6:1 (Hawksworth and Rossman 1987).

### 3.3.1 Endomyco Diversity in *Adenocalymma alliaceum* Miers

*A. alliaceum,* is commonly known as garlic creeper or *lahsun lata* plant. It is a member of the family Bignoniaceae, a highly medicinal, evergreen tropical shrubby vine plant that is native to the Amazon rainforest. In the absence of garlic, its leaf can be used as a substitute for cooking purposes. Every part of the plant is well used by the indigenous people of the Amazon as folk medicine for curing various disorders. Despite having several compounds, it is considered analgesic, anti-inflammatory, depurative, purgative and widely used against arthritis, rheumatism, body aches, muscle pain, cholesterol and injuries. Its leaves are also used to cure flu, pneumonia, cough, fever and headache. Kharwar et al. (2011) reported the isolation of total 149 fungal endophytic isolates belonging to 17 fungal taxa from 270 segments of leaf, stem and petiole (90 segments of each tissue). Collectively, among the total isolates recovered, hyphomycetes were more frequent (74.47%) followed by mycelia sterilia (10.07%), ascomycetes (8.05%) and coelomycetes (4.03%) (Table 3.1, Fig. 3.1). Among all tissues studied, leaves showed greater colonization of endomycobiota (72.22%) compared to stem (67.78%) and petiole (25.54%). *A. alternata* (6.30%), *A. niger* (5.93%), *Stenella agalis* (5.20%), *Fusarium oxysporum* (5.18%), *C. lunata* (4.18%) and *Fusarium roseum* (4.07%) were recovered as the dominant genera. However, *Penicillium* sp. and *Rhizoctonia* sp. were the least frequent with equal CF of 1.85%. Out of 17 taxa, *Penicillium* sp., *C. globosum* and *Rhizoctonia* sp. were only restricted to stem tissue, and as per authors this may be because of displacement of their spores from root and substrate specificity supported by stem.

**Table 3.1** Endophytic fungal diversity among eight different medicinal plants

| Host plants | *A. alliaceum* | *A. marmelos* | *A. indica* | *C. roseus* | *C. camphora* | *E. citriodora* | *N. arbortristis* | *T. cordifolia* | Total | |
|---|---|---|---|---|---|---|---|---|---|---|
| Total segments plottedt | 270 | 550 | 600 | 300 | 105 | 600 | 800 | 7200 | 10425 | |
| Endophytic fungi | 17 | 32 | 44 | 19 | 26 | 32 | 34 | 29 | | Total CF |
| *Acremonium acutatum* | | | 34 | | | | | | 34 | 0.329 |
| *Acromonium* sp. | | | 15 | | | | 40 | 6 | 61 | 0.585 |
| *Alternaria alternata* | 17 | 29 | 22 | 16 | | 25 | 59 | 78 | 246 | 2.35 |
| *Alternaria chlamydospora* | | | 12 | | | | | | 12 | 0.115 |
| *Alternaria cinerariae* | | | | | 10 | | | | 10 | 0.095 |
| *Alternaria dennsii* | | | 19 | | | | | | 19 | 0.182 |
| *Alternaria longipes* | | | 19 | 4 | | | | | 23 | 0.220 |
| *Alternaria* sp. | 5 | | | | | 6 | | | 11 | 0.105 |
| *Arthrinium* sp. | | | | | 1 | | | | 1 | 0.009 |
| *Arthrobotrys* sp. | | | | | 6 | | | | 6 | 0.057 |
| *Aspergillus flavus* | | | 18 | | | | | 52 | 70 | 0.671 |
| *Aspergillus fumigatus* | 8 | 14 | 2 | 10 | 7 | 19 | 20 | | 80 | 0.767 |
| *Aspergillus niger* | 16 | 29 | 38 | 10 | 17 | 8 | 25 | 61 | 204 | 1.956 |
| *Aspergillus oryzae* | | | 7 | | | | | | 7 | 0.067 |
| *Aspergillus sydowii* | | | | | | | | 36 | 36 | 0.345 |
| *Aspergillus terreus* | | | | 6 | | 10 | | 28 | 44 | 0.422 |
| *Aspergillus tubingensis* | | | | | | | | 20 | 20 | 0.191 |
| *Aureobasidium pullulans* | | | | | | 22 | 24 | | 46 | 0.441 |
| *Aureobasidium* sp. | | 63 | | | | | | | 63 | 0.604 |
| *Basidiobotrys* sp. | | | | | | 8 | | | 8 | 0.076 |
| *Bipolaris* sp. | | | | 12 | | | | | 12 | 0.115 |
| *Botryosphaeria rhodina* | | | | | | | | 38 | 38 | 0.364 |
| *Botrytis cinerea* | | | | | | 21 | | | 21 | 0.201 |
| *Botrytis* sp. | | | | | | | | 16 | 16 | 0.153 |
| *Cercinella mucoroides* | | | 1 | | | | | | 1 | 0.009 |
| *Chaetomium crispatum* | | | 2 | | | | | | 2 | 0.019 |
| *Chaetomium globosum* | 6 | 37 | 9 | 3 | | 20 | 25 | 93 | 193 | 1.85 |
| *Chaetomium* sp. | 6 | | | | 3 | | | | 9 | 0.086 |
| *Chaetophoma* sp. | | | | | 1 | | | | 1 | 0.009 |
| *Chloridium virescenc* | | | 1 | 9 | | | | | 10 | 0.095 |
| *Cladosporiella* sp. | | | | | | 7 | | | 7 | 0.067 |
| *Cladosporium acaciicola* | | | 5 | | | | | | 5 | 0.047 |
| *Cladosporium apicale* | | | | | | | | 16 | 16 | 0.153 |
| *Cladosporium cladosporioides* | | 53 | 39 | 16 | | 65 | 94 | 87 | 354 | 3.39 |
| *Cladosporium* sp. | | | | | | | 13 | | 13 | 0.124 |
| *Cladosporium tennuissimum* | | | | | 8 | | | | 8 | 0.076 |
| *Colletotrichum crassipes* | | | | | | | | 27 | 27 | 0.258 |
| *Colletotrichum dematium* | | 12 | | | | | 33 | 77 | 122 | 1.17 |
| *Colletotrichum gloeosporioides* | | 18 | | | | 10 | | | 28 | 0.268 |
| *Colletotrichum linicola* | | | | | | | | 58 | 58 | 0.556 |
| *Colletotrichum* sp. | | | 3 | 6 | | | | | 9 | 0.086 |
| *Corynespora* sp. | | 1 | | | | | 13 | | 14 | 0.134 |
| *Curvularia catanulata* | | | 3 | | | | | | 3 | 0.028 |
| *Curvularia fallax* | | | | | | | 3 | | 3 | 0.028 |

**Table 3.1** (continued)

| Host plants | *A. allia-ceum* | *A. marme-los* | *A. indica* | *C. roseus* | *C. cam-phora* | *E. citrio-dora* | *N. arbor-tristis* | *T. cordi-folia* | Total | |
|---|---|---|---|---|---|---|---|---|---|---|
| Total segments plottedt | 270 | 550 | 600 | 300 | 105 | 600 | 800 | 7200 | 10425 | |
| Endophytic fungi | 17 | 32 | 44 | 19 | 26 | 32 | 34 | 29 | | Total CF |
| *Curvularia intermedia* | | | | | | | | 31 | 31 | 0.297 |
| *Curvularia lunata* | 13 | 32 | 21 | 5 | 1 | 39 | 56 | 56 | 223 | 2.13 |
| *Curvularia oryzae* | | | | | | | 17 | | 17 | 0.163 |
| *Diatrype* sp. | | | | | | | 14 | | 14 | 0.134 |
| *Drechslera* sp. | | | 1 | 11 | 4 | | | | 16 | 0.153 |
| *Drechslera ellisii* | | 12 | | | | | 25 | | 37 | 0.354 |
| *Drechslera graminea* | | | | | | | | 25 | 25 | 0.239 |
| *Drechslera rostrata* | | | 2 | | | 15 | | | 17 | 0.163 |
| *Emericella nidulans* | | | | | | | | 31 | 31 | 0.297 |
| *Emericella* sp. | | 3 | | | | | | | 3 | 0.028 |
| *Fusarium chlaydosporum* | | | 12 | | | | | | 12 | 0.115 |
| *Fusarium moniliformae* | | | 4 | 3 | | | | | 7 | 0.067 |
| *Fusarium oxysporum* | 14 | 4 | 21 | | | 8 | 8 | 35 | 90 | 0.863 |
| *Fusarium roseum* | 11 | 8 | | 15 | | | | | 34 | 0.326 |
| *Fusarium soloni* | | | 3 | | | | | | 3 | 0.028 |
| *Fusarium* sp. | | | 9 | | | | | | 9 | 0.086 |
| *Gliomastix* sp. | | | 1 | | 7 | | | | 8 | 0.076 |
| *Guignardia* sp. | | | | | | | | 6 | 6 | 0.057 |
| *Helicosporum* sp. | | | | | | | 4 | | 4 | 0.038 |
| *Humicola grisea* | | | 2 | | 6 | 7 | 19 | | 34 | 0.326 |
| *Humicola* sp. | | 14 | | 9 | | 20 | | 23 | 66 | 0.633 |
| *Macrophoma* sp. | | | | | | | 16 | | 16 | 0.153 |
| *Melanconium* sp. | | | | | | 20 | 37 | | 57 | 0.546 |
| *Monilia* sp. | | | | | | | | 10 | 10 | 0.095 |
| Morphospecies 1 | | | | | 6 | | | | 6 | 0.057 |
| Morphospecies 2 | | | | | 5 | | | | 5 | 0.0479 |
| Morphospecies 3 | | | | | 7 | | | | 7 | 0.067 |
| Morphospecies 4 | | | | | 7 | | | | 7 | 0.067 |
| Mycelia sterilia | | 2 | 18 | | | | | | 20 | 0.191 |
| Mycelia sterilia | 4 | | | | | | | | 4 | 0.038 |
| Mycelia sterilia | 6 | | | | | | | | 6 | 0.057 |
| Mycelia sterilia | 5 | | | | | | | | 5 | 0.047 |
| Mycelia sterilia | | 9 | | | | | | | 9 | 0.086 |
| Mycelia sterilia | | 3 | | | | | | | 3 | 0.028 |
| Mycelia sterilia | | 1 | | | | | | | 1 | 0.009 |
| Mycelia sterilia | | 10 | | | | | | | 10 | 0.095 |
| Mycelia sterilia | | | | | | 14 | | | 14 | 0.134 |
| Mycelia sterilia | | | | | | 8 | | | 8 | 0.076 |
| Mycelia sterilia | | | | | | 7 | | | 7 | 0.067 |
| Mycelia sterilia | | | | | | | 7 | | 7 | 0.067 |
| Mycelia sterilia | | | | | | | 4 | | 4 | 0.038 |
| Mycelia sterilia | | | | | | | 35 | | 35 | 0.335 |
| Mycelia sterilia | | | | | | | 12 | | 12 | 0.115 |
| Mycelia sterilia | | | | | | | 7 | | 7 | 0.067 |
| *Nigrospora oryzae* | | 13 | 18 | 6 | 6 | 12 | 39 | 70 | 164 | 1.57 |

**Table 3.1** (continued)

| Host plants | *A. alliaceum* | *A. marmelos* | *A. indica* | *C. roseus* | *C. camphora* | *E. citriodora* | *N. arbortristis* | *T. cordifolia* | Total | |
|---|---|---|---|---|---|---|---|---|---|---|
| Total segments plottedt | 270 | 550 | 600 | 300 | 105 | 600 | 800 | 7200 | 10425 | |
| Endophytic fungi | 17 | 32 | 44 | 19 | 26 | 32 | 34 | 29 | | Total CF |
| *Oidiodendron clamydosporum* | | | | | | 13 | | | 13 | 0.124 |
| *Penicillium citrinum* | | | | 16 | | | | | 16 | 0.153 |
| *Penicillium cristata* | | | 7 | | | 10 | | | 17 | 0.163 |
| *Penicillium crysogenum* | | | | 11 | | | | | 11 | 0.105 |
| *Penicillium* sp. | 5 | 2 | 1 | | 8 | | 9 | | 25 | 0.239 |
| *Penicillium* sp. 1 | | | | | | | | 87 | 87 | 0.834 |
| *Penicillium* sp. 2 | | | | | | | | 58 | 58 | 0.556 |
| *Periconia* sp. | | | 1 | | 5 | | | | 6 | 0.057 |
| *Periconia tirupatiensis* | | | | | | 12 | | | 12 | 0.115 |
| *Pestalotia macrotricha* | | 28 | 2 | | | | | | 30 | 0.287 |
| *Pestalotiopsis* sp. | | | 34 | | 10 | | | | 44 | 0.422 |
| *Pestelotia* sp. | | | | | | 9 | | | 9 | 0.086 |
| *Phacidium* sp. | | | | | 2 | | | | 2 | 0.019 |
| *Phaeotrichoconis* sp. | | | | | | | 7 | | 7 | 0.067 |
| *Phoma eupyrena* | | | 18 | | | | | | 18 | 0.172 |
| *Phoma glomerata* | | 25 | | | | 58 | 31 | | 114 | 1.09 |
| *Phoma herbarum* | | 20 | | | | | | | 20 | 0.191 |
| *Phoma* sp. | | 15 | | | | | | | 15 | 0.143 |
| *Phomopsis helianthi* | | | | | | | 50 | | 50 | 0.479 |
| *Phomopsis oblonga* | | | 27 | | | | | | 27 | 0.258 |
| *Phomopsis* sp. | 6 | 28 | | | 4 | 24 | 19 | | 81 | 0.776 |
| *Phyllosticta minima* | | | 2 | | | | | | 2 | 0.019 |
| *Phyllosticta nobilis* | | | | | 11 | | | | 11 | 0.105 |
| *Pseudofusicoccum violaceum* | | | | | | | | 8 | 8 | 0.076 |
| *Rhizoctonia* sp. | 5 | 16 | | | | | 22 | | 43 | 0.412 |
| *Scytalidium* sp. | | | 2 | | | | 7 | | 9 | 0.086 |
| *Stachybotrys* | | | | | 4 | | 5 | | 9 | 0.086 |
| *Stenella agalis* | 14 | 4 | | | | | | | 18 | 0.172 |
| *Stenella* sp. | | | 2 | | | | | | 2 | 0.019 |
| *Trichoderma harzianum* | | | | | 11 | | | | 11 | 0.105 |
| *Trichoderma* sp. | 8 | | | | | | | | 8 | 0.076 |
| *Trichoderma viride* | | 1 | 27 | | | 23 | | 8 | 59 | 0.565 |
| *Ulocladium chlamydosporum* | | | 7 | | | | | | 7 | 0.067 |
| Mycelia sterilia/unidentified | | 4 | | | | | | | 4 | 0.038 |
| Mycelia sterilia/unidentified | | | | 15 | 5 | | | | 20 | 0.191 |
| Mycelia sterilia/unidentified 1 | | | | | | 11 | | | 11 | 0.105 |
| Mycelia sterilia/unidentified 2 | | | | | | 7 | | | 7 | 0.067 |
| Mycelia sterilia/unidentified 5 | | | | | | 14 | | | 14 | 0.134 |
| *Veronaea musae* | | | | | | | | 10 | 10 | 0.095 |
| *Verticillium albo-atrum* | | | 2 | | | | | | 2 | 0.019 |
| *Verticillium* sp. | | | | | | | | | 1 | 0.009 |
| *Verticillium tenuissimum* | | | 2 | | | | | | 2 | 0.019 |
| Total isolates | 149 | 511 | 495 | 183 | 162 | 552 | 799 | 1151 | 4002 | 38.38 |

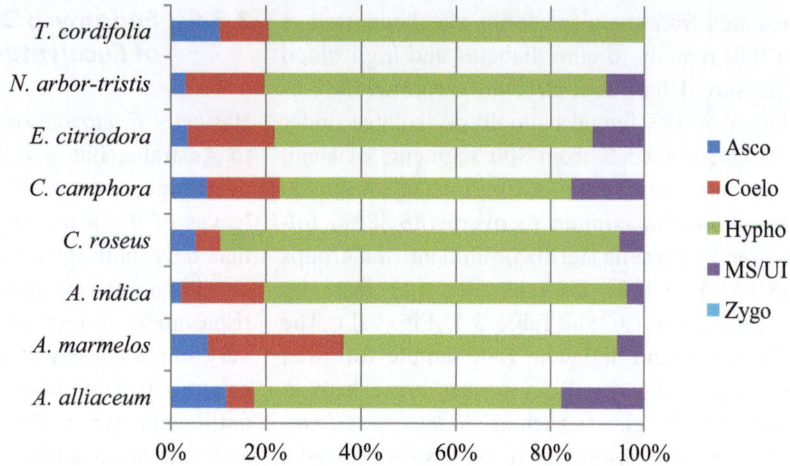


**Fig. 3.1** Per cent recovery of different classes of endophytic fungi

### 3.3.2 Endomyco Diversity in *Aegle marmelos*

*A. marmelos* is an Indian plant having medicinal and religious importance as well. The plant is used in Indian system of ayurvedic medicine against variety of diseases including diarrhoea, dysentery and dyspeptic symptoms. Green leaves of the plant are used for lowering blood sugar level. The plant was also reported to possess antifungal and antibacterial properties. Gond et al. (2007, 2011) isolated total of 511 endophytic fungal isolates representing 32 endophytic fungal taxa from 550 segments of bark, leaf and root. In the study, bark was found to harbour greater number of endophytic fungi followed by leaf and root. Among total taxa recovered, the *Aureobasidium* sp. (11.45 % CF) was found to be the highly dominated taxon. Among different endophytic classes, hyphomycetes showed maximum colonization 57.92 % followed by 28.57 % coelomycetes, 7.82 % ascomycetes and 5.67 % mycelia sterilia (Table 3.1, Fig. 3.1).

### 3.3.3 Endomyco Diversity in *Azadirachta indica*

*A. indica* is native to India and one of the most effective and popular medicinal plant, commonly known as neem, belongs to family Meliaceae. Different parts or extracts of the plant are used as antibacterial, antiretroviral, antiarthritic, anti-inflammatory and antiulcer. Over 400 bioactive compounds from neem plant and 32 from its endophytes have been reported so far. Verma et al. (2007) isolated 495 endophytic fungal isolates from 600 segments of leaf, stem and bark, root and fruit of neem collected from Varanasi region. The total endophytic fungal isolates recovered belonged to 44 fungal species including mycelia sterilia. In whole of the study, hyphomycetes dominated with 76.56 % followed by 17.37 % coelomycetes, 3.63 % mycelia sterilia, 2.22 % ascomycetes and interestingly only a single isolate (0.02 %) of zygomycete (Table 3.1, Fig. 3.1). However, genera like *Cladosporium, Aspergillus, Acremonium, Pestalotiopsis, Phomopsis, Curvularia and Trichoderma* were observed as dominant fungi. Among 495 isolates, 223 isolates were recovered from 200 segments of leaf, bark and stem while 272 isolates were isolated from 400 segments of root and fruit (Verma et al. 2007, 2011; Verma 2009).

### 3.3.4 Endomyco Diversity of *Catharanthus roseus*

*C. roseus* is commonly known as Madagascar periwinkle or sadabahar belonging to family Apocynaceae. A number of anticancer vinca alkaloids such as vincristine, vindesine, vinorelbine, vinblastin and vinflunine have been isolated from the plant. It has also been used as a folk remedy to cure diabetes and high blood pressure. Kharwar et al. (2008) reported the isolation of 183 fungal endophytic isolates under 19 fungal species from 300 segments of stem, leaf and root of Varanasi region. Hyphomycetes showed maximum recovery (86.88 %) followed by mycelia sterilia or unidentified groups (8.19 %), 3.27 % coelomycetes and least by ascomycetes 1.63 % (Table 3.1, Fig. 3.1). The CF was found higher in root sample followed by leaf and stem. Root tissues were heavily colonized by genera such as *Alternaria, Cladosporium* and *Aspergillus*. Leaf tissues showed a greater diversity of endophytes and *Drechslera, Curvularia, Bipolaris, Alternaria* and *Aspergillus* spp. were the dominant fungi.

### 3.3.5 Endomyco Diversity in *Cinnamomum camphora*

*C. camphora* is commonly known as camphor or *kapoor* plant. The plant belongs to the family Lauraceae and is native to Taiwan, southern Japan, Southeast China and Indochina. The oil of camphor is used as an anti-inflammatory, antiseptic, a cardiac, carminative, diuretic, febrifuge, an insecticide, a laxative, rubefacient, stimulant and vulnerary agent. Kharwar et al. (2012) claimed the isolation of 162 endophytic fungal isolates belonging to 26 species from more than 100 segments of leaf, stem and petiole. Among isolates recovered, hyphomycetes ranked first with 62.96 % isolation frequency (IF) followed by coelomycetes 16.66 %, mycelia sterilia 15.43 % and least from ascomycetes 4.93 % (Table 3.1, Fig. 3.1). Among all the segments studied, leaf harbour maximum (40.44 %) endomyco isolates followed by stem (29.04 %) and petiole (30.24 %). Among all the species observed, *A. niger* (10.49 %) was found to be most dominated followed by *Phyllosticta nobilis* and *Trichoderma harzianum* with equal IF of 6.79 % while *Arthrinium* sp. and *C. lunata* were recorded as rare isolates with IF value of 0.61 %.

### 3.3.6 Endomyco Diversity of *Eucalyptus citriodora*

Basically *E. citriodora* is a long tree and native to Australia, but it is frequently grown in the northeastern states of India. The bluish-green leaves of the plant contain fragrant volatile oil that have antiseptic, expectorant, antibacterial, anti-inflammatory, deodorant, diuretic and antispasmodic properties. Commonly used and a very important essential oil, it is known as eucalyptol, isolated from the leaves and used as an anti-cough syrup, for aromatherapy, dentistry, and to treat bronchitis, sinusitis, chronic rhinitis and asthma, etc. (Gond et al. 2010; Gond 2011). A total of 552 fungal endophytic isolates belonging to 32 fungal species from 600 segments of leaf and stem at Varanasi and Sonbhadra regions were isolated. Hyphomycetes was found to be the highly dominated group (67.02 %) followed by coelomycetes (18.29 %), mycelia sterilia or unidentified taxa (11.05 %) while ascomycetes represented the least IF (3.62 %; Table 3.1, Fig. 3.1). *Cladosporum cladosporioides* with an IF of 11.77 % was the most dominant taxon followed by *P. glomerata* at 10.50 %.

### 3.3.7 Endomyco Diversity of *Nyctanthes arbor-tristis*

*N. arbor-tristis* is a well-known medicinal plant native to the Indian subcontinent and grows abundantly in all parts of the country. It is commonly known as Harsinghar, Parijata, or night jasmine and belongs to the family Oleaceae. The flowers and leaves of *N. arbor-tristis* are well known for their interesting antibacterial, antifungal, antileishmanial and cytotoxic activity. Gond 2011 described the endomyco diversity of leaf and stem of *N. arbor-tristis* collected from Varanasi and Sonbhadra regions. From 800 segments (400 segments for each tissue) of leaf and stem, the author reported the isolation of 799 endophytic isolates. In this study, the recovery of hyphomycetes was found maximum with 72.09 % followed by coelomycetes 16.64 %, mycelia sterilia 8.13 %

and least from ascomycetes 3.12% (Table 3.1, Fig. 3.1). A total of 34 endophytic fungal species were observed from both tissues collectively. Among the total 34 species recorded, 32 were isolated from the leaves while only 19 species from the stem. *C. cladosporioides* (11.63%), *A. alternata* (7.38%), *Phomopsis helianthi* (6.25%) were observed as dominated taxa. *C. cladosporioides, C. lunata, C. dematium, Drechslera ellisii, Acremonium* sp., *N. oryzae, Phomopsis* sp. and *Rhizoctonia* sp. were isolated as common species for both tissues; *Aspergillus fumigates, A. niger, Helicosporium* sp. *Scytilidium* sp. and *Stachybotrys* sp. were only isolated from the leaf segments while isolate NAH3 only reported from stem segments; however, these results are a fine example of tissue specificity of endophytic fungi. Gond 2011 concluded that leaves harbour a higher number and high diversity of endophytic fungi in comparison to the stem, and this may be due to the large surface area of leaves exposed to the outer environment and the presence of stomata providing passage to the entry of fungal mycelia.

### 3.3.8 Endomyco Diversity in *Tinospora cordifolia* Miers

*T. cordifolia* is a widely used medicinal plant in the Indian Ayurvedic system of medicine. It is commonly known as Guduchi, Gurch, Giloe or Amrita, having a large, glabrous, deciduous, shade-loving climbing shrub belonging to family Menispermiaceae. A number of chemical constituents such as alkaloids, diterpenoids, lactones, phenolics, glycosides, aliphatic compounds and steroids have been isolated from *T. cordifolia.* It is used as an anti-inflammatory, antiperiodic, antifever, antidyspepsia, antiarthritic, anti-allergic and antidiabetic agent. The plant is also used to cure scorpion stings, and its watery extract used in febrifuge which is called 'Indian quinine' (Chopra et al. 1982; Singh and Panda 2005). The plant contains a polyclonal B cell mitogen with antioxidant activity which can be used as an immunomodulator (Venna et al. 2002). Mishra et al. (2012a) isolated 1,151 endophytic fungal isolates representing 29 taxa from 7,200 segments of leaf, stem, petiole and root (1800 segments of each tissue) collected at three locations of Varanasi district in three different seasons (winter, summer and monsoon). The IF of hyphomycetes (74.80%) was found greater followed by coelomycetes(14.07%) and ascomycetes (11.12%; Table 3.1, Fig. 3.1). Leaf tissues harbour maximum endophytes (29.38% of the isolates), followed by stem (18.16%), petiole (10.11%) and root segments (6.27%). The leaf segments harbour greater species (29) followed by stem (26), petiole (23) and root (18). CF was maximal during monsoon (23.23%) followed by winter (15.35%) and minimal during summer (8.85%). Among the isolates, *Penicillium* spp. were dominant (12.62% of all isolates), followed by *Colletotrichum* spp. (11.75%), *Cladosporium* spp. (8.93%), *C. globosum* (8.06%), *Curvularia* spp. (7.55%) and *A. alternata* (6.75%). *Trichoderma viride, Monilia* sp., *Acremonium* sp. and *Guignardia* sp. were rare (0.69%, 0.86%, 0.52% and 0.52%). The paper suggested that some endophytes are season specific for example *Colletotrichum linicola* occurred almost exclusively in winter and *F. oxysporum* only in winter and summer but never during monsoon while *C. lunata* was found only in winter and during monsoon but never in summer. It was concluded that the effect of season and tissue type on CF and species diversity was much more pronounced than the effect of the location.

## 3.4 Biopotential of Endophytic Fungi

Microbes have played an important role in the discovery of novel and effective drugs. More than 22,000 secondary metabolites from natural sources are reported with various bioactive properties, but not more than 200 compounds could reach the market which certainly is a discouraging figure (Bérdy 2005). Due to the rising resistance ability in pathogens against existing antibiotics and ingress of newer diseases in society, there is an urgent need to discover the novel and potent antimicrobials. For this, one should go with a novel alternative source. This is the reason that endophytic fungi are getting attention from the scientific community for their ability to produce novel natural metabolites. As the literature suggests, the microbes residing in special niches may be able to produce novel and potent compounds as well. After the discovery of taxol from an endophytic fungi, *T. andreanae* isolated from the Pacific yew (Stierle et al. 1993), the endophytic research came to light as an alternative source and till today over 100 (57 % novel and 43 % known) anticancer compounds have been isolated and characterized from endophytic fungi (Kharwar et al. 2011a). Not only anticancer but a number of effective and potential antibacterial and antifungal compounds were also isolated from endophytic fungi against a range of Gram +ve and Gram −ve strains. Some of them are colletotric acid isolated from *Colletotrichum gloesporioides,* an endophytic fungus of *Artimisia mongolica* active against *Bacillus subtilis, Staphylococcus aureus* and *Sarcina leutea* (Zou et al. 2000); Javanacin isolated from endophytic fungus *Chloridium* sp. resident of *A. indica* showed strong antibacterial property against *Bacillus* sp., *Escherichia coli, Pseudomonas fluorescens* and *Pseudomonas aeruginosa.* The compound was also active against several fungal pathogens (Kharwar et al. 2009). Cryptocandin isolated from endophytic *Cryptosporiopsis quercina* showed avtivity against human pathogenic fungi *Trichophyton rubrum* (ATCC 28188), *Trichophyton mentagrophytes* (ATCC 28185), *Candida albicans* (ATCC 90028), *Candida parapsilosis and Histoplasma capsulatum* (Strobel et al. 1999). Excluding other diseases, malaria alone is responsible to kill about 1 million people throughout the world every year. Endophytic fungi produce several antimalarial compounds such as phomoxanthones A and B from an endophyte *Phomopsis* sp., which are known to display antimalarial activity against *Plasmodium falciparum* K1 (Isaka et al. 2001). Pestacin and isopestacin, obtained from endophytic *Pestalotiopsis microspora* from the interior of *Terminalia morobensis,* displayed an antioxidant activity (Strobel et al. 2002). Cytonic acids A and B are novel protease inhibitors, isolated from *Cytonaema* sp., an endophyte of *Quercus* sp. against human cytomegalovirus (hCMV) (Guo et al. 2000). L-783 and 281 are nonpeptidal fungal metabolites isolated from endophytic *Pseudomassaria* sp. The compound acts as an insulin mimetic, but without destroying the digestive tract (Zhang et al. 1999). Subglutinol A and B are immunosuppressive, noncytotoxic diterpene pyrones isolated and characterized from an endophytic fungus *Fusarium subglutinans* of *Tripterygium wilfordii* (Lee et al. 1995). Nodulisporic acid A is a potential insecticide obtained from an endophytic fungus *Nodulisporium* sp. of *Bontia daphnoides* (Ondeyka et al. 1997). 3-Hydroxypropionic acid was isolated from *Phomopsis phaseoli* endophytically present in *Betula pendula and Betula pubescens* showed selective nematicidal activity against the plant-parasitic nematode *Meloidogyne incognita* (Schwarz et al. 2004). In addition to endophytic fungal diversity of eight plants, this chapter also covers the biopotential of the endophytic diversity. Literatures reveal that of the total endophytic community reported to have bioactive potential, 35 % belong to medicinal plants, followed by crops at 29 %, and the rest is equally divided between plants with special niches and other plants, each at 18 % (Selim et al. 2012).

### 3.4.1 Biopotential of Endophytic Fungi of *Adenocalymma alliaceum*

Out of 17 endophytic taxa, only 12 taxa were tested for their antibacterial activity against five human bacterial pathogens. Among 12 endophytic taxa, nine were found to be active against at least one bacterial pathogen. *A. alternata, C. globosum, C. lunata and Penicillium* sp. were active against 4 of 5 tested pathogens. *Salmonella enteritidis* (IMS/GN3) was found to be the most susceptible pathogen (Kharwar et al. 2011b).

### 3.4.2 Biopotential of Endophytic Fungi of *Aegle marmelos*

#### 3.4.2.1 *Antibacterial Activity*

Seventeen endophytic fungi isolated from *A. marmelos* were tested for antibacterial activity against human pathogenic bacteria. Fifteen (88.23 %) endophytic fungi showed antibacterial activity against one or more pathogenic bacteria. Out of 17 endophytic fungi, four were active against five bacteria (*Shigella flexnii, Shigella boydii, S. enteritidis, Salmonella paratyphi* and *P. aeruginosa*). *Phoma herbarum* had exhibited an impressive antibacterial activity against seven of eight bacteria tested. The extract of *P. herbarum* showed strongest activity (inhibition zone 23 mm) against *S. boydii*. Among the endophytes of *A. marmelos, P. herbarum* gave least minimum inhibitory concentration (MIC; 40 µg/ml) against *S. flexnii* and *S. boydii. S. boydii* was found to be most susceptible followed by *P. aeruginosa* towards the extract of endophytic fungi. Fifteen endophytic fungal extracts were active against *S. boydii* and 13 against *P. aeruginosa* (Gond 2011).

#### 3.4.2.2 *Antifungal Activity*

Seventeen endophytic fungi of *A. marmelos* were also tested against eight pathogenic fungi by dual culture assay. Out of 17 endophytic fungi, 10 were found to be active against one or more fungal pathogens. *P. herbarum* was most active that inhibited growth of five out of eight fungal pathogens. It inhibited 54.47 % growth of *C. lunata. Pestalotia macrotricha* was most active against *C. cladosporioides* showing 47.03 % growth inhibition, while *C. globosum* showed 24.03, 26.90, 27.07 and 39.13 % growth inhibition against *C. cladosporioides, F. oxysporum, Fusarium udum* and *C. lunata,* respectively. *Colletotrichum dematium* showed activity against *C. cladosporioides, F. oxysporum, F. udum* and *C. lunata* with 29.03, 32.7, 33.73 and 43.00 % inhibition, respectively. The endophytic *Phomopsis* sp. showed inhibitory activity against *A. alternata* (40.73 %), *C. cladosporioides* (30.03 %) and *C. lunata* (38.89 %). The pathogenic *C. lunata* was found most susceptible whereas *Microsporium gypseum* was resistant against all endophytic fungi tested (Gond 2011).

#### 3.4.2.3 *Antimarial Activity*

*P. herbarum* (*A. marmelos*) was assessed for antimalarial activity against 3D7 strain of *P. falciparum*. The extract of *P. herbarum* gave only 55 % Schizont maturation inhibition of 3D7 strain of *P. falciparum* at the concentration of 50 µg/ml (Gond 2011).

#### 3.4.2.4 *Antioxidant Activity*

The free radical-scavenging activity of fungal extract was carried out by using 2,2-diphenyl-1-picrylhydrazyl (DPPH). The $IC_{50}$ of *P. herbarum* isolated from *A. marmelos* was 125.63 µg/ml (Gond 2011), which could further be studied for detail and precise activity.

#### 3.4.2.5 Extracellular Enzyme Production

Out of 32 endophytic fungi, only *A. alternata, C. globosum, P. herbarum, C.olletotrichum dematium, T. viride* of *A. marmelos* were tested for extracellular production of amylase, xylanase and phosphate solubilization. All five endophytic fungi were found to produce amylase, while only *P. herbarum, C. dematium* secreted xylanase whereas no fungi were observed for solubilizing the phosphate in solid media (Gond 2011).

### 3.4.3 Biopotential of Endophytic Fungi of *A. indica*

Among endophytic isolates of *A. indica,* six endophytic fungi (*Alternaria* sp., *Colletotrichum* sp., *Chloridium* sp., *Nigrospora* sp., *Pestalotiopsis* sp., *Scytalidim* sp.) were evaluated for their anti-dermatophyte activity. Among the six endophytic taxa, ethyl-acetate-extracted *Pestalotiopsis* metabolite was found more effective against dermatophytes at MIC 80 µg/ml while acetone-extracted *Scytalidium* sp. exhibited least activity with 400 µg/ml. Javanicin, a napthaqunone isolated from *Chloridum* sp., an endophytic fungus resident of neem tree root (Kharwar et al. 2009), showed antibacterial as well as antifungal activity. Among all tested pathogens, *P. fluorescens* and *P. aerugenosa* were observed more sensitive at MIC 2 µg/ml followed by *Cercospora arachidicola* at 5 µg/ml. At the rate of 10 µg/ml, the compound inhibited the growth of *Rhizoctonia solani* and *Verticillium dahalae,* and *F. oxysporum* at 20 µg/ml, whereas the suppression of *Bacillus* sp., *E. coli* and *C. albicans* were observed at 40 µg/ml (Kharwar et al. 2009). The isolation of azadirachtin was previously only known from *A. indica* but Kusari et al. (2012) described the isolation and characterization of azadirachtin A and B from *Eupenicillium parvum* isolated from *A. indica*.

### 3.4.4 Biopotential of Endophytic Fungi of *C. roseus*

The endophytic fungi isolated from *C. roseus* collected in China are known to produce vinka alkaloids. Endophytic *Alternaria* sp. and *F. oxysporum* isolated from the phloem of *C. roseus* were able to produce vinblastine and vincristine. These alkaloids have anticancer property (Guo et al. 1998; Zhang et al. 2000).

### 3.4.5 Biopotential of Endophytic Fungi of *Cinnamomum camphora*

Five out of 26 endophytic taxa were tested against 11 fungal (five human and five phytopathogens) and single bacterial pathogens. *Pestalotiopsis* sp. showed significant inhibition against *Phytophthora cryptogea* (57.7 %), *Pythium aphanidermatum* (54.5 %), *Microsporum nanum* (51.4 %), *T. rubrum* (49.7 %), *Microsporum gypseum* (48.5 %) and *P. fluorescence* (47.1 %), while *Phomopsis* sp. showed significant inhibition only to *P. aphanidermatum* (50.6 %) (Kharwar et al. 2012).

### 3.4.6 Biopotential of Endophytic Fungi of *Eucalyptus citriodora*

#### 3.4.6.1 *Antibacterial Activity*

Thirteen (72.22 %) out of 18 endophytic fungi were found active against one or more human bacterial pathogens. *C. globosum, Rhizoctonia* sp., *P. glomerata* and *T. viride* were found to be active against four bacteria. *Pestalotia* sp. was most active against *S. flexnii* and *S. boydii* with an inhibition zone of 16.33 mm and 16.00 mm, respectively. *Periconia* sp. showed the activity only against *S. enteritidis*. *S. paratyphi* showed most susceptibility against *Rhizoctonia* sp. with 10 mm diameter of inhibition zone. The extract of seven endophytic fungal species inhibited *P. aeruginosa.* An unidentified species ECB2 (mycelia sterilia) gave maximum inhibition to *P. aeruginosa* with 12 mm diameter. *Citrobacter freundii* was only inhibited by extract of *Pestalotia* sp. *P. vulgaris* was inhibited by *P. glomerata and T. viride.* However, *Morganella morganii* was resistant against all the fungal extracts (Kharwar et al. 2010).

#### 3.4.6.2 *Antifungal Activity*

Out of 18 endophytic fungi, eight were found active against at least one phytopathogenic fungus. *Phomopsis* sp. was the most active taxon against *C. lunata* followed by an unidentified fungus ECB1 and with 48.88 and 47.1 % radial growth inhibition, respectively. *C. globosum* also inhibited 32.87 % growth of *T. rubrum*. *Phomopsis* sp. and ECB1 inhibited growth of four pathogenic fungi out of the eight tested. Pathogenic *A. alternata* was inhibited only by endophytic *F. oxysporum* (34.57 %) while *F. udum* was inhibited (30.7 %) only by *Phomopsis* sp. (Kharwar et al. 2010).

#### 3.4.6.3 Extracellular Enzyme Production

Eight endophytic fungi of *E. citriodora* were tested for amylase, xylanase and phosphate solubilization activity. Except *C. globosum,* all seven were found to produce amylase. Among them, *Periconia* sp. gave maximum zone of amylase production on solid agar medium. Only *Colletotrichum gloeosporioides* and *Aspergillus terreus* were observed to produce xylanase. Like *A. marmelos,* none of the endophytic fungus of *E. citriodora* had exhibited phosphate-solubilization activity (Gond et al. 2012).

### 3.4.7 Biopotential of Endophytic Fungi of *Nyctanthes arbor-tristis*

#### 3.4.7.1 Antibacterial Activity

Sixteen endophytic fungi isolated from *N. arbor-tristis* were tested for antibacterial activity by disc diffusion assay against eight clinical isolates of human pathogenic bacteria (*S. flexnii, S. boydii, S. enteritidis, S. paratyphi, P. aeruginosa, C. freundii, M. morganii* and *P. vulgaris*). Among them, 12 (75%) were found active at a rate of 5 mg/disc. *C. dematium* and *Chaetomiun globosum* exhibited antibacterial activity against five pathogens with inhibition ranged from 6.00 to 14.00 mm while *Nigrospora* sp. inhibited the growth of four bacterial pathogens, i.e. *S. paratyphi* (22.00 mm), *S. flexnii* (15.00 mm), *S. boydii* (18.00 mm) and *P. aeruginosa* (15.66 mm). In a study, *C. freundii, M. morganii* and *Proteus vulgaris* were found resistant against all the fungal extracts. *S. boydii* was found most susceptible and was inhibited by ten endophytic fungal extracts (Gond et al. 2012).

#### 3.4.7.2 Antifungal Activity

Nine out of 16 endophytic fungi exhibited antifungal activity. *C. dematium* displayed the inhibitory activity against five phytopathogens, however its maximum activity was pronounced against *C. lunata* producing 55.87% radial growth inhibition in dual culture. *Acremonium* sp. and *N. oryzae* inhibited the growth of three of eight pathogenic fungi. *C. cladosporioides* was found to be the most susceptibile species that was inhibited by *Aspergillus fumigatus,* and *F. oxysporum* with 39.66%, 39.57% while 31.60% by *Dreschlera rostrata.* The growth of *F. oxysporum* was restricted by a single unidentified fungus MS/NAB2 up to 38.47%.

#### 3.4.7.3 Antimalarial Activity

*N. oryzae* isolated from *N. arbor-tristis* showed 100% Schizont maturation inhibition of a malarial parasite 3D7 strain of *P. falciparum* at the concentration of 50 µg/ml (Gond 2011).

#### 3.4.7.4 Antioxidant Activity

The free radical-scavenging activity of fungal extract of *N. oryzae* was carried out by using DPPH . The $IC_{50}$ for *N. oryzae* was found at 265.53 µg/ml (Gond 2011), which was quite higher than *P. herbarum* isolated from *A. marmelos.*

#### 3.4.7.5 Extracellular Enzyme Production

Nine endophytic fungi of *N. arbor-tristis* were tested for amylase, xylanase and phosphate-solubilization activity. Only *N. oryzae, Helicosporium* sp., *Diatrype* sp., *Macrophoma* sp. were found to produce amylase. Except *N. oryzae,* none were observed to produce xylanase. Unlike *A. marmelos* and *E. citriodora,* three endophytic fungi of *N. arbor-tristis*, i.e. *P. glomerata, Scytilidium* sp. and *Diatrype* sp. were able to solubilize phosphate (Gond 2011).

### 3.4.8 Biopotential of Endophytic Fungi of *Tinospora cordifolia*

Twenty nine endophytic taxa were tested for their antibacterial activity against eight human bacterial pathogens. More than 50% (15 out of 29) of the endophytic taxa exhibited antimicrobial activity. *Botryosphaeria rhodina* (JQ031157) and *C. globosum* showed activity against all bacterial human pathogens tested, with the former showing higher activity than the latter. *B. rhodina* (JQ031157) exhibited strongest activity against *C. freundii* (IMS/GN5) producing an inhibition zone of 45.66±0.33 mm whereas lowest against *M. morganii* (IMS/GN6) with an inhibition zone of 12.83±0.16 mm at the rate of 5 mg/ml. *C. linicola, A. alternata, C. cladosporioides, N. oryzae* and *Pseudofusicoccum violaceum* (JQ031159) were active against a single pathogen. *S. flexnii* IMS/GN1 was observed to be the most susceptible bacterial pathogen, inhibited by 11 endophytic taxa followed by *E. coli* ATCC 25922, inhibited by six endophytic taxa, *S. paratyphi* and *P. vulgaris* inhibited by five endophytic taxa whereas *S. enteritidis* IMS/GN3, *P. aeruginosa* ATCC 27853, *C. freundii* IMS/GN5, *M. morganii* IMS/GN6 were found to be the least susceptible and were only inhibited by three endophytic taxa (Mishra et al. 2012a).

## 3.5 Future Prospective

Endophytic fungi are relatively less studied, unexploited and hidden microbes of the microbial community. All the plants studied to date for their endophytic fungi were found to harbour either at least a single or plethora of fungi. Hawksworth and Rossman (1987) estimated there may be as many as 1.5 million different fungal species, while only about 100,000 have been described, and this study raises the question, Where are the rest of the fungi? Are they in form of endophytes or somewhere else? These are some of the basic questions regarding the diversity of endophytic fungi that require more endophytic research which may help in the isolation and characterization of new fungal species and/or bioactive compounds. Since a considerable number of novel fungal genera and species have been reported from this relatively hidden (inside healthy plant tissues) source that may be a good repertoire for filling the gap between reported and estimated fungal diversity. Literatures suggest that endophytes enhance resistance in their hosts against herbivores, pathogenic fungi, bacteria, viruses, insects, nematodes illness, reduced seed production, temperature and salinity and also against drought and minerals (Mishra et al. 2012b). Today a major problem in the front of scientists is the development of resistance in pathogenic microbes (bacteria, fungi and other microbes, malarial parasite, viruses, etc.), pests and weeds that have become a serious trouble for humans, animals and agriculture. To overcome this problem, there is an urgent need for a novel and alternative source for drug discovery. Endophytic fungi can serve as a good alternative because a number of antibacterial, antidiabetic, antifungal, antimalarial, antioxidant, antiviral and other bioactive compounds exhibited promising activity isolated and characterized from this source. The endophytic fungi isolated from the above described eight medicinal plants are under the process of isolation and characterization of antibacterial, antifungal, antioxidant and antimalarial compounds. Suryanarayanan and his colleagues found endophytic fungi as a prolific source for production of extracellular amlyases, cellulases, chitinases, chitosonases, laccases, lipases, pectinases and proteases (Suryanarayanan et al. 2012).

The isolation of several fungi and their isolates (Alternaria, Phomopsis, Chaetomium, Chollelotrichum, Fusarium, etc) from these medicinal plants indicate that some endophytic fungi may be probable candidates to produce some cytotoxic compounds. Another interesting aspect of fungal endophytes is to produce antimicrobial volatile organic compounds (VOCs) reported from mitosporic xylariales fungus *Muscodor albus* and *Muscodor vitigenus* isolated from *Cinamomum zeylanicum* (Strobel et al. 2001). Recently, some hydrocarbon derivatives as major constituents of diesel fuel (Mycodiesel) were reported for the first time from a fungal endophyte *Gliocladium roseum,* NRRL 50072 (Strobel et al. 2008); however, some genuine technical questions on mycodiesel production were raised (Stadler and Schulz 2009). Nevertheless, the successful trial of running a Honda (100 cc) motorbike using eucalyptol, a better and safe alternative of gasoline received from a fungal endophyte of *Eucalyptus* sp. (Tomsheck et al. 2010) furthered our understanding towards other interesting aspects.

Very recently, some interesting works have been published in order to enhance the production of cryptic and known bioactive compounds through epigenetic modulations, and these works may point the way in future that can reduce the stigma of reduced yield of fungal endophytes in successive generations (Sun et al. 2012; Hassan et al. 2012). A huge diversity of endophytic fungi isolated from these plants and significant antimicrobial and biochemical activity of crude extracts provide us potential fungal endophyte pools to isolate pure and novel bioactive compounds. In future, epigenetic modulation may play a very crucial role in isolating the cryptic secondary metabolites which are not either known, or it may enhance multifold production of known compounds from fungal source which may serve the need of society.

**Acknowledgment** The authors are thankful to the Head of the Department of Botany, Banaras Hindu University, Varanasi India, for providing the necessary facilities. The authors also extend their thanks to the CSIR/UGC/DST, New Delhi, for providing financial assistance in the form of JRF/SRF. RNK expresses his appreciation to the DST, New Delhi for providing financial assistance (File No SR/SO/PS-78-2009, dt-10-5-2010).

## References


Backman PA, Sikora RA (2008) Endophytes: an emerging tool for biological control. Biol Control 46:1–3

Bacon CW, White JF (2000) Microbial endophytes. Marcel Deker, New York

Bashyal BP, Wijeratne EMK, Faeth SH, Gunatilaka AAL (2005) J Nat Prod 68:724–728

Bérdy J (2005) Bioactive microbial metabolites: a personal view. J Antibiotics 58:1–26

Brem D, Leuchtmann A (2001) Epichloë grass endophytes increase herbivore resistance in the woodland grass Brachypodium sylvaticum. Oecologia 126:522–530

Chambers SM, Williams PG, Seppelt RD, Cairney JWG (1999) Molecular identification of Hymenoscyphus sp. from the rhizoids of the leafy liverwort Cephaloziella exiliflora in Australia and Antarctica. Mycol Res 103:286–288

Chopra RN, Chopra LC, Handa KD, Kapur LD (1982) Indigenous drugs of India. 2nd ed. M/S Dhar VN & Sons, KolKata

Clay K (1990) Fungal endophytes of grasses. Ann Rev Ecol Syst 21:255–297

De Bary A (1866) Morphologie und Physiologie der Pilze, Flechtem, und Myxomycelen. Holfmeister's Handbook of Physiological Botany. Vol 2. Leipzig

Fisher PJ, Graf F, Petrini LE, Sutton BC, Wookey PA (1995) Fungal endophytes of Dryas octopetala from a high arctic polar semi-desert and from the Swiss Alps. Mycologia 87:319–323

Gond SK (2011) Study of endophytic mycoflora of some medicinal plants from eastern Uttar Pradesh. Thesis, Banaras Hindu University

Gond SK, Verma VC, Kumar A, Kumar V, Kharwar RN (2007) Study of endophytic fungal community from different parts of Aegle marmelos Correae (Rutaceae) from Varanasi (India). World J Microbiolo and. Biotech 23:1371–1375

Gond SK, Verma VC, Mishra A, Kumar A, Kharwar RN (2010) Role of fungal endophytes in plant protection. In: Arya A, Perello AE (eds) Management of fungal plant pathogens. CAB, London, pp 183–197

Gond SK, Mishra A, Sharma VK, Verma SK, Kumar J, Kharwar RN, Kumar A (2012) Diversity and antimicrobial activity of endophytic fungi isolated from Nyctanthes arbor-tristis, a well-known medicinal plant of India. Mycoscience 53:113–112

Guo B, Li H, Zhang L (1998) Isolation of the fungus producing vinblastine. J Yunnan University (Natural Science Edition) 20:214–215

Guo B, Dai J, Ng S, Huang Y, Leong C, Ong W, Carte BK (2000) Cytonic acids A and B: novel tridepside inhibitors of hCMV protease from the endophytic fungus Cytonaema species. J Nat Prod 63:602–604

Hawksworth DC, Rossman AY (1987) Where are the undescribed fungi? Phytopathology 87:888–891

Hassan SRU, Strobel G, Booth E, Kingston B, Floerchinger C, Sears J (2012) Epigenetic modulation of volatile organic compound formation in the mycodiesel producing endophyte- Hypoxylon sp. CI-4. Microbiology 158:465–473

Huang C, Wang Y (2011) New Records of endophytic fungi associated with the Araucariaceae in Taiwan. Coll Res 24:87–95

Hyde KD, Soytong K (2008) The fungal endophyte dilemma. Fungal Divers 33:163–173

Isaka M, Jaturapat A, Rukseree K, Danwisetkanjana K, Tanticharoen M, Thebtaranonth Y (2001) Phomoxanthones A and B, novel xanthone dimers from the endophytic fungus Phomopsis species. J Nat Prod 64:1015–1018

Kharwar RN, Verma VC, Strobel G, Ezra D (2008) The endophytic fungal complex of Catharanthus roseus (L.) G. Don. Curr Sci 95:228–235

Kharwar RN, Verma VC, Kumar A, Gond SK, Strobel G (2009) Javanicin, an antibacterial naphthaquinone from an endophytic fungus of neem—Chloridium sp. Curr Microbiol 58:233–238

Kharwar RN, Gond SK, Kumar A, Mishra A (2010) A comparative study of endophytic and epiphytic fungal association with leaf of Eucalyptus citriodora Hook. and their antimicrobial activity. World J Microbiol Biotechnol 26:1941–1948

Kharwar RN, Mishra A, Gond SK, Stierle A, Stierle D (2011a) Anticancer compounds derived from fungal endophytes: their importance and future challenges. Nat Prod Rep 28:1208–1228

Kharwar RN, Verma SK, & Mishra A, Gond SK, Sharma VK, Afreen T, Kumar A (2011b) Assessment of diversity, distribution and antibacterial activity of endophytic fungi isolated from a medicinal plant Adenocalymma alliaceum Miers. Symbiosis 55:39–46

Kharwar RN, Maurya AL, Verma VC, Kumar A, Gond SK, Mishra A (2012) Diversity and antimicrobial activity of endophytic fungal community isolated from medicinal plant Cinnamomum camphora. Proc Natl Acad Sci India, Sect B Biol Sci. doi:10.1007/s40011-012-0063-8

Kusari S, Verma VC, Lamshoeft M, Spiteller M (2012) An endophytic fungus from Azadirachta indica A. Juss.that produces azadirachtin. World J Microbiol Biotechnol 28:1287–1294

Lee JC, Lobkovsky E, Pliam NB, Strobel G, Clardy J (1995) Subglutinols A and B: immunosuppressive compounds from the endophytic fungus Fusarium subglutinans. Org Chem 60:7076–7077

Li H, Wei D, Shen M, Zhou Z (2012) Endophytes and their role in hytoremediation. Fungal Divers 54:11–18

Malinowski D, Leuchtmannv A, Schmidt D, Nösberger J (1997) Agronom J 89:673–678

Maria GL, Sridhar KR, Raviraja NS (2005) Antimicrobial and enzyme activity of mangrove endophytic fungi of southwest coast of India. J Agricultural Tech 1:67–80

Mishra A, Gond SK, Kumar A, Sharma VK, Verma SK, Kharwar RN, Sieber TN (2012a) Season and tissue type affect fungal endophyte communities of the Indian medicinal plant Tinospora cordifolia more strongly than geographic location. Microb Eco 64:388–398

Mishra A, Gond SK, Kumar A, Sharma VK, Verma SK, Kharwar RN (2012b) Sourcing the fungal endophytes: a beneficial transaction of biodiversity, bioactive natural products, plant protection and nanotechnology. In: Satyanarayana T, Johri BN, Prakesh A (eds) Microorganisms in sustainable agriculture and biotechnology. Springer, pp 581–612

Ondeyka JG, Helms GL, Hensens OD, Goetz MA, Zink DL, Tsipouras A, Shoop WL, Slayton L, Dombrowski AW, Polishook JD, Ostlind DA, Tsou NN, Ball RG, Singh SB (1997) Nodulisporic acid A, a novel and potent insecticide from a Nodulisporium Sp. isolation, structure determination, and chemical transformations. J Am Chem Soc 119:8809–8816

Østergaard LH, Olsen HS (2010) Industrial applications of fungal enzymes. In: Hofrichter XM (ed) The mycota. Springer, Berlin, pp 269–290

Redman RS, Sheehan KB, Stout RG, Rodriguez RJ, Henson JM (2002) Thermotolerance onferred to plant host and fungal endophyte during mutualistic symbiosis. Science 298:1581

Rice JS, Pinkerton BW, Stringer WC, Undersander DJ (1990) Crop Sci 30:1303–1305

Rosa LH, Vaz ABM, Caligiorne RB, Campolina S, Rosa CA (2009) Endophytic fungi associated with the Antarctic grass Deschampsia antarctica Desv. Polar Biol 32:161–167

Schmid E, Oberwinkler F (1995) A light and electron-microscopic study on a vesicular-arbuscular host-fungus interaction in gametophytes and young sporophytes of the Gleicheniaceae (Filicales). New Phytol 129:317–324

Schwarz M, Kopcke B, Weber RWS, Sterner O, Anke H (2004) 3-Hydroxypropionic acid as a nematicidal principle in endophytic fungi. Phytochemistry 65:2239–2245

Selim KA, El-Beih AA, AbdEl-Rahman TM, El-Diwany AI (2012) Biology of endophytic fungi. Curr Res Environ App Mycol. doi:10.5943/cream/2/1/3

Shweta S, Zuehlke S, Ramesha BT, Priti V, Mohana Kumar P, Ravikanth G, Spiteller M, Vasudeva R, Uma Shaanker R (2010) Endophytic fungal strains of Fusarium solani, from Apodytes dimidiata E. Mey. ex Arn (Icacinaceae) produce camptothecin, 10-hydroxycamptothecin and 9-methoxycamptothecin. Phytochemistry 71:117–122

Singh MP, Panda H (2005) Medicinal herbs with their formulation. Vol 2. Daya Publishing House, Delhi, pp 834–835

Stadler M, Schulz B (2009) High energy biofuel from endophytic fungi? Trends Plant Sci 14:353–355

Stierle A, Strobel GA, Stierle D (1993) Taxol and taxane production by Taxomyces andreanae. Science 260:214–216

Strobel GA (2002) Microbial gifts from rain forests. Can J Plant Pathol 24:14–20

Strobel GA, Miller RV, Miller C, Condron M, Teplow DB, Hess WM (1999) Cryptocandin, a potent antimycotic from the endophytic fungus Cryptosporiopsis cf. quercina. Microbiology 145:1919–1926

Strobel GA, Dirkse E, Sears J, Markworth C (2001) Volatile antimicrobials from Muscodor albus, a novel endophytic fungus. Microbiology 147:2943–2950

Strobel GA, Ford E, Worapong JJ, Harper K, Arif AM, Grant DM, Fung PCW, Chan K (2002) Ispoestacin, an isobenzofuranone from Pestalotiopsis microspora, possessing antifungal and antioxidant activities. Phytochemistry 60:179–183

Strobel G, Knighton B, Kluck K, Ren Y, Livinghouse T, Griffen M, Spakowicz D, Sears J (2008) The production of mycodiesel hydrocarbons and their derivatives by the endophytic fungus Gliocladium roseum. Microbiology 154:3319–3328

Sun J, Awakawa T, Noguchi H, Abe I (2012) Induced production of mycotoxins in an endophytic fungus from the medicinal plant Datura stramonium L. Bioorg Med Chem Lett 22:6397–6400

Suryanarayanan TS, Venkatesan G, Murali TS (2003) Endophytic fungal communities in leaves of tropical forest trees: diversity and distribution patterns. Curr Sci 85:489–493

Suryanarayanan TS, Thirunavukkarasu N, Govindarajulu MB, Venkatesan G (2012) Fungal endophytes: an untapped source of biocatalysts. Fungal Divers 54:19–30

Tomsheck AR, Strobel GA, Booth E, Geary B, Spakowicz D, Knighton B, Floerchinger C, Sears J, Liarzi O, Ezra D (2010) Hypoxylon sp., an endophyte of Persea indica, producing 1,8-Cineole and other bioactive volatiles with fuel potential. Microbial Ecol 60:903–914

Tung CY, Yang DB, Gou M (2002) A preliminary study on the condition of the culture and isolate of endophytic fungus producing vincristine. J Chuxiong Normal University 6:39–41

Venna RD, Kamat JP, Sainis KB (2002) An immunomodulator from Tinospora cordifolia with antioxidant activity in cell free system. Proc Indian Acad Sci 6:713–719

Verma SK, Gond SK, Mishra A, Sharma VK, Kumar J, Singh DK, Kumar A, Goutam J, Kharwar RN (2013) Impact of environmental variables on isolation, diversity and antibacterial activity of endophytic fungal community from Madhuca indica Gmel at different locations of India. Ann Microbiol. doi:10.1007/s13213-013-0707-9)

Verma VC (2009) Screening and assessment of endophytic fungi and actinomycetes from Azadirachta indica A. Juss. Thesis, Banaras Hindu University

Verma VC, Gond SK, Kumar A, Kharwar RN, Strobel GA (2007) Endophytic mycoflora from leaf, bark, and stem of Azadirachta indica A Juss. from Varanasi India. Microbial Eco 54:119–125

Verma VC, Gond SK, Kumar A, Kharwar RN, Boulanger LA, Strobel G (2011) Endophytic fungal flora from Roots and Fruits of an Indian neem plant Azadirachta indica A. Juss., and impact of culture media on their isolation. Indian J Microbiol 51:469–476

Verma VC, Lobkovsky E, Gange AC, Singh SK, Prakash S (2011) Piperine production by endophytic fungus

Periconia sp isolated from Piper longum sp. J Antibiot 64:427–431

Wang S, Li XM, Teuscher F, Li DL, Diesel A, Ebel R, Proksch P, Wang BG (2006) Chaetopyranin, a benzaldehyde derivative, and other related metabolites from Chaetomium globosum, an endophytic fungus derived from the marine red alga Polysiphonia urceolata. J Nat Prod 69:1622–1625

Yang RY, Li CY, Lin YC, Peng GT, She ZG, Zhou SN (2006) Lactones from a brown alga endophytic fungus (No. ZZF36) from the South China Sea and their antimicrobial activities. Bioorganic Medicinals Chem Lett 16:4205–4208

Zhang B, Salituro G, Szalkowski D, Li Z, Zhang Y, Royo I, Vilella D, Dez M, Pelaez F, Ruby C, Kendall RL, Mao X, Griffin P, Calaycay J, Zierath JR, Heck JV, Smith RG, Moller DE (1999) Discovery of a small molecule insulin mimetic with antidiabetic activity in mice. Science 284:974–977

Zhang L, Guo B, Li H, Zeng S, Shao H, Gu S, Wei R (2000) Preliminary study on the isolation of endophytic fungus of Catharanthus roseus and its fermentation to produce products of therapeutic value. Chinese Trad Herbal Drugs 31:805–807

Zou WX, Meng JC, Lu H, Chen GX, Shi GX, Zhang TY, Tan RX (2000) Metabolites of Colletotrichum gloeosporioides, an endophytic fungus in Artemisia mongolica. J Nat Prod 63:1529–1530

# Unlocking the Myriad Benefits of Endophytes: An Overview

4

Sanjana Kaul, Maroof Ahmed, Tanwi Sharma and Manoj K. Dhar

**Abstract**

Endophytes are the organisms that exist within the living tissues of most plant species without inducing any pathogenic symptoms in the host. Their association with the host plant may be symbiotic, mutualistic, commensalistic, trophobiotic and their interaction with them may be biochemical and genetic as well. It is their enormous diversity and specialized biotopes that make them a stimulating field of study. The fact that endophytes can mimic some of the phytochemicals originally characteristic of the host has changed the scope of endophytic biology. As a result of this fact, various classes of natural products that have been discovered from endophytes which include antitumour, antibacterial, antifungal, antiviral and various different industrial enzymes. Even some unusual molecules that act as immunomodulatory, antidiabetic, insecticidal, herbicidal, etc. have been isolated from them. Endophyte–plant relationship can be exploited to promote plant health as it also plays an important role in low-input sustainable agricultural applications for both food and nonfood crops. Recent studies on endophytes have shown the beneficial roles of host-endophyte associations as protection against mammals, resistance to pathogenic fungi and other insect herbivores, increased growth and development, nutrient uptake and stress tolerance in various plants including agriculturally important crops. The main objective of this review is to explore the potential of endophytes in varied fields that can prove beneficial to mankind. As a result, the role of endophytes has been discussed in some of the key fields of interest such as biotransformation, biodegradation, phytoremediation, seed production, seed predation, plant growth promotion, plant stress relievers, biocatalysis, biofuel production, biocontrol agents, agricultural importance, source of novel natural products etc. The present

S. Kaul (✉) · M. Ahmed · T. Sharma · M. K. Dhar
School of Biotechnology, University of Jammu, Jammu
180006, India
e-mail: sanrozie@rediffmail.com

R. N. Kharwar et al. (eds.), *Microbial Diversity and Biotechnology in Food Security*,
DOI 10.1007/978-81-322-1801-2_4, © Springer India 2014

review is an attempt to unlock the enormous benefits that the microbial endophytes can offer.

**Keywords**

Endophytes · Natural products · Biocontrol · Biofuel · Microbes

## 4.1 Endophytes

Endophytes have over the period of time evolved from being defined just as the microbes living within plants indicating not only their location but also the type of association that they have with the host. They are the organisms that live asymptomatically within the internal tissues of the plant. Endophytes live for its life or part of it without causing any apparent disease to the host plant. The word endophytes (Greek endo = within + phytes = plants) has also been defined as the microorganisms inhabiting inside of healthy plant tissues. Petrini (1991) described them as all organisms that inhabit or colonize internal plant tissues at some part of their life without any immediate deleterious effect on their host. They are now considered as ubiquitous symbionts of plants from their common detection to many plant species.

Endophytes have now been considered as an important component of microbial biodiversity, as they are found in virtually every plant on earth ranging from terrestrial to aquatic. Common endophytes include a variety of bacteria, fungi and actinomycetes which can be isolated from wild (Brooks et al. 1994) or cultivated crops (Liu and Tang 1996) of either the monocotyledonous (Fisher et al. 1992) or dicotyledonous plant groups (El-Shanshoury et al. 1996). A single plant can harbour several to hundreds of endophyte species, some of which may also show host specificity. They exhibit a variety of relationships with their host ranging from symbiotic to pathogenic. The observed biodiversity of endophytes suggests that they can also be aggressive saprophytes or opportunistic pathogens. Endophytes usually inhabit above-ground plant tissues (leaves, stems, bark, petioles and reproductive structures) which distinguish them from better known mycorrhizal symbionts. The distinction is not so firm, because endophytes may also inhabit root tissues.

The type of symbiosis exhibited between plant and endophyte is often related to the reproductive mode of the endophyte. Most of the endophytes are horizontally transmitted. These have sexual mode of transmission via spores and can be spread by wind, insects and vectors; so, these can evolve a more pathogenic lifestyle as they can escape their host plant, whereas vertically transmitted endophytes are asexual and transmit via fungal hyphae penetrating the host's seed (e.g. *Neotyphodium*).

Endophytes have a very intimate and also a coevolutionary relationship with their host. Therefore, they have the potential to influence the physiology of the plant. They confer enhanced fitness to the host plants by producing functional metabolites in lieu of the nutrition and protection from the host. They offer a plethora of benefits to the plant ranging from resistance against diseases to insect herbivory.

## 4.2 Endophytes as Source of Novel Bioactive Metabolites

Endophytes can be considered as potential, reliable and promising source of novel organic natural metabolites exhibiting a variety of biological activities viz. antimicrobial, antiparasitic, cytotoxic, neuroprotective, antioxidant, insulin mimetic, enzyme inhibitive, immunosuppressive and many more. A broad variety of bioactive secondary metabolites with unique structures, including alkaloids, benzopyranones, chinones, flavonoids, phenolic acids, quinones, steroids, terpenoids, tetralones, xanthones and many more, are the members of the diverse metabolite bank of endophytes. Many reports witness the various usages of endophytic fungal metabolites in different human diseases due to their highly functionalized activity. Various fungal endophytes from selected medicinal plants of western Himalayas have been reported to show antimicrobial and immunomodulatory activities (Qadri et al. 2013). Antibacterial activity of four fungal endophytes (*Cladosporium* sp., *Aspergillus flavus*, *Aspergillus* sp., *Curvularia lunata*) isolated from *Kigelia africana* has been reported by Idris et al. (2013). Fungal endophytes isolated from the plant *Aralia elata* have also been reported as potential antibacterial agents (Wu et al. 2012). Metabolites of endophytes isolated from *Artemisia annua* have been reported to show anti-acetylcholinesterase activity and can therefore be used in the treatment of Alzheimer's disease (Aly et al. 2011). Some of the endophytes have the potential to produce the same bioactive metabolites for which the host plant is known, hypericin, camptothecin, podophyllotoxin, paclitaxel and diosgenin (Zhao et al. 2010). There is an increasing interest in studying fungal endophytes as alternative potential source of functional metabolites. So far, about 19 genera from different plant sources have been screened for the production of paclitaxel and its analogues. *Cephalotheca faveolata,* a fungal endophyte isolated from *Eugenia jambolina* has been reported to exhibit anticancerous activity (Giridharan et al. 2012). Therefore, keen studies on endophytes are required for the commercial and economical production of taxol and other host origin chemicals from endophytic microbes, as their production from host plant is comparatively costlier, time consuming and exerts biodiversity pressure.

Endophytic fungi isolated from six species of Indonesian medicinal plants Mahoni (*Swietenia mahagoni* Jacq.), Sambiloto (*Andrographis paniculata* Ness.), Kumis kucing (*Orthosiphon spicatus BBS*), Mengkudu (*Morinda citrifolia* L.), Sirih merah (*Piper crocatum* L.) and Sirih hitam (*Piper ornatum* sp.) have the potential to produce antidiabetic agents (Dompeipen 2011). The foliar endophytes of conifers have been reported to synthesize anti-insect toxins (Sumarah et al. 2009). So, strategic use of specific endophytes or their metabolites can be used in plant–insect/pest management programmes.

Some of the important bioactive metabolites isolated recently from endophytes and their bioactive functions are mentioned in detail in Table 4.1. For more information readers are suggested to read reviews by Tan and Zou (2001); Schulz et al. (2002); Strobel and Daisy (2003); Strobel et al. (2004); Joseph and Priya (2011); Kharwar et al. (2011); Premjanu and Jayanthy (2012).

## 4.3 Seed Production and Seed Predation

Presence or absence of endophytes influences the production and predation of seeds among different plants, especially the grasses. Saari et al. (2010) studied the grass populations of tall fescue and meadow fescue for the effect of presence or absence of endophytes on production and predation of seeds. Endophyte infected (E+) and endophyte free (E−) tall fescue plants did not show any difference in the mean seed mass or production of seeds, but seeds from E+ plants showed significantly lower proportions of seed predation compared to seeds from E− plants. In contrast, endophyte infected (E+) meadow fescue plants produced marginally more seeds than endophyte free (E−) plants. Seed predation of meadow fescue plants by cocksfoot moth did not differ between endophyte infected and endophyte free plants. These results suggest that endophyte infection benefits tall fescue grasses by enhancing host resistance to seed predators rather than through enhanced reproductive effort. On the other hand, it also suggests that endophyte infection benefits meadow fescue grasses by enhancing their capabilities of seed production rather than enhancement of host resistance to seed predators. The possible reasons for this variability in the effects of endophytes is firstly, the high genetic variability among wild grasses; secondly, different species of endophytes infecting

**Table 4.1** Important bioactive metabolites isolated from endophytes and their bioactive function

| Bioactive metabolites | Endophyte | Host plant | Bioactive function | Reference |
|---|---|---|---|---|
| Camptothecine | *Fomitopsis* sp. | *Miquelia dentata* | Anticancer | Singh et al. 2013 |
| 9-methoxy camptothecine | *Alternaria alternata* | | | |
| 10-hydroxy camptothecine | *Phomopsis* sp. | | | |
| (4R,8R)-foedanolide | *Pestalotiopsis foedan* | *Bruguiera sexangula* | Anticancer | Yang and Li 2013 |
| (4S,8S)-foedanolide | | | | |
| Isobenzofuranone derivatives | *Cephalosporium* sp. | *Sinarundinaria nitida* | Antioxidant | Huang et al. 2012 |
| 3-hydroxyfumiquinazolin A | *Aspergillus fumigates* | *Melia azedarach* | Antifungal | Li et al. 2012 |
| 5-hydroxyramulosin | *Phoma* sp. | *Cinnamomum mollissimum* | | Santiago et al. 2012 |
| α-tetralone derivatives, cercosporamide, β-sitosterol and trichodermin | | *Arisaema erubescens* | Antifungal and antibacterial | Wang et al. 2012 |
| Cajaninstilbene acid | *Fusarium* sp. | *Cajanus cajan* | Antioxidant | Zhao et al. 2012 |
| 16α-hydroxy-5N-acetylardeemin | *Aspergillus terreus* | *Artemisia annua* | Anti-acetylcholinesterase | Aly et al. 2011 |
| Nodulisporic acid A | *Nodulisporium* sp. | *Bontia daphnoides* | Insecticidal | |
| Palmarumycin | *Edenia* sp. | *Petrea volubilis* | Growth inhibitors of *Leishmania donovani* | |
| Macrosporin | *Stemphylium globuliferum* | *Mentha pulegium* | Protein kinase inhibitors | |
| Pestalotheol | *Pestalotiopsis theae* | Unknown | HIV inhibitor | |
| Phomoarcherins A-C | *Phomopsis archeri* | *Vanilla albindia* blume | Antimalarial | Hemtasin et al. 2011 |
| Kakadumycin | *Streptomyces* sp. | *Grevillea pteridifolia* | Antibiotic | Pimentel et al. 2010 |
| Pestacin | *Pestalotiopsis microspora* | *Terminalia morobensis* | Antioxidant | |
| Oblongolide Y | *Phomopsis* sp. | *Musa acuminate* | Anticancer | Bunyapaiboonsri et al. 2010 |
| Eremophilanolides | *Xylaria* sp. | *Licuala spinosa* | | Isaka et al. 2010 |
| Terpenoids | *Phomopsis* sp. | *Plumeria acutifolia* Poiret | Antibacterial | Nithya et al. 2010 |
| Anthracenedione | *Halorosellinia* sp. and *Guignardia* sp. | Mangrove plant | Anticancer | Zhang et al. 2010 |
| Paclitaxel | *Pestalotiopsis microspora* | *Taxus wallachiana* | | Zhao et al. 2010 |
| Podophyllotoxin | *Alternaria neesex* | *Sinopodophyllum hexandrum* | Anticancer, antiviral, antioxidant, antibacterial, immuno stimulation, antirheumatic | |
| Camptothecin | *Fusarium solani* | *Camptotheca acuminate* | Antineoplastic | |
| Cochliodinol, Isocochliodinol | *Chaetomium* sp. | *Salvia officinalis* | Cytotoxic | Debbab et al. 2009 |
| Sesquiterpene aryl esters | *Armillaria mellea* | *Gastrodia elata* | Antimicrobial | Gao et al. 2009 |
| 5-methoxy-7-hydroxyphthalide | *Ascomycete* | *Meliotus dentatus* | Antifungal | Hussain et al. 2009 |

**Table 4.1** (continued)

| Bioactive metabolites | Endophyte | Host plant | Bioactive function | Reference |
|---|---|---|---|---|
| Cochlioquinone-A, isocochlioquinone-A | *Cochliobolus* sp. | *Piptadenia adiantoides* | Antiparasitic | Campos et al. 2008 |
| Javinicin | *Chloridium* sp. | *Azadirachta indica* | Antibacterial | Kharwar et al. 2009 |
| Taxol | *Phyllosticta* sp. | *Ocimum basilicum* | Anticancer | Gangadevi 2007 |
| Pestasol | *Drechslera gigantea* | *Agropyron repens* | Insect repellant | Bunkers et al. 1991 |
| Podophyllotoxin | *Podophyllum hexandrum* | *Trametes hirsute* | Anticancer | Puri et al. 2006 |
| Isoflavonoids | *Phomopsis* sp. | *Erythrina crista-galli* | Antimicrobial | Redko et al. 2006 |
| Ergosterol, cerevesterol | 2L-5 (unidentified) | *Ocimum basilicum* | | Haque et al. 2005 |
| Graphislactone-A | *Cephalosporium* sp. | *Trachelospermum jasminoides* | Antioxidant | Song et al. 2005 |

meadow fescue and tall fescue grasses may have different effects on various invertebrate and vertebrate herbivores (Cheplick and Faeth 2009); thirdly, the production of mycotoxins by endophytes, especially alkaloids, is conditional on environmental factors such as weather conditions or soil nutrients (Saikkonen et al. 2006; Saikkonen et al. 2010) and the genotype of the plant and endophyte (Cheplick and Faeth 2009).

Endophytes also aid in the seed production in particular types of grasses such as bird deterrent grasses or grass swards used at airports (Pyke et al. 2010). These grasses can deter birds either by reducing insect populations for insectivorous birds or by deterring the feeding of herbivorous birds causing them to shift to some other areas for feeding. These deterrent properties of the grasses are provided by endophytic fungi specifically selected for these characteristic features. Endophytic fungi grow within the host plant into the developing seed and when these seeds are sown, they develop into the new grass seedlings that confer bird deterrence to the developing sward (sward grass).

## 4.4 Phytoremediation

Benefits of industrialization, modernization and urbanization are always associated with some harmful effects such as accumulation of a wide variety of anthropogenic chemicals which includes halogenated hydrocarbons, petroleum hydrocarbons, polycyclic aromatic hydrocarbons, pesticides, heavy metals, solvents, etc. in the environment. This has caused havoc by enhancing the environmental problems (Meagher 2000; Rajkumar et al. 2010; Ma et al. 2011a, b). Phytoremediation by endophytes has been found to be a promising technology for the remediation of contaminated soils (Li et al. 2012). For the phytoremediation of heavy metals, the endophytes resistant to heavy metals can decrease metal phytotoxicity, enhance plant growth and affect metal accumulation and translocation in plants. Endophytes have application in the phytoremediation of organic contaminants produce various enzymes that degrade these contaminants thereby reducing both the evapotranspiration and phytotoxicity of volatile compounds. Li et al. (2012) have also discussed about various endophytes (bacteria and fungi) that aid in phytoremediation of heavy metals and organic contaminants. Various metal resistant bacterial endophytes belong to the taxa *Bacillus, Clostridium, Arthrobacter, Enterobacter, Leifsonia, Microbacterium, Paenibacillus, Pseudomonas, Xanthomonadaceae, Staphylococcus*, etc. Likewise, various reports of metal-resistant fungal endophytes include species of *Alternaria, Microsphaeropsis, Steganosporium, Mucor, Phoma, Peyronellaea* and *Aspergillus*. Microbial endophytes that are able to degrade organic contaminants include *Pseudomonas rhodesiae, Bacillus megaterium,*

*Paenibacillus amylolyticus, Burkholderia macroides, Achromobacter xylosoxidans, Pseudomonas tolaasii* and *Pseudomonas putida*. Two main groups of contaminants that are being degraded by endophyte-assisted phytoremediation include heavy metals (Ni, Pb, Cu, Cd, Cr, Zn, Co, etc.) and organic contaminants (trichloroethylene, naphthalene, catechol, phenol, alkanes, BTEX (benzene, toluene, ethylbenzene and xylene), etc.).

Engineered bacterial endophytes are being incorporated into the plants in order to improve the phytoremediation of contaminants particularly organic ones, so that they can efficiently degrade these organic contaminants (Taghavi et al. 2011). These engineered bacterial endophytes have also been tested to improve the phytoremediation of heavy metals but this study is still in its infancy (Lodewyckx et al. 2001).

Phytoremediation of organic compounds especially benzene, toluene, ethylbenzene, and xylene (BTEX) or trichloroethylene (TCE) brings about the evaporation of a significant amount of metabolites from these contaminants into the atmosphere thereby increasing their levels much above the optimal amounts. Therefore, to prevent the environment from such hazards, there is a need for certain alternative means by which these contaminants can be minimized. One of the best and safe ways is to inoculate plants with engineered bacterial endophytes so that they can optimally metabolize the contaminants of interest, thereby improving the overall process of phytoremediation.

## 4.5 Biotransformation

Biotransformation may simply be defined as any change or conversion of any substance by or in a biological system. It has been found that endophytes have excellent biotransformation abilities as they adapt so well with their host plants thereby having a strong tolerance towards the unique metabolites including toxins of their hosts (Wang and Dai 2011).

To control the excessive growth of endophytes, plants sometimes produce certain toxins against them. Endophytes have gradually formed a variety of tolerance mechanisms towards hosts' metabolites over their long period of coevolution, and have acquired a powerful ability to transform complex compounds. These endophytes may degrade the hosts' macromolecules into simpler ones or convert toxins to other substances thereby increasing their adaptability. In addition to plants, sometimes even endophytes become aggressive towards their host plants thereby producing mycotoxins or exoenzymes in order to colonize a particular host (Schulz et al. 2002). Several cases of endophytic virulence have also been reported (Peters et al. 1998; Costa et al. 2000). Likewise endophytes, plants also have defense mechanisms against the virulence effects of endophytes. Thus, from these studies, it is concluded that both the host plant and the endophyte produce unique metabolites and enzymes against each other and even in their defence reactions thereby making this host–endophyte relationship much more complex (Saikkonen et al. 2004).

The process of biotransformation mediated by endophytes has various applications, some of which are discussed as under:

### 4.5.1 Detoxification of Toxic Metabolites Produced by Host Plants

One of the important abilities of the endophytes is the detoxification of highly bioactive defence metabolites of host plant and converting them to simpler compounds to be easily metabolized. For example, plants usually produce unique class of compounds like benzoxazinones for defence against pests such as fungi, bacteria and insects. Once the plant defence response is triggered on attack of pests, hydroxamic acids are chemically transformed into toxic benzoxazolinon 2-benzoxazolinone (BOA) or the less toxic 2-hydroxy-1,4-benzoxazin-3-(2H)-one (HBOA) and some other toxic compounds. These toxins are detoxified by the biotransformation abilities of endophytes as reported by Zikmundova et al. (2002). They isolated four fungal endophytes, among which *Fusarium sambucinum* detoxified (BOA) to N-(2- hydroxyphenyl) malonamic acid and the rest detoxified (HBOA) to 2-Hydroxy-N-(2-hydroxyphenyl) acetamide, N-(2-hydroxyphenyl) acetamide and some other structural analogues.

### 4.5.2 Stereoselective Biotransformation Mediated by Endophytes

Endophytes are being exploited to obtain more active metabolites because fungal endophytes as a result of their effective biotransformation enzymes have been employed to change the three-dimensional conformation of compounds, e.g. thioridazine (THD), a phenothiazine neuroleptic drug, is commercially available as a racemic mixture of the enantiomers. Borges et al. (2008) isolated four fungal endophytes that showed stereoselective biotransformation potential, thereby transforming THD into pharmacologically active metabolites. Other examples of endophyte mediated stereoselective biotransformation includes transformation of flavans, i.e. stereoselective oxidation of flavans at C-4 by the endophytic fungus *Diaporthe* sp. isolated from a tea plant *Camelia sinensis* (Agusta et al. 2005) and biotransformation of tetrahydrofuran lignans by the endophytic fungus *Phomopsis* sp. inhabiting *Viguiera arenaria* (Verza et al. 2009).

### 4.5.3 *In Vivo* Simulation of Drug Metabolism in Mammals by Endophytes

It has been found that sometimes, the endophyte-mediated enzymatic reactions are similar to metabolic responses in mammals. Thus, these metabolic pathways and the active metabolites involved therein could be very much useful in studies on metabolism in mammals, e.g. two naturally occurring triterpenes: betulinic acid and betulonic acid found in many plants, exhibit important biological properties like antiviral and antineuroblastomic activity. Bastos et al. (2007) isolated four fungal endophytes *Arthrobotrys* sp., *Chaetophoma* sp., *Dematium* sp. and *Colletotrichum* sp. that showed certain oxidation reactions thereby converting betulinic acid to many oxygenated derivatives and some of these oxidizing reactions also occur in mammals.

## 4.6 A Potential Tool for Biodegradation

A large number of extracellular enzymes are being produced by endophytes including proteinases, cellulases, lipoidases, pectinases, phenoloxidases and lignin catabolic enzymes (Tan and Zou 2001; Oses et al. 2006; Bischoff et al. 2009). Such broad spectrum of enzyme production helps the endophytes in degradation and decomposition of plant debris. Four fungal endophytes from Chilean tree showed a nonselective white-rot wood decay pattern (Oses et al. 2006). Jordaan et al. (2006) isolated three different fungal endophytes (*Alternaria* sp., *Phoma* sp. and *Phomopsis* sp.) from surface-sterilized pods of *Colophospermum mopane* showing lignocellulolytic enzyme activity that helps them to accelerate the decay of pods. By doing this, they allowed the germination of seeds more effectively under favourable conditions. Endophytic fungi are now being exploited for the degradation of environmental pollutants like polycyclic aromatic hydrocarbons (PAHs). Dai et al. (2010) isolated a novel endophytic fungus *Ceratobasidum stevensii* from the euphorbiaceae plant that metabolized 89.51% of environmental pollutant phenanthrene. Likewise, various other novel fungal endophytes harboured by different plants are being studied for the degradation of various energetic compounds like 2,4,6-trinitrotoluene (TNT), octahydro-1,3,5,7-tetranitro-1,3,5,7-tetrazocine (HMX), hexahydro-1,3,5-trinitro-1,3,5-triazine (RDX), etc. (Aken et al. 2004).

## 4.7 Endophytes as a Source of Biocatalysts

It has been found that the endophytes produce a large number of enzymes that aid in the catabolism of a wide variety of complex organic molecules such as proteins, cellulose, chitin, etc. This is due to the absorptive mode of nutrition of these fungi and a wide variety of substrates which they exploit for their growth and thus produce a diverse array of extracellular enzymes like cellulases, chitinases, amylases, lipases and proteases, etc. (Suryanarayanan et al. 2012). Ostergaard and Olsen (2010), concluded that around 60 % of the industrial enzymes used currently are of fungal origin with a wide variety of applications, e.g. enzymes produced by endophytes are being used for meat processing, fermentation of coffee beans, baking, manufacturing corn syrup, hydrolysing milk proteins, removing stains, separating racemic mixtures of amino acids, bioremediation and biosensing. Hawksworth (1991) found that there are only five genera of fungi (*Penicillium, Rhizopus, Aspergillus, Humicola* and *Trichoderma*) out of 1.5 million estimated members which account for almost three-quarters of the 60 % industrial enzymes produced by endophytic fungi (Ostergaard and Olsen 2010). This leads to the immediacy for screening of novel and more efficient biocatalysts from different ecological groups of fungi (Peterson et al. 2011). There are a wide variety of industrial enzymes derived from endophytes acting as biocatalysts with diverse applications. Some of the common enzymes isolated from endophytes are as follows:

### 4.7.1 Alkaline Protease

Suryanarayanan et al. (2012) isolated several fungal endophytes from the leaves of dicotyledonous trees which showed positive results for alkaline protease activity using agar plate assay. It has been found that many fungal endophytes belonging to the genera *Nodulisporium, Corynespora, Colletotrichum, Xylaria, Curvularia* and *Robillarda* produce alkaline proteases using dot-blot assay (Thirunavukkarasu and Suryanarayanan unpublished). This industrial enzyme can be exploited for production of protein hydrolysates, mediation of drug release, brewing, baking, detergent additive, denture cleaners, tanning processes, cosmetics, animal feeds, waste water treatment, etc. (Gupta et al. 2002).

### 4.7.2 Chitinase and Chitosanase

Suryanarayanan et al. (2012) isolated foliar fungal endophytes inhabiting the trees of different forests in the Western Ghats mountain ranges (in southern India) that produced various extracellular enzymes including chitinases and chitosanases. These enzymes can be used in drug preparations for treating cancer, diabetes, asthma, inflammation, immunity modulation, anti-inflammatory, bacterial/fungal infections, wounds, sialic acid production, etc. (Hartl et al. 2012).

### 4.7.3 Laccase

Kumaresan et al. (2002) isolated two fungal endophytes *Pestalotiopsis* sp. and *Glomerella* sp. inhabiting *Rhizophora apiculata* and screened them for the production of extracellular enzymes such as laccase and cellulase. Promputtha et al. (2010) also reported the production of laccase from *Corynespora cassiicola*, endophytic in *Magnolia liliifera*. This enzyme can be exploited for processing of wine, fruit juice and beer, bioremediation of phenolic compounds, delignification, bio-bleaching, etc. (Kunamneni et al. 2008).

### 4.7.4 Chitin Deacetylase

Suryanarayanan et al. (2012) observed the production of chitin deacetylase from the fungal endophytes *Pestalotiopsis* sp. and *Sordaria* sp. inhabiting the tropical forest trees. This enzyme is used for biological control of insect pests, human and plant pathogens and can even be exploited for preparation of chitosan from chitin (Zhao et al. 2010).

### 4.7.5 Acidic Protease

Many foliar fungal endophytes, i.e. *Nodulisporium, Xylaria, Colletotrichum, Corynespora* and *Curvularia* show positive results for the production of acidic proteases (Suryanarayanan et al. 2012). This enzyme aids in preparation of fermented foods and seasoning material as well as in digestion (Rao et al. 1998).

### 4.7.6 β-glucosidase

Suryanarayanan et al. (2012) have observed the production of β-glucosidases from the endophytic *Phomopsis* sp. Chauve et al. (2010) reported that this enzyme is mainly used for biofuel production from lignocellulosic biomass.

### 4.7.7 Tannase

Production of tannase enzyme from the endophytic *Phomopsis* sp. has been reported by Suryanarayanan et al. (2012). This enzyme is used as a clarifying agent in instant tea, fruit juices and wine (Ramirez-Coronel et al. 2003).

## 4.8 Agricultural Importance of Endophytes

In order to feed the growing human population, the use of man-made fertilizers (particularly N and P fertilizers) have increased manifold to increase the crop production (Vance 2001). People are mainly interested in large-scale crop production, thereby using wide varieties of pesticides, fungicides and bactericides without caring about their harmful effects on ground and surface water (Cook 1992).

One of the best alternative methods involves the use of endophytes as biofertilizers in agriculture. In the recent times, use of endophytes in agriculture have received a substantial attention due to their broad host range, tolerance to a vast range of environmental conditions and promotion of vegetative growth including seed production of various economically important plants (Singh et al. 2000; Shende et al. 2006). Endophytes have become a valuable tool for agriculture due to their capability to colonize the internal tissues of the host and to contribute towards plant growth.

Over the last 40 years, nutrient use efficiency (NUE) in crop production has decreased considerably (Hirel et al. 2007). Due to this and rapid depletion of mines mineral deposits (particularly Potassium and phosphorus) along with their increasing costs have raised threat to current crop production and future food security (Tilman et al. 2002; Cordell et al. 2009). Nutrient use efficiency in crop plants may significantly be increased by root-colonizing endophytes. These endophytes help the crop plants in two ways:

a. Direct mode of action through nitrogen fixation, phosphorus and micronutrient uptake, etc.
b. Indirect mode of action through antagonism against soil- and seed-borne pathogens, induction of plant resistance, etc.

Nowadays, a strongly growing field of land use is desert agriculture, but the problems caused by soil-borne pathogens limit the yield. However, endophytic biological control agents (BCAs) are the promising candidates for plant protection that can suppress these soil-borne pathogens. But for desert agriculture, only specific and drought resistant strains of BCAs will prove beneficial (Koberl et al. 2011). Khan et al. (2012) reported an endophyte, *Paraconiothyrium* sp. as a potential producer of phytotoxin. So, it can be exploited for weed control strategies.

Among the most important microbes beneficial for agriculture are the different species of genus *Trichoderma,* being used worldwide as active ingredients of biopesticides and biofertilizers (Harman et al. 2004). Specific species of this genus can act endophytically in the vascular system by colonizing the root tissues thereby, establishing a physiological interaction with the host plant based on molecular crosstalk. This results in the promotion of plant growth and development along with resistance to diseases, degradation of toxic compounds, promotion of water and nutrient uptake and finally brings about the production of bioactive *Trichoderma* metabolites.

## 4.9 Endophytes as Nanoparticle Biosynthesizers

Endophytes can be used as potential biosynthesizers for production of various economically and industrially important compounds. Various endophytes can be exploited for the extracellular or intracellular synthesis of different indigenous or nonindigenous compounds. *Pestalotia* sp. isolated from the healthy leaves of *Syzygium* has been used for the extracellular synthesis of silver nanoparticles which exhibited good antimicrobial potential (Raheman et al. 2011). Similarly, *Aspergillus clavatus,* isolated from the sterilized stem tissue of *Azadirachta indica,* was also reported to be capable of synthesizing extracellular silver nanoparticles (Verma et al. 2010). The same endophyte was found to produce diverse range of gold nanoparticles as well, especially nanotriangles in the size range from 20 to 35 nm. These nanotriangles possess special optical and physical properties which can be exploited for vital applications such as biomedicine, optics, electronics, catalysis and many more (Verma et al. 2010). There is a big thrust in the area of synthesizing different metal nanoparticles using biological systems these days because of their potential application in nanomedicines (Panneerselvam et al. 2011). There are wide applications of nanoparticles in diverse fields as catalysts, sensors, medicines, etc. that largely depend on the different physical and optical properties of the particles. The synthesis of nanoparticles using biological systems provides new routes to develop nanoparticles with desired properties for making their exploitation possible in diverse fields (Pugazhenthiran et al. 2009). The nonpathogenic and ecofriendly behaviours of endophytes make them as good candidates for biosynthesizers.

## 4.10 Endophytes as Host Plant Growth Promoters

Endophytes might prove to be important biological agents in near future and revolutionize the agriculture industry without harming the nature. By artificially inoculating specific endophytes in non-host plants, the property of microbes to improve host plant growth potential can be transferred to new and economically important crops. It has been observed that seed bacterization of bhendi plant with some selected endophytic isolates promote plant growth (Vetrivelkalai et al. 2010). Another endophyte *Piriformospora indica* has great potential as plant growth-promoting agent and is sometimes termed as plant probiotic. Phosphorous (P) is one of the most essential mineral nutrients for proper plant growth and development. *P. indica* has important role in P transfer to the plant particularly under P-deprived conditions (Kumar et al. 2011). Some endophytes are reported to have nitrogen fixation ability that can also promote the host plant growth (Hurek et al. 2003). *Enterobacter* sp. 638, isolated from poplar plant, has been reported as a plant growth-promoting gamma-proteobacterium (Taghavi et al. 2010). Poplar plant is one of the potentially important biofuel feed stock plant. Combined genome, transcriptome and metabolome analysis revealed synergistic interaction between poplar and its growth-promoting endophyte. Nitrogen fixation, P solubilization, indole acetic acid (IAA) production and siderophore secretion are some of the commonly followed mechanisms through which endophytes are able to promote the host plant growth. Improved plant growth, higher nutrient content, resistance to insect pests and herbivores, resistance or tolerance to diseases, increased competitiveness, enhanced tolerance to stressful factors such as heavy metals, low pH, high salinity, etc., are the other benefits from the endophytic interaction (Kumar et al. 2011). Endophytes can be used as biofertilizers for plant growth promotion (Pandya et al. 2010). The plant growth-promoting potential of endophytes can also be exploited for the improved and sustainable production of important crops on marginal and nonagricultural soils (Table 4.2).

## 4.11 Endophytes as Host Plant Stress Relievers

Different kinds of abiotic factors such as drought, flood, high soil salinity, heat, cold, oxidative stress, heavy metal toxicity and nutrient deficiency, etc. lead to stress in the plant resulting in lower or no yield. Plant growth-promoting

**Table 4.2** Endophytes characterized for their plant growth promoting potential

| Endophyte | Host plant | Plant growth-promoting activity | Reference |
|---|---|---|---|
| *Paecilomyces Formosus* | Cucumber | Gibberellin synthesis | Khan et al. 2012 |
| *Methylobacterium* spp. | *Citrus* sp. | Siderophore secretion | Vendan et al. 2010 |
| *Micrococcus luteus* | *Ginseng* sp. | IAA synthesis | |
| *Lysinibacillus fusiformis* | | Phosphate solubilisation | |
| *Stenotrophomonas Maltophilia* | | Nitrogen fixation | |
| *Penicillium citrinum* | *Ixeris repens* | Phytohormone synthesis | Khan et al. 2008 |
| *Pantoea agglomerans* | *Oryza sativa* | Nitrogen fixation | Feng et al. 2006 |

**Table 4.3** Important endophytes used as plant abiotic stress relievers

| Endophyte | Abiotic stress | Host | Reference |
|---|---|---|---|
| *Fusarium culmorum* | Drought | *Oryza sativa* | Redman et al. 2011 |
| *Piriformospora indica* | Drought | *Brassica campestris* | Sun et al. 2010 |
| *Trichoderma hamatum* | Drought | *Theobroma cacao* | Bae et al. 2009 |
| *Curvularia protuberate* | Drought | *Oryza sativa* | Rodriguez et al. 2008 |
| *Curvularia protuberate* | Heat | *Lycopersicon esculentum* | |
| *Fusarium culmorum* | Salinity | *Oryza sativa, Lycopersicon esculentum* | |
| *Piriformospora indica* | Salinity | *Hordeum vulgare* | Waller et al. 2005 |
| *Neotyphodium* | Drought | *Festuca arizonica* | Morse et al. 2002 |

potential of endophytes can be exploited to overcome the limitations of crop productivity caused by different abiotic stress factors and it may prove to be a promising alternative strategy (Table 4.3). For the sustainable existence of plants, endophytes have done and are still doing a good job as symbionts (Singh et al. 2011). *P. indica* and *Sebacina vermifera* are common root endophytes found in wide variety of plant species. Both organisms have the potential to promote their host plant growth, and to induce resistance against soil-borne fungal pathogens (Ghahfarokhi et al. 2010). Agricultural productivity under saline conditions can be improved by exploiting the potential of endophytic fungi as they have been reported to produce important stress-relieving phytohormones (Khan et al. 2012). The ability of fungal endophytes to confer stress tolerance to plants may provide a novel strategy for mitigating the impacts of global climate change on agricultural and native plant communities (Rodriguez et al. 2008).

## 4.12 Endophytes as a Source of Biofuel

Constantly increasing energy consumption, energy demand and environmental concerns are constantly heralding the need of an alternative biofuel resource. Many endophytes possess enzymes of interest that have the ability to degrade available carbon sources into compounds that can be used as biofuels (Suryanarayanan et al. 2012). Studies say that endophytes can be exploited for the synthesis of fuel-based compounds utilizing cellulose-based medium from host plant. Solid-phase micro-extraction (SPME)-gas chromatography/mass spectrometry (GC/MS) analysis of an endophytic fungus, *Gliocladium roseum* revealed the production of a series of volatile hydrocarbons and their derivatives under controlled conditions (Strobel et al. 2008). The hydrocarbons characterized were named as myco-diesel as some of them had properties similar to diesel fuel. Some of the endophytes isolated from different species of tropical plants may prove to be potential biofuel sources as they contain high concentrations of methyl esters which are effective biofuel precursors (Santos et al. 2011).

## 4.13 Endophytes as Biocontrol Agents

Many plant endophytes are reported as potential biocontrol agents against plant pathogens. The biocontrol activity imposed by endophytes may be due to competition for space and nutrients, secretion of chitinolytic enzymes, myco-parasitism and production of inhibitory compounds. Some root endophytic fungi like *Piroformospora indica, Sebacina vermifera* and *Trichoderma* species have been reported as potential biocontrol agents against plant pathogens such as *Gaeumannomyces graminis,* causative agent of Take-all disease of wheat; *Fusarium oxysporum,* causative agent of fusarium wilt of lentil and many more (Ghahfarokhi et al. 2010; Dolatabadi et al. 2012). Shalini and Kotasthane (2007) reported strains of *Trichoderma* including *Trichoderma harzianum, Trichoderma viride* and *Trichoderma aureoviride* as growth inhibitors of *Rhizoctonia solani,* a potent plant pathogen. Some endophytes, if not completely eliminate, reduce the severity of the host plant disease to an efficient extent. *P. indica*-colonized host plants were reported to be affected less severely by *Pseudocercosporella herpotrichoides* disease (Pandya et al. 2010). It has been reported that *P. indica* has the ability to improve the production of tomato by acting antagonistically against *Verticillium dahliae* (Fakhro et al. 2010). The host plant growth promotion by *P. indica* was not only observed in soil but also in hydroponic cultures. *Muscodor albus,* an endophytic fungal isolate from the plant *Cinnamomum zeylanicum* was reported to produce bioactive volatile organic compounds (VOCs). The mixture of VOCs produced by the isolate was found to be effective against various human and plant pathogenic fungi and bacteria (Strobel et al. 2006). Some of the fungal endophytes isolated from different vegetable crops were reported to exhibit anti-oomycete property (Kim et al. 2007). In dual culture tests, the fungal isolates were shown to inhibit the growth of various oomycete plant pathogens, viz. *Pythium ultimum, Phytophthora infestans* and *Phytophthora capsici.* So, endophytes can be considered as potent anti-oomycete agents. More intensive research is required to optimize the application of endophytes as biocontrol agents so that the quality of crops in terms of taste and health should not be negatively affected.

### 4.13.1 Concluding Statement and Future Perspectives

In today's world, man has become almost inadequate to cope up with the medical problems due to the emergence of new diseases, increased incidences of fungal infections and drug resistance by microorganisms. All this speaks about the need for new and useful bioactive compounds to provide relief and assistance to human sufferings. Endophytes are considered as a mine for bioprospecting and thereby, help in various aspects to overcome many problems. It is proposed that many interesting cases are yet to be discovered even after the exploitation of number of new endophytes and the metabolites thereof reported since last century. Endophytes have not only benefited animals, but have been exploited to protect various plants from a wide range of pathogens and insect pests, etc. Some among them function by antagonism, inducing host defenses and also help in improving plants health and nutrition.

Among the various endophytic microbes, fungi have particularly proved to be a promising and largely untapped reservoir of bioactive metabolites. The natural products have been optimized by evolutionary, environmental and ecological factors yielding useful and effective bioactive secondary metabolites.

Majority of the drugs that are now being used against infective organisms, agronomic pests, parasites of livestock and humans and diseases such as tuberculosis, cancer, malaria, diabetes, etc., are mainly composed of a diverse range of bioactive plant secondary metabolites.

To achieve a range of applications associated with plant–endophyte relationship, there is a need for understanding the mechanisms behind the interaction of endophytes with their host plant so that the biotechnological potential of such a relationship can be fully elucidated. A deeper understanding at the molecular and genetic levels of this relationship may prove beneficial for inducing and optimizing secondary metabolite production in vitro to yield potential bioactive metabolites. As a result of genetic engineering and metabolic regulation, the content and the yield of active ingredients in certain known strains can be accordingly increased for large-scale production. The present review has unlocked the myriad benefits of endophytes which can be utilized in different ways for the improvement of plant growth, development and productivity. Endophytes, hence is an important and interesting complex occupying a unique niche worth for exploration.

## References

Agusta A, Maehara S, Ohashi K, Simanjuntak P, Shibuya H (2005) Stereoselective oxidation at C-4 of flavans by the endophytic fungus *Diaporthe* sp. isolated from a tea plant. Chem Pharm Bull 53:1565–1569

Aken VB, Yoon JM, Schnoor JL (2004) Biodegradation of nitro-substituted explosives 2, 4, 6-trinitrotoluene, hexahydro-1, 3, 5-trinitro-1, 3, 5- triazine, an octahydro-1, 3, 5, 7-tetranitro-1, 3, 5-tetrazocine by a phytosymbiotic *Methylobacterium* sp. associated with poplar tissues (*Populus deltoides* x nigra DN34). Appl Environ Microbiol 70:508–517

Aly AH, Debbab A, Proksch P (2011) Fungal endophytes: unique plant inhabitants with great promises. Appl Microbiol Biotechnol 90:1829–1845

Bae H, Sicher RC, Kim MS, Kim SH, Strem MD, Rachel L, Melnick RL, Bailey BA (2009) The beneficial endophyte *Trichoderma hamatum* isolate DIS 219b promotes growth and delays the onset of the drought response in *Theobroma cacao*. J Exp Bot 60(11):3279–3295

Bastos DZL, Idac P, de Jesus DA, de Oliveira BH (2007) Biotransformation of betulinic and betulonic acids by fungi. Phytochemisty 68:834–839

Bischoff KM, Wicklow DT, Jordan DB, de Rezende ST, Liu SQ, Hughes SR, Rich JO (2009) Extracellular hemicellulolytic enzymes from the maize endophyte *Acremonium zeae*. Current Microbiology 58:499–503

Borges KB, Borges WDS, Pupo MT, Bonato PS (2008) Stereoselective analysis of thioridazine-2-sulfoxide and thioridazine-5- sulfoxide: an investigation of rac-thioridazine biotransformation by some endophytic fungi. J Pharmaceut Biomed 46:945–952

Brooks DS, Gonzalez CF, Appel DN, Filer TH (1994) Evaluation of endophytic bacteria as potential biological control agents for oak wilt. Biol Contr 4:373–381

Bunkers GF, Kenfield D, Strobel GA (1991) Production of petasol by *Drechslera gigantia* in liquid culture. Mycological Research 95:347–351

Bunyapaiboonsri T, Yoiprommarat S, Srikitikulchai P, Srichomthong K, Lumyong S (2010) Oblongolides from the endophytic fungus *Phomopsis* sp. BCC 9789. J Nat Prod 73(1):55–59

Campos FF, Rosa LH, Cota BB, Caligiorne RB, Rabello ALT, Alves TMA, Rosa CA, Zani CL (2008) Leishmanicidal metabolites from *Cochliobolus* sp., an endophytic fungus isolated from Piptadenia adiantoides (Fabaceae). PLoS Negl Trop Dis 2(12):e348

Chauve M, Mathias H, Huc D, Casanave D, Monot F, Ferreira (2010) Comparative kinetic analysis of two fungal β-glucosidases. Biotechnol Biofuels. doi:10.1186/1754–6834-3–3

Cheplick GP, Faeth SH (2009) Ecology and evolution of the grass-endophyte symbiosis. Oxford University Press, Oxford. doi:10.1093/acprof:oso/9780195308082.001.0001

Cook GW (1992) Fertilizing for maximum yield, 3rd edn. MacMillan, New York

Cordell D et al (2009) The story of phosphorus: global food security and food for thought. Global Environ Change 19:292–305

Costa LSR, Azevedo JL, Pereira JO, Carneiro ML, Labate CA (2000) Symptomless infection of banana and maize by endophytic fungi impairs photosynthetic efficiency. New Phytol 147:609–615

Dai CC, Tian LS, Zhao YT, Chen Y, Xie H (2010) Degradation of phenanthrene by the endophytic fungus *Ceratobasidum stevensii* found in *Bischofia polycarpa*. Biodegradation 21:244–255

Debbab A, Aly AH, Edrada-Ebel RA, Muller WEG, Mosaddak M, Hakiki A, Ebel R, Proksch P (2009) Bioactive secondary metabolites from the endophytic fungus *Chaetomium* sp. isolated from *Salvia officinalis* growing in Morocco. Biotechnol Agron Soc Environ 13(2):229–234

Dolatabadi HK, Goltapeh EM, Mohammadi N, Rabiey M, Rohani N, Varma A (2012) Biocontrol Potential of Root Endophytic Fungi and *Trichoderma* Species Against *Fusarium* Wilt of Lentil Under In vitro and Greenhouse Conditions. J Agr Sci Tech 14:407–420

Dompeipen EJ, Srikandance Y, Suharso WP, cahyana H, Simanjuntak P (2011) Potential endophytic microbes selection for antidiabetic bioactive compounds production. Asian J Biochem 6(6):465–471

El-Shanshoury AR, El-Sououd SMA, Awadalla OA, El-Bandy NB (1996) Effects of *Streptomyces corchorusii*, *Streptomyces mutabilis*, pendimethalin and metribuzin on the control of bacterial and *Fusarium* wilt of tomato. Can J Bot Rev 74:1016–1022

Fakhro A, Linares DRA, Bargen SV, Bandte M, Buttner C, Grosch R, Schwarz D, Franken P (2010) Impact of *Piriformospora indica* on tomato growth and on interaction with fungal and viral pathogens. Mycorrhiza 20:191–200

Feng Y, Shen D, Song W (2006) Rice endophyte *Pantoea agglomerans* YS19 promotes host plant growth and affects allocations of host photosynthates. J Appl Microbiol 100(5):938–945

Fisher PJ, Petrini O, Lazpin SHM (1992) The distribution of some fungal and bacterial endophytes in maize (*Zea mays* L.). New Phytol 122:299–305

Gangadevi V, Muthumary J (2007) Endophytic fungal diversity from young, mature and senescent leaves of *Ocimum basilicum* L. with special reference to Taxol production. Ind J Sci Tech 1(1):1–12

Gao LW, Li WY, Zhao YL, Wang JW (2009) The cultivation, bioactive components and pharmacological effects of *Armillaria mellea*. Afr J Biotech 8(25):7383–7390

Ghahfarokhi RM, Goltapeh ME (2010) Potential of the root endophytic fungus *Piriformospora indica*; *Sebacina vermifera* and *Trichoderma* species in biocontrol of take-all disease of wheat *Gaeumannomyces graminis* var. *tritici in vitro*. J Agr Tech 6(1):11–18

Giridharan P, Verekar SA, Khanna A, Mishra PD, Deshmukh SK (2012) Anticancer activity of Sclerotiorin isolated from an endophytic fungus Cephalotheca faveoleta Yaguchi, Nishim & Udagawa. Ind J Exp Biol 50:464–468

Gupta R, Beg QK, Lorenz P (2002) Bacterial alkaline proteases: molecular approaches and industrial applications. Appl Microbiol Biotechnol 59:15–32

Haque MA, Hossain MS, Rahman MZ, Rahman MR, Hossain MS, Mosihuzzaman M, Nahar N, Khan SI (2005) Isolation of bioactive secondary metabolites from the endophytic fungus of *Ocimum basilicum*. J Pharm Sci 4(2):127–130

Harman GE, Howell CR et al (2004) *Trichoderm a* species-Opportunistic, avirulent plant symbionts. Nat Rev Microbiol 2(1):43–56

Hartl L, Zach S, Seidl-Seiboth V (2012) Fungal chitinases: diversity, mechanistic properties and biotechnological potential. Appl Microbiol Biotechnol 93:533–543

Hawksworth DL (1991) The fungal dimension of biodiversity: magnitude, significance, and conservation. Mycological Research 95:641–655

Hemtasin C, Kanokmedhakul S, Kanokmedhakul K, Hahnvajanawong C, Soytong K, Prabpai S, Kongsaeree P (2011) Cytotoxic Pentacyclic and tetracyclic aromatic sesquiterpenes from *Phomopsis archeri*. J Nat Prod 74(4):609–613

Hirel B et al (2007) The challenge of improving nitrogen use efficiency in crop plants: towards a more central role for genetic variability and quantitative genetics within integrated approaches. J Exp. Bot 58:2369–2401

Huang XZ, Zhu Y, Guan XL, Tian K, Guo JM, Wang HB, Fu GM (2012) A novel Antioxidant Isobenzofuranone Derivative from fungus *Cephalosporium* sp. AL031. Molecules 17:4219–4224

Hurek T, Hurek BR (2003) *Azoarcus* sp. strain BH72 as a model for nitrogen-fixing grass endophytes. J Biotech 106:169–178

Hussain H, Krohn K, Draeger S, Meier K, Schulz B (2009) Bioactive chemical constituents of a sterile endophytic Fungus from *Meliotus dentatus*. Rec Nat Prod 3(2):114–117

Idris A, Al-tahir I, Idris E (2013) Antibacterial activity of endophytic fungi extracts from the medicinal plant Kigelia africana. Egypt Acad J Biolog Sci 5(1):1–9

Isaka M, Chinthanom P, Boonruangprapa T, Rungjindamai N, Pinruan U (2010) Eremophilanetype sesquiterpenes from the fungus *Xylaria* sp. BCC 21097. J Nat Prod 73:683–687

Jordaan A, Taylor JE, Rossenkhan R (2006) Occurrence and possible role of endophytic fungi associated with seed pods of *Colophospermum mopane* (Fabaceae) in Botswana. South Afr J Bot 72:245–255

Joseph B, Priya RM (2011) Bioactive compounds from endophytes and their potential in pharmaceutical effect: a review. Am J Biochem Mol Biol 1:291–309

Khan SA, Hamayun M, Yoon H, Kim HY, Suh SJ, Hwang SK et al (2008) Plant growth promotion and *Penicillium citrinum*. BMC Microbiology. doi:10.1186/1471–2180-8–231

Khan AL, Hamayun M, Kang SM, Kim YH, Jung HY, Lee JH, Lee IJ (2012) Endophytic fungal association via gibberellins and indole acetic acid can improve plant growth under abiotic stress: an example of *Paecilomyces formosus* LHL10. BMC Microbiology 12(3):1–14

Kharwar RN, Verma VC, Kumar A, Gond SK, Harper JK, Hess WM, Lobkovosky E, Ma C, Ren Y, Strobel GA (2009) Javanicin, an antibacterial Naphthaquinone from an endophytic fungus of Neem, *Chloridium* sp. Current Microbiology 58:233–238

Kharwar RN, Mishra A, Gond SK, Stierle A, Stierle D (2011) Anticancer compounds derived from fungal endophytes: their importance and future challenges. Nat Prod Rep 28:1208–1228

Kim HY, Choi GJ, Lee HB, Lee SW (2007) Some fungal endophytes from vegetable crops and their anti-oomycete activities against tomato late blight. Lett Appl Microbiol 44:332–337

Koberl M, Muller H, Ramadan EM, Berg G (2011) Desert farming benefits from microbial potential in arid soils and promotes diversity and plant health. PLOS ONE 6(9)

Kumar M, Yadav V, Kumar H, Sharma R, Singh A, Tuteja N, Johri AK (2011) *Piriformospora indica* enhances plant growth by transferring phosphate. Plant Signal Behav 6(5):723–725

Kumaresan V, Suryanarayanan TS (2002) Endophyte assemblages in young, mature and senescent leaves of *Rhizophora apiculata*: evidence for the role of endophytes in mangrove litter degradation. Fungal Diversity 9:81–91

Kunamneni A, Camarero S, Garcia-Burgos C, Plou FJ, Ballesteros A, Alcalde M (2008) Engineering and applications of fungal laccases for organic synthesis. Microb Cell Factories. doi:10.1186/1475–2859-7–32

Li HY, Wei DQ, Shen M, Zhou ZP (2012) Endophytes and their role in phytoremediation. Fungal Diversity 54:11–18

Li XJ, Zhang Q, Zhang AL, Gao JM (2012) Metabolites from *Aspergillus fumigatus*, an endophytic fungus associated with *Melia azedarach*, and their antifungal, antifeedant, and toxic activities. J Agric Food Chem 60:3424–3431

Liu SF, Tang WH (1996) The study on endophytic streptomyces of cotton. In: Tang WH, Cook RJ, Rovira A (eds) Advances in biological control of plant diseases. China Agricultural University, China, pp 212–213

Lodewyckx C, Taghavi S, Mergeay M et al (2001) The effect of recombinant heavy metal resistant endophytic bacteria in heavy metal uptake by their host plant. Int J Phytoremediation 3:173–187

Ma Y, Prasad MNV, Rajkumar M, Freitas H (2011b) Plant growth promoting rhizobacteria and endophytes accelerate phytoremediation of metalliferous soils. Biotechnol Advances 29:248–258

Ma Y, Rajkumar M, Luo Y, Freitas H (2011a) Inoculation of endophytic bacteria on host and non-host plants-effects on plant growth and Ni uptake. J Hazard Mater 195:230–237

Meagher RB (2000) Phytoremediation of toxic elemental and organic pollutants. Curr Opin Plant Biol 3:153–162

Morse LJ, Day TA, Faeth SH (2002) Effect of *Neotyphodium* endophyte infection on growth and leaf gas exchange of Arizona fescue under contrasting water availability regimes. Environ Exp Bot 48:257–268

Nithya K, Muthumary J (2010) Secondary metabolite from *Phomopsis* sp. isolated from *Plumeria acutifolia* Poiret. Recent Res Sci Tech 2(4):99–103

Oses R, Valenzuela S, Freer J, Baeza J, Rodriguez J (2006) Evaluation of fungal endophytes for lignocellulolytic enzyme production and wood biodegradation. Int Biodeterior Biodegrad 57:129–135

Ostergaard LH, Olsen HS (2010) Industrial applications of fungal enzymes. In: Hofrichter XM (ed) The mycota. Springer, Berlin, pp 269–290

Pandya U, Saraf M (2010) Application of fungi as a biocontrol agent and their biofertilizer potential in agriculture. J Adv Dev Res 1(1):90–99

Panneerselvam C, Ponarulselvam S, Murugan K (2011) Potential anti-plasmodial activity of synthesized silver nanoparticle using *Andrographis paniculata* Nees (Acanthaceae). Arch Appl Sci Res 3(6):208–217

Peters S, Dammeyer B, Schulz B (1998) Endophyte-host interactions. I. Plant defense reactions to endophytic and pathogenic fungi. Symbiosis 25:193–211

Peterson R, Grinyer J, Nevalainen H (2011) Extracellular hydrolase profiles of fungi isolated from koala faeces invite biotechnological interest. Mycol Prog 10:207–218

Petrini O (1991) Fungal endophytes of tree leaves. In: Andrews JH, Hirano SS (eds) Microbial ecology of leaves. Springer Verlag, New York, pp 179–197

Pimentel MR, Molina G, Isio APD, Marostica Junior MR, Pastore GM (2010) The use of endophytes to obtain bioactive compounds and their application in biotransformation Process. Biotechnol Res Int. doi:10.4061/2011/576286

Premjanu N, Jayanthy C (2012) Endophytic fungi a repository of bioactive compounds—a review. Int J Inst Pharm Life Sci 2(1) :135–162

Promputtha I, Hyde KD, McKenzie EHC, Peberdy JF, Lumyong S (2010) Can leaf degrading enzymes provide evidence that endophytic fungi becoming saprobes? Fungal Diversity 41:89–99

Pugazhenthiran N, Anandan S, Kathiravan G, Prakash NKU, Crawford S, Kumar MA (2009) Microbial synthesis of silver nanoparticles by *Bacillus* sp. J Nanopart Res 11:1811–1815

Puri SC, Nazir A, Chawla R, Arora R, Hasan SR, Amna T, Ahmed B, Verma V, Singh S, et al (2006) The endophytic fungus *Trametes hirsuta* as a novel alternative source of podophyllotoxin and related aryl tetralin lignans. J Biotechnol 122:494–510

Pyke N, Rolston P, Chynoweth R, Kelly M, Pennell (2010) Seed production of bird deterrent grass for use at airports. 29th Meeting of the International Bird Strike Committee, Cairns (Australia) 2010

Qadri M, Johri S, Shah BA, Khajuria A, Sidiq T, Lattoo SK et al (2013) Identification and bioactive potential of endophytic fungi isolated from selected plants of the Western Himalayas. SpringerPlus 2:8

Raheman F, Deshmukh S, Ingle A, Gade A, Rai M (2011) Silver nanoparticles: novel antimicrobial agent synthesized from a endophytic fungus *Pestalotia* sp. isolated from leaves of *Syzygium cumini* (L.). Nano Biomed Eng 1(3):174–178

Rajkumar M, Ae N, Prasad MNV, Freitas H (2010) Potential of siderophore-producing bacteria for improving heavy metal phytoextraction. Trends Biotech 28:142–149

Ramirez-Coronel MA, Viniegra-González G, Darvil A, Augur C (2003) A novel tannase from *Aspergillus niger* with β-glucosidase activity. Microbiology 149:2941–2946

Rao MB, Tanksale AM, Ghatge MS, Deshpande V (1998) Molecular and biotechnology aspects of microbial proteases. Microbiol Mol Biol Rev 62:597–635

Redko F, Clavin M, Weber D, Anke T, Martino V (2006) Search for active metabolites of *Erythrina crista-galli* and its endophyte *Phomopsis* sp. Mol Medi Chem 10:24–26

Redman RS, Kim YO, Claire JDA, Woodward CJDA, Greer C, Espino L, Doty SL, Rodriguez RJ (2011) Increased fitness of rice plants to abiotic stress via habitat adapted symbiosis: a strategy for mitigating impacts of climate change. PLOS ONE 6(7):1–10

Rodriguez RJ, Henson J, Volkenburgh EV, Hoy M, Wright L, Beckwith F, Yong-Ok Kim YO, Redman RS (2008) Stress tolerance in plants via habitat-adapted symbiosis. The ISME Journal 2:404–416

Saari S, Helander M, Faeth SH, Saikkonen K (2010) The effects of endophytes on seed production and seed predation of tall fescue and meadow fescue. Microb. Ecol 60:928–934

Saikkonen K, Wali P, Helander M, Faeth SH (2004) Evolution of endophyte-plant symbioses. Trends Plant Sci 9:275–280

Saikkonen K, Lehtonen P, Helander M, Koricheva J, Faeth SH (2006) Model systems in ecology: dissecting the endophyte-grass literature. Trends Plant Sci 11:428–433

Saikkonen K, Saari S, Helander M (2010) Defensive mutualism between plants and endophytic fungi? Fungal Diversity 41:101–113

Santiago C, Fitchett C, Murno MHG, Jalil J, Santhanam J (2012) Cytotoxic and antifungal activities of 5-Hydroxyramulosin, a compound produced by an endophytic fungus isolated from *Cinnamomum mollisimum*. Hindawi publishing corporation. doi:10.1155/2012/689310

Santos F, Florisvaldo C, Taicia PF, Joanita N, Marcos RM, Edson RF (2011) Endophytic fungi as a source of biofuel precursors. J Microbiol Biotechnol 21(7):728–733

Schulz B, Boyle C, Draeger S, Rommert AK, Kronh K (2002) Endophytic fungi: a source of novel biologically active secondary metabolites. Mycological Research 106:996–1004

Shalini S, Kotasthane AS (2007) Parasitism of *Rhizoctonia solani* by strain of *Trichoderma* spp. EJEAFChe 6(8):2272–2281

Shende S, Bhagwat K, Wadegaonkar P, Rai M, Varma A (2006) *Piriformospora indica* as a new and emerging mycofertilizer and biotizer: potentials and prospects in sustainable agriculture. In: Rai M (ed) Handbook of microbial biofertilizers. Food products press, New York, pp 477–496

Singh R, Adholeya A, Mukerji KG (2000) Mycorrhiza in control of soil-borne pathogens. In: Mukerji KG, Chamola BP, Singh J (eds) Mycorrhizal biology. Kluwer, New York, pp 173–196

Singh LP, Gill SS, Tuteja N (2011) Unraveling the role of fungal symbionts in plant abiotic stress tolerance. Plant Signaling Behav 6(2):175–191

Singh S, Bukkambudhi RG, Gudasalamani R, Uma Shaanker R, Manchanahally BS (2013) Endophytic fungi from Miquelia dentata Bedd., produce the anti-cancer alkaloid, Camptothecine. Phytomedicine 20:337–342

Song YC, Huang WY, Sun C, Wang FW, Tan RX (2005) Characterization of graphislactone a as the antioxidant and free radical-scavenging substance from the culture of *Cephalosporium* sp. IFB-E001, an endophytic fungus in *Trachelospermum jasminoides*. Biol Pharm Bull 28(3):506–509

Strobel G (2006) Harnessing endophytes for industrial microbiology. Curr Opin Microbiol 9:240–244

Strobel G, Daisy B (2003) Bioprospecting for microbial endophytes and their natural products. Microbial Mol Biol Rev 67(4):491–502

Strobel GA, Daisy B, Castillo U, Harper J (2004) Natural products from endophytic microorganisms. J Nat Prod 67(2):257–268

Strobel GA, Knighton B, Kluck K, Ren Y, Livinghouse T, Griffin M, Spakowich D, Sears J (2008) The production of myco-diesel hydrocarbons and their derivatives by the endophytic fungus *Gliocladium roseum* (NRRL 50072). Microbiology 154:3319–3328

Sumarah MW, Miller JD (2009) Anti-insect secondarys metabolites from fungal endophytes of conifer trees. Nat Prod Commun 4(11):1497–1504

Sun C, Johnsona JM, Caib D, Sherametia I, Oelmüllera R, Louc B (2010) *Piriformospora indica* confers drought tolerance in Chinese cabbage leaves by stimulating antioxidant enzymes, the expression of drought-related genes and the plastid-localized CAS protein. J Plant Physiol 167:1009–1017

Suryanarayanan TS, Thirunavukkarasu N, Govindarajulu MB, Gopalan V (2012) Fungal endophytes: an untapped source of biocatalysts. Fungal Diversity 54:19–30

Taghavi S, Lelie DV, Hoffman A, Zhang YB, Walla MD, Vangronsveld J, Newman L, Monchy S (2010) Genome sequence of the plant growth promoting endophytic bacterium *Enterobacter* sp. 638. PLOS Genet 6(5):1–15

Taghavi S, Weyens N, Vangronsveld J, van der Lelie D (2011) Improved phytoremediation of organic contaminants through engineering of bacterial endophytes of trees. endophytes of forest trees: biology and applications. Forestry Sci 80(4):205–216

Tan RX, Zou WX (2001) Endophytes: a rich source of functional metabolites. Nat Prod Rep 18:448–459

Tilman D et al (2002) Agricultural sustainability and intensive production practices. Nature 418:671–677

Vance CP (2001) Symbiotic nitrogen fixation and phosphorus acquisition. Plant nutrition in a world of declining renewable resources. Plant Physiology 127:390–397

Vendan RT, Yu YJ, Lee SH, Rhee YH (2010) Diversity of endophytic bacteria in *Ginseng* and their potential for plant growth promotion. J Microbiol 48(5):559–565

Verma VC, Kharwar RN, Gange AC (2010) Biosynthesis of antimicrobial silver nanoparticles by the endophytic fungus *Aspergillus clavatus*. Nanomedicine 5(1):33–40

Verza M, Arakawa NS, Lope NP, Kato MJ, Pupo MT, Said S, Carvalho I (2009) Biotransformation of a tetrahydrofuran lignin by the endophytic fungus *Phomopsis* sp. J Braz Chem Soc 20:195–200

Vetrivelkalai P, Sivakumar M, Jonathan EI (2010) Biocontrol potential of endophytic bacteria on *Meloidogyne incognita* and its effect on plant growth in bhendi. J Biopest 3(2):452–457

Waller F, Achatz B, Baltruschat H, Fodor J, Becker K, Fischer M, Heier T, Huckelhoven R, Neumann C, von Wettstein D, Franken P, Kogel KH (2005) The endophytic fungus *Piriformospora indica* reprograms barley to salt stress tolerance, disease resistance and higher yield. PNAS 102:13386–13391

Wang Y, Dai CC (2011) Endophytes: a potential resource for biosynthesis, biotransformation, and biodegradation. Ann Microbiol 61:207–215

Wang LW, Xu BG, Wang JY, Su ZZ, Lin FC, Zhang CL, Kubicek CP (2012) Bioactive metabolites from *Phoma* species, an endophytic fungus from the Chinese medicinal plant *Arisaema erubescens*. Appl Microbiol Biotechnol 93:1231–1239

Wu H, Yang H, You X, Li Y (2012) Isolation and characterization of Saponin-producing fungal endophytes from Aralia elata in Northeast China. Int J Mol Sci 13:16255–16266

Yang XL, Li ZZ (2013) New Spiral γ-Lactone Enantiomers from the plant endophytic fungus Pestalotiopsis foedan. Molecules 18:2236–2242

Zhang JY, Tao LY, Liang YJ, Chen LM, Mi YJ, Zheng LS, Wang F, She ZG, Lin YC, To KKW, Fu LW (2010) Anthracenedione derivatives as anticancer agents isolated from secondary metabolites of the mangrove endophytic fungi. Drugs 8(4):1469–1481

Zhao J, Zhou L, Wang J, Shan T, Zhong L, Liu X, Gao X (2010) Endophytic fungi for producing bioactive compounds originally from their host plants. Current research, technology and education topics in applied microbiology microbial biotechnology 567–576

Zhao Y, Park RD, Muzzarelli RAA (2010) Chitin deacetylases: properties and applications. Marine Drugs 8:24–46

Zhao JT, Fu YJ, Luo M, Zu YG, Wang W, Zhao CJ, Gu CB (2012) Endophytic fungi from Pigeon Pea [*Cajanus cajan* (L.) Mill sp.] produce Antioxidant Cajaninstilbene Acid. J Agric Food Chem 60:4314–4319

Zikmundova M, Drandarov K, Bigler L, Hesse M, Werner C (2002) Biotransformation of 2-Benzoxazolinone and 2-Hydroxy-1, 4- Benzoxazin-3-one by endophytic fungi isolated from *Aphelandra tetragona*. Appl Environ Microbiol 10:4863–4870

# Fungal Endophytes: An Amazing and Hidden Source of Cytotoxic Compounds

# 5

Sunil Kumar Deshmukh and Shilpa A. Verekar

**Abstract**

This review covers substantially the cytotoxic compounds isolated from endophytic fungi from terrestrial and mangrove plants during 2009–2012. Endophytes living asymptomatically within plant tissues have been found in almost all plants studied to date. Many of the compounds reported here were originally isolated from plants, then from endophytic fungi, whereas some are exclusively isolated from endophytic fungi. The anticancer activities in this review are from the published cytotoxicity against specific cancer cell lines. Development of these natural compounds is based on their cytotoxic activity profiles, chemical structures, and potential structure–activity relationship deduced from the biochemical and cytotoxic studies.

**Keywords**

Endophytic fungi · Cytotoxic compounds · Anticancer · Mangrove plants · Medicinal plants

## 5.1 Introduction

Based on the data reported by the World Health Organization's Cancer Control Program, (http://www.who.int/cancer), on a worldwide scale more than 11 million people are diagnosed with cancer every year. So, it is estimated that there will be 16 million new cases of cancer every year by 2020. Cancer is a chronic disease that is caused by defective genome-surveillance and signal-transduction mechanisms (Davoodi et al. 2012). It is characterized by out-of-control cell growth leading to genomic instability. The development of tumors is a multistep process involving various environmental and genetic factors (Li et al. 2011). Increasing insight into the genetics and molecular biology of cancer has resulted in the identification of an increasing number of potential molecular targets for anticancer drug discovery and development (Hanahan and Weinberg 2011).

Emerging evidences indicate that the anticancer activities of certain natural products and their chemically modified derivatives are due to their involvement in the induction of apoptosis, which

S. K. Deshmukh (✉) · S. A. Verekar
Department of Natural Products, Piramal Enterprises Limited, 1, Nirlon Complex, Off Western Express Highway, Goregaon (East), 400063, Mumbai, India
e-mail: sunil.deshmukh@piramal.com

R. N. Kharwar et al. (eds.), *Microbial Diversity and Biotechnology in Food Security*,
DOI 10.1007/978-81-322-1801-2_5, © Springer India 2014

**Table 5.1** Cytotoxic compounds reported from endophytic fungi

| Sr. No. | Fungus | Plant source | Compound | Reference |
|---|---|---|---|---|
| 1 | *Taxomyces andreanae* | *Taxus brevifolia* | Paclitaxel (Taxol) (**1**) | Stierle et al. 1993 |
| 2 | Unidentified YF1 | *Taxus yunnanensis* | Taxol (**1**) | Qiu et al. 1994 |
| 3 | *Alternaria* sp. Ja-69 | *Taxus cuspidata* | Taxol (**1**) | Strobel et al. 1996 |
| 4 | *Pestalotia bicilia* Tbx-2 | *Taxus baccata* | Taxol (**1**) | Strobel et al. 1996 |
| 5 | *Pestalotiopsis microspora* Ja-73 | *Taxus cuspidata* | Taxol (**1**) | Strobel et al. 1996 |
| 6 | *Pestalotiopsis microspora* Ne-32 | *Taxus wallachiana* | Taxol (**1**) | Strobel et al. 1996 |
| 7 | *Pestalotiopsis microspora* No. 1040 | *Taxus wallachiana* | Taxol (**1**) | Strobel et al. 1996 |
| 8 | *Fusarium lateritium* Tbp-9 | *Taxus baccata* | Taxol (**1**) | Strobel et al. 1996 |
| 9 | *Monochaetia* sp. Tbp-2 | *Taxus baccata* | Taxol (**1**) | Strobel et al. 1996 |
| 10 | *Pithomyces* sp. P-96 | *Taxus sumatrana* | Taxol (**1**) | Strobel et al. 1996 |
| 11 | *Pestalotiopsis microspora* Cp-4 | *Taxodium distichum* | Taxol (**1**) | Li et al. 1996 |
| 12 | *Pestalotiopsis* sp. W-x-3 | *Wollemia nobilis* | Taxol (**1**) | Strobel et al. 1997 |
| 13 | *Pestalotiopsis* sp. W-1f-1 | *Wollemia nobilis* | Taxol (**1**) | Strobel et al. 1997 |
| 14 | *Pestalotiopsis guepinii* W-1f-2 | *Wollemia nobilis* | Taxol (**1**) | Strobel et al. 1997 |
| 15 | *Periconia* sp. No. 202 | *Torreya grandifolia* | Taxol (**1**) | Li et al. 1998a |
| 16 | *Pestalotiopsis microspora* Ne 32 | *Taxus wallachiana* | Taxol (**1**) | Li et al. 1998b |
| 17 | *Tubercularia* sp. TF5 | *Taxus chinensis* var. *mairei* | Taxol (**1**) | Wang et al. 2000 |
| 18 | *Alternaria* sp. | *Ginkgo biloba* | Taxol (**1**) | Kim et al. 1999 |
| 19 | *Taxomyces* sp. | *Taxus yunnanensis* | Taxol (**1**) | Wang et al. 2001 |
| 20 | *Ectostroma* sp. XT5 | *Taxus chinensis* var. *mairei* | Taxol (**1**) | Hu et al. 2006 |
| 21 | *Botrytis* sp. XT2 | *Taxus chinensis* var. *mairei* | Taxol (**1**) | Hu et al. 2006 |
| 22 | *Papulaspora* sp. XT17 | *Taxus chinensis* var. *mairei* | Taxol (**1**) | Hu et al. 2006 |
| 23 | *Alternaria alternata* TPF6 | *Taxus chinensis* var. *mairei* | Taxol (**1**) | Tian et al. 2006 |
| 24 | *Ozonium* sp. BT2 | *Taxus chinensis* var. *mairei* | Baccatin III (**2**), Taxol (**1**) | Guo et al. 2006 |
| 25 | *Fusarium mairei* Y1117 | *Taxus chinensis* var. *mairei* | Taxol (**1**) | Cheng et al. 2007 |
| 26 | *Aspergillus* sp. NSZJ043 | *Taxus yurmanensis* | Baccatin III (**2**) | Yang et al. 2007 |
| 27 | *Botrytis* sp. HD181–23 | *Taxus cuspidata* | Taxol (**1**) | Zhao et al. 2008 |
| 28 | *Fusarium arthrosporioides* F-40 | *Taxus cuspidata* | Taxol (**1**) | Li et al. 2008 |
| 29 | *Fusarium mairei* UH23 | *Taxus chinensis* var. *mairei* | Taxol (**1**) | Dai and Tao, 2008 |
| 30 | *Fusarium solani* | *Taxus celebica* | Taxol (**1**) | Chakravarthi et al. 2008 |
| 31 | *Pestalotiopsis pauciseta* CHP-11 | *Cardiospermum helicacabum* | Taxol (**1**) | Gangadevi et al. 2008 |
| 32 | *Phyllosticta citricarpa* No. 598 | *Citrus medica* | Taxol (**1**) | Kumaran et al. 2008a |

**Table 5.1** (continued)

| Sr. No. | Fungus | Plant source | Compound | Reference |
|---|---|---|---|---|
| 33 | *Phyllosticta melochiae* Yates | *Melochia corchorifolia* | Taxol (**1**) | Kumaran et al. 2008b |
| 34 | *Phyllosticta spinarum* No. 625 | *Cupressus* sp. | Taxol (**1**) | Kumaran et al. 2008c |
| 35 | *Aspergillus fumigatus* EPTP-1 | *Podocarpus* sp. | Taxol (**1**) | Sun et al. 2008 |
| 36 | *Botryodiplodia theobromae* BT115 | *Taxus baccata* | Taxol (**1**) | Venkatachalam et al. 2008 |
| 37 | *Bartalinia robillardoides* | *Aegle marmelos* | Taxol (**1**) | Gangadevi and Muthumary 2008 |
| 38 | *Chaetomella raphiger* TAC-15 | *Terminalia arjuna* | Taxol (**1**) | Gangadevi and Muthumary 2009a |
| 39 | *Pestalotiopsis terminaliae* | *Terminalia arjuna* | Taxol (**1**) | Gangadevi and Muthumary 2009b |
| 40 | *Aspergillus niger* var. *taxi* HD86–9 | *Taxus cuspidata* | Taxol (**1**) | Zhao et al. 2009 |
| 41 | *Cladosporium cladosporioides* MD2 | *Taxus media* | Taxol (**1**), 10-deacetylbaccatin III. (**3**) | Zhang et al. 2009 |
| 42 | *Phomopsis* BKH 27 | *Taxus cuspidata, Ginkgo biloba, Larix leptolepis* | Taxol (**1**) | Kumaran et al. 2009a |
| 43 | *Metarhizium anisopliae* H-27 | *Taxus chinensis* | Taxol (**1**) | Liu et al. 2009 |
| 44 | *Phyllosticta dioscoreae* No. 605 | *Hibiscus rosa-sinensis* | Taxol (**1**) | Kumaran et al. 2009b |
| 45 | *Fusarium solani* Tax-3 | *Taxus chinensis* | Taxol (**1**) | Deng et al. 2009 |
| 46 | *Gliocladium* sp. | *Taxus baccata* | Taxol (**1**), 10-deacetyl baccatin III (**3**) | Sreekanth et al. 2009 |
| 47 | *Mucor rouxianus* sp., *DA10* | *Taxus chinensis* | Baccatin III (**2**), 10-deacetyl baccatin III (**3**) | Miao et al. 2009 |
| 48 | *Fusarium* sp. LNUF014 | *Taxus chinensis* var. *mairei* | Taxol (**1**) | Ai et al. 2010 |
| 49 | *Pestalotiopsis pauciseta* | *Tabebuia pentaphylla* | Taxol (**1**) | Vennila, et al. 2010 |
| 50 | *Lasiodiplodia theobromae* | *Morinda citrifolia* | Taxol (**1**) | Pandi, et al. 2011 |
| 51 | *Colletotrichum capsici* | *Capsicum annuum* | Taxol (**1**) | Kumaran et al. 2011 |
| 52 | *Alternaria* sp. | *Catharanthus roseus* | Vinblastine (**4**) | Guo et al. 1998 |
| 53 | *Fusarium oxysporum* | *Catharanthus roseus* | Vincristine (**5**) | Zhang, et al. 2000 |
| 54 | Unidentified fungus | *Catharanthus roseus* | Vincristine (**5**) | Yang et al. 2004 |
| 55 | *Entrophosphora infrequens* | *Nothapodytes foetida* | Camptothecin (**6**) | Puri et al. 2005 |
| 56 | *Neurospora crassa* | *Camptotheca acuminate* | Camptothecin (**6**) | Rehman et al. 2008 |
| 57 | *Nodulisporium* sp. | *Nothapodytes foetida* | Camptothecin (**6**) | Rehman et al. 2009 |
| 58 | *Botryosphaeria parva* | *Nothapodytes nimmoniana* | Camptothecin (**6**) | Gurudatt et al. 2010 |
| 59 | *Fusarium solani* | *Apodytes dimidiate* | Camptothecin (**6**) | Shweta et al. 2010 |
| 60 | *Fusarium solani* | *Camptotheca acuminate* | 9-methoxycamptothecin (**7**) | Kusari et al. 2009a |
| 61 | *Fusarium solani* | *Camptotheca acuminate* | 10-hydroxycamptothecin (**8**) | Kusari et al. 2009a |
| 62 | *Xylaria* sp. M20 | *Camptotheca acuminate* | 10-hydroxycamptothecin (**8**) | Liu et al. 2010 |

**Table 5.1** (continued)

| Sr. No. | Fungus | Plant source | Compound | Reference |
|---|---|---|---|---|
| 63 | Unidentified endophytic fungal strain XK001 | *Camptotheca acuminata* | 10-hydroxycamptothecin (**8**) | Min and Wang 2009 |
| 64 | *Fusarium solani* | *Apodytes dimidiate* | 9-methoxycamptothecin (**7**), 10-hydroxycamptothecin (**8**) | Shweta et al. 2010 |
| 65 | *Trametes hirsuta* | *Podophyllum hexandrum* | Podophyllotoxin (**9**) | Puri et al. 2006 |
| 66 | *Phialocephala fortinii* | *Podophyllum peltatum* | Podophyllotoxin (**9**) | Eyberger et al. 2006 |
| 67 | *Fusarium oxysporum* | *Juniperus recurva* | Podophyllotoxin (**9**) | Kour et al. 2008 |
| 68 | *Aspergillus fumigatus* | *Juniperus communis* | Deoxypodophyllotoxin (**13**) | Kusari et al. 2009b |
| 69 | *Thielavia subthermophila* | *Hypericum perforatum* | Hypericin (**14**), emodin (**15**) | Kusari et al. 2008 |
| 70 | *Fusarium proliferatum* (MTCC 9690) | *Dysoxylum binectariferum* | Rohitukine (**16**) | Mohanakumara et al. 2012 |
| 71 | *Annulohypoxylon squamulosum* BCRC 34022 | *Cinnamomum* sp. | Annulosquamulin (**18**), (3 S)-7-hydroxymellein (**19**) | Cheng et al. 2012 |
| 72 | *Penicillium* sp. (CR1642D) | Unidentified plant | Dicerandrol B (**20**) | Cao et al. 2012 |
| 73 | *Cephalotheca faveolata* | *Eugenia jambolana* | Sclerotiorin (**21**) | Giridharan et al. 2012 |
| 74 | *Phoma* species ZJWCF006 | *Arisaema erubescens.* | Cercosporamide (**22**) | Wang et al. 2012 |
| 75 | *Phoma* sp. | *Cinnamomum mollissimum* | 5-hydroxyramulosin (**23**) | Santiago et al. 2012 |
| 76 | *Fusarium* sp. BCC14842 | Bamboo leaf | 5-methoxydihydrofusarubin B (**24**), 5-hydroxydihydrofusarubin C (**25**), javanicin (**26**), 3-O-methylfusarubin (**27**) | Kornsakulkarn et al. 2011 |
| 77 | *Paraconiothyrium* sp. MY-42. | Stem of a beech branch | 19-(2-acetamido-2-deoxy-α-D-glucopyranosyloxy) isopimara-7,15-dien-3β-ol (**28**), 19-(α-D-glucopyranosyloxy) isopimara-7,15-dien-3-one (**29**) | Shiono et al. 2011 |
| 78 | *Phomopsis archeri* | *Vanilla albidia* | Phomoarcherins B-C (**30–31**) | Hemtasin et al. 2011 |
| 79 | *Mycosphaerella* sp. F2140 | *Psychotria horizontalis* | Cercosporin (**32**) | Moreno et al. 2011 |
| 80 | *Pestalotiopsis* sp. | *Podocarpus macrophyllus* | Pestaloquinols A (**33**), B (**34**) | Ding et al. 2011 |
| 81 | *Phomopsis* sp. CMU-LMA | *Alpinia malaccensis* | LMA-P1, (**35**), Sch-642305 (**36**), benquoine (**37**) | Adelin et al. 2011 |
| 82 | *Cytospora* sp. | *Ilex canariensis* | Cytospolides B and E (**38–39**) | Lu et al. 2011 |
| 83 | *Massrison* sp. | *Rehmannia glutinosa* | Massarigenin D (**40**), spiromassaritone (**41**), paecilospirone (**42**) | Sun et al. 2011 |

**Table 5.1** (continued)

| Sr. No. | Fungus | Plant source | Compound | Reference |
|---|---|---|---|---|
| 84 | *Eutypella* sp. BCC 13199 | *Etlingera littoralis* | Diaporthein B (**43**), scopararane A (**44**) | Isaka et al. 2011 |
| 85 | *Myrothecium roridum* IFB-E091 | *Artemisia annua* | Roritoxin E (**45**) | Shen et al. 2010 |
| 86 | *Fusarium oxysporum.* | *Cinnamomum kanehirae* | Beauvercin (**46**) | Wang et al. 2011 |
| 87 | *Chaetomium globosum* IFB-E041 | *Artemisia annulata* | Chaetoglobosin V (**47**), W (**48**), A (**49**), Fex (**50**), C(**51**), F (**52**), chaetoglobosin E (**53**), chaetoglobosin G (**54**) | Zhang et al. 2010 |
| 88 | *Chaetomium globosum* | *Ginkgo biloba* | Chaetomugilin D (**55**), chaetomugilin A (**56**), chaetoglobosin A (**49**), C (**51**) | Qin et al. 2009 |
| 89 | *Botryosphaeria rhodina* | *Bidens pilosa* | Botryorhodines A-B (**57–58**) | Abdou et al. 2010 |
| 90 | *Nigrospora sphaerica* (SS67) | *Smallanthus sonchifolius* | Aphidicolin (**59**), (22E,24R)-ergosta-4,6,8(14),22-tetraen-3-one (**60**) | Gallo et al. 2009 |
| 91 | *Phoma betae* | *Smallanthus sonchifolius* | (22E,24R)-8,14-epoxyergosta-4,22-diene-3,6-dione (**61**) | Gallo et al. 2010 |
| 92 | *Xylaria* sp. BCC 21097 | *Licuala spinosa* | Eremophilanolide 1 (**62**), 2 (**63**), 3 (**64**) | Isaka et al. 2010 |
| 93 | *Pestalotiopsis fici* | *Camellia sinensis* | Chloropupukeananin (**65**) | Liu et al. 2008 |
| 94 | *Pestalotiopsis fici* | *Camellia sinensis* | Chloropestolide A (**66**) | Liu et al. 2009a |
| 95 | *Pestalotiopsis fici* | *Camellia sinensis* | Chloropupukeanolide A (**67**) | Liu et al. 2010 |
| 96 | *Pestalotiopsis fici* | *Camellia sinensis* | Pestalofíciol L (**68**) | Liu et al. 2009b |
| 97 | Ascomycetous endophyte PM0651480 | *Mimosops elengi* | Ergoflavin (**69**) | Deshmukh et al. 2009 |
| 98 | *Dothideomycete* sp. LRUB20 | *Leea rubra* Blume ex Spreng | Dothideopyrone D (**70**) | Chomcheon et al. 2009 |
| 99 | *Preussia* sp. | *Aquilaria sinensis* | Spiropreussione A (**71**) | Chen et al. 2009 |
| 100 | *Pestalotiopsis photiniae* | *Roystonea regia* | Photinides A–F (**72–77**) | Ding et al. 2009 |
| 101 | *Eutypella* sp. BCC 13199 | *Etlingera littoralis* | Eutypellin A (**78**) | Isaka et al. 2009a |
| 102 | *Aspergillus fumigatus* | *Cynodon dactylon* | 9-deacetoxyfumigaclavine C (**79**) | Ge et al. 2009 |
| 103 | *Diaporthe* sp. XZ-07 | *Camptotheca acuminate.* | Phomopsolide B (**80**) | Yuan et al. 2009 |
| 104 | *Chaetomium* sp. | *Salvia officinalis* | Cochliodinol (**81**), isocochliodinol (**82**) | Debbab et al. 2009 |
| 105 | *Edenia* sp. | *Petrea volubilis* | Preussomerin EG1 (**83**), palmarumycin CP2 (**84**), CJ-12,371 (**85**) | Martinez-Luis et al. 2008 Martinez-Luis et al. 2009 |
| 106 | *Endothia gyrosa* IFB-E023 | *Vatica mangachapo* | Cytochalasin H (**86**), cytochalasin J (**87**), epoxycytochalasin H (**88**), cytochalasin Z10 (**89**), Z11 (**90**) | Xu et al. 2009a |

**Table 5.2** Cytotoxic compounds reported from mangrove fungi

| Sr. No. | Fungus | Plant source | Compound | Reference |
|---|---|---|---|---|
| 1 | Mangrove endophytic fungus No. 1403 | Unidentified mangrove plant | 1403P-3 (**91**) | Zhang et al. 2007, Yuan et al. 2011 |
| 2 | *Alternaria* sp. ZJ9–6B, | *Aegiceras corniculatum* | Alterporriol K (**92**), L (**93**) | Huang, et al. 2011 |
| 3 | Mangrove endophytic fungus No. 1403. | Unidentified mangrove plant | Bostrycin (**94**) | Xu et al. 2010 |
| 4 | *Nigrospora* sp.(No. 1403) | *Kandelia candel* | Bostrycin (**94**), deoxybostrycin (**95**) | Xia et al. 2011 |
| 5 | *Talaromyces flavus* | Unidentified mangrove plant | Talaperoxide B(**96**) and D (**97**) | Li et al. 2011 |
| 6 | *Fusarium* sp. ZZF60 | Unidentified mangrove plant | 6,8-dimethoxy-1-methyl-2-(3-oxobutyl) anthrakunthone (**98**) | Huang et al. 2010a |
| 7 | Endophytic fungus No. 1893 | Unidentified mangrove plant | Mycoepoxydiene (**99**) | Chen et al. 2005, Chen et al. 2006, Wang et al. 2010 |
| 8 | *Phomopsis* sp. ZSU-H26 | Unidentified mangrove plant | 5-hydroxy-6,8-dimethoxy-2-benzyl-4 H-naphtho[2,3-b]-pyran-4-one (**100**) | Huang et al. 2010b |
| 9 | Mangrove endophytic fungus No. ZH19 | Unidentified mangrove plant | 1,7-dihydroxy-2-methoxy-3-(3-methylbut-2-enyl)-9 H-xanthen-9-one (**101**), 1-hydroxy-4,7-dimethoxy-6-(3-oxobutyl)-9 H-xanthen-9-one (**102**) | Huang et al. 2010c |
| 10 | Mangrove endophytic fungus | *Xylocarpus granatum* | Merulin A (**103**), Merulin C (**104**) | Chokpaiboon et al. (2010) |
| 11 | *Talaromyces* sp. ZH-154 | *Kandelia candel* | 7-epiaustdiol(**105**), 8-O-methylepiaustdiol (**106**), stemphyperylenol (**107**), Skyrin (**108**), secalonic acid A (**109**), emodin (**15**), norlichexanthone (**110**) | Liu et al. 2010 |
| 12 | *Penicillium* sp. | *Acanthus ilicifolius* | Penicinoline (**111**) | Shao et al. 2010a |
| 13 | *Fusarium* sp. (No. b77) | Unidentified mangrove plant | Anhydrofusarubin (**112**) | Shao et al. 2010b |
| 14 | *Halorosellinia* sp. and *Guignardia* sp. | Unidentified mangrove plant | 9,10-anthracenedione (**113**) | Zhang et al. 2010c |
| 15 | Mangrove endophytic fungus Zh6-B1 | Unidentified mangrove plant | 3R,5R-sonnerlactone (**114**), 3R,5 S-sonnerlactone (**115**) | Li et al. 2010b |
| 16 | Mixed broth of two mangrove fungi (strain no. K38 and E33) | Unidentified mangrove plant | (−)-byssochlamic acid bisdiimide (**116**) | Li et al. 2010a |
| 17 | *Penicillium expansum* 091006 | *Excoecaria agallocha* | ExpansolA (**117**) and B (**118**) | Lu et al. 2010 |
| 18 | Endophytic fungus No. 1403 | Unidentified mangrove plant | SZ-685 C (**119**) | She et al. 2008, Xie et al. 2010 |
| 19 | *Phomopsis* sp. ZSU-H76 | Unidentified mangrove plant | 2-(7' -hydroxyoxooctyl)-3-hydroxy-5-methoxybenzeneacetic acid Et ester (**120**) | Huang et al. 2009a |
| 20 | Mangrove endophytic fungus ZSU-H19 | Unidentified mangrove plant | Ophioglonin (**121**) Ophioglonin-7-O-β-D-glucopyranoside (**122**) | Huang et al. 2009b |

**Table 5.2** (continued)

| Sr. No. | Fungus | Plant source | Compound | Reference |
|---|---|---|---|---|
| 21 | Mangrove endophytic fungus No. ZSU44. | Unidentified mangrove plant | Secalonic acid D (**123**) | Zhang et al. 2009 |
| 22 | *Aspergillus flavipes* | *Acanthus ilicifolius* | Cytochalasins- Z16 (**124**), Z17 (**125**), Z19 (**126**), rosellichalasin (**127**) | Lin et al. 2009 |
| 23 | *Pestalotiopsis* sp. | *Rhizophora mucronata* | Pestalotiopsone F (**128**) | Xu et al. 2009b |
| 24 | *Penicillium* sp. 091402 | Unidentified mangrove plant | (3R*,4S*)-6,8-dihydroxy-3,4,7-trimethylisocoumarin (**129**) | Han et al. 2009a |
| 25 | *Penicillium* sp. HK13–8 | *Rhizophora stylosa* | S-curvularin (**130**) | Han et al. 2009b |
| 26 | *Aigialus parvus* BCC 5311 | Unidentified mangrove wood | Hypothemycin (**131**), 4-O-demethylhypothemycin (**132**) | Isaka et al. 2009b |
| 27 | *Paecilomyces* sp. (tree 1–7) | Unidentified mangrove Plant | Secalonic acid A (**109**), penicillixanthone A (**133**), paecilin A (**134**) | Wen et al. 2009 |
| 28 | *Fusarium* sp. (ZZF41) | Unidentified mangrove plant | 5-O-methyl-2' -methoxy-3' -methylalpinumisoflavone (**135**) | Huang et al. 2010d |

is regarded as the preferred way to manage cancer. Understanding the molecular mechanisms of actions of these compounds and their effects on cellular signaling processes as well as their structure–activity relationships is necessary for the development of new compounds with more favorable profiles in chemopreventive activities. Efficient and effective applications of natural products will improve the drug discovery process by reducing the side effects (Kuo et al. 2005).

To date, no fungal metabolite or its derivative has been approved as an anticancer drug; however, there are several compounds in clinical trials. For example, irofulven, a chemical derivative of illudin S, first isolated from the basidiomycetes *Omphalotus illudens* (formerly *Clitocybe illudens*) is undergoing phase I and II trials for various types of cancer and phase II trials in combination with approved chemotherapeutics (Greve et al. 2010). Another example is plinabulin (NPI-2358), a chemical derivative based on the diketopiperazine phenylahistin (also known as halimide), which was isolated from cultures of the marine fungus *Aspergillus ustus* (Kanoh et al. 1997). In combination with docetaxel, plinabulin is at present in phase II clinical studies against advanced non-small cell lung cancer (Greve et al. 2010).

Endophytes are biodiverse microorganisms with inherent properties to colonize and survive in internal tissues of all plant species, creating a huge biodiversity with unknown novel natural products with the possibilities to become the frontiers of drug discovery. After the discovery of Taxol, a billion dollar anticancer drug, the research on endophytic microorganisms has yielded several potential drug lead compounds with antibacterial, antiviral, antioxidant, antineurodegenerative, immunosuppressant, anti-inflammatory, antidiabetic, and anticancer properties (Zhang et al. 1999; Strobel and Daisy 2003; Gunatilaka 2006; Deshmukh and Verekar 2009; Kharwar et al. 2011; Deshmukh and Verekar 2012; Mishra et al. 2013). Endophytes play a major role in the physiological activities of host plants influencing enhancement of stress, insect, nematode, and disease resistance (Carroll 1988; Hallmann and Sikora 1996; Sturz and Nowak 2000; Azevedo and Araujo 2007).

The collection of plants for isolation of endophytes should be done with a definite strategy in mind. The biogeography of ecosystem along with the medicinal property of the plant selected should be taken into consideration for the collection program. Plants already known for production of notable drugs, used in traditional medicine, can be taken for the isolation of endophytes. It is also important to know which season traditional healers are using for collection of these plants. These endophytes can be screened for several disease areas irrespective of what the traditional healer is using it for. For example, the *Catharanthus roseus* plant was initially used for treating diabetes, but the alkaloids isolated from this plant are in use for treating cancer. This will give us a broader aspect of the medicinal potential of these fungi.

In this review, we have attempted to cover metabolites discovered from endophytic fungi isolated from terrestrial plants and mangroves during 2009–2012 and their potential as anticancer agents. Brief descriptions of the cytotoxic activity of these compounds against selected cancer cell lines are given. Many of these compounds are shown in Tables 5.1 and 5.2.

## 5.2 Compounds from Endophytic Fungi of Terrestrial Plants

Paclitaxel (Taxol) (**1**), a well-known and highly functionalized tetracyclic diterpenoid bioactive compound, was originally isolated from the bark of *Taxus brevifolia* (Wani et al. 1971). Taxol is found in extremely low amounts in the needles, bark, and roots of yews (*Taxus* sp.). It is specifically targeted to treat prostate, ovarian, breast, and lung cancers (Rowinsky 1997). Taxol stabilizes plus end dynamic instability of microtubules both in vitro and in vivo (Jordan et al. 1993, Yvon et al. 1999). The discovery of a paclitaxel-producing endophytic fungus *Taxomyces andreanae* from the Pacific yew (*T. brevifolia*) by Stierle et al. (1993) generated interest in the scientific community for the isolation of endophytic fungi, which produce not only Taxol but other active metabolites also. Over the last 19 years, there have been numerous publications of fungi from yews (Stierle et al. 1993; Strobel et al. 1996; Yuan et al. 2006; Guo et al. 2006; Ruiz-Sanchez et al. 2010) and other plant species (Kumaran et al. 2008a, b, 2009a, b, 2011, Pandi et al. 2011).

To date, at least 20 genera of endophytic fungi (*Alternaria, Aspergillus, Botryodiplodia, Botrytis, Cladosporium, Ectostroma, Fusarium, Metarhizium, Monochaetia, Mucor, Nigrospora, Ozonium, Papulaspora, Periconia, Pestalotia, Pestalotiopsis, Phyllosticta, Pithomyces, Taxomyces,* and *Tubercularia*) have been reported to produce paclitaxel and its analogues such as baccatin III (**2**) and 10-deacetylbaccatin III (**3**) (Table 5.1). The hosts of paclitaxel-producing fungi mainly include *Taxus* species (i.e., *T. baccata, T. cuspidata, T. media,* and *T. yunnanensis*) that belong to the family Taxaceae, and non-*Taxus* species such as *Cardiospermum halicacabum* (Sapindaceae), *Citrus medica* (Rutaceae), *Cupressus* sp. (Cupressaceae), *Ginkgo biloba* (Ginkgoaceae), *Hibiscus rosa-sinensis* (Malvaceae), *Podocarpus* sp. (Podocarpaceae), *Taxodium distichum* (Taxodiaceae), *Terminalia arjuna* (Combretaceae), *Torreya grandifolia* (Taxaceae), and *Wollemia nobilis* (Araucariaceae). Such a large number and wide host range of plants imply that both paclitaxel-producing fungi and their hosts have a broad biological diversity (Zhao et al. 2011).

Vinca alkaloids, vinblastine (**4**) and vincristine (**5**), were isolated from the Madagascar periwinkle, *C. roseus* (Gueritte and Fahy 2005) and recent semisynthetic analogues of these agents, viz. vinolrelbine, vindesine, and vinflunine, have been reported (Cragg and Newman 2005; Okouneva et al. 2003; Simoens et al. 2008). These agents act through the inhibition of tubulin polymerization and are used in combination with other cancer chemotherapeutics for the treatment of a variety of cancers including leukemias, lymphomas, advanced testicular cancer, breast cancer, lung cancer, and Kaposi's sarcoma (Cragg and Newman 2009). These agents can also be isolated from endophytic fungi of *C. roseus*. Vinblastine was reported from *Alternaria* sp. associated with *C. roseus* (Guo et al. 1998). Later vincristine was reported from *Fusarium oxysporum* and an unidentified endophytic fungus from *C. roseus* (Zhang et al. 2000; Yang et al. 2004).

Camptothecin (**6**) was initially isolated from the bark of *Camptotheca acuminata* (Nyssaceae) (Wall et al. 1966) followed by *Nothapodytes foetida* (Fulzele et al. 2001). Later on camptothecin was reported from another endophytic fungus *Entrophosphora infrequens* of *N. foetida* (Puri et al. 2005). It was also reported from *Neurospora crassa,* a seed endophyte of *C. acuminata* (Rehman et al. 2008), from *Nodulisporium* sp. isolated from *N. foetida* (Rehman et al. 2009), *Botryosphaeria parva* isolated from *Nothapodytes nimmoniana* (Icacinaceae) (Gurudatt et al. 2010), and *Fusarium solani* strains MTCC9667 and MTCC9668 isolated from *Apodytes dimidiate* (Shweta et al. 2010). Camptothecin acts through inhibition of topoisomerase I, an important enzyme involved in the replication pathway of DNA during cell-cycle progression and, to date, remains the most important class of topoisomerase I inhibitors so far (Cragg and Newman 2004).

Two of camptothecin analogues, 9-methoxycamptothecin (**7**) and 10-hydroxycamptothecin (**8**) were isolated from *F. solani,* an endophytic fungus of *C. acuminata* (Kusari et al. 2009a). 10-hydroxycamptothecin was also isolated from an endophytic fungus *Xylaria* sp. M20, from *C. acuminata* (Liu et al. 2010) and an unidentified endophytic fungal strain XK001 from *C. acuminata* (Min and Wang 2009). *F. solani* strain MTCC9668 isolated from *Apodytes dimidiate* (Icacinaceae) from the Western Ghats, India produces 9-methoxycamptothecin and 10-hydroxycamptothecin (Shweta et al. 2010).

Podophyllotoxin (**9**) is an important natural product which was originally isolated in 1950 from the higher plant *Podophyllum emodi* (Leiter et al. 1950). Podophyllotoxin is currently used as a treatment for genital warts, but its greater value is its role as the precursor to three anticancer drugs, the topoisomerase I inhibitors etoposide (**10**), teniposide (**11**), and etoposide phosphate (**12**) (Puri et al. 2006; Eyberger et al. 2006). An alternative source of podophyllotoxin is an endophytic fungus *Trametes hirsuta* obtained from the dried rhizomes of *Podophyllum hexandrum* collected from the northwestern Himalayan region of India (Puri et al. 2006). Simultaneously, two different strains of *Phialocephala fortinii* from rhizomes of *Podophyllum peltatum* yielded podophyllotoxin (Eyberger et al. 2006). Podophyllotoxin is also produced by an endophytic fungus *F. oxysporum* isolated from *Juniperus recurva* (Kour et al. 2008). Anticancer prodrug deoxypodophyllotoxin (**13**) was isolated from an endophytic fungus *Aspergillus fumigatus* isolated from *Juniperus communis* L. Horstmann (Kusari et al. 2009b).

Hypericin (**14**), a naphthodianthrone derivative, originally isolated from the herb *Hypericum perforatum* (St. John's Wort) has been used since ancient times to treat depression and other ailments (Brockmann et al. 1939; Nahrstedt and Butterweck 1997). Several in vitro studies have revealed that the multifaceted cytotoxic activity of hypericin is a result of photodynamic activity (Hadjur et al. 1996; Delaey et al. 2001; Kamuhabwa et al. 2001). For the first time hypericin (**14**), along with emodin (**15**), was isolated from an endophytic fungus *Thielavia subthermophila* from *H. perforatum* from India (Kusari et al. 2008) and exhibited photodynamic cytotoxicity against the human acute monocytic leukemia cell line (THP-1) in two different assays. THP-1 cells were exposed to varying concentrations of the fungal extract in the dark and after the extract had been irradiated with visible light for 20 min. In the resazurin-based assay, dark versus light cell viability was 92.7 versus 4.9 %, and in the ATP-lite assay, dark versus light cell viability was 91.1 versus 1.0 % (Kusari et al. 2009c).

Rohitukine (**16**), a chromane alkaloid, was first reported from *Amoora rohituka* (Harmon et al. 1979) and then from *Dysoxylum binectariferum* (Naik et al. 1988), both from the family Meliaceae. Rohitukine exhibits both anti-inflammatory as well as immunomodulatory properties besides acting as an anticancer compound (Naik et al. 1988). Rohitukine is an important precursor for the semisynthetic derivative, flavopiridol (**17**) (Carlson et al. 1996; Sedlacek et al. 1996). Flavopiridol inhibits several cyclin-dependent kinases (CDKs), a family of kinases which govern progression of cells through the cell cycle, and displays unique anticancer properties (Sausville et al. 1999). In addition to directly inhibiting CDKs, flavopiridol is also known to selectively induce apoptotic cell death as well as exhibiting some antiangiogenic properties (Takada and Aggarwal 2003). Rohitukine exhibits cytotoxicity against SKOV3, T47D, MDAMB 273, NCI/ADR-RES, and MCF-7 cell lines with $IC_{50}$ value of 20, 50, 3, 2.8, and 15 µg/ml, respectively (Mohanakumara et al. 2010). Recently, it has also been reported from an endophytic fungus *Fusarium proliferatum* (MTCC 9690) from the inner bark tissue of *D. binectariferum* (Mohanakumara et al. 2012).

Annulosquamulin (**18**) and (3 S)-7-hydroxymellein (**19**) were isolated from *Annulohypoxylon squamulosum* BCRC 34022, an endophyte of *Cinnamomum* sp. Compounds **18** and **19** possessed moderate cytotoxicity against MCF-7, NCI-H460, and SF-268 cancer cell lines with $IC_{50}$ values of 3.19, 3.38, 2.46, and 2.78, 3.17, 2.38 µg/ml. The positive control Actinomycin D possessed cytotoxicity against MCF-7, NCI-H460, and SF-268 cancer cell lines with $IC_{50}$ values of 0.96, 0.56, and 0.52 µg/ml, respectively (Cheng et al. 2012).

Dicerandrol B (**20**) was isolated from an endophytic isolate of *Penicillium* sp. (CR1642D), collected from the Costa Rican rainforest. Dicerandrol B exhibits moderate activity against Dox40, Farage, H929, HT, OPM2, and RPMI8226 in the presence of stromal cells with $IC_{50}$ values of 2.3, 1.3, 3.4, 1.3, 1.5, and 1.2 µM, respectively. The activity of dicerandrol B against cancer cell lines RPMI8226 and H929 is doubled or tripled in the presence of stromal cells ($IC_{50}$ 1.2 µM, but 2.4 µM without stromal cells) (Cao et al. 2012).

Sclerotiorin (**21**) was isolated from an endophytic fungus *Cephalotheca faveolata*, obtained from the leaf petiole of *Eugenia jambolana.* Sclerotiorin exhibited cytotoxicity against ACHN, Panc-1, Calu-1, HCT-116, and H460 cell lines with $IC_{50}$ value of 1.2, 1.6, 2.1, 0.63, and 1.6 µM, respectively, while it showed an $IC_{50} > 10$ µM in normal breast epithelium cells (MCF10A). It was also found that sclerotiorin induced apoptosis in colon cancer (HCT116) cells via the activation of BAX and downregulation of Bcl-2, which lead to activation of cleaved caspase-3 thereby causing apoptosis of cancer cells (Giridharan et al. 2012).

Cercosporamide (**22**) was isolated from an endophytic fungus *Phoma* sp. ZJWCF006 in *Arisaema erubescens*. Cercosporamide exhibited cytotoxic activity against HT-29, SMMC-772, MCF-7, HL-60, MGC80–3, and P388 cell lines with $IC_{50}$ values of 9.3, 27.87, 48.79, 37.57, 27.83, and 30.37 μM, respectively. The positive control cisplatin was cytotoxic against HT-29, SMMC-772, MCF-7, HL-60, MGC80–3, and P388 cancer cell lines with $IC_{50}$ values of 25, 71.2, 73.6, 3.5, 12.3, and 2.8 μg/ml, respectively (Wang et al. 2012).

A polyketide compound, 5-hydroxyramulosin (**23**) was isolated from the endophytic fungus *Phoma* sp., from the plant *Cinnamomum mollissimum*. The compound was cytotoxic against murine leukemia cells with an $IC_{50}$ of 2.10 μg/mL (Santiago et al. 2012).

5-methoxydihydrofusarubin B (**24**) and 5-hydroxydihydrofusarubin C (**25**), Javanicin (**26**) and 3-O-methylfusarubin (**27**) were isolated from the endophytic fungus *Fusarium* sp. BCC14842, isolated from bamboo leaves. 5-methoxydihydrofusarubin B was cytotoxic against KB cells, MCF-7 cells, NCIeH187 cells, and Vero cells with $IC_{50}$ values of 13.47, 28.26, 11.89, and >50 μg/mL, respectively. 5-hydroxydihydrofusarubin C was cytotoxic against KB cells, MCF-7 cells, NCIeH187 cells, Vero cells with $IC_{50}$ values of 9.28, 8.50, 5.38, 12.77 μg/mL, respectively. Javanicin was cytotoxic against KB cells, MCF-7 cells, NCIeH187 cells, Vero cells with $IC_{50}$ values of 1.62, 3.40, 1.91, 6.98 μg/mL, respectively. 3-O-methylfusarubin exhibited cytotoxic activity against KB cells, MCF-7 cells, NCIeH187 cells, Vero cells with $IC_{50}$ values of 9.28, 8.50, 5.38, and 12.77 μg/mL, respectively (Kornsakulkarn et al. 2011).

Isopimarane diterpenes, 19-(2-acetamido-2-deoxy-α-D-glucopyranosyloxy)isopimara-7,15-dien-3β-ol (**28**), and 19-(α-D-glucopyranosyloxy) isopimara-7,15-dien-3-one (**29**) were isolated from the endophytic fungus *Paraconiothyrium* sp. MY-42 of a beech branch collected in October 2008 at Mt. Gassan, Yamagata, Japan. Compounds (**28**) and (**29**) showed moderate cytotoxicity against the human promyelocytic leukemia cell line HL60 with an $IC_{50}$ value of 6.7 μM and 9.8 μM, respectively, while camptothecin showed an $IC_{50}$ of 0.01 μM (Shiono et al. 2011).

Phomoarcherins B (**30**), C (**31**) were isolated from the endophytic fungus *Phomopsis archeri* of the stem of *Vanilla albidia*. Phomoarcherins B exhibited in vitro cytotoxic activity against KKU-100, KKU-M139, KKU-M156, KKU-M213, KKU-M214, and KB cell lines with $IC_{50}$ values of 8.0, 0.1, 2.0, 20, 5.0, and 9.4 μg/ml, respectively. Phomoarcherins C exhibited in vitro cytotoxic activity against KKU-100, KKU-M139, KKU-M156, KKU-M213, and KKU-M214 cell lines with $IC_{50}$ values 8.9, 8.9, 18.0, 15.4, and 18.8 μg/ml, respectively (Hemtasin et al. 2011).

Cercosporin (**32**) was isolated from an endophytic fungus, *Mycosphaerella* sp. strain F2140, associated with the plant *Psychotria horizontalis* (Rubiaceae). Cercosporin showed an $IC_{50}$ of 4.68 μM against MCF7 cancer cell lines, whereas camptothecin showed an $IC_{50}$ value of 0.01 μM (Moreno et al. 2011).

Pestaloquinols A (**33**) and B (**34**) were isolated from an endophytic fungus *Pestalotiopsis* sp. isolated from *Podocarpus macrophyllus*. When both were tested for cytotoxicity against HeLa (cervical epithelium) cells, both showed $IC_{50}$ values of 8.8 μM (the positive controls VP-16 and D-24851 showed $IC_{50}$ values of 1.63 and 0.88 μM, respectively (Ding et al. 2011).

LMA-P1 (**35**), Sch-642305 (**36**), and benquoine (**37**) were isolated from the cultivation broth of *Phomopsis* sp. CMU-LMA was isolated from *Alpinia malaccensis*. LMA-P1 exhibited the highest cytotoxic activity against the colonic epithelial cancer cell line HCT-116 cells with an $IC_{50}$ of 41 nM, benquoine and Sch-642305 had lower activity at 210 and 140 nM, respectively (Adelin et al. 2011).

Cytospolides B and E (**38, 39**) were isolated from the endophytic fungus *Cytospora* sp. isolated from *Ilex canariensis*. Cytospolides B and E displayed strong cytotoxicity against A-549 cell line with $IC_{50}$ values of 5.15 and 7.09 μg/mL, respectively (Lu et al. 2011).

Massarigenin D (**40**), Spiromassaritone (**41**), and Paecilospirone (**42**) were isolated from an endophytic fungus *Massrison* sp. isolated from wild *Rehmannia glutinosa*. Spiromassaritone exhibited cytotoxicity against L-02, HepG-2, MCF-7, and A-549 cell lines with an $IC_{50}$ value of 7.2, 5.6, 6.8, and 9.8 µg/ml, respectively, while paecilospirone exhibited cytotoxicity against L-02, HepG-2, MCF-7, and A-549 cell lines with an $IC_{50}$ value of 12.4, 10.4, 7.6, and 6.8 µg/ml, respectively. Massarigenin D exhibited cytotoxicity against L-02, HepG-2, MCF-7, and A-549 cell lines with an $IC_{50}$ value of 19.6, 20.8, 11.2, and 14.4 µg/ml, respectively. Mitomycin, a positive control, exhibited cytotoxicity against L-02, HepG-2, MCF-7, and A-549 cell lines with an $IC_{50}$ value of 13.6, 6.8, 3.6, and 3.2 µg/ml, respectively (Sun et al. 2011).

Diaporthein B (**43**) and scopararane A (**44**) were isolated from the endophytic fungus *Eutypella* sp. BCC 13199 was isolated from *Etlingera littoralis*. Diaporthein B exhibited significant cytotoxic activities against NCI-H187, MCF-7, KB, and Vero cell lines with an $IC_{50}$ of 0.15, 1.2, 3.2, and 2.8 µM, respectively. Scopararane A exhibited significant cytotoxic activities against NCI-H187, MCF-7, KB, and Vero cell lines with an $IC_{50}$ value of 0.024, 5.7, 9.3, and 2.6 µM, respectively (Isaka et al. 2011).

Roritoxin E (**45**) was isolated from an endophytic fungus *Myrothecium roridum* IFB-E091 residing inside *Artemisia annua* root. Roritoxin E exhibited cytotoxicity against the gastric carcinoma SGC-7901 and hepatocarcinoma SMMC-7721 cell lines, with $IC_{50}$ values of 0.26 and 10.54 µg/mL, respectively. 5-fluorouracil co-assayed as a positive control had an $IC_{50}$ value of 6.66 µg/mL against SGC-7901 cells, and it demonstrated only a 9.98 % growth inhibition against SMMC-7721 cells at 10 µg/mL (Shen et al. 2010).

Beauvercin (**46**) was isolated from *F. oxysporum,* isolated from the bark of *Cinnamomum kanehirae,* an endemic plant of Taiwan. Beauvericin showed cytotoxicity against PC-3, Panc-1, and A549 with $IC_{50}$ values of 49.5, 47.2, and 10.4 µM, respectively (Wang et al. 2011).

Chaetoglobosin V (**47**), W (**48**), A (**49**), Fex (**50**), C (**51**), F (**52**), E (**53**), and G (**54**) were isolated from an endophytic fungus *Chaetomium globosum* IFB-E041, isolated from *A. annulata.* Compound **47–49** and **51–54** showed moderate cytotoxic activity against KB cell lines with $IC_{50}$ values in the range of 20–30 µg/ml. Compound **50** and **52–54** exhibited considerable cytotoxic activity against the K562 cell lines with $IC_{50}$ values of 19.25, 18.89, 20.90, and 25.40 µg/ml, respectively. Chaetoglobosin V displayed moderate cytotoxic activity against MCF-7 cell lines with $IC_{50}$ values of 27.86 µg/ml and chaetoglobosin W exhibited moderate cytotoxic activity against HepG2 cell lines with $IC_{50}$ values of 27.87 µg/ml. Doxorubicin hydrochloride, a positive control, exhibited cytotoxic activities against four human cancer cell lines, i.e., KB, K562, MCF-7, and HepG2 with $IC_{50}$ values of 0.12, 0.68, 0.78, and 0.76 µg/mL, respectively (Zhang et al. 2010a).

Chaetomugilin D (**55**), chaetomugilin A (**56**), chaetoglobosin A (**49**), and C (**51**) were isolated from an endophytic fungus *C. globosum,* isolated from *G. biloba.* Compounds **55, 56, 49**, and **51** displayed significant growth inhibitory activity against the brine shrimp (*Artemia salina*) and toxicity at a concentration of 10 µg/ml, with mortality rates (%) of 75.2, 78.3, 83.4, and 75.3, respectively (Qin et al. 2009).

Botryorhodines A (**57**) and B (**58**) were isolated from an endophytic fungus *Botryosphaeria rhodina,* isolated from *Bidens pilosa.* Compounds **57** and **58** exhibited potent antiproliferative activity against HUVEC cell line ($GI_{50}$ value 1.67 and 0.07 µM) and K-562 ($GI_{50}$ of 0.84 and 0.003 µM). Botryorhodine A and B show moderate to weak cytotoxic activities against HeLa cell lines with a $CC_{50}$ of 96.97 and 36.41 µM, respectively (Abdou et al. 2010).

Aphidicolin (**59**) and (22E,24R)-ergosta-4,6,8(14),22-tetraen-3-one (**60**) were isolated from the endophytic fungus *Nigrospora sphaerica* (SS67) and *Phoma betae* isolated from plant *Smallanthus sonchifolius* (Asteraceae). Aphidicolin exhibited cytotoxicity against HCT-8, MDA-MB435, SF295, and HL-60 cell lines with $IC_{50}$ values of 0.05, 0.20, 0.16, and 0.09 µg/ml, respectively. Compound **60** exhibited cytotoxicity against HCT-8, MDA-MB435, SF295, and HL-60 cell lines with $IC_{50}$ values of 6.24, 14.11,

17.03, and 5.29 μg/ml, respectively (Gallo et al. 2009). (22E, 24R)-8,14-epoxyergosta-4,22-diene-3,6-dione (**61**) was isolated from an endophytic fungus *Papulaspora immerse,* isolated from *S. sonchifolius*. Compound **61** showed the highest cytotoxic activity against human tumor cell lines MDA-MB435, HCT-8, SF295, and HL-60 with $IC_{50}$ values of 3.3, 14.7, 5.0, and 1.6 μM, respectively (Gallo et al. 2010).

Three eremophilane-type sesquiterpenes (**62–64**) were isolated from the endophyte *Xylaria* sp. BCC 21097 associated with *Licuala spinosa*. The three compounds, eremophilanolide **62, 63**, and **64** exhibited moderate cytotoxic activity with $IC_{50}$ values of 3.8–21 μM against cancer cell lines KB, MCF-7, and NCI-H187 (Isaka et al. 2010).

Chloropupukeananin (**65**), chloropestolide A (**66**), chloropupukeanolide A (**67**), pestalofiol L (**68**) were isolated from the endophytic fungus *Pestalotiopsis fici,* isolated from the branches of *Camellia sinensis* (Liu 2011). Chloropupukeananin displayed cytotoxic effect against HeLa and HT29 cells, showing $IC_{50}$ values of 1.4 and 6.7 μM, respectively (Liu et al. 2008). Chloropestolide A showed significant cytotoxicity against HeLa and HT29 cells, with $IC_{50}$ values of 0.7 and 4.2 μM, respectively (Liu et al. 2009a). Chloropupukeanolide A showed significant cytotoxicity against the human tumor cell lines, MDA-MB-231, HeLa, and MCF-7, with $IC_{50}$ values of 16.9, 15.5, and 15.9 μM, respectively (Liu et al. 2010). Pestalofiol L displayed cytotoxic activity against HeLa and MCF-7 cells, with $IC_{50}$ values of 8.7 and 17.4 μM against positive control 5-fluorouracil with $IC_{50}$ values of 10.0 and 15.0 μM, respectively (Liu et al. 2009b).

Ergoflavin (**69**) has been isolated from a leaf ascomycetous endophyte of *Mimosops elengi* designated PM0651480. Ergoflavin exhibited cytotoxicity against ACHN, H460, Panc1, HCT116, and Calu1 cancer cell lines, with $IC_{50}$ values of 1.2, 4.0, 2.4, 8.0, and 1.5 μM, respectively. Flavopiridol, a positive control, exhibited cytotoxicity against ACHN, H460, Panc-1, HCT116, and Calu1 cell lines, with $IC_{50}$ value of 0.84, 0.38, 0.23, 0.25, and 0.41 μM, respectively (Deshmukh et al. 2009).

Dothideopyrone D (**70**) was isolated from the endophytic fungus *Dothideomycete* sp. LRUB20, isolated from *Leea rubra* (Family Leeaceae). Dothideopyrone D exhibited cytotoxic activity against HeLa, HuCCA-1, HepG2, T47D, MDA–MB231, S102, A549, HL60, and MOLT3 cancer cell lines with $IC_{50}$ of 23, 19, 21, 21, 20, 24, 25, 16, and 13.8 μg/ml, respectively (Chomcheon et al. 2009).

Spiropreussione A (**71**) was isolated from an endophytic fungus *Preussia* sp., isolated from the stem of *Aquilaria sinensis*. Spiropreussione A showed cytotoxicity toward A2780 and BEL-7404 cells with $IC_{50}$ values of 2.4 and 3.0 μM, respectively. It was inactive ($IC_{50} > 10$ μM) against HCT-8, BGC-823, and A-549 human cancer cell lines (Chen et al. 2009).

Photinides A–F (**72–77**) were isolated from an endophytic fungus *Pestalotiopsis photiniae,* isolated from *Roystonea regia*. Compounds **72–77** showed modest but selective cytotoxicity against MDA-MB-2311, with inhibitory rates in the range of 23.1–24.6% when tested at 10 μg/mL, whereas none of these compounds displayed cytotoxicity against HeLa cells at the same concentration (Ding et al. 2009).

Eutypellin A (**78**) was isolated from the endophytic fungus *Eutypella* sp. BCC 13199, an endophyte of *E. littoralis*. Eutypellin A is a γ-lactone that exhibited cytotoxic activity against NCI-H187, MCF-7, KB, and nonmalignant Vero cells with $IC_{50}$ values of 12, 84, 38, and 88 μM, respectively. Ellipticine, a positive control, exhibited cytotoxic activity against NCI-H187, KB, and nonmalignant Vero cells with $IC_{50}$ values of 3.6, 2.5, and 5.5 μM, respectively (Isaka et al. 2009a).

9-deacetoxyfumigaclavine C (**79**) was isolated from an endophyte *A. fumigatus,* which was obtained from stem of *Cynodon dactylon*. It exhibited potent cytotoxicity against human leukemia cells (K562) with an $IC_{50}$ value of 3.1 μM, while doxorubicin hydrochloride, a drug which is currently used for the treatment of leukemia at 1.2 μM (Ge et al. 2009).

Phomopsolide B (**80**) was isolated from the endophytic fungal strain *Diaporthe* sp. XZ-07 of *C. acuminata*. Phomopsolide B significantly inhibited the growth of human-tumor HeLa cells with an $IC_{50}$ of 5.7 μg/mL and $IC_{50}$ value of the positive control Cisplatin was 3.5 μg/ml (Yuan et al. 2009).

Cochliodinol (**81**) and isocochliodinol (**82**) were isolated from an endophytic fungus *Chaetomium* sp. of *Salvia officinalis*. These compounds were tested for cytotoxicity against L5178Y mouse lymphoma cells. Cochliodinol was an order of magnitude more potent than its isomer, with an $EC_{50}$ of 7.0 mg/mL, compared to 71.5 mg/mL for isocochliodinol (Debbab et al. 2009).

Preussomerin EG1 (**83**), palmarumycin CP2 (**84**), and CJ-12,371 (**85**) were isolated from the endophytic fungus *Edenia* sp., isolated from leaves of *Petrea volubilis* (Martinez-Luis et al. 2008). Compounds **83**–**85** showed moderate activity against MCF-7 cancer cell lines with $IC_{50}$ value of 23.5, 25.8, and 18.3 μM (Martinez-Luis et al. 2009).

Cytochalasin H (**86**), cytochalasin J (**87**), epoxycytochalasin H (**88**), cytochalasin Z10 (**89**), and Z11 (**90**) were isolated from an endophytic fungi *Endothia gyrosa* IFB-E023, from the plant *Vatica mangachapo* (Dipterocarpaceae). Metabolites **86**–**90** were demonstrated to be substantially cytotoxic to the human leukemia K562 cell line with $IC_{50}$ values of 10.1, 1.5, 24.5, 28.3, and 24.4 μM, respectively, which are comparable to that of 5-fluorouracil (33.0 μM), co-assayed as the positive reference (Xu et al. 2009a).

## 5.3 Compounds from Endophytic Fungi of Mangroves

Anthracenedione derivative, 1403P-3 (**91**) was isolated from the mangrove endophytic fungus No. 1403. 1403P-3 exihibited cytotoxicity against human epidermoid carcinoma drug-sensitive parental KB cells and multidrug-resistant KBv200 cells with $IC_{50}$ values of 19.66 and 19.27 μM, respectively (Zhang et al. 2007). This compound also exhibited cytotoxicity against human breast cancer cells MCF-7 and MDA-MB-435 with $IC_{50}$ values of 9.7 and 7.6 μM, respectively. 1403P-3 induced breast cancer cell apoptosis by blocking Akt activation (Yuan et al. 2011).

Bianthraquinone derivatives, alterporriol K (**92**) and L (**93**) were isolated from the endophytic fungus *Alternaria* sp. ZJ9–6B, isolated from the mangrove *Aegiceras corniculatum*. Alterporriol K exhibited cytotoxicity against MDA-MB-435 and MCF-7 cells with $IC_{50}$ values of 26.97 and 29.11 μM, respectively, and alterporriol L exhibited cytotoxicity against MDA-MB-435 and MCF-7 cells with $IC_{50}$ values of 13.11 and 20.04 μM, respectively (Huang et al. 2011).

Bostrycin (**94**) was isolated from the mangrove endophytic fungus No. 1403. In the yeast *Saccharomyces cerevisiae* as a model, bostrycin inhibits cell proliferation by blocking the cell cycle at G1 phase and ultimately leads to cell death in a time- and dose-dependent manner. Bostrycin-induced lethal cytotoxicity is accompanied with increased levels of intracellular ROS and hallmarks of apoptosis such as chromatin condensation, DNA fragmentation, and externalization of phosphatidylserine. Bostrycin decreases mitochondrial membrane electric potential and causes mitochondrial destruction during the progression of cell death. Bostrycin-induced cell death was promoted in YCA1 null yeast strain but was partially rescued in AIF1 null mutant both in fermentative and respiratory media, strongly indicating that bostrycin induces apoptosis in yeast cells through a mitochondria-mediated but caspase-independent pathway (Xu et al. 2010). Bostrycin (**94**) and deoxybostrycin (**95**) were isolated from the marine endophytic fungus *Nigrospora* sp. (No. 1403) resident of *Kandelia candel* wood. Bostrycin exhibited cytotoxicity against A549, Hep-2, Hep G2, KB, MCF-7, and Adr with MCF-7 with $IC_{50}$ values of 2.64, 5.39, 5.90, 4.19, 6.13, and 6.68 μM/ml, respectively. Deoxybostrycin was also cytotoxic against all the cell lines tested, with $IC_{50}$ values of 2.44, 3.15, 4.41, 3.15, 4.76, and 5.46 μM/ml, respectively (Xia et al. 2011).

Talaperoxides B (**96**) and D (**97**) have been isolated from a mangrove endophytic fungus, *Talaromyces flavus*. Talaperoxide B exhibited in vitro cytotoxic activity against MCF-7, MDA-MB-435, HepG2, HeLa, and PC-3 cells lines with $IC_{50}$ values of 1.33, 2.78, 1.29, 1.73, and 0.89 μg/mL, respectively. Talaperoxides D exhibited in vitro cytotoxic activity against MCF-7, MDA-MB-435, HepG2, HeLa, and PC-3 cells lines with $IC_{50}$ values of 1.92, 0.91, 0.90, 1.31, and 0.70 μg/mL, respectively (Li et al. 2011).

An anthraquinone derivative, 6, 8-dimethoxy-1-methyl-2-(3-oxobutyl) anthrakunthone (**98**) was isolated from the marine mangrove endophytic fungus *Fusarium* sp. ZZF60 from the South China Sea. It showed cytotoxicity toward Hep2 and HepG2 with an $IC_{50}$ of 16 and 23 μM/L, respectively (Huang et al. 2010a).

Mycoepoxydiene (**99**), a polyketide was isolated from a mangrove endophytic fungus No. 1893 (Chen et al. 2005) and showed antitumour activity with an $IC_{50}$ of 5.5 μg/mL against HeLa cell lines (Chen et al. 2006). Mycoepoxydiene inhibited the growth of HeLa cells by inducing apoptosis accompanied with cytoskeletal rearrangement and cell-cycle arrest at G2/M phase. The cytoskeleton reorganization in actively growing HeLa cells is by promoting formation of actin stress fiber and inhibiting polymerization of tubulin. Mycoepoxydiene induced G2/M arrest by increasing p21 and cyclin B1 expression and decreasing the expression of cdc2, GADD45, wee1, myt1, chk1, and chk2. Apoptosis in HeLa cells was induced by decrease of the Bcl-2 level and increase of the Bad and Bak levels resulting in release of cytochrome C from mitochondria that subsequently activated caspases and the cleavage of PARP. Additionally, mycoepoxydiene also activated p38, MAPK, and JNK (Wang et al. 2010).

A naphtho-γ-pyrone, 5-hydroxy-6,8-dimethoxy-2-benzyl-4 H-naphtho[2,3-b]-pyran-4-one (**100**) was isolated from the mangrove endophytic fungus *Phomopsis* ZSU-H26, isolated from the stem of the *Excoecaria agallocha*. It exhibited cytotoxicity against Hep2 and HepG2 cells with an $IC_{50}$ of 10 and 8 μg/mL, respectively (Huang et al. 2010b).

Two xanthone derivatives 1,7-dihydroxy-2-methoxy-3-(3-methylbut-2-enyl)-9 H-xanthen-9-one (**101**) and 1-hydroxy-4,7-dimethoxy-6-(3-oxobutyl)-9 H-xanthen-9-one (**102**) were isolated from the mangrove endophytic fungus (No. ZH19). Compounds **101** and **102** inhibited KB cells with an $IC_{50}$ value of 20 and 35 μM/mL and KBv200 cells with an $IC_{50}$ value of 30 and 41 μM/mL, respectively (Huang et al. 2010c).

Merulin A (**103**) and C (**104**) were isolated from an endophytic fungus of *Xylocarpus granatum*. Merulin A and C showed cytotoxicity against human breast cancer (BT474) cell lines with $IC_{50}$ values of 4.98 and 1.57 μg/mL and also against colon cancer (SW620) cell lines with $IC_{50}$ values of 4.84 and 4.11 μg/mL, respectively (Chokpaiboon et al. 2010).

7-epiaustdiol (**105**), 8-O-methylepiaustdiol (**106**), stemphyperylenol (**107**), skyrin (**108**), secalonic acid A (**109**), norlichexanthone (**110**), and emodin (**15**) were isolated from the mangrove endophytic fungus *Talaromyces* sp. ZH-154, from the stem bark of *K. candel*. Compounds **105–110** and **15** exhibited cytotoxic activity against KB cells with $IC_{50}$ values in the range of 0.63–20.38 μg/ml and against KBv200 cells with $IC_{50}$ values in the range of 1.05–44.35 μg/ml (Liu et al. 2010).

A pyrrolyl 4-quinolinone alkaloid, penicinoline (**111**) was isolated from *Penicillium* sp., isolated from the bark of the mangrove plant *Acanthus ilicifolius* collected from the South China Sea. It exhibited potent in vitro cytotoxicity against 95-D and HepG2 cell lines with $IC_{50}$ values of 0.57 and 6.5 μg/mL, respectively (Shao et al. 2010a). Anhydrofusarubin (**112**) was isolated from the mangrove endophytic fungus *Fusarium* sp. (No. b77). It showed strong inhibitory effects on the growth of HepG2 and Hep2 cells, with $IC_{50}$ values of 1.0 and 2.5 μg/mL, respectively (Shao et al. 2010b).

9,10-anthracenedione (**113**) was isolated from the mangrove endophytic fungi *Halorosellinia* sp. and *Guignardia* sp. It displayed strong cytotoxicity against KB and KBv200 cells with an $IC_{50}$ of 3.17 and 3.21 μM, respectively. The mechanism involved in the apoptosis induced by compound **113** is probably related to mitochondrial dysfunction (Zhang et al. 2010c). 3R,5R-sonnerlactone (**114**) and 3R,5 S-sonnerlactone (**115**) were isolated from the mangrove endophytic fungus Zh6-B1. Compounds **114** and **115** inhibited the multidrug-resistant human oral floor carcinoma cells KV/MDR growth by 42.4% and 41.6%, respectively at 100 μM (Li et al. 2010b).

A new diimide derivative named (−)-byssochlamic acid bisdiimide (**116**) was isolated from the mixed culture of two mangrove fungi (strain no. K38 and E33) from the South China Sea coast. It exhibited weak cytotoxicity against Hep2 and HepG2 cells with an $IC_{50}$ of 45 and 51 µg/mL, respectively (Li et al. 2010a).

Expansols A (**117**) and B (**118**) were isolated from *Penicillium expansum* 091006 of mangrove plant *E. agallocha*. Expansol A exhibited moderate cytotoxicity against the HL-60 cell line with an $IC_{50}$ value of 15.7 µM, and expansol B inhibited the proliferation of A549 and HL-60 cells with $IC_{50}$ values of 1.9 and 5.4 µM, respectively (Lu et al. 2010). SZ-685 C (**119**), an anthracycline analogue, was isolated from the mangrove endophytic fungus No. 1403 collected from the South China Sea (She et al. 2008). SZ-685 C suppressed the proliferation of six cancer cell lines derived from human breast cancer, prostate cancer, glioma, and hepatoma ($IC_{50}$ values ranged from 3.0 to 9.6 µM) and the growth of breast cancer xenografts in mice. SZ-685 C had a direct apoptosis-inducing effect through both extrinsic and intrinsic apoptotic pathways, by activation of caspase-8 and 9 as well as effector caspase-3 and poly (ADP-ribose) polymerase. Phosphorylation of Akt and its downstream effectors, forkhead box protein O1 and forkhead box protein O3a were downregulated in SZ-685 C-treated cancer cells (Xie et al. 2010). A polyketide, 2-(7'-hydroxyoxooctyl)-3-hydroxy-5-methoxybenzeneacetic acid Et ester (**120**) was isolated from the mangrove endophytic fungus *Phomopsis* sp. ZSU-H76 obtained from the South China Sea. It exhibited cytotoxicity against Hep2 and HepG2 cells with $IC_{50}$ values of 25 and 30 µg/mL, respectively (Huang et al. 2009a).

Ophioglonin (**121**) and ophioglonin-7-O-β-D-glucopyranoside (**122**) were isolated from the mangrove endophytic fungus ZSU-H19 from the South China Sea. Ophioglonin and ophioglonin-7-O-β-D-glucopyranoside showed weak inhibitory activity on the KB cell line with $IC_{50}$ values of $3.0 \times 10^{-2}$ and $1.0 \times 10^{-2}$ g/L, respectively (Huang et al. 2009b). Secalonic acid D (**123**) was isolated from the mangrove endophytic fungus, *Paecilomyces* sp. (tree 1–7). It showed cytotoxicity toward KB cells with an $IC_{50} < 1$ µg/ml and inhibited human topoisomerase I with an $IC_{50}$ of 0.16 µM/mL (Guo et al. 2007). Secalonic acid D was also isolated from the mangrove endophytic fungus No. ZSU44. It showed potent cytotoxicity to HL60 and K562 cells, and the $IC_{50}$ values were 0.38 and 0.43 µM/L, respectively. Annexin V-FITC/PI assay and western blot indicated that secalonic acid D induced apoptosis in HL60 and K562 cells. In addition, secalonic acid D also induced cell-cycle arrest of G1 phase related to down regulation of c-Myc. The downregulation of c-Myc and cell-cycle arrest of G1 phase were caused not by formation of G-quadruplex structures but by activation of GSK-3β followed by degradation of β-catenin (Zhang et al. 2009). Cytotoxic cytochalasins namely, cytochalasin Z16 (**124**), Z17 (**125**), Z19 (**126**), and rosellichalasin (**127**) were isolated from *Aspergillus flavipes,* an endophytic fungus associated with *A. ilicifolius*. Compounds **124–127** showed cytotoxic activities against A-549 cell lines with $IC_{50}$ values of 19.5, 5.6, 17.4, and 7.9 µM (Lin et al. 2009).

Pestalotiopsone F (**128**) was isolated from the mangrove endophytic fungus *Pestalotiopsis* sp., associated to Chinese mangrove plant *Rhizophora mucronata*. It exhibited relatively impressive cytotoxicity against the murine cancer cell line L5178Y with an $EC_{50}$ value of 8.93 µg/mL (Xu et al. 2009b). (3R*,4S*)-6,8-dihydroxy-3,4,7-trimethylisocoumarin (**129**) was isolated from the mangrove endophytic fungus *Penicillium* sp. 091402. It exhibited moderate cytotoxicity against K562 cell lines with an $IC_{50}$ value of 18.9 µg/ml (Han et al. 2009a). S-curvularin (**130**) was isolated from the mangrove endophytic fungus *Penicillium* sp. HK13–8, obtained from *Rhizophora stylosa*. S-curvularin exhibited significant cytotoxicity against HL-60 cells with an $IC_{50}$ value of 2.56 µM (Han et al. 2009b).

Two cytotoxic compounds hypothemycin (**131**) and 4-O-demethylhypothemycin (**132**) were isolated from the mangrove fungus, *Aigialus parvus* BCC 5311. Hypothemycin exhibited cytotoxicity with an $IC_{50}$ of 2.0 and 2.1 µg/mL against NCI-H187 and Vero cell lines, respectively. 4-O-demethylhypothemycin exhibited cytotoxicity with an $IC_{50}$ of 2.6, 3.6, and 0.77 µg/

mL against BC1, NCI-H187, and Vero cell lines, respectively (Isaka et al. 2009b). Secalonic acid A (**109**), penicillixanthone A (**133**), and paecilin A (**134**) were isolated from mangrove endophytic fungus *Paecilomyces* sp. (tree 1–7) from the Taiwan Strait. Interestingly, secalonic acid A inhibited KB and KBv cells at $IC_{50}$ values of less than 1.57 nmol/mL, whereas penicillixanthone A inhibited KB and KBv cells at $IC_{50}$ values of less than 1.22 nmol/mL. Paecilin A inhibited KB and KBv cells at $IC_{50}$ values of 40, 50 nmol/mL (Wen et al. 2009), respectively. A new isoflavone, 5-O-methyl-2' -methoxy-3' -methylalpinumisoflavone (**135**) was isolated from the mangrove endophytic fungus, *Fusarium* sp. (ZZF41). Compound **135** inhibited Hep-2 and HepG2 cells with $IC_{50}$ values of 4 and 11 μM, respectively (Huang et al. 2010d).

## 5.4 Outlook

It is evident from the extensive data quoted here by the authors on anticancer compounds that endophytes are a very good source for the discovery of novel molecules. The existence of endophytes has been known for more than 100 years (Freeman 1904); however, they have been studied for their potential as novel sources of effective new drugs for only the last two decades. Endophytes represent a huge diversity of microbial community that have developed in special and sequestered environments, and their diversity in specialized habitat make them an exciting field of study in the search for new medicines. To fulfil this objective, all the 12 hot spots of plant diversity should be sampled for isolating endophytes which may help in exploring the chemical diversity. The medicinal values of the plants should be kept in mind while selecting the plants. The endophytes from the same medicinal plant should be investigated from different parts of the globe for bioactive metabolites and may be correlated with the activity (Owen and Hundley 2004). Taxonomic identification of the plants should be taken into account so as to correlate the metabolites of host plant and the metabolites of endophytes existing in the same plant. Endophytic fungi should be screened for their secondary metabolite spectrum under different growth conditions so as to induce the synthesis of several new metabolites that can be useful in different therapeutic indications (Bode et al. 2002). The synthesis of metabolites produced by these fungi should be correlated with the genes present in the plants/endophytes along with the biosynthetic pathway. Simultaneous metabolite profiling of plants and endophytes will help in understanding the origin of compounds and not hypothesis.

Mangroves can be another source of endophytic fungi. They are a group of woody plants growing in coasts, estuaries, intertidal zones of tropical and subtropical climates and constitute the second most important ecosystem among the marine ecosystems. Mangroves are adapted to anaerobic conditions of both salt- and freshwater environment. Although there are several studies regarding different ecological groups such as epiphytes, saprophytes, superficial biotrophs, and pathogens associated with mangroves (Goh and Yipp 1996), little is reported on endophytes of mangroves (Suryanarayanan and Johnson 2005).

Other possible areas might include endophytes from plants growing in Antarctica, high alpine regions, desert areas, wetlands, marine weeds, etc. as they are struggling to survive in harsh conditions. The less explored endophytes are fungi associated with lichens, mosses, bryophytes, and algae along with tropical endemic plants.

The Indian mycoflora is explored for diversity but less explored for bioactive metabolites. The reports are very scanty (Puri et al. 2005; Kusari et al. 2008; Kharwar et al. 2009; Deshmukh et al. 2009; Periyasamy et al. 2012). There is a great need of integration of mycologists, chemists, pharmacologists, and toxicologists to form teams and explore these fungi for pharmaceutical uses. These fungi can also be a good source of industrial enzymes (Suryanarayanan et al. 2012) and also to get nanoparticles (Verma et al. 2010; Raheman et al. 2011). So far no anticancer drugs are developed from fungi but endophytes are proven as alternative source of anticancer drugs which were originally produced by plants. This will help us in getting these drugs by fermentation in place of plant tissue culture or by cutting the plants.

As most of the plant diversity is found in tropical countries, there will be a great need of fungal culture collections from such plants. The collection will help in getting bioactive metabolites; some of which can be leads for chemical modification for various pharmaceutical applications and similarly for agricultural applications. Collections of fungi will be valuable to obtain selective enzymes for biotransformation and novel products such as chitosans.

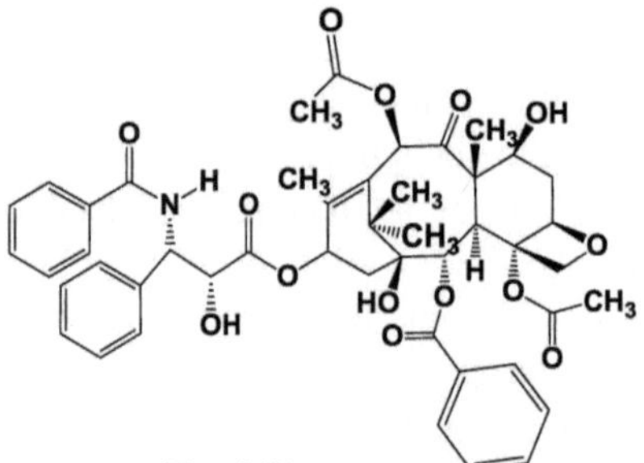

**Taxol (1)**

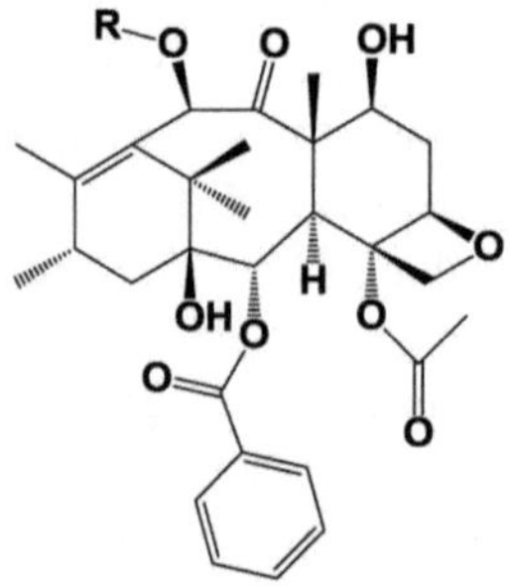

**R= $COCH_3$ Baccatin III (2)**
**R=H 10-deacetylbaccatin III (3)**

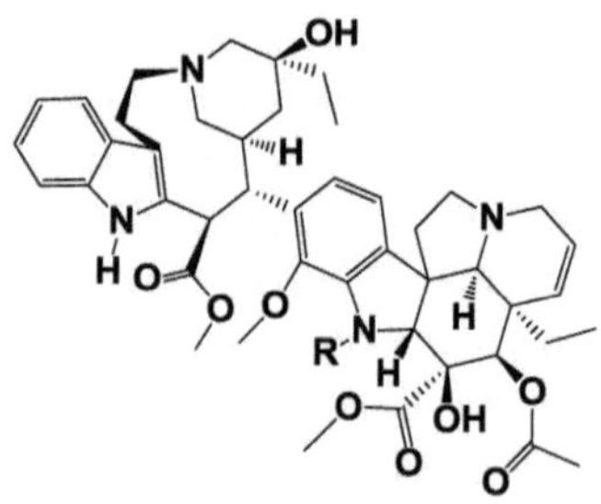

**R= $CH_3$ Vinblastine (4)**
**R= CHO Vincristine (5)**

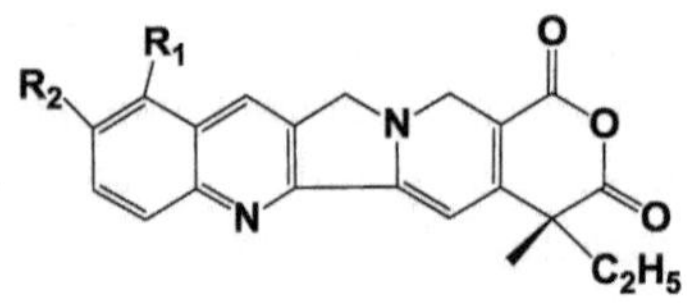

**R1=R2= H Camptothecin (6)**
**R1= $OCH_3$, R2=H 9-methoxycampothecin (7)**
**R1=H, R2=OH 10-hydroxycampothecin (8)**

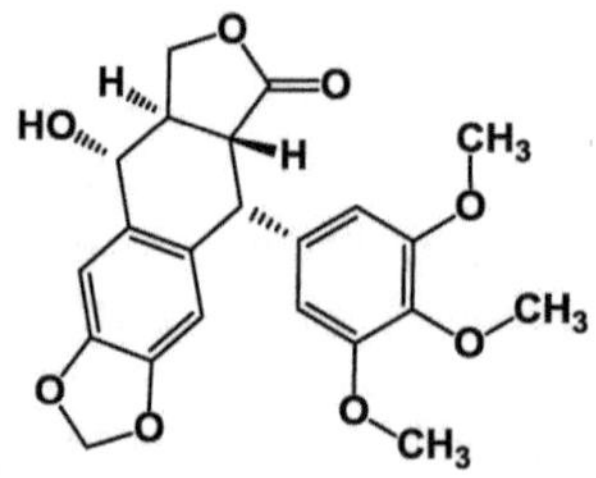

**Podophyllotoxin (9)**

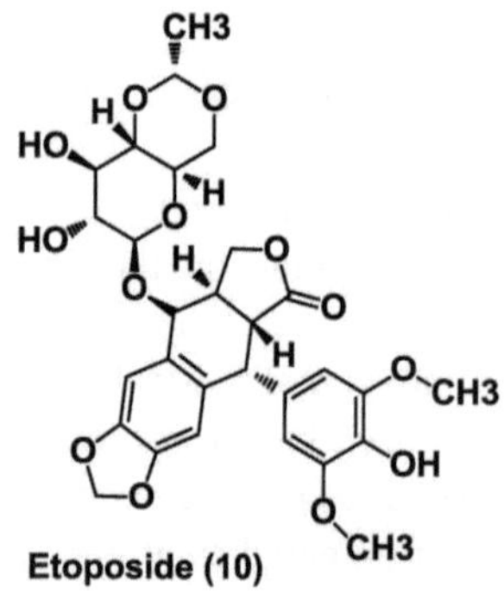

**Etoposide (10)**

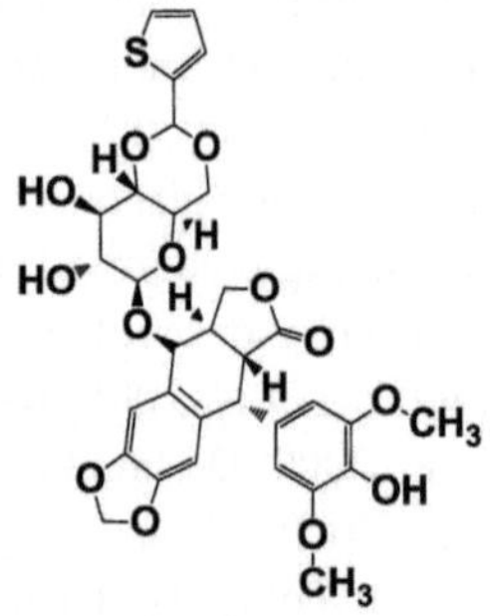

**Teniposide (11)**

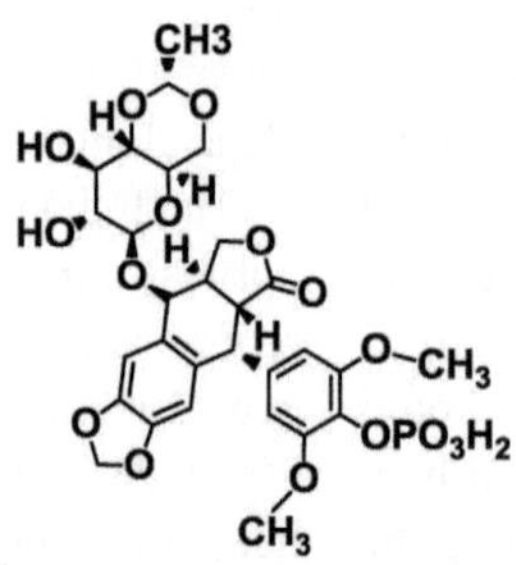

**Etoposide phosphate (12)**

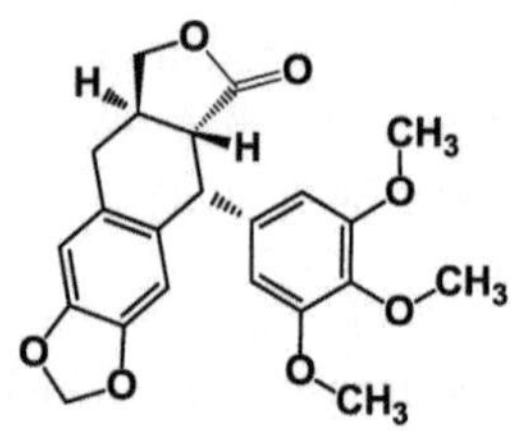

**Deoxypodophyllotoxin (13)**

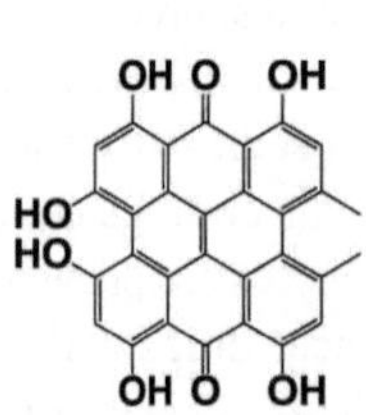

**Hypericin (14)**

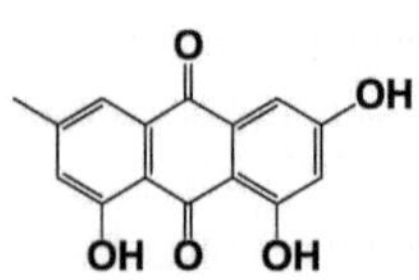

**Emodin (15)**

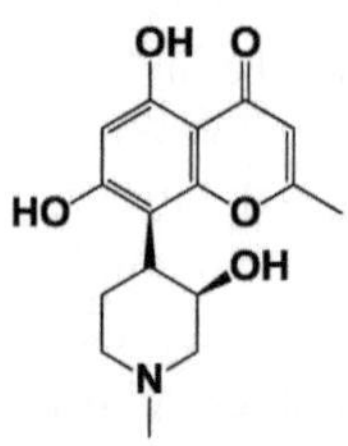

**Rohitukine (16)**

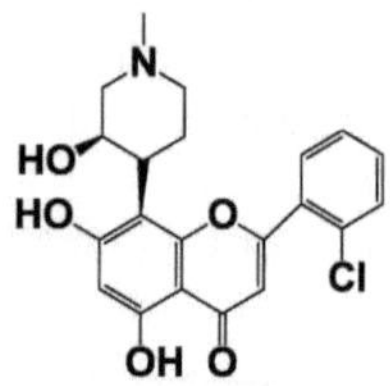

**Flavopiridol (17)**

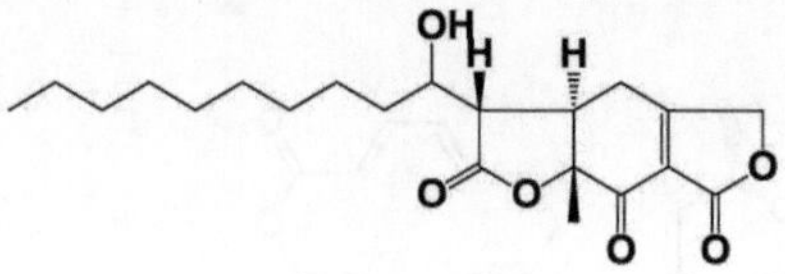

Annulosquamulin (18)

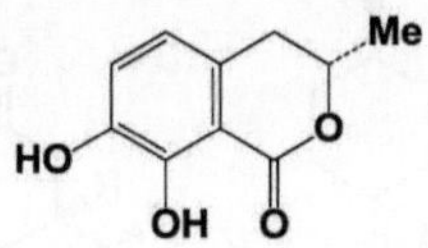

(3S)-7-hydroxymellein (19)

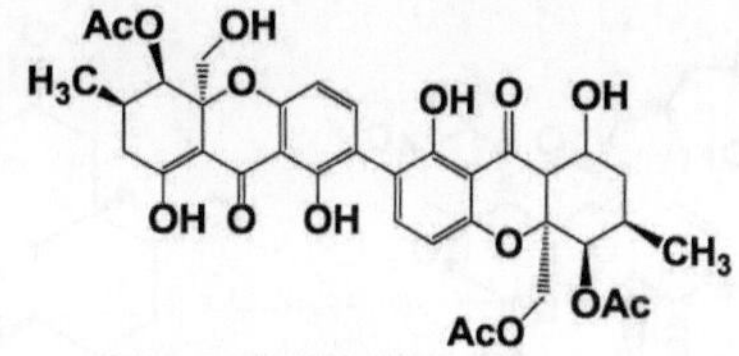

Dicerandrol B (20)

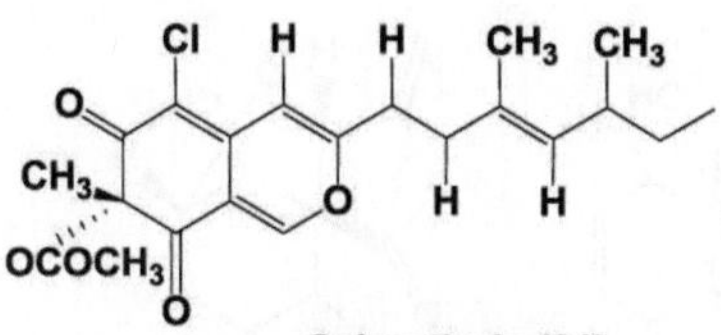

Sclerotiorin (21)

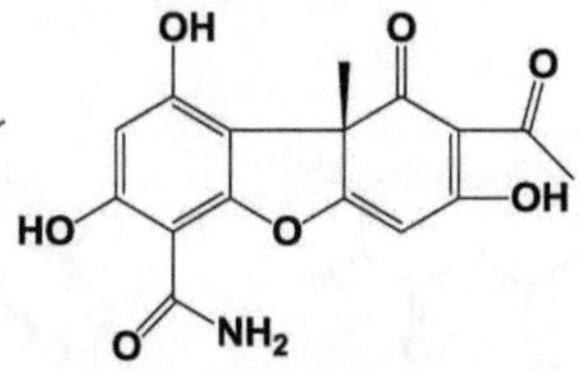

Cercosporamide (22)

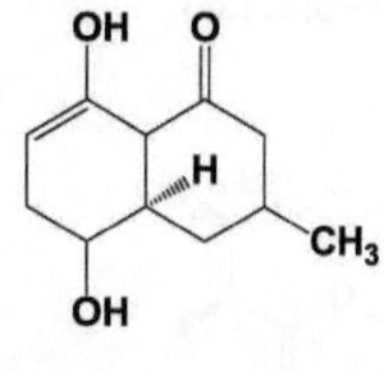

5-hydroxyramulosin (23)

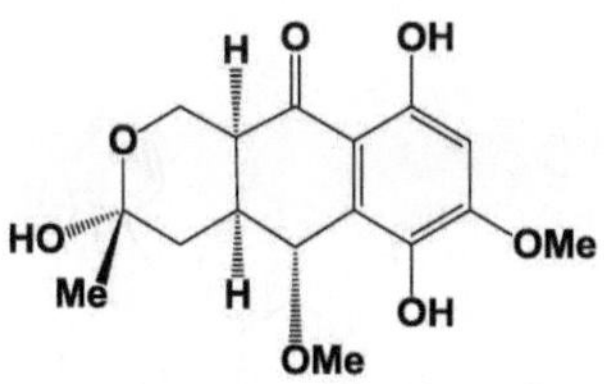

5-methoxydihydrofusarubin B (24)

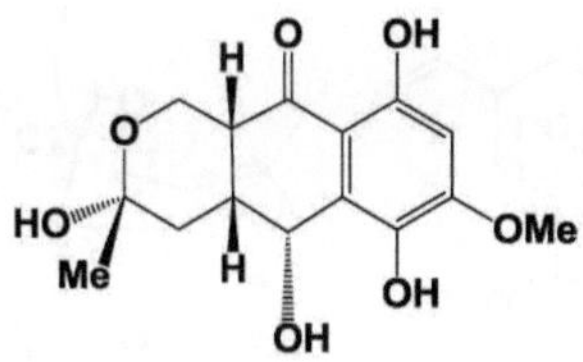

5-hydroxydihydrofusarubin C (25)

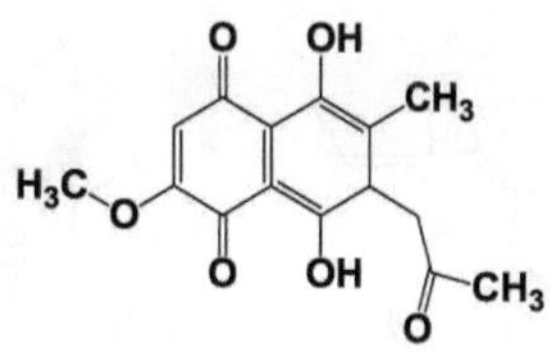

Javanicin (26)

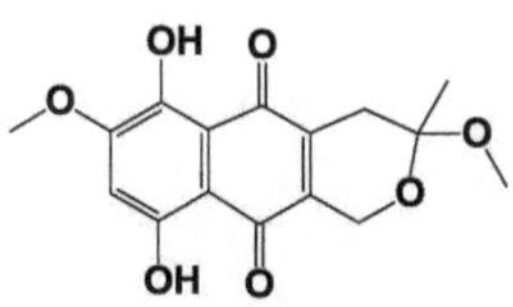

3-O-Methylfusarubin (27)

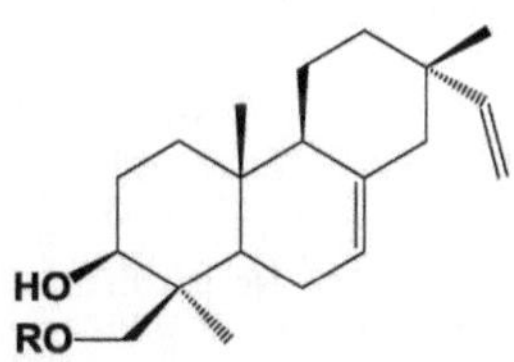

R= 2-acetamido-2-deoxy-α-D-glucopyranosyl
19-(2-acetamido-2-deoxy-α-D-glucopyranosyloxy) isopimara-7,15-diene-3β-ol (28)

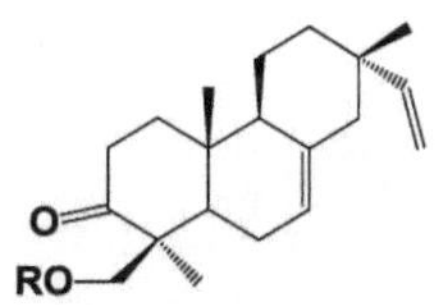

R= α-D-glucopyranosyl
19-(α-D-glucopyranosyloxy) isopimara-7,15-diene-3-one (29)

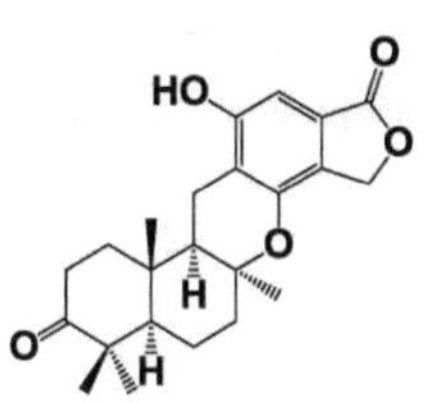

Phomoarcherin B (30)

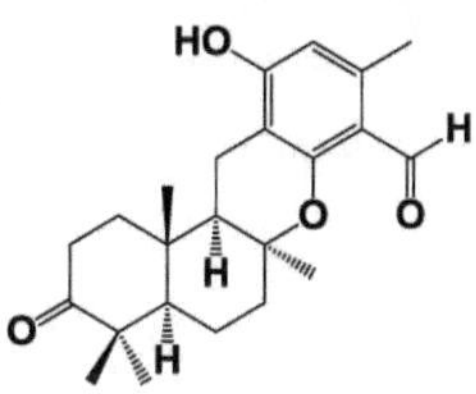

Phomoarcherin C (31)

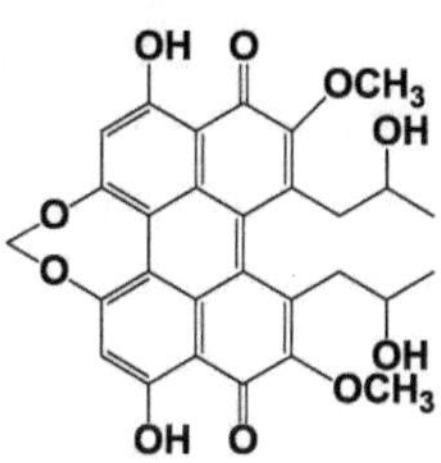

Cercosporin (32)

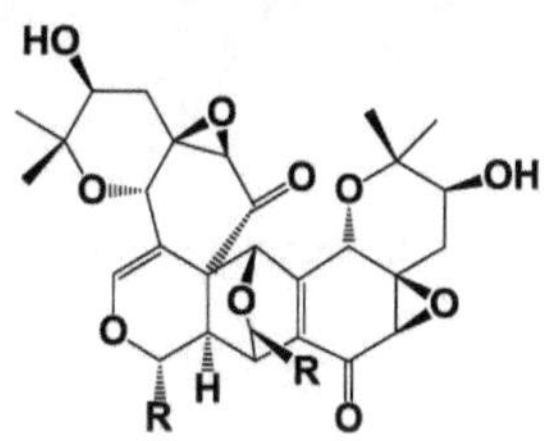

R= Pentyl

Pestaloquinol A (33)

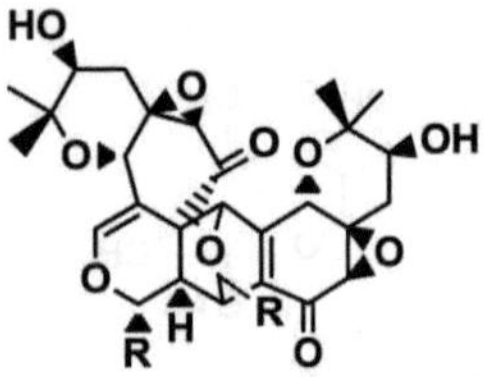

R= Pentyl

Pestaloquinol B (34)

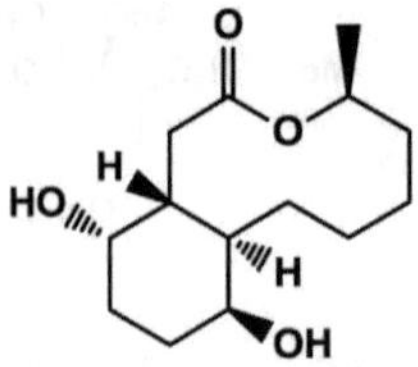

LMA-P1 (35)

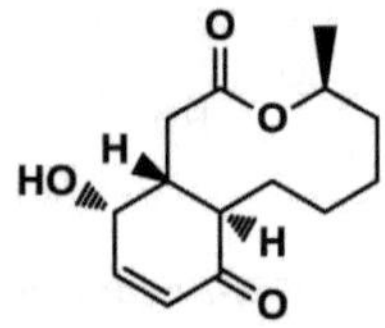

Sch-642305 (36)

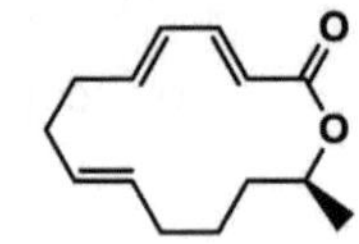

Benzoquinone (37)

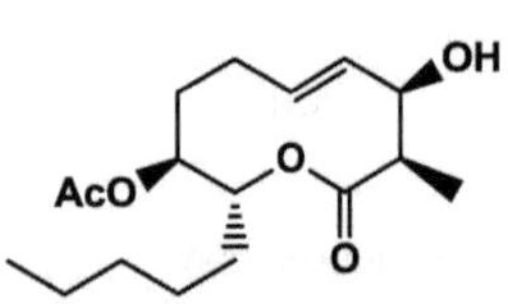

Cytospolide B(38)

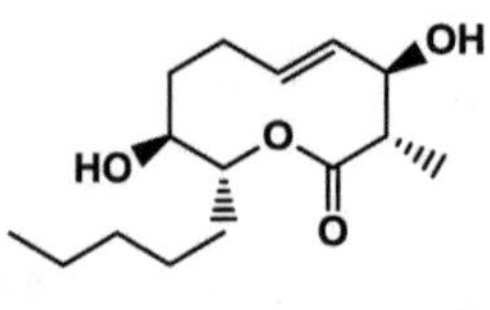

Cytospolide E (39)

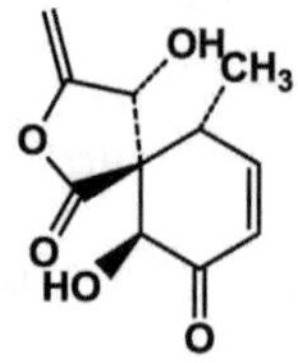

Massarigenin D (40)

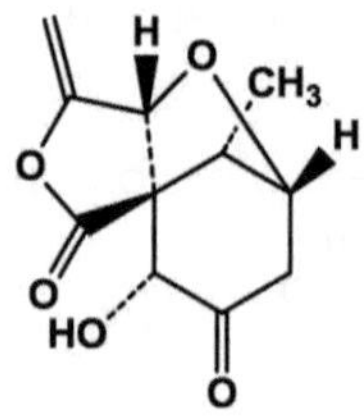

Spiromassaritone (41)

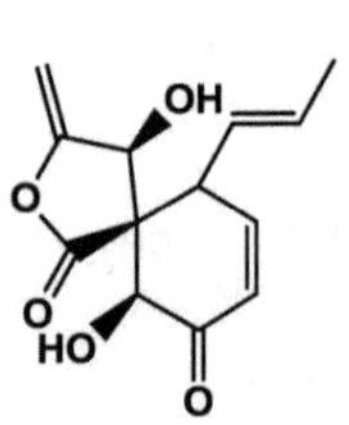

Paecilospirone (42)

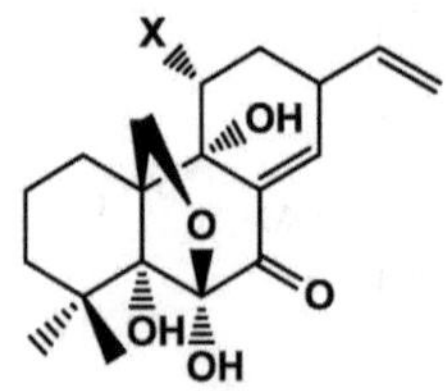

X= OH Diaporthein B (43)
X=H Scopararane A (44)

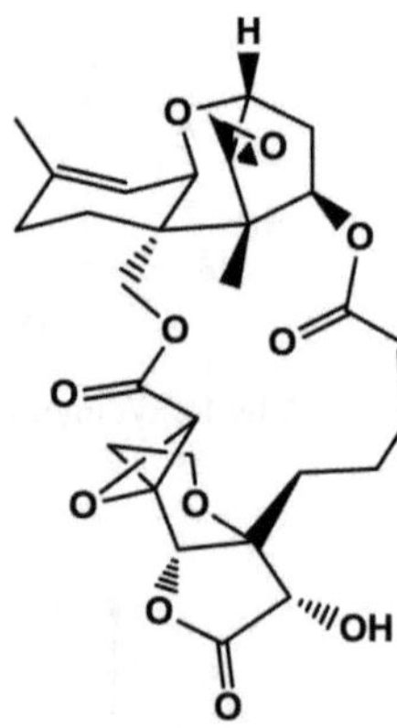

Roritoxin E (45)

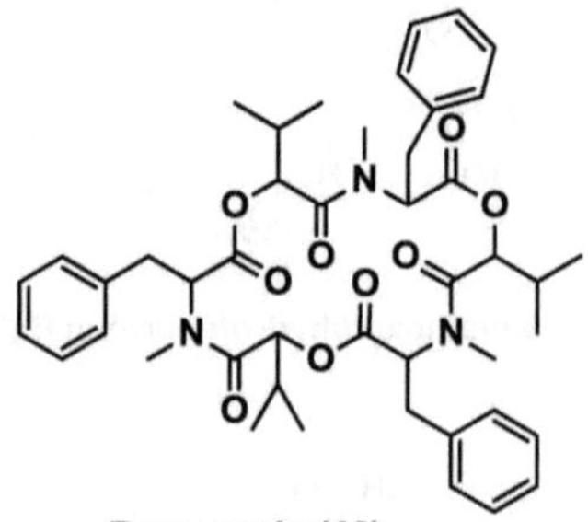

Beauvercin (46)

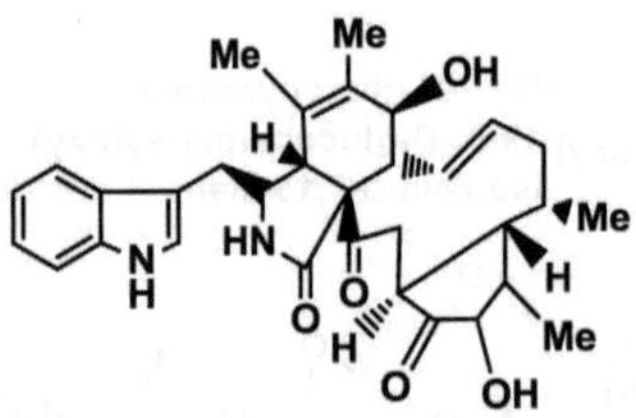

Chaetoglobosin V (47)

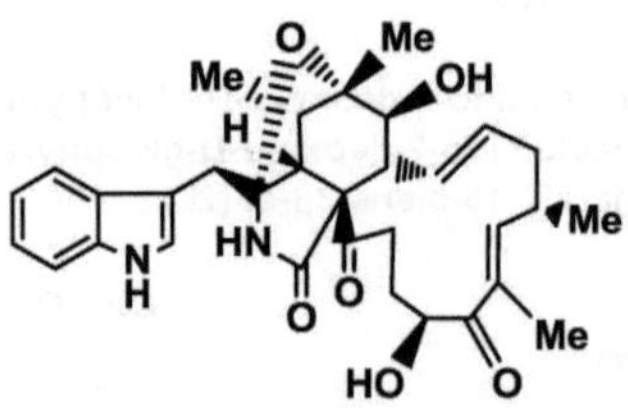

Chaetoglobosin W (48)

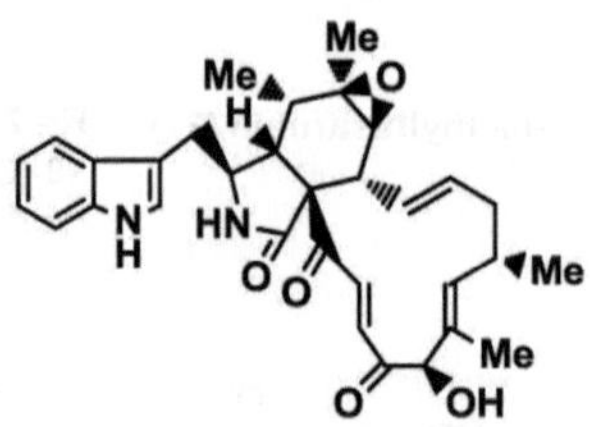

Chaetoglobosin A (49)

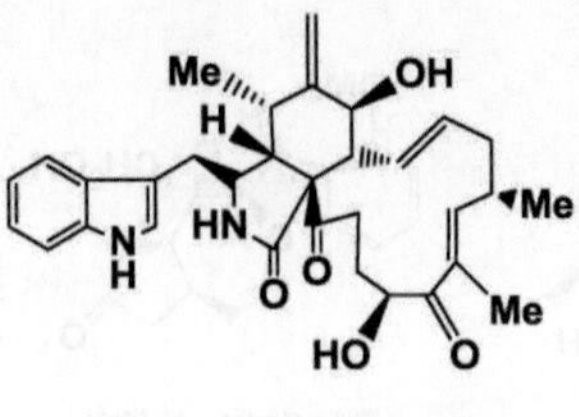

Chaetoglobosin Fex (50)

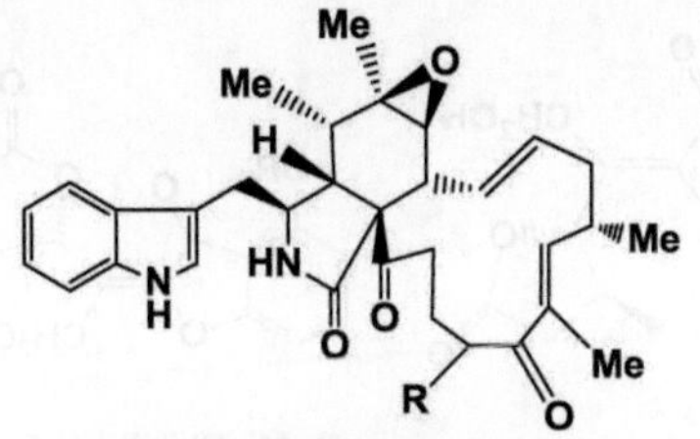

R=O Chaetoglobosin C (51)
R=OH Chaetoglobosin F (52)

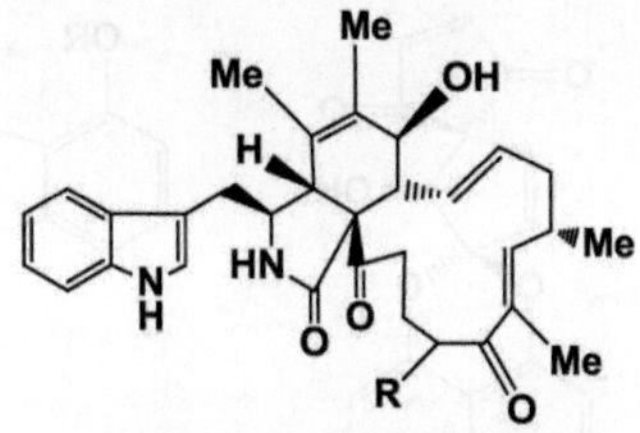

R=OH Chaetoglobosin E (53)
R=O Chaetoglobosin G (54)

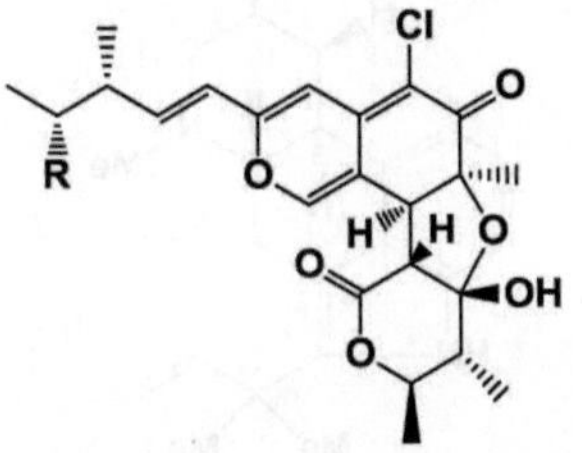

R=H Chaetomugilin D (55)
R=OH Chaetomugilin A (56)

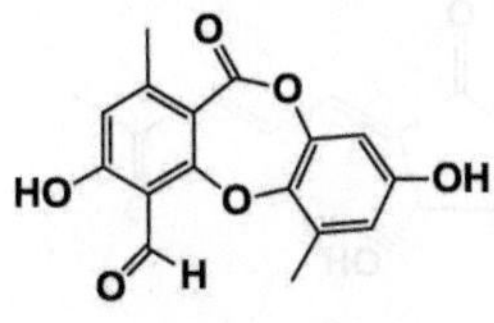

Botryorhodine A (57)

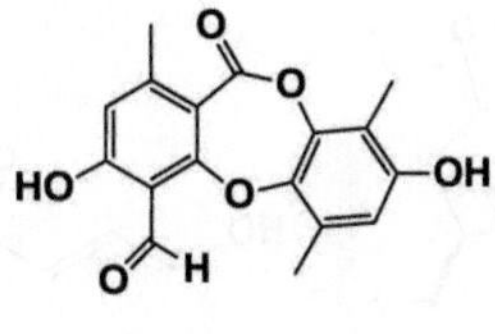

Botryorhodine B (58)

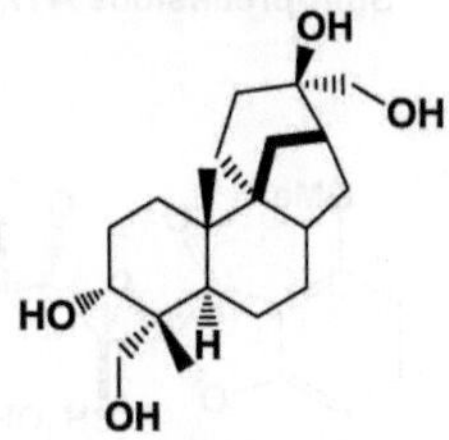

Aphidicolin (59)

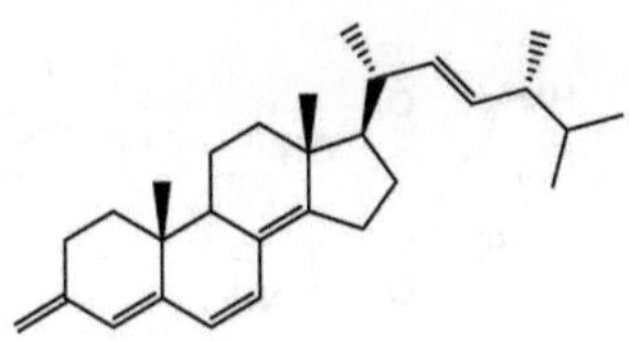

(22E,24R)-ergosta-4,6,8(14), 22-tetraen-3-one (60)

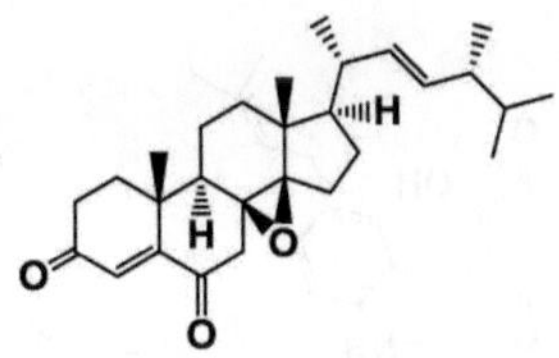

(22E,24R)-8,14 epoxyergosta-4, 22-diene-3,6-dione (61)

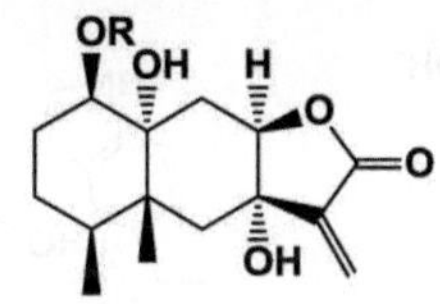

R=H Eremophilanolide 1 (62)
R=$CH_3$ Eremophilanolide 2 (63)

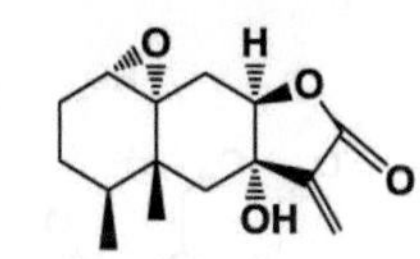

Eremophilanolide 3 (64)

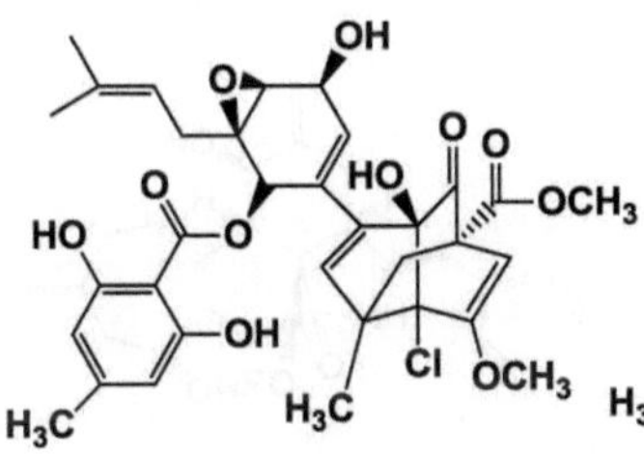

Chloropupukeananin (65)

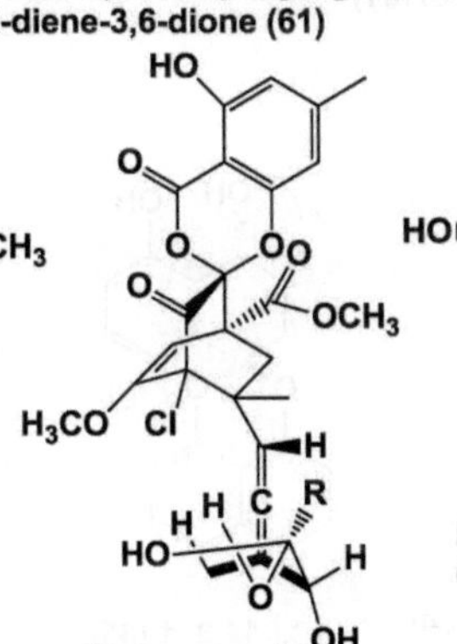

R= Isoprenyl
Chloropestolide A (66)

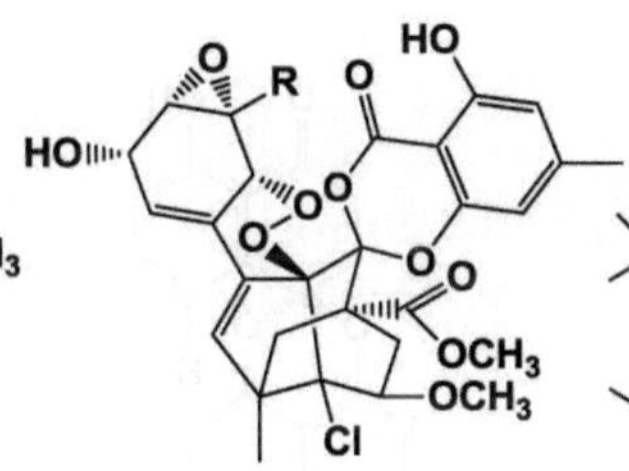

R= Isoprenyl
Chloropupukeanolide A (67)

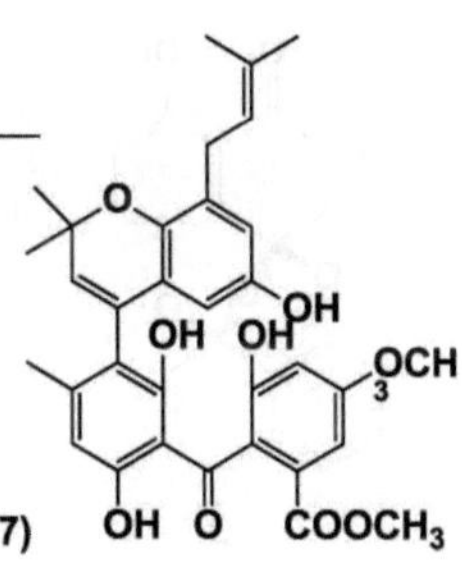

Pestaloficiol L (68)

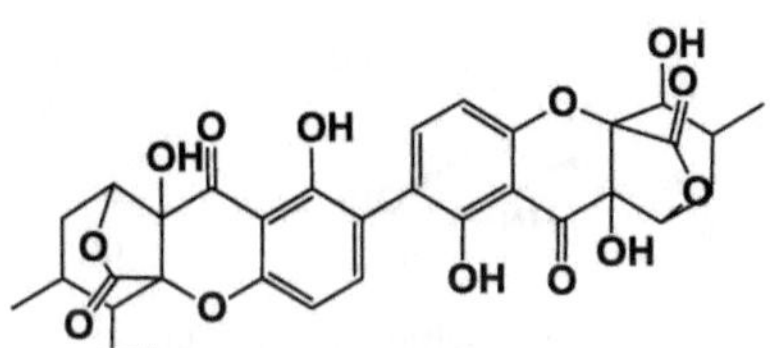

Ergoflavin (69)

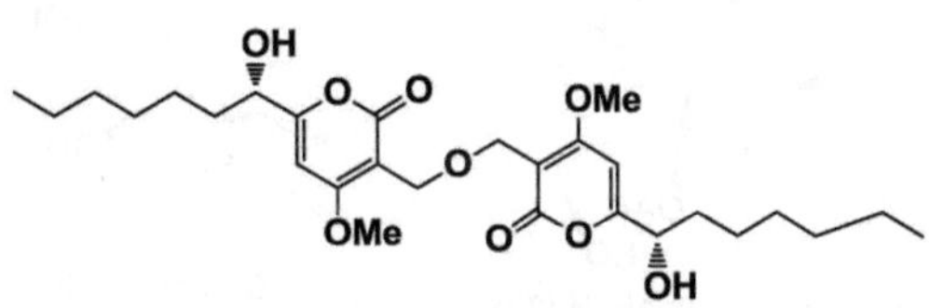

Dothideopyrone D (70)

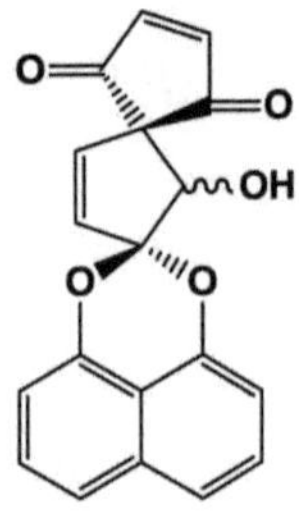

Spiropreussione A (71)

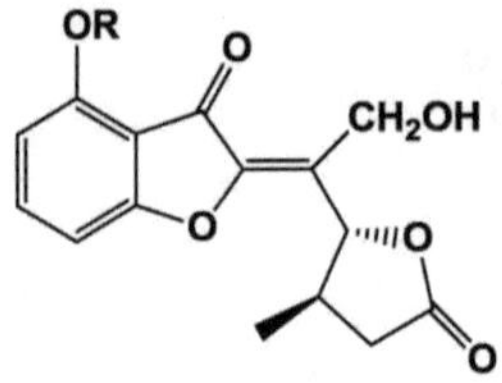

R=Me Photinide A (72)
R= H Photinide C (74)

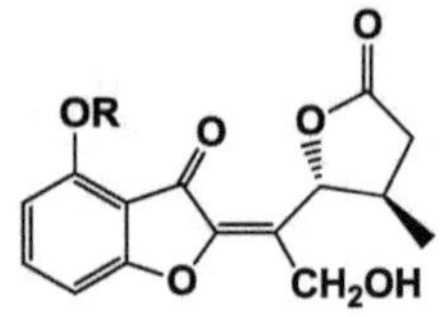

R=Me Photinide B (73)
R= H Photinide D (75)

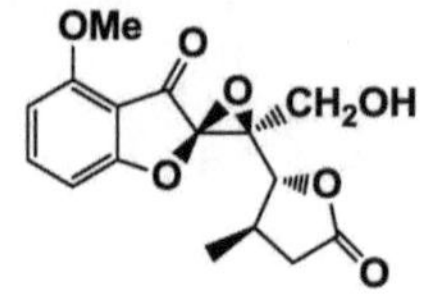

Photinide E (76)

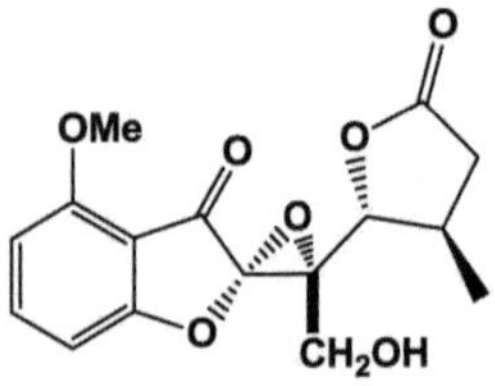

Photinide F (77)

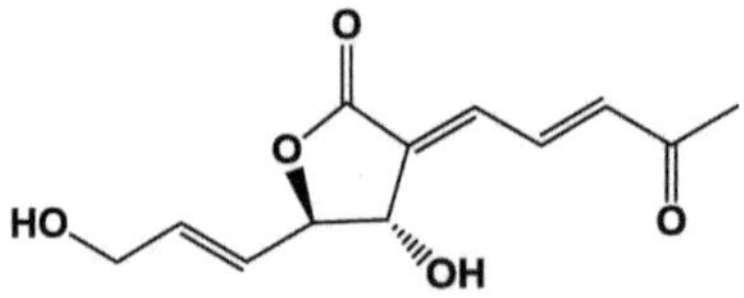

Eutypellin A (78)

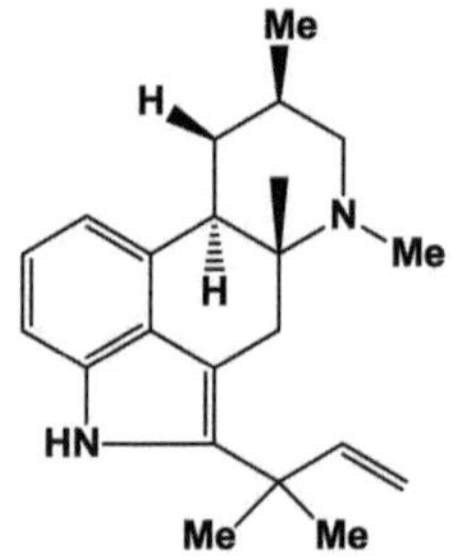

9-deacetoxyfumigaclavine C (79)

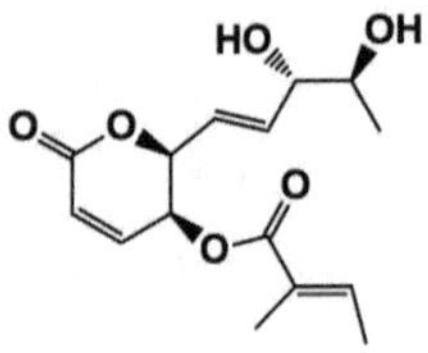

Phomopsolide B (80)

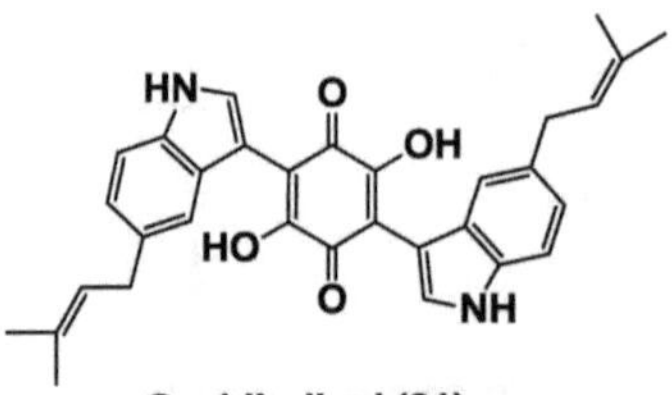

Cochliodinol (81)

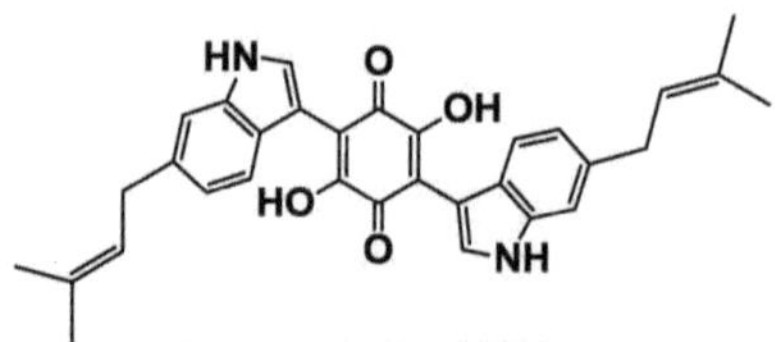

Isocochliodinol (82)

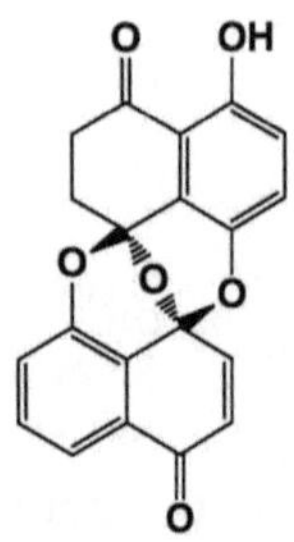

Pressomerin EG1 (83)

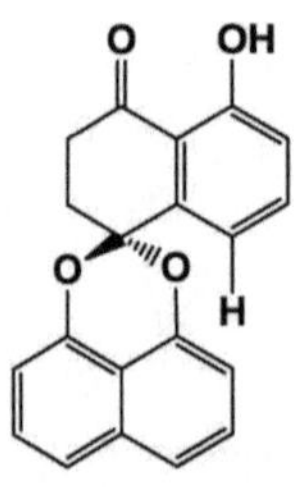

Palmarumycin CP2 (84)

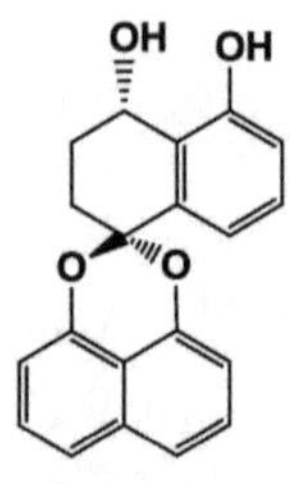

CJ-12,371 (85)

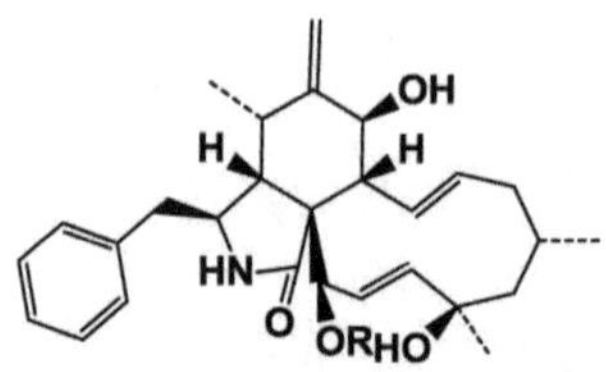

R= Ac Cytochalasin H (86)
R= H Cytochalasin J (87)

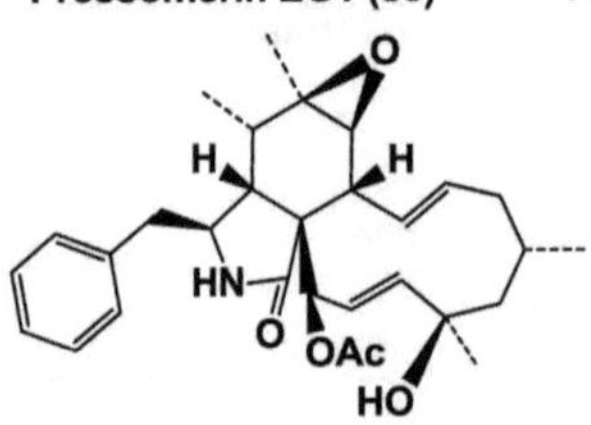

Epoxycytochalasin H (88)

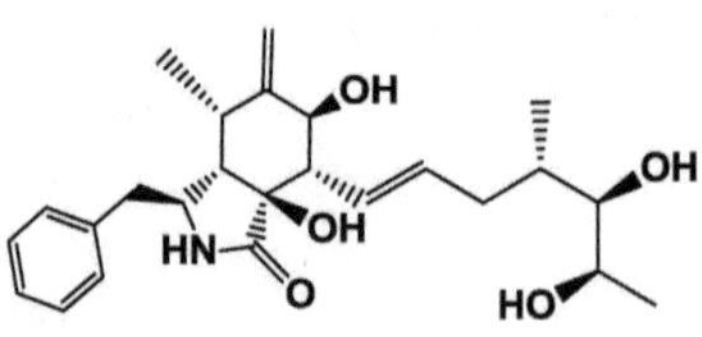

Cytochalasin Z10 (89)

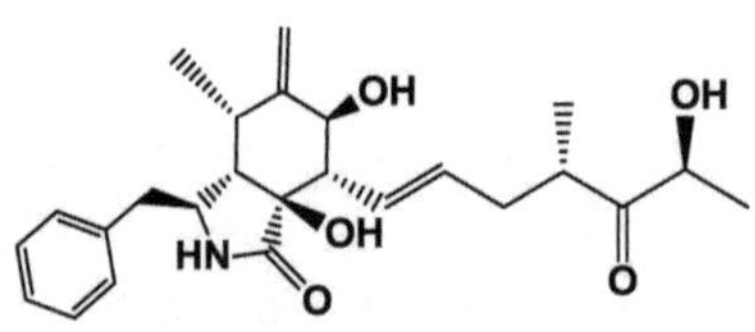

Cytochalasin Z11 (90)

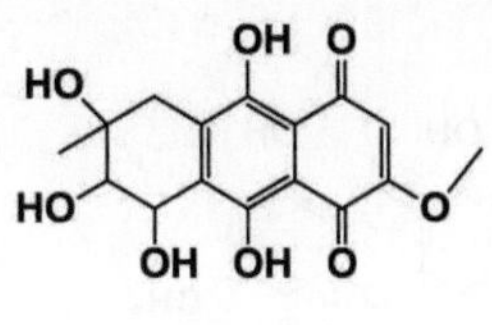

1403P-3 (91)

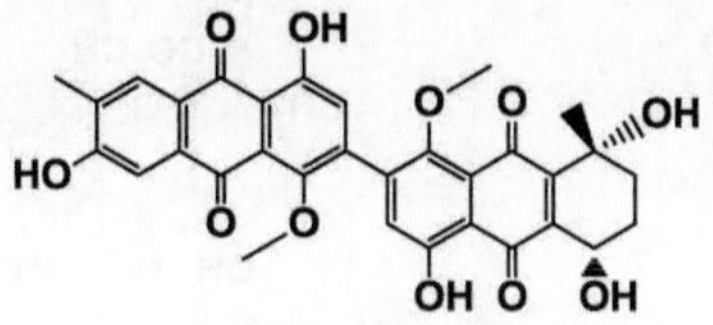

Alterporriol K (92)

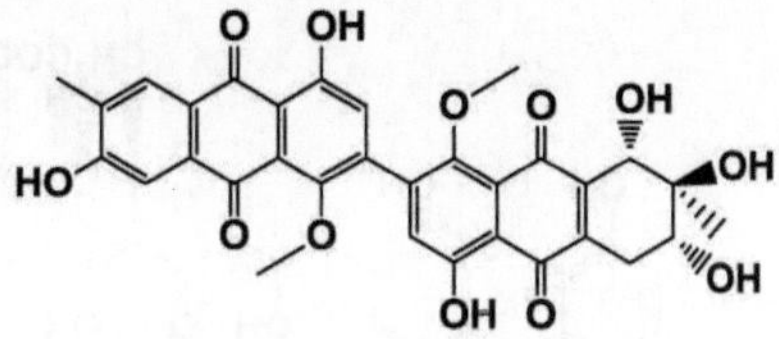

Alterporriol L (93)

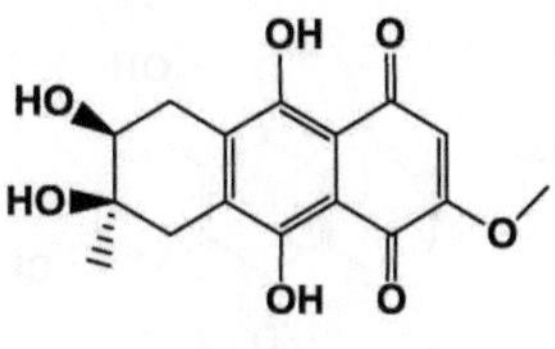

Bostrycin (94)

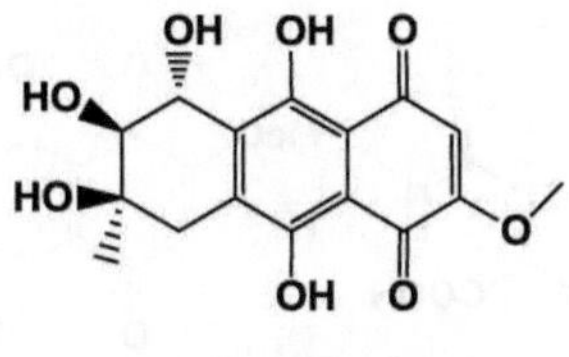

Deoxybostrycin (95)

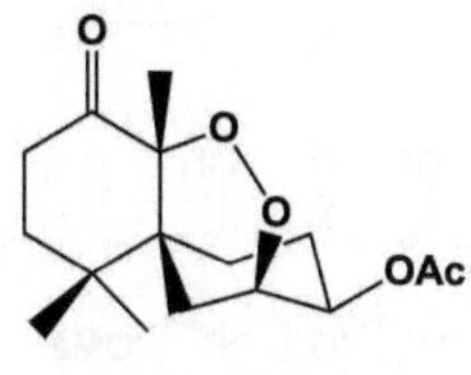

Talaperoxide B (96)

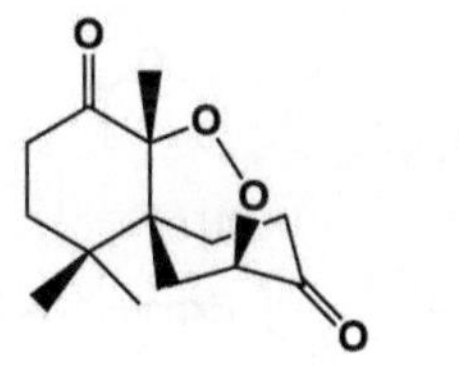

Talaperoxide D (97)

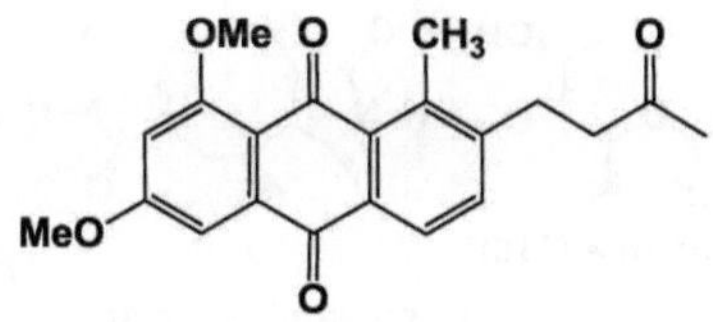

6,8-dimethoxy-1-methyl-2-(3-oxobutyl) anthrakunthone (98)

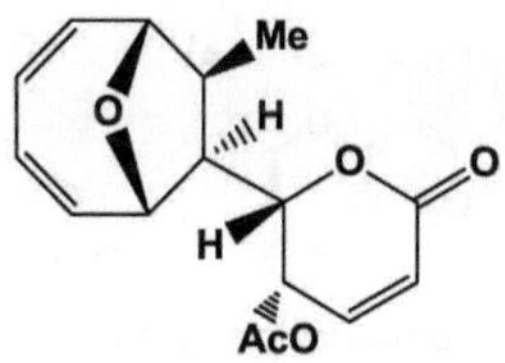

Mycoepoxydiene (99)

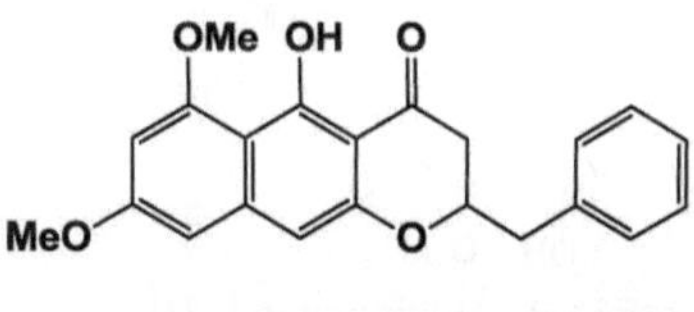

5-hydroxy-6,8-dimethoxy-2-benzyl-4H -naphtho[2,3-b]-pyran-4-one (100)

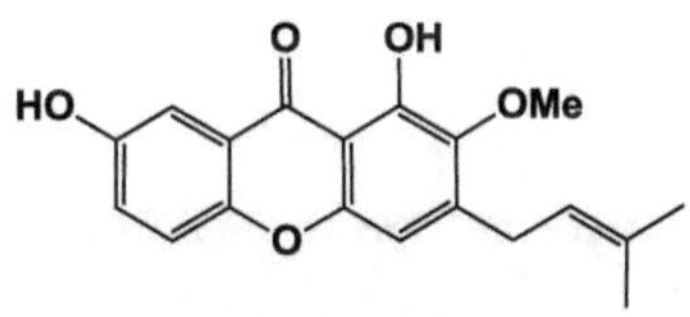

1,7-dihydroxy-2-methoxy-3-(3-methylbut-2-enyl)-9H-xanthen -9-one (101)

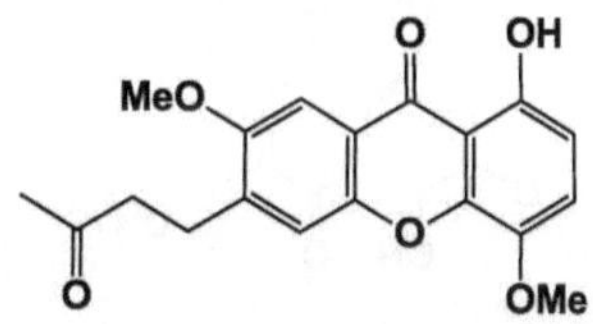

1-hydroxy-4,7-dimethoxy-6 -(3-oxobutyl)-9H-xanthen-9-one (102)

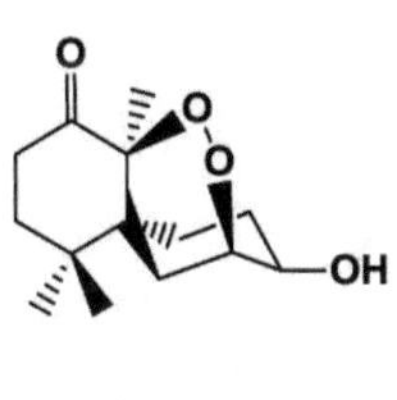

Merulin A (103)

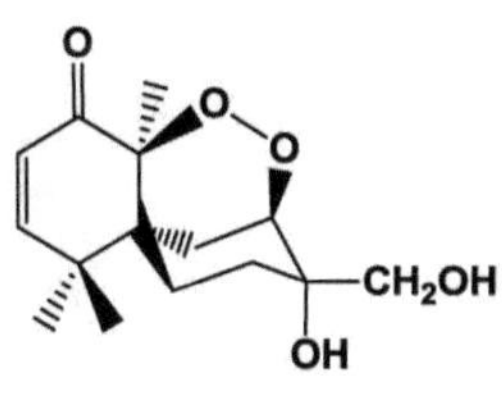

Merulin C (104)

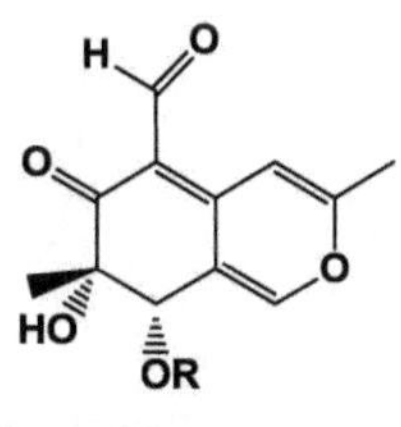

R=H 7-epiaustdiol (105)
R= $CH_3$ 8-O-methylepiaustdiol (106)

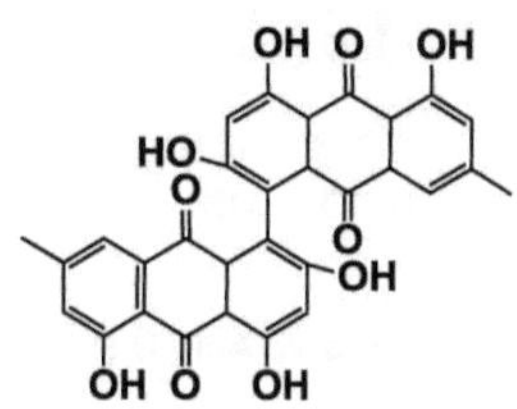

Stemphyperylenol (107)

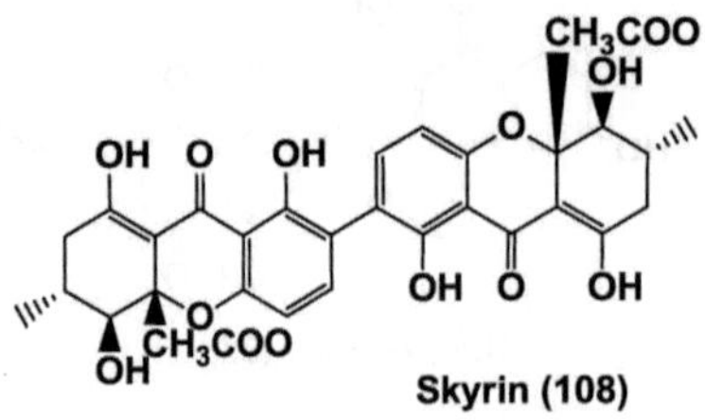

Skyrin (108)

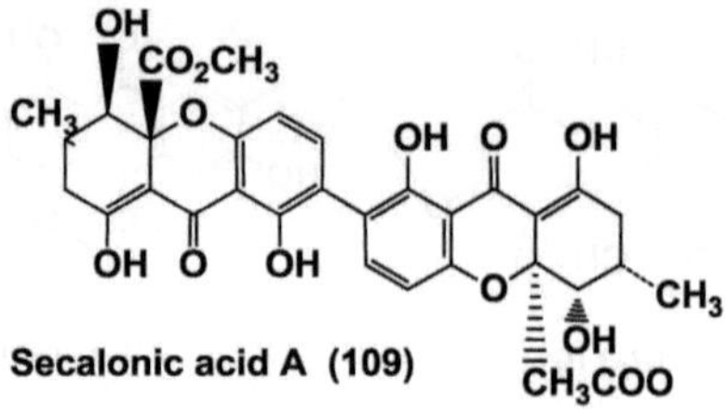

Secalonic acid A (109)

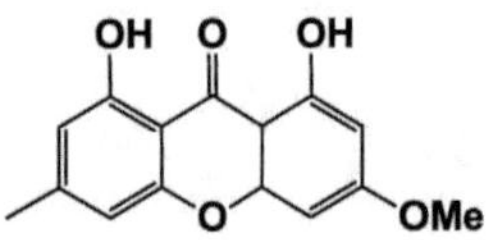

Norlichexanthone (110)

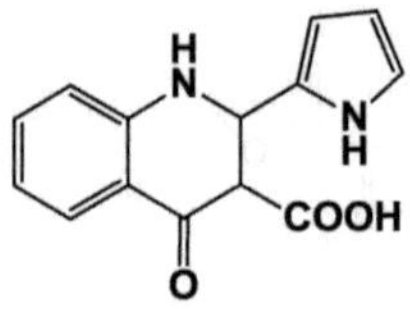

Penicinoline (111)

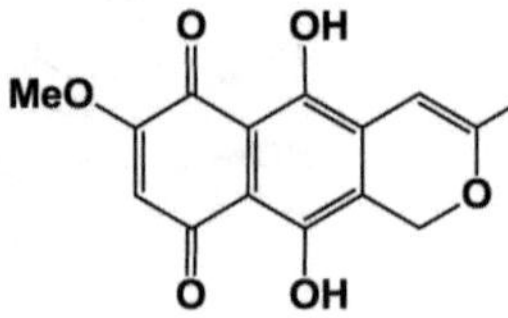

Anhydrofusarubin (112)

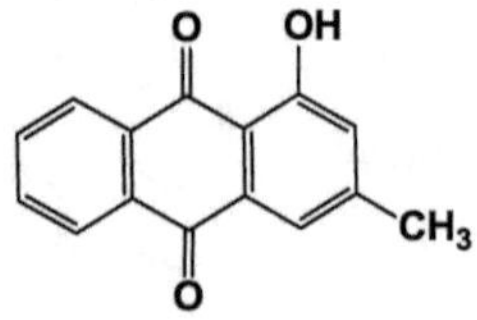

9,10 -anthracenedione (113)

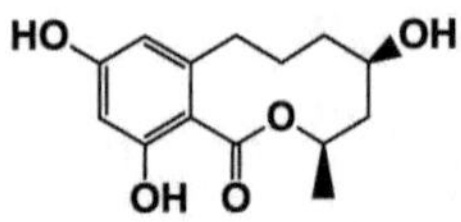

3R,5R-sonnerlactone (114)

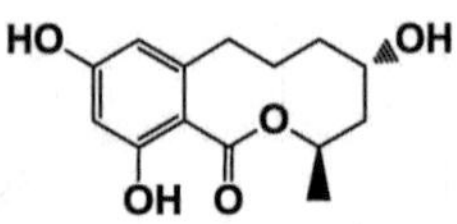

3R,5S-sonnerlactone (115)

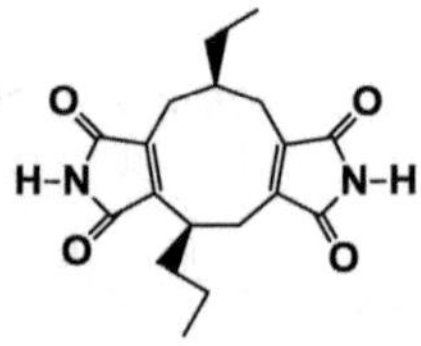

(-)-byssochlamic acid bisdiimide (116)

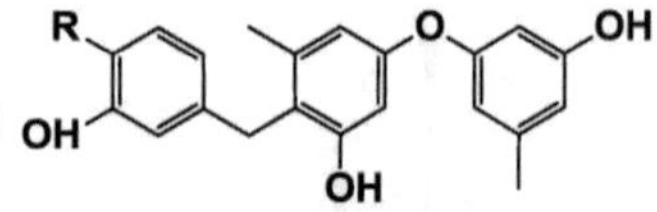

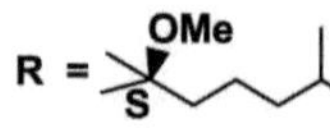

Expansol A (117)

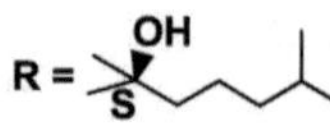

Expansol B (118)

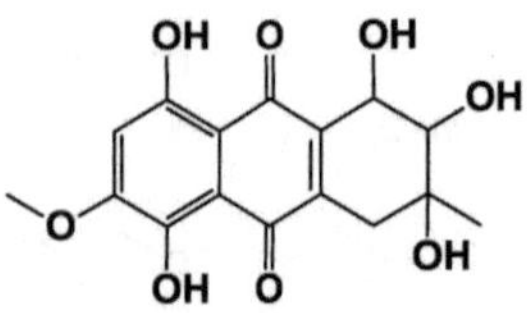

SZ-685C (48) (119)

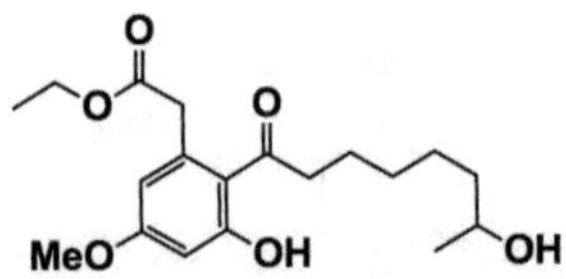

2-(7'-hydroxyoxooctyl)-3-hydroxy-5-methoxybenzeneacetic acid Et ester (120)

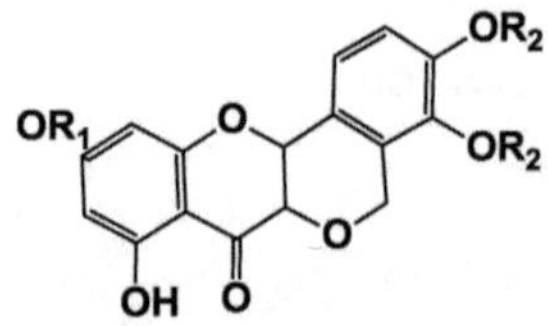

R1=R2= H Ophioglonin (121)
R1=b-D-glucopyranosyl, R2= H Ophioglonin-7-O-b-D-glucopyranoside (122)

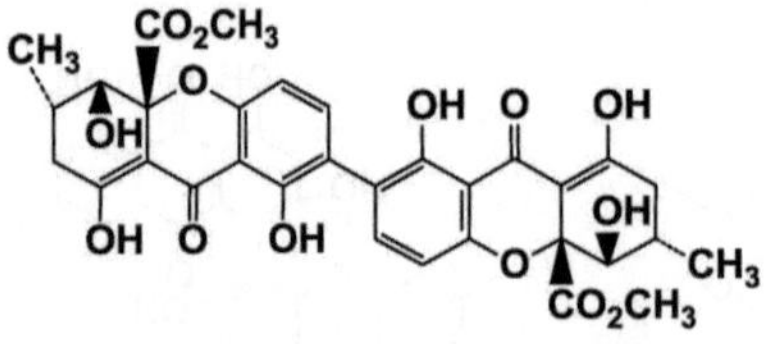

Secalonic Acid D (123)

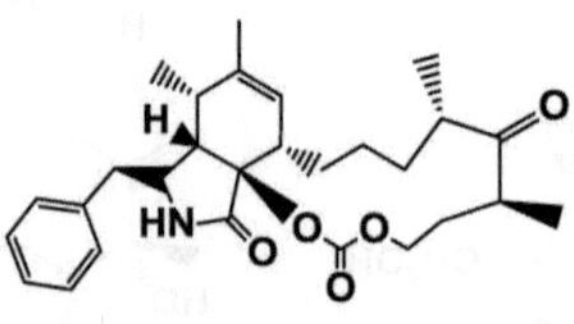

Cytochalasin Z16 (124)

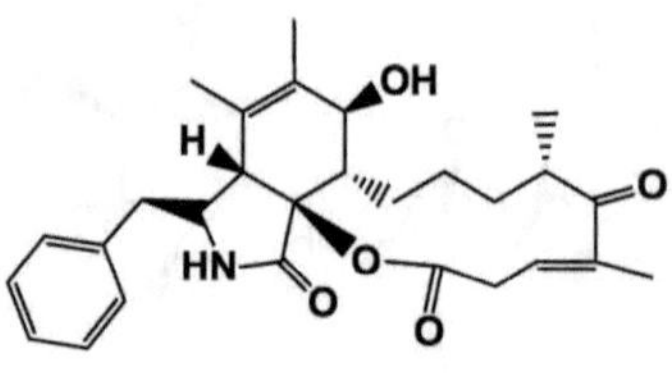

Cytochalasin Z17(125)

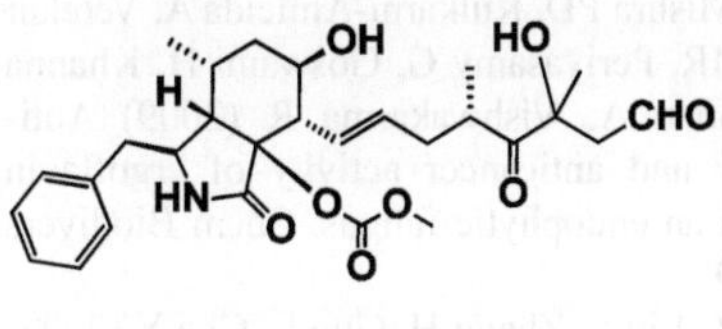

**Cytochalasin Z19 (126)**

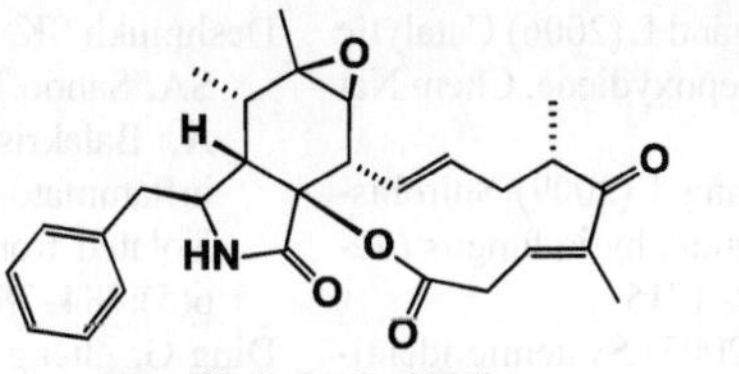

**Rosellichalasin (127)**

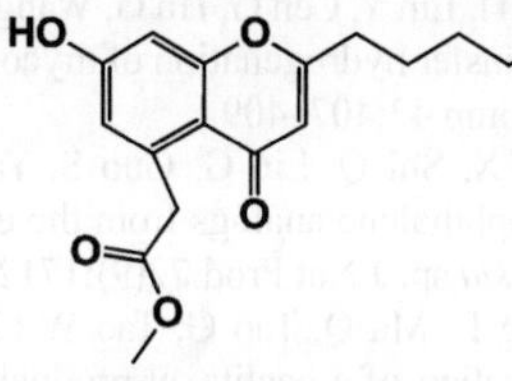

**Pestalotiopsone F (128)**

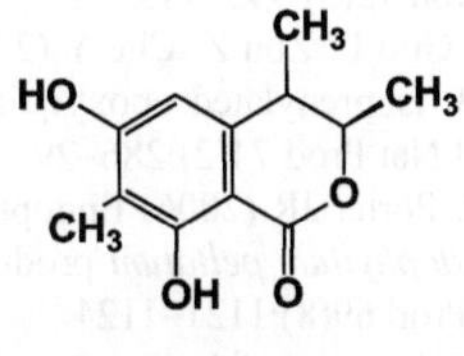

**(3R*,4S*)-6,8-dihydroxy-3,4,7-trimethylisocoumarin (129)**

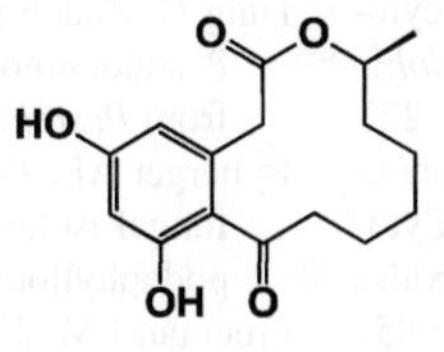

**S-Curvularin (130)**

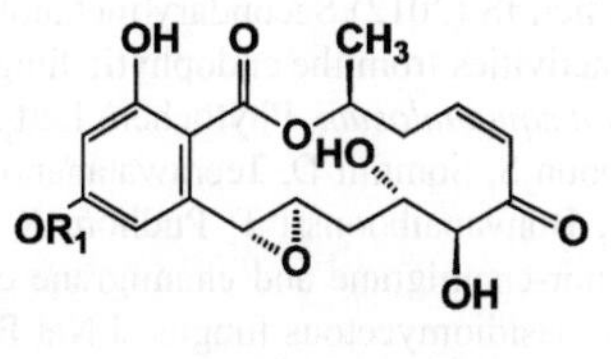

**R1= $CH_3$ Hypothemycin (131)**
**R1= H 4-O-demethylhypothemycin (132)**

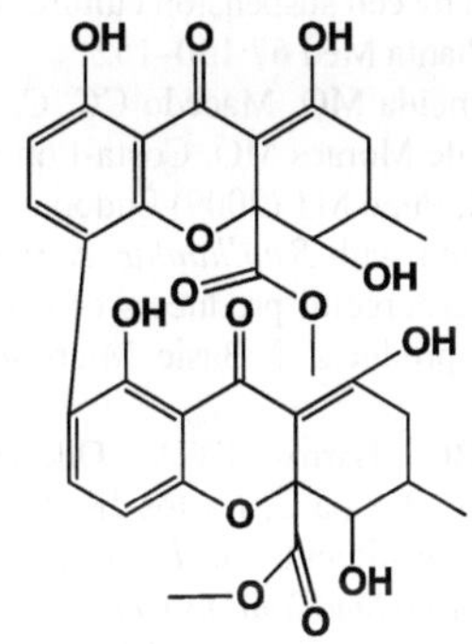

**Penicillixanthone A (133)**

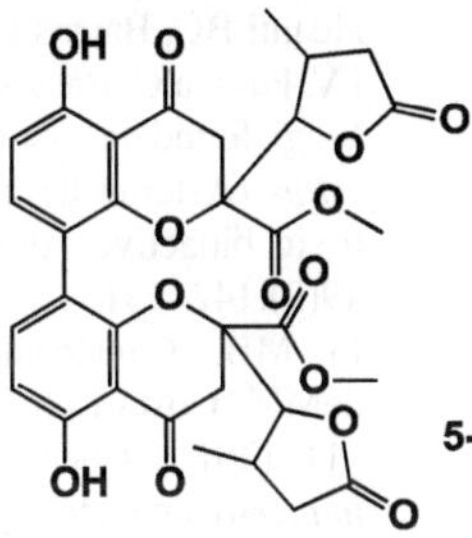

**Paecillin A (134)**

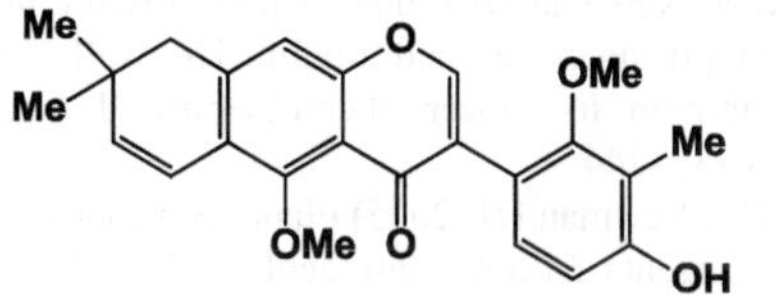

**5-O-methyl-2'-methoxy-3'-methylalpinumisoflavone (135)**

## References


Abdou R, Scherlach K, Dahse HM, Sattler I, Hertweck C (2010) Botryorhodines A-D, antifungal and cytotoxic depsidones from *Botryosphaeria rhodina*, an endophyte of the medicinal plant *Bidens pilosa*. Phytochem 71:110–116

Adelin E, Servy C, Cortial S et al (2011) Isolation, structure elucidation and biological activity of metabolites from Sch-642305-producing endophytic fungus *Phomopsis* sp. CMU-LMA. Phytochem 72:2406–2412

Ai H, Fen Y, Zh C, Zhen F, Xi R, Wan X, Din G, Li T, Zha H, Li H (2010) Isolation and identification of a taxol-producing endophytic fungus LNUF014. Weishengwuxue Zazhi 30(4):58–62

Azevedo JL, Araujo WL (2007) Diversity and applications of endophytic fungi isolated from tropical plants. In: Ganguli BN, Deshmukh SK (eds) Fungi: multifaceted microbes. CRC press, Boca Raton, pp 189–207

Bode HB, Bethe B, Höfs R, Zeek A (2002) Big effects from small changes: possible ways to explore nature's chemical diversity. Chem Bio Chem 3:619–627

Brockmann H, Haschad MN, Maier K, Pohl F (1939) Hypericin, the photodynamically active pigment from *Hypericum perforatum*. Naturwissenschaften 32:550–555

Cao S, McMillin DW, Tamayo G, Delmore J, Mitsiades CS, Clardy J (2012) Inhibition of tumor cells interacting with stromal cells by xanthones isolated from a Costa Rican *Penicillium* sp. J Nat Prod 75(4):793–797

Carlson BA, Dubay MM, Sausville EA, Brizuela L, Worland PJ (1996) Flavopiridol induces G1 arrest with inhibition of cyclin-dependent kinase CDK2 and CDK4 in human breast carcinoma cells. Cancer Res 56:2973–2978

Carroll G (1988) Fungal endophytes in stems and leaves: from latent pathogens to mutualistic symbionts. Ecology 69:2–9

Chakravarthi BVSK, Das P, Surendranath K, Karande AA, Jayabaskaran C (2008) Production of paclitaxel by *Fusarium solani* isolated from *Taxus celebica*. J Biosci 33:259–267

Chen H, Qi P, Lin YC, Chen GY, Wang L, Vrijoed LLP (2005) Preparation of hexahydromycoepoxydiene by catalytic hydrogenation of mycoepoxydiene. Zhongshan Daxue Xuebao. Ziran Kexueban 44(3):122–123

Chen H, Lin Y, Cen G, Hu G, Wangand L (2006) Catalytic transfer hydrogenation of mycoepoxydiene. Chem Nat Comp 42:407–409

Chen X, Shi Q, Lin G, Guo S, Yang J (2009) Spirobisnaphthalene analogs from the endophytic fungus *Preussia* sp. J Nat Prod 72(9):1712–1715

Cheng L, Ma Q, Tao G, Tao W (2007) Systemic identification of a paclitaxel-producing endophytic fungus. Ind Microbiol 37:23–30

Cheng MJ, Wu MD, Yuan GF, Chen YL, Su YS, Hsieh MT, Chen IS (2012) Secondary metabolites and cytotoxic activities from the endophytic fungus *Annulohypoxylon squamulosum*. Phytochem Lett 5(1):219–223

Chokpaiboon S, Sommit D, Teerawatananond T, Muangsin N, Bunyapaiboonsri T, Pudhom K (2010) Cytotoxic nor-chamigrane and chamigrane endoperoxides from a basidiomycetous fungus. J Nat Prod 73:1005–1007

Chomcheon P, Wiyakrutta S, Sriubolmas N, Ngamrojanavanich N, Mahidol C, Ruchirawat S, Kittakoop P (2009) Metabolites from the endophytic mitosporic *Dothideomycete* sp. LRUB20. Phytochemistry 70(1):121–127

Cragg GM, Newman DJ (2004) A tale of two tumor targets: topoisomerase I and tubulin. The Wall and Wani contribution to cancer chemotherapy. J Nat Prod 67(2):232–244

Cragg GM, Newman DJ (2005) Plants as a source of anticancer agents. J Ethnopharmacol 100:72–79

Cragg GM, Newman DJ (2009) Nature: a vital source of leads for anticancer drug development. Phytochem Rev 8:313–331

Dai W, Tao W (2008) Preliminarly study on fermentation conditions of taxol-producing endophytic fungus. Chem Ind Eng Progress 27:883–886

Davoodi H, Hashemi SR, Seow HF (2012) Increased NFk-B activity in Hct116 colorectal cancer cell line harboring *TLR4* Asp299Gly *polymorphism*. Iran J Allergy Asthma Immunol 11(2):121–132

Debbab A, Aly AH, Edrada-Ebel RA, Muller WEG, Mosaddak M, Hakiki A, Ebel R, Proksch P (2009) Bioactive secondary metabolites from the endophytic fungus *Chaetomium* sp. isolated from *Salvia officinalis* growing in Morocco. Biotechnol Agron Soc Environ 13:229–234

Delaey EM, Obermueller R, Zupko I, De Vos D, Falk H, de Witte PA (2001) *In vitro* study of the photocytotoxicity of some hypericin analogs on different cell lines. Photochem Photobiol 74:164–171

Deng BW, Liu KH, Chen WQ, Ding XW, Xie XC (2009) *Fusarium solani*, Tax-3, a new endophytic taxol-producing fungus from *Taxus chinensis*. World J Microbiol Biotechnol 25:139–143

Deshmukh SK, Verekar SA (2009) Fungal Endophytes: a potential source of anticancer compounds. In: Carpinella MC, Rai MK (eds) Novel therapeutic agents from plants: progress and future perspectives. Science Publishers Inc., Enfield, pp. 175–206

Deshmukh SK, Verekar SA (2012) Fungal endophytes: a potential source of antifungal compounds. Front Biosci (Elite Ed) 4:2045–2070

Deshmukh SK, Mishra PD, Kulkarni-Almeida A, Verekar SA, Sahoo MR, Periyasamy G, Goswami H, Khanna A, Balakrishnan A, Vishwakarma R (2009) Anti-inflammatory and anticancer activity of ergoflavin isolated from an endophytic fungus. Chem Biodivers 6(5):784–789

Ding G, Zheng Z, Liu S, Zhang H, Guo L, Che Y (2009) Photinides A–F, cytotoxic benzofuranone-derived γ-lactones from the plant endophytic fungus *Pestalotiopsis photiniae*. J Nat Prod 72(5):942–945

Ding G, Zhang F, Chen H, Guo L, Zou Z, Che Y (2011) Pestaloquinols A and B, isoprenylated epoxyquinols from *Pestalotiopsis* sp. J Nat Prod 74(2):286–291

Eyberger AL, Dondapati R, Porter JR (2006) Endophyte fungal isolates from *Podophyllum peltatum* produces podophyllotoxin. J Nat Prod 69(8):1121–1124

Freeman EM (1904) The seed fungus of *Lolium temulentum* L. Phil Trans R Soc Lond (Biol) 196:1–27

Fulzele DP, Satdive RK, Pol BB (2001) Growth and production of camptothecin by cell suspension cultures of *Nothapodytes foetida*. Planta Med 67:150–152

Gallo MB, Chagas FO, Almeida MO, Macedo CC, Cavalcanti BC, Barros FW, de Moraes MO, Costa-Lotufo LV, Pessoa C, Bastos JK, Pupo MT (2009) Endophytic fungi found in association with *Smallanthus sonchifolius* (Asteraceae) as resourceful producers of cytotoxic bioactive natural products. J Basic Microbiol 49(2):142–151

Gallo MB, Cavalcanti BC, Barros FWA, Odorico deMM, Costa-Lotufo LV, Pessoa C, Bastos JK, Pupo MT (2010) Chemical constituents of *Papulaspora immersa*, an endophyte from *Smallanthus sonchifolius* (Asteraceae) and their cytotoxic Activity. Chem Biodivers 7(12):2941–2950

Gangadevi V, Muthumary J (2008) Taxol, an anticancer drug produced by an endophytic fungus *Bartalinia robillardoides* Tassi, isolated from a medicinal plant, *Aegle marmelos* Correa ex Roxb. World J Microbiol Biotechnol 24(5):717–724

Gangadevi V, Muthumary J (2009a) A novel endophytic Taxol-producing fungus *Chaetomella raphigera* isolated from a medicinal plant, *Terminalia arjuna*. Appl Biochem Biotechnol 158(3):675–684

Gangadevi V, Muthumary J (2009b) Taxol production by *Pestalotiopsis terminaliae*, an endophytic fungus of *Terminalia arjuna* (arjun tree). Biotechnol Appl Biochem 52:9–15

Gangadevi V, Murugan M, Muthumary J (2008) Taxol determination from *Pestalotiopsis pauciseta*, a fungal endophyte of a medicinal plant. Chin J Biotechnol 24:1433–1438

Ge HM, Yu ZG, Zhang J, Wu JH, Tan RX (2009) Bioactive alkaloids from endophytic *Aspergillus fumigatus*. J Nat Prod 72(4):753–755

Giridharan P, Verekar S, Khanna A, Mishra PD, Deshmukh SK (2012) Anticancer activity of Sclerotiorin isolated from an endophytic fungus *Cephalotheca faveolata* Yaguchi, Nishim, Udagawa. Indian J Exp Biol 50:464–468

Goh TK, Yipp MW (1996) *In vivo* and *in vitro* studies of three new species of *Trimmatostroma* associated with sooty spots of the mangrove *Aegiceras corniculatum* in Hong Kong. Mycol Res 100(12):1489–1497

Greve H, Mohamed IE, Pontius A, Kehraus S, Gross H, Konig GM (2010) Fungal metabolites: structural diversity as incentive for anticancer drug development. Phytochem Rev 9:537–545

Gunatilaka AAL (2006) Natural products from plant-associated microorganisms: distribution, structural diversity, bioactivity and implications of their occurrence. J Nat Prod 69:509–526

Gueritte F, Fahy J (2005) The vinca alkaloids. In: Cragg GM, Kingston DGI, Newman DJ (eds) Anticancer agents from natural products. Taylor and Francis, Boca Raton, pp 123–136

Guo B, Li H, Zhang L (1998) Isolation of the fungus producing vinblastine. J Yunnan Univ (Nat Sci Edn) 20:214–215

Guo BH, Wang YC, Zhou XW, Hu K, Tan F, Miao ZQ, Tang KX (2006) An endophytic taxol producing fungus BT2 isolated from *Taxus chinensis* var. mairei. Afr J Biotechnol 5:875–877

Guo Z, She Z, Shao C, Wen L, Liu F, Zheng Z, Lin Y (2007) Spectral assignments and reference data; 1H and 13C NMR signal assignments of paecilin A and B, two new chromone derivatives from mangrove endophytic fungus *Paecilomyces* sp. (tree 1–7). Magn Reson Chem 45(9):777–780

Gurudatt PS, Priti V, Shweta S, Ramesha BT, Ravikanth G, Vasudeva R, Amna T, Deepika S, Ganeshaiah KN, Shaanker RU, Puri S, Qazi GN (2010) Attenuation of camptothecin production and negative relation between hyphal biomass and camptothecin content in endophytic fungal strains isolated from *Nothapodytes nimmoniana* Grahm (Icacinaceae). Curr Sci 98(8):1006–1010

Hadjur C, Richard MJ, Parat MO, Jardon P, Favier A (1996) Photodynamic effect of Hypericin on lipid peroxidation and antioxidant status in melanoma cells. Photochem Photobiol 64:375–381

Hallmann J, Sikora RA (1996) Toxicity of fungal endophytic secondary metabolites to plant parasitic nematodes and soil borne plant pathogenic fungi. Eur J Plant Pathol 102:155–162

Han Z, Mei W, Zhao Y, Deng Y, Dai H (2009a) A new cytotoxic isocoumarin from endophytic fungus *Penicillium* SP. 091402 of the mangrove plant *Bruguiera sexangula*. Chem Nat Comp 45(6):805–807

Han X, Lin Z, Tao H, Liu P, Wang Y, Zhu W (2009b) Cytotoxic metabolites from symbiotic fungus *Penicillium* sp. HK13–8 with *Rhizophora stylosa*. Zhongguo Haiyang Yaowu 28(5):11–16

Hanahan D, Weinberg RA (2011) Hallmarks of cancer: the next generation. Cell 144(5):646–674

Harmon AD, Weiss U, Silverton JV (1979) The structure of rohitukine, the main alkaloid of *Amoora rohituka* (syn. *Aphanamixis polystachya*) (Meliaceae). Tetrahedron Lett 8:721–724

Hemtasin C, Kanokmedhakul S, Kanokmedhakul K, Hahnvajanawong C, Soytong K, Prabpai S, Kongsaeree P (2011) Cytotoxic pentacyclic and tetracyclic aromatic sesquiterpenes from *Phomopsis archeri*. J Nat Prod 74(4):609–613

Hu K, Tan F, Tang K, Zhu S, Wang W (2006) Isolation and screening of endophytic fungi synthesizing taxol from *Taxus chinensis* var. *mairei*. J Southwest China Normal Univ (Nat Sci Edn) 31:134–137

Huang Z, Guo Z, Yang R, Yin X, Li X, Luo W, She Z, Lin Y (2009a) Chemistry and cytotoxic activities of polyketides produced by the mangrove endophytic fungus *Phomopsis* SP. ZSU-H76. Chem Nat Comp 45(5):625–628

Huang Z, Yang R, Su G, She Z, Lin Y (2009b) Study on metabolites of mangrove endophytic fungus ZSU-H19 from South China Sea. Guangxi Shifan Daxue Xuebao. Ziran Kexueban 27(4):57–60

Huang Z, Yang RY, Guo ZY, She ZG, Lin YC (2010a) New anthraquinone derivative produced by cultivation of mangrove endophytic fungus *Fusarium* sp. ZZF60 from the South China Sea. Yingyong Huaxue 27(4):394–397

Huang Z, Yang R, Guo Z, She Z, Lin Y (2010b) A new naphtho- γ- -pyrone from mangrove endophytic fungus ZSU-H26. Chem Nat Comp 46(1):15–18

Huang Z, Yang R, Yin X, She Z, Lin Y (2010c) Structure elucidation and NMR assignments for two xanthone derivatives from a mangrove endophytic fungus (No. ZH19). Magn Reson Chem 48(1):80–82

Huang Z, Yang J, She Z, Lin Y (2010d) Isoflavones from the mangrove endophytic fungus *Fusarium* sp. (ZZF41). Nat Prod Commun 5(11):1771–1773

Huang CH, Pan JH, Chen B, Yu M, Huang HB, Zhu X, Lu YJ, She ZG, Lin YC (2011) Three bianthraquinone derivatives from the mangrove endophytic fungus *Alternaria* sp. ZJ9–6B from the South China Sea. Mar Drugs 9:832–843

Isaka M, Palasarn S, Lapanun S, Chanthaket R, Boonyuen N, Lumyong S (2009a) Gamma-lactones and ent-eudesmane sesquiterpenes from the endophytic fungus *Eutypella* sp. BCC 13199. J Nat Prod 72(9):1720–1722

Isaka M, Yangchum A, Intamas S, Kocharin K, Jones EBG, Kongsaeree P, Prabpai S (2009b) Aigialomycins and related polyketide metabolites from the mangrove fungus *Aigialus parvus* BCC 5311. Tetrahedron 65:4396–4403

Isaka M, Chinthanom P, Boonruangprapa T, Rungjindamai N, Pinruan U (2010) Eremophilane-type sesquiterpenes from the fungus *Xylaria* sp. BCC 21097. J Nat Prod 73(4):683–687

Isaka M, Palasarn S, Prathumpai W, Laksanacharoen P (2011) Pimarane diterpenes from the endophytic fungus *Eutypella* sp. BCC 13199. Chem Pharm Bull 59(9):1157–1159

Jordan MA, Toso RJ, Thrower D, Wilson L (1993) Mechanism of mitotic block and inhibition of cell proliferation by taxol at low concentrations. Proc Natl Acad Sci USA 90:9552–9556

Kamuhabwa AR, Agostinis PM, D'Hallewin MA, Baert L, de Witte PA (2001) Cellular photo destruction induced by hypericin in AY-27 rat bladder carcinoma cells. Photochem Photobiol 74(2):126–132

Kanoh K, Kohno S, Asari T, Harada T, Katada J, Muramatsu M, Kawashima H, Sekiya H, Uno I (1997) (-)-Phenylahistin: a new mammalian cell cycle inhibitor produced by *Aspergillus ustus*. Bioorg Med Chem Lett 7:2847–2852

Kharwar RN, Verma VC, Gond SK, Kumar A, Strobel G (2009) Javanicin, an antibacterial naphthaquinone from an endophytic fungus of Neem- *Chloridium* sp. Curr Microbiol 58:233–238

Kharwar RN, Mishra A, Gond SK, Stierle A, Stierle D (2011) Anticancer compounds derived from fungal endophytes: their importance and future challenges. Nat Prod Rep 28(7):1208–1228

Kim SU, Strobel GA, Ford E (1999) Screening of taxol-producing endophytic fungi from *Ginkgo biloba* and *Taxus cuspidata* in Korea. Agric Chem Biotechnol 42:97–99

Kornsakulkarn J, Dolsophon K, Boonyuen N, Boonruangprapa T, Rachtawee P, Prabpai S, Kongsaeree P, Thongpanchang C (2011) Dihydronaphthalenones from endophytic fungus *Fusarium* sp. BCC14842. Tetrahedron 67(39):7540–7547

Kour A, Shawl AS, Rehman S, Sultan P, Qazi PH, Suden P, Khajuria RK, Verma V (2008) Isolation and identification of an endophytic strain of *Fusarium oxysporum* producing podophyllotoxin from *Juniperus recurva*. World J Microbiol Biotechnol 24(7):1115–1121

Kumaran RS, Hur BK (2009a) Screening of species of the endophytic fungus *Phomopsis* for the production of the anticancer drug taxol. Biotechnol Appl Biochem 54(1):21–30

Kumaran RS, Muthumary J, Hur BK (2008a) Taxol from *Phyllosticta citricarpa*, a leaf spot fungus of the Angiosperm *Citrus medica*. J Biosci Bioeng 106:103–106

Kumaran RS, Muthumary J, Hur BK (2008b) Isolation and identification of taxol, an anticancer drug from *Phyllosticta melochiae* Yates, an endophytic fungus of *Melochia corchorifolia* L. Food Sci Biotechnol 17(6):1246–1253

Kumaran RS, Muthumary J, Hur BK (2008c) Production of taxol from *Phyllosticta spinarum*, an endophytic fungus of *Cupressus* sp. Eng Life Sci 8:438–446

Kumaran RS, Muthumary J, Kim EK, Hur BK (2009b) Production of taxol from *Phyllosticta dioscoreae*, a leaf spot fungus isolated from *Hibiscus rosa-sinensis*. Biotechnol Bioproc Eng 14:76–83

Kumaran RS, Jung H, Kim HJ (2011) *In vitro* screening of taxol, an anticancer drug produced by the fungus, *Colletotrichum capsici*. Eng Life Sci 11:264–271

Kuo PL, Hsu YL, Lin CC (2005) The chemopreventive effects of natural products against human cancer cells. Int J Appl Sci Eng 3:203–214

Kusari S, Lamshöft M, Zuhlke S, Spiteller M (2008) An endophytic fungus from *Hypericum perforatum* that produces Hypericin. J Nat Prod 71:159–162

Kusari S, Zuhlke S, Spiteller M (2009a) An endophytic fungus from *Camptotheca acuminata* that produces camptothecin and analogs. J Nat Prod 72(1):2–7

Kusari S, Lamshoeft M, Spiteller M (2009b) *Aspergillus fumigatus* Fresenius, an endophytic fungus from *Juniperus communis* L. Horstmann as a novel source of the anticancer pro-drug deoxypodophyllotoxin. J Appl Microbiol 107(3):1019–1030

Kusari S, Zuhlke S, Kosuth J, Cellarova E, Spiteller M (2009c) Light-independent metabolomics of endophytic *Thielavia subthermophila* provides insight into microbial hypericin biosynthesis. J Nat Prod 72(10):1825–1835

Leiter J, Downing V, Hartwell JL, Shear MJ (1950) Damage induced in sarcoma 37 with podophyllin, podophyllotoxin alpha-peltatin, beta-peltatin, and quercetin. J Natl Cancer Inst 10(6):1273–1293

Li JY, Strobel GA, Sidhu R, Hess WM, Ford EJ (1996) Endophytic taxol-producing fungi from bald cypress, *Taxodium distichum*. Microbiology 142:2223–2226

Li JY, Sidhu RS, Ford EJ, Long DM, Hess WM, Strobel GA (1998a) The induction of taxol production in the endophytic fungus-*Periconia* sp. from *Torreya grandifolia*. J Ind Microbiol Biot 20:259–264

Li JY, Sidhu RS, Bollon A, Strobel GA (1998b) Stimulation of taxol production in liquid cultures of *Pestalotiopsis microspora*. Mycol Res 102:461–464

Li CT, Li Y, Wang QJ, Sung CK (2008) Taxol production by *Fusarium arthrosporioides* isolated from yew, *Taxus cuspidata*. J Med Biochem 27:454–458

Li CY, Ding WJ, Shao CL, She ZG, Lin YC (2010a) A new diimide derivative from the co-culture broth of two mangrove fungi (strain no. E33 and K38). J Asian Nat Prod Res 12(9):809–813

Li KK, Lu YJ, Song XH, She ZG, Wu XW, An LK, Ye CX, Lin YC (2010b) The metabolites of mangrove endophytic fungus Zh6-B1 from the South China sea. Bioorg Med Chem Lett 20(11):3326–3328

Li G, Xiao Z, Liu J, Li C, Li F, Chen Z (2011a) Cancer: a proteomic disease. Sci China Life Sci 54(5):403–408

Li H, Huang H, Shao C, Huang H, Jiang J, Zhu X, Liu Y, Liu L, Lu Y, Li M, Lin Y, She Z (2011b) Cytotoxic norsesquiterpene peroxides from the endophytic fungus *Talaromyces flavus* isolated from the mangrove plant *Sonneratia apetala*. J Nat Prod 74(5):1230–1235

Lin ZJ, Zhang GJ, Zhu TJ, Liu R, Wei HJ, Gu QQ (2009) Bioactive cytochalasins from *Aspergillus flavipes*, an endophytic fungus associated with the mangrove plant *Acanthus ilicifolius*. Helv Chim Acta 92(8):1538–1544

Liu L (2011) Bioactive metabolites from the plant endophyte *Pestalotiopsis fici*. Mycology 2(1):37–45

Liu L, Liu SC, Jiang LH, Chen XL, Guo LD, Che YS (2008) Chloropupukenananin, the first chlorinated pupukeanane derivative, its precursors from *Pestalotiopsis fici*. Org Lett 10:1397–1400

Liu K, Ding X, Deng B, Chen W (2009) Isolation and characterization of endophytic taxol-producing fungi from *Taxus chinensis*. J Indus Microbiol Biotechnol 36:1171–1177

Liu L, Li Y, Liu SC, Zheng ZH, Chen XL, Guo LD, Che YS (2009a) Chloropestolide A, an antitumor metabolite with an unprecedented spiroketal skeleton from *Pestalotiopsis fici*. Org Lett 11:2836–2839

Liu L, Liu SC, Niu SB, Guo LD, Chen XL, Che YS (2009b) Isoprenylated chromone derivatives from the plant endophytic fungus *Pestalotiopsis fici*. J Nat Prod 72:1482–1486

Liu F, Cai XL, Yang H, Xia XK, Guo ZY, Yuan J, Li MF, She ZG, Lin YC (2010a) The bioactive metabolites of the mangrove endophytic fungus *Talaromyces* sp. ZH-154 isolated from *Kandelia candel* (L.) Druce. Planta Medica 76(2):185–189

Liu K, Ding X, Deng B, Chen W (2010b) 10-Hydroxycamptothecin produced by a new endophytic *Xylaria* sp., M20, from *Camptotheca acuminata*. Biotechnol Lett 32(5):689–693

Liu L, Niu S, Lu X, Chen X, Zhang H, Guo L, Che Y (2010c) Unique metabolites of *Pestalotiopsis fici* suggest a biosynthetic hypothesis involving a Diels-Alder reaction and then mechanistic diversification. Chem Commun (Cambridge, UK) 46(3):460–462

Lu Z, Zhu H, Fu P, Wang Y, Zhang Z, Lin H, Liu P, Zhuang Y, Hong K, Zhu W (2010) Cytotoxic polyphenols from the marine-derived fungus *Penicillium expansum*. J Nat Prod 73:911-914

Lu S, Kurtan T, Yang G, Sun P, Mandi A, Krohn K, Draeger S, Schulz B, Yi Y, Li L, Zhang W (2011) Cytospolides A-E, new nonanolides from an endophytic fungus, *Cytospora* sp. Eur J Org Chem 2011(28):5452–5459

Martinez-Luis S, Cherigo L, Spadafora C, Gerwick WH, Cubilla-Rios L (2009) Additional anti-leishmanial constituents of the panamanian endophytic fungus *Edenia* sp. Rev Latinoam Quím 37(2):104–114

Martínez-Luis S, Della-Togna G, Coley PD, Kursar TA, Gerwick WH, Cubilla-Rios L (2008) Antileishmanial constituents of the Panamanian endophytic fungus *Edenia* sp. J Nat Prod 71(12):2011–2014

Miao Z, Wang Y, Yu X, Guo B, Tang K (2009) A new endophytic taxane production fungus from *Taxus chinensis*. Appl Biochem Microbiol 45:81–86

Min C, Wang X (2009) Isolation and identification of the 10- hydroxycamptothecin-producing endophytic fungi from *Camptotheca acuminata* Decne. Acta Bot Boreali-Occidential Sinica 29:614–617

Mishra PD, Deshmukh SK, Kulkarni-Almeida A, Roy S, Jain S, Verekar SA, Balakrishnan A, Vishwakarma R (2013) Anti-inflammatory and anti-diabetic naphthoquinones from an endophytic fungus. Indian J Chem, Sect B 52B:pp (Accepted for Publication)

Mohana Kumara P, Sreejayan N, Priti V, Ramesha BT, Ravikanth G, Ganeshaiah KN, Vasudeva R, Mohan J, Santhoshkumar TR, Mishra PD, Viswakarma R, Shaanker RU (2010) *Dysoxylum binectariferum* Hook. f (Meliaceae), a rich source of rohitukine. Fitoterapia 81(2):145–148

Mohana Kumara P, Zuehlke S, Priti V, Ramesha BT, Shweta S, Ravikanth G, Vasudeva R, Santhoshkumar TR, Spiteller M, Uma Shaanker R (2012) *Fusarium proliferatum*, an endophytic fungus from *Dysoxylum binectariferum* Hook.f, produces rohitukine, a chromane alkaloid possessing anti-cancer activity. Antonie Van Leeuwenhoek 101(2):323–329

Moreno E, Varughese T, Spadafora C, Arnold AE, Coley PD, Kursar TA, Gerwick WH, Cubilla-Rios L (2011) Chemical constituents of the new endophytic fungus *Mycosphaerella* sp. nov. and their anti-parasitic activity. Nat Prod Commun 6(6):835–840

Nahrstedt A, Butterweck V (1997) Biologically active and other chemical constituents of the herb of *Hypericum perforatum* L. Pharmacopsychiatry 30(Suppl 2):129–134

Naik RG, Kattige SL, Bhat SV, Alreja B, de Souza NJ, Rupp RH (1988) An anti-inflammatory cum immunomodulatory piperidinylbenzopyranone from *Dysoxylum binectariferum*: isolation, structure and total synthesis. Tetrahedron 44:2081–2086

Okouneva T, Hill BT, Wilson L, Jordan MA (2003) The effects of vinflunine, vinorelbine, and vinblastine on centromere dynamics. Mol Cancer Ther 2:427–436

Owen NL, Hundley N (2004) Endophytes—the chemical synthesizers inside plants. Sci Prog 87(2):79–99

Pandi M, Kumaran RS, Choi YK, Kim HJ, Muthumary J (2011) Isolation and detection of taxol, an anticancer drug produced from *Lasiodiplodia theobromae*, an endophytic fungus of the medicinal plant *Morinda citrifolia*. Afr J Biotechnol 10(8):1428–1435

Puri SC, Verma V, Amna T, Qazi GN, Spiteller M (2005) An endophytic fungus from *Nothapodytes foetida* that produces camptothecin. J Nat Prod 68:1717–1719

Puri SC, Nazir A, Chawla R, Arora R, Riyaz-ul-Hasan S, Amna T, Ahmed B, Verma V, Singh S, Sagar R, Sharma A, Kumar R, Sharma RK, Qazi GN (2006) The endophytic fungus *Trametes hirsuta* as a novel alternative source of Podophyllotoxin and related aryl tetralin lignans. J Biotech 122(4):494–510

Qin JC, Zhang YM, Gao JM, Bai MS, Yang SX, Laatsch H, Zhang AL (2009) Bioactive metabolites produced by *Chaetomium globosum*, an endophytic fungus isolated from *Ginkgo biloba*. Bioorg Med Chem Lett 19(6):1572–1574

Qiu D, Huang M, Fang X, Zhe C (1994) Isolation of an endophytic fungus associated with *Taxus yunnanensis* et L.K.Fu. Acta Mycol Sinica 13:314–316

Rehman S, Shawl AS, Kour A, Andrabi R, Sudan P, Sultan P, Verma V, Qazi GN (2008) An endophytic *Neurospora* sp. from *Nothapodytes foetida* producing camptothecin. Appl Biochem Microbiol 44:203–209

Rehman S, Shawl AS, Kour A, Sultan P, Ahmad K, Khajuria R, Qazi GN (2009) Comparative studies and identification of camptothecin produced by an endophyte at shake flask and bioreactor. Nat Prod Res 23(11):1050–1057

Raheman F, Deshmukh S, Ingle A, Gade A, Rai MK (2011) Silver nanoparticles: novel antimicrobial agent synthesized from an endophytic fungus *Pestalotia* sp. isolated from leaves of *Syzygium cumini* (L). Nano Biomed Eng 3(3):174–178

Rowinsky E (1997) The development and clinical utility of the taxane class of antimicrotubule chemotherapy agents. Annu Rev Med 48:353–374

Ruiz-Sanchez J, Flores-Bustamante ZR, Dendooven L, Favela-Torres E, Soca-Chafre G, Galindez-Mayer J,

Flores-Cotera LB (2010) A comparative study of taxol production in liquid and solid-state fermentation with *Nigrospora sp.* a fungus isolated from *Taxus globosa*. J Appl Microbiol 109:2144–2150

Santiago C, Fitchett C, Munro MH, Jalil J, Santhanam J (2012) Cytotoxic and antifungal activities of 5-Hydroxyramulosin, a compound produced by an endophytic fungus isolated from *Cinnamomum mollisimum*. Evid Based Complement Alternat Med 2012:689310

Sausville EA, Zaharevitz D, Gussio R, Meijer L, Louarn-Leost M, Kunick C, Schultz R, Lahusen T, Headlee D, Stinson S, Arbuck SG, Senderowicz A (1999) Cyclin-dependent kinases: initial approaches to exploit a novel therapeutic target. Pharmacol Therapeut 82:285–292

Sedlacek HH, Czech J, Naik R, Kaur G, Worland P, Losiewicz M, Parker B, Carlson B, Smith A, Senderowicz AM, Sausville EA (1996) Flavopiridol (L86 8275; NSC 649890), a new kinase inhibitor for tumor therapy. Int J Oncol 9:1143–1168

Shao CL, Wang CY, Gu YC, Wei MY, Pan JH, Deng DS, She ZG, Lin YC (2010a) Penicinoline, a new pyrrolyl 4-quinolinone alkaloid with an unprecedented ring system from an endophytic fungus *Penicillium* sp. Bioorg Med Chem Lett 20(11):3284–3286

Shao CL, Wang C, Zheng C, She Z, Gu Y, Lin Y (2010b) A new anthraquinone derivative from the marine endophytic fungus *Fusarium* sp. (No. b77). Nat Prod Res, Part A: Struct Synth 24(1):81–85

She ZG, Chen SP, Lin YC, Yuan J, Pang JY, Li MF et al (2008) SZ-685C preparation method and antitumor application. Application No: 00810028628.3, Aplication Date: 2008.6.6

Shen L, Wang JS, Shen HJ, Song YC, Tan RX (2010) A new cytotoxic trichothecene macrolide from the endophyte *Myrothecium roridum*. Planta Med 76(10):1004–1006

Shiono Y, Kikuchi M, Koseki T, Murayama T, Kwon E, Aburai N, Kimura K (2011) Isopimarane diterpene glycosides, isolated from endophytic fungus *Paraconiothyrium* sp. MY-42. Phytochem 72(11–12):1400–1405

Shweta S, Zuehlke S, Ramesha BT, Priti V, Kumar PM, Ravikanth G, Spiteller M, Vasudeva R, Shaanker RU (2010) Endophytic fungal strains of *Fusarium solani*, from *Apodytes dimidiate* E. Mey. ex Arn (Icacinaceae) produce camptothecin, 10- hydroxycamptothecin and 9-methoxycamptothecin. Phytochem 71:117–122

Simoens C, Lardon F, Pauwels B, De Pooter CMJ, Lambrechts HAJ, Pattyn GGO, Breillout F, Vermorken JB (2008) Comparative study of the radiosensitising and cell cycle effects of vinflunine and vinorelbine, in-vitro. BMC Cancer 8:65

Sreekanth D, Syed A, Sarkar S, Sarkar D, Santhakumari B, Ahmad A, Khan MI (2009) Production, purification, and characterization of taxol and 10-DABIII from a new endophytic fungus *Gliocladium* sp. isolated from the Indian yew tree, *Taxus baccata*. J Microbiol Biotechnol 19(11):1342–1347

Stierle A, Strobel GA, Stierle D (1993) Taxol and taxane production by *Taxomyces andreanae*, an endophytic fungus of Pacific yew. Science 260:214–216

Strobel GA, Daisy B (2003) Bioprospecting for microbial endophytes and their natural products. Microbiol Mol Bio Rev 67:491–502

Strobel GA, Hess WM, Ford E, Sidhu RS, Yang X (1996) Taxol from fungal endophytes and issue of biodiversity. J Ind Microbiol 17:417–423

Strobel GA, Hess WM, Li JY, Ford E, Sears J, Sidhu RS, Summerell B (1997) *Pestalotiopsis guepinii*, a taxol-producing endophyte of the Wollemi pine, *Wollemia nobilis*. Austral J Bot 45:1073–1082

Sturz AV, Nowak J (2000) Endophytic communities of rhizobacteria and the strategies required to create yield enhancing associations with crops. Appl Soil Ecol 15:183–190

Sun D, Ran X, Wang J (2008) Isolation and identification of a taxol-producing endophytic fungus from *Podocrapus*. Acta Microbiol Sin 48:589–595

Sun ZL, Zhang M, Zhang JF, Feng J (2011) Antifungal and cytotoxic activities of the secondary metabolites from endophytic fungus *Massrison* sp. Phytomedicine 18(10):859–862

Suryanarayanan TS, Johnson JA (2005) Fungal endophytes of tropical plants: a critical review. In: Satyanarayana T, Johri BN (eds) Microbial diversity: current perspectives and potential application. I K International Private Limited, New Delhi, pp 207–224

Suryanarayanan TS, Thirunavukkarasu N, Govindarajulu MB, Gopalan V (2012) Fungal endophytes: an untapped source of biocatalysts. Fungal Divers 54(1):19–30

Takada Y, Aggarwal BB (2003) Genetic deletion of the tumor necrosis factor receptor p60 or p80 sensitizes macrophages to lipopolysaccharide-induced nuclear factor-kappa B, mitogen-activated protein kinases, and apoptosis. J Biol Chem 278:23390–23397

Tian R, Yang Q, Zhou G, Tan J, Zhang L, Fang C (2006) Taxonomic study on a taxol producing fungus isolated from bark of *Taxus chinensis* var. *mairei*. J Wuhan Bot Res 24:541–545

Venkatachalam R, Subban K, Paul MJ (2008) Taxol from *Botryodiplodia theobromae* (BT 115)-an endophytic fungus of *Taxus baccata*. J Biotechnol 136:S189–S190

Vennila R, Thirunavukkarasu SV, Muthumarya J (2010) *In-vivo* studies on anticancer activity of taxol isolated from an endophytic fungus *Pestalotiopsis pauciseta* Sacc. VM1. Asian J Pharm Clin Res 3(4):30–34

Verma VC, Kharwar RN, Gange AC (2010) Biosynthesis of antimicrobial silver nanoparticles by the endophytic fungus *Aspergillus clavatus*. Nanomedicine 5(1):33–40

Wall ME, Wani MC, Cook CE, Palmer KH, McPhail AT, Sim GA (1966) Plant antitumor agents. 1. The isolation and structure of camptothecin, a novel alkaloidal leukemia and tumor inhibitor from *Camptotheca acuminata*. J Am Chem Soc 88:3888–3890

Wang J, Li G, Lu H, Zheng Z, Huang Y, Su W (2000) Taxol from *Tubercularia* sp. Strain TF5, an endo-

phytic fungus of *Taxus mairei*. FEMS Microbiol Lett 193(2):249–253
Wang B, Li A, Wang X (2001) An endophytic fungus for producing taxol. Sci China (Series C) 31:271–274
Wang J, Zhao B, Zhang W, Wu X, Wang R, Huang Y, Chen D, Park K, Weimer BC, Shen Y (2010) Myco-epoxydiene, a fungal polyketide, induces cell cycle arrest at the G2/M phase and apoptosis in HeLa cells. Bioorg Med Chem Lett 20(23):7054–7058
Wang QX, Li SF, Zhao F, Dai HQ, Bao L, Ding R, Gao H, Zhang LX, Wen HA, Liu HW (2011) Chemical constituents from endophytic fungus *Fusarium oxysporum*. Fitoterapia 82(5):777–781
Wang LW, Xu BG, Wang JY, Su ZZ, Lin FC, Zhang CL, Kubicek CP (2012) Bioactive metabolites from *Phoma* species, an endophytic fungus from the Chinese medicinal plant *Arisaema erubescens*. Appl Microbiol Biotechnol 93(3):1231–1239
Wani MC, Taylor HL, Wall ME, Coggon P, McPhail AT (1971) Plant antitumor agents VI: the isolation and structure of taxol, a novel antilekemic and antitumor agent from *Taxus brevifolia*. J Am Chem Soc 93:2325–2327
Wen L, Guo Z, Liu F, Wan Q, Yu Z, Lin Y, Fu L (2009) Studies on the secondary metabolites and bioactivity of mangrove endophytic fungus *Paecilomyces* sp. (tree 1–7). Huaxue Yanjiu Yu Yingyong 21(2):198–202
Xia X, Li Q, Li J, Shao C, Zhang J, Zhang Y, Liu X, Lin Y, Liu C, She Z (2011) Two new derivatives of griseofulvin from the mangrove endophytic fungus *Nigrospora* sp. (strain No. 1403) from *Kandelia candel* (L.) Druce. Planta Med 77(15):1735–1738
Xie G, Zhu X, Li Q, Gu M, He Z, Wu J, Li J, Lin Y, Li M, She Z, Yuan J (2010) SZ-685C, a marine anthraquinone, is a potent inducer of apoptosis with anticancer activity by suppression of the Akt/FOXO pathway. Br J Pharmacol 159(3):689–697
Xu S, Ge HM, Song YC, Shen Y, Ding H, Tan RX (2009a) Cytotoxic cytochalasin metabolites of endophytic *Endothia gyrosa*. Chem Biodivers 6(5):739–745
Xu J, Kjer J, Sendker J, Wray V, Guan H, Edrada RA, Lin W, Wu J, Proksch P (2009b) Chromones from the endophytic fungus *Pestalotiopsis* sp. isolated from the Chinese mangrove plant *Rhizophora mucronata*. J Nat Prod 72(4):662–665
Xu C, Wang J, Gao Y, Lin H, Du L, Yang S, Long S, She Z, Cai X, Zhou S, Lu Y (2010) The anthracenedione compound bostrycin induces mitochondria-mediated apoptosis in the yeast *Saccharomyces cerevisiae*. FEMS Yeast Res 10(3):297–308
Yang X, Zhang L, Guo B, Guo S (2004) Preliminary study of a vincristine- producing endophytic fungus isolated from leaves of *Catharanthus roseus*. Chin Tradit Herb Drugs 35:79–81
Yang L, Liu J, Yang D, Zhu W (2007) Screening of endophytic fungi producing baccatin III from *Taxus yunnanensis* and preliminary optimization of the culture media. Xiandai Shengwuyixue Jinzhan 7(5):692–695
Yuan J, Jian-Nan B, Bing Y, Xu-Dong Z (2006) Taxol-producing fungi: a new approach to industrial production of taxol. Chin J Biotechnol 22:1–6
Yuan L, Lin X, Zhao PJ, Ma J, Huang YJ, Shen YM (2009) New Polyketides from Endophytic *Diaporthe* sp. XZ-07. Helv Chim Acta 92(6):1184–1190
Yuan J, He Z, Wu J, Lin Y, Zhu X (2011) A novel adriamycin analogue derived from marine microbes induces apoptosis by blocking Akt activation in human breast cancer cells. Mol Med Report 4(2):261–265
Yvon AM, Wadsworth P (1997) Non-centrosomal microtubule formation and measurement of minus end microtubule dynamics in A498 cells. J Cell Sci 110:2391–2401
Zhang B, Salituro G, Szalkowski D, Li Z, Zhang Y, Royo I, Vilella D, Díez MT, Pelaez F, Ruby C, Kendall RL, Mao X, Griffin P, Calaycay J, Zierath JR, Heck JV, Smith RG, Moller DE (1999) Discovery of a small molecule insulin mimetic with antidiabetic activity in mice. Science 284:974–977
Zhang L, Guo B, Li H, Zeng S, Shao H, Gu S, Wei R (2000) Preliminary study on the isolation of endophytic fungus of *Catharanthus roseus* and its fermentation to produce products of therapeutic value. Chin Tradit Herb Drugs 31:805–807
Zhang JY, Wu HY, Xia XK, Liang YJ, Yan YY, She ZG, Lin YC, Fu LW (2007) Anthracenedione derivative 1403P-3 induces apoptosis in KB and KBv200 cells via reactive oxygen species-independent mitochondrial pathway and death receptor pathway. Cancer Biol Ther 6(9):1413–1421
Zhang JY, Tao LY, Liang YJ, Yan YY, Dai CL, Xia XK, She ZG, Lin YC, Fu LW (2009a) Secalonic acid D induced leukemia cell apoptosis and cell cycle arrest of G1 with involvement of GSK-3beta/beta -catenin/c-Myc pathway. Cell Cycle 8(15):2444–2450
Zhang P, Zhou PP, Yu LJ (2009b) An endophytic Taxol-producing fungus from *Taxus media, Cladosporium cladosporioides* MD2. Curr Microbiol 59(3):227–232
Zhang J, Ge HM, Jiao RH, Li J, Peng H, Wang YR, Wu JH, Song YC, Tan RX (2010a) Cytotoxic chaetoglobosins from the endophyte *Chaetomium globosum*. Planta Med 76(16):1910–1914
Zhang JY, Tao LY, Liang YJ, Chen LM, Mi YJ, Zheng LS, Wang F, She ZG, Lin YC, To KKW, Fu LW (2010c) Anthracenedione derivatives as anticancer agents isolated from secondary metabolites of the mangrove endophytic fungi. Mar Drugs 8:1469–1481
Zhao K, Zhao L, Jin Y, Wei H, Ping W, Zhou D (2008) Isolation of a taxol-producing endophytic fungus and inhibiting effect of the fungus metabolites on HeLa cell. Mycosystema 27:735–744
Zhao K, Ping W, Li Q, Hao S, Zhao L, Gao T, Zhou D (2009) *Aspergillus niger* var. *taxi*, a new species variant of taxol-producing fungus isolated from *Taxus cuspidata* in China. J Appl Microbiol 107:1202–1207
Zhao J, Shan T, Mou Y, Zhou L (2011) Plant-derived bioactive compounds produced by endophytic Fungi. Mini-Rev Med Chem 11:159–168

# Diversity and Bioactivity of Endophytic Fungi from *Nothapodyte foetida* (Wt.) Sleumer and *Hypericum mysorense* Heyne

6

Pradeepa V. Samaga and V. Ravishankar Rai

**Abstract**

Endophytic fungi were isolated from the medicinal plants *Nothapodyte foetida* and *Hypericum mysorense*, collected from the cold peaks of the Western Ghats of India, which is considered as one among ten important biodiversity hotspots of the world. The diversity of foliar endophytic biota in both plants was assessed using various indices. Endophytes were screened for the antimicrobial and free radical-scavenging activity. Ethyl acetate extracts of the culture broths of two isolates *Bionectria ochroleuca* NOTL33 from *N. foetida* and *Chaetomium globosum* HYML55 from *H. mysorense* were further characterized for their composition and minimum inhibitory concentrations (MIC) based on their significant activities during primary screening. MIC of both extracts ranged from 78 to 625 μg/ml, with the exception of *Pseudomonas aeruginosa* (5 mg/ml). 1,1-diphenyl-2-picrylhydrazyl (DPPH) and 2, 2'-azinobis-3-ethylbenzothiazoline-6-sulfonic acid (ABTS) free radical quenching by the extracts were dose dependent. 2-heptanone, furfural, methyl furfural, benzyl pentyl ketone, benzyl decanoate, amyl isovalerate and palmitic acid were identified in NOTL33 extract by gas chromatography–mass spectrometry (GC-MS) analysis. A yellow amorphous powder (HYML 8A) was purified from HYML55 extract and further characterized for MIC and minimum cidal concentrations against the test bacteria. The yield of the compound in different media was estimated.

**Keywords**

*Bionectria* · Endophytic fungi · Antimicrobial · Free radical · *Nothapodyte* · *Hypericum*

V. R. Rai (✉) · P. V. Samaga
Department of Studies in Microbiology,
University of Mysore, Manasagangotri,
Mysore, Karnataka 570006, India
e-mail: raivittal@gmail.com

R. N. Kharwar et al. (eds.), *Microbial Diversity and Biotechnology in Food Security*,
DOI 10.1007/978-81-322-1801-2_6, © Springer India 2014

## 6.1 Introduction

Plants have served as source for the antimicrobial chemicals for millennia. But, the exhaustive extraction procedures have threatened the diversity of the medicinal plants, which has diverted the search for bioactive metabolites to the endophytic fungi. Endophytes are microorganisms, which spend the whole or part of their life cycle inside the healthy tissues of the host plants, typically causing no apparent symptoms of diseases (Zhao et al. 2010). A myriad of metabolites belonging to different chemical categories such as alkaloids, steroids, terpenoids, isocoumarin derivatives, quinones, flavonoids, cytochalasines, furandiones, phenylpropanoids, lignans, peptides, phenol, phenolic acids, aliphatic compounds and chlorinated metabolites (Verma et al. 2009) have been reported from endophytes, which are known for their antibiotic, antiviral, volatile antibiotic, anticancer, antioxidant, insecticidal, antidiabetic and immunosuppressive properties. Surprisingly, over 100 anticancer compounds are reported only from fungal endophytes after 1993 (Kharwar et al. 2011). Hence, the endophytes are currently considered to be a wellspring of novel secondary metabolites offering the potential for medical, agricultural and/or industrial exploitation (Strobel and Daisy 2003).

*Nothapodytes foetida* (Wt.) Sleumer (*Mappia foetida* Miers), Icacinaceae, is a medium size tree, grows wild in the forests of the Western Ghats. The tree is a major source of important antineoplastic alkaloids such as camptothecin, 9-methoxycamptothecin (Govindachari and Viswanathan 1972), 9-methoxy-20-O-acetylcamptothecin (Srinivas and Das 2003) exhibiting antimicrobial (Kumar et al. 2002), antiviral (Liu et al. 2010) and anticancer properties. *Hypericum mysorense,* a perennial shrub growing luxuriantly in the cold peaks of the Western Ghats is known to produce many xanthone derivatives (Gunatilaka et al. 1982) having antiviral and antioxidant activities (Vijayan et al. 2004). The Western Ghats of India are considered as one among ten important biodiversity hotspots of the world. Endophytic fungi from medicinal plants and biodiversity hotspots are expected to yield a diverse and unexplored biota with significant bioactivities. Therefore, in the present study, diversity and bioactivities of endophytes from *N. foetida* and *H.mysorense* were investigated.

## 6.2 Materials and Methods

### 6.2.1 Sample Collection and Processing

Leaves from average sized *N. foetida* trees were collected from Aloor region of Hassan, Karnataka, India. *H. mysorense* plants were collected from Talacauvery region, Madikeri district, Karnataka, India. The samples were carried to the laboratory in sterile polypropylene bags and processed within 24 h of collection.

### 6.2.2 Isolation of Endophytic Fungi

The samples were washed with distilled water and surface sterilized using the following immersion sequences: 0.1 % $HgCl_2$ solution for 1 min, thorough sterile water wash, 90 % ethanol for 2 min and sterile water wash. The samples were then cut into pieces of approximately 0.5 $cm^2$ using a sterile scalpel and the leaf discs were then placed onto sterile water agar (pH 4.8) and incubated at an ambient temperature in the dark. The plates were frequently observed for the emergence of the fungal hyphae through the cut ends of the tissues and the emerging hyphal tissues were transferred onto potato dextrose agar (PDA; Himedia, India). The cultures were then maintained on PDA slants at 4 °C for the further studies (Lu et al. 2012).

### 6.2.3 Identification of the Fungal Isolates and Diversity Study

The endophytic fungal isolates were identified based on the morphological features using the standard mycological manuals (Fig. 6.1; Ellis 1971; Ellis 1976; Gilman 1957). The lead isolates were identified based on sequencing

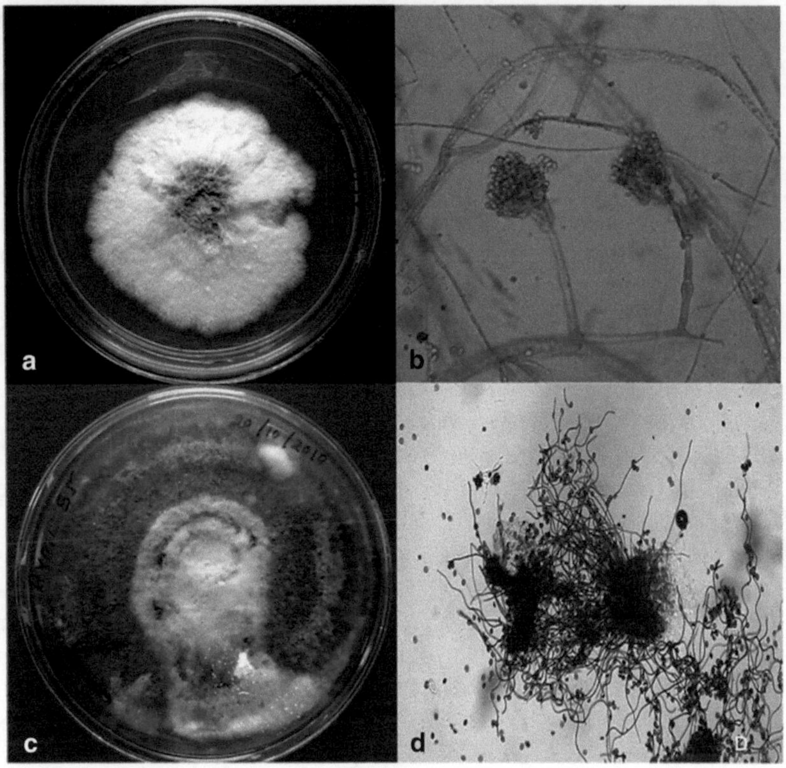


**Fig. 6.1** The colony morphology and microscopic features of the endophytic isolates *Bionectria ochroleuca* NOTL33 (**a, b**) and *Chaetomium globosum* HYML55 (**c, d**), respectively

of ITS1, 5.8S rRNA and ITS2 regions. The primers used were ITS1 (forward): 5′-TCCGTAGGTGAACCTGCG-3′ and ITS4 (reverse): 5′-TCCTCCGCTTATTGATATG-3′. The reaction volume was 25 µl, with 12.5 µl 2× PCR master mix (Genei, Bangalore), 1 µl each of forward and reverse primers and template DNA, 9.5 µl water. The amplification conditions were initial denaturation at 95 °C for 5 min, followed by 30 cycles of 95 °C for 1 min, 55 °C for 1 min, 72 °C for 1 min and final extension at 72 °C for 5 min.

The diversity was studied using the various diversity indices.

### 6.2.4 Fermentation

The agar discs (6 mm diameter) with actively growing hyphae of test organisms were transferred to 45 ml fermentation medium (Potato dextrose broth (Himedia, India) 20 g/l, Soytone (Himedia, India) 5 g/l) in wide-mouth tissue culture bottles of 200 ml capacity. Each bottle was inoculated with three agar blocks. The flasks were kept undisturbed for 20 days at ambient (28 °C) temperature.

### 6.2.5 Extraction

After incubation, the fermentation broth was filtered through four layers of destarched and sterilized cheese cloth. The mycelial mass was finely ground in a mixer grinder. The filtrate and the mycelial slurry were extracted thrice with equal volume of ethyl acetate (SD Fine chemicals, India). The fractions were pooled together, moisture removed using sodium sulphate and the solvent evaporated using a vacuum rotary evaporator. The residue was dissolved in minimal amount of ethyl acetate and stored at −20 °C till further use.

### 6.2.6 Evaluation of Antimicrobial Activity

#### 6.2.6.1 Test Strains

The microbial strains used in the study were obtained from microbial type culture collection (MTCC) of Institute of Microbial Technology (IMTECH), Chandigarh. The bacterial cultures used were *Pseudomonas aeruginosa* (MTCC 7093), *Escherichia coli* (MTCC 40), *Enterobacter aerogenes* (MTCC 111), *Klebsiella pneumoniae* (MTCC 661), *Shigella flexneri* (MTCC 1457), *Bacillus subtilis* (MTCC 121), *Salmonella enterica* ser. Typhi (MTCC 733), *Staphylococcus aureus* (MTCC 7443), methycillin-resistant *S. aureus* (clinical isolate) and *Streptococcus pyogens* (MTCC 1925). The dermatophytes used were *Candida albicans* (MTCC 183), *Microsporum gypseum* (MTCC 2830), *Microsporum canis* (MTCC 2831) and *Trichophyton rubrum* (MTCC 296). The fungal strains used were *Aspergillus nidulans* (MTCC 803), *Aspergillus terreus* (MTCC 2580), *Aspergillus flaviceps* (MTCC 1990), *Aspergillus fumigatus* (MTCC 3008), *Aspergillus clavatus* (MTCC 1323), *Aspergillus parasiticus* (MTCC 2797) and *Penicillium citrinum* (MTCC 1784).

#### 6.2.6.2 Disc Diffusion Assay

The antimicrobial assay was carried out using paper disc diffusion method (Bauer et al. 1966). Briefly, sterile filter paper discs (6 mm diameter, Hi Media, India) impregnated with ethyl acetate extracts (40 µl or 400 µg per disc) were placed on the pre-seeded petri plates. The discs loaded with solvent alone (40 µl) were used as the negative control. Chloramphenicol discs (HiMedia, 30 µg per disc) were used as the positive control for bacteria and nystatin discs served as positive controls for dermatophytes. After incubation for 18 h, inhibition zones around the discs were measured using the zone scale (HiMedia, India) and the mean of the triplicate readings were recorded.

#### 6.2.6.3 Minimum Inhibitory Concentration (MIC) and Minimum Microbicidal Concentration (MMC)

The minimum inhibitory concentration was determined by broth micro-dilution method (Eloff 1998) with slight modifications. The assay was carried out in 96-well microtitre plates. The assay mixture consisted of 90 µl of standardized inoculum and 10 µl extract of various concentrations in methanol to yield a twofold decrease in the concentration from 5 to 0.0195 mg/ml. The plates were incubated for 24 h at $35 \pm 1$ °C and triphenyl tetrazolium chloride (TTC) was used as microbial growth indicator. The assay was carried out in triplicates, with streptomycin (for bacteria) and nystatin (for dermatophytes) dilutions as positive controls and broth with only methanol as a negative control (growth controls). The least concentration with no visible growth indicated the MIC. The broth from the well of MIC and above concentrations was streaked onto the agar medium to test the cell viability. The least concentration with no viable cells represents the minimum microbicidal concentration.

#### 6.2.6.4 Test for Volatile Antimicrobials

The ability of endophytic *Bionectria ochroleuca* (NOTL33) to produce volatile antifungal chemicals was determined according to Ting et al. (2011). Briefly, the endophyte was point inoculated and incubated for 5 days. The test fungal strains were point inoculated on PDA plates. The lids of the plates were removed and the pre-incubated endophytic plates were inverted over the test fungi and sealed airtight with parafilm. The point inoculated test fungi alone served as the control. The percentage inhibition of growth diameter of test fungi over the control was calculated using the formula:

$$\text{Growth inhibition}\left(\%\right) = \left[\left(D_{\text{control}} - D_{\text{test}}\right)/D_{\text{control}}\right] \times 100$$

where $D$ is diameter of colony.

The experiment was repeated thrice and the mean inhibition was expressed. The significance inhibition was calculated by one tailed paired *t*-test (using Microsoft-Excel function) between control and experimental pairs.

### 6.2.7 Free Radical-Scavenging Activity

#### 6.2.7.1 DPPH Radical-Scavenging Activity

The free radical-scavenging activity of the extract was measured by using stable 1,1-diphenyl-2-picrylhydrazyl (DPPH) (Sigma, Germany) free radicals in 96-well microtitre plates (Flat bottom, Tarsons, India) with modifications (Predes 2011). The activity of the extract was compared with that of quercetin (Himedia, India). In brief, 100 µl aliquots of different concentrations of the extract (twofold dilutions from 5000 µg/ml to 39.06 µg/ml in methanol) were mixed with 100 µl of DPPH solution (40 µmol/l) and incubated for 30 min at 25 °C. An appropriate control was maintained by adding 100 µl of DPPH to 100 µl methanol. The decrease in absorbance (*A*) was measured at 517 nm.

$$\text{DPPH scavenging activity }(\%) = \left[\left(A_{\text{control}} - A_{\text{sample}}\right)/A_{\text{control}}\right]\times 100$$

The data was presented as mean of triplicate. The concentration required for 50 % reduction of DPPH radical ($IC_{50}$) was determined graphically.

#### 6.2.7.2 ABTS Radical-Scavenging Activity

The free radical-scavenging activity of the crude extracts was also determined by using 2, 2'-azinobis-3-ethylbenzothiazoline-6-sulfonic acid (ABTS) stable cationic free radicals (Osman et al. 2011). The assay was carried out in a 96-well flat bottom microtiter plate (Tarsons, India). The ABTS free radical solution was prepared by mixing 3.75 mM ABTS diammonium salt with 1.225 mM potassium persulphate solution. After the incubation of the mixture overnight at 30 °C for completion of the reaction, the ABTS cationic free radical solution was adjusted for the absorbance of 0.6 ± 0.05 as measured in the microtiter plate at 734 nm. 200 µl of the standardized ABTS solution was mixed with various concentrations of the extract in 10 µl aliquots (5 mg/ml to 0.039 mg/ml in two-fold dilutions in methanol) and the absorbance (*A*) was read at 734 nm for every 5 min up to 60 min. The ABTS-quenching activity was calculated by the formula:

$$\text{ABTS scavenging activity }(\%) = \left[\left(A_{\text{control}} - A_{\text{sample}}\right)/A_{\text{control}}\right]\times 100$$

The data were presented as mean of triplicate. The concentration required for 50 % reduction of ABTS radical ($IC_{50}$) was determined graphically.

### 6.2.8 GC-MS Analysis of NOTL33 Extract

Shimadzu table top gas chromatography–mass spectrometry (GC-MS; GC 17A ver. 3) equipped with DB-1 capillary column (30 m, ID = 0.25 mm, film thickness = 0.15 mm) was used for the analysis of NOTL33 ethyl acetate extracts. The conditions during the analysis were as follows: initial injection temperature was 250 °C, interface temperature was 230 °C, oven temperature was raised from 60 to 250 °C over a period of 5–30 min. Control mode was kept as split with the split ratio and column inlet pressure 64.5 kPa. The column flow of nitrogen carrier gas was set at a linear velocity of 45.9 cm/s with a flow rate of 1.3 ml/min (total flow 38.9 ml/min). Mass spectrometer (GCMS-QP5050A) detector with scan acquisition mode was set at m/z from 40 to 500 with scan speed of 1,000 over 30 min time, interface temperature 230 °C, detector gain 1.3 kV. The mass peaks were interpreted tentatively based on the mass peak split pattern in the mass spectrum.

### 6.2.9 Purification of the Active Compound

HYML55 extract (500 mg) was loaded onto a silica gel column and fractionated by using the solvents of increasing polarity. Fractions with similar UV–visible absorption spectra were pooled together and the antimicrobial activity of individual fractions was noted. The fraction with desired activity was dried by evaporation and further purified by solvent–solvent extraction.

### 6.2.10 Effect of the Media on the Production of the Active Compound

HYML55 was grown in yeast extract sucrose broth (YESB), potato dextrose broth (PDB), Sabaraud's dextrose broth (SDB) and M1D broth (40 ml each in triplicates). They were fermented for 20 days and individually extracted using ethyl acetate, evaporated under vacuum and the active compound in the extract was quantified by high performance liquid chromatography (HPLC) using a purified compound as the reference. HPLC was performed using analytical HPLC system (Agilent Technologies assembled 1100 and 1200 series) equipped with quaternary pump and UV–Visible detector. Reversed phase chromatographic analyses were carried out in isocratic conditions using RP-$C_{18}$ column (4.6 mm × 250 mm) packed with 5 μm diameter particles. The separation conditions were as follows: The mobile phase was water–acetonitrile–acetic acid (80:20:3, v/v/v) filtered through a membrane filter (0.45 μm) and degassed by ultrasonication; flow rate 0.8 ml/min; injection volume 10 μl and detection was done at 280 nm. The reference compound was injected at 1,000 ppm and the residues at 500 ppm each.

## 6.3 Results

The isolation frequency for the foliar endophytes was 79.56 % in *N. foetida* and 96 % in *H. mysorense*. The major genera isolated were *Gliocladium, Phomopsis, Xylaria, Aspergillus, Penicillium, Phoma, Cylindrotrichum, Monodictys, Chaetomium, Helminthosporium, Pestalotiopsis, Fusarium, Cladosporium, Colletotrichum, Curvularia, Trichoderma, Lasiodiplodia, Xylaria, Verticillium, Nectria, Bionectria, Botryosphaerea* and *Mycelia sterilia* including some unidentified genera. The diversity indices are given in Table 6.1. The isolates were screened for the antimicrobial and free radical-scavenging activities (data not shown). Based on the screening, two endophytic isolates, *B. ochroleuca* (NOTL33 anamorph: *Gliocladium roseum*) from *N. foetida* and *Chaetomium globosum* HYML55 from *H. mysorense* were selected for further characterization. Both isolates were identified based on rDNA sequencing. The rDNA sequences for *B. ochroleuca* (JQ794833.1) and *C. globosum* (JX500352.1) were submitted to Genbank.

### 6.3.1 Antimicrobial Activity

Ethyl acetate extracts from NOTL33 and HYML55 inhibited the tested bacteria and fungi. Antibacterial activity was of broad spectrum as the extracts inhibited both Gram-negative and Gram-positive bacteria. The diameter of inhibition zones of the extracts and standard antibiotics were compared (Table 6.2). The insignificant *t*-statistics ($P \geq 0.05$) suggests that the activity of NOTL33 and HYML55 extracts were comparable with that of the standard antibiotics used. The MIC and MMC of the extracts are given in Table 6.3. The MIC of the extracts ranged from 78 to 625 μg/ml with the exception of *P. aeruginosa* against which MIC was in the range of 2.5–5 mg/ml. Streptomycin was used as the positive control. The MIC and MMC of the antibiotic was much lower than the extracts. However, the antibiotic was only bacteriostatic for *P. aeruginosa* and growth was resumed on fresh agar, even at the highest concentration tested.

### 6.3.2 Test for Volatile Antimicrobials

The volatile compounds from the endophyte *B. ochroleuca* NOTL33 significantly inhibited the growth of *A. terreus* (MTCC 2580), *A. parasiticus* (MTCC 2792) and *F. oxysporum* (Fig. 6.2). The inhibition of growth was 57.13 % ($P \leq 0.01$) for *A. terreus*, 72.75 % ($P \leq 0.05$) for *A. parasiticus* and 44.03 % ($P \leq 0.01$) for *F. oxysporum*.

**Table 6.1** Diversity of endophytes from *N. foetida* and *H. mysorense*

| Sl. No. | Diversity indices | *N. foetida* | *H. mysorense* |
|---|---|---|---|
| 1. | Simpson index | 0.04187 | 0.07658 |
| 2. | Simpson's diversity index | 0.9581 | 0.9234 |
| 3. | Reciprocal Simpson index | 23.88 | 13.06 |
| 4. | Shannon–Weiner index (log) | 3.952 | 3.635 |
| 5. | Species richness | 18 | 16 |
| *Comparative indices* | | | |
| 6. | Sorenson's similarity index | 0.529 | |
| 7. | Jaccard coefficient | 0.36 | |
| 8. | Jaccard distance | 0.64 | |

**Table 6.2** Antibacterial and antidermatophytic activity of ethyl acetate extracts of endophytic *Bionectria ochroleuca* NOTL33 and *Chaetomium globosum* HYML55

| Test organisms | Diameter of inhibition zones[a] (in mm) | |
|---|---|---|
| | *Bionectria ochroleuca* NOTL33 | *Chaetomium globosum* HYML55 |
| *Escherichia coli* (MTCC 40) | 31.5±0.70 | ND |
| *Enterobacter aerogens* (MTCC 111) | 21.0±2.82 | 22.0±0.0 |
| *Bacillus subtilis* (MTCC 121) | 30.5±0.70 | 21.5±0.70 |
| *Klebsiella pneumonia* (MTCC 661) | 25.0±0.00 | ND |
| *Shigella flexneri* (MTCC 1457) | 34.0±0.00 | 22.5±0.70 |
| *Salmonella enterica* ser. Typhi (MTCC 733) | 23.34±1.15 | 23.0±0.00 |
| *Staphylococcus aureus* (MTCC 7443) | 15.34±0.57 | ND |
| *Pseudomonas aeruginosa* (MTCC 7083) | NA | 23.0±1.41 |
| Methicillin resistant *Staphylococcus aureus* | 19.0±0.00 | ND |
| *Streptococcus pyogens* (MTCC 1925) | ND | 20.5±0.70 |
| *Candida albicans* (MTCC 183) | ND | 16.5±0.07 |
| *Microsporum canis* (MTCC 2831) | ND | 29.5±0.70 |
| *Microsporum gypseum* (MTCC 2830) | ND | 25.5±0.70 |
| *Aspergillus nidulans* (MTCC 803) | 25.5±0.70 | ND |
| *A. terreus* (MTCC 2580) | 25.0±0.00 | ND |
| *A. flaviceps* (MTCC 1990) | 39.0±1.41 | ND |
| *A. fumigatus* (MTCC 3008) | 19.5±0.70 | ND |
| *A. clavatus* (MTCC 1323) | 39.5±0.70 | ND |
| *Penicillium citrinum* (MTCC 1784) | 33.0±0.00 | ND |
| *A. parasiticus* (MTCC 2797) | 20.0±0.00 | ND |

[a] Mean value±SD (in mm) of triplicate values, including the diameter of the disc (6 mm)
*ND* not done, *NA* no activity

### 6.3.3 Free Radical-Scavenging Activity

*B. ochroleuca* NOTL33 and *C. globosum* HYML55 extracts effectively scavenged both DPPH and ABTS free radicals. The kinetics and dose dependency of the free radical scavenging was studied (Fig. 6.3). Free radical-scavenging activity of extracts was dose dependent. A maximum of 46.17 and 98.65 % of DPPH free radicals were quenched by NOTL33 and HYML55, respectively, at 5 mg/ml concentration. The ABTS cation radical was quenched up to 54.13 % by HYML55 and 72.16 % by NOTL33. Total antioxidant capacity of HYML55 and NOTL33 was 13.62 mg and 18.17 mg per gram of the extracts, respectively, based on ascorbic acid stan-

**Table 6.3** Minimum inhibitory concentration (MIC) and minimum microbicidal concentration (MMC) of ethyl acetate extracts of *Bionectria ochroleuca* NOTL33 and *Chaetomium globosum* HYML55

| Test organisms | NOTL 33, (*Bionectria* sp.) | | HYML55, (*Chaetomium* sp.) | |
|---|---|---|---|---|
| | MIC in mg/ml | MMC in mg/ml | MIC in mg/ml | MMC in mg/ml |
| *Shigella flexneri* (MTCC 1457) | 0.078125 | 0.15625 | 0.078125 | 0.15625 |
| *Salmonella enterica* ser. *Typhi* (MTCC 733) | 0.31250 | 0.625 | 0.15625 | 0.3125 |
| *Staphylococcus aureus* (MTCC 7443) | 0.15625 | 0.3125 | 0.15625 | 0.3125 |
| *Pseudomonas aeruginosa* (MTCC 7083) | 5.0 | – | 2.5 | 5.0 |
| *Candida albicans* (MTCC 183) | 0.15625 | 0.3125 | 0.15625 | 0.31250 |
| *Microsporum canis* (MTCC 2831) | 0.31250 | 0.625 | 0.078125 | 0.15625 |
| *Microsporum gypseum* (MTCC 2830) | 0.6250 | 0.625 | 0.15625 | 0.31250 |

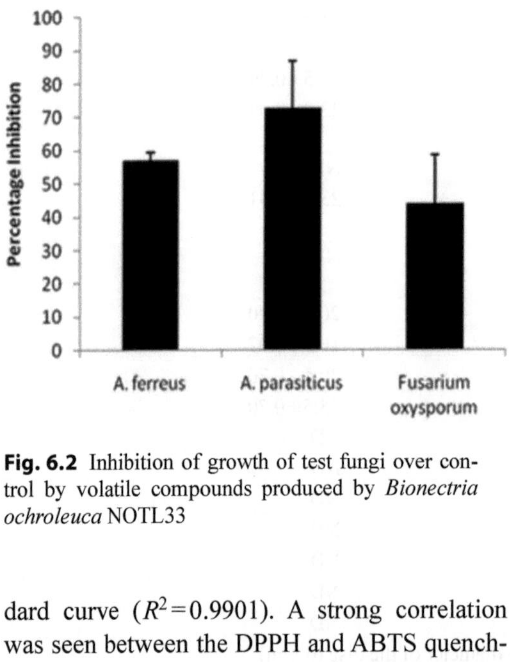


**Fig. 6.2** Inhibition of growth of test fungi over control by volatile compounds produced by *Bionectria ochroleuca* NOTL33

dard curve ($R^2$=0.9901). A strong correlation was seen between the DPPH and ABTS quenching activity of HYML55 ($R^2$=0.94472) and NOTL33 ($R^2$=0.99327) as determined by linear correlation analysis.

### 6.3.4 GC-MS Analysis

Since the NOTL33 extract produced volatile antimicrobials, the ethyl acetate extract of the same was subjected to GC-MS analysis. By the mass peak fragmentation pattern, eight compounds were tentatively identified (Table 6.4).

### 6.3.5 Compound Isolation and Effect of Media on the Production of the Active Compound

In total, nine different fractions were obtained after fractionation, out of which only the eighth fraction had the antimicrobial activity. The evaporation of which followed by crystallization using hexane yielded a yellow amorphous powder with significant antibacterial activity. The residue was purified by repeated recrystallization in hexane. The purity was confirmed by HPLC as 88 %. The MIC of the pure compound (HYML55–8A) is given in Table 6.5. The yield of the compound and the biomass in different media are given in Fig. 6.4.

## 6.4 Discussion

Plants growing in unique geographical areas and those with known medicinal properties are particularly selected for the endophytic isolation (Qadri et al. 2013). Previous studies showed that the endophytic fungi were phylogenetically distinct from that of free living forms (Syed et al. 2009)

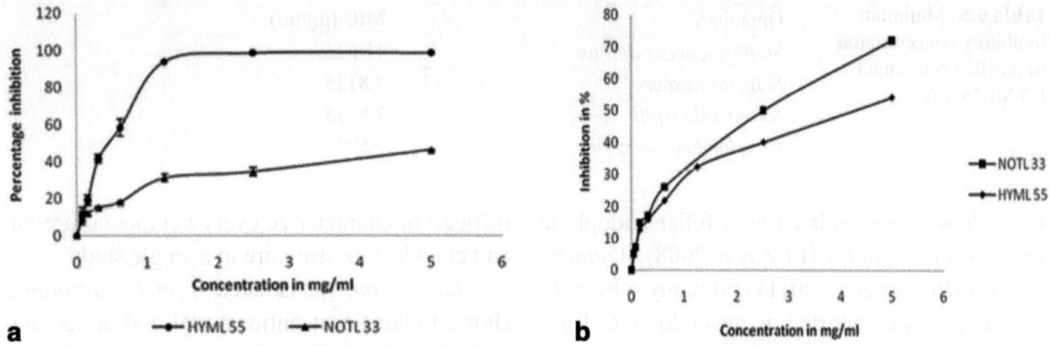


**Fig. 6.3** Dose dependent radical DPPH (a) and ABTS (b) radical scavenging activity of *Chaetomium globosum* HYML55 and *Bionectria ochroleuca* NOTL33

**Table 6.4** GC-MS analysis of the NOTL 33 ethyl acetate extract

| Peak numbers | Retention time (min) | Area (%) | Mass | Major mass peaks | Identified molecule |
|---|---|---|---|---|---|
| 1 | 11.972 | 7.66 | 114 | 71, 57, 43 | Heptanone |
| 2 | 15.865 | 2.13 | 96 | 96, 95, 67, 42 | Furfural |
| 3 | 16.597 | 14.77 | 172 | 115, 85, 57, 43 | Amyl-isovalerate |
| 4 | 19.765 | 0.38 | 168 | 168, 151, 125, 111, 97, 83, 69, 55, 41 | Decanoic acid |
| 5 | 22.030 | 18.17 | 110 | 110, 109, 95, 81 | Methyl furfural |
| 6 | 22.775 | 1.51 | 262 | 262, 261, 219, 191, 163, 135, 91, 41 | Benzyl decanoate |
| 7 | 23.119 | 16.86 | 256 | 256, 239, 211, 57, 43 | Palmitic acid |
| 8 | 27.646 | 1.19 | 190 | 190, 119, 91, 64, 57, 43, 38 | Benzyl pentyl ketone |

and they represent unique biotopes due to them occupying unique niches. Since, the metabolites that microorganisms produce seem to be characteristic of certain biotopes (Strobel and Daisy 2003), this study was aimed at the isolation and screening of the endophytic fungi from two medicinally important plants *N. foetida* and *H. mysorense* growing in the Western Ghats of Karnataka, India, which is an important biodiversity hotspot.

Isolation frequency and diversity indices (Table 6.1) reflect that the selected plants comprised diverse endophytic biota. Higher humidity and cold temperature of the Western Ghats may be the reason for this observation. Simpson's diversity index and species richness quantifies the endophytic diversity as more in *N. foetida* than *H. mysorense*. Relatively higher Shannon–Weiner index of *N. foetida* than *H. mysorense* suggests that the former harbours richer and evenly distributed endophytic biota than the latter. Larger surface area of *N. foetida* leaves than *H. mysorense* offering maximum endophytic fungal infestations may be the reason for this observation (Gond et al. 2012). Sorenson's similarity index indicated a similarity of 0.529 between two host plants.

Endophytic association in plants may be vertical or horizontal. The horizontal transmission of the endophyte is widespread in nature. Most of the foliar endophytes growing between the cells of the leaf are of this category (Syed et al. 2009). The genera *Gliocladium, Phomopsis, Xylaria, Aspergillus, Penicillium, Phoma, Cylindrotrichum, Monodictys, Chaetomium, Helminthosporium, Pestalotiopsis, Fusarium, Cladosporium, Colletotrichum, Curvularia, Trichoderma, Lasiodiplodia, Xylaria, Verticillium, Nectria, Bionectria, Botryosphaerea* and *Mycelia sterilia* isolated in the present study were previously reported as endophytes (Rubini et al. 2005). *Chaetomium* sp. and *Bionectria* sp. have previously been reported as foliar endophytes.

**Table 6.5** Minimum inhibitory concentration of purified compound HYML55–8A

| Organism | MIC (μg/ml) |
|---|---|
| *Staphylococcus aureus* | 15.625 |
| *Shigella flexneri* | 7.8125 |
| *Salmonella typhi* | 7.8125 |
| *Pseudomonas aeruginosa* | >500 |

*C. globosum* was isolated as a foliar endophyte from *Glinus lotoides* (El-Zayat 2008), *Viguiera robusta* (Borges et al. 2011) and many others. *B. ochroleuca* was reported as an endophyte from *Dendrobium* sp. (Chen et al. 2011). Nevertheles, this study is the first report on *Chaetomium* sp. in *H. mysorense* and *B. ochroleuca* in *N. foetida.*

*B. ochroleuca* NOTL33 (Anamorph: *Gliocladium roseum*) from *N. foetida* exhibited remarkable antimicrobial property through production of antimicrobial volatile compounds. Stinson et al. (2003) reported the volatile antimicrobial compounds from an endophytic *Gliocladium* sp. from *Eucryphia cordifolia*, which inhibited the growth of *Aspergillus ochraceous* by 65.3 %, but failed to inhibit *F. oxysporum. B. ochroleuca* NOTL33 volatile compounds showed better activity by inhibiting *A. parasiticus* and *F. oxysporum* by 72.75 and 44.03 %, respectively. Antibacterial activity of ethyl acetate extract was broad spectrum as it inhibited both Gram-positive and Gram-negative bacteria tested. However, the extract failed to inhibit *Pseudomonas*. Liouane et al. (2010) reported the broad-spectrum antibacterial and antifungal activity of *Gliocladium* extract, where *P. aeruginosa* was inhibited but failed to inhibit *C. albicans*. Antifungal activity of the tested extract was significant against both filamentous fungi and dermatophytes including *C. albicans.* The GC-MS analysis of the ethyl acetate extract of *B. ochroleuca* revealed the presence of many active compounds. The antimicrobial activity of heptanone and its derivatives such as (Mendonça et al. 2009) furfural (Rossmoore 2001) and decanoic acid (Isaacs 2001) are well documented. Amyl isovalerate was reported as one of the major components of many plant essential oils with antimicrobial property (Morris et al. 1979). Because of the diversity and complexity of the natural mixtures of bioactive compounds in the crude plant extract and fungal cultures, it is rather difficult to characterize every compound present and elucidate its structure in a single study.

*Chaetomium* sp. isolated from *H. mysorense* showed significant antibacterial and antidermatophytic activity. The MIC values ranged from 78 to 156 μg/ml against the tested bacteria. The antibacterial and antifungal activity of endophytic *Chaetomium* sp. have been reported earlier and attributed to the presence of bioactive metabolites such as musanahol (Marwah et al. 2007), chaetomugilin A, chaetoglobosins A and C, chaetomugilin D (Qin et al. 2009). Previous studies on other *Chaetomium* isolates exhibited similar MIC values of 80 μg/ml–1.25 mg/ml (Huang et al. 2007). Pure antibacterial compound purified from HYML55 was effective against bacterial pathogens with significant MIC values from 7.8 to15.625 μg/ml. The yield was high when grown in YESB compared to others. Previous reports also substantiate the higher yield of bioactive compounds in YESB (Tong et al. 2011).

In living organisms, the oxidative stress created by reactive oxygen species (ROS) resulting from metabolism, in the form of superoxide anion ($O^{2-}$), hydroxyl radical (OH), hydrogen peroxide ($H_2O_2$) and nitric oxide (NO) leads to conditions such as cancer, stroke, myocardial infarctions, diabetes, sepsis and hemorrhagic shock, neurodegenerative diseases by inducing biomolecular oxidations (Cavas and Yurdakoc 2005; Chew et al. 2008). Therefore, food and pharmaceutical industries need effective free radical-scavenging molecules. NOTL33 and HYML55 extracts exhibited free radical-scavenging activity comparable with that of BHA. The free radical-scavenging activity of the endophytic extracts could be attributed to the presence of phenolic compounds (Huang et al. 2007).

This study is the first report on the diversity of endophytes from *H. mysorense* and endophytic occurrence of *B. ochroleuca* in *N. foetida.* The study shows that the medicinal plants from the

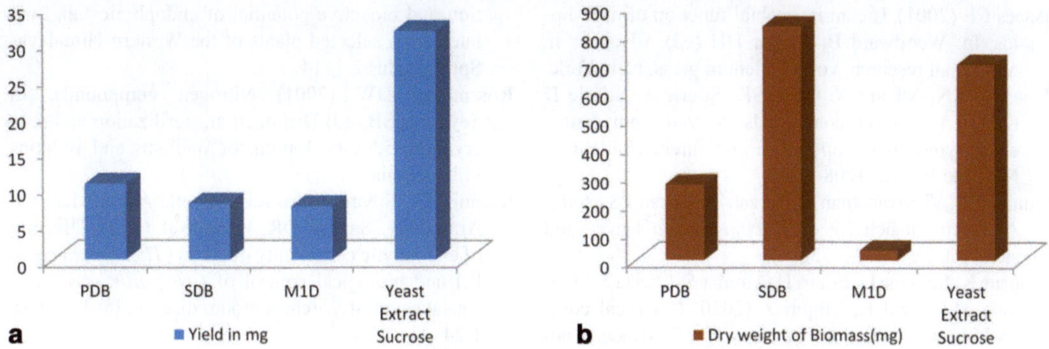


**Fig. 6.4** Dry weight (a) and yield of Compound HYML 55-8A (b) in different media

Western Ghats are a rich source of endophytic fungi, the careful evaluation of which may lead to the discovery of novel fungal strains as well as novel natural compounds with bioactive potential. Two endophytic isolates studied here exhibited significant antimicrobial and antiradical activity proving them to be a potential source for bioactive compounds. However, further investigation is essential for the purification and characterization of the active metabolites from these two endophytic fungal isolates, which may lead to the discovery of novel bioactive molecules of industrial and pharmaceutical importance.

**Acknowledgments** We are grateful to Dr. K. R. Prabhu, Senior Scientific Officer, Department of Organic Chemistry, IISc, Bangalore for providing the GC-MS facility. We are also grateful to the director of the Central Food Technological Research Institute (CFTRI), India; Dr. Varadaraj, Head, Human Resource Development (HRD) Department, CFTRI, India and Dr. Joseph, Scientist CFTRI, India for the HPLC facility. We are thankful to Prof. K. M. Lokanath Rai, Department of Studies in Chemistry, University of Mysore, India for his valuable suggestions.

## References

Bauer AW, Kirby WMM, Scherris JC, Turck M (1966) Antibiotic susceptibility testing by a standardized single disk method. Am J Clin Pathol 45:493–496

Borges WS, Mancilla G, Guimarães DO, Durán-Patrón R, Collado IG, Pupo MT (2011) Azaphilones from the endophyte *Chaetomium globosum*. J Nat Prod 74:1182–1187

Cavas L, Yurdakoc K (2005) An investigation on the antioxidant status of the invasive alga *Caulerpa racemosa* var. *cylindracea* (Sonder) Verlaque, Huisman, et Boudoresque (Caulerpales, Chlorophyta). J Exp Mar Biol Ecol 325:189–200

Chen J, Hu KX, Hou XQ, Guo SX (2011) Endophytic fungal assemblages from 10 *Dendrobium* medicinal plants (Orchidaceae). World J Microbiol Biotechnol 27:1009–1016

Chew YL, Lim YY, Omar M, Khoo KS (2008) Antioxidant activity of three edible seaweeds from two areas in South East Asia. LWT—Food Sci Technol 41:1067–1072

Ellis MB (1971) Demataceous hyphomycetes. Commonwealth mycological institute, Kew

Ellis MB (1976) More demataceous hyphomycetes. Commonwealth mycological institute, Kew

Eloff JN (1998) A sensitive and quick microplate method to determine the minimal inhibitory concentration of the plant extract for bacteria. Planta Med 64:711–713

El-Zayat SA (2008) Preliminary studies on laccase production by *Chaetomium globosum* an endophytic fungus in *Glinus lotoides*. Am Euresian J Agric Environ Sci 3:86–90

Gilman J (1957) A manual of soil fungi. Iowa State University Press, Iowa

Gond SK, Mishra A, Sharma VK, Verma SK, Kumar J, Kharwar RN, Kumar A (2012) Diversity and antimicrobial activity of endophytic fungi isolated from *Nyctanthes arbour-tristis*, a well known medicinal plant of India. Mycosci 53:113–121

Govindachari TR, Viswanathan N (1972) Alkaloids of *Mappia foetida*. Phytochem 11:3529–3531

Gunatilaka LAA, Jasmin De Silva AMY, Sotheeswaran S (1982) Minor xanthones of *Hypericum mysorense*. Phytochem 21:1751–1753

Huang WY, Cai YZ, Hyde KD, Corke H, Sun M (2007) Endophytic fungi from *Nerium oleander* L (Apocynaceae): main constituents and antioxidant activity. World J Microbiol Biotechnol 23:1253–1263

Isaacs CE (2001) The antimicrobial function of milk lipids. In: Woodward B, Draper HH (ed) Advances in nutritional research, vol 10. Plenum press, New York

Kharwar RN, Mishra A, Gond SK, Stierle A, Stierle D (2011) Anticancer compounds derived from fungal endophytes: their importance and future challenges. Nat Prod Rep 28:1208–1228

Kumar NR, Vishwanathan H, Suresh T, Mohan PS (2002) Antibacterial activity of *Mappia foetida* leaves and stem. Fitoter 73:734–736

Liouane K, Saidana D, Edziri H, Ammar S, Chriaa J, Mahjoub MA, Said K, Mighri Z (2010) Chemical composition and antimicrobial activity of extracts from *Gliocladium* sp. growing wild in Tunisia. Med Chem Res 19:743–756. doi:10.1007/s00044–009-9227–3

Liu YQ, Liu ZL, Tian X, Yang L (2010) Anti-HSV activity of camptothecin analogues. Nat Prod Res 24:509–514. doi:10.1080/14786410802270779

Lu Y, Chen C, Chen H, Zhang J, Chen W (2012) Isolation and identification of endophytic fungi from *Actinidia macrosperma* and investigation of their bioactivities. Evid Based Complement Alternat Med. doi:10.1155/2012/382742

Marwah RG, Fatope MO, Deadman ML, Al-Maqbali YM, Husband J (2007) Musanahol: a new aureonitol-related metabolite from a *Chaetomium* sp. Tetrahedron 63:8174–8180

Mendonça AL, Silva CE, Mesquita FL, Campos RS, Nascimento RR, Ximenes EC, Sant'Ana AE (2009) Antimicrobial activities of components of the glandular secretions of leaf cutting ants of the genus *Atta*. Antonie Van Leeuwenhoek 95:295–303

Morris JA, Khettry A, Seitz EW (1979) Antimicrobial activity of aroma chemicals and essential oils. J Am Oil Chem Soc 56:595–603

Osman H, Rahim AA, Isa NM, Bakhir NM (2011) Antioxidant activity and phenolic content of *Paederia foetida* and *Syzygium aqueum*. Molecules 14:970–978

Predes SF, Ruiz ALTG, Carvalho JE, Foglio MA, Dolder H (2011) Antioxidative and *in vitro* antiproliferative activityof *Arctium lappa* root extracts. BMC Complement Alternat Med 11:25

Qin JC, Zhang YM, Gao JM, Bai MS, Yang SX, Laatsch H, Zhang AL (2009) Bioactive metabolites produced by *Chaetomium globosum*, an endophytic fungus isolated from *Ginkgo biloba*. Bioorgan Med Chem Lett 19:1572–1574

Quadri M, Johri S, Shah BA, Khajuria A, Sidiq T, Lattoo SK, Abdin MZ, Riyaz-Ul-Hassa S (2013) Identification and bioactive potential of endophytic fungi isolated from selected plants of the Western Himalayas. Springerplus 2:1–14

Rossmoore HW (2001) Nitrogen compounds. In: Seymour SB (ed) Disinfection, sterilization and preservation, 5th edn. Lippinicot Wiiliams and Wilikins, Philadelphia

Rubini MR, Silva-Ribeiro RT, Pomella AWV, Maki CS, Araujo WL, Santose DR, Azevedo JL (2005) Diversity of endophytic community of cacao (*Theobroma cacao* L.) and biological control of *Crinipellis pernisiosa*, causal agent of Witche's broom disease. Int J Biol Sci 1:24–33

Srinivas KVNS, Das B (2003) 9-Methoxy-20-O-acetyl-camptothecin, a minor new alkaloid from *Nothapodites foetida*. Biochem Syst Ecol 31:85–87

Stinson M, Ezra D, Hess WM, Sears J, Strobel G (2003) An endophytic *Gliocladium* sp. of *Eucryphia cordifolia* producing selective volatile antimicrobial compounds. Plant Sci 165:913–922

Strobel G, Daisy D (2003) Bioprospecting for microbial endophytes and their natural products. Microbiol Mol Biol Rev 491–502. doi:10.1128/MMBR.67.4.491–502.2003

Syed NA, Midgley DJ, Ly PKC, Saleeba JA, McGee PA (2009) Do plant endophytic and free-living *Chaetomium* species differ? Australas Mycol 28:51–55

Ting ASW, Mah SW, Tee CS (2011) Detection of potential volatile inhibitory compounds produced by endobacteria with biocontrol properties towards *Fusarium oxysporum* f. sp. *cubense* race 4. World J Microbiol Biotechnol 27:229–235

Tong WY, Darah I, Latiffah Z (2011) Antimicrobial activities of endophytic fungal isolates from medicinal herb Orthosiphon stamineus Benth. J Med Plants Res 5:831–836

Verma VC, Kharwar RN, Strobel GA (2009) Chemical and functional diversity of natural products from plant associated endophytic fungi. Nat Prod Commun 4:1511–1532

Vijayan P, Raghu C, Ashok G, Dhanaraj SA, Suresh B (2004) Antiviral activity of medicinal plants of Nilgiris. Indian J Med Res 120:24–29

Zhao J, Zhou L, Wang J, Shan T, Zhong L, Liu X, Gao X (2010) Endophytic fungi for producing bioactive compounds originally from their host plants. In: Vilas MA (ed) Current research, technology and education topics in applied microbiology and microbial biotechnology. Formatex Research Center

# 7 Diversity of Arbuscular Mycorrhizal Fungi in Field and Trap Cultures from Rhizosphere Soils of *Flemingia vestita* Benth. ex Baker

L. S. Songachan and H. Kayang

**Abstract**

Diversity of arbuscular mycorrhizal fungi (AMF) was investigated from natural and cultivated sites of *Flemingia vestita* (Fabaceae). Sampling was done at monthly intervals for two crop cycles. Root samples were evaluated for AMF colonization in the form of arbuscules, vesicles and hyphae. AMF colonization and spore density was higher in natural site as compared to cultivated site. A total of 68 AMF species were isolated belonging to six genera: *Acaulospora, Ambispora, Gigaspora, Glomus, Pacispora* and *Scutellospora*. Trap cultures of AMF were also established with four different host plants (*Oryza sativa* L., *Paspalum notatum* Flüggé, *Trifolium ripens* L. and *Zea mays* L.) using rhizosphere soils from natural and cultivated sites. A total of 33 AMF species were isolated from trap cultures, out of which three species were not detected in *F. vestita* rhizosphere soil. Thus, trap cultures allow development and sporulation of AMF species which otherwise were not recovered from field soil.

**Keywords**

Arbuscular mycorrhizal fungi · *Flemingia vestita* · Species

## 7.1 Introduction

Arbuscular mycorrhizal fungi (AMF) are obligate biotrophic symbionts associated with roots of over 80% terrestrial plant species (Smith and Read 1997). AMF benefit from this association by obtaining carbon compounds which are necessary for their growth and in return, they have diverse, beneficial impacts on plants and soils (Brundrett et al. 1999; Li et al. 2006). AMF increases resistance to environmental stresses, enhances plant nutrient acquisition, water relations, disease resistance and improves soil quality (Smith and Read 2008). Soil microorganisms that form mutually beneficial relationships with plant roots have become a target of increasing interest in agricultural research and development because they offer a biological alternative to promote plant growth and reduce chemical inputs in

L. S. Songachan (✉) · H. Kayang
Microbial Ecology Laboratory, Department of Botany, North Eastern Hill University, Shillong 793 022, India
e-mail: rayskybl@yahoo.co.in

R. N. Kharwar et al. (eds.), *Microbial Diversity and Biotechnology in Food Security*,
DOI 10.1007/978-81-322-1801-2_7, © Springer India 2014

sustainable cropping systems (Hart and Trevors 2005).

Distribution and diversity of AMF species is influenced by the individual plant species and plant communities in natural and farming systems through differential effects on hyphal growth and sporulation (Eom et al. 2000; Jefwa et al. 2006). It is also highly influenced by several environmental factors including climatic conditions, soil physicochemical status, age and variety of host plant. AMF species richness reduces in agricultural systems compared to plant species-rich natural sites (Börstler 2010).

In all parts of Meghalaya, many plants have been investigated for their botanical aspects and medicinal values. However, mycorrhizal investigations have been rarely conducted. *Flemingia vestita* Benth. ex Baker is an indigenous plant of Meghalaya. It is a weak climber that produces an edible root tuber, which is somewhat juicy, sweet and nut-like in flavour and is eaten raw and has a high local market value. In addition, its root-tuber peel is used as curative against worm infection in traditional medicine among the natives of Meghalaya. Anthelmintic efficacy of this plant-derived material has provided evidences that support and authenticate the usage of the tuberous root of this plant as vermifuge and vermicide (Das et al. 2004). Data on the occurrence and diversity of AMF association in *F. vestita* are virtually, non-existent. Thus, the present study was aimed to investigate AMF diversity in field and trap cultures from rhizosphere soils of *F. vestita*.

## 7.2 Materials and Methods

### 7.2.1 Site Description and Field Sampling

Sampling was done at monthly intervals for two crop cycle (March to December, 2008 and 2009) from natural (MF) and cultivated sites (TF) in East Khasi Hills, Meghalaya, northeast India. Rhizosphere soils and root samples of *F. vestita* were collected with ten replicates per site, kept in a sterilized plastic bag and transported to the laboratory for analysis.

### 7.2.2 Analysis of AMF Colonization

Roots were washed thoroughly in tap water and cut into approximately 1 cm segments. The roots were then cleared in 10 % (w/v) KOH by heating at 90 °C for 1 h. It was then washed and stained with Trypan blue (Phillips and Hayman 1970). The stained root samples were mounted on microscope slides and examined for AMF structure under light microscope. The colonization of root length with arbuscules (RLA), vesicles (RLV) and hyphae (RLH) per sample were quantified by the magnified intersection method of McGonigle et al. (1990) and expressed it in percentage.

### 7.2.3 AMF Spore Isolation, Enumeration and Identification

AMF spores were extracted by wet sieving and decanting method of Gerdemann and Nicolson (1963). Suspension of 25 g soil sample in water was decanted through a series of 710–737 μm sieves. The residues in the sieves were washed into the beaker and the sievates were dispersed in water and filtered through filter papers and the spores were counted using a dissection microscope at 40× magnification. Sporocarps and spore clusters were considered as one unit. AMF spores were picked up using a needle and mounted in polyvinyl alcohol-lactic acid-glycerol (PVLG) with Meltzer's reagent on a glass slide for identification based on morphological descriptions published by International Culture Collection of Vesicular and Arbuscular Mycorrhizal Fungi (http://invam.caf.wvu.edu), AMF phylogeny (www.amf-phylogeny.com), Oehl and Sieverding (2004) and Goto et al. (2008).

### 7.2.4 Trap Culture Setup

Trap cultures of AMF were established with four different host plants (*Oryza sativa* L., *Paspalum notatum* Flüggé, *Trifolium ripens* L. and *Zea mays* L.). Field soils and root fragments of *F. vestita* from MF and TF served as inocula for AMF trap cultures. Approximately, 40–50 seeds

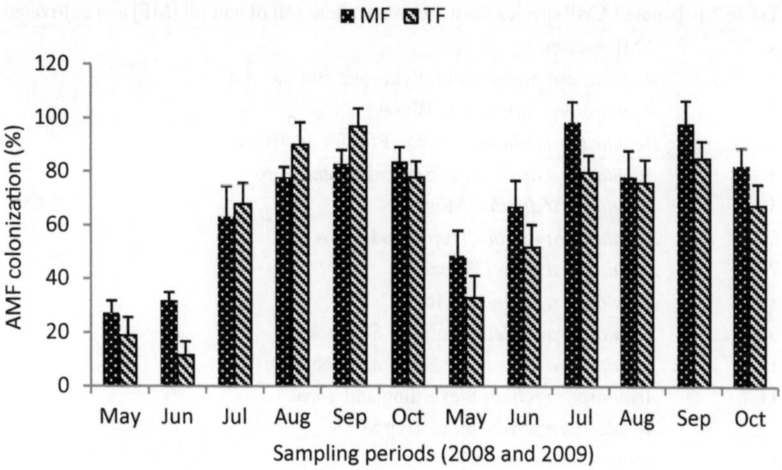


**Fig. 7.1** Arbuscular mycorrhizal fungi (*AMF*) colonization in natural (*MF*) and cultivated site (*TF*) of *Flemingia vestita*

of each plant species were evenly sown on 25 cm diameter plastic pots containing 1.5 kg of inoculum and autoclaved coarse sand (1:1 v/v) and monitored in greenhouse condition. It was watered whenever required. After 5 months, trap plants were checked out for AMF colonization, spores density and species composition.

### 7.2.5 Statistical Analysis

Spore density and species richness were expressed as number of AMF spores and numbers of AMF species in 25 g soil samples, respectively. Relative abundance, isolation frequency, Shannon-Wiener index of diversity (H') and Simpson's index of dominance (D) were calculated (Dandan and Zhiwei 2007). Data were statistically analysed using one-way analysis of variance (ANOVA). Pearson correlation coefficient was employed to determine the relationships between mycorrhizal structural colonization and soil physicochemical properties. Standard errors of means were calculated.

## 7.3 Results and Discussion

The roots of *F. vestita* were colonized by AMF throughout the crop cycle in both sites. AMF colonization in the form of arbuscules, vesicles and hyphae, and occasionally, intraradical spores were observed. The monthly variation in total AMF colonization is given in Fig. 7.1. Total AMF colonization ranged from 27.13 to 98.03 % in natural site and 11.38 to 96.76 % in cultivated site. No tillage condition in natural site could have stimulated mycorrhizal activity in soil of *F. vestita*. This finding is also in accordance with Dodd et al. (2000). Low colonization in cultivated site could be due to the agricultural practices such as uses of agricultural machineries and conventional tillage which might have reduced the hyphal network in the soil.

High AMF spore density was observed in rhizosphere soil of *F. vestita* in both study sites. The spore numbers ranged from 574 to 3,074 in 25 g soil sample in natural site, while in cultivated site, it ranged from 552 to 1,854 in 25 g soil sample. The rhizosphere soils of *F. vestita* revealed that natural site had higher number of AMF species as compared to the cultivated site. A total of 68 AMF species belonging to six genera (*Glomus, Acaulospora, Scutellospora, Gigaspora, Ambispora* and *Pacispora*) were isolated from *F. vestita* rhizosphere soils (Table 7.1). Some of the isolated AMF species are given in Fig. 7.2. In this investigation, *Glomus* was the most widely distributed AMF species and it is considered as cosmopolitan AMF species in many ecosystems (Sýkorová et al. 2007). It was followed by *Acaulospora, Scutellospora* and *Gigaspora* in both study sites. In natural site, the dominant AMF species was *Glomus*. Klironomos and Hart (2002) suggested that mycelium is of major importance as propagule for some *Glomus* species.

**Table 7.1** Isolated AMF species from the rhizosphere soil of natural (MF) and cultivated sites (TF) of *Flemingia vestita*

| Sl. No. | AMF species | MF | TF |
|---|---|---|---|
| 1. | *Acaulospora alpina* Oehl, Sykorova and Sieverd | + | + |
| 2. | *Acaulospora capsiculata* Blaszk | + | + |
| 3. | *Acaulospora delicata* Walker, Pfeiffer and Bloss | + | + |
| 4. | *Acaulospora denticulata* Sieverding and Toro | + | + |
| 5. | *Acaulospora dilatata* Morton | − | + |
| 6. | *Acaulospora foveata* Trappe and Janos | + | − |
| 7. | *Acaulospora koskei* Blaszk | + | + |
| 8. | *Acaulospora lacunosa* Morton | + | + |
| 9. | *Acaulospora mellea* Spain and Schenck | + | + |
| 10. | *Acaulospora morrowiae* Spain and Schenck | + | + |
| 11. | *Acaulospora rehmii* Sieverding and Toro | − | + |
| 12. | *Acaulospora scrobiculata* Trappe | + | + |
| 13. | *Acaulospora spinosa* Walker and Trappe | + | + |
| 14. | *Ambispora brasiliensis* Goto, Maia and Oehl | + | + |
| 15. | *Gigaspora albida* Schenck and Smith | + | + |
| 16. | *Gigaspora decipiens* Hall and Abbott | − | + |
| 17. | *Gigaspora margarita* Becker and Hall | − | + |
| 18. | *Gigaspora rosea* Nicolson and Schenck | + | + |
| 19. | *Glomus aggregatum* Schenck and Smith | + | + |
| 20. | *Glomus ambisporum* Smith and Schenck | + | + |
| 21. | *Glomus aurantium* Błaszk., Blanke, Renker and Buscot | + | + |
| 22. | *Glomus badium* sp. nov. Oehl, Redecker and Sieverd | + | + |
| 23. | *Glomus caledonium* Nicolson and Gerdemann | + | − |
| 24. | *Glomus claroideum* (Schenck and Smith emend. Walker and Vestberg) | − | + |
| 25. | *Glomus clavisporum* (Trappe) Almeida and Schenck | + | − |
| 26. | *Glomus convolutum* Gerdemann and Trappe | + | − |
| 27. | *Glomus coronatum* Giovann | + | + |
| 28. | *Glomus corymbiforme* Blaszkowski | + | + |
| 29. | *Glomus eburneum* Kenn., Stutz and Morton | + | + |
| 30. | *Glomus etunicatum* Becker and Gerdemann | + | + |
| 31. | *Glomus fasciculatum* (Thaxter) Gerdemann and Trappe | + | + |
| 32. | *Glomus fistulosum* Skuo and Jakobsen | + | + |
| 33. | *Glomus fuegianum* (Spegazzini) Trappe and Gerdemann | + | − |
| 34. | *Glomus geosporum* (Nicol. and Gerd.) Walker | + | + |
| 35. | *Glomus gibbosum* Błaszk | − | + |
| 36. | *Glomus glomerulatum* Sieverding | + | + |
| 37. | *Glomus heterosporum* Smith and Schenck | − | + |
| 38. | *Glomus hoi* Berch and Trappe | + | + |
| 39. | *Glomus intraradices* Schenck and Smith | + | + |
| 40. | *Glomus lamellosum* Dalpe, Koske and Tews | + | − |
| 41. | *Glomus luteum* Kenn., Stutz and Morton | + | + |
| 42. | *Glomus macrocarpum* Tul. and Tul | + | + |
| 43. | *Glomus manihotis* Howeler, Sieverding and Schenck | + | − |
| 44. | *Glomus melanosporus* Gerd. and Trappe | − | + |
| 45. | *Glomus microaggregatum* Koske, Gemma and Olexia | + | − |
| 46. | *Glomus microcarpum* Tul. and Tul | + | + |
| 47. | *Glomus minutum* Blaszkowski, Tadych et Madej, sp. Nov. | + | − |
| 48. | *Glomus mosseae* (Nicol. and Gerd.) Gerdemann and Trappe | + | − |

**Table 7.1** (continued)

| Sl. No. | AMF species | MF | TF |
|---|---|---|---|
| 49. | *Glomus rubiforme* Gerdemann and Trappe | + | + |
| 50. | *Glomus sinuosum* (Gerd. and Bakshi) Almeida and Schenck | + | + |
| 51. | *Glomus spinosum* Hu | + | – |
| 52. | *Glomus tenebrosum* (Thaxter) Berch | + | – |
| 53. | *Glomus tortuosum* Schenck and Smith | + | + |
| 54. | *Glomus verruculosum* Blaszkowski and Tadych | + | + |
| 55. | *Glomus versiforme* (Karsten) Berch | + | + |
| 56. | *Glomus viscosum* Nicolson | + | + |
| 57. | *Pacispora robigina* Oehl and Sieverd | – | + |
| 58. | *Scutellospora calospora* Walker and Sanders | + | – |
| 59. | *Scutellospora cerradensis* Spain and Miranda | + | + |
| 60. | *Scutellospora coralloidea* Koske and Walker | – | + |
| 61. | *Scutellospora erythropa* (Koske and Walker) Walker and Sanders | – | + |
| 62. | *Scutellospora fulgida* Koske and Walker | + | – |
| 63. | *Scutellospora heterogama* (Nicolson and Gerd.) Walker and Sanders | + | + |
| 64. | *Scutellospora pellucida* (Nicolson and Schenck) Walker and Sanders | + | + |
| 65. | *Scutellospora pernambucana* Oehl, Silva, Freitas and Maia | + | + |
| 66. | *Scutellospora rubra* Sturmer and Morton | + | + |
| 67. | *Scutellospora scutata* Walker et Diederichs | + | + |
| 68. | Unidentified species | + | + |

'+' indicates presence of species and '–' absence of species

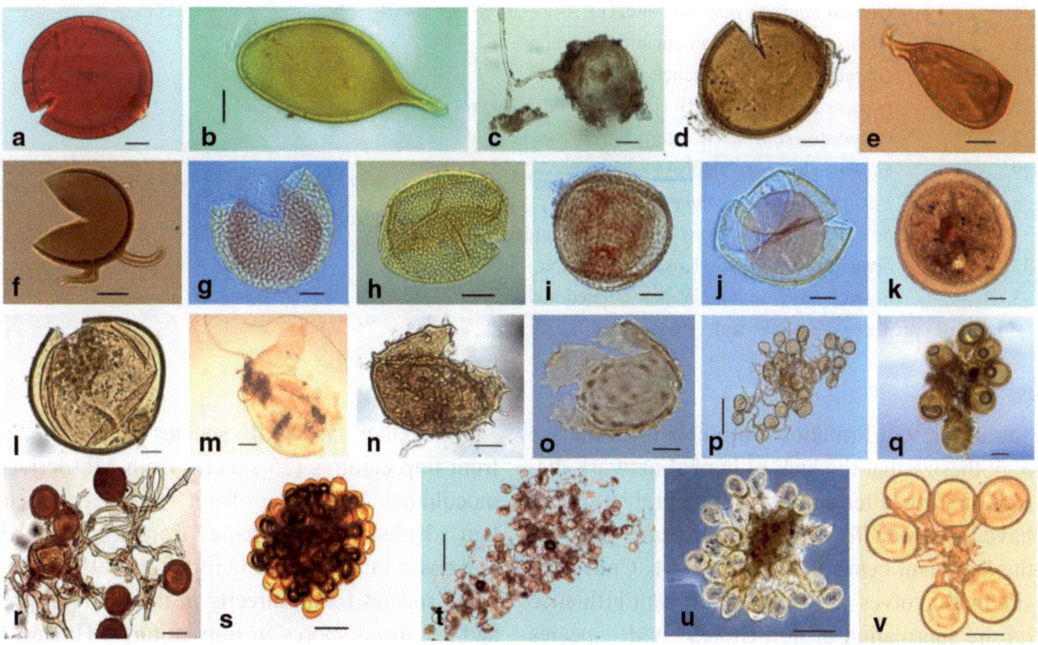


**Fig. 7.2** AMF spores and sporocarps isolated from rhizosphere soil of *F. vestita*. **a–f** *Glomus* species—*G. badium, G. indraradices, G. tortuosum, G. indraradices, G. clavisporum* and *G. glomerulatum*. **g–k** *Acaulospora* species—*A. scrobiculata, A. denticulata, A. delicata* and *A. foveata*. **l–m** *Scutellospora* species. **n–o** Unidentified species. **p–v** Sporocarpic species. (*Scale bar:* **a–o** = 50 μm, **p** = 200 μm, **q–r** = 100 μm, **s–u** = 100 μm, **v** = 60 μm)

**Table 7.2** AMF species isolated from four trap plants set up with natural (MF) and cultivated (TF) rhizosphere soils of *F. vestita*

| Sl. No. | AMF species | MF | TF |
|---|---|---|---|
| 1. | *Acaulospora capsiculata* Blaszk | + | − |
| 2. | *Acaulospora delicata* Walker, Pfeiffer and Bloss | + | + |
| 3. | *Acaulospora koskei* Blaszk | + | − |
| 4. | *Acaulospora mellea* Spain and Schenck | + | − |
| 5. | *Acaulospora morrowiae* Spain and Schenck | − | + |
| 6. | *Acaulospora rehmii* Sieverding and Toro | − | + |
| 7. | *Acaulospora scrobiculata* Trappe | + | + |
| 8. | *Gigaspora rosea* Nicolson and Schenck | + | + |
| 9. | *Gigaspora margarita* Becker and Hall | − | + |
| 10. | *Glomus aggregatum* Schenck and Smith | + | + |
| 11. | *Glomus badium* sp. nov. Oehl, Redecker and Sieverd | + | + |
| 12. | *Glomus caledonium* Nicolson and Gerdemann | − | + |
| 13. | *Glomus clavisporum* (Trappe) Almeida and Schenck | + | − |
| 14. | *Glomus etunicatum* Becker and Gerdemann | + | − |
| 15. | *Glomus eburneum* Kenn., Stutz and Morton | − | + |
| 16. | *Glomus fasciculatum* (Thaxter) Gerdemann and Trappe | + | + |
| 17. | *Glomus fistulosum* Skuo and Jakobsen | − | + |
| 18. | *Glomus fuegianum* (Spegazzini) Trappe and Gerdemann | + | − |
| 19. | *Glomus geosporum* (Nicol. and Gerd.) Walker | + | + |
| 20. | *Glomus glomerulatum* Sieverding | + | + |
| 21. | *Glomus intraradices* Schenck and Smith | + | + |
| 22. | *Glomus luteum* Kenn., Stutz and Morton | + | + |
| 23. | *Glomus macrocarpum* Tul. and Tul. | + | − |
| 24. | *Glomus microcarpum* Tul. and Tul. | + | − |
| 25. | *Glomus rubiforme* Gerdemann and Trappe | − | + |
| 26. | *Glomus tortuosum* Schenck and Smith | + | − |
| 27. | *Glomus verruculosum* Blaszkowski and Tadych | + | − |
| 28. | *Glomus versiforme* (Karsten) Berch | + | − |
| 29. | *Glomus viscosum* Nicolson | − | + |
| 30. | *Paraglomus brasilianum* | + | − |
| 31. | *Scutellospora calospora* Walker and Sanders | + | − |
| 32. | *Scutellospora fulgida* Koske and Walker | + | − |
| 33. | *Scutellospora scutata* Walker et Diederichs | + | − |

'+' indicates presence of species and '−' absence of species

Taking these affirmations into consideration we hypothesize that the lack of hyphal network disruption in no-tilled condition in natural site could have favoured *Glomus* species. *Acaulospora* was the dominant genus in cultivated site. Cultivated site that involves tillage has significant influence on the sporulation of non-*Glomus* AMF species (Castillo et al. 2006), and thus it reflects the relative abundance and the dominant characteristics of *Acaulospora* species under cultivated site.

Altogether, 33 AMF species were isolated from trap cultures (25 species from MF-derived inoculum and 18 species from TF-derived inoculum). Isolated AMF species from trap cultures are given in Table 7.2. Only about 49 % of the AMF species found directly in the field samples had produced spores in trap cultures. However, it helps in recovery of one additional AMF species (*Paraglomus brasilianum*) from MF-derived trap cultures and two additional AMF species (*Glomus caledonium* and *Glomus glomerulatum*) from TF-derived trap cultures that were not isolated from field soils. Occurrence of additional AMF species in the traps is a well-documented phenomenon, justifying the use of trap cultures for more complete AMF surveys than direct isolation of spores from the field soils (Brundrett et al. 1999; Jansa et al. 2002; Oehl et al. 2004). Given the different environmental conditions in trap pots in comparison to the fields, some of the AMF rarely sporulating in the field soil, might start forming spores in the pots. Thus, this approach should be incorporated in the analysis of AMF species diversity as it revealed additional AMF species.

## 7.4 Conclusions

AMF are ecologically important root symbionts of most terrestrial plants, and their roles in natural and agricultural ecosystems are increasingly being recognized. The important observation of the present investigation is that *F. vestita* harbours a relatively high AMF community, supporting the view that the representatives of Fabaceae have a high mycorrhizal dependency (Duponnois et al. 2001). This study also suggested that AMF diversity is higher in natural site as compared to the cultivated site of *F. vestita* and it could be affected by different agricultural management practices.

Many plant species are in high demand for their medicinal properties, and for other various purposes. Owing to the multiple beneficial effects on plant performance and soil health, recognition of mycorrhizal status and application of mycorrhizal technology which is cost-effective and environmental friendly, are potentially important in increasing plant productivity. Therefore, appropriate management of ecosystem services rendered by AMF will have an impact on the net gain for human society.

**Acknowledgment** The authors are thankful to Head, Department of Botany, North Eastern Hill University for providing laboratory facilities. The first author is also grateful to the Rajiv Gandhi National Fellowship, New Delhi, for financial support in the form of research fellowship.

## References

Börstler B (2010) Diversity of cultured isolates and field populations of the arbuscular mycorrhizal fungus Glomus intraradices: development and application of molecular detection methods for mitochondrial haplotypes. Ph.D Thesis, Universität Basel

Brundrett MC, Abbott LK, Jasper DA (1999) Glomalean mycorrhizal fungi from tropical Australia I. Comparison of the effectiveness and specificity of different isolation procedures. Mycorrhiza 8:305–314

Castillo CG, Rubio R, Rouanet JL, Borie F (2006) Early effects of tillage and crop rotation on arbuscular mycorrhizal fungal propagules in an Ultisol. Biol Fertil Soils 43:83–92

Dandan Z, Zhiwei Z (2007) Biodiversity of arbuscular mycorrhizal fungi in the hot-dry valley of the Jinsha river, southwest China. Appl Soil Ecol 37:118–128

Das B, Tandon V, Saha N (2004) Anthelmintic efficacy of *Flemingia vestita* (Fabaceae): alterations in glucose metabolism of the cestode, *Raillietina echinobothrida*. Parasitol Int 53:345–350

Dodd JC, Boddington CL, Rodriguez A, Gonzalez-Chavez C, Mansur I (2000) Mycelium of arbuscular mycorrhizal fungi (AMF) from different genera: form, function and detection. Plant Soil 226:131–151

Duponnois R, Plenchette C, Thioulouse J, Cadet P (2001) The mycorrhizal soil infectivity and arbuscular mycorrhizal fungal spore communities in soils of different aged fallows in Senegal. Appl Soil Ecol 17:239–251

Eom AH, Hartnett DC, Wilson GWT (2000) Host plant species effects on arbuscular mycorrhizal fungal communities in tallgrass prairie. Oecologia 122:435–444

Gerdemann JW, Nicolson TH (1963) Spores of mycorrhizal Endogone species extracted from soil by wet sieving and decanting. Trans Brit Mycol Soc 46:235–244

Goto BT, Maia LC, Oehl F (2008) *Ambispora brasiliensis*, a new ornamented species in the arbuscular mycorrhiza-forming Glomeromycetes. Mycotaxon 105:11–18

Hart MM, Trevors JT (2005) Microbe management: application of mycorrhizal fungi in sustainable agriculture. Frontiers in Ecol Environ 3:533–539

Jansa J, Mozafar A, Anken T, Ruh R, Sanders IR, Frossard E (2002) Diversity and structure of AMF communities as affected by tillage in a temperate soil. Mycorrhiza 12:225–234

Jefwa JM, Sinclair R, Maghembe JA (2006) Diversity of Glomale mycorrhizal fungi in maize/sesbania intercrops and maize monocrop systems in southern Malawi. Agroforestry Syst 67:107–114

Klironomos JN, Hart MM (2002) Colonization of roots by arbuscular mycorrhizal fungi using different sources of inoculum. Mycorrhiza 12:181–184

Li H, Smith SE, Holloway RE, Zhu Y, Smith FA (2006) Arbuscular mycorrhizal fungi contribute to phosphorous uptake by wheat grown in a phosphorous-fixing soil even in the absence of positive growth responses. New Phytol 172:536–543

McGonigle TP, Miller MH, Evans DG, Fairchild GL, Swan JA (1990) A new method which gives an objective measure of colonization of roots by vesicular-arbuscular mycorrhizal fungi. New Phytol 115:495–501
Oehl F, Sieverding E (2004) *Pacispora*, a new vesicular arbuscular mycorrhizal fungal genus in the Glomeromycetes. J Appl Bot 78:72–82
Oehl F, Sieverding E, Mäder P, Dubois D, Ineichen K, Boller T, Wiemken A (2004) Impact of long-term conventional and organic farming on the diversity of arbuscular mycorrhizal fungi. Oecologia 138:574–583
Phillips JM, Hayman DS (1970) Improved procedures for clearing roots and staining parasitic and vesicular–arbuscular mycorrhizal fungi. Trans Brit Mycol Soc 55:158–160
Smith SE, Read DJ (1997) Mycorrhizal symbiosis, 2nd edn. Academic Press Ltd., London
Smith SE, Read DJ (2008) Mycorrhizal Symbiosis, 3rd ed. Academic Press, San Diego
Sýkorová Z, Ineichen K, Wiemken A, Redecker D (2007) The cultivation bias: different communities of arbuscular mycorrhizal fungi detected in roots from the field, from bait plants transplanted to the field, and from a greenhouse trap experiment. Mycorrhiza 18:1–14

# 8 Molecular Identification and Characterization of the Taxol-Producing *Colletotrichum gloeosporioides* from *Moringa oleifera* Linn.

K. Gokul Raj, P. Rajapriya, J. Muthumary and M. Pandi


**Abstract**

Cancer is the uncontrolled proliferation of abnormal cells whose diagnosis and treatment continues to be an Achilles heel to the medical fraternity. In spite of several breakthroughs in cancer research, economically viable, safe treatment is still uncommon to the commoners. The unwarranted side effects of the available drugs and their high cost necessitate the search for novel, safer and cheaper bioactive molecules for the treatment. Taxol, a diterpenoid, alkaloid first isolated from the bark of the Pacific yew tree, *Taxus brevifolia,* is one of the better known anticancer drugs. The increased demand for taxol, coupled with its limited availability from the protected Pacific yew, led the researchers scrambling for alternate sources. Our present study details the isolation and molecular identification of the taxol-producing endophytic fungus *Colletotrichum gloeosporioides* from *Moringa oleifera*. Morphotyping and ITS-based identification were employed to confirm the identity of our organisms, while their potential for taxol production was evaluated at the genetic level using DBAT (10-deactylbaccatin III-10-O-acetyl transferase) and BAPT (C-13 phenylpropanoid side chain-CoA acyl transferase) genes. The fungal taxol produced in M1D medium was extracted, partially purified and confirmed using different spectral and analytical methods.



**Keywords**

Fungal taxol · Endophytic fungi · DBAT and BAPT genes



M. Pandi (✉) · K. G. Raj
Department of Molecular Microbiology, School of Biotechnology, Madurai Kamaraj University, Madurai 625021, Tamil Nadu, India
e-mail: an_pandi@rediffmail.com; pandimohan81@gmail.com

P. Rajapriya
Department of Microbial Technology, School of Biological Sciences, Madurai Kamaraj University, Madurai 625021, Tamil Nadu, India


## 8.1 Introduction

The process of carcinogenicity presents a major challenge to scientists and provides limited tools for its control. For more than 50 years, the war on cancer is being fought with three tools—surgery (cut), radiation therapy (burn) and chemotherapy (chemical; Lam 2003). In common, all


R. N. Kharwar et al. (eds.), *Microbial Diversity and Biotechnology in Food Security,*
DOI 10.1007/978-81-322-1801-2_8, © Springer India 2014

cancers are characterized by the uncontrolled proliferation of abnormal cells in the body. Increase in global failure (or decrease in the curing ability) of currently available medicine for many diseases, especially cancer, diabetes, malaria, leishmania, AIDS, etc., undermines our ability to overcome the problem (Culver et al. 1998). While new drugs and drug sources continue to be explored, endophytic fungi continue to be a pivotal group producing novel bioactive metabolites (Strobel and Daisy 2003; Joseph and Minipriya 2011; Xing et al. 2011).

Endophytic fungi are those which live within plant tissues for all or part of theirlife cycle and cause no apparent harm to the host plant; thus, they are also called symptomless parasitic fungi (Maheshwari 2006). Secondary metabolites produced by endophytes have been documented to vary according to the host and their environmental condition. Taxol, a diterpenoid natural product, first isolated from the bark of the Pacific yew tree, *Taxus brevifolia*, is highly functionalized and is the world's first billion dollar anticancer drug (Stierle et al. 1993; Strobel 2002). Isolation of taxol-producing endophytic fungi had greatly alleviated the pressure on yew trees and is considered to bridge the gap in meeting the rising demand. Several endophytic isolates from varied host plants have been reported to produce taxol and their derivatives, through laborious conventional isolation and screening procedures. As an alternative method for screening such endophytes, genes encoding for taxol biosynthetic enzymes have been used as a molecular marker for screening taxol-producing fungal endophytes (Kumaran and Hur 2009).

As a part of a major research project, we attempted to document the diversity of endophytic fungal community from *Moringa oleifera,* a deciduous tree with edible leaves, flowers and fruits, of rich medicinal utility. *M. oleifera* is used as an antibacterial, antioxidant, antiseptic, anti-anaemic, antihypertensive, antitumoural, radio protective and external cancer preventive measure, and to treat sores, prostate, diabetes/hypoglycaemia, hypocholesterolaemia, bronchitis, energy, protein, vitamin/mineral deficiency, thyroid, hepatorenal, colitis, diarrhoea, dysentery, lactation, scurvy, gastritis, rheumatism, tonic, etc. (Rajangam et al. 2001; Thurber and Fahey 2005). In this study, we report the isolation of *Colletotrichum gloeosporioides,* an endophytic fungus from *M. oleifera* and its identification using morphotypic parameters such as growth pattern, spore morphology with internal transcribed spacer (ITS) region-specific primers as molecular markers. Erstwhile, the morphological species recognition (MSR) concept held good for genus-level identification in the common fungal community; recently, ITS region-based phylogeny has been widely used because of its ease and advantage to resolve qualms at species level. *C. gloeosporioides,* isolated from various hosts growing in different environments, had been reported to produce taxol (Gangadevi and Muthumary 2008; Nithya and Muthumary 2009; Kumaran et al. 2008, 2010; Kumaran and Hur 2009). We screened our isolate for the presence of DBAT (10-deactylbaccatin III-10-O-acetyl transferase) and BAPT (C-13 phenylpropanoid side chain-CoA acyl transferase) genes and intermediate genetic elements involved in taxol biosynthesis. Also, production of fungal taxol was further confirmed using ultraviolet (UV) spectroscopy, infrared (IR) spectroscopy and high performance liquid chromatography (HPLC) analysis (Strobel et al. 1996).

### 8.1.1 Sample Collection and Isolation of Endophytic Fungi

*M. oleifera* leaf samples were collected from four different locations across Chennai, and were separately processed within 24 h of collection for endophytic fungi isolation as detailed by Strobel (2002) and Strobel and Daisy (2003).

J. Muthumary
Centre for Advanced Studies in Botany, University of Madras, Guindy Campus, Chennai 600025, Tamil Nadu, India

### 8.1.2 Isolation and Identification of Endophytic Fungi from *M. oleifera*

The phylloplane (epiphytic) fungal propagules adhering onto the leaf surfaces were killed with surface sterilants using the modified procedure described by Dobranic et al. (1995) to facilitate endophyte isolation. Specifically, the leaves were washed with running tap water, sterilized with ethanol (75% V/V) for 1 min and sodium hypochlorite (2.5% V/V) for 5 min, then rinsed in sterile water three times and cut into 1 cm long segments. Plant segments transferred onto potato-dextrose-agar (PDA) plates amended with ampicillin (200 µg/ml) and streptomycin (200 µg/ml) were incubated for 12 h of light/dark condition at 25 °C and observed for emergence of hyphae. Axenic cultures were raised using hyphal tip method (Huang et al. 2001) and maintained in PDA. Fungal identification methods were based on the morphology of the fungal culture, the mechanism of spore production and the characteristics of the spores (Sutton 1980; Nagraj 1993).

### 8.1.3 Molecular Identification and Screening of Taxol-Producing Fungi

#### 8.1.3.1 DNA Extraction

Genomic DNA was extracted from the endophytic fungi using modified method (Azmat et al. 2012). *C. gloeosporioides* isolate was cultured in 50 mL potato dextrose broth (PDB) medium for a period of 7 days to obtain fungal mycelial mat. The mycelium was harvested and filtered through Whatman No.1 filter paper and was later lyophilized. A total of 50 mg of mycelium, grounded in liquid nitrogen, was dissolved in 500 µL TES (100 mM Tris, pH 8.0, 10 mM ethylenediaminetetraacetic acid (EDTA), 2% sodium dodecyl sulfate (SDS)) and incubated for 60 min at 55–60 °C with addition of 50 µg of proteinase K from an appropriate stock solution. The salt concentration was adjusted to 1.4 M with 5 M NaCl (140 µL) and 1/10 volume (65 µL) of 10% cetyl trimethylammonium bromide (CTAB) was added and incubated for 10 min at 65 °C. A total of 700 µL of Phenol: Chloroform (24:1) (SEVAG) buffer phenol/chloroform/isoamyl alcohol was added gently, incubated for 30 min at 0 °C and then centrifuged for 10 min (4 °C) at 10,000 rpm. The supernatant was collected in a 1.5 mL tube, and 225 µL of 3 M sodium acetate was added, mixed gently and incubated in ice for 30 min and centrifuged again at 10,000 × rpm. The supernatant was transferred to a fresh tube and 550 µLof isopropanol was added to precipitate DNA, and was centrifuged immediately for 10 min at 10,000 × rpm. The supernatant was aspirated and the pellet was washed twice with cold 70% ethanol, air-dried and resuspended in 50 µL of RNase-free water.

#### 8.1.3.2 Amplification of Internal Transcribed Spacer (ITS) Sequences

The ITS gene from the isolated endophytic fungus was amplified using universal primer by polymerase chain reaction (PCR). The reaction was carried out by using the forward primer ITS1f (5′TCC-GTA-GGT-GAA-CCT-GCG-G 3′) and reverse primer ITS4r (5′TCC-TCC-GCT-TAT-TGA-TAT-GC 3′; Sim et al. 2010). The PCR amplification was performed in an L196G-GD Model Peltier Thermal Cycler Version 2.0 with a total 25 µl reaction that comprised 20 ng of genomic DNA, 2.5 µl 10× buffer with 25 mM $MgCl_2$, 0.5 µl of 10 mM DNTPs, 2 U of Taq DNA polymerase and 10 pmol of each primer (all molecular chemicals were purchased from Sigma Aldrich). The cycling conditions were: 4 min at 94 °C for initial denaturation, 30 cycles each of 30 s at 94 °C for denaturation, 1 min at 58.2 °C for annealing, 2 min at 72 °C for extension and a final extension at 72 °C for 7 min. The amplified DNA fragments were analysed by 1% agarose gel electrophoresis along with the 100 bp ladder (New England Biolabs), and the amplicons were visualized using a gel documentation system (Uvitech). A non-template control was included in each run. PCR products were purified using mini columns (PCR Preps DNA purification System, Sigma) according to the manufacturer's protocol.

#### 8.1.3.3 ITS Region Sequencing and Phylogenetic Analysis

The purified PCR products were subsequently sequenced in Applied Biosystems 3730xl sequencer, USA (Eurofins Private Limited, Bangalore, India). The samples (mixed with sequencing buffer, dye (BigDye Terminator) labelled dNTPs and universal primers for ITS) were amplified with the following PCR conditions: 25 cycles, each of 96 °C for 10 s, 55 °C for 5 s and 60 °C for 4 min. The sequencing run was completed using a standard programme. The extension products were purified with ethanol and EDTA, separated by capillary electrophoresis and interpreted with an inbuilt software programme. The resultant nucleotide sequences were analysed using n-BLAST (http://blast.ncbi.nlm.nih.gov) and deposited in GenBank database. A phylogenetic tree was constructed using a MEGA 5.1 software.

#### 8.1.3.4 Screening for DBAT and BAPT Genes

The presence of DBAT and BAPT genes were screened with gene-specific primers. The PCR conditions for *dbat* gene are initial denaturation at 94 °C for 4 min before 30 cycles of amplification and denaturation at 94 °C for 45 s, annealing at 56.9 °C for 30 s, extension at 72 °C for 30 s and final extension at 72 °C for 6 min. The PCR conditions for *bapt* were initial denaturation at 94 °C for 4 min before 30 cycles of amplification and denaturation at 94 °C for 50 s, annealing at 56 °C for 50 s, extension at 72 °C for 30 s and finally at 72 °C for 6 min. The amplified DNA fragments were analysed by agarose gel electrophoresis along with the 100 bp ladder (New England Biolabs) and the amplicons were visualized under a gel documentation system. All the PCR amplifications were performed in an L196GGD Model Peltier Thermal Cycler Version 2.0 (Zhang et al. 2008).

### 8.1.4 Extraction of Fungal Taxol

*C. gloeosporioides* isolate was inoculated into 2 L of M1D broth and incubated for 21 days at 23–24 °C for taxol production with 12 h of light and 12 h of dark condition. At the end of the incubation period, culture supernatant separated by filtering through cheese cloth was added with 0.25 g of $Na_2CO_3$ to clarify the solution. Subsequently, the filtrate was mixed with dimethyl sulpoxide (DMSO) in a 1:2 ratio to extract the fatty acid-free organic phase, which was further concentrated under controlled pressure at 35 °C using rotary vacuum evaporator. The dry solid residue was redissolved in 1 ml of methanol and screened for the presence of taxol-like compounds on pre-coated silica gel plates (Merck) along with authentic taxol as a positive control. The plates were developed by the solvent system as reported (Strobel et al. 1996b), and fractions with the same mobility as that of standard taxol were carefully separated on preparative thin layer chromatographic (TLC) plates and were utilized for further analysis.

### 8.1.5 Characterization of Fungal Taxol Using UV, IR Analyses and HPLC Analysis

The fractionated metabolite's UV-vis absorption spectrum was recorded in a Beckman *DU-40 spectrophotometer followed by IR analyses (Shimadzu). Furthermore, the metabolite was analysed by HPLC (Shimadzu 9 A model) using a reverse phase $C_{18}$ column with a UV detector. A total of 20 µl of the metabolite was injected along with methanol, acetonitrile and water (the ratio was 25:35:40) as mobile phase at a flow rate of 1.0 ml $min^{-1}$ and was detected at 232 nm. The sample and the mobile phase were filtered through 0.2 µm PVDF filter before injecting into the column (Pandi et al. 2011; Kumaran and Hur 2009; Kumaran et al. 2010).

## 8.2 Results

Eighteen isolates were recorded from *M. oleifera* leaves collected across various sites in Chennai (Fig. 8.1). Colonies grew well on PDA and good sporulation was observed after 10 days of incubation. The isolate was white coloured mycelium

Fig. 8.1 Plant of *Moringa oleifera*

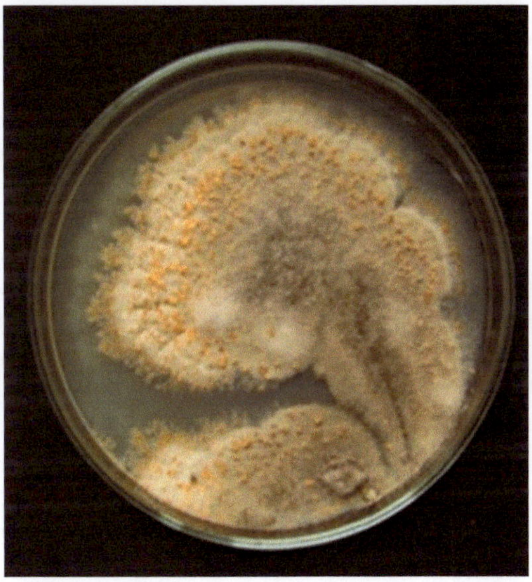

Fig. 8.2 Morphology of *C. gloeosporioides*

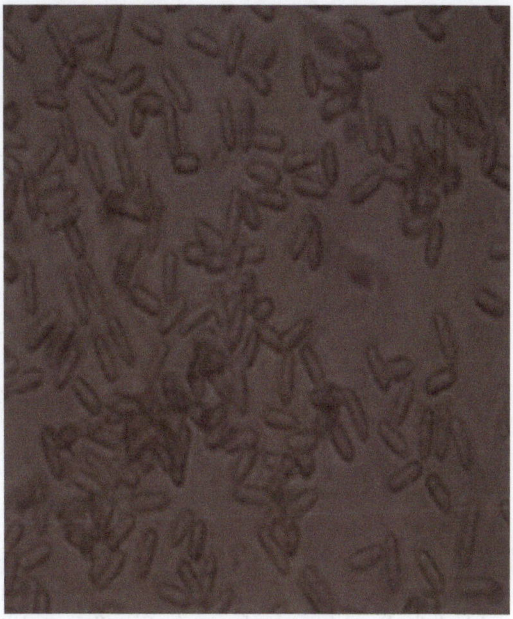

Fig. 8.3 Spore photo *C. gloeosporioides* (40×)

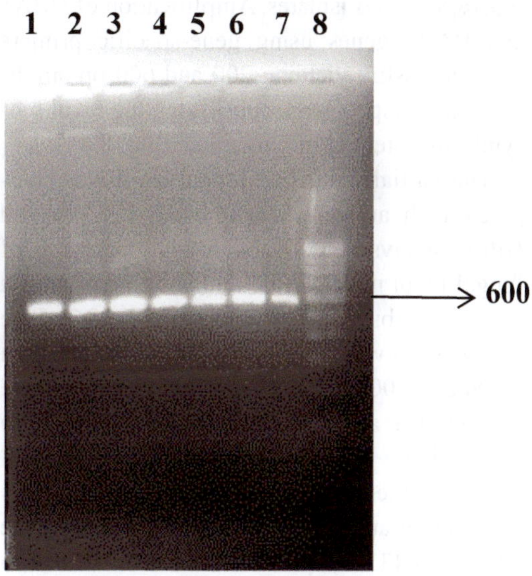


Fig. 8.4 Representative PCR analysis for the presence of ITS region *Lane 2*—PCR product; *Lane 8*—100 bp ladder; 3 μL of sample + 2 μL of loading dye

(Fig. 8.2) which turned grey on sporulation; the spore (Fig. 8.3) as observed under the light microscope (40×) was orange in colour, cylindrical in shape and was identified to be *C. gloeosporioides,* using standard monographs. Amplification of fungal ITS region using universal primers was successful, which was used to establish the molecular identity of our isolate. The amplified fungal (18 S rRNA) ITS region (Fig. 8.4) was sequenced and the product size ranged up to 504 bp. The sequence information was submitted in GenBank database for universal accession (JQ33884). Furthermore, an n-BLAST search for our sequence showed higher homology with ITS sequences of *C. gloeosporioides* isolates from NCBI's GenBank database. A phylogenetic tree

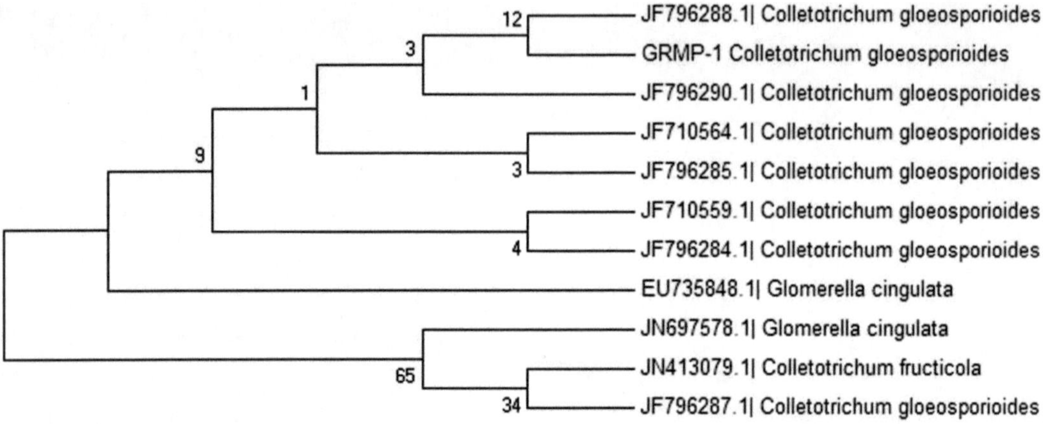


**Fig. 8.5** Phylogenetic tree of GRMP-1 *Colletotrichum gloeosporioides* inferred from the nearest neighbour of fungal endophytes in this study

drawn with neighbour-joining (NJ) algorithm with default settings is shown in Fig. 8.5, and it depicts the placement of our isolate within *Colletotrichum* clade and among earlier reported *C. gloeosporioides* isolates. Amplification of DBAT and BAPT genes using gene-specific primers was successful, yielding 200 and 600 bp amplicons size, respectively, confirming the taxol biosynthesis potential of our isolate (Fig. 8.6).

The partially purified fungal taxol was compared with authentic taxol using UV, IR and HPLC analysis. UV absorption spectrum of fungal taxol matched the authentic taxol with a maximum absorption at 235 nm (Fig. 8.7). The IR spectrum was recorded in the region between 4,000 and 500 $cm^{-1}$ and corroborated the above finding. The appearance of bands in IR spectra convincingly illustrates the identical chemical nature of the extracted taxol from the fungus with that of authentic taxol. A broad peak in the range of 3,431–3,347 $cm^{-1}$ observed was due to hydroxyl (OH) and amide (NH) groups' stretch. The aromatic ring (C=C) stretching frequency was observed in the range of 1,590–1,735 $cm^{-1}$. The peak observed in the range of 1,040–1,120 $cm^{-1}$ was due to the presence of aromatic C, H bends (Fig. 8.8). The fungal taxol analysed using HPLC technique exhibited a peak with similar retention time as that of the authentic taxol (Fig. 8.9).

## 8.3 Discussion

Taxol, or 'paclitaxel' (generic name), is a mitotic inhibitor which has been used in chemotherapy for many types of cancers since 1970s. It is known to be produced by a considerable number of endophytic fungi such as *Taxomyces andreanae, Taxomyces* sp., *Trichoderma, Tubercularia* sp., *Monochaetia* sp., *Fusarium lateritium, Pestalotiopsis microspora, Pestalotiopsis guepinii, Pithomyces* sp., *Pestalotia bicilia, Papulaspora* sp.1, *Pseudomonas aureofaciens, Pleurocytospora taxi, Cephalosporium* spp., *Chaetomium, Martensiomyces* spp., *Mycelia sterilia, Nodulisporium sylviforme, Rhizoctonia* sp., *Penicillium, Alternaria* sp., *Alternaria taxi, Ectostroma* sp., *Botrytis* sp., *Alternaria alternate and Botrytis* taxi (Pandi et al. 2011; Kumaran and Hur 2009; Kumaran et al. 2010; Zhao et al. 2009).

The present study is the first report of taxol-producing *C. gloeosporioides* associated with the leaves of *M. oleifera*. While more recent methods, such as dilution-to-extinction plating techniques (Unterseher and Schnittler 2010) and pyrosequencing-based techniques (Cordier et al. 2012), have started getting wide acceptance, conventional protocols (Strobel et al. 1996) still continue to be used. Our isolate was identified using MSR and phylogenetic species recognition (PSR) concepts (Gazis et al. 2011), as molecular identification has been increasingly used as a

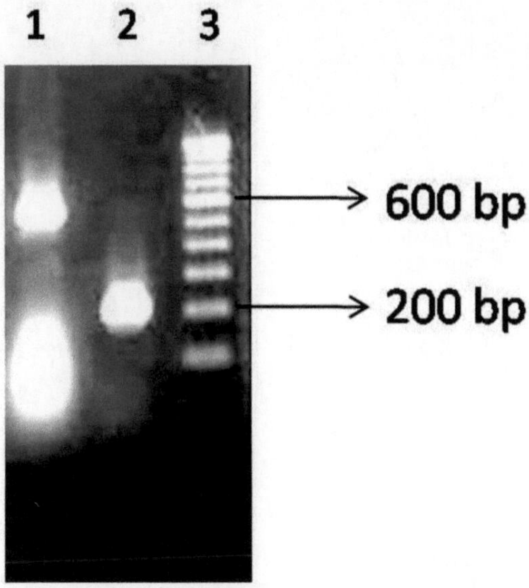


**Fig. 8.6** Representative PCR analysis for the presence of DBAT and BAPT gene. Lane: 1 *bapt* gene and Lane: 2 *dpat* gene Lane 3-100 bp ladder; 3 μL of sample + 2 μL of loading dye

supplementary tool for the traditional systematic classification. ITS region, β-tubulin and cytochrome oxidase II are the most commonly employed molecular markers for phylogenetic analysis of the isolates. ITS region, a non-functional RNA sequence of 500–600 bp length, located in the structural ribosomal RNA is of high taxonomic utility, because of its high copy number, conserved 5.8 S region in the centre and highly variable ITS1 and ITS2 regions on either side (Won and Renner 2005; Promputtha et al. 2007; U'ren et al. 2009; Sim et al. 2010).

Taxol, a diterpenoid of high commercial value, had been reported to be produced from *C. gloeosporioides* isolated from *Justicia gendarussa* (Gangadevi and Muthumary 2008), *Plumeria acutifolia Poiret* (Nithya and Muthumary 2009) and *Salacia chinensis* (Sharma et al. 2011). We employed a two-stage (genetic level and fermentation) screening to evaluate the taxol production potential of our isolate. Two genes *dbat* and *bapt*, part of the taxol biosynthesis pathway, had been screened with gene-specific primer in the present study. The *dbat* gene codes for enzyme that converts DBAT to baccatin III, which is the immediate diterpenoid precursor of taxol, while the *bapt* gene codes for enzymes catalysing the 13-O-acylation of baccatin III with beta phenylalanoyl-CoA as the acyl donor to form *N*-debenzoyl-2′-deoxytaxol; it catalyzes the attachment of an important biological taxol precursor side chain. Recent reports on genetic screening of taxol bio-

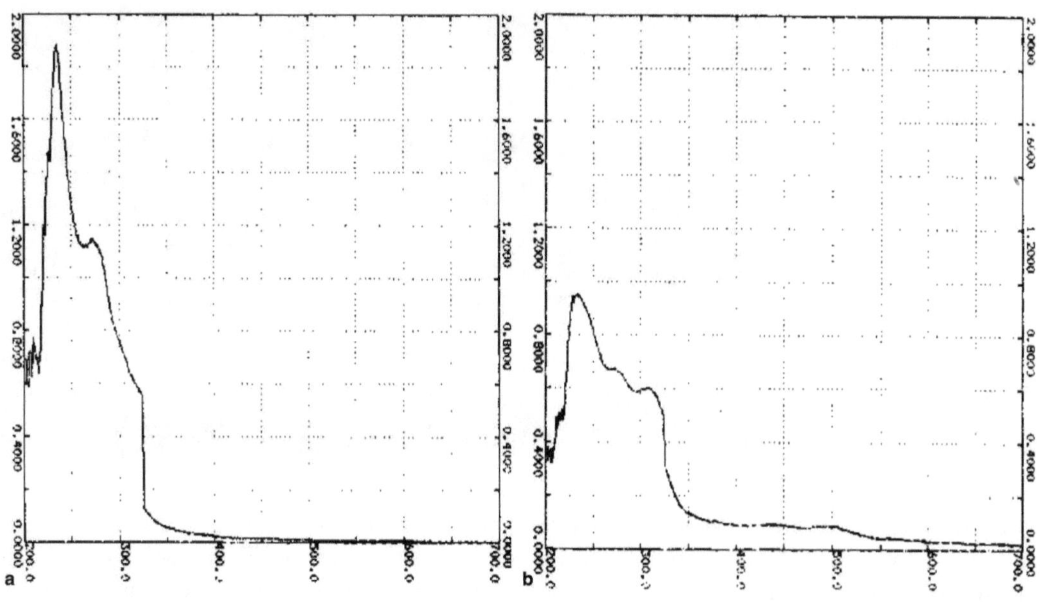


**Fig. 8.7** UV spectroscopic analysis of **a** authentic taxol and **b** fungal taxol

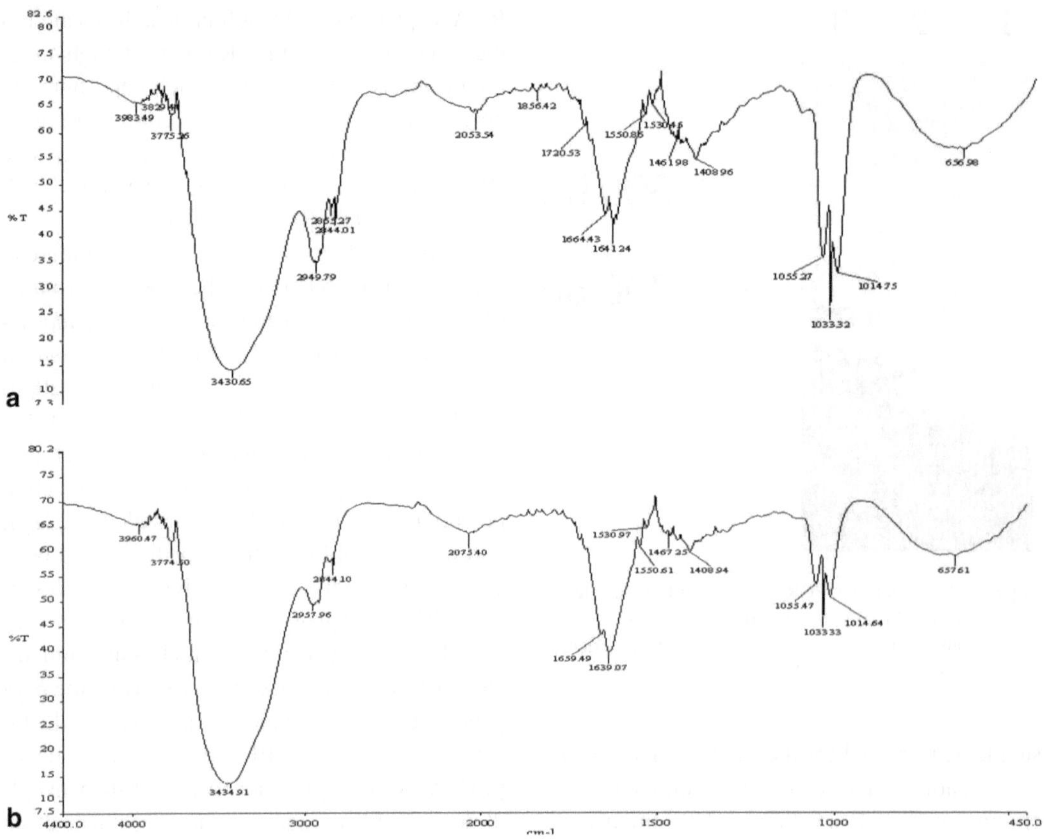


**Fig. 8.8** IR spectroscopic analysis of **a** authentic taxol and **b** fungal taxol

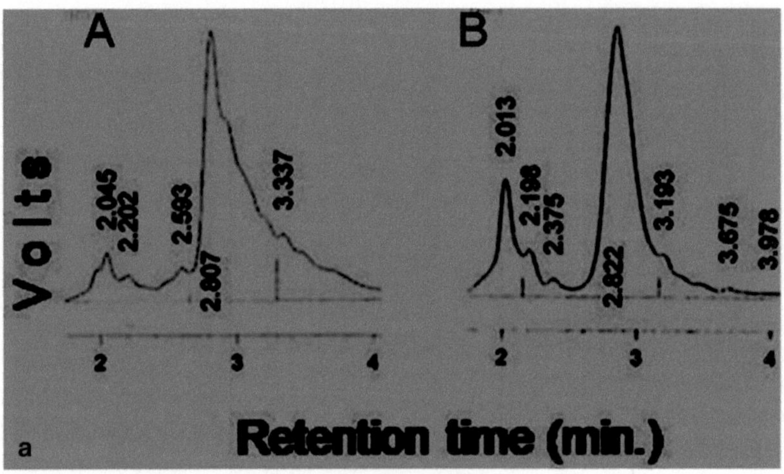


**Fig. 8.9** HPLC analysis of **a** authentic taxol and **b** fungal taxol

synthetic genes have presented scintillating results. It is now evident that the taxol biosynthetic pathways in plants and endophytes are different (Heinig and Jennewein 2009). Nevertheless, of the limitations of our basic understanding of the evolution of taxol biosynthetic genes, some fungi appear to have sequences similar to those of *dbat* and *bapt* genes of *Taxus*, which made it distinctly possible that the species are taxol-producing fungi (Zhang et al. 2008). Our isolate had both the genes; hence, we hypothesised it to have whole gene construct for taxol biosynthesis and probed for production of taxol in M1D broth under optimized conditions.

Techniques like UV, TLC, IR, HPLC, the high resolution $^{1}$H and $^{13}$C NMR and mass spectroscopy (MS) are the tools applied in the confirmation test for the antitumour compound taxol isolated from fungi and are supported by many workers (Strobel et al. 1993, 1996, 1997). Sreekanth et al. (2009) also reported the presence of taxol from *Gliocladium sp*. Kumaran and Hur (2009; Kumaran et al. 2010) also reported the presence of taxol from different endophytic fungi isolated from different plant hosts. Zhao et al. (2009) showed the presence of fungal taxol from *Aspergillus niger* var. On the TLC plate, the metabolite (fungal taxol) had a pinkish colour on spraying vanillin sulphuric acid reagent and subsequently turned grey. Based on the results of UV, IR and HPLC analysis of the partially purified metabolite, we conclude it to be fungal taxol. The wide occurrence of taxol-producing endophytes in hosts of different environments (Zhao et al. 2012) underscores the possibility of important ecological function for this compound. Our future work focuses on elucidating its structure, biosynthetic pathway and enhancing taxol production. The isolation and identification of many endophytic fungi able to produce taxol has created a possibility for a cheaper and more widely available product to be eventually produced via industrial fermentation. The fungus also has potential for genetic engineering to improve taxol production.

**Acknowledgment** The authors thank the University Grants Commission, Delhi for financial support and Madurai Kamaraj University, Madurai, Tamil Nadu, India for providing the laboratory facilities.

## References

Azmat MA, Khan IA, Cheema HMN, Rajwana IA, Khan AS, Khan AA (2012) Extraction of DNA suitable for PCR applications from mature leaves of *Mangifera indica* L. J Zhejiang University-Science B 13(4):239–243

Cordier T, Robin C, Capdevielle X, Desprez-loustau ML, Vacher C (2012) Spatial variability of phyllosphere fungal assemblages: genetic distance predominates over geographic distance in a European beech stand (*Fagus sylvatica*). Fungal Ecol 5:509–520

Culver JB, Hull J, Levy-Lahad E, Daly M, Burke W (1998) Breast cancer genetics: an overview. GeneReviews at GeneTests: Medical Genetics Information Resource (database online). University of Washington, Seattle 2003

Dobranic JK, Johnson JA, Alikhan QR (1995) Isolation of endophytic fungi from eastern larch (*Larix laricina*) leaves from New Brunswick Canada. Can J Microbiol 41:194–198

Gangadevi V, Muthumary J (2008) Isolation of *Colletotrichum gloeosporioides*, a novel endophytic taxol-producing fungus from the leaves of a medicinal plant, *Justicia gendarussa*. Mycol Balcanica 5:1–4

Gazis R, Rehner S, Chaverri P (2011) Species delimitation in fungal endophyte diversity studies and its implications in ecological and biogeographic inferences. Mol Ecol 20:3001–3013

Heinig U, Jennewein S (2009) Taxol: a complex diterpenoid natural product with an evolutionarily obscure origin. Afr J Biotechnol 8 (8):1370–1385

Huang Y, Wang J, Li G, Zheng Z, Su W (2001) Antitumor and antifungal activities in endophytic fungi isolated from pharmaceutical plants *Taxus mairei, Cephalataxus fortunei* and *Torreya grandis*. FEMS Immunol Med Microbiol 31:163–167

Joseph B, Minipriya R (2011) Bioactive compounds from endophytes and their potential in pharmaceutical effect: a review. Am J Biochem Mol Biol 1(3):291–309

Kumaran RS, Muthumary J, Hur BK (2008) Isolation and Identification of Taxol, an Anticancer Drug from Phyllosticta melochiae Yates, an Endophytic Fungus of Melochia *corchorifolia* L. Food Sci Biotechnol 17(6): 1246–1253

Kumaran RS, Hur BK (2009) Screening of species of the endophytic fungus *Phomopsis* for the production of the anticancer drug Taxol. Biotechnol Appl Biochem 54:21–30

Kumaran RS, Hyung JK, Hur BK (2010) Taxol promising fungal endophyte, *Pestalotiopsis* species isolated from *Taxus cuspidate*. J Biosci Bioeng 110:(5) 541–546

Lam M (2003) Beating cancer with natural Medicine. Author House Pub USA 35–48

Maheshwari R (2006) What is an endophytic fungus? Current science 90(10):1309

Nagraj TR (1993) Coelomycetous anamorphs with appendage-bearing conidia. Mycologue publications, Ann Arbor.

Nithya K, Muthumary J (2009) Growth studies of *Colletotrichum gloeosporioides* (Penz.) (Penz.) (Penz.) Sacc.—a taxol producing endophytic fungus from axol producing endophytic fungus from *Plumeria acutifolia*. Ind J Sci Tech 2(11):14–19

Pandi M, Kumaran RS, Choi YK, Kim HJ, Muthumary J (2011) Isolation and detection of taxol, an anticancer drug produced from *Lasiodiplodia theobromae*, an endophytic fungus of the medicinal plant *Morinda citrifolia*. Afr J Biotechnol 10(8):1428–1435

Promputtha I, Lumyong P, Dhanasekaran V, McKenzie EHC, Hyde KD, Jeewon R (2007) A phylogenetic evaluation of whether endophytes become saprotrophs at host senescence. Micro Ecol 53:579–590

Rajangam J, Azahakia Manavalan RS, Thangaraj T, Vijayakumar A, Muthukrishan N (2001) Status of production and utilisation of Moringa in Southern India. In: Fuglie LJ (ed) The miracle tree/the multiple attributes of Moringa. CTA, USA

Sharma KR, Bhagya N, Sheik S, Samhitha M (2011) Isolation of Endophytic *Colletotrichum gloeosporioides* Penz. from *Salacia chinensis* and its Antifungal Sensitivity. J Phytol 3(6):20–22

Sim JH, Khoo CH, Lee LH, Cheah YQ (2010) Molecular diversity of fungal endophytes isolated from *Garcinia mangostana* and *Garcinia parvifolia*. J Microbiol Biotechnol 20(4):651–658

Sreekanth D, Syed A, Sarkar S, Sarkar D, Santhakumari B, Ahmad A, Khan MI (2009) Production, Purification, and Characterization of Taxol and 10-DABIII from a new Endophytic Fungus *Gliocladium sp*. Isolated from the Indian Yew Tree, *Taxus baccata*. J Microbiol Biotechnol 19(11):1342–1347

Stierle A, Strobel GA, Stierle D (1993) Taxol and taxane production by *Taxomyces andreanae*, an endophytic fungus of Pacific Yew. Science 260:214–216

Strobel GA, Stierle A, Stierle D, Hess WM (1993) Taxomyces andreanae a proposed new taxon for a bulbilliferous hyphomycete associated with Pacific yew. Mycotaxon 47:71–78

Strobel GA (2002) Microbial gifts from the rainforest. Can J Phytopathol 24:14–20

Strobel GA, Daisy B (2003) Bioprospecting for microbial endophytes and their natural products. Microbiol Mol Biol Rev 67:491–502

Strobel GA, Xianshu Y, Sears J, Kramer R, Sidhu RS, Hess WM (1996) Taxol from *Pestalotiopsis microspora*, an endophytic fungus of *Taxus wallachiana*. Microbiol. 142:435–440

Strobel, GA, Hess WM, Ford E, Sidhu RS, and Yang X (1996b) Taxol from fungal endophytes and the issue of biodiversity. J Ind Microbiol 17:417–423

Strobel GA, Hess WM, Li JY (1997) *Pestalotiopsis guepinii*, a Taxol producing endophyte of the wollemi pine, *Wollemia robilis*, Aust J Biotechnol 45:1073–1082

Sutton TC (1980) The Coelomycetes—fungi imperfecti with pycnidia, acervuli and stromata. Commonwealth Mycological Institute, Kew

Thurber MD, Fahey JW (2009) Adoption of *Moringa oleifera* to combat under-nutrition viewed through the lens of the "diffusion of innovations" theory. Ecol Food Nutr 48(3):212–225

U'ren JM, Dalling JW, Gallery RE (2009) Diversity and evolutionary origins of fungi associated with seeds of a Neotropical pioneer tree: a case study for analysing fungal environmental samples. Mycol Res 113:432–449

Unterseher M, Schnittler M (2010) Species richness analysis and ITS rDNA phylogeny revealed the majority of cultivable foliar endophytes from beech (*Fagus sylvatica*). Fungal Ecol 3(4):366–378

Won H, Renner SS (2005) The internal transcribed spacer of nuclear ribosomal DNA in the gymnosperm *Gnetum*. Mol Phylogenet Evol 36:581–597

Xing JG, Deng HY, Luo DQ (2011) Two new compounds from an endophytic fungus *Pestalotiopsis heterocornis*. J Asian Nat Prod Res 13(12):1069–1073

Zhang P, Zhou PP, Jiang C, Yu H, Yu LJ (2008) Screening of taxol-producing fungi based on PCR amplification from *Taxus*. Biotechnol Lett 30:2119–2123

Zhao J, Zhou L, Wang J, Shan T, Zhong L, Liu X, Gao X (2012) Endophytic fungi for producing bioactive compounds originally from their host plants. Current research, technology and education topics in applied microbiology and microbial technology. A mendez villas (ed); Formatex Research centre: Badajoz, Spain, 1:567–576

Zhao K, Ping W, Li Q, Hao S, Zhao L, Gao T, Zhou D (2009) *Aspergillus niger* var. taxi, a new species variant of Taxol-producing fungus isolated from *Taxus cuspidate* in China. J Appl Microbiol 107:1202–1207

# Linking Mycorrhizal Technology with Medicinal Plant Secondary Metabolites

9

Richa Raghuwanshi and Shilpam Sinha

**Abstract**

Mycorrhizae are symbiotic associations formed between the roots of most plant species and fungi. Symbiosis is characterized by bidirectional movement of nutrients where carbon flows to the fungus from the plant and inorganic nutrients move to the plant, thereby providing a critical linkage between the plant, root and soil. Symbiosis of arbuscular mycorrhizal fungi (AMF) with higher plants results in growth promotion and accumulation of secondary metabolites like alkaloids, flavonoids and terpenoids, which have pharmacological characteristics. A detailed study revealed that mutualistic symbiosis has great applications in cultivation of medicinal/endangered plants for obtaining a high level of bioactive compounds. Recent work suggest that mycorrhization not only has a positive effect on various plant growth parameters but also enhances the essential oil concentrations, alkaloids, flavanoids and other secondary metabolites in a number of plants from different plant families. Endangered plant species like *Plantago atrata, Pulsatilla slavica* (IUCN Red list) and *Senecio umbrosus* have wide possibilities in improving their *ex-situ* conservation and enhancement of metabolic content through mycorrhizal technology. Mycorrhizal technology may be used in the enhancement of plant's secondary metabolites efficacious as antioxidant, antibacterial, antifungal content and also in conservation of endangered plant species.

**Keywords**

Arbuscular mycorrhizal fungi (AMF) · Bioactive compounds · Mycorrhizal technology · Plant secondary metabolites · Symbiosis

S. Sinha (✉) · R. Raghuwanshi
Department of Botany, Mahila Mahavidyalaya,
Banaras Hindu University, Varanasi 221005, India
e-mail: nuttyshilpam17@gmail.com

R. N. Kharwar et al. (eds.), *Microbial Diversity and Biotechnology in Food Security*,
DOI 10.1007/978-81-322-1801-2_9, © Springer India 2014

## 9.1 Introduction

Plants may be considered as a chemical factory for biosynthesis of a huge array of secondary metabolites. Many of these chemicals are utilized as medicine, scent, dyes and pesticides and are of great commercial importance. Medicinal plants are the richest bio-resources of drugs of traditional medicinal systems, modern medicines, nutraceuticals, food supplements, folk medicines, pharmaceuticals, intermediate and chemical entitled for synthetic drugs (Hammer et al. 1999). These properties are due to the presence of bioactive compounds termed secondary metabolites. The demand for medicinal plants is increasing in both developing and developed countries due to the growing recognition of natural products, being nontoxic, having no side effects and affordable prices.

It has been universally accepted that plant secondary metabolites actively participate in plant–microbe interaction including the plant-symbiotic fungi associations. Enhancement of secondary products accumulation in plants is of great importance in medicinal plants cultivation industry. There is growing concern about diminishing populations, loss of genetic diversity, local extinctions and habitat degradation. Therefore, co-culture system is assumed to be a meaningful and effective tool to biotic elicitation of secondary metabolite production in plants upon symbiotic fungi infection. Symbiotic fungi usually perform compatible and friendly interactions with host plants, which contribute to growth promotion and secondary metabolites accumulation simultaneously, such as alkaloids and terpenoids with pharmacological characteristics (Zhilin Y et al. 2007). The current chapter covers the various aspects of arbuscular mycorrhizal fungus (AMF) inoculation on the levels of secondary metabolites of medicinal plants.

## 9.2 Plant Secondary Metabolites

Secondary metabolites (idiolites) are special metabolites usually possessing bizarre chemical structures and although not essential for the organism's growth in pure culture, they have survival functions in nature (Demain 1986). Every plant at its earlier stage synthesizes primary metabolites—"the building blocks"—which are the basic needs of the plant to grow and develop. These include carbohydrate, protein and nucleotides. The plant starts synthesizing secondary metabolites to defend ifself from abiotic and biotic stress situations at later stage of its life cycle. These are complex substances derived from primary metabolites.

The secondary metabolites are known to play a major role in the adaptation of plants to their environment and also represent an important source of pharmaceuticals (Siahsar et al. 2011). Secondary metabolites are natural products that often have an ecological role in regulating the interactions between plants and their environment. They can be defensive substances, such as phytoalexins and phytoanticipins, anti-feedants, attractants and pheromones (Hanson 2003). It regulates physiology and growth, buffer environmental extremes, deters pest attacks, respond to damage by pests, repairs leaf and fruit injuries, ultraviolet light protection and molecular signal to promote colonization by mycorrhizae and rhizobia (Wink 1999b).

### 9.2.1 Types of Secondary Metabolites and their Effects

Broadly secondary metabolites can be divided into three distinct groups:

1. Nitrogen containing compounds—Terpenoids, Phenols, Flavonoids, Steroids and Saponins
2. No Nitrogen containing compounds—Alkaloids, Amines, Cyanogenic gycosides, Glucosinulates and nonprotein amino acids
3. Sulphur containing compounds—GSH (organic S in the soluble fraction), GSL (low molecular mass N and S containing plant glucosides), Phytoalexins, Thionins, Defensins and Lectins

Plants have enormous ability to synthesize aromatic secondary metabolites, most of which are phenols or their oxygen-substituted derivatives

(Geissman 1963). The important subclasses in this group of compounds include phenols, phenolic acids, quinones, flavones, flavonoids, flavonols, tannins and coumarins. Fragrance of a plant is carried by essential oil fractions which are secondary metabolites and highly enriched in isoprene structure-based compounds. They are called terpenes but when the compound contains an additional element as oxygen they are termed as terpenoids. Essential oils possess strong antimicrobial properties. These groups of compounds show antimicrobial effect and serve as plant defence mechanisms against pathogenic microorganisms. Simple phenols and phenolic acids are bioactive phytochemicals consisting of a single substituted phenolic ring. Quinones are characteristically highly reactive, coloured compounds with two ketone substitutions in aromatic ring. Flavones, flavonoids and flavonols are phenolic structures with one carbonyl group. They are synthesized by plants in response to microbial infection (Dixon et al.1983) and are often found to be effective under in vitro as antimicrobial substance against a wide array of microorganisms (Bennet et al. 1994). Tannins are polymeric phenolic substances possessing the astringent property. These compounds are soluble in water, alcohol and acetone and give precipitates with proteins (Basri and Fan 2005). Tannins are also known to have antimicrobial, antihelminthic, antimutagenic, anti-inflammatory and antioxidant properties (Makkar and Becker 2009). Coumarins are phenolic substances made of fused benzene and pyrene rings (O'Kennedy and Thornes 1997). They have a characteristic odour and several of them have antimicrobial properties. Saponins help in enhancement of immunity, reduction in blood glucose and other antidiabetic effects, and reduction in blood cholesterol (Francis et al. 2002).

## 9.3 Plant Mycorrhizal Interaction

AMF are a unique group of soil fungi that form symbiotic association with the higher plants and facilitate the uptake of diffusion limited plant nutrition, such as P, Zn, Cu, Fe and Mn (Tinker 1984). These fungi show a preferential colonization to the hosts, and thereby, the extent to which a host is benefited depends on the fungal species involved in the symbiosis (Smith and Read 1997).

AMF inoculation can be applied in agriculture since it benefits plants by enhancing nutrient acquisition, increasing resistance to environmental stresses and improving soil quality. Moreover, it can improve yield of medicinal plants by influencing the content of bioactive compounds. Arbuscular mycorrhizas are associations where Glomeromycete fungi produce arbuscules, hyphae and vesicles within root cortex cells. These associations are defined by the presence of arbuscules. AMF colonization of root cortical cells produces several cytological and metabolic changes, such as a marked proliferation of plastids during arbuscule formation (Fester et al. 2001; Hans et al. 2004; Lohse et al. 2005). The activation of plastid biosynthetic pathways and Krebs cycle leads to increased metabolic activity and higher production of amino acids, fatty acids and apocarotenoids, linked to the formation of symbiotic structures (Lohse et al. 2005).

Mycorrhizae have the ability to form a network between the roots of a plant and with the soil around them, which allows the fungus to uptake nutrients from a wider area thereby increasing the surface area of the roots (Sylvia et al. 2005) and also increasing the availability of water for the plants growth. The symbiotic partnership provides benefit to the fungus as well. While aiding the plant in the uptake of nutrients and water, the plant in turn provide 10–20%of the carbon they fix from photosynthesis to the fungus (Allen et al. 2011). Overall, this is a small price for the plant to pay given that the fungus is providing nutrients and water that will allow it to prosper even in stressful environment. Mycorrhization, i.e. inoculation of mycorrhizal fungi into roots of plants should be done at an early stage of plant growth, by layering inoculums below seeds or mixing inoculum into the growth substrate for selected plants.

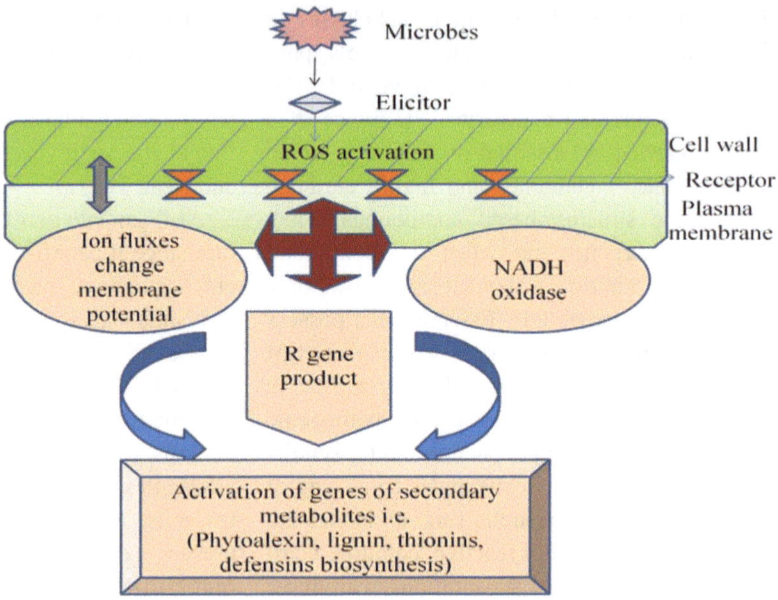


**Fig. 9.1** Elicitation of secondary metabolites in plants by microbes

### 9.3.1 Effect of Mycorrhization on Plant Secondary Metabolites

Medicinal plant contains phytochemical contents which have antioxidant, antibacterial, antifungal, antiviral and anticarcinogenic properties. The synthesis of secondary metabolites is also dependent on plant age, developmental stage (Maffei et al. 1989) and the environmental conditions. Enhancement of secondary metabolites accumulation in medicinal plant is of great importance to the medicinal plant cultivation industry. As the upraised cases of drug resistant microorganism against antibiotics are opening new challenges, the wonderful sources of medicinal herbs could be used to make tablets, injections and syrups to meet the consumer demands in an environment friendly way.

AMF showing symbiotic phenomenon with most of medicinal plants may be exploited as a bioinoculant to improve the essential oil concentration, alkaloids, flavanoids, triterpenoids, phenols and saponins of the medicinal plants (Zhi-lin Y et al. 2007). AM symbiosis influences primary and secondary metabolism of host plants (Schliemann et al. 2008). It induces important changes both in enzymatic activities (i.e. superoxide dismutase and catalase) (Ruiz-Lozano et al. 1996; Marin et al. 2002) and in physiological mechanisms leading to the accumulation of secondary metabolites, such as carotenoids and polyphenols (Walter et al. 2000; Lambais et al. 2003; Fester et al. 2001; Marulanda et al. 2007; Toussaint et al. 2007). Higher levels of reactive oxygen species (ROS) in colonized roots suggest that mycorrhizal plants respond to oxidative stresses by the accumulation of antioxidative enzymes and carotenoids (Fester and Hause 2005).

Plant–microbe interactions occur at phyllosphere (aerial portion), rhizosphere (root portion) and endosphere (internal portion).

The microbes are responsible for activating defence responses in which mycorrhizal fungi contributes in attenuating responses (Garcìa-Garrido and Ocampo 2002). The Fig. 9.1 shows the production of secondary metabolites and has been associated with increased levels of ROS which are the byproducts of abiotic or biotic stress (Chaudhary et al. 2008). Abiotic and biotic stresses cause fluxes between plant primary and secondary metabolism resulting in a diversion of available resources from growth to defence, which can result in excessive production of secondary metabolites (Szakiel et al. 2010). For example, ROS triggered by stress in *Artemisia* are scavenged by the intermediate product dihydroartemisinic acid finally converting into the end product artemisinin (Wallaart et al. 2000). ROS accumulation is reported in AM roots (Fester and Hause 2005). Artemisinin concentration may be considered as a defence response to AMF colonization. Thus, AMF may alter signaling pathways thereby altering the secondary metabolites level in plants.

Studies reveled that AMF may respond to plant's allocation to defence in different ways, i.e. by modulating nutritional status of the host plant. Fungal partners receive up to 19 times more carbohydrates from their hosts than normal leakage of the root system through symbiosis resulting in a strong carbohydrate demand of infected roots and as a consequence, a more efficient plant photosynthesis in return providing more nutrients in the form of soluble minerals like phosphate and others to the plants. Greater nutrient availability could lead to an increase in primary productivity that provides more resources for the plant to use in the biosynthesis of indirect defensive metabolites, such as IGs (Iridoid glycosides) or VOCs (Volatile organic compounds) in *Plantago lanceolata* L. (Fontana et al. 2009). Some secondary metabolites derived from root exudation act as signal molecules, which induce the spore germination and hypha branching in mycorrhizal fungi.

Tremendous researches have revealed the molecular basis and principles of the plant–microbe interactive mechanism (Lugtenberg et al. 2002), which indicates that plant secondary products usually act as signal molecules or respond to pathogen and symbiont colonization. Although the roles of secondary metabolites in plant–pathogen interactions have been well documented (Field et al. 2006; Hahlbrock et al. 2003; Saunders and O'neill 2004; Grayer and Kokubun 2001), only limited information is available from published studies about the significance of host secondary products involved with plant-mutualistic fungi associations.

Just like the well-studied plant–pathogen interaction, phytoalexins accumulation during mycorrhizal infection has also been investigated (Morandi 1996). Raising the possibility that signal perception and transduction proceed via similar pathways between the symbiosis and pathogenesis of plants (Garcia-Garrido and Ocampo 2002; Baron and Zambryski 1995) leads to an increased level of secondary metabolites. However, the defence response in plant-mycorrhizal association is probably weak and transient once the symbiosis becomes established.

Alkaloids are also constitutive defence-related secondary metabolites in plants. These include trigonelline, castanospermine and camptothecin which get enhanced by AMF inoculation (Wei and Wang 1989; Abu-Zeyad et al. 1999; Rojas-Andrade et al. 2003). Using in vitro co-culture system, plant growth effect and dendrobine (pseudoalkaloid or sesquiterpene alkaloid) production was found to be promoted to a certain extent in *Dendrobium* sp. (Chen and Guo 2005).

There has been extensive research devoted to studying the terpenoids metabolism in mycorrhizal infected plants. Isoprenoid metabolism in arbuscular mycorrhizal roots has been reviewed (Strack and Fester 2006). It was shown that some gramineous plant roots accumulated mycorradicin, so-called "yellow pigment" compounds upon mycorrhization. Another category of mycorrhiza-induced secondary metabolite is blumenin. Chemical analysis have identified that they are carotenoid origin of cyclohexenone derivatives (Strack et al. 2003). AMF induced the accumulation of mycorradicin via non-mevalonate methylerythritol phosphate pathway (MEP pathway). cDNA encoding two enzymes central to this pathway, 1-deoxy-D-xylulose 5-phosphate synthase (DXS) and 1-deoxy-D-xylulose 5-phosphate reductoisomerase (DXR) have been cloned from plants. Strong induction of transcript levels of DXS and DXR in mycorrhizal plants has been investigated (Walter et al. 2000).

Saponins, another diverse group of secondary metabolites characterized by their structure containing a steroidal or triterpene aglycone and one or more sugar chains (Hostettmann and Marston 1995; Osbourn 2003; Güçlü-Üstündağ and Mazza 2007; Vincken et al. 2007), occur constitutively in many plants and may be considered as part of the plant defence system (Potter and Kimmerer 1989; Hoagland et al. 1996; Osbourn 1996; Agrell et al. 2003). Saponin content in plants seems to be dynamic, responding to many external factors including various biotic stimuli connected to herbivory attack, pathogenic infection, as well as involved in plant mutualistic symbioses with rhizobial bacteria and mycorrhizal fungi.

Metabolite kinetics during colonization of *Medicago truncatula* with *Glomus intraradices* showed clear mycorrhiza-dependent differences in level of primary and secondary metabolites (Harrison and Dixon 1993). *M. truncatula* is treated as a model legume for studies on symbiosis. Mycorrhization of *M. truncatula* roots resulted in reduction of saponin malonylation, and the difference between saponin content in mycorrhizal and control plants was not quantitative, but qualitative (Schliemann et al. 2008). By contrast, significant increase in the level of triterpene saponin, glycyrrhizin in roots of *Glycyrrhiza uralensis* Fisch inoculated with AMF *Glomus mosseae* and *Glomus versiforme*, was reported.

Mycorrhizal treatment also resulted in significant increase of the content of bioactive constituents including saponins in *Plectranthus amboinicus* (Lour) Spreng, known as Indian borage (also as Cuban oregano, Spanish thyme, Mexican mint), a perennial succulent herb native to South and East Africa but naturalized and widely cultivated on other continents for its leaves with a strong flavour of mixed herbs. Inoculation of *P. amboinicus* roots with seven different indigenous AMF influenced the plant growth and the amount of phenols, alkaloids, tannins and saponins. Application of symbiotic fungi was also reported to enhance the growth and the saponin content of *Bacopa monieri* L., Indian water hyssop, commonly known as Brahmi, an important medicinal plant used in mental disorders treatment and for memory improvement. Reports on potential use of mycorrhiza in enhancing plant secondary metabolites are listed in Table 9.1.

However, inoculating AMF in the plant does not always lead to enhancement of metabolites in the plant. If the outflow of photosynthates to the fungal symbionts is greater than the increase in productivity due to enhanced nutrient supply, there may be a net decrease in carbon supply that could lead to a decline in defence metabolites production, i.e. secondary metabolites (Fontana et al. 2009).

### 9.3.2 Acclimatization of *In Vitro* Propagated Plants Through AMF and Their Secondary Metabolites

Commercial production of high-value phytochemicals by tissue culture and in vitro system of growing plant organs, explants, tissues, cells or protoplasts has a long history. However, acclimatization of in vitro cultured plants and their field transfer has always been a big challenge. AMF are applied into commercial transplant production systems in the greenhouse and are of importance for field growth and survival. The in vitro culture of plant tissue to develop into the whole plant with the help of growth regulator is called micropropagation. Endangered and lower-yield plants are micropropagated, and AMF is introduced during field transfer of these plants since it is beneficial to plant growth in multiple ways. The root colonization with AM improves plant nutritional status, water management and disease resistance (Fortin et al. 2002). The benefit of mycorrhization depends on several factors like the growing medium, plant, mycorrhizal species and the degree of root colonization. AM-stimulated growth of plants is frequently expressed only after acclimatization. Based on reports, it is suggested to develop a co-culture system where the mycorrhizal fungi are introduced in vitro during the rooting stage (Wang et al. 1993) of the plant. In strawberry, mycorrhized plants had a better establishment rate and produced more runners than non-mycorrhized controls (Murphy et al. 1997). In garlic, improved growth was observed after post vitro transplant inoculation (Lubraco et al. 2000). Increased rooting and reduction in weaning stress have been reported in the medicinal plant, *Babtista tinctoria* (L.) R. BR. (Grotkass et al. 2000). AMF-inoculated plantlets exhibited a larger root system, better shoot growth and a higher (more negative) osmotic potential when compared to non-mycorrhizal controls. It is suggested that the enhancement of osmotic potential is important in the pre-adaptation step prior to full acclimation of plantlets for transplanting. In vitro propagation of root colonized by AMF increased sesquiterpenic acid concentrations

**Table 9.1** Reports on enhanced plant secondary metabolites through mycorrhizal inoculation

| Plant name | Family | Secondary metabolite | Mycorrhizal fungi applied | References |
|---|---|---|---|---|
| *Andrographis paniculata* Nees | Acanthaceae | Andrographolite | *Glomus mosseae, Glomus fasciculatum* | Tejavathi et al. 2011 |
| *Anethum graveolens* L. | Apiaceae | Essential oils | *Glomus macrocarpus* | Kapoor et al. 2002a |
| *Arnica montana* L. | Asteraceae | Phenols | *Glomus intraradices* | Jurkiewicz et al. 2010 |
| *Artemisia annua* L. | Asteraceae | Artemisinin | *Glomus mosseae, Glomus aggregatum, Glomus fasiculatum, Glomus intraradices* | Kapoor et al. 2007, Awasthi et al. 2011 |
| *Artemisia annua* L. | Asteraceae | Artemisinin, essential oils | *Glomus macrocarpum, Glomus fasciculatum* | Chaudhary et al. 2008 |
| *Azadirachta indica* A. Juss. | Meliaceae | Azadirachtin | *Glomus mosseae, Glomus fasciculatum* | Venkateswarlu et al. 2008 |
| *Bacopa monieri* L. | Scrophulariaceae | Saponins | *Glomus mosseae, Glomus fasiculatum* | Sowmya et al. 2004 |
| *Begonia malabarica* Lam. | Begoniaceae | Phenols, tannins, flavonoids and alkaloids | *Glomus mosseae* | Selvaraj et al. 2008 |
| *Catharanthus roseus* (L.) G Don | Apocynaceae | Ajmalicine | *Glomus mosseae* | Karthikeyan et al. 2008 |
| *Chlorophytum borivilianum* L. | Liliaceae | Saponins | *Glomus mosseae, Glomus intraradices, Glomus fasiculatum* | Dave et al. 2011 |
| *Coriandrum sativum* L. | Apiaceae | Essential oils | *Glomus macrocarpum* | Kapoor et al. 2002b |
| *Cynara cardunculus* L. | Asteraceae | Phenols | *Glomus intraradices, Glomus mosseae* | Ceccarelli et al. 2010 |
| *Foeniculum vulgare* Mill. | Apiaceae | Essential oils | *Glomus macrocarpum* | Kapoor et al. 2004 |
| *Glycine max* (L.) Merr. | Fabaceae | Flavonoids | *Glomus mosseae* | Xie et al. 1995 |
| *Hypericum perforatum* L. | Hypericaceae | Anthraquinone | *Glomus mosseae, Glomus constrictum, Glomus geosporum, Glomus intraradices* | Zubek et al. 2011 |
| *Medicago sativa* L. | Papilionaceae | Flavonoids | *Glomus intraradix, Glomus mosseae, Gigaspora rosea* | Larose et al. 2002, Catford et al. 2006 |
| *Mentha piperata L. Mentha spicata L., Mentha virdis* (L.) L | Lamiaceae | Essential oils | *Glomus etunicatum* | Karagiannidis et al. 2011, 2012 |
| *Nicotiana tabacum* L. | Solanaceae | Terpenoids | *Glomus intraradices* | Maier et al. 1999 |
| *Ocimum bascilicum* L.var Genovese | Lamiaceae | Essential oils | *Glomus etuonicatum, Glomus intraradices, Glomus fasiculatum* | Copetta et al. 2006, Rasouli-Sadaghiani et al. 2010 |
| *Oreganum onites* L., *Oreganum vulgare* L. | Lamiaceae | Essential oils | *Glomus etunicatum* | Karagiannidis et al. 2011, 2012 |
| *Plectranthus amboinicus* (Lour) Spreng. | Lamiaceae | Phenols, alkaloids, flavonoids, saponins tannins | *Acaulospora birecticulata, Acaulospora scrobiculata, Gigaspora margarita, Glomus aggregatum, Glomus geosporum, Glomus mosseae, Scutellospora heterogama* | Rajesh kumar et al. 2008 |

**Table 9.1** (continued)

| Plant name | Family | Secondary metabolite | Mycorrhizal fungi applied | References |
|---|---|---|---|---|
| *Pogostemon patchouli* Pellet | Lamiaceae | Phenols | *Acaulospora scrobiculata, Gigaspora margarita, Glomus aggregatum, Glomus geosporum, Glomus mosseae, Sclerocystis pakistanika, Scutellospora heterogama* | Selvaraj et al. 2009 |
| *Prosopis laevigata* (Willd.) M. Johnston | Fabaceae | Alkaloids | *Glomus rosea* | Rojas-Andrade et al. 2003 |
| *Salvia officinalis* L. | Lamiaceae | Essential oil, rosmarinic acid, phenols | *Glomus mosseae, Glomus intraradices* | Nell et al. 2009 |
| *Trifolium pratense* L. | Fabaceae | Flavonoids | *Glomus mosseae* | Khaosaad et al. 2008 |
| *Trifolium repens* L. | Fabaceae | Flavonoids | *Glomus intraradices* | Ponce et al. 2004 |
| *Valeriana officinalis* L. | Valerianaceae | Terpenes | *Glomus mosseae, Glomus intraradices* | Nell et al. 2010 |

in *Valeriana officinalis* L. (Nell et al. 2010). Vesicular- Arbuscular Mycorrhiza (VAM) improves establishment of micropropagated *Leucaena leucocephala* plantlets (Puthur et al. 1998). Few studies have been performed to check the effect of AMF on secondary metabolites of micropropagated plants as in *Andrographis paniculata*, used for the treatment of autoimmune diseases (Tejavathi et al. 2011) where the level of andrographolide was found to increase. In case of *Catharanthus roseus*, Vinblastine and Vincristine metabolites increase (Collin 2001), and in case of *V. officinalis* L. used as a sedative, there was an increase in sesquiterpenic acid (Nell et al. 2009). Endangered plant species like *Plantago atrata, Pulsatilla slavica* (IUCN Red list) and *Senecio umbrosus*, getting extinct in wild, also have wide possibility in improving their ex-situ conservation and enhancement of metabolic content through mycorrhizal technology (Zubek et al. 2009). Increasing the production of active phytochemicals constituents is a well-established target for genetic manipulation through biotechnology, but present some severe challenges as the metabolic pathways by which active compounds are biosynthesized are mostly poorly understood, and relatively few genes for key enzymatic or regulatory steps have been isolated.

## 9.4 Conclusion

Plant-symbiotic fungi interactions result in reprogramming the host cell's metabolic state which results in an increase in the secondary metabolite content of the plant as revealed by most of the studies, and this can definitely impel the development of attractive strategies to bring medicinal plants cultivation into a new era for pharmaceutical purpose. Since the production of secondary metabolites within the plants is under diverse physiological, biochemical, metabolic and genetic regulation, which can be optimized by altering the growing conditions, the mycorrhizal technology can be promising in enhancing the secondary metabolite content of medicinal plants and bringing the unproductive land under cultivation with an economic profit.

## References

Abu-Zeyad R, Khana G, Khoo C (1999) Occurrence of arbuscular mycorrhiza in *Castanospermum australe and* effects on growth and production of castanospermine. Mycorrhiza 9:111–117

Agrell J, Oleszek W, Stochmal A, Olsen M, Anderson P (2003) Herbivore-induced responses in alfalfa (*Medicago sativa*). J Chem Ecol 29:303–320

Allen M, Swenson W, Querejeta JI, Egerton-Warburton LM, Treseder KK (2011) Ecology of Mycorrhizae: a conceptual framework for complex interactions among plants and fungi. The Ann Rev Phytopathol 41:271–300

Awasthi A, Bharti N, Nair P, Singh R, Shukla AK, Gupta MM, Darokar MP, Kalra A (2011) Synergistic effect of *Glomus mosseae* and nitrogen fixing *Bacillus subtilis* strain Daz26 on artemisinin content in *Artemisia annua* L. Appl Soil Ecol 49:125–130

Baron C, Zambryski PC (1995) The plant response in pathogenesis, symbiosis, and wounding: variations on a common theme? Ann Rev Genet 29:107–129

Basri DF, Fan SH (2005) The potential of aqueous and acetone extracts of galls of *Quercus infectoria* as antibacterial agents. Ind J Pharmacol 37:26–29

Bennett RN, Wallsgrove RM (1994) Secondary metabolites in plant defense mechanisms. Phytother Res 271:617–633

Catford JG, Staehelin C, Larose G, Pich´e YVierheiligH (2006) Systemically suppressed isoflavonoids and their stimulating effects on nodulation and mycorrhization in alfalfa split-root systems. Pt Soil 285:257–266

Ceccarelli N, Curadi M, Martelloni L, Sbrana C, Picciarelli P, Giovannetti M (2010) Mycorrhizal colonization impacts on phenolic content and antioxidant properties of artichoke leaves and flower heads two years after field transplant. Pt Soil 335:311–323

Chaudhary V, Kapoor R, Bhatnagar AK (2008) Effectiveness of two arbuscular mycorrhizal fungi on concentrations of essential oil and artemisinin in three accessions of *Artemisia annua* L. Appl Soil Ecol 40:174–181

Chen XM, Guo SX (2005) Effects of four species of endophytic fungi on the growth and polysaccharide and alkaloid contents of *Dendrobium nobile*. China J Chin Mater Medica 30:253–257

Collin HA (2001) Secondary product formation in plant tissue cultures. Plant Growth Regul 34:119–134

Copetta A, Lingua G, Berta G (2006) Effects of the three AM fungi on growth, distribution of glandular hairs and essential oil production in *Ocimum basilicum* L. var. Genovese. Mycorrhiza 16:485–494

Dave S, Dasb J, Tarafdarc JC (2011) Effect of vesicular arbuscular mycorrhizae on growth and saponin accumulation in *Chlorophytum borivilianum*. Science Asia 37:165–169

Demain AL (1986) Regulation of secondary metabolism in fungi. Pure Appl Chem 58:219–226

Dixon R, Dey P, Lamb C (1983) Phytoalexins: enzymology and molecular biology. Adv Enzymol 55:1–136

Fester T, Strack D, Hause G (2001) Reorganization of tobacco root plastids during arbuscule developement. Planta 213:864–868

Fester T, Hause G (2005) Accumulation of reactive oxygen species in arbuscular mycorrhizal roots. Mycorrhiza 15:373–379

Field B, Jordan F, Osbourn A (2006) First encounters—deployment of defense-related natural products by plants. New Phytol 172:193–207

Fontana A, Reichelt M, Hempel S, Gershenzon J, Unsicker SB (2009) The effects of arbuscular mycorrhizal fungi on direct and indirect defense metabolites of *Plantago lanceolata* L. J Chem Ecol 35:833–843

Fortin JA, Bécard G, Declerck S, Dalpé Y, St-Arnaud M, Coughlan AP, Piché Y (2002) Arbuscular mycorrhiza on root-organ cultures. Can J Bot 80:1–20

Francis G, Kerem Z, Makkar HPS, Becker K (2002) The biological action of saponins in animal systems–a review. Bri J Nutr 88:587–605

Garcia-Garrido JM, Ocampo JA (2002) Regulation of the plant defense response in arbuscular mycorrhizal symbiosis. J Exp Bot 53:1377–1386

Geissman TA (1963) Flavonoid compounds, tannins, lignins and related compounds. In: Florkin M, Stotz EH (eds) p. 265, Pyrrole pigments, isoprenoid compounds and phenolic plant constituents, vol 9. Elsevier, New York

Grayer RJ, Kokubun T (2001) Plant-fungal interactions: the search for phytoalexins and other antifungal compounds from higher plants. Phytochem 56:253–263

Grotkass C, Hutter I, Feldmann F (2000) Use of arbuscular mycorrhizal fungi to reduce weaning stress of micropropagated *Baptisia tinctoria* (L.) R.Br. Acta Hort 530:305–312

Güçlü-Üstündağ O, Mazza G (2007) Saponins: properties, applications and processing. Crit Rev Food Sci Nutr 47:231–258

Hahlbrock K, Bednarek P, Ciolkowski I, Hamberger B, Heise A, Liedgens H, Logemann E, Nurnberger T, Schmelzer E, Somssich IE, Tan J (2003) Non-self recognition, transcriptional reprogramming, and secondary metabolite accumulation during plant/pathogen interactions. Proc Natl Acad Sci 100:14569–14576

Hammer KA, Carson CF, Riley TV (1999) Antimicrobial activity of essential oils and other plant extracts. J Appl Microbiol 86:985–990

Hans J, Hause BD, Strack MH (2004) Walter Cloning, characterization and immunolocalisation of a mycorrhiza-inducible 1-deoxy-d-xylulose 5-phosphate reductoisomerase in arbuscule-containing cells of maize. Pt Physiol 134:614–624

Hanson JR 2003 Natural products: the secondary metabolites Cambridge: Royal Society of Chemistry; ISBN 0–85404-490-6

Harrison MJ, Dixon RA (1993) Isoflavonoid accumulation and expression of defense gene transcripts during the establishment of vesicular-arbuscular mycorrhizal associations in roots of *Medicago truncatula*. Mol Pt Microbe Interact 6:643–654

Hoagland RA, Zablotowicz RM, Renny KN (1996) Studies on the phytotoxicity of saponins on weed and crop plants. Adv Exp Med Biol 405:57–73

Hostettmann KA, Marston A (1995) Chemistry and pharmacology of natural products Saponins Cambridge University Press. Cambridge, United Kingdom, pp 1–2

Jurkiewicz A, Ryszka P, Anielska T, Waligorski P, Bialonska D, Goralska K, Tsimilli-Michael M, Turnau K (2010) Optimization of culture conditions of *Arnica Montana* L.: effects of mycorrhizal fungi & competing plants. Mycorrhiza 20:293–306

Kapoor R, Giri B, Mukerij KG (2002a) *Glomus macrocarpus*: a potential bioinoculant to improve essential oil quality and concentration in Dill (*Anethum graveolens L.* ) and Carum (*Trachyspermum ammi* (Linn.) Sprague). World J Microbiol Biotechnol 18:459–463

Kapoor R, Giri B, Mukerji KG (2002b) Mycorrhization of coriander (*Coriandrum sativum* L.) to enhance the concentration and quality of essential oil. J Sci Food Agricult 82:339–342

Kapoor R, Giri B, Mukerji KG (2004) Improved growth and essential oil yield and quality in *Foeniculum vulgare* mill on mycorrhizal inoculation supplemented with P-fertilizer. Bioresour Technol 93:307–311

Kapoor R, Chaudhary V, Bhatnagar AK (2007) Effect of arbuscular mycorrhiza and phosphorus application on artemisinin concentration in *Artemisia annua* L. Mycorrhiza 17:581–587

Karagiannidis N, Thomidis T, Lazari D, Panou-Filotheou E, Karagiannidou C (2011) Effect of three Greek arbuscular mycorrhizal fungi in improving the growth, nutrient concentration, and production of essential oils of oregano and mint plants. Sci Horticult 129:329–334

Karagiannidis N, Thomidis T, Lazari D, Panou-Filotheou E, Karagiannidou C (2012) Response of three mint and two oregano species to *Glomus etunicatum* inoculation. Aust J Crop Sci 6:164–169

Karthikeyan B, Jaleel CA, Changxing Z, Joe MM, Srimannarayan J, Deiveekasundaram M (2008) The effect of AM fungi and phosphorous level on the biomass yield and ajmalicine production in *Catharanthus roseus*. Eur Asia J BioSci 2:26–33

Khaosaad T, Krenn L, Medjakovic S, Ranner A, Lossl A, Nell M (2008) Effect of mycorrhization on the isoflavone concentration and the phytoestrogen activity of red clover. J Plant Physiol 165:1161–1167

Lambais MR, Rios-Ruiz WF, Andrade RM (2003) Antioxidant responses in bean (*Phaseolus vulgaris*) roots colonized by arbuscular mycorrhizal fungi. New Phytol 160:421–428

Larose G, Chênevert R, Moutoglis P, Gagné S, Piché Y, Vierheilig H (2002) Flavonoid levels in roots of *Medicago sativa* are modulated by the developmental stage of the symbiosis and the root colonizing arbuscular mycorrhizal fungus. J Plant Physiol 159:1329–1339

Lohse S, Schliemann W, Ammer C, Kopka J, Strack D, Fester T (2005) Organisation and metabolism of plastids and mitochondria in arbuscular mycorrhizal roots of *Medicago truncatula*. Pt Physiol 139:329–340
Lubraco G, Schubert A, Previati A (2000) Micropropagation and mycorrhization of *Allium sativum*. Acta Hort 530:339–344
Lugtenberg BJ, Chin-A-Woeng TF, Bloemberg GV (2002) Microbe–plant interactions: principles and mechanisms. Anton van Leeuwen 81:373–383
Maffei M, Chialva F, Sacco T (1989) Glandular trichomes and essential oils in developing peppermint leaves. New Phytol 111:707–716
Maier W, Schmidt J, Wray V, Walter M, Strack D (1999) The arbuscular mycorrhizal fungus *Glomus intraradices* induces the accumulation of cyclohexenone derivatives in tobacco roots. Planta 207:620–623
Makkar HPS, Becker K (2009) The biological action of saponins in animal systems – a review. British J Nutr 88:587–605
Marin M, Ybarra M, Garcia-Ferriz F, Garcia-Ferriz L (2002) Effect of arbuscular mycorrhizal fungi and pesticides on *Cynara cardunculus* growth. Agr Food Sci 11:245–251
Marulanda A, Porcel R, Barea JM, Azcon R (2007) Drought tolerance and antioxidant activities in lavander plants colonized by native drought-tolerant of drought-sensitive *Glomus* species. Microb Ecol 54:543–552
Morandi D (1996) Occurrence of phytoalexins and phenolic compounds in endomycorrhizal interactions and their potential role in biological control. Pt Soil 185:241–251
Murphy J, Mark L, Periappuram C, Walsh C, Cassells AC (1997) Microbial characterization and preparation of inoculum for *in vitro* mycorrhization of strawberry in autotrophic culture. In: Pathogen and microbial contamination management in micropropagation. Cassells AL (ed) Kluwer Acad Pub., Dordrecht, pp 345–350
Nell M, Votsch M, Vierheilig H, Steinkellner S, Zitterl-Eglseer K, Franz C, Novak J (2009) Effect of phosphorus uptake on growth and secondary metabolites of garden sage (*Salvia officinalis* L.). J Sci Food Agric 89:1090–1096
Nell M, Wawrosch C, Steinkellner S, Vierheilig H, Kopp B, Lossi A, Franz C, Novak J, Zitterl Eglseer K (2010) Root colonization by symbiotic Arbuscular Mycorrhizal fungi increases sesquiterpenic acid concentrations in *Valeriana officinalis* L. Planta Med 76:393–398
O'Kennedy R, Thornes RD (1997) Coumarins: biology, applications and mode of action. John Wiley & Sons, Inc., New York
Osbourn A (1996) Saponins and plant defense—a soap story. Trends in. Plant Sci 1:4–9
Osbourn AE (2003) Saponins in cereals. Phytochem 62:1–4
Ponce MA, Scervino JM, Erra Balsells R, Ocampo JA, Godeas A (2004) Flavonoids from shoots and roots of *Trifolium Repens* (white clover) grown in presence or absence of the arbuscular mycorrhizal fungus *Glomus intraradices*. Phytochem 65:1925–1930
Potter DA, Kimmerer TW (1989) Inhibition of herbivory on young holly leaves: evidence for the defensive role of saponins. Oecologia 78:322–329
Puthur JT, Prasad KVSK, Sharmila P, Saradhi PP (1998) Vesicular arbuscular mycorrhizal fungi improves establishment of micropropagated *Leucaena leucocephala* plantlets. P Cell. Tiss Org Cul 53:41–47
Rajesh Kumar S, Nisha MC, Selvaraj T (2008) Variability in growth, nutrition and phytochemical constituents of *Plectranthus amboinicus* (Lour) Spreng. as influenced by indigenous arbuscular mycorrhizal fungi. Mj Int J Sci Tech 2:431–439
Rasouli-Sadaghiani MH, Hassani A, Barin M, Danesh YR, Sefidkon F (2010) Effects of arbuscular mycorrhizal (AM) fungi on growth, essential oil production and nutrients uptake in basil. J Med Pts Res 4:2222–2228
Rojas-Andrade R, Cerda-Garcia-Rojas CM, Frias-Hernandez JT, Dendooven L, Olalde-Portugal V, Ramos-Valdivia AC (2003) Changes in the concentration of trigonelline in a semi-arid leguminous plant (*Prosopis laevigata*) induced by an arbuscular mycorrhizal fungus during the presymbiotic phase. Mycorrhiza 13:49–52
Ruiz-Lozano JM, Azcon R, Palma JM (1996) Superoxide dismutase activity in arbuscular mycorrizal *Lactuca sativa* plants subjected to drought stress. New Phytol 134:327–333
Saunders JA, O'neill NR (2004) The characterization of defense response to fungal infection in alfalfa. Bio Control 49:715–728
Schliemann W, Ammer C, Strack D (2008) Metabolite profiling of mycorrhizal roots of *Medicago truncatula*. Phytochem 69:112–146
Selvaraj T, Rajeshkumar S, Nisha MC, Wondimu L, Tesso M (2008) Effect of *Glomus mosseae* and plant growth promoting rhizomicroorganisms (PGPR's) on growth, nutrients and content of secondary metabolites in *Begonia malabarica* Lam. Mj Int J Sci Tech 2:516–525
Selvaraj T, Nisha MC, Rajesh Kumar S (2009) Effect of indigenous arbuscular mycorrhizal fungi on some growth parameters and phytochemical constituents of *Pogostemon patchouli* Pellet. Maejo Int J Sci Technol 3:222–234
Siahsar B, Raissi AS, Tavassoli A, Rahimi M (2011) Pattern of gene and enzyme in secondary pathways of medicinal plants. J Med Plants Res 5:5953–5957
Smith SE, Read DJ (1997) Mycorrhizal symbiosis. 2nd ed. Academic Press, London, 605 pp
Sowmya R, Tejavathi DH, Sukuda M (2004) Utilization of VA mycorrhizal fungi and *Trichoderma viride* on plant growth and drug content of micropropagated *Bacopa monnieri*(L.) Pennell. Ecoport article online publication, FAO, USA. http:// ecoport.org/ ep? Search Type =earticle View and Earticled =145 and page = -2 # section 1900

Strack D, Fester T, Hause B, Schliemann W, Walter MH (2003) Arbuscular mycorrhiza, biological, chemical and molecular aspects. J Chem Ecol 29:1955–1979

Strack D, Fester T (2006) Isoprenoid metabolism and plastid reorganization in arbuscular mycorrhizal roots. New Phytol 172:22–34

Sylvia D, Fuhrmann J, Hartel P, Zuberer D (2005) Principles and applications of soil microbiology. Pearson, Upper Saddle River

Szakiel A, Paczkowski C, Henry M (2010) Influence of environmental biotic factors on the content of saponins in plants. Phytochem Rev 10:493–502

Tejavathi DH, Anitha P, Murthy SM, Nijagunaiah R (2011) Effect of AM fungal association with normal and micropropagated plants of *Andrographis paniculata* Nees on biomass, primary and secondary metabolites. Int Res J Pt Sci 2:338–348

Tinker PB (1984) The role of microorganisms in mediating and facilitating the uptake of plant nutrient from soil. Pt Soil 76:77–91

Toussaint JP, Smith FA, Smith SE (2007) Arbuscular mycorrhizal fungi can induce the production of phytochemicals in sweet basil irrespective of phosphorus nutrition. Mycorrhiza 17:291–297

Venkateswarlu B, Pirat M, Kishore N, Rasul A (2008) Mycorrhizal inoculation in neem (*Azadirachta indica*) enhances azadirachtin content in seed kernels. World J Microbiol Biotechnol 24:1243–1247

Vincken JP, Heng L, de Groot A, Gruppen H (2007) Saponins, classification and occurrence in the plant kingdom. Phytochem 68:275–297

Walter MH, Fester T, Strack D (2000) Arbuscular mycorrhizal fungi induce the non-mevalonate methylerythritol phosphate pathway of isoprenoid biosynthesis correlated with accumulation of the 'yellow pigment' and other apocarotenoids. Plant J 21:571–578

Wallaart TE, Pras N, Beekman AC, Quax WJ (2000) Seasonal variation of artemisinin and its biosynthetic precursors in plants of *Artemisia annua* of different geographical origin: proof for the existence of chemotypes. Planta Med 66:57–62

Wang H, Parent S, Gosselin A, Desjardins Y (1993) Vesicular arbuscular mycorrhizal peat- based substrates enhances symbiosis establishment and growth of three micropropagated species. J Am Soc Hort Sci 118:896–901

Wei GT, Wang HG (1989) Effects of VA mycorrhizal fungi on growth, nutrient uptake and effective compounds in Chinese medicinal herb *Datura stramonium* L. Sci Agri Sinica 22:56–61

Wink M (1999b) Function of plant secondary metabolites and their exploitation in biotechnology. Sheffield Academic Press and CRC Press. Ann Plant Rev 3:362

Xie ZP, Staehelin C, Vierheilig H, Wiemken A, Jabbouri S, Broughton WJ, Vogeli-Lange R, Boller T (1995) Rhizobial nodulation factors stimulate mycorrhizal colonization of nodulating and non-nodulating soybeans. Plant Physiol 108:1519–1525

Zubek S, Turnau K, Tsimilli-Michael M, Strasser RJ (2009) Response of endangered plant species to inoculation with arbuscular mycorrhizal fungi and soil bacteria. Mycorrhiza 19:113–123

Zubek S, Mielcarek S, Turnau K (2011) Hypericin and Pseudohypericin concentrations of a valuable medicinal plant *Hypericum perforatum* L. are enhanced by arbuscular mycorrhizal fungi. Mycorrhiza 22:149–156

Zhi-lin Y, Chuan-chao D, Lian-qing C (2007) Regulation and accumulation of secondary metabolites in plant-fungus symbiotic system. Afr J Biotech 6:1266–1271

# Ecology of Arbuscular Mycorrhizal Fungi

**10**

D. J. Bagyaraj

**Abstract**

It is a well-established fact that arbuscular mycorrhizal (AM) fungi improve plant growth. The main effect of AM fungi in improving plant growth is through improved uptake of nutrients, especially phosphorus, which is due to exploration by the external hyphae of the soil beyond the root hair and phosphorus depletion zone. Fungal hyphae are also known to absorb phosphorus from lower concentrations compared to non-mycorrhizal roots. They also improve the uptake of minor elements like Zn, Cu, etc., and water. They also produce plant hormones, increase the activity of beneficial soil organisms in the root zone and reduce the severity of disease caused by root pathogens. Thus the benefits the plant derives from mycorrhizal inoculation seem to be enormous.

The ecology of these fungi, in tropics, is not fully understood. These fungi are geographically ubiquitous. An explanation for their remarkably wide spread distribution is that these fungi were disseminated inter-continentally prior to continental drift. Further, these fungi can grow actively, to a limited extent, spending their own energy. Passive dissemination can occur through biotic agents like earthworms, ants, wasps, etc. and abiotic agents like wind, water, etc. The various agricultural practices are known to influence the occurrence of these fungi qualitatively and quantitatively. The approaches in understanding the occurrence, dissemination, survival and persistence of these fungi in tropics is discussed.

**Keywords**

Arbuscular mycorrhizal (AM) fungi · Diversity · Ecology · Plant growth promoting rhizomicroorganisms

D. J. Bagyaraj (✉)
Center for Natural Biological Resources and Community Development (CNBRCD), 41, RBI Colony, Anand Nagar, 560 024 Bangalore, India
e-mail: djbagyaraj@gmail.com

R. N. Kharwar et al. (eds.), *Microbial Diversity and Biotechnology in Food Security*,
DOI 10.1007/978-81-322-1801-2_10, © Springer India 2014

## 10.1 Introduction

Plant roots provide an ecological niche for many of the microorganisms that abound in soil. In (Frank 1885), German Botanist Albert Bernard Frank introduced the Greek word *mycorrhiza,* which literally means "fungus root", to scientific terminology. In natural ecosystems much of the root system can be colonized by mycorrhizal fungi. Colonization is restricted to the root cortex and does not enter the vascular cylinder. The symbiosis is so well balanced that, although many of the host cells are invaded by the fungal endophyte, there is no visible tissue damage, and under certain conditions it enhances the growth and vigor of the host plant. These associations are grouped, based on morphological and anatomical characters, as ectomycorrhizae and endomycorrhizae. Endomycorrhiza include arbutoid, monotropoid, ericoid, orchid and arbuscular mycorrhizal (AM) forms. Arbuscular mycorrhizae are the most common and widely occurring of all the mycorrhizal associations and have great economic significance. They cannot be cultured on laboratory media. As most economically important plants form AM, the subject is currently attracting much attention in agricultural, horticultural, and forestry research.

AM fungi are said to establish a mutualistic relationship with 90% of vascular plants (Trappe 1977; Gianinazzi and Gianinazzi-Pearson 1986; Wang and Qiu 2006, Smith and Read 2008). Plants that rarely form AM fungal association include members of Caryophyllaceae, Brassicaceae, Chenopodiaceae and Cyperaceae (Hirrel et al. 1978). In addition to their widespread distribution throughout the plant kingdom, AM fungi are ubiquitous and occur in plants grown in arctic, temperate and tropical regions (Mosse et al. 1981; Allen et al. 2003; Bagyaraj 2011). They have been reported to be associated with plants grown in sand dunes, coal mines (Khan 1978) and aquatic environments (Bagyaraj et al. 1979). Blaszkowski (1994) observed variations in AM fungal diversity with the changes in plant species. Plants of a particular family are colonized by specific types of AM fungi, and a few AM fungal genera were found only in the plants of a particular family. AM fungi have the widest host range and distribution of all the mycorrhizal associations. AM fungi have been observed in 1,000 genera of plants representing some 200 families. There are at least 300,000 receptive hosts in the world flora, and there are about 220 species of AM fungi. If the hosts are divided up evenly among the fungi, with no overlap in host range, each fungus would have more than 1,360 potential partners. We know that the host range overlaps extensively, suggesting that some individual AM fungi may well have access to thousands of host (Gianinazzi et al. 2010; Bagyaraj 2011).

## 10.2 Arbuscular Mycorrhizal Fungi: Classification, Importance and Occurrence

AM fungi are obligate symbionts and cannot be cultured on synthetic media. Their penetration takes place through root hairs or epidermal cells and then grows intercellularly or intracellularly in the root cortex, ultimately developing short haustoria like structures called arbuscules within the cortical cells. These arbuscules function as sites of nutrient exchange between the fungus and host roots. Vesicles are formed in the cortical cells, which are thin walled structures of various sizes and shapes and function as storage organs. The presence of vesicles and arbuscules is the criteria for identifying AM fungus in the roots.

AM fungi belong to the phylum Glomeromycota, which has three classes (Glomeromycetes, Archaeosporomycetes and Paraglomeromycetes) with five orders (Glomerales, Diversisporales, Gigasporales, Paraglomerales and Archaeosporales), 14 families and 26 genera (Sturmer 2012). The commonly occurring genera of AM fungi are *Glomus, Gigaspora, Scutellospora, Acaulospora* and *Entrophospora*.

Improved plant growth due to inoculation of soil with AM fungi has been demonstrated especially under phosphorus (P) deficient conditions (Mosse 1977; Hodge et al. 2010). The growth improvement is mainly because of enhanced P uptake. AM fungi can also enhance tolerance or resistance to root pathogens (Borowicz 2001) and abiotic stresses such as drought and metal toxicity (Meharg and Cairney 2000). AM fungi play a role in the formation of stable soil aggregates, build up a macroporous structure of soil that allows penetration of water and air and prevents erosion (Miller and Jastrow 1992). There is well documented evidence that AM fungi have important effects on plant P uptake. Greater soil exploration by mycorrhizal roots as a means of increasing phosphate uptake is well established. In phosphate deficient soils immobile phosphate ions develop a phosphate depletion zone around the roots. The hyphae spread beyond this zone and directly translocate nutrients from the soil to the root cortex (Hayman 1983). Experiments with $^{32}P$ labeled phosphate indicate that AM fungal hyphae obtain their extra phosphate from the labile pool rather than by accessing insoluble phosphate by solubilizing it (Raj et al. 1981). Sparingly soluble rock phosphate is better utilized by the hyphae by closer physical contact with the ions dissociating at the particle surface (Bagyaraj 1991).

The increased growth of plants inoculated with AM fungi is not only attributed to improved phosphate uptake but also to better availability of other elements like Zn, Cu, K, Al, Mn, Fe etc. AM fungi affect the levels of plant hormones. Allen (1991) measured levels of plant hormones like cytokinins- and gibberellin-like substances. AM fungi can tolerate a wide range of soil water regimes and also improve water relationships of many plants. It is still unclear whether the observed effects are directly due to the fungus itself or indirectly due to some alteration in host physiology as a result of improved P nutrition. Anatomical and other physiological studies have brought out that mycorrhizal plants have increased rates of respiration, photosynthesis and increased amounts of sugars, amino acids, RNA etc. and larger and/or more number of chloroplasts, mitochondria, xylem vessels, motor cells etc. Changes in the root exudations and altered rhizosphere microorganisms (which also affect plant growth) may result because of colonization of roots by AM fungi (Machado and Bagyaraj 1995).

Mycorrhizal colonization may also allow introduced populations of beneficial soil organisms like *Azotobacter, Azospirillum* and phosphate solubilizing bacteria to be maintained in high numbers than around non-mycorrhizal plants and to exert synergistic effects on plant growth. It is apparent from the investigations on AM fungi–plant pathogen interaction that AM fungi can usually (though not always) deter or reduce the severity of disease caused by soil-borne pathogens. All these studies bring out that AM fungi help the host plant in more than one way and that AM fungal inoculation helps plants growth (Hodge 2000).

AM fungi, in addition to their widespread distribution throughout the plant kingdom, are also geographically ubiquitous and occur in plants growing in arctic, temperate and tropical regions (Mosse 1981). As an explanation for their remarkably widespread distribution, Trappe (1977) proposed that AM fungi were disseminated intercontinentally prior to the continental drift. The super continent Gondwanaland is thought to have begun to break apart and drift north about 125 million years ago. In general, AM fungal population is more in cultivated soil, and their numbers decrease markedly below the top 15 cm (Redhead 1977). They are normally not found in depths beyond the normal root range of plants (Mosse 1981). Although AM fungi are ubiquitous in soils, the patterns or distribution of individual species have not been fully understood. Studies on the distribution of species have either covered large geographical areas (Hall 1977) or smaller regions (Abbott and Robson 1977). The distribution of species of AM fungi varies with climatic and edaphic environment as well as with land use. For example, *Acaulospora laevis* is common in western Australia (Abbott and Robson 1977) but occurs less frequently in soils of eastern Australia while *Glomus* spp. appears to have the widest distribution. *Gigaspora* and *Sclerocystis* spp. are more common in tropical soils. *Acaulospora* seems to be better adapted to soils with $pH < 5.0$. Infact, certain AM fungi have been linked to particular kind of soil: *Glomus mosseae* with fine textured, fertile high pH soils; *A. laevis* with coarse textured, acid soils; and *Gigaspora* species with sand dune soils (Kendrick and Berch 1985; Bagyaraj 1991).

## 10.3 Dissemination

AM fungi are indigenous to soil throughout the world. In fact, many species are represented on most continents. As an explanation for their remarkably wide distribution, Trappe (1977) proposed that AM fungi were disseminated intercontinentally prior to continental drift. Majority of land plants form mycorrhizal associations, many plants colonizing disturbed sites are non-mycorrhizal. The AM fungi on these distributed sites are dependent upon a variety of dispersal agents. Dispersal of AM fungi is usually by spread from one living root to another through mycorrhizal propagules including mycelia and spores which can be moved by biotic and abiotic agents. Dispersal of spores and other propagules over greater distances are dependent upon wind and water that are probably important dispersing agents, especially in arid environments. Animal dispersal of AM spores is well documented in many cases and can occur through ingestion of spores (Bagyaraj and Ravindra 1997). These can be divided into two groups:

1. Active dissemination (growth of mycelium through soil)
2. Passive dissemination, where AM fungi are moved by wind, water and animal vectors

### 10.3.1 Active Dissemination

AM fungi may be disseminated in a variety of ways. Active dispersal occurs as mycelia grow through soil, although it is effective over a limited range. Infective mycelia can be dispersed in one of the three general forms: (a) infected plant parts, (b) mycelial fragments and (c) mycelial extension between plants.

The first two of these could be mediated by either animal or abiotic factors. The third requires only suitable medium and a short enough plant-to-plant distance. Powell (1979) determined that an efficient mycorrhizal fungus would advance 65 m in 150 years or 0.43 m per year under green house conditions. Powell (1979) demonstrated that AM fungal species differed in their rate of spread and in their ability to retain possession of colonized parts by other AM fungal species. It has been shown that plant species and root density may significantly influence the rate of AM fungus spread (Warner and Mosse 1982). In clover, the rate of spread of *Glomus fasciculatum* was 1 cm per week, whereas in fescue *G. fasciculatum* spread at a rate of only 0.7 cm per week, and it was concluded that root density and plant species significantly influence the rate of AM fungus spread. In fact supra-optimal root density was achieved in fescue (a grass with an extensive root system), and rate of fungal spread was reduced as size of the plant increased. Similarly, supra-optimal root density was not achieved in clover because of less extensive root system.

The above experiments were conducted in fumigated soil in a green house, as it is difficult to project the rate of mycelial growth and spread of AM fungus through field soils. Mosse et al. (1982) demonstrated that *Glomus caledonicum* was able to spread 7–13 cm from an inoculation point after 13 weeks. No correlation was observed between rate and plant size but the spread rate was greater in non-sterilized plots than in those receiving formalin treatments. Higher rates of fungal spread (1.5–3.4 m per year) have been reported for certain soil-borne plant pathogenic fungi in non-sterile soils (Wallace 1978), although factors such as soil fertility, seasonal fluctuations in moisture, temperature and microbial activity influence the rate of spread of AM fungi.

The active spread of AM hyphae from mycorrhizal *Leucaena* plant through a compartment of soil with no roots and then to colonize uninoculated plants was studied by Harinikumar and Bagyaraj (1995). Core samples taken from different distance at periodical intervals showed that the hyphae from the mycorrhizal plant travelled through soil (no root zone) to a distance of 300 mm in 180 days. The presence of AM colonization in the root system of uninoculated plants after 180 days was an evidence of active spread of AM fungi. AM hyphae travelled a distance of 300 mm in 180 days. It was deduced that AM hyphae can spread actively through soil in the absence of roots at the rate of 1.66 mm per day. Powell's (1979) experiments which are more comparable to the present experiment brought out that AM fungal hyphae spread at the rate of 5.6 mm per day in the presence of root to root contact.

Whether AM fungi grow in a directed way, i.e. towards a root stimulus or randomly in soil has been debated. Directed growth would most likely make optimum use of energy supplies in the spore and would increase the number of infective hyphae or strands which reach a host (Wallace 1978). Powell (1976) demonstrated buried slide technique in partially sterilized soil and showed that there was no attraction of AM hyphae to root until random contact occurred, except with hyphae from honey coloured spores (*A. laevis*) which frequently grew towards the roots. Chemotactic attraction of hyphae of *Gigaspora margarita* to host roots *in vitro* was demonstrated by Koske (1981). Since hyphae would pass through the air to reach the host roots suspended above germinating spores, the attractant is probably a volatile substance. Whether such chemotactic substances are produced under field conditions and can direct mycelial growth in the field has not been studied.

### 10.3.2 Passive Dissemination

Many soil borne fungi have developed highly specialized methods of dispersal. This is particularly true of the hypogeous fungi which fruit in sporocarps below the ground. Mature spores of these hypogeous fungi frequently emit an odour which, by becoming increasingly strong, attracts rodents. These rodents eat the sporocarps, digest the peridium or glebal mycelial constituents and defecate the spores which remain intact. In addition, adaption to dispersal by rodents may increase the probability that spores will be deposited on or near the roots of susceptible host plant (Trappe and Masser 1976).

Passive dissemination can also occur through other biotic agents like worms, insects and birds or through abiotic agents like wind and water. A wide variety of animals are known to have AM fungal spores in their gut tracts or faeces. Animals that ingest digestion-resistant plant parts might pass them through their faeces and inoculate soil. Allen and Allen (1980) found AM fungal mycelia in the faeces of Jack rabbit (*Lepus californicus*). As early as 1922, mycorrhizal spores were observed in the digestive tracts of millipedes (Thaxter 1992), and more recently they have been found in crickets and grasshoppers (Hansen and Uckert 1970). We found that earthworm casts give rise to typical AM colonization, and AM propagules can survive in earthworm cast stored for a period of 11 months (Harinikumar and Bagyaraj 1994). In this group, millipedes, earthworms and wasps are obvious potential vectors because they handle soil in one way or another. Grasshoppers and crickets, are less likely candidates since, they get spores from plant surfaces.

The major contribution of some of these vectors may be that soil containing spores and spores themselves are brought to the soil surface, thus favouring further dispersal by wind, if that occurs (McIveen and Cole 1976). Soils containing spores can also be brought to soil surface by activity of mud dauber wasps, robins or sparrows in whose nests AM spores have been found. Spores from swallow nests were also able to initiate typical AM colonization. In Kuwait, birds have been observed to feed on sporocarps of certain desert truffles and may be important spore vectors. AM spores were found in termitaria (Harinikumar and Bagyaraj 1994). Their numbers were more in fresh live termitaria compared to old dry termitaria. The soil from termite mound contained 7.32 spores/g compared to 7.6 spores/g in adjacent check soil. The AM spores in termitaria were nonviable and loss of viability was attributed to the probable presence of inhibitory chemicals present in saliva of termites (Delinge et al. 1981).

Infectivity and viability of AM fungal propagules in faecal pellet of millipedes was tested by Harinikumar and Bagyaraj (1994). AM propagules were viable in faecal pellet only for 4 days. The Low acidic pH (4.5–5.0) of the faecal pellet was probably unfavourable for AM activity. Of the 42 samples tested, 22 samples constituting 66 % only, were positive for AM colonization. The digestive enzymes present in alimentary tract perhaps inactivated AM spores.

Wasp (*Sceliphron madraspatnum*) nests had 310 infective propagules/g while those of

*S. spinoli* and *Eumenes conica* had 190 and 92 I.P/g, respectively. *Rhynchium* sp. is known to secrete a gummy substance from its body which is used in the construction of nests, and this gummy substance was found to be inhibitory to AM fungi (Spradbery 1973). Harinikumar and Bagyaraj (1994) observed that AM spores in ant nests collected from the vicinity of plants harboured more mycorrhizal propagules. Ants have been estimated to bring out soil to surface annually at the rate of 0.11 kg $m^{-2}$ of dry earth (Baxter and Hole 1966).

Fogel and Trappe (1978) observed a relationship between mammal size and the size of sporocarps ingested, it is not surprising that sporocarps generally smaller (1–10 mm diameter) than those formed by Ascomycetes and Basidiomycetes are ingested by smaller mammals such as jumping mice, mice, rats, lemmings, voles and pikas. Despite the large spore size, these spores pass through rodent digestive system and still remain viable (Trappe and Masser 1976). These defecated spores are capable of initiating typical AM fungal infections (Rothwell and Holt 1978). The authors felt that only known dispersal method for spores of AM fungi is mycophagy.

Any bulk movement of soil by an animal might move spores. The faecal pellets of worms, millipeds, grasshoppers, crickets and the mud nest materials of wasps might be examples of this phenomenon. The worms, millipedes and ants probably, are limited to vertical movement from subsoil to surface. Movement of spores over greater lateral distance by birds, crickets, grasshoppers and mud wasps has a higher potential. The importance of digging and burrowing animals is not to ingest fungi but to bring soil to the surface (Hetrick 1984). It is reported that jumping mouse (*Zapus hudsonicus*) consumes sporocarp, and they may be dusted with spores which adhere to their bodies. This mode of dispersal, rarely observed, may ultimately have significance. There is no requirement of digestion resistant spores in this case. Additionally, spores may be deposited over a broad area rather than in discrete faecal packages. AM mycorrhizal propagules brought to the soil surface by a variety of vectors can be disseminated by abiotic agents like wind, water, etc.

Ponder (1980) reported the presence of AM fungal spores in grasshoppers and rabbit droppings. While occasional bits of soil might be ingested by these animals, they being primarily leaf feeders. The presence of AM fungal spores in their digestive tracts or faeces, therefore, implies that spores were present on leaves prior to feeding, probably as a result of wind dispersal. The presence of viable AM fungal spores in rabbit dropping certainly implies that the spores can survive wind dissemination (Ponder 1980). Taber (1982) observed AM fungal spores in *Portulaca* seed capsules, which are oriented on the plant towards the wind. AM fungal spores appear to be trapped on the mucilaginous surface of seeds with the capsules and were probably wind disseminated. Airborne AM spores were collected from modified sticky traps thus giving direct evidence of wind dispersal of AM spores (MacMohan and Warner 1984). Wind dispersal of spores up to 2 km has been demonstrated (Warner et al. 1987).

The turbulence and wind velocity is necessary to disseminate spores of AM fungi and the distance they could be transported have not been studied. The unusually large size of AM fungal spores makes comparison with dissemination of other fungal spores difficult. Tommerup and Carter (1982) devised a method to separate spores from soil and demonstrated that velocities of 0.10–0.55 m/sec would transport spores. The maximum velocity used was that necessary to move 100 μm quartz particles. Spores exposed to these wind velocities were also demonstrated to be viable.

Surface flow of water that would cause mass flow of soil might also move spores. Rain flow as surface sheet erosion is ideal for movement of spores at or near the soil surface. Powell's study dealt with the mycorrhizal infectivity of eroded soils in which spore density decreased with increasing soil erosion, especially in open environments like deserts and tundras with sparse vegetation (Powell 1980). Wind and water may be more important dispersal agents than animals.

Although our understanding of the ecology of AM fungi is increasing, there is still much to learn. The composite life cycle of these AM fungi is becoming clearer, but the ecological adaptation of the various fungal species has received little attention. It is clear that there are numerous potential avenues for the dispersal of AM spores. It has been postulated that for all processes involved in the phenomenon of succession, significant differences in the relative importance of animals as influents in these processes could occur and that such variations were characteristic of various biomes. We believe that the dispersal of spores is another example of a more general phenomenon. That is in physically rigorous environments (those with extreme values of environment variables and/or where such variables are unpredictable), abiotic factors will be mainly responsible for spore dispersal. In contrast, equable environments would be dominated by biotically mediated dispersal.

## 10.4 Effect of Agricultural Practices on AM Fungal Population and Diversity

Modern agricultural practices to enhance food production to meet the needs of increasing human population are posing problems to AM fungi. The agricultural intensification declined the AM abundance and effectiveness with respect to root colonization and plant growth promotion. The important agricultural practices commonly followed are cropping pattern, crop rotation, tillage operations, organic amendments, season, fertilizer application, weeding, crop protection, etc (Lakshmipathy et al. 2007).

Monocropping with a particular crop results in the development of a predominant AM fungus in soil. Continuous cultivation of maize in Philadelphia, USA, for 3 years resulted in the development of *Gigaspora gigantia* in a soil (Schenck and Kinloch 1980). Mixed cropping is common practice in the tropics. In an experiment, the effect of mono and mixed cropping with soybean and maize on AM fungal population in soil showed that mixed cropping stimulated the proliferation of AM fungi, compared with monocropping with maize or soybean (Harinikumar et al. 1990). Soybean being a legume, possibly provides nitrogen to maize through AM fungi. Sieverding and Leihner (1984) found that mycorrhizal root infection of cassava increased by intercropping with legumes. One reason for the higher propagule density under mixed cropping may be the more intensively rooted soil in the mixed system. Additionally, through higher plant density, nutrients are extracted faster from the soil, thereby stimulating AM fungal reproduction. Mixed cropping is a common practice in the tropics.

Graminaceous and leguminous crops are generally believed to increase AM fungal population, while non-mycotrophic plants decrease the population of mycorrhizal fungi (Sieverding and Lihner 1984). Taking non-mycorrhizal hosts like mustard or leaving the land fallow will reduce the propagules of AM fungi in soil (Harinikumar and Bagyaraj 1988b). In contrast, use of a crop which is strongly mycorrhizal will increase their numbers. An experiment was carried out in a P deficient soil, where in first season, finger millet was grown in all the plots and in the second season, a mycorrhizal host (cowpea) was grown in two thirds of the plot and rest was left fallow. In the third season, cowpea was grown in all the plots. The results showed reduction in mycorrhizal propagules in the soil left fallow (Harinikumar and Bagyaraj 1988b). Growing a non-mycorrhizal host significantly reduced the native mycorrhiza, but the reduction was not as bad as when the land was left fallow. A mycorrhizal host taken in the third season resulted in a slow build up of mycorrhizal population in the soil. However, at the end of third season it did not reach the same level as in plots cropped continuously with mycorrhizal host. This suggests the reduction in mycorrhizal population caused by leaving the land fallow or growing a non-mycorrhizal host. It may take at least two cropping seasons with a mycorrhizal host to rebuild the reduced mycorrhizal population to the original level. Therefore, it is best to grow a variety of crops in rotation. Further, some plants do not become colonized by AM fungi and therefore will depress populations of these fungi.

Weeds can act as a kind of an instantaneous crop rotation. Since the diversity of the AM fungus community can be proportional to the associated plant community, strict and complete weed control decreases the diversity and efficacy to the indigenous community of AM fungi. Earlier studies have shown that less tillage of soil is better for the buildup of mycorrhizal populations (McGonigle and Miller 1993). Less disturbance of the soil with ridge tillage resulted in more mycorrhizal population, compared with mouldboard ploughing. Studies have shown that less tillage of soil is better for the fill up of mycorrhizal population (McGonigle and Miller 1993). Reduced disturbance of soil with ridge tillage resulted in more mycorrhizal population compared with mould-board ploughing. Further, a cover crop of hairy vetch planted after harvest of winter wheat became significantly more colonized by AM fungi in untilled soil, where hyphal networks were intact, than in plots subjected to mould-board plough tillage. Untilled soils may have more mycorrhizal spores in the top soil, while tilled soils may have more at the 8–15 cm depth (Abbott and Robson 1991). The mycorrhizal hyphae in the soil act as the nutrient-absorbing organ of the mycorrhiza and the way in which new roots are colonized. Tillage disrupts both of these functions. On the other hand, seedlings grown in untilled soils become colonized by AM fungi more rapidly and have greater P status than those grown in tilled soils.

Another practice that has negative impacts on AM fungi is over winter bare fallow. This removes potential host roots, from which the fungi can receive sugar during mild fall and spring weather, thereby decreasing viability and ability of the fungi to colonize the next crop. An over winter cover crop may not only be useful for the mycorrhizal fungi but will also boost the amount of AM fungi in the soil. Studies conducted in temperate countries reveal that AM fungi sporulate during summer with higher temperature and longer day length (Hayman 1974; Furlan and Fortin 1977). An experiment conducted by Harinikumar and Bagyaraj (1988a) and Mallesha and Bagyaraj (1991) revealed that AM fungi sporulate during winter in the tropics. This is probably because the optimum soil temperature for sporulation occurs during winter in the tropics while it occurs during summer in the temperate countries. The optimum temperature for sporulation by mycorrhizal fungi appears to be around 25 °C.

In tropical soils, application of organic matter either in the form of farm yard manure, compost or organic amendments stimulates proliferation of AM fungi (Harinikumar and Bagyaraj 1989). This is probably because of the low organic matter content in tropical soils. Harinikumar and Bagyaraj (1988a) found that addition of organic amendments such as paddy straw, maize straw and pongamia leaf increased the mycorrhizal activity. Of the three amendments studied, the addition of pongamia leaf encouraged AM fungi to the maximum, followed by maize straw.

For modern agriculture, fertilizer application is an essential and often the most promising method to increase crop production in infertile soils. In tropical soil, P is one of the most limiting elements for crop production. High P availability is reported to be negatively correlated with AM fungal activity (Krishna and Bagyaraj 1982). Apparently, the internal P content of plants regulate AM fungal infection and reproduction (Menge et al. 1978). Tissue P concentration is not always a good estimate for mycorrhizal colonization, because the mycorrhiza themselves influence the factor. It is likely that P influences AM colonization by affecting concentrations of root carbohydrates or the amount of root exudates. The percentage of P in plants at the time of AM colonization is the best indicator to identify a soil, which provides good AM colonization (Jasper et al. 1979). The rock phosphate applied at 100 ppm P level resulted in more infective propagules of *G. fasciculatum* (Sreenivasa and Bagyaraj 1989). Clarke and Mosse (1981) observed that rock phosphate encourages better proliferation of AM fungi, compared to bone meal and super phosphate. Research findings show that years of P fertilization can lead to very high soil P levels. Plants that are able to absorb sufficient P *via* their roots alone in high nutrient soils inhibit the spread of colonization by the fungus. This reduces the flow of sugars to the fungus, which lessens the amount of AM fungi in the soil. Hence, low or no P fertilization is necessary in such soils. Addition of phosphatic fertilizer decreased the AM fungal formation compared to no added P control and P supplied through organic matter. Even disturbance of soil pre-established with AM fungi affected its further establishment (Boddington and Dodd 2000).

Application of heavy doses of nitrogen fertilizers (188 kg N/ha per year) can have a large negative effect on AM population (Hayman 1975), and nitrate as a source of nitrogen has been shown to be more inhibitory to AM development than ammonium salts (Menge 1984; Sreenivasa and Bagyaraj 1990). Menge (1984) noted that regular fertilization of citrus with more than 100 ppm N as a mixture of $NO_3$ and $NH_4$ retarded mycorrhizal development. Among the salts, calcium nitrate, urea and calcium ammonium nitrate at different levels were compared, and it was found that calcium ammonium nitrate applied at 80 ppm N level to soilrite:perlite mix substrate with negligible N produced maximum number of infective propagules of *G. fasciculatum,* in association with Rhodes grass (Sreenivasa and Bagyaraj 1988). Levels greater than 80 ppm decreased the number of infective propagules. These results indicate that nitrogen content in soils could greatly influence the distribution and abundance of AM fungi. Potassium content of acidic tropical soils is generally low, and sustainable crop yields depend on K application. Cassava has a very high K demand. About 5.8 kg has to be applied for each ton of cassava yield (Howeler 1980). It was found that increasing K application levels up to 200 kg K/ha increased mycorrhizal root infection ratings as well as tuber yield in Columbia.

Most pesticides inhibit colonization and development of AM fungi in plants (Ocampo and Hayman 1980) although the majority of pesticides tested adversely affect the symbiosis (Menge 1982) others do not appear to damage mycorrhizal fungi. Some may even increase mycorrhizal colonization (Trappe et al. 1984). Fumigation of soil with biocides such as methyl bromide, chloropicrin, etc. effectively kills endophytes. However, AM fungi can reinvade most fumigated soils within several years (Trappe et al. 1984). The systemic fungicides like thiobendazole, benomyl and triademefon are most toxic to these fungi (Nemec 1980) Pentachloronitrobenzene which is not systemic is also toxic (Nemec and Tucker 1983). Fungicides like Captan and Rilon applied at half the recommended levels had no adverse effects on AM fungi (Praveen Kumar and Bagyaraj 1999). The insecticides metasystox and aldrin differ in their activity on AM fungi, the former being less toxic while most nematicides exihibit slight inhibitory effects on AM fungi. Interestingly, some nematicides such 1,2-dibromo-3-chloropropane (DBCP) can stimulate root infection in host plant (Menge et al. 1978). This response could be due to control of competitive pathogenic microflora and possible stimulation of root exudates of host plants or other factors (Trappe et al. 1984).

## 10.5 Effect of Land Use Patterns on the Arbuscular Mycorrhizal Fungal Population and Diversity

Land use pattern is the way/purpose/extent to which a particular area of land is being utilized. Land use patterns are mainly classified into two types, *viz.* cultivated and uncultivated. In cultivated type different land use patterns are forests, grasslands, plantations, agricultural lands, etc. In uncultivated type different land use patterns are human dwellings, industrial areas, roads, barren lands, etc. Studies have been carried out in cultivated areas. Many studies were done to know AM fungal population and diversity because of change in land use intensities/patterns, *i.e.* conversion of forest to agricultural lands or to pasture, pastures to agriculture lands, barren lands to forest plantations etc. Studies were also done regarding comparison of AM fungal population diversity in different land used intensities or land use patterns, *e.g.* between natural forest, plantations, grasslands and within the plantations between different crops, etc.

Oehl et al. (2003) studied the impact of land use intensity on the diversity of AM fungi at eight different sites; three sites with low input grass lands, two sites with low to moderate input farming with seven year crop rotation and three sites with high input continuous maize monocropping. They observed a decrease in AM fungal diversity from low input grasslands to high input monocropping farming system. As the land use intensification increased, the AM fungal species diversity decreased. Apart from this they also observed more spore abundance in low input grasslands than in low to medium input farming system and high input farming system. Even the AM fungal species diversity index was more in grasslands. Among two different farming systems, AM fungal species diversity index was more in sites with crop rotation rather than sites with monocropping. Galvez et al. (2001) studied the populations of spores of AM fungi, mycorrhizal formation and nutrient utilization of maize in mould-board ploughed, chisel disked or untilled soil under conventional and low input agricultural systems. Soils under low input management had higher AM fungal spore populations than soils under conventional management. Spore load and colonization of maize roots by AM fungi were higher in non-tilled than in mould-board ploughed and chisel disked soil.

Glomalin, an arbuscular mycorrhizal fungal soil protein playing an important role in soil aggregation was significantly affected by land use pattern. Glomalin concentrations were highest in native forest soils, moderate in afforested soils and lowest in agricultural lands. Soil C and N were highly correlated with glomalin across all soils and within each land use type, indicating that some glomalin may be under similar controls as soil C. These results also show that glomalin may be useful as an indicator of land use change (Rilling et al. 2003).

Carpenter et al. (2001) studied the spore density and diversity of AM fungi in different land uses and across different gradients of erosion. They found that the diversity and composition of AM fungi changed due to change in land use types, and AM fungal diversity decreased due to erosion. Picone (2000) compared spore abundance and diversity of AM fungi in soils of lowland evergreen forests and pastures in Nicaragua. Species composition, dominance-diversity curves and Simpson's diversity indices were similar for both forest and pasture soils. Of 28 distinct fungus morphospecies, 11 produced more spores in pasture, while only one produced more spores in forest. According to species-accumulation curves, local AM fungal species richness did not significantly decline following conversion of forest to pasture. Because pastures contained a surprising abundance and diversity of AM fungal spores compared to native forest, a lack of mycorrhizal fungi is unlikely to limit plant succession, restoration or reforestation in the pastures studied. In addition to these trends in diversity, species that sporulated more in pasture tended to have small spores, while the species that sporulated more in forest had the largest spores. Similarly, only large-spored fungi (300 μm) showed seasonal variation in spore abundance, being more common in the wet season. Wolfe (2002) compared species diversity of AM fungi in three pairs of forest fragment and adjacent pasture sites in a coffee farming community in Coto Brus, Southern Costa Rica. They determined species diversity of AM fungi by both direct assessment of field-collected soil samples and estimates from trap cultures in the greenhouse. Results suggested that conversion of forest to pasture produced shifts in abundance of many AM fungi species rather than general declines. Species richness of AM fungi was similar in pasture and forest sites, despite the dipartite nature of the pasture plant community relative to that of the forest. While some AM fungi species were common in both forest and pasture sites, others were abundant in one vegetation type and rare or absent in the other. These results suggested that pasture plants support AM fungi community that may be as species rich as that of forested sites, but differing in its composition.

Lovelock et al. (2003) assessed the spatial and temporal distribution of the AM fungal community in a wet tropical rainforest in Costa Rica. Host tree species differed in their associated AM fungal communities, but differences in the AM community between the hosts could not be generalized over life history groupings of the hosts. Changes in the relative abundance of a few common AM fungal species were the cause of differences in AM fungal communities for different host tree species instead of differences in the presence and absence of AM fungal species. Thus, AM fungal communities were spatially distinguishable in the forest, even though all species were widespread. Soil fertility ranging between 5 and 9 mg/ha P did not affect composition of AM fungal communities. However, sporulation was more abundant in lower fertility soils. Sampling soils over seasons revealed that some AM fungal species sporulated profusely in the dry season compared to the rainy season. On one host tree species sampled at two sites with vastly different rainfall, relative abundance of spores from *Acaulospora* was lower and that of *Glomus* was relatively higher at the site with lower and more seasonal rainfall. Johnson and Wedin (1997) found that diversity of mycorrhizal spore communities (measured by Sorenson's similarity index) was lower in the grassland plots than in the forest plots, indicating that grass invasion had caused some convergence. However, total spore diversity and alpha diversity of mycorrhizal spore communities (measured by species richness and Simpson's diversity index) were not altered by wildfires and grass invasion. These results suggest that persistence and regeneration of forest plant species in the grasslands may not be constrained to a significant degree by the lack of mycorrhizal symbionts. Stutz (2003) studied species richness, composition, spore density and diversity of AM fungi in four different land use types, *viz*. urban residential, urban non-residential, agriculture and desert. The agricultural sites were associated with decreased spore densities and decreased species richness. Spores of *G. microaggregatum* were most abundant in urban sites, while those of *G. eburneum* were most abundant in desert and agricultural sites. A recent study suggests that a particular AM fungal community may be better matched ecologically to its local habitat than communities taken from other locations (Ji et al. 2010). More investigations are needed in this aspect.

## 10.6 Conclusion

It is now established that AM fungi improve plant growth, mainly through P nutrition; other beneficial effects are in the biological control of root pathogens, biological nitrogen fixation, hormone production and greater ability to withstand water stress. Because of their unique ability to increase the uptake of P by plants, mycorrhizal fungi can be utilized as practical substitutes for phosphatic fertilizers. Several workers have stressed the potential for commercial utilization of AM fungi in crop production.

Man is trying to use these beneficial fungi to increase the productivity of food, fuel and fiber. Principally there could be two strategies for managing AM fungi. The first strategy is to develop inoculation techniques with efficient AM fungi, adapted to the crop plant and the environment. Considerable work is being undertaken at present on this aspect. The second strategy is to manage the indigenous AM fungi by agricultural practices in such a way that efficient native fungi are enhanced and inefficient fungi are depressed. A good deal of research is necessary before we formulate some general recommendations for this management method for each soil and crop. Therefore, manipulation of this symbiotic association to attain its full ecological and economic potential should be the goal for future studies. Without a clear understanding of the ecology of AM fungi, man's ability to manipulate the mycorrhizal symbiosis for maximizing productivity will be severely limited.

## References

Abbott LK, Robson AD (1977) The distribution and abundance of vesicular endophytes in some western Australian soils. Aust J Bot 25:515–522

Abbott LK, Robson AD (1991) Field management of mycorrhizal fungi. In: Kelster DL, Cregan PB (eds) The rhizosphere and plant growth. Kluwer Academic, Dordrecht, pp 355–362

Allen MF (1991) The Ecology of Mycorrhizae. Cambridge University Press, Cambridge

Allen EB, Allen MF (1980) Natural re-establishment of vesicular-arbuscular mycorrhizae following strip mine reclamation in Wyoming. J Appl Ecol 17:139–147

Allen MF, Swenson W, Querejeta JI, Egerton-Warburton LM, Treseder KK (2003) Ecology of mycorrhizae: a conceptual frame work for complex interactions among plants and fungi. Ann Rev Phytopathol 41:271–303

Bagyaraj DJ (1991) Ecology of vesicular-arbuscular mycorrhizae. In: Arora DK, Rai B, Mukerji KG, Knudsen GR (eds) Handbook of applied mycology. Marcel Decker, New York, pp 3–34

Bagyaraj DJ (2011) Microbial biotechnology for sustainable agriculture, horticulture and forestry. New India Publishing Agency, New Delhi

Bagyaraj DJ, Ravindra TP (1997) Distribution and dissemination of VA mycorrhizal fungi. In: Tiwari JP, Saxena G, Mittal N, Tewari I, Chamola BP (eds) New approaches in microbial ecology. Aditya Books, New Delhi, pp 167–182

Bagyaraj DJ, Manjunath A, Patil RB (1979) Occurrence of vesicular arbuscular infection in some tropical aquatic plants. Trans Br Mycol Soc 73:164–167

Baxter FP, Hole FD (1966) The ant that ploughed the Praire. Crop Soil 19:11–13

Blaszkowski J (1994) Comparative studies on the occurrence of arbuscular fungi and mycorrhizae (Glomales) in cultivated and uncultivated soils of Poland. Acta Mycol 28:93–140

Boddington CL, Dodd JC (2000) The effect of agriculture practices on the development of indigenous arbuscular mycorrhizal fungi-II. Studies in experimental microcosms. Plant Soil 218:145–157

Borowicz VA (2001) Do arbuscular mycorrhizal fungi alter plant-pathogen relations? Ecology 82:3057–3068

Carpenter FL, Palacioss S, Gonzalez E, Schroeder M (2001) Land-use and erosion of a Costa Rican Ultisol affects soil chemistry, mycorrhizal fungi and early regeneration. For Ecol Manag 144:1–17

Clarke C, Mosse B (1981) Plant growth response to vesicular arbuscular mycorrhiza-XII. Field inoculation responses of barley at two soil P levels. New Phytol 87:695–703

Delinge J, Queenedey A, Blum MS (1981) The enemies and defense mechanisms of termites. In: Herman HR (ed) Social insects. Academic, London, pp 1–76

Fogel R, Trappe JM (1978) Fungal consumption (Mycophagy) by small animals. Northwest Sci 52:1–13

Frank AB (1885) Uber die out Wurzetsymbiose berohende ernahausing gewisser Baume durch utnerimdinche. Berdent Bot Gessel 3:128–145

Furlan V, Fortin JA (1977) Effects of light intensity on the formation of vesicular arbuscular endomycorrhizas on *Allium cepa* by *Gigaspora calospora*. New Phytol 79:335–340

Galvez L, Douds DD Jr, Drinkwater LE, Wagoner P (2001) Effect of tillage and farming system upon VAM fungus populations and mycorrhizas and nutrient uptake of maize. Plant Soil 228:299–308

Gianinazzi S, Gianinazzi-Pearson V (1986) Progress and headaches in endomycorrhizae biotechnology. Symbiosis 2:139–149

Gianinazzi S, Gollotte A, Binet MN, van Tuinen D, Redecker D, Wipf D (2010) Agroecology: the key role of arbuscular mycorrhizas in ecosystem services. Mycorrhiza 20:519–530

Hall IRS (1977) Species and mycorrhizal infections of New Zealand endogonaceae. Trans Br Mycol Soc 68:341–356

Hansen RM, Uckert PN (1970) Dietary similarity of some primary consumers. Ecology 51:641–680

Harinikumar KM, Bagyaraj DJ (1988a) The effect of season on VA mycorrhiza of leucaena and mango in semiarid tropic. Arid Soil Res Rehabil 7:139–143

Harinikumar KM, Bagyaraj DJ (1988b) Effect of crop rotation on native VA mycorrhizal propagules in soil. Plant Soil 110:77–80

Harinikumar KM, Bagyaraj DJ (1989) Effect of cropping sequences, fertilizers and FYM on VA mycorrhizal fungi. Biol Fertil Soils 7:173–175

Harinikumar KM, Bagyaraj DJ (1994) Potential of earthworms, ants, millipedes and termites in dissemination of vesicular arbuscular mycorrhizal fungi in soil. Biol Fertil Soils 18:115–118

Harinikumar KM, Bagyaraj DJ (1995) Spread of vesicular arbuscular mycorrhizal hyphae in soil. Microbiol Res 150:77–80

Harinikumar KM, Bagyaraj DJ, Mallesha BC (1990) Effect of intercropping and organic soil amendments on native VA mycorrhiza in semi arid tropics. Arid Soil Res Rehabil 4:193–197

Hayman DS (1974) Plant growth responses to vesicular arbuscular mycorrhiza-VI. Effect of light and temperature. New Phytol 73:71–80

Hayman DS (1975) The occurrence of mycorrhiza in crops as affected by soil fertility. In: Sanders FE, Mosse B, Tinker PS (eds) Endomycorrhizas. Academic, London, pp 495–509

Hayman DS (1983) The physiology of vesicular arbuscular endomycorrhizal symbiosis. Can J Bot 61:944–963

Hetrick BDA (1984) Ecology of VA mycorrhizal fungi. In: Powell CL, Bagyaraj DJ (eds) VA mycorrhizae. CRC, Boca Raton, pp 35–55

Hirrel MC, Mehravaran H, Gerdemann JW (1978) Vesicular arbuscular mycorrhizae in Chenopodiaceae and Cruciferaceae: do they occur? Can J Bot 56:2813–2817

Hodge A (2000) Microbial ecology of arbuscular mycorrhiza. FEMS Microbiol Ecol 32:91–96

Hodge A, Helgasson T, Fitter AH (2010) Nutritional ecology of arbuscular mycorrhizal fungi. Fungal Ecol 3:267–273

Howeler RH (1980) Soil related cultural practices for cassava. In: Webes EJ, Toro MJC, Graham M (eds) Cassava cultural practices, Proc. of a workshop held at Salvador, Bahia, March 18–21, CAB Int, pp 159–169

Jasper PA, Robson AD, Abbott LB (1979) Phosphorus and the formation of vesicular arbuscular mycorrhiza on the growth and metabolism of sweet orange. New Phytol 90:665–670

Ji B, Bentivenga SP, Casper BB (2010) Evidence for ecological matching of whole fungal communities to the local plant-soil environment. Ecology 91:3037–3046

Johnson NC, Wedin DA (1997) Soil carbon, nutrients, and mycorrhizae during conversion of dry tropical forest to grassland. Ecol Appl 7:171–182

Kendrick B, Berch S (1985) Mycorrhizae: applications in agriculture and forestry. In: Robinson CW (ed) Comprehensive biotechnology, vol 4. Pergamon, Oxford, pp 109–150

Khan AG (1978) Vesicular arbuscular mycorrhizas in plants colonizing black wastes from bituminous coal mining in the Illawarra region of New South Wales. New Phytol 81:53–63

Koske RE (1981) A preliminary study of interactions between species of VA fungi in a sand dune. Trans Br Mycol Soc 76:411–416

Krishna KR, Bagyaraj DJ (1982) Effect of vesicular arbuscular mycorrhiza and soluble phosphate on *Abelmoscus esculentus* (L.) Moench. Plant Soil 64:209–213

Lakshmipathy R, Bagyaraj DJ, Balakrishna AN (2007) Can agricultural practices and land use patterns affect arbuscular mycorrhizal fungal population and diversity? In: Ganguly BN, Deshmukh SK (eds) Fungi: multifaceted microbes. Anamaya, New Delhi, pp 304–315

Lovelock CE, Andersen K, Morton JB (2003) Arbuscular mycorrhizal communities in tropical forests are affected by host tree species and environment. Oecologia 132:268–279

Machado C, Bagyaraj DJ (1995) Mycorrhization bacteria and its influence on growth of cowpea. In: Adholeya A, Sujan S (eds) Mycorrhizae, Biofertilizers for the Future. Tata Energy Research Institute Pub, New Delhi, pp 192–196

MacMohan JA, Warner A (1984) Dispersal of mycorrhizal fungi: processes and agents. In: Williams SE, Allen MF (eds) VA mycorrhizae and reclamation of arid and semiarid lands. University of Wyoming Publications, Wyoming, pp 24–41

Mallesha BC, Bagyaraj DJ (1991) Season favouring sporulation of VA mycorrhizal fungi in cardamom plantations. J Soil Biol Ecol 11:75–78

McGonigle TP, Miller MH (1993) Mycorrhizal development and phosphorus adsorption in maize under conventional and reduced tillage. Soil Sci Soc Am J 57:1002–1006

McIveen WD, Cole H (1976) Spore dispersal of Endogonaceae by worms, ants, wasps and birds. Can J Bot 54:1486–1489

Meharg AA, Cairney JWG (2000) Co-evaluation of mycorrhizal symbionts and their hosts to metal-contaminated environments. Adv Ecol Res 30:69–112

Menge JA (1982) Effect of soil fumigants and fungicides on vesicular arbuscular fungi. Phytopathology 72:1125–1132

Menge JA (1984) Inoculum production. In: Powell CL, Bagyaraj DJ (eds) VA mycorrhizae. CRC, Boca Raton, pp 188–189

Menge JA, Steirle D, Bagyaraj DJ, Jhonson ELV, Leonard TT (1978) Phosphorus concentrations in plants responsible for inhibition of mycorrhizal infection. New Phytol 80:575–578

Miller RM, Jastrow JD (1992) The application of VA mycorrhizae to ecosystem restoration. In: Allen MF (ed) Mycorrhizal Functioning. Chapman and Hall, London, pp 438–467

Mosse B (1977) Plant growth responses of vesicular-arbuscular mycorrhiza-IV. Soil given additional phosphorus. New Phytol 72:127–136

Mosse B (1981) Vesicular Arbuscular Mycorrhizal Research for Tropical Agriculture. Honolulu University of Hawaii Press, Hawaii, p 54

Mosse B, Stribley DP, Le Tacon F (1981) Ecology of mycorrhizae and mycorrhizal fungi. Adv Microb Ecol 5:137–209

Mosse B, Warner A, Clarke CA (1982) Plant growth responses of vesicular-arbuscular mycorrhiza-XIII. Spread of an introduced VA endophyte in the field and residual growth effects of inoculation in the second year. New Phytol 90:521–528

Nemec S (1980) Effects of eleven fungicides on endomycorrhizal development in sour oranges. Cad J Bot 58:522–527

Nemec S, Tucker D (1983) Effects of herbicides on endomycorrhizal fungi in Florida citrus (Citrus sp) soils. Weed Sci 31:417–431

Ocampo IA, Hayman DS (1980) Effects of pesticides on mycorrhiza in field grown barley, maize and potatoes. Trans Br Mycol Soc 74:413–416

Oehl F, Sieverding E, Ineichen L, Mader P, Boller T, Wiemken A (2003) Impact of land use intensity on the species diversity of arbuscular mycorrhizal fungi in agroecosystems of central Europe. Appl Environ Microbiol 69:2816–2824

Picone CM (2000) Diversity and abundance of arbuscular mycorrhizal fungus spores in tropical forest and pasture. Biotropica 32:734–750

Ponder F (1980) Rabbits and grasshoppers: vectors of endomycorrhizal fungi of new coal mine spoil. Research Note NC- 250, USDA, North Central Forest Experiment Station, Newtown Square, Pennsylvania, USA

Powell CL (1976) Development of mycorrhizal infections from Endogone spores and infected root segments. Trans Br Mycol Soc 66:439–445

Powell CL (1979) Spread of mycorrhizal fungi through soil. N Z J Agric Res 22:335–339

Powell CL (1980) Mycorrhizal infectivity of eroded soils. Soil Biol Biochem 12:247–251

Praveen Kumar KA, Bagyaraj DJ (1999) Mass production of arbuscular mycorrhiza as influenced by some agrochemicals. Proc Nat Acad Sci India 69:61–66

Raj J, Bagyaraj DJ, Manjunath A (1981) Influence of soil inoculation with arbuscular mycorrhiza and a phosphate dissolving bacterium on plant growth and $^{32}P$-uptake. Soil Biol Biochem 13:105–108

Redhead JF (1977) Endotrophic mycorrhizas in Nigeria: species of the Endogonaceae and their distribution. Trans Br Mycol Soc 69:275–280

Rilling MC, Ramsey PW, Morris S, Paul EA (2003) Glomalin, an arbuscular mycorrhizal fungal soil protein, response to land use changes. Plant Soil 253:293–299

Rothwell FM, Holt C (1978) Vesicular-arbuscular mycorrhizae established with *Glomus fasciculatum* spores isolated from the faeces of cricetine mice. Research Note NC- 259, USDA, North Central Forest Experiment Station, Newtown Square, Pennsylvania, USA

Schenck NC, Kinloch RA (1980) Incidence of mycorrhizal fungi on six field crops in mono culture on a newly cleared woodland site. Mycologia 72:445–455

Sieverding E, Leihner DE (1984) Influence of crop rotation and intercropping of cassava with legume on VA mycorrhizal symbiosis of cassava. Plant Soil 80:143–146

Smith SE, Read DJ (2008) Mycorrhizal symbiosis, 3rd edn. Academic, London

Spradbery JP (1973) Wasps. An account of the biology and natural history of solitary and social wasps. Sidgwicks and Jackson, London

Sreenivasa MN, Bagyaraj DJ (1988) *Chloris gayana* (Rhodes grass) a better host for mass production of *Glomus fasciculatum* inoculum. Plant Soil 106:109–112

Sreenivasa MN, Bagyaraj DJ (1989) Suitable form and level of phosphorus for mass production of the VA mycorrhizal fungus, *Glomus fasciculatum*. Zentralbl Mikrobiol 144:34–36

Sreenivasa MN, Bagyaraj DJ (1990) Suitable source and level of nitrogen for mass production of VA mycorrhizal fungi. In: Jalalli BL, Chand A (eds) Current trends in mycorrhizal research. Haryana Agril. Univ. Press, Hissar, pp 35–36

Sturmer SL (2012) A history of the taxonomy and systematics of arbuscular mycorrhizal fungi belonging to the phylum Glomeromycota. Mycorrhiza 22:247–258

Stutz JC (2003) Preliminary assessment of arbuscular mycorrhizal fungal diversity and community structure in an urban ecosystem. Mycorrhiza 13:319–326

Taber RA (1982) Occurrence of *Glomus* spores in weed seeds in soil. Mycologia 74:515–520

Thaxter R (1992) A revision of Endogonaceae. Proc Am Acad Arts Sci 57:293–341

Tommerup IC, Carter DJ (1982) Dry separation of microorganisms from soil. Soil Biol Biochem 14:69–71

Trappe JM (1977) Biogeography of hypogeous fungi: trees, mammals and continental drift. In: Bigelow HE, Simmons EG (eds) Abstracts 2nd International Mycology Congress. University of South Florida, Tampa, pp 675

Trappee JM, Masser C (1976) Germination of *Glomus marcrocarpus* (Endogonaceae) after passage through a rodent digestive tract. Mycologia 68:433–436

Trappe JM, Monila R, Castellano M (1984) Reactions of mycorrhizal fungi and mycorrhiza formation to pesticides. Ann Rev Phytopathol 22:331–359

Wallace HR (1978) Dispersal in time and space: soil pathogens. In: Horsefall JS, Cowling EB (eds) Plant disease: an advanced treatise, vol 2. Academic, New York, pp 181–202

Wang GM, Qiu YL (2006) Phylogenetic distribution and evolution of mycorrhizas in land plants. Mycorrhiza 16:299–363

Warner A, Mosse B (1982) Factors affecting the spread of vesicular-arbuscular mycorrhizal fungi in soil-I Root density. New Phytol 90:529–536

Warner MJ, Allen MF, MacMohan JA (1987) Dispersal agents of VA mycorrhizal fungi in disturbed arid ecosystem. Mycologia 79:721–730

Wolfe AL (2002) Species diversity and community composition of arbuscular mycorrhizal fungi in tropical forest fragments and adjacent pastures. 87th Annual Meeting of the Ecological Society of America and Annual International Conference of the Society for Restoration, Tucson, 4–9 Aug, pp 62

# Part II
# Microbial Diversity and Plant Protection

# Screening of *Brassica rapa* L. var. Yellow Sarson Genotypes Against Downy Mildew and *Alternaria* blight

11

Kamlesh Kumar Prajapati, O. P. Verma, Prakash Singh, Sanjeev Singh and Dhirendra K. Singh

**Abstract**

The use of resistant varieties is considered to be the best method of disease control. Therefore, the study was carried out to find out the sources of resistance against downy mildew (caused by *Peronospora brassicae* (Pers. ex. Fr.)) and *Alternaria* blight (*Alternaria brassicae* (Berk.) Sacc. and *Alternaria brassicicola* (Schw.) Wiltshire) in *Brassica rapa*. The pathogen is greatly influenced by weather with the highest disease incidence in wet seasons and in areas with relatively high rainfall. *A. brassicae* can affect host species at all stages of growth, including seed. Symptoms of the disease are characterized by formation of spots on leaves, stem and siliquae. This investigation was conducted under three different environments in randomized block design with three replications. A total of 20 genotypes of yellow sarson were screened for varietal reactions against the diseases under field conditions during rabi 2010 and 2011. None of the genotypes was found free from both the disease, in case of downy mildew disease; only one genotype named Jagrati was found moderately resistant. For *Alternaria* blight, seven genotypes named, Benoy, Pusa Gold, YSK 09–1, PYS 2008–2, NRCYS-05–02, YSWB-2012 and NDYS-427 were found moderately resistant.

**Keywords**

Yellow sarson · Downy mildew · *Alternaria* blight · Resistance

## 11.1 Introduction

The genus *Brassica* is an important member of the Brassicaceae family. It comprises several (*Brassica nigra, Brassica oleraceae, Brassica campestris, Brassica carinata, Brassica juncea, Brassica napus*) economically important species (Table 11.1) which yield edible roots, stems, leaves, buds, flowers and seed condiment. Among them, the first three species are elementary and diploids with $2n = 16$, 18 and 20 chromosomes

P. Singh (✉) · D. K. Singh
Department of Genetics and Plant Breeding, Institute of Agricultural Sciences, Banaras Hindu University, Varanasi, Uttar Pradesh 221005, India
e-mail: prakash201288@gmail.com

K. K. Prajapati · O. P. Verma
Department of Genetics and Plant Breeding, Narendra Deva University of Agriculture and Technology, Faizabad 224229, India

S. Singh
Department of Agricultural Botany, Udai Pratap Autonomous College, Varanasi, Uttar Pradesh 221002, India

R. N. Kharwar et al. (eds.), *Microbial Diversity and Biotechnology in Food Security*,
DOI 10.1007/978-81-322-1801-2_11, © Springer India 2014

**Table 11.1** Classification and origin of rapeseed–mustard. (Source: National Research Centre on Rapeseed-Mustard, Sewar, Bharatpur, Rajasthan)

| Sl.no. | English name | Vernacular name | Botanical name | Origin | Identification characteristics of seeds |
|---|---|---|---|---|---|
| 1. | Indian mustard/brown mustard | Rai, ryada, raya, laha, lahta, sasve, herbo | *Brassica juncea* (L.) Czern. and Coss. | Originated from China to India via northeastern India and spread to Afghanistan via Punjab | Seeds are medium sized, round and dark brown or black in colour |
| 2. | Indian rape/rapeseed/toria | Toria, tori, lahi | *Brassica rapa L.* var. *toria* (*syn. B. campestris* L. var. toria) | It is a native of India and spread to parts of Pakistan | Seeds are dark brown, bold and large sized |
| 3. | Brown sarson/rapeseed | Brown sarson, bhoori sarson | *Brassica rapa* L. var. brown sarson (*syn. B. campestris* L. var. brown sarson) | Eastern Afghanistan and adjoining parts of India and Pakistan | Seeds are light reddish in colour, bold, large sized |
| 4. | Yellow sarson/colza/rapeseed | Yellow sarson, pilli sarson | *Brassica rapa* L. var. yellow sarson (*syn. B. campestris* L. var. yellow sarson) | Eastern part of India | Seeds are slightly smaller than sarson, ovoid in shape, yellow in colour |
| 5. | Rapeseed | Gobhi sarson | *Brassica napus L.* | It is a native of Europe | Seeds are brownish black and large sized |
| 6. | Abyssinian mustard/Ethiopian mustard | Karan rai | *Brassica carinata A. Br.* | Introduced in India though it is a native of southern Europe and North America | Seeds are small, round and reddish brown in colour |
| 7. | Rocket Salad | Duan, tera, tara, saundh, taramira | *Eruca sativa* Mill. | - | Seeds are light reddish brown coloured and distinctly ovoid shape |

and other three are tetraploids with chromosome members $2n = 34$, 36 and 38. The oil content in rapeseed–mustard is ranging from 36.2 to 41.9% and the edible oil is obtained mainly from *B. napus, B. juncea* and *B. campestris* (NRC-RM Sewar 2010). Rapeseed–mustard is a group of crops that contributes 32% of the total oilseed production in India, and it is the second largest indigenous oilseed crop (Meena et al. 2010).

### 11.1.1 Global Scenario

The area and production of rapeseed–mustard seed have been increasing consistently. Being an important source of oil and protein meal, mustard seed is grown across the world. The area of rapeseed–mustard seed has risen from 24.68 to 30.06 million ha with a compounded annual growth of 3.02% while the production has increased from 37.32 to 55.97 million t with an annual growth rate of 5.70% during 2000–2011 (USDA 2011). India ranks second in acreage and third in rapeseed–mustard production in the world after China and Canada, and contributed around 26.1% of the total oilseed production in the world (FAO 2010; Meena et al. 2011).

### 11.1.2 Domestic Scenario

In India, during 2009–2010, the rapeseed–mustard crop production was 6.40 million t from an area of 6.45 million ha with an average productivity of 990 kg/ha. The projected demand for oilseeds in India by 2020 is around 34 million t which are to be met by rapeseed–mustard (Anonymous 2010). It is estimated that about 90 % of domestic production of rapeseed–mustard is crushed for extracting edible oil, which is mostly traded and consumed in northern, northeastern, eastern and central India. Out of 57,856 thousand t of rapeseed–mustard seed produced over 30,308 thousand ha in the world, India produced 5,833 thousand t from 5,750 thousand ha (FAO 2010). In India, rapeseed and mustard are being grown largely in Uttar Pradesh, Rajasthan, Haryana, Assam, Gujarat, Punjab, West Bengal and Madhya Pradesh. However, in Uttar Pradesh it is grown on 820 thousands ha with production of 900 thousand tons. The average productivity of Uttar Pradesh is 11.41 t/ha which is more than the national average productivity (Anonymous 2010).

### 11.1.3 Symptoms of Downy Mildew and *Alternaria* Blight

Downy mildew disease is caused by *Peronospora brassicae* (Pers. ex. Fr.). In disease affected plants, yellow irregular spots appear on the upper surface (dorsal surface) of the leaves and white growth is visible on the lower surface of the spots. If the attack is severe, inflorescence is also affected; the affected inflorescence is malformed, twisted and covered with a white powder. No pods are produced on such inflorescence (Dange et al. 2002; Agrios 2005). However, *Alternaria* blight disease caused by *Alternaria brassicae* (Berk.) Sacc. and *A. brassicicola* (Schw.) Wiltshire is one among the important diseases of rapeseed–mustard, which has been reported from all the continents of the world, depending on the prevailing environmental factors particularly at critical stages. There is no proven source of resistance against the disease reported till date in any of the host (Kumar and Chauhan 2005; Chattopadhyay 2008). In *Brassica* species, resistance to downy mildew is controlled by single dominant allele and resistance of *Alternaria* leaf blight is governed by a single dominant gene or nuclear genes with partial dominance (Nashaat et al. 1996). Severe infection causes substantial yield loss as a result of early defoliation, flower bud abortion, premature ripening, siliquae dehiscence and seed shrivelling (Seidle et al. 1995).

### 11.1.4 Environmental Factors Affecting the Disease

The effect of environmental factors, temperature (15–25°C), relative humidity (>80 %) and leaf wetness (rain or dew) for 4–24 h, for the disease development and its epidemics have been reported in some countries like Canada (Degenhardt et al. 1982), the UK (Humpherson-Jones and Phelps 1989; Mridha and Wheeler 1993; Hong and Fitt 1995) and India (Ansari et al. 1989; Yadav et al. 2010).

### 11.1.5 Annual Losses Due to Downy Mildew and *Alternaria* Blight

*Alternaria* blight and downy mildew are causing 10–70 % yield losses depending on the crop species, being high in *Brassica rapa,* with 35–40 % in Indian mustard (Chattopadhyay 2008). Kolte et al. (1987) reported the losses in 1,000-seed weight of yellow sarson and mustard of 23 % and 24 %, respectively. Yield losses of 20–30 % were recorded in Canada (Degenhardt et al. 1974; Conn et al. 1990). In India, losses of yield due to diseases were reported 15–71 % (Kumar 1986; Ram and Chauhan 1998; Chattopadhyay 2008). Dange et al. (2002) reported about 24 % loss in Gujarat, whereas, on an average, loss of about 25 % on different varieties of rapeseed–mustard have been reported from Haryana.

Therefore, the present investigation was carried out to find out the sources of resistance against the downy mildew and *Alternaria* blight for use in breeding programme to develop the resistant variety.

**Table 11.2** Name of the entries and their place of origin

| Sl.no. | Name of genotypes | Source |
|---|---|---|
| 1. | RAUDYS 07–76 | RAU, Dholi (Bihar) |
| 2. | YSK 09–2 | CSAUA &T, Kanpur (UP) |
| 3. | Benoy | PORS, Berhampore (WB) |
| 4. | Pusa Gold | IARI, New Delhi |
| 5. | NDYS 107–1 | NDUA &T, Kumarganj, Faizabad (UP) |
| 6. | RYS-1 | ARS, Navgaon |
| 7. | YSK 09–1 | CSAUA & T, Kanpur (UP) |
| 8. | DRMRYS-09–99 | DRMR, Bharatpur (Raj.) |
| 9. | YSWB-2009 | PORS, Bharatpur (Raj.) |
| 10. | PYS 2007–7 | GBPUA & T, Pantnagar |
| 11. | YSB-2010 | CCS, HAU, Haryana |
| 12. | PYS 2008–2 | GBPUA & T, Pantnagar |
| 13. | NRCYS-05–02 | DRMR, Bharatpur (Raj.) |
| 14. | YSWB-2012 | PORS, Bharatpur (Raj.) |
| 15. | NDYS-427 | NDUA & T, Faizabad (UP) |
| 16. | DRMRYS-09–103 | DRMR, Bharatpur (Raj.) |
| 17. | NDYS 141–3 | NDUA & T, Kumarganj, Faizabad (UP) |
| 18. | NDYS-425 | NDUA & T, Kumarganj, Faizabad(UP) |
| 19. | Jagrati | NDUA & T, Kumarganj, Faizabad (UP) |
| 20. | Type-151 | CSAUA & T, Kanpur (UP) |

## 11.2 Materials and Methods

### 11.2.1 Experimental Site and Climatic Condition

The present investigation was conducted under three different environments, i.e. irrigated timely sown ($E_1$), rainfed timely sown ($E_2$) and under irrigated late sown ($E_3$) conditions at Research Farm of Genetics and Plant Breeding, Narendra Deva University of Agriculture and Technology, Kumarganj, Faizabad during rabi season of (from the fourth week of October to the last week of March during 2009–2010 and from the third week of October to the fourth week of March in 2010–2011) both the years. The experimental site is situated between 26°47′ N latitude, 82°12′ E longitude and at an altitude of 113 m above the mean sea level in the Gangetic plain of eastern Uttar Pradesh. The climate of the district Faizabad was semi-arid with hot summer and cold winter. Nearly 80 % of total rainfall is during the monsoon (only up to September) with a few showers in the winter. The soil types of the experimental plot were silt loam.

### 11.2.2 Experimental Materials and Statistical Design

The experimental material consisted of 20 different genotypes of *B. rapa* (L.) var. yellow sarson collected from different locations (Table 11.2), and these materials were sown in randomized block design with three replications from 25th October to 29th March, in 2009–2010 and from 19th October to 26th March, in 2010–2011.

### 11.2.3 Cultural Practices and Fertilizer Application

Each entry was sown in single row of 3 m length spaced at 45 cm apart and the distance between plant to plant (15 cm) was maintained by thinning after 15 days of sowing. The fertilizers were applied at 50 kg N, 30 kg $P_2O_5$ and 30 kg $K_2O$/ha under irrigated and rainfed condition. Half of N and full dose of $P_2O_5$ and $K_2O$ as basal were given in irrigated, and full dose of nitrogen, $P_2O_5$ and $K_2O$ were given as basal

**Table 11.3** Disease scoring scale (0–9 scale)

| Grade | Reaction | Severity |
|---|---|---|
| 0 | Immune | No lesions |
| 1 | Highly resistant | Non-sporulation pinpoint size or small necrotic spots, less than 5 % leaf area covered by lesions |
| 3 | Resistant | Small, roundish slightly sporulation, larger necrotic spots about 1–2 mm in diameter with a distinct margin or yellow halo, 5–10 % leaf area covered by lesions |
| 5 | Moderately resistant | Moderately sporulating, non-coalescing larger spots, about 2–4 mm in diameter with a distinct margin or yellow halo, 11–25 % leaf area covered by lesions |
| 7 | Susceptible | Moderately sporulating, coalescing larger spots about 4–5 mm in diameter 26–50 % leaf area covered by the lesions |
| 9 | Highly susceptible | Profusely sporulating, rapidly coalescing spots measuring more than 6 mm diameter without margins covering more than 50 % leaf area |

in rainfed condition. The remaining half of nitrogen was applied after first irrigation. All the recommended cultural practices were followed for raising the crop under irrigated and rainfed conditions.

Coordinated Research Project on Rapeseed–Mustard (Directorate of Rapeseed and Mustard, Sewar, Bharatpur, Rajasthan, Planning and Review Session 2010–2011 (AICRP-RM 2010)), which is stated below (Table 11.3).

Average disease severity (ASC) was recorded with the help of the formula given below:

$$\text{ASC} = \frac{(N-1\times 0)+(N-2\times 1)+(N-3\times 3)+(N-4\times 5)+(N-5\times 7)+(N-6\times 9)}{\text{Number of leaf samples}}\times 100.$$

Per cent disease intensity (PDI) was also calculated with the help of the following formula:

$$\text{PDI} = \frac{(N-1\times 0)+(N-2\times 1)+(N-3\times 3)+(N-4\times 5)+(N-5\times 7)+(N-6\times 9)}{\text{Number of leaf samples}\times 9}\times 100,$$

where,

$N-1$ to $N-6$ represents frequency of leaves in the respective score.

### 11.2.4 Disease Screening Techniques and Statistical Analysis

After germination, the crop was regularly watched for appearance of major diseases. The observations on disease severity were recorded on ten plants randomly selected from each genotype. The disease severity was recorded at maximum disease occurrence. The disease was scored on the basis of maximum severity following 0–9 scale as suggested in the proceeding of All India

## 11.3 Results and Discussion

The use of resistant varieties is considered to be the best method of disease control. Therefore, the study was carried out to find the sources of resistance against downy mildew and *Alternaria* blight. A total of 20 genotypes of yellow sarson were

**Table 11.4** Reactions of yellow sarson genotypes against downy mildew and *Alternaria* blight

| Grade | Reaction | Name of genotypes | | | |
|---|---|---|---|---|---|
| | | Downy mildew | No. of entries | *Alternaria* blight | No. of entries |
| 0 | Immune | - | Nil | - | Nil |
| 1 | Highly resistant | - | Nil | - | Nil |
| 3 | Resistant | - | Nil | - | Nil |
| 5 | Moderately resistant | Jagrati | 1 | Benoy, Pusa Gold, YSK 09-1, PYS 2008-2, NRCYS-05-02, YSWB-2012 and NDYS-427 | 7 |
| 7 | Susceptible | RAUDYS 07-76, YSK 09-2, Benoy, Pusa Gold, NDYS 107-1, RYS-1, YSK 09-1, DRMRYS-09-99, YSWB-2009, PYS 2007-7, YSB-2010, PYS 2008-2, NRCYS-05-02, YSWB-2012, NDYS-427, DRMRYS-09-103, NDYS 141-3, NDYS-425 and Type-151 | 19 | RAUDYS 07-76, YSK 09-2, NDYS 107-1, RYS-1, YSK 09-1, DRMRYS-09-99, YSWB-2009, PYS 2007-7, YSB-2010, DRM-RYS-09-103, NDYS 141-3, NDYS-425 and Type-151 | 13 |
| 9 | Highly susceptible | - | Nil | - | Nil |

screened for varietal reactions against the diseases under field conditions during rabi 2009–2010 and 2010–2011, and the results obtained are presented in Table 11.4. It indicated the per cent disease severity of various yellow sarson genotypes recorded obtained during 2009–2011 crop season.

### 11.3.1 Downy Mildew Screening

Out of 20 genotypes screened against downy mildew none of the genotypes was found free from the disease, only one genotype named Jagrati was found moderately resistant, (11–25% leaf area covered by the lesions moderately sporulating, non-coalescing larger spots, about 2–4 mm in diameter with a distinct margin or yellow halo) and 19 genotypes named RAUDYS 07-76, YSK 09-2, Benoy, Pusa Gold, NDYS 107-1, RYS-1, YSK 09-1, DRMRYS-09–99, YSWB-2009, PYS 2007-7, YSB-2010, PYS 2008-2, NRCYS-05-02, YSWB-2012, NDYS-427, DRMRYS-09-103, NDYS 141-3, NDYS-425 and Type-151 were found susceptible (26–50% leaf area covered by the lesions moderately sporulating, coalescing larger spots about 4–5 mm in diameter). Out of the 20 genotypes screened against downy mildew, none of the genotypes was found highly susceptible for this disease. Singh and Bhajan (2006) also reported 11 lines, namely NDYS-2, NDYS-8, NDYS-121, NDYS-132, NDYS-135, NDYS-136, NDYS-139, NDYS-140, YSC-4-1, YSC-5, YSC-24-1 as resistant and 50 lines as moderately resistant against the downy mildew disease.

### 11.3.2 *Alternaria* Blight Screening

For *Alternaria* blight, out of the 20 genotypes screened against this disease, none of the genotypes was found highly susceptible for this disease, seven genotypes named, Benoy, Pusa Gold, YSK 09-1, PYS 2008-2, NRCYS-05-02, YSWB-2012 and NDYS-427 were found moderately resistant (11–25% leaf area covered by the lesions moderately sporulating, non-coalescing larger spots, about 2–4 mm in diameter with a distinct margin or yellow halo) and 13 genotypes, namely RAUDYS 07-76, YSK 09-2, NDYS 107-1, RYS-1, YSK 09-1, DRMRYS-09–99, YSWB-2009, PYS 2007-7, YSB-2010, DRMRYS-09-103, NDYS 141-3, NDYS-425 and Type-151 were found susceptible (26–50% leaf area covered by the lesions moderately sporulating, coalescing larger spots about 4–5 mm in diameter). The crop is attacked by *A. brassicae* causing *Alternaria* leaf blight all over the country, but the severity of the disease is much higher in *B. rapa* L. var. *toria* (Shrestha and Shrestha 1992; Meena et al. 2004). Our assessments show that the critical period for disease development in the foliage was from 2nd week of December to the 3rd week of January. During this period, the disease appeared in severe form in both years causing heavy damage to the plants in the unsprayed plots. At first, the disease was observed as small greyish-brown necrotic spots on leaves, which later coalesced each other causing leaf blight (Singh et al. 2008). Kumar et al. (2009) also identified the potential source of resistance against *Alternaria* blight. They found one genotype as resistant, six genotypes as moderately resistant and 25 genotypes as moderately susceptible.

### 11.3.3 Per se Performance of Cultivars Against Downy Mildew and *Alternaria* blight

The per se performance of top-performing cultivars out of the 20 lines under different environmental conditions such as irrigated-timely sown, rainfed-timely sown and irrigated-late sown are provided in Table 11.5. The performance of those lines in terms of other traits is also given in Table 11.5 for three different climatic conditions. The highly susceptible top four yielding mustard cultivars with their average performances are discussed in this section. The mean disease severity in downy mildew of different cultivars varied from 24.69 (Jagrati) to 34.14 (YSB-2010) with the population mean of (28.89 ± 2.63). Four genotypes named Benoy (32.22), YSK 09–1 (33.15), YSWB-2009 (32.77) and YSB-2010 (34.19) showed significantly higher disease severity than population mean. Cultivar YSK 09-1 recorded maximum yield/plant (7.15 g) followed by NDYS-427 (6.32 g) and DRMRYS-09-103 (6.05) as compared to the check. The high yield of the cultivars is due to a good number of silique/plant, seed to silique ratio and 1,000-seed weight. Cultivar PYS 2007–7, followed by NDYS-427 and NDYS-141-3 had the highest oil content than other cultivars in all environments. The mean disease severity in *Alternaria* blight of different genotypes varied from 22.09 (YSWB-2012) to 32.24 (Jagrati) with the population mean of (25.62 ± 2.52). Three genotypes named RYS-1, NDYS-141-3 and Jagrati showed significantly higher disease severity (26–50 % leaf area covered by the lesions moderately sporulating, coalescing larger spots about 4–5 mm in diameter) than population mean. Several workers (Dang et al. 1995; Gupta et al. 2001; Kumar et al. 2009; Yadav et al. 2010) have realized since long, the importance of G × E interactions in influencing the performance of genotypes under environmental conditions. Alternatively, the genotypes of an individual may be such that it can develop different phenotypes in different environments such that each phenotype being better adapted for the specific environment.

## 11.4 Conclusion and Future Application

Based on the above results, none of the genotypes was found to be completely resistant (free from the diseases) against downy mildew disease; while only one genotype Jagrati was found moderately resistant. However, in case of *Alternaria* blight, seven genotypes, namely Benoy, Pusa Gold, YSK 09-1, PYS 2008-2, NRCYS-05-02, YSWB-2012 and NDYS-427 were found moderately resistant. The sources of resistance identified above against the downy mildew and *Alternaria* blight will be used further in breeding programme for developing the resistant variety of rapeseed–mustard. The above study recommended that for using disease-free seeds for sowing, select the least susceptible or moderately resistant cultivars and spray them with effective fungicide (if needed) at critical period during the disease development. In future, the developed resistant (or moderately resistant) varieties will be important in an integrated disease management programme in India.

**Table 11.5** Mean performance of genotypes in respect to 11 characters in yellow sarson

| Sl. No | Genotypes | Days to maturity | Plant height (cm) | Primary branches per plant | Siliquae per plant | Seeds per siliquae | 1,000-seed weight (g) | Seed yield per plant (g) | Oil content (%) | Downy mildew | *Alternaria* blight | Seed siliqua ratio |
|---|---|---|---|---|---|---|---|---|---|---|---|---|
| 1. | RAUDYS 07-76 | 116.3[b] | 120.34[b] | 3.31 | 51.68 | 33.716[b] | 3.59 | 5.75 | 43.61 | 27.62 | 25.32 | 49.25 |
| 2. | YSK 09-2 | 109.7 | 93.38 | 3.51 | 48.57 | 32.909[a] | 3.69 | 5.66 | 44.53 | 28.53 | 25.49 | 51.40 |
| 3. | Benoy | 109.4 | 86.13 | 4.07[a] | 55.41 | 24.45 | 3.52 | 4.57 | 44.63 | 32.22[a] | 23.96 | 52.58 |
| 4. | Pusa Gold | 110.6[b] | 110.00 | 2.94 | 37.81 | 30.70 | 4.11[b] | 5.66 | 44.22 | 30.31 | 23.74 | 51.29 |
| 5. | NDYS 107-1 | 108.7 | 92.66 | 2.91 | 35.86 | 29.69 | 4.18[b] | 4.24 | 44.57 | 28.93 | 25.13 | 51.64 |
| 6. | RYS-1 | 133.1[b] | 131.25 | 2.45 | 43.09 | 30.56 | 4.34[b] | 5.45 | 43.86 | 27.53 | 29.04[a] | 52.10 |
| 7. | YSK 09-1 | 110.6[b] | 115.60[b] | 3.43 | 68.66[b] | 29.49 | 3.96 | 7.15[b] | 44.16 | 33.15[b] | 23.97 | 54.68[a] |
| 8. | DRMRYS-09-99 | 110.9[b] | 109.89 | 3.00 | 51.12[a] | 29.98 | 3.58 | 5.75 | 44.10 | 29.10 | 25.59 | 48.05 |
| 9. | YSWB-2009 | 111.6[b] | 103.51 | 4.59 | 70.47[b] | 18.97 | 3.74 | 4.78 | 43.73 | 32.77[a] | 28.20 | 48.69 |
| 10. | PYS 2007-7 | 110.3[b] | 99.23 | 3.24 | 38.76 | 38.093[b] | 3.93 | 5.58 | 45.29[b] | 30.64 | 25.64 | 52.14 |
| 11. | YSB-2010 | 131.7[b] | 120.53[b] | 3.94 | 50.35 | 20.80 | 4.36[b] | 4.56 | 44.50 | 34.19[b] | 25.96 | 56.61[b] |
| 12. | PYS 2008-2 | 108.0 | 84.62 | 3.05 | 36.52 | 30.97 | 3.25 | 3.87 | 44.47 | 25.92 | 22.18 | 54.54[a] |
| 13. | NRCYS-05-02 | 111.2[b] | 96.68 | 3.30 | 44.86 | 32.03 | 3.76 | 5.84 | 44.51 | 25.49 | 22.59 | 52.61 |
| 14. | YSWB-2012 | 107.9 | 89.40 | 3.88 | 62.55[b] | 23.17 | 3.64 | 5.07 | 44.64 | 27.20 | 22.09 | 54.34 |
| 15. | NDYS-427 | 108.2 | 95.77 | 3.22 | 42.11 | 34.908[b] | 4.31[b] | 6.32[a] | 44.98[b] | 27.21 | 24.48 | 48.18 |
| 16. | DRM-RYS-09-103 | 109.3 | 108.40 | 3.49 | 46.41 | 29.47 | 4.46 | 6.05 | 44.21 | 29.37 | 27.15 | 52.80 |
| 17. | NDYS 141-3 | 110.8[b] | 103.54 | 2.82 | 45.86 | 32.52 | 3.68 | 5.18 | 44.95[b] | 27.42 | 29.91[b] | 48.35 |
| 18. | NDYS-425 | 112.6[b] | 115.52[b] | 2.63 | 50.06 | 30.02 | 3.81 | 5.53 | 44.06 | 26.29 | 23.87 | 49.80 |
| 19. | Jagrati | 107.3 | 118.65[b] | 3.37 | 45.64 | 31.36 | 3.68 | 5.73 | 43.99 | 24.69 | 32.24[b] | 52.77 |
| 20. | Type-151 | 110.2[b] | 116.17[b] | 3.36 | 52.58 | 21.67 | 3.95 | 4.34 | 44.18 | 29.11 | 25.79 | 53.08 |
| Population Mean | | 112.42 | 105.56 | 3.24 | 48.92 | 29.29 | 3.88 | 5.30 | 44.36 | 28.89 | 25.62 | 51.74 |
| Sem± | | 1.100 | 4.300 | 0.362 | 5.297 | 2.094 | 0.124 | 0.681 | 0.380 | 2.633 | 2.520 | 1.762 |
| CD at 5% | | 1.912 | 6.519 | 0.669 | 7.244 | 3.057 | 0.168 | 0.801 | 0.425 | 3.011 | 2.885 | 2.618 |

a Significant at 5% probability level
b Significant at 1% probability level

## References


Agrios GN (2005) Plant pathology, 5th edn. Elsevier Academic, California

AICRP-RM (2010) All India co-ordinated research project on Rapeseed-mustard. Directorate of Rapeseed-mustard. Sewar, Planning and review report, 2010–2011

Anonymous (2010) Director's report presented at the 17th Annual Group Meeting of Rapeseed-mustard Research Workers, held at RVSKVV, Gwalior

Ansari NA, Khan MW, Muheet A (1989) Effect of some factors on growth on sporulation of *Alternaria brasicae* causing *Alternaria* blight of rapeseed and mustard. Acta Bot Indic 17:49–53

Chattopadhyay C (2008) Management of diseases of rapeseed-mustard with special reference to Indian conditions. In: Kumar A, Chauhan JS, Chattopadhyay C (eds) Sustainable production of oilseeds: rapeseed mustard technology. Agrotech, Udaipur, pp 364–388

Conn KL, Tiwari JP, Awasthi RP (1990) A disease assessment key for *Alternaria* blackspot in rapeseed and mustard. Can Plant Dis Surv 70:19–22

Dang JK, Kaushik CD, Sangwan MS (1995) Quantitative relationship between *Alternaria* leaf blight of rapeseed-mustard and weather variables. Indian J Mycol Plant Pathol 25:184–188

Dange SRS, Patel RL, Patel SI, Patel KK (2002) Assessment of losses in yield due to powdery and downy mildew disease in mustard under north Gujarat conditions. J Mycol Plant Pathol 32:249–250

Degenhardt KJ, Petrie GA, Morrall RAA (1982) Effects of temperature on spore germination and infection of rapeseed by *Alternaria brassicae, A. brassicicola and A. raphani*. Canadian J Plant Pathology 4:115–118

Degenhardt KJ, Skoropad WP, Kondra ZP (1974) Effects of *Alternaria* blackspot on yield, oil content and protein content of rapeseed. Can J Plant Sci 54:795–799

FAO (2010) Food and agricultural commodities production. In: http://www.faostat.fao.org/site/339/default.aspx downloaded on 12 Apr 2010, 1226 hrs IST (2010)

Gupta K, Saharan GS, Singh D, Gupta K (2001) Sources of resistance in Indian mustard against white rust and *Alternaria* blight. Cruciferae Newsletter 23:59–60

Hong CX, Fitt BD (1995) Effect of inoculum concentration, leaf age and wetness period on the development of dark leaf and pod spot (*Alternaria brassicae*) on oil seed rape (*Brassicae napus*). Ann Appl Biol 127(2):283–295

Humpherson-Jones FM, Phelps K (1989) Climatic factors influencing spore production in *Alternaria brassicae* and *Alternaria brassicicola*. Ann Appl Biol 114:449–458

Kolte SJ, Awasthi RP, Vishwanath (1987) Assessment of yield losses due to *Alternaria* blight in rapeseed and mustard. Indian Phytopathol 40:209–211

Kumar A, Chauhan JS (2005) Strategies and future thrust areas of rapeseed -mustard research in India. Indian J Agric Sci 75:621–635

Kumar PR (1986) Rapeseed—mustard research in India. In: Proceedings of the SAARC Member Countries Counterpart Scientists Meeting for Multi-location Trial on Rapeseed-Mustard, Kathmandu, 9–11 Dec

Kumar S, Singh RB, Singh RN (2009) Fungicides and genotypes for the management of foliar disease of Rapeseed-Mustard. Proc Nat Acad Sci India 79(2):189–193

Meena PD, Meena RL, Chattopadhyay C, Kumar A (2004) Identification of critical stage of disease development and biocontrol of *Alternaria* blight of Indian mustard (*Brassica juncea*). J Phytopathol 152:204–209

Meena PD, Awasthi RP, Chattopadhyay C, Kolte SJ, Kumar A (2010) *Alternaria* blight: a chronic disease in rapeseed-mustard. J Oilseed Brassica 1(1):1–11

Meena PD, Chattopadhyay C, Kumar A, Awasthi RP, Singh R, Kaur S, Thomas L, Goyal P, Chand P (2011) Comparative study on the effect of chemicals on Alternaria blight in Indian mustard – A multi-location study in Indian J Environ Biol 32:375–379

Mridha MAU, Wheeler BEJ (1993) In vitro effects of temperature and wet periods on infection of oilseed rape by *brassicae*. Plantt Pathol 42:671–675

Nashaat NI, Heram A, Mitchell S (1996) Resistance to downy mildew and its interaction with in rapeseed-mustard. Proceeding of 2nd Int Crop Sci Congress New Delhi, pp 199

NRC-RM Sewar (2010) Package of practices and contingency plan for enhancing production of rapeseed-mustard. National Research Centre on Rapeseed-Mustard, Sewar, p 39

Ram RS, Chauhan VB (1998) Assessment of yield loss due to *Alternaria* leaf spot in various cultivars of mustard and rapeseed. J Mycopathol Res 36(2):109–111

Seidle E, Rude S, Petrie A (1995) The effect of Alternaria black spot of canola on seed quality and seed yield and studies on disease control. Agriculture and Agri-Food Canada Saskatoon, Canada, p 41

Shrestha K, Shrestha SK (1992) Major diseases of rapeseed-mustard and their control measures. In: Proc. of the First HMG-DANIDA-FAO Nat. Train. Course of Seed Production of Rapeseed-Mustard Crops, Nawalpur, pp 131–160

Singh HK, Singh RB, Kumar K, Verma OP (2008) Screening of *Brassica* genotypes against *Alternaria* blight of rapeseed-mustard caused by *Alternaria brassicae* (Berk.) sacc. Cruciferae Newsletter 27:35–36

Singh RB, Bhajan R (2006) Status of downy mildew (*Peronospora parasitoca*) and genetic variability for resistance in yellow *sarson* (*Brassica campestris* L. var. *yellow sarson*). J oilseeds Res 23(2):361–362

USDA (2011) The U.S. Department of Agriculture: agriculture production 2010. http://www.cnpp.usda.gov/dgas2011 downloaded on 27 May 2011

Yadava DK, Giri SC, Yadav S, Vasudev AK, Dass B, Raje RS, Vignesh M, Singh R, Mohapatra T, Prabhu KV (2010) Stability analysis in Indian mustard (*Brassica juncea*) varieties. Indian J Agric Sci 80(9):761–765

# Mycofloristics of Some Forest Localities in Khammam: Some New Additions to the Fungi of Andhra Pradesh, India

12

D. N. Nagaraju, I. K. Kunwar and C. Manoharachary

**Abstract**

India with its varied topography, climate, forest types, soil types, altitudes and specialized ecological niches possesses a much diversified mycoflora. India has been the cradle for fungi and one-third of global fungal diversity exists in India. The forests of Bhadrachalam, Kothagudem and Paloncha of Khammam District, AP, India which were not explored earlier were surveyed (2007–2010) for micro- and macrofungi colonizing litter, fruit, bark, humid soils, wood and dung. It is interesting to note that the survey has resulted in the critical evaluation of 11 fungi which form new additions to the fungi of A.P. India.

**Keywords**

Diversity · Forest · Fungi · Litter

## 12.1 Introduction

The estimated fungi in the world are around 1.5 million (Hawksworth 1991). The number of fungi identified in the world are 97,861 (Kirk et al. 2008) and in India it exceeds 29,000 species (Manoharachary et al. 2005). The fungal diversity that the mycologists have unraveled through conventional techniques is only a fraction of the amazing real diversity.

C. Manoharachary (✉) · D. N. Nagaraju · I. K. Kunwar
Mycology and Molecular Plant Pathology Laboratory, Dept. of Botany, Osmania University, Hyderabad, A.P. 500007, India
e-mail: cmchary@rediffmail.com

From India, fungal diversity has been worked out by many researchers. Recently available literature on biodiversity and taxonomy of fungi indicates the wealth of fungal diversity (Arya et al. 2008; Prasher et al. 2008; Pratibha and Bhat 2008, 2010; Kaviyarasan et al. 2009; Singh et al. 2009; Sureshkumar et al. 2006, 2009; Wahegaonkar 2009; Bhat 2010; Bhosle et al. 2010; Gawas and Bhat 2010; Manoharachary and Kunwar 2010a, 2010b, 2010c; Kumar et al. 2010; Kumar and Kaviyarasan 2011; Patil et al. 2011; Sharma et al. 2010; Tiwari et al. 2010).

Contributions on mitosporic fungi and macrofungi colonizing various substrates from Andhra Pradesh (AP) include that of Rao (1988) and Gopal (1990) on Basidiomycetous fungi,

R. N. Kharwar et al. (eds.), *Microbial Diversity and Biotechnology in Food Security*,
DOI 10.1007/978-81-322-1801-2_12, © Springer India 2014

and on aquatic fungi by Rao (1980) and Galaiah (1985). Significant and extensive work is available on Hyphomycetes fungi from AP (Rao and Rao 1964a, b, c, d, e, f, g, h, i, j, k, l; Manoharachary and Ramarao 1971, 1972, 1973, 1974a, b; Manoharachary et al. 1971, 1975a, b, 2006a, b, c, d, 2007a, b, 2009; Ramarao 1975; Rao and Chandravathi 1982; Rao et al. 1982; Rao and de Hoog 1986; Gopal et al. 1991; Manoharachary and Gopal 1991; Reddy et al. 1999; Rao et al. 2004, 2005; Sureshkumar et al. 2006, 2009; Bagyanarayana et al. 2009; Nagaraju et al. 2009, 2011a, b).

In spite of all such surveys conducted from time to time, there is a huge hidden wealth of fungi available in various forests of AP. We surveyed (2007–2010) the forests of Bhadrachalam, Kothagudem and Paloncha in Khammam District of AP for micro- and macrofungi from various habitats. The present study has revealed 11 new additions to the mycofloristics of AP, India. Therefore all the 11 fungi are described.

## 12.2 Information About Collection Site

Khammam District in AP is located between 16°45′ and 18°35′ N latitude and 79°47′ and 80°47′ E latitude. The district is surrounded by Chhattisgarh state to the north, Odisha state to the northeast, in AP east and west Godavari Districts to the east, Krishna District to the south, Nalgonda District to the southwest, and Warangal District to the west. Khammam District has a total area of 16,029 $km^2$ out of which forest cover is of 7945.35 $km^2$ (approximatly 49 %, as against 23.41 % for AP, and 22.80 % for India). It has five territorial and one wildlife division. The Godavari river passes through the eastern part of the district and harbors the richest biodiversity in the state; other tributaries of Godavari like Taliperu, Kinnerasani and Sabari also help in bringing prosperity to the area. The district has a total population of 2,578,927. The present investigation is confined to the forests of Bhadrachalam, Kothagudem and Paloncha of Khammam District, AP, India.

The district receives 1,096 mm rainfall annually and is relatively less drought prone. It also has two extreme temperatures in the state, as it touches 10 °C during winter and 50 °C during summer. The forest areas covered in the present study in Khammam District were of natural and mixed teak forests of good quality, but slowly the percentage of teak came down due to various reasons. The standing forest consists of primarily angiospermic plant species.

## 12.3 Materials and Methods

### 12.3.1 Collection

The fungal materials were collected from the forest localities of Khammam District for a period of 3 years (Aug 2007–July 2010). Collections were made throughout the year, with an interval of 15 days. The collection included dead twigs, barks, stems, decaying wood, pods, fruits, dead leaves, decaying leaf litter, fresh water foam, macrofungi, etc. Field data containing place of collection, date and other relevant information were immediately recorded and maintained. If possible, the name of the host or substrate on which fungal material was present was also noted in the field notes. The materials were then carried to the laboratory by placing them carefully in fresh polythene bags, for systematic observation and study of fungi.

### 12.3.2 Isolation of Microfungi

Microfungi colonizing litter, bark, twigs, senescent plant parts, water, etc. were isolated by the following methods:

1. Direct observation from plant litter/other substrates.
2. Moist chamber incubation method.
3. Isolation by particle plating method.

#### 12.3.2.1 Direct Observation from Plant Litter/Other Substrates

The sample, say a decaying leaf, litter, nut or bark, was scanned under a stereomicroscope to locate a fungal colony. A small portion of the fungal material was picked by a fine-tipped needle and placed in distilled water or lactophenol mountant and was examined under microscope. The detailed study of morpho-taxonomic characteristics of the fungus was done using a light transmitted microscope.

### 12.3.3 Moist Chamber Incubation

A thin layer of absorbent cotton superimposed by a circular piece of blotting paper was placed in a Petri dish (20 cm diameter) and drenched with distilled water. Two slides were placed crisscrossing on the filter paper. The plates were sterilized at 121 °C and 15 lbs/cm³ pressure in an autoclave for 20 min. The sample was thoroughly washed in sterile distilled water, placed on the sterile slides in moist plates and incubated at room temperature. Beginning from the 3rd day, the incubated samples were scanned daily under a stereomicroscope for growth of the fungi. The fungal colony was picked up and mounted on a slide containing a drop of distilled water or lactophenol or lactophenol cotton blue for microscopic examination.

### 12.3.4 Isolation by Particle-Plating

Decaying leaves, twigs or bark were cut into small pieces and grounded to fine particles in an electric blender. The particles were filtered through three superimposed metal sieves with mesh size of 1,000, 250 and 100 µm. Fine particles of size between 100 and 250 µm, trapped in the lower sieves were repeatedly washed in sterile distilled water, diluted to suitable concentration and plated onto malt extract agar (MEA) medium incorporated with a mixture of antibiotics (bacitracin 0.02 g, neomycin 0.02 g, penicillin G 0.02 g, polymyxin 0.02 g, streptomycin 0.02 g and tetramycin 0.04 g dissolved in 10 ml of distilled water and added to 1 L of MEA medium). The fungal hyphae arising from the particles were aseptically and individually transferred to fresh MEA slants (Bills and Polishook 1994). Some fungi sporulated in culture after several days/weeks of incubation. These were examined under the microscope, isolated and identified.

### 12.3.5 Drying and Preservation

The materials brought to the laboratory were taken out of the package and necessary drying was done. The materials were then fumigated with the help of 0.1 % mercuric chloride, 1 % silver nitrate or 4 % formalin vapour (Domsch et al. 1980) to prevent contamination by other microorganisms. The materials were trimmed properly to have only the parts where suspected colonies of fungi could be found, and then colonies were marked with indelible ink after being observed under the low power of a binocular dissection microscope.

Stiff and quality paper folders of standard size (15 × 10.5 cm) were used for the preservation of the trimmed material. Before preservation, each material was wrapped in a clean and soft tissue paper. Labels of CABI standard size (12.6 × 7.69 cm) with relevant details, such as: name of the collector, date of collection and accession number, etc. were prepared before affixing them to the folder, accession number given in abbreviation being OUFH/DNR/No (OUFH standing for Osmania University Fungal Herbarium and DNR for collector's name in abbreviation). Each of the paper packets or folder was placed in a polythene cover to prevent desiccation and a piece of naphthalene ball was placed inside the polythene cover to prevent infestation of mites/insects, etc. All these folders were arranged alphabetically, taxa wise in suitable containers.

### 12.3.6 Mountant and Stains

Aman's lactophenol (Dade and Gunnel 1969) (phenol crystals—20 g, lactic acid—20 g, glycerol—40 g, water—20 g) was prepared for microscopic study. For the observation of hyaline fungal material, 0.05 g of cotton blue was added to the preparation. To observe dematiaceous nature of the material, mounts were also prepared in distilled water.

### 12.3.7 Identification and Deposition of Fungal Herbaria

Identification of fungi was done with the help of available manuals such as Subramanian (1971), Ellis (1971, 1976), Matsushima (1975), Domsch et al. (1980); Ellis and Ellis (1998), Nagamani et al. (2006), etc. besides taking help from relevant mycological papers from CABI, UK and other taxonomic literature published in different national and international journals. Some novel and interesting taxa were deposited in HCIO, IARI, New Delhi and received the accession numbers.

### 12.3.8 Study of Macrofungi

The methods given by Kaviyarasan et al. (2009) have been used in the present study. Fresh materials were collected in polythene bags and subjected for further study on the day of collection. Field data, viz. shape, size, colour, texture were recorded besides taking field photograph.

The laboratory observations included morphology and thin sections of gill tissue, vegetative mycelia, pileus, annulus, sporeprint, basidospores, etc. The collected material was dried, packed, descriptions and identifications were made as the per procedures of Kaviyarasan et al. (2009). Photographs of macrofungi were taken, and photomicrographs were taken with the help of Leica microscope with attached camera.

## 12.4 Results and Discussion

Fungal diversity of several parts of the world is known and huge literature is available about the fungi from soil, litter, dung, water, wood, bark and other substrates (Barron 1968; Subramanian 1971; Ellis 1971, 1976; Ingold 1975; Matsushima 1975; Dix and Webster 1995; Kirk et al 2008). About one third (around 29,000) of the fungal diversity of the globe exists in India (Manoharachary et al. 2005). Bhat et al. (2009) reported that Western Ghats and Eastern Himalayas are very rich in fungal diversity. The present investigation has shown that varied substrates under different climatic conditions, altitudes and forest vegetation along with edaphic factors help in the colonization of fungi, indicating that there is remarkable hidden wealth of fungi in different geographic parts of the country.

The mycofloristic survey revealed the association of 131 fungal species of which 96 were microfungi and 35 being macrofungi. The percentage occurrence of fungal groups was studied, and 5 % belonged to Zygomycotina, 13 % to Ascomycotina, 26 % to Basidiomycotina and 56 % of fungi belonged to anamorphic fungi. Critical evaluation of fungi and survey of literature (Butler and Bisby 1960; Bilgrami et al. 1979, 1981; Sarbhoy et al. 1986, 1996; Jamaluddin et al. 2004) indicates that 11 fungi form new additions to the fungi of AP, India, hence described.

*Glomerularia corni* (Peck) D.M. Hend. (1961) Notes. R. Bot. Gdn Edinb. 23(4):500 (Fig. 12.1a, b).

Conidiophores borne in groups in spots in living leaves, mostly short, simple or divided into branches, conidiophores up to 126 µm long; conidia (aleuriospores) globose, in short chains, somewhat unequally clustered forming few spored heads, 1-celled, hyaline, 4.8–7.4 µm.

Material on litter from Bhadrachalam Forest of Khammam District, AP, collected on 25th Dec 2009, Coll. DNR, OUFH. No. 741.

*Gyrothrix hughesii* Pirozynski (1962) Mycol. Pap. 84 (Fig. 12.2).

Colonies effuse, mouse gray to brown, velvety, irregular, up to 5 mm diameter, scattered or coalescing; mycelium superficial, composed of a loose network of branched and anastomizing, smooth walled, septate, subhyaline to dilute olivaceous hyphae bearing sporogenous cells, here and there becoming thickened and dark brown at the point of origin of the setae; setae erect,

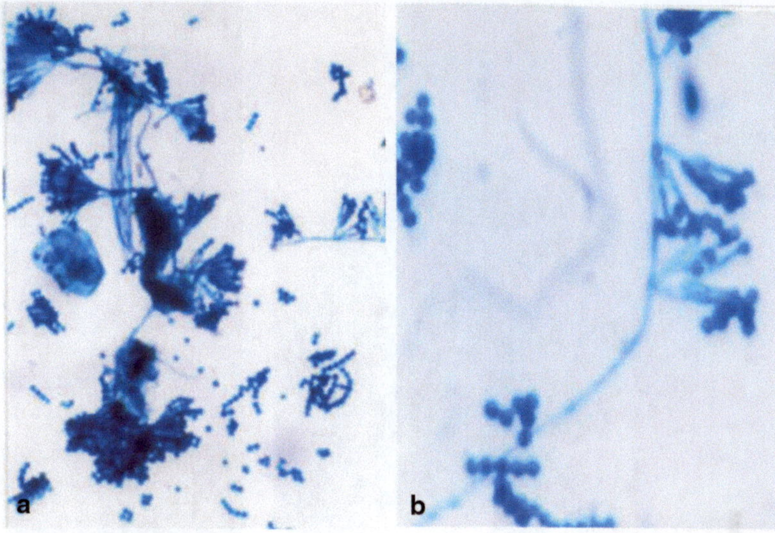


**Fig. 12.1** New additions to the fungi of AP *Glomerularia corni.* **a** =100x. **b** =400x

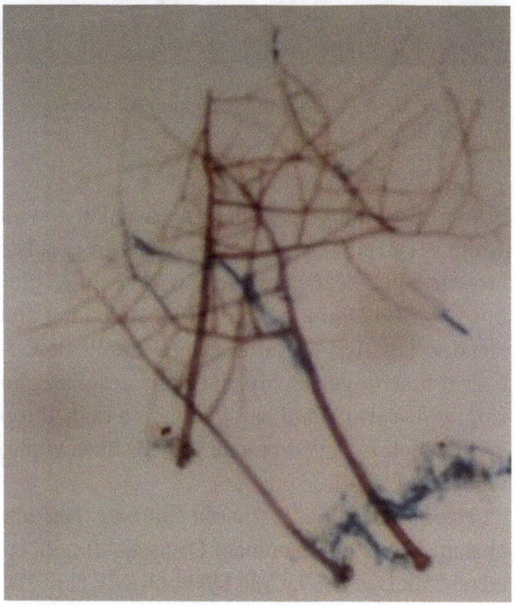

**Fig. 12.2** New additions to the fungi of AP *Gyrothrix hughesii.* 100x

distinctly septate, thick walled, brown, translucent, smooth; two or three times branched, 70–120 µm high, 3–4 µm wide at the base, 0.5 µm at the apices of the branches, branches arising in one or two whorls, along the vertical axis, erect, curling inwards, paler in colour towards the apices; sporogenous cells born laterally on the superficial hyphae, not crowded, obclavate to lageniform, subhyaline, 7–12 µm high, 3–5 µm broad below, 1 µm or less above; conidia aggregated into a whitish layer at the bases of setae, cylindrical to fusiform, continuous, hyaline, 10–15 × 1.5–2 µm.

Material on Eucalyptus leaf collected from Bhadrachalam Forest, Khammam District, AP, 14th Nov 2008. Coll. DNR, OUFH. No. 745.

The present isolate differs from type description in having slightly bigger conidia.

*Libertella betulina* Desm. (1830), Annls. Sci. Nat., Bot., se´r. 1 19: 276 (Fig. 12.3).

Acervuli small or expanded, at first pinkish, covered by the elevated periderm, then divided into many tortuous chambers, rich golden yellow; conidiophores crowded, branched, acicular, nearly straight, about as long as the conidia; conidia yellow in mass, fusoid, faintly curved, pointed at both ends, issuing in golden-yellow tendrils, 13–16 × 0.75–1 µm. *Libertella betulina* is the pycnidial stage of Diatrype stigma.

Collected on *Sesbania sp.* fruit litter from Bhadrachalam Forest of Khammam District, AP, on 26th July, 2009. Coll. DNR, OUFH No. 752.

The present fungus is variable in many respects. However, for the present, it is accommodated in *L. betulina.*

*Amanita fulva* Fr. (1815) Observ. mycol. (Havniae) 1: 2 (Fig. 12.4).

Cap 4–7 cm, semi-ovate at first, soon expending, often slightly depressed with umbo. Splitting at margin which is noticeably striate. Orange

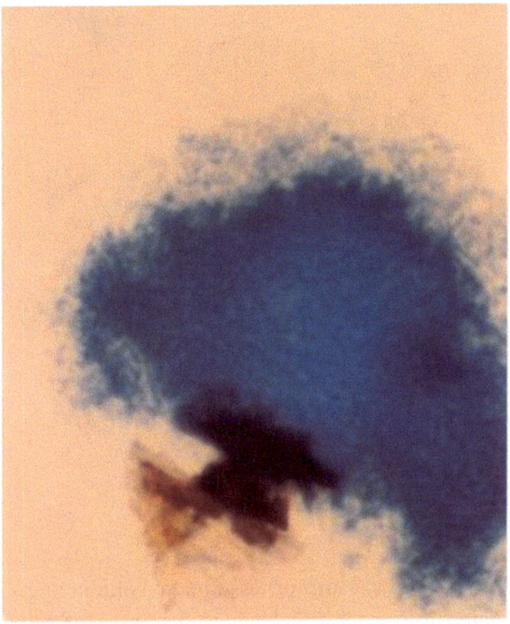

**Fig. 12.3** New additions to the fungi of AP *Libertella betulina*. 200×

**Fig. 12.5** New additions to the fungi of AP *Clavulinopsis dichotoma*

**Fig. 12.4** New additions to the fungi of AP *Amanita fulva*

brown becoming browner, occasionally with white velar remains; gills pure white, free, rather crowded but neatly spaced; spores white, globose, 9–11(12) μm, non-amyloid; stipe long and slender, tapering upwards, white, thinly floccullose at first, tinged tawny especially near base, no ring; volva white and bag like with a ragged top edge which is tinged orange brown; flesh white, tender and fragile.

Habitat: Material on humid soil from Bhadrachalam Forest of Khammam District, AP, collected on 25th Dec 2009, Coll. DNR, OUMF No. 303.

The present collection has got smaller spores and bigger fruitbodies than type description.

*Clavulinopsis dichotoma* (Godey) Corner (1950), Ann. Bot. Mem. 1:365 (Fig. 12.5).

Fruit bodies up to 4 cm high, up to 2 cm broad, branching dichotomously throughout; stipe up to 2 cm long, up to 2 mm thick, equal or tapering slightly downward, white, spreading up to 1.5 mm, bone white, terete; axils narrowly angled below, acutely angled to rounded in ultimate rank, often decurrent by a depressed line, apices swollen, irregularly lobed or cusped to subturbinate, up to 2 mm broad.

**Fig. 12.6** New additions to the fungi of AP *Coprinus cinereus*

**Fig. 12.7** New additions to the fungi of AP *Lentinus cladopus*

Material on humid soil from Bhadrachalam Forest of Khammam District, AP, collected on 25th Dec 2009, Coll. DNR, OUMF No. 307.

The measurments are similar to the type description.

*Coprinus cinereus* (Schaeff.) Gray (1821), Nat. Arr. Brit. Pl. (London) 1:634 (Fig. 12.6).

Cap 2.5–5 cm across when expanded and 1–3 cm high, oval or campanulate, then soon conico expanded, later splitting radially before auto-digesting, at first it is covered with a dense dirty-white wooly coating, which breaks up into patches and eventually falls away leaving the cap shiny, the centre brownish gray and the margin dark grey and striate; gills crowded and free, white but very quickly dissolving into a black liquid; spores violaceous black, elliptical and smooth with a germ pore, 10–12 × 6–6.5 μm.

Material on dung heap from Bayyaram Forest of Khammam District, AP, collected on 25th Dec 2009, Coll. DNR, OUMF No. 308.

Spore colour and germspores are more clear than the type description.

*Lentinus cladopus* Le'v in (1844), Ann. Sci. Nat. Bot., 32: 174 (Fig. 12.7).

Pileus up to 100 mm broad, white, convex, depressed to infundibuliform, margin striate, sometimes recurved; pileus surface smooth in the centre with squarrose scales in the margin; lamellae decurrent, white to cream, up to 1 mm wide; lamellulae present; stipe up to 75 mm long and 10 mm wide, cylindrical, solid, white, smooth, often stipes two or more; basidiocarp arising from a single base; pileus surface made up of interwoven hyphae, hymenophoral trama regular; context white, thin, consisting of dimitic hyphal system with generative and binding hyphae; generative hyphae up to 3 mm diameter, hyaline, thin walled and with prominent clamp connections, binding hyphae up to 8 mm diameter; hyaline, thick walled, and with tapering lateral branches; spore print colour white; spores ellipsoid hyaline, thin walled, 4.2–5.6 × 2.5–3.5 μm; basidia clavate, four spored 12.5–140 × 4.2–5.6 μm; pleurocystidia and cheilocystidia absent; hyphal pegs up to 50 μm long and 5 μm wide, consisting of thin walled cylindric hyphae.

**Fig. 12.8** New additions to the fungi of AP *Marasmius silvicola*

On soil, Kothagudem Forest, Khammam District, AP, 7th July 2008, Coll. DNR, OUMF No. 319.

Basidia and basidiospores are marginally smaller than the type description.

*Marasmius silvicola* Singer and Digilio (1953), Lilloa 25 199 (Fig. 12.8).

Pileus 2.5–7.5 cm diameter, broadly convex to plane, umbonate, centre brown, rest light brown, dry, margin entire, smooth to crenate, decurved to plane, plicate-striate to pellucid, lamellae adnexa-adnate, subdistant, grayish orange; lamellulae of 2–3 different lengths seen; stipe 5.8–12(16) × 0.5–1 cm, cylindrical, cartilaginous, smooth, hollow, slightly tapering towards apex, apex grayish orange, base light brown; spores hyaline, cylindrical, guttulate, in amyloid, thin walled.

On soil humus from Bhadrachalam Forest, Khammam District, AP, 14th Sep 2008. Coll. DNR, OUMF No. 324.

The descriptive account is comparable to type description.

*Mycena leptocephala* (Pers.) Gillet (1876), *Hymenomycetes* (Alencon) 267 (Fig. 12.9).

Cap 1–3 cm, conic-campanulate, striate, dull grey or grey brown but can be olive tinted or nearly black; gills subdistant, adnate, linear or ventricose, whitish then brownish or dark grey, edges white; spores white, elliptic cylindric; basidia 4-spored, 8–12 × 4.5–6 μm, amyloid; stipe 5–8 cm long, slender and rigid, smooth and shining with a villose base, more or less concolorous with cap; flesh whitish and thin, has a mild taste and nitrous smell.

**Fig. 12.9** New additions to the fungi of AP *Mycena leptocephala*

Habitat: Material on soil from Yellandu Forest of Khammam District, AP, collected on 25th Dec 2009, Coll. DNR, OUMF No. 326.

Basidia are marginally bigger in size than type description.

*Polyporus grammocephalus* Berk, Hooker's (1953), London J. Bot. 1:184 (Fig. 12.10).

Sporophore annual, stipitate or nearly sessile, solitary or imbricate, soft and fleshy, on drying rigid, fan shaped, obovate or reniform, 30–170 mm across, 3 mm thick, stipe lateral, expanding into pileus above, up to 20 μm long and broad, upper surface white when fresh, ochraceous or reddish brown when dry, usually with fine striations, smooth or minutely scaly, evident under hand lens, margin smooth, entire or wavy; context light buff, corky, up to 2 mm

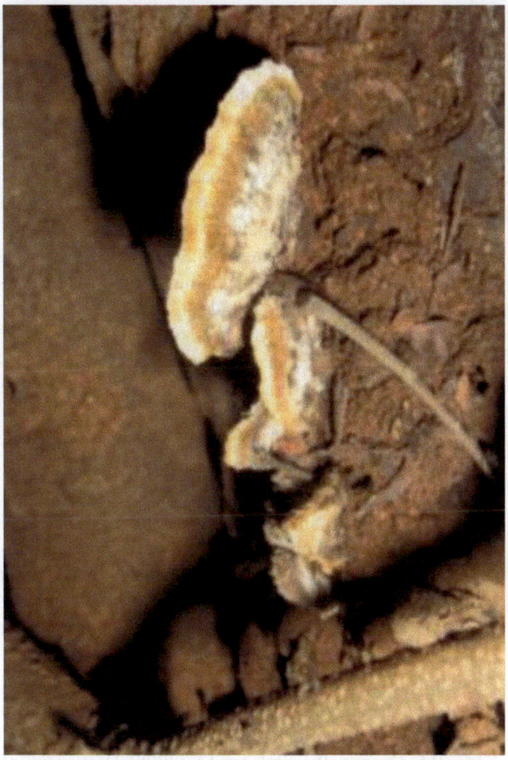

**Fig. 12.10** New additions to the fungi of AP *Polyporus grammocephalus*

thick; hymenial surface white when fresh, yellow to brown, usually with a silky sheen, margin fertile; pores round, regular or irregular, sometimes the adjacent ones join to form large ones, 4–5 per mm, pore wall thin, pore tube concolorous up to 1 mm long; basidia broadly clavate, 12–14 × 5.7–6.4 µm; basidiospores round, hyaline 4–7.5 × 2.2–3 µm diameter; hyphae, hyaline, thick walled with narrow or little lumen, flexuous, unbranched, 3–5 µm broad; hyaline, thin walled, branched with occasional clamp connections, often collapsing, 2–3.5 µm broad, less common.

On dead wood from Bhadrachalam Forest, Khammam District, AP, 3rd Aug 2008. Coll. DNR, OUMF No. 330.

Sporophores and clamp connections are more prominent than original description.

*Scleroderma citrinum* Persoon ex Soothill and Fairhurst (1978), The New Field Guide to fungi: 168 (Fig. 12.11).

**Fig. 12.11** New additions to the fungi of AP *Scleroderma citrinum*

Fruit balls about 5–10 cm in diameter, growing on the ground most often on base soils in the woods and waste lands, peridium is slightly coloured, the fruit body arises from cord like mycelium coming from the soil, the mycelium smooth, septate, with clamp connections or sometimes forming mycorhiza with surrounding trees, grasses and plants; fruit bodies or basidiocarp vertical section shows outer thick peridium made up of mycelium and skeletal cells, central mass known as gleba with a system of sterile veins there are no columella, capillitium, the basidiospores arise on the basidia which are sessile and autodigested and the basidiospores are freely suspended, globose or spherical, 10–15 µm in diameter, verrucose or reticulate, pinkish or blackish brown with purple tinge. Soothill and Fairhurst (1978), recorded that the basidiocarps are visible from August–December whereas we have observed them only from the month of November onwards, the peridium ruptures irregularly to release the spores.

Material on humid soil Bhadrachalam Forest of Khammam District, AP, collected on 5th Apr 2009, Coll. DNR, OUMF No. 332.

Mostly sessile basidiospores are observed while in type description it is mentioned that they are mostly attached.

The above data clearly indicate that there is a vast potential of unexplored tropical fungi in India. Many of the habitats and substrates in the tropical climate are now realized to be rich source and natural repositories of useful fungal genomes and their biotechnological utility.

**Acknowledgments** The authors are grateful to Ministry of Environment and Forests for the financial support.

## References

Arya A, Albert S, Nagadas PK (2008) New and interesting records of Basidiomycetous fungi from Ratanmahal Wildlife Sanctuary, Gujarat, India. J Mycol Plant Pathol 38(2):221–226

Bagyanarayana G, Krishna Rao N, Kunwar IK (2009) *Manoharachariella*, a new dematiaceous hyphomycetous genus from India. Mycotaxon 109:301–30

Barron GL (1968) The genera of Hyphomycetes from soil. The Williams & Wilkins Co., Baltimore

Bhat DJ (2010) Fascinating microfungi (Hyphomycetes) of the Western Ghats-India. Broadway Book Centre, Goa

Bhat DJ, Pratibha J, Gawas P, Sarita KY, Swapnaja D (2009) Diversity of microfungi in the forests of Western Ghats in Goa and surrounding regions. In: Krishnan S, Bhat DJ (eds) Plant and fungal biodiversity and bioprospecting. Broadway Book Centre, Panaji, pp 117–133

Bhosle S, Ranadeve K, Bapat G, Garad S, Deshande G, Vaidya J (2010) Taxonomy and diversity of *Ganoderma* from the Western Ghats of Maharashtra (India). Mycosphere 1:3

Bills GF, Polishook JD (1994) Abundance and diversity of microfungi in leaf litter of a lowland rain forest in Costa Rica. Mycologia 86:187–198

Bilgrami KS, Jamaluddin, Rizvi MA (1979) The fungi of India Part I (List and reference). Today and Tomorrow Printers and Publication, New Delhi

Bilgrami KS, Jamaluddin, Rizvi MA (1981) The fungi of India Part II (Host Index and Addenda). Today and Tomorrow Printers and Publication, New Delhi

Butler EJ, Bisby G (1960) The fungi of India (Revised by RS Vasudeva). ICAR Publ, New Delhi

Dade HA, Gunnel J (1969) Class work with fungi: notes for teachers, 2nd edn. CMI, Kew

Dix NJ, Webster J (1995) Biodiversity of fungi: inventorying and monitoring methods. Elsevier Academic Press, New York, pp 128–172

Domsch KH, Gams W, Anderson TH (1980) Compendium of soil fungi, vol I, II. Academic Press, London

Ellis MB (1971) Dematiaceous Hyphomycetes. CMI, Kew

Ellis MB (1976) More Dematiaceous Hyphomycetes. CMI, Kew

Ellis MB, Ellis JP (1998) Microfungi on miscellaneous substrates: an identification hand book. The Richmond Publishing Co Ltd, London

Galaiah K (1985) Taxo-ecological studies on fungi of submerged leaves and forest soil profiles from Ananthagiri hills, Andhra Pradesh, India. PhD Thesis, Osmania University, Hyderabad

Gawas PS, Bhat DJ (2010) Mycoflora associated with *Aegle marmelos*, a medicinal plant of forests of Western Ghats, India. Kavaka 37/38:37–46

Gopal KV (1990) Studies on higher fungi of few forest localities of Andhra Pradesh, India. PhD Thesis, Osmania University, Hyderabad, pp 154

Gopal KV, Manoharachary C, Rao NSS (1991) Agaricales from Andhra Pradesh, India. J Indian Bot Soc 70:130–160

Hawksworth DL (1991) The fungal dimension of biodiversity: magnitude, significance, and conservation. Mycol Res 95:641–655

Ingold CT (1975) An illustrated guide to aquatic and water-borne hyphomycetes with notes on their biology. Freshw Biol Assoc (Scientific Publ No. 30) 61(2):270

Jamaluddin, Goswami MG, Ojha (2004) Fungi of India (1989–2001) Scientific Publishers, India

Kaviyarasan V, Ravindran C, Senthilarasu G, Narayanan K, Kumaresan V, Kumar M (2009) A field guide to South Indian Agaricales. Madras University, Chennai

Kirk PM, Cannon PF, Stalpers JA (2008) Dictionary of the fungi, 10th edn. CABI, UK

Kumar M, Jagadeesh R, Kaviyarasan V (2010) A new species of *Volvariella* from Tamilnadu, India. Kavaka 37/38:21–23

Kumar M, Kaviyarasan V (2011) A rare agaric (Agaricomycetes: Agaricaceae) from a sacred grove of eastern ghats, India. J Threat Taxa 3(5):1778–1781

Manoharachary C, Kunwar IK (2010a) Taxonomy of Eriocercosporaceous fungi. In: K.G. Mukerji, C. Manoharachary (eds) Taxonomy and Ecology of Indian fungi. IK Internat Pvt Ltd, New Delhi, pp 9–12

Manoharachary C, Kunwar IK (2010b) *Spegazzinia* species from India. In: K.G. Mukerji, C. Manoharachary (eds) Taxonomy and Ecology of Indian fungi. IK Internat Pvt Ltd, New Delhi, pp 13–18

Manoharachary C, Kunwar IK (2010c) *Helicosporium* species from India. In: K.G. Mukerji, C. Manoharachary (eds) Taxonomy and Ecology of Indian fungi. IK Internat Pvt Ltd, New Delhi, pp 1–7

Manoharachary C, Gopal KV (1991) Mycofloristics of Agaricales from Andhra Pradesh and some aspects of mushroom cultivation in India. Mushrooms JMS, 3–5.

Manoharachary C, Ramarao P (1971) A new *Annellophorella* on *Zizyphus*. Curr Sci 40:471–472

Manoharachary C, Ramarao P (1972) *Stemphyliomma terricola* sp nov from pond mud. Curr Sci 41:718–719

Manoharachary C, Ramarao P (1973) *Thielavia boothi* sp nov from pond mud. Trans Br mycol Soc 61:196–198

Manoharachary C, Ramarao P (1974a) Studies on soil fungi VI. Notes on some interesting fungi from soils of Hyderabad, India. Indian J Mycol Pl Pathol 41:89–92

Manoharachary C, Ramarao P (1974b) Studies on soil fungi VIII. Notes on some fungi from Hyderabad. Indian Phytopathol 27(4):653–656

Manoharachary C, Raghuveer Rao P, Ramarao P (1971) A new *Trimmatostroma* from soil. Curr Sci 46(22):788

Manoharachary C, Karan D, Raghuveer Rao P (1975a) Ecological distribution of certain fungi Imperfecti. New Bot 3/4:154–155

Manoharachary C, Raghuveer Rao P, Rehana AR, Ramarao P (1975b) Notes on microfungi from Andhra Pradesh—I. A new species of *Graphium* from soil. Nova Hedwig 25:473–476.

Manoharachary C, Agarwal DK, Suresh Kumar G, Kunwar IK, Sharath Babu K (2006a) *Memnoniella mohanramii* sp nov and *Zygosporium anupamvermae* sp nov from India. Indian Phytopathol 59(4):489–491

Manoharachary C, Rao NK, Kunwar IK, Agarwal DK (2006b) Two new species of *Trichocladium* Herz. from India. Indian Phytopathol 59(3):356–358

Manoharachary C, Rao NK, Agarwal DK, Kunwar IK (2006c) Two new species of *Acrodictys* M.B. Ellis from India. Indian Phytopathol 59(1):91–93

Manoharachary C, Rao NK, Kunwar IK, Agarwal DK (2006d) *Cheiromyces ananthgiriensis* sp nov and *Vanibandha* gen nov from India. Indian Phytopathol 59(2):210–214

Manoharachary C, Bagyanarayana G, Rao NK, Kunwar IK (2009) *Chasakopama*, a new dematiaceous hyphomycetous genus from India. Mycotaxon 110:459–464

Manoharachary C, Kunwar IK, Rao NK (2007a) Two new species of *Dictyosporium* from India. Indian Phytopathol 60(3):341–344

Manoharachary C, Kunwar IK, Rao NK, Agarwal DK (2007b) *Deightoniella mayeei* sp nov and *Dischloridium gangawanei* sp nov from A.P., India. Indian Phytopathol 60(1):88–91

Manoharachary C, Sridhar K, Singh R, Adholeya A, Suryanarayanan TS, Rawat S, Johri BN (2005) Fungal biodiversity: distribution, conservation and prospecting of fungi from India. Curr Sci 89(1):58–71

Matsushima T (1975) Icones Microfungorum a Matsushima Lectorum. Kobe, Japan

Nagamani A, Kunwar IK, Manoharachary C (2006) Handbook of soil fungi. IK Internat Publication House Pvt Ltd, New Delhi

Nagaraju D, Kunwar IK, Manoharachary C Agarwal DK (2009) *Ulocladium gpagarwalii* and *U. lignicola* two new sp nov from Andhra Pradesh. Indian Phytopath 62(2):237–239

Nagaraju D, Kunwar IK, Sureshkumar G, Manoharachary C (2011a) *Custingophora lignicola* sp nov and *Chaetopsina indica* sp nov from India. J Mycol Plant Pathol 41(1):6–10

Nagaraju D, Kunwar IK, Sureshkumar G, Manoharachary C (2011b) A new synnematous hyphomycetous fungus—*Bhadradriella* gen nov from India. J Mycol Plant Pathol 41(2):238–240

Patil VRSY, Patil SY, Nemade LC, Borse BD, Naik VS (2011) Aquatic fungi from Buldhana district. Curr Bot 2(1):56–58

Prasher IB, Manoharachary C, Kunwar IK, Agarwal DK (2008) New species of *Dicranidion* Harkn from India. Indian Phytopath 61(3):367–368

Pratibha J, Bhat DJ (2008) New and unusual hyphomycetes from Mahabaleshwar, India. Mycotaxon 105:423–432

Prathibha J, Bhat DJ (2010) *Cercospora* spp from Goa and neighboring areas of Karnataka, Maharashtra. Kavaka 37/38:69–78

Ramarao P (1975) Notes on microfungi from AP—I: a new species of *Graphium* from soil. Nova Hedwigia 26:473–476

Rao MM (1980) Studies on the myco-ecology of certain aquatic, semi-aquatic and terrestrial ecosystems of Mannanur Forest, A.P, India. PhD Thesis, Osmania University, Hyderabad

Rao D, Rao PR (1964) *Vrikshopama*—a new genus of Stillbaceae. Mycopath Mycol Appl 23(4):287–290

Rao D, Rao PR (1964b) A new species of *Edmundmasonia* Subram from Hyderabad. Mycopath Mycol Appl 22(4):242–244

Rao D, Rao PR (1964c) Some allied Dematiaceae *Amerosporae* from India. Trans Am Micros Soc 83(4):399–406

Rao D, Rao PR (1964d) *Sporoschisma* Berk & Br from India. Mycopath Mycol Appl 24(2):81–84

Rao PR, Rao D (1964e) *Berkleasmium* Zobel from India. Mycopath Mycol Appl. 22(4):311–314

Rao PR, Rao D (1964f) *Diplorhinotrichum* Hohn from India. Mycopathol Mycol Appl 23:291–293

Rao PR, Rao D (1964g) *Piricauda* Bubak from India. Nature 204:200

Rao PR, Rao D (1964h) Some allied Dematiaceae Dictyosporae from India. Mycopathol Mycol Appl 23(1):23–28

Rao PR, Rao D (1964i) Some *Helicosporae* from Hyderabad. Mycopathol Mycol Appl 22(1):47–54

Rao PR, Rao D (1964j) Some species of *Camposporium* Harkn from India. Antonie von Leeuwenhoek 30:60–64

Rao PR, Rao D (1964k) The genus *Periconia* from India. Mycopathol Mycol Appl 22(4):285–310

Rao PR, Rao D (1964l) *Kumanasamuha* a new genus of Dematiaceae. Mycopathol Mycol Appl 22:291–293

Rao V, de Hoog GS (1986) New or critical Hyphomycetes from India. Stud Mycol 28:1–84

Rao NK, Manoharachary C, Kunwar IK, Agarwal DK (2004) *Hawksworthia* gen. nov. from India. Indian Phytopath 57(4):499–500

Rao NK, Agarwal DK, Kunwar IK (2005) *Manoharachariomyces*, a new hyphomycetous fungus from India. Indian Phytopath 58(1):96–99

Rao NSSS (1988) Studies on macrofungi of few forest localities of Andhra Pradesh, India. PhD Thesis, Osmania University, Hyderabad

Rao PR, Chandravathi CL (1982) A new species of *Passalora* Fries from India. Biol Bull India 4(3):195–196

Rao PR, Manoharachary C, Ramarao P (1982) *Eriocercospora websteri* sp. nov. and related species. Curr Sci 51:1155–1156

Reddy BS, Manoharachary C, Rao V (1999) Two new Hyphomycetous fungal species from India. J Indian Bot Soc 76:173–175

Sarbhoy AK, Varshney JL, Agarwal DK (1986) Fungi of India (1977–81). Associated Publ Co, New Delhi

Sarbhoy AK, Varshney JL, Agarwal DK (1996) Fungi of India (1982–92). CBS Publ & Distributions, New Delhi

Sharma K, Luka LS, Deo S (2010) Soil mycoflora of Lachung, Sikkim, India. Kavaka 37/38:67–68

Singh SK, Gaikwad VP, Waingarkar VM (2009) A new endophytic ascomycete from healthy leaves of *Pongamia pinnata* Merr. Indian Phytopath 62(1):124–125

Soothill E, Fairhurst A (1978) The new field guide to fungi. Michael Joseph, London, pp 191

Subramanian CV (1971) Hyphomycetes, an account of Indian species, except Cercosporae. ICAR, New Delhi

Sureshkumar G, Manoharachary C, Kunwar IK (2009) *Bhadradriomyces*, a new Dematiaceous Hyphomycetous genus from India. J Mycol Plant Pathol 39(2):238–240

Sureshkumar G, Sharathbabu K, Kunwar IK, Manoharachary C, Prasad V (2006) New records of fungi from India. J Mycol Plant Pathol 36(1):8–10

Tiwari CK, Parihar J, Verma RK (2010) Additions to wood decaying fungi of India. J Threat Taxa 2(6):970–973

Wahegaonkar N, Salunkhe SM, Palsingankar PL, Shinde SY (2009) Fungal diversity in soil samples from cultivated, barren and garden lands. J Mycol Plant Pathol 39(3):462–467

# The *Gomphus* Paradox of Meghalaya: Wild Edible Fungus or a Poisonous Mushroom?

13

Polashree Khaund and S. R. Joshi


**Abstract**

Wild edible mushrooms form an integral part of the local cuisine among the ethnic Khasi tribe of Meghalaya. Edible macrofungi collected from their wild habitats are sold extensively in the traditional rural markets of the east Khasi hills of Meghalaya. Among the mushroom species consumed as local delicacy, the consumption of *Gomphus* is particularly interesting owing to the fact that members of this genus have been reported to be poisonous. Molecular characterization of the mushroom specimens collected from the local markets showing morphological resemblance to *Gomphus* spp. using the fungi-specific primers nu-SSU-0817–5 and nu-SSU-1536–3 revealed them to be related to *Gomphus floccosus*. Consumption of this species is not recommended owing to the presence of norcaperatic acid which is indicated in the development of gastrointestinal disorders with delayed onset of typical symptoms like nausea, vomiting and diarrhea. The consumption of this poisonous mushroom by the ethnic population without any apparent harm may be hypothesized to reduced toxicity of the specific local strain available in the region, indigenous cooking practices and methodology or even to genetic resistance to the mushroom toxin among the local population. Efforts are currently being made to document the norcaperatic acid levels, traditional knowledge relating to the usage of this species and also to develop molecular tools for proper taxonomic identification of the regional strains collected from the traditional markets and natural habitats. The chapter deals with collection of *Gomphus* spp. from the traditional markets of Meghalaya, to document the morphological characteristics of the collected specimens, to evaluate the antimicrobial activity of crude metabolites against selected pathogenic bacterial strains and to carry out their molecular characterization.



S. R. Joshi (✉) · P. Khaund
Microbiology Laboratory, Department of Biotechnology & Bioinformatics, North-Eastern Hill University, Shillong, Meghalaya 793022, India
e-mail: srjoshi2006@yahoo.co.in



R. N. Kharwar et al. (eds.), *Microbial Diversity and Biotechnology in Food Security*, DOI 10.1007/978-81-322-1801-2_13, © Springer India 2014

**Keywords**

Wild edible mushroom · Khasi tribe · Traditional rural markets · *Gomphus floccosus* · Poisonous mushroom · Meghalaya · India

## 13.1 Introduction

Wild edible fungi (WEF) are an important source of food and income in both developing and developed countries (Hosford et al. 1997; Wong et al. 2001; Boa 2004). The state of Meghalaya in the North-eastern region of India is rich in a variety of natural forest resources including mushrooms. Edible mushrooms are collected from these forests on the basis of ethnomycological knowledge and are sold extensively in the traditional rural markets of the region. Wild edible mushrooms form an integral part of the local cuisine among the ethnic tribes of Meghalaya. The discrimination of the poisonous nature for the collected mushrooms is based on the traditional knowledge of the tribes.

Among the mushroom species consumed as local delicacy, the consumption of *Gomphus* is particularly interesting owing to the fact that members of this genus have been reported to be poisonous (Henry and Sullivan 1969). There are a few reports on wild edible mushrooms of Meghalaya and their nutritional values (Barua et al. 1998; Agrahar-Murugkar and Subbulakshmi 2005) but a comprehensive evaluation on the *Gomphus* poisonous mushroom from the region is lacking. The aim of the present study was, therefore, to collect *Gomphus* spp. from the traditional markets of Meghalaya, to document the morphological characteristics of the collected specimens, to evaluate the antimicrobial activity of crude metabolites against selected pathogenic bacterial strains and to carry out their molecular characterization.

## 13.2 Materials and Methods

### 13.2.1 Collection of *Gomphus* spp. from the Traditional Markets

Samples were collected from the local markets in the east Khasi hills district of Meghalaya for their documentation and study. The samples were wrapped in aluminium foil and stored in sterile containers prior to their transport to the laboratory for preservation and identification. Care was taken to avoid distortion of the fleshy fungi.

### 13.2.2 Preservation and Storage

The collected specimens were stored in the in-house culture collection of the Department of Biotechnology and Bioinformatics, North-Eastern Hill University, Shillong in 4 % formaldehyde as per standard procedures (Tanti et al. 2011). Alternatively, the specimens were also oven dried at 80 °C for 5 days, wrapped in aluminium foil and placed in labelled containers for further analysis. The containers were then stored at −20 °C. Some part of the dried sample was used for the preparation of extract.

### 13.2.3 Identification of the Edible Mushroom Specimens

The morphological characters of the fruiting bodies were noted down after comparing them with standard manuals for mushroom identification (Purkayastha and Chandra 1985; Svrcek and Coxon 1975) and mycokeys available at www.mushroomexpert.com and www.mycokey.com.

### 13.2.4 PCR Amplification and 18S rDNA Sequencing

Genomic DNA was extracted using fungal genomic DNA miniprep purification spin kit (QIAGEN, Germany). The universal 18S rDNA primers, viz forward nu-SSU-0817–5 (TTAGCATGGAATAATRRAATAGGA) and reverse nu-SSU-1536–3 (ATTGCAATGCYCTATCCCCA), were used for the amplification of the 18S rRNA gene (Borneman and Hartin 2000). Amplification of DNA was carried out with a 9700 Gold thermal cycler (Applied Biosystems, UK) under the following conditions: initial denaturation at 94 °C for 2 min, 35 cycles of denaturation at 94 °C for 0 s, annealing at 56 °C for 10 s, extension at 72 °C for 30 s and a final extension at 72 °C for 2 min. The amplified PCR product was analyzed on an agarose gel, amplified DNA was purified using QIA quick® gel extraction kit and sequenced using BigDye terminator protocol (Applied Biosystems, UK).

### 13.2.5 Sample Preparation

Dried samples were grounded into coarse powder. 10 g of the powdered samples was extracted by stirring with 100 mL ethanol at 30 °C at 150 rpm for 24 h and filtered through Whatman No. 4 filter paper. The residue was then extracted with two additional volumes of 100 mL of ethanol as described above. The combined ethanolic extracts were then rotary evaporated at 40 °C to dryness and stored at 4 °C for further use (Gezer et al. 2006).

### 13.2.6 Antimicrobial Assay

The antimicrobial potential of the extracted crude metabolites was tested against a total of five microbial strains comprising of four bacteria—*Bacillus cereus* MTCC 430, *Salmonella enterica* ser. *paratyphi* MTCC 735, *Escherichia coli* MTCC 730, *Staphylococcus aureus* subsp. *aureus* MTCC 96—and one yeast, *Candida albicans* MTCC 183, obtained from the Microbial Type Culture Collection (MTCC), Chandigarh, India.

Antibacterial and antifungal activity of the macrofungal extract was tested using well diffusion method (Bauer et al. 1996). All the bacteria mentioned above were incubated at 37 ± 0.1 °C for 24 h by inoculation into Brain Heart Infusion broth. The yeast was inoculated into Potato Dextrose Broth and incubated at 28 ± 0.1 °C for 48 h. The prepared cultured plates (Mueller Hinton Agar for the bacteria and PDA for the yeast) were then inoculated with different bacteria and yeast by using plate method (100 μL). Wells were made on the agar surface with 6 mm cork borer. The dried mushroom extract was dissolved in dimethylsulfoxide (DMSO) to a final concentration of 2 mg $mL^{-1}$ and sterilized by filtration through a 0.22 μm membrane filter (Tepe et al. 2005). 50 μL of the extracts were poured directly into the well. The plates were incubated at 37 ± 2 °C for 24 h for bacterial activity and at 28 ± 0.1 °C for 48 h for yeast activity. The plates were then observed for the zone formation around the wells and the inhibition zones formed on the medium were evaluated in mm.

## 13.3 Results and Discussion

### 13.3.1 Morphological Characteristics

The most obvious distinguishing characters of the genus are the vase or funnel shape of the cap and recurrent forked gills that in most species are so thick that they appear more like ridges than true gills (Table 13.1, Fig. 13.1a, b).

### 13.3.2 Molecular Characterization

The product of PCR amplification using the primers nu-SSU-0817–59 and nu-SSU-1536–39 was approximately 762 bp (Fig. 13.2). Multiple sequence alignments were carried out using 18S rDNA sequences of 13 type strains retrieved from the NCBI database (http://www.ncbi.nlm.nih.gov/) and the multiple sequence alignment was achieved using ClustalW. A phylogenetic tree was constructed by the neighbour joining (NJ) method using MEGA 4 software. Multiple alignments and the phylogenetic tree showed that the sample (Gen Bank accession no: KJ411941) had maximum similarity with *Gomphus floccosus* gbAF026637.1 (Fig. 13.3).

### 13.3.3 Antimicrobial Assays

The mushroom extract showed antimicrobial activity against selected pathogenic strains (Table 13.2). Prominent zone of inhibition was

**Table 13.1** Morphological features of the samples collected

| | Size | Characteristics |
|---|---|---|
| Cap | 5–8 cm wide | Vase or funnel shaped, at first conical or almost cylindrical with a flat top, later shallow funnel shaped, yellow to pale orange, margin curved downward or rolled inward |
| Gills | 8–12 cm, 1–2 mm wide,1 mm thick | Flesh white, ridge like, long decurrent but ending rather abruptly on the stem, frequently forked and joined, yellow to reddish yellow |
| Stem | Stem 3–6 cm long, 1–2 cm thick | Uniform in diameter, solid, pale yellow |

**Fig. 13.1** *Gomphus floccosus*. **a** Growing in the wild and, **b** Sample collected from the market (PKSR1)

observed against *B. cereus* and *S. aureus* subsp. *aureus* (Fig. 13.4a, b). However, there was very little inhibition against *E. coli* and *C. albicans* and no inhibition was observed against *S. enterica* ser. *paratyphi*.

## 13.4 Conclusion

*G. floccosus* has been described by some authors as an inedible mushroom (Henry and Sullivan 1969). Consumption of this particular species in the northern USA and in Canada has been reported to cause severe gastrointestinal disorders (Ammirati et al. 1985).

Based on the 18S rRNA gene phylogeny using the nu-SSU-0817–5 and nu-SSU-1536–3 universal primers, the local strain examined in

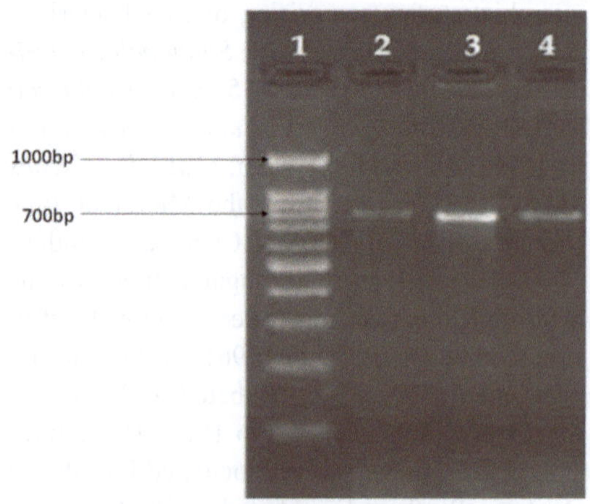


**Fig. 13.2** Gel picture of the amplified 18S rDNA. Lane 1: 100 bp DNA ladder; lane 2, 3 and 4: PKSR1

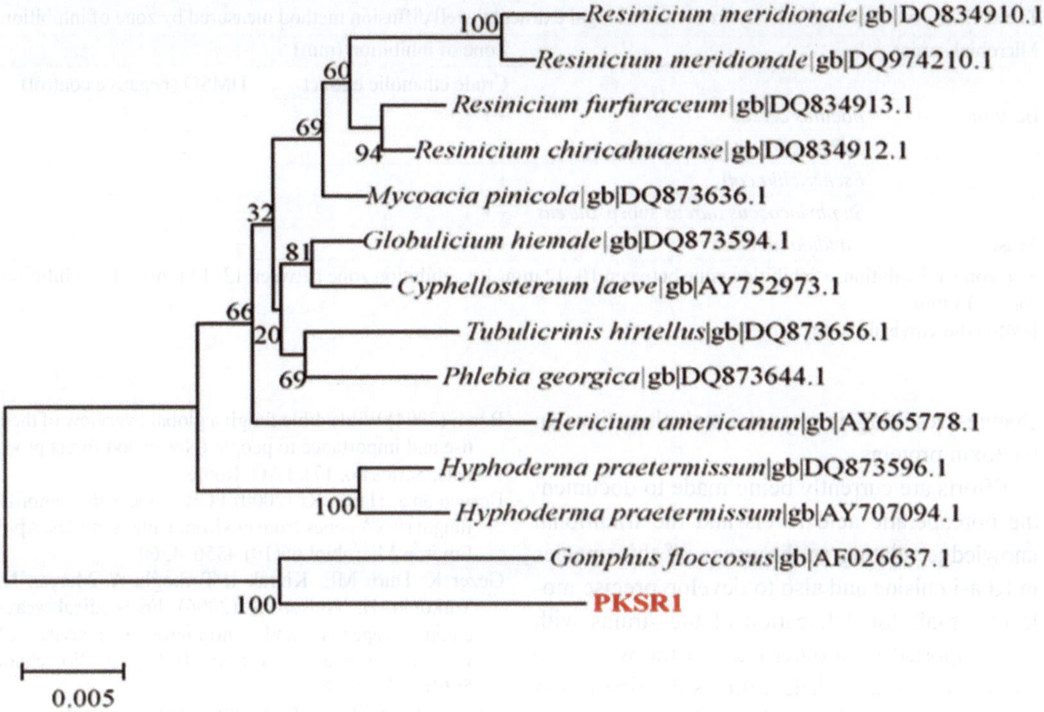


**Fig. 13.3** Phylogenetic relationship of PKSR1 with closest match

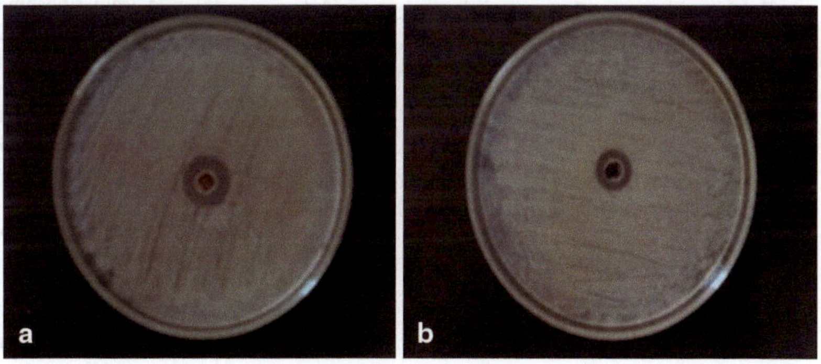


**Fig. 13.4** Antimicrobial activity of the crude ethanolic extract of the mushroom against **a** *Bacillus cereus,* and **b** *Staphylococcus aureus* subsp. *aureus*

the current study showed high similarity with *G. floccosus* Genbank accession no: AF026637.1 (Fig. 13.3) which was reported by Hibbett et al. in 1997. This indicates that from an evolutionary genetic perspective the strains of the present study are similar to those reported from North America. In light of this finding, the relative lack of severe gastrointestinal disorders in the ethnic population of Meghalaya that has been traditionally using this mushroom species in their daily diet can possibly be attributed to the reduced toxicity of the local strains owing to lesser production of the toxin, i.e. norcaperatic acid, and to the time-tested indigenous culinary practices of the ethnic tribes of Meghalaya that may be responsible for inactivating the toxin during food preparation itself. Further, there is also a possibility that the ethnic population is genetically resistant or has developed resistance to the mushroom, which is otherwise reported to be toxic in literature. The other possibility could be that the mushroom has undergone significant mutational

**Table 13.2** Antimicrobial activity from macrofungal extract by well diffusion method measured by zone of inhibition

| Microbial strains | | Zone of inhibition (mm) | |
|---|---|---|---|
| | | Crude ethanolic extract | DMSO (negative control) |
| Bacteria | *Bacillus cereus* | +++ | – |
| | *Salmonella enterica* ser. *paratyphi* | – | – |
| | *Escherichia coli* | + | – |
| | *Staphylococcus aureus* subsp. *aureus* | ++ | – |
| Yeast | *Candida albicans* | + | – |

– no zone of inhibition, + inhibition zone between 10–12 mm, ++ inhibition zone between 12–14 mm, +++ inhibition zone >14 mm
DMSO dimethylsulfoxide

changes resulting in expressional alterations in the toxin proteins.

Efforts are currently being made to document the norcaperatic acid levels and the traditional knowledge relating to the usage of this species in Khasi cuisine and also to develop precise molecular tools for delineation of the strains with those reported from other parts of the world. To the best of our knowledge, this is the first report of a phylogenetic study on the *Gomphus* spp. of Meghalaya that aimed at probing into the evolutionary genetic basis of the *Gomphus* Paradox. We believe that as more information becomes available with time, the remaining pieces of the jigsaw puzzle that we have started to scientifically explain, the *Gomphus* Paradox, will be finally put together.

**Acknowledgement** The study formed a part of the work under research project sanctioned by DST, Govt of India.

## References

Agrahar-Murugkar D, Subbulakshmi G (2005) Nutritional value of edible wild mushrooms collected from the Khasi hills of Meghalaya. Food Chem 89:599–603

Ammirati JF, Traquair JA, Horgen PA (1985) Poisonous mushrooms of the United States and Canada. University of Minnesota Press (2037 University Avenue South east), Minneapolis

Barua P, Adhikary RK, Kalita P, Bordoloi D, Gogol P, Singh RS, Ghosh AC (1998) Wild edible mushrooms of Meghalaya. Anc Sci Life 3(17):190–193

Bauer HW, Kirby WMM, Slerris JC, Truck M (1996) Antibiotic susceptibility testing by a standardized single disc method. Am J Clin Path 45:493–496

Boa E (2004) Wild edible fungi: a global overview of their use and importance to people (Non-wood forest products, Series no. 17). FAO, Rome

Borneman J, Hartin RJ (2000) PCR primers that amplify fungal rRNA genes from environmental samples. Appl Environ Microbiol 66(10):4356–4360

Gezer K, Duru ME, Kivrak I, Turkoglu A, Mercan N, Turkoglu H, Gulcan S (2006) Free-radical scavenging capacity and antimicrobial activity of wild edible mushroom from Turkey. J Biotechnol 5(20):1924–1928

Henry ED, Sullivan G (1969) Phytochemical evaluation of some cantharelloid fungi. J Pharm Sci 58(12):1497–1500

Hibbett DS, Pine EM, Langer E, Langer G, Donoghue MJ (1997) Evolution of gilled mushrooms and puffballs inferred from ribosomal DNA sequences. Proc Natl Acad Sci USA 94:12002–12006

Hosford D, Pilz D, Molina M, Amaranthus M (1997) Ecology and management of the commercially harvested American Matsutake mushroom. General technical report PNW-GTR-412, Portland (US Department of Agriculture, Forest Service, Pacific Northwest Research Station)

Purkayastha RP, Chandra A (1985) Manual of Indian edible mushrooms. Today and tomorrow's publication, New Delhi

Svrcek M, Coxon D (1975) A color guide to familiar mushrooms and fungi. Octopus Books Limited, London

Tanti B, Gurung L, Sarma GC (2011) Wild edible fungal resources used by ethnic tribes of Nagaland, India. Indian J Tradit Knowl 10(3):512–515

Tepe B, Daferera D, Sokmen A, Sokmen M, Polissiou M (2005) Antimicrobial and antioxidant activities of the essential oil and various extracts of *Salvia tomentosa* Miller (Lamiaceae). Food Chem 90:333–340

Wong JLG, Thornber K, Baker N (2001) Resource assessment of non-wood forest products. Experience and biometric principles (Non-wood forest products Series no. 13). FAO, Rome

# Identification of Tomato Leaf Curl Virus Infecting *Acalypha indica*: An Ethnomedicinal Weed in North-Eastern Uttar Pradesh

14

Smriti Mall, Swapna Gupta and P. P. Upadhyaya

**Abstract**

During an extensive survey of viral disease on medicinal plants in the summer season of 2011–2012 in different localities of north-eastern Uttar Pradesh, a variety of symptoms were recorded on *Acalypha indica* plants. The infected plants showed yellow-green mosaic symptoms followed by yellow patches intermingled with normal green. Suspected infected plants were examined by electron microscopy and nested polymerase chain reaction (PCR) by using CRv301 and CRc1152 primer pairs. The electron micrographs prepared with a leaf-dip method revealed the presence of geminate particles of size 20 nm in diameter. PCR products of the expected size ~870 bp, were obtained from *Acalypha* samples. The results of the PCR investigation revealed that the *Acalypha* plant showing yellow mosiac was positive for *Tomato leaf curl virus* (ToLCV). The virus isolate was identified as ToLCV of genus *Begomovirus* and family Geminiviridae. This is the first report of occurrence of ToLCV on *A. indica* from India on the basis of symptomatology, particle morphology and PCR analysis.

**Keywords**

*Tomato leaf curl virus* (ToLCV) · North-eastern Uttar Pradesh · Symptoms · Nested PCR

## 14.1 Introduction

*Acalypha indica* is one of the most extensively investigated medicinal plant and well known for its medicinal importance. *A. indica* is an erect herb of the family Euphorbiaceae having ethnomedicinal importance. It occurs throughout tropical Africa and South Africa, in India and Sri Lanka as well as in Yemen and Pakistan. It has possibly been introduced elsewhere as a weed (Schmelzer and Gurib-Fakim 2007). The active principles identified from these plants are acalyphine, cyanogenetic glucoside and triacetonamine which are extremely poisonous to rabbit, causes discolouration of blood and gastrointestinal irritations (Chopra et al. 1956).

S. Mall (✉)
Department of Botany, St. Andrew's College, Gorakhpur, UP 273001, India
e-mail: smriti.mall@rediffmail.com

S. Gupta · P. P. Upadhyaya
Department of Botany, DDU Gorakhpur University, Gorakhpur, UP 273009, India

R. N. Kharwar et al. (eds.), *Microbial Diversity and Biotechnology in Food Security*,
DOI 10.1007/978-81-322-1801-2_14, © Springer India 2014

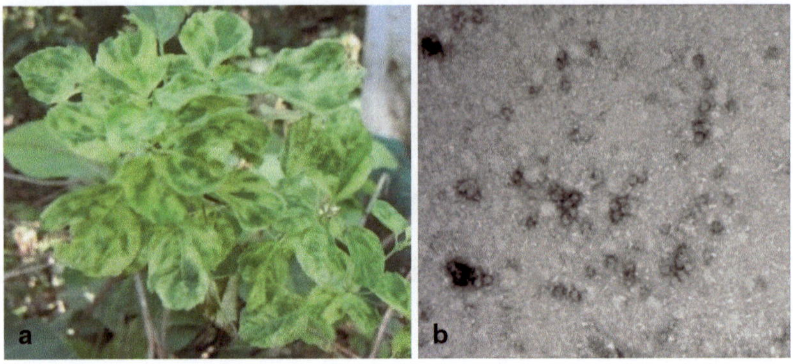

**Fig. 14.1** **a** Virus symptoms of yellow mosaic on naturally grown plants at Gorakhpur. **b** Electron micrograph showing geminate particles of size 20 nm in *Acalypha indica* plant with green-yellow mosaic disease

The *A. indica* plant is diuretic, cathartic and expectorant, used as a substitute for senegal, useful in bronchitis, pneumonia and asthma (Chandramohan et al. 2012). Root is cathartic; however, the leaves are laxative, used in scabies and in snakebite. Fresh leaf juice is useful in rheumatoid arthritis and skin infection. Juice with salt is applied on eczema. For joint pains and eczema a handful of leaves along with small amount of salt, made into a paste, can be applied externally (Sandhya et al. 2006). The leaf juice when mixed with neem oil and applied to the inner part of children's tongue with the help of a quill, induces vomiting and acts as an expectorant (Muthaliar 1988).

So far six viruses, viz. *Bitter gourd yellow mosaic virus* (Rajinimala et al. 2009), *Tomato leaf curl virus* (ToLCV) (Dafalla and Sidig 1997), *Acalypha yellow mosaic virus* (Raj et al. 1996), *Yellow vein mosaic virus* (Muniyappa et al. 2003), *Mungbean yellow mosaic virus* (Green and Kim 1992), *Hibiscus chlorotic ringspot virus* (Doan et al. 2003), has been reported on *A. indica*. But there is no report of any virus on the survey area, hence, in the present study an attempt has been made to identify the causal virus on this ethnomedicinal weed from Gorakhpur during the survey of different geographical area of north-eastern Uttar Pradesh (2006 and 2007). *A. indica* plants were characterized by yellow-green mosaic symptoms at University campus, Gorakhpur, India. The present study describes identification of ToLCV infecting *A. indica*.

## 14.2 Materials and Methods

### 14.2.1 Survey and Symptomatology

Extensive surveys were conducted during the summer season of 2011–2012 in the different geographical regions of north-eastern Uttar Pradesh, to study the incidence of any virus infection on medicinally important plants. During the survey, yellow-green mosaic symptoms were observed from this locality (Fig. 14.1a). There was a slight reduction of leaf lamina of viral symptoms on some potential medicinal plant genera. The leaves showed yellow patches intermingled with normal green. The infected samples were collected and examined for the etiology of the virus.

### 14.2.2 Electron Microscopy

Electron microscopy by leaf-dip preparation was carried out as described by Brandes (1964). Electron micrographs of well-separated *Acalypha* virus were taken at different magnification on a plate film in transmission electron microscopy. (TEM)-1011. The negatives were magnified five times the original magnification and the measurements of the length of 100 particles (in nm) were taken in the prints. The electron micrographs prepared with leaf-dip method revealed the presence of geminate particles.

### 14.2.3 DNA Extraction and PCR Amplification

The total DNA was extracted from 100 mg leaf tissue of infected as well as healthy leaf samples, using the method described earlier by Dellaporta et al. (1983). The DNA pellet was suspended in 20 ml Tris-ethylenediaminetetraacetic acid (TE) buffer. The quality and quantity of the genomic DNA was checked on 1 % agarose gel and stored at −20 °C till further use. Total DNA was extracted from infected leaf samples and polymerase chain reaction (PCR) was performed using primers specific to amplify the coat protein gene of Indian isolates of ToLCV viz. ATGKCSAAGC-GWCCRGCAGA (CRv301) and TTWARAAT-GTAAWWKGAGCAG (CRc1152) (Reddy et al. 2005).

| | | | |
|---|---|---|---|
| 94°C | - | 5 Minute | 1 Cycle |
| 94°C | - | 30 Sec | 30 Cycles |
| 47°C | - | 30 Sec | |
| 72°C | - | 40 Sec | |
| 72°C | - | 5 Minute | 1 Cycle |

PCR reactions were carried out in a total of 50 µl volume containing 1 µl (20 ng) template DNA, 5.0 µl (10 ×) PCR buffer, 1.0 µl (10 mM of each) dNTPs, 3 µl (25 mM) $MgCl_2$, 1.0 µl (25 pmole) of each forward and reverse primers and 1.0 µl (3U) *Taq* DNA polymerase. Amplifications were performed in a Peltier thermal cycler under the following conditions: initial denaturation at 94 °C for 5 min followed by 30 cycles of denaturation at 94 °C for 30 s, primers annealing at 47 °C for 30 s and extension at 72 °C for 40 s. Later, a final extension was given at 72 °C for 5 min. The amplified products were electrophoresed with DNA 100 bp marker in 1 % agarose.

## 14.3 Results and Discussion

During an extensive survey of viral disease on medicinal plants in the summer season of 2011–2012, in different localities of north-eastern Uttar Pradesh, India, a variety of symptoms were recorded on *A. indica* plants at the university campus of Gorakhpur. The infected plants showed yellow-green mosaic symptoms followed by yellow patches intermingled with normal green. There was a slight reduction of leaf lamina (Fig. 14.1a). Slight reduction of leaf lamina was also noticed in affected plants. The electron micrographs prepared with leaf-dip method revealed the presence of geminate particles of size 20 nm in diameter (Fig. 14.1b). The presence of virus was demonstrated by electron microscopy and electrophoresing the PCR products on agarose gels (Fig. 14.2).

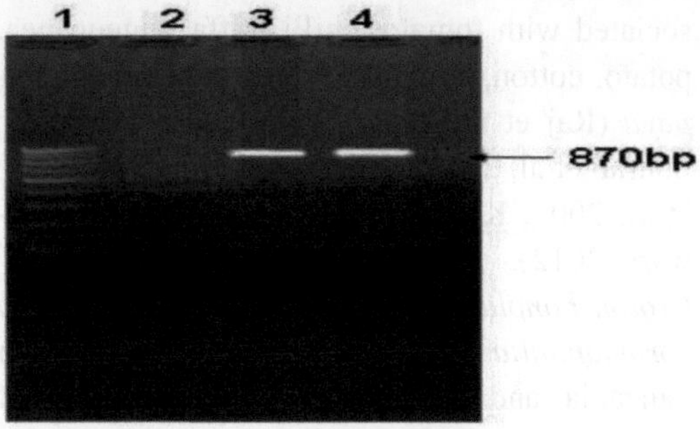


**Fig. 14.2** Amplification of coat protein gene of geminivirus isolate using CRv301 and Crc1152 primer pair specific to coat protein gene of *Tomato leaf curl virus* (*Lane 1*: 100 bp ladder; *Lane 2*: healthy control; *Lane 3* and *4*: infected)

PCR was carried out using ATGKCSAAGC-GWCCRGCAGA (CRv301) and TTWARAAT-GTAAWWKGAGCAG (CRc1152) (Reddy et al. 2005) primers specific to amplify coat protein gene. PCR products of the expected size ~870 bp were obtained from *Acalypha* samples with symptoms but not from healthy ones. The results of the PCR investigation revealed that the *Acalypha* samples were positive for ToLCV.

From available literature, it was found that ToLCV has been reported from Pakistan on various hosts such as tomato, chilli, watermelon, bitter gourd, croton, *Eclipta* and potato (Mansoor et al. 2000; Hussain et al. 2004, 2005; Tahir and Haider 2005; Haider et al. 2006) and in Thailand ToLCV was associated only with luffa. In India, ToLCV have been reported from northern and southern regions and mainly associated with tomato, chilli, luffa, pigeon pea, potato, cotton, cowpea, *Croton* and *Zinnia, Papaya* (Raj et al. 2008; Srivastava et al. 1995; Sohrab et al. 2003; Usharani et al. 2004; Reddy et al. 2005; Khan et al. 2006; Panday and Tiwari 2012). Various weed species, namely *Croton bonplandianum, A. indica, Malvastrum coromandalianum, Eclipta* were reported from Tanzania, and a study showed the presence of *alba, Ageratum conyzoides, Coccinia grandis, Corchorus olitorius, Nicotiana plumbaginifolia, Parthenium hysterophorus, Solanum nigrum, Sonchus oleraceus* and *Trigonella corniculata* which grow naturally in or nearby the fields of cultivated crops in many places of India and are proven to be the alternate host or reservoir hosts of one or many *begomoviruses* (Raj et al. 2011; Panday and Tiwari 2012).

However, only six viruses, viz. *Bitter gourd yellow mosaic virus* (Rajinimala et al. 2009), ToLCV (Dafalla and Sidig 1997), *Acalypha yellow mosaic virus* (Raj et al. 1996), *Yellow vein mosaic virus* (Muniyappa et al. 2003), *Mungbean yellow mosaic virus* (Green and Kim 1992) and *Hibiscus chlorotic ringspot virus* (Doan et al. 2003) have been reported on *A. indica* from different parts of the world. From earlier reports it was clear that ToLCV has been reported mostly on vegetable crops and only on two wild species, *Croton* and *Eclipta* from across the world, among which *Croton* is reported from India. Literature survey confirmed that very little information is available on the occurrence of ToLCV on *A. indica* from different parts of the world.

On the basis of this study, it was concluded that only one ToLCV was reported from Sudan on *A. indica* (Dafalla and Sidig 1997) and there is no report of any virus on the survey area. Hence, ToLCV reported on *A. indica* in the present study is the first report from India on the basis of symptomatology, particle morphology and PCR analysis. The virus isolate was identified as isolate of ToLCV of genus *Begomovirus* and family Geminiviridae. Among wild plants, after *Croton* only *A. indica* is found to be associated with ToLCV in India.

**Acknowledgments** Author is thankful to SERB FAST TRACK PROJECT New Delhi for providing fund.

## References

Brandes J (1964) Identifizierugvon Getreckten pflanzenpathogenen viren auf morphologischer grundlage. Mitt Boil Bundesanstalt Land Forstwirtsch 110:1–130 (Berlin- Dahlem)

Chandramohan S, ElayaRaja R, Thiripura Salini S, Elavarasan A, Senthil Kumar R (2012) Synthesis, characterization and biological evaluation of some novel thiazolidinones derivatives. Int J Pharma Sci Res 3:1516–1519

Chopra RN, Nayar SL, Chopra IC (1956) Glossary of Indian medicinal plants. National Institute of Science Communication and Information Resources (CSIR), New Delhi

Dafalla GA, Sidig SA (1997) Management of the whitefly virus complex in Sudan. Food and Agricultural Organization (FAO) Plant Product and protection paper 143. Rome, IT: p 214

Dellaporta SL, Wood J, Hicks JB (1983) A plant minipreparation: version II. Plant Mol Biol Rep 1:19–21

Doan DN, Lee KC, Laurinmaki P, Butcher S, Wong SM, Dokland T (2003) Three dimensional reconstruction of Hibiscus chlorotic ringspot virus. J Struct Biol 144:253–261

Green SK, Kim D (1992) Mungbean yellow mosaic disease: proceeding of an international workshop, 2–3 July 1991, Bangkok. Thailand, Issue 92. AVRDC, Shanhua, pp 373–379

Haider MS, Tahir M, Latif S, Briddon RW (2006) First report of *tomato leaf curl New Delhi virus* infecting *Eclipta prostrata* in Pakistan. J Phytopathol 55:285–285

Hussain M, Mansoor S, Iram S, Zafar Y, Briddon RW (2004) First report of *tomato leaf curl New Delhi virus* affecting chilli pepper in Pakistan. Plant Pathol 53(6):794

Hussain M, Mansoor S, Iram S, Naureen A, Zafar Y (2005) The nuclear shuttle protein of *tomato leaf curl New Delhi virus* is a pathogenecity determinant. J Virol 79(7):4434–4439

Khan MS, Raj SK, Singh R (2006) First report of *tomato leaf curl* New Delhi virus infecting chilli in India. Plant Pathol 55:289

Mansoor S, Khan SH, Hussain M, Mushtag N, Zafar Y, Malik KA (2000) Evidence that water melon leaf curl disease in Pakistan is associated with *tomato leaf curl virus*- India, a bipartite *Begomovirus*. Plant Dis 84:102

Muniyappa V, Maruthi MN, Babithu CR, Colvin J, Briddon RW, Rangaswamy KT (2003) Characterisation of pumpkin yellow vein mosaic virus from India. Ann Appl Biol 142:323

Muthaliar M (1988) Materia Medica (Vegetable section), Volume I, 4th edn. Tamilnadu Siddha Medical Council, Chennai, p 359

Panday N, Tiwari AK (2012) Identification of *Zinnia leaf curl virus* infecting *Zinnia elegans* in India/. J Biotechnol Bioinforma 2(1):6–10. doi: 10.5897/ISAAB-JBB12.001

Raj SK, Srivastava M, Singh BP (1996) Evidence from nucleic acid hybridization tests for Geminivirus infection of ornamental crotons in India. Eur J Plant Pathol 102:201–203

Raj SK, Snehi SK, Tiwari AK, Rao GP (2011) First molecular characterization of *Ageratum enation virus* associated with mosaic disease of pointed gourd (*Trichosanthes dioica* Roxb.) in India. Phytoparasitica. doi: 10.1007/s12600-011-0182-4

Raj SK, Snehi SK, Khan MS, Singh R and Khan AA (2008) Molecular evidence for association of *Tomato leaf curl New Delhi virus* with leaf curl disease of Papaya (*Carica papaya* L.) in India. Aust. Plant Dis. Notes 3:152–155

Rajinimala N, Rabindran R, Ramaiah M (2009) Host range and purification of bitter gourd yellow mosaic virus (BGYMV) in bitter gourd (*Momordica charantia*). Arch Phytopathol Plant Prot 42:499–507

Reddy RV, Colvin C, Muniyappa JV, Seal S (2005) Diversity and distribution of *Begomovirus* infecting tomato in India. Arch Virol 150:845–867

Sandhya B, Thomos S, Isabel W, Shenbagarathai R (2006) Ethnomedicinal plants used by the valaiyan community of piranmalai hills (Reserved forests), Tamilnadu, India-A Pilot Study. Afr J Trad CAM 3:101–114

Schmelzer GH, Gurib-Fakim A (2007) *Acalypha indica* L. ((Internet) Record from Protabase). *PROTA* (Plant Resources of Tropical Africa/Ressources végétales de l'Afrique tropicale). Prota Foundation, Wageningen

Sohrab SS, Mandal B, Pant RP, Varma A (2003) First report of association of *Tomato leaf curl virus-New Delhi* with Yellow Mosaic Disease of *Luffa cylindrica* in India. Dis Notes 87:1148

Srivastava KM, Hallan V, Raizad RK, Govinda C, Singh BD, Sane PV (1995) Molecular cloning of Indian *Tomato leaf curl virus* genome following a simple method of concentrating the supercoiled replicative form of DNA. J Virol Meth 51:297–304

Tahir M, Haider MS (2005) First report of *tomato leaf curl New Delhi virus* infecting bitter gourd in Pakistan. Plant Pathol 54:807

Usharani KS, Surendranath P, Paul-Khurana SM, Garg ID, Malathi VG (2004) Potato leaf curl- a new disease of potato in northern India caused by a strain of *tomato leaf curl New Delhi virus*. Plant Pathol 53:23

# Occurrence of Antiviral Systemic Resistance Inducer in *Pseuderanthemum bicolor* Radlk., Its Mode of Action and Biophysico-Chemical Properties

15

M. M. Abid Ali Khan, S. Rais Haider, M. Zahid Rizvi and S. Arshad Hasan Rizvi

**Abstract**

Out of several medicinal and ornamental plants screened for anti-tobamovirus activity, the leaf extract of *Pseuderanthemum bicolor Radlk.* plant (Family: Acanthaceae) showed 94–100 % antiviral activity on *Cyamopsis tetragonoloba* Taub and *Nicotiana tabacum var* Np 31 test hosts. *Pseuderanthemum bicolor* systemic resistance inducer (PBSRI) occurring in the leaves of *P. bicolor* plant was most effective, and pre-inoculation sprays of SRI completely prevented the infection and to some extent multiplication of tobamo and potex viruses when sprayed over the test hosts reacting hypersensitively or systemically.

The degree of resistance induced varied slightly in different host–virus combinations and increased with time and could be reversed by simultaneous application of actinomycin-D (20 μg/ml) (Merck, Sharp and Dhome, U.S.A), which indicated that induction of resistance was host mediated and associated with synthesis of some virus-inhibitory agent (VIA) in treated host. The treated plants developed antiviral state due to alterations in metabolic state after 18–24 h of PBSRI treatment which lasted for about 15 days.

PBSRI was highly thermostable at higher dilution and was active even after prolonged storage at pH 6–8. The active principle occurring in PBSRI when mixed was absorbed by charcoal and celite, was not sedimented by ultracentrifugation at 120,000 g for 90 min, and thus appears to be smaller than the smallest virus, is non-dialyzable, actively precipitated by protein precipitants, i.e. ammonium sulfate (BDH), absolute alcohol and lost its activity when incubated with proteolytic enzymes like trypsin, chymotrypsin, pronase and papain (Serva Fine Biochemica, Germany).

The PBSRI showed broad spectrum antiviral activity against different host–virus combinations.

M. M. A. A. Khan (✉) · S. R. Haider · M. Z. Rizvi · S. A. H. Rizvi
Department of Botany, Shia P. G. College, Lucknow 226020, India
e-mail: mmabidalikhan265@gmail.com

R. N. Kharwar et al. (eds.), *Microbial Diversity and Biotechnology in Food Security*,
DOI 10.1007/978-81-322-1801-2_15, © Springer India 2014

**Keywords**

*Pseuderanthemum bicolor* Systemic Resistance Inducer (PBSRI) · Actinomycin-D · Proteolytic enzymes and antiviral

## 15.1 Introduction

Recent work published on virus inhibitory activity (VIA) of plants and plant products displayed that few of them contained strong antiviral agent(s) of virus infection and multiplication (Gianinazzi 1982; Govier and Jacqueline 1995; Jain et al. 1990; Khan et al. 1991; Khan and Zaim 1992; Murty and Nagarajan 1996; Verma and Khan 1984; Verma 1986). Besides local protection (Verma and Baranwal 1983), some of them were also capable of inducing systemic resistance in treated plants (Khan et al 2011a, d; Khan and Zaim 1992; Verma and Prasad 1983; Verma et al. 1984, 1985; Khan et al. 2013).

During present investigations, a broad spectrum antiviral systemic resistance inducer occurring in the leaves of *Pseuderanthemum bicolor Radlk.* (PBSRI) was studied for its effect on different host–virus combinations and biophysicochemical properties.

## 15.2 Materials and Methods

### 15.2.1 Virus Cultures

The cultures of tobacco mosaic virus (TMV), sunnhemp rosette virus (SRV), tomato yellow mottle mosaic virus (TYMMV), potato virus-x (PVX) and cucumber green mottle mosaic virus (CGMMV) were maintained on their respective systemic hosts, viz. *Nicotiana tabacum var. Np31, Crotalaria juncea L., Lycopersicon esculentum Mill., Solanum tuberosum L.* and *Lagenaria siceraria standl.*

Virus cultures with severe systemic symptoms were maintained in glass house conditions. SRV caused severe mosaic and rosetting of leaves, reduced flower size and number; seeds were smaller and deformed in *C. juncea* L. plants. TYMMV caused severe dark green mosaic with reduced leaf lamina leading to fern leaf symptoms in *L. esculentum* Mill plants. CGMMV turned young leaves light yellow, showed downward curling and mosaic mottle later on all the leaves, showed dark green mosaic symptoms in *L. siceraria*, Mol. Standal plants. PVX infected young leaves displayed dark green mosaic symptoms along with downward curling, and necrosis in *S. tuberosum* L.

### 15.2.2 Test Plants

The test plants were raised in a glass house, in unsterilized compost soil in earthen pots. They were used for the experiment when they grew to 4–5 leaves stage.

### 15.2.3 Preparation of Virus Inocula

The young leaves showing severe disease symptoms were ground in a sterilized pestle and mortar using distilled water as the diluent (1 g/5 ml). The pulp obtained was squeezed through two-folds of muslin cloth; the filtrate obtained was centrifuged at 5,000 rpm for 15 min. The supernatant was diluted suitably with distilled water and used as viral inocula.

### 15.2.4 Extraction of PBSRI

Fresh or air-dried leaves of *P. bicolor* were ground in a sterilized mortar, suitable amount of distilled water/0.05 M phosphate buffer at pH 7.0 was added, and the pulp was squeezed through a muslin cloth. The filtrate was centrifuged at 5,000 rpm for 15 min, clarified with solvent ether, precipitated with 75 % ammonium sulfate, centrifuged at 5,000 rpm for 15 min. The precipitate thus obtained was dissolved in minimum amount of distilled water. The solution was centrifuged again at 3,000 rpm for 10 min, and the supernatant was dialyzed in a cellophane bag against running water. The non-dialyzable fraction was collected and centrifuged at 3,000 rpm for 10 min and was used for further experiments.

### 15.2.5 Antiviral Testing

To detect the antiviral activity, PBSRI was applied on two lower leaves of test hosts, having four or five healthy leaves. Similarly, in control plants, two lower leaves were rubbed with distilled water. After 1, 3 and 6 days, the treated leaves were washed with distilled water dusted evenly with 600 mesh carborundum powder and inoculated with virus. Four fifths of the leaves in hypersensitive hosts and two lower leaves in systemic hosts were inoculated with virus. The experiments were performed in a glass house.

Local lesions were counted 2–6 days after virus challenge in local lesion hosts, and symptoms were observed in systemic hosts after 1 or 2 weeks; the active viral titre in control and treated sets of test plants was estimated by inoculating the crude sap from infected leaves on suitable local lesion hosts.

Per cent reduction in viral titre was calculated by using the formula, $IP = (1 - T / C) \times 100$, where C is the number of lesions on control leaves and T is the number of lesions on treated leaves. Details of the analysis of data were same as described earlier (Snedecor 1961).

### 15.2.6 Biophysico-Chemical Properties

The tests were performed on local lesions-producing host *Cyamopsis tetragonoloba* plants against SRV, and all the properties were studied in crude leaf extract.

### 15.2.7 Dilution

The crude leaf extract was diluted from 1/5, 1/10, 1/20, 1/40, 1/50, 1/100, 1/200 w/v with distilled water, and antiviral activity of the diluted samples was tested.

### 15.2.8 Thermal Stability

A total of 5 ml of 1/5 w/v leaf extract was taken in separate test tubes and heated at 40, 50, 60, 70, 80, 90 and 98 °C for 10 min respectively in a temperature controlled water bath. Samples were cooled, centrifuged at 3,000 rpm for 10 min and antiviral activity of the samples was tested separately.

### 15.2.9 Storage

Leaf extract was stored at laboratory temperature in the stoppered sterilized test tubes. Antiviral activity of the sample was tested regularly at an interval of 5 days.

### 15.2.10 Dialyzability

Leaf extract was dialyzed in a cellophane bag (Serva, F.B.C. Germany) under running water for 24 h at room temperature. The bag content was tested, and the undialyzed sample was kept as control.

### 15.2.11 Sedimentability

Leaf extract (20 ml) was centrifuged first at 3,000 rpm for 15 min, and then supernatant was centrifuged at 120,000 g for 2 h in a Beckman model L5–50B ultracentrifuge. The antiviral activity of pellets (in 5 ml distilled water) and supernatant was tested separately.

### 15.2.12 Adsorption

Leaf extract was mixed with activated charcoal and celite separately for 30 min and was centrifuged at 3,000 rpm for 15 min. The supernatant was filtered through Whatman No.1 filter paper.

**Table 15.1** Antiviral screening of different plant extracts

| S. no. | Names of plants | Family | Per cent decrease in virus titre | |
|---|---|---|---|---|
| | | | Leaf | Root |
| 1 | *Andrographis paniculata* Nees | Acanthaceae | 59* | 40** |
| 2 | *Anageissus latifolia* | Combretaceae | 67* | 72* |
| 3 | *Berrintome acutanqule* | Lecythidaceae | 64* | 50* |
| 4 | *Calotropis procera* Br. (L.) | Asclepiadaceae | 73* | 61* |
| 5 | *Curcuma amada* Roxb | Zingibaraceae | 13 | 26** |
| 6 | *Clitorea ternatae* L. | Leguminosae | 8 | 7 |
| 7 | *Desmodium gangeticum* | Leguminosae | 59* | 62* |
| 8 | *Ixora parviflora* Vahl | Rubiaceae | 44** | 41** |
| 9 | *Lantana amara* L | Lantanaceae | 21** | 36** |
| 10 | *Paederia foetida* L. | Rubiaceae | 61* | 42** |
| 11 | *Pseuderanthemum bicolor* Radlk | Acanthaceae | 100* | 100* |
| 12 | *Psoralea corylifolia* L | Leguminosae | 82* | 53* |
| 13 | *Raulfia serpentine* Benth | Apocynaceae | 74* | 58* |
| 14 | *Ricinus communis* L | Euphorbiaceae | 12 | 58* |
| 15 | *Sambucus nigra* L | Caparifoliaceae | 9 | 26** |
| 16 | *Santalum album* L | Santalaceae | 59* | – |
| 17 | *Woodfordia fruticosa* Kurtz | Lythraceae | 70* | 51* |
| 18 | *Zingiber officinal* | *Zingiberaceae* | 8 | 39** |

Extracts were tested on *C. tetragonoloba* plants against sunnhemp rosette virus (SRV) infection
*Data significant at 1 % level; **Data significant at 5 % level

The clear supernatants were tested for antiviral activity.

Reversal experiment with actinomycin-D (AD)

Following PBSRI treatment on to the test host, AD (Merck and Dhome, U.S.A.), 20 µg/ml, was applied at different intervals to the same leaves that had earlier been treated with inhibitor. An equal number of identical leaves in control sets was treated with AD alone, PBSRI alone, or distilled water. In each case, the SRV was challenge-inoculated for 24 h after PBSRI treatment on *Cyamopsis tetragonolobus* (L.) taub, 5 days after the inoculation, from upper (untreated) and basal (treated) leaves separately.

## 15.3 Results

### 15.3.1 Screening of Plants

Antiviral screening of higher plants revealed that *P. bicolor* leaves contained strong antiviral agent (Table 15.1). The PBSRI when applied a couple of hours before virus challenge more or less completely prevented infection and to some extent multiplication of tobamo and potex viruses in their susceptible host reacting hypersensitively and systemically (Table 15.2 and 15.3).

The degree of protection varied in different host–virus combinations. The inhibitory response of PBSRI was not pronounced in *Chenopodium amaranticolor, Nicotiana tabacum var. xanthi nc, N. tabacum var. samsun NN and N. glutinosa* plants. The PBSRI showed a broad spectrum and most promising antiviral action, and the maximum prevention of infection observed after 24–72 h of treatment later on gradually decreased in several other host–virus combinations, viz. hypersensitive hosts *C. tetragonoloba, Datura stramonium, D. metel, D. innoxia, Nicotiana rustica, Vigna sinensis, Gomphrena globosa* and systemic hosts *N. tabacum var. Np31, N. tabacum var. white burley, Lycopersicum esculentum, C. juncea, Cucumis melo* and *S. tuberosum* against their respective viruses (Table 15.2 and 15.3).

**Table 15.2** Anti-tobamovirus activity of PBSRI in their hypersensitive and systemic hosts

| Host used | Per cent decrease in virus infectivity/titre time interval between SRI application and virus challenge in hours | | | | | | | | | | | |
|---|---|---|---|---|---|---|---|---|---|---|---|---|
| | TMV | | | SRV | | | Tm YMMV | | | CGMMV | | |
| *Datura stramonium* L | 100* | 96* | 83* | – | - | - | 100* | 91* | 72* | – | – | – |
| *D. metel* | 100* | 70* | 66* | – | – | – | – | – | – | – | – | – |
| *D. innoxia* | 100* | 84* | 71* | – | – | – | – | – | – | – | – | – |
| *Nicotiana glutinosa* | 70* | 73* | 52* | – | – | – | 81* | 65* | 53* | – | – | – |
| *N. rustica* | 98* | 83* | 65* | – | – | – | – | – | – | – | – | – |
| *N. tabacum* var Samsunn NN | 54* | 48** | 43** | – | – | – | – | – | – | – | – | – |
| *N. tabacum* var *Xanthi nc* | 60* | 52* | 37** | – | – | – | – | – | – | – | – | – |
| *C. amaranticolor* Coste and Ryne | 71* | 85* | 52* | 80* | 84* | 86* | 72* | 55* | 49* | 70* | 95* | 88* |
| *N. tabacum* var. Np 31[a] | 100* | 100* | 97* | – | – | – | - | – | – | – | – | – |
| *N. tabacum* var. white burley [a] | 100* | 100* | 93* | – | – | – | – | – | – | – | – | – |
| *N. tabacum* var. Ky 58 | 100* | 68* | 57* | – | – | – | – | – | – | – | – | – |
| *L. esulentum* Mill[a] | 100* | 100* | 93* | – | – | – | 100* | 100* | 94* | – | – | – |
| C. tetragonoloba Taub | – | – | – | 100* | 94* | 81* | – | – | – | – | – | – |
| *Crotalaria juncea* L.[a] | – | – | – | 98* | 87* | 63* | – | – | – | – | – | – |
| *Vigna sinensis* L | – | – | – | – | – | – | – | – | – | – | – | – |
| *Cucumis melo* L.[a] | – | – | – | – | – | – | – | – | – | 100* | 100* | 91* |
| *L. siceraria* Mohl and Standl.[a] | – | – | – | – | – | – | – | – | – | 100* | 96* | 93* |

*Significant at 1 % level; **Significant at 5 % level
Not done/non host
[a]Systemic hosts

**Table 15.3** Anti-potex virus (PVX) activity of PBSRI in the hosts reacting hypersensitively and systemically

| Host used | Per cent decrease in virus infectivity/titre | | |
|---|---|---|---|
| | Time interval between SRI application and virus challenge in hours | | |
| | *24* | *72* | *144* |
| *Gomphrena globosa* L | 100* | 90* | 65* |
| [a]*Solanum tuberosum* L. | 100* | 100* | 61* |

The differences due to treatment are significant
*Significant at 1 % level
[a]Systemic hosts

**Table 15.4** Demonstration of antiviral systemic resistance induced at top untreated leaves (at remote site) of plants, when basal leaves treated with PBSRI

| Host | Virus | Per cent reduction in local lesions number | |
|---|---|---|---|
| *C. tetragonoloba* | TMV | 100* | 100* |
| *D. stramonium* | SRV | 100* | 100* |
| *G. globosa* | PVX | 96* | 98* |
| *D. stramonium* | Tm YMMV | 97* | 100* |

*Data significant at 1 % level

### 15.3.2 Mode of Action

#### 15.3.2.1 Development of Antiviral State in Treated Plants

Treatment in the lower leaves provided complete protection to the entire plant against subsequent viral infection. Lesion production was completely prevented not only in the leaves which had been treated but also in the top leaves on which no treatment was given; thus, it would appear that the treatment induced systemic inhibitory response of a very high order in the test plants and made them completely refractory to viral infection (Table 15.4).

#### 15.3.2.2 Reversal of Induced Resistance

Induced systemic resistance was completely reversed, when AD (20 μg/ml) was given along with PBSRI. It clearly displayed that perhaps some metabolic alterations take place within treated host (Table 15.5).

#### 15.3.2.3 Biophysico-Chemical properties

Studies were conducted on the test plants *C. tetragonoloba* against SRV.

A range of 80–100 % protection was observed against viral infection on diluting the PBSRI up to 1/20 w/v; however, no VIA was recorded at 1/200 w/v dilution. PBSRI was highly thermostable; however, antiviral action was completely lost when heated up to 90 °C for 10 min, on storing up to 2 months at room temperature in a closed vial at pH 6–8.

The active principle was adsorbed on charcoal and celite, is non-dialyzable, not sedimented on high speed ultracentrifugation up to 120,000 g for 120 min, and on incubation with proteolytic enzymes, viz. trypsin, chymotrypsin, pronase and papain (20 ug/ml), lost the antiviral action. The active principle can be precipitated following the addition of ammonium sulfate and ethanol; however, addition of tricarboxylic acid (TCA) inactivated it (Table 15.5).

## 15.4 Discussion

The study concludes that PBSRI not only interferes with the initial stages of viral infection but also suppresses the infectivity of the viruses within the cell; therefore, intracellular virus synthesis was interfered by altering the host susceptibility. Antivirals from several non-host plants which alter the susceptibility of the host rather than affecting the virus particles directly have been reported earlier (Awasthi 1981; Govier and Jacqueline 1995; Khan and Zaim 1992; Khan et al. 1990, 1996; Verma and Khan 1984; Verma and Prasad

**Table 15.5** Biophysico-chemical properties of PBSRI

| Properties | Results |
|---|---|
| Dilution end point | Active up to 1/200 w/v |
| Thermal stability | Inactivated on heating at 90 °C for 10 min |
| Storage | Active on storing up to 2 months at room temperature in a closed vial |
| Sedimentability | Not sedimented by ultracentrifugation up to 120,000 g for 120 min |
| Dialyzability | Non-dialyzable |
| Adsorption | Adsorbed on charcoal and celite |
| pH sensitivity | Active between pH 6 and 8 |
| Sensitivity to proteolytic enzymes | Antiviral activity was lost on incubation with trypsin, chymotrypsin, pronase and papain |
| Precipitation with protein precipitants | Actively precipitated by ammonium sulphate and ethanol however TCA inactivated the PBSRI |
| Reversal of inhibition | Reversed by simultaneous application of actinomycin-D |

1983). Like antivirals reported earlier (Awasthi 1981; Verma and Prasad 1983; Simons and Ross 1971; Khan and Verma 1990), the PBSRI also induced systemic resistance which reduced the susceptibility of treated and non-treated parts of plants, this type of interference differed from the competitive mechanism because a time interval was needed and its development was sensitive to metabolic inhibitor like AD (Awasthi 1981; Gianinazzi 1982; Khan and Zaim 1992; Verma and Khan 1985; Verma et al. 1985) suggesting that resistance-inducing agent or protective substance(s) synthesized in the treated areas were translocable to the entire plant (Verma and Khan 1984; Verma 1986). Antiviral activity was not observed in *Boerhaavia diffusa, Clerodendrum* sp., *Bougainvillea spectabilis* after simultaneous application of PBSRI and AD (Khan et al. 2011b, d; Khan and Zaim 1992; Verma 1986; Verma and Prasad 1987, 1988; Verma et al. 1985). Interference by AD was completely checked when applied up to 6 h after PBSRI treatment. However, when AD was given 18 h after PBSRI treatment, when such protective agents probably must have been produced, the interference by AD was not observed in treated test plants.

There might be some analogy between interference mediated by interferon in animal system and interference induced by plant antivirals (Gianinazzi 1982; Khan and Verma 1990; Verma 1986) and PBSRI because the induced resistance in all the cases was not virus specific and was sensitive to AD. Since the viral inhibition induced by PBSRI was observed at the site of application and at the remote site, it leads to the conclusion that PBSRI treatment at the epidermal surface initiates a type of biochemical reaction, resulting in the alteration of cell metabolism and production of protective substance(s) (Gianinazzi 1982; Khan et al. 2011c; Khan and Zaim 1992; Khan and Verma 1990; Verma 1986).

Since PBSRI could not pass through cellophane membrane, it is adsorbed on charcoal and celite, precipitated with protein precipitants, active at pH 6–8, not sedimentable on ultracentrifugation, thermostable, active up to higher dilutions and showed a proteinaceous nature (Khan et al. 2011d); these properties more or less resemble those of other antivirals described earlier (Khan et al. 2011b, d; Khan and Zaim 1992; Verma 1986; Verma et al. 1985), but differences still exist as far as storage, dilution and thermal stability are concerned. The PBSRI showed highly promising results over other antivirals.

A clear-cut analogy exists between the action of the antiviral agents in plants and the interferon in animal system (Gianinazzi 1982; Khan and Verma 1990; Khan and Zaim 1992). Properties shown by an interferon are (a) antiviral protein (b) lack of virus specificity (c) association with DNA-dependent protein synthesis, (d) acts through cellular metabolic processes (e) labile to pH and heat (f) induce the production of an antiviral protein in adjacent cells. Similar to other plant antivirals (Gianinazzi 1982; Khan and Verma 1990), PBSRI also resembles interferon as it is also a protein and is small enough to remain in solution at high speed ultracentrifugation but is large enough to remain with tubing dialysis. PBSRI acts at cellular transcription and translation level, hence it is better to call it an interferon-like antiviral.

## 15.5 Conclusion

The aqueous extract from several higher plants has shown the broad spectrum and highly significant antiviral activity in both systemic and hypersensitive host when applied prior to viral inoculation which resulted in gradual increase in virus inhibition with the increase in time interval between treatment and virus challenge up to 72 h. Thereafter, antiviral resistance in treated host gradually decreased; however, the maximum inhibition of viruses was achieved during treatment with PBSRI obtained from *P. bicolor* leaves. The biophysical properties studied clearly displayed that it was a comparatively high potential antiviral systemic resistance inducer which can be exploited in the prophylactic management of several viral diseases in economically important crops. A world survey cumulative data have reported that heavy losses in crop production due to onset of plant viral diseases estimated to US$ 60 billion loss in crop yields worldwide each year (http://en.wikipedia.org/wiki/Plant_virus). Thus, it provides a clear-cut food security problem worldwide, and so further attempts can be made to exploit such natural antiviral isolated from higher plants for the management of plant viral diseases for global food security.

**Acknowledgments** Dr. M. M. Abid Ali Khan is thankful to the Council of Scientific and Industrial Research, New Delhi, Government of India for providing financial assistance during the tenure of present investigations at the Plant Virus Laboratory, Department of Botany, Lucknow University, Lucknow (India). The work presented in this research paper is a part of PhD thesis of Dr. M. M. Abid Ali Khan awarded by the Lucknow University, Lucknow, India.

## References

Awasthi LP (1981) The purification and nature of an antiviral protein from *Cuscuta reflexa* plants. Arch Virol 70:215–223

Gianinazzi S (1982) Antiviral agents and inducers of virus resistance: analogies with interferon. In: Wood RKS (ed) Active defense mechanism in plants. Plenum Publishing Corp., New York, pp 275–296

Govier DA, Jacqueline W (1995) An inhibitor in faba bean sap suppressing infection of *Chenopodium quinoa* by bean yellow mosaic virus. Ann Appl Biol 110:657–660

Jain DC, Khan MM Abid Ali, Zaim M, Thakur RS (1990) Antiviral evaluations of some steroids and their glycosides: a new report. Nat Acad Sci Lett 13:41–42

Khan MM Abid Ali, Verma HN (1990) Partial characterization of an induced virus inhibitory protein, associated with systemic resistance in *Cyamopsis tetragonoloba* (L) Taub plants. Ann Appl Biol 117:617–623

Khan MM Abid Ali, Zaim M (1992) Physico-Chemical properties and mode of action of inhibitors of plants virus replication present in *Operculina turpethum* L. and *Scilla indica* baker. Z Pflanzenkrankh Pflanzenschutz 99:71–79

Khan MM Abid Ali, Jain DC, Bhakuni RS, Zaim M, Thakur RS (1991) Occurrence of some antiviral sterols in *Artemisia annua*. Plant Sci 75:161–165

Khan MM Abid Ali, Singh N, Dhawan KN (1996) Occurrence and identification of a new antiviral saponin from *Lawsonia alba* Lam. Fruits. Nat Acad Sci Lett 19:145–148

Khan MM Abid Ali, Zaidi SNH, Musanna SA (2011a) Isolation and identification of new virus inhibitory diosgenin saponin from *Scilla indica* Baker plant. Presented in XXXIV All India Botanical Conference held at Department of Botany, Luckow University, Lucknow, India, on October 10–12, pp 100

Khan MM Abid Ali, Zaidi SNH, Abidi AB (2011b) Production of new anti tobamo virus protein by Operculina turpethum L. Cells in suspension culture. Presented at National conference on Frontiers in Biological Sciences held at faculty of science, Veer Bahadur Singh Purvanchal University, Jaunpur, UP, India. On December 4–5:45

Khan MM Abid Ali, Rizvi AH, Rizvi MZ (2011c) Oxido reductases alterations associated with host mediated systemic resistance. Proceedings of the national conference on frontiers in biological sciences, at Veer Bahadur Singh Purvanchal University, Jaunpur, U P, India, December 4–5

Khan MM Abid Ali, Haider SR, Kazmi SHA (2011d) Purification of a low molecular weight broad spectrum anti viral protein from *Pseuderanthenum bicolor* Radlk leves presented in XXXIV All India botanical conference held at Department of Botany, Luckow University, Lucknow, India, on October 10–12, pp. 80

Khan MM Abid Ali, Abidi AB, Hasnain S (2013) Total proteins and enzymes alterations associated with induction of antiviral resistance in treated plants. Natl Acad Sci Lett 36(2):147–149. doi:10.1007/s40009-013-0111-9

Murty NS, Nagarajan K (1996) Role of plant extracts in control of TMV infection in nursery and field grown tobacco. Indian Phytopathol 39:98–100

Simons TJ, Ross AF (1971) Metabolic changes associated with systemic induced resistance to tobacco mosaic virus in Samsun NN tobacco. Phytopathol Z 61:293–300

Snedecor GW (1961) Statistical methods. Allied Pacific Pvt. Ltd, Bombay

Verma HN (1986) Interferon like antiviral agent from plants. In: Verma A, Verma JP (eds) Vistas in plant pathology. Malhotra Publishing House, New Delhi, pp 481–490

Verma HN, Baranwal VK (1983) Antiviral activity and the physical properties of the leaf extract of *Chenopodium ambroseides* L. Proc Indian Acad Sci (Plant Sci) 92:461–465

Verma HN, Khan MM Abid Ali (1984) Management of plant virus diseases by *Pseuderanthemum bicolor* leaf extract. Z Pflanzenkrankh Pflanzenschutz 91:266–272

Verma HN, Khan MM Abid Ali (1985) Occurrence of a strong virus interfering agent in susceptible plants sprayed with *Pseuderanthemum atroprupureum* leaf extract. Indian J Virol 01:26–34

Verma HN, Prasad V (1983) Inhibitors of viruses systemic resistance inducers from higher plants.In: Husain A, Singh K, Singh BP, Agnihotri VP (eds) Recent advances in plant pathology. Print House, Lucknow pp 311–324

Verma HN, Prasad V (1987) Systemic induced antiviral resistance by plant extract alters physiology of susceptible test host. Ind J Plant Pathol 5:69–72

Verma HN, Prasad V (1988) Metabolic alterations associated with host mediated systemic antiviral resistance. Indian Phytopathol 41:332–335

Verma HN, Chowdhury B, Rastogi P (1984) Antiviral activity in leaf extracts of different *Clerodendrum* L. species. Z Pflanzenkrankh Pflanzenschutz 88:228–234

Verma HN, Khan MM Abid Ali, Dwivedi SD (1985) Biophysical properties of highly antiviral agents present in *Pseuderanthemum atropurpureum* and *Bougainvillea spectabilis* extracts. Indian J Plant Pathol 3:13–20

# An Impact of Seed Priming on Disease Resistance: A Review

16

Sananda Mondal and Bandana Bose

**Abstract**

Seed priming is basically a physiological seed quality enhancement method which offers a hydration treatment that allows controlled imbibition and induction of the pre-germinative metabolism (activation), but radicle emergence is prevented. The beneficial effects of this technology are greater cellular membrane integrity, counter action of lipid peroxidation, antipathogenic effects, repair of biochemical lesions by the cellular enzymatic repair system and the metabolic removal of toxic substances. Disease is the disorder of the structure or function of a particular system, caused mainly with the aid of fungi, bacteria, viruses and nematodes. Wilt, blight, blast, rust, canker, decay, root diseases, etc. represent the common diseases of plants that cause a reduction in their yield. To get rid of plant diseases a number of fungicides, bactericides, etc. are in use, which impose their residual effects on the users of the plant parts. Consequently, a number of hazardous effects are also noted in human beings. To avoid this kind of hazardous effects on mankind, originated from the residual effects of different kinds of pesticide, an alternative measure must be taken into consideration. In this respect, different kinds of seed priming like hydro-priming, bio priming, osmo-priming, matrix priming and halo-priming can be adopted. Collar rot (*Sclerotium rolfsii*) in chickpea, yellow mosaic virus of mung bean and downy mildew of pearl millet are found to decrease with the use of hydro-primed seeds whereas the use of *Trichoderma* in the form of bio priming has controlled the cowpea root rot pathogens. Salicylic acid alone or in combination with magnesium nitrate ($Mg(NO_3)_2$) induced resistance in groundnut and mustard plants against *Alternaria alternate* and *Alternaria brassicae*. Hence, the review frames a norm to work out the effects of seed priming towards the pathogen-related defence mechanism because seed-priming technology offers a number of benefits and helps to minimize pollution in different ways.

B. Bose (✉) · S. Mondal
Seed Physiology Laboratory, Department of Plant Physiology, Institute of Agricultural Sciences, Banaras Hindu University, Varanasi 221005, India
e-mail: bbosebhu@gmail.com

R. N. Kharwar et al. (eds.), *Microbial Diversity and Biotechnology in Food Security*, DOI 10.1007/978-81-322-1801-2_16, © Springer India 2014

**Keywords**

Seed · Priming · Disease resistance

## 16.1 Introduction

Seeds are the best delivery system in agriculture. High-quality seeds show synchronized germination, and the formation of potential seedlings makes them able to attain the optimum level of their genetic potential. Hence, always, the good quality seeds attracted researchers to get proper production potential of a particular crop. Quality seeds, nowadays, can be attained by various means; the basic and applied knowledge of plant physiology, genetics and seed technology, all are integrated to improve the quality of seeds. With these efforts, a technology now has come into existence known as seed priming, which has many fold beneficial effects on various growth phases of plants' life in general, and it also develops a qualitative change in growing plants by improving their stress-tolerance capacity towards heat, temperature, light, water, nutrition as well as diseases. The present review is focussed on the influence of seed priming on disease resistance.

## 16.2 Seed and Seed-Treatment Technology

Seed is the ripened ovule that consists of an embryo and stored food for germination and contains a protective covering, or one can also say that a seed is a small embryonic plant enclosed in a covering called seed coat along with some stored food. The term seed also has a general meaning, i.e. a seed represents any living material that can be sown which gives rise to a functional plant, e. g. seed potato is a part of a tuber and the sets of sugarcane are parts of the stem. High-quality seed leads to excellent seedling performance in the field. It is the ultimate basis of successful companies that breed crop plants for seed production. Seed quality is a complex trait that is determined by interactions between multiple genetic factors and environmental conditions. Modern approaches to improve seed quality therefore combine classical genetics, plant molecular biology and a variety of seed technologies. These 'seed biotechnologies' enhance the physiological quality, vigour and synchronicity of seeds to establish a crop in the field under diverse environmental conditions.

Seed-treatment technology is an important entrepreneurial link between seed producers and crop production industry. Its aim is to allow the seed-treatment product to be used in such a form that represents the highest quality in the market. It is also referred as seed enhancement. Seed-treatment technology includes priming, pre-sowing hydration treatments which include noncontrolled water uptake systems (methods in which water is freely available and not restricted by the environment) and controlled systems (methods that regulate seed moisture content preventing the completion of germination) (Taylor et al. 1998), pelleting (which adds thicker artificial coverings to seeds, which can be used to cover irregular seed shapes and adds chemicals to the pellet matrix, e.g. of sugar beet or vegetable seeds; the pellet matrix consists of filling materials and glue and is also used to increase the size of very small horticultural seeds), coating (film-coating methods allow the chemicals to be applied in a form of synthetic polymer that is sprayed onto the seeds and provide a solid, thin coat covering on them; the advantage of the polymers is that they adhere tightly to the seed and prevent loss of active materials like fungicides, nutrients, colorants or plant hormones), artificial seeds and other novel seed-treatment methods of applied seed biology. Our basic and applied seed research projects focus on embryo growth and on the different seed covering layers (e.g. testa, endosperm, pericarp), which are determinants of seed quality and exhibit the biodiversity of seed structures. Seed germination is controlled by environmental factors (light, temperature, water) and plant hormones as endogenous regulators (gibberellins, abscisic acid, ethylene, auxin, cytokinins and brassinosteroids). The utilization of plant hormones and inhibitors as well as their biosynthesis and action in seed treatment technologies affects seed germination and seedling emergence. The genes, enzymes, signalling components and downstream targets of some plant hormones provide molecular marker for seed quality and seedling performance.

Important methods have been developed to enhance seed quality and seedling performance through the addition of chemicals to protect the seed from pathogens and/or to improve germination.

Seed priming is an innovative concept of treating seeds using various solvents including water, which activates physiological processes of seeds. Generally, osmoregulators like polyethylene glycol (PEG), mannitol, glycerol, etc. are being extensively used in seed treatment for various purposes. It is hydration treatment that allows controlled imbibition and induction of the pre-germinative metabolism ('activation'), but radicle emergence is prevented. The hydration treatment is stopped before desiccation tolerance is lost. An important problem is to stop the priming process at the right moment; this right time depends on the species and the seed batch. Molecular markers can be used to control the priming process. Priming solutions can be supplemented with plant hormones or beneficial microorganisms. The seeds can be dried back for storage, distribution and planting. Priming can induce the germination speed and synchronisation (Bose and Tandon 1991); it can improve seed vigour which requires very short or no activation time during germination. It may introduce a wider range of temperature for germination (Anaytullah and Bose 2007), and can break the dormancy or may shorten the time of emergence with improved seedling vigour (Mondal et al. 2011). This leads to better crop stands and higher yields. The pretreatment of seeds with priming agents facilitates the active absorption of ionic molecules with greater adenosine triphosphate (ATP) availability and repair of deteriorated seed parts for reducing leakage of metabolites leading to faster embryo growth (Dahal et al. 1990).

The beneficial effects of these priming treatments were reflected in greater cellular membrane integrity, counter action of lipid peroxidation and free radical chain reaction often found to be directly correlated with the maintenance of viability and reduced moisture uptake by hydrated–dehydrated seeds (Dollypan and Basu 1985), antipathogenic effects (Powell and Mathews 1986), repair of biochemical lesions by the cellular enzymatic repair system (Villiers and Edgcumbe 1975) and metabolic removal of toxic substances (Basu et al. 1973) and counteraction of free radical and lipid peroxidation reactions (Rudrapal and Basu 1982).

Any cellular disorder or distinction resulting due to the presence of some biotic factors like fungi, bacteria, viruses and nematodes can come under the persistence of disease. Wilt, blight, blast, rust, canker, decay, root diseases, etc. represent the common diseases of plants, causing a reduction in their yield. To get rid of plant diseases, a number of fungicides, bactericides, etc. are in use which may impose their residual effects on the users of the plant parts and soil (Roger et al. 1994; Hart 1995). More than 98 % of sprayed insecticides and 95 % of herbicides reach a destination other than their target species, including nontargeted species, air, water, bottom sediments and food (Miller 2004). Though there can be benefits using pesticides, inappropriate use can counterproductively increase pest resistance and kill the natural enemies of pests. Many users are inadequately informed about potential short- and long-term risks, and the necessary precautions in the correct application of such toxic chemicals are not always made (Damalas and Eleftherohorinos 2011). Pesticides can contaminate unintended land and water when they are sprayed aerially or allowed to run off fields, or when they escape from production sites and storage tanks or are inappropriately discarded (Tashkent 1998). Consequently, a number of hazardous effects are also noted in human beings. To avoid this kind of hazardous effects on mankind originated from the residual effects of different kinds of pesticide, an alternative measure can be taken into consideration. In this respect, different kinds of seed priming like hydro-priming, bio priming, osmo-priming, matrix priming and halo-priming are taken into consideration.

## 16.3 Hydro-Priming with Respect to Disease Resistance

Hydro-priming is achieved by continuous or successive addition of a limited amount of water to the seeds. Pre-hydration of seeds is an important approach to enhance germination and emergence in drought-affected areas, where adequate moisture is not available for proper crop establishment. Afzal et al. (2002) reported significant grain yield improvement in double-cross hybrid maize using hydro-primed seeds. Similarly, hydro-priming showed significant improvement in percentage and mean time of emergence on sorghum (Moradi and Younesi 2009).

Indian farmers reported that primed chickpea suffered less damage from pod borers (Harris et al. 1999), and the damage in Bangladesh was much reduced, but the apparent difference was not statistically significant (Musa et al. 2001). However, damage in farmers' trials caused by collar rot (*Sclerotium rolfsii*) in Bangladesh was significantly reduced by priming seeds overnight, by 45 % in 1998–1999 (30 trials) and by 30 % in 1999–2000 (35 trials) (Musa et al. 2001).

An on-station trial in Peshawar, Pakistan in 2002 (Rashid et al. 2004a) showed that priming seeds of mung bean cv. NM 92 for 8 h in water resulted in a significant fivefold increase in grain yield relative to a non-primed crop. This was associated with a large difference in the severity of symptoms of mung bean yellow mosaic virus (MYMV) assessed using a visual scoring index. More than 70 % of the non-primed plants had severe or lethal symptoms whereas only 14 % of the primed plants were similarly affected. Only 9 % of non-primed plants showed no disease symptoms in contrast to 32 % of primed plants. Rashid et al. (2004b) also observed similar differences in MYMV infection in other on-station mung bean priming trials.

Downy mildew disease caused by the obligate biotroph *Sclerospora graminicola* (Sacc.) Schroet. is a major constraint to pearl millet yields. A standard greenhouse screening method (Jones et al. 1995) was used to investigate the effect of seed priming on the disease resistance of pearl millet. Priming seeds in water for 8 h before sowing significantly reduced the incidence of downy mildew disease in seedlings of a highly susceptible cultivar from about 80 % to less than 60 % (Harris et al. 2005). They also stated that the effect was confirmed in additional glasshouse experiments (unpublished). Although the screen would not allow plants to be assessed at later stages of growth, there is a high degree of correlation between performance of cultivars in the screen and their resistance to downy mildew in the field (Jones et al. 2002). However, a preliminary trial at International Crops Research Institute for the Semi-Arid-Tropics (ICRISAT), India in 2006 failed to confirm this effect in the field.

### 16.3.1 Bio Priming with Respect to Disease Resistance

Bio priming involves coating of seeds with a bacterial biocontrol agent like *Pseudomonas aureofaciens* and hydrating for 20 h under warm conditions (23 °C) in moist vermiculite or on moist germination blotters in a self-sealing plastic bag. The seeds are taken out from the solution before radical emergence (Callan et al. 1990). It integrates the biological and physiological aspects of disease control and was recently used as an alternative method for controlling many seed- and soil-borne pathogens.

Controlling soil-borne pathogens depends mainly on fungicidal applications that cause hazards to the human health and environment. Soil amendment and bio priming seed treatment are gaining importance in management of many plant pathogens as another alternative to chemical fungicides in recent times. Seed coating with biocontrol agents is an effective treatment for controlling root rot diseases (Callan et al. 1991). Nemec et al. (1996) noted that amended planting mixes with formulation of commercial biocontrol agents such as *Trichoderma harzianum, Bacillus subtilis, Gilocladium virens* and *Strepomyces* sp. reduced root rot and crown rot diseases on tomato, bell pepper, celery and citrus.

EI-Mohamedy et al. (2006) evaluated the efficacy of soil amendment with *T. harzianum* formulated on sugarcane bagasse and/or bio-priming seed treatment in controlling cowpea root rot pathogens under greenhouse and field conditions. The percentage of root rot diseases caused by *Fusarium solani, Rhizoctonia solani and Macrophomina phaseolinae* were reduced significantly. The most effective treatments were bagasse + *T. harzianum* (10 %), bio priming and bagasse + *T. harzianum* (5 %); they reduced *Fusarium* root rot by 73.9, 60.8 and 56.5 %, *Rhizoctonia* root rot by 78.6, 75, 71.4 % and *M. phaseolinae* charcoal rot by 70.8, 62.5 and 62.5 %, respectively.

Nayaka et al. (2010) attempted the use of *T. harzianum* as a seed treatment for controlling maize ear rot and managing fumonisin (synthesized by *Fusarium verticillioides*) accumulation in maize seeds. Seed treatments with *T. harzianum* improve seed germination and emergence, vigour index, plant height, yield and 1,000 seed weight of maize. Again, it reduces the incidence of ear rot disease and the level of fumonisins in maize samples.

Rao et al. (2009) did an experiment to test the efficacy of integrated seed treatment options for the management of *Alternaria* blight of sunflower. Seeds of hybrid sunflower were treated with carbendazim + iprodione (quintal) at 0.3 % along with different organic solvents as priming agents such as PEG 6000, acetone, dichloromethane (DCM) and glycerol, and compared with a treatment with water. *Pseudomonas fluorescens* Migula was used for bio priming of seeds with priming agents such as vermiculite, jelly, moist blotters, salicylic acid (SA) and was compared with direct seed treatment. A total of 100 g of sunflower seeds were soaked for 24 h at 25 ± 2 °C in 500 ml solutions of acetone, glycerol, dichloromethane, PEG (30 % v/v) and water containing 0.3 % of quintal separately and were dried under shade. In bio priming, 100 g of seeds were treated with the bacterial biocontrol agent *P. fluorescens* using priming agents such as vermiculite, jelly and moist blotter. Results showed that the seed treatment with carbendazim + iprodione at 0.3 % in PEG alone with foliar spray of hexaconazole recorded the least per cent disease indexes of 19.24, 28.86 and 37.74 % at 45, 60 and 75 disease activity score (DAS), respectively, and this treatment also recorded the highest yield of 17.12 q/ha with test weight and head diameter of 5.51 g and 25 cm, respectively. Again, Moeinzadeh et al. (2010) reported that bio priming of sunflower seeds with *P. fluorescens* improve seed invigoration and seedling growth.

Nagaraju et al. (2012) reported that seven plant growth-promoting fungi (PGPF) native to the rhizospheric soil were screened for their potential to promote growth and to induce resistance in sunflower against *Plasmopara halstedii,* an incitant of sunflower downy mildew disease. The effect of seed priming with each of these seven PGPF isolates on seed quality parameters and resistance induction were studied under in vitro and greenhouse conditions. Root colonization ability of these PGPF was also studied under greenhouse conditions. The results showed that seed priming with conidial suspension of PGPF at 1 × 108 CFU $ml^{-1}$ significantly increased seed germination and seedling vigour compared to non-primed control. Plants raised with primed seeds demonstrated a significant reduction in downy mildew disease severity and provided a maximum of 61 % protection under greenhouse conditions when compared to the untreated control. The experimental results have proven that the tested PGPF, promoted growth and induced systemic resistance (ISR) in sunflower plants against the downy mildew disease caused by *P. halstedii,* due to their growth-promoting and biocontrol abilities.

Abuamsha et al. (2012) observed that the efficacy of a seed treatment of oilseed rape (OSR) (*Brassica napus*) with the rhizobacteria *Serratia plymuthica* (strain HRO-C48) and *Pseudomonas chlororaphis* (strain MA 342) applied alone or in combination against the blackleg disease caused by *Leptosphaeria maculans* was tested with different cultivars. Seeds were soaked in bacterial suspensions (bio-priming) to obtain $\log_{10}6$–7 CFU $seed^{-1}$. Cotyledons were inoculated with a 10 µl droplet of *L. maculans* spore suspension of $\log_{10}7$ spores $ml^{-1}$, and the disease index (size of lesions) was evaluated 14 days later. A mean disease reduction of 71.6 % was recorded for *S. plymuthica* and of 54 % for *P. chlororaphis*. The combined treatment was not superior to the treatment with *S. plymuthica* alone. The reduction of the disease caused by *S. plymuthica* was independent of the cultivar's susceptibility, whereas the control effect recorded with *P. chlororaphis* increased with decreasing cultivar resistance to blackleg disease. The bacterial colonization of OSR was restricted to the roots and hypocotyls. No significant difference in bacterial colonization of the rhizosphere was observed between different cultivars, or between single or combined bacterial seed treatments.

### 16.3.2 Halo-Priming with Respect to Disease Resistance

Halo-priming is a pre-sowing soaking of seeds in salt solution, which enhance germination and seedling emergence uniformly under adverse environmental conditions (Bose and Mishra 1999; Ashraf et al. 2003; Basra et al. 2005a). The adverse and depressive effects of salinity and water stress on germination can be alleviated by various seed-priming treatments. Cayuela et al. (1996) showed that the higher-salt tolerance of plants raised from primed seeds is the result of higher capacity for osmotic adjustment since plants from primed seeds have more $Na^{+}$ and $Cl^{-}$ in roots and more sugars and organic acids in leaves than plants from non-primed seeds.

SA is a phenolic phytohormone and is found in plants with roles in plant growth and development, photosynthesis, transpiration, ion uptake and transport. SA also induces specific changes in leaf anatomy and chloroplast structure. SA is involved in endogenous signalling, mediating in plant defence against pathogens (Hayat and Ahmad 2007). It plays a role in the resistance to pathogens by inducing the production of pathogenesis-related proteins (Huijsduijnen 2009). It is involved in the systemic-acquired resistance (SAR) in which a pathogenic attack on one part of the plant induces resistance in other parts. The signal can also move to nearby plants by SA being converted to the volatile ester, methyl salicylate. Ratnam et al. (2004) observed that exogenous application of SA at 1.5 mM and benzothiadiazole (Bion at 5 mM) on sunflower (cv. Modern) leaves induced systemic resistance against *Alternaria helianthi*. SA was found in inducing resistance in groundnut plants against *Alternaria alternate.* Foliar application of SA at 1 mM significantly reduced leaf blight disease intensity and increased pod yield under glasshouse conditions. The changes in the activities of phenylalanine ammonia lyase (PAL), chitinase, beta-1,3 glucanase and in phenolic content on groundnut after application of SA and inoculation with *A. alternate* were also studied and in SA-treated leaves an increase in phenolic content was observed after 5 days (Chitra et al. 2008). Kuril (2010) reported that the plants obtained from hardened seeds ($Mg(NO_3)_2$ (magnesium nitrate is a hygroscopic salt, used in the ceramics, printing, chemical and agriculture industries), SA and $Mg(NO_3)_2$ + SA) has less percent disease index (PDI (%)) as compared to non-hardened control seeds. $Mg(NO_3)_2$ + SA treatment was found to have the lowest PDI (%) towards *Alternaria brassicae* as identified by the method of Ellis (1971); $Mg(NO_3)_2$ alone can also lower down the PDI (%) in respect to SA treatment where the latter has less PDI (%) than control. He also reported that $Mg(NO_3)_2$ and SA either alone or in combination were able to improve the yield potential (seed as well as oil content) and disease (reduced PDI of *A. brassicae*)-resistance capability in hardened seeds.

Nitric oxide (NO) is a signalling molecule that takes part in pathophysiological and developmental processes and acts mainly against oxidative stress and also plays a role in plant–pathogen interactions. It has been shown that NO plays a prominent role in the activation of defence-associated responses in several plants against various phytopathogen infections. NO donors nitroso-R-salt, 2-nitroso-1-naphthol and sodium nitroprusside (SNP) were evaluated for their effectiveness in protecting pearl millet ((*Pennisetum glaucum* L.) R. Br.) plants against downy mildew disease caused by *S. graminicola* ((Sacc). Schroet). Optimization experiments with NO donors showed no adverse effect either on the host or pathogen. Aqueous SNP seed treatment with or without PEG priming was most effective in inducing the host resistance against downy mildew both under greenhouse and field conditions. Potassium ferrocyanide, a structural analog of NO donor lacking NO moiety, failed to protect the pearl millet plants from downy mildew indicating a role for NO in induced host resistance reported by Manjunatha et al. (2008a). Treatment of wheat leaves with NO donor, SNP, elevated the level of defence enzyme PAL and disease protection was recorded against *Puccinia striformis* reported by Guo et al. (2004). Again, Modolo et al. (2002) stated that NO treatment-induced phytoalexins like isoflavanoids and pterocarpons accumulation in soybean in response to stem canker pathogen *Diaporthe phaseolurum* f. sp. *meridionalis*. Potato plants, treated with the NO-releasing compound 1-hydroxy-2-oxo-3,3-bis(2-aminoethyl)-1-triazene (NOC18) induced the accumulation of rishitin (a phytoalexin) in response to *Phytopthora infestans* that is known to confer disease resistance in several solanaceous plants. Treatment of NO-releasing compounds to tobacco plants lead to a state of induced resistance and significant reduction of lesion size caused by tobacco mosaic virus (TMV) (Noritake et al. 1996; Song and Goodman 2001).

Chitosan is a linear polysaccharide composed of randomly distributed β-(1–4)-linked D-glucosamine (deacetylated unit) and N-acetyl-D-glucosamine (acetylated unit). It has a number of commercial and possible biomedical uses. In agriculture, chitosan is used primarily as a natural seed treatment and plant growth enhancer, and as an eco-friendly biopesticide that boosts the innate ability of plants to defend themselves against fungal infections. A natural biocontrol active ingredients, chitin/chitosan, is found in the shells of crustaceans, such as lobsters, crabs and shrimp, and many other organisms, including insects and fungi. Chitosan active biopesticides represent a new tier of cost-effective biological control of crops for agriculture and horticulture (Goosen 1996). The biocontrol mode of action of chitosan elicits natural innate defence responses within plants to resist insects, pathogens and soil-borne diseases when applied to foliage or the soil (Linden and Stoner 2005). Chitosan increases photosynthesis, promotes and enhances plant growth, stimulates nutrient uptake, increases germination and sprouting, and boosts plant vigour. When used as seed treatment or seed coating on cotton, corn, seed potatoes, soybeans, sugar beets, tomatoes, wheat and many other seeds, it elicits an innate immunity response in developing roots which destroys parasitic cyst nematodes without harming beneficial nematodes and organisms (Smiley et al. 2002, Stoner and Linden 2006). Manjunatha et al. (2008b) observed that when pearl millet seeds were primed with chitosan at different concentrations: 0.5, 1.5, 2.5 and 3 g $kg^{-1}$ seed, then 2.5 g $kg^{-1}$ was found to be optimum, with enhanced seed germination of 99 % and seedling vigour of 1,782, whereas the untreated control recorded values of 87 % and 1,465, respectively. At optimum concentration, chitosan did not inhibit sporulation and release of zoospores from sporangia. Furthermore, pearl millet seedlings raised after seed treatment with chitosan showed an increased level of the defence-related enzymes chitosanase and **peroxidase** as compared with the untreated pearl millet seedlings on downy mildew pathogen inoculation. The effect of chitosan in reducing downy mildew incidence was evaluated in both greenhouse and field conditions, in which 79.08 and 75.8 % disease protection was obtained, respectively.

Manjunatha et al. (2009) observed that chitosan seed priming elevated NO accumulation in pearl millet seedlings, beginning from 2 h post-inoculation, and it was found to be involved in the activation of early defence reactions such as hypersensitive reaction, callose deposition and PR-1 protein expression. Pretreatment with NO scavenger 2-4-carboxyphenyl-4,4,5,5-tetramethylimidazoline-1-oxyl-3-oxide (C-PTIO) and nitric oxide synthase (NOS) inhibitor L-NG-nitroarginine methyl ester (L-NAME) before pathogen inoculation reduced the disease-protecting ability of chitosan, and defence reactions were also downregulated, which indicated a possible role for NO in chitosan-induced resistance.

Shailasree et al. (2001) reported that beta-aminobutyric acid (BABA) treatment of pearl millet (*Pennisetum glaucum* (L) R Br) seeds influenced seedling vigour and protected the seedlings from downy mildew disease caused by the oomycetous biotropic fungus *S. graminicola* (Sacc) Schroet.

Of the different concentrations of BABA tested, viz. 25, 50, 75 and 100 mM, seeds treated with 50 mM for 6 h resulted in the maximum of 1,428 seedling vigour and showed 23 % disease incidence in comparison with the control which recorded a seedling vigour of 1,260 and 98 % disease incidence, i.e. 75 % protection from disease.

Worrall et al. (2012) observed that priming of defence is a strategy employed by plants exposed to stress to enhance resistance against future stress episodes with minimal associated costs on growth. They tested the hypothesis that application of priming agents to seeds can result in plants with primed defences. They measured resistance to arthropod herbivores and disease in tomato (*Solanum lycopersicum*) plants grown from seed treated with jasmonic acid (JA) and/or BABA. Plants grown from JA-treated seeds showed increased resistance against herbivory by spider mites, caterpillars and aphids, and against the necrotrophic fungal pathogen, *Botrytis cinerea*. BABA seed treatment provided primed defence against powdery mildew disease caused by the biotrophic fungal pathogen, *Oidium neolycopersici*. Priming responses were long lasting, with significant increases in resistance sustained in plants grown from treated seed for at least 8 weeks, and were associated with enhanced defence gene expression during pathogen attack. There was no significant antagonism between different forms of defence in plants grown from seeds treated with a combination of JA and BABA. Long-term defence priming by seed treatments was not accompanied by reductions in growth, and may therefore be suitable for commercial exploitation.

Nakaune et al. (2012) demonstrated that the low salt (sodium chloride) seed priming is effective in the promotion of germination and seedling emergence and subsequent growth of tomato plants and an increased tolerance to *Ralstonia solanacearum,* the causative agent of bacterial wilt of tomato, compared with the hydro-primed and non-primed seedling.

The osmo-conditioning effect on pearl millet seeds was tested with different concentrations among which 1 % showed an optimum effect. Seeds osmo-primed with mannitol for 3 h followed by overnight drying offered maximum germination of 99 % and seedling vigour of 1,465 followed by PEG and glycerol. All the three osmo-priming agents recorded a significant increase on growth parameters like the height of the plant, leaf area, the number of tillers per plant. The results indicated that mannitol offered a maximum positive effect followed by PEG and glycerol. Notably, osmo-priming has advanced 5 days of flowering and also has a positive effect on the number and length of the ear heads. The osmo-conditioning agent increased the 1,000 seed weight significantly over the untreated control. When the osmo-primed seeds were germinated and inoculated with downy mildew pathogen, the maximum protection of 61.76 % was observed in mannitol treatment followed by PEG and glycerol under greenhouse conditions reported by Roopa et al. (2009).

L-ascorbic acid (AA, vitamin C) is a multifaceted molecule with diverse physiological functions in plants (Smirnoff 1996) and animals. It is best characterized as an antioxidant that detoxifies reactive oxygen species (ROS), particularly hydrogen peroxide ($H_2O_2$) (Smirnoff 2000). Mukherjee et al. (2010) stated that the AA-deficient *Arabidopsis thaliana vtc1–1* mutant exhibits increased resistance to the virulent bacterial pathogen *Pseudomonas syringae*. This response correlates with heightened levels of SA, which induces antimicrobial pathogenesis-related (PR) proteins.

Ahn et al. (2007) observed that thiamine confers SAR on susceptible plants through priming, leading to rapid counterattack against pathogen invasion and perturbation of disease progress. Priming reduces the metabolic cost required for constitutive expression of acquired resistance. To investigate the effects of priming by thiamine on defence-related responses, *A. thaliana* was treated with thiamine and the effects of pathogen challenge on the production of active oxygen species, callose deposition, hypersensitive cell death and pathogenesis-related 1 (PR1)/PAL1 gene expression were analysed. Thiamine did not induce cellular and molecular defence responses except for transient expression of PR1 per se; however, subsequent *P. syringae* pv tomato challenge triggered pronounced cellular defence responses and advanced activation of PR1/PAL1 gene transcription.

## 16.4 Future Prospects

For the crop growers, handling seeds is easier in respect to growing crops in the field. To eradicate any disease, large amounts of pesticides are required, and the available literature suggests that a good percentage of the used pesticide either remains within the plants in residual form or goes to the soil/water/atmosphere creating hazardous effects or pollution. It has been realized that the seed priming technology, used for increasing the yield/production potential of crops, can also improve stress tolerance characters including disease resistance in the treated plants. This may open the window for researchers to explore ways for minimizing the use of pesticide and curing the diseases in growing plants by using seed-priming technology as a practice. The chemicals used during seed priming are easily available and required in low amounts, so they are economical and eco-friendly in nature. Besides this, our planet now has more than 6 billion population which will double in the next 50 years. About 12 % of global population is suffering from hunger and living without secure access to food. Hence, seed priming will be the most prominent technology in the coming days and can be a step towards food security as well as sustainable agriculture.

## References


Abuamsha R, Salman M, Ehlers RU (2012) Effect of seed priming with *Serratia plymuthica* and *Pseudomonas chlororaphis* to control *Leptosphaeria maculans* in different oilseed rape cultivars. Eur J Plant Pathol 130:287–295

Afzal I, Ahmad N, Basra SMA, Ahmad R, Iqbal A (2002) Effect of different seed vigour enhancement techniques on hybrid maize (*Zea mays* L.). Pak J Agric Sci 39:109–112

Ahn Il-P, Kim S, Lee Y-H, Suh S-C (2007) Vitamin B1-induced priming is dependent on hydrogen peroxide and the npr1 gene in arabidopsis1. Plant Physiol 143:838–848

Anaytullah, Bose B (2007) Nitrate-hardened seeds increase germination, amylase activity and proline content in wheat seedlings at low temperature. Physiol Mol Biol Plants 13:199–207

Ashraf M, Kausar A, Ashraf MY (2003) Alleviation of salt stress in pearl millet (*Pennisetum glaucum* (L.) R. Br.) through seed treatments. Agronomie 23:227–234

Basra SMA, Afzal I, Anwar S, Shafique M, Haq A, Majeed K (2005a) Effect of different seed invigoration techniques on wheat (*Triticum aestivum* L.) seeds sown under saline and non-saline conditions. J Seed Technol 28:36–45

Basu RN, Chattopadhyay K, Pal P (1973) Maintenance of seed viability in rice (*Oryza sativa* L.) and Jute (*Corchorus capsularis* L. and *C. olitorius* L.). Ind Agr 18:76–79

Bose B, Mishra T (1999) Influence of pre-sowing soaking treatment in *Brassica juncea* seeds with Mg salt on growth, nitrate reductase activity total protein content and yield responces. Physiol Mol Biol Plant 5:83–88

Bose B, Tandon A (1991) Effect of magnesium nitrate on metabolism in germinating maize seeds. Indian J Plant Physiol 34:69–71

Callan NW, Mathre DE, Miller TB (1990) Bio-priming seed treatment for biological control of *Pythium ultinum* pre-emergence damping-off in the sweet corn. Plant Dis 74:368–371

Callan NW, Mathre DT, Miller JB (1991) Yield performance of sweet corn seed bio-primed and coated with *Pseudomonas flurescence* AB 254. Hort Sci 26:1163–1165

Cayuela E, Perez-Alfocea F, Caro M, Bolarin MC (1996) Priming of seeds with NaCl induces physiological changes in tomato plants grown under salt stress. Physiol Pt 96:231–236

Chitra K, Ragupathi N, Dhanalakshmi k, Mareeshwari P, Indra N, Kamalakannan A, Sankaralingam A, Rabindran R (2008) Salicylic acid induced systemic resistant on peanut against *Alternaria alternate*. Arch Phytopathol Plant Prot 41:50–56

Dahal P, Bradford KJ, Jones RA (1990) Effects of priming and endosperm integrity on seed germination rates of tomato genotypes II. J Exp Bot 41:1441–1453

Damalas CA, Eleftherohorinos IG (2011) Pesticide exposure safety issues and risk assessment indicators. Int J Environ Res Public Health 8(5):1402–1419 (Web of Science)

Dollypan, Basu RN (1985) Mid-storage and pre-sowing seed treatments for lettuce and carrot. Scientia Hort 33:1026–1027

El-Mohamedy RSR, Abd Alla MA, Badia RI (2006) Soil amendment and seed bio-priming treatments as alternative fungicides for controlling root rot diseases on cowpea plants in Nobaria province. Res J Agr Biol Sci 2:391–398

Ellis MB (1971) Dematiaceous hyphomycetes, 1st edn. Commonwealth Mycological Institute, Kew, pp 608

Goosen MFA (1996) Applications of chitin and chitosan. CRC Press. pp 132–139

Guo P, Cao Y, Li Z, Zhao B (2004) Role of an endogenous nitric oxide burst in the resistance of wheat to stripe rust. Plant Cell Environ 27:473–477

Harris D, Joshi A, Khan PA, Gothkar P, Sodhi PS (1999) On-farm seed priming in semi-arid agriculture development and evaluation in maize, rice and chickpea in India using participatory methods. Exp Agr 35:15–29

Harris D, Breese WA, Kumar Rao JVDK (2005) The improvement of crop yield in marginal environments using 'on-farm' seed priming nodulation nitrogen fixation and disease resistance. Aust J Agr Res 56:1211–1218

Hart M (1995) Effects of pesticide on the soil microbial biomass and microbial activity. Thesis University of Nottingham, pp 1–223

Hayat S, Ahmad A (2007) Salicylic acid—a plant hormone. Springer, ISBN 1402051832

Huijsduijnen H Van (2009) Induction by salicylic acid of pathogenesis-related proteins and resistance to alfalfa mosaic virus infection in various plant species. http://vir.sgmjournals.org/cgi/reprint/67/10/2135.pdf. Accessed 28 May 2009

Jones ES, Liu CJ, Gale MD, Hash CT, Witcombe JR (1995) Mapping quantitative trait loci for downy mildew resistance in pearl millet. Theo Appl Genetics 91:448–456

Jones ES, Breese WA, Liu CJ, Singh SD, Shaw DS, Witcombe JR (2002) Mapping quantitative trait loci for resistance to downy mildew in pearl millet: field and glasshouse screens detect the same QTL. Crop Sci 42:1316–1323

Kuril SK (2010) Influence of seed hardening with $Mg(NO_3)_2$ and salicylic acid on timely and late sown mustard (*Brassica juncea* L. Czern and Coss.) varieties. Ph. D. Thesis, I Ag Sc, BHU, Varanasi, India

Linden JC, Stoner RJ (2005) Proprietary elicitor affects seed germination and delays fruit senescence. J Food, Agr Environ. http://www.yeacrops.com/Elicitor%20-%20Ethylene%20Reduction.pdf

Manjunatha G, Raj SN, Shetty NP, Shetty HS (2008a) Nitric oxide donor seed priming enhances defense responses and induces resistance against pearl millet downy mildew disease. Pesticide Biochem Physiol 91:1–11

Manjunatha G, Roopa KS, Prashanth GN, Shetty HS (2008b) Chitosan enhances disease resistance in pearl millet against downy mildew caused by *Sclerospora graminicola* and defence-related enzyme activation. Pest Manag Sci 64:1250–1257

Manjunatha G, Niranjan-Raj S, Prashanth GN, Deepak S, Amruthesh KN, Shetty HS (2009) Nitric oxide is involved in chitosan-induced systemic resistance in pearl millet against downy mildew disease. Pest Manag Sci 65:737–743

Miller GT (2004) Sustaining the earth, 6th edn. Thompson Learning, Inc. Pacific Grove, California, pp 211–216 (Chapter 9)

Modolo LV, Cunha FQ, Braga MR, Salgado I (2002) Nitric oxide synthase-mediated phytoalexin accumulation in soybean cotyledons in response to the *Diaporthe phaseolorum* f. sp. *meridionalis* elicitor. Plant Physiol 130:1288–1297

Moeinzadeh A, Sharif-Zadeh F, Ahmadzadeh M, Tajabadi FH (2010) Bio-priming of sunflower (*Helianthus annuus* L.) seed with *Pseudomonas fluorescens* for improvement of seed invigoration and seedling growth. Aust J Crop Sci 4:564–570

Mondal S, Vijai P, Bose B (2011) Role of seed hardening in rice variety Swarna (MTU 7029). Res J Seed Sci 4:157–165

Moradi A, Younesi O (2009) Effects of osmo and hydro-priming on seed parameters of grain sorhgum. Aust J Basic Applied Sci 3:1696–1700

Mukherjee M, Larrimore KE, Ahmed NJ, Bedick TS, Barghouthi NT, Traw MB, Barth C (2010) Ascorbic acid deficiency in *Arabidopsis* induces constitutive priming that is dependent on hydrogen peroxide, salicylic acid, and the *npr1* gene. Mol Plant Microbe Inter 23:340–351

Musa AM, Harris D, Johansen C, Kumar J (2001) Short duration chickpea to replace fallow after aman rice: the role of on-farm seed priming in the High Barind Tract of Bangladesh. Exp Agric 37:509–521

Nagaraju A, Murali M, Sudisha J, Amruthesh KN, Murthy MS (2012) Beneficial microbes promote plant growth and induce systemic resistance in sunflower against downy mildew disease caused by *Plasmopara halstedii*. Current Botany 3:12–18

Nakaune M, Tsukazawa K, Uga H, Asamizu E, Imanishi S, Matsukura C, Ezura H (2012) Low sodium chloride priming increases seedling vigor and stress tolerance to *Ralstonia solanacearum* in tomato. Plant Biotechnol 29:9–18

Nayaka SC, Niranjana SR, Shankar ACU, Raj SN, Reddy MS, Prakash HS, and Mortensen CANP (2010) Seed biopriming with novel strain of *Trichoderma harzianum* for the control of toxigenic *Fusarium verticillioides* and fumonisins in maize. Arch phytopathol and plant protection, vol 43, nr. 3, s. 264–282., http://dx.doi.org/10.1080/03235400701803879

Nemec S, Datnoff LE, Strondbery T (1996) Efficacy of bio-control agents in planting mixes to colonize plant roots and control root diseases of vegetable and citrus. Crop Prot 15:735–743

Noritake T, Kawatika K, Doke N (1996) Nitric oxide induces phytoalexin accumulation in potato tuber tissues. Plant Cell Physiol 37:113–116

Pesticides, Wikipedia. http://en.wikipedia.org/wiki/Pesticides, section entitled 'The public'

Powell AA, Mathews S (1986) Cell membranes and seed leachate conductivity in relation to the quality of seeds for sowing. J Seed Tech 10:81–100

Rao MSL, Kulkarni S, Lingaraju S, Nadaf HL (2009) Bio-priming of seeds: a potential tool in the integrated management of *Alternaria blight* of sunflower. Helia 32:107–114

Rashid A, Harris D, Hollington PA, Ali S (2004a) On-farm seed priming reduces yield losses of mungbean (*Vigna radiata*) associated with mungbean yellow mosaic virus in the North West Frontier Province of Pakistan. Crop Protec 23:1119–1124

Rashid A, Harris D, Hollington PA, Rafiq M (2004b) Improving the yield of mungbean (*Vigna radiata)* in the North West Frontier Province of Pakistan using on-farm seed priming. Expl Agr 40:233–244

Ratnam MVS, Reddy PN, Rao SC, Raju RB (2004) Systemic induced resistance in sunflower to *Alternaria* leaf blight by foliar application of SA and Bion. J Oilseeds Res 21:104–107

Roger PA, Simpson I, Oficial R, Aradales S, Jimenez R (1994) Effects of pesticide on soil and water microflora and mesofauna in wetland ricefields: a summary of current knowledge and extrepolation to temperate environments. Aust J Exp Agric 34:1057–1068

Roopa KS, Geetha NP, Sharathchandra RG, Pushpalatha HG, Sudisha J, Amruthesh KN, Prakash HS, Shetty HS (2009) Osmopriming enhances pearl millet growth and induces downy mildew disease resistance. Natl Acad Sci 42:979–987

Rudrapal AB, Basu RN (1982) Use of chlorine and bromine in controlling mustard seed deterioration. Seed Res 9:188–191

Shailasree S, Sarosh BR, Vasanthi NS, Shetty HS (2001) Seed treatment with beta-aminobutyric acid protects *Pennisetum glaucum* systemically from *Sclerospora graminicola*. Pest Manag Sci 57:721–728

Smiley R, Cook RJ, Pauliz T (2002) Seed treatment for sample cereal grains Oregon State University, EM 8797.http://extension.oregonstate.edu/catalog/pdf/em/em8797.pdf#search=%22YEA!%20Seed%20treatment%22

Smirnoff N (1996) The function and metabolism of ascorbic acid in plants. Ann Bot 78:661–669

Smirnoff N (2000) Ascorbic acid: Metabolism and functions of a multifaceted molecule. Curr. Opin Plant Biol 3:229–235

Song F, Goodman RM (2001) Activity of nitric oxide is dependent on, but is partially required for function of, salicylic acid in the signalling pathway in tobacco systemic acquired resistance. Mol Plant Microbe Interact 14:1458–1462

Stoner R, Linden J (2006) Micronutrient elicitor for treating nematodes in field crops, Patent Pending, Pub. no.: US 2008/0072494 A1. http://www.google.com/patents?id=XMeqAAAAEBAJ&dq=micronutrients+nematode+suppression

Tashkent (1998) Part 1. Conditions and provisions for developing a national strategy for biodiversity conservation. Biodiversity conservation national strategy and action plan of republic of Uzbekistan. Prepared by the National Biodiversity Strategy Project Steering Committee with the financial assistance of The Global Environmental Facility (GEF) and technical assistance of United Nations Development Programme (UNDP). Retrieved on September 17, 2007

Taylor AG, Allen PS, Bennett, MA, Bradford KJ, Burris JS, and Misra, MK (1998) Seed enhancements. Seed Sci. Res. 8:245–256

Villiers TA, Edgcumbe DJ (1975) On the cause of seed determination in dry storage. Seed Sci Tech 3:761–774

Worrall D, Holroyd GH, Moore JP, Glowacz M, Croft P, Taylor JE, Paul ND, Roberts MR (2012) Treating seeds with activators of plant defence generates long-lasting priming of resistance to pests and pathogens. New Phytol 193:770–778

# 17 Occurrence of Stone Fruit Yellows Phytoplasma Disease (*Candidatus* Phytoplasma prunorum) in Hungary and Central Europe

Gábor Tarcali, György J. Kövics and Emese Kiss

**Abstract**

Plant diseases caused by phytoplasmas have an increasing importance all over the world for fruit growers. Lately, phytoplasma diseases occur on many fruit species and are responsible for serious losses both in quality and quantity for fruit production. Apricot phytoplasma disease (*Candidatus* Phytoplasma prunorum) was reported first from France in Europe in 1924. Then, the pathogen spread in all European apricot-growing areas. In 1992, the disease was identified in Hungary first. Based on the growers' signals, serious damage of *Ca.* Phytoplasma prunorum (Seemüller and Schneider, International Journal of Systematic and Evolutionary Microbiology, 54, 2004, 1217–1226; formerly: European stone fruit yellows (ESFY) phytoplasma) could be observed in several stone fruit orchards in the famous apricot-growing area nearby the town of Gönc, northern Hungary. Field examinations were started in 2009 in the infested stone fruit plantations in Borsod-Abaúj-Zemplén County mainly in Gönc region, which is one of the most important apricot-growing areas in Hungary, named "Gönc Apricot-growing area". Our goals were to diagnose the occurrence of *Ca.* Phytoplasma prunorum on stone fruits (especially on apricot) in the North Hungarian growing areas by visual diagnostics and to confirm data by laboratory polymerase chain reaction (PCR)-based examinations. All the 40 collected samples were tested in laboratory trials and in 22 samples from apricot, peach, cherry, sour cherry and wild plum were confirmed the presence of phytoplasma (ESFY). Field investigations were done in a western Romanian apricot plantation, and the presence of

G. Tarcali (✉) · G. J. Kövics
Institute of Plant Protection, University of Debrecen,
Böszörményi str. 138, 4032 Debrecen, Hungary
e-mail: tarcali@agr.unideb.hu

E. Kiss
Department of Biotechnology, Plant Protection Institute,
Hungarian Academy of Sciences, Herman O. str. 15,
1022 Budapest, Hungary

R. N. Kharwar et al. (eds.), *Microbial Diversity and Biotechnology in Food Security,*
DOI 10.1007/978-81-322-1801-2_17, © Springer India 2014

apricot phytoplasma disease was confirmed. It was the first finding that *Ca.* Phytoplasma prunorum occurred in the western part of Romania. On the basis of these observations, it seems evident that the notable losses caused by *Ca.* Phytoplasma prunorum is a new plant health challenge for fruit growers to manage, especially for apricot producers in Hungary and other central European countries.

**Keywords**

Phytoplasma · *Ca.* Phytoplasma prunorum European stone fruit yellows phytoplasma · Gönc apricot-growing area · Stone fruits · Apricot · Peach · Cherry · *Cacopsylla pruni*

## 17.1 Introduction

### 17.1.1 Importance of Phytoplasma Diseases

Plant diseases caused by phytoplasmas have increasing importance for fruit growers. Phytoplasma diseases occur on several crops throughout the world, and these pathogens cause serious economic losses in cultivated fruit and various field crops. In the long run, these diseases cause the destruction of the fruit producing trees.

Until 1967, plant diseases known as "yellows diseases" were thought to be caused by viruses. In 1967, Japanese researchers (Doi et al. 1967) found microorganism by electron microscope in yellows diseased plants. This new class of plant such a disease agents was named "mycoplasma-like organism" (Welliver 1999). Mycoplasma-like organisms (MLOs) are nonculturable, parasitic prokaryotes of the class Mollicutes associated with diseases of several hundred plant species (McCoy et al. 1989). Until recently, their differentiation and characterization was mainly based on host range and the symptoms induced in natural hosts and in the experimental host *Catharanthus roseus* (periwinkle) (Marwitz 1990). The need for more reliable and specific traits to classify MLOs has resulted in the development of MLO specific serological and DNA hybridization assays (Kiske et al. 1991). In 1992, characterization of the organisms associated with yellows diseases had progressed to a point where they were recognized as unique and so were given their own name: phytoplasma (ICSB 1993; Gundersen et al. 1994).

Phytoplasmas are single-celled organisms that are similar to bacteria but lack a rigid cell wall. Phytoplasmas are obligate parasites. They grow and reproduce in the cytoplasm of host cells, both in insect vectors and in plants. Phytoplasmas are very small agents. They look like amorphous sacks or blobs, ranging from 70–1.000 nm in diameter. Phytoplasmas reproduce asexually by budding. Phytoplasmas reside in the phloem tissues of the plants, and are transmitted by phloem-feeding insect vectors. Phytoplasmas cannot be transmitted mechanically.

Phytoplasmas are serious pathogens of several important plants, including coconut, sugarcane, rice, sandal wood in tropical and subtropical regions of the world, causing a wide variety of symptoms that range from mild yellowing to the death of infected plants. Phytoplasmas also cause very serious diseases on several important crops and fruits in the temperate zone. Stolbur disease of potato and tomato (*Ca.* Phytoplasma solani) is one of the most common plant diseases caused by phytoplasmas. Apple and pear also have phytoplasma-originated diseases (*Candidatus* Phytoplasma mali, *Candidatus* Phytoplasma pyri), but grape and maize are also endangered by these pathogens. Phytoplasmas require a vector to be transmitted from plant to plant, which normally takes the form of sap sucking insects such as leaf hoppers, in which they are also able to replicate.

### 17.1.2 Importance of Apricot in the World and in Hungary

Apricot (*Prunus armeniaca* L.) is one of the most favourite fruits in Hungary and in Europe. Products made from apricot are also popular in Europe. The most important apricot-growing countries are in southern Europe and in the Middle East. The Mediterranean coastal area has the most advantageous climatic conditions for apricot growing. Turkey is the greatest apricot producer in the world with more than 500.000 t yield per year. In Europe, Italy is the main apricot producer country, while France, Spain, Greece are also important growers. Apricot is also a notable fruit crop of mid-hill and dry-temperature regions of India, especially in Himachal Pradesh, Jammu and Kashmir and Uttar Pradesh. According to historians: "The origin of the apricot's domestication was in the Chinese region, but another source says the apricot was first cultivated in India in about 3000 BC" (Huxley 1992).

Hungary also belongs to the main European apricot-producer countries. It has several excellent apricot-growing sites, and apricot cultivation has a long history in Hungary. The best apricot yields were in the 1960s with more than 130.000 t per year. Although Hungarian apricots are easy to sell because of their excellent flavour, nowadays there is not enough quantity in the country. The apricot production has decreased considerably in the Hungarian apricot-growing sites over the last 20 years. The reasons of losses in quantity of apricot production in Hungary are as follows: the uncertain weather conditions, the ageing of several plantations, the unfavourable economical situation for fruit growers and several plant diseases.

### 17.1.3 Phytoplasma Disease on Apricot and Other Stone Fruits

Currently, one of the most important reasons of apricot yield loss in Hungary is the phytoplasma disease. Until 2004, it was named European stone fruit yellows phytoplasma (ESFY) (Kövics 2009). Since 2004, the new official name of the pathogen has been *Ca.* Phytoplasma prunorum (Seemüller and Schneider 2004). Although the disease is well known in many European countries (Lederer and Seemüller 1992), it is a relatively new and serious pathogen for the apricot cultivation in Hungary and the neighbouring central European countries in the Carpathian basin. Moreover, the pathogen is able to infect other stone fruit species (e.g. peach, cherry, and sour cherry).

Apricot phytoplasma disease (named that time as "apoplexy") was reported first in Europe in 1924, from France, (Chabrolin 1924) and then the pathogen was observed in several other European apricot-growing countries (Lederer and Seemüller 1992). In many European countries the disease has been identified as one of the most prevalent problems threatening apricot trees (Jarausch et al. 2001; Navratil et al. 2001; Torres et al. 2004). In 1992, the disease was also observed on apricot in Hungary (Süle, unpublished) although its symptoms had been suspected before too. Later on (Viczián et al. 1997; Süle et al. 1997), the occurrence of the ESFY was confirmed by molecular biological examinations in Hungary. After that, similar symptoms were also observed on other stone fruits (Mergenthaler 2004). The occurrence of phytoplasma was observed in peach (Németh et al. 2001), Japanese cherry (*Prunus serrulata*) (Lorenz et al. 1994), Mahaleb cherry (*P. mahaleb cv.*) Cemany (Varga et al. 2001) and Blackthorne (*P. spinosa*) (Jarausch et al. 2001) as well. According to earlier natural experiences and examinations based on artificial inoculation, several kinds of plum are tolerant to *Ca.* Phytoplasma prunorum. European plums have been determined to be tolerant to *Ca.* Phytoplasma prunorum, whereas Japanese plums (*P. salicina*) are highly susceptible (Carraro et al. 1998; Mona et al. 2008). The pathogen is able to propagate in the tolerant host plum trees without any typical symptoms, therefore these trees obviously have an important role in the spreading of the pathogen (Morvan 1977; Carraro et al. 1998). The psyllid *Cacopsylla pruni* (Scopoli Fig. 17.1) was described as the main vector of *Ca.* Phytoplasma prunorum (Carraro et al. 2001; Fialová et al. 2007). Occurrence of phytoplasma disease on cherry was reported in France as "Molieres-disease" (Bernhard et al. 1977), but several ex-

**Fig. 17.1** The vector of *Ca.* Phytoplasma prunorum–*Cacopsylla pruni.* (Source: Dr Wolfgang Jarausch, Agroscience)

**Table 17.1** Scale of infection index (*Ii*) classification system. (Tarcali and Kövics 2009)

| Infection degrees | Symptoms |
|---|---|
| I | Healthy tree |
| II | Symptoms on leaves, on one branch |
| III | General yellowing or drying, symptoms on several branches |
| IV | One dead branch |
| V | Dead or felled tree |

perts thought that cherry is resistant to phytoplasma infection (Jarausch et al. 2000).

Symptoms of *Ca.* Phytoplasma prunorum disease are visible on the leaves, the branches and the bark. Yellow colour change and rolling of leaves can be seen. General yellowing and "scalding-like" dryings are visible on the branches. There is an orange to light brown colour change in the phloem. Finally, the disease causes a "sudden death" of apricot trees, similar to the destruction of another important apricot disease, the apricot die-back (apoplexy) caused by fungal (*Cytospora* spp.), bacterial (*Pseudomonas syringae*) pathogens and abiotic reasons (e.g. frost damage). However, there is a significant difference between the destructions of these diseases. In the case of phytoplasma disease, there is no secretion of resin in contrast with apoplexy.

## 17.2 Materials and Methods

In Hungary, there are eight great apricot-growing areas (Balaton, Mecsek, Lake-Velencei, Buda, Pest-Gödöllö, Mátra-Bükkalja, Area between Danube and Tisza Rivers and Gönc). One of the most important sites is the Gönc Apricot-growing area, situated in Borsod-Abaúj-Zemplén County, northern Hungary, which was the main location of our field observations. Field examinations were done on several stone fruit orchards on Gönc Apricot-growing area between 2009 and 2011 (12 apricot, 1 peach, 4 sour cherry, and 1 cherry plantations). That place is among the most favoured growing areas for stone fruits, especially for apricot in central Europe. Besides, field examinations were also done in Hajdúdorog, close to the Gönc region (in a neighbouring county, Hajdú-Bihar) and in another region near Oradea city, which belongs to Romania (close to the Hungarian border) in 2010–2011.

The main objectives of the research were to:

- Visually check of *Ca.* Phytoplasma prunorum infection on stone fruits (especially on apricot) on the examined northern Hungarian and Romanian growing areas.
- Calculate the infection ratio (*I*%) and infection index (*Ii*) (according to a classification system (Table 17.1)) in the various stone fruit plantations based on the visible symptoms of the disease caused by *Ca.* Phytoplasma prunorum.

The classification system contains five infection degrees, and the symptoms get more heavy from the first degree to the fifth degree. Infection degrees were classified on the basis of the following symptoms:

- On leaves: yellow colour change and rolling of leaves to its abaxial surface
- On branches: general yellowing or "scalding-like" drying
- In the trunk: having striped the bark of tree, orange or light brown colour change is visible in the phloem
- On trees: general yellowing on several branches or general drying; withered, dead or felled tree, and there is no secretion of resin
- At the plantation: infections and destruction of trees starting in a circular direction around the infected tree.

**Table 17.2** Used sequences and programmes on laboratory examinations

| Name of primer (1) | Sequences (5′–3′) (2) | Position (bp) (3) | Programme (4) |
|---|---|---|---|
| P1 | AAGAGTTTGATCCTG-GCTCAGGATT | 6–28 | 94 °C-5 min; 94 °C-1 min 55 °C-1 min 72 °C-2 min (35 cycle); 72 °C-10 min |
| P7 | TTCTCGGC-TACTTCCTGC | 1,818–1,836 | |
| fU5 | CGGCAATGGAG-GAAACT | 370–387 | 95 °C-3 min; 95 °C-1 min 55 °C-1 min 72 °C-1 min (35 cycle); 72 °C-5 min |
| rU3 | TTCAGCTACTCTTTG-TAACA | 1,230–1,250 | |
| ECA1 | AATAATCAAGAA-CAAGAAGT | | 95 °C-1 min; 95 °C-30 s 55 °C-30 s 72 °C-30 s (35 cycle); 72 °C-3 min |
| ECA2 | GTTTATAAAAATTAAT-GACTC | | |
| fO1 | CGGAAACTTT-TAGTTTCAGT | 61–81 | 94 °C-3 min; 94 °C-1 min 55 °C-1 min 72 °C-1 min (35 cycle); 72 °C-7 min |
| rO1 | AAGTGCCCAACTAAAT-GAT | 1,115–1,135 | |

Around 100 trees were examined on a researched site (except the smaller fruit gardens), out of which 10 fruit trees of a circle were randomly selected for examination from 10 circles.

Plant samples were collected from the supposedly infected trees, based on the visible symptoms (living leaves, pieces of branches and pieces of roots) by a sharp sampling knife for further laboratory examinations. The identification of phytoplasma is only possible from living plant parts. Identification is not possible from dead plant samples because of the life of the pathogen twit to the living phloem of the plant (Mergenthaler 2004). In the laboratory, molecular biological examinations were applied (PCR) to confirm the presence of the phytoplasma. The primers, sequences and programmes which were applied on the PCR examination in order to identify the phytoplasma are shown in Table 17.2.

The venue of the laboratory examinations was at the molecular biology laboratory of the Plant Protection Institute, University of Debrecen, in Debrecen, Hungary, and in the Sequence Laboratories Göttingen GmbH, in Göttingen, Germany. First, universal primers and sequences (fP1/rP7, fU5/rU3) were used to prove the phytoplasma infection in the examined plant samples (Kirkpatrick et al. 1994). Then, group-specific primers (fO1/rO1, ECA1/ECA2) were applied to identify species of phytoplasma (Kirkpatrick et al. 1994; Jarausch et al. 1998).

## 17.3 Results and Discussion

### 17.3.1 Results of the Field Examinations

Our examination on the research of *Ca.* Phytoplasma prunorum began on 2nd October 2009, when we visited the first infested stone fruit orchard in the village of Bekecs (near Szerencs city). The first visual experience was the view of a very depressing situation of the apricot plantations in the Gönc apricot-growing area. Several apricot plantations were heavily destructed by phytoplasma disease, and a great number of apricot trees were dead or felled in the orchards (Fig. 17.2). Yellowing and rolling leaves on the apricot branches and several drying branches were found on the apricot trees (Fig. 17.3). A similar situation was visible on peach (Fig. 17.4), and the same situation was experienced on cherry and sour cherry trees (Fig. 17.5).

A comparatively new, only 4-year-old apricot plantation was examined first. Most trees were healthy, but there were a few trees (2 %) infected by *Ca.* Phytoplasma prunorum (Table 17.3). According to the description of Süle et al. (2003),

**Fig. 17.2** Stumps of destructed and felled apricot trees

**Fig. 17.3** Yellowing leaves on apricot trees

**Fig. 17.4** Yellowing branches on peach

the first symptoms of the pathogen can be observed from the age of 3 or 4, and this thesis was justified in the visited apricot orchard.

The second apricot plantation was about 8 or 9 years old. At first glance, it was clearly visible that there is a very serious destruction on apricot trees caused by phytoplasma. More than 50 % of

**Fig. 17.5** A dried and felled sour cherry tree

the trees were infected and 40 % of apricot trees were dead. Most of the killed trees were felled (about 35 %). The owner of the plantation said that in the previous year only one to two trees showed the symptoms of the disease. The general drying has begun in the year 2009 at the end of the blooming of apricot trees (in the first half of May), and then destruction has progressed fast. There was another interesting fact to observe. As there were a few old plum trees among the apricots in the fruit garden, the presence of *Cacopsylla pruni* on the plum trees was evident, which is the main vector in the transmission of the pathogen (Fig. 17.1).

The varieties of apricot grown on the plantations were the following ones: Ceglédi Óriás (Cegléd Giant), Ceglédi Arany (Cegléd Gold) and Magyar Kajszi (Hungarian Apricot). The Cegléd varieties (a new local one in Hungary) are more susceptible to phytoplasma disease than the old variety, the Hungarian Apricot.

A serious destruction of 85 % was experienced in the third apricot orchard on 3 ha. It was the most heavy infested and destructed apricot population among the examined fields. A very serious infection ratio was measured; out of the 100 sample examined trees as many as 85 were infected, and according to the classification system (Table 17.1), 65 were found dead or felled, as shown in column V. of Table 17.3. The sight was similar on the fourth examined plantation, where 12 to 13-year-old apricot trees are grown; 25 % of apricot trees were dead, and another 35 % were in the phase of fast destruction.

**Table 17.3** Phytoplasma infection data on the examined fruit plantations (results of field examination)

| No. | Date of field examination | Tree species | Age (year) | Area (ha) | Number of trees | Degree of infection | | | | | Ii | I% |
|---|---|---|---|---|---|---|---|---|---|---|---|---|
| | | | | | | I | II | III | IV | V | | |
| 1 | 02.10.2009. | Apricot | 4 | 20 | 100 | 98 | 1 | 1 | – | – | 1.03 | 2 |
| 2 | 02.10.2009. | Apricot | 8–9 | 5 | 100 | 45 | 4 | 6 | 5 | 40 | 2.91 | 55 |
| 3 | 02.10.2009. | *Apricot* | *8* | *3* | *100* | *15* | *7* | *7* | *6* | *65* | *3.99* | *85* |
| 4 | 02.10.2009. | Apricot | 12–13 | 10 | 100 | 30 | 6 | 4 | 35 | 25 | 3.21 | 70 |
| 5 | 02.10.2009. | Peach | 8 | 6 | 100 | 79 | 7 | 2 | 2 | 10 | 1.57 | 21 |
| 6 | 02.10.2009. | Cherry | 10 | 22 | 100 | 70 | 9 | 4 | 6 | 11 | 1.79 | 30 |
| 7 | 02.10.2009. | *Sour cherry* | *8–9* | *5* | *100* | *38* | *14* | *10* | *8* | *30* | *2.78* | *62* |
| 8 | 02.10.2009. | Sour cherry | 7 | 5 | 100 | 91 | 3 | 1 | 1 | 4 | 1.24 | 9 |
| 9 | 02.10.2009. | Sour cherry | 30 | 8 | 100 | 64 | 6 | 9 | 13 | 8 | 1.95 | 36 |
| 10 | 07.09.2010. | *Apricot* | *13* | *22,6* | *70* | *11* | *12* | *2* | *10* | *35* | *3.66* | *84* |
| 11 | 07.09.2010. | Apricot | 13 | 22,6 | 78 | 17 | 6 | 3 | 11 | 41 | 3.68 | 78 |
| 12 | 07.09.2010. | Sour cherry | 7 | 5 | 104 | 43 | 7 | 12 | 12 | 30 | 2.78 | 59 |
| 13 | 07.10.2010. | Apricot | 21 | 50 | 100 | 41 | 10 | 9 | 11 | 28 | 2.72 | 59 |
| 14 | 07.10.2010. | Apricot | 4 | 5 | 54 | 34 | 4 | 4 | 3 | 9 | 2.06 | 37 |
| 15 | 07.10.2010. | Apricot | ~12 | 6 | 50 | 46 | 1 | 2 | 1 | – | 1.16 | 8 |
| 16 | 07.10.2010. | Apricot | ~25 | 15 | 100 | 23 | 24 | 12 | 21 | 26 | 3.21 | 77 |
| 17 | 07.10.2010. | Apricot | ~15 | 10 | 50 | 45 | 3 | 1 | 1 | – | 1.16 | 10 |
| 18 | 14.10.2010. | Apricot | 25 | 6 | 100 | 97 | 2 | 1 | – | – | 1.04 | 3 |
| 19 (18) | 06.08.2011. | Apricot | 26 | 6 | 100 | 87 | 6 | 3 | 3 | 1 | 1.25 | 13 |
| 20 (2) | 05.06.2011. | Apricot | 10–11 | 5 | 100 | 35 | 7 | 8 | 7 | 43 | 3.16 | 65 |
| 21 (5) | 05.06.2011. | Peach | 10 | 6 | 100 | 81 | 6 | – | 1 | 12 | 1.57 | 19 |
| 22 (6) | 05.06.2011. | Cherry | 12 | 22 | 100 | 64 | 11 | 6 | 4 | 15 | 1.95 | 36 |
| 23 (9) | 05.06.2011. | Sour cherry | 32 | 8 | 100 | 55 | 11 | 10 | 15 | 9 | 2.22 | 45 |
| 24 | 25.05.2011. | Sour cherry | 12 | 11 | 100 | 100 | – | – | – | – | – | – |

Spots of the field examinations: 1–9: Bekecs; 10–12: Bükkaranyos; 13: Rátka; 14: Göncruszka; 15: Vizsoly; 16: Boldogkőváralja; 17: Abaújkér; 18–19: Biharpüspöki (Romania); 20–23: Bekecs-2, -5, -6, -9, 24: Hajdúdorog

Other stone fruit species were also examined during the field investigations. A more moderate infection was observed in a 12 to 13-year-old peach orchard where phytoplasma infection with a rate of 21 % was observed. The destruction rate on peach was not so high as it was on apricot, but the problem with *Ca.* Phytoplasma prunorum seemed to be evident. Three sour cherry and one cherry plantation were examined as well. On the first examined sour cherry plantation, the infection rate was very high (62 %) and there were several withered or felled trees. It was easy to realize that sour cherry and cherry are also endangered by *Ca.* Phytoplasma prunorum infection.

Further field research was done during September and October 2010 in the Hungarian stone fruit orchards. Seven apricot and one sour cherry plantations were investigated in Borsod-Abaúj-Zemplén County, and one apricot orchard in another region in the village of Biharpüspöki, near the city of Oradea (western part of Romania). Very high infection rates were measured in Bükkaranyos on two apricot orchards (infection rate: 7–84 %) and one sour cherry plantation (infection rate: 9 %) as shown in Table 17.3. The rate of the totally destructed or felled trees was also very high. No high infection rates were measured in the examined Romanian apricot orchard, but the presence of apricot phytoplasma disease was evident. It was the first time that *Ca.* Phytoplama prunorum had been identified in the western part of Romania. Plant samples of the supposedly infected trees were collected for further laboratory research during each field examination.

Summarizing the results of the field experiences and the degree of infection, we can say that the plant health conditions of stone fruit plantations on the visited areas are rather bad (illustrated by the photos in Figs. 17.2, 17.3, 17.4, 17.5, 17.6, 17.7, 17.8 and 17.9).

**Fig. 17.6** Dead apricot tree

**Fig. 17.7** Dried branches on apricot tree

### 17.3.2 Results of the Laboratory Examinations

As many as 40 plant samples were collected on the fields, which were examined in laboratory by PCR. The presence of *Ca.* Phytoplasma prunorum was detected on 22 samples (Table 17.4). Phytoplasma infection was detected from the collected plant samples on all examined stone fruit species (apricot, peach, cherry, sour cherry, wild plum). The results of phytoplasma identifications using universal primers and group specific primers are shown on Figs. 17.10, 17.11 and 17.12. The presence of the pathogen without any typical symptoms in tolerant wild plum was also detected, and it is obvious that wild plum may have an important part in the spreading of the pathogen.

It was confirmed by the laboratory results that *Ca.* Phytoplasma prunorum is a rather serious danger for stone fruit plantations in Hungary.

### 17.3.3 Control Measures to Phytoplasmas

A promising strategy to avoid phytoplasma disease is the identification or development of resistant plant varieties (Welliver 1999). But management and control have to focus mainly on the clean stock programmes, eliminating sources of the phytoplasma and controlling vectors as follows:

- Propagates from phytoplasma-free plants
- Elimination of perennial and biennial weed hosts
- Avoiding planting susceptible plants next to plant-harbouring phytoplasma
- Controlling the vector in the plants and nearby weeds early in the season
- Planting varieties that are more resistant to the disease

The ecology of phytoplasmas is complex, and affected by the host range and geographic distribution of both phytoplasma and the insects that

**Fig. 17.8** A destructed cherry tree

**Fig. 17.9** Leaf rolling symptoms on peach

**Table 17.4** Rates of examined and DNA-isolated samples of different fruit trees and the results of phytoplasma detection

| Fruit tree species | Number of examined samples | Number of positive samples | Identified phytoplasma |
|---|---|---|---|
| Apricot (*Prunus armeniaca*) | 21 | 12 | ESFY |
| Peach (*Prunus persica*) | 6 | 2 | ESFY |
| Cherry (*Prunus avium*) | 2 | 2 | ESFY |
| Sour cherry (*Prunus cerasus*) | 10 | 5 | ESFY |
| Wild plum (*Prunus cerasifera*) | 1 | 1 | ESFY |

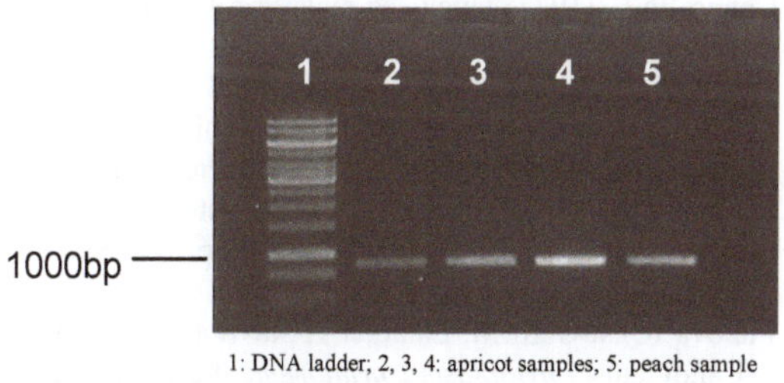


**Fig. 17.10** DNA fragments amplified by FU5/rU3 primers in 1 % agarose gel

transmit them, and is strongly affected by weather conditions. As more is learned about relationships among causal agent, vectors and hosts, surprising ecological niches have been uncovered, and theories of how disease may have evolved have been developed. These suggestions may be important in choosing management strategies for disease, and in forecasting where new disease outbreaks may occur.

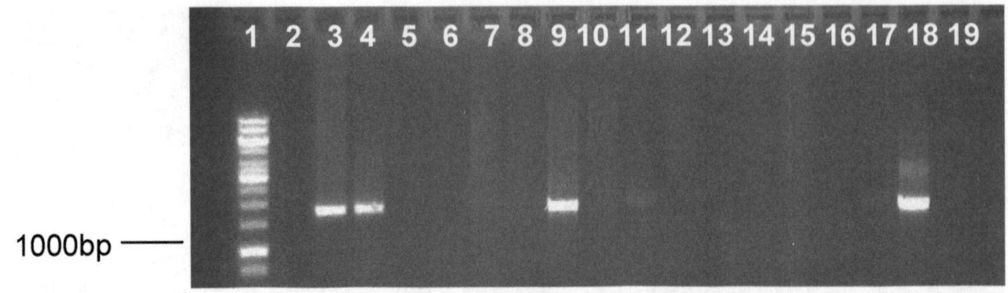


1: DNA ladder; 2,5,6,8,10,12,13,14,15,16: negative samples; 18: positive ESFY control; 19: negative control; 3, 4, 7: apricot samples infected by phytoplasma; 9: infected wild plum sample; 11, 17: infected sour cherry and cherry samples

**Fig. 17.11** DNA fragments amplified by P1/P7 primers in 1 % agarose gel

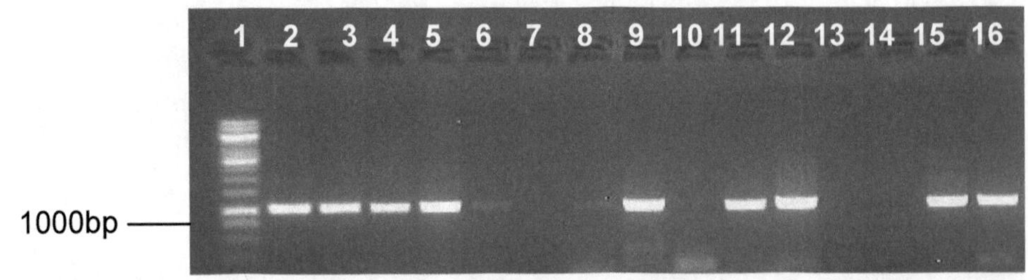


1: DNA ladder; 2,9: direct PCR; 2, 3, 4: infected apricot samples; 5: infected wild plum sample; 6, 7, 8: negative sour cherry and peach samples; 9: positive ESFY control; 10-16: nested PCR: 10: negative control; 16: positive ESFY control; 11: infected apricot sample; 12, 15: infected sour cherry samples; 13,14: negative sour cherry and peach samples

**Fig. 17.12** DNA fragments amplified by FO1/rO1 group-specific primers in 1 % agarose gel

## 17.4 Conclusions

The presence of *Ca.* Phytoplasma prunorum in the examined northern Hungarian apricot-growing sites is confirmed by field examinations and laboratory PCR tests. It can be easily seen that the problem is very serious, and *Ca.* Phytoplasma prunorum endangers almost every stone fruit plantation in that area. The disease caused by *Ca.* Phytoplasma prunorum is an increasing and relatively new problem for fruit growers in Hungary, and it seriously threatens the Hungarian apricot cultivation mainly, and other stone fruits as well. The pathogen causes serious destruction of the apricot trees on the examined Hungarian fruit orchards in the Gönc region. The main control methods to apply against the disease: to propagate phytoplasma-free plants and to control vector (*Cacopsylla pruni).* Our experiences resulting from our investigation show that we have to pay attention to the increasing phytoplasma problem in stone fruits, and have to develop new and effective management strategies.

## References


Bernhard R, Marenaud C, Eymet J, Sechet J, Fos A, Moutous G (1977) Une maladie complex de certain *Prunus*: Le dépérissement de Moliéres. CR Acad Agric 2(2):178–189

Carraro L, Loi N, Ermacora P, Osler R (1998) High tolerance of European plum varieties to plum leptonecrosis. Eur J Pt Pathol 104:141–145

Carraro L, Loi N, Ermacora P (2001) Transmission characteristics of the European stone fruit yellows phytoplasma and its vector *Cacopsylla pruni*. Eur J Plant Pathol 107:695–700

Chabrolin C (1924) Quelques maladies des arbres fruitiéres de la vallée du Rhone. Ann Epiphyties 10:265–333

Doi Y, Teranaka M, Yora K, Asuyama H (1967) Mycoplasma or PLT group-like microorganisms found in the phloem elements of plants infected with mulberry dwarf, potato witches broom, aster yellows, or paulownia witches broom. Ann Phytopathol Soc Jpn 33:259–266

Fialová R, Navrátil M, Lauterer P, Navrkalová V (2007) Candidatus *Phytoplasma prunorum*: the phytoplasma infection of *Cacopsylla pruni* from apricot orchards and from overwintering habitats in Moravia (Czech Republic). Bull Insectology 60(2):183–184

Gundersen DE, Lee IM, Rehner SA, Davis RE, Kingsbury DT (1994) Phylogeny of mycoplasma organisms (phytoplasmas): a base for their classification. J Bacteriol 176:5244–5254

Huxley A et al (1992) New RHS dictionary of gardening 1:203–205. Macmillan ISBN 0-333-47494-5

International Committee on Systematic Bacteriology Subcommittee on the Taxonomy of Mollicutes (1993) Minutes of the interim meetings, 1 and 2 August 1992, Ames, Iowa. Int J Syst Bacteriol 43:394–397

Jarausch W, Lansac M, Saillard C, Broquaire JM, Dosba F (1998) PCR assay for specific detection of European stone fruit yellows phytoplasmas and its use for epidemiological studies in France. Eur J Plant Pathol 104:17–27

Jarausch W, Eyquard JP, Lansac M, Mohns M, Dosba F (2000) Susceptibility and tolerance of new French *Prunus domestica* cultivars to European stone fruit yellows phytoplasma. J Phytopathol 148(7):489–493

Jarausch W, Jarausch-Wehrheim B, Danet JL, Broquaire JM, Dosba F, Saillard C, Garnier M (2001) Detection and identification of European stone fruit yellows and other phytoplasmas in wild plants in the surroundings of apricot chlorotic leaf roll-affected orchards in southern France. Eur J Plant Pathol 107:209–217

Kirkpatrick BC (1989) Strategies for characterizing plant pathogenic mycoplasma-like organisms and their effects on plants. In: Kosuge T, Nester EW (eds) Plant-microbe interactions, molecular and genetic perspectives. vol. 3. McGraw-Hill, New York, pp 241–293

Kirkpatrick BC (1991) Mycoplasma-like organisms. Plant and invertebrate pathogens. In: Balows A, Triiper GH, Dworkin M, Harder W, Schliefer KH (eds) The prokaryotes vol. 2. Springer, New York.

Kirkpatrick BC, Smart CD, Gardner S et al (1994) Phylogenetic relationship of plant pathogenic MLO-s established by 16/23 S rDNA spacer sequences. IOM Lett 3:228–229

Kiske CR, Kirkpatrick BC, Seemüller E (1991) Differentiation of virescence MLOs using western aster yellows mycoplasma-like organism chromosomal DNA probes and restriction fragment length polymorphism analysis. J Gen Microbiol 137:153–159

Kövics G (2009) Növénykórtani vademecum. NOFKA Debrecen pp 470

Lederer M, Seemüller E (1992) Demonstration of mycoplasmas in *Prunus* species in Germany. J Phytopathol 134:89–96

Lorenz KH, Dosba F, Poggi Pollini C, Llacer G, Seemüller E (1994) Detection of the apple proliferation and pear decline phytoplasmas by PCR amplification of ribosomal and nonribosomal DNA. Phytopathol 85:771–776

Marwitz R (1990) Diversity of yellows disease agents in plant infections. Zentralblatt für Bakteriologie Suppl 20(43):1–434

McCoy RE, Caudwell A, Chang CJ, Chen TA, Chiykowski IN, Cousin MT, Dale JL, de Leeuw GTN, Golino DA, Hackett KJ, Kirkpatric BC, Marwitz R, Petzhold H, Sinha RC, Sugiura M, Whitecomb F, Young IL, Zhu BM, Seemüller E (1989) Plant diseases associated with mycoplasma-like organisms. In: Whitcomb RF, Tully JG (eds) The Mycoplasmas vol V. Academic Press, San Diego, pp 545–640

Mergenthaler E (2004) Fitoplazmás betegségek Magyarországon: Korszerü diagnosztikai módszerek fejlesztése. Doktori értekezés. Budapesti Közgazdaságtudományi és Államigazgatási Egyetem Kertészettudományi Kar, Budapest pp 164

Mona G, Kadriye C, Cigdem US, Levent S (2008) Evaluations of apricot trees infected by *Candidatus* Phytoplasma prunorum for horticultural characteristics. Romanian biotechnological letters, Bucharest University. Rom Soc Biol Sci 14(1):4123–4129

Morvan G (1977) Apricot clorotic leaf roll. *EPP* Bull 7:37–55

Navratil M, Valova P, Fialova R, Patrova K (2001) Survey for stone fruit phytoplasmas in the Czech Republic. Acta Hortic 550:377–382

Németh M (1986) Virus, mycoplasma and rickettsia diseases of fruit trees. Martinus Nijhoff Publishers, the Netherlands and Akadémiai Kiadó, Budapest, pp 840

Németh M, Ember I, Krizbai L, Kölber M, Hangyál R, Bozsics G (2001) Detection and identification of phytoplasmas in peach based on woody indexing and molecular methods. Int J Hortic Sci 7:37–41

Seemüller E, Schneider B (2004) Candidatus *Phytoplasma mali*, Candidatus *Phytoplasma pyri* and Candidatus *Phytoplasma prunorum*, the causal agents of apple proliferation, pear decline and European stone fruit yellows, respectively. Int J Syst Evol Microbiol 54:1217–1226

Süle S (2003) A kajszi baktériumos és fitoplazmás betegségei. In: Pénzes B, Szalay L (eds) Kajszi. Mezögazda Kiadó, Budapest. pp 282–291

Süle S, Viczián O, Pénzes B (1997) A kajszi fitoplazmás pusztulása. Kertészet és Szölészet 45:8–11

Tarcali G, Kövics GJ (2009) Occurrence of stone fruit yellows phytoplasma disease in Gönc region, northern-Hungary. 5th international plant protection symposium at University of Debrecen, 20–22 October 2009, Debrecen, Hungary. Journal of Agricultural Sciences/ Acta Agraria Debreceniensis, University of Debrecen 38:69–74

Torres E, Martin MP, Paltrinier S, Vila A, Masalles R, Bertaccini A (2004) Spreading of EFSY phytoplasmas in stone fruit in Catalonia (Spain). J Phytopathology 152:432–437

Varga K, Kölber M, Németh M, Ember I, Erdös Z, Bíró E, Paltrinieri S, Martini M, Bertaccini A (2001) Identification of phytoplasmas infecting sour cherry in Hungary. XVIII international symposium on virus and virus-like diseases of temperate fruit crops—top fruit diseases. ISHS Acta Horticulturae 550:383–388

Viczián O, Süle S, Pénzes B, Seemüller E (1997) A kajszi fitoplazmás pusztulása Magyarországon. Új Kertgazd 1:48–51

Welliver R (1999) Diseases Caused by phytoplasmas. Regulatory Horticulture. Plant Pathol Circ 42:17–22

# Prevention of Virus Infection and Multiplication by Inducing Virus-Interfering Agent(s) in Treated Crop Plants Under Field Conditions

18

M. M. Abid Ali Khan, S. N. H. Zaidi, S. H. A. Kazmi and S. A. Musanna

**Abstract**

Natural virus inhibitors from higher plants are known to inhibit the infection of several plant viruses in hypersensitive and systemic test hosts. During the present investigation, the clarified sap from dried leaves of the medicinal plant *Operculina turpethum* L. (Family: Convolvulaceae) was sprayed on the leaves of several economically important crop plants, viz. *Lycopersicon esculentum, Lagenaria siceraria* (Mol). Standl., *Crotalaria juncea* L. and *Solanum tuberosum* L. in microplots under field conditions which prevented the infection and multiplication of tomato yellow mottle mosaic virus (TmYMV), cucumber green mottle mosaic virus (CGMMV), sunn-hemp rosette virus (SRV) and potato virus-x (PVX). Almost complete protection was observed and treated plants showed no disease symptoms if six sprays were given prior to virus challenge and two postinoculation sprays were given at weekly intervals.

The *O. turpethum*-clarified sap was completely safe and non-phytotoxic. The treated plants were healthy and showed no systemic symptoms of virus infection.

The resistant leaves' sap extracted from treated crop plants and assayed under in vitro conditions with their respective viruses on their hypersensitive hosts showed the ability to inhibit virus infectivity. The induction of systemic antiviral state in treated plants presumably involved the production of some highly active induced virus-interfering agent(s), such agent(s) were absent in nontreated control crop plants which showed severe disease symptoms.

**Keywords**

*Operculina turpethum* L. · Virus inhibitor · Virus interfering agent · Systemic host

M. M. A. A. Khan (✉) · S. N. H. Zaidi · S. H. A. Kazmi ·
S. A. Musanna
Department of Botany, Shia P. G. College,
Lucknow 226020, India
e-mail: mmabidalikhan265@gmail.com

R. N. Kharwar et al. (eds.), *Microbial Diversity and Biotechnology in Food Security,*
DOI 10.1007/978-81-322-1801-2_18, © Springer India 2014

## 18.1 Introduction

The infection due to plant viruses causes an estimated US$ 60 billion loss in crop yields worldwide each year (http://en.wikipedia.org/wiki/Plant_virus). Investigations on prevention of viral diseases in several economically important plants showed that strong virus preventive agents (VPAs) occur in a very few number of plants (Verma and Dwivedi 1983; Verma and Khan 1984; Ostermann et al. 1987; Zaidi et al. 1988; Khan et al. 1991; Khan and Zaim 1992; Khan et al. 2011, 2013). However, not much effort has been made to exploit the natural VPAs to manage virus diseases of crop plants sown in compost unsterilized soil under field conditions. Therefore, during present investigations, the VPAs occurring in a medicinal plant *Operculina turpethum* L. (Family: Convolvulaceae) leaf extract (OTE) were partially clarified and used for prevention of virus diseases in several economically important crops under field conditions.

## 18.2 Materials and Methods

### 18.2.1 Virus Cultures

The cultures of sunn-hemp rosette virus (SRV), tomato yellow mottle mosaic virus (TmYMV), cucumber green mottle mosaic virus (CGMMV), potato virus-x (PVX) were maintained on their respective systemic hosts, viz. *Crotalaria juncea* L., *Lycopersicon esculentum* Mill., *Lageneria siceraria* (Mol). Standl. and *Solanum tuberusum*, respectively, under glasshouse conditions.

Plant virus cultures with systemic symptoms were maintained in glass house conditions.

SRV

Severe mosaic and rosetting of leaves, reduced flower size and number; seeds were smaller and deformed in *C. juncea* L. plants.

TmYMV

Severe dark green mosaic with reduced leaf lamina leading to fern leaf symptoms in *L. esculentum* Mill. plants.

CGMMV, young leaves turned light yellow, showed downward curling and mosaic mottle later on all the leaves showed dark green mosaic symptoms in *L. siceraria* (Mol). Standl. plants.

PVX

Young leaves turned dark green, mosaic showed downward curling and necrosis in *S. tuberosum* L.

### 18.2.2 Virus Inocula and Inoculation

The young diseased leaves of systemically maintained virus cultures from their respective infected hosts were ground in a pestle-mortar separately, and distilled water was added as a diluent. The pulp obtained was squeezed through muslin cloth, centrifuged at 5,000 rpm for 10 min. The supernatant was diluted to 1:20 w/v. The viral suspensions were prepared separately for SRV, TmYMV, CGMMV and PVX and used as virus inocula.

The inoculation was done by forefinger using carborundum powder (600 meshes) as an abrasive.

### 18.2.3 Extraction of Natural Virus-Preventive Agent(s) (VPAs)

A total of 100 g of fresh leaves of *O. turpethum* L. were ground in a mixer with 100 ml of distilled water. The pulp obtained was squeezed through a muslin cloth. The filtrate was centrifuged at 3,000 rpm for 10 min, partially clarified with ether as a solvent, make up the concentration to 1:20 w/v by adding distilled water and was used for experiments under field conditions.

### 18.2.4 Raising of Systemic Test Crops Under Field Conditions

The systemic test crops, viz. *C. juncea* L., *L. esculentum* Mill., *S. tuberosum* L. and *L. siceraria* (Mol). Standl., were grown under field conditions in four microplots (size 2 m × 2 m) provided with compost soil. Healthy plants with five to six leaves were selected and used for experimental purposes.

### 18.2.5 Field Trails

Hosts with uniform growth were selected in all the four microplots and labelled as mp-1, mp-2 and mp-3 which served as treated plants and were given two preinoculation sprays of OTE per day up to 3 days and mp-4 which served as control was given two preinoculation sprays of distilled water daily for 3 days. After 1, 3 and 6 days of last preinoculation sprays all the treated and control plants raising in microplots were mechanically inoculated with their respective virus thereafter, two postinoculation sprays were given weekly. After 15, 30 and 45 days the test leaves were collected and inocula prepared separately as before.

The infectivity was tested on their local lesion test hosts by counting the local lesions. The number of local lesions was counted after 3–4 days and per cent decrease in virus titre was calculated using the formula: $C - T/C \times 100$, where $C$ and $T$ are the number of local lesions produced by sap of control and treated plants, respectively.

Observations for development of disease symptoms, growth, flowering and fruiting in all cases were taken after 30 or 45 days of virus challenge.

### 18.2.6 Production of Virus-Interfering Agent(s) (VIAs) in OTE-Treated and Test Plants

In another experiment, test hosts, viz. *C. juncea, L. esculentum, L. siceraria* and *S. tuberosum* separately were sprayed with OTE (1:2 w/v) twice a day for 3 days and at an interval of 1, 3 and 6 days leaves were removed from treated hosts and saps were prepared using distilled water and designated as R-juices (1:2 w/v). The virus-interfering activity of different saps extracted from the plants listed in table 18.5 was estimated by mixing equal volume of sap with their respective virus inoculums. After 30 min of incubation, mixtures were tested for virus inhibitory activity by infectivity assay on local lesion hosts *Cyamopsis tetragonoloba, Nicotiana glutinosa* and *Chenopodium amaranticolor.* Control in each crop under study consisted of leaf sap extracted from distilled water-sprayed plots and mixed with their respective viruses prior to infectivity assay as before.

The data for significance were statistically analysed by Snedecor 1961 method.

## 18.3 Results

Six preinoculation sprays of VPAs occurring in the leaves of OTE when given to *C. juncea, L. esculentum, L. siceraria* and *S. tuberosum* crops followed by SRV, TmYMV, CGMMV and PVX challenge after 1 and 3 days of treatment followed by two preinoculation sprays weekly in microplots under field conditions prevented 96–100 % virus infection and multiplications as no disease symptoms appeared. However, when viruses were challenged in their respective crops after 6 days of OTE treatment the percentage of virus infection decreased from 58.5 to 100 % after 15, 30 and 45 days of inoculation. Results clearly displayed that OTE had most significant virus infection and multiplication-preventing property although this property gradually decreased after 30 days of virus challenge and treated plants showed mild disease symptoms. (Tables 18.1, 18.2, 18.3 and 18.4).

The clarified OTE sap was completely safe, devoid of any phytotoxicity (not tested in the study). The treated plants were healthy with normal growth and flowering; however, after 30–45 days of virus challenge, mild disease symptoms gradually appeared without affecting normal growth and average flowering of treated crops.

It was also observed that the control experimental crops showed severe disease symptoms, abnormal growth and development with poor flowering (Tables 18.1, 18.2, 18.3 and 18.4).

It was observed that development of antiviral state in treated crop plants was associated with the formation of some VIAs in treated crop plants.

**Table 18.1** Prevention of sunn-hemp rosette virus (SRV) infection in sunn hemp (*C.juncea*) by *O. turpethum* leaf extract

| Treatment (six sprays with distilled water or extract before virus challenge and two sprays per week after virus challenge) | Symptoms and severity | Per cent decrease in virus titre | | Average after 45 days | |
|---|---|---|---|---|---|
| | | After 15 days | After 30 days | Height in cm | No. of flowers |
| Control sprayed with distilled water | Severe mosaic, rosetting, reduced flower size and number; seeds smaller and deformed | Nil | Nil | 38.5 | 10 |
| Treatment interval between last preinoculation sprays with OTE and virus challenge (in days) | | | | | |
| 1 | No symptoms | 100[a] | 100[a] | 76 | 21 |
| 3 | No symptoms | 90.0[a] | 76.5[a] | 69.6 | 18 |
| 6 | Only mild symptoms after 45 days | 78.5[a] | 58.8[a] | 58.7 | 15 |

Active virus assayed on *Cyamopsis tetragonoloba* L.
[a] Data significant at 1 % level

**Table 18.2** Prevention of tomato yellow mottle mosaic virus (TmYMV) infection in *Lycopersicon esculentum* by *O. turpethum* leaf extract

| Treatment (six sprays with distilled water or extract before virus challenge and two sprays per week after virus challenge) | Symptoms and severity | Per cent decrease in virus titre | | | Average after 45 days | |
|---|---|---|---|---|---|---|
| | | After 15 days | After 30 days | After 45 days | Height in cm | No. of flowers |
| Control sprayed with distilled water | Severe dark green mosaic, reduced leaf lamina leading to fern leaf symptoms | Nil | Nil | Nil | 40.5 | 7.2 |
| Treatment interval between last preinoculation sprays with OTE and virus challenge (in days) | | | | | | |
| 1 | No symptoms | 100[a] | 82[a] | 70[a] | 48.4 | 16 |
| 3 | No symptoms | 94[a] | 71[a] | 60[a] | 48.2 | 15.6 |
| 6 | Mild symptoms | 72[a] | 60[a] | 54[a] | 44.6 | 12.8 |

Active virus assayed on *N. glutinosa* L.
[a] Data significant at 1 % level

The resistant leaf sap extracted from treated crop plants and assayed under in vitro conditions with their respective viruses on their hypersensitive hosts showed the ability to inhibit virus infectivity.

The induction of systemic antiviral state in treated plants presumably involved the production of some highly active induced VIAs, which was absent in nontreated control crop plants which showed severe disease symptoms (Table 18.5).

## 18.4 Discussion

Preinoculation sprays with extracts from several higher plants have been reported to prevent plant virus diseases of economically important crops (Verma and Kumar 1979, 1980; Awasthi and Mukerjee 1980; Verma et al. 1982). These extracts induced systematic resistance in plants (Verma et al. 1982, 1984, 1985; Verma and Prasad 1983; Verma and Khan 1985).

**Table 18.3** Prevention of cucumber green mottle mosaic virus (CGMMV) infection in *Lagenaria siceraria* by *O. turpethum* leaf extract

| Treatment (six sprays with distilled water or extract before virus challenge and two sprays per week after virus challenge) | Symptoms and severity | Percent decrease in virus titre | | | Average after 45 days | | |
|---|---|---|---|---|---|---|---|
| | | After 15 days | After 30 days | After 45 days | Length of leaves cm | Breadth of leaf cm | No. of flowers (average) |
| Control sprayed with distilled water | Young leaves turned light yellow, showed downward curling and mosaic mottle later on all the leaves, showed dark green mosaic symptoms | Nil | Nil | Nil | 6.8 | 5.2 | 2.5 |
| Treatment interval preinoculation sprays and virus challenge (in days) | | | | | | | |
| 1 | No symptoms | 100[a] | 96[a] | 78[a] | 8.9 | 7.6 | 20.3 |
| 3 | No symptoms | 100[a] | 90[a] | 74[a] | 8.6 | 7.2 | 18.3 |
| 6 | Mild symptoms | 98.5[a] | 83[a] | 65[a] | 8.5 | 7.2 | 19.2 |

Active virus assayed on *Chenopodium amaranticolor*
[a] Data significant at 1 % level

**Table 18.4** Prevention of potato virus x (PVX) in *Solanum tuberosum* L. by *O. turpethum* leaf extract

| Treatment (six sprays with distilled water or extract before virus challenge and two sprays per week after virus challenge) | Symptoms and severity | Percent decrease in virus titre | | | Average after 45 days | | |
|---|---|---|---|---|---|---|---|
| | | After 15 days | After 30 days | After 45 days | Length of leaves in cm | Breadth of leaf in cm | No. of flowers (average) |
| Control sprayed with distilled water | Young leaves turned dark green, showed downward curling and necrosis later on all the leaves showed symptoms | Nil | Nil | Nil | 3.8 | 3.2 | 2.5 |
| Treatment interval preinoculation sprays and virus challenge (in days) | | | | | | | |
| 1 | No symptoms | 100[a] | 93[a] | 88[a] | 3.9 | 3.6 | 20.3 |
| 3 | No symptoms | 98[a] | 92[a] | 76[a] | 3.6 | 3.2 | 18.3 |
| 6 | Mild symptoms | 90.5[a] | 81.5[a] | 70.5[a] | 3.5 | 3.2 | 19.2 |

Active virus assayed on Nicotiana glutinosa
[a] Data significant at 1 % level

**Table 18.5** Production of virus interfering agent (VIA) in OTE treated and nontreated plants

| Source of plant VIA | Average number of local lesions/leaf ± SEM[a] | | | |
|---|---|---|---|---|
| | NR juice + PVX[c] (treated leaves with distilled water) | R juice + PVX[c] (leaves sprayed with distilled water) | NR juice + SRV[b] (leaves sprayed with OTE) | R-juice + SRV[b] (from leaves sprayed with OTE) |
| *Crotalaria juncea* | 172 ± 6.75 | 0 ± 0.0 | 159 ± 9.2 | 0 ± 0.0 |
| *Lycoperisicon esculentum* | 203 ± 4.5 | 0 ± 0.0 | 198 ± 6.6 | 0 ± 0.0 |
| *Solanum tuberosum* | 176 ± 9.85 | 0 ± 0.0 | 178 ± 11.5 | 0 ± 0.0 |
| *Lagenaria siceraria L.* | 108 ± 8.6 | 0 ± 0.0 | 113 ± 2.55 | 0 ± 0.0 |

[a] SEM Standard error mean

[b] Virus assayed on C. tetragonoloba

[c] Virus assayed on *N. glutinosa*

In the present investigations, results showed that treatment with OTE in different crops provided complete protection against the viruses belonging to the tobamo- and potexvirus group. Total suppression of disease symptoms and virus titre was presumably due to production of some VIAs in the treated crops showing systemic induced resistance.

Total resistance developed after 24–72 h of OTE application. The degree of induced resistance, however, varied in different crop virus combinations and it was not specific to plant species or viruses. Prophylactic treatment with OTE was highly significant and was not associated with any phytotoxicity symptoms on the treated crops. Thus, OTE can be an effective and practical means for the management of many plant virus diseases of economically important crops.

A brief contact between the extract and hosts or even a single spray was sufficient enough to provide significant resistance to the plants against virus infection and multiplication. A few sprays for inducing complete protection might indicate that the development of the VIA was gradual.

Several virus inhibitors have been purified earlier, which were responsible for the prevention of plant virus infection, were protein (Khan et al. 2011) and glycoprotein in nature (Awasthi et al. 1984). Some were antiviral sterol in nature. (Khan et al. 1991). Induction of systemic antiviral resistance by leaf extracts of several plants have been reported earlier. The induced antiviral resistance was due to the synthesis of new virus inhibitory protein with molecular weight 15,500 Daltons in treated *C. tetragonoloba* Taub. Test host had been isolated by Khan and Verma (1990). The increase in protein content, catalase and peroxidase enzymes in *C. tetragonoloba* Taub host as a consequence of systemic resistance inducers application appears evident that enhancement may be either due to new enzymes synthesis signifying a genetic outcome or may be due to the augmented activity of existing enzymes on account of a catalytic influence. Increase in the activity of the above-mentioned enzymes have also been recorded earlier (Prasad et al. 1989; Khan et al. 2013). It may be assumed that the enhanced enzyme activity may lead to the production of new virus inhibitory proteins, which could hamper virus replication in treated susceptible test plants. Enhancement in total proteins and enzymes activity as a consequence of inducers application indicated that host mediation is required, similar changes have also been reported in case of resistance induced by systemic resistance inducers isolated from *Clerodendrum aculeatum* leaves in tobacco cv. Samsunn NN test plant (Prasad 1986), also by systemic resistance inducers from *Cerodendrum fragrans* and *Boerhaavia diffusa* in treated plants (Verma and Prasad 1988). The alterations of total proteins content and enzymes activity in *Pseuderanthemum bicolor* extract treated *C. tetragonoloba* Taub plants host indicated that host mediation is required, which make the plant refractory against virus infection (Khan et al. 2013).

## 18.5 Conclusions

There is no direct control for management of viral diseases in crop plants which is responsible for heavy losses in crop production worldwide, therefore an indirect method must be applied to manage plant viral diseases and several investigators have reported prophylactic effect of several natural inhibitors occurring in medical and wild plants. These inhibitors provided protection against virus infection in treated and untreated parts of plants thus the effect was systemic in nature, the present chapter reported the significant protective effect of *O. turpethum* L. extract when sprayed on the leaves of several economically important crop plants, for example, *L. esculentum, L. siceraria* (Mol). Standl., *C. juncea* L. and *S. tuberosum* L. in microplots under field conditions which prevented the infection and multiplication of TmYMV, CGMMV, SRV and PVX. Almost complete protection was observed and treated plants showed no diseases symptoms.

The OTE-clarified sap was completely safe and non-phytotoxic and treated plants were healthy and devoid of any symptoms of virus infection.

The effect of OTE inhibitor effectively prevented virus infection and multiplication by inducing antiviral resistance to whole of the treated plants. The systemic induced resistance was due to the development of certain new induced VIAs in treated plants. However, the prevention of viral diseases requires prolonged treatment with OTE inhibitor. The virus disease management is important to reduce the heavy losses in crop production and to combat the problems of population explosion and food security.

**Acknowledgments** Dr. M. M. Abid Ali Khan is highly grateful to the Director, CIMAP-CSIR, Lucknow for appointment as a Research Associate (CSIR) during the tenure of the present investigations in this research paper.

## References

Awasthi LP, Mukerjee K (1980) Protection of potato virus X-infection by plant extracts. Biol Plantarum (Praha) 22:205–209

Awasthi LP, Chowdhury B, Verma HN (1984) Prevention of plant virus disease by *Boerhavia diffusa* inhibitor. Int J Trop Plant Dis 2:41–44

Khan MMAA, Verma HN (1990) Partial characterization of an induced virus inhibitory protein, associated with systemic resistance in *Cyamopsis tetragonoloba* (L.) Taub plants. Ann Appl Biol 117:617–623

Khan MMAA, Zaim M (1992) Physico-chemical properties and mode of action of inhibitors of plants virus replication present in *Operculina turpethum* L. and *Scilla indica* baker. Z Pflkrankh Pflschutz 99:71–79

Khan MMAA, Jain DC, Bhakuni RS, Zaim M, Thakur RS (1991) Occurrence of some antiviral sterols in *Artemisia annua*. Plant Sci 75:161–165

Khan MMAA, Zaidi SNH, Abidi AB (2011) Production of new anti tobamo virus protein by *Operculina turpethum* L. Cells in suspension culture. Presented at National conference on frontiers in biological sciences held at faculty of science, Veer Bahadur Singh Purvanchal University, Jaunpur, 4–5 December, p 45

Khan MMAA, Abidi AB, Hasnain S (2013) Total proteins and enzymes alterations associated with induction of antiviral resistance in treated plants. Natl Acad Sci Lett 36(2):147–149. doi:10.1007/s40009–013-0111–9

Ostermann WD, Meyer U, Leiser PM (1987) Induction of plant virus resistance: 2. Leaf extract from carnation plants (*Dianthus caryophyllus* L.) as inducer of resistance. Zentralbl Mikrobiol 142(3):229–238

Prasad V (1986) Alteration in enzyme activity during induced antiviral state by leaf extract. J Indian Bot Soc 65:90–94

Prasad V, Srivastava P, Dwivedi K, Verma HN (1989) Changes in phenol-oxidizing enzymes upon induction of systemic antiviral resistance in tomato plants. Indian Phytopathol 42:279–281

Snedecor GW (1961) Statistical methods. Allied Pacific Pvt. Ltd., Mumbai

Verma HN, Dwivedi SD (1983) Prevention of plant virus disease in some economically important plants by *Bounainvillea* leaf extract. Indian J Plant Pathol 1(1):97–100

Verma HN, Khan MMAA (1984) Management of plant virus diseases by *Pseuderanthemum bicolor* leaf extract. Z Pflkrankh Pflschutz 91:266–272

Verma HN, Khan MMAA (1985) Occurrence of a strong virus interfering agent in susceptible plants sprayed with *Pseuderanthemum atroprupureum* leaf extract. Indian J Virol 1:26–34

Verma HN, Kumar V (1979) Prevention of potato plants from viruses and insect vectors. J Indian Potato Assoc 6:157–161

Verma HN, Kumar V (1980) Yellow mottle disease of tomato caused by a strain of tobacco mosaic virus. New Botanist 7:31–36

Verma HN, Prasad V (1983) Inhibitors of viruses systemic resistance inducers from higher plants. In: Husain A, Singh K, Singh BP, Agnihotri VP (eds) Recent advances in plant pathology. Print House, Lucknow, pp 311–324

Verma HN, Prasad V (1988) Metabolic alterations associated with host mediated systemic antiviral resistance. Indian Phytopathol 41:332–335

Verma HN, Awasthi LP, Mukerjee K (1982) Characteristics and mode of action of natural inhibitors of virus infection. In: Bilgirami KS et al (eds) Advancing frontiers of mycology and plant pathology. Today and tomorrow's Printers and Publishers, New Delhi, pp 255–264

Verma HN, Chowdhury B, Rastogi P (1984) Antiviral activity in leaf extracts of different *Clerodendrum* species. Z PflKrankh PflSchutz 91:34–41

Verma HN, Khan MMAA, Dwivedi SD (1985) Biophysical properties of highly antiviral agents present in *Pseuderanthemum atropurpureum* and *Bougainvillea spectabilis* extracts. Indian J Plant Pathol 3:13–20

Zaidi ZB, Gupta VP, Samad A, Naqvi QA (1988) Inhibition of spinach mosaic virus by extract of some medicinal plants. Curr Sci 57(3):151–152

# Biocontrol of Phytopathogenic Fungi of Rice Crop Using Plant Growth-Promoting Rhizobacteria

19

Mohamed A. Gad, Manab Deka, Naglaa A. Ibrahim, Sherif S. Mahmoud, R. N. Kharwar and Tarun C. Bora

**Abstract**

Rice is widely affected by quite a number of diseases caused by fungi, bacteria, viruses, and mycoplasma that result in high yield loss. Among the fungal diseases, aggregate sheath spot caused by *Rhizoctonia oryzae-sativae* is an important disease affecting the rice production. Application of beneficial bacteria as seed coat or seedling root dip to protect from these diseases may be an alternative strategy to chemical control. In the present study, out of 200 bacterial strains isolated from plant rhizosphere of cereal crops, one bacterial strain *Pseudomonas aurogenosa* "NEIST 003" was selected for its antagonistic ability against different rice fungal pathogens under *in vitro* conditions. The per cent inhibition of mycelial growth of different rice fungal pathogens namely "*R. oryzae-sativae, Fusarium moniliforme, Rhizoctonia solani, Curvularia oryzae, Fusarium oxysporium,* and *Pyricularia grisea* by *P. aurogenosa* "NEIST 003" was 35.11, 21.33, 33.11, 25, 34.22, and 41.67 %, respectively. Under greenhouse conditions, *P. aurogenosa* "NEIST 003" promoted plant growth besides inhibiting rice fungal pathogen *R. oryzae-sativae* and gave highly significant growth improvement compared to the other treatments for all parameters including shoot height, root length, tiller number, number of leaves, shoot fresh

M. A. Gad (✉) · T. C. Bora
Biotechnology Division, North East Institute of Science and Technology (NEIST), Council of Scientific and Industrial Research (CSIR), Jorhat, Assam 785006, India
e-mail: mohamedabo2002@yahoo.com

M. Deka
Biotechnology Division, Gauhati University, Guwahati, Assam, India

N. A. Ibrahim · S. S. Mahmoud
Department of Biology, Alexandria University, Alexandria, Egypt

R. N. Kharwar
Centre of Advanced Study in Botany, Banaras Hindu University, Varanasi 221005, India

R. N. Kharwar et al. (eds.), *Microbial Diversity and Biotechnology in Food Security*, DOI 10.1007/978-81-322-1801-2_19, © Springer India 2014

weight, shoot dry weight, root fresh weight, root dry weight, moisture content, chlorophyll content, and disease incidence at different days of seedlings infection. *P. aurogenosa* "NEIST 003" has the ability to promote plant growth and suppressed most of the rice fungal pathogens.

**Keywords**

Rice · Biocontrol · *Rhizoctonia oryzae-sativae* · *Pseudomonas aurogenosa*

## 19.1 Introduction

Rice has been under cultivation since time immemorial, being grown under varying climatic conditions in different parts of the world, including India. It is affected by a number of diseases that result in high yield loss (Ou 1985). Aggregate sheath spot (AgSS) disease is one of the important diseases affecting the rice production which is caused by *Rhizoctonia oryzae-sativae* (telemorph: *Ceratorhiza oryzae-sativae*) (Moore 1989; Seint et al. 2009). Initially, AgSS lesions appear on the lower leaf sheaths at the waterline following infection from *R. oryzae-sativae* sclerotia floating on the water. The disease progresses to the upper leaf sheaths and, under favorable conditions, can spread to the flag leaf and cause yield loss by reducing the photosynthetic area. Bordered sheath spot and brown sclerotium disease caused by *Rhizoctonia oryzae* Ryker et Gooch (*Waitea circinata* f. sp. *oryzae*, WAG-O) and *R. oryzae-sativae* (Sawada) Mordue (*Ceratobasidium oryzae-sativae* Gunnell & Webster), respectively, occur worldwide in rice-growing areas of Southeast Asia, India, Japan, the USA, and Brazil (Hashioka and Makino 1969; Ou 1984). In Australia, yield losses caused by *R. oryzae* and *R. oryzae-sativae* were as high as 10 and 20 %, respectively (Lanoiselet et al. 2005). A recent survey documented the occurrence of *R. oryzae-sativae* from diseased rice sheaths in Myanmar (Aye et al. 2009). The management of rice fungal diseases is done through fungicides, growing resistant varieties, application of organic amendments, balanced nutrition, biological agents, and resistance-inducing chemicals. Chemical fungicides pollute the environment and disturb the ecological balance for all living microorganisms and cause harmful effects to beneficial microorganisms. The development of biological products based on beneficial microorganisms can extend the range of options for maintaining the health and yield of crops. PGPR are known to improve growth in different crops and suppression of diseases by induced resistance in plants (Kumar 1999). Many strains of rhizobacteria, especially fluorescent *Pseudomonads*, are reported to enhance plant growth/yield and suppress disease when applied as seed inoculants (Schippers 1993). Some of the PGPR strains, especially fluorescent *Pseudomonads* and *Bacillus subtilis*, have shown promising results as biological control agents by reducing crop damage caused by major plant pathogens (Utkhede and Smith 1992, 1993; Schippers 1993).

## 19.2 Materials and Methods

### 19.2.1 Collection of Plant Samples and Isolation of Fungal Pathogens

Infected plant samples showing typical AgSS symptoms were collected from farmer's fields of Assam state in sterile polythene bags and kept airtight. It was then brought to the laboratory and stored at 4 °C until isolation.

The diseased plant parts were washed in running tap water followed by sterile distilled water to remove impurities. The diseased sample was cut with a sterile blade in 1–2 cm length ensuring that fresh plant tissues were also included. It was then immersed in 0.2 % $HgCl_2$ solution for 2 min followed by two to three times washings with sterile distilled water. After that the plant part was blotted dry with sterile filter paper. The thin sections were inoculated on potato dextrose agar medium (PDA, HiMedia, India), supplemented with 100 mg streptomycin sulphate/L. Petri plates were incubated at 25±2 °C for 2–4 days or until colony appears. The fungal colony was repeatedly subcultured on PDA medium, followed by microscopic examination of the colony characters till the pure culture was obtained. The colonies thus obtained were then cultured in PDA slants and stored at 4 °C until further use (Gad 2012).

### 19.2.2 The Disease Symptoms

AgSS disease caused by *R. oryzae-sativae* (Moore 1989) is similar to the early stages of sheath blight and is caused by a related fungus. Small sclerotia initiate the disease on the outermost leaf sheath. A reddish-brown lesion develops on the plant near the waterline. During internode elongation, the sheath spot may be pushed up above the waterline because of this new growth. With age, the lesion enlarges slightly to assume an elliptical to irregular shape (1/2 to 1 inch long) with a distinct purple-brown border and a tan-to-straw colored center. The sheath spot fungus does not develop on the leaf blades as is the case with sheath blight. Rather, it remains on the outer sheath only. Sometimes, the fungus will cause yellowing of the leaf attached to the sheath it has infected. A reduction in yield occurs in tillers that are infected.

### 19.2.3 Isolation of Bacteria From Rhizosphere Samples

Rhizosphere samples were collected from different locations of Northeast India. The plantlets were uprooted from the field, placed in polyethylene boxes with covers and secured tightly with rubber bands. These samples were used for the isolation of rhizobacteria. Samples were kept in the cold room at 4 °C for isolation.

Gram from each rhizosphere sample was shaking in 10 ml of sterile distilled water for 30 min. The samples were serially diluted up to $10^{-5}$ dilution. A volume of 100 µl of the suspension was taken from dilution $10^{-3}$ and $10^{-4}$ and plated on King's B (KB) agar medium and the plates were incubated at 28 °C for 48 h. Isolated single colonies were selected at random from the samples and were further streaked on KB agar medium to obtain pure cultures. The purified strains were stored in sterile 50 % glycerol for further use (Karthikeyan and Gnanamanickam 2007).

### 19.2.4 Identification of Bacterial Strains

The bacterial antagonist showing the highest antifungal activity was identified through the study of the morphological, physiological, and biochemical characteristics of the isolates according to Bergy's Manual of Systematic Bacteriology (Collins 1964; Holding and Collee 1971; Goodfellow 1989). Final confirmatory identification of the potential bacterial strains was done from, Microbial Type Culture Collection and Gene Bank Division of Institute of Microbial Technology, Council of Scientific and Industrial Research "IMTECH" Chandigarh, India.

### 19.2.5 *In Vitro* Antifungal Activity

Three potential bacterial strains isolated from plant rhizosphere of cereal crops and identified as *Pseudomonas aeruginosa* "NEIST 003," *Brevibacillus laterosporus* "NEIST 041," and *Bacillus safensis* "NEIST 050" were selected out of 200 rhizobacterial strains on the basis of the antagonism against rice fungal pathogens (*Pyricularia grisea, Rhizoctonia solani, R. oryzae-sativae, Fusarium moniliforme, Fusarium oxysporium,* and *C. oryzae*) under *in vitro* conditions.

Antagonism against phytopathogenic fungi was assayed by dual culture (Yuan and Crawford 1995):

PDA was used for examining the antagonism. For this, an actively growing mycelial disc of fungal culture (approximately 9 $mm^2$) was placed at one side of the petri plate, 2 cm inside the periphery and a loopful of the rhizobacterial strain was streaked in a line on the opposite side at a distance of 5 cm from the mycelial disc. The plates were incubated at $25 \pm 2\,°C$ until fungal mycelium completely covered the agar surface in control plate. Inhibition zone was measured as the distance (in cm) between the respective rhizobacterial test antagonist and fungal pathogen.

Percent inhibition was measured by the formula:

$$\text{Inhibition\%} = (A - B/A) \times 100$$

Where, *A* is the diameter of fungal growth in control plate and *B* is the diameter of fungal growth in experimental plate.

### 19.2.6 Greenhouse Assays

Greenhouse experiments were conducted at the experimental greenhouse of NEIST, Jorhat, Assam, during rice growing season at 25–30 °C under natural daylight conditions. Popular local rice variety "Mahsuri" susceptible to *Ceratorhiza* spp. was obtained from, Rice Research Station, Assam Agricultural University, Titabor, Jorhat, Assam.

### 19.2.7 Treatments

All the laboratory as well as greenhouse experiments were done with the following treatments throughout the entire period of investigation:

(1) *R. oryzae-sativae* alone, (2) *P. aeruginosa* alone, (3) *R. oryzae sativae* +*P. Aeruginosa*, and (4) Control (no bacteria, no fungus).

### 19.2.8 Inoculation

*Ceratorhiza* spp. bioassays were performed essentially as described in Rodrigues (2003). Four-week-old plants were challenged by placing a 1 cm toothpick colonized by *Ceratorhiza* spp. inside the sheath of the second youngest fully expanded leaf. Inoculated plants were maintained inside humid inoculation chambers ($\geq 92\%$ relative humidity; $30 \pm 4\,°C$) for 72 h and thereafter, transferred to greenhouse conditions. Five days after the challenge infection, disease incidence was assessed by observation of water-soaked lesions.

### 19.2.9 Growth Promotion and Disease Suppression Studies Under Greenhouse Condition

Seed bacterization was done according to Kumar and Bezbaruah (1996). Seeds of commercial rice (*Oryza sativa* L.) were surface disinfected with 1 % sodium hypochlorite for 2 min and rinsed three times with sterile distilled water prior to sowing. For bacterization, bacteria grown on KB medium for 48 h were scrapped with a sterile glass rod and mixed with moist sterile soil to prepare a paste (approximately $1.0 \times 10^7$ cells/g soil). The rice seedlings were placed in this paste for 12 h and 5 g of the soil paste was attached around the root portion of the seedling before planting. Seedlings treated with moist soil without any bacteria served as control. Both treated and control seedlings were planted in pots containing sterile field soil and maintained under greenhouse conditions. The experiment was laid out in completely randomized design as with three set of replication. Under each replication three seedlings were treated as per treatment. Proper hygienic conditions under the greenhouse was maintained by weeding, light forking, and irrigation as and when required maintaining the healthy condition of seedlings and minimizing the interference of external factors other than desired. After planting in pots, seedlings were again treated with respective PGPR strain at 15 days interval. At 20, 40, and 60 days of infection, 3 numbers of representative seedlings were selected randomly and data on growth promotion in terms of increase in shoot height, number of leaves were taken before uprooting of seedlings. These plants were then uprooted from the pots with utmost care to keep the roots intact and washed gently under running tap water to remove the adhering soil particles. Data on root length, chlorophyll contents of leaves, fresh and dry weight of shoot, root were recorded. Disease symptoms were recorded for aggregate sheath blight spot on plant and after 20, 40, and 60 days of treatment per cent disease incidence was calculated.

**Table 19.1** *In vitro* antagonistic activity of potential bacterial strains against rice fungal pathogens

| Test organisms | Inhibition over control (%) | | |
|---|---|---|---|
| | *P. aeruginosa* (NEIST 003) | *B. laterosporus* (NEIST 041) | *B. safensis* (NEIST 050) |
| *Rhizoctonia sativae* | 35.11 | 6.44 | 21.89 |
| *Fusarium moniliforme* | 21.33 | 10 | 15.22 |
| *Rhizoctonia solani* | 33.11 | 10.89 | 17 |
| *Curvularia oryzae* | 25 | 19.33 | 18.56 |
| *Fusarium oxysporium* | 34.22 | 6.89 | 15.78 |
| *Pyricularia grisea* | 41.67 | 30.89 | 45.78 |
| Control | 0.00 | 0.00 | 0.00 |

### 19.2.10 Data Analysis

Data was subjected to analysis of variance (ANOVA) and critical difference used to compare the means for all the variables within the experiment at ($P = 0.05$).

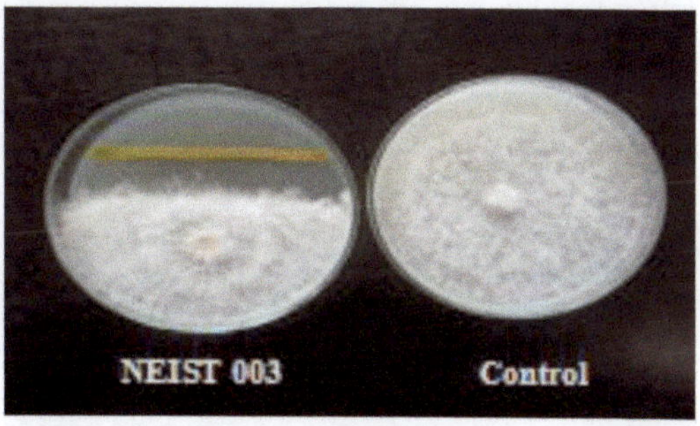


**Fig. 19.1** *In vitro* antagonism of bacterial strain NEIST 003 against *R. oryzae*

## 19.3 Results

### 19.3.1 Antagonistic Activity of Phytopathogenic Fungi

Three potential bacterial strains were evaluated for their antifungal activity *in vitro* against rice fungal pathogens: *P. grisea, R. solani, R. oryzae-sativae, F. moniliforme, F. oxysporium*, and *C. oryzae* as shown in Table 19.1.

From the result, *P. aeruginosa* "NEIST 003"was found to show maximum inhibition zone for all rice fungal pathogens. This suggested that *P. aeruginosa* "NEIST 003" was able to suppress the growth of all the selected rice fungal pathogens under laboratory conditions (Figs. 19.1, 19.2, 19.3, 19.4, 19.5 and 19.6).

### 19.3.2 *In Vivo* Greenhouse Studies

#### 19.3.2.1 Effect of Bacterization on Occurrence of Disease Incidence

Under greenhouse conditions, the non-bacterized seedlings challenged with fungus showed 87, 90, and 93.5% disease incidence and 41, 35, and 39% disease incidence was recorded in seedlings bacterized with *P. aeruginosa* "NEIST 003" and challenged with fungus at 20, 40, and 60 days, respectively, as shown in (Fig. 19.7). Higher per cent disease incidence (41%) was recorded in treatment with *R. oryzae sativae* +*P. aeruginosa* and lowest (35%) at 20 and 40, respectively. Seedlings treated with the pathogen alone showed per cent disease incidence (93.5) after 60 days of treatment (Fig. 19.8). The treatment in which the pathogen was inoculated with bacterized rice seedlings, disease symptoms were observed after 20 days followed by 40 days and 60 days in *R. oryzae sativae* +*P. aeruginosa. P. aeruginosa* treated seedlings remained healthy up to the last day (60th day) of observation along with the control (Table 19.2).

#### 19.3.2.2 Effect of Bacterization on Plant Growth Promotion

Seed bacterization had a significant effect on increase of shoot height, root length, tiller number, and number of leaves over the control under greenhouse conditions. Highest increase in shoot height (56, 59, 69 cm), root length (26, 34, 36 cm), tiller number (13, 15, 14), and number of

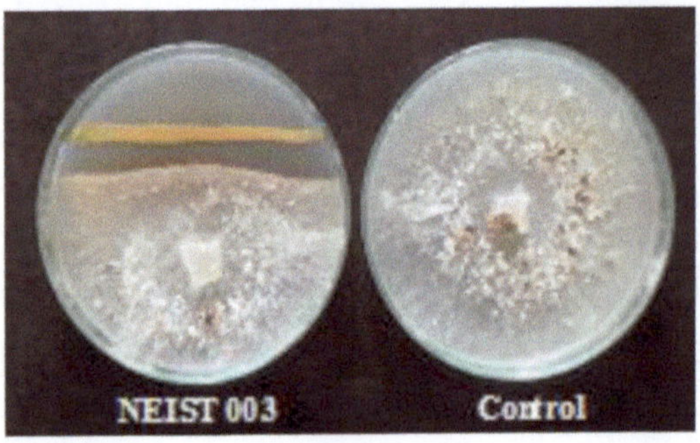


**Fig. 19.2** *In vitro* antagonism of bacterial strain NEIST 003 against *R. solani*

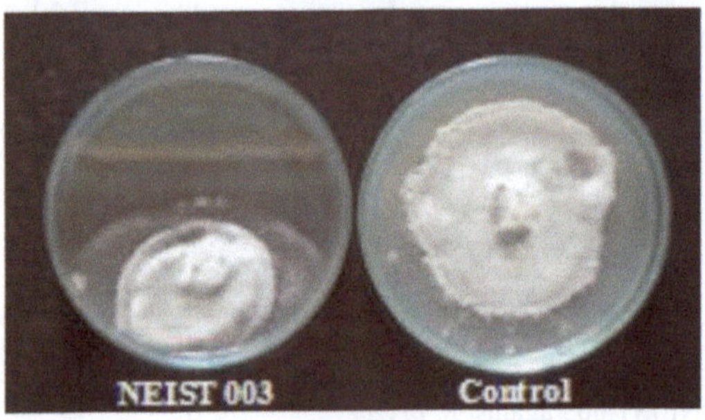


**Fig. 19.3** *In vitro* antagonism of bacterial strain NEIST 003 against *P. grisea*

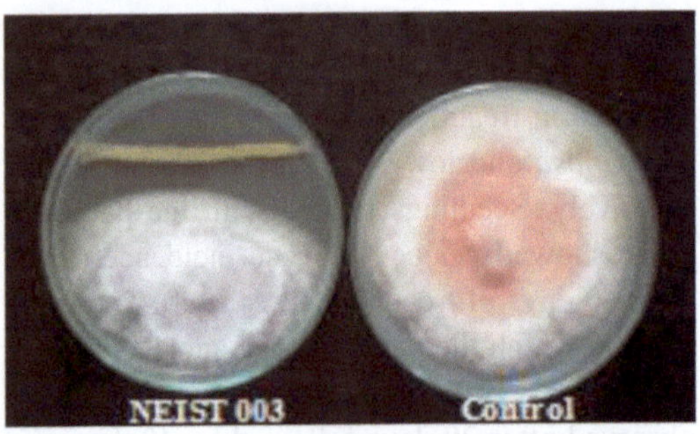


**Fig. 19.4** *In vitro* antagonism of bacterial strain NEIST 003 against *F. oxysporium*

leaves (77, 83, 78) were recorded in NEIST 003 at 20, 40, and 60 days of treatment, respectively. *R. oryzae-sativae* infected seedlings showed decrease in shoot height, root length, tiller number, and number of leaves over control. Among the treatments, where pathogen and PGPR were applied, highest increase in shoot height (51, 52, 60 cm), root length (21, 30, 31 cm), tiller number (11, 13, 13), and number of leaves (59, 62, 62) were recorded in treatment *R. oryzae-sativae* + NEIST 003 over the control at 20, 40, and 60 days of treatment, respectively.

**Fig. 19.5** *In vitro* antagonism of bacterial strain NEIST 003 against *F. moniliforme*

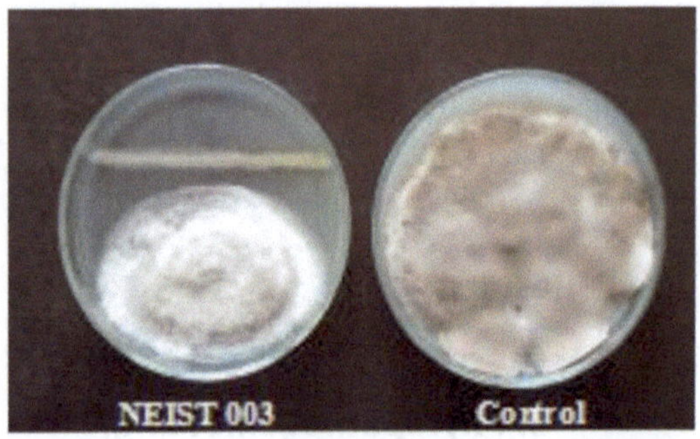


**Fig. 19.6** *In vitro* antagonism of bacterial strain NEIST 003 against *C. oryzae*

## 19.4 Discussion

Sheath diseases of rice caused by *R. solani, R. oryzae, R. oryzae-sativae,* and *Sclerotium hydrophilum* are important phytopathogens distributed worldwide and cause yield losses in rice-growing countries (Matsumoto 2003). Rice is grown as an important cereal crop all over the world, but mostly in Southeast Asian countries. Among different diseases that attack rice crops, sheath diseases have become one of the dominant diseases causing significant reduction in rice yield. In addition, soil-borne diseases can contaminate the soil by establishing its inocula permanently in the soil. Therefore, it is important

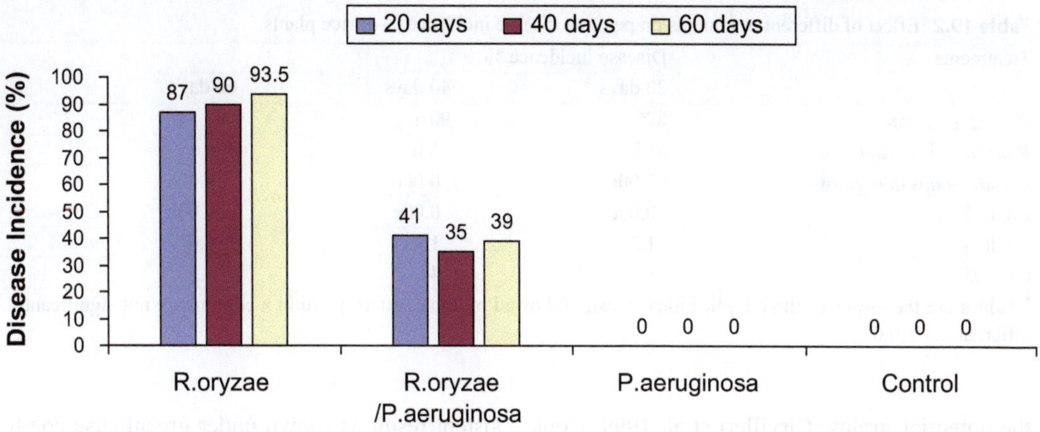


**Fig. 19.7** Effect of bacterization on per cent disease incidence

**Fig. 19.8** Symptoms of aggregate sheath spot disease in rice

to find out the control measures that are environmentally safe to reduce the incidence of these pathogens. The aim of the present study was to investigate the effect of PGPR to suppress rice fungal pathogen *R. oryzae-sativae* and promote plant growth. Seint and Masaru (2011) found that the phytoextracts, cloves, neem leaf, rosemary, and pelargonium are potential phytoextracts to control phytopathogens of rice, such as *R. solani, R. oryzae, R. oryzae-sativae,* and *S. hydrophilum* under *in vitro* conditions. Linquist et al. (2006) found that the overall severity of AgSS was most pronounced at lower nitrogen (N) levels. AgSS, is one of the major rice (*Oryza sativa* L.) diseases in California. It has been reported in Asia (Taheri et al. 2007), Australia (Lanoiselet et al. 2005a), the Middle East (Rahimian 1989), South America (Cedeno et al. 1998), and the USA (Gunnell and Webster 1984). In California, where rice is grown on over 200,000 ha, it has been observed since the late 1960s and is one of the prevalent rice diseases (Gunnell and Webster 1984). All cultivars currently grown in California are somewhat susceptible to AgSS (Miller and Webster 2001). Research in Australia has shown that AgSS can result in yield declines of up to 20% (Lanoiselet et al. 2005). Two hundred bacterial strains belonging to *Bacillus, Proteus,* and *Pseudomonas* genera were screened for their ability to inhibit the growth of rice fungal pathogen in PDA. The *P. aeruginosa* strain “NEIST 003” showed best inhibition against the test pathogens followed by *B. safensis* “NEIST 050,” and *B. laterosporus* “NEIST 041.” These strains were further selected for detailed studies. Selection of potential strains for crop improvement and/or control of plant pathogens through *in vitro* antibiosis have been done by several workers (Chakraborty et al. 2005b; Kumar and Bezbaruah 1997; Dekaboruah and Kumar 2002; Utkhede and Rahe 1983; Gad 2007). Although the *in vitro* antibiosis test does not always co-relate with suppression of soil-borne plant disease, but because of the magnitude of the rhizosphere population and the lack of a more reliable method, *in vitro* screening of organisms has been a valuable tool to select

**Table 19.2** Effect of different treatments on per cent disease incidence % of rice plants

| Treatments | Disease incidence % | | |
|---|---|---|---|
| | 20 days | 40 days | 60 days |
| *R. oryzae sativae* | 87[a] | 90 a | 93.5a |
| *R.oryzae +P. aeruginosa* | 41 b | 35 b | 39b |
| *Pseudomonas aeruginosa* | 0.00c | 0.00c | 0.00c |
| Control | 0.00c | 0.00c | 0.00c |
| S.Ed(±) | 1.2 | 1.7 | 0.6 |
| CD*0.05* | 3 | 4 | 1.5 |

* Values are the means of three replications. Means followed by same letter(s) within a column are not significantly different at $P \leq 0.05$

the potential strains (Cirvilleri et al. 1999; Cook 1993). In the present study, bacterization of rice seedlings with "NEIST 003" improved the plant growth leading to plant biomass improvement over the pathogen treated as well as control plants under greenhouse conditions. There are several reports that PGPR promote growth and control many diseases in several crops (Singh et al. 2002; Wei et al. 1991). Application of bacterial strains to the rhizosphere of tea plants for enhanced growth with increased number of leaves and lateral branches are reported earlier (Chakraborty et al. 2005a, 2005b). Seed bacterization with fluorescent *Pseudomonas* has been a potential method for enhancement of plant growth as well as suppression of plant pathogenic fungi (Folders et al. 2001; Rao et al. 1999). Bacterization of rice seedlings with "NEIST 003" resulted in highest increase of shoot height, root length, number of leaves, fresh and dry weight of root and leaf. Increased shoot height and root length ranging from 16–18% and enhanced grain yield by 22.61% were also recorded by Nautiyal (1997), when chickpea seeds were bacterized with a *Pseudomonas fluorescens* strain. *Bacillus* species particularly *B. subtilis* are reported to have growth-promotion and disease-suppression ability (Harish et al. 1998; Podile 1994; Podile and Prakash 1996; Utkhede and Rahe 1983a).

## 19.5 Conclusion

From the results, it is concluded that the PGPR strain was able to suppress rice fungal pathogens under *in vitro* conditions, and also showed a consistent result as shown under greenhouse conditions by inducing disease resistance in the host. This positive effect suggests the possible use of *Pseudomonas* strain NEIST 003 as a biocontrol agent against these destructive fungi. Therefore, the biological method of plant disease management seems to be a better alternative to chemical fungicides that resulted in environmental pollution and ill health to biotic community as a whole and for that it plays an important role in food security.

**Acknowledgments** The authors are grateful to Dr. P. G. Rao, Director, CSIR-NEIST, Jorhat for constant support and encouragement. MAG gratefully acknowledges DBT, Department of Biotechnology, Govt. of India and TWAS—The Academy of Sciences for the Developing World, Italy for PhD. fellowship and financial support and also thankful to the authority of Plant Pathology Research Institute, Agriculture Research Centre, Giza, Egypt for granting study leave during the work.

## References

Aye SS, Myint YY, Lwin T, Matsumoto M (2009) *Rhizoctonia oryzae-sativae*, causal agent of aggregate sheath spot disease of rice in Myanmar. New Disease Reports, Vol. 19. http://www.bspp.org.uk/publications/new-disease-reports/volumes.php

Cedeno L, Nass H, Carrero C, Cardona R, Rodriguez H, Aleman L (1998) *Rhizoctonia orizae-sativae*, causal agent of the aggregated stain of rice in Venezuela. Interciencia 23:248–251

Collins CH (1964) Microbiological Methods, 1st edn. Butterworth and Co Ltd., London

Chakraborty U, Basnet M, Chowdhury RP, Chakraborty BN (2005a) Interactions of *Bacillus pumilus* with tea root rot pathogens. In: Chakraborty U, Chakraborty.

BN (eds) Stress Biology, Narosa Publishing House, New Delhi, pp 252–256

Chakraborty U, Chakraborty BN, Basnet M, Bhutia PL (2005b) PGPR-mediated induction of resistance in tea plants for management of root rot diseases. In: Gnanamanickam SS, Balasbramanian R, Anand N (eds) Emerging trends in plant-microbe interactions. Proceedings of the Asian conference on emerging trends in plant-microbe interactions. University of Madras, Chennai, pp 42–50, 8–18 Dec 2005

Cirvilleri G, Catara V, Bella P, Marchese E (1999) Isolation and characterization of rhizosphere bacteria of potential interest for biological control of soil borne pathogens. Phytopathol Suppl 8:79–87

Cook RJ (1993) Making greater use of introduced microorganisms for biological control of plant pathogens. Ann Rev Phytopathol 31:53–80

Dekaboruah HP, Kumar BS (2002) Biological activity of the secondary metabolites produced by pseudomonas styrain RRL 008. Folia Microbiol 47:359–363

Folders J, Algra J, Roelfs MS, Van Loon LC, Tommassen J, Bitter W (2001) Characterization of *Pseudomonas aeruginosa* chitinase.a gradually secretion protein. J Bacteriol 183:7044–7052

Gad MA (2007) Microbiological and serological studies on wheat wilt and root-rot diseases. MSc. Dissertation, Benha University, Botany Department, Egypt, pp 81

Gad MA (2012) Molecular characterization of potent phytopathogenic fungal strains and their biocontrol through plant growth promoting rhizobacteria. Ph. D. Dissertation, Gauhati University, India, pp 307

Goodfellow M (1989) Bergy's manual of systematic bacteriology. Wiliams & Wilkins Company, Baltimore

Gunnell PS, Webster RK (1984) Aggregate sheath spot of rice in California. Plant Dis 68:529–531

Harish S, Manjula K, Podile AR (1998) *Fusarium udum* is resistant to the mycolytic activity of a biocontrol strain of *Bacillus subtillis* AF.J. FEMS Microbiol Ecol 25:385–390

Hashioka Y, Makino M (1969) *Rhizoctonia* group causing the rice sheath spots in the temperate and tropical regions, with special reference to *Pellicularia sasakii* and *Rhizoctnoia oryzae*. Res Bull Fac Agric Gifu Univ 28:51–63

Holding AJ, Collee JG (1971) Routine biochemical tests. In: Norris JR, Ribbons DW (eds) Methods in Microbiology, Vol. 6A. Academic press, York, pp 1–34

Karthikeyan V, Gnanamanickam SS (2007) Biological control of *Setaria* blast (*Magnaporthe grisea*) with bacterial strains. Crop Prot 27:263–267

Kumar BS (1999) Fusarial wilt suppression and crop improvement through two rhizobacterial strains in chick pea growing in soil infested with *Fusarium oxyspiorium* f.sp.*ciceris*. Biol Fert Soils 29:87–91

Kumar BS, Bezbaruah B (1996) Antibiosis and plant growth promotion by a pseudomonas strain isolated from soil under tea cultivation. Ind J Microbiol 36:45–48

Kumar BS, Bezbaruah B (1997) Plant growth promotion and fungal pest control through an antibiotic and siderophore producing *fluorescent pseudomonas* strain from tea (*camellia sinensis*(L.)O. Kuntze) plantations. Ind J Exp Biol 35:289–292

Lanoiselet VM, Cother EJ, Ash GJ, Harper JDI (2005) Yield loss in rice caused by *Rhizoctonia oryzae* and *R. oryzae-sativae* in Australia. Australas Plant Pathol 34:175–179

Lanoiselet VM, Cother EJ, Ash GJ, Hind-Lanoiselet TL, Murray GM, Harper JDI (2005a) Prevalence and survival, with emphasis on stubble burning, of *Rhizoctonia* spp., causal agents of sheath diseases of rice in Australia. Australas Plant Pathol 34:135–142

Linquist BA, Brouder SM, Hill JE (2006) Winter straw and water management effects on soil nitrogen dynamics in California rice systems. Agron J 98:1050–1059

Matsumoto M (2003) A qualitative baiting technique for selective isolation and DNA diagnosis of *Rhizoctonia* spp., causal agents of rice sheath diseases from soil. J Fac Agric Kyushu Univ 48:13–20

Miller TC, Webster RK (2001) Soil sampling techniques for determining the effect of cultural practices on *Rhizoctonia oryzae-sativae* inoculum in rice field soils. Plant Dis 85:967–972

Moore RT (1989) *Ceratorhiza oryzae-sativae*, a new combination for the anamorph of *Cerabasidium setariae*. Antonie Van Leeuwenhoek 55(4):393–395

Nautiyal CS (1997) Selection of chickpea–rhizosphere-competent. *Pseudomonas fluroscens* NBR11303 antagnostic to *Fusarium oxysporium* f.sp. *ciceris, Rhizoctonia bataticola* and *Pythium* sp. Curr Microbiol 35:52–58

Ou SH (1984) Fungus disease-diseases of stem, leaf sheath and root. Commonwealth Mycological Institute, Kew, pp 247–300. (In: Rice Diseases)

Ou SH (1985) Rice diseases. Common Wealth Agricultural Bureaux, Wallingford

Podile AR (1994) Survival of Bacillus subtilis AF 1 in the bacterized peanut rhizosphere and its influence on native microflora and seedling growth. World J Microbiol Biotechnol 10:700–703

Podile AR, Prakash AP (1996) Lysis and biological control of *Asperigillus niger* by *Bacillus subtilis* AF1. Can J Microbiol 42:533–538

Rahimian H (1989) Occurrence of aggregate sheath spot of rice in Iran. J Phytopathol 125:41–46

Rao CVS, Sachan LP, Johri BN (1999) Influence of *Pseudomonas* on growth and nodulation of lentil (Lens esculentus) in *Fusarium* infested soil. Ind J Microbiol 39:23–29

Rodrigues FA, Vale FXR, Datnoff LE, Prabhu AS, Korndorfer GH (2003) Effect of rice growth stages and silicon on sheath blight development. Phytopathol 93:256–261

Schippers B (1993) Exploitation of microbial mechanisms to promote plant health and growth. Phytoparasitica 21:275–279

Seint SA, Masaru M (2011) Effect of some plant extracts on *Rhizoctonia* spp. and *Sclerotium hydrophilum*. J Med Plants Res 5(16):3751–3757

Seint SA, Myint YY, Thein L, Masaru M (2009) *Rhizoctonia oryzae-sativae*, causal agent of aggregate sheath spot disease of rice, found in Myanmar. N Dis Rep 19:22–24

Singh UP, Prithviraj P, Singh KP, Sarma BK (2002) Control of powdery mildew (*Erysiphe pisi*) of pea (*Pisum sativum*) by combined application of plant growth promoting rhizobacteria and neemazaltim. J Plant Dis Prot 107:59–66

Taheri P, Gnanamanickam S, Hofte M (2007) Characterization, genetic structure, and pathogenicity of *Rhizoctonia* spp. associated with rice sheath diseases in India. Phytopathol 97:373–383

Utkhede RS, Rahe RE (1983) Interaction of antagonist and and pathogen in biological control of white rot. Phytopathol 73:890–893

Utkhede RS, Rahe RE (1983a) Effect of Bacillus subtilis on growth and protection of onion against white rot. Phytopathol Z 106:199–203

Utkhede RS, Smith EM (1992) Promotion of apple tree growth and fruit production by the EBW-4 strain of *Bacillus subtilis* in apple replant disease soil. Can J Microbiol 38:1270–1273

Utkhede RS, Smith EM (1993) Biological treatment for planting apple tree in soil previously planted with cherry trees in the Kootenay valley of British Columbia. Soil Boil Biochem 25:1689–1692

Wei L, Kloepper JW, Tuzun S (1991) Induction of systemic resistance of cucumber to *Colletotrichum orbiculare* by selected strains of plant growth promoting rhizobacteria. Phytopathol 81:1508–1512

Yuan WM, Crawford DL (1995) Characterization of *Streptomyces lydicus* WYEC 108 as a potential biocontrol agent against fungal root and seed rots. Appl Env Microbiol 61:3119–3128

# Incidence and Preliminary Control of Blast Disease of Rice in Southwest Nigeria

20

David B. Olufolaji

**Abstract**

Studies were carried out on the occurrence of a rice blast disease in southwest Nigeria. Blast disease caused by the fungal pathogen *Pyricularia oryzae* which was not in existence in southwest region of Nigeria in the past is now assuming an economic dimension. Pathogenicity tests were carried out and assessment of the extent of damage in incidence and severity which caused considerable loss to the cultivation of rice were investigated and ascertained. While 35–65 % incidence was obtained on the farm land, the severity ranged from 3.2 to 5.4. This was very devastating in rice production. From the studies conducted, 30–50 % of yield loss in rice is recorded and this has caused untold hardship to the rice researchers and farmers in the agro-ecological region of Nigeria. A weed, *Panicum maximum*, was identified as the alternative host of *P. oryzae* which may also hinder the pathogen's control. As a preliminary control measure, aqueous extracts of three botanicals, namely *Ocimum gratissimum, Gmelina arborea* and *Chromolaena odorata* were employed at concentrations 20, 40, 60 and 80 % on the growth and sporulation of *P. oryzae,* the causal organism of rice blast disease. *G. arborea* at 60 % and *C. odorata* at 80 % show some promise in reducing mycelial growth and sporulation of the causal organism. This served as a preliminary control strategy for rice blast. A single isolate is still being observed but molecular studies will be carried out to determine their variation.

**Keywords**

*Pyricularia oryzae* · Incidence · Severity · Yield loss · Molecular · Pathogenicity · Blast

D. B. Olufolaji (✉)
Department of Crop, Soil and Pest Management, The Federal University of Technology, Akure, Nigeria
e-mail: tundefolaji022@gmail.com

## 20.1 Introduction

Rice (*Oryza sativa)* which is grown widely in most ecological locations in the world is the major food source for about 40 % of the world's

R. N. Kharwar et al. (eds.), *Microbial Diversity and Biotechnology in Food Security,*
DOI 10.1007/978-81-322-1801-2_20, © Springer India 2014

human population (WARDA 1993). According to WARDA (1993), there are five major rice-growing ecologies which include the rain-fed upland, irrigated lowland, rain-fed lowland, mangrove and the deep water or floating rice ecologies. Approximately, the lowlands provide 75 % of the world's rice while 10 % is grown on the uplands.

Rice has become a major staple cereal in most countries of Africa especially West Africa where it accounted for more than 25 % of the cereals consumed (Africarice 2004). Nigeria is the largest producer of rice in West African subregion, producing about 45 % of the total production (Africarice 2004). The importance of rice in the economy of Nigeria and the availability of varying rice-growing ecologies and diverse production systems led to the special focus placed on rice production in the country by Africarice (2004). In Nigeria, the importance of rice in the diet of the people is steadily on the increase. The annual consumption of the staple for an average Nigerian is as high as 24.8 kg of rice which represents 9 % of the total calorie intake (Akpokodje et al. 2001). In spite of the efforts made in increasing rice area under cultivation, yields remained very low; thus, the production has not been able to meet the consumption level of the growing population. Rice production in Africa is seriously affected by diseases. Some of the common diseases of rice in Africa are blast, bacterial blight, brown spot, leaf scald, sheath blight, sheath rot and Rice yellow mottle virus (RYMV) (Ou 1985).

Blast disease caused by the fungal pathogen *P. oryzae* which was not in existence in southwestern Nigeria in the past is now assuming an economic dimension as rice is becoming increasingly important in Nigeria (Akpokodje et al. 2001). For this reason, the government is investing heavily in the development of the domestic rice sector. However, the expensive efforts to increase rice production, by the development of irrigation schemes where water and water management are available, allowing double cropping and promotion of productive varieties from Asia might be hindered by the occurrence of rice blast disease due to its projected impact on the crop (Abo et al. 1998).

This work was therefore directed to the investigation of the sudden occurrence of rice blast disease in southwestern Nigeria and preliminary control measures to check its epidemics.

## 20.2 Materials and Methods

### 20.2.1 Experimental Observatory Locations

Observatory personnel of The Teaching and Research Farm of The Federal University of Technology, Akure, Nigeria brought to the Crop Protection Diagnostic Laboratory samples of rice blast-diseased plants, found from some locations as reported by rice farmers in the southwestern rice-farming locations of Nigeria. The rice samples infected with blast were assessed in the laboratory.

Then, each state of the region was visited to assess the disease incidence and severity on the rice fields. The states visited were Ekiti, Kwara, Lagos, Ondo, Ogun, Osun and Oyo.

### 20.2.2 Disease Incidence and Severity

Disease incidence was determined by taking a count of infected plants and expressing it as a percentage of the total plants assessed:

$$\text{Disease incidence} = \frac{\text{No of plants affected}}{\text{Total number of plants}} \times \frac{100}{1}.$$

The disease severity ratings were recorded on a 1–6 point scale similar to that of Allen et al. (1981):

1 No disease symptoms
2 1–40 % leaves and 1–20 % petioles affected
3 41–60 % leaves and 21–50 % petioles and panicles affected
4 61–80 % leaves, 80 % petioles and panicles affected
5 Over 80 % leaves, petioles and panicles affected
6 Plant completely killed

The mean data of the farms were recorded for each state's location.

Diseased plants were collected and taken to the laboratory for symptoms description, mycoflora isolation and identification. The isolated organism was plated in 14 petri dishes of potato dextrose agar (PDA) (two per samples from each state) incubated at room temperature (28–30 °C) and observed daily for signs of fungal growth. Fungal mycelia from the plates were subcultured on freshly prepared PDA. At exactly 7 days after the culturing when spore would have ensued from the cultures, microscopic examinations of mycelia and spores produced were carried out.

### 20.2.3 Pathogenicity Test

Pathogenicity tests were carried out on freshly established rice plants in the screen house to ascertain the role of *P. oryzae* as the causal organism of rice blast disease.

Twenty 10-L plastic pots three-quarters filled with sterilized loamy-sandy soil were utilized to raise 4-week-old rice seedlings (two plants per stand). Ten potted plants were for FARO-2 (susceptible variety) while ten were for NERICA-8 (resistant variety).

Spore suspension ($5^{10}$ $ml^{-1}$ water) of the isolated *P. oryzae* was prepared and used as the inoculum for the experiment. The potted rice plants were sprayed until runoff and humidity was maintained by covering the potted plants, with polyethylene bags during the first 3 days with daily aeration for the growth and development of the pathogen to aid infection process. The plants were observed daily for the appearance of disease symptoms. The observations were also carried out till maturity of the rice.

### 20.2.4 Preliminary Control with Botanicals

As a preliminary control measure, botanicals were utilized since the epidemics need to be prevented.

Botanicals used were *Chromolaena odorata, Gmelina arborea* and *Ocimum gratissimum*.

#### 20.2.4.1 Botanical Extracts

Leaf samples of each of *C. odorata, G. arborea* and *O. gratissimum* were collected from the Teaching and Research Farm of The Federal University of Technology, Akure, Ondo state in the southwest of Nigeria. Southwest region of Nigeria falls into the rain forest area with thick forest ecosystem, a little above the equator and generally has 7–8 months of rain in a year. It has a considerable collection of biodiversity with many trees and shrubs from which many varieties of plants are available.

Extracts of the botanicals were prepared by first air drying the leaves and pulverizing them into powdered form which was kept for use. Cold extraction was produced by dissolving 20, 40, 60 and 80 g each of air-dried ground leaf samples of the plants in 100 ml distilled water and was left for 48 h. After soaking, they were filtered through four layers of sterile cheese cloth and the filtrate was kept for further bioassay (Olufolaji 2006). The cold aqueous extracts of the filtrates were passed through membrane/millipore filter of pore size 0.2 μm to avoid any microbial contamination, and each of the extract's concentrations were kept sterile in a 150 ml conical flask. The content in the flasks were exposed to UV light for further sterilization.

#### 20.2.4.2 *In Vitro* Assay

The bioassay of each of the plant's extracts at concentrations 20, 40, 60 and 80 % was carried out in determining the effect of concentrations and various plants extracts on the growth and sporulation of *P. oryzae*, as described by Nene and Thapliyal (1971) and Olufolaji (2006). The PDA crude extracts were prepared by spreading 1 ml of the extract, separately on the surface of the molten PDA in the petri dishes and was allowed to dissolve into the PDA before solidifying. The control was PDA with distilled water spread on the surface. With the aid of 5 mm sterile cork borer, disks were cut from the 7-day-old culture grown on PDA and placed at the centre of the petri dish containing the PDA crude extract. The set up was carried out for each of the species *C. odorata* and *G. arborea.* The whole set up which was in completely randomized design was replicated three times. The experiment was carried out under a temperature of 27 ± 2 °C and terminated at exactly 7 days, when the my-

**Table 20.1** Impact of blast disease and varietal differences on rice cultivation in the states of southwestern Nigeria

| S/N | Location states | Variety | Disease incidence (%) | Disease severity | Cultivation type | Yield t/ha (paddy) | % Yield loss |
|---|---|---|---|---|---|---|---|
| 1 | Lagos | NERICA-8 | 35 | 2.2 | Lowland | 2,9 | 29.5 |
| | | FARO-2 | 36 | 2.3 | Upland | 2.2 | 34.5 |
| 2 | Ogun | NERICA-8 | 40 | 4.1 | Lowland | 2.5 | 45.6 |
| | | FARO-2 | 44 | 4.2 | Upland | 2.1 | 35.6 |
| 3 | Oyo | NERICA-8 | 50 | 5.0 | Lowland | 2.3 | 47.5 |
| | | FARO-2 | 53 | 4.8 | Upland | 1.8 | 50.5 |
| 4 | Osun | NERICA-8 | 53 | 5.2 | Lowland | 2.1 | 53.7 |
| | | FARO-2 | 54 | 5.0 | Upland | 1.7 | 54.7 |
| 5 | Ondo | NERICA-8 | 52 | 4.9 | Lowland | 1.9 | 48.8 |
| | | FARO-2 | 51 | 5.3 | Upland | 1.7 | 45.8 |
| 6 | Ekiti | NERICA-8 | 42 | 4.5 | Lowland | 1.8 | 43.7 |
| | | FARO-2 | 39 | 4.7 | Upland | 1.7 | 43.6 |
| 7 | Kwara | NERICA-8 | 65 | 5.3 | Lowland | 1.4 | 50.4 |
| | | FARO-2 | 66 | 5.4 | Upland | 1.2 | 52.4 |

celia in the control would have covered the entire petri dish. Radial growth and sporulation was calculated (Olufolaji 2006). Statistical analysis was carried out for the data and mean separation was determined using standard error of the mean (SEM).

#### 20.2.4.3 *In Vivo* Assay

The set up of the screen-house experiments was a randomized complete block design having seven treatments including the control. They were replicated three times. The treatments comprised the three plant extracts in aqueous extraction and in concentrations 60 and 80 %. Thus, the total number of 10-L plastic buckets used in the screen house were 7 × 3 replicates totalling 21. The set up was a preventive mode of application. Thus, the plant extracts were applied by spraying until runoff on to the rice seedlings before inoculation with the spore suspension of the pathogen (*P. orazae*).

## 20.3 Results

### 20.3.1 Disease Survey

The survey revealed that the disease is present at a very considerable proportion in the rice field of southwestern Nigeria. It was observed that the blast disease have wide epidemics in both upland and lowland ecosystems of southwestern Nigeria. Most of the farms visited had the presence of blast disease with varying degree of incidence and severity (Table 20.1). The early stage of the disease was manifested as small circular, dark brown, water-soaked lesions about 0.5 and 2 cm in diameter (Fig. 20.1). Infection begins on the lobes and sides of the leaf where water often seems to collect (Fig. 20.1) as explained by Africarice (2004).

The lesions enlarged and developed characteristic yellow halos. A delicate white fungal fuss was consistently observed around the lesion on both sides of the leaf giving older lesions zonation appearance (Fig. 20.2). A clear light-yellowish fluid which turned bright yellow/golden when dry was observed from the centre of some of the lesions. These exudates eventually became crusty and dark brown on drying out (Fig. 20.2). Leaf lamina, petioles as well as stems were also affected (Fig. 20.1). Irregularly shaped lesions completely covered the leaf blade and dried it up, but these dried infected leaves did not drop from the petioles (Figs. 20.1 and 20.2). These symptoms observed in the fields and samples brought to the laboratory were similar to those described by other researchers (Ou 1985). The panicles were covered with brown patches characteristic of the blast disease.

**Fig. 20.1** Blast lesions on foliar parts (leaves and stems) of rice

**Fig. 20.2** Blast lesions on foliar parts (leaves, petioles and stems) of rice

#### 20.3.1.1 Disease Incidence and Severity

While 35–65 % incidence was obtained on the farm land, the severity ranged from 3.2 to 5.4. This was very devastating in rice production. From the studies conducted, 30–50 % of yield loss in rice is recorded and this has caused untold hardship to the rice researchers and farmers in the agro-ecological region of Nigeria. A weed, *Panicum maximum*, was identified as the alternative host of *P. oryzae* which may also hinder the pathogen's control. As a preliminary control measure, aqueous extracts of three botanicals, namely *O. gratissimum, G. arborea* and *C. dorata* were employed at concentrations 20, 40, 60 and 80 % on the growth and sporulation of *P. oryzae*, the causal organism of rice blast disease. *G. arborea* at 60 % and *C. odorata* at 80 % show some promise in reducing mycelia growth and sporulation of the causal organism.

### 20.3.2 Preliminary Blast Disease Control with the Botanicals

#### 20.3.2.1 *In Vitro* Studies

The three plant extracts (*C. odorata, G. arborea* and *O. gratissimum)* significantly inhibited both mycelial growth and sporulation of *P. oryzae,* the causal organism of blast disease of rice (Figs. 20.3 and 20.4). Furthermore, the inhibitory effects of the plant extracts on mycelial growth and sporulation increased as their concentrations increased with 80 % concentrations generally as the best.

However, *C. odorata* and *G. arborea* under cold aqueous extracts did not differ significantly in mycelial growth inhibition while *O. gratissimum* was significantly the least in inhibiting mycelial growth of the fungus. While *G. arborea* at 60 and 80 % concentrations gave 71.4 and 75.6 %, *C. odorata* gave 68.5 and 71.3 % mycelial growth inhibition, respectively (Fig. 20.3). The effects of the plant extracts on sporulation of *P oryzae* followed the same trend as the mycelial growth inhibition. Among the cold aqueous extracts, *G. arborea* was the best in that at 80 % concentration it yielded 76.3 % as the highest sporulation inhibition level (Fig. 20.4). This was not significantly different from that of 80 % concentration of *C. odorata* which had 70.6 % sporulation inhibition. The least performance was observed in *O. gratissimum* with 41.2 % of sporulation inhibition under 80 % extract concentration (Fig. 20.2).

It was observed from the in vivo studies that the plant extracts had a significant reduction in the impact of the blast disease on rice (Fig. 20.5). In the impact on disease incidence, cold extracts of the three plants at 80 % concentration was the best with 56, 42 and 70 % inhibition for *C. odorata, G. arborea* and *O. gratissimum,* respectively. However, each of them was not significantly higher than those of 60 % concentration (Fig. 20.5). On the severity of the blast disease, 80 % extracts concentration of all the cold aqueous extracts were significantly the best among

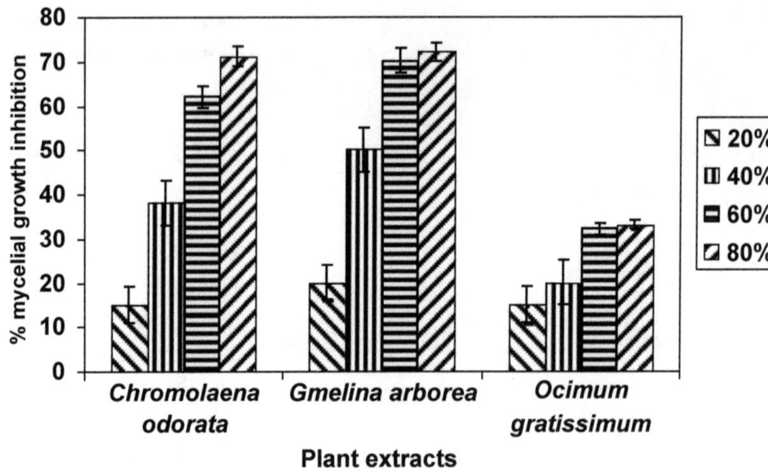


**Fig. 20.3** Effect of aqueous plant extracts on mycelial growth inhibition of *Pyricularia oryzae, I*=standard error

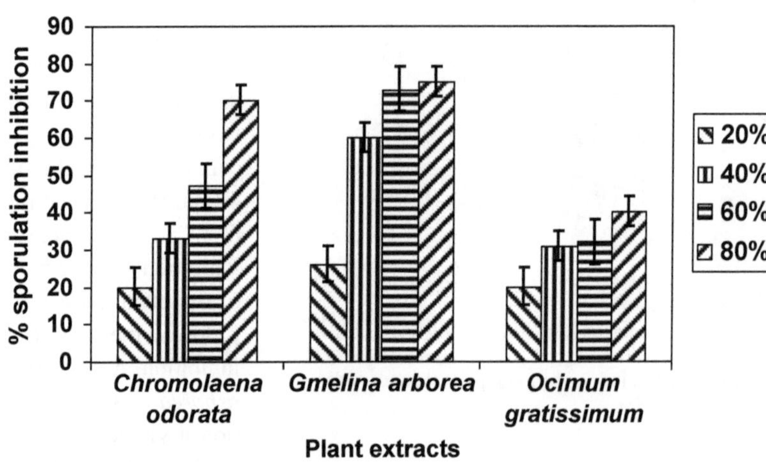


**Fig. 20.4** Effect of aqueous plant extracts on mycelial growth inhibition of *Pyricularia oryzae, I*=standard error

all the tested concentrations. The three botanical extracts at 80% concentration gave 3, 2.5 and 5.4% inhibition for *C. odorata, G. arborea* and *O. gratissimum,* respectively, while at 60% extracts concentrations, they yielded severity levels of 2.8, 3 and 5.8% inhibition for *C. odorata, G. arborea* and *O. gratissimum,* respectively.

## 20.4 Discussion

It is quite evident from the survey, and the impact of disease incidence and severity, that rice blast could cause a great deal of loss on the all important crop. However, its occurrence in the rice field shows a sign of threat to rice production in southwestern Nigeria in which rice is a very important staple food.

The use of the botanicals as fungicides for plant disease control has been explored in recent times by plant protectionists worldwide (Olufolaji 2008; Srivastava et al. 2009; Olufolaji 2011).

Reduction of mycelial growth and sporulation of the rice blast disease fungus (*P. oryzae)* is an indication of the presence of toxic principles inside the various plant extracts utilized in this study. This also conforms with the findings of some previous workers who studied the effects of some related plant extracts on the pathogens of some crop plants (Srivastava et al. 1999; Olufolaji 2006, 2008, Shcheribakova 2011).

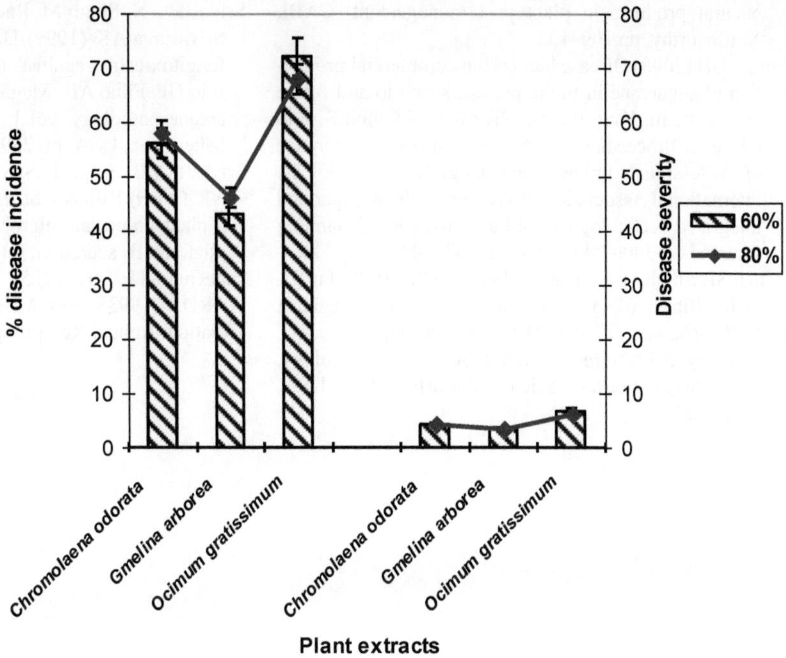


**Fig. 20.5** Effect of aqueous plant extracts on incidence and severity of blast disease on rice variety NERICA-8 Nigeria, *I*=standard error

The negative impact of rice diseases has caused a great yield reduction in rice and resultant loss in rice production (Olufolaji 1999). However, the reduction of this negative impact by the botanicals has demonstrated that the use of botanicals is a promising way of controlling the disease and increasing the rice production. It is thus evident from this study that botanicals are coming up to replace the synthetic fungicides, if we care for safe food and pollution-free environment.

## References


Abo ME, Sy AA, Alegbejo MD (1998) Rice yellow mottle virus (RYMV) in Africa: evolution, distribution and economic significance on sustainable rice production and management strategies. J Sustain Agric 11:85–111

Africarice (2004) For the future. Africa focus bulletin. htpp://www.africafocus.org/docs04/rice0401.php. Accessed Feb 2004

Akpokodje G, Lancon F, Erestein O (2001) Nigeria's rice economy: state of the art. West African Rice Development Agency (WARDA), Cote d'Ivoire, pp 65–78

Allen DJ, Emechebe AM, Ndimande B (1981) Identification of cowpea to diseases of African Savannas. Trop Agr (Trinidad) 58:267–279

Nene VI, Thapliyal PN (1971) Fungicides in plant disease control. Oxford and IBHI Public, New Delhi, pp 404–425

Olufolaji DB (1999) Effects of *Colletotrichum falcatum* (red-rot fungus) on the morphology and yield attributes of sugarcane in Nigeria. In: Rao GP, Filho AB, Magarey RC, Autrey JC (eds) Sugarcane pathology, vol. I: fungal disease. Science Publishers Inc., USA, pp 155–163

Olufolaji DB (2006) Effects of crude extracts of *Eichhornia crassipes* and *Chromolaena odorata* on the control of red-rot disease of sugar-cane in Nigeria. In: Yang-Rui Li, Solomon, S (eds) Proceedings of the International Conference of Professionals in sugar and integrated technologies. International Association of Professionals in Sugar and Integrated Technologies 174 East Daxue Road Nanning Guangxi—530007 P R China. pp 337–342

Olufolaji DB (2008) Synergistic effect of aqueous extracts of *Chromolaena odorata* and *Ocimum gratissimum* in the control of *Colletotrichum falcatum*, the sugarcane red-rot disease causal organism. Proceedings of the 5th International Congress of Plant Pathology, Torino Italy. 24–29th August 2008, pp 687–689

Olufolaji DB (2011) Prospects of large-scale use of natural products as alternatives to synthetic pesticides in developing countries. In: Dubey NK (ed) Natural products in plant pest management. CABI, Northworthy, pp 191–204

Ou SH (1985) Rice diseases 2nd edn. Commonwealth Mycological Institute, Kew, 368 p

Shcheribakova LA (2011) Some natural proteinaceous and polyketide compounds in plant protection and their potential in green consumerization. In: Dubey NK (ed)

Natural products in plant pest management. CABI, Northworthy, pp 109–133

Singh HB (2008) Disease hampering commercial production of sugarcane in India: present scenario and future prospects. In: Yang-Rui Li, Nasr MI, Solomon S, Rao GP (eds) Proceedings of the International Conference of Professionals in sugar and integrated technologies. International Association of Professionals in Sugar and Integrated Technologies 174 East Daxue Road Nanning Guangxi—530007, PR China. pp 431–433

Singh SP, Singh M, Rao GP, Uphadyaya PP (1999) Fungitoxic efficacy of some volatile plants products against *Colletotrichum falcatum* Went. In: Rao GP, Filho AB, Magarey RC, Autrey JC (eds) Sugarcane Pathology, vol I: fungal disease. Science Publishers Inc., USA, pp 263–277

Srivastava N, Singh M, Rao GP, Tripathi SM, Singh HB, Srivastava AK (1999) Development of herbal volatile fungitoxicants against *Colletotrichum falcatum*. In: Rao GP, Filho AB, Magarey RC, Autrey JC (eds) Sugarcane pathology, vol I: fungal disease. Science Publishers Inc., USA, pp 239–262

Srivastava B, Singh P, Srivastava AK, Shukla R, Dubey NK (2009) Efficacy of *Artabotrys odoratissimus* oil as a plant based antimicrobial against storage fungi and aflatoxin $B_1$ secretion. Int J Food Sci Technol Microbiol Technol 44:1909–1915

WARDA (1993) West African Rice Development Association, Annual Report, pp 20–25

# 21 Evaluation of Biotic and Abiotic Factors for Production of Healthy Apple *(Malus × domestica)* Seedling

Jitender K. Verma, D. K. Kishore, S. K. Sharma and Asha Sharma

**Abstract**

Cultivation of apple has become the main occupation and major source of economy to the farming community of Himachal Pradesh, India. The production and supply of healthy apple nursery is key factor in the establishment of healthy orchard. The nursery production can be carried out at lower elevation at an altitude of 1,200–2,500 m. One-year-old seedlings embrace the price tag of ₹ 30–70. It can be a good prospective for small farmers with limited input. The biocontrol agent *Trichoderma viride* (at the rate 100 g/m$^2$) and its combined use with other soil amendment, viz. *Vitex* leaves (at the rate 1 kg/m$^2$), soil heating by burning and soil solarization, influences the soil microflora and growth parameters of seedling in apple nursery. The experiment was conducted during the year 2009–2010 to compare effects of various nonchemical methods and recommended chemical methods on the growth of seedlings raised through seeds in Karsog valley of Himachal Pradesh at farmer's field. Apple saplings growth parameters, health, vigorous, population of soil micro flora, pH, electrical conductivity, and organic carbon/matter contents were affected by soil amendments in nursery at all experimental sites.

**Keywords**

*Trichoderma* · Biocontrol · *Vitex nigundo* · Soil micro flora · Plant protection

## 21.1 Introduction

The own production of apple nursery helps in the rejuvenation of old apple orchard and establishment of new apple orchard. It reduces the dependency of orchardist on other agency for procurement of apple saplings. Most of the apple nurseries are at the altitude between 1,200 and

J. K. Verma (✉) · D. K. Kishore
IARI, Regional Station (CHC), Amartara Cottage, Shimla-4, Himachal Pradesh, 171004, India
e-mail: jatin_k_verma@yahoo.com

S. K. Sharma
YSP UH&F, R.& E. Station, Sharbo, Kinnaur, Himachal Pradesh, India

A. Sharma
Department of Botany, MDU, Rohtak, Haryana 124001, India

R. N. Kharwar et al. (eds.), *Microbial Diversity and Biotechnology in Food Security*,
DOI 10.1007/978-81-322-1801-2_21, © Springer India 2014

2,700 m above mean sea level in the Himalaya ranges. Apple cultivation requires 1,000–1,500 h of chilling below 7 °C during winter to break the rest period. Soil depth, drainage and pH determine the suitability of soil types. Loamy soils, rich in organic matter having a pH of 5.5–7.5 with gentle to moderate slope, proper drainage, and good aeration are most suitable. The average summer temperature should be around 21–24 °C during active growth period. In winter, the average temperature should not exceed the chilling requirement i.e., below 7 °C. The rainfall of 100–125 cm throughout the growing season is favorable (Kishore et al. 2006). Verma et al. (2009), investigated that the improvement in soil moisture availability, pH, organic carbon, and nutrient status of the soil was significant after input of organic manure. Growth parameters, fruit characteristics, and yield were recorded maximum with the application of organic manure. Soil solarization for 12 weeks resulted in maximum reduction in the population of fungi, bacteria, and actinomycetes, at 5, 10, and 15 cm depths with 19.56, 35.00, and 40.00 % survival of *Dematophora necatrix* propagules at these depths, respectively (Sharma and Sharma 2005). Soil solarization carried out for longer durations can be utilized for soil-borne disease management in temperate regions (Sharma et al. 2005). Effect of integration of cultural, chemical, and biocontrol methods was studied on the incidence of white root rot of apple caused by *Dematophora necatrix* to devise a suitable management strategy in nursery. In pot culture, all the combination gave cent per cent disease control, and no disease incidence was recorded even after 60 days of inoculation. However, under nursery conditions, maximum disease control was recorded in a combination of deodar needles+ neem cake +carbendazaim (0.1 %) + *Trichoderma* treatment. All the combinations were effective in improving the plant health except the individual treatments in pot culture. In addition, all treatments were effective in improving the plant health under nursery conditions (Rana et al. 2010). Sanchez et al. (2006), evaluated the effects of soil management on yield, growth, and soil fertility in an organic apple orchard cv. Royal Gala/EM 26 and cover crop treatments were applied to the inter-row spaces planted at 4 × 2 m. Soil organic matter increased in the topsoil especially with permanent cover crops, but decreased due to both annual soil tillage and less input of groundcover biomass. It is concluded that tree growth and yield are affected by soil management. Perennial cover crops perform better than annual common vetch. Disking is not a recommended practice because it may decrease the content of soil organic matter and lead to poor tree vigor that corresponds to low fruit bearing potential. However, even with the use of permanent cover crops the addition of organic fertilizers is necessary in order to sustain good yields and proper tree vigor. Measurement of pH, electrical conductivity (EC) parameters provides valuable information for assessing soil condition for plant growth, nutrient cycling, and biological activity. Soil and crop management practices having significant effect on pH and EC are considered good indicators of change. Elevated pH values indicate potential losses of nitrate and subsequent water contamination. The tendencies for soil acidification suggest insufficient use of ammoniac fertilizers and increased leaching losses (Smith and Doran 1996). Hassan et al. (2009), inoculated 1-year-old apple seedling of cv. "Red Delicious" with biocontrol agents, viz., *Glomus mosseae* (an endomycorrihza), *Laccaria laccata* (an ectomycorrhiza) and *Trichoderma harzianum* (a fungal antagonist) in presence of root rot pathogen, *Pythium ultimum*, to assess their impact on seeding growth and disease control. All the biocontrol agents significantly enhanced plant height, seedling diameter, root length, and plant biomass as compared to uninoculated or pathogen-inoculated controls. Molin and Rabello (2011), in their study clearly indicated that EC relates with soil texture and moisture, and may represent an important and low price tool for collecting data and characterizing soil physical properties. The objectives of the present study was, firstly to use easy, cost free, and eco-friendly practices in the production of healthy apple nurseries for orchards, biodynamically. Secondly, recycling of organic waste of fields, energy input along with soil amendments to make nursery production practice familiar and

**Fig. 21.1** Experimental field of apple (*Malus × domestica*) nursery

recognizing its economic importance along with use of biocontrol agents to regulate soil health and the production of healthy and vigorous standard rootstocks for grafting with scion of desired variety.

## 21.2 Materials and Methods

The present investigation was carried out at IARI, Regional Station (CHC), Shimla-4 and data was recorded from experimental field (Fig. 21.1) at Churag, in Mandi district of Himachal Pradesh (India) situated at an elevation, above mean sea level of 1,740 m 31°20′ N latitude 77°08′ E longitudes. The present study was aimed for the production of healthy apple sapling in nurseries and evaluation for growth parameters, viz., plant height, stem girth, fresh root weight, dry root weight, and leaf area against soil characteristics i.e., pH, EC, and organic carbon/matter after application of soil amendment treatments, viz., soil solarization (SS), soil heating by burning (SHB), *Trichoderma viride* (TV), *Vitex* leaves (*vitex nigundo*) (VL) at farmer's field during 2009–2010. The treatments were applied during growing season while growth parameters were recorded during dormancy period in winter. The soil characteristics were recorded consequently during rainy and winter season of same year also.

The overall details of the treatments at experimental field were as follow:

- T1—Application of TV at the rate 100 g/m$^2$
- T2—Soil amendments i.e. *Vitex*/garlic leaves at the rate 1 kg/m$^2$
- T3—SHB straw/SS
- T4—T1+T3
- T5—T1+T2+T3
- T6—T1+T2+T3+T8 (if required)
- T7—T1+T2
- T8—Application of chemicals
- T9—SS (polythene sheet, i.e., 25 μm thick)
- T10—Control

The soil samples were drawn from experimental fields consequently during same growth season to determine the soil characteristics, viz., pH, EC, microbial population, organic carbon and matter as presented in Table 21.1. The pH and EC was determined by "Lab Quest" 1:2 soil:water suspension. The soil samples were further analyzed for the estimation of organic carbon and organic matter collected from experimental nurseries. The chromic acid titration/rapid digestion method was used for its estimation (Walkley 1947). Plant growth characteristics such as plant height, girth, leaf area, fresh and dry root weight were recorded for each treatment at experimental sites and presented in Table 21.2. The plants were selected randomly and plant height was recorded from stem base to the tip of the seedlings. Plant girth was measured with the help of vernier caliper and recorded in millimeter. Leaf area measurements were recorded with the help of leaf area meter and expressed in square centimeter. Ten leaves were taken from each treatment. Root weight was recorded after uprooting three plants at random from each treatment. The root portion was cut from the stem base, washed in tap water, air dried, and fresh root weight was recorded in grams per plant. After that the roots were dried in oven at 50 °C for 5–6 days till constant dry weight was achieved. The data so obtained for different characters was averaged and analyzed statistically.

## 21.3 Results and Discussion

Soil moisture, pH, organic carbon, and available nitrogen (N), phosphorous (P), and potassium (K) were recorded maximum after the application of commercial organic manure at the rate 20 kg tree$^{-1}$ treatment and farmyard manure at

**Table 21.1** Status of microbial population, organic carbon/matter, pH, and electrical conductivity of soil at experimental nursery site

| Treatments | pH | Average microbial population | | | Organic carbon (%) | Organic matter (%) | Electrical conductivity in µs/cm Range(0–2,000) |
|---|---|---|---|---|---|---|---|
| | | Fungi $10^3$ cfu/g | Bacteria $10^4$ cfu/g | Actinomycetes $10^4$ cfu/g | | | |
| T1 (Trichoderma viride) | 6.89 | 1.33 | 7.00 | 6.33 | 1.15 | 1.98 | 485 |
| T2 (soil amendments) | 6.79 | 0.33 | 8.33 | 5.33 | 1.07 | 1.84 | 428 |
| T3 (soil burning) | 6.68 | 0.66 | 10.33 | 7.00 | 0.98 | 1.68 | 462 |
| T4 (T1+T3) | 6.80 | 0.33 | 9.00 | 4.66 | 1.00 | 1.72 | 470 |
| T5 (T1+T2+T3) | 6.75 | 0.33 | 5.66 | 5.00 | 1.27 | 2.18 | 454 |
| T6 (T1+T2+T3+T8 (if required) | 6.81 | 0.33 | 8.33 | 4.66 | 1.20 | 2.06 | 471 |
| T7 (T1+T2) | 6.78 | 1.00 | 7.66 | 6.00 | 1.23 | 2.12 | 405 |
| T8 (application of chemicals) | 6.48 | 0.66 | 6.00 | 5.00 | 1.13 | 1.94 | 395 |
| T9 (soil solarization) | 6.63 | 0.33 | 5.33 | 5.66 | 1.10 | 1.89 | 472 |
| T10 (control) | 6.60 | 0.33 | 9.66 | 9.66 | 0.95 | 1.63 | 480 |

**Table 21.2** Effect of different treatments on plant growth characteristics at nursery trial site

| Treatment No | Average seedling height (cm) | Average fresh root weight (g) | Average dry roots weight (g) | Average leaf area ($cm^2$) | Average seedling girth (mm) |
|---|---|---|---|---|---|
| T1—Trichoderma viride | 73.50 | 2.30 | 1.19 | 19.75 | 5.44 |
| T2—Soil amendment—Vitex leaves | 91.80 | 3.80 | 1.52 | 22.08 | 6.77 |
| T3—Soil heating by burning stubbles | 85.60 | 2.28 | 1.47 | 23.65 | 6.65 |
| T4—(T1+T3) | 88.19 | 1.97 | 0.85 | 25.94 | 6.68 |
| T5—(T1+T2+T3) | 107.10 | 1.69 | 0.75 | 23.99 | 7.18 |
| T6—(T5+T8 if required) | 105.00 | 3.25 | 1.28 | 22.10 | 6.87 |
| T7—(T1+T2) | 98.10 | 2.06 | 0.93 | 21.99 | 6.67 |
| T8—(Bavistin at the rate 0.1%) | 91.70 | 2.23 | 0.95 | 21.82 | 5.99 |
| T9—Soil solarization | 94.50 | 2.07 | 0.84 | 20.30 | 6.07 |
| T10—Control | 95.40 | 1.27 | 0.53 | 23.71 | 5.93 |

the rate 100 kg tree$^{-1}$. Better quality apple fruits were also obtained under the application of commercial organic manure at the rate 20 kg tree$^{-1}$ and farmyard manure at the rate 100 kg tree$^{-1}$ during the years of experimentation (Verma et al. 2010). The study to assess the effects of integrated, organic, and conventional apple production systems on horticultural performance, soil quality, and orchard profitability indicated that the integrated and organic production systems maintained higher soil quality than did the conventional system (Glover et al. 2000). The application of composted apple pruning promotes the growth of apple nursery trees and is an effective method of improving the chemical properties of soil (Sakamoto et al. 2010). The growth parameters of apple saplings and soil characteristics, viz., pH, EC, organic carbon, and organic matter were evaluated to find out best treatments at trial sites. The soil samples were analyzed for microbial population, pH, EC, organic carbon/matter status, and compared with the growth characteristics as follow.

### 21.3.1 pH

The pH of all samples ranged from 6.48(T8) to 6.89(T1). Increase in the pH value toward neutral i.e., pH 7 increased the plant height and girth value. The fresh root weight (3.80 g) and dry root weight (1.52 g) had maximum value at pH 6.79 (T2). The maximum leaf area (25.94 $cm^2$) was observed at pH 6.80 (T4). Shoot growth increased with soil pH. The highest yield (boxes/tree) was obtained at pH 6.0–6.5 and the largest fruits were produced at pH 5.5–6.9. Low soil pH was associated with high concentrations of manganese (Mn) in the leaves and fruit and a reduction in fruit red skin color (Raese 1995).

### 21.3.2 Electrical Conductivity (EC)

The EC (0–2,000 μs/cm) ranged from 395μs/cm (T8) to 485 μs/cm (T1). The maximum plant height and girth was observed on EC value 454.00 μS/cm (T5). The maximum value of fresh and dry root weight was observed at EC value 428.00 μS/cm (T2).The maximum leaf area was observed at EC value 470.00 μS/cm (T4). It is clearly indicated that EC relates with soil texture and moisture, and may represent an important and low price tool for collecting data and characterizing soil physical properties (Molin and Rabello 2011).

### 21.3.3 Organic Carbon/Matter Contents

Organic matter (OM) and organic carbon (OC) ranged from 1.63% (T10) to 2.18% (T5) and 0.95% (T10) to 1.27% (T5), respectively. The maximum plant height and girth were observed at the OC value of 1.27% and OM value of 2.18% in T5. The maximum fresh and dry root weight were observed at the OC value of 1.07% (T2) having OM value 1.84%. The maximum leaf area was observed at the OC of 1.00% and OM 1.72% in T4. Organic fertilization and soil management can interfere directly on the yield and growth of apple trees, especially in shallow soils. The application of litter poultry promoted the recovery of plant growth and in a more efficient way when herbicide was used as weed control (Nava 2010).

### 21.3.4 Microbial Population

Microbial count showed that the fungal population ranged from $0.33 \times 10^3$ (T10) to $1.33 \times 10^3$ (T1) and the bacterial population ranged from $5.33 \times 10^4$ (T9) to $10.33 \times 10^4$ (T3). Similarly, actinomycetes population ranged from $4.66 \times 10^4$ (T4) to $9.66 \times 10^4$ (control).The microbial population of bacteria, fungi, and actinomycetes were maximum in T3, T1, and control, respectively. The minimum values for the same were observed in T9, T10, and T4, respectively (Fig. 21.2). The higher organic content was associated with higher fungal diversity and population levels and quality of poor soil can be altered by enriching and modifying the microbial flora of the soil (Wahegaonkar et al. 2009).

### 21.3.5 Growth Parameters

Maximum average plant height was 107.10 cm in T5 followed by T6 (105.00 cm), and minimum average height was 73.50 cm (T1). The seedling girth was ranged from 5.44 mm (T1) to 7.18 mm (T5). The fresh root weight ranged from 1.27 g (control) to 3.80 g (T2), and dry weight ranged from 0.53 g (control) to 1.52 g (T2). The average leaf area ranged from 19.75 $cm^2$ (T1) to 25.94 $cm^2$ (T4).

The soil characteristics of above treatments gave maximum growth parameter values in T2 and T5. The fresh root weight (3.80 g) and dry root weight (1.52 g) had maximum value at pH 6.79 (T2). The maximum leaf area (25.94 $cm^2$) was observed at pH 6.80 (T4). The maximum plant height and girth was observed at EC value 454.00 μS/cm (T5) and pH 6.75 (T5). The maximum value of fresh and dry root weight was observed at EC value 428.00 μS/cm (T2). The maximum leaf area was observed at EC value 470.00 μS/cm (T4).The maximum plant height and girth were observed on OC value 1.27% and OM value 2.18% in T5. The maximum fresh and dry root weight were observed on OC value 1.07% (T2). The OM value for the same was 1.84% (T2). The maximum leaf

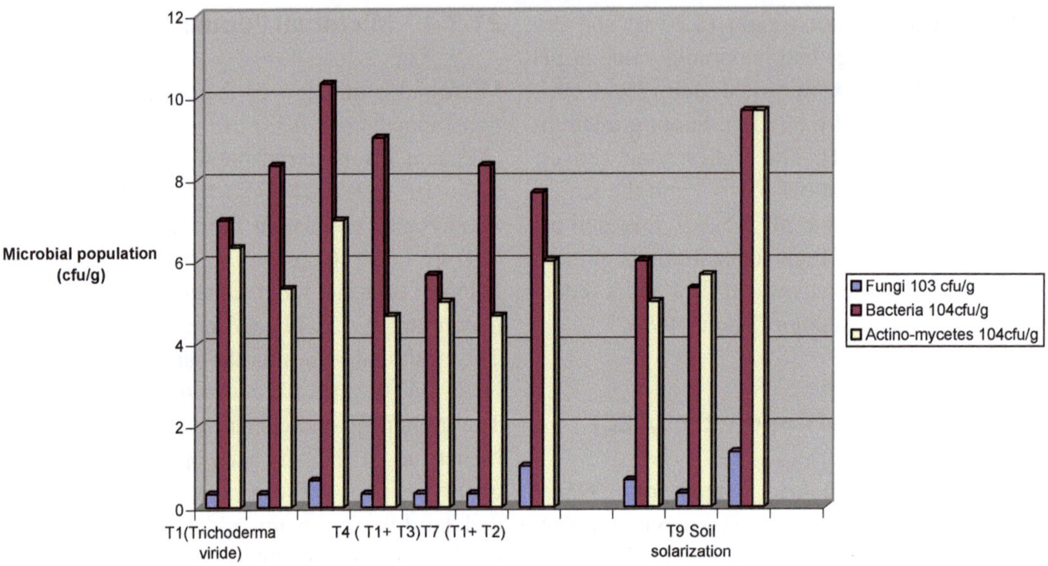


**Fig. 21.2** Treatment wise average microbial population

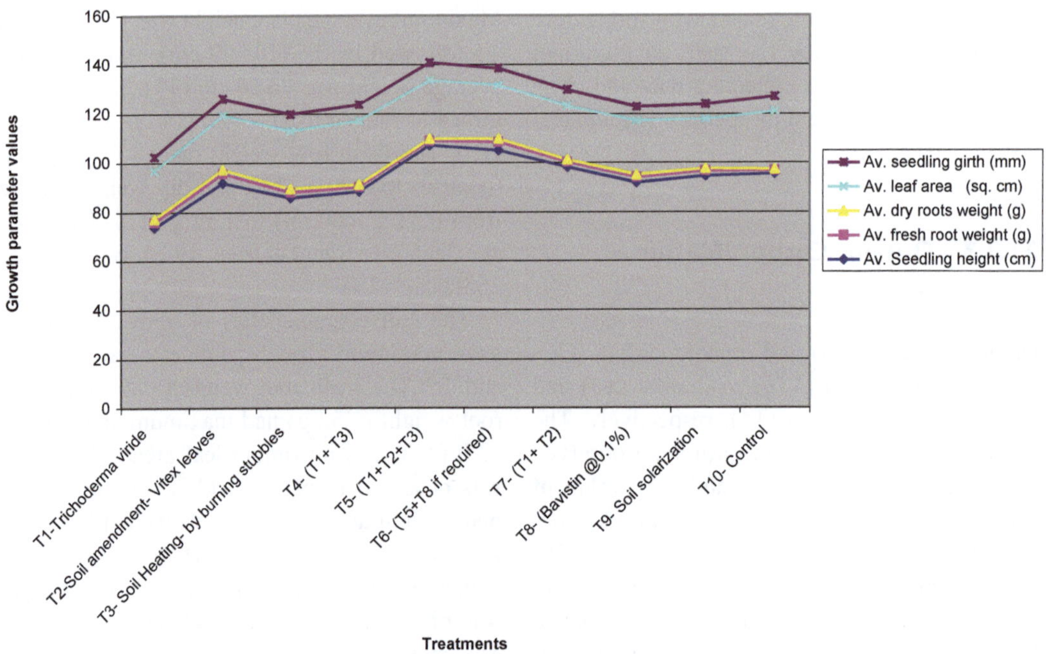


**Fig. 21.3** Effect of treatments on growth characteristics of apple (*Malus* × *domestica*) saplings

area was observed on value OC 1.00% and OM 1.72% in T4. The microbial population of bacteria, fungi, and actinomycetes was examined maximum in T3, T1, and control, respectively. The treatment wise growth characteristics are presented in Fig. 21.3. These soil amendment treatments can overcome the drawbacks of conventional method of apple nursery production and ensure organic nursery production with desired traits.

## 21.4 Conclusions

The application of biocontrol agent TV at the rate 100 g/m$^2$ mixed with FYM and its combined use with other soil amendments, viz., VL at the rate 1 kg/m$^2$, SHB i.e., TV+VL+SHB influenced the soil characteristics and growth parameters at experimental field. The effect of treatments on biotic and abiotic factors of apple nursery revealed that pH of treated nursery soil ranged from 6.48 (chemicals) and 6.89 (TV). The fungal population ranged from $1.33 \times 10^3$ (TV) to $0.33 \times 10^3$ (control). The bacterial population ranged from $5.33 \times 10^4$ (SS) to $10.33 \times 10^4$ (SHB). The actinomycetes population ranged from $4.66 \times 10^4$ (TV and SHB) to $9.66 \times 10^4$ (control). The organic matter and carbon ranged from 1.63% (control) to 2.18% (TV, VL, and SHB) and 0.95% (control) to 1.27% (TV, VL, and SHB), respectively. The average plant height was maximum 107.10 cm (TV+VL+SHB) and minimum 73 cm (control) whereas plant girth ranged from 5.44 mm (TV) to 7.18 mm (TV+VL+SHB). The fresh root weight ranged from 1.27 g (control) to 3.80 g (VL). The dry root weight ranged from 0.53 g (control) to 1.52 g (VL). The leaf area ranged from 25.94 cm$^2$ (TV and SHB) to 19.75 cm$^2$ (TV). It indicates that health and vigorousity was affected by the soil amendments in nursery at the present geographical experimental sites. The use of biocontrol agents and chopped leaves of the medicinal plant VL along with SHB overcomes the drawbacks of conventional method of apple nursery production and ensures the production of an organic nursery with vigorous and healthy saplings. The less input along with organic waste recycling in the cultivation of apple nursery impact soil health and growth parameters of apple sapling. It regulated the population of soil microflora and increased the organic carbon or matter contents also.

## References


Glover JD, Reganold JP, Andrews PK (2000) Systematic method for rating soil quality of conventional, organic, and integrated apple orchards in Washington State. Agric Ecosyst Environ 80:29–45

Hassan DHG, Banday S, Beig AM, Fatima N (2009) Efficacy of biocontrol agents in improving plant growth and control of *Pythium* root rot in apple. Plant Dis Res 24(2):142–145

Kishore DK, Sharma SK, Paramanick KK (2006) A text book on current scenario of temperate fruits in H.P. New India Publishing Agency, Pitam Pura, pp 1–5

Molin JP, Rabello LM (2011) Studies about soil electrical conductivity measurements. J Braz Ass Agric Engg 31(1):90–101

Nava G (2010) Organic fertilization and weed control effects on yield and growth of 'Fuji' apple. (Portuguese). Revista Brasileira de Fruticultura 32(4):1231–1237

Raese T (1995) Tree fruit nutrition II: soil pH and apple tree growth. Good Fruit Grow 46(5):42–44

Rana S, Sharma SK, Sharma NJ (2010) Integrationn of cultural, chemical and biocontrol methods in managements of white root rot of apple. Indian Phytopathol 63(2):207–211

Sakamoto K, Aoyama M, Sakamoto K, Aoyama M (2010) Effects of applying composted of apple prunings on the growth of apple (Malus pumila) nursery trees and physico-chemical properties of soils. Horticultural Research (Japan) 9(2):153–158

Sanchez EE, Cichon LI, Fernandez D (2006) Effects of soil management on yield, growth and soil fertility in an organic apple orchard. Acta Hortic 72(1):49–53

Sharma A, Sharma SK (2005) Effect of soil solarization on soil borne pathogens and soil microbial population in apple nurseries. Plant Dis Res 20(2):138–142

Sharma SK, Kishore DK, Pramanick KK (2005) Effect of soil solarization on soil microflora and survival of *Dematophora necatrix* in temperate fruit nurseries. Acta Hortic (ISHS) 696:381–386

Smith JL, Doran JW (1996) Measurement and use of pH and electrical conductivity for soil quality analysis. Methods for assessing soil quality. Soil Science Society of America inc (SSSA spl Publisher), pp 169–185

Verma ML, Singh C, Bhardwaj SP (2009) Effects of biofertilizers on soil moisture, nutrient status and fruit productivity under organic cultivation of apple in Himachal Pradesh. Indian J Soil Conserv 37(3):201–205

Verma ML, Sharma R, Singh C, Rathore AC (2010) Influence of organic manuring on apple performance and soil properties in temperate zone of Himachal Pradesh. Indian J Soil Conserv 38(3):212–216

Wahegaonkar N, Salunkhe SM, Palsingankar PL, Shinde SY (2009) Fungal diversty in soil samples from cultivated, barren and gardens lands. J Mycol Plant Pathol 39(3):462–467

Walkley A (1947) A critical examination of a rapid method for determination of organic carbon in soils—effect of variations in digestion conditions and of inorganic soil constituents. Soil Sci 63:251–257

## 21.4 Conclusions

The application of biocontrol agent TV at the rate 100 g/m² mixed with FYM and its combination with other soil amendments, viz., VL at the rate 1 kg per m² (i.e., TV + VL + SH3) influenced the soil characteristics and growth parameters at experimental field. The effect of treatments on biotic and abiotic factors of apple nursery revealed that pH of treated nursery soil ranged from 6.48 (treatments) and 6.89 (TV). The fungal population ranged from 1.33 × 10⁶ (TV) to 0.34 × 10⁶ (control). The bacterial population ranged from 5.31 × 10⁷ (SS) to 10.33 × 10⁷ (SHB). The actinomycetes population ranged from 4.66 × 10⁵ (TV and SHB) to 9.66 × 10⁵ (control). The organic matter and carbon ranged from 1.63 % (control) to 2.18 % (TV, VL and SHB) and 0.95 % (control) to 1.27 % (TV, VL and SHB), respectively. The average plant height was maximum 15.10 cm (TV+VL+SHB) and minimum 7.3 cm (control) whereas plant girth ranged from 5.44 mm (TV) to 7.18 mm (TV+VL+SHB). The fresh root weight ranged from 1.27 g (control) to 3.80 g (TV), dry root weight ranged from 0.53 g (control) to 1.52 g (TV). The leaf area ranged from 13.94 cm² (TV and SHB) to 19.75 cm² (TV). It indicates that health and vigorousity was affected by the soil amendments in nursery at the present geographical experimental sites. The use of biocontrol agents and chopped leaves of the medicinal plant VL along with SHB overcomes the drawbacks of conventional method of apple nursery production and ensures the production of an organic nursery with vigorous and healthy saplings. This less input along with organic waste recycling in the cultivation of apple nursery impact soil health and growth parameters of apple sapling. It regulated the population of soil microflora and increased the organic carbon or matter contents also.

## References


Glover JD, Reganold JP, Andrews PK (2000) Systematic method for rating soil quality of conventional, organic, and integrated apple orchards in Washington State. Agric Ecosyst Environ 80:29–45

Hassan DHG, Bhaday S, Baig AM, Jonson N (2009) Efficacy of biocontrol agents in improving plant growth and control of *Pythium* root rot in apple. Plant Dis Res 24(2):142–145

Kishore DK, Sharma SK, Pramanick KK (2006) Recent outlook on current and future temperate fruits in H.P. (Eds.) in: Temperate Agroecosystems. Mohit Lom, pp 1–5

Mahn H, Ranchl EK (2011) Studies about electrical soil conductivity measurement. Fet Ask Ass Mater Eng 19(1):99–105

Neto D (2010) Organic fertilization and weed control effects on yield and growth of Fuji apple (Portuguese). Revista Brasileira de Fruticultura 32(4):1171–1235

Rudolf (1995) Leaf fruit surfaces – the need for apple tree growth. Good Fruit Grower 46(5):12–13

Rana SK, Sharma SK, Sharma N (2009) Integrated management of white root rot of apple with chemical and biocontrol methods in nursery. Indian Phytopathol 62(2):207–211

Sakamoto K, Awasthi I, Nakamura S, Aoyama M (2000) Effects of applying composted apple pomace on the growth of apple (Malus pumila) seedlings and physicochemical properties of soil. Horticultural Research (Japan) 69(2):152–158

Sanchez EE, Giehoi LB, Laurenteniz D (2006) Effects of soil management on yield, growth and soil fertility in an organic apple orchard. Acta Hortic 721:49–54

Sharma A, Sharma SK (2009) Effect of soil solarization on soil borne pathogens and soil microbial population in apple nurseries. Plant Dis Res 24(2):138–143

Sharma SK, Kishore DK, Pramanick KK (2011) Effect of soil solarization on soil microflora and survival of *Dematophora necatrix* in temperate fruit nurseries. Acta Hortic (ISHS) 896:381–386

Smith JL, Doran JW (1996) Measurement and use of pH and electrical conductivity for soil quality analysis. Methods for assessing soil quality. Soil Science Society of America Inc (SSSA Spl Published), pp 169–185

Verma ML, Singh C, Bhardwaj SP (2009) Effects of biofertilizers on soil moisture, nutrient status and fruit productivity under organic cultivation of apple in Himachal Pradesh. Indian J Soil Conserv 37(3):201–205

Verma ML, Sharma R, Singh C, Rathore AC (2010) Influence of organic manuring on apple performance and soil properties in temperate zone of Himachal Pradesh. Indian J Soil Conserv 38(3):212–216

Wakegaonkar N, Salunkhe SM, Pathingankar P, Shinde SY (2009) Fungal diversity in soil samples from cultivated, barren and gardens lands. J Mycol Plant Pathol 39(3):462–467

Walkley A (1947) A critical examination of a rapid method for determination of organic carbon in soils—effect of variations in digestion conditions and of inorganic soil constituents. Soil Sci 63:251–257

# 22 Evaluation of Antifungal Activity of *Metarhizium anisopliae* Against Plant Phytopathogenic Fungi

K. Ravindran, S. Chitra, A. Wilson and S. Sivaramakrishnan

**Abstract**

An indigenous insect pathogenic fungi *Metarhizium anisopliae* (TK09) was isolated and evaluated for antifungal activity against phytopathogenic fungi *Fusarium oxysporum*, *Cladosporium herbarum* and *Curvularia clavata*. The dichloromethane (DCM) solvent was used for the extraction of fungal secondary metabolites and was assessed against plant pathogenic fungi at various concentrations ranging 500–1,200 µg/ml by agar disk diffusion method. As a result, the fungal crude metabolite showed the highest inhibitory activity against *C. herbarum* and lowest to *F. oxysporum*. Thus, the overall results suggest that isolate TK09 may be used as a control for plant pathogens and insects pest as well.

**Keywords**

*M. anisopliae* · Insecticidal activity · Antifungal activity · Plant pathogens

## 22.1 Introduction

Fungal phytopathogens are the major causative agent of many plant diseases with subsequent loss of crop yields, especially in subtropical and tropical regions (Brimner and Boland 2003). In the present scenario, numerous reports are available on potential usages of biocontrol agents (Bacteria, Fungi), as a novel alternative of agrochemicals (Yang et al. 2008). Microbial assemblage present in the soil is an essential part of soil ecosystem for sustainable agriculture, for controlling pest population and suppressing the soil-borne pathogens. The soil environment represents an important reservoir for a diversity of entomopathogenic fungi (EPF) which can contribute significantly to the regulation of insect population. Many species belonging to Hypocreales inhabit the soil for a significant part of their life cycle as dormant conidia in the vicinity of dead host cadaver with limited saprobic growth. The infective propagules cause muscardine disease in living insect host for a relatively short period of time. *Beauveria* and *Metarhizium* spp., were the two most widely used EPF to combat

K. Ravindran (✉) · S. Chitra · A. Wilson · S. Sivaramakrishnan
Department of Biotechnology, School of Life Sciences, Bharathidasan University, Tiruchirappalli, Tamil Nadu 620024, India
e-mail: ravindmbs@gmail.com

R. N. Kharwar et al. (eds.), *Microbial Diversity and Biotechnology in Food Security*, DOI 10.1007/978-81-322-1801-2_22, © Springer India 2014

insect population which has been exploited to develop as a biocontrol agent for many important insect and plant pathogens. This fungus has been deemed safe, environmentally acceptable alternative to chemical pesticides (Zimmermann 1986).

Selection of indigenous *M. anisopliae* may be the best-suited control agent to local pest in that area by producing variety of bioactive compounds for the development of an efficient and eco-friendly formulation of biopesticides. Therefore, the study was focused on the isolation of EPFs from agricultural soils and evaluation of its crude metabolites against phytopathogens.

## 22.2 Material and Methods

### 22.2.1 Isolation of EPF

Entomopathogenic fungi were isolated from soil samples using the "Galleria bait method" (GBM) (Zimmermann 1986). After isolation, the isolate was evaluated for virulence against model insect. Further, the isolate TK09 was cultivated on oatmeal agar medium and incubated for 7 days at 27 °C or until sporulation. For long-term storage, TK09 conidial suspensions were prepared in 20–30 % glycerol solutions and stored at − 80 °C as a stock culture which will be subcultured for further analysis (Zimmermann 1986).

### 22.2.2 Preparation of Crude Extract of TK09

The aerial conidium of TK09 was harvested from 15-day-old sporulating cultures by scraping the surface with a spatula and suspended the conidia in sterile 0.05 % aqueous (w/v) Triton X-100. The isolate was further cultivated in 250 ml Erlenmeyer flasks containing 100 ml of Czapek Dox agar medium with 0.5 % (w/v) bacto peptone, pH 7.0. The flasks were inoculated with 1 ml of conidial suspension containing 107 conidia/ml and incubated at 25 ± 1 °C on an orbital shaker at 150 rpm for 9–12 days. The mycelium and impurities were removed by centrifugation (8,000 g, 30 min, 4 °C). In addition, the culture filtrate was filtered through a Buchner funnel lined with Whatman filter paper No. 1 to ensure complete removal of conidia and hyphen debris. The crude culture filtrate was mixed with dichloromethane solvent (3 × 500) for 2 h followed by cheesecloth filtration to remove the biomass. Then the solvent was saturated with rotary vacuum evaporator for solvent-free extracts. The extracted metabolites were further stored at − 4 °C for further study.

### 22.2.3 Antimicrobial Assay

The extracted metabolite of TK09 was further mixed with dimethyl sulfoxide (DMSO) for antifungal activity against plant pathogenic fungal *F. oxysporum, C. herbarum and C. clavata.* The test pathogens were procured from the department of Biotechnology and maintained on potato-dextrose-agar (PDA) plates. Different concentration of DMSO-mixed extract was prepared and evaluated against pathogens by disk diffusion method and plane DMSO was used as a control. Plates were monitored for the development of zones of inhibition every 24 h for 4 days. Three replicates were used for each pathogen.

## 22.3 Results

### 22.3.1 Pathogenicity of EPF on Preliminary Screening

EPF was evaluated for its efficacy and pathogenicity using a model insect *Galleria mollenella* larva (Koch's postulates). The *M. anisopliae* TK09 isolate was assayed for its virulence and infectivity rate ranges from 30 to 100 % within 3–5 days. The TK09 isolate produced 100 % mortality rate on fourth day and exhibited exposed green mycelial development in cadavers.

### 22.3.2 Growth Inhibitory Action

The potential strain of *M. anisopliea* (TK09) fungal crude extracts was tested against plant patho-

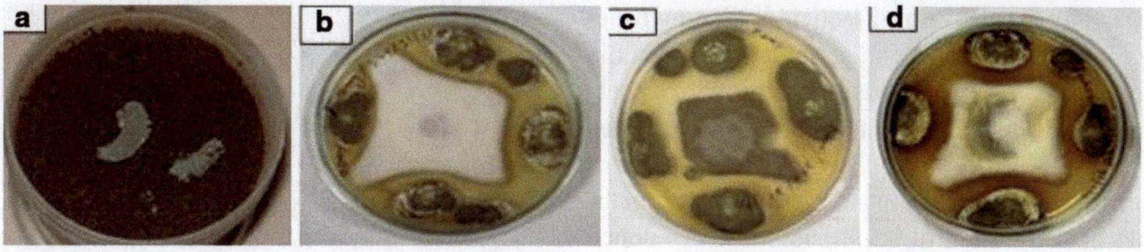


**Fig. 22.1** *Galleria* biting technique (**a**), Antifungal activity against three plant pathogenic fungi on *Fusarium oxysporum* (**b**) *Curvularia clavata* (**c**) and *Cladosporium herbarum* (**d**) plates after 7 days at 25 ± 1 °C

**Table 22.1** Antifungal activity of *M. anisopliae* against plant pathogenic fungi

| Test organisms | Concentraction (μg/ml)/zone of inhibition | | | | | | | |
|---|---|---|---|---|---|---|---|---|
| | 500 | 600 | 700 | 800 | 900 | 1,000 | 1,100 | 1,200 |
| *F. oxysporum* | – | 2.9 ± 0.3 | 3.9 ± 0.9 | 6.5 ± 0.3 | 6.7 ± 0.2 | 7.9 ± 0.5 | 12 ± 1.1 | 17.6 ± 2.0 |
| *C. herbarum* | – | 2.1 ± 0.2 | 4.6 ± 0.2 | 5.3 ± 0.5 | 9.0 ± 1.5 | 11 ± 1.5 | 14 ± 1.1 | 20.6 ± 2.5 |
| *C. clavata* | – | 4.3 ± 1.5 | 7.0 ± 0.5 | 8.6 ± 0.5 | 10.3 ± 0.5 | 10.6 ± 1.1 | 16 ± 2.0 | 19.6 ± 2.0 |
| DMSO | – | – | – | – | – | – | – | – |

[a] DMSO - Negative control
[b] Values are represent mean± SE of three experiments

gens such as, *F. oxysporum, C. herbarum and C. clavata* (Fig. 22.1b). Entomopathogenic fungus of *M. anisopliea* was found to be very effective for fungal pathogens when checked by a well-diffusion method. The minimum inhibitory concentration was determined according to the different concentrations (500–1,200 μg/ml) by isolated compounds. It showed highest activity against *C. herbarum* (20.6 ± 2.5/1,200 μg/ml) and lowest activity against *F. oxysporum* (17.2 ± 2/1,200 μg/ml) (Table 22.1) when the zone of inhibition was measured. In DMSO, no activity was observed. The dichloromethane crude extraction of *M. anisopliea* showed a significant variation in concentration between 700 and 1,200 μg/ml against plant pathogens.

## 22.4 Discussion

Biological control agents are perceived to have specific advantages over synthetic fungicides, including fewer nontarget and environmental effects, efficacy against fungicide-resistant pathogens and reduced probability of resistance development (Harman 2000; Cook 1988; Anonymous 1999; Tsror et al. 2001). Insecticides have a potential to affect the various developmental stages of entomopathogenic fungi. All tested insecticides displayed varying degree of potential to inhibit growth and conidial germination of both entomopathogenic fungi which corroborate with previous findings (Mietkiewski and Gorski 1995; Gupta et al. 1999). They observed variations in toxicity response of entomopathogenic fungi from synergistic, antagonistic or neutral to insecticides. Hassan and Charnley (1989) also reported inconsistent interaction between fungus and insecticides. Fungitoxic effects of insecticides vary as a function of the chemical nature of the products and interacting microbial species (Antonio et al. 2001).

A given insecticide may have different fungitoxic effects on various developmental stages of the fungus (Li and Holdom 1994). The potential inhibitory effects of pesticides on germination and mycelial growth of entomopathogenic fungi vary among taxa and strains (Vanninen and Hokkanen 1988). Effect of insecticides on conidial germination is the most important aspect to evaluate fungus–insecticide compatibility (Neves et al. 2001; Hirose et al. 2001). This happens because the fungi infect insects through the conidial germination which is the first step of the infection process (Oliveria et al. 2003). Our results suggested that conidial germination was more sensitive to insecticides than myceliel growth of the fungi. The germination of conidia is more severely affected than the growth of entomopathogenic fungi in the presence of pesticides (Hall 1981).

Low-cost production and field dispersal method of the biocontrol agent should be developed to improve production, effectiveness and sustainability. *M. anisopliae* continues to remain as the most potential biological control agent, for plant pathogens and pest attacking, to plant. The fungi infects and inhibits pathogenic growth and therefore has the potential to provide longer-term suppression of subsequent pest generations and to develope suitable agro ecosystem.

## References

Adrio JL, Demain AL (2003) Fungal biotechnology. Internat Microbiol 6(3):191–199

Ali-Shtayeh MS, Abdel-Basit M, Jamous R (2002) Distribution, occurrence and characterization of entomopathogenic fungi in agricultural soil in the Palestinian area. Mycopathologia 156:235–244

Amiri B, Ibrahim L, Butt TM (1999) Antifeedant properties of destruxins and their use with the entomogenous fungus *Metarhizium anisopliae* for improved control of crucifer pests. Biocontrol Sci Technol 9:487–498

Amiri-Besheli B, Khambay B, Cameron S, Deadman M, Butt TM (2000) Inter- and intra-specific variation in destruxin production by the insect pathogenic fungus, *Metarhizium*, and its significance to pathogenesis. Mycoll Res 104:447–452

Anonymous (1999) Organic agriculture. Canadian General Standards Board, Ottawa, Ontario, Canada

Ansari MA, Tirry L, Vestergaard S, Moens M (2009) Selection of a highly virulent fungal isolate *Metarhizium anisopliae* CLO 53 for controlling *Hopliaphilanthus*. J Invertebr Pathol 85(2):89–96

Antonio BF, Almeida JEM, Clovis L (2001) Effect of Thiamethoxam on entomopathogenic microorganisms. Neotropic Entomol 30:437–447

Bandani AR, Khambay BPS, Faull J, Newton R, Deadman M, Butt TM (2000) Production of efrapeptins by *Tolypocladium* species (Deuteromycotina: hyphomycetes) and evaluation of their insecticidal and antimicrobial properties. Mycol Res 104:537–544

Brimner TA, Boland GJ (2003) A review of the non-target effects of fungi used to biologically control plant diseases. Agric Ecosyst Environ 100:3–16

Cook RJ (1988) Biological control and holistic plant-health care in agriculture. Am J Altern Agric 3:51–62

Demain AL (1999) Pharmaceutically Active Secondary Metabolites of Microorganisms. Appl Microbiol Biotechnol 52(4):455–463

Domsch KH, Gams W, Anderson TH (1993) Compendium of soil fungi. IHW Verlag, Eching, pp 458–459

Espada A, Dreyfuss MM (1997) Effect of the cyclopeptolide 90–215 on the production of destruxins and helvolic acid by *Metarhizium anisopliae*. J Ind Microbiol Biotechnol 19:7–11

Georgianna, DR, Fedorova, ND, Burroughs JL, Dolezal AL, Bok JW (2010) Beyond aflatoxin: four distinct expression patterns and functional roles associated with *Aspergillus flavus* secondary metabolism gene clusters. Mol Plant Pathol 11:213–226

Goettel MS, Inglis GD, Wraight SP (2000) Fungi. In: Lacey LA, Kaya HK (eds) Field manual of techniques in invertebrate pathology, application and evaluation of pathogens for control of insects and other invertebrate pests. Kluwer Academic Publishers, Dordrecht, pp 255–279

Gupta P, Paul MS, Sharma SN (1999) Studies on compatibility of white muscardine fungus *Beauveria bassinia* with neem products. Indian Phytopathol 52(3):278–280

Hall RA (1981) Laboratory studies on the effects of fungicides, acaricides and insecticides on the entomopathogenic fungus, *Verticillium lecanii*. Entomol Exp Appl 29(1):39–48

Harman GE (2000) Myths and dogmas of biocontrol: changes in perceptions derived from research on *Trichoderma harzianum* T-22. Plant Dis 84:377–393

Hassan AEM, Charnely AK (1989) Ultrastructural study of the penetration by *Metarhizium anisopliae* through dimilin affected cuticle of Mandusasexta. J Invertebr Pathol 54(1):117–124

Hatta R, Ito K, Hosaki Y, Tanaka T, Tanaka A, Yamamoto M, Akimitsu K, Tsuge T (2002) A conditionally dispensable chromosome controls host-specific pathogenicity in the fungal plant pathogen *Alternaria alternate*. Genetics 161:59–70

Hirose E, Neves PMOJ, Zequi JAC, Martins LH, Peralta CH, Moino A Jr (2001) Effect of biofertilizers and Neem oil on the entomopathogenic fungi *Beauveria bassiana* (Bals.) Vuill. and *Metarhizium anisopliae* (Metsch) Sorok. Braz Arch Biol Technol 44(4):419–423

Hernandez-Crespo P, Santiago-Alvarez C (1997) Entomopathogenic fungi associated with natural populations of the Moroccan locust *Dociostaurus maroccanus* (Thunberg) (Orthoptera: Gomphocerinae) and other acridoidea in Spain. Biocontrol Sci Technol 7:357–363

Johnson L (2008) Iron and siderophores in fungal–host interactions. Mycol Res 112:170–183

Keller C, Maillard M, Keller J, Hostettmann K (2002) Screening of European fungi for antibacterial, antifungal, larvicidal, molluscicidal, antioxidant and free-radical scavenging activities and subsequent isolation of bioactive compounds. Pharm Boil 40:518–525

Kershaw MJ, Moorhouse ER, Bateman R, Reynolds SE, Charnley AK (1999) The role of destruxins in the pathogenicity of *Metarhizium anisopliae* for three species of insects. J Invertebr Pathol 74:213–223

Kodaira Y (1961) Biochemical studies on the muscardine fungi in the silkworm, *Bombyxmori*. J Fac Text Sci Technol Shinshu Univ Ser E 5:1–68

Li DP, Holdom DG (1994) Effects of pesticides on growth and sporulation of *Metarhizium anisopliae* (Deuteromycotina: Hyphomycets). J Invertebr Pathol 63:209–211

Liu BL, Chen JW, Tzeng YM (2000) Production of cyclodepsipeptides destruxin A and B from *Metarhizium anisopliae*. Biotechnol Prog 16:993–999

Miyazawa M, Honjo Y, Kameoka H (1998) Biotransformation of the sesquiterpenoid ( + )- 1- gurjunene using a plant pathogenic fungus *Glomerella cingulata* as a biocatalyst. Phytochemistry 38(4):172–174

Meitkiewski R, Gorski R (1995) Growth of selected entomopathogenic fungi species and isolates on media containing insecticides. Acta Mycol 30(1):27–33

Molnar I, Gibson DM, Krasnoff SB (2010) Secondary metabolites from entomopathogenic Hypocrealean fungi. Nat Prod Rep 27(9):1241–1275

Neves PMOJ, Hirose E, Techujo PT, Moino A Jr (2001) Compatibility of entomopathogenic fungi with Neonicotinoid insecticides. Neotropical Entomol 3:263–268

Oliveria GN, Neves PMOJ, Kawazoe LS (2003) Compatibility between the entomopathogenic fungus *Beauveria bassiana* and insecticides used in coffee plantations. Sci Agric 60(4):663–667

Ownley BH, Windham MT (2007) Biological control of plant pathogens. In: Trigiano RN, Windham MT, Windham AS (eds) Plant pathology concepts and laboratory exercises, 2nd edn. CRC Press, New York, pp 423–436

Ownley BH, Gwinn KD, Vega FE (2009) Endophytic fungal entomopathogens with activity against plant pathogens: ecology and evolution. BioControl 55:113–128

Pedras MSC, Irina ZL, Ward DE (2002) The destruxins: synthesis, biosynthesis, biotransformation, and biological activity. Phytochemistry 59:579–596

Roh KB, Kim CH, Lee H, Kwon HM, Park JW (2009) Proteolyticcascade for the activation of the insect Toll pathway induced by the fungal cell wall component. J Biol Chem 284:19474–19481

Shah PA, Pell JK (2003) Entomopathogenic fungi as biological control agents. Appl Microbiol Biotechnol 61:413–423

St Leger RJ, Wang C (2010) Genetic engineering of fungal biocontrol agents toachieve greater efficacy against insect pests. Appl Microbiol Biotechnol 85:901–907

Sumarah MW, Adams GW, Berghout J, Slack GJ, Wilson AM, Miller JD (2008) Spread and persistence of a rugulosin-producing endophyte in Picea glaucaseedlings. Mycol Res 112:731–736

Tsror L, Barak R, Sneh B (2001) Biological control of black scurf on potato under organic management. Crop Prot 20(2):145–150

Vanninen I, Hokkanen H (1988) Effects of pesticides on four species of entomopathogenic fungi. Ann Agri Fenn 27:345–353

Vega FE, Simpkins A, Aime MC, Posada F, Peterson SW, Rehner SA, Infante F, Castillo A, Arnold AE (2010) Fungal endophyte diversity in coffee plants from Colombia, Hawai'i, Mexico, and Puerto Rico. Fungal Ecol 3:122–138

Vey A, Hoagland RE, Butt TM (2001) Toxic metabolites of fungal biocontrol agents. In: Butt TM, Jackson C, Magan N (eds) Fungi as biocontrol agents: Progress, problems and potential. CAB International, Bristol, pp 311–346

Wang C, Skrobek A, Butt TM (2004) Investigations on the destruxin production of the entomopathogenic fungus *Metarhizium anisopliae*. J Invertebr Pathol 85:168–174

Woodrin JL, Kaya HK (1988) Steinernematid and heterorhabditid nematodes: a handbook of biology and techniques. Arkansas Agricultural Experiment Station, Fayetteville

Yang L, Xie JT, Jiang DH, Fu YP, Li GQ (2008) Antifungal substances produced by Penicillium oxalicum strain PY-1—potential antibiotics against plant pathogenic fungi. World J Microbiol Biotechnol 24:909–915

Zimmermann G (1986) The "Galleria bait method" for detection of entomopathogenic fungi in soil. Z Angew Entomol 102:213–215

# Antifungal Activity of Plant Growth Promoting Rhizobacteria Against *Fusarium oxysporum* and *Phoma* sp. of Cucurbitaceae

23

T. S. Avinash and Ravishankar V. Rai

**Abstract**

Cucurbitaceae is the largest group of summer vegetables which includes about 118 genera and 825 species and most of which are known to possess medicinal values. Several major pathogens are known to infect cucurbits and cause yield loss up to 70–80 %. Among them, *Fusarium* wilt and black rot are prevalent in and around Mysore, Karnataka, India. Plant growth promoting bacteria (PGPR) in the rhizosphere, enhances plant growth by exerting their beneficial effects through metabolites that directly or indirectly influence the plant growth. In the present study, seven bacterial isolates were isolated from the rhizosphere soils of Cucurbitaceae crops near Mysore, India. The bacterial isolates were identified based on the morphological and biochemical tests. The root rot pathogen *Fusarium oxysporum* and black rot causing *Phoma* sp. were isolated from Cucurbitaceae crops. The antifungal activity of PGPR isolates against *F. oxysporum* and *Phoma* sp. was studied by dual culture assay.

All the isolates were screened for their plant growth promoting traits such as phosphate solubilisation and ability to produce—indole acetic acid (IAA), siderophore, hydrogen cyanide (HCN) and enzymes. The PGPR strains showing more than two traits also showed good antifungal activity against *F. oxysporum* and *Phoma* sp. An increase in the growth rate accompanied by decreased disease intensity was noticed in the plants treated with selected PGPR.

**Keywords**

PGPR. · *Fusarium oxysporum* · Antagonism

## 23.1 Introduction

The increasing population worldwide necessitated the practices to ensure food security to combat against the population growth-related problems

R. V. Rai (✉) · T. S. Avinash
Department of Studies in Microbiology,
University of Mysore, Manasagangothri,
Mysore, Karnataka 570 006, India
e-mail: raivittal@gmail.com

R. N. Kharwar et al. (eds.), *Microbial Diversity and Biotechnology in Food Security*,
DOI 10.1007/978-81-322-1801-2_23, © Springer India 2014

such as malnutrition and energy deficiency. The need for food security also emphasised by other factors such as climate change, desertification, lack of water and increased resistance in pathogens. Infectious diseases have long been a major threat to food security, directly relevant to a world population that has been growing at an annual rate. The modernization of agriculture emphasises on excessive use of chemicals and change in traditional cultivation practices. But this has resulted in the deterioration of physical, chemical and biological health of the soil (Cummings 2009; Mader et al. 2011). This scenario will exert an increased demand on the available agricultural land for more production to meet the requirement. In the developing world, to control or minimize the use of chemicals in agriculture and to increase the nutrient availability for the plants, plant growth promoting rhizobacteria (PGPR) is being used from past three decades. They replenish micronutrients to soil, promote the plant growth as well as increase the crop yield and help in disease control by being associated with the plant rhizosphere (Rana et al. 2011; Glick 2012; Singh 2013). The worldwide harvest of watermelons, cucumbers, melons, squashes and pumpkins was approximately 213,000,000 metric tons with 33 % growth in 2010. Watermelon, muskmelon, cantaloupe along with other melons constitute 126,000,000 metric tons with 12 % growth in 2010 (Egel and Martyn 2007; FAOSTAT 2011).

Many microorganisms that inhabit the rhizosphere can be grouped into bacteria, fungi, actinomycetes, algae and protozoa. *Agrobacterium, Azospirillum, Azotobacter, Bacillus, Burkholderia, Pseudomonas, Serratia, Streptomyces, Rhizobia* are some of the bacterial strains used as biofertilizers as well as biocontrol agents to increase the yield of the crops. There are many direct and indirect mechanisms to enhance the growth of plants such as the production of hormones like indole acetic acid (IAA) (Zhao 2012) and gibberellic acid (GA) (Liu et al. 2013), production of cell wall-degrading enzymes like chitinase and β-1,3-glucanase (Kumar et al. 2012), production of hydrogen cyanide (HCN) (Jayaprakashvel et al. 2010), iron-chelating siderophores (Yu et al. 2011), mineralization or solubilisation of phosphorous (Gupta et al. 2011) and antagonism (Kumar et al. 2010). All these multiple plant growth-promoting traits represents the potential of PGPR. In this study, consideration is given to control cucurbit diseases and promote plant growth using PGPR.

## 23.2 Materials and Methods

### 23.2.1 Sample Collection and Microbial Isolation

Rhizosphere soil was collected from six different Cucurbitaceae fields located in Mysore and Mandya. The bacteria were isolated from soil by plating the serially diluted soil samples on nutrient agar medium (NA). The PGPR strains were confirmed by the rapid screening bioassay (Silva et al. 2003).

The fungal cultures of *Fusarium oxysporum* and *Phoma* sp. were isolated from cucurbit samples with disease symptoms from the native fields whose pathogenicity was confirmed by reinoculation. These pathogens were used for in vitro and in vivo experiments.

### 23.2.2 Standard PGPR Strains

*Pseudomonas aeruginosa* MTCC2581, *Bacillus coagulans* MTCC3543, *Bacillus subtilis* MTCC2763 and *Bacillus circulans* MTCC7194 were procured from Microbial Type Culture Collection (MTCC) Chandigarh and used as standard PGPR strains.

### 23.2.3 Confirmation of PGPR Traits

Rhizobacterial samples were characterised on the basis of biochemical tests as described by Cuppuccino and Sherman (2008). The PGPR traits were confirmed by following standard tests.

### 23.2.4 Phosphate Solubilisation

The rhizobacterial isolates were inoculated on the Pikovskaya's agar media and the plates were incubated for 4 days. A clear zone around the bacterial colonies confirms phosphate solubilisation (Laslo et al. 2012).

### 23.2.5 HCN Production

Glycine (4.4 g/l) and iron chloride hexahydrate ($FeCl_3 \cdot 6H_2O$, 0.3 mM) were amended with 10% trypticase soy agar slants. Whatman no.1 filter papers were cut into uniform strips saturated with alkaline picrate solution and placed in the sides of the agar slants. The relative quantification of HCN was done spectrophotometrically at 625 nm as described by Nagarajkumar et al. (2004).

### 23.2.6 IAA Production

The fermented broth of bacterial isolates were subjected to qualitative screening for IAA using Salkowski's reagent (2% of 0.5 M $FeCl_3$ in 35% $HClO_4$) by spectrophotometer (Rana et al. 2011).

### 23.2.7 Siderophore Production

Siderophore production was tested based on the method of Hu and Xu (2011) using chrome azurol S, a blue indicator dye. Bacterial isolates exhibiting an orange halo after 3 days of incubation at 37±2 °C were considered positive for the production of siderophores.

### 23.2.8 Production of Enzymes

PGPR isolates were tested further for their ability to produce the following enzymes.

#### 23.2.8.1 Amylase

Bacterial isolates were streaked onto starch medium and incubated for 2–3 days at 37 °C. The plates were flooded with Lugol's iodine solution and observed for zone of hydrolysis. Appearance of clear zone of hydrolysis around and under the bacterial growth indicates hydrolysis of starch (Cuppuccino and Sherman 2008).

#### 23.2.8.2 Cellulase

Production of cellulase was determined in basal medium supplemented with carboxymethyl cellulose (CMC) (10 g/l) incubated at 37 °C for 24 h (Kavamura et al. 2013).

#### 23.2.8.3 Protease

The test bacteria were inoculated onto the sterile skimmed milk agar plates and incubated for 48 h at 28 °C. The agar plates were observed for formation of clear zones around the colonies (Suresh et al. 2010).

#### 23.2.8.4 Chitinase

The rhizobacterial strains were inoculated on the colloidal chitin agar plates incubated at 37±2 °C for 4 days. The plates were stained with 0.1% Congo red solution after incubation to detect the production of chitinase indicated by halo zones around the colony (Shanmugaiah et al. 2008).

#### 23.2.8.5 β-1,3-glucanase

β-1,3-glucanase is detected using the sterilized media amended with laminarin and incubated for 2–3 days at 37±2 °C. The bacteria forms halo zone by producing laminarin-degrading β-1,3-glucanase (Kumar et al. 2012).

### 23.2.9 *In Vitro* Antagonism Assay

Antagonistic nature of all selected rhizobacterial isolates against two phytopathogens *F. oxysporum* and *Phoma* sp. isolated from the cucurbitaceous field was determined by employing a dual culture technique. Bacterial isolates were seeded at the edges of a 90 mm petri plate containing PDA and incubated for 36 h at 28±2 °C. A 9 mm diameter plug of fungus was placed on the centre of the circle. Plates were incubated at 28±2 °C for 7 days. The radii of the fungal colony towards and away from the bacterial colony were measured. The percentage of growth inhibition was calculated using the formula given by (Pastor et al. 2012).

$$\% \text{ Inhibition} = (R - r)/R \times 100$$

where $r$ is the radius of the fungal colony opposite to the bacterial colony, and $R$ is the maximum radius of the fungal colony away from the bacterial colony.

Each of the rhizobacterial isolates was maintained in triplicate and all the experiments were carried out in triplicates.

## 23.3 Results and Discussion

PGPR in the rhizosphere enhances plant growth by exerting their beneficial effects through metabolites that directly or indirectly influence the plant growth. PGPR strains were reported to antagonize against the plant pathogens by siderophore production, antibiotics and regulation of the ethylene level to protect plants, and they also managed to promote the plant growth (Beneduzi et al. 2012). Mader et al. (2011) reported after an inoculation of the PGPR an increase in micronutrient was seen in wheat, rice and black gram soils. It increases yield by 41 % and protects soil nutrients for future crops in sustainable agriculture. The main aim of the study was to screen the potential PGPR from the cucurbitaceous fields against two main cucurbit pathogens *F. oxysporum* and *Phoma* sp. of Cucurbitaceae.

Along with eight PGPR strains isolated from rhizosphere soil, another four PGPR strains from MTCC were used as standards. Among the isolates, two were Gram negative and ten were Gram positive (Table 23.1). All the isolates were rods, tolerant to salt and high temperature. Isolate MIC 1, MIC 6, MTCC 2581 and MTCC 2763 helped in phosphate solubilisation and increased the soil fertility. Hariprasad and Niranjana (2009) reported the decrease in the incidence of *F. oxysporum* wilt of tomato after the field treatment with the phosphate solubilising microorganisms. The PGPR secretes organic acids which solubilise complex phosphorous molecules to the phosphorous (Singh and Satyanarayana 2011).

Microbes produce a row of enzymes tryptophan transaminase and decarboxylase to control the L-tryptophan-dependent pathway in the plant growth promotion (Idris et al. 2007). All PGPR isolates used in the study produced IAA by using tryptophan as substrate, out of which the production was significantly high in MIC 6 isolate (212 mg/ml). Seedlings inoculated with PGPR showed a correlation between length and number of lateral and hair roots as ethylene biosynthesis in plant is mediated by bacterial IAA increasing surface area of the root (Bhattacharyya and Jha 2012). Ahmed et al. (2008) confirmed the enhanced root growth due to enhanced IAA production by *Bacillus, Pseudomonas* and *Azotobacter* in the presence of increased concentration of the tryptophan.

HCN is a volatile antibiotic produced by PGPR isolates against the microbes. In our tests only the isolates MIC 4 and MTCC 3543 produced HCN. George et al. (2013) reported that HCN produced by microbes is an important antifungal trait to control root infecting fungi. Blom et al. (2011) summarized 43 strains of PGPR with different types of volatiles increase the growth of *Arabidopsis thaliana.*

Siderophores are metal chelating molecules produced by many PGPR microorganisms which increase the iron near the rhizosphere and enhance the growth of the plant. Siderophore assay was based on competition for iron between ferric complex of an indicator dye—CAS and chelator or siderophore produced by microorganisms to remove iron from the CAS. Four isolates MIC 1, MIC 6, MTCC 2581 and MTCC 2763 produced siderophores. Hu and Xu (2011) confirmed production of siderophore by *Bacillus* sp. in various media. In *Oryza sativa,* siderophore producing bacteria showed strong biological control against phytopathogens like *F. oxysporum, Pyricularia oryzae, Sclerotium* sp. and *Pseudomonas aureofaciens* (Chaiharn et al. 2009). Yu et al. 2011 reported that *Bacillus subtilis* induced systemic resistance by siderophore production to give protection against the *Fusarium* wilt of the pepper.

PGPR microbes produce many enzymes such as chitinase, β-1,3 glucanase, protease, cellu-

**Table 23.1** Morphological, physiological and biochemical characteristics of bacterial isolates from Cucurbitaceae roots

| Characteristics | | Bacterial isolates | | | | | | | | | | |
|---|---|---|---|---|---|---|---|---|---|---|---|---|
| | | MIC 1 | MIC 2 | MIC 3 | MIC 4 | MIC 5 | MIC 6 | MIC 7 | MTCC 2581 | MTCC 2763 | MTCC 3543 | MTCC 7194 |
| Gram's test | | −ve | −ve | +ve | −ve | +ve | +ve | +ve | −ve | +ve | +ve | +ve |
| Shape | | Rod | Rod | Rod | Rod | Rod | Rod | Rod | Rod | Rod | Rod | Rod |
| Endospore | | + | + | + | + | + | − | + | + | − | + | + |
| Growth at 40 °C | | + | + | + | + | + | + | + | + | + | + | + |
| KOH test | | + | + | − | + | − | + | − | + | + | + | + |
| Urease test | | − | − | − | − | + | − | − | − | − | − | − |
| Catalase test | | + | − | + | + | − | + | − | + | + | + | − |
| Gelatin hydrolysis | | + | − | − | − | + | + | − | + | − | − | − |
| Oxidase test | | − | − | − | − | − | + | − | − | − | − | − |
| Lipase medium | | − | − | − | − | − | − | + | − | − | − | − |
| $H_2S$ production | | − | − | − | − | − | − | − | − | − | − | − |
| Indole production | | + | + | − | − | − | + | − | + | + | + | - |
| Citrate utilization | | + | − | − | − | − | + | − | + | − | + | + |
| MR test | | + | − | − | − | − | − | − | − | − | − | − |
| VP test | | + | − | + | + | − | − | + | + | + | + | + |
| 2% NaCl tolerance | | + | + | + | + | + | + | + | + | + | + | − |
| *Carbohydrate utilization* | | | | | | | | | | | | |
| Glucose | Gas | − | − | − | − | − | − | − | − | − | + | + |
| | Acid | + | + | + | + | + | + | + | − | + | + | − |
| Lactose | Gas | − | − | − | − | − | − | − | − | − | − | − |
| | Acid | − | − | − | − | − | − | + | − | − | − | − |
| Sucrose | Gas | − | − | − | − | − | − | − | − | − | + | − |
| | Acid | + | + | + | + | + | + | + | − | + | + | − |

For each bacterial isolates all the tests were repeated thrice
+ Positive reaction, − Negative reaction, *A* Acid formation, *G* Glucose formation

lase and lipase, which have the capacity to lyse or degrade the cell walls of fungi. The enzyme production capacity of PGPR strains used in the study is tabulated in Table 23.2. The strains mainly produced chitinase, β-1,3 glucanase and protease enzymes. Compant et al. (2005) reported that the biocontrol capacity of the PGPR strains is because of their capacity to produce chitinase and β-1,3-glucanase. Production of chitinase, cellulase, pectinase, amylase, β-1,3 glucanase, protease and lipase was reported in *Pseudomonas* sp. gave protection to tomato plants from fungal pathogens (Pastor et al. 2012; Solanki et al. 2012).

PGPR isolates inhibited the fungal mycelial growth, when tested in vitro by dual culture assay. Isolates MIC 3 and MIC 5 showed good antagonistic activity against *F. oxysporum* and *Phoma* sp (Fig. 23.1). In the case of *Fusarium,* not only the mycelial growth was inhibited but the colour of the mycelia also changed from white to reddish brown. This shows that they were not only inhibited by antibiosis but also by other antifungal metabolites such as siderophores, hydrogen ions and gaseous products including ethylene, HCN and ammonia (Idris et al. 2007). Laslo et al. (2012) reported 47 strains of the PGPR tested against the *Alternaria* and *Fusarium* given the positive antagonism. *Bacillus* and *Pseudomonas* were reported to be most effective against the wilt pathogens, by showing more than 50 % inhibition

**Table 23.2** Different Traits and Enzyme Production by PGPR

| Isolates | PGPR traits | | | | | | | | | |
|---|---|---|---|---|---|---|---|---|---|---|
| | IAA | HCN | Siderophore | Phosphate | Cellulase | Starch | Pectinase | Chitinase | Protease | β-1,3-glucanase |
| MIC 1 | + | + | + | + | − | − | − | + | − | − |
| MIC 2 | + | + | − | − | + | + | + | − | + | − |
| MIC 3 | + | − | − | − | + | + | + | + | + | − |
| MIC 4 | + | − | − | − | + | + | + | − | + | − |
| MIC 5 | + | + | − | − | + | + | − | + | − | − |
| MIC 6 | *ND* | + | + | + | + | − | − | − | + | + |
| MIC 7 | *ND* | − | − | − | − | − | + | − | − | − |
| MTCC 2581 | + | + | + | + | − | − | − | + | − | + |
| MTCC 2763 | + | + | + | + | − | + | + | − | − | − |
| MTCC 3543 | + | − | − | − | − | + | − | − | − | − |
| MTCC 7194 | + | + | − | − | − | + | − | − | − | − |

For each bacterial isolates all the tests were repeated thrice
+ Positive reaction, − Negative reaction, *ND* Not done

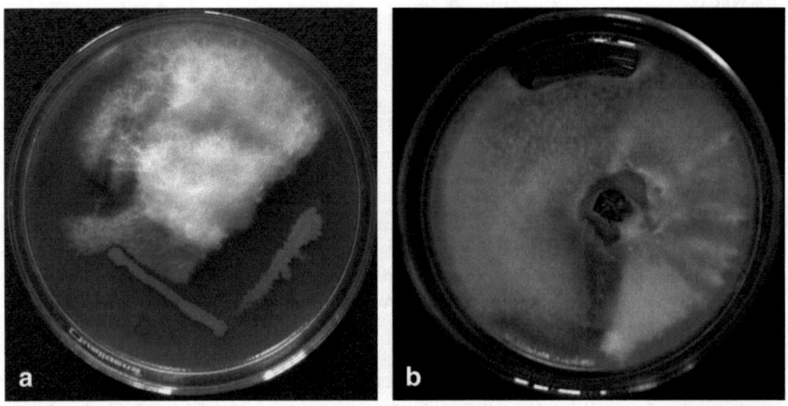


**Fig. 23.1** Antagonist activity of PGPR against **a** MIC 3 with *F. oxysporum* and **b** MIC 5 with *Phoma* sp.

to the pathogen and also by increasing the yield of the crop (Solanki et al. 2012; Gul et al. 2013).

## 23.4 Conclusion

The PGPR produced different types of enzymes, siderophores and gaseous products including ethylene, HCN as well as ammonia by inhibiting the pathogen against *F. oxysporum* and *Phoma* sp. in vitro. PGPR strains MIC 3 and MIC 5 exhibited good antagonistic activity against *F. oxysporum* and *Phoma* sp. confirmed by the dual culture assay and can be used as an effective biocontrol agent. They produce increased IAA in vitro compared to others. PGPR was very effective as well as has potential for enriching the soil fertility and enhancing the agricultural yield. Current and future progress in our understanding of PGPR diversity, colonization ability, mechanisms of action and application could facilitate their development. PGPR supplies phosphorous, minor nutrients and minerals to the barren region and improves the yield as reliable components in the management of sustainable agricultural systems.

## References


Ahmad F, Ahmad I, Khan MS (2008) Screening of free-living rhizospheric bacteria for their multiple plant growth promoting activities. Microbiol Res 163:173–181

Beneduzi A, Ambrosini A, Passaglia LMP (2012) Plant growth-promoting rhizobacteria (PGPR): Their potential as antagonists and biocontrol agents. Genet Mol Biol 35:1044–1051

Bhattacharyya JA, Jha DK (2012) Plant growth promoting rhizobacteria (PGPR): emergence in agriculture. World J Microbiol Biotechnol 28:1327–1350

Blom D, Fabbri C, Connor EC, Schiestl FP, Klauser DR, Boller T, Eberl L, Weisskopf L (2011) Production of plant growth modulating volatiles is widespread among rhizosphere bacteria and strongly depends on culture conditions. Environ Microbiol 13:3047–3058

Chaiharn M, Chunhaleuchanon S, Lumyong S (2009) Screening siderophore producing bacteria as potential biological control agent for fungal rice pathogens in Thailand. World J Microbiol Biotechnol 25:1919–1928

Compant S, Duffy B, Nowak J, Clement C, Barka AE (2005) Use of plant growth-promoting bacteria for biocontrol of plant diseases: principles, mechanisms of action, and future prospects. Appl Environ Microbiol 71:4951–4959

Cummings SP (2009) The application of plant growth promoting rhizobacteria (PGPR) in low input and organic cultivation of graminaceous crops; potential and problems. Environ Biotechnol 5:43–50

Cuppuccino JG, Sherman N (2008) Microbiology a laboratory manual. Pearson, Singapore

Egel DS, Martyn RD (2007) Fusarium wilt of watermelon and other cucurbits. Plant Health Instr. doi:10.1094/PHI-I- 2007-0122-01

FAOSTAT (2011) Food and Agriculture Organization of the United Nations—Rome, Italy. http://faostat.fao.org/site/567/DesktopDefault.aspx?PageID=567#ancor. Accessed 18 Apr 2013

George P, Gupta A, Gopal M, Thomas L, Thomas GV (2013) Multifarious beneficial traits and plant growth promoting potential of Serratia marcescens KiSII and Enterobacter sp. RNF 267 isolated from the rhizosphere of coconut palms (Cocos nucifera L.). World J Microbiol Biotechnol 29:109–117

Glick BR (2012) Plant growth-promoting bacteria: mechanisms and applications. Scientifica 2012:1–15. http://dx.doi.org/10.6064/2012/963401

Gul A, Ozaktan H, Kidoglu F, Tuzel Y (2013) Rhizobacteria promoted yield of cucumber plants grown in perlite under Fusarium wilt stress. Sci Hortic 153:22–25

Gupta M, Bisht S, Singh B, Gulati A, Tewari R (2011) Enhanced biomass and steviol glycosides in Stevia rebaudiana treated with phosphate-solubilizing bacteria and rock phosphate. Plant Growth Regul 65:449–457

Hariprasad P, Niranjana SR (2009) Isolation and characterization of the phosphate solubilising rhizobacteria to improve plant health of tomato. Plant Soil 316:13–24

Hu Q, Xu J (2011) A simple double-layered chrome azurol S agar (SD-CASA) plate assay to optimize the production of siderophores by a potential biocontrol agent *Bacillus*. Afr J Microbiol Res 5:4321–4327

Idris HA, Labuschagne N, Korsten L (2007) Screening rhizobacteria for biological control of Fusarium root and crown rot of sorghum in Ethiopia. Biol Control 40:97–106

Jayaprakashvel M, Muthezhilan R, Srinivasan R, Hussain AJ, Gobalakrishnan S, Bhagat J, Kaarthikeyan C, Muthulakshmi R (2010) Hydrogen cyanide mediated biocontrol potential of *Pseudomonas* sp. AMET1055 isolated from the rhizosphere of coastal sand dune vegetation. J Adv Biotechnol 9:38–42

Kavamura VN, Santosa SN, da Silva JL, Parmaa MM, Avilaa LA, Viscontia A, Zucchia TD, Taketania RG, Andreoteb FD, de Meloa IS (2013) Screening of Brazilian cacti rhizobacteria for plant growth promotion under drought. Microbiol Res 168:183–191

Kumar H, Bajpai VK, Dubey RC, Maheshwari DK, Kang SC (2010) Wilt disease management and enhancement of growth and yield of Cajanus cajan (L) var. Manak by bacterial combinations amended with chemical fertilizer. Crop Prot 29:591–598

Kumar P, Dubey RC, Maheshwari DK (2012) Bacillus strains isolated from rhizosphere showed plant growth promoting and antagonistic activity against phytopathogens. Microbiol Res 167: 493–499

Laslo E, Gyorgy E, Mara G, Tamas E, Abraham B, Lanyi S (2012) Screening of plant growth promoting rhizobacteria as potential microbial inoculants. Crop Prot 40:43–48

Liu Y, Sun Y, He S, Zhu Y, Ao M, Li J, Cao Y (2013) Synthesis and characterization of gibberellin–chitosan conjugate for controlled-release applications. Int J Biol Macromol. http://dx.doi.org/10.1016/j.ijbiomac.2013.03.024

Mader P, Kaiser F, Adholeya A, Singh R, Uppal HS, Sharma AK, Srivastava R, Sahai V, Aragno M, Wiemken A, Johri BN, Fried PM (2011) Inoculation of root microorganisms for sustainable wheat-rice and wheat-black gram rotations in India. Soil Biol Biochem 43:609–619

Nagarajkumar M, Bhaskaran R, Velazhahan R (2004) Involvement of secondary metabolites and extracellular lytic enzymes produced by *Pseudomonas fluoresces* in inhibition of *Rhizoctonia solani* the rice sheath blight pathogen. Microbiol Res 159:73–81

Pastor N, Carlier E, Andrés J, Rosas SB, Rovera M (2012) Characterization of rhizosphere bacteria for control of phytopathogenic fungi of tomato. J Environ Manage 95:S332–S337

Rana A, Joshi M, Prasanna R, Nain L (2011) Identification of multi-trait PGPR isolates and evaluating their potential as inoculants for wheat. Ann Microbiol 61:893–900

Shanmugaiah V, Mathivanan N, Balasubramanian N, Manoharan PT (2008) Optimization of cultural conditions for production of chitinase by *Bacillus laterosporous* MML2270 isolated from rice rhizosphere soil. Afr J Biotechnol 7(15):2562–2568

Singh JS (2013) Plant growth promoting rhizobacteria -potential microbes for sustainable agriculture.

Resonance 276–281. www.ias.ac.in/resonance/March2013/p275–281.pdf
Singh B, Satyanarayana T (2011) Microbial phytases in phosphorus acquisition and plant growth promotion. Physiol Mol Biol Plants 17:93–103
Silva HSA, da Silva RR, Mounteer AJ (2003) Development of a root colonization bioassay for rapid screening of rhizobacteria for potential biocontrol agents. J Phytopathol 151:42–46
Solanki MK, Kumar S, Pandey AK, Srivastava S, Singh RK, Kashyap PL, Srivastava AK, Arora DK (2012) Diversity and antagonistic potential of Bacillus spp. associated to the rhizosphere of tomato for the management of Rhizoctonia solani. Biocontrol Sci Technol 22:203–217. http://dx.doi.org/10.1080/09583157.2011.649713
Suresh A, Pallavi P, Srinivas P, Kumar VP, Chandra SJ, Reddy SR (2010) Plant growth promoting activities of fluorescent pseudomonads associated with some crop plants. Afr J Microbiol Res 4:1491–1494
Yu X, Ai C, Xin L, Zhou G (2011) The siderophore-producing bacterium, Bacillus subtilis CAS15, has a biocontrol effect on Fusarium wilt and promotes the growth of pepper. Eur J Soil Biol 47:138–145
Zhao Y (2012) Auxin biosynthesis: a simple two-step pathway converts tryptophan to indole-3-acetic acid in plants. Mol Plant 5:334–338

# 24 Integrated Management of Web Blight (*Rhizoctonia solani* Kühn) of French Bean

R. P. Gupta, B. C. Yadav, S. K. Singh and S. P. Singh

**Abstract**

Experiment was conducted in laboratory and field to find out the comparative performance of few biocontrol agents and fungicides and their integration for management of web blight (*Rhizoctonia solani* Kühn) in French bean. Efficacy of four bioagents, *Trichoderma viride, Trichoderma harzianum, Trichoderma virens* and *Aspergillus niger* was tested *in vitro.* All the bioagents significantly inhibited the mycelial growth as well as number of sclerotia of *R. solani.* The maximum mycelial growth inhibition (62.12%) was observed by *T. harzianum* followed by *T. viride* (57.28%), *T. virens* (51.38%) and *A. niger* (46.72%), respectively. The per cent reduction in the number of sclerotia was also maximum by *T. harzianum* (69.76%) followed by *T. viride* (65.28%), *T. virens* (60.05%) and *A. niger* (52.82%) after 168 h of incubation in lab.

**Keywords**

*Phaseolus vulgaris* · Web blight · *Rhizoctonia solani* · Bioagents · Fungicides and management

## 24.1 Introduction

French bean (*Phaseolus vulgaris* L) is an important pulse crop and plays important role in the supply of the proteins, vitamins and minerals to vegetarian population. In 2010, total world production of dry beans was 23 million metric t. The area harvested for dry beans was over 30 million hectares. World production of green beans in 2010 was 17.7 million tons, harvested in 15.1 million ha (FAO 2011). It suffers from a number of diseases. Among them, web blight caused by *Rhizoctonia solani* Kühn is gradually becoming a serious menace in cultivation of French bean. Losses from web blight vary generally in different localities from year to year. Several workers have reported varying degree of losses due to web blight in various parts of the world. Deighton (1928) reported huge losses of French bean in Sierra Lione. Weber (1939) also

S. K. Singh (✉) · R. P. Gupta · B. C. Yadav · S. P. Singh
Department of Plant Pathology, N.D. University of Agriculture & Technology, Kumarganj, Faizabad, U.P., India
e-mail: sks_nduat@rediffmail.com

S. P. Singh
e-mail: singhsamir.ppnd@gmail.com

R. N. Kharwar et al. (eds.), *Microbial Diversity and Biotechnology in Food Security,*
DOI 10.1007/978-81-322-1801-2_24, 

reported serious losses of snap bean in Florida over a period of several years. Rajnauth (1987) reported 25–100 % yield losses during rainy season in Trinidad. Sanchez and Cardenas (1988) observed that the pathogen affected the plant population from emergence to the first 30 days of crop growth. In Brazil, under favourable environmental conditions, the bean crop was destroyed within 48–72 h (Sartorato 1988). Galvez et al. (1989) reported that web blight causes mild to severe foliage blight resulting in bean seed yield losses up to 90 % in Central America. In India, Sharma and Sohi (1980) reported that yield losses of green pod at different stages of plant growth varied from 8.45 to 64.68 %.

The main symptoms of disease are appearance of greyish brown spots on leaves, leaf stalks and stem of affected plants. Smaller water-soaked spots on leaves, stem, pods and foliage of much lighter colour were also reported by various workers (Weber 1935; Zaumeyer and Thomas 1957; Rajnauth 1987; Sartorato 1988). Mathew and Gupta (1996) and Upmanyu (2002) reported that defoliation occurred in advanced stage of infection and tan brown to radish brown discolouration was also observed on infected seed, located immediately below the spots on the affected pods.

Being soil and seed borne in nature, the management of disease is difficult. Frequent use of synthetic fungicides has led to the emergence of several problems like, environmental pollution, residual effect, killing of beneficial organisms and development of fungicides-resistant strains of pathogen. Among the various control methods, integrated management with bioagents and fungicides is cheap, safe and ecofriendly. Therefore, keeping in view the importance and seriousness of the disease, present study was undertaken to find out the most effective combination of bioagents and fungicides for managing the disease in laboratory and field conditions.

## 24.2 Materials and Methods

*R. solani* causing web blight of French bean was isolated from infected leaf, stem and pods on potato dextrose agar (PDA). French bean leaves and pods showing characteristic web blight symptoms were collected from Student's Instructional Farm of Narendra Deva University of Agriculture and Technology, Kumarganj, Faizabad, U.P., India and brought to the laboratory for isolation of the pathogen. Infected leaves were washed with sterilized water. Smaller pieces having half healthy and half diseased portion were cut down. Thus, obtained pieces were surface sterilized with 0.1 % mercuric chloride solution followed by thoroughly washing thrice with sterilized water. Excess water was removed by placing on the folds of sterilized blotting paper. These pieces were subsequently transferred to petri dishes and PDA slants under aseptic conditions. Petri dishes were properly marked with glass marker and incubated at $26 \pm 1$ °C in biological oxygen demand (BOD) incubator.

The culture of *R. solani* was purified by using hyphal tip method. Water suspension of mycelium (1.0 ml) was poured aseptically over molten but still warm plain agar (2 %) petri dishes to form a very thin layer. The growth of fungus was allowed on plain agar for 24–48 h and was critically observed under the microscope. The areas having hyphal tip were marked with a glass pencil on the back of petri dish. The hyphal tips along with medium was scooped out and transferred to slants to obtain a single hyphal tip culture. After proper growth of fungus was obtained by hyphal tip, regular subculturing was done to check contamination at 15 days interval. These PDA slants having *R. solani* were kept in refrigerator at 6–8 °C for further studies.

### 24.2.1 Efficacy of Bioagents Against *R. solani In Vitro*

Four bioagents, viz. *Trichoderma viride, Trichoderma harzianum, Trichoderma virens* and *Aspergillus niger* were obtained from the Bio-agent laboratory, Department of Plant Pathology of G. B. Pant University of Agriculture and Technology, Pantnagar to test their efficacy against *R. solani* by using dual culture technique (Upadhyay and Rai 1987). About 20 ml of sterilized melted PDA was aseptically poured in sterilized petri dishes (90 mm) and allowed to solidify. Then, 5 mm disc of each antagonist and test pathogen was cut with the help of sterilized cork borer from the edge of 3-day-old culture and placed opposite to each other, 60 mm apart on PDA in four replications. Petri dishes only with *R. solani* bits served as check (control). The petri dishes were incubated at 25 ± 1 °C in incubator.

Observations were recorded on colony growth and number of sclerotia formation of *R. solani* at 72 and 168 h of incubation, respectively, in dual culture and control. Percent inhibition of mycelial growth was calculated as per the formula (Vincent 1947):

$$I = \frac{C - T}{C} \times 100$$

Where, I = Percent inhibition; C = Colony diameter in control (mm); T = Colony diameter in different treatments (mm)

### 24.2.2 *In Vivo* Effect of Fungicides, Bioagent and Their Integration

#### 24.2.2.1 Inoculum Preparation and Spraying

Potato dextrose broth medium was used for preparation of the inoculum in large quantity. Two hundred gram peeled potatoes were cut in 12 mm cubes. Two hundred gram of potatoes cubes were rinsed in water and boiled for 20 min in 500 ml water. Potato broth was filtered through cheese cloth and filled in the measuring cylinder. Twenty gram dextrose was added in it. The final volume was made up to 1,000 ml by adding distilled water. The pH was adjusted to 7.0.

Potato dextrose broth medium was collected in conical flasks and sterilized at 15 psi for 20 min in an autoclave. Three discs of pure culture of *R. solani* grown in petri dishes were cut by 5 mm cork borer and then transferred to each flasks containing potato dextrose broth after cooling. These flasks were incubated at 26 ± 1 °C for 3 days and were shaken daily by hand to achieve high growth of *R. solani*. For artificial inoculation, contains of each flask were ground by pestle and mortar after 5 days of incubation and filtered through cheese cloth. This mycelial suspension were used for artificial inoculation of the plants.

#### 24.2.2.2 Spraying of Bioagents

Spore suspension of bioagents was prepared by suspending one full plate of 7-day-old culture growth in 1 L of water and filtered through muslin cloth to prepare the spore suspension of approximately $10^7$ spores per ml concentration. A total of 10 L suspension was used to spray all the plants for treatment with the help of hand sprayer after 30 days of planting.

Among bioagents, *T. harzianum* was selected for field study based on its best performance in vitro. Efficacy of fungicides and bioagent used as seed dresser, foliar sprays alone or in combination against web blight in French bean was studied in field conditions. Field trail was conducted during *rabi* 2005–2006 and 2006–2007, crop seasons, in randomized block design with three replication using highly susceptible cultivar, contender. Three different fungicides, viz. Carbendazim, Bitertanol (Baycor) and Hexaconazole and a biocontrol agent *T. harzianum* were used. The seed treatment with Carbendazim (0.2 %) and *T. harzianum* was done 24 h prior to sowing. Seeds without any treatment served as check. The plot size was 2.7 × 2.7 $m^2$ with inter and intra-row spacing of 45 and 10 cm, respectively. Recommended agronomical practices were followed for raising good crop.

The fungicides, viz. Carbendazim (0.1 %), Bitertanol (0.1 %), Hexaconazole (0.1 %) and *T. harzianum* (0.5 %) were sprayed thrice at 15 days interval starting from first appearance of disease. Control plot was sprayed with water. Details of the treatments are as follows:

$T_1$ (seed treatment with Carbendazim at the rate 0.2 %), $T_2$ (seed treatment with *T. harzianum* at the rate 0.5 %), $T_3$ (foliar spray of Carbendazim at the rate 0.1 %), $T_4$ (foliar spray of Bitertanol at the rate 0.1 %), $T_5$ (foliar spray of Hexaconazole at the rate 0.1 %), $T_6$ (foliar spray of *T. harzianum* at the rate 0.5 %), $T_7$ ($T_1 + T_3$), $T_8$ ($T_1 + T_4$), $T_9$ ($T_1 + T_5$), $T_{10}$ ($T_1 + T_6$), $T_{11}$ ($T_2 + T_3$), $T_{12}$ ($T_2 + T_4$), $T_{13}$ ($T_2 + T_5$), $T_{14}$ ($T_2 + T_6$) and $T_{15}$ (control)

The percent disease intensity (PDI) was calculated as follow:

**Table 24.1** Effect of bioagent on per cent inhibition on mycelial growth and reduction in number of sclerotia against *R. solani in vitro*

| Bioagents | Mycelial growth inhibition (%) after 72 h | Reduction in number of sclerotia (%) after 168 h |
|---|---|---|
| *Trichoderma viride* | 57.28 (49.20) | 65.28 (53.91) |
| *T. harzianum* | 62.12 (52.00) | 69.76 (56.60) |
| *T. virens* | 51.38 (45.80) | 60.05 (50.77) |
| *Aspergillus niger* | 46.72 (43.11) | 52.82 (46.61) |
| Control | 0.00 (0.00) | 0.00 (0.00) |
| SEM ± | 0.98 | 0.89 |
| CD at 5% | 1.92 | 1.76 |

Figures in parentheses are arcsine-transformed value

$$\text{PDI} = \frac{\text{Sum of all numerical rating}}{\text{Total number of leaves examined} \times \text{maximum grade}} \times 100$$

The percent disease control (PDC) was determined by using the following formula:

$$\text{PDC} = \frac{\text{PDI in unprotected plot} \times \text{PDI in protected plot}}{\text{PDI in unprotected plot}} \times 100$$

The per cent avoidable loss was calculated as follows:

$$\text{Percent avoidable loss} = \frac{\text{Yield in protected plot} - \text{Yield in unprotected plot}}{\text{Yield in protected plot}} \times 100$$

## 24.3 Results and Discussion

### 24.3.1 *In Vitro* Efficacy of Bioagents Against *R. solani*

All the bioagents significantly inhibited the mycelial growth as well as number of sclerotia of *R. solani*. Maximum percent inhibition in mycelial growth (62.12%) was observed for *T. harzianum* followed by *T. viride* (57.28%), *T. virens* (51.38%) *and A. niger* (46.72%), respectively. The inhibition differed significantly from each other after 72 h of incubation (Table 24.1).

The percent reduction in number of sclerotia was maximum by *T. harzianum* (69.76%) followed by *T. viride* (65.28%), *T. virens* (60.05%) and *A. niger* (52.82%) after 168 h of incubation. The reduction in the number of sclerotia differs significantly from each other (Table 24.1).

The inhibitory effect of these bioagents against *R. solani* has also been reported by several workers (Hadar et al. 1979; Hazarika and Das 1998; Mathew and Gupta 1998). Kim and Roh (1987) and Sen (2000) also reported the antagonistic activity of *T. harzianum, T. viride, T. virens* and *A. niger* against *R. solani*. These are in agreement with the present findings.

### 24.3.2 Effect of Fungicides, Bioagent and Their Integration in Field

#### 24.3.2.1 Effect on Seed Germination

Maximum seed germination (91.05%) was recorded in $T_{14}$ (seed treatment with *T. harzianum* + foliar application of *T. harzianum*) followed by $T_{11}$ (88.89%), $T_{12}$ (87.96%), $T_{13}$ (87.20%), $T_2$ (86.78%), $T_{10}$ (85.03%), $T_7$ (84.26%), $T_8$ (83.32%), $T_9$ (82.55%), $T_1$ (82.23%), $T_3$ (73.46%), $T_6$ (73.43%), $T_4$ (72.69%), $T_5$ (72.22%) and $T_{15}$ (71.30%), respectively. Seed germination was significantly higher in $T_{14}$ treatment. However, it was at par with $T_{11}$, $T_{12}$, $T_{13}$ and $T_2$ but showed significantly higher seed germination in comparison to $T_{10}$. The treatment $T_{10}$, $T_7$ and $T_8$ were non-significant to each other (Table 24.2). The present finding is in agreement with the report of Hadar et al. (1979). Mathew and Gupta (1998) and Upmanyu (2002) also reported the seed treatment with *T. harzianum* + foliar application of *T. harzianum* significantly increased seed germination. It seems that higher seed germination was obtained when bioagent was used as seed treatment along with foliar application of *T. harzianum*. This might be due to suppression of *R. solani* present in soil and seed since early

**Table 24.2** Effect of bioagent and fungicides on seed germination, disease intensity and yield of French bean (pooled data for 2005–2006 and 2006–2007)

| Treatment | Dose (%) | | Seed germination (%) | Disease intensity (%) | PDC | Green pod yield (q/ha) | Increase green pod yield (%) | Dry seed yield (q/ha) | Increase dry seed yield (%) |
|---|---|---|---|---|---|---|---|---|---|
| | Seed treatment | Foliar spray | | | | | | | |
| $T_1$-Carbendazim | 0.2 | – | 82.23 (65.06) | 40.94 (39.78) | 40.18 (39.34) | 55.69 | 35.58 (36.62) | 9.83 | 32.53 (34.77) |
| $T_2$-*T. harzianum* | 0.5 | – | 86.78 (68.68) | 44.83 (42.03) | 34.50 (35.97) | 49.46 | 27.48 (31.60) | 8.83 | 24.87 (29.90) |
| $T_3$-Carbendazim | – | 0.1 | 73.46 (58.98) | 28.55 (32.30) | 58.28 (49.76) | 76.12 | 52.87 (46.64) | 12.75 | 48.00 (43.85) |
| $T_4$-Bitertanol (Baycor) | – | 0.1 | 72.69 (58.59) | 32.66 (34.85) | 52.29 (46.32) | 69.84 | 48.63 (44.21) | 11.82 | 43.89 (41.50) |
| $T_5$-Hexaconazole | – | 0.1 | 72.22 (58.20) | 36.27 (37.03) | 47.01 (43.28) | 63.26 | 43.30 (41.15) | 10.90 | 39.15 (38.73) |
| $T_6$-*T. harzianum* | – | 0.5 | 73.43 (58.98) | 48.50 (44.14) | 29.14 (32.67) | 44.16 | 18.80 (25.70) | 8.03 | 17.43 (24.66) |
| $T_7$-$T_1+T_3$ | 0.2 | 0.1 | 84.26 (66.62) | 11.72 (20.01) | 82.88 (65.56) | 106.10 | 66.19 (54.45) | 16.58 | 60.00 (50.77) |
| $T_8$-$T_1+T_4$ | 0.2 | 0.1 | 83.32 (65.89) | 15.39 (23.11) | 77.52 (61.69) | 100.68 | 64.36 (53.34) | 15.85 | 58.16 (49.69) |
| $T_9$-$T_1+T_5$ | 0.2 | 0.1 | 82.55 (65.31) | 19.27 (26.04) | 71.84 (57.94) | 94.44 | 62.02 (52.01) | 14.98 | 55.75 (48.30) |
| $T_{10}$-$T_1+T_6$ | 0.2 | 0.5 | 85.03 (67.22) | 36.77 (37.32) | 46.28 (42.86) | 62.19 | 42.32 (40.57) | 10.74 | 38.26 (38.20) |
| $T_{11}$-$T_2+T_3$ | 0.5 | 0.1 | 88.89 (70.54) | 21.61 (27.69) | 68.42 (55.81) | 87.93 | 59.25 (50.33) | 14.17 | 53.22 (46.84) |
| $T_{12}$-$T_2+T_4$ | 0.5 | 0.1 | 87.96 (69.70) | 25.16 (30.10) | 63.23 (52.66) | 82.11 | 56.31 (48.62) | 13.42 | 50.59 (45.34) |
| $T_{13}$-$T_2+T_5$ | 0.5 | 0.1 | 87.20 (69.04) | 28.39 (32.20) | 58.52 (49.90) | 76.22 | 52.92 (46.67) | 12.76 | 48.06 (43.88) |
| $T_{14}$-$T_2+T_6$ | 0.5 | 0.5 | 91.05 (72.58) | 40.55 (39.55) | 40.75 (39.68) | 56.37 | 36.37 (37.08) | 9.92 | 33.17 (35.15) |
| $T_{15}$-Control | – | – | 71.30 (57.61) | 68.44 (55.82 | 0.00 (0.00) | 35.87 | 0.00 (0.00) | 6.63 | 0.00 (0.00) |
| SEM± | – | – | 0.86 | 0.63 | 0.78 | 0.91 | 0.69 | 0.36 | 0.67 |
| CD at 5% | – | – | 1.76 | 1.30 | 1.60 | 1.85 | 1.40 | 0.74 | 1.38 |

Figures in parentheses are arcsine-transformed value. *SEM* Standard Error Mean, *CD* Critical Difference

stages of the crop. Furthermore, spraying of *T. harzianum* on standing crop, also reduced disease intensity by inhibiting *R. solani* as it was clearly indicated by *in vitro* test.

#### 24.3.2.2 Effect on Disease Intensity

The first appearance of disease was noticed 5–7 days after inoculation. Minimum disease intensity (11.72 %) was recorded in $T_7$ (seed treatment with Carbendazim + foliar application of Carbendazim) followed by $T_8$ (15.39 %), $T_9$ (19.27 %), $T_{11}$ (21.61 %), $T_{12}$ (25.16 %), $T_{13}$ (28.39 %), $T_3$ (28.55 %), $T_4$ (32.66 %), $T_5$ (36.27 %), $T_{10}$ (36.77 %), $T_{14}$ (40.55 %), $T_1$ (40.94 %), $T_2$ (44.83 %), $T_6$ (48.50 %) and $T_{15}$ (68.44 %), respectively. The PDI significantly differed in $T_7$, $T_8$, $T_9$, $T_{11}$, $T_{12}$, and $T_{13}$. However, no significant difference was observed in between $T_{13}$ and $T_3$, $T_5$ and $T_{10}$, $T_{14}$ and $T_1$ (Table 24.2).

#### 24.3.2.3 Effect on Percent Disease Control (PDC)

The PDC is calculated from the data of PDI with the formula described in materials and methods. Maximum disease control (82.88 %) was recorded in $T_7$ (seed treatment with Carbendazim + foliar application of Carbendazim) followed by $T_8$ (77.52 %), $T_9$ (71.84 %), $T_{11}$ (68.42 %), $T_{12}$ (63.23 %), $T_{13}$ (58.52 %), $T_3$ (58.28 %), $T_4$ (52.29 %), $T_5$ (47.01 %), $T_{10}$ (46.28 %), $T_{14}$ (40.75 %), $T_1$ (40.18 %), $T_2$ (34.50 %) and $T_6$ (29.14 %), respectively. The PDC differed significantly in $T_7$, $T_8$, $T_9$, $T_{11}$, $T_{12}$, and $T_{13}$. However, PDC was statistically similar in between $T_{13}$ and $T_3$, $T_5$ and $T_{10}$, $T_{14}$ and $T_1$ (Table 24.2). Hence the PDC was higher in those treatments where disease intensity was lower. Similar results were also reported by various workers (Mathew and Gupta 1996; Tiwari and Tiwari 1998; Gupta and Singh 2002).

#### 24.3.2.4 Effect on Green Pod Yield

Maximum green pod yield (106.10 q/ha) was recorded in $T_7$ (seed treatment with Carbendazim + foliar application of Carbendazim) followed by $T_8$ (100.68 q/ha) $T_9$ (94.44 q/ha), $T_{11}$ (87.93 q/ha), $T_{12}$ (82.11 q/ha), $T_{13}$ (76.22 q/ha), $T_3$ (76.12 q/ha), $T_4$ (69.84 q/ha), $T_5$ (63.26 q/ha), $T_{10}$ (62.19 q/ha), $T_{14}$ (56.37 q/ha), $T_1$ (55.69 q/ha), $T_2$ (49.46 q/ha), $T_6$ (44.16 q/ha) and $T_{15}$ (35.87 q/ha), respectively. The green pod yield significantly differs in $T_7$, $T_8$, $T_9$, $T_{11}$, $T_{12}$, and $T_{13}$. However, green pod yield was at par in between $T_{13}$ and $T_3$, $T_5$ and $T_{10}$, $T_{14}$ and $T_1$ (Table 24.2).

#### 24.3.2.5 Effect on Dry Seed Yield

Maximum dry seed yield (16.58 q/ha) was recorded in $T_7$ (seed treatment with Carbendazim + foliar application of Carbendazim) followed by $T_8$ (15.85 q/ha) $T_9$ (14.98 q/ha), $T_{11}$ (14.17 q/ha), $T_{12}$ (13.42 q/ha), $T_{13}$ (12.76 q/ha), $T_3$ (12.75 q/ha), $T_4$ (11.82 q/ha), $T_5$ (10.90 q/ha), $T_{10}$ (10.74 q/ha), $T_{14}$ (9.92 q/ha), $T_1$ (9.83 q/ha), $T_2$ (8.83 q/ha), $T_6$ (8.03 q/ha) and $T_{15}$ (6.63 q/ha), respectively. The dry seed yield significantly differs in $T_7$, $T_8$, $T_9$, $T_{11}$, $T_{12}$, and $T_{13}$. However, it was statistically similar in between $T_{13}$ and $T_3$, $T_5$ and $T_{10}$, $T_{14}$ and $T_1$ (Table 24.2).

Thakur et al. (1991), Mathew and Gupta (1996), Upmanyu et al. (2002) also reported that highest yield of green pods and dry seeds were obtained in pre-seed treatment with Carbendazim followed by foliar spray of Carbendazim. The present findings were similar to the above reports.

## 24.4 Conclusion

French bean (*P. vulgaris* L) is one of the most important pulse crops in the world. It plays a vital role in providing nutritional security to human health as it is the rich source of the proteins, nutrients, vitamins, minerals, antioxidants and other growth promoting phytochemicals. Shifting from a non-vegetarian diet to vegetarian, global recognition of importance of the pulses for human health, and their medicinal and nutritional value have contributed to be a steady upward trend in pulse production. The productivity per unit area of French bean in developing countries is much lower as compared to developed countries due to different diseases. Disease pressures in the standing crop from the seedling stage to harvest are the main constraints in total French bean production. Among diseases, web blight caused by *R. solani* Kühn is the main reason for the extensive damage of French bean, especially in tropical and subtropical countries.

The management of web blight of French bean is difficult because of soil-borne nature of the pathogen, wide host range of the pathogen, long survivability of sclerotia and unstable source of resistance due to high level of variability in pathogen. Though, web blight has been managed through use of fungicides but due to its high cost and detrimental effect on environment, soil and also by killing of non-target organism(s), its use should be minimized. Hence, losses caused by this disease could be minimized by integrated disease management by using fungicide and bio-agents alone or in combination.

Maximum disease control (82.88 %) was recorded in $T_7$ (seed treatment with Carbendazim + foliar application of Carbendazim) followed by $T_8$ ($T_1$+$T_4$) (77.52 %), $T_9$ ($T_1$+$T_5$) (71.84 %), $T_{11}$ ($T_2$+$T_3$) (68.42 %), $T_{12}$ ($T_2$+$T_4$) (63.23 %), $T_{13}$ ($T_2$+$T_5$) (58.52 %), $T_3$ (three foliar spray of Carbendazim at the rate 0.1 % at fortnight intervals) (58.28 %), $T_4$ (three foliar spray of Baycor at the rate 0.1 % at fortnight intervals) (52.29 %), $T_5$ (three foliar spray of Hexaconazole at the rate 0.1 % at fortnight intervals) (47.01 %), $T_{10}$ ($T_1$+$T_6$) (46.28 %), $T_{14}$ ($T_2$+$T_6$) (40.75 %), $T_1$ (seed treatment with Carbendazim at the rate 0.2 %) (40.18 %), $T_2$ (Seed treatment with *T. harzianum* at the rate 5 g/kg seed) (34.50 %) and $T_6$ (three foliar spray of *T. harzianum* at the rate 5 g/L water at fortnight intervals) (29.14 %), respectively, in field and seed germination and yield were also higher.

## References


FAO (2011) UN Food & Agriculture Organisation. https://en.wikipedia.org

Deighton FC (1928) Mycological section. Annual Report of Land and Forests Deptartment, Sierre Leone for the year 1927. pp 13–17

Galvez CF, Mora B, Pastor Corrales MA (1989) Web blight. In: Schwartz HF, Pastor corrales MA (eds) Bean production proble in Tropics. CIAT, Columbia, pp 195–209

Gupta RP, Singh RV (2002) Fungicidal management of web blight of mungbean. J Mycol Pt Pathol 32(1):141–142

Hadar Y, Chet I, Henis Y (1979) Biological control of *Rhizoctonia solani* damping off with wheat bran culture of *Trichoderma harzianum*. Phytopathology 69:64–68

Hazarika DK, Das KK (1998) Biological management of root rot of French bean (Phaseolus vulgaris L.) caused by *Rhizoctonia solani*. Plant Dis Res 13:101–105

Kim HK, Roh MJ (1987) Isolation, identification and evaluation of biocontrol potential of rhizosphere and antagonists to *Rhizoctonia solani*. Korean J Plant Prot 26: 87

Mathew KA, Gupta SK (1996) Studies on web blight of French bean caused by *Rhizoctonia solani* and its management. Indian J Mycol Plant Pathol 26(2):171–177

Mathew KA, Gupta SK (1998) Biological control of root rot of French bean caused by *Rhizoctonia solani*. J Mycol Plant Pathol 28:202–205

Rajnauth G (1987) Web blight—an important disease of bean and pakchoi in Trinidad. Trop Agr 64(4):356–358

Sanchez AJH, Cardenas AM (1988) Etiology and damage of root rot disease in beans *Phaseolus vulgaris* in the state of Durango. Rev Chapingo 12:43–49

Sartorato A (1988) Web blight. In Cultura do Feijoero Factores que afetam a productividade (Zimmerman MJ de O, Rocha M and Yamada T, Eds.). Piracicaba Brazil, Associacao Brasileira para Perquisa de Potassa e do Fosfato. pp 503–520

Sen B (2000) Biological control: a success story. Indian Phytopathol 53:243–249

Sharma SR, Sohi HS (1980) Assessment of losses in French bean caused by *Rhizoctonia solani* Kuhn. Indian Phytopathol 33:366–369

Thakur RS, Sugha SK, Singh BM (1991) Evaluation of systemic fungicides for control of *Rhizoctonia solani* under glasshouse condition. Indian J Agri Sci 61:230–232

Tiwari A and Tiwari A (1998) Incidence of web blight disease of Kharif legume crops in Madhya Pradesh a study. *Bhartiya Krishi Anusandhan Patrika,* 13:(3-4): 117–121

Upadhyay RS, Rai B (1987) Studies on antagonism between *Fusarium udum* Butler and root region microflora of pigeonpea. Plant Soil 101:79–93

Upmanyu S (2002) Epidemiology and management of root rot web blight of Frenchbean caused by Rhizoctonia solani Kuhn. Ph. D. Thesis, Dr. Y S Parmar University of Horticulture and Forestry, Solan, India, 134 p

Upmanyu S, Gupta SK, Shyam KR (2002) Innovative approaches for the management of root rot and web blight (*Rhizoctonia solani)* of French bean. J Mycol Plant Pathol 32(3):317–331

Vincent JM (1947) Distortion of fungal hyphae in the presence of certain inhibitors. Nature 159:850–853

Weber GF (1935) An aerial Rhizoctonia on bean. Phytopathology 25:38

Weber GF (1939) Web blight, a disease of beans caused by *Corticium microsclerotia*. Phytopathology 29:575–599

Zaumeyer WJ, Thomas HR (1957) A monographic study of bean diseases and methods for their control. USDA Tech Bull No. 868:255

# Role of Antagonistic Microbes in Management of Phytopathogenic Fungi of Some Important Crops

**25**

S. K. Dwivedi and Sangeeta

**Abstract**

Biocontrol involves harnessing disease-suppressive microorganisms to improve plant health. Biocontrol agents are successfully being used to control plant diseases and are now commercially being used to control the phytopathogens. Mostly they are used to control soil-borne diseases. Some of the most important biological control agents which are being used to control phytopathogenic fungi of some important crops are reviewed in this chapter.

**Keywords**

Biocontrol · Phytopathogens · Fungal pathogens · Antagonistic activity · Biopesticides

## 25.1 Introduction

Agriculture has increased many folds after the green revolution. Nowadays, agricultural practices totally depend on the chemical fertilizers and pesticides to enhance the productivity of various crops. These chemical fertilizers to some extent have increased the crop productivity but have posed devastating effects on the fertility of the soil and also various pests have become resistant to these pesticides. That is why sustainable agriculture is the main need of our modern society. We should focus on such agricultural practices which are more ecofriendly. One of the most appropriate approaches for getting sustainable agriculture is the use of microbial technology. Plants are found to be susceptible to different types of pathogens which are responsible for destruction of plants. These phytopathogens reduce the plant growth or may cause the death of the plant (Gupta and Sharma 2008). Many phytopathogens are most complicated and challenging ones and causes economic loss (Naik et al. 2008). Biological control of these phytopathogens can be brought about by antagonistic microorganisms (Vittal Rao and Rao 1966). The biological control agents have enormous antimicrobial potential. They are very effective in curing infectious diseases; therefore, biological control agents can

Sangeeta (✉) · S. K. Dwivedi
Department of Environmental Science, Babasaheb Bhimrao Ambedkar (A Central) University, 226025 Lucknow, Uttar Pradesh, India
email: sangibhushan7184@gmail.com

R. N. Kharwar et al. (eds.), *Microbial Diversity and Biotechnology in Food Security*,
DOI 10.1007/978-81-322-1801-2_25, 

be successfully exploited as an agricultural method for phytopathogens (Papavizas and Lumsden 1980). The present review is an effort to draw attention on biological control of phytopathogenic fungi of some important crops.

## 25.2 Biological Control

Biocontrol or biological control is another alternative source of pathogen control, involving disease-suppressive microorganisms to improve plant health. Disease suppression by biocontrol agents is the sustained manifestation of interactions among the plant, the pathogen, the biocontrol agent, the microbial community on and around the plant and the physical environment (Handelsman and Stabb 1996). Biocontrol of soil-borne diseases is particularly complex because these diseases occur in the dynamic environment at the interface of root and soil (Rovira 1965, 1969, 1991; Hawes 1991; Waisel et al. 1991)

## 25.3 History of Biological Control

The study on biological control started in 1921 with the introduction of 12 isolates of saprophytic fungi and one bacterium in nursery bed for the control of damping off of pine seed caused by *Pythium debaryanum* (Hartley 1921). Interest in biological control aroused after finding of parasitization of hyphae of *Rhizoctonia* sp. and production of antifungal compound gliotoxin by *Trichoderma* sp. (Weindling 1932). Least attention on biocontrol agents was paid during 1936–1970 for their ecofriendly approach of disease management. Later on, species of *Trichoderma* were reported to show high antagonistic potentiality against chickpea wilt pathogen (*Fusarium oxysporum* f.sp. *ciceri*) under cultural conditions (Padwick 1941). Certain inhibitory substances such as geodin, terricin, terric acid, aspergillic acid, dermadin etc. are produced by antagonists to inhibit the growth of plant pathogen (Brian et al. 1945). The fungistasis of soil have been reported by many investigators (Lingappa and Lockwood 1961; Lockwood 1977). It was found that fungistasis of soil caused inhibition of the plant pathogens. The antagonistic action of *Trichoderma* sp. against *F. oxysporum* f.sp.*ciceri* was studied and reported that the *Trichoderma* species released antibiotic substances aerially and in nutrient media that killed *F.oxysporum* f.sp. *ciceri*; causal organism of chickpea wilt (Khodzhayan 1970). *Trichoderma viride, Trichoderma harzianum, Aspergillus, F. oxysporum* and *Rhizoctonia bataticola* were also reported to be antagonistic against *Sclerotium rolfsii* causing root rot of sugar beet (Mathur and Sarbhoy 1971).Effectiveness of antibiotics varies according to the nature, quality and quantity of the antibiotics secreted by the antagonists (Dennis and Webster 1971; Skidmore and Dickinson 1976).

*Trichoderma koningii* was found to be strongly inhibitory against *F.oxysporum* and *Fusarium culmorum* but less effective to *Fusarium solani*. The inoculation of *T. koningii* at the time of sowing increased the seed germination and reduced the number of infected plants at the seedling, flowering and at maturity stages by 20.3, 29.6 and 34.7 %, respectively, both under in vitro and natural conditions (Kirik and Steblyuk 1974). The activity of *T. harzianum* and *Bacillus subtilis* were demonstrated to be antagonistic against *S. rolfsii* causing collar rot of lentil (Agarwal et al. 1975). Seed treatment of *T. harzianum* with carboxin constantly showed the best performance in minimizing wilt incidence and enhancing seed germination and grain yield under sick field conditions. The co-inoculation of *Trichoderma* and carboxin was found to be superior over any one treatment alone, due to combined effect of the treatments and variations in the mode of action of the fungicides and bioagents. The pathogen as well as soil microflora were weakened by the chemical and are therefore, better controlled by *Trichoderma* (Henis et al. 1978; Henis and Papavizas 1982). The species of *Trichoderma* were reported to be highly antagonistic towards the soil-borne pathogens (Papavizas and Lumsden 1980). The antagonistic activity of isolates of *T. viride,* three of *T. harzianum* and *Myrothecium* sp. against *Fusarium moniliformae, Rhizoctonia solani, Rhizoctonia fragariae, Phythium ultimum* and *Verticillium dahlia* were demonstrated (D'Ercole et al. 1983).

*Trichoderma* species were reported to be highly antagonistic towards different plant pathogenic fungi. Under in vitro condition, most of *Trichoderma* spp. overgrew the growth of pathogens, and the metabolite in the culture filtrate of *Trichoderma* spp. inhibited the linear growth of pathogens. *Trichoderma* spp. is now the most common fungal biological control agents that have been extensively researched and deployed throughout the world (Pande 1985). The primary mechanism of antagonism in *Trichoderma* is mycoparasitism (Chet 1987). Lytic activity is the key feature responsible for the expression of mycoparasitism against several fungal pathogens. Biological control agent can be easily combined with fungicide seed dressing to enhance the efficacy for controlling diseases (Cook 2000; Harman and Taylor 1990).

*Trichoderma* and *Gliocladium* spp. were observed to be the most potent ones in inhibiting the *Phytophthora* root and crown rot of apple (Smith et al. 1990). Under field conditions, *Gliocladium virens* and *T. harzianum* have been recognized as the most effective antagonists for biological control of several plant pathogens. *G. virens* influences the infectivity of sclerotia of *S. rolfsii* (Papavizas and Collins 1990). The relationship between suppressiveness to *Fusarium* wilt of flax (caused by *F. oxysporum* f.sp.*lini*) and indigenous microflora of the soil with special emphasis on nonpathogenic *Fusarium* was studied and reported that *F. oxysporum* is more important in the mechanism of suppression than *F. solani* and *Fusarium roseum* (Tamietti and Pramotton 1990). Nine strains of *F. oxysporum* and *F. solani* were compared for their specific ability to colonize sterilized soil and the cortex of flax roots growing in the soil precolonized by these strains, observed that the two nonpathogenic strains that were best in reducing the percentages of plants infected by the pathogen were also the best root colonizers. However, a relationship between the ability of the strains to colonize roots and to induce biological control of wilts was not confirmed (Nagao et al. 1990). The population of *T. harzianum* and *B. subtilis* in the soybean spermosphere gradually increased after sixth day of sowing, and *R. bataticola* (*Macrophomina phaseolina*) and *S.* (*Corticium*) *rolfsii* associations decreased. It is suggested that these increases and decreases are due to the antagonistic activity of *T. harzianum* and *B. subtilis* on *M. phaseolina* and *C. rolfsii* (Kumar and Khare 1990). Different *Fusarium* strains decrease vascular fusariosis of flax. The most competent strains reduced crop losses by 80%, while the least competent ones had virtually no effect. When the abilities of strains to colonize disinfected soil were estimated by enumeration on agar plates, *F. oxysporum* strains were proven superior to *F. solani* due to production of more numerous conidia and therefore greater numbers of colonies, without significantly increasing biomass (Amir 1991). Seventy-four strains of fluorescent *Pseudomonas* were tested for their ability to reduce *Fusarium* wilt of flax (caused by *F. oxysporum* f.sp. *lini*) when applied alone or in association with one preselected nonpathogenic strain of *F. oxysporum* (Fo47). Based on the effect of bacteria on disease severity on their own or in association with Fo47, four classes were established and was found that most of the strains did not modify the percentage of wilted plants, but 10.8% of them, although having no effect on their own, significantly improved control due to Fo47 (Lemanceau and Alabouvette 1991). Twelve fungi were isolated from *Sclerotinia*-infested soils of pea fields out of which seven fungi viz. *Penicillium cyclopium, Paecilomyces lilacinus, Aspergillus niger, Aspergillus fumigatus, Acremonium implicatum, Penicillium sheari* and *Trichothecium roseum* were antagonistic to *Sclerotinia sclerotium*, the causal agent of white mould of peas. *T. harzianum* destroyed sclerotia within 15 days. *T. roseum* gave 100% inhibition of sclerotial germination after 30 days of co-incubation in soil (Singh 1991). *T. harzianum* and *T. koningii* was found to be strongly antagonistic to *R. solani,* the causal agent of adzuki bean root rot disease. In order to improve biocontrol efficacy and survival in the field, benomyl resistance was induced in the structure of *T. koningii* without affecting the mycoparasitic lytic and antibiotic activity against *R. solani* (Liu 1991).

## 25.4 Microbial Biocontrol Agents

The integrated control of chickpea wilt complex by *Trichoderma* and chemical methods and reported that chickpea wilt complex was effectively controlled by *T. harzianum* alone or in combination with fungicides. The soil application of *T. harzianum* showed a decrease of 53.5–85.7% in the disease under glasshouse conditions (Mukhopadhyay et al. 1992). Seed treatment with vitavox-200 (carboxin + thiram) and ziram resulted in 29.9% disease control which increased to 63.3% with the addition of *T. harzianum* (Kaur and Mukhopadhyay 1992). The culture filtrate of *Trichoderma lignorum* and *T. viride* inhibited growth of *Fusarium oxysporum* f.sp. *psidii* by 70 and 71.4%, respectively, and *T. harzianum* by 60 and 64% of *F. solani* and *Fusarium longipes* (Dwivedi 1992). *T. harzianum* was reported to increase the root development in maize and several other crop plants both under greenhouse or field conditions (Harman 2000). Soil fungistasis plays an important role in the biological control of *Fusarium udum* in the soil. The inoculation of the antagonists might have increased the fungistasic activity of the treated soil through time due to the production of certain substances of antibiotic/toxic nature in soil (Upadhyay 1992). The combination of Captafol and *G. virens* controls the *F. oxysporum* f.sp. *ciceri* causing wilt in chickpea very successfully (Singh et al. 1993). *A. niger, A. terreus, G. virens, Penicillium citrinum, T. harzianum* and species of *Bacillus* control the soil-borne diseases (Whipps and McQuilken 1993). Antagonistic activity of *T. harzianum* was studied under in vitro condition on artificially inoculated plants of cucumber cv. straight eight and bean (*Phaseolus vulgaris*) cv. *Porrillo sintetico* (Elias et al. 1993). The efficacy of two species of *Trichoderma* (*T. harzianum* and *T. viride*) as a biological control against *Rhizoctonia solanikühn* isolated from string bean root rot in Italy was studied. The two species of *Trichoderma* were found to be more effective than other antagonistic fungi. *T. viride* produced more effective nonvolatile compounds than *T. harzianum* when grown on cellophane membrane (Roberti et al. 1993).

Out of 17 fungi isolated from the rhizosphere of *Vigna radiata* cultivars SML-32 and ML-5 during spring (March–May) and kharif (July–October) seasons, *Epicoccum purpurascens* (*E. nigrum*), *T. harzianum* and *T. viride* were antagonistic to *M. phaseolina* (Singh et al. 1993). *E. nigrum* and *T. viride* showed antibiosis, while *T. harzianum* exhibited hyperparasitism as the mechanism of pathogen inhibition. *Aspergillus flavus, A. niger* and *T. viride* amended in soil suppressed the growth of *F. oxysporum* f.sp. *ciceri* and exhibited strong fungistatic activity against germination of conidia of test pathogen (Bashar and Rai 1994). Two nonpathogenic isolates of *F. oxysporum* were examined for their ability to counteract *Fusarium solani* f.sp. *pisi* which causes foot and dry rot in peas. Both nonpathogenic *F. oxysporum* isolates reduced disease severity and prevented the plant weight losses caused by *F. solani* f.sp. *pisi* in sterilized soil. Precolonization of sterilized soil with nonpathogenic isolates increased the antagonistic effect (Oyarzun et al. 1994). The antagonistic potential of 14 *Trichoderma* isolates collected from seven beans (*Phaseolus vulgaris*) and cowpea growing areas were tested against *C. rolfsii*. Inhibition of mycelial growth and production of sclerotia, hyperparasitic activity and reduction of preemergence damping off in *Proteus vulgaris* seedlings grown under greenhouse conditions were evaluated. Isolates TN-50, TN-21 and TN-52 gave the best inhibition of mycelia growth; TN-21, TN-50 and TR-1 gave the best control of sclerotia production and isolates TN-21 and TN-16 showed the greatest hyperparasitic activity towards sclerotia of the pathogen. Overall, TN-21 isolate gave the best control of *S. rolfsii* (Silveira et al. 1994).

*Penicillium chrysogenum* and its antifungal extracts were evaluated as potential biological control agents against *Botrytis fabae* on faba beans. Antagonistic fungal (*P. chrysogenum*) isolates (AP1.S20, AP2.R16, Z1.S23 and AP2.R19), from root and soil samples of faba bean crops, inhibited mycelia growth of *B. fabae* in vitro by up to 44.9 %, producing clear inhibition zones (Jackson et al. 1994). The paired cultures and cellophane paper technique was used to select *Trichoderma* spp. with better antagonistic performances against *Colletotrichum lindemuthianum* races in vitro. *T. harzianum* T25, *T. koningii* T15, *T. pseudokoningii* T26, *T. aureoviride* T10 and *T. viride* Tr2 were used against races Ba2 and Ba10 of *C. lindemuthianum*. It was observed that five *Trichoderma* spp. caused morphological changes in the hyphae of the two phytopathogenic races indicating an antagonistic reaction. In case of Ba10, *T. viride* being the most efficient whereas *T. pseudokoningii* did not cause a significant decrease in growth of race Ba10 (Barros et al. 1995). The effects of *T. harzianum* and *T. viride* on cowpea stem and root rot fungus *M. phaseolina* in vitro was studied and reported to inhibit *M. phaseolina* growth on potato dextrose agar (PDA), but *T. viride* caused greater inhibition (Singh and Majumdar 1995). Integrated control of soil-borne plant pathogens by solar heating and antagonistic microorganisms was observed. The application of polyethylene mulching alone allowed significant control of *R. solani* on bean (*Phaseolus vulgaris*), *Pythium ultimum* on cucumber (*Cucumis sativus*) and *Fusarium oxysporum* f.sp. *basilicum* on basil (*Ocimum basilicum*). The biological control agents were effective on *P. vulgaris,* but not on cucumber or basil when sown in non-solarized soils (Minuto et al. 1995).

The potential antagonist of *F. solani* f.sp. *pisi* isolate 48(Fs48) in increasingly complex system which were selected from soil samples with varying degrees of receptivity to this pathogen. Most species tested in vitro was able to antagonize Fs48. In soils naturally infested with pea root rot pathogens, which were stored in humid conditions at 4 °C for a period > 1 year, various isolates of *Fusarium, Gliocladium* and *Penicillium* spp. were able to reduce root rot (Castejon-Munoz and Oyarzun 1995).

The effect of *Arachniotus* sp. along with soil amendments for the control of *F. oxysporum* f.sp. *ciceri* was assessed and found that the activity of *F. oxysporum* f.sp. *ciceri* was suppressed by the addition of antagonist *Arachniotus* sp. The antagonistic activity of *Arachniotus* sp. was enhanced by the addition of wheat straw and further increased by addition of glucose and urea (Ansar et al. 1996a). Out of five different organic substrates viz., wheat straw, rice husk, rice shell, berseem (*Trifolium alexandrium* L.) straw and chickpea stalk along with *Arachniotus* sp., wheat straw showed better results in the control of chickpea wilt caused by *F. oxysporum* f.sp. *ciceri* (Ansar et al. 1996b). The bioagents like *T. viride, T. harzianum* and *Trichoderma hamatum* were effective in controlling pigeon pea wilt caused by *F. oxysporum* f.sp. *udum* (Somasekhara et al. 1996). The antagonistic fungi viz., *T. harzianum* ($T_1$ and $T_2$), *T. viride* ($T_3$ and $T_4$), *T. koningii* ($T_5$) and *G. virens* ($G_1$ and $G_2$) were tested to control chickpea root rot caused by *F. solani*. The result showed that the antagonistic fungi decreased root rot by 40, 56, 69, 44, 64, 36 and 32 %, respectively. The antagonists in order efficacy were $T_3 > T_5 > T_2 > T_4 > T_1 > G_1 > G_2$ (Okhovat and Karampour 1996). The study on biological control of chickpea seed rot and damping off caused by *P. ultimum* using an antagonistic fungus *T. harzianum, T. viride* and *G. virens* was done. The results indicated that *G. virens, T. viride* + metalaxyl treatments gave significant control of *P. ultimum* (Shahriary et al. 1996).

The antagonistic activity of three *Trichoderma* species (*T. viride, T. harzianum* and *T. koningii*) against *F. udum,* the causal agent of wilt of pigeon pea, at different temperature; pH and C/N ratio and found that the antagonistic potential of *Trichoderma* spp. was not much altered by changing the environmental conditions suggesting that *Trichoderma* spp. can withstand a wide range of variations in C/N ratio without losing their antagonistic vigour. However, *Trichoderma* spp. showed maximum antagonistic potential against *F. udum* at 35 + or −2 °C and pH 6.5 (Bhatnagar 1996).

The effects of biocontrol agents viz., *T. viride, T. harzianum* and *B. subtilis* on the growth of *M. phaseolina,* the causal agent of leaf blight of *Vigna aconitifolia* (moth bean) were observed and found that of the three antagonists *T. harzianum* caused maximum growth inhibition of *M. phaseolina* (Majumdar et al. 1996). *T. viride, T. harzianum* and *G. virens* were highly effective in the management of chickpea wilt (Gurha et al. 1997). Three fungal antagonists viz., two isolates of *Trichoderma* and one isolate of *E. nigrum* mycelia strain affects the growth of chickpea plants and wilt caused by *F. oxysporum* f.sp. *ciceri*. In sterilized and unsterilized soil inoculated with *T. harzianum* and *F. oxysporum* f.sp. *ciceri,* 80 and 60% of chickpea plants remain healthy, respectively. Similarly 33.3% and 40% of plants in sterilized soil and 20 and 40% of plants in unsterilized soil were healthy in the presence of *T. viride* and *E. nigrum* respectively (Singh et al. 1997).

The antagonistic property of *Penicillium aurantiogriseum* to the soybean stem canker agent *Diaporthe phaseolorum* f.sp. *meridionalis* was tested in vitro using the dual culture technique on PDA medium and it was proved that *P. aurantiogriseum* was a good biocontrol agent, inhibiting the mycelia growth of the pathogen and performing well at 25°C and 20°C (Arias et al. 1997).

An isolate of *Gliocladium roseum* was proven to be highly antagonistic to *Botrytis cinerea.* Sporulation of *B. cinerea* on chickpea seed naturally infected or inoculated with *B. cinerea* was suppressed by seed treatment with conidial suspensions of *G. roseum* at 107 and 108 conidia/ml, respectively (Burgess et al. 1997). Isolation of *T. harzianum* from nine soil samples was carried out using *Trichoderma* specific medium and found that only isolate T1 was an effective antagonist against *F. oxysporum* f.sp.*udum* (*F. udum*), the wilt pathogen of pigeon pea, as it overgrew mycelial growth of the pathogen by up to 50%. Except for isolate T9, all the isolates overgrew mycelial growth of *R. solani* by more than 50%. Isolate T2 and T5 gave the highest degree of overgrowth of up to 89.1 and 90.2%, respectively. It was also found that isolate T5 was least antagonistic to *Fusarium* and most antagonistic to *Rhizoctonia* (Biswas 1999). *T. harzianum* has been proven as a potential bioagent of soil-borne plant pathogens (Dubey 1998, 2000, 2003) and was also found effective against *F. oxysporum* f.sp. *ciceri.* It can be used alone or in combination with carboxin as a seed treatment for the management of the disease. Chickpea wilt was effectively controlled with seed treatment by *Rhizobium, T. viride, T. harzianum* and *Azotobacter* sp. (Kolte et al. 1998). The concept of active defense mechanism in plants was studied and reported that some soil inhabiting microorganisms can reduce damage caused from disease by inducing host resistance (Hutchinson 1998). The biological control of brown collar rot induced by *R. solani* in many cropped plants has been studied among which annual pulses were studied (Sesan et al. 1998). Several *T. viride* isolates were tested against *R. solani* to control *Rhizoctonia* disease in bean (*Phaseolus vulgaris*). A number of the 16 isolates tested were proven to be highly antagonistic to this pathogen. Application of *T. viride* to soil and seed, and seven mutants originating from the isolate Td5 (TdC, TdE, TdI, TdK, TdM, TdN) as seed dressing provided protection of bean seedling. The best efficacy was shown by the mutant TdK, followed by TdN, TdM, TdJ, TdI under both greenhouse and field (Uzlina and Tulcea, Romania).

The antagonistic activity of microflora isolated from soybean root nodules was investigated toward *Bradyrhizobium japonicum* in Udaipur, Rajasthan, India. Cultures of *A. flavus, A. niger, F. solani, M. phaseolina* and *R. solani* were antagonistic to *B. japonicum* (Singh and Lodha 1998).

The efficiency of microbiological dressing of pea (*Pisum stivum* L.) against pathogenic soil-borne fungi was tested. Seed dressing prepared from individual strains of antagonistic microorganisms (*Bacillus* spp.; *Pseudomonas* spp.; *Trichoderma* spp.; *Gliocladium* spp.; and *Serratia mercescens* 9) and also mixtures of bacteria and fungus (*Bacillus* sp. 100 + *Trichoderma koningii* 41 and *Pseudomonas fluorescens* 4 + *T. koningii* 41) were tested against plant pathogenic fungi on 6-year old pea monocultures. Mixtures of *Bacillus* sp. 100 + *T. koningii* 41 and *P. fluorescens* 4 + *T. koningii* 41 gave the best protection against infection by soil fungi. *Serratia mercescens* 9 was the least effective treatment. Seed dressings prepared from individual strains of antagonistic microorganisms (*Bacillus* spp., *Pseudomonas* spp., *Trichoderma* spp. and *Gliocladium* spp.) and also mixtures of bacteria and fungi (*Bacillus* sp. S1 + *Trichoderma koningii* 51 and *P. fluorescens* 7+ *T. koningii* 51) was to be antagonistic against plant pathogenic fungi on 8-year old soybean (*Glycinemax* (L.) Merrill) monoculture. The most effective treatment was *Bacillus* sp. S1 + *T. koningii* 51 and *P. fluorescens* 7+ *T. koningii* 51, and the least effective were *Bacillus mycoids* 27 and *Gliocladium fimbriatum* 31 (Pieta et al. 1998). Field trials were done to determine protective ability of antagonistic microorganism's viz. cells of *B. subtilis* 7 and *P. fluorescens* 17 and the spores of *Trichoderma koningii* 31, *T. viride* 27, *Gliocladium catenulatum* 19 and *G. roseum* 43 against infection by soil-borne pathogenic fungi. These antagonists were used as a seed treatment of soybean, pea and common bean (*Phaseolus vulgaris*). *B. subtilis* 7, *T. koningii* 31 and *T. viride* 27 provided the best disease control (Pieta 1998).

The effects of commercial and indigenous microorganisms of *Fusarium* wilt development in chickpea were observed. Three antagonistic microorganisms, namely *B. subtilis,* nonpathogenic *F. oxysporum* isolate Fo 90105 and *T. harzianum* were applied alone or in combination to chickpea genotypes ICCV4 and PV61 with differing levels of resistance to *Fusarium* wilt (*F. oxysporum* f.sp. *ciceri*) to determine if they could effectively suppress disease development caused by the highly virulent *F. oxysporum* f.sp. *ciceri* race 5. All three antagonists effectively colonized the roots of both chickpea genotypes whether alone or in combination, and suppression was higher and more consistent in PV61 than in ICCV4 (Hervas et al. 1998). Two biological control agents, nonpathogenic Fo47 and *Pseudomonas putida* WCS 358 were evaluated for suppression of *Fusarium* wilt of flax grown in nutrient solution and for suppression of the population density and metabolic activity of the causal organism *F. oxysporum* f.sp. *lini* strain Foln 3GUS on root surfaces. It was found that suppression of *Fusarium* wilt of flax by Fo47 is related to reductions in the population density and metabolic activity of the pathogen on the root surface; WCS 358 can enhance the biological control activity of Fo47, but this enhancement depends on the population of Fo47 relative to the pathogen; and *Pseudobactin* contributes to suppression of *Fusarium* wilt by the combination of Fo47 and WCS 358 on roots (Duijff et al. 1999).

The biocontrol agents viz., six species of *Trichoderma* (*T. hamatum, T. harzianum, T. koningii, T. pseudokoningii, T. longibrachiatum* and T. viride), *G. virens, B. subtilis* and *P. fluorescens* were tested for their antagonistic activity against seed-borne *C. lindemuthianum* in *Phaseolus vulgaris. T. viride* recorded the maximum inhibition of mycelia growth followed by *P. fluorescens* and *T. harzianum* in a dual culture technique. The culture filtrate of T. viride exerted the maximum inhibition of pathogen spore germination and mycelia growth (Ravi et al. 1999). The antagonistic effect of saprophytic microorganisms (*T. koningii, T. viride, Gliocladium catenulatum, G. roseum, Bacillus, Pseudomonas and Erwinia*) was studied on pathogenic fungi (*Botrytis cinerea, Fusarium culmorum, F. oxysporum f.sp. glycines, F. solani, Phoma exigua, Pythiumdebarianum, R. solani and Sclerotinia sclerotiorum*) isolated from affected soybean roots. All isolates of *T. koningii* and *T. viride* were antagonistic to the studied pathogenic fungi. This effect, in case of genus *Gliocladium* was noted on average in 70% of the tested isolates. However, in the genus *Bacillus* and *Pseudomonas* the antagonistic effect was observed in 65 and 41% of the examined colonies, respectively (Pastucha 1999).

*T. viride* exhibited an antagonistic effect against the chickpea grey mould pathogen, *Botrytis cinerea.* In dual culture the hyphal growth of the pathogen was inhibited at the zone of contact with the hyphae of the antagonist. Microscopic examination revealed that the hyphal tips of *B. cinerea* swelled and became curved. Seed treatment with the spore mass of the antagonist at 200 and 500 mg/kg seed increased percentage of seed germination, roots and shoot length (Agarwal and Tripathi 1999). Bacterial bioagent (*B. subtilis*) and kalisena TMa commercial formulation of *A. niger,* earlier found effective against wilt but they were not effective under sick field conditions (Sen 2000). The rhizospheric interactions and the exploitation of microbial agents for the biological control of plant-pathogenic fungi was studied and reported that some soil microorganisms can reduce damage from disease by inducing resistance against pathogens (Kerry 2000). Some microorganisms, particularly those in soil, can reduce damage from diseases by promoting plant growth or by inducing host resistance against a myriad of pathogens (Cook 2000). The rhizosphere of healthy pigeon pea (cv. Bihar) plants was heavily colonized by *A. niger* and *Penicillium* sp., while *F. udum* dominated the rhizosphere of diseased plants. There was a mixed population of *Aspergillus* spp., *Penicillium* sp. and *F. udum* in non-rhizosphere soil of pigeon pea. Eleven fungi and four bacterial isolates were screened for their antagonism to *F. udum* using dual culture technique. Isolates of *T. harzianum, G. virens* and *T. viride* exhibited strong antagonism by inhibiting hyphal growth of *F. udum*. None of the bacterial isolates were antagonistic to the pathogen (Pandey and Upadhyay 2000).

The effect of five biocontrol strains of *Trichoderma* was assessed on pea growth and their antagonistic activity against large *P. ultimum* inocula. In the absence of *P. ultimum, Trichoderma* strain N47 significantly increased pea wet shoot weight by 15 %, but did not significantly affect dry weight, while T4 and N47 significantly increased root weight by 22 and 8 % respectively. TH1 and N47 resulted in significantly greater root lengths (Naseby et al. 2000). Seven different formulations of *G. virens* were evaluated for their efficacy in preventing chickpea wilt complex in the glass house and in the field and found that out of seven, six were effective in controlling the disease (Tiwari and Mukhopadhyay 2001). The volatile compounds produced by *T. virens* (T9) were very effective and showed maximum inhibition of growth of *F. solani* f.sp.*pisi* (Kumar and Dubey 2001). The fungal antagonists were evaluated against *Thanatephorus cucumeris* causing web blight urad (*Vigna mungo*) and mung bean (*Vigna radiata*).It was proven that *T harzianum* is an effective antagonist against several soil and seed-borne diseases (Dubey and Patel 2001).

*T. viride* and *T. harzianum* were reported to be best antagonists for growth inhibition of many soil and seed-borne plant pathogens. Both the *Trichoderma* spp. were highly effective as antagonist against collar rot of French bean (Dubey 2002). Two antagonistic fungi viz., *T. harzianum* (PDBCTH-10) and *T. viride* (PDBCTV) were observed against wilt (*F. oxysporum* f.sp.*ciceri*) and wet root rot (*R. solani*) of chickpea in field and reported that soil application of *T. harzianum* and *T. viride* 1 week before sowing as more effective than seed treatment in reducing wilt and wet root rot of chickpea (Prasad et al. 2002). Biological control of *F. udum* causing wilt disease of pigeon pea was studied in vitro, as well as, in vivo. *A. flavus, A. niger, Bacillus licheniformis* (strain-2042), *G. virens, Penicillium citrinum* and *T. harzianum,* were found to be the most potent ones in inhibiting the test pathogen. The maximum reduction of the wilt disease was caused by *G. virens* both in pots and in fields (Singh et al. 2002). The efficacy of some antagonists (*T. viride, T. harzianum, B. subtilis* and *P. fluorescens*) was evaluated, singly or in combination with fungicides (carboxin, carbendazin, Topsin M-70 (thiophanatemethyl) and thiram) against *F. oxysporum* f.sp. *ciceri,* (chickpea cultivars (JG 62 and Ujjain 21)). The fungal antagonists did not reduce wilt incidence in JG 62, but significantly reduce disease incidence in Ujjain 21. The lowest disease incidence (6.5 %) was recorded for *P. fluorescens, T. harzianum* + *P. fluorescens* and *B. subtilis* + *P. fluorescens*. In JG 62, wilt incidence was reduced only when *T. viride* and *T. harzianum* were applied with the fungicides (Agarwal et al. 2002). The improved performance of linseed due to inoculation of phosphate solubilizing fungi was studied. Thirty phosphate-solubilizing fungi (PSF) collected from agriculture fields were evaluated for their potential to solubilize tricalcium phosphate under in vitro condition. *Fusarium* sp., *Aspergillus awamori* and *A. niger* were reported of high potential. Their efficacy in improving the performance of the linseed in terms of growth and yield under greenhouse conditions was evaluated. In general, all three PSF improved the performance of the crop but only *A. niger* showed an all-around versatility in improving root/shoot biomass, P contents in shoots and yields (Kehri et al. 2002). Two isolates of *T. viride* 1 and 2, and one each of *T. harzianum, G. virens, Chaetomium globosum* and *B. subtilis* as biological control agent of pigeon pea wilt. The biological control agents tested in vitro by dual culture method were antagonistic to *F. udum*. Inhibition in growth of *F. udum* was highest (38.3 %) with *T. viride* 1 followed by *T. viride* 2 (35.3 %). *C. globosum* was the least effective in controlling *F. udum* (Singh et al. 2002).

The bioagents were tested against *F. oxysporum* f.sp. *ciceri* causing chickpea wilt. The efficacy of *T. harzianum, T. viride, T. hamatum, G. virens, P. fluorescens and B. subtilis* in controlling *F. oxysporum* f.sp. ciceri causing wilt in chickpea was determined *in vitro* and in field experiments. *T. harzianum* recorded the highest control of the pathogen both *in vitro* and under field conditions (Singh et al. 2003). A biological control agent colonizes the rhizosphere, the site requiring protection and leaves no toxic residues as opposed to chemicals. The first requirement of biological control is the identification and deployment of highly effective strains. The filamentous fungi, Trichoderma have attracted the attention because of their multipronged action against various plant pathogens (Harman et al. 2004). The application of *T. harzianum* mutants and carbendazina were studied to manage chickpea wilt (*F. oxysporum* f.sp. *ciceri*). They reported that *T. harzianum* decreased wilt incidence of chickpea (Poddar et al. 2004). The antagonistic activity of *T. viride, T. harzianum, T. virens (G. virens), A. niger, A. flavus* and *Penicillium citrinum* were evaluated against *F. oxysporum* f.sp.*ciceri*. All biological agents inhibited the growth of the pathogen in dual culture technique and increased chickpea germination percentage. Growth inhibition after 120 h ranged from 14.8 % with *T. virens* to 43.4 % with *P. citrinum* and *T. harzianum* (Gangwar et al. 2004). An experiment was conducted to study the effect of *T. harzianum* and *P. fluorescens*, alone and in combination on the seeds of chickpea (*Cicer arietinum*) to control wilt, *F. oxysporum* f.sp. *ciceri*. Both biofungicides suppressed wilt severity, the most effective being *T. harzianum* + *P. fluorescens* (66 %). On chickpea inoculated with the wilt, yield increased by 39 % with *P. fluorescens*, by 33 % with *T. harzianum* + *P. fluorescens*, by 44 % with *T. harzinum* (Khan et al. 2004). The most effective fungal biocontrol agents for the management of chickpea wilt were *T. viride, G. virens and T. harzianum; T. viride* was very effective and gave 77.8 % control after seed treatment (Harichand and Singh 2005). Some important microbial (mainly fungal) antagonists are listed in Table 25.1.

## 25.5 Conclusion and Future Aspects

More than 50 % of loss in crop yield occurs due to the diseases caused by microbes inhabiting the soil. Fungal pathogens are the most important factors which are responsible for the serious damages and plant losses (Brimner and Boland 2003). Chemical pesticides have been used for the control of several phytopathogens but it has created serious environmental and health problems in the world. It affects the environment and non-target organisms. To overcome these problems a new sustainable method such as use of microbial antagonists for the control of phytopathogenic fungi can be used (Spadaw and Gullino 2005). Biological control by using fungal antagonists is an important alternative to the chemical pesticides for the control of the diseases of various crops and has no hazardous effect on human health and the environment (Khan et al. 2012). Because of much recent interest in biological control of soil-borne diseases it is difficult to find commercial exploitation of fungal antagonists this may be due to the inappropriate laboratory facilities. However, in order to enhance the marketing of these antagonistic microfungi as biological control agents, feasible commercial production processes are of greatest importance (Verma et al. 2007).

**Table 25.1** List of some microbial biocontrol agents

| Crop | Pathogen/disease | Bioagent | Reference |
|---|---|---|---|
| Potato | *Ralstonia solanacearum* | *Bacillus subtilis, T. album* and *T. hamatum* | Abd-El-Khair and El-Nasr (2012) |
| Sugar beet | *Fusarium lateritium* (Nees), *Fusarium xylarioides* (Steyaert) and *Fusarium camptocearas* (Wollenw and Reinking) | *Trichoderma harzianum* | Abo-Elnaga and Heidi (2012) |
| Banana | *Mycosphaerella fijiensis* and *Cordane musae* (Black sigatoka and Cordana leaf spot) | *Trichoderma harzianum* DGA01 and *Bacillus anylolique faciens* DG14 | Alvindia (2012) |
| Tomato | *Fusarium oxysporum* f.sp. *lycopersici* | *A. niger, P. citrinum, T. harzianum* and *Penicillium* sp. | Alwathnani and Perveen (2012) |
| Wheat | *Sclerotium rolfsii* | *Talaromyces flavus* (NAIMCC-F-01948) | Chakraborty et al. (2012) |
| Pea | *Sclerotium rolfsii* | *Talaromyces flavus* (NAIMCC-F-01948) | Chakraborty et al. (2012) |
| Bean | *Sclerotium rolfsii* | *Talaromyces flavus* (NAIMCC-F-01948) | Chakraborty et al. (2012) |
| Tomato | *Fusarium oxysporum* f.sp. *lycopersici* | *Trichoderma* spp. | Devi et al. (2012) |
| Guava | *Fusarium solani* | *Aspergillus flavus, Aspergillus luchuensis, Penicilliun citrinum P. chrysogenum* and *Trichoderma viride* | Dwivedi and Dwivedi (2012) |
| Bean | Root rot | *Trichoderma harzianum* and *T. viride* | Hameed et al. (2012) |
| Apple | *Botrytis cinerea* (grey mould), *Colletotrichum acutatum* (bitter rot) or *Penicillium expansum* (blue mould) | *Aureobasidium pullulans,* strains L1 and L8 | Mari et al. (2012) |
| Brinjal | *Fusarium solani* | *Glomus fasciculatum* and Salicylic acid (0.5 and 1.0 mM) | Ojha et al. (2012) |
| Potato | *Fusarium solani* | *Trichoderma* (*T. brevicompactum* (T1), *T. longibrachiatum* (T5) and *T. asperellum* (T2)) | Ommati and Zaker (2012) |
| Capsicum | *Fusarium pallidoroseum* (*Fusarium* wilt) | *Trichoderma harzianum* and *Trichoderma viride* | Pandey and Namdeo (2012) |
| Kinnow | *Penicillium digitatum* (Green mould rot) | *Trichoderma hamatum* (HP-20), *T. harzianum* (TG-1), *T. viride*- 1, *T. viride*- 2, *Gliocladium deliquescens, G. virens* and *Chaetomium globosum* (HP-29) | Sharma et al. (2012) |
| Tomato | *Ralstonia solanacearum* (bacterial wilt) | *Trichoderma harzianum* and *T. viride* | Sharma et al. (2012) |
| Groundnut | *Aspergillus niger, Apergillus flavus, Sclerotium rolfsii, Thievaliopsis basicola, Rhizoctonia solani* and *Pythium phanidermatum* | *Trichoderma harzianum* (Th3) | Sharma et al. (2012) |
| Safflower | *Fusarium oxysporum* f.sp. *carthami* | *Trichoderma harzianum* and *T. viride* | Shinde and Hallale (2012) |
| Sugar beet | *Sclerotium rolfsii* | *Pseudomonas fluorescens* strain pf1 and *T. asperellum* strain TTH1 or *Bacillus subtilis* strain EPCO-16 | Thilagavathi et al. (2012) |
| Maize | *Fusarium moniliforme* (Post flowering stalk rot) | *Trichoderma viride* | Thori et al. (2012) |
| Bean | Root rot/*Fusarium solani, Rhizoctonia solani, Sclerotium rolfsii* and *Macrophomina phaseolina* | Essential oils and *Trichoderma harzianum* | Abdel-Kader et al. (2011) |

**Table 25.1** (continued)

| Crop | Pathogen/disease | Bioagent | Reference |
|---|---|---|---|
| Chickpea | *Fusarium oxysporum* f.sp. *ciceri*; *F. solani* and *Rhizoctonia solani* | *Trichoderma virens* and *T. viride* | Ansari et al. (2011) |
| Okra | Powdery Mildew/*Erysiphe pumcichoracerum* | *Epicoccum nigrum*, *E. minitans*, *Epicoccum* sp., *Trichoderma harzianum*, *T. viride* and *Bacillus pumilus* with or without Penconazole | Derbalah et al. (2011) |
| Squash | Powdery Mildew | *Epicoccum nigrum*, *E. minitans*, *Epicoccum* sp., *Trichoderma viride* and *Bacillus pumilus* | Elkot and Derbalah (2011) |
| Tomato | Bacterial spot and early blight/*Xanthomonas euvesicatoria* and *Alternaria solani* | *Trichoderma* spp. | Fontenelle et al. (2011) |
| Brinjal | *Verticillium dahliae* | Nonpathogenic *Fusarium oxysporum* F2 strain | Gizi et al. (2011) |
| Safflower | Wilt/*Fusarium Oxysporum* f.sp. *carthami* | *Trichoderma harzianum*, *Bacillus subtilis* and *Pseudomonas fluorescens* | Govindappa et al. (2011) |
| – | *Fusarium oxysporum*, *F. solani*, *Macrophomina phaseolina*, *Aspergillus japonicum* var *aculeatus* and *Cladosporium cladosporioides* | *Penicillium italicum* and *P. simplissimum* | Khokhar et al. (2011) |
| Cotton | *Verticillium* wilt | *Glomus etunicatum* | Kobra et al. (2011) |
| – | *Rhizoctonia solani*, *Macrophomina* sp. *Sclerotium rolfsii* and *Pythium aphanidermatum* | *Trichoderma* spp., *T. koningii* and *T. viride* | Kumar et al. (2011) |
| Sesame | *Alternaria* blight, White rust, Powdery mildew and *Sclerotinia* rot | Garlic bulb extract, *Trichoderma harzianum* and *Pseudomonas fluorescens* | Meena et al. (2011) |
| Capsicum | Damping off/*Pythium aphanidermatum* | *Bacillus licheniformis*, *T. harzianum* | Mehetre and Kale (2011) |
| Brinjal | Wilt/*Fusarium solani* f.sp. *melongenae* | *Gliocladium roseum*, *Paecilomyces varioti*, *Trichothecium roseum* and *Aspergillus flavus*, *Trichoderma viride* (isolate-I&II); *Trichoderma harzianum* (isolate-I &II); *Pseudomonas fluorescens* | Najar et al. (2011) |
| Bean | Anthracnose/*Colletotrichum lindemuthianum* | *Trichoderma viride*, *T. harzianum*, *T. hamatum* and *Gliocladium virens* | Padder and Sharma (2011) |
| Chickpea | *Sclerotinia sclerotiorum* | *Trichoderma virens*, *T.harzianum* and*Pseudomonas fluorescens* | Pandey et al. (2011) |
| Tomato | *Fusarium* wilt/*F. oxysporum* f.sp. *lycopersici* | Nonpathogenic *Fusarium moniliforme* (Fu3, Fu7 and Fu24), *Fusarium oxysporum* (Fu2, Fu4), *F. solani* (Fu 25) and *F. merismoides* (Fu1) | Patil et al. (2011) |
| Pigeon pea | Wilt/*Fusarium udum* | *Trichoderma viride*, *Pseudomonas fluorescens* and *P. aeruginosa* | Ram and Pandey (2011) |
| Tomato | *Fusarium oxysporum* f.sp. *lycopersici race* 1CU1 (Fol) | *Penicillium* sp.EU0013 | Sartaj et al. (2011) |
| Bean | Root rot/*Macrophomina phaseolina* | *Burkholderia* sp. strain TNAU-1 | Satya et al. (2011) |
| Guava | *Fusarium oxysporum* f.sp. *psidii* | *Trichoderma* spp., *Aspergillus niger*, *Penicillium* spp. | Srivastava et al. (2011) |
| Sesame | Wilt and root rot/*Fusarium oxysporum* f.sp. *sesame* (Zap.) cast and *Macrophomina phaseolina* (Moubl) Ashby | *Glomus* spp. and *Lums* spp. *Trichoderma viride* or *Bacillus subtilis* | Ziedan et al. (2011) |

**Table 25.1** (continued)

| Crop | Pathogen/disease | Bioagent | Reference |
|---|---|---|---|
| Wheat | Spot blotch/*Bipolaris sorokiniana* | *Chaetomium globosum* (Cg1, Cg5, Cg6, Cg7 and Cg8) | Agarwal et al. (2010) |
| Bean | Root rot/*Fusarium solani* | Arbuscular Mycorrhizal Fungi (AMF) *Glomus mosseae, G. intraradices, G. clarum, Gigaspora gigantea* and *Gigaspora margarita* | Al-Askar and Rashad (2010) |
| Apple | *Penicillium expansum* | *Trichoderma virens* | Bordbar et al. (2010) |
| Safflower | Root rot/*Macrophomina phaseolina* | *Trichoderma harzianum, Pseudomonas fluorescens, Bacillus subtilis* | Govindappa et al. (2010) |
| Guava | *Fusarium* wilt/*Fusarium oxysporum* f.sp. *psidii* and *F. solani* | *Tricoderma* spp. (*T. virens and T. viride*) | Gupta et al. (2010) |
| Banana | *Fusarium* wilt/*F. oxysporum* f.sp. *cubense* | Nonpathogenic, endophytic *Fusarium oxysporum* strains, *Trichoderma harzianum* Eco-T®, silicon and mulching using macadamia husks | Kidane and Laing (2010) |
| Apple | Stem brown canker or *Botryosphaeria* canker | Arbuscular mycorrhizal fungi | Krishna et al. (2010) |
| Tomato | *Verticillium albo-atrum* | *Talaromyces flavus* | Naraghi et al. (2010) |
| Capsicum | Fruit rot/*Colletotrichum gloeosporioides* | *T. viride* and *Pseudomonas fluorescens* | Ngullie et al. (2010) |
| Bengal gram | *M. phaseolina* | *Trichoderma* spp. | Pan and Jash (2010) |
| Potato | Black scurf/*Rhizoctonia solani* | *Trichoderma viride, Bacillus cereus* strain B4 and *B. subtilis* strain B5 | Somani and Arora (2010) |
| Tomato | *Fusarium* wilt/*Fusarium oxysporum* f.sp. *lycopersici* | *Fluorescent Pseudomonas, Trichoderma harzianum* and *Glomus intraradices* | Srivastava et al. (2010) |
| Paddy | *Rhizoctonia solani* | *Aspergillus* spp.*(A. ochraceous, A. niger, A. fumigatous, A. flavus and A. terreus) T. virens* | Vibha (2010) |
| Brinjal | Damping off/*Rhizoctonia solani* | *Bacillus subtilis* CA32 and *Trichoderma harzianum* RU01 | Abeysinghe (2009) |
| Capsicum | Damping off/*Rhizoctonia solani* | *Bacillus subtilis* CA32 and *Trichoderma harzianum* RU01 | Abeysinghe (2009) |
| Bean | *Sclerotinia sclerotiorum* (Lib.) de Bary | Arbuscular Mycorrhizal Fungi (AMF) *Glomus mosseae* (Gm), *Glomus fasciculatum* (Gf) and *Rhizobium leguminosarum* biovar *phaseoli* (Rlp) | Aysan and Demir (2009) |
| Brinjal | Wilt/eggplant (*Solanum melongena*) | *Trichoderma harzianum* and *T. viride* | Chakraborty et al. (2009) |
| Potato | *Rhizoctonia solani* | *Trichoderma harzianum* Rifai MUCL 29707 | Gallou et al. (2009) |
| Banana | Post-harvest crown rot/*Lasiodiplodia theobromae* and *Colletotrichum musae* | *T. viride, T. harzianum* and *T.koningii* | Ganesan et al. (2009) |
| Guava | *Fusarium* oxysporum f.sp. *psidii* | *Aspergillus niger, Trichoderma* sp., *Penicillium citrinum* | Gupta et al. (2009) |
| Brinjal | Collar rot/*Sclerotium rolfsii* Sacc. | *Trichoderma viride* (T5) | Jadon (2009) |
| Sesame | Stem rot/*Sclerotinia sclerotiorum* | *T. harzianum*-8, *T. atroviride* PTCC5220 and *T. longibrachiatum* PTCC5140 | Matroudi et al. (2009) |
| Sesame | *Sclerotinia sclerotiorum*/*Sclerotinia* rot | *T. harzianum* isolate GR, FYM of *T. harzianum* isolate SI-02, garlic bulb aqueous extract | Meena et al. (2009) |

**Table 25.1** (continued)

| Crop | Pathogen/disease | Bioagent | Reference |
|---|---|---|---|
| Guava | *F. oxysporum* f sp. *psidii* and *F. solani* | *Aspergillus niger*,*Trichoderma* sp. (*T. virens, T. harzianum* and *T. viride*) and *Penicillium citrinum* | Misra and Gupta (2009) |
| Tomato | *Fusarium oxysporum* f.sp. *radicis-lycopersici* | *Trichoderma koningiopsis* (Th003) | Moreno et al. (2009) |
| Cumin | *Fusarium oxysporum* f.sp. *cumini* | *Trichoderma harzianum, T. viride, Pseudomonas fluorescens* and *Bacillus subtilis* | Nitin and Gangopad-hyay (2009) |
| Tomato | *Fusarium* wilt/*Fusarium oxysporum* | *Trichoderma, T. harzianum, T. viride* and *T. hamatum* | Ojha and Chatterjee (2009) |
| Capsicum | Wilt/*Fusarium solani* | *Trichoderna viride*-16, *T. harzia-num*-10, *Pseudomonas fluorescens* (Pf-1) | Rani et al. (2009) |
| Tomato | *Ralstonia* wilt/*Ralstonia solanacearum* | *Glomua mosseae, Trichoderma viride* and *Azotobacter + Phosphobactrin* | Sharma and Kumar (2009) |
| Maize | *Fusarium verticillioides* | *T. harzianum* (strain 1:/IMI 380934; strain 2: IMI 380935; strain 3:IMI 380938) | Sobowale et al. (2009) |
| Brinjal | Root-knot-wilt/*Meloidogyne incognita* and *Fusarium solani* | *Aspergillus niger, Trichoderma har-zianum, Paecilomyces lilacinus* and *Pseudomonas fluorescens* | Vipin et al. (2009) |
| Brinjal | *Sclerotinia sclerotiorum* | *Trichoderma harzianum* and *Amylo liquefaciens* | Abdullah et al. (2008) |
| Banana | *Lasiodiplodia theobromae*/Banana crown rot | *Clonostachys byssicola, Curvularia pallescens, Penicillium oxalicum, Trichoderma harzianum* | Alvindia and Natsuaki (2008) |
| Tomato | Pre- and postemergence rots and foliar disease | *Trichoderma viride, T. harzianum, Pseudomonas fluorescens* and *Aspergillus niger* | Hooda et al. (2008) |
| Tomato | *Fusarium* crown and root-rot/*Fusarium oxysporum* f.sp. *radicis-lycopersici* | *Fusarium equiseti* GF191 and biode-gradable pots (BPs) | Horinouchi et al. (2008) |
| Chickpea | Charcoal rot/*Macrophomina phaseolina* | *Trichoderma harzianum* 25–92 | Jyotsna et al. (2008) |
| Sesame | *Alternaria* blight | *Allium sativum* and *Trichoderma harzianum* | Meena et al. (2008) |
| Linseed | *F. oxysporum* f.sp. *lini* | *Trichoderma viride, T. harzianum*; *T. viride* + , *T. harzianum* + thiram; *T. viride* + thiram; *T. harzianum* + thiram, thiram; *T. harzianum* + *T. viride* and FYM (farmyard manure) | Singh et al. (2008) |
| Linseed | *Sclerotium rolfsii* (*Corticium rolfsii*)/Collar rot | *Trichoderma* spp. | Bhosale et al. (2007) |
| Bean | Dry root rot/*Macrophomina phaseolina* | AM fungi viz., *Glomus fasciculatum, G. mosseae, G. aggregatum, G. claroideum, G. macrocarpum,* and *G. multicaule* | Chandra et al. (2007) |
| Brinjal | Damping off/*Fusarium equiseti* | *Trichoderma harzianum* and *Aspergil-lus niger* | Datar (2007) |
| Tomato | *Fusarium oxysporum* f.sp. *lycopersici* | *Fluorescent pseudomonas*, nonpatho-genic *Fusarium* strain and *Tricho-derma harzianum* T-22 | Fahri Yigit and Dikili-tas (2007) |
| Pigeonpea | Wilt/*Fusarium udum* | *Trichoderma harzianum* | Gade et al. (2007) |

**Table 25.1** (continued)

| Crop | Pathogen/disease | Bioagent | Reference |
|---|---|---|---|
| Potato | Black scurf/*Rhizoctonia solani,* Late blight/*Phytophthora infestans,* Bacterial brown spot/*Fusarium* and *Phoma* sp. | *Trichoderma* spp. | Gogoi et al. (2007) |
| Flax | *Fusarium* wilt | Endophytic symbiont *Acremonium* | Grunewaldt-Stocker et al. (2007) |
| Tomato | *Fusarium* crown and root rot/*Fusarium oxysporum* f.sp. *radicis-lycopersici* | Plant growth-promoting fungi, *Fusarium equiseti* GF191 | Horinouchi et al. (2007) |
| Chickpea | Wilt/*F. oxysporum* f.sp. *ciceri* | Nonpathogenic *Fusarium oxysporum,* Fo52, Fo47 and Fo47b10 | Kaur and Singh (2007) |
| Chickpea | *Fusarium oxysporum* f.sp.*ciceri* | Nonpathogenic *Fusarium oxysporum* and *Pseudomonas fluorescent* | Kaur et al. (2007) |
| Brinjal | *Rhizoctonia solani* | *Aspergillus niger* | Khan and Anwer (2007) |
| Tomato | Damping off/*Pythium aphanidermatum* | *Trichoderma* spp. | Kumar and Hooda (2007) |
| Tomato | *Fusarium* wilt/*F. oxysporum* f.sp. *lycopersici* | *Fusarium oxysporum* strain CS-20 | Panina et al. (2007) |
| Tomato | *Fusarium oxysporum* f.sp. *lycopersici* | *Trichoderma* spp. | Singh (2007) |
| Banana | *Lasiodiplodia theobromae*/crown rot | *T. pseudokoningii, T. viride* S7, *T. viride* RT1, *T. viride* S17, and *Pseudomonas aeruginosa* and *P. viridiflava* | Thangavelu et al. (2007) |
| Banana | Wilt/*Fusarium oxysporum* f.sp. *cubense* | Nonpathogenic *Fusarium oxysporum* isolate upm31p1 and UPM39B3 | Ting et al. (2007) |
| Fiber flax | *Fusarium avenaceum* (Fr.) Sacc., *Fusarium oxysporum* (Schlecht.) Snyd. et Hans., *Alternaria alternata* (Fr.) Keissl., *Botrytis cinerea* Pers., *Rhizoctonia solani* Kühn., *Mucor, Aspergillus niger* Tiegh., *Penicillium* spp. | *Trichoderma lignorum* T 13–82 | Pristchepa et al. (2006) |

## References


Abdel-Kader M, El-Mougy N, Lashin S (2011) Essential oils and *Trichoderma harzianum* as integrated control measure against *Faba Bean* Root Rot pathogens. J Plant Prot Res 51(3):306–313

Abd-El-Khair H, El-Nasr HIS (2012) Applications of *Bacillus subtilis* and *Trichoderma* spp. for controlling the potato brown rot in field. Arch Phytopathol Plant Prot 45(1):1–15

Abdullah MT, Ali NY, Suleman P (2008) Biological control of *Sclerotinia sclerotiorum* (Lib.) de Bary with *Trichoderma harzianum* and *Bacillus amyloliquefaciens*. Crop Prot 27:1354–1359

Abeysinghe S (2009) Effect of combined use of *Bacillus subtilis* CA32 and *Trichoderma harzianum* RU01 on biological control of *Rhizoctonia solani* on *Solanum melongena* and *Capsicum annum*. Plant Pathol J 8(1):9–16

Abo-Elnaga, Heidi IG (2012) Biological control of damping off and root rot of wheat and sugar beet with *Trichoderma harzianum*. Plant Pathol J 11:25–31

Agarwal A, Tripathi HS (1999) Biological and chemical control of *Botrytis* gray mould of chickpea. J Mycol Plant Pathol 29(1):52–56

Agarwal SC, Khare MN, Agarwal PS (1975) Biological control of *Sclerotium rolfsii* causing collar rot of lentil. Indian Phytopathol 30:176–178

Agarwal SC, Sharma S, Prasad KVV (2002) Efficacy of biological compents in wilt management of chickpea. Indian J Pulses Res 15(2):177–178

Agarwal R, Renu, Srinivas P, Malathi VG (2010) Assessment of genetic diversity in *Chaetomium globosum*, a potential biocontrol agent by amplified fragment length polymorphism. Indian Phytopathol 63(1):2-5

Al-Askar AA, Rashad YM (2010) Arbuscular Mycorrhizal Fungi: A biocontrol agent common bean *Fusarium* root rot disease. Plant Pathol J 9(1):31–38

Alvindia DG (2012) Inhibitory influence of biocontrol agents, plant oils and an inorganic salt on *Mycosphae-*

*rella fijiensis* and *Cordanemusae*, the causal pathogen of black sigatoka and leaf spot of banana. Afr J Microbiol Res 6(19):4179–4184

Alvindia DG, Natsuaki KT (2008) Evaluation of fungal epiphytes isolated from banana fruit surfaces for biocontrol of banana crown rot disease. Crop Prot 27:1200–1207

Alwathnani HA, Perveen K (2012) Biological control of *Fusarium* wilt of tomato by antagonist fungi and cyanobacteria. Afr J Biotechnol 11(5):1100–1105

Amir H (1991) Correlations between the ability of different *Fusarium* strains to decrease vascular fusariosis of flax, their respiratory activity and saprophytic development in disinfected soil. Can J Microbiol 37(12):889–896

Ansar M, Akhtar CM, Ahmad R, Alam SS (1996a) Effect of *Arachniotus* sp. along with soil amendments in the control of chickpea wilt caused by *Fusarium oxysporum* f. sp. *ciceri*. Pak J Phytopathol 8(1):37–39

Ansar M, Akhtar CM, Ahmad R, Alam SS (1996b) Effect of *Arachniotus* sp. along with soil amendments in the control of chickpea wilt caused by *Fusarium oxysporum* f. sp. *ciceri*. Pak J Phytopathol 8(1):40–42

Ansari M, Vaid A, Razdan V (2011) Evaluation of different measures to control wilt causing pathogens in chickpea. J Plant Prot Res 51:55–59

Arias SMS, Machado Jda C, Arias ERA (1997) Evaluation of the antagonistic property of *Penicillium*. Fitopatologia. Brasileira 22(3):437–440

Aysan E, Demir S (2009) Using Arbuscular Mycorrhizal Fungi and *Rhizobium leguminosarum* Biovar *phaseoli* against *Sclerotinia sclerotiorum* (Lib.) de Bary in the Common Bean (*Phaseolus vulgaris* L.). Plant Pathol J 8:74–78

Barros ST, Oliveira NT, Bastos STG (1995) *Trichoderma* spp. in the biological contol of *Colletotrichum lindemuthianum* (Sacc.et Magn.) Scribb, agent of the bean (*Phaseolus vulgaris* L.) anthracnose. Bol Micol 10(1/2):5–11

Bashar MA, Rai B (1994) Antagonistic potential of root region microflora of chickpea against*Fusarium oxysporum* f. sp. *ciceri*. Bangladesh J Bot 23:13–19

Bhatnagar H (1996) Influence of environmental conditions on antagonistic activity of *Trichoderma* spp. against *Fusarium udum*. Indian. J Mycol Plant Pathol 26(1):58–63

Bhosale PM, Verma KP, Zape AS, Nichal SS (2007) Antagonism by different isolates of *Trichoderma* spp. against the collar rot pathogen of linseed. J Plant Dis Sci 2(2):247

Biswas KK, (1999) Screening of isolates of *Trichoderma harzianum* Rifai for their relative biocontrol efficacy against *Fusarium oxysporum* f.sp. *udum* and *Rhizoctonia solani Kühn*. Ann Plant Prot Sci 7(2):125–130

Bordbar F, Etebarian H, Sahebani N, Rohani H (2010) Control of postharvest decay of apple fruit with *Trichoderma virens* isolates and induction of Defence Responses. J Plant Prot Res 50:146–152

Brian PW, Hemming HG, McGowan JC (1945) Origin of toxicity in warhen health soil. Nature 155:637

Brimner TA, Boland GJ (2003) A review of the non-target effects of fungi used to biologically control plant diseases. Agric Ecosyst Environ 100(1):3–16

Burgess DR, Bretag T, Keane PJ (1997) Biocontrol of seed-borne *Botrytis cinerea* in chickpea with *Gliocladium roseum*. Plant Pathol 46(3):298–305

Castejon-Munoz M, Oyarzun PJ (1995) Soil receptivity to *Fusarium solani* f.sp. *pisi* and biological control of root rot of pea. Eur J Plant Pathol 101(1):35–49

Chakraborty MR, Chatterjee NC, Quimio TH (2009) Integrated management of *Fusarial* wilt of eggplant (*Solanum melongena*) with soil solarization. Micol Apl Int 21(1):25–36

Chakraborty BN, Chakraborty U, Sunar K, Dey PL (2012) Evaluation of plant growth promoting and antifungal activities of *Talaromyces flavus* (NAIMCC-F-01948) against *Sclerotium rolfsii*. Indian Phytopathol 65(3):258–263

Chandra S, Khare V, Kehri H K (2007) Evaluation of arbuscular mycorrhizal fungi against *Macrophomina phaseolina* causing dry root-rot of urd and mung bean. Indian Phytopathol 60(1):42–47

Chet I, (1987) *Trichoderma*-application, mode of action and potential as a biological control agent of soil-borne plant pathogenic fungi. In: Chet I (ed) Innovative approaches to plant diseases control. Wiley, New York, pp 137–160

Cook RJ (2000) Advances in plant health management in the 20th century. Annu Rev Phytopathol 38:95–116

D'Ercole N, Sportelli M, Nipoti P (1983) *In vitro* trials of antagonistic fungi. Plant Pathol IntrestInformatore Fitopatol 33:55–58

Datar VV (2007) Investigations on pre- and post- emergence mortality in eggplant (*Solanum melongena* L.). Indian Phytopathol 60(2):156–161

Dennis C, Webster J (1971) Antagonistic properties of species group of *Trichoderma* I. Production of volatile antibiotics. Trans Brit Mycol Soc 57:41–48

Derbalah AS, El Kot GA, Hamza AM (2011) Control of powdery mildew in okra using cultural filtrates of certain bioagents alone and mixed with penconazole. Arch Phytopathol Plant Prot 44(20):2012–2023

Devi SS, Sreenivasulu Y, Saritha S, Kumar MR, Kumar KP, Sudhakar P (2012) Molecular diversity of native *Trichoderma* isolates against *Fuasrium oxysporum* f.sp.*lycopersici* (Sacc.). A causal agent of *Fusarium* wilt in tomato (*Lycopersicon esculentum* Mill.). Arch Phytopathol Plant Prot 45(6):686–698

Dubey SC (1998) Evaluation of fungal Antagonists of *Thanatephorus cucumeris* web blight of horse gram. J Mycol Plant Pathol 28 PP:15–17

Dubey SC (2000) Biological management of web blight of groundnut (*Rhizoctonia solani*). J Mycol Plant Pathol 30 PP:89–90

Dubey SC (2002) Bio-agent based integrated management of collar root of French bean. Indian Phytopathol 55:230–231

Dubey SC (2003) Integrated management of web blight of urd/mung bean by bio-seed treatment. Indian Phytopathol 56:34–38

Dubey SC, Patel B (2001) Evaluation of fungal antagonist against *Thanatephorus cucumeris* causing web blight urd and mung bean. Indian Phytopathol 54:206–209

Duijff BJ, Recorbet G, Bakker PAHM, Loper JE, Lemanceau P (1999) Microbial antagonism at the root level is involved in the suppression of *Fusarium* wilt by the combination of nonpathogenic *Fusarium oxysporum* Fo47 and *Pseudomonas putida* WCS 358. Phytopathology 89(11):1073–1079

Dwivedi SK (1992) Effect of culture filtrate of soil microbes on pathogen inciting wilt disease of guava (*Psidium guajaver*) under *in vitro* conditions. Natl Acad Sci Lett 15:33–35

Dwivedi SK, Dwivedi N (2012) *In vitro* bioefficacy of some selected fungal antagonists against guava wilt pathogen. IOSR J Eng 2(5):1217–1223

Elias R, Arcos O, Arbelaez G (1993) A study of antagonism of some species of *Trichoderma harzianum* from Colombian soils in control of *Fusarium oxysporum* and *Rhizoctonia solani*. Agron-colomb 10(1):52–61

Elkot G, Derbalah A (2011) Use of cultural filtrates of certain microbial isolates for powdery mildew control in squash. J Plant Prot Res 51(3):252–260

Fontenelle ADB, Guzzo SD, Lucon CMM, Harakava R (2011) Growth promotion and induction of resistance in tomato plant against *Xanthomonas euvesicatoria* and *Alternaria solani* by *Trichoderma* spp. Crop Prot 30:1492–1500

Gade RM, Zote KK, Mayee CD (2007) Integrated management of pigeon pea wilt using fungicide and bioagent. Indian Phytopathol 60(1):24–30

Gallou A, Cranenbrouck S, Decterck S (2009) *Trichoderma harzianum* elicits defence response genes in roots of potato plantlets challenged by *Rhizoctonia solani*. Eur J Plant Pathol 124(2):219–230

Ganesan S, Swaminathan U, Arjunan M (2009) Biocontrol with *Trichoderma* species for the management of post-harvest crown rot of banana. Phytopathol Mediterr 48:214–225

Gangwar RK, Prajapati RK, Kumar K (2004) Evaluation of fungal antagonists against *Fusarium oxysporum* f.sp. *ciceri*. Ann Plant Prot Sci 12(2):444–445

Gizi D, Stringlis IA, Tjamos SE, Paplomatas EJ (2011) Seedling vaccination by stem injecting a conidial suspension of F2, a non-pathogenic *Fusarium oxysporum* strain, suppresses *Verticillium* wilt of eggplant. Biol Control 58:387–392

Gogoi R, Saikia M, Helim R, Ullah Z (2007) Management of potato diseases using *Trichoderma viride* formulations. J Mycol Plant Pathol 37:227–230

Govindappa M, Lokesh S, Rai VR, Naik VR, Raju SG (2010) Induction of systemic resistance and management of safflower *Macrophomina phaseolina* root-rot disease by biocontrol agents. Arch Phytopathol Plant Prot 43(1):26–40

Govindappa M, Rai VR, Lokesh S (2011) *In vitro* and *In vivo* responses of different treating agents against wilt disease of safflower. J Cereals Oilseeds 2(1):16–25

Grunewaldt-Stocker G, Riediger N, Dietrich C (2007) Suitability of GFP-transformed isolates of the fungal root endophyte *Acremonium strictum* W. Gams for studies on induced *Fusarium*-wilt resistance in flax. Plant Root 1:46–56

Gupta VK, Sharma RC (2008) Integrated management of soil-borne diseases. Advances in soil-borne plant diseases. New India Publishing Agency, New Delhi, pp 415–427

Gupta VK, Misra AK, Pandey BK, Ram RA, Mishra SP, Chauhan UK (2009) Eco-friendly management of *Fusarium* sp. causing wilt disease of guava (*Psidium guajava* L.) by bio-dynamic antagonists. J Eco-Friendly Agric 4:77–79

Gupta V, Misra A, Gupta A, Pandey B, Gaur R (2010) RAPD-PCR of *Trichoderma* isolates and *in vitro* Antagonism against *Fuasrium* wilt pathogens of *Psidium guajaval*. J Plant Prot Res 50:256–262

Gurha SN, Singh RA, Vishwadhar (1997) Prospects of wilt Management in chickpea. Paper presented in International conference on integrated plant disease management for sustainable agriculture, New Delhi, 10–15 Nov

Hameed ER, El-Gamal NG, El-shami AR (2012) Efficacy of formulation and storage on rice straw waste on the activation of bioagents against root-rot diseases of bean plants. Arch Phytopathol Plant Prot 45(1):22–32

Handelsman Jo, Stabb Eric V (1996) Biocontrol of soil-borne plant pathogens. Plant Cell 8:1855–1869

Harichand, Singh S (2005) Control of chickpea wilt (*Fusarium oxysporum* f.sp. *ciceri*) using bioagents and plant extracts. Indian J Agric Sci 75:115–116

Harman GE (2000) Myths and dogmas of biocontrol: changes in perceptions derived from research on *Trichoderma harzianum* T22. Plant Dis 84:377–393

Harman GE, Taylor AG (1990) Development of an effective biological seed treatment system. In: Horn D (ed) Biological control of soil-borne plant pathogens. CAB International, Wallingford, pp 415–426

Harman GE, Howell CR, Viterbo A, Chet I, Lorito M (2004) *Trichoderma* species-opportunistic, avirulent plant symbionts. Nat Rev 2:43–56

Hartley C (1921) Damping off in forest nurseries, U.S. Dep Agric Bull 934:1–99

Hawes MC (1991) Living plant cells released from the root cap: a regulator of microbial populations in the rhizosphere? In: Keister DL, Cregan PB (eds) The rhizosphere and plant growth. Kluwer Academic Publishers, Boston, pp 51–59

Henis Y, Papavizas GC (1982) Factors affecting susceptibility of *Sclerotium rolfsii* sclerotia to *Trichoderma harzianum* in natural soil. Phytopathol 72:1010

Henis Y, Elad Y, Chet I, Hadar Y, Hadar E (1978) Integrated control of *Rhizoctonia solani* damping off radish: effect of successive planting, PCNB and *Trichoderma harzianum* on pathogen and disease. Phytopathol 68:900–907

Hervas A, Landa B, Datnoff LE, Jimenez, Diaz RM (1998) Effects of commercial and indigenous microorganisms of *Fusarium* wilt development in chickpea. Biol Control 13(3):166–176

Hooda KS, Bhatt JC, Joshi D, Sushil SN, Gupta HS (2008) Biocontrol agents vis-à-vis fungicides in managing various diseases of tomato (*Lycopersicon esculentum* Mill.) in hills of Uttarakhand. Indian Phytopathol 61(3):331–336

Horinouchi H, Muslim A, Suzuki T, Hyakumachi M (2007) *Fusarium equiseti* GF191 as an effective biocontrol agent against *Fusarium* crown and root rot of tomato in rock wool systems. Crop Prot 26:1514–1523

Horinouchi H, Katsuyamaa N, Taguchi Y, Hyakumachi M (2008) Control of *Fusarium* crown and root-rot of tomato in a soil system by combination of a plant growth-promoting fungus, *Fusarium equiseti*, and biodegradable pots. Crop Prot 27:859–864

Hutchinson SW (1998) Current concepts of active defense in plants. Annu Rev Phytopathol 36:59–90

Jackson AJ, Walters DR, Marshall G (1994) Evaluation of *Penicillium chrysogenum* and its antifungal extracts as potential biological control agents against *Botrytis fabae* on faba beans. Mycol Res 98(10):1117–1126

Jadon KS (2009) Eco-friendly management of brinjal collar rot caused by *Sclerotium rolfsii* Sacc. Indian Phytopathol 62(3):345–347

Jyotsna, Srivastava A, Singh RP, Srivastava AK, Saxena AK, Arora DK (2008) Growth promotion and charcoal rot management in chickpea *Trichoderma harzianum*. J Plant Prot Res 48(1):81–92

Kaur NP, Mukhopadhyay AN (1992) Integrated control of chickpea wilt complex by *Trichoderma* and chemical methods in India. Trop Pest Manag 38(4):372–375

Kaur R, Singh RS (2007) Study of induced systemic resistance in *Cicer arietinum* L. due to non-pathogenic *Fusarium oxysporum* using a modified split root technique. J Phytopathol 155(11–12):694–698

Kaur R, Kaur J, Singh RS, Alabouvette C (2007) Biological control of *Fusarium oxysporum* f. sp. *ciceri* by non-pathogenic *Fusarium* and *Fluorescent Pseudomonas*. Int J Bot 3:114–117

Kehri SS, Chandra S, Kehri HK (2002) Improved performance of linseed due to inoculation of phosphate solubilizing fungi. Front microb biotechnol Plant Pathol 2002:193–198

Kerry BR (2000) Rhizosphere interactions and the exploitation of microbial agents for the biological control of plant-pathogenic fungi. Annu Rev Phytopathol 38:423–441

Khan MR, Anwer MA (2007) Molecular and biochemical characterization of soil isolates of *Aspergillus niger* aggregate and an assessment of their antagonism against *Rhizoctonia solani*. Phytopathol Mediterr 46(3):304–315

Khan MR, Khan SM, Mohiddin FA (2004) Biological control of *Fusarium*wilt of chickpea through seed treatment with the commercial formulation of *Trichoderma harzianum* and/or *Pseudomonas fluorescens*. Phytopathol Mediterr 43(1):20–25

Khan S, Guo L, Maimaiti Y, Mijit M, Qiu D (2012) Entomopathogenic fungi as microbial biocontrol agent. Mol Plant Breed 3(7):63–79

Khodzhayan EA (1970) A study on the antagonistic action of *Trichoderma* sp. on *Fusarium oxysporum Trud 11: 2* Asheh-Rest. Arm-ssr 1:263–268

Khokhar I, Mukhtar I, Mushtaq S (2011) Antifungal effect of *Penicillium* metabolites against some fungi. Arch Phytopathol Plant Prot 44(14):1347–1351

Kidane EG, Laing MD (2010) Integrated Control of *Fusarium* Wilt of Banana (*Musa* Spp.). Acta Hort (ISHS) 879:315–321. http://www.actahort.org/books/879/879_32.htm

Kirik NN, Steblyuk NI (1974) Assessment of the effectiveness of *Trichoderma konongii* oud. out. In biological control of Fusariosis of Pea. Mycol-i-Fitopathologia 8:108–112

Kobra N, Jalil K, Youbert G (2011) *Arbuscular mycorrhizal* fungi and biological control of *Verticillium* wilted cotton plants. Arch Phytopathol Plant Prot 44(10):933–942

Kolte SO, Thakre KG, Gupta M, Lokhande VV (1998) Biocontrol of *Fusarium* wilt of chickpea (*Cicer arietinum*) under wilt sick field condition. Paper submitted, ISOPP at National Symposium on management of soil and soil-borne diseases. 9–10 Feb 1998, p 22

Krishna H, Das B, Attri BL, Grover M, Ahmed N (2010) Suppression of *Botryosphaeria* canker of apple by arbuscular mycorrhizal fungi. Crop Prot 29:1049–1054

Kumar D, Dubey SC (2001) Management of collar rot of Pea by the integration of biological and chemical methods. Indian Phytopathol 57:62–66

Kumar R, Hooda I (2007) Evaluation of antagonistic properties of *Trichoderma* species against *Pythium aphanidermatum* causing damping off of tomato. J Mycol Plant Pathol 37:240–243

Kumar SM, Khare MN (1990) Studies on the antagonistic relationship of soybean spermosphere microflora with *Rhizoctonia bataticola* and *Sclerotium rolfsii*. J Biol Control 4(1):72–74

Kumar K, Amaresan N, Bhagat S, Madhuri K, Udhayaraj P, Chandra R (2011) Genetic and physiological relatedness of antagonistic *Trichoderma* isolates against soil borne plant pathogenic fungi. Arch Phytopathol Plant Prot 44(14):1399–1409

Lemanceau P, Alabouvette C (1991) Biological control of *Fusarium* diseases by the association of *fluorescent Pseudomonas* and non-pathogenic *Fusarium*. Bull SROP 14(8):45–50

Lingappa BT, Lockwood JL (1961) The nature of wide spread soil fungistasis. J Gen Microbiol 26:473–475

Liu SD (1991) Biological control of adzuki bean root rot disease caused by *Rhizoctonia solani*. Plant Prot Bull Taipei 33(1):63–71

Lockwood JL (1977) Fungistasis in soils. Boil Rev 52:1–43

Majumdar VL, Jat JR, Gour HN (1996) Effects of biocontrol agents on the growth of *Macrophomina phaseolina* the incitant of blight of moth bean. Indian J Mycol Plant Pathol 26(2):202–203

Mari M, Martini C, Spadoni A, Rouissi W, Bertolini P (2012) Biocontrol of apple postharvest decay by

*Aureobasidium pullulans*. Postharvest Biol Technol 73:56–62

Mathur SB, Sarbhoy AK (1971) Biological control of *Sclerotium* root rot of sugarbeet. Indian Phytopathol 31:365–367

Matroudi S, Zamani MR, Motallebi M (2009) Antagonistic effects of three species of *Trichoderma* sp. on *Sclerotinia sclerotiorum*, the causal agent of canola stem rot. Egypt J Biol 11:37–44

Meena PD, Chattopadhyay C, Meena RL (2008) Eco-friendly management of *Alternaria* Blight in *Brassica juncea*. Indian Phytopathol 61(1):65–69

Meena PD, Kumar A, Chattopadhyay C, Sharma P (2009) Eco-friendly management of *Sclerotinia* rot in Indian mustard (*Brassica juncea*). 16th Australian Research Assembly on Brassicas. Ballarat Victoria

Meena PD, Awasthi RP, Godika S, Gupta JC, Kumar A, Sandhu PS, Sharma P, Rai PK, Singh YP, Rathi AS, Prasad R, Rai D, Kolte SJ (2011) Eco-friendly approaches managing major diseases of Indian mustard. World Appl Sci J 12(8):1192–1195

Mehetre ST, Kale SP (2011) Comparative efficacy of thermophilic bacterium, *Bacillus licheniformis* (NR1005) and antagonistic fungi, *Trichoderma harzianum* to control *Pythium aphanidermatum* induced damping-off in chilli (*Capsicum annum* L.). Arch Phytopathol Plant Prot 44(11):1068–1074

Minuto A, Migheli Q, Garibaldi A (1995) Integrated control of soil-borne plant pathogens by solar heating and antagonistic microorganisms. Acta Hortic 382:138–143

Misra AK, Gupta VK (2009) *Trichoderma*: Biology, biodiversity and biotechnology. J Eco-Friendly Agric 4:99–117

Moreno CA, Castillo F, Gonza´lez A, Bernal D, Jaimes Y, Chaparro M, Gonza´ lezC, Rodriguez F, Restrepo S, Cotes AM (2009) Biological and molecular characterization of the response of tomato plants treated with *Trichoderma koningiopsis*. Physiol Mol Plant Pathol 74:111–120

Mukhopadhyay AN, Shrestha SM, Mukherjee PK (1992) Biological seed treatment for control of soil-borne plant pathogens.*F.A.O.* Plant Prot Bull 40(1–2):21–30

Nagao H, Couteaudier Y, Alabouvette C (1990) Colonization of sterilized soil and flax roots by strains of *Fusarium oxysporum* and *Fusarium solani*. Symbiosis Reho 9(1–3):343–354

Naik MK, Devika Rani GS, Prasad RD, Patil MB, Sen B (2008) An over-view of soil borne plant pathogens. Advances in soil borne plant diseases. New India Publishing Agency, New Delhi, pp 1–31

Najar AG, Anwar A, Masoodi L, Khar MS (2011) Evaluation of native biocontrol agents against *Fusarium solani* f.sp. *melongenae* causing Wilt disease of brinjal in Kashmir. J Phytol 3(6):31–34

Naraghi L, Heydari A, Rezaee S, Razavi M, Jahanifar H, Khaledi E (2010) Biological control of Tomato *Verticillium* wilt disease of *Talaromyces flavus*. J Plant Prot Res 50:360–365

Narendra S, Ram P, Udit N, Mathuria OP (2008) Efficacy of bio-agents in management of linseed wilt. Ann Plant Prot Sci 16(1):245–246

Naseby DC, Pascual JA, Lynch JM (2000) Effect of biocontrol strains of *Trichoderma* on plant growth, *Pythium ultimum* populations, soil microbial communities and soil enzyme activities. J Appl Microbiol 88(1):161–169

Ngullie M, Daiho L, Upadhyay D (2010) Biological management of fruit rot in the world's hottest chilli (*Capsicum chinense* Jacq.). J Plant Prot Res 50(3):269–273

Nitin C, Gangopadhyay S (2009) Integration of organic amendments and bioagents in suppressing cumin wilt caused by *Fusarium oxysporum* f. sp. *cumini*. Indian Phytopathol 62(2):209–216

Ojha S, Chatterjee NC (2009) Mycoparasitism of *Trichoderma* spp. in biocontrol of *Fusarial* wilt of tomato. Arch Phytopathol Plant Prot 44(8):771–782

Ojha S, Chakraborty M, Chatterjee NC (2012) Influence of salicylic acid and *Glomus fasciculatum* on *Fusarial* wilt of tomato and brinjal. Arch Phytopathol Plant Prot 45(13):1599–1609

Okhovat M, Karampour F (1996) Effect of some isolates of antagonistic fungi on the control of chickpea black root rot caused by *Fusarium solani* under greenhouse conditions. Iran J Agric Sci 27(2):37–43

Ommati F, Zaker M (2012) In vitro and greenhouse evaluations of *Trichoderma* isolates for biological control of potato wilt disease (*Fusarium solani*). Arch Phytopathol Plant Prot 45(14):1715–1723

Oyarzun PJ, Postma J, Luttikholt AJG, Hoogland AE (1994) Biological control of foot and root rot in pea caused by *Fusarium solani* with non-pathogenic *Fusarium oxysporum* isolates. Can J Bot 72(6):843–852

Padder BA, Sharma PN (2011) *In vitro* and *in vivo* antagonism of biocontrol agents against *Colletotrichum lindemuthianum* causing bean anthracnose. Arch Phytopathol Plant Prot 44(10):961–969

Padwick GC (1941) Report of the Imperial Mycologist. Sci Rep Agric Res Inst, New Delhi 1939–1940:94–101

Pan S, Jash S (2010) Variability in induction of defense response in Bengal gram against *Trichoderma* species. Indian Phytopathol 63(1)

Pande A (1985) Biocontrol characteristics of some moulds. Biovigyanam 11:14–18

Pandey A, Namdeo A (2012) Antagonistic behavior of two different species of *Trichoderma* Against *Fusarium pallidoroseum*, Pathogenic to *Capsicum frutescens*. Mycology, plant pathlogy, microbiology & nanoscience. Proc. 99th Indian Science Congress, Part II: Abstracts of Oral/Poster Presentations 48

Pandey KK, Upadhyay JP (2000) Microbial population from rhizosphere and non-rhizosphere soil of pigeonpea: Screening for resident antagonist and mode of mycoparasitism. J Mycol Plant Pathol 30(1):7–10

Pandey P, Kumar R, Mishra P (2011) Integrated approach for the management of *Sclerotinia sclerotiorum* (Lib.) de Bary, causing stem rot of chickpea. Indian Phytopathol 64(1):37–40

Panina Y, Fravel DR, Baker CJ, Shcherbakova LA (2007) Biocontrol and plant pathogenic *Fusarium oxysporum*-induced changes in phenolic compounds in tomato leaves and roots. J Phytopathol 155(7-8):475–481

Papavizas GC, Collins DJ (1990) Influence of *Gliocladium virens* on infectivity of sclerotia of *Sclerotium rolfsii*. Phytopathology 80:827–830

Papavizas GC, Lumsden RD (1980) Biological control of soil borne fungal propagules. Annu Rev Phytopathol 18:389–413

Pastucha A (1999) Protective effect of antagonistic microorganisms in limiting soybean (*Glycine max* (L.) Merrill) root diseases. Ann Univ Mariae Curie Skodowska Sectio EEE Hortic 7:119–135

Patil S, Sriram S, Savitha MJ, Arulmani N (2011) Induced systemic resistance in tomato by nonpathogenic *Fusarium* species for the management of *Fusarium* wilt. Arch Phytopathol Plant Prot 44(16):1621–1634

Pieta D (1998) The role of antagonistic microorganisms in the control of Plant diseases. *Roczniki* Akademii Rolniczejw Poznaniu, Ogrodnictwo 1998(27):221–227

Pieta D, Pastucha A, Patkowska E (1998a) The efficiency of microbiological dressing of soybean seed (*Glycine max* (L.) Merrill) against root and stem base diseases. Ann Agric Sci, Ser E, Plant Prot 27(1/2):103–109

Pieta D, Patkowska E, Pastucha A (1998b) The efficiency of microbiological dressing of Pea (*Pisum sativum* L.) against pathogenic soil-borne fungi. Ann Agric Sci, Ser E, Plant Prot 27(1/2):81–89

Poddar RK, Singh DV, Dubey SC (2004) Integrated application of *Trichoderma harzianum* mutants and carbendazina to manage chickpea wilt (*Fusarium oxysporum* f.sp. *ciceri*). Indian J Agric Sci 74:346–348

Prasad RD, Rangeshwran R, Anuroop CP, Rashni HJ (2002) Biological control of wilt and root rot of chickpea under field conditions. Ann Plant Prot Sci 10(1):72–75

Pristchepa L, Voitka D, Kasperovich E, Stepanova N (2006) Influence of Trichodermin-BL on the decrease of fiber flax infection by diseases and the improvement of ITS production quality. J Plant Prot Res 46(1):97–102

Ram H, Pandey RN (2011) Efficacy of bio-control agents and fungicides in the management of wilt of pigeon pea. Indian Phytopathol 64(3):267–271

Rani GSD, Naik MK, Patil MB, Prasad PS (2009) Biological control of *Fusarium solani* causing wilt of chilli. Indian Phytopathol 62(2):190–198

Ravi S, Doraiswamy S, Valluvaparidasan V, Jeyalakshmi C (1999) Effect of biocontrol agents on seed-borne *Colletotrichum* in French bean. Plant Dis Res 14(2):146–151

Roberti R, Ghisellini L, Pisi A, Flori P, Filippini G (1993) Efficacy of two species of *Trichoderma* as a biological control against *Rhizoctonia solani Kühn* isolated from string bean root rot in Italy. Adv Hortic Sci 7(1):19–25

Rovira AD (1965) Interactions between plant roots and soil microorganisms. Annu Rev Microbiol 19:241–266

Rovira AD (1969) Plant root exudates. Bot Rev 35:35–57

Rovira AD (1991) Rhizosphere research-85 years of progress and frustration. In Keister DL, Cregan PB (eds) The rhizosphere and plant growth. Kluwer Academic Publishers, Boston, pp 3–13

Sartaj AY, Sakamoto K, Inubushi K (2011) Effect of *Penicillium* sp. EU0013 inoculation on tomato growth and *Fusarium* wilt. Hortic Res 65:69–73

Satya V, Vijayasamundeeswari A, Paranidharan V, Velazhahan R (2011) *Burkholderia* sp. strain TNAU-1 for Biological control of Root Rot in Mung Bean (*Vigna radiate* L.) caused by *Macrophomina phaseolina*. J Plant Prot Res 51(3):273–278

Sen B (2000) Biological control: a success story. Ind Phytopathol 53:243–249

Sesan TE, Baicu T, Gogoaso C (1998) Biological control of brown collar rot (*Rhizoctonia solani Kühn*) in annual pulses. Rom Agric Res 9/10:49–53

Shahriary D, Okhovat M, Rouhani H (1996) Biological control of *Pythium ultimum*, Trow, the causal agent of chickpea seed-root and damping-off disease by antagonistic fungi. Iran J Agric Sci 27(3):1-7(Pc), 8(En)

Sharma JP, Kumar S (2009) Management of *Ralstonia* wilt of tomato through microbes, plant extract and combination of cake and chemicals. Indian Phytopathol 62(4):417–423

Sharma JP, Kumar S, Das B (2012a) Soil application of *Trichoderma harzianum* and *T. viride* on biochemical constituents in bacterial wilt resistant and susceptible cultivars of tomato. Indian Phytopathol 65 (3)

Sharma RN, Maharshii, Gaur RB (2012b) Biocontrol of post-harvest green mould rot (*Penicillium digitatum*) of kinnow fruits using microbial antagonists. Indian Phytopathol 65(3)

Shinde AB, Hallale BV (2012) Biochemical management of *Fusarium Wilt*. DAV. Int J Sci 1(2):87–90

Silveira NSS, Michereffi SJ, Menezes M, campos-Takaki GM (1994) Potential of *Trichoderma* spp. isolates on the control of *Sclerotium rolfsii* on beans. Summa-Phytopathol 20(1):22–25

Singh D (1991) Biocontrol of *Sclerotinia sclerotiorum* (Lib.) de Bary by *Trichoderma harzianum*. Trop Pest Manag 37(4):374–378

Singh D (2007) Role of fungicides and biocontrol agents in the management of fusarial wilt of chilli. J Mycol Plant Pathol 37:361–362

Singh J, Lodha PC (1998) Antagonistic activity of microflora isolated from soybean root nodules towards *Bradyrhizobium japonicum*. Plant Dis Res 13(2):195–197

Singh M, Majumdar VL (1995) Antagonistic activity of *Trichoderma* sp. to *Macrophomina phaseolina* (Tassi) Goid *in vitro*. Environ Ecol 13(2):481–482

Singh RN, Upadhyay JP, Ojha KL (1993a) Management of chickpea wilt by fungicides and*Gliocladium*. J Appl Biol 46–51

Singh RS, Singh N, Kang MS (1993b) Rhizosphere mycoflora of mung bean and their interaction with *Macrophomina phaseolina*. Plant Dis Res 8(1):25–28

Singh B, Mane SS, Pal M (1997a) Management of chickpea wilt paper presented in international conference of integrated plant disease management for sustainable Agriculture, New Delhi, 10–15 Nov

Singh RS, Singh D, Singh HV (1997b) Effect of fungal antagonists on the growth of chickpea plants and wilt caused by *Fusarium oxysporum* f.sp.*ciceri*. Plant Dis Res 12(2):103–107

Singh R, Singh BK, Upadhyay RS, Rai B, Lee Y (2002a) Biological control of *Fusarium* wilt disease of Pigeonpea. Plant Pathol J 18(5):279–283

Singh SK, Singh RH, Dutta S (2002b) Integrated management of Pigeonpea wilt by biotic agents and biopesticides. Ann Plant Prot Sci 10(2):323–326

Singh BK, Srivastava M, Narain U (2003) Evaluation of bioagents against *Fusarium oxysporum* f.sp.*ciceri* causing chickpea wilt

Singh N, Palat R, Narain U, Mathuria OP (2008) Efficacy of bio-agents in management of linseed wilt. Ann Plant Prot Sci 16(1):245–246

Skidmore AM, Dickinson CH (1976) Colony interactions and hyphal interference between *Septoria nodorum* and Phyloplane fungi. Trans Br Mycol Soc 66:57–64

Smith VL, Wilcox WF, Harman GE (1990) Potential for biological control of Phytopthora root and crown rot of apple by *Trichoderma* and *Gliocladium* spp. Phytopathology 80:880–885

Sobowale AA, Odebode AC, Cardwell KF, Bandyopadhyay R (2009) Suppression of Growth of *Fusarium Verticillioides* Niren. Using strains of *Trichoderma harzianum* from Maize (*Zea Mays*) plant parts and its rhizosphere. J Plant Prot Res 49(4):452–459

Somani AK, Arora RK (2010) Field efficacy of *Trichoderma viride*, *Bacillus subtilis* and *Bacillus cereus* in consortium for control of *Rhizoctonia solani* causing black scurf disease of potato. Indian Phytopathol 63(1)

Somasekhara YM, Anilkumar TB, Siddarad AH (1996) Biocontrol of pigeonpea wilt *Fusarium udum*. Mysore J Agric 30:159–163

Spadaw D, Gullino ML (2005) Improving the efficacy of biocontrol agents against soil-borne pathogens. Crop Prot 24:601–613

Srivastava R, Khalid A, Singh US, Sharma AK (2010) Evaluation of arbuscular mycorrhizal fungus, *fluorescent Pseudomonas* and *Trichoderma harzianum* formulation against *Fusarium oxysporum* f. sp. *lycopersici* for the management of tomato wilt. Biol Control 53:24–31

Srivastava S, Singh VP, Kumar R, Srivastava M, Sinha A, Simon S (2011) *In vitro* Evaluation of Carbendazim 50 % WP, antagonists and botanicals against *Fusarium oxysporum* f. sp. *psidii* associated with rhizosphere soil of guava. Asian J Plant Pathol 5:46–53

Tamietti G, Pramotton R (1990) Soil suppressiveness to *Fusarium* wilt: relationship between suppressiveness and indigenous microflora of the soil with special emphasis on non-pathogenic *Fusarium*. Agronomie 10(2):69–76

Thangavelu R, Sangeetha G, Mustaffa MM (2007) Cross-infection potential of crown rot pathogen (*Lasiodiplodia theobromae*) isolates and their management using potential native bioagents in banana. Australas Plant Pathol 36(6):595–605

Thilagavathi R, Rajendran L, Nakkeeran S, Raguchander T, Balakrishnan A, Samiyappan R (2012) Vermicompost based bioformulation for the management of sugar beet root rot caused by *Sclerotium rolfsii*. Arch Phytopathol Plant Prot 45(18):2243–2250

Thori HR, Bunker RN, Mathur K, Sharma SS (2012) Integrated management of post flowering stalk rot of Maize caused by *Fusarium moniliforme*. Indian Phytopathol 65(2)

Ting ASY, Sariah M, Kadir J, Gurmit S (2007) Field evaluation of non-pathogenic *Fusarium oxysporum* isolates upm31p1 and UPM39B3 for the control of *Fusarium* wilt in '*Pisang Berangan*'(MUSA, AAA). Proceedings of the International Symposium on Recent Advances in Banana Crop Protection for Sustainable Production and Improved Livelihoods, Sept. 2007 ISHS Acta Horticulturae, pp 139–144

Ting ASY, Maha SW, Tee CS (2012) Evaluating the feasibility of induced host resistance by endophytic isolate *Penicillium citrinum* BTF08 as a control mechanism for *Fusarium* wilt in banana plantlets. Biol Control 61:155–159

Tiwari AK, Mukhopadhyay AN (2001) Testing of different formulations of *Gliocladium virens* against chickpea wilt compex. Indian Phytopathol 54(1):67–71

Upadhyay RS (1992) Ecology and biological control of *Fusarium udum* in relation to soil fungistasis of antagonistic microorganisms. J Plant Prot Trop 9:1–9

Verma M, Brar SK, Tyagi RD, Surampalli RY, Valero JR (2007) Antagonistic fungi, *Trichoderma* spp. panoply of biological control. Biochem Eng J 37:1–20

Vibha (2010) Effect of fungal metabolites and amendments on mycelial growth of *Rhizoctonia solani*. J Plant Prot Res 50:93–97

Vipin K, Akhtar H, Anita S (2009) Integrated management of *Meloidogyne incognita-Fusarium solani* disease complex of brinjal cv. Pusa Kranti. Ann Plant Prot Sci 17(1):192–194

Vittal Rao M, Rao AS (1966) A study of the effects of antagonistic microorganisms on soil-borne plant pathogenic fungi. Indian Phytopathol 19:251–256

Waisel Y, Eshel A, Kafkafi U (1991) Plant Roots: the Hidden Half. Marcel Dekker, New York

Weindling R (1932) *Trichoderma lignorum* as a parasite of other soil fungi. Phytopathology 22:837–845

Whipps JM, McQuilken MP (1993) Aspects of biocontrol of plant pathogens. In: Jones DG (ed) Exploitation of Microorganisms. Chapman and Hall, London, pp 45–68

Yigit F, Dikilitas M (2007) Control of *Fusarium* wilt of tomato by combination of *Fluorescent Pseudomonas*, non-pathogen *Fusarium* and *Trichoderma harzianum* T-22 in greenhouse conditions. Plant Pathol J 6:159–163

Ziedan E, Sayed EI, Mostafa M, Sahab A (2011) Application of Mycorrhizae for controlling root diseases of sesame. J Plant Prot Res 51(4):355–361

# *In Vitro* Evaluation of PGPR Strains for Their Biocontrol Potential Against Fungal Pathogens

26

Urja Pandya and Meenu Saraf

**Abstract**

Crop protection has become a basic requirement of the sustainable agriculture to ensure increased crop production. Biological control has been actively practiced as a crop protection measure for more than five decades and the history of biocontrol, its successes and failures, have been extensively reviewed. Plant growth promoting rhizobacteria (PGPR) are an important group of microorganisms, which play a major role in the biocontrol of plant pathogens. All plant-associated microenvironments, especially the rhizosphere, are colonized in high abundances by antagonistic microbes. Between 1 and 35 % of the microbial inhabitants show antagonistic capacity to inhibit the growth of pathogens *in vitro* by various biocontrol mechanisms that include production of antibiotics, siderophores, lytic enzymes, HCN and induced systemic resistance. In recent years, the popularity of biocontrol agents has increased substantially, as extensive and systematic research has enhanced their effectiveness and consistency.

**Keywords**

Biocontrol · Plant growth promoting rhizobacteria (PGPR) · Siderophores · HCN · Induced systemic resistance

## 26.1 Introduction

Soil-borne fungal diseases cause worldwide economically significant diseases that affect important crops. Soil-borne fungal diseases are the limiting factors in the productivity of many crop systems. Controlling soil-borne fungal pathogens has always been very difficult. Fungicide drenches are expensive and impractical, and they have undesirable effects on the environment. Physical methods, such as soil sterilization by heat and soil solarisation, can sometimes be useful for reducing the pathogen inoculums. However, these techniques can have undesirable side effects on the plants and on beneficial microflora. Among the major soil-borne fungal pathogens, *Rhizoctonia* and the oomycete *Pythium,* are especially problematic in disinfected soil and soilless

U. Pandya (✉) · M. Saraf
Department of Microbiology, University School of Sciences, Gujarat University, Ahmedabad, Gujarat 380009, India
e-mail: urjapandya1@yahoo.co.in

R. N. Kharwar et al. (eds.), *Microbial Diversity and Biotechnology in Food Security,*
DOI 10.1007/978-81-322-1801-2_26, © Springer India 2014

substrates, where the microbial diversity and biological buffering present in natural soils are lacking. Moreover, sclerotia-producing fungi such as *Botrytis, Sclerotinia* and *Sclerotium* are very difficult to eradicate from the soil by existing methods. In this scenario, the use of antagonistic rhizospheric microorganisms would be an effective alternative means of control. In fact, the biocontrol of soil-borne fungal pathogens has probably been the subject of more research than any other form of plant disease biological control (Perez-Garcia et al. 2011).

These fungal pathogens need to be controlled by improving crop protection strategies, to prevent damage, to ensure a substantial contribution, the food security and economy of people. The use of chemical fertilizers and pesticides in agriculture had showed spectacular improvements in crop productivity over the past decades. However, the hazardous effects such as degradation of soil, undesirable effects on no target organisms, environmental pollution, etc. were seen. Furthermore, the growing cost of pesticides, particularly in less affluent regions of the world and consumer demand for pesticide-free food and ecofriendly management strategies has led to search for a substitute for these agricultural inputs (Tapadar and Jha 2013).

Biological control of plant diseases has been considered a viable alternative method to manage plant diseases. Biological control is the inhibition of growth, infection or reproduction of one organism using another organism. Biocontrol is environmentally safe, and in some cases is the only option available to protect plants against pathogens (Heydari and Pessarakli 2010). Total 14 bacteria and 12 fungi were registered by Fravel (2005) with the US Environmental protection agency for control of plant diseases. Among the 14 registered bacterial biocontrol agents (BCAs), 6 are based on *Bacillus,* 5 on *Pseudomonas,* 2 on *Agrobacterium* and 1 on *Streptomyces*. *Pseudomonas* spp. are particularly suited as a BCAs because they can use many exudates as nutrient source, are abundantly present in natural soils, especially in the rhizosphere, have a high growth rate, can be directly plant-growth promoting and have the ability to control diseases by a variety of mechanisms. They are also the most extensively studied group of bacterial BCAs, since *Pseudomonas* bacteria are amenable to mutation and modification using genetic tools (Hofte and Altier 2010).

Beneficial rhizobacteria capable of aggressively colonizing the rhizosphere and facilitating plant growth are often termed as plant growth-promoting rhizobacteria (PGPR) (Kloepper and Schroth 1978). Plant growth promoting rhizobacteria (PGPR) are free-living soil bacteria that can either directly or indirectly facilitate rooting and growth of plants. PGPR indirectly enhance plant growth via suppression of phytopathogens by a variety of mechanisms. These include the ability to produce siderophores that chelate iron, making it unavailable to pathogens; the ability to synthesize antifungal metabolites such as antibiotics, fungal cell wall-lysing enzymes or production of volatiles such as hydrogen cyanide, which suppress the growth of fungal pathogens; the ability to successfully compete with pathogens for nutrients or specific niches on the root; and the ability to induce systemic resistance (ISR; Saraf et al. 2010). Bacteria belonging to the genera *Agrobacterium, Bacillus, Burkholderia, Enterobacter, Erwinia, Lysobacter, Pseudomonas* and *Serratia* are successfully used as BCAs against many plant diseases. Some of the fungal biocontrol genera used are *Ampelomyces, Aspergillus, Coniothyrium, Gliocladium, Laetisaria, Penicillium, Phlebiopsis, Sporodesmia, Talaromyces, Tilletiopsis, Trichoderma* and *Trichothecium* (Mathivanan and Manibhushanrao 2004; Mathivanan et al. 2006). In addition, several species of actinomycetes belonging to the genera *Streptomyces, Actinoplanes, Actinomadura, Micromonospora, Streptosporangium, Streptoverticillium* and *Spirillospora* are used as BCAs. Interestingly, they produce biologically active secondary metabolites that have a potential in controlling plant pathogens (Doumbou et al. 2002; El-Tarabily and Sivasithamparam 2006; Prabavathy et al. 2008; Ramesh 2009).

Pertaining to massive world population pressure, increase in food grain production is a troublesome task in today's world. The need of the day is sustainable agriculture without harming the balance of soil ecology as well as unlocking the mystery of biota influencing plant growth by using PGPR. PGPR are nowadays applied in a wide array of agro and allied industries in the form of inoculants (biofertilizer and BCA) in a range of agro-economically important plants including leguminous and nonleguminous crops, trees and plants of forest, horticulture, sericulture, medicinal, fodder, oilseed and cash crops for enhancing their growth and productivity (Aeron et al. 2011).

PGPR, as BCAs have certain advantages over conventional chemical control compounds (Labuschagne et al. 2010). They are mentioned below:

1. PGPR are beneficial, naturally occurring microorganisms, which are environmentally friendly and nontoxic.
2. As per ecological perspective, their application is sustainable (long term).
3. Another advantage of PGPR is the fact that they possess a diverse range of modes of action including antibiosis, production of siderophore, cell wall degrading enzymes, bio-surfactants and volatiles, and also induces systemic resistance in plants.
4. The fact that some PGPR by definition directly enhance the growth of plants is an additional advantage.

In biocontrol research, screening is a critical step in the development of BCAs, and the ultimate success of biocontrol depends on how well the searching and screening process is done (Fravel 2005). The places to look for potential control agents must be selected carefully and the control agent eventually selected must be able to survive and grow in the environment in which it is expected to show the biocontrol potential (Campbell 1989). Isolates from culture collections rarely prove useful organisms for the field because they are usually adapted to the high nutrient levels in common media. However, culture collections are important in biocontrol research because reference strains are needed in taxonomic and phylogenetic studies to identify the newly isolated BCAs and to study their genotypic and phenotypic diversity. Some authors separate biocontrol strategies into two broad categories. One strategy, which could be considered preventive, follows a fundamentally ecological approach. This biocontrol strategy reaches a long-time plant protection against the pathogen, and it is mainly based on induction or improvement of suppressive soils. On the other hand, a second strategy, which could be considered curative, uses microorganisms as biopesticides and resembles, in some important respects, the approach of chemical pesticide treatment, which aims for control at a limited period of time (Knudsen et al. 1997). These differences in control strategy should influence the choice of isolation as well as the screening method (Kohl 2009).

The mode of action of PGPR as biocontrol is discussed with reference to the production of antibiotics, siderophore, and cell wall degrading enzymes as well as induction of systemic resistance, root colonization and rhizosphere competence. The aim of this chapter is to point out the status of biocontrol, related with its mode of action and to draw the focus on research strategies for the development of better inoculants.

## 26.2 PGPR as a Biocontrol Agent Against Soil-Borne Pathogens

During interaction process of PGPR with phytopathogens, the former produce certain antibiotics, siderophore, HCN or induce systemic resistance, and release of such metabolites decides the fate of the pathogen.

### 26.2.1 Antibiotics Production

Antibiotic production by biocontrol-PGPR is the most powerful mechanism of biocontrol against phytopathogens. Many different types of antibiotics produced and effective under laboratory conditions, may not necessarily be under field conditions. As the genes involved in the production of some antibiotics are known, it is possible to enhance antibiotic activity, and hence enhance suppression of phytopathogens (Bashan and Bashan 2005). These antibiotics may be antitumor, antiviral, antimicrobial, antihelmenthic and cytotoxic (Fernando et al. 2005). Among various BCAs, fluorescent pseudomonads (FPs) are found to be the prolific producers of a wide variety of metabolites such as phenazines, pyrrolnitrin, pyoluteorin, oomycin A, viscosinamide and hydrogen cyanide (Dwivedi and Johri 2003). *Bacillus* spp. were found to produce many antibiotics such as zwittermycin A, kanosamine, rhizocticin C, iturins, fungicin and saltavalin and they are also capable of producing thermostable antimicrobial peptides (Emmert and Handelsman 1999; Kavitha et al. 2005). Among fungal BCAs, *Trichoderma* spp. produce a range of antibiotic metabolites such as trichodermin, peptaibols, pyrones, etc. (Mathivanan et al. 2008). Other bacteria such as *Agrobacterium radiobacter, Burkholderia cepacia, Lysobacter* sp. and *Pantoea agglomerans* have been reported to produce various antibiotic compounds such as agrocin 84, pyrrolnitrin, pseudane, xanthobaccin A and herbicolin (Kerr 1980; Homma et al. 1989; Islam et al. 2005; Sandra et al. 2001; Zhao et al. 2010).

Genetic analysis of many biocontrol strains of *Pseudomonas* indicated that there is a positive correlation between disease suppression and antibiotic production. It was demonstrated that with increasing populations of *Pseudomonas* spp., which produce the antibiotic DAPG, there was a rapid decline in Take-all disease in wheat caused by the fungus *Gaeumanomyces graminis* var. *tritici* (Labuschagne et al. 2010). Zhou et al. (2012) studied the main antimicrobial compound of *Pseudomonas brassicacearum* J12 which strongly inhibited the growth of *Ralstonia solanacearum* as 2,4-diacetylphloroglucinol (2,4- DAPG) by HPLC–ESI-MS analysis. The gene cluster phlACBD, which is responsible for 2,4-DAPG production, was identified and expressed in the bacterial strain *Escherichia coli* DH5 α. *Bacillus subtilis* strain PRBS-1 and AP-3 inhibited five soybean seed pathogenic fungi, viz. *Rhizoctonia solani, Colletotrichum truncatum, Sclerotinia sclerotium, Macrophomina phaseolina,* and *Phomopsis* spp. under in vitro conditions (Araujo et al. 2005). Zhou et al. (2008) isolated *Paenibacillus* strain HT16 from locusts, which showed strong inhibition to *Penicillium expansum* and produced antifungal protein with the molecular weight of 4,517 Da. Pueyo et al. (2009) showed a large group of lipopeptides produced by soil bacterium *B. megaterium* and their antagonistic activity similar to surfactins, lichenysins, iturin A, and fengycins. The antifungal metabolite produced by *Paenibacillus polymyxa* strain HKA-15 showed strong antagonism against *Rhizoctonia bataticola* causing charcoal rot disease in soybean. Two bioactive fractions collected from the culture filtrate of *Paenibacillus polymyxa* strain HKA-15 by preparative HPLC were characterized as cyclic peptide and depsipeptide (Senthilkumar et al. 2007). *Paenibacillus lentimorbus* strain WJ5, a soil isolate, produced antifungal metabolite, which was extracted with n-butanol. The FT-IR spectrum of the antifungal metabolite confirmed the presence of the peptide and glycosidic bonds (Lee et al. 2008).

Romero et al. (2007) showed the involvement of iturin and fengycin antibiotics from four *B. subtilis* strains UMAF6614, UMAF6616, UMAF6639 and UMAF8561 in the suppression of powdery mildew of cucurbits caused by *Podosphaera fusca*. The culture supernatant could successively inhibit the powdery mildew at levels previously reported for vegetative cells (Romero et al. 2004). The chemical analysis of culture filtrate together with the recovery of inhibitory components (surfactin, fengycin, and iturin A or bacillomycin) from the melon leaves treated with two strains (UMAF6614 and UMAF6639) strongly supported the evidence of in situ production of these antimicrobials. Bais et al. (2004) demonstrated the protective action of surfactin produced by *B. subtilis* against the infection caused by *Pseudomonas syringae* in *Arabidopsis thaliana* and suggested that surfactin was not only necessary for root colonization but also provided protection against the pathogen. The disease suppression was correlated with inhibitory concentrations of surfactin produced by the organism on roots. Moyne et al. (2001) identified *B. subtilis* strain AU195 capable of producing antifungal peptides showing similarity with bacillomycin (group iturin A).

### 26.2.2 Siderophore Production

PGPRs also exert their antagonistic activity against plant pathogens by means of secretion of siderophores. These low molecular weight compounds (400–1, 500 Da) preferentially chelate iron ($Fe^{+++}$) and transport it into the cell across the cell membrane. The siderophores bind most of the $Fe^{+3}$ in the rhizosphere and effectively prevent the proliferation of fungal pathogens by depriving them of available iron. Suppression of the pathogens arises because iron deficiency causes growth inhibition, decrease in nucleic acid synthesis, inhibition of sporulation and causes changes in cell morphology (Labuschagne et al. 2010). The production of siderophores by plant-associated bacteria has received major attention because of their role in both biological control of diseases and in virulence of plant pathogens (Neilands and Leong 1986; Loper and Buyer 1991). This is particularly true for siderophores of *Pseudomonas* spp., which are produced in a large variety to sustain survival and growth of bacterial cells under iron-limiting conditions. Possibly, production and utilization of siderophores are evolutionary responses to the diverse and often adverse habitats in which these bacteria live (Ishimaru and Loper 1993). Pyoverdines (or pseudobactins) are the prevalent class of siderophores produced by fluorescent *Pseudomonas* spp. They are yellow–green water-soluble chromopeptides, fluorescent under ultraviolet irradiation ($k = 366$ nm), and with a rather complex structure compared to that of most of the microbial siderophores described. They have both catechol and hydroxamate groups that chelate iron (Leong 1986). Their molecular structures, gene clusters responsible for biosynthesis, excretion and uptake, and their regulation have been extensively studied and reviewed (Crosa 1997; Meyer 2000; Ravel and Cornelis 2003). Fluorescent siderophores production was observed as a mechanism of biocontrol of bacterial wilt disease in the fluorescent pseudomonads RBL 101 and RSI 125 (Jagadeesh et al. 2001) while Akhtar and Siddiqui (2009) reported that siderophore producing *Pseudomonas* strains significantly reduced the root-rot disease in chickpea. Press et al. (2001) reported the catechol siderophore biosynthesis gene in *Serratia marcescens* 90–166 is associated with induced resistance in cucumber against anthracnose, while *P. fluorescens* inhibited the growth of *Fusarium culmorum* in vitro (Kurek and Jaroszuk-Scisel 2003). Enhanced plant growth caused by pseudomonad strains was often accompanied by the reduction in pathogen populations on the roots. There is convincing evidence to support a direct role of siderophore-mediated iron competition in the biocontrol activity exhibited by such isolates (Leong 1986; Loper and Buyer 1991).

The antagonism depends on the amount of iron available in the medium; siderophores produced by a BCA and sensitivity of target pathogens (Kloepper et al. 1980; Weger et al. 1988). Battu and Reddy (2009) reported 10 strains of *P. fluorescens* based on preliminary screening of all these isolates for antifungal activity against rice fungal pathogens (*P.oryzae* and *R.solani*); inhibited the growth of rice fungal pathogens in Fe-deficient King's B medium that varied from 3 to 58 % inhibition. Among these, Pf 003 strain completely inhibited the mycelial growth of two rice pathogens (*P.oryzae* and *R.solani*) both in presence and absence of $FeCl_3$ which indicated the siderophore mediation along with antifungal metabolites. Patel et al. (2011) reported hydroxymate types of siderophores that ranged from 11–50 mM under saline conditions.

### 26.2.3 HCN Production

HCN is an effective inhibitor of cytochrome c oxidase (Knowles 1976) and other metalloenzymes (Blumer and Haas 2000). Its production by PGPR is implicated in biological control of black root rot of tobacco (Voisard et al. 1989), root rot of tomato and *Pythium* damping off of cucumber (Ramette et al. 2003). Cyanogenesis by *Pseudomonas* spp. is the product of an HCN synthase, encoded by three biosynthetic genes (hcnABC) (Laville et al. 1998). These three genes are sufficient for HCN biosynthesis from glycine, the primary precursor (Blumer and Haas 2000). Multiple mechanisms contribute to HCN resistance in *P. aeruginosa.*

These include rhdA, which encodes a rhodonase that detoxifies HCN (Cipollone et al. 2007); the cioAB gene pair, which encode a cyanide insensitive terminal oxidase (Cooper et al. 2003); and the secretion of α-ketoglutarate, which detoxifies cyanide (Blumer and Haas 2000).

### 26.2.4 Induced Systemic Resistance

The increased level of resistance using external agents, without modifying the genome of the plant, is known as induced or acquired resistance. The expression of induced resistance can be local or systemic when it is expressed at sites not directly exposed to the inducers agent (Stadnik 2000). Several PGPR that colonize root systems by seed applications and protect plants against foliar disease include *Pseudomonas fluorescens, P. putida, Bacillus pumilus* and *Serratia marcescens*. Fewer published accounts of ISR by *Bacillus* spp. are available that show specific strains of the species *B. amyloliquefaciens, B. subtilis, B. pasteurii, B. cereus, B. pumilus, B. mycoides* and *B. sphaericus* can elicit significant reductions in the incidence or severity of various diseases in diverse hosts (Choudhary and Johri 2009). PGPR elicit ISR in plants by increasing the physical and mechanical strength of the cell wall as well as changing the physiological and biochemical reactions of the host. This results in the synthesis of defence chemicals such as chitinase-, peroxidase- and pathogenesis-related proteins (Ramamoorthy et al. 2001; Nandakumar et al. 2001; Silva et al. 2004).

The difference between ISR and SAR is that ISR is induced by nonpathogenic rhizobacteria, while SAR is induced systemically after inoculation with necrotizing pathogens. Moreover, ISR is independent of salicyclic acid but involves jasmonic acid and ethylene signalling, while SAR requires salicyclic acid as a signalling molecule in plants. ISR is accompanied by the expression of sets of genes distinct from the PR genes whereas SAR is accompanied by the induction of pathogenesis-related proteins. Both ISR and SAR are effective against a broad spectrum of plant pathogens (Kuc 1982; van Loon et al. 1998). Fluorescent pseudomonads are also involved in the induction of plant resistance against insects and nematodes and specific metabolites produced by these bacteria have been involved in elicitation of defence reactions of the host plants. Similar to classical induced resistance, PGPR–mediated resistance is known to produce broad spectrum resistance, which is correlated with increased amounts of pathogenesis-related (PR) proteins, peroxidase, chitinases and β 1, 3-glucanase in plant tissue (Pathma et al. 2011). A salicylate overproducing recombinant of *P.fluorescens* strain P3 affords enhanced protection to tobacco against tobacco necrosis virus compared with the wild type P3, which indicates that salicylate might also stimulate defence. In another *Pseudomonas* biocontrol strain, a combination of pyocyanin and pyochelin seems to be most effective for inducing resistance in tomatoes (Audenaert et al. 2002). Fluorescent pseudomonads (SE21 and RD 41) are resistance inducers (chitin and salicyclic acid) for plant growth promotion and biological control of damping off of pepper caused by *Rhizoctonia solani*. *P. fluorescens* strains CHA0 and Pf1 were investigated by Rajkumar et al. (2008) for their biocontrol efficacy against Banana bunchy top virus (BBTV) in bananas (*Musa spp.*) alone and in combination with chitin under glasshouse and field conditions.

Increased accumulation of oxidative enzymes, peroxidase (PO), polyphenol oxidase (PPO), phenylalanine ammonia lyase (PAL), pathogenesis-related (PR) proteins, chitinase, β-1,3-glucanase and phenolics were observed in CHA0 bioformulation amended with chitin-treated plants challenged with BBTV under glasshouse conditions (Kavino et al. 2007). *Bacillus subtilis* AF1 isolated from soils are suppressive to pigeon pea (*Cajanus cajan*) wilt caused by *Fusarium* sp. caused lysis of *A. niger* by stimulating the production of PAL and peroxidase by the plant thereby eliciting induction of systemic resistance (Kloepper et al. 2004). Protection resulting from ISR elicited by *Bacillus* spp. has been reported against leaf spotting fungal and bacterial pathogens, systemic viruses, a crown rotting fungal pathogen, root-knot nematodes and a stem blight fungal pathogen as well as damping off blue mould and late blight diseases (Govindasamy et al. 2010).

## 26.3 Molecular Studies on PGPR as Biocontrol Agents

Genomic information describes the analysis of the mode of action, detailed investigations of interactions as well as optimization of fermentation and formulation processes for biocontrol products (Gross and Loper 2009). De Bruijn et al. (2007) used genome mining to discover unknown gene clusters and their traits that are highly relevant and significant in the lifestyle of *P. fluorescens* SBW25. Proteomic and transcriptomic studies are interesting to study the function of BCAs. Garbeva et al. (2011) studied transcriptional and antagonistic responses of *P. fluorescens* Pf0–1 to phylogenetically different bacterial competitors (*Bacillus, Brevundimonas* and *Pedobacter*), which demonstrated that Pf0–1 shows a species-specific response to bacterial competitors. In another transcriptomic study published by Hassan et al. (2010), a whole genome oligonucleotide microarray was developed for *P. fluorescens* Pf-5 and used to assess the consequences of a gacA mutation: GacA significantly influenced transcript levels of 10 % of the 6,147 annotated genes in the Pf-5 genome including genes involved in the production of hydrogen cyanide, pyoluteorin and the extracellular protease. Transcriptomic studies can also lead to new insights into plant responses on BCAs: *Pseudomonas*-primed barley genes indicated that, as is the case in dicots, jasmonic acid plays a role in host responses (Petti et al. 2010).

Metabolomics studies are important for the analysis of metabolites in situ and for registration procedures of products. Frimmersdorf et al. (2010) used a metabolomic approach to show how *P. aeruginosa* adapts to various environments. In addition, analysis of the mobilome of strains can result in interesting findings for biocontrol research as shown for *P. fluorescens* Pf-5 by Mavrodi et al. (2009), in which mobile genetic elements contain determinants that contribute to Pf-5's ability to adapt to changing environmental conditions and/or colonize new ecological niches. Studying the colonization of plants has been greatly reported by the application of fluorescent proteins which are used as vital markers and reporter genes (Bloemberg 2007). New methods for the in situ analysis of antifungal gene expression using flow cytometry combined with green fluorescent protein (GFP)-based reporter fusions (de Werra et al. 2008), barcode pyrosequencing (Gomes et al. 2010), and ultra-deep sequencing (Velicer et al. 2006) are studied for biocontrol mechanisms.

Evolution of molecular tools has permitted the development of new screening strategies for BCA, such as the development of sequence-based T-RFLP-derived molecular markers to direct the identification and isolation of novel bacteria. This technique provides a cost-effective approach to finding generalist populations that consistently contribute to suppression across environments. T-RFLP analyses compare the bacterial community structure in soils differing in their disease-suppressive capacities, revealing the positive association of multiple bacterial populations (marked with different TRFs) with disease suppression (Pliego et al. 2011). Whole-genome sequences are now available for two biocontrol strains belonging to the *P. fluorescens* lineage (strain SBW 25) or a closely related species, as well as non biocontrol *P. fluorescens* Pf01. Comparative genomics, gene array-based expression studies and integrated, *in situ* molecular analyses of microbe–host interactions have started to provide advanced knowledge on plant protection properties and rhizosphere competence of these BCAs (Couillerot et al. 2009).

## 26.4 Global Productions of Biocontrol Products Derived from PGPR

The process of developing biological control begins with in vitro and in vivo screenings that continues with the study of mechanisms of control such as competition, antibiosis, siderophore, HCN and induced systemic resistance under laboratory experiments. The next stage is the production of large amounts of efficient biomass at a low cost, which requires studies of microbial physiology and the use of biotechnological processes. Adequate formulations and application methods have to be designed so that the microbial biomass will attain a high level of biocontrol activity (Schisler et al. 2004). The selection of best antagonistic strain is carried out by screening for antimicrobial action against different soil-borne pathogens apart from the target pathogen. The plant, pathogen and antagonists are then co exposed to controlled environmental conditions. Promising antagonists are further tested for their efficacy in field trials along with standard recommended fungicides (Pengnoo et al. 2000). Mass production is achieved through liquid (Manjula and Podile 2001), semisolid and solid fermentation techniques (Lewis 1991).

It is also important to focus on the critical stages of commercialization of BCAs. Screening of new agents should consider the biology and ecology of the pathosystem, as well as agricultural practices associated with the crop (Fravel 2007). This knowledge will help prevent variation in field performance which is responsible for lack of wider adoption of biocontrol for disease management. The formulation stage aims to deliver the BCA in a physiologically active state for the needed control. The formulation must be economical with a good shelf life and in a suitable form for shipping, storage and application. Risk assessment of human health and to the environment are needed before releasing the new product, and early in the screening, even microorganisms with good biocontrol potential but capable of growing at human body temperature should be eliminated (Fravel 2007).

In the USA, organisms currently registered for biocontrol and active compounds isolated from plants or other organisms are listed at http://www.epa.gov/oppbppd1/biopesticides/ingredients/index.htm. A few examples of PGPR and their commercialised biocontrol products (Figueiredo et al. 2010) are: *Agrobacterium radiobacter* K1026 (Nogall®), *Bacillus pumilus* QST 2808 (Sonata® TM), *B. pumilus* GB34 (YieldShield®), *B. subtilis* GBO3(Kodiak®), *Pantoea agglomerans* C9–1 (BlightBan C9–1®), *P. agglomerans* E325 (Bloomtime®), *Pseudomonas aureofaciens* Tx-1(Spot-Less®T), *P. syringae* ESC-10 and ESC-11 (Bio-save®), *P. fluorescens* A506 (BlightBan®), *P. chlororaphis* MA 342 (Cedomon®), *Streptomyces griseoviridis* K61 (Mycostop®), and *S. lydicus* WYEC 108 (Actinovate®). Some of the important PGPR strains as biocontrol product along with their commercial products are listed in Table 26.1.

Companies in India manufacturing biocontrol products for their commercialisation as listed in Table 26.2. In China, PGPRs have been successfully applied for over two decades in an area of 20 million ha in different crop plants for commercial development. Owing to the potentiality of *Bacillus* spp., more than 20 different commercial products of *Bacillus* origin are sold in China to mitigate soil-borne diseases (Backman et al. 1997). Besides, *Bacillus* spp., certain other PGPR strains belonging to the genera such as *Agrobacterium, Azospirillum, Bulkholderia, Pseudomonas* and *Streptomyces* are also used for the production of several commercial products, which are generally being applied against several target pathogens like *Botrytis cinerea, Penicillium* spp., *Mucor pyroformis, Geotrichum candidum, Erwinia amylovora,* russet-inducing bacteria, *Fusarium* sp., *Rhizoctonia* sp., *Pythium* sp., *Fusarium* sp., *Phytophthora* sp., and *P. tolassii* (Nakkeeran et al. 2005).

Droby et al. (2009) has well-documented commercial antagonistic microorganisms available in the global market for postharvest control of decays of fruits and vegetable. These are Bio-save (*Pseudomonas syringae* Van Hall), which are registered in the USA and used mostly for the control of sweet potato and potato diseases (Stockwell and Stack 2007), and "Shemer" (*Metschnikowia fructicola* Kurtzman and Droby) registered in Israel and used commercially for the control of sweet potato and carrot storage diseases (Kurtzman and Droby 2001; Blachinsky et al. 2007). The two yeast-based products,

**Table 26.1** Commercial products developed using different PGPR strains. (Adapted Modified from Bhattacharya and Jha 2012)

| PGPR | Products | Intended crop |
|---|---|---|
| *Agrobacterium radiobacter* | Diegall, Galltrol-A, Nogall, Norbac 84 C | Fruit, nut, ornamental nursery stock and trees |
| *Azospirillum brasilense* | Azo-Green | Turf and forage crops |
| *Bacillus subtilis* | Epic, HiStick N/T, Kodiak, Rhizo-Plus, Serenade, Subtilex | Barley, beans, cotton, legumes peanut, pea, rice and soybean |
| *B. amyloliquefaciens* GB99 | Quantum 4000 | Broccoli, cabbage, cantaloupe, cauliflower, celery, cucumber, lettuce, ornamentals, peppers, tomato and watermelon |
| *Burlkholderia cepacia* | Blue Circle, Deny, Intercept | Alfalfa, barley, beans, clover, cotton, maize, peas, sorghum, vegetables and wheat |
| *Pseudomonas fluorescens* | BlightBan A506, Conquer, Victus | Almond, apple, cherry, mushroom, peach, pear, potato, Strawberry and tomato |
| *P. syringae* | Bio-save10 | Citrus and pome fruit |
| *Streptomyces griseovirdis* K61 | Mycostop | Field, ornamental and vegetable crops |
| *T. harzianum* | TRICHODEX | Strawberry, tomato, rice |
| *T. harzianum* Rifai ATCC20476 | Binab-TF-WP | Cotton and other agricultural crops and vegetables |
| *T. polysporum* Rifai ATCC20475 | Binab-TF-WP-Konc | |

**Table 26.2** List of Indian companies manufacturing the biocontrol products. (DBT India 2010)

| Company name | Biocontrol agent | Biocontrol product | Sources |
|---|---|---|---|
| Ajay Bio-Tech (I) Limited, Pune | *T. viride*<br>*B. thuringiensis var. kurstaki* | Trichoguard-WP, Trichoguard-L<br>Bio-Dart | Web: http://www.ajaybio.com |
| Bio-Control Research Laboratories, Karnataka | *Trichogramma*<br>*T. viride*<br>*T. harzianum* | Tricho-Card<br>NIPROT<br>NIPROT | E-mail: bcrl@vsnl.com/jayanthk@vsnl.com |
| Biotech International Limited, New Delhi | *B. thuringiensis var. kurstaki*<br>*T. viride*<br>*P. fluorescens*<br>*Trichogramma spp.* | Biolep<br>Bioderma<br>Biomonas<br>Biogramma | E-mail: info@biotech-int.com<br>web: http://www.biotech-int.com |
| Bioved Research and Communication Centre, Uttar Pradesh | *Trichoderma viride*<br>*T. harzianum*<br>*Aspergillus niger*<br>*Saccharomyces cervisiae* | Biovidi<br>Biozim<br>Bionizer<br>Sachcer | E-mail: bioved2003d@yahoo.com |
| Esvin Advanced Technologies Limited, Tamil Nadu | *P. fluorescens*<br>*Beauveria bassiana*<br>*Beauveria bassiana*<br>*Verticillium lecanii*<br>*Trichoderma viride* | Esvin Pseudo<br>Biolarvex<br>Biogrubex<br>Biosappex<br>Eswin Tricho | E-mail: admin@esvintech.com/tsv@vsnl.com |
| International Panaacea Limited, New Delhi | *Trichoderma* sp.<br>*Pseudomonas* sp.<br>*Beauveria* sp. | Sanjeemni, WP<br>P-Suraksha, WP<br>Daman, WP | E-mail: info@iplbiotech.com<br>Web: http://www.iplbiotech.com/ |

AspireTM (Ecogen, US) and Yield Plus (Anchor Yeast, South Africa) developed in the USA and South Africa are no longer available (Droby et al. 2009). Currently, BioNext (Belgium) and Leasaffre International (France) have developed a commercial product, based on the same yeast used in AspireTM, *Candida oleophila*. A similar yeast-based product, *Candida saitoana* was developed by Neova Technologies (Abbotsford, British Columbia, Canada). Additionally, Spain has also developed a commercial formulation of *Candida sake* for use on pome fruit under the name "Candifruit".

## 26.5 Conclusions

The inconsistent performance of BCAs in the field developed thus far has plagued efforts to exploit them for commercial applications. There is a compelling need to identify efficient and dependable BCAs to be used singly or as mixtures, so as to ensure consistent performance in the farmer's field. Biological control offers exciting possibilities for the future. Opportunities for creating superior strains of BCAs and transgenic crops which express microbial secondary metabolites such as an antibacterial antibiotic and other antifungal proteins are endless. The choice of the right microbial candidates is one of the most important factors governing the success of biocontrol programs on a commercial basis. Ideal BCAs would reduce the severity of more than one pathogen, as this will make their application cost-effective. It needs to be remembered that most of the world's farmers, who live in Asia, are resource poor. Therefore, only cost-effective formulations of BCAs that perform consistently in the field, either by themselves or as part of an integrated disease management package, will benefit low-income agricultural crop growers. In this lies the key to the ultimate success of biocontrol research for integrated disease management.

**Acknowledgments** We gratefully acknowledge the financial support by the Department of Science and Technology (DST), New Delhi under Women Scientist Scheme (WOS-A).

## References

Aeron A, Kumar S, Pandey P, Maheshwari DK (2011) Emerging role of plant growth promoting rhizobacteria in agrobiology. In: Maheshwari DK (ed) Bacteria in agrobiology: crop ecosystem. Springer-Verlag, Berlin

Akhtar MS, Siddiqui ZA (2009) Use of plant growth promoting rhizobacteria for the biocontrol of root rot disease complex of chickpea. Australas Plant Pathol 38:44–50

Araujo FF, Henning AA, Hungria M (2005) Phytohormones and antibiotics produced by *Bacillus subtilis* and their effects on seed pathogenic fungi and on soybean root development. World J Microbiol Biotechnol 21:1639–1645

Audenaert K, Pattery T, Cornelis P, Hofte M (2002) Induction of systemic resistance to *Botrytis cinerea* in tomato by *P. aeruginosa* 7NSK2: role of salicyclic acid, pyochelin and pyocyanin. Mol Plant Microbe Interact 11:1147–1156

Backman PA, Wilson M, Murphy JF (1997) Bacteria for biological control of plant diseases. In: Rechcigl NA, Rechecigl JE (eds) Environmentally safe approaches to crop disease control. Lewis Publishers, Boca Raton

Bais HP, Park SW, Weir TL, Callaway RM, Vivanco JM (2004) How plants communicate using the underground information superhighway. Trends Plant Sci 9:26–32

Bashan Y, de-Bashan LE (2005) Bacteria/plant growth promotion. In: Hillel D (ed) Encyclopedia of soils in the environment. Elsevier, Oxford

Battu PR, Reddy MS (2009) Siderophore mediated antibiosis of rhizobacterial fluorescent pseudomonads against rice fungal pathogens. Int J PharmTech Res 1:227–229

Bhattacharya PN, Jha DK (2012) Plant growth promoting rhizobacteria: emergence in agriculture. World J Microbiol Biotechnol 28:127–1350

Blachinsky D, Antonov J, Bercovitz A, Elad B, Feldman K, Husid A, Lazare M, Marcov N, Shamai I, Keren-Zur M, Droby S (2007) Commercial applications of "Shemer" for the control of pre- and postharvest diseases. IOBCWPRS Bull 30:75–78

Bloemberg GV (2007) Microscopic analysis of plant bacteria interactions using auto fluorescent proteins. Eur J Plant Pathol 119:301–309

Blumer C, Haas D (2000) Mechanism, regulation, and ecological role of bacterial cyanide biosynthesis. Arch Microbiol 173:170–177

Campbell R (1989) Biological control of microbial plant pathogens. Cambridge University Press

Choudhary D, Johri BN (2009) Interactions of *Bacillus* spp. and plants with special reference to induced systemic resistance (ISR). Microbiol Res 164:493–513

Cipollone R, Frangipani E, Tiburzi F, Imperi F, Ascenzi P, Visca P (2007) Involvement of *Pseudomonas aeruginosa* rhodanese in protection from cyanide toxicity. Appl Environ Microbiol 73:390–398

Cooper M, Tavankar GR, Williams HD (2003) Regulation of expression of the cyanide- insensitive terminal oxidase in *Pseudomonas aeruginosa*. Microbiology 149:1275–1284

Couillerot O, Prigent-Combaret C, Caballero-Mellano J, Moenne-Loccoz Y (2009) *Pseudomonas fluorescens* and closely related fluorescent pseudomonads as biocontrol agents of soil borne phytopathogens. Lett Appl Microbiol 48:505–512

Crosa JH (1997) Signal transduction and transcriptional and posttranscriptional control of iron-regulated genes in bacteria. Microbiol Mol Biol Rev 61:319–336

de Bruijn I, de Kock MJ, Yang M, de Waard P, van Beek TA, Raaijmakers JM (2007) Genome-based discovery, structure prediction and functional analysis of cyclic lipopeptide antibiotics in *Pseudomonas* species. Mol Microbiol 63:417–428

de Werra P, Baehler E, Huser A, Keel C, Maurhofer M (2008) Detection of plantmodulated alterations in antifungal gene expression in *Pseudomonas fluorescens* CHA0 on roots by flow cytometry. Appl Environ Microbiol 74:1339–1349

Doumbou CL, Hamby Salove MK, Crawford DL, Beaulieu C (2002) Actinomycetes, promising tools to control plant diseases and to promote plant growth. Phytoprotection 82:85–102

Droby S, Wisniewski M, Macarisinb D, Wilson C (2009) Twenty years of postharvest biocontrol research: is it time for a new paradigm? Postharvest Biol Technol 52:137–145

Dwivedi D, Johri BN (2003) Antifungals from fluorescent pseudomonads: biosynthesis and regulation. Curr Sci 12:1693–1703

El-Tarabily KA, Sivasithamparam K (2006) Non-streptomycete actinomycetes as biocontrol agents of soil-borne fungal plant pathogens and as plant growth promoters. Soil Biol Biochem 38:1505–1520

Emmert EAB, Handelsman J (1999) Biocontrol of plant disease: a (gram-) positive perspective. FEMS Microbiol Lett 171:1–9

Fernando WGD, Nakkeeran S, Zhang Y (2005) Biosynthesis of antibiotics by PGPR and its relation in biocontrol of plant diseases. In: Siddiqui ZA (ed) PGPR: biocontrol and biofertilization. Springer, Dordrecht, pp 111–142

Figueiredo MV, Seldin L, de Araujo F, Mariano RLR (2010) Plant growth promoting rhizobacteria: fundamentals and applications. In: Maheshwari DK (ed) Plant growth and health promoting bacteria, microbiology monographs. Springer Verlag, berlin, pp 21–44

Fravel D (2007) Commercialization of biocontrol agents for use against plant pathogens. In: IX Reunia˜o Brasileira sobre Controle Biolo´gico de Doenc¸as de Plantas, Campinas, S. Paulo, Brasil, CD-ROM, pp 1–2

Fravel DR (2005) Commercialization and implementation of biocontrol. Ann Rev Phytopathol 43:337–359

Frimmersdorf E, Horatzek S, Pelnikevich A, Wiehlmann L, Schomburg D (2010) How *Pseudomonas aeruginosa* adapts to various environments: a metabolomic approach. Environ Microbiol 12:1734–1747

Garbeva P, Silby MW, Raaijmakers JM, Levy SB, Boer WD (2011) Transcriptional and antagonistic responses of *Pseudomonas fluorescens* Pf0-1 to phylogenetically different bacterial competitors. ISME J 5(6):973–985

Gomes NC, Cleary DF, Pinto FN, Egas C, Almeida A, Cunha A, Mendonça-Hagler LC, Smalla K (2010) Taking root: enduring effect of rhizosphere bacterial colonization in mangroves. PLoS ONE 5:14065

Govindasamy V, Senthilkumar M, Magheshwaran V, Kumar U, Bose P, Sharma V, Annapurna K (2010) *Bacillus* and *Paenibacillus* spp.: potential PGPR for sustainable agriculture. In: Maheshwari DK (ed) Plant growth and health promoting bacteria, microbiology monographs 18. doi:10.1007/978-3-642-13612-2_15

Gross H, Loper JE (2009) Genomics of secondary metabolite production by *Pseudomonas* spp. Nat Prod Rep 26:1408–1446

Hassan KA, Johnson A, Shaffer BT, Ren Q, Kidarsa TA, Elbourne LD, Hartney S, Heydari A, Pessarakli M (2010) A review on biological control of fungal plant pathogens using microbial antagonists. J Biol Sci 10:273–290

Heydari A, Pessarakli M (2010) A review on biological control of fungal plant pathogens using microbial antagonists. J Biol Sci 10:273–290

Hofte M, Altier N (2010) Fluorescent pseudomonads as biocontrol agents for sustainable agricultural systems. Res Microbiol 161:464–471

Homma Y, Sato Z, Hirayama F, Konno K, Shirahama H, Suzui T (1989) Production of antibiotics by *Pseudomonas cepacia* as an agent for biological control of soil borne pathogens. Soil Biol Biochem 21:723–728

Ishimaru CA, Loper JE (1993) Biochemical and genetic analysis of siderophores produced by plant-associated *Pseudomonas* and *Erwinia* species. In: Barton LL, Hemming BC (eds) Iron chelation in plants and soil microorganisms. Academic Press, San Diego

Islam MT, Hashidoko Y, Deora A, Ito T, Tahara S (2005) Suppression of damping-off disease in host plants by the rhizoplane bacterium *Lysobacter* sp. strain SB-K88 is linked to plant colonization and antibiosis against soil-borne peronosporomycetes. Appl Environ Microbiol 71:3786–3796

Jagadeesh KS, Kulkarni JH, Krisharaj PU (2001) Evaluation of role of fluorescent siderophore in the biological control of bacterial wilt in tomato using Tn5 mutants of fluorescent *Pseudomanas* sp. Curr Sci 81:882–883

Kavino M, Harish S, Kumar N, Saravanakumar D, Damodaran T, Soorianathasundaram K, Samiyappan R (2007) Rhizosphere and endophytic bacteria for induction of systemic resistance of banana plantlets against bunchy top virus. Soil Biol Biochem 39:1087–1098

Kavitha S, Senthilkumar S, Gnanamanickam SS, Inayathullah M, Jayakumar J (2005) Isolation and partial characterization of antifungal protein from *Bacillus polymyxa* strain VLB16. Process Biochem 40:3236–3243

Kerr A (1980) Biological control of crown gall through production of agrocin 84. Plant Diseases 64:25–30

Kloepper JW, Schroth MN (1978) Plant growth promoting rhizobacteria on radishes. In: Proceeding of the 4th international conference on plant pathogenic bacteria. Vol. 2, Station de Pathologie Vegetale et Phytobacteriologie, INRA, Angers, France, pp 879–882

Kloepper JW, Leong J, Teintze M, Schroth MN (1980) *Pseudomonas* siderophores: a mechanism explaining disease suppressive soils. Curr Microbiol 4:317–320

Kloepper JW, Ryu CM, Zhang S (2004) Induced systemic resistance and promotion of plant growth by *Bacillus* spp. Phytopathology 94:1259–1266

Knowles CJ (1976) Microorganisms and cynide. Bacteriol Rev 40:652–680

Knudsen IMB, Hockenhull J, Jensen DF, Gerhardson B, Hökeberg M, Tahvonen R, Teperi E, Sundheim L, Henriksen B (1997) Selection of biological control agents for controlling soil and seed-borne diseases in the field. Eur J Plant Pathol 103:775–784

Kohl J (2009) Screening of biocontrol agents for control of foliar diseases. In: Gisi et al (eds) Recent developments in management of plant diseases. doi:10.1007/978-1-4020-8804-9_9

Kuc J (1982) Induced immunity to plant disease. Bioscience 32:854–860

Kurek E, Jaroszuk-Scisel J (2003) Rye (Secale cereale) growth promotion by *Pseudomonas fluorescens* strains and their interactions with *Fusarium culmorum* under various soil conditions. Biol Control 26:48–56

Kurtzman CP, Droby S (2001) *Metschnikowia fructicola*, new ascosporic yeast with potential for biocontrol of postharvest fruit rots. Syst Appl Microbiol 24:395–399

Labuschagne N, Pretorius T, Idris AH (2010) Plant growth promoting rhizobacteria as biocontrol agents against soil borne plant diseases. In: Maheshwari DK (ed) Plant growth and health promoting bacteria, Microbiology monographs. Springer Verlag, berlin, pp 211–230

Laville J, Blumer C, Von Schroetter C, Gaia V, Defago G, Keel C, Haas D (1998) Characterization of the hcnABC gene cluster encoding hydrogen cyanide synthase and anaerobic regulation by ANR in the strictly aerobic biocontrol agent *Pseudomonas fluorescens* CHA0. J Bacteriol 180:3187–3196

Lee H, Churey JJ, Worobo RW (2008) Purification and structural characterization of bacillomycin F produced by a bacterial honey isolate active against *Byssochlamys fulva* H25. J Appl Microbiol 105:663–673

Leong J (1986) Siderophores: their biochemistry and possible role in the biocontrol of plant pathogens. Ann Rev Phytopathol 24:187–209

Lewis JA (1991) Formulation and delivery system of biocontrol agents with emphasis on fungi *Beltsville symposia* in agricultural research. In: Keister DL, Cregan PB (eds) The rhizosphere and plant growth. pp 279–287

Loper JE, Buyer JS (1991) Siderophores in microbial interactions on plant surfaces. Mol Plant-Microbe Interact 4:5–13

Manjula K, Podile AR (2001) Chitin supplemented formulations improve biocontrol and plant growth promoting efficiency of *Bacillus subtilis* AF1. Can J Microbiol 47:618–625

Mathivanan N, Manibhushanrao K (2004) An overview of current strategies on biological control of soil-borne pathogens. In: Prakash HS, Niranjana RS (eds) Vistas in applied botany. Department of Applied Botany and Biotechnology, University of Mysore, Mysore, pp 119–148

Mathivanan N, Manibhushanrao K, Murugesan K (2006) Biological control of plant pathogens. In: Anand N (ed) Recent trends in botanical research. University of Madras, Chennai, pp 275–323

Mathivanan N, Prabavathy VR, Vijayanandraj VR (2008) The effect of fungal secondary metabolites on bacterial and fungal pathogens. In: Karlovsky P (ed) Secondary metabolites in soil ecology. Springer, Berlin, pp 129–140

Mavrodi DV, Loper JE, Paulsen IT, Thomashow LS (2009) Mobile genetic elements in the genome of the beneficial rhizobacterium *Pseudomonas fluorescens* Pf-5. BMC Microbiol. doi:10.1186/1471-2180-9-8

Meyer JM (2000) Pyoverdines: pigments, siderophores and potential taxonomic markers of *fluorescent Pseudomonas* species. Archi Microbiol 174:135–142

Moyne AL, Shelby R, Cleveland TE, Tuzun S (2001) Bacillomycin D: an iturin with antifungal activity against *Aspergillus flavus*. J Appl Microbiol 90:622–629

Nakkeeran S, Renukadevi P, Marimuthu T (2005) Antagonistic potentiality of *Trichoderma viride* and assessment of its efficacy for the management of cotton root rot. Arch. Phytopathol Plant Prot 38:209–225

Nandakumar R, Babu S, Viswanathan R, Raguchander T, Samiyappan R (2001) Induction of systemic resistance in rice against sheath blight disease by *Pseudomonas fluorescens*. Soil Biol Biochem 33:603–612

Neilands JB, Leong SA (1986) Siderophores in relation to plant growth and disease. Annu Rev Plant Physiol 37:187–208

Patel D, Jha CK, Tank N, Saraf M (2011) Growth enhancement of chickpea in saline soils using plant growth promoting rhizobacteria. J Plant Growth Regul doi:10.1007/s00344-011-9219-7

Pathma J, Kennedy K, Sakthivel N (2011) Mechanisms of fluorescent Pseudomonads that mediate biological control of phytopathogens and plant growth promotion of crop plants. In: Maheshwari DK (ed) Bacteria in agrobiology: plant growth responses. doi:10.1007/978-3-642-20332-9_4, Springer-Verlag, Berlin, pp 77–104

Pengnoo A, Kusonwiriyawong C, Nilratana L, Kanjanamaneesathian M (2000) Greenhouse and field trials of the bacterial antagonists in pellet formulations to suppress sheath blight of rice caused by *Rhizoctonia solani*. Biocontrol 45:245–256

Perez-Garcia A, Romero D, Zeriouh H, de Vicente A (2011) Biological control of phytopathogenic fungi by aerobic endospore formers. In: Logan NA, De Vos P (eds) Endospore forming soil bacteria. Springer-Verlag, Berlin, pp 157–180

Petti C, Khan M, Doohan F (2010) Lipid transfer proteins and protease inhibitors as key factors in the priming of barley responses to *Fusarium* head blight disease by a biocontrol strain of *Pseudomonas fluorescens*. Funct Integr Genomics 10:619–627

Pliego C, Ramos C, de Vicente A, Cazorla FM (2011) Screening for candidate bacterial biocontrol agents against soilborne fungal plant pathogens. Plant Soil 340:505–520

Prabavathy VR, Vajayanandraj VR, Malarvizhi K, Mathivanan N, Mohan N, Murugesan K (2008) Role of actinomycetes and their metabolites in crop protection. In: Khachatourian GC, Arora DK, Rajendran TP, Srivastava AK (eds) Agriculturally important microorganisms. Academic World International, Bhopal, pp 243–255

Pueyo MT, Jr CB, Carmona-Ribeiro AM, Mascio P (2009) Lipopetides produced by a soil *Bacillus megatarium* strain. Microbiol Ecol 57:367–378

Press CM, Lopper JE, Kloepper JW (2001) Role of iron in rhizobacteria mediated induced systemic resistance of cucumber. Phytopathology 91:593–598

Rajkumar M, Lee KJ, Freitas H (2008) Effects of chitin and salicyclic acid on biological control of *Pseudomonas* spp. against damping off of pepper. S Afr J Bot 74:268–273

Ramamoorthy V, Viswanathan R, Raguchander T, Prakasam V, Samayapan R (2001) Induction of systemic resistance by plant growth promoting rhizobacteria in crop plants against pests and diseases. Crop Protect 20:1–11

Ramesh S (2009) Marine actinomycetes diversity in Bay of Bengal, India: isolation and characterization of bioactive compounds from *Streptomyces fungicidicus* MML1614. Ph. D. thesis, University of Madras, Madras, India

Ramette A, Frapolli M, Defago G, Moenne-Loccoz Y (2003) Phylogeny of HCN synthase-encoding hcnBC genes in biocontrol fluorescent pseudomonads and its relationship with host plant species and HCN synthesis ability. Mol Plant Microbe Interact 16:525–535

Ravel J, Cornelis P (2003) Genomics of pyoverdine-mediated iron uptake in pseudomonads. Trends in Microbiol 11:195–200

Romero D, Pe´rez-Garcia A, Rivera ME, Cazorla FM, de Vicente A (2004) Isolation and evaluation of antagonistic bacteria towards the cucurbit powdery mildew fungus *Podosphaera fusca*. Appl Microbiol Biotechnol 64:263–269

Romero D, de Vicente A, Rakotoal RH, Dufour SE, Veening JW, Arrebola E, Cazorla FM, Kuipers OP, Paquot M, Perez-Garcia A (2007) The iturin and fengycin families of lipopeptides are key factors in antagonism of *Bacillus subtilis* towards *Podosphaera fusca*. Mol Plant Microbe Interact 20:430–440

Sandra AI, Wright CH, Zumoff LS, Steven VB (2001) Pantoea agglomerans strain EH318 produces two antibiotics that inhibit *Erwinia amylovora* in vitro. Appl Env Microbiol 67:282–292

Saraf M, Jha CK, Patel D (2010) The role of ACC deaminase producing PGPR in sustainable agriculture. In: Maheshwari DK (ed) Plant growth and health promoting bacteria, microbiology monographs. Springer Verlag, berlin, pp 365–386

Schisler DA, Slininger PJ, Behle RW, Jackson MA (2004) Formulation of *Bacillus* spp. for biological control of plant diseases. Phytopathology 94:1267–1271

Senthilkumar M, Govindasamy V, Dureja P, Annapurna K (2007) Purification and partial characterization of antifungal peptides from soybean endophyte- *Paenibacillus* sp strain HKA-15. J Plant Biochem Biotechnol 16:131–134

Silva HSA, de Silva RR, Macagnan D, de Almeda Halfeld-VieraB, Pereira MCB, Mounteer A (2004) Rhizobacterial induction of systemic resistance in tomato plants: non specific protection and increase in enzyme activities. Biol Control 29:288–295

Stadnik MJ (2000) Induc¸a˜o de resiste^ncia a Oı´dios. Summa Phytopath 26:175–177

Stockwell VO, Stack JP (2007) Using *Pseudomonas* spp. for integrated biological control. Phytopathology 97:244–249

Tapadar SA, Jha DK (2013) Disease management in staple crops: a bacteriological approach. In: Maheshwari DK (ed) Bacteria in agrobiology: disease management. Springer-Verlag, Berlin

van Loon LC, Bakker PAHM, Pieterse CMJ (1998) Systemic resistance induced by rhizosphere bacteria. Ann Rev Phytopath 36:453–483

Velicer GJ, Raddatz G, Keller H, Deiss S, Lanz C, Dinkelacker I, Schuster SC (2006) Comprehensive mutation identification in an evolved bacterial cooperator and its cheating ancestor. Proc Natl Acad Sci U S A 103:8107–8112

Voisard C, Keel C, Haas D, Defago G (1989) Cyanide production by *Pseudomonas fluorescens* helps suppress black root rot of tobacco under gnotobiotic conditions. EMBO J 8:351–358

Weger LA, Arendonk JJ, Recourt K, Hofstad GA, Weisbeek PJ, Lugtenberg B (1988) Siderophore-mediated uptake of $Fe^{3+}$ by the plant growth-stimulating *Pseudomonas putida* strain WCS358 and by other rhizosphere microorganisms. J Bacteriol 170:4693–4698

Zhao Z, Wang Q, Wang K, Brian K, Liu C, Gu Y (2010) Study of the antifungal activity of *Bacillus vallismortis* ZZ185 in vitro and identification of its antifungal components. Biores Technol 101:292–297

Zhou WW, Huang JX, Niu TG (2008) Isolation of an antifungal *Paenibacillus* strain HT 16 from locusts and purification of its medium-dependent antagonistic compound. J Appl Microbiol 105:912–919

Zhou T, Chen D, Li C, Sun Q, Li L, Liu F, Shen Q, Shen B (2012) Isolation and characterization of *Pseudomonas brassicacearum* J12 as an antagonist against *Ralstonia solanacearum* and identification of its antimicrobial components. Microbiol Res. doi:10.1016/j.micres.2012.01.003

# Pathogenicity, Ecology and Genetic Diversity of the *Fusarium* spp. Associated with an Emerging Bakanae Disease of Rice (*Oryza sativa* L.) in India

27

B. M. Bashyal, Rashmi Aggarwal, Sagar Banerjee, Sangeeta Gupta and Sapna Sharma

**Abstract**

Bakanae disease is one of the emerging diseases of rice (*Oryza sativa* L.). *Gibberella fujikuroi* species complex were detected in popularly grown rice varieties of India with infection percentage ranging from 1 to 24 %. Pathogenicity test of *Fusarium* spp. was performed in susceptible rice variety Pusa 1121, which showed reduced seed germination and possessed varying ability to cause symptoms. On the basis of internal transcribed spacer (ITS) and Translation Elongation Factor (TEF), three *Fusarium* spp., viz. *F. verticillioides, F. fujikuroi* and *F. proliferatum* were found associated with bakanae disease of rice in India. Maximum numbers of slender and chlorotic leaves were produced by *F. fujikuroi* (90 %), whereas crown rot and stem rot was produced by *F. verticillioides* (50 %). *F. proliferatum* produced both elongation and rotting symptoms. Information on the bakanae disease, its distribution, characterization and identification in India could be helpful for the development of management strategies.

**Keywords**

Bakanae disease · Biocontrol · *Fusarium* · Plant pathogen · Genetic diversity

## 27.1 Introduction

Rice (*Oryza sativa* L.) is one of the three major food crops of the world. Being grown worldwide, it is the staple food for more than one and a half of the world's population. India is the largest rice-growing country accounting for about one third of the world acreage under the crop. It is grown in almost all states of India, covering more than 30 % of the total cultivated area. India is the largest producer and exporter of basmati rice in the world. During the year 2011, the rice variety Pusa Basmati 1121 was grown by the farmers on 1.35 million ha area, producing 5.8 million ton of basmati paddy worth 11,600 crores. However, bakanae disease

B. M. Bashyal (✉) · R. Aggarwal · S. Banerjee · S. Gupta · S. Sharma
Division of Plant Pathology, Indian Agricultural Research Institute, New Delhi 110012, India
e-mail: bishnumayabashyal@gmail.com

R. N. Kharwar et al. (eds.), *Microbial Diversity and Biotechnology in Food Security*,
DOI 10.1007/978-81-322-1801-2_27, © Springer India 2014

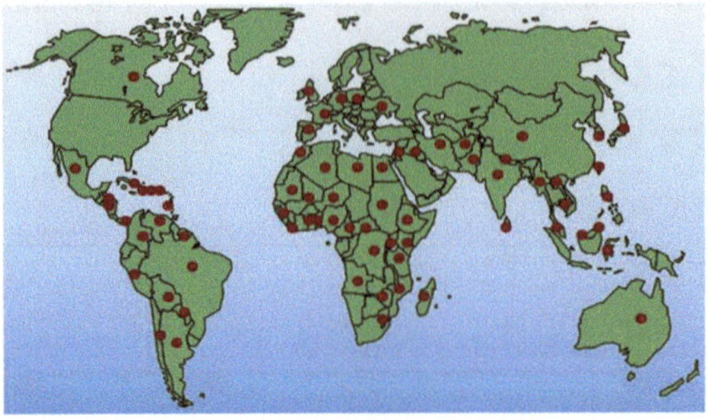

**Fig. 27.1** Worldwide distribution of bakanae disease of rice

is major constraint for this highly popular variety. Bakanae disease incidence was reported high from Punjab, Haryana, Uttar Pradesh and Uttarakhand. In the year 2011, up to 40% disease incidence was reported from Kapurthala, Ropar, Patiala, Ludhiana, Amritsar, Gurudaspur and Hoshiarpur district of Punjab.

The disease is said to be known in Japan since 1828 and it was described and called foot rot in India by Thomas in 1931. Bakanae disease is one of the emerging diseases of rice (*Oryza sativa* L.). This disease has been reported from the rice tracts of South Asia, European countries and the USA (Fig. 27.1). It is emerging as a potential threat in Japan, Taiwan, Thailand and India (Webster and Gunnell 1992; Kini et al. 2002; Saremi 2005; Anonymous 2007). The name bakanae means 'bad' or 'foolish' seedlings in Japanese, referring to the elongation symptoms specific for the disease, and caused by gibberellins production by the pathogen upon infection of the host. Bakanae is traditionally associated with rice, but water grass plants such as *Echinochloa* spp., with classic symptoms of bakanae were also observed in California in 2002 (Carter et al. 2008). Although the bakanae disease usually causes dieback or sterility of rice, mycotoxin contamination also poses a concern since the pathogen is seed borne.

The disease is said to have known in Japan since 1828 (Ito and Kimura 1931). It was first described by Hori (1808) who identified the causal organism as *Fusarium heterosporum* Nees. Fujikuro found the telomorph which was described as *Lisea fujikuroi* by Sawada (1917). It was later put on the genera *G. fujikuroi* (Sawada) Ito (Ito and Kimura 1931), with *Fusarium moniliforme* Sheld as its anamorph. Kurosawa 1926 demonstrated hypertrophic or bakanae effect of the fungus on its hosts. This unique phenomenon attracted the attention of the biochemists and plant physiologists and led to the isolation of gibberellins and other growth regulators by Yabuta et al. 1934. The disease was described and called foot rot in India by Thomas (1931).

## 27.2 Yield Losses

Ito and Kimura (1931), reported up to 20% loss in Hokkaido (Japan). Pavgi and Singh (1964) stated that losses of 15% occurred in eastern districts of Uttar Pradesh (India) and Kanjanasoon (1965) found 3.7–14.7% loss in northern and central Thailand. However, recent reports reported 40% losses in rice (Ou 1987).

## 27.3 Pathogen

Although bakanae disease was first described more than 100 years ago in Japan, it is still not clear which *Fusarium* species are associated with different symptoms. Early work in Japan identified the pathogen as *F. moniliforme* in a broad sense (Ou 1985); however, this taxon comprises a number of distinct species, now collectively termed the *Gibberella fujikuroi* species complex. Sun and Snyder (1981), produced perithecia by crossing strains of *F. fujikuroi* in the laboratory; then four reproductively isolated groups of *G. fujikuroi* were designated as groups A, B, C and D, with genetically interfertile strains from rice designated as mating group C (Hseieh et al. 1977; Kuhlman 1982). Additional genetic studies have identified nine biological species or mating populations, designated A–I, within the *G. fujikuroi* species complex (Leslie 1995; Viljoen et al. 1997; Leslie and Summerell 2006). *G. fujikuroi* species complex is generally designated as section Liseola, comprising nine biological species (Table 27.1).

Three mating populations of section Liseola (A, C and D) of the *G. fujikuroi* complex

**Table 27.1** Mating populations of *Gibberella fujikuroi* species complex

| S. No. | Anamorph | Telomorph | Mating population |
|---|---|---|---|
| 1. | *F. verticillioides* | *G. fujikuroi* | MP-A |
| 2. | *F. sacchari* | *G. fujikuroi* | MP-B |
| 3. | *F. fujikuroi* | *G. fujikuroi* | MP-C |
| 4. | *F. proliferatum* | *G. fujikuroi* | MP-D |
| 5. | *F. subglutinans* | *G. fujikuroi* | MP-E |
| 6. | *F. thapsinum* | *G. fujikuroi* | MP-F |
| 7. | *F. nygamai* | *G. fujikuroi* | MP-G |
| 8. | *F. circinatum* | *G. fujikuroi* | MP-H |
| 9. | *F. konzum* | *G. fujikuroi* | MP-I |

**Table 27.2** Rice varieties infected with *Gibberella fujikuroi* species complex from India

| S. No. | Rice variety | Infection (%) |
|---|---|---|
| 1. | Pusa Basmati 1121 | 24 |
| 2. | Pusa Basmati-1 | 4 |
| 3. | Tarori Basmati | 20 |
| 4. | Samba Mahsuri | 2 |
| 5. | Jaya | 10 |
| 6. | Ajaya | 4 |
| 7. | Vikramaraya | 6 |
| 9. | Swarnadhan | 1 |
| 10. | IR 50 | 4 |
| 11. | IR 64 | 4 |
| 12. | MTU 1010 | 4 |

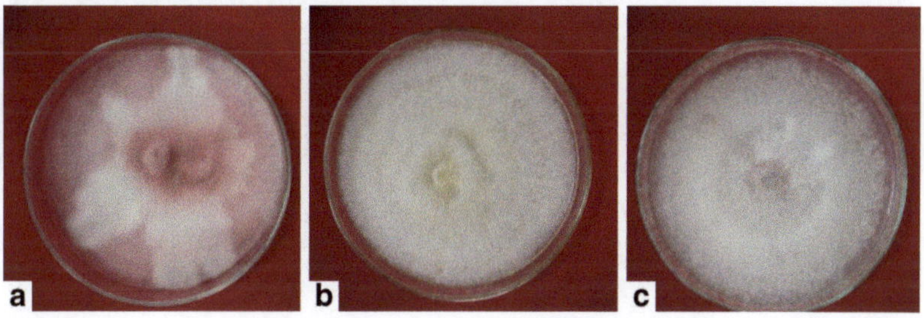


**Fig. 27.2** Bakanae disease pathogens **a** *Fusarium fujikuroi*. **b** *Fusarium proliferatum*. **c** *Fusarium verticillioides*

have been associated with bakanae disease of rice. Mating population C (MP-C) (anamorph, *Fusarium fujikuroi*; Nirenberg 1976), was first identified in 1977 among strains from rice from Taiwan (Hseieh et al. 1977). It has been found responsible for bakanae disease in Italy (Amatulli et al. 2010). Mating population A (MP-A) (anamorph, *Fusarium verticillioides* (synonym, *F. moniliforme*)) and mating population D (MP-D) (anamorph, *Fusarium proliferatum*) have been isolated from rice from Asia, and MP-D has been isolated from rice from Africa, Australia and the USA (Desjardins et al. 1997; Amoah et al. 1996; Voigt et al. 1995). Thus, more than one species of *Fusarium* may be able to infect rice and cause symptoms of bakanae disease.

Wulff et al. (2010) isolated and characterized African and Asian populations of *Fusarium* spp. (*G. fujikuroi* species complex) associated with bakanae of rice (*Oryzae sativa* L.) with respect to ecology, phylogenetics, pathogenicity and mycotoxin production. Independent of the origin, *Fusarium* spp. were detected in the different rice seed samples with varied infection rate ranging from 0.25 to 9%. Four *Fusaria* (*F. andiyazi, F. fujikuroi, F. proliferatum* and *F. verticillioides*) were found associated with bakanae disease of rice. While three of the *Fusaria* were found in both African and Asian seed samples, *F. fujikuroi* was only detected in seed samples from Asia. Phylogenetic studies showed a broad genetic variation among the strains that were distributed into four different genetic clades. Pathogenicity tests showed that all strains reduced seed germination and possessed varying ability to cause symptoms of bakanae on rice, some species (i.e. *F. fujikuroi*) being more pathogenic than others. *G. fujikuroi* species complex were detected in popularly grown rice varieties with infection percentage ranging from 1 to 24% in India (Bashyal and Aggarwal 2013), (Table 27.2). Three *Fusarium* spp. viz., *F. verticillioides, F. fujikuroi* and *F. proliferatum* were found associated with bakanae disease of rice in India (Fig. 27.2). Maximum incidence of slender and chlorotic leaves were produced by *F. fujikuroi* (90%) while maximum incidence of crown rot and stem rot was produced by *F. verticillioides* (50%).

**Fig. 27.3** Bakanae disease of rice in field showing **a** Elongation. **b** Rotting. **c** Mycelial growth. **d** Foot rot and adventitious root formation symptoms

## 27.4 Disease Symptoms

Yamanaka and Honkura (1978), classified five types of symptoms (a) elongation, (b) elongation and normal growth, (c) elongation then stunted growth, (d) stunted growth and (e) no growth. The amount of each type of the symptom varied with the isolate. Sun and Synder (1978) reported that varying inoculums potential produced different types of symptoms. Sasaki (1973), reported lesion formation on rice leaves but Sun (1975), considered that no vegetative parts above ground are sites for the infection. Sasaki (1976) reported the presence of elongation symptoms in ratoon plants in Japan.

The typical symptoms of bakanae are slender, chlorotic and abnormally elongated primary leaves, however, not all infected seedlings show these symptoms, as crown rot is also seen, resulting in stunted rice plants. In crops reaching maturity, infected plants show tall lanky tillers bearing pale green flag leaves which are conspicuous above the general level of the crop. Infected plants usually have small numbers of tillers and leaves dry up one after another from below and die in few weeks. Occasionally infected plants survive until maturity but bear empty panicles. A white or pink mycelial growth may be noticed in lower parts of the infected plants (Fig. 27.3).

## 27.5 Disease Cycle

The disease is seed borne. Seeds are infected at the flowering stage. In moderately diseased rice fields, 100 % of seed yielded the pathogen and 30 % of them showed bakanae symptoms when planted. Hino and Furuta (1968), reported the average embryo infection of 8.85% in July, 8.1 % in August and 0.03 % in September. When severely infected, the kernels develop a reddish discolouration due to the presence of the conidia of the pathogen. Seto (1937) determined, that the most favourable stage for the development of seed infection is the flowering time.

Kanjanasoon (1965), in Thailand showed that artificially inoculated soil caused 93 % infection immediately after inoculation. Infection decreased

**Table 27.3** Incidence of bakanae disease in different rice varieties. (except Pusa 1121)

| States | Districts | Disease incidence (%) | Variety grown | Symptoms |
|---|---|---|---|---|
| Haryana | Karnal | 2 | CSR 30 | Elongation and rotting |
| Punjab | Fatehgarh | 10 | 1401 | Elongation and rotting |
| Uttar Pradesh | Aligarh | 15 | 1401 | Elongation and rotting |
| | Buladshaher | 2 | 2511 | Elongation and rotting |
| | Gautam Budh Nagar | 3 | 1509 | Elongation and rotting |
| Uttarakhand | Haridwar | 2 | Dehradun Basmati | Rotting |
| | Udham Singh Nagar | 2 | Pakistani Basmati | Rotting |

with the passage of time and only 0.7% infection occurred after 90 days of the inoculation and no infection resulted after 180 days. This indicated that the fungus does not survive in soil for long in tropics. According to Sun (1975), fungus survives in soil for about 4 months in the form of thick-walled hyphae or macroconidia.

Nishikado and Kimura 1941 found microconidia and mycelium of the pathogen could be observed in vascular bundles especially in the large-pitted vessels and the lacunae of the xylem. The phloem and parenchyma were not invaded to any extent. The pathogen was found to be discontinuously distributed. The pathogen survives in the winter (summer in the tropics) in infected seeds or other parts of diseased plants. Kanjansoon (1965) found that the fungus is viable in seeds and other parts of the diseased plants up to 4–10 months in room temperature and more than 3 years in cold storage at 7 °C.

## 27.6 Disease Epidemiology

It is generally observed that bakanae plants are few or not observed at all when the temperature is low. The temperature of 35 °C is most favourable for seedling growth and also for infection. At the temperature of 25 °C, bakanae plants could still be found but at 20 °C they failed to appear. However, fungus was isolated from these outwardly healthy looking plants. The optimum temperature for the fungus is 27–30 °C; however, optimum temperature for the disease is 35 °C. It has been observed that damp soil conditions are favourable for elongated symptoms and dry soil conditions are favourable for stunt growth. Kanjanasoon (1965) obtained higher percentage of infection when dry seed were sown than pre-soaked or germinated seeds.

## 27.7 Host Resistance

Although the disease attacks both scented and unscented groups of paddy, it is more common in high yielding dwarf and aromatic tall varieties (export quality). Rice variety Pusa 1121 was observed highly susceptible to the bakanae. However, the disease was observed in other basmati varieties too. In the year 2012, rice varieties Pusa 1401, Pusa 2511, CSR 30, Dehradun basmati and Pakistani basmati were infected by the disease (Table 27.3). Some of the varieties identified with different degree of resistance in India and Pakistan against bakanae disease are listed below (Table 27.4).

## 27.8 Molecular Detection of the Pathogen

Molecular methods are among the most precise tools for differentiation between species and identification of new strains/isolates collected from infected samples. They differ regarding discriminatory power, reproducibility, ease of use and interpretation. DNA fingerprinting of *Fusarium* has been successfully used for characterization of individual isolates and grouping them into standard racial classes and groups. This is particularly useful when any unknown fungal sample is to be identified. A comparison at the DNA se-

**Table 27.4** Genotypes showing different degree of resistance to bakanae disease of rice

| Genotypes | Disease ratings[a] |
|---|---|
| BR 1067-84-1-3-2-1, BR 1257-31-1-1,BR 4367-8-11-4-9, IR 58109-109-1-1-3, IR-6, DR-82,ADT-40, ADT-44,ADT-41, ASD-16, Amulya, Sabita, Ereimaphou, Prasanna, GR-4, IR-64, Akutphou, MTU-1010 | HR |
| AS 34011, BG 936, CNA 3886, HKR86-104, IR 39464-54-1-3-2-1-3, PR 106, C 4-64 (green base), Karjat x 13-21, IR-8, DR-83, KS-282, DM-15-1-95 | R |
| BR 802-118-4-2, BR 827-35-2-1-HR, 6, HKR 42, HKR 46, HKR 126, HKR 91-104, HKR 91-108, HKR 91-110, HKR 91-112, HKR 91-119, HKR 91-120, IR 51673-50-2-1, IR 57301-195-3-3, Jaya, PR 110, RP 2235-113-85-20, RP 2240-52-4-8, RP 49455-20-2-2-3-3-1, 52616, Guinata, Hansraj, HKR 91-417, Kamod, Lua Nhe, Macunting, Milpal, S-1732 | MR |

[a] Disease incidence of 0, <1, 1–5 was rated as *HR* highly resistant, *R* resistant, *MR* moderately resistant, respectively

quence level provides accurate classification of fungal species and is beginning to elucidate the evolutionary and ecological relationships among diverse species (Mule et al. 2005).

The sequences most commonly used to distinguish *Fusarium* spp. are portions of the genomic sequences encoding the translocation elongation factor 1-α (TEF) (Wulff et al. 2010), β-tubulin (tub2) (O'Donnell et al. 1998), calmodulin (O'Donnell et al. 2000), internally transcribed spacer regions in the ribosomal repeat region (ITS1 and ITS2) (Waalwijk et al. 1996; O'Donnell and Cigelnik 1997) and the intergenic spacer region (IGS) (Yli-Mattila and Gagkaeva 2010). Other molecular techniques such as RAPDs (Du-Teau and Leslie 1991; Mitter et al. 2002; Voigt et al. 1995), mitochondrial RFLPs (Correll et al. 1992), AFLPs (Chulze et al. 2000; Zeller et al. 2003) and CHEF-gel karyotypes (Xu et al. 1995) have been also used to differentiate members of the *G. fujikuroi* species complex. Based on the results of these analyses, the *G. fujikuroi* complex has been delineated into three lineages, designated as the African, Asian and American clades (O'Donnell et al. 1998). Not all sequences work equally well for all species. *TEF1* gene (primer ef1 (5′-ATGGGTAAGGA (A/G) GACAAGA C-3′) and primer ef2 (5′-GGA (G/A) GTACCAGT (G/C) ATCATGTT-3′) the most widely accepted across the genus. Phylogenetic analysis of ITS region produced dendogram consisted of three different clades in India (Bashyal and Aggarwal 2013). Clade I comprised of *F. verticillioides* isolates. Clade II consisted of *F. fujikuroi* and clade III was formed by *F. proliferatum* strains with 100% bootstrap support (Fig. 27.4).

## 27.9 Management

Currently the most common management practice for bakanae disease is seed treatment with fungicides. Seed treatment with benomyl, thiram and benomyl combination, and carbendazim are effective. Dry seed coating with the fungicide benomyl or benomyl-T, 1–2% seed weight and seed soaking in 1:1,000 for 1 h and 1:2,000 for 5 h gave good control. Some workers have reported that treatment is more effective in pre-germinated seeds when the shoot is 1 mm long.

## 27.10 Conclusion

More than one pathogen of *G. fujikuroi* spp. complex (mostly *Fusarium fujikuroi, Fusarium verticilloides, F. proliferatum*) has been associated with bakanae disease of rice. Only *Fusarium fujikuroi* was found associated with typical symptoms of bakanae disease, i.e. slender and elongated leaves and abnormal growth of the plant. In the last decade, the organization of *Fusarium* spp. into well-defined lineages and their mapping to specific geographic locations have been achieved by analysis of genes involved in mycotoxin biosynthesis or other metabolic processes to study the pathogen populations prevalent in those regions. Knowledge of the distribution and aggressiveness pattern of *Fusarium* spp. is very

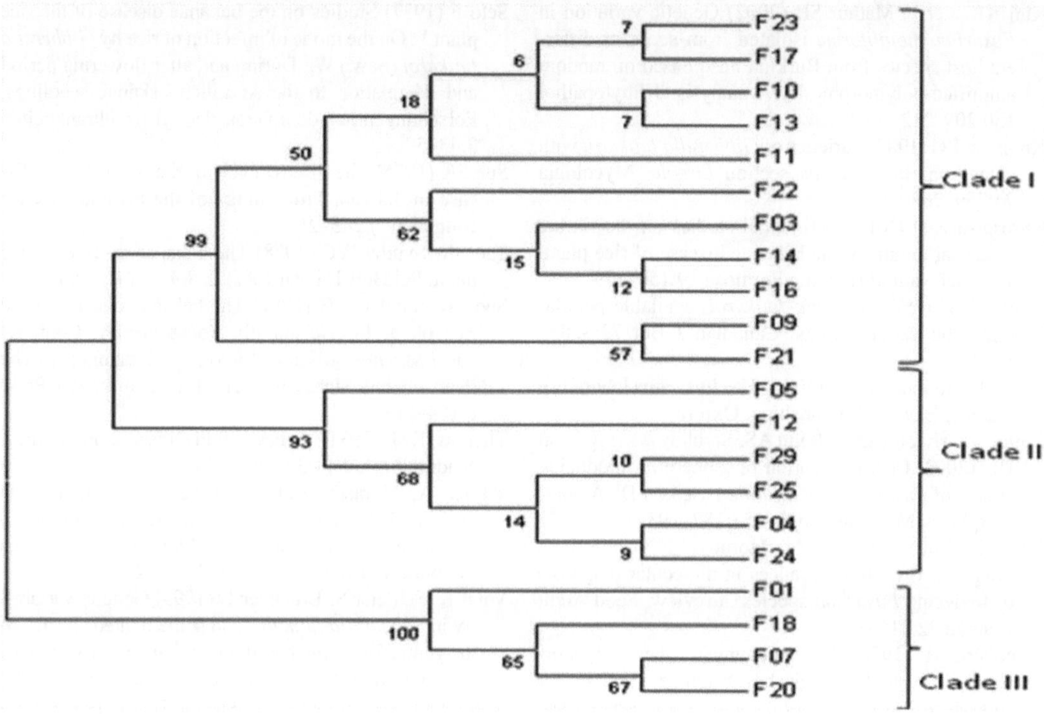


**Fig. 27.4** Dendogram constituted by neighbour-joining method based on ITS sequences of *Fusarium* spp. Bootstrap values based on 1,000 replications are indicated as percentages in the internodes when replication frequencies exceed 50 %

useful for understanding and developing strategies to control the bakanae disease of rice.

## References

Amatulli MT, Spadaro D, Gullino ML, Garibaldi A (2010) Molecular identification of *Fusarium* spp. associated with bakanae disease of rice in Italy and assessment of their pathogenicity. Plant Pathol 59(5):839–844

Amoah BK, Macdonald MV, Rezanoor N, Nicholson P (1996) The use of the random amplified polymorphic DNA technique to identify mating groups in the *Fusarium* section *Liseola*. Plant Pathol 45:115–125

Anonymous (2007) Bakanae disease an emerging problem of Basmati rice. NCIPM Newsletter 13:4

Bashyal BM, Aggarwal R (2013) Molecular identification of *Fusarium* spp. associated with bakanae disease of rice in India. Indian J Agr Sci 83(1):72–77

Carter LLA, Leslie FJ, Webster RK (2008) Population Structure of *Fusarium fujikuroi* from California rice and water grass. Phytopathol 9:992–998

Chulze SN, Ramirez ML, Torres A, Leslie JF (2000) Genetic variation in *Fusarium* section Liseola from no-till maize in Argentina. Appl Environ Microbiol 66:5312–5315

Correll JC, Gordon TR, McCain AH (1992) Genetic diversity in California and Florida populations of the pitsch canker fungus *Fusarium subglutinans f. sp. pini*. Phytopathology 82:415–420

Desjardins AE, Plattner RD, Nelson PE (1997) Production of fumonisin B1 and moniliformin by *Gibberella fujikuroi* from rice from various geographic areas. Appl Environ Microbiol 63:1838–1842

Du Teau NM, Leslie JF (1991) RAPD markers for *Gibberella fujikuroi* (Fusariu section Liseola). Fun Genet Newsl 38:37

Hino T, Furuta T (1968) Studies on the control of rice plant disease caused by *Gibberella fujikuroi*-2. Influence of flowering season of rice plant on seed transmissibility through flower infection. Bull Chigoku-Shikoku Agric Exp Stn E-2:97–102

Hori S (1808) Researches on bakanae disease of rice plant II, on the infection of rice by *Lisea fujikuroi* Sawada and *Gibberella saubinetii* (Mont.) Sacc. Forschungen aus dem Gebit der Pflanzenkrankheiten 1:99–100

Hseieh WH, Smith SN, Snyder WC (1977) Mating groups in *Fusarium moniliforme*. Phytopathology 67:1041–1043

Ito S, Kimura J (1931) Studies on the bakanae disease of the rice plant. Rep Hokkaido Agr Exp Stn 27:1–95

Kanjanasoon P (1965) Studies on the bakanae disease of rice in Thailand. Doc Agr Thesis, Tokyo University, Japan

Kini KR, Let V, Mathur SB (2002) Genetic variation in *Fusarium moniliforme* isolated from seeds of different host species from Burkina Faso based on random amplified polymorphic DNA analysis. J Phytopathol 150:209–212

Kuhlman EG (1982) Varieties of *Gibberella fujikuroi* with anamorphs in *Fusarium* section *Liseola*. Mycologia 74:759–768

Kurosawa E (1926) Experimental studies on the filtrate of causal fungus of the bakanae disease of rice plant. Transact Natural Hist Soc Formosa 19:150–179

Leslie JF (1995) *Gibberella fujikuroi*: available populations and variable traits. Canadian J Bot 73:S282–S291

Leslie JF, Summerell BA (2006) The *Fusarium* laboratory manual. Blackwell Publishing, Oxford

Mitter N, Srivastava AC, Renu AS, Sarbhoy AK, Agarwal DK (2002) Characterization of gibberellin producing strains of *Fusarium moniliforme* based on DNA polymorphism. Mycopathologia 153:187–193

Mule G, Gonzalez-Jaen MT, Hornok L, Nicholson P, Waalwijk C (2005) Advances in molecular diagnosis of toxigenic *Fusarium* species: a review. Food Addit Contam 22:316–323

Nirenberg H (1976) Untersuchungen uber die morphologische und biologische Differenzierung in der *Fusarium*-Sektion *Liseola*. Mitt Biol Bundesanst Land-Forstwirtsch Berlin-Dahlem 169:1–117

Nishikado Y, Kimura K (1941) A contribution to the pathological anatomy of rice plants affected by *Gibberella fujikuroi* (Saw.) Wolllenweber I. Des ohara Instituts. Fur Landwirtschaftlic Forschungen. Okayama Universitat 8:421–428

O'Donnell K, Cigelnik E (1997) Two divergent intragenomic rDNA ITS2 types within a monophyletic lineage of the fungus *Fusarium* are nonorthologous. Mol Phylo Evol 7:103–116

O'Donnell K, Cigelnik E, Nirenberg HI (1998) Molecular systematics and phylogeography of the *Gibberella fujikuroi* species complex. Mycologia 90:465–493

O'Donnell K, Nirenberg HI, Aoki T, Cigelnik E (2000) A multigene phylogeny of the *Gibberella fujikuroi* species complex: detection of additional phylogenetically distinct species. Mycoscience 41:61–78

Ou SH (1985) Rice diseases. Commonwealth Mycological Institue, Kew

Ou SH (1987) Rice diseases. 2nd edn. CAB International, Slough

Pavgi MS, Singh J (1964) Bakanae and foot rot of rice in Uttar Pradesh, India. Plant Dis Rep 48:340–342

Saremi H (2005) *Fusarium*, biology, ecology and taxonomy, 1st edn. Jehad Daneshgahi Press University of Mashhad, Iran, pp 153

Sasaki T (1973) Lesion formation on rice leaves by *Fusarium moniliforme* Sheldon. Ibid 39:435–437

Sasaki T (1976) Elongation of ratoon in rice plants inoculated with *Fusarium moniliforme* Sheldon. Ibid 42:606–608

Sawada K (1917) Beitrage uber Formosas-Pilze no. 14. Transac Natural Hist Soc Formosa 3:31–133

Seto F (1937) Studies on the bakanae disease of the rice plant V. On the mode of infection of rice by *Gibberella fujikuroi* (Saw.) Wr. During and after flowering period and its relation to the so called bakanae seedlings. Forschungen aus dem Gebit der Pflanzenkrankheiten 3:43–57

Sun SK (1975) The disease cycle of the rice bakanae disease in Taiwan. Proceedings of the national science council 8(2):245–256

Sun SK, Synder WC (1978) The bakanae disease of rice plant. Sci Bull Taiwan 10(7):2, 8(4), 9 (4), 10(4)

Sun SK, Snyder WC (1981) The bakanae disease of the rice plant. In: Nelson PE, Toussoun TA, Cook RJ eds *Fusarium*: disease, biology and taxonomy. The Pennsylvania State University Press, University Park, p 104–113

Thomas KM (1931) A new paddy disease in Madras. Madras Agric J 19:34–36

Viljoen A, Marasas WFO, Wingfield MJ, Viljoen CD (1997) Characterization of *Fusarium subglutinans* f. sp. *pini* causing root disease of *Pinus patula* seedlings in South Africa. Mycol Res 101:437–445

Voigt K, Schleier S, Bruckner B (1995) Genetic variability in *Gibberella fujikuroi* and some related species of the genus *Fusarium* based on random amplification of polymorphic DNA (RAPD). Curr Genet 27:528–535

Waalwijk C, de Koning JRA, Baayen RP, Gams W (1996) Discordant groupings of *Fusarium* spp. from sections *elegans*, *liseola*, and *dlaminia* based on ribosomal ITS1 and ITS2 sequences. Mycologia 88:361–368

Webster RK, Gunnell PS (1992) Compendium of rice diseases. The American Phytopathological Society Press, St. Paul

Wulff ED, Sorensen JL, Lubeck M, Nlelson KF, Thrane U, Torp J (2010) *Fusarium* spp. associated with rice bakanae: ecology, genetic diversity, pathogenicity abd toxigenicity. Environ Microbiol 12(3):649–657

Xu JR, Yan K, Dickman MB, Leslie JF (1995) Electrophoretic karyotypes distinguish the biological species of *Gibberella fujikuroi* (*Fusarium* section Liseola). Mol Plant-Microbe Interact 8:74–84

Yabuta T, Kobe K, Hayashi T (1934) Biochemical studies of the bakanae fungus of rice. I. Fusaric acid, a new product of the bakanae fungus. Ibid 10:1059

Yamanaka S, Honkura R (1978) Symptoms on rice seedlings inoculated with 'Bakanae' disease fungus, *Fusarium moniliforme* Sheldon. Ann Phytopathol Soc Japan 44:57–58

Yli-Mattila T, Gagkaeva T (2010) Molecular chemotyping of *Fusarium gramineaum, F. culmorum* and *F. cerealis* isolates from Finland and Russia. In: Gherbawy Y, Voigt K (eds) Molecular identification of fungi. Springer, Berlin

Zeller KA, Summerell BA, Bullock S, Leslie JF (2003) *Gibberella konza* (*Fusarium konzum*) sp nov from prairie grasses, a new species in the *Gibberella fujikuroi* species complex. Mycologia 95:943–954

# Development of Novel Molecules for the Control of Plant Pathogenic Fungi in Agriculture

28

Santosh G. Tupe, Preeti M. Chaudhary, Sunita R. Deshpande and Mukund V. Deshpande

**Abstract**

More than 100 different fungicide molecules are in use with global market value of more than US$ 10 billion. These molecules were selected by random screening and might be with little knowledge about their mode of action on the fungi. A variety of technologies have given us different molecules like Triazoles, Imidazole, Morpholines, etc. to combat fungal pathogens. However, rapid development of resistance by fungi to these classes of molecules has generated further quest for new agrochemical fungicides. The obvious choice of new target is based on criteria like: (a) should be present in several pathogenic fungi, (b) should be essential for either growth or virulence of fungi, and (c) should be exclusive for fungi, i.e. preferably absent in plants or humans. The nontoxic nature of the molecules to the nontargeted organisms is a major concern too. One of the approaches can be the identification of targets through a combination of bioinformatics, comparative genomics and pathway analysis. The strategies for the identification of the targets, development of novel high throughput screening methods, generation of chemical libraries, performance in the field of lead molecules, etc. will be discussed.

**Keywords**

Fungicide· Phytopathogenic fungi · Antifungal agents · Crop protection agents

## 28.1 Introduction

A large variety of crops are cultivated in various parts of the world throughout the year, including rice, cereals, pulses, oil seeds, potato, sugar beet, fruits, vegetables and others. Due to their intensive cultivation, they are prone to serious fungal diseases which, if not managed lead to severe crop losses. It is most striking, when the epidemic

M. V. Deshpande (✉) · S. G. Tupe · P. M. Chaudhary
Biochemical Sciences Division, National Chemical Laboratory, Pune 411008, India
e-mail: mv.deshpande@ncl.res.in

S. R. Deshpande
Organic Chemistry Division, National Chemical Laboratory, Pune 411008, India

R. N. Kharwar et al. (eds.), *Microbial Diversity and Biotechnology in Food Security*,
DOI 10.1007/978-81-322-1801-2_28, © Springer India 2014

**Table 28.1** Fungal pathogens affecting crops of economic importance

| Crop | Disease | Causative agent |
|---|---|---|
| Arecanut | Koleroga | *Phytophthora arecae* |
| Beans | Anthracnose | *Colletotrichum lindemuthiamum* |
| Black pepper | Foot rot and root rot | *Phytophthora capsici* |
| Cardamom | Azhukal capsule rot | *Phytophthora parasitica* |
| Chillies | Powdery mildew | *Leveillula taurica* |
| | Die-back, fruit rot | *Colletotrichum capsici* |
| Citrus | Gummosis | *Phytophthora citrophthora* |
| | Powdery mildew | *Oidium tingitanium* |
| Coconut | Bud rot | *Phytophthora palmivora* |
| Cucurbits | Downy mildew | *Pseudoperonospora cubensis* |
| | Powdery mildew | *Sphaerotheca fulginea* |
| | Anthracnose | *Colletotrichum lagenarium* |
| Cumin | Powdery mildew | *Erysiphe polygoni* |
| | Blight | *Alternaria burnsii* |
| Grapes | Downey mildew | *Plasmopara viticola* |
| | Powdery mildew | *Uncinula necator* |
| | Anthracnose | *Gloeosporium ampelophagum* |
| Green peas | Powdery mildew | *Erysiphe polygoni* |
| Groundnut | Early and late leaf spots | *Cercospora* spp. |
| Maize | Downy mildew | *Sclerophthora raysii* |
| | Corn smut | *Ustilago maydis* |
| Mango | Powdery mildew | *Oidium mangiferae* |
| | Anthracnose | *Colletotrichum gloeosporioides* |
| Potato | Late blight | *Phytophthora infestans* |
| Rice | Sheath blight | *Rhizoctonia solani* |
| | Blast | *Pyricularia oryzae/Magnaporthe grisea* |
| | Brown leaf spot | *Drechslera oryzae* |
| Rubber | Abnormal leaf fall | *Phyyophthora meadii* |
| Soya bean | Soya bean rust | *Phakospora pachyrhizi* |
| Tomato | Fruit rot | *Phytophthora infestans* |
| | Early blight | *Alternaria solani* |
| Wheat | Stem rust | *Puccinia graminis* |

is intensively spread over large areas and if all the plants are more or less susceptible to the devastating pathogens. Major fungal pathogens affecting crops of economic importance are listed in Table 28.1. A recent review published in Nature by Fisher et al. (2012) highlights jeopardizing of food security by plant disease epidemics. Based on the global harvests for the year 2009/2010 for five major crops, rice, wheat, maize, potato and soya bean, the estimated loss due to rice blast (10–35 %), soya bean rust (10–80 %), stem rust in wheat (10–70 %), corn smut in maize (2–20 %) and late blight in potatoes (5–78 %) is at least 125 million t. Fisher et al. (2012) further suggested that this could have fed 596–4287 million mouths for a year, i.e. 8.5–61.2 % of the world's population. The only solution for the problem is effective prevention and timely control of the fungal diseases. Judicious use of available fungicides and development of pathogen-specific new fungicides will be an important arsenal in achieving this goal.

## 28.2 Fungicides

The modern fungicide development started in the mid-seventeenth century with the observation that wheat seed recovered from sea was free of bunt. Farmers started brining of grain with salt

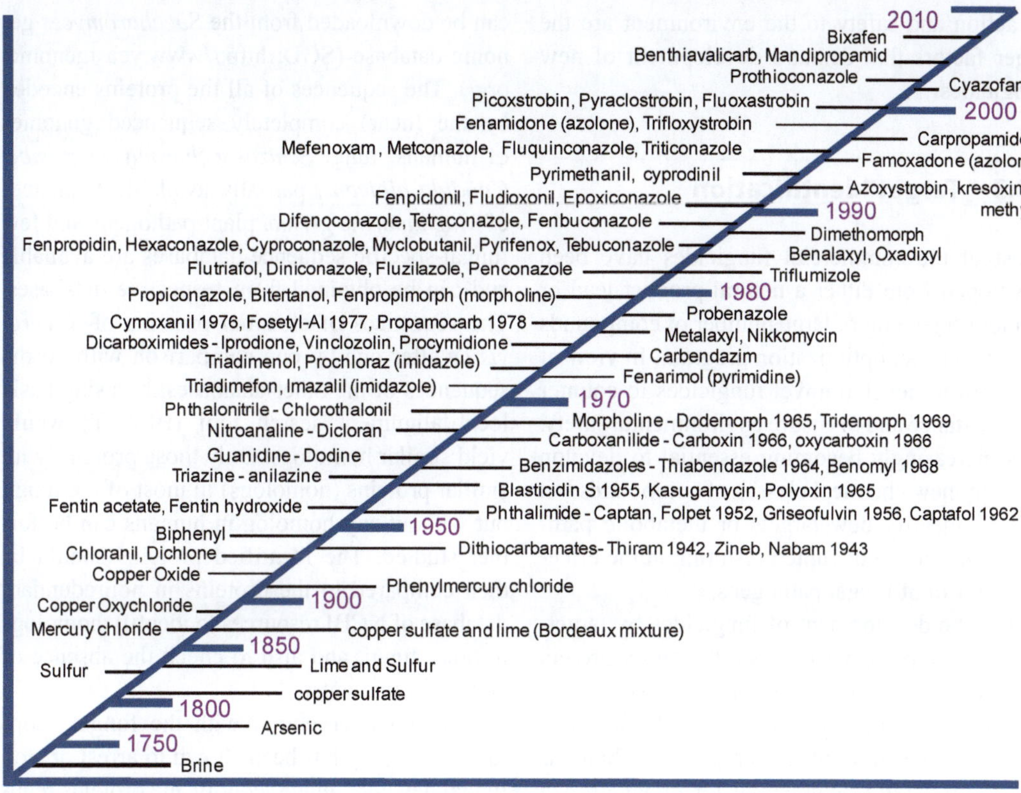


**Fig. 28.1** Timeline for fungicide development. (Compiled from Klittich 2008, Morton and Staub 2008)

water and then liming to control bunt. In 1755, Tillet demonstrated that seed-borne fungi (*Tilletia tritici, T. laevis*) caused bunt of wheat and that it could be controlled by seed treatment with lime, or lime and salt. First chemical fungicide, copper sulphate was discovered in 1807 by Prevost based on his serendipitous observation that the growth of the wheat bunt causal agent was prevented when its spore suspension was held in a copper vessel. After eight decades, the famous Bordeaux mixture, i.e. a mixture of copper sulphate and lime was discovered by Millardet. A farmer in the Bordeaux region of France had used it to discourage thievery of grapes. Millardet noticed that the sprayed vines retained their leaves, whereas unsprayed plants were defoliated by downy mildew. It was developed as the first foliar fungicide and even today many copper-based foliar fungicides are used to control a variety of fungal diseases, particularly on fruits and vegetables (Klittich 2008; Morton and Staub 2008). Thereafter, several fungicides were discovered and currently > 150 different compounds are in use with a global end user market value of US$ 10 billion. The fungicide development progress is summarized in Fig. 28.1.

Most of the chemicals mentioned in Fig. 28.1 were identified by random screening and serendipity with little knowledge about their mode of action on the fungi. A variety of technologies have given us different molecules like Triazoles, Imidazole, Morpholines, etc. to combat fungal diseases. However, rapid development of resistance by fungi to these classes of molecules has generated further quest for new agrochemical fungicides. For instance, *Botrytis cinerea* is one of the most serious plant pathogens, infecting more than 200 crop plants in the field as well as in storage. Chemical control of *Botrytis* infection is limited, mainly due to the development of fungicide-resistant strains. Apart from resistance, optimum level of persistence, extended duration of action and safety to the environment are the other factors that demand development of new fungicides.

## 28.3 Target Identification

Most of the commercial fungicides have been developed from either a natural product lead or random screening of large number of compounds followed by an optimization program. In view of the need to develop novel fungicides to counter the resistance and to ensure environmental safety, it is increasingly becoming essential to develop not only new chemical classes of compounds but also to identify new targets or metabolic pathways that can be disrupted resulting in the effective control of fungal pathogens.

For the development of fungicides by targeting specific proteins/enzymes, the target protein must have qualities such as: (1) it should be present in several fungi, (2) it should not preferably be present in plants, animals and humans and (3) it should be essential for either growth or virulence of fungi. By choosing such targets, one can ensure that the fungicides developed will be broad spectrum (i.e. active against most fungal pathogens), but at the same time less toxic to plants, livestock and humans.

Table 28.2 lists commercial fungicides based on their mode of action and target site. Best example of fungicide development based on a novel mode of biochemical action is the ergosterol biosynthesis inhibitor fungicides. More than 20 agricultural fungicides were developed based on *N*-substituted 1,2,4-triazoles and imidazoles. However, resistance is now reported in various fungal pathogens. Moreover, resistance and toxicity have been reported for most of the mechanisms which have prompted researchers to identify new targets. Bioinformatics approach to identify novel antifungal targets present in most fungi, while absent in plants and humans, and which are essential for growth/viability of fungi can be employed. Winzeler et al. (1999) and Giaever et al. (2002) have experimentally demonstrated that ~ 1,000 proteins are essential for yeast *Saccharomyces cerevisiae*. The sequences of these proteins can be downloaded from the *Saccharomyces* genome database (SGD; http://www.yeastgenome.org/). The sequences of all the proteins encoded by the (near) completely sequenced genomes of humans, fungi *Schizosaccharomyces pombe, Candida albicans,* partially available sequences of *Magnaporthe grisea,* plant pathogens and few fungal-specific sequence databases are available and can be obtained from respective databases. The sequences of essential proteins of *S. cerevisiae* after one by one comparison with all the sequences of the other databases, by using basic local alignment search tool (BLAST) would yield similar homologs. Only those proteins with similar proteins (homologs) in most of the fungi, but without any homolog in humans can be further studied. The identified proteins should be then compared to the proteins in nonredundant database of NCBI resources to identify homologs in other fungi, and also to check the absence of homologs in plants.

External intervention with the fungal apoptotic machinery has been shown to arrest or stop fungal development. Secondly, the proteins regulating fungal apoptosis are adequately different from their mammalian or plant counterparts. Hence, development of novel antifungal drugs and fungicides that activates natural fungal cell death pathways is a feasible and attractive strategy. Apoptosis inducers will affect proteins that are regulators of fungal apoptosis, thereby causing activation of global suicide response resulting in killing of the fungus (Sharon 2009).

After identification of a novel biochemical or molecular target, subsequent rational design and directed synthesis will lead to effective fungicides.

## 28.4 Screening for Fungicides

An in vivo fungicide testing assay developed by Boyce Thompson Institute using plants growing in glasshouse/greenhouse or controlled environment room was mostly employed by companies till recently and still is the cornerstone for screening of fungicides (McCallan et al. 1943; Shephard 1987). However, due to recent advances in

**Table 28.2** Classification of currently used fungicides based on the mode of action. (Adapted and modified from Fungicide Resistance Action Committee (FRAC) code list)

| Site of action | Target site | Example | Resistance |
|---|---|---|---|
| Nucleic acids synthesis | RNA polymerase I | Metalaxyl, furalaxyl | Well-known resistance and cross resistance in oomycetes |
| | Adenosin-deaminase | Bupirimate, dimethirimol | Medium-risk resistance and cross resistance in powdery mildews |
| | DNA/RNA synthesis | Hymexazole, octhilinone | Resistance not reported |
| | DNA topoisomerase type II (gyrase) | Oxolinic acid | Resistance not reported in fungi |
| Mitosis and cell division | ß-tubulin assembly in mitosis | Benomyl, carbendazim | Resistance common in many fungi |
| | Cell division | Pencycuron | Resistance not reported |
| | Delocalisation of spectrin-like proteins | Fluopicolide | Resistance not reported |
| Respiration | Complex I: NADH oxidoreductase | Diflumetorim | Resistance not reported |
| | Complex II: succinate-dehydrogenase | Boscalid, bixafen | Resistance known in various fungal species |
| | Complex III: cytochrome bc1 (ubiquinol oxidase) at Qo site (*cyt b* gene) | Azoxystrobin, coumoxystrobin, enoxastrobin | Resistance known in various fungal species |
| | Complex III: cytochrome bc1(ubiquinone reductase) at Qi site | Cyazofamid, amisulbrom | Resistance risk unknown |
| | Uncouplers of oxidative phosphorylation | Binapacryl, meptyldinocap | Resistance risk unknown |
| | Inhibitors of oxidative phosphorylation, ATP synthase | Fentin acetate, fentin chloride | Low- to medium-risk resistance |
| | ATP production | Silthiofam | Low-risk resistance reported |
| | Complex III: cytochrome bc1 (ubiquinone reductase) at Q x (unknown) site | Ametoctradin | Medium- to high-risk resistance |
| Amino acids and protein synthesis | Methionine biosynthesis (proposed) (*cgs* gene) | Cyprodinil, mepanipyrim | Resistance known in *Botrytis* and *Venturia* |
| | Protein synthesis | Blasticidin-S, kasugamycin | Low- to high-risk resistance |
| Signal transduction | Signal transduction | Quinoxyfen, proquinazid | Resistance to quinoxyfen known |
| | MAP/Histidine-Kinase in osmotic signal transduction (*os-2, HOG1*) | Fenpiclonil, fludioxonil | Resistance found sporadically |
| | MAP/Histidine-Kinase in osmotic signal transduction (*os-1, Daf1*) | Chlozolinate, iprodione | Resistance common in *Botrytis* and some other pathogens |
| Lipids and membrane synthesis | Phospholipid biosynthesis, methyltransferase | Edifenphos, iprobenfos | Resistance known in specific fungi |
| | Lipid peroxidation | Biphenyl, chloroneb, dicloran | Resistance known in some fungi |
| | Cell membrane permeability | Iodocarb, propamocarb | Low- to medium-risk resistance |
| | Microbial disrupters of pathogen cell membranes | *Bacillus subtilis* strain QST 713 | No resistance reported |

**Table 28.2** (continued)

| Site of action | Target site | Example | Resistance |
|---|---|---|---|
| Sterol biosynthesis in membranes | C14-demethylase in sterol biosynthesis (*erg11/cyp51*) | Myclobutanil, azaconazole, ipconazole | Resistance is known in various fungal species |
| | Δ14-reductase and Δ8→Δ7-isomerase in sterol biosynthesis (*erg24, erg2*) | Aldimorph, dodemorph | Decreased sensitivity for powdery mildews |
| | 3-keto reductase, C4-demethylation (*erg27*) | Fenhexamid | Low to medium risk resistance |
| | Squalene epoxidase in sterol biosynthesis (*erg1*) | Pyributicarb, terbinafine | Resistance not known |
| Cell wall biosynthesis | Trehalase and inositol biosynthesis | Validamycin | Resistance not known |
| | Chitin synthase | Polyoxin | Medium-risk resistance |
| | Cellulose synthase | Dimethomorph, mandipropamid | Low- to medium-risk resistance |
| Melanin synthesis in cell wall | Reductase in melanin biosynthesis | Fthalide pyroquilon tricyclazole | Resistance not known |
| | Dehydratase in melanin biosynthesis | Carpropamid, diclocymet, fenoxanil | Medium risk resistance |
| Host plant defence induction | Salicylic acid pathway | Acibenzolar-S-methyl | Resistance not known |

automation and combinatorial chemistry, a large number of compounds can be generated within a short time which cater to the need for in vitro high throughput screens prior to the greenhouse or field studies. The potential of fungicide is mainly adjudged by growth inhibition or spore germination inhibition assays. Cell-based bioassays like disc diffusion method or broth microdilution technique according to Clinical and Laboratory Standards Institute (CLSI) guidelines employing the target organism are regularly used for screening. In industries, automated screens for whole cell/specific target inhibition based on microtiter plate format are widely used which allow screening of > 10,000 compounds per week (Knight et al. 1997). Screening for antifungal agents based on inhibition of specific enzymes such as chitin synthase will offer the greatest degree of specificity for screening and will lead to safer and target-specific fungicide development. Most of the phytopathogens are filamentous and the hyphae grow apically. Usually hypal tip is plastic, thin walled and a site for the deposition of nascent chitin. In the presence of specific chitin synthase/glucan synthase inhibitor in osmotically stabilized medium, the hyphal tip bursting can be observed within few minutes under the microscope (Patil et al. 2001). Similarly, inhibition of a germ tube formation, yeast to hypha transition in a dimorphic fungus, appresorium formation can be used for initial screening of fungicides.

Disruption and specific deactivation of fungal genes will be helpful in identification of novel, validated target sites. For instance, mutants in which apoptosis is blocked are expected to be less sensitive to apoptosis inducers and can be used as genetic screens. Another emerging screening strategy which employs mutants is haploinsufficiency assay.

### 28.4.1 Haploinsufficiency Assay

Giaever et al. (1999) developed a yeast genome-wide drug induced haploinsufficiency screen using *S. cerevisiae* as a model. Removal of one gene copy from two copies of a diploid yeast results in a heterozygote. The resulting mutant strain becomes sensitive to any drug that acts on the product of this gene as compared to wild-type strain. This haploinsufficient phenotype thereby identifies the gene product of the heterozygous locus as the likely antifungal target (Baetz et al. 2004). For instance, diploid wild-type *S. cerevisiae* strain and its mutant (haploid for the targeted gene *Chs2*–chitin synthase 2 gene) are used for the screening of the CS inhibitors. As the mutant has only one copy of *Chs2,* the effect of chitin synthase inhibitors on growth will be more pronounced on it as compared to the wild type, whereas both the strains will respond similarly to other non-target compounds. *S. cerevisiae* genes, essential for growth and with no homologs in humans may be selected as specific antifungal target. Similarly, haploinsufficient strains can be generated for specific pathogens and used for screening. For instance, Xu et al. (2007) used haploinsufficiency in *C. albicans* (termed as CaFT—*C. albicans* Fitness Test) and screened 45% of the *C. albicans* genome for the molecular targets of growth inhibitory compounds.

For the development of target-specific fungicide, above and few other methods/strategies are useful and as an example work on development of chitin synthase inhibitors ongoing at National Chemical Laboratory (NCL) is presented in the following section.

## 28.5 Fungicide and Antifungal Agents Development Program at NCL

Chitin synthase enzyme satisfies all the aforementioned criteria for a good target. It is responsible for the formation of β-1,4-glycosidic linkages between chitin oligomer and GlcNAc, and is an enzyme absent in both plants and animals although it is present in insects. The Peptidyl-nucleoside antibiotic inhibitors, nikkomycins and polyoxins, isolated from the culture filtrates of some *Streptomyces* spp. are known inhibitors for this enzyme for several years, but have not been sufficiently exploited. Hydrolytic lability and decreased fungal permeability are major obstacles in effective in vivo use against human pathogenic fungi (Zhang and Miller 1999), but they were found to be effective against plant pathogenic fungi. The prohibitive cost of synthesis hampers commercial exploitation to a significant extent.

Most of the efforts to synthetically modify nikkomycin/polyoxin to overcome drawbacks ended up with comparable or less antifungal activity (Chaudhary et al. 2013). The 1,2,3-triazole substitution at 5′ position of uridine nucleoside (general structure of polyoxin/nikkomycin) was selected. The design was based on replacing the peptide bond of polyoxin with triazole moiety as 1,2,3-triazole unit was considered a surrogate for a peptide bond due to similar atom placement and electronic properties. The 1,2,3-triazoles are transported easily inside the cells and are stable in the cell environment. The azole cores interact strongly with biological targets through hydrogen bonding to nitrogen atoms and their large dipole moments. Using click chemistry approach, 7 aryl ether linked 1,2,3 triazolyl uridine, 7 aryl ester linked 1,2,3 triazolyl uridine (Chaudhary et al. 2009) and 8 cyano uridine derivatives were synthesized (Fig. 28.2). The 1,2,3 triazolyl uridine and cyano uridine derivatives showed good antifungal activity against plant pathogens *Fusarium oxysporum, Drechslera oryzae, Magnaporthe grisea* and *Colletotrichum capsici* (Table 28.3 and 28.4). However, SC139, a 1,2,3 triazole uridine with hydrophobic alkyl chain (Fig. 28.2) was most potent with minimum inhibitory concentration (MIC) values of 32, 32 and 16 μg/ml for *M. grisea, D. oryzae* and *F. oxysporum,* respectively.

### 28.5.1 Dimeric and Hybrid Inhibitors as Novel Fungicides

Based on the alternating orientation of the GlcNAc residues within the chitin chain, it was proposed that CS possesses two active sites (Saxena 1995). Yeager and Finey (2004a, 2004b) tested the hypothesis of presence of two active sites in CS by synthesizing dimeric nucleoside inhibitors with the view that if two adjacent active sites are present in close proximity, then dimeric inhibitors should show bivalent inhibition. Dimeric inhibitors of 5′-deoxy-5′-aminouridine were synthesized by joining two units through carbamate linkers of different lengths. Efficacy of the dimeric inhibitors was dependent on the length of the spacer used, with shorter spacer (~14 Å) showing better efficacy. Similarly, uridine analogues joined by tartarate amide spacer showed significant inhibition of CS activity. There are also many reports wherein dimerization has been

**Table 28.3** Antifungal activity of the 1,2,3 triazolyl uridine derivatives (100 μg/ml) against plant pathogens by disc diffusion assay

| Compound | | Substituents (R) | | Zone of inhibition (diameter in mm) | | | |
|---|---|---|---|---|---|---|---|
| | | | | *Fusarium oxysporum* | *Drechslera oryzae* | *Magnaporthe grisea* | *Colletotrichum capsici* |
| Aryl ethers 1,2,3-triazolyl linked uridine derivatives | PC 115 | 4-$NO_2$ | | 6 | 10 | 10 | 11 |
| | PC 107 | H | | – | 6 | – | 10 |
| | PC 109 | 3-Me | | 10 | 4 | – | 14 |
| | PC 121 | 4-F | | 6 | 6 | – | 12 |
| | SC 058 | 4-Me | | 8 | 6 | 11 | 2 |
| Aryl esters 1,2,3-triazolyl linked uridine derivatives | SC 062 | R | X | 7 | 4 | 10 | 15 |
| | | 2,4-Cl | $CH_2$ | | | | |

**Table 28.4** Antifungal activity of the cyano uridine derivatives against plant pathogens by disc diffusion assay

| Compound | Substituents | Inhibition zone diameter in mm (Concentration used 100 μg/ml) | | | |
|---|---|---|---|---|---|
| | R | *Fusarium oxysporum* | *Drechslera oryzae* | *Magnaporthe grisea* | *Colletotrichum capsici* |
| PC 204b | Piperazine | 6 | 5 | 6 | 10 |
| PC 045 | Methyl-2-amino acetate | 6 | 4 | 5 | 12 |
| PC 076 | Boc-protected $NH_3$ | 4 | 2 | 4 | 10 |
| PC 086 | Ethanolamine | 8 | 4 | 10 | 14 |
| SC 027 | Cyclohexyl amine | 12 | 6 | 10 | 11 |
| SC 034 | Piperidine | 10 | 4 | 10 | 14 |
| SC 009 | N-Boc ethylenediamine | 5 | 8 | 15 | 11 |

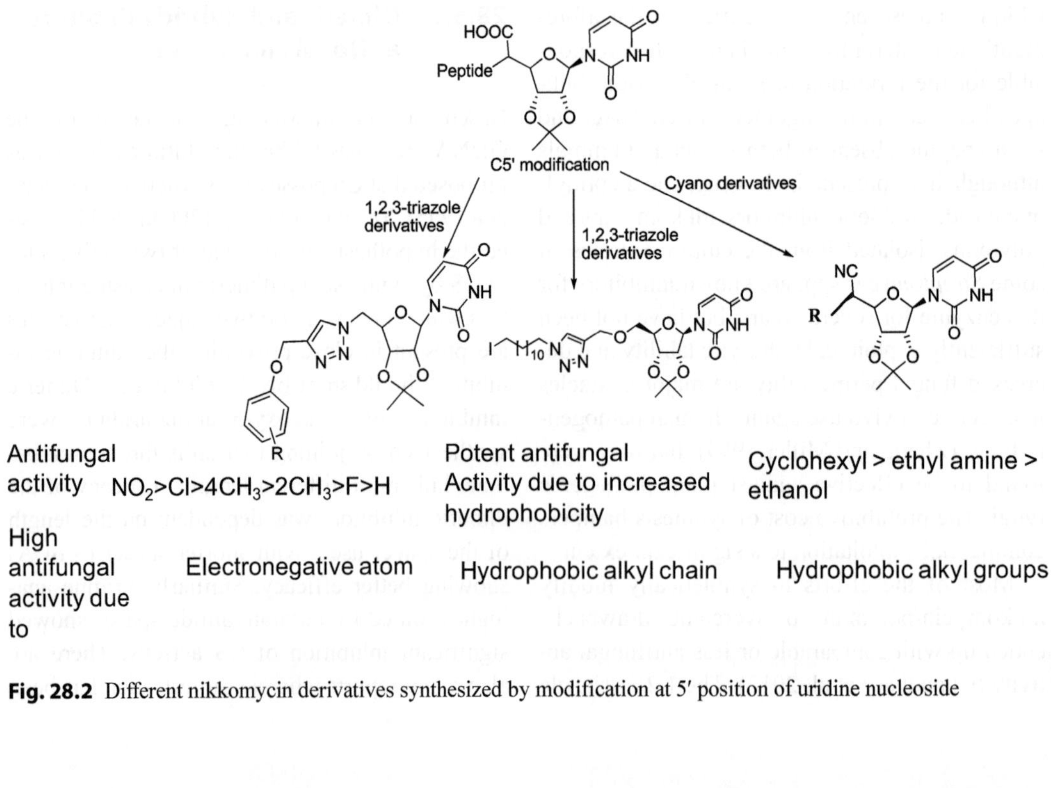


**Fig. 28.2** Different nikkomycin derivatives synthesized by modification at 5′ position of uridine nucleoside

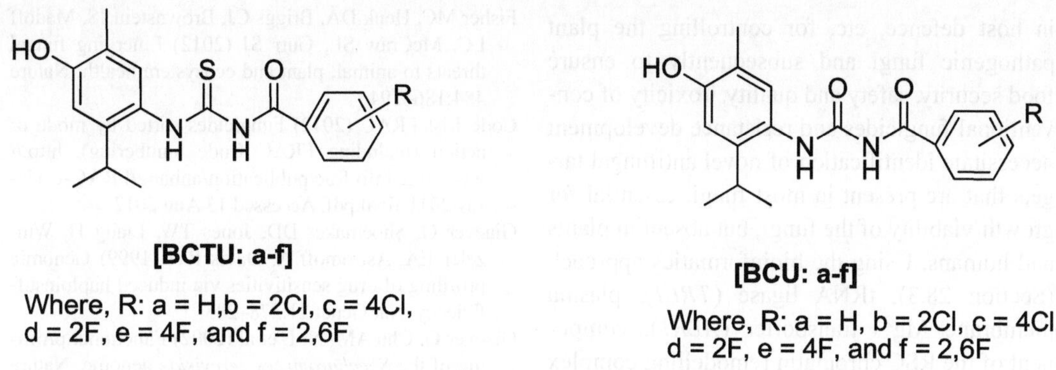


**Fig. 28.3** Hybrid inhibitors of benzoylphenyl urea and carvacrol

shown to result in improved biological activity. The strategy can be used for synthesis of dimeric molecules of CS or other enzyme inhibiting monomers using different linkers for improved antifungal activity.

Synergistic action of two different antifungal compounds like Nikkomycin Z and caspofungin (Sandovsky-Losica 2008) suggests that formation of hybrid inhibitors with dual specificity may be a good strategy, in which the properties of one molecule may improve the efficacy of the other. For instance, above two inhibitors can be joined by hydrolysable spacers. This may tackle the problem of resistance development against one mechanism. This approach was used for the synthesis of novel agrochemicals. Benzoyl phenyl urea (commercial insecticide), a class of insect growth regulator's inhibits chitin synthesis. Carvacrol, a naturally occurring monoterpenoid is an effective antifungal agent. Carvacrol was modified structurally by introducing benzoylphenyl urea linkage. Two series of benzoylcarvacryl thiourea (BCTU) and benzoylcarvacryl urea (BCU) derivatives were synthesized (Fig. 28.3). Most of the new molecules showed comparable insecticidal activity with the standard lufenuron against cotton red bug, *Dysdercus koenigii* as well as potent antifungal activity against phytopathogenic fungi, viz. *M. grisae, F. oxysporum, D. oryzae* and food spoilage yeasts, viz. *Debaromyces hansenii, Pichia membranifaciens* (Pete et al. 2012).

β-Methoxyacrylate group is an important active principle of commercially used strobilurin fungicides. Whereas, triazoles are well-known antifungal compounds. These two groups were brought together in 1,2,4-triazole thiol and 1,2,3-triazole substituted strobilurin derivatives. 1,2,4-Triazole thiol substituted strobilurin derivatives were found to inhibit the growth of plant pathogens such as *F. oxysporum, M. grisea* and *D. oryzae* with MIC in the range of 16–256 µg/ml (unpublished data).

Li et al. (2011) used combinatorial approach for the generation of hybrid CS inhibitors. The structures of nikkomycin and polyoxin are similar. They introduced gene responsible for the biosynthesis of dipeptide of polyoxin from *Streptomyces cacaoi* in the *Streptomyces ansochromogenes* mutant, which produced nucleoside moiety of nikkonycin X. This resulted in production of two hybrid antibiotics by the new mutant, out of which one was identified as polyoxin N and another named as polynik A. These two hybrid molecules were stable at neutral and alkaline conditions and showed better antifungal activity against five different phytopathogenic fungi including *Alternaria kikuchiana* and *Botrytis cinerea* as compared to polyoxin B.

## 28.6 Conclusions

Present situation demands for development of newer fungicides, derivatization of current fungicides for better activity, random screening of natural compounds, chemical libraries, development of high throughput screens, improvement in host defence, etc. for controlling the plant pathogenic fungi and subsequently to ensure food security, safety and quality. Toxicity of conventional fungicides and resistance development necessitate identification of novel antifungal targets that are present in most fungi, essential for growth/viability of the fungi, but absent in plants and humans. Using the bioinformatics approach (Section 28.3), tRNA ligase (*TRL1*), plasma membrane $Mg^{2+}$ transporter (*ALR1*), component of the RSC chromatin remodelling complex (*RSC9*), osmo-sensory signalling pathway protein (*YPD1*), lanosterol 14-α demethylase (*ERG11*) and chitin synthase II (*CHS2*) were identified as ideal targets (Chaudhary et al. 2013). Inhibitors against the products of these genes will be target specific, without any side effects to other eukaryotes. Haploinsufficiency assay (Section 28.4.1) will aid in faster screening for the target-specific compounds. At NCL, the assay using different haploinsufficient mutants including mutants for (*CHS2* and *ERG11*) are routinely employed for screening of compounds. In near future, rational design and synthesis of monomeric and dimeric chitin synthase inhibitors, hybrid inhibitors with dual mode of action, derivatization of existing inhibitors might lead to safer and better fungicides for increased crop productivity.

**Acknowledgements** SGT thanks the Council of Scientific and Industrial Research (CSIR), India for Research Associateship. The fungicide development program at NCL was funded by the CSIR-New Millennium Indian Technology Leadership Initiative.

## References

Baetz K, McHardy L, Gable K, Tarling T, Reberioux D, Bryan J, Andersen RJ, Dunn T, Hieter P, Roberge M (2004) Yeast genome-wide drug-induced haploinsufficiency screen to determine drug mode of action. Proc Natl Acad Sci U S A 101:4525–4530

Chaudhary PM, Chavan SR, Shirazi F, Razdan M, Nimkar P, Maybhate SP, Likhite AP, Gonnade R, Hazara BG, Deshpande MV, Deshpande SR (2009) Exploration of click reaction for the synthesis of modified nucleosides as chitin synthase inhibitors. Bioorg Med Chem 17:2433–2440

Chaudhary PM, Tupe SG, Deshpande MV (2013) Chitin synthase inhibitors as antifungal agents. Minirev Med Chem 13:222–236

Fisher MC, Henk DA, Briggs CJ, Brownstein JS, Madoff LC, McCraw SL, Gurr SJ (2012) Emerging fungal threats to animal, plant and ecosystem health. Nature 484:186–194

Code List FRAC (2012) Fungicides sorted by mode of action (including FRAC Code numbering). http://www.frac.info/frac/publication/anhang/FRAC-Code-List2011-final.pdf. Accessed 13 Aug 2012

Giaever G, Shoemaker DD, Jones TW, Liang H, Winzeler EA, Astromoff A, Davis RW (1999) Genomic profiling of drug sensitivities via induced haploinsufficiency. Nat Genet 21:278–283

Giaever G, Chu AM, Ni L et al (2002) Functional profiling of the *Saccharomyces cerevisiae* genome. Nature 418:387–391

Klittich CJ (2008) Milestones in fungicide discovery: chemistry that changed agriculture. Plant Health Prog doi:10.1094/PHP-2008-0418-01-RV

Knight SC, Anthony VM, Brady AM, Greenland AJ, Heaney SP, Murray DC, Powell KA, Schulz MA, Spinks CA, Worthington PA, Youle D (1997) Rationale and perspectives on the development of fungicides. Ann Rev Phytopathol 35:349–372

Li J, Li L, Tian Y, Niu G, Tan H (2011) Hybrid antibiotics with the nikkomycin nucleoside and polyoxin peptidyl moieties. Metab Eng 13:336–344

McCallan SEA, Wellman RH (1943) A greenhouse method of evaluating fungicides by means of tomato foliage diseases. Contributions from Boyce Thompson Institute 13:93–134

Morton V, Staub T (2008) A short history of fungicides. APSnet Features. doi: 10.1094/APSnetFeature-2008-0308

Patil RS, Deshpande AM, Natu AA, Nahar P, Chitnis M, Ghormade V, Laxman RS, Rokade S, Deshpande MV (2001) Biocontrol of root infecting plant pathogenic fungus, *Sclerotium rolfsii* using mycolytic enzymes and chitin metabolism inhibitors singly and in combination. J Biol Control 15:157–164

Pete UD, Zade CM, Bhosale JD, Tupe SG, Chaudhary PM, Dikundwar AG, Bendre RS (2012) Hybrid molecules of carvacrol and benzoyl urea/thiourea with potential applications in agriculture and medicine. Bioorg Med Chem Lett 22:5550–5554

Sandovsky-Losica H, Shwartzman R, Lahat Y, Segal E (2008) Antifungal activity against *Candida albicans* of nikkomycin Z in combination with caspofungin, voriconazole or amphotericin B. J Antimicrob Chemother 62:635–637

Saxena IM, Brown RM Jr, Fevre M, Geremia RA, Henrissat B (1995) Multidomain architecture of beta-glycosyl transferases: implications for mechanism of action. J Bacteriol 177:1419–1424

Sharon A, Finkelstein A, Shlezinger N, Hatam I (2009) Fungal apoptosis: function, genes and gene function. FEMS Microbiol Rev 33:833–854

Shephard MC (1987) Screening for fungicides. Ann Rev Phytopathol 25:189–206

Winzeler EA, Shoemaker DD, Astromoff A et al (1999) Functional characterization of the *S. cerevisiae* genome by gene deletion and parallel analysis. Science 285:901–906

Xu D, Jiang B, Ketela T, Lemieux S, Veillette K, Martel N, Davison J, Sillaots S, Trosok S, Bachewich C, Bussey H, Youngman P, Roemer T (2007) Genome-wide fitness test and mechanism-of-action studies of inhibitory compounds in *Candida albicans*. PLoS Pathog 3:e92

Yeager AR, Finney NS (2004a) The first direct evaluation of the two-active site mechanism for chitin synthase. J Org Chem 69:613–618

Yeager AR, Finney NS (2004b) Second-generation dimeric inhibitors of chitin synthase. Bioorg Med Chem 12:6451–6460

Zhang D, Miller D (1999) Polyoxins and nikkomycins: progress in synthetic and biological studies. Curr Pharm Design 5:73–99

# Sustainable Agriculture and Plant Growth Promoting Rhizobacteria

29

S. K. Dwivedi and Ram Gopal

**Abstract**

Plant growth-promoting rhizobacteria (PGPR) are naturally occurring soil bacteria that colonize the roots of the plant and promote the plants in terms of their growth. Biomass of the crops is enhanced if the PGPR are supplied to the crop plants at a very early stage of their growth. Inoculation of crops with PGPR may result in multiple effects such as seedling germination, plant height, root and shoot fresh and dry weight, nutrient content and increased nodulation in soybean crop. They help in increasing nitrogen fixation in soybean crop and also help in promoting free-living nitrogen-fixing bacteria; increase supply of other nutrients such as phosphorus and produce plant hormones. They control fungal and bacterial diseases.

**Keywords**

Plant growth-promoting rhizobacteria (PGPR) · Nitrogen fixation · IAA · Siderophores · Phosphate solublization · HCN

## 29.1 Introduction

In the last few decades, crop yield has increased many folds due to use of inorganic fertilizers and pesticides, but their excessive and indiscriminate use has led to serious environmental problems such as ground water pollution, decrease in soil fertility and development of resistance in pathogens. It is very difficult to feed such a huge population without damaging the environment (Reddy and Wang 2011). Sustainable agriculture is the vital need of today's world so that we can meet our needs. Sustainable agriculture means full utilization of environmental resources with no harm to it (Singh et al. 2011). The concept of sustainable development is the management of natural resources for human use. One of the best ways for sustainable agriculture and to decrease the negative impacts of pesticides and fertilizers on environment is the use of PGPR (Adesemoye et al. 2009). They are the heterogeneous group of bacteria that are found in the

R. Gopal (✉) · S. K. Dwivedi
Department of Environmental Science, Babasaheb Bhimrao Ambedkar (A Central) University, Lucknow, Uttar Pradesh 226025, India
e-mail: gopal6754@gmail.com

R. N. Kharwar et al. (eds.), *Microbial Diversity and Biotechnology in Food Security*,
DOI 10.1007/978-81-322-1801-2_29, © Springer India 2014

rhizosphere, at the root surface and in association with plant roots (Ahmad et al. 2008). The rhizosphere is the region of soil influenced by plant roots and/or in association with root hairs, and plant-produced materials (Bringhurst et al. 2001). This space includes soil bound by plant roots, often extending a few millimetre from the root surface and can include the plant root epidermal layer (Mahaffee and Kloepper 1997). Plant roots modify the rhizosphere (Chen et al. 2001). Plant exudates which are released in the rhizosphere includes amino acids and sugars which provide energy and nutrients for PGPR resulting in greater bacterial populations in the rhizosphere than that of non-rhizosphere. Similarly, the soil aggregates affect the diversity of microbial communities (Vadakattu et al. 2006). The bacteria which colonize the plant root systems and thus promote the plant growth are referred to as plant growth-promoting rhizobacteria (PGPR) (Kloepper and Schroth 1978).

PGPR enhance the plant growth by following different mechanisms such as P-solubilization, nitrogen ($N_2$) fixation and by controlling effects of phtyopathogenic microorganisms (Laslo et al. 2012). PGPR was divided into two classes, i.e. biocontrol: PGPR and PGPR (Bashan and Holguin 1998), but this classification does not seem to be widely accepted. Later on, it was classified into four categories, i.e. (i) biofertilizers (increasing availability of nutrients to plants), (ii) phytostimulators (plant growth promoting hormones), (iii) rhizoremediators (degrading organic pollutants) and (iv) biopesticides (controlling diseases and pests) (Somers et al. 2004). On the basis of their relationship with their host plant, they are broadly divided into two major groups, i.e. (i) symbiotic bacteria (bacteria that exist inside plant cells forming nodules), (ii) free-living bacteria (bacteria that are found in rhizosphere soil) (Khan 2005; Hayat et al. 2010). However, on the basis of their localization, they are classified into two groups, i.e. (i) intracellular PGPR (iPGPR): bacteria residing inside plant cells, producing nodules and being localized inside those specialized structures (e.g. nodules) and (ii) extracellular PGPR (ePGPR): those bacteria living outside plant cells and not producing nodules but enhancing plant growth through production of signal compounds that directly stimulate plant growth, improve plant disease resistance or improve nutrient status of soil. The ePGPR has further been subdivided into three types based on the degree of association with plant roots (i) those living near but not in contact with the roots, (ii) those colonizing the root surface and (iii) those living in the space between cells of the root cortex (Gray and Smith 2005). Of these PGPR, the iPGPR are mostly Gram-negative and rod shaped. Generally, iPGPR includes genera *Rhizobium, Bradyrhizobium, Sinorhizobium, Azorhizobium, Mesorhizobium* and *Allorhizobium,* which belong to the family Rhizobiaceae, and are capable of forming nodules on the root system on leguminous plants (Martinez-Romero and Wang 2000) whereas ePGPR do not form nodules. It includes some agronomically important genera, such as *Bacillus* (Ryder et al. 1999), *Pseudomonas* (De Freitas and Germida 1991), *Erwinia* (Nelson 1998), *Enterobacter, Flavobacterium, Actinobacter* sp., *Caulobacter, Serratia* (Zhang et al. 1996), *Aeromonas caviae* (Inbar and Chet 1991), *Agrobacterium* (Ryder and Jones 1990), *Alcaligenes* sp. (Yuen et al. 1985), *Phyllobacterium* sp. (Lambert et al. 1990), and *Bacillus thuringiensis* (Bai et al. 2002), *Hyphomicrobium, Azotobacter, Azospirillum* and *Acetobacter* (Prithiviraj et al. 2003).

## 29.2 How Plant Growth-Promoting Rhizobacteria Promote Growth?

Plant growth is either indirectly or directly influenced by the plant growth-promoting rhizobacteria (Beauchamp 1993; Kloepper 1993; Glick 1995; Kapulnik 1996; Lazarovits and Nowak 1997; Gupta et al. 2000; Antoun and Prevost 2005). The indirect promotion of plant growth is due to the inhibition or prevention of the deleterious and phytopathogenic organisms, whereas direct promotion of plant growth involves the synthesis of a compound or uptake of certain nutrients from the environment which are facilitated by the PGPR, and involves nitrogen fixation,

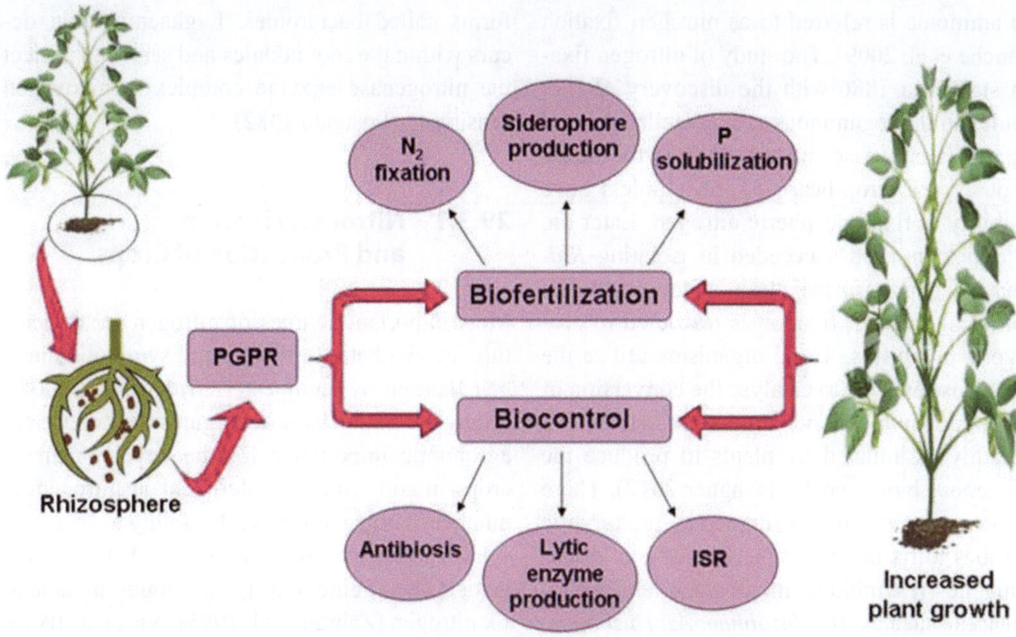


**Fig. 29.1** Schematic illustration of mechanisms for plant growth promotion by PGPR. (Kumar et al. 2011)

phosphate solubilization, production of phytohormones and production of siderophore (Ghosh et al. 2003). PGPR enhance the plant growth by following a wide variety of mechanisms (Ashrafuzzaman et al. 2009; Bhattacharyya and Jha 2012) such as phosphate solubilization, siderophore production, biological nitrogen fixation (Bowen and Rovira 1999; Bhattacharyya and Jha 2012), rhizosphere engineering, production of 1-aminocyclopropane-1-carboxylate deaminase (ACC), quorum sensing (QS) signal interference and inhibition of biofilm formation, phytohormone production (Bowen and Rovira 1999; Benizri et al. 2001; Bhattacharyya and Jha 2012), exhibiting antifungal activity, production of volatile organic compounds (VOCs), induction of systemic resistance, promoting beneficial plant–microbe symbiosis, interference with pathogen toxin production, etc. (Bhattacharyya and Jha 2012). Out of all these mechanisms, direct mechanisms include, the production of stimulatory bacterial VOCs and phytohormones, lowering of ethylene level in plants, improvement of the plant nutrient status and stimulation of disease-resistance mechanisms (Antoun and Prevost 2005), whereas indirect method includes PGPR as a biocontrol agent that reduce diseases, stimulates other beneficial symbiosis and protects the plant by degrading xenobiotics in inhibitory contaminated soils (Jacobsen 1997). PGPR increase the germination rate, tolerance to drought, root and shoot weight, yield and plant growth under salt stress condition (Kokalis-Burelle et al. 2006). They have potentiality to produce antibacterial compounds that are effective against certain plant pathogens (Herman et al. 2008; Fig. 29.1).

## 29.3 Symbionts and Nitrogen Fixation

Nitrogen is considered as one of the major limiting nutrient in plant growth (Graham and Vance 2000). It is a major component of chlorophyll and the most important pigment for photosynthesis as well as amino acids. Even though it is the most abundant element found in the environment, plants are unable to utilize it because it is in the form of nitrogen gas, and plants can utilize only the reduced form of this element (Vance 2001). The biological process which is responsible for the reduction of molecular nitrogen into ammonia is referred to as nitrogen fixation (Franche et al. 2009). The study of nitrogen fixation started in 1886 with the discovery of root nodules in the leguminous crops. Hellriegel and Wilfarth (German scientists) reported that leguminous crops (crop-bearing root nodules) have capability to fix atmospheric nitrogen. Later on, Beijerinck in 1888 succeeded in isolating *Rhizobium leguminosarum* strain (Stewart 1969). Biological nitrogen fixation is restricted to prokaryotic organisms. These organisms utilize the nitrogenase enzyme to catalyse the conversion of atmospheric nitrogen into ammonia ($NH_3$) which is readily assimilated by plants to produce the nitrogenous biomolecules (Wagner 2012). There are three principal modes adopted by prokaryotic microbes to fix the atmospheric nitrogen to ammonia, i.e. (i) symbiotic (members of family Rhizobeaceae such as *Allorhizobium, Azorhizobium, Bradyrhizobium, Mesorhizobium, Rhizobium* and *Sinorhizobium*) form symbiosis with leguminous host (Vessey 2003); (ii) non-symbiotic (a large number of bacteria and cyanobacteria are capable of non-symbiotic nitrogen fixation. *Azotobacter, Azomonas, Azotococcus, Mycobacterium* spp., *Methylosinus trichosporium, Thiobacillus ferrooxidans, Chlorobiumthiosulfatophilum, Chromatiumvinosum, C. minutissimum, Bacillus polymixa, B. macerans, Enterobacteraerogenes (Aerobacter aerogenes), Escherichia intermedia, Escherichia coli(E. coli), Klebsiella* spp., *Rhodospirillumrubrum, Rhodomicrobium, Rhodopseudomonas, Clostridium* spp., *Desulfovibrio* spp. are examples of free living nitrogen-fixing bacteria) (iii) associative symbiotic (no nodules are formed; Singh 2005). Associative symbiotic bacteria grow in close association with roots and sometimes invade the outer cortical region of the roots and fix atmospheric nitrogen.

In symbiotic nitrogen fixation, bacteria invade the root hairs of host plants where they multiply and stimulate the formation of root nodules. Bacteria inside the root nodules convert nitrogen to nitrates, which is utilized by the plant for its growth and development (www.britannica.com/EBchecked/topic/967311/nitrogen-fixing-bacteria). Within the resulting root nodules, the bacteria exist as highly pleomorphic $N_2$-fixing forms called bacteroides. Leghaemoglobin occurs within the root nodules and serves to protect the nitrogenase enzyme complex from oxygen sensitivity (Postgate 1982).

### 29.3.1 Nitrogen Fixation and Production of Crops

Most important sources of nitrogen for agriculture are synthetic fertilizers and symbiotic nitrogen fixation by legumes (Herridge et al. 2008). Nitrogen fixation by microorganisms is of greater agronomic importance for the development of crops in soil which are deficient in nitrogen. A number of organisms have the ability to fix atmospheric nitrogen. However, only a very few organisms belonging to different groups are able to fix nitrogen (Zahran et al. 1995). Nitrogen-fixing bacteria include, *Achromobacter, Acetobacter, Alcaligenes, Arthrobacter, Azospirillum, Azotobacter, Azomonas, Bacillus, Beijerinckia, Clostridium, Corynebacterium, Derxia, Enterobacter, Herbaspirillum, Klebsiella, Pseudomonas, Rhodospirillum, Rhodopseudomonas* and *Xanthobacter* (Bashan and de Bashan 2005). The atmospheric nitrogen is converted to usable forms of nitrogen (i.e. $NH_4^+$) through nitrogenase which is most sensitive to oxygen. However, each group of nitrogen-fixing organism possess a mechanism to protect nitrogenase from oxygen toxicity.

The most common nitrogenase (Mo- or conventional nitrogenase) contains a prosthetic group with molybdenum FeMoCo. Some bacteria such as *Azotobacter* and several photosynthetic nitrogen fixers (including some cyanobacteria) carry additional forms of nitrogenase whose cofactor contains vanadium (V-nitrogenase) or only iron (Fe-nitrogenase; Newton 2007). The leguminous crops form symbiotic relationship with root-nodule bacteria collectively called as rhizobia (Doyle and Lucknow 2003). The rhizobia are Gram-negative and belong to *Proteobacteria* division. The alpha-proteobacterial genera *Agrobacterium, Allorhizobium, Azorhizobium, Bradyrhizobium, Mesorhizobium, Rhizobium, Sinorhizobium, Devosia, Methylobacterium, Ochrobactrum* and *Phyllobacterium* all harbour nodule-forming bacteria, and so do the beta-proteobacterial *Burkholderia* and *Cupriavidus* (Lindström and Martínez-Romero 2007). PGPR can fix atmospheric $N_2$ and enhance crop yield, for example *Azospirillum* species enhance rice (Pedraza et al. 2009; Ashrafuzzaman et al. 2009), maize (Montanez et al. 2009) and wheat (Sala et al. 2007) yield by BNF mechanisms. Co-inoculation of *Bradyrhizobium* and certain PGPRs positively affected the symbiotic nitrogen fixation by enhancing both root nodule number and mass (Yahalom et al. 1987) and increasing the nitrogenase activity (Alagawadi and Gaur 1988). Co-inoculation of PGPRs with *Bradyrhizobium* increased nodulation and nitrogen fixation in *Glycine max* at low root zone temperature (Dashti et al. 1998).

## 29.4 Plant Growth Promoters

Plant growth regulators (PGR) or growth regulators or plant hormones are chemicals that alter the growth of the plant or plant part (http://edis.ifas.ufl.edu). PGR is an organic compound, either natural or synthetic, that modifies or controls one or more specific physiological processes within a plant (Lemaux 1999). Plant hormone is the compound produced within the plant. According to the Environmental Protection Agency, PGRs are "any substance or mixture of substances intended, through physiological action, to accelerate or retard the rate of growth or maturation, or otherwise alter the behavior of plants or their produce. Additionally, plant regulators are characterized by their low rates of application; high application rates of the same compounds often are considered herbicidal." PGR can be used as an enhancer for a wide range of plants such as tea, tobacco, banana, carambola, dragon fruit, mango, citrus, grapes, guava, strawberry, beans, capsicum, carrot, cabbages, Chinese cabbage, celery, cucurbits, tomato, spinach, eggplant, cut flowers, roses, chrysanthemum, rice, etc. (http://www.nufarm.com/MY/PlantGrowthRegulators).

PGRs are the chemicals that physiologically alter particular areas of plant growth. They can accelerate or retard growth rate, promote flowering, enhance the colour of fruit, stimulate the healthy root growth and also promote or reduce stem elongation (http://chameleongrowsystems.com/Chameleon_Grow_Systems/Plant_Growth_Regulators_(PGR)). The phytohormones and other compounds synthesized by PGPR affecting plant growth are discussed as under.

### 29.4.1 Plant Growth-Promoting Substances

#### 29.4.1.1 Indole Acetic Acid

Auxins are compounds that influence the cell enlargement, bud formation and root initiation. They also control the growth of stems, roots and fruits (Osborne and McManus 2005). Indole acetic acid (IAA) is the most common naturally occurring plant hormone of the class auxin found in plants (Simon and Petrášek 2011). Chemically, IAA is a carboxylic acid (http://en.wikipedia.org/wiki/Indole-3-acetic_acid). IAA is a phytohormone and it may function as an important signal molecule in the regulation of plant development (Ashrafuzzaman et al. 2009). It is mainly produced in the cells of the young apex (bud) and very young leaves of plants (Zhao 2010). In plants, IAA is produced from tryptophan through indole-3-pyruvic acid (Won et al. 2011) and through indole-3-acetaldoxime in *Arabidopsis* (Sugawara et al. 2009). The discovery of IAA as a PGR coincided with the first indication of the molecular mechanisms involved in tumour genesis induced by *Agrobacterium*. *Agrobacterium*-induced tumours were shown to be sources of IAA (Link and Eggers 1941) and capable of growth in the absence of PGRs, which are normally, required to incite growth of callus from sterile plant tissues (White and Braun 1941). It was later found that not only plants but also microorganisms including bacteria and fungi are able to synthesize IAA (Costacurta et al. 1994).

IAA (phytohormone) is the most native auxin, it functions as a signal molecule in the regulation of plant development including organogenesis, tropic responses, cellular responses such as cell expansion, division and differentiation and gene regulation (Ryu and Patten 2008). IAA produced by bacteria improves plant growth by increasing root hairs and lateral roots. In soil, the microbial synthesis of IAA is increased by tryptophan from root hairs and decaying cells (Frankenberger and Arshad 1991). The endogenous level of IAA in plants is important for successful rooting. Other indolic compounds such as indole-pyruvic, indole-acetamide and indole-carboxylic-acid can be involved in root formation (Costacurta and Vanderleyden 1995). Diazotrophic PGPR promotes plant growth by producing phytohormone IAA and by fixing nitrogen biologically. Its ability to produce IAA appears to be widespread in nature. *Azospirillum, Burkholderia, Herbaspirillum* and *Pseudomonas* produce IAA. Diazotrophic PGPR improves nutrient cycling by increasing root surface and potentially delays pathogenic spread (Aziz et al. 2012).

#### 29.4.1.2 Indole Acetic Acid Production by Rhizobacteria

IAA is commonly produced by PGPR (Lynch 1985). When plant tissues were analysed for the presence of bacteria then IAA-producing PGPR were being detected (Rosenblueth and Martinez-Romero 2006). In plants, IAA affects many important physiological processes including cell enlargement and division, tissue differentiation and responses to light and gravity (Teale et al. 2006). Some species of bacteria such as *Pseudomonas* sp., *Bacillus* sp., *Klebsiella* sp., *Azospirillum* sp., *Enterobacter* and *Serratia* sp. produce IAA (Frankenberger and Arshad 1995). IAA was detected by colorimetric method (Dey Vay et al. 1968). The main precursor for IAA biosynthesis pathways in bacteria is tryptophan. Five different pathways for the production of IAA using tryptophan are indole-3-acetamide (IAM) pathway, indole-3-pyruvate pathway, tryptamine pathway, tryptophan side-chainoxidase pathway and indole-3-acetonitrile pathway (Spaepen et al. 2007). Rhizobia are exclusively known for nitrogen fixation, but many of them are reported to produce IAA. It has been reported that auxins play a key role in creating nodule in leguminous plants and establishing a symbiotic association with rhizobia (Neeru et al. 2000). The species of *Brayrhizobium, Rhizobium* and *Mesorhizobium* produced a substantial amount of IAA under in vitro conditions (Ahmad et al. 2008). Amongst other PGPR strains, *Pseudomonas, Bacillus, Agrobacterium* sp., *Alcaligenes piechaudii* and two strains of *Comamonas acidovorans* secreted IAA at lower levels as compared to deleterious bacteria (Rajkumar et al. 2006). Root treatment with *Azotobacter paspali* secreted IAA and significantly increased dry weight of leaves and roots of several plant species. Inoculation of seeds of canola with *Pseudomonas putida* GR 12–2 significantly increased the root length 2–3-fold (Caron et al. 1995). *Azospirillum brazilance* GR12–2 increased the number and length of lateral roots of wheat (Barbieri et al. 1986). *Pseudomonas* and *Agrobacterium* use a tryptophan-2-monooxygeanse (iaaM) to convert tryptophan to IAM which is hydrolysed into IAA by a hydrolase iaaH (Camilleri and Jouanin 1991). It has been known that the iaaM/ iaaH pathway is the only known complete trp-dependent IAA biosynthesis pathway but plants do not use the iaaM/ iaaH pathway to make IAA. However, IAM exists in plant extracts and has been suggested as a key intermediate in converting indole-3-acetaldoxime (IAOx) to IAA (Sugawara et al. 2009).

#### 29.4.1.3 Siderophores

Siderophores are low molecular weight (500–1,000 Da) ferric ion chelating agents with remarkable chemical properties which are excreted by microorganisms under iron deficiency conditions. Earlier, siderophores have been called siderochromes or sideramines (Raymond et al. 1984). Iron is one of the major limiting factors and is very essential for microbial life (Saha et al. 2012). It plays an important role in various physiological (respiration, photosynthetic transport, nitrate reduction, chlorophyll synthesis and nitrogen fixation) and biochemical processes (Robinson and Postgate 1980). Despite its high abundance in the earth's crust (1–6 %), it is not readily available in the required form to plants and microorganisms. Therefore, microorganisms produce small, high affinity iron-chelating molecules, i.e. siderophores for its acquisition.

They have the ability to produce a wide variety of siderophores controlled at the molecular level by different genes to accumulate, mobilize and transport iron for metabolism. Siderophores also play a critical role in the expression of virulence and development of biofilms by different microbes. Apart from maintaining microbial life, siderophores can be harnessed for the sustainability of human, animals and plants (Saha et al. 2012). Siderophores are amongst the strongest soluble $Fe^{3+}$binding agents known. The inadequacy of iron can affect the microorganisms in many ways such as inhibit growth, decrease genetic material and inhibit sporulation and can also change cell morphology (Crowley et al. 1987). Under iron-deficient condition, many bacteria excrete siderophores exhibiting phenolate or 2,3-dihydroxybenzoate (DHB) iron-binding groups. In 1970, the first tricatechol siderophore, i.e. enterobactin (also known as enterochelin) was isolated from culture fluids of *Escherichia coli* (*E. coli*), *Aerobacter aerogenes* and *Salmonella typhimurium* (Pollack and Neilands 1970). Despite the considerable variation of structure of siderophore, their common feature is to form six-coordinate octahedral complexes with ferric ion of great thermodynamic stability. The ferrichromes, fusarinines and ferrioxamines are typical trihydroxamate siderophores, while enterobactin is a cyclic tricatecholate siderophore (Harris et al. 1979).

#### 29.4.1.4 Siderophore Production by PGPR

Siderophores serve as a vehicle for the transport of Fe(III) into the microbial cell. The availability of iron in the rhizosphere for microbial assimilation is extremely limited (Loper and Buyer 1991). To survive in such an iron stress condition, organisms secrete iron-binding ligands called siderophores that have capability bind the ferric ion and make it available to the host organisms (Gupta and Gopal 2008). They promote growth of several annual crops through uptake of iron through siderophores (Sindhu et al. 1997). They have greater binding potentials (Schippers et al. 1988). Siderophore-producing bacteria are good for plant growth promotion especially in neutral and alkaline soil. The two PGPR strains $DKC_2$ and $DKM_5$ were found to produce siderophore on chrome azurol S (CAS) agar plates. The nature of siderophore produced by $DKC_2$ strain was catecholate in nature (Ladwal et al. 2012). Rhizobacteria contribute antibiosis by producing siderophores and antibiotics (Maksi-mov et al. 2011). In soil, siderophore production plays a central role in determining the ability of different microorganisms to improve plant growth and development. Microbial siderophores enhance iron uptake by plants that are able to recognize the bacterial ferric-siderophore complex (Dimkpa et al. 2009) and are also important in the iron uptake by plants in the presence of other metals such as nickel and cadmium (Dimkpa et al. 2008). Bacterial siderophores produced by pseudomonads are known for their high affinity to the ferric ion. The potent siderophore, pyoverdin can inhibit the growth of bacteria and fungi (Kloepper et al. 1980a). Pseudobactin siderophore produced by *P. putida* B10 strain was also able to suppress *Fusarium oxysporum* in soil deficient in iron but this suppression was lost when the soil was replenished with iron (Kloepper et al. 1980b). *Pseudomonas* sp. (Loper and Henkels 1999), *Azotobacter* (Husen 2003), *Bacillus megaterium* (16) play the vital role in stimulating plant growth and controlling several plant diseases (Lemanceau and Albouvette 1993). They function as biocontrol agents by depriving the pathogen from iron nutrition, thus resulting in increased yield of crop (O'Sullivan and O'Gara 1992).

#### 29.4.1.5 Phosphate Solubilizers and Crop Productivity

Phosphorus (P) is the second most plant nutrient available in soil after nitrogen. It is present in high amounts in soil, but most of the P is present in insoluble form in the soil as iron and aluminium phosphates in acidic soils and calcium phosphates in alkaline soils (Hariprasad and Niranjan 2009). Globally, it is the major yield-limiting nutrient and it is estimated that world's P reserve will end by 2050 (Shah et al. 2011).

P is one of the major nutrients whose deficiency restricts crop yields severely. Most of the P is present in soil in the form of insoluble phosphates and cannot be utilized by plants (Pradhan and Sukla 2006). However, PGPR can be utilized to solubilize precipitated phosphates and enhance phosphate availability to plant. It has been reported that phosphate-solubilizing bacteria (PSB) is commonly found in the rhizosphere as compared to non-rhizosphere (Raghu and Mac Rae 1966). It can be readily isolated from rhizospheric soil plants (deFreitas et al. 1997). The use of phosphate-solubilizing rhizobacteria as inoculants will simultaneously increase the P uptake by plants. Mostly, strains of genera *Pseudomonas, Bacillus* and *Rhizobium* are known to be strong phosphate solubilizers (Rodríguez and Fraga 1999). P is one of the major macronutrient and it is present in soil at the level of 400–1,200 mg/kg soil (Fernández and Novo 1988). But the concentration of soluble P in soil is about or less than 1 ppm (Goldstein 1994). Many P compounds which are present in soil are of high molecular weight which should be first bio-converted to either soluble ionic phosphate form or low molecular weight organic phosphate to get absorbed by the cell (Goldstein 1994). The cells are able to absorb the P in many forms but they mostly absorb in the form of $HPO_4^{2-}$ or $H_2PO_4^-$(Beever and Burns 1980). P has the ability to increase the strength of cereal straw, promote flower formation and fruit production, stimulate root development and is also essential for seed formation. Adequate P fertilization may improve the quality of fruits, vegetables and grain crops and increase their resistance to diseases and adverse conditions. It is essential for the development of meristematic tissues, in stimulation of early root growth (Bisen and Verma 1996).

PSB have the ability to release metabolites such as organic acid which through their hydroxyl and carboxyl groups chelate the cation bound to phosphate, latter being converted to soluble forms. Phosphate solubilization also takes place through various microbial processes (Sharma et al. 2011). The population of PSB totally depends on the physical and chemical properties, organic matter and P content of the soil. Larger populations of PSB are found in agricultural and rangeland soils (Sharma et al. 2011).

#### 29.4.1.6 Antibiotic and HCN Production by Plant Growth-Promoting Rhizobacteria

Antibiotic and hydrogen cyanide (HCN) production is the primary and one of the most effective mechanisms through which PGPR can control the growth of phytopathogens (Weller and Thomashow 1993). Mainly fluorescent *Pseudomonas* and *Bacillus* species play an active role in the suppression of phytopathogenic organisms by synthesizing extracellular metabolites such as antibiotics and HCN that inhibit the growth of the pathogen even at a very low concentration. These antibiotics may be antitumour, antiviral, antimicrobial, antihelminthic and cytotoxic (Fernando et al. 2005). For example, *Bacillus subtilis* produces antibiotic which inhibit the growth of *Fusarium oxysporum* f.sp. *ciceri,* the causal agent of *Fusarium* wilt in chickpea (Kumar 1999). In the plant-promoting bacterium *Pseudomonas fluorescens* CHAO, the GacS/GacA system is essential for the production of antibiotic compounds and hence for the biological control of root pathogenic fungi (Kay et al. 2005). Antibiotic synthesis is totally dependent on the metabolic status of the cell (Thomashow 1996) which in turn is dependent on nutrient availability and other environmental factors such as pH, temperature, carbon content, major and minor minerals, etc. (Nielsen and Sørensen 2003). Antibiotics are produced as/by microbial biocontrol agents as a secondary metabolites for hampering the phytopathogens (Deshwal et al. 2003). The antibiotics contribute microbial competitiveness besides their role in suppressing the growth of plant root pathogens. The antibiotic-producing PGPR strains are of considerable interest as a biological control agent (Thomshow et al. 2003). Several antimicrobial compounds belonging to polypeptides, heterocyclic nitrogenous compounds and lipopeptides groups active against phytopathogens have been reported (Thomshaw and Webler 1995; Table 29.1).

**Table 29.1** Plant growth regulators

| Growth regulators | Organisms | References |
|---|---|---|
| IAA (Indole acetic acid) | *Pseudomonas aeruginosa* PS 1 | Ahemad and Khan (2010a) |
| | *P. aeruginosa*<br>*P. putida*<br>*P. polymyxa*<br>*Bacillus subtillis*<br>*B. boronophillus* | Yadav et al. (2010) |
| | *Mesorhizobium* sp. | Ahemad and Khan (2010b) |
| | *Rhizobium* sp. | Ahemad and Khan (2010c) |
| | *Klebsiella* sp. | Ahemad and Khan (2010d) |
| | *Enterobacterasburiae* | Ahemad and Khan (2010e) |
| | *Rhizobium leguminosarum* | Ahemad and Khan (2009a) |
| | *Azotobacter*<br>*Fluorescent pseudomonas*<br>*Bacillus* | Ahmad et al. (2008) |
| | *Pseudomonas* sp. | Poonguzhali et al. (2008) |
| | *Bacillus subtilis* | Singh et al. (2008) |
| | *Serratiamarcescens* | Selvakumar et al. (2008) |
| | *Acinetobacter* sp., *Pseudomonas* sp. | Indiragandhi et al. (2008) |
| | *Enterobacter* sp. | Kumar et al. (2008) |
| | *Mesorhizobium*sp. | Ahmad et al. (2008); Wani et al. (2008) |
| | *Burkhoideria* | Jiang et al. (2008) |
| | *Pseudomonas jessenii* | Rajkumar and Freitas (2008) |
| | *Pseudomonas aeruginosa* | Ganesan (2008) |
| | *Pseudomonas* spp., *Bacillus* spp.,*Azotobacter* spp., | Joseph et al. (2007) |
| | *Klebsiellaoxytoca* | Jha and Kumar (2007) |
| | *Pseudomonas, Azotobacterchroococcum* | Wani et al. (2007c) |
| | *Bacillus* spp. | Wani et al. (2007a, b, c) |
| Siderophore | *Pseudomonas aeruginosa* PS 1 | Ahemad and Khan (2010a) |
| | *Klebsiella* sp. | Ahemad and Khan (2010d) |
| | *Enterobacter asburiae* | Ahemad and Khan (2010e) |
| | *Mesorhizobium* sp. | Ahemad and Khan (2010b) |
| | *Rhizobium* sp. | Ahemad and Khan (2010c) |
| | *Proteus vulgaris* | Rani et al. (2009) |
| | *Rhizobium leguminosarum* | Ahemad and Khan (2009a) |
| | *Azotobacter,*<br>*Fluorescent Pseudomonas,*<br>*Bacillus* | Ahmad et al. (2008) |
| | *Pseudomonas* sp. | Poonguzhali et al. (2008) |
| | *Bacillus subtilis* | Singh et al. (2008) |
| | *Serratiamarcescens* | Selvakumar et al. (2008) |
| | *Enterobacter* sp. | Kumar et al. (2008) |
| | *Burkhoideria* | Jiang et al. (2008) |
| | *Pseudomonas jessenii* | Rajkumar and Freitas (2008) |
| | *Pseudomonas aeruginosa* | Ganesan (2008) |
| | *Brevibacillusbrevis*<br>*Enterobacter* sp.<br>*Pseudomonas* sp.<br>*P. fluorescens*<br>*Azospirillumbrasilense* | Gupta and Gopal (2008) |

**Table 29.1** (continued)

| Growth regulators | Organisms | References |
|---|---|---|
| | *Pseudomonas, Azotobacterchroococcum, Bacillus*spp. | Wani et al. (2007c) |
| HCN (hydrogen cyanide) | *Pseudomonas aeruginosa* PS 1 | Ahemad and Khan (2010a) |
| | *Klebsiella* sp. | Ahemad and Khan (2010d) |
| | *Enterobacterasburiae* | Ahemad and Khan (2010e) |
| | *Mesorhizobium* sp. | Ahemad and Khan (2010b) |
| | *Rhizobium* sp. | Ahemad and Khan (2010c) |
| | *Rhizobium leguminosarum* | Ahemad and Khan (2009a) |
| | *Azotobacter*<br>*Fluorescent pseudomonas*<br>*Bacillus* | Ahmad et al. (2008) |
| | *Serratiamarcescens* | Selvakumar et al. (2008) |
| | *Azotobacter* sp.,*Mesorhizobium* sp.<br>*Pseudomonas* sp., *Bacillus* sp. | Ahmad et al. (2008) |
| | *Bacillus* spp. | Wani et al. (2007a, b, c) |
| Ammonia | *Pseudomonas aeruginosa* PS 1 | Ahemad and Khan (2010a) |
| | *P. aeruginosa*<br>*P. putida*<br>*P. polymyxa*<br>*Bacillus subtillis*<br>*B. boronophillus* | Yadav et al. (2010) |
| | *Klebsiella* sp. | Ahemad and Khan (2010d) |
| | *Enterobacterasburiae* | Ahemad and Khan (2010e) |
| | *Mesorhizobium* sp. | Ahemad and Khan (2010b) |
| | *Rhizobium* sp. | Ahemad and Khan (2010c) |
| | *Rhizobium leguminosarum* | Ahemad and Khan (2009a) |
| | *Azotobacter*<br>*Fluorescent Pseudomonas*<br>*Bacillus* | Ahmad et al. (2008) |
| | *Pseudomonas* spp., *Bacillus* spp.,*Azotobacter* spp., | Joseph et al. (2007) |
| | *Bacillus* spp. | Wani et al. (2007a, b, c) |

## 29.5 Conclusion and Future Applications

The above discussion clearly demonstrates that plant growth promoting bacteria can enhance and promote plant growth and development in different ways (Laslo et al. 2012). It is known to increase plant growth, vigour and plant nutrient contents (Lenin and Jayanthi 2012). The mechanisms of these rhizobacteria to promote plant growth are an environmentally friendly tool for sustainable agriculture. They may directly or indirectly benefit the crop. Indirectly they prevent the growth of plant pathogens through competition for space and nutrients, antibiosis, production of hydrolytic enzymes, inhibition of pathogen by producing enzymes or toxins and through induction of plant defence mechanisms (Weyens et al. 2009).

In order to meet the challenges of providing food to increasing population, there is an urgent need to increase the crop production. To meet the increasing demand of the increasing population, the excessive usage of chemical fertilizers and pesticides are taking place. Although we are achieving the satisfactory results with the use of chemical pesticides and fertilizers, they are damaging the environment. Excessive use of chemical fertilizers and pesticides are polluting the large water resources, decreasing the soil fertility, destroying the beneficial microorganisms and making the pathogen resistant to pesticides. So, to overcome all these problems, in recent years scientists have diverted their mind towards plant growth promoting rhizobacteria for sustainable agriculture (Lwin et al. 2012). Sustainable agriculture practice by the use of PGPR provides a better option for increasing the soil fertility and decreasing the environmental damage as well as a better option for food production (Verma et al. 2013). Hence, the next step should be to explore the PGPR, singly or in combinations (with multiple plant growth promoting properties) as biofertilizers and commercialized it.

## References

Adesemoye AO, Torbert HA, Kloepper JW (2009) Plant growth promoting rhizobacteria allow reduced application rates of chemical fertilizers. Microb Ecol 58:921–929

Ahemad M, Khan MS (2009a) effect of insecticide- tolerant and plant growth-promoting *Mesorhizobium* on the performance of chickpea growth in insecticide stressed alluvial soils. J Crop Sci Biotechnol 12:213–222

Ahemad M, Khan MS (2010a) Ameliorative effects of *Mesorhizobium* sp. MRC4 on chickpea yield and yield components under different doses of herbicide stress. Pestic Biochem Physiol 98:183–190

Ahemad M, Khan MS (2010b) Comparative toxicity of selected insecticides to pea plants and growth promotion in response to insecticide- tolerant and plant growth promoting rhizobium leguminosarum. Crop Prot 29:325–329

Ahemad M, Khan MS (2010c) Growth promotion and protection of lintil (*Lens esculenta*) against herbicide stress by rhizobium species. Ann Microbiol 60:735–745

Ahemad M, Khan MS (2010d) Phosphate solubilizing and plant growth promting *Pseudomonas aeruginosa* PS1 improves greengram performance in quizalafop-p-ethyl and clodinafop amended soil. Arch Environ Contam Toxicol 58:361–372

Ahemad M, Khan MS (2010e) Plant growth promoting activities of phosphate-solubilizing Enterobacter-asburiae as influenced by fungicides. Eur Ashian J Biosci 4:88–95

Ahmad F, Ahmad I, Khan MS (2008) Screening of free-living rhizospheric bacteria for their multiple plant growth promoting activities. Microbiol Res 163:173–181

Alagawadi AR, Gaur AC (1988) Associative effect of rhizobium and phosphate-solubilizing bacteria on the yield and nutrient uptake of chickpea. Plant Soil 105:241–246

Antoun H, Prevost D (2005) Ecology of plant growth promoting Rhizobacteria. Chapter 1. In: Siddiqui ZA (ed) PGPR: Biocontrol and Biofertilization. Springer, Dordrecht, pp 1–38

Ashrafuzzaman M, Hossen FA, Ismail MR, Hoque MA, Islam MZ, Shahidullah SM, Meon S (2009) Efficiency of plant growth-promoting rhizobacteria (PGPR) for the enhancement of rice growth. Afr J Biotechnol 8(7):1247–1252

Aziz ZFA, Saud HM, Rahim KA, Ahmed OH (2012) Variable responses on early development of shallot (*Allium ascalonicum*) and mustard (*Brassica juncea*) plants to *Bacillus cereus* inoculation. Malays J Microbiol 8(1):47–50

Bai Y, Souleimanov A, Smith DL (2002) An inducible activator produced by *Serratia proteamaculans* strain and its soybean growth promoting activity under greenhouse conditions. J Exp Bot 53:1495–1502

Barbieri P, Zanelli T, Galli E et al (1986) Wheat inoculation with *Azospirillum brasilence* Sp6 and some mutants altered in nitrogen fixation and indole-3-acetic acid. FEMS Microbiol Lett 36:87–90

Bashan Y, de-Bahan LE (2005) Bacteria. In: Hillel D (ed) Encyclopedia of soils in the environment, vol 1. Elsevier, UK, pp 103–115

Bashan Y, Holguin G (1998) Proposal for the division of plant growth-promoting rhizobacteria into two classifications: biocontrol-PGPB (plant growth-promoting bacteria) and PGPB. Soil Biol. Biochem 30:1225–1228

Beauchamp CJ (1993) Mode of action of plant growth-promoting rhizobacteria and their potential use as biological control agents. Phytoprotection 71:19–27

Beever RE, Burns DJW (1980) Phosphorus uptake, storage and utilization by fungi. Adv Bot Res 8:127–219

Benizri E, Baudoin E, Guckert A (2001) Root colonization by inoculated plant growth-promoting Rhizobacteria. Biocontrol Sci Technol 11(5):557–574

Bhattacharyya PN, Jha DK (2012) Plant growth-promoting rhizobacteria (PGPR): emergence in agriculture. World J Microbiol Biotechnol 28:1327–1350

Bisen PS, Verma K (1996) Handbook of microbiology. CBS, New Delhi

Bowen GD, Rovira AD (1999) Therhizosphere and its management to improve plant growth. Adv Agron 66:1–102

Bringhurst RM, Cardon ZG, Gage DJ (2001) Galactosides in the rhizosphere: utilization by *Sinorhizobium meliloti* and development of a biosensor. PNAS 98:4540–4545

Camilleri C, Jouanin L (1991) The TR-DNA region carrying the auxin synthesis genes of the Agrobacterium rhizogenes agropine-type plasmid pRiA4: nucleotide sequence analysis and introduction into tobacco plants. Mol Plant Microbe Interact 4:155–162

Caron M, Patten CL, Ghosh S et al (1995) Effects of plant growth promoting rhizobacteria *Pseudomonas putida* GR-122 on the physiology of canolla roots. In: Green

DW(ed) Proceedings of the plant growth regulation society of America, 22nd proceeding, 18-20 July

Chen CC, Wang MK, Chiu CY, Huang PM, King HB (2001) Determination of low molecular weight dicarboxylic acids and organic functional groups in rhizosphere and bulk soils of Tsuga and Yushania in a temperate rain forest. Plant Soil 231:37–44

Costacurta A, Vanderleyden J (1995) Synthesis of phytohormones by plant-associated bacteria. Crit Rev Microbiol 21:1–18

Costacurta A, Keijers V, Vanderleyden J (1994) Molecular cloning and sequence analysis of an *Azospirillum brasilense* indole-3-pyruvate decarboxylase gene. Mol Gen Genet 243:463–472

Crowley DE, Reidd CPP, Szaniszlo PJ (1987) Mirobial siderophores as iron sources for plants. In: Winkelmann GD, Van der H, Neilands JB (eds) Iron transport in animals, plants and microorganisms. VCH Chemie, Weinheim

Dashti N, Zhang F, Hynes R, Smith DL (1998) Plant growth promoting rhizobacteria accelerate nodulation and increase nitrogen fixation activity by field grown soybean (Glycine max (L) Merr) under short season condition. Plant Soil 200:205–213

De Freitas JR, Germida JJ (1991) *Pseudomonas cepacia* and *Pseudomonas putida* as winter wheat inoculants for biocontrol of *Rhizobium solani*. Can J Microbiol 37:780–789

De Freitas JR, Banerjee MR, Germida JJ (1997) Phosphate-solubilizing rhizobacteria enhance the growth and yield but not phosphorus uptake of canola (Brassica napus L.). Biol Fertil Soils 24(4):358–364

Deshwal VK, Dubey RC, Maheshwari DK (2003) Isolation of plant growth promoting strains of *Bradyrhizobium* (*Arachis*) sp. with biocontrol potential against *Macrophomina phaseolina* causing charcol rot of peanut. Curr Sci 84:443–444

Dey Vay JE, Lukezic FL, Sinden SL, English H, Coplin DL (1968) A biocide produced by pathogenic isolates of *Pseudomonas syringae* and its possible role in bacterial canker disease of peach trees. Phytopathology 58:95–101

Dimkpa C, Svatos A, Merten D, Büchel G, Kothe E (2008) Hydroxamate siderophores produced by *Streptomyces acidiscabies* E13 bind nickel and promote growth in cowpea (*Vigna unguiculata* L.) under nickel stress. Can J Microbiol 54:163–172

Dimkpa CO, Merten D, Svatos A, Büchel G, Kothe E (2009) Siderophores mediate reduced and increased uptake of cadmium by *Streptomyces tendae* F4 and sunflower (*Helianthus annuus*), respectively. J Appl Microbiol 107:1687–1696

Doyle JJ, Luckow MA (2003) The rest of the iceberg. Legume diversity and evolution in a phylogenetic context. Plant Physiol 1331:900–910

Fernández C, Novo R, (1988) Vida Microbiana en el Suelo, II. La Habana: Editorial Pueblo y Educación, p 220

Fernando WGD, Nakkeeran S, Zhang Y (2005) Biosynthesis of antibiotics by PGPR and its relation in biocontrol of plant diseases. In: Siddiqui ZA (ed) PGPR: Biocontrol and biofertilization. Springer, Dordrecht, pp 111–142

Franche C, Lindström K, Elmerich C (2009) Nitrogen-fixing bacteria associated with leguminous and non-leguminous plants. Plant Soil 321:35–59

Frankenberger WT Jr, Arshad M (1991) Microbial production of plant growth regulating substances in soil. In: Keel C, Koller B, Defago G (eds) Plant growth-promoting rhizobacteria, progress and prospects. The Second International Workshop on PGPR. Interlaken, p 162-171, 14–19 Oct 1990

Frankenberger WT Jr, Arshad M (1995) Phytohormones in soils. Microbial Production and Function. Marcel Dekker, New York, pp 5–40

Ganesan V (2008) Rhizoremediation of cadmium soil using a cadmium-resistant plant growth-promoting rhizopseudomonad. Curr Mcobiol 56:403–407

Ghosh D, Bal B, Kashyap VK, Pal S (2003) Molecular phylogenetic exploration of bacteria diversity in a Bakreshwar (India) hot spring and culture of Shewanella related thermophiles. Appl Environ Microbiol 69:4332–4336

Glick BR (1995) The enhancement of plant growth by free living bacteria. Can. J Microbiol 41:109–250

Goldstein AH (1994) Involvement of the quinoprotein glucose dehydrogenase in the solubilization of exogenous phosphates by gram-negative bacteria. In: Torriani-Gorini A, Yagil E, Silver S (eds) Phosphate in microorganisms: cellular and molecular biology. ASM, Washington DC, pp 197–203

Graham PH, Vance CP (2000) Nitrogen fixation in perspective: an overview of research and extension needs. Field Crop Res 65:93–106

Gray EJ, Smith DL (2005) Intracellular and extracellular PGPR: commonalities and distinctions in the plant-bacterium signaling processes. Soil Biol Biochem 37:5–412

Gupta A, Gopal M (2008) Siderophore production by plant growth promoting rhizobacteria. Indian J Agric Res 42(2):153–156

Gupta A, Gopal M, Tilak KV (2000) Mechanism of plant growth promotion by rhizobacteria. Indian J Exp Biol 38:856–862

Hariprasad P, Niranjan SR (2009) Isolation and characterization of phosphate solubilizing rhizobacteria to improve plant health of tomato. Plant Soil 316:13–24

Harris WR, Carrano CJ, Raymond KN (1979) Microbial Iron Acquisition: Marine and Terrestrial Siderophores. J Am Chem Soc 101:2722

Hayat R, Ali S, Amara U, Khalid R, Ahmed I (2010) Soil beneficial bacteria and their role in plant growth promotion: a review. Ann Microbiol 60:579–598

Herman MAB, Nault BA, Smart CD (2008) Effects of plant growth promoting rhizobacteria on bell pepper production and green peach aphid infestations in New York. Crop Protect 27:996–1002

Herridge DF, Peoples MB, Boddey RM (2008) Global inputs of biological nitrogen fixation to agricultural systems. Plant Soil 311:1–18

Husen E (2003) Screening of soil bacteria for plant growth promotion activities *in vitro*. Indones J Agric Sci 4(1):27–31

Inbar J, Chet I (1991) Evidence that chitinase produced by *Aeromonas caviae* is involved in the biological control of soil-borne plant pathogens by bacterium. Soil Biol Biochem 23:974–978

Indiragandhi P, Anandham R, Madhaiyan M, Sa TM (2008) Characterization of plant growth-promoting traits of bacteria isolated from larval guts of diamondback moth Plutella xylostella (Lepidoptera: Plutellidae). Curr Microbiol 56:327–333

Jacobsen CS (1997) Plant protection and rhizosphere colonization of barley by seed inoculated herbicide degrading *Burkholderia* (*Pseudomonas*) *cepacia* DBO1(pRO101) in 2,4-D contaminated soil. Plant Soil 189:139–144

Jha PN, Kumar A (2007) Endophytic colonization of *Typhaaustralis* by a plant growth-promoting bacteria *Klebsiella oxytoca* strain GR-3. J Applied Micobiol 103:1311–1320

Jiang C, Sheng X, Qian M, Wang Q (2008) Isolation and characterization of a heavy metal- resistant *Burkholderia* sp. from heavy metal-contaminated paddy field soil and its potential in promoting plant growth and heavy metal accumulation in metal-polluted soil. Chemosphere 72:157–164

Joseph B, Patra RR, Lawerence R (2007) Characterization of plant growth promoting rhizobacteria associated with chickpea (*Cicer arietinum* L.). Int J Plant Prod 2:141–152

Kapulnik Y (1996) Plant growth promoting rhizosphere bacteria. In: Waisel Y, Eshel A, Kafkafi U (eds) Plant roots the hidden half. Marcel Dekker, NY, pp 769–781

Kay E, Dubuis C, Haas D (2005) Three small RNAs jointly ensure secondary metabolism and biocontrol in *Pseudomonas fluorescens* CHA0. Proc Natl Acad Sci 102(47):17136–17141

Khan AG (2005) Role of soil microbes in the rhizospheres of plants growing on trace metal contaminated soils in phytoremediation. J Trace Elem Med Biol 18:355–364

Kloepper JW (1993) Plant-growth-promoting rhizobacteria as biological control agents. In: Metting FB Jr (ed) Soil Microbial Ecology. Marcel Dekker, New York, pp 255–273

Kloepper JW, Schroth MN (1978) Plant growth-promoting rhizobacteria on radishes. Fourth international conference on plant pathogen bacteria, vol 2. Angers, pp 879–882

Kloepper JW, Leong J, Teintze M, Schroth MN (1980a) Enhancing plant growth by siderophores produced by plant growth-promoting rhizobacteria. Nature 286:885–886

Kloepper JW, Leong J, Teintze M, Schroth MN (1980b) *Pseudomonas* siderophores: a mechanism explaining disease-suppressive soils. Curr Microbiol 4:317–320

Kokalis-Burelle N, Kloepper JW, Reddy MS (2006) Plant growth promoting rhizobacteria as transplant amendments and their effects on indigenous rhizosphere microorganisms. Appl Soil Ecol 31:91–100

Kumar A, Prakash A, Johri BN (2011) Bacillus as PGPR in Crop Ecosystem. In: Maheshwari DK (ed) Bacteria in agrobiology: crop ecosystems, pp 37–59

Kumar BSD (1999) Fusarial wilt suppression and crop improvement through two rhizobacterial strains in chick pea growing in soils infested with *Fusarium oxysporum* f.sp. *ciceris*, Biol Fert Soils 29:87–91

Kumar KV, Singh N, Behl HM, Srivastava S (2008) Influence of plant growth promoting bacteria and its mutant on heavy metal toxicity in *Brassica juncea* grown n fly ash amended soil. Chemosphere 72:678–683

Ladwal A, Bhatia D, Malik DK (2012) Effect of coinoculation of *Mesorhizobium cicer* with PGPR on *Cicer arietinum*. Aust J Basic Appl Sci 6(9):183–187

Lambert B, Joos H, Dierick S, Vantomme R, Swings J, Kerters K, Van Montagu M (1990) Identification and plant interaction of *Phyllobacterium* sp. a predominant rhizobacterium of young sugar beet. Appl Environ Microbiol 56:1093–1102

Laslo E, György E, Mara G, Tamás É, Ábrahám B, Lányi S (2012) Screening of plant growth promoting rhizobacteria as potential microbial inoculants. Crop Prot 40:43–48

Lazarovits G, Nowak J (1997) Rhizobacteria for improvement of plant growth and establishment. Hort Sci 32:188–192

Lemanceau P, Albouvette C (1993) Suppression of *Fusarium* wilts by fluorescent *Pseudomonads*: mechanisms and applications. Biocontrol Sci Technol 3:219–234

Lemaux PG (1999) Plant growth regulators and biotechnology. Western plant growth regulator society presentation, Anaheim

Lenin G, Jayanthi M (2012) Efficiency of Plant Growth Promoting Rhizobacteria (PGPR) on enhancement of growth, yield and nutrient content of Catharanthus roseus. Int J Res Pure Appl Microbiol 2(4):37–42

Lindström K, Martínez-Romero E (2007) International committee on systematics of prokaryotes subcommittee on the taxonomy of Agrobacterium and Rhizobium: minutes of the meeting, 23-24 July 2006, Århus. Int J Syst Evol Microbiol 57:1365–1366

Link GKK, Eggers V (1941) Hyperauxiny in crown gall of tomato. Bot Gaz 103:87–106

Loper JE, Buyer JS (1991) Molecular Pl. Microb Interact 4(1):5–13

Loper JE, Henkels MD (1999) Utilization of heterologous siderophores enhances levels of iron available to Pseudomonas putida in the rhizosphere. Appl Environ Microbiol 65(12):5357–5363

Lwin KM, Myint MM, Tar T, Aung WZM (2012) Isolation of plant hormone (indole-3-acetic acid-IAA) producing rhizobacteria and study on their effects on maize seedling. Eng J 16(5):137–144

Lynch JM (1985) Origin, nature and biological activity of aliphatic substances and growth hormones found in soil. In: Vaughan D, Malcom RE, Martinus N (eds) Soil organic matter and biological activity. Dr. W. Junk Publishers, Dordrecht, pp 151–174

Mahaffee WF, Kloepper JW (1997) Temporal changes in the bacterial communities of soil, rhizosphere, and endorhiza associated with fieldgrown cucumber (*Cucumis sativus* L.). Microb Ecol 34:210–223

Maksi-mov IV, Abizgil'dina RR, Pusenkova LI (2011) Plant growth promoting rhizobacteria as alternative to chemical crop protectors from pathogens (Review). Appl Biochem Microbiol 47:333–345

Martinez-Romero E, Wang ET, (2000) Sesbania herbacea-Rhizobium huautlense nodulation in flooded soils and comparative characterization of S. herbaces nodulating rhizobia in different environments. Microb Ecol 41:25–32

Montanez A, Abreu C, Gill PR, Hardarson G, Sicardi M (2009) Biological nitrogen fixation in maize (*Zea mays* L.) by N-15 isotope-dilution and identification of associated culturable diazotrophs. Biol Fertil of Soil 45:253–263

Neeru N, Vivek K, Rishi K, Wolfgancy M (2000) Effect of P-solubilizing *Azotobacter chroococcum* on N, P, K uptake in p-responsive genotypes grown under greenhouse condition. J Plant Nutr Soil Sci 163:393–398

Nelson EB (1998) Biological control of *pythium* seed rot and preemergence damping-off of cotton with *Enterobacter cloacae* and *Ervinis herbicola* applied as seed treatments. Plant Dis 72:140–142

Newton WE (2007) Physiology, biochemistry and molecular biology of nitrogen fixation. In: Bothe H, Ferguson SJ, Newton WE (eds) Biology of the nitrogen cycle. Elsevier, Amsterdam, pp 109–130

Nielsen TH, Sørensen J (2003) Production of cyclic lipopeptides by *Pseudomonas fluorescens* strains in bulk soil and in the sugar beet rhizosphere. Appl Environ Microbiol 69:861–868

O'Sullivan DJ, O'Gara F (1992) Traits of fluorescent *Pseudomonads* species involved in suppression of plant root pathogens. Microbiol Rev 56:662–676

Osborne DJ, McManus MT (2005) Hormones, signals and target cells in plant development. Cambridge University Press p 158. ISBN 978-0-521-33076–33083

Pedraza RO, Bellone CH, Bellone de S, Sorte PMB, Teixeira KRD (2009) *Azospirillum* inoculation and nitrogen fertilization effect on grain yield and on the diversity of endophytic bacteria in the phyllosphere of rice rainfed crop. Eur J Soil Biol 45:36–43

Pollack JR, Neilands JB (1970) Enterobactin, an iron transport compound from Salmonella typhimurium. Biochem Biophys Res Commun 38:989

Poonguzhali S, Madhaiyan M, Sa TM (2008) Isolation and identification of phosphate solubilizing bacteria from Chinese cabbage and their effect on growth and phosphorus utilization of plants. J Microbiol Biotechnol 18:773–777

Postgate JR (1982) The fundamentals of nitrogen fixation. Cambridge University Press, Cambridge United Kingdom p 271

Pradhan N, Sukla LB (2006) Solubilization of inorganic phosphates by fungi isolated from agriculture soil. Afr J Biotechnol 5:850–854

Prithiviraj B, Zhou X, Souleimanov A, Kahn W, Smith DL (2003) A host-specific bacteria-to-plant signal molecule (Nod factor) enhances germination and early growth of diverse crop plants. Planta 216:437–445

Raghu K, MacRae IC (1966) Occurrence of phosphate-dissolving microorganisms in the rhizosphere of rice plants and in submerged soils. J Appl Bacteriol 29:582–586

Rajkumar M, Freitas H (2008) Influence of metal resistant-plant growth promoting bacteria on the growth-promoting bacteria on the growth of *Ricinus communis* in soil contaminated with heavy metals. Chemosphere 71:834–842

Rajkumar M, Nagendran R, Kui JL, Wang HL, Sung ZK (2006) Influence of plant growth promoting bacteria and Cr (VI) on the growth of Indian mustard. Chemosphere 62:741–748

Rani A, Souche YS, Goel R (2009) Comparative assessment of *in situ* bioremediation potential of cadmium resistant acidophilic *pseudomonas putida* 62BN and alkalophilic *pseudomonas monteilli* 97AN strains on soybean. Int Biodeterior Biodegrad 63:62–66

Raymond KN, Müller G, Matzanke BF (1984) Complexation of Iron by siderophores review of their solution and structural chemistry and biological function. Top Curr Chem 123:49–102

Reddy MS, Wang Q (2011) Plant growth promoting rhizobacteria (PGPR) for sustainable agriculture. Proceedings of the 2nd Asian PGPR conference August 21-24, Beijing, P.R. China.

Robinson RL, Postgate JR (1980) Oxygen and nitrogen in biological nitrogen fixation. Ann Rev Microbiol 34:182–207

Rodríguez H, Fraga R (1999) Phosphate solubilizing bacteria and their role in plant growth promotion. Biotechnol Adv 17:319–339

Rosenblueth M, Martinez-Romero E (2006).Bacterial endophytes and their interactions with hosts. Mol Plant-Microb Interact 19:827–837

Ryder MH, Jones DA (1990) Biological control of crown gall. In: Hornby D, Cook RJ, Henis Y (eds) Biological control of soil-borne plant pathogens. CAB International, Oxford, UK, pp 45–63

Ryder MH, Yan Z, Terrace TE, Rovira AD, Tang W, Correll RL (1999) Use of strains of *Bacillus* isolated in China to suppress take-all and rhizoctonia root rot, and promote seedling growth of glasshouse grown wheat in Australian soils. Soil Biol Biochem 31:19–29

Ryu R, Patten CL (2008) Aromatic amino acid-dependent expression of indole-3-pyruvate decarboxylase is regulated by 4 TyrR in *Enterobacter cloacae* UW5. Am Soc Microbiol 190(21):1–35

Saha R, Saha N, Donofrio RS, Bestervelt LL (2013) Microbial siderophores: a mini review. J Basic Microbiol 53, 303–317. Doi: 10. 1002/jobm. 201100552

Sala VMR, Cardoso E, Freitas de JG, Silveira da APD (2007) Wheat genotypes response to inoculation of diazotrophic bacteria in field conditions. Pesquisa Agropecuária Brasileira 42:833–842

Schippers et al (1987) Interactions of deleterious and beneficial rhizosphere microorganisms and the effect of cropping practices. Ann Rev Phytopathol 25:339–358

Selvakumar G, Mohan M, Kundu S, Gupta AD, Joshi P, Nazim S, Gupta HS (2008) Cold tolerance and plant growth promotion potential of *serratia marcescens* strain SRM (MTCC 8708) Isolated from flowers of summer squash (*Cucurbita pepo).* Lett Appl Microbiol 46:171–175

Shah Z, Shah AN, Ansari ST (2011) Utilizing rhizobacteria for sustainable wheat production: role of phosphate solubilizing ACC-deaminase rhizobacteria under phosphorus deficiency stress. Lap Lambert Academic Publishing

Sharma S, Kumar V, Tripathi RB (2011) Isolation of phosphate solubilizing microorganism (PSMs) from soil. J Microbiol Biotech Res 1(2):90–95

Simon S, Petrášek J (2011) Why plants need more than one type of auxin. Plant Sci 180(3):454–460

Sindhu SS et al (1997) Biotechnological approaches in soil microorganisms for sustainable crop production. Scientific Publishers, Jodhpur, pp 149–170

Singh JS, Pandey VC, Singh DP (2011) Efficient soil microorganisms: a new dimension for sustainable agriculture and environmental development. Agric Ecosyst Environ 140:339–353

Singh N, Pandey P, Dubey RC, Maheshwari DK (2008) Biological contro of root rot fungus *Macrophomina phaseolina* and growth enhancement of *Pinus roxburghii* by rhizosphere competent *Bacillus subtilis* BN1. World J Microbiol Biotechnol 24:1669–1679

Singh RP (2005) Biological nitrogen fixation. Microbiology Kalyani Publishers, Head office B-1/292, Rajinder Nagar, Ludhiana

Somers E, Vanderleyden J, Srinivasan M (2004) Rhizosphere bacterial signalling: a love parade beneath our feet. Crit Rev Microbiol 30:205–240

Spaepen S, Vanderleyden J, Remans R, (2007) Indole-3-acetic acid in microbial and microorganism-plant signaling. FEMS Microbiol Rev 31:425–448

Stewart WDP (1969) Biological and ecological aspects of nitrogen fixation by free-living microorganisms. Proc Roy Soc B (London) 172:367–388

Sugawara S, Hishiyama S, Jikumaru Y, Hanada A, Nishimura T, Koshiba T, Zhao Y, Kamiya Y, Kasahara H, (2009) Biochemical Analyses of Indole-3-acetaldoxime-dependent auxin biosynthesis in Arabidopsis. Proc Natl Acad Sci U S A 106(13):5430–5435

Teale WD, Paponov IA, Palme K (2006) Auxin in action: signaling, transport and the control of plant growth and development. Nat Rev Mol Cell Biol 7:847–859

Thomashow LS (1996) Biological control of plant root pathogens. CurrOpin Biotechnol 7:343–347

Thomshaw LS, Webler DM (1995) Current concepts in the use of introduced bacteria for biological disease control of *Gaeumamomyces graminis* var *tritici*. J Bacteriol 170:3499–3508

Thomshow LS, Bonsal RF, Weller DM (2003) Detection of antibiotics production by soil and rhizosphere microbes *in situ*. Thomashow lab methods: 1–13

Vadakattu GVSR, Kasper ML, Jankovic-Karasoulos T, Elliott ET (2006) Macroaggregate environment influences the composition and activity of associated microbiota communities. 18th World Congress of Soil Science Philadelphia Pennsylvania USA

Vance C (2001) Symbiotic nitrogen fixation and phosphorus acquisition. Plant nutrition in a world of declining renewable resources. Plant Physiol 127:391–397

Verma JP, Yadav J, Tiwaric KN, Kumar A (2013) Effect of indigenous Mesorhizobium spp. and plant growth promoting rhizobacteria on yields and nutrients uptake of chickpea (Cicer arietinum L.) under sustainable agriculture. Ecol Eng 51:282–286

Vessey JK (2003) Plant growth-promoting rhizobacteria as biofertilizers. Plant Soil 255:571–586

Wagner SC (2012) Biological nitrogen fixation. Nat Educ Knowl 3(10):15

Wani PA, Khan MS, Zaidi A (2007a) Co inoculation of nitrogen fixing and phosphate solubilizing bacteria to promote growth, yield and nutrient uptake in chickpea. Acta Agron Hung 55:315–323

Wani PA, Khan MS, Zaidi A (2007b) Effect of metal tolerant plant growth promoting *Bradyrhizobium* sp. (Vigna) on growth symbiosis seed yield and metal uptake by Greengram plants. Chemosphere 70:36–45

Wani PA, Khan MS, Zaidi A (2007c) Synergistic effects of the inoculation with nitrogen fixing and phosphate solubilizing rhizospheria on the performance of yield grown chickpea. J Plant Nutr Soil Sci 170:283–287

Wani PA, Khan MS, Zaidi A (2008) Chromium reducing and plant growth-promoting *Mesorhizobium* improves chicpea growth in chromium-amended soil. Biotechnol Lett 30:159–163

Weller DM, Thomashow LS (1993) Use of rhizobacteria for biocontrol. Curr Opin Biotechnol, 4:306–311

Weyens N, van der Lelie D, Taghavi S, Newman L, Vangronsveld J (2009) Exploiting plantemicrobe partnerships to improve biomass production and remediation. Trends Biotechnol 27:591–598

White PR, Braun AC (1941) Crown gall productions by bacteria free tumor tissues. Science 93:239–241

Won C, Shen X, Mashiguchi K, Zheng Z, Dai X, Cheng Y, Kasahara H, Kamiya Y (2011) Conversion of tryptophan to indole-3-acetic acid by Tryptophan Aminotransferases of Arabidopsis and Yuccas in *Arabidopsis*. Proc Natl Acad Sci U S A 108(45):18518–18523

Yadav J, Verma JP, Tiwari KN (2010) Effect of plant growth promoting rhizobacteria on seed germination and plant growth chickpea (*Cicer arietinum* L.) under in vitro conditions. Biol Forum Int J 2(2):15–18

Yahalom E, Okon Y, Dovrat A (1987) *Azosprillium* effect on susceptibility to *rhizobium* nodulation and on nitrogen fixation of several forage Can. J Microbiol 33:510–514

Yuen GY, Schroth MN, McCain AH (1985) Reduction of *Fusarium* wilt of carnation with suppressive soils and antagonistic bacteria. Plant Dis 69:1071–1075

Zahran HH, Ahmed MS, Afkar EA (1995) Isolation and characterization of nitrogen-fixing moderate halophilic bacteria from saline soils of Egypt. Basic Microbiol 35:269–275

Zhang F, Dhasti N, Hynes R, Smith DL (1996) Plant growth promoting rhizobacteria and soybean (*Glycinemax* (L.)*Merr.*) nodulation and nitrogen fixation at sub-optimal root zone temperatures. Ann Bot 77: 453–459

Zhao Y (2010) Auxin biosynthesis and its role in plant development. Annu Rev Plant Biol 61:49–64

# Biochemical Activity of *Ocimum gratissimum* Essential Oil Against Fruit-Rotting Fungi *Penicillium expansum* and *Penicillium digitatum*

30

Arshad H. Rizvi, M. M. Abid Ali Khan, Praveen C. Verma and Gauri Saxena

**Abstract**

The fruit-rotting fungi *Penicillium expansum* and *Penicillium digitatum* are the primary cause of postharvest losses as they cause blue mold of apples and green mold of oranges, respectively. In addition to rotting, they also contaminate food with their highly toxic chemical known as patulin, which is well known for its carcinogenic effect. In the present study, the essential oil of *Ocimum gratissimum* obtained by hydro-distillation was tested for its antifungal assay against *P. expansum* and *P. digitatum* in *in vitro* culture conditions. 500 ppm MIC of the essential oil completely inhibited the mycelial growth of both the test fungi *in vitro*. The inhibitory effect of the essential oil was also observed directly on the fruits, apples and oranges inoculated with *P. expansum* and *P. digitatum* spore suspensions *in vivo*. They were fumigated with the essential oil of *O. gratissimum* for 10 days and the results obtained showed potent biochemical activity up to 93 % against blue mold rot of apples and up to 75 % against green mold rot of oranges. The present study suggests use of *Ocimum* oil as a safe, effective and alternative means to control fruit rotting fungi.

A. H. Rizvi (✉)
Department of Botany, Shia PG College, Lucknow, India
e-mail: a_hrizvi@yahoo.co.in

M. M. Abid Ali Khan
Department of Botany, Shia PG College, Lucknow, India
e-mail: mmabidalikhan265@gmail.com

P. C. Verma
National Botanical Research Institute (CSIR),
Rana Pratap Marg,
Lucknow, India

G. Saxena
Department of Botany, University of Lucknow,
Lucknow, India

R. N. Kharwar et al. (eds.), *Microbial Diversity and Biotechnology in Food Security*,
DOI 10.1007/978-81-322-1801-2_30, © Springer India 2014

**Keywords**

Blue mold · Green mold · *Ocimum gratissimum* · Essential oil · Post harvest losses · Antifungal

## 30.1 Introduction

*Ocimum* is an extremely versatile plant, having 160 species belonging to family Lamiaceae. It has long been used as a medicinal and aromatic plant in many countries like India, Egypt, Greece, Italy and Morocco. The genus is cosmopolitan in distribution being present in tropical, subtropical and temperate regions. *Ocimum gratissimum,* a wildly growing species in India, contains an essential oil in leaves, which is used for its aromatic values as flavouring in foods and beverages and as fragrance in pharmaceutical and industrial products. It has been well documented that the essential oil obtained from *O. gratissimum* shows antifungal property (Reuveni et al. 1984; Rashmi and Yadav 1999; Dubey et al. 2000a, b). Studies on the composition of *O. gratissimum* essential oil from Madhya Pradesh India, showed that the eugenol was the main component of the essential oil, and the other constituents being β-caryophyllene, farnesene, methyl eugenol, linalool, limonene and in trace amount α-terpineol, β-selinene, methyl isoeugenol, bisabolol, geraniol, α-pinene, *p*-Cymene, cubebene, camphene, t-cadinol, γ-eudesmol, sabinene, myrcene, β-bisaboline, α-humulene and β-elemene (Pandey and Chaudhary 2001).

The fruit-rotting fungi frequently infects unripe fruits and causes relatively minor damage until ripening, when they may cause extensive decay. Such quiescent infections have been observed in fruits of tropical, subtropical and deciduous plants. The resultant decays have great economic loss as they reduce the shelflife of fruits during storage and transport (Prusky and Keen 1993). In India, tropical and temperate fruits like *Citrus* and *Malus* have been well cultivated, transported and exported to different noncultivated regions. Fruit-rotting fungi represent one of the main causes of postharvest losses. In addition to rotting, they also contaminate fruits with highly toxic chemicals “patulin” a mycotoxins (Richard et al. 1989), well known for its carcinogenic and mutagenic effects on man and animals (Lancaster et al. 1961; Newberne et al. 1968; Hayes 1980; Natrajan 1989). Medicinal plants produce varieties of chemicals which are being explored as indigenous antifungal agents and, nontoxic and renewable sources for plant disease control. Phytochemists and biologists are trying to find out the potentiality of secondary metabolites of medicinal plants, as alternative source of chemicals in control of plant diseases caused by fungi, nematodes, bacteria and insects. Several plant parts and their constituents have recently proven their effect in providing lesser phytotoxic, more systemic, easily biodegradable and host metabolism stimulating fungicides (Fawcett and Spencer 1970; Jacobson and Crosby 1971; Beye 1978; Dixit et al. 1978; Yang and Tong 1988).

The present report deals with the gas chromatography analysis of essential oil obtained from leaves of *Ocimum gratissimum* and its potentiality as ecofriendly, natural fungi-toxicant against fungal deterioration of apple and orange during storage and transportation.

## 30.2 Materials and Methods

### 30.2.1 Plant Material

The crop of *O. gratissimum* was grown in Lucknow, India situated at 120 m altitude. Confirmation of identity of the plant was done with the help of authentic herbarium specimens lodged in Central Institute of Medicinal and Aromatic Plants (CIMAP), Lucknow, India. Fresh aerial parts of the plant were harvested before the flowering stage in the month of September.

### 30.2.2 Isolation of Oil

The fresh plant material was hydro-distilled for 3 h using a clevenger-type apparatus and the isolated oil was dried over anhydrous sodium sulphate.

### 30.2.3 Gas Chromatography

Gas chromatography (GC) was performed on a Perkin Elmer Autosystem *XL*, with flame ionization detector (FID) attachment under the following conditions: injector and detector temperatures, 220 °C and 250 °C respectively, using a capillary column PE-5, (5% phenyl and 95% dimethyl-polysiloxane) 50 mm × 0.32 mm, film thickness 0.25 µm. Oven temperature programmed from 100 to 220 °C at the flow rate of 30 °C/min with initial hold of 2 min. Helium ($H_2$) was used as carrier gas with column head pressure 10 psi. The area percentage of constituent was obtained through Total Chrome software. Gas chromatography–mass spectrometry (GC-MS) data were obtained on Perkin Elmer Turbomass spectrometer coupled with computer library (NIST and WILEY) using a PE-5 column (50 × 0.32 mm), film thickness 0.25 µm. The carrier gas was $H_2$ and the temperature programming was 3 min hold time at 100 °C, rising at the rate of 3 °C/min to 280 °C.

### 30.2.4 Identification of Constituents

The identity of the components was assigned by comparing their GC retention time with those of authentic samples, as well as with the comparison of the fragmentation patterns of the mass spectra with those reported in the literature.

### 30.2.5 Antifungal Assay

#### 30.2.5.1 Culture of Test Fungus

*Penicillium expansum* and *Penicillium digitatum* was isolated from the decaying apple and orange respectively. Isolate from infected apple and orange fruit were obtained and cultures were maintained by growing the organism on potato dextrose agar (PDA, Difco, USA) slants for 7 days at 25 ± 2 °C, at the end of which the entire slant surface was covered with spores of the molds. The fungal pathogenicity and virulence were maintained by inoculating fruits with the fungus and re-isolating it. Apple and oranges utilized in this work had not been treated with synthetic fungicides.

#### 30.2.5.2 Preparation of Spore Suspension

Spore suspensions of *P. expansum* and *P. digitatum* were prepared following Mishra and Dubey (1994). The fungal spores were harvested from a 7-day-old culture of each fungus and suspended in 100 ml of distilled water containing 0.05% Tween80 (polyoxyethylene sorbitan mono-oleate) and the tubes were shaken gently to disperse the spores. The spore suspension was filtered through two layers of sterile cheese cloth to remove mycelia and is diluted with sterile water to obtain a density of spore suspension of $10^4$ spores/ml. The number of spores was determined using haemocytometer.

#### 30.2.5.3 Minimum Inhibitory Concentration (MIC)

To find out the minimum concentration at which the oils showed absolute fungi-toxicity, experiments were carried out by the usual poisoned food technique. Different concentrations of the oil *viz.* 100, 250, 500, 1,000 and 2,000 ppm were prepared by dissolving its requisite amount in 0.5 ml acetone and then mixing with 9.5 ml PDA medium. Control sets contained requisite amount of sterilized water dissolved in 0.5 ml acetone in place of the oil. The plates were inoculated aseptically with the assay disc of the test fungus and incubated for 6 days.

**Table 30.1** Constituents of essential oil of *O. gratissimum*

| S No. | Compounds | Area % | Kovats indices | Identification |
|---|---|---|---|---|
| 1 | Cis-3-Hexenol | 00.07 | 0851 | KI, MS |
| 2 | α-Pinene | 00.31 | 0939 | KI, MS |
| 3 | Sabinene | 00.45 | 0976 | KI, MS |
| 4 | β-Pinene | 00.26 | 0980 | KI, MS |
| 5 | Limonene | 00.03 | 1,031 | KI, MS |
| 6 | 1,8-Cineole | 11.23 | 1,033 | KI, MS |
| 7 | β-Ocimene | 00.54 | 1,040 | KI, MS |
| 8 | Linalool | 00.48 | 1,098 | KI, MS |
| 9 | Terpinen-4-ol | 00.43 | 1,177 | KI, MS |
| 10 | Methylchavicol | 00.41 | 1,195 | KI, MS |
| 11 | Eugenol | 77.81 | 1,356 | KI, MS |
| 12 | Methyl cinnamate (trans) | 00.06 | 1,301 | KI, MS |
| 13 | β-Elemene | 00.30 | 1,391 | KI, MS |
| 14 | β-caryophyllene | 01.30 | 1,404 | KI, MS |
| 15 | Germacrene-D | 04.07 | 1,480 | KI, MS |

Elution order on PE-5 capillary column

The observation was made on the seventh day and percentage mycelial inhibition was calculated.

### 30.2.6 Efficacy of Essential Oil Against Mold Rot of Apple and Oranges

Testing *O. gratissimum* oils as fumigants for protection of apples (*Malus pumila* Mill) from blue mold and oranges (*Citrus reticulata* Blanco) from green mold was done by the method adopted by Mishra and Dubey (1994).

Mature medium sized and healthy fruits were selected and washed in running water and then disinfected by immersing in 1.0% sodium hypochlorite solution, rinsed with sterile water and dried in a sterile chamber. The pathogenicity of respective fungi was tested on respective fruits following Garcha and Singh (1980).

An injury (3 mm wide and 1 mm deep) was made on the outer surface of fruits with the help of sterilized needle. The apples and oranges were inoculated by spore suspension ($10^4$ spores/ml) of *P. expansum* and *P. digitatum*, respectively, with the help of atomizer on the scratched peel at the stem end of each lot of fruit samples. Eight fruits of each set were then placed separately in plastic containers. Requisite amount of oil of *O. gratissimum* was introduced in these containers by soaking in cotton pieces so as to get required concentrations (v/v). The control set contained two sets: un-inoculated controls and inoculated controls. In the un-inoculated controls, the fruit samples were stored as such in plastic containers whereas in inoculated controls each fruit sample was inoculated with normal spore suspension of test fungi as usual to treated sets. The plastic containers were kept for 10 days in a biochemical oxygen demand (BOD) incubator set at 25±2 °C. Three replicates were kept for treatment and control sets. The percentage loss of fruit tissue was calculated in the form of disease index (0–10) scale which was based on average value following Garcha and Singh (1980). Percent disease control was calculated by following formula:

$$\text{Disease control} = \frac{dc - dt}{dc} \times 100.$$

Where, $dc$=% disease index in inoculated control, $dt$=% disease index in treated sets.

## 30.3 Results and Discussion

Essential oil of *O. gratissimum* was examined by GC & GC-MS. Fifteen constituents were identified in *O. gratissimum*, representing 97.75% of the oil (Table 30.1). The major constituents of *O. gratissimum* oil were eugenol (77.81%) and

**Table 30.2** Minimum inhibitory concentration of *O. gratissimum* against *P. expansum* and *P. digitatum*

| Cocentrations in ppm. | % Mycelial inhibition of test fungi | |
|---|---|---|
| | *P. expansum* | *P. digitatum* |
| 100 | 30 | 30 |
| 250 | 60 | 70 |
| 500 | 100 | 100 |
| 1,000 | 100 | 100 |
| 2,000 | 100 | 100 |

**Table 30.3** Efficacy of *O. gratissimum* essential oil against mold rot disease

| Groups | 500 ppm (MIC) | | | | 1,000 ppm (MIC) | | | |
|---|---|---|---|---|---|---|---|---|
| | Blue mold | | Green mold | | Blue mold | | Green mold | |
| | DI | PC | DI | PC | DI | PC | DI | PC |
| Uninoculated control | 0.2 | – | 0.1 | – | 0.1 | – | 0.0 | – |
| Inoculated control | 100 | – | 99 | – | 100 | – | 100 | – |
| Inoculated and treated | 0.5 | 93 | 0.62 | 75.0 | 0.0 | 100 | 0.0 | 100 |

*DI* average value of disease index, *PC* percentage of disease control

1,8-cineole (11.23 %). The composition of *O. gratissimum* oils mentioned in the present study was found to be different from those obtained from the cultivars of *O. gratissimum* grown in Madhya Pradesh, India. The variation in oil constituents in the present material may call for an interpretation premised not only upon geographic divergence but also upon other factors, such as the uniqueness of the cultivars, age of plants, time of harvesting and other relevant variables occurring under varying ecological conditions.

Results clearly reveal that the constituents of essential oil possess fungi-toxic properties. It is evident from Table 30.2 that at 500 ppm and above, the *O. gratissimum* oil completely inhibited the mycelial growth of both the test fungi. Therefore, the MIC of the *O. gratissimum* oil was assigned to be 500 ppm. It is clear from Table 30.3 that *O. gratissimum* oils have controlled blue mold rot of apples caused by *P. expansum* on their respective MIC(s) by 93 %. While, it was 75 % against green mold rot of oranges caused by *P. digitatum.* Moreover, the oils exhibited 100 % control of blue and green mold rot above their respective MIC(s). Thus the oil exhibited potency to control green mold of oranges and blue mold apples.

**Acknowledgments** The authors are grateful to the Central Laboratory Facility, Central Institute of Medicinal and Aromatic Plants, Lucknow, for GC and GC-MS spectra.

## References

Beye F (1978) Insecticides from the vegetable kingdom. Plant Res Dev 7:13–31

Dixit SN, Tripathi NN, Tripathi SC (1978) Fungitoxicity of some seed extract. Nat Acad Sci Lett 1:287–288

Dubey NK, Tiwari TN, Mandin D, Andriamboavonjy H, Chaumout JP (2000a) Antifungal properties of O. *gratissimum* essential oil. Fitoterapi 71:567–569

Dubey NK, Tripathi P, Singh HB, (2000b) Prospects of some essential oils as antifungal agents. J Med Aromat Plant Sci 22:350–354

Fawcett CH, Spencer DM (1970) Plant chemotherapy with natural products. Annual Rev Phytopath 8:403–418

Garcha HC, Singh V (1980) Post-harvest diseases of fruits in Punjab. Ind Phytopath 33:42–47

Hayes AW (1980) Mycotoxin: a review of biological effects and their role in human disease. Clin Toxicol 17:45

Jacobson M, Crosby DG (1971) Naturally occurring insecticides. Marcel Dekke, New York

Lancaster MC, Jenkins FP, Philip MLJ (1961) Toxicity Associated with certain samples of groundnuts. Nature 192:1095

Mishra AK, Dubey NK (1994). Evaluation of some essential oils for their toxicity against fungi causing deterioration of stored food commodities. Appl Environ Microbial 60(4):1101–1105

Natrajan KR (1989) Mycotoxin and human health. Biol Edu 6:23–27

Newberne PM, rogers AE, Wagan GN (1968) Hepatorenal lesions in rats fed low lipotrope diet and exposed to aflatoxin. J Nutri XCIV:331–342

Pandey AK, Chaudhary AR, (2001) Composition of the essential oil of Ocimum *gratissimum* grown in Madhya Pradesh. JMAP Sci 22 & 23:26–28

Prusky D, Keen NT (1993) Involvement of performed antifungal compound in the resistance of subtropical fruits to fungal decay. Pt Dis 77:114–119

Rashmi, Yadav BP (1999) A comparative efficacy of fungicides and plant extracts on radial growth and biomass production of *Alternaria alternata*. J App Biol 9:73–76

Reuveni R, Fleisher A, Putievsky E (1984) Fungistatic activity of essential oils from *Ocimum basilicum* chemotypes. Phytopathology 110:20–22

Richard JL, Cole RJ, Archibald SO (1989) Mycotoxin, economic and health risks. Council of Agricultural Science and Technology Report, p 116

Yang RZ, Tong CS (1988) Plants used for pest control in China. Econ Bot 42:376–406

# Seed Quality Status of Polymer-Coated Bt Cotton (*Gossypium* sp.) During Storage Under Coastal Environment

31

C. Rettinassababady and T. Ramanadane

**Abstract**

An experiment was conducted during 2009 at Karaikal Dist., Puducherry (UT) to evaluate the seed quality status and storability of polymer-coated seeds of Bt cotton NHH 44. The seeds were coated with synthetic polymer (Polykote at the rate 3 ml/kg diluted in 5 ml of water $kg^{-1}$ seed) alone or in combination with fungicide (Flowable Thiram (Royal flow)) at the rate 2.5 g $kg^{-1}$ seed) or insecticide (Imidachloprid at the rate 6 ml $kg^{-1}$ of seed) or both and stored in 700 gauge polythene bags and cloth bags for 4 months. Besides the above mentioned, Vitavax 200 at the rate 2 g $kg^{-1}$ seed and seeds primed and dried to original moisture content and primed seed treated with polymer along with Flowable Thiram and Imidachloprid were also included as treatments. Observations on percentage seed germination, seed moisture content and seed infection were recorded at bimonthly intervals up to 4 months. The results of standard germination test on treated and untreated stored seeds indicated that seeds treated with Royal flow + Imidachloprid (90 %) were on par with Vitavax 200 (88.5 %) and Flowable Thiram alone (89 %) excelled other treatments irrespective of containers and period of storage. The increase in seed moisture content was gradual in the case of polythene bag when compared to cloth bag-stored seeds. The results on seed health status indicated that among the treatments, Polykote-coated seeds were highly susceptible to seed infection. Seed coating with Polykote coupled with Royal flow and Imidachloprid and seeds coated with Vitavax 200 effectively contained the seed infection. Between containers, seeds stored in 700 gauge polythene bags excelled cloth bag by maintaining higher germination and less seed infection during storage.

**Keywords**

Polymer coating · Bt cotton · Seed moisture · Seed infection · Coastal environment

C. Rettinassababady (✉) · T. Ramanadane
Pandit Jawaharlal Nehru College of Agriculture and Research Institute, Karaikal 609 603, Union Territory of Puducherry, India
e-mail: crsvaisu@yahoo.co.in

R. N. Kharwar et al. (eds.), *Microbial Diversity and Biotechnology in Food Security*,
DOI 10.1007/978-81-322-1801-2_31, © Springer India 2014

## 31.1 Introduction

Cotton is an important fibre crop known as "King of fibre" and in recent years is known as "White Gold". It is one of the most important commercial crops grown in about 80 countries in the world. Cotton belongs to Malvaceae family and its seed loses viability and vigour rapidly during storage, being a poor storer. The *linted* cotton seed harbours many pathogens and insect pests during storage which reduces the seed quality. Cotton seeds carry destructive pathogens that often take heavy toll by causing severe diseases on crops raised from them. Hence, seed coating with synthetic polymer in combination with fungicides and insecticides may be a potent tool for quality improvement and effective disease management against seed and soil-borne pathogens. Film coating technology is a sophisticated process of applying precise amount of active ingredients along with a liquid material, directly on to the seed surface without obscuring its shape; total seed weight may increase up to 1–2 % (Vijaykumar et al. 2007; Shakuntala et al. 2010). The film formulations consists of a mixture of polymer, plasticizer and colourants (Robani 1994) that are commercially available as ready to use liquids or as dry powders (Ni 1997). Seed coating provides an opportunity to package effective quantities of materials so that they can improve the germination and seedling growth. The polymer film coating may act as a physical barrier, which has been reported to reduce the leaching of inhibitors from the seed coverings and may restrict oxygen diffusion to the embryo (Vanangamudi et al. 2003). Hence, an attempt was made to study the seed quality status of polymer-coated Bt cotton during storage under coastal environment.

## 31.2 Materials and Methods

A laboratory experiment was conducted under National Seed Project (crops), at Pandit Jawaharlal Nehru College of Agriculture and Research Institute, Karaikal, Puducherry (UT) during 2009 to evaluate the seed health status and storability of polymer-coated seeds of Bt cotton NHH 44 with various seed treatment combinations. The seeds were coated with synthetic polymer (Polykote at the rate 3 ml/kg diluted in 5 ml of water $kg^{-1}$ seed) alone (T1) or in combination with fungicide (Flowable Thiram (Royal flow)) at the rate 2.5 g $kg^{-1}$ seed) (T2) or insecticide (Imidachloprid at the rate 6 ml $kg^{-1}$ of seed) (T3) or both as treatment T4 (Polykote + fungicide + insecticide) and stored in 700 gauge polythene bags and cloth bags for 4 months. Besides the above, Polykote along with Vitavax 200 at the rate 2 g $kg^{-1}$ seed (T5) and seeds primed and dried to original moisture content (T6) and primed seeds treated with polymer along with Flowable Thiram and Imidachloprid (T7) and untreated control (T0) were also included as treatments. The experiment was designed as completely randomised block design with three replications. The seeds were packed in a cloth bag and 700 gauge polythene bags and stored in ambient conditions of Karaikal district. Observations on per cent seed moisture content (ISTA 1999), seed germination per cent (Anonymous 1996) and percentage seed infection (ISTA 1999) were recorded at bimonthly intervals up to 4 months. The statistical analysis was done as per the procedure described by Panse and Sukhatme (1985).

## 31.3 Results and Discussion

### 31.3.1 Effect of Polykote Treatment on Seed Germination

The results of seed germination test on treated and stored seeds indicated that seed germination declined with progress of storage period (Table 31.1). Among the treatments, seeds treated with Polykote along with Royal flow and Imidachloprid (T4) registered the highest germination of 90 %, which was on par with Vitavax 200 (88.5 %) (T5) and Flowable Thiram alone (89 %) (T2) excelled other treatments irrespective of containers and period of storage. The percentage germination gradually decreased from 85 to 79 and from 85 to 82 in cloth and polythene bags, respectively and it was above minimum seed certification standards at the end of 4 months of storage.

**Table 31.1** Effect of Polykote seed coating on seed germination (%) during storage in Bt cotton NHH 44

| Treatments | Cloth bag ($C_1$) | | | | Polythene bag ($C_2$) | | | |
|---|---|---|---|---|---|---|---|---|
| | $P_0$ | $P_1$ | $P_2$ | Mean | $P_0$ | $P_1$ | $P_2$ | Mean |
| $T_0$: Untreated control | 74 | 70 | 70 | 71 | 74 | 73 | 70 | 72 |
| $T_1$: Polykote at the rate 3 ml/kg dilute with 5 ml water | 89 | 83 | 76 | 83 | 89 | 87 | 80 | 85 |
| $T_2$: $T_1$ + Flowable Thiram at the rate 2.5 g/kg | 89 | 89 | 88 | 89 | 89 | 89 | 89 | 89 |
| $T_3$: $T_1$ + Imidachloprid at the rate 6 ml/kg | 83 | 82 | 80 | 82 | 83 | 82 | 80 | 82 |
| $T_4$: $T_1$ + Royal flow 40 SC at the rate 2.4 ml/kg + Imidachloprid at the rate 6 ml/kg | 91 | 90 | 90 | 90 | 91 | 90 | 90 | 90 |
| $T_5$: $T_1$ + Vitavax 200 at the rate 2 g/kg | 91 | 90 | 80 | 87 | 91 | 90 | 90 | 90 |
| $T_6$: Primed and dried to original moisture content | 78 | 75 | 66 | 73 | 78 | 78 | 76 | 77 |
| $T_7$: $T_6$ + Polykote + Thiram + Imidachloprid | 85 | 85 | 83 | 84 | 85 | 85 | 84 | 85 |
| Mean | 85 | 83 | 79 | | 85 | 84 | 82 | |
| | C | P | T | CxP | PxT | CxT | CxPxT | |
| SEd. | 0.73 | 0.89 | 1.45 | NS | NS | NS | NS | |
| CD ($P$ = 0.05) | 1.44 | 1.76 | 2.88 | | | | | |

*Mean Table*

| Treatments (T) | | | | Period of storage (P) | | Containers (C) | |
|---|---|---|---|---|---|---|---|
| $T_0$ | 71.5 | $T_4$ | 90.0 | $P_0$ | 85.0 | $C_1$ | 82.3 |
| $T_1$ | 84.0 | $T_5$ | 88.5 | $P_1$ | 83.5 | $C_2$ | 83.7 |
| $T_2$ | 89.0 | $T_6$ | 78.0 | $P_2$ | 80.5 | | |
| $T_3$ | 82.0 | $T_7$ | 84.5 | | | | |

$P_0$ Initial, $P_1$ 2 months after storage, $P_2$ 4 months after storage

In the case of cloth bag storage, the effect of seed coating with chemicals on germination was concerned and all the treatments recorded significantly higher seed germination up to 4 months of storage as compared to control. The primed seeds (T6) without any seed treatment had lesser germination of 75 % next to untreated control during the entire period of storage. The decline in percentage germination might be attributed to ageing effect, leading to depletion of food reserves and decline in synthetic activities of embryo apart from death of seed because of fungal invasion, insect damage and storage conditions (Vijaykumar et al. 2007; Manjunatha et al. 2008). Similarly, decrease in germination with increase in storage period was reported in cotton, (Vijaykumar et al. 2007) soya bean (Kurdikeri et al. 1996) and also due to dye treatment in sorghum (Tonapi 1989). Thiram acts as a protective agent against seed deterioration due to fungal invasion and physiological ageing as a result of which the seed viability was maintained for a comparatively longer period of time (Savitri et al. 1994). The film formed around the seed acts as a physical barrier, which has been reported to reduce leaching of inhibitors from the seed coverings and may restrict oxygen diffusion to the embryo (Duan and Burris 1997). The higher germination per cent observed in the present study was due to increase in the rate of imbibition, where the fine particles in the coating act as a "wick" or a moisture-attracting material to improve germination. Similar observations were made by Dadlani and Vashisht (2006) in soya bean and Rettinassababady and Ramanadane (2012) in hybrid rice.

### 31.3.2 Effect of Polykote Treatment on Seed Moisture

Amount of moisture in seeds is probably the most important factor influencing seed viability during storage (Table 31.2). After 4 months of storage, the seed moisture content was found to increase in both uncoated and coated seeds irrespective of containers. The increase was gradual in the case of polythene-stored seeds when compared to cloth bag-stored seeds. An initial mean moisture content of 7.67 % was increased to 9.99 % after a period of 4 months of storage under ambient condition irrespective of containers. Among the treatments, T5 registered the mean minimum seed moisture content of 8.15 % during storage, irrespective of the containers since Vitavax 200 was used as a dry dressing. The fluctuation in the moisture content was higher in cloth bag than polythene bag storage. Moisture content in polythene bag-stored seeds recorded the lower moisture content (8.32 %) when compared to cloth bag (8.93 %) at the end of the storage. Therefore, the viability of seeds in cloth bag declined rapidly than the seeds stored in polythene bag. This underlines the suitability of seed coatings as a barrier to the entry of moisture strengthening them against the fungal infection as experimentally demonstrated by West et al. (1985), Pham and Rame (2007) in hybrid rice, Jitendra et al. (2007) in soya bean and Vijaykumar et al. (2007) in cotton.

### 31.3.3 Effect of Polykote Treatment on Seed Infection

During seed storage, the fungi accelerate the seed deterioration which is reflected in reduced seed germination and increased seed moisture content. The inoculum of fungi is universally present and its proliferation during storage is governed by factors such as seed moisture, seed temperature, relative humidity and chemicals used for treating seeds, etc. The results on seed health status indicated that about fivefold increase in seed infection was noticed after 4 months of storage period (Table 31.3). Among the treatments, Polykote alone coated seeds (T1) were highly susceptible to seed infection. Seed coating with Polykote coupled with Royal flow and Imidachloprid (T4) and seeds coated with Vitavax 200(T5) effectively contained the seed infection. Between containers, seeds stored in 700 gauge polythene bags (2.8 %) excelled cloth bag (5.5 %) by maintaining higher germination and lesser seed infection during storage. Storage fungi produce spores in large numbers under favourable conditions and their appearance on stored seed is a sign

**Table 31.2** Effect of Polykote seed coating on seed moisture (%) during storage in Bt cotton NHH 44

| Treatments | Cloth bag ($C_1$) | | | | Polythene bag ($C_2$) | | | |
|---|---|---|---|---|---|---|---|---|
| | $P_0$ | $P_1$ | $P_2$ | Mean | $P_0$ | $P_1$ | $P_2$ | Mean |
| $T_0$: Untreated control | 7.85 | 8.34 | 10.92 | 9.04 | 7.85 | 8.19 | 8.76 | 8.27 |
| $T_1$: Polykote at the rate 3 ml/kg dilute with 5 ml water | 7.95 | 8.85 | 11.12 | 9.31 | 7.95 | 8.29 | 8.98 | 8.41 |
| $T_2$: $T_1$+Flowable Thiram at the rate 2.5 g/kg | 7.28 | 8.84 | 11.27 | 9.13 | 7.28 | 8.03 | 8.91 | 8.07 |
| $T_3$: $T_1$+Imidachloprid at the rate 6 ml/kg | 7.56 | 8.22 | 10.71 | 8.83 | 7.56 | 7.57 | 8.8 | 7.98 |
| $T_4$: $T_1$+Royal flow 40 SC at the rate 2.4 ml/kg+Imidachloprid at the rate 6 ml/kg | 7.55 | 8.63 | 11.10 | 9.09 | 7.55 | 7.78 | 9.56 | 8.3 |
| $T_5$: $T_1$+Vitavax 200 at the rate 2 g/kg | 7.22 | 7.95 | 10.23 | 8.47 | 7.22 | 7.41 | 8.84 | 7.82 |
| $T_6$: Primed and dried to original moisture content | 8.03 | 8.94 | 10.61 | 9.19 | 8.03 | 8.21 | 9.29 | 8.51 |
| $T_7$: $T_6$+Polykote+Thiram+Imidachloprid | 7.91 | 8.11 | 10.70 | 8.91 | 7.91 | 7.93 | 10.06 | 8.63 |
| Mean | 7.67 | 8.28 | 10.83 | | 7.67 | 8.13 | 9.15 | |
| | C | P | T | CxP | PxT | CxT | CxPxT | |
| SEd. | 0.04 | 0.04 | 0.07 | 0.06 | 0.12 | 0.1 | 0.17 | |
| CD ($P$=0.05) | 0.07 | 0.09 | 0.14 | 0.12 | 0.24 | 0.2 | 0.34 | |

*Mean Table*

| Treatments (T) | | | | Period of storage (P) | | Containers(C) | |
|---|---|---|---|---|---|---|---|
| $T_0$ | 8.66 | $T_4$ | 8.70 | $P_0$ | 7.67 | $C_1$ | 8.93 |
| $T_1$ | 8.86 | $T_5$ | 8.15 | $P_1$ | 8.21 | $C_2$ | 8.32 |
| $T_2$ | 8.60 | $T_6$ | 8.85 | $P_2$ | 9.99 | | |
| $T_3$ | 8.41 | $T_7$ | 8.77 | | | | |

$P_0$ Initial, $P_1$ 2 months after storage, $P_2$ 4 months after storage

**Table 31.3** Effect of Polykote seed coating on seed infection (%) during storage in Bt cotton NHH 44

| Treatments | Cloth bag ($C_1$) | | | | Polythene bag ($C_2$) | | | |
|---|---|---|---|---|---|---|---|---|
| | $P_0$ | $P_1$ | $P_2$ | Mean | $P_0$ | $P_1$ | $P_2$ | Mean |
| $T_0$: Untreated control | 3.3 | 13.3 | 13.3 | 10.0 | 3.3 | 3.3 | 13.3 | 6.6 |
| $T_1$: Polykote at the rate 3 ml/kg dilute with 5 ml water | 3.3 | 3.3 | 23.0 | 9.9 | 3.3 | 3.3 | 10.0 | 5.5 |
| $T_2$: $T_1$ + Flowable Thiram at the rate 2.5 g/kg | 0.0 | 3.3 | 3.3 | 2.2 | 0.0 | 0.0 | 0.0 | 0.0 |
| $T_3$: $T_1$ + Imidachloprid at the rate 6 ml/kg | 3.3 | 3.3 | 16.7 | 7.8 | 3.3 | 3.3 | 13.3 | 6.6 |
| $T_4$: $T_1$ + Royal flow 40 SC at the rate 2.4 ml/kg + Imidachloprid at the rate 6 ml/kg | 0.0 | 0.0 | 0.0 | 0.0 | 0.0 | 0.0 | 0.0 | 0.0 |
| $T_5$: $T_1$ + Vitavax 200 at the rate 2 g/kg | 0.0 | 3.3 | 6.7 | 3.3 | 0.0 | 0.0 | 0.0 | 0.0 |
| $T_6$: Primed and dried to original moisture content | 3.3 | 3.3 | 16.7 | 5.6 | 3.3 | 3.3 | 3.3 | 3.3 |
| $T_7$: $T_6$ + Polykote + Thiram + Imidachloprid | 0.0 | 0.0 | 10.0 | 3.3 | 0.0 | 0.0 | 0.0 | 0.0 |
| Mean | 1.7 | 3.7 | 11.2 | | 1.7 | 1.7 | 5.0 | |
| | C | P | T | CxP | PxT | CxT | CxPxT | |
| SEd. | 0.4 | 0.5 | 0.8 | 0.7 | 1.38 | NS | 1.95 | |
| CD ($P$=0.05) | 0.8 | 1.0 | 1.6 | 1.4 | 2.73 | | 3.87 | |

*Mean Table*

| Treatments (T) | | | | Period of storage (P) | | Containers(C) | |
|---|---|---|---|---|---|---|---|
| $T_0$ | 8.3 | $T_4$ | 0.0 | $P_0$ | 1.7 | $C_1$ | 5.5 |
| $T_1$ | 7.7 | $T_5$ | 1.7 | $P_1$ | 2.7 | $C_2$ | 2.8 |
| $T_2$ | 1.1 | $T_6$ | 5.5 | $P_2$ | 8.1 | | |
| $T_3$ | 7.2 | $T_7$ | 1.7 | | | | |

$P_0$ Initial, $P_1$ 2 months after storage, $P_2$ 4 months after storage

of deterioration in seed quality (Williams and McDonald 1983; Jitendra et al. 2007; Manjunatha et al. 2008). In the present investigation, the seed microflora, predominantly found in most of the treatments under storage, were *Aspergillus spp.*, *Curvularia sp.* and *Penicillium* sp. Among the treatments, Polykote seed coating with Imidachloprid (T3) did not contain the seed infection significantly. However, in combination with Royal flow thiram (T4) registered no seed infection even after 4 months of storage. Likewise, primed and dried seeds treated with Polykote + Thiram + Imidachloprid (T6) and stored in polythene bags showed better germination with no seed infection. With respect to liquid formulated polymer treatments, seeds coated with polymer have to be necessarily treated with fungicides as prestorage seed treatment to avoid seed infection being caused by hydrophobic nature of polymers.

## 31.4 Conclusion

From the foregoing discussions, it could be concluded that seeds coated with different combinations of Polykote treatments in general deteriorate at slower pace as manifested in higher percentage germination with less or no seed infection over the control. However, Polykote alone coated seeds were highly susceptible to seed infection and have to be necessarily treated with fungicides. Seed coating with Polykote coupled with Royal flow and Imidachloprid or Vitavax 200 effectively contained the seed infection. Between containers, seeds stored in 700 gauge polythene bags excelled cloth bag by maintaining higher germination and lesser seed infection during storage.

## References

Anonymous (1996) International Rules for seed testing. Seed Sci Tech 13:299–355

Dadlani M, Vashisht V (2006) Prolonged storability of soybean seed. Abstr XII National seed Seminar, 24–26, February 2006, ANGRAU Hyderabad, p 73

Duan X, Burris JS (1997) Film coating impairs leaching of germination inhibitors in sugarbeet seeds. Crop Sci 37:515–520

ISTA (1999) International rules for seed testing. International Seed Testing Association, Basserdorf

Jitendra Kumar, Nisar K, Arun Kumar MB, Suresh Walia, Shakil NA, Rajender Prasad, Parmer BS (2007) Development of polymeric seed coats for seed quality enhancement of soyabean (*Glycine max*). Ind J Agr Sci 77(11):738–743

Kurdikeri MB, Basavaraj GT, Hiremath MV, Aswathanarayan SC (1996) Storability of soyabean (*Glycine max* (L.) Merill) seed under ambient condition. Karnataka J Agr Sci 9:552–554

Manjunatha SN, Ravi Hunje BS, Vyakaranhal BS, Kalappanavar IK (2008) Effect of seed coating with polymers and fungicides on seed quality of chilli during storage. Seed Res 36(1):42–46

Ni BR (1997) Seed coating, film coating and pelleting. In: Seed industry and Agricultural development, Chinese Association of Agricultural Sciences, DOA, Ministry of Agriculture. Agriculture Press, Beijing, pp 737–747

Panse VG, Sukhatme PV (1985) Statistical methods for agricultural workers. ICAR, New Delhi, pp 327–340

Pham LG, Rame G (2007) Influence of seed coating with synthetic polymers and chemicals on seed quality and storability of hybrid rice (*Oryza sativa* L.). Omonrice 15:68–74

Rettinassababady C, Ramanadane T (2012) Role of polymer seed coating on seed health status of hybrid rice during storage. Symposium Papers Vol. 2, International Symposium on 100 years of rice science and looking beyond, TNAU, Coimbatore, 9–12, January, 2012, p 378

Robani H (1994) Film coating in Horticultural seed. Horticultural Tech 4:104–105

Savitri H, Sugunakar Reddy M, Muralimohan Reddy B (1994) Effect of seed treatment with fungicides and insecticides on seed-borne fungi, storage insect pests and seed viability and seedling vigour in sorghum. Seed Res 22(2):146–155

Shakuntala NM, Vyakaranhal BS, Shekargouda I, Deshpande VK, Pujari BT, Nadaf HL (2010) Effect of seed polymer coating on growth and yield of sunflower hybrid RSFH-130. Karnataka J Agr Sci 23(5):708–711

Tonapi VA (1989) Longevity and storability of sorghum seeds in relation to stage of harvest and position of seed on earhead alongwith seed treatment and storage containers. Ph. D. Thesis, TNAU, Coimbatore, India

Vanangamudi K, Srimathi P, Natarajan N, Bhaskaran M (2003) Current scenario of seed coating polymer. In proc. of ICAR—short course on seed hardening and pelleting technologies for rainfed/garden land ecosystems, New Delhi pp 80–100

Vijaykumar K, Ravi H, NK BP, Vyakarnhal BS (2007) Effect of seed coating with polymer, Fungicide and insecticide on seed quality in cotton during storage. Karnataka J Agric Sci 20(1):137–139

West SH, Loffin SKW, M Batich CD, Beatty CL (1985) Polymers are Moisture barriers to maintain seed quality. Crop Sci 25:941–944

Williams RJ, McDonald B (1983) Grain moulds in the tropics: Problems and importance. Ann Rev Phytopath 21:153–178

# Viruses Infecting *Cucurbita pepo*: Current Status and Management

# 32

A. K. Tiwari and G. P. Rao

**Abstract**

*Cucurbita pepo* has been known since the dawn of time and presently widely cultivated particularly by small land-holding farmers for business purposes. This particular crop is threatened by several viral diseases resulting in losses through reduction in growth and yield. The viruses cause striking damages by distortion and mottling of fruit, making the produces unmarketable. Viruses are gaining international importance because of yield losses and adverse quality. Moreover, most of the causal viruses induced unspecific symptoms and this make their diagnosis difficult. The most important viruses on *C. pepo* are *Cucumber mosaic virus, Watermelon mosaic virus, Zucchini yellow mosaic virus, Papaya ringspot virus* and *Begomovirus* (*es*). These viruses have been reported from different regions of the world. In this chapter, collective information on occurrence, symptomatology, characterization, transmission and management approaches of *C.pepo* viruses has been discussed.

**Keywords**

Viruses · *C. pepo* · Symptomatology · Characterization · Management

**Abbreviations**

| | |
|---|---|
| CMV | *Cucumber mosaic virus* |
| WMV | *Watermelon mosaic virus* |
| ZYMV | *Zucchini yellow mosaic virus* |
| PRSV | *Papaya ringspot virus* |
| ZYMV | *Zuccini yellow mosaic potyvirus* |

G. P. Rao (✉)
Division of Plant Pathology, Indian Agriculture Research Institute, Pusa Campus-110012, New Delhi, India
e-mail: gprao_gor@rediffmail.com

A. K. Tiwari
Central Lab, U P Council of Sugarcane Research, Shahjahnapur 242001, UP, India

R. N. Kharwar et al. (eds.), *Microbial Diversity and Biotechnology in Food Security*, DOI 10.1007/978-81-322-1801-2_32, © Springer India 2014

| | |
|---|---|
| CGMMTV | *Cucumber green mottle mosaic tobamovirus* |
| TSWV | *Tomato spotted wilt virus* |
| CYSDV | *Cucurbit yellow stunting disorder virus* |
| CABYV | *Cucurbit aphid-borne yellows virus* |
| SLCBV | *Squash leaf curl bigeminivirus* |
| BMV | *Bryonia mottle virus* |
| CYVV | *Clover yellow vein virus* |
| BPYV | *Beet pseudo-yellows virus* |
| Begomovirus | *Bean golden mosaic virus* |
| TMPV | *Trichosanthes mottle potyvirus* |
| MWMPV | *Moroccan watermelon mosaic potyvirus* |

## 32.1 Introduction

Pumpkins and squash (*Cucurbita* L. spp fam; Cucurbitaceae) are major vegetable crops and are grown in almost all the regions, from cool temperature to tropical. *Cucurbita pepo* is by far most widely cultivated and has high monetary value (Paris and Maynard 2008). *C. pepo* has been known since the dawn of time, and it is today widely cultivated as food and for decorative purpose in all warm and temperate parts of the globe. Cooked, mature fruit flesh also has nutritive value. The seeds are consumed in some regions and the oil pressed from the seeds is highly valued in central Europe (Andres 2000). Mature fruits are also valued for autumn decoration in North America and Europe, witnessed by the large and ever-expanding pumpkin-growing industry. It is used mostly for the consumption of its cooked immature fruits, which are known as summer squash (Paris et al. 2012). Crops assigned as *C. pepo* include squash, zucchini and pumpkin. Their main nutrients are lutein, and both alpha- and beta-carotene. Their purpose is to generate vitamin A in the body (Chakravarty 1982). The seeds have been used in traditional medicine as an anthelmintic and a taenicide, demulcent, diuretic and tonic. A tea made from the seeds has been used as a remedy for hypertrophy of the prostate gland. Today, the pumpkin seeds are utilized in the treatment of urological symptoms associated with benign prostatic hyperplasia (BPH) (Bombardelli and Morazzoni 1997).

Viruses are the most common causes of diseases affecting cucurbits in India and other cucurbits-growing countries. These diseases result in losses through reduction in growth and yield and are responsible for distortion and mottling of fruit, making the product unmarketable. French and Elder (1999) reported that the important viruses which cause serious damages on cucurbitaceous crops are *Cucumber mosaic cucumovirus, Tomato bushy stunt tombusvirus, Cucumber necrosis tombusvirus, Melon necrotic spot carmovirus, Tomato ringspot nepovirus, Tobacco ringspot nepovirus, Squash mosaic comovirus, Zuccini yellow mosaic potyvirus*, *Papaya ringspot mosaic virus, Cucumber green mottle mosaic tobamovirus* and *Tomato spotted wilt virus*. Out of these viruses, the disease caused by potyvirus, begomovirus, tospovirus and ilarivirus are gaining international importance because of serious losses and epidemiology in several Asian countries, mainly in India (Raj et al. 2008).

## 32.2 Detailed Descriptions of Symptomatology, Host Range, Transmission, Genome Organization of Important Viruses Affecting *C. pepo*

### 32.2.1 Cucumber Mosaic Virus (CMV)

CMV is a type member of genus *Cucumovirus*, family *Bromoviridae*. CMV is economically important because of its capability to infect a large number of plant species and has the widest host range among all known plant viruses (Raj et al. 2008).

*Cucumoviruses* have icosahedra particles of 29 nm in diameter with 180 capsid protein subunits. The molecular weight of CMV falls in the range of 5.8–6.7 million and consists of about 18 % RNA and the remaining 82 % protein. The RNA is tightly packed by the protein shell, leaving a hollow core of about 110 Å along the threshold axes. The genome of *Cucumoviruses* consists of three single-stranded (ss) messenger sense RNAs designated as 1, 2 and 3 in order of decreasing size. RNAs 1 and 2 are encapsulated separately whereas RNA 3 and sub genomic RNA 4 are probably encapsidated in the same particle (Raj et al. 2008).

CMV causes a variety of symptoms and significant losses to various economically important vegetable food crops. CMV seldom attacks the seedlings of cucurbits and causes mosaic. Infected leaves become mottled, distorted and wrinkled. The subsequent growth of the infected plants is reduced and they appear dwarfed, with shorter stem internodes and petioles, and underdeveloped leaves. Infected plants produce few runners and also few flowers and fruit. Older leaves develop chlorotic and then necrotic areas along the margins which later spread over the entire leaf.

In early 1963, Reddy and Narriani (1963) confirmed the first-time presence of CMV on *C. pepo*. In 1985, CMV affecting *C. pepo* was identified by enzyme-linked immunosorbent assay (ELISA) in Mexico (Delgadillo et al. 1988). Association of PRSV, ZYMV with CMV was also reported through ELISA technique from the same location. In 1991, Fernandes et al. (1991) from the USA reported presence of CMV. This was the first report of CMV viruses infecting *C. pepo* plant in Louisiana.

In France, epidemiology of CMV on *C. pepo* and courgette was described by Lecoq (1992) and Desbiez et al. (1996) through the serological and molecular techniques. In Serbia, study was carried out by Branka et al. (2002) in order to identify the major viruses infecting pumpkins (*C. pepo*) grown in Serbia. Virus-infected plants showed mild mosaic, yellowing, spotting and mottling to deformation of the leaf lamina. The most frequent symptoms were the deformation of leaf lamina, yellow–green mosaic of different intensity and blistering of leaf lamina. The infected samples were tested by the biotest, as well as by two serological methods, ELISA and EBIA and the presence of CMV infection was observed in more than 58 % collected leaf samples (Branka et al. 2002). In Egypt, mixed infection by ZYMV and CMV of the cucurbits host *C. pepo* resulted in a synergistic effect where symptoms in infected plants were more pronounced than plants singly infected by either virus. ELISA values revealed that the concentration of CMV in doubly infected Zucchini squash plants was significantly higher than in singly infected plants (Fattouh 2003).

CMV was also reported from Iran by Farhangi et al. (2004) with the help of double antibody sandwich (DAS)-ELISA technique on *C pepo* plants. At the same time, in Vietnam, Revill et al. (2004) collected the cucurbits samples in a survey and after molecular analysis found the existence of CMV in *C. pepo* with other collected cucurbits samples. In Australia, presence of CMV was confirmed by Coutts and Jones (2005) and in Cyprus by Papayiannis et al. (2005) who found infection of CMV in all the tested plants of *C. pepo* through polymerase chain reaction (PCR)-based technique. From Turkey, Yardumici and Ozgonen (2007) through serological technique found infection of CMV on *C. pepo* plants.

Recently, CMV infection in pumpkin seedlings was identified through DAS-ELISA and reverse transcription PCR (RT-PCR). The PCR detection assays were based on CMV-specific primers. The CMV primers amplify an approximately 910-bp fragment including the complete 656-bp coat protein gene in all the collected infected samples (Tobias et al. 2008).

### 32.2.2 Watermelon Mosaic Virus (WMV)

*C. pepo* is the natural host of WmMV2 (Purcifull et al. 1984). WmMV2 is distributed worldwide and poses a serious problem in cucurbit production. It causes economic damage in watermelon and other susceptible cucurbits, and is reflected mostly as reduced plant growth, yield losses and decrease in fruit quality (Raj et al. 2008).

The WmMV2 filamentous particles of 760 nm consist of a coat protein subunit of 34 KDa and RNA that sediment at 39S (Purcifull et al. 1984). Isolates differ based on host range, aphid transmissibility and the degree of serological relatedness. WmMV2 has good antigenicity. WmMV2 was described separately to distinguish it from the earlier described WMV 1, which is now designated as a strain of PRSV (Purcifull and Hibert 1979; Raj et al. 2008).

The initial symptom of WmMV2 is mild chlorosis at leaves, followed by mosaic and leaf distortion. Green mosaic along the veins and/or green bubble-like protuberances on chlorotic interveinal leaf parts occur as the disease develops. WmMV2 infection early in the season causes poor development of young leaves, internode shortening, show plant growth and reduced fructification (Raj et al. 2008).

In the USA, survey for viruses infecting *C. pepo,* cucumber and watermelon was conducted during spring seasons of 1988–1989 and WMV was detected by Fernandes et al. (1991). In Iran, Delgadillo et al. (1988) confirmed the presence of WMV on *C. pepo* through a serological technique. Later, Farhangi et al. (2004) reported through DAS-ELISA and the existence of other viruses on *C. pepo* from Iran. To determine the distribution of CMV, ZYMV and WMV, a survey was made by them in 2002–2003 and 466 samples were collected from the squash field. The frequency of ZYMV, WMV and CMV was 36, 27 and 25 %, respectively. Triple infection was found in 7 % of samples. This was the first report of WMV on *C. pepo* in Teharan, Iran. From France, Lecoq (1992) and Desbiez et al. (1996) confirmed the existence of WMV through serological and molecular techniques, respectively. In Jordan, Mansoor et al. (1997) reported WMV with the association of ZYMV in *C. pepo* plants. From Australia, Coutts and Jones (2005) reported WMV through molecular study and also confirmed the existence of other viruses on *C. pepo* from the same location. In Turkey, presence of WMV was reported by Yardumici and Ozgonen (2007) with the help of ELISA technique.

### 32.2.3 Papaya Ringspot Virus-W (PRSV)

Until a few years ago, PRSV-W was known as WMV 1. A number of studies, however, determined that WMV-1 and PRSV are identical in many respects. Presently, two pathotypes of this potyvirus are recognized: PRSV_P (papaya strain), which infects *Carica papaya* and most of the *Cucurbitaceae,* and PRSV-W (watermelon strain), which infects all *Cucurbitaceae* but not papaya (Freitas and Rezende 2008). PRSV-type W is one of the most important viruses that may cause damage to zucchini squash (*C. pepo*) throughout the world. This virus, responsible for the zucchini common mosaic, belongs to the family *Potyviridae* and has the shape of a cylindrical and flexuous filament about 780 × 12 nm, consisting of a ss, positive RNA with molecular weight of approximately 10.3 kb. It is transmitted in the field in a nonpersistent, noncirculative manner by several aphid species (*Aphididae*). It can also be transmitted by mechanical inoculation, but not via the seeds. The PRSV-W symptoms in *C. pepo* consist of mosaic, presence of blisters and deformations on leaves, fruit malformation with changed colour, and plant stunting. This virus has been quite difficult to control in the field (Freitas and Rezende 2008).

In Iran, Delgadillo et al. (1988) reported the presence of PRSV on *C. pepo* through serological technique. Later, Fernandes et al. (1991) from the USA reported the existence of this virus in Louisiana on *C. pepo* through serological technique. In China, PRSV was reported first time on *C. pepo* in 1994 by PCR analysis (Xiao and Fan 1994). In France, PRSV on melon and courgette is described by Lecoq (1992) through serological technique and, through molecular technique it was confirmed by Desbiez et al. (1996). Revill et al. (2004) conducted a survey between November 1998 and May 2001 in Vietnam with cucurbit samples. Virus symptoms including mosaics, yellowing, mottling and vein clearing were most commonly observed in pumpkin (*Cucurbita* sp.) and cucumber (*Cucumis sativa*). In Zucchini (*C. pepo*), the only crop examined, stunting, twisting and plant death were observed. PRSV were detected in 47% cucurbit samples. In Australia, Coutts and Jones (2005) and in Cyprus, Papayiannis et al. (2005) also reported presence of PRSV through PCR-based analysis. In Brazil, PRSV–type W (PRSV-W) is the most prevalent virus in cucurbit crops and is responsible for frequent yield losses. Diseases caused by these viruses are difficult to control. Many reports have been made by different researchers of Brazil on the presence of viruses on *C. pepo*. Bonilha et al. (2009) again established the presence of PRSV and WMV on *C. pepo* and also evaluated the effects of mild strains PRSV-W-1 and ZYMV-M on the yield of *C. pepo* L. in greenhouse and field conditions. Plants infected with ZYMV-M and grown in a plastic greenhouse did not exhibit typical leaf symptoms or significant alterations in quantitative and qualitative fruit yield. However, when infected with PRSV-W-1, or PRSV-W-1+ZYMV-M, the plants exhibited severe leaf mosaic symptoms and reduced fruit quality, although there were no changes in the number and mean fruit weight harvested from these plants. When these plants were infected with PRSV-W-1 and studied simultaneously in the field and plastic greenhouse, intensification of symptoms in the fruits and leaves was more pronounced under the greenhouse conditions. Quantitative yield did not change. Environmental factors seem to influence symptoms induced by PRSV-W-1(Bonilha et al. 2009).

### 32.2.4 Zucchini Yellow Mosaic Virus (ZYMV)

ZYMV was first reported almost simultaneously in 1981 from Italy (Lisa et al. 1981) and France, where it was named *Muskmelon yellow stunt virus* (Lisa and Dellavalle 1981). In less than 5 years, it was reported in 20 countries of 5 continents, often causing devastating epidemics. Virus particles are flexuous, 750 nm long and contain a ssRNA. Although circumstantial evidence points to seed transmission of this virus, demonstrating this avenue of spread has been very difficult. Two reports have indicated a low level of transmission in some varieties of *C. pepo* (Schrijnwerkers et al. 1991). Unquestionably, this is one of the most destructive viruses occurring in cultivated cucurbits. The virus incites very prominent foliage mosaic, severe malformation and plant stunting. Fruits often develop knobby areas, resembling those caused by PRSV-W, thus, differentiating the symptoms caused by these two viruses is often very difficult. Serologically, ZYMV is related to WMV. A few strains and pathotypes of ZYMV have already been identified (Lecoq and Pitrat 1984; Lisa and Lecoq 1984; Provvident et al. 1984; Raj et al. 2008).

A survey for viruses infecting *C. pepo* was conducted in southeastern Louisiana, USA, during spring seasons of 1988–1989 and ZYMV was detected by Fernandes et al. (1991). A greenhouse-grown cucumber virus isolate induced symptoms on *C. pepo* resembling those incited by ZYMV. This isolate has similar host range as ZYMV original isolate, except that it did not infect *Gomphrena globosa*. The complete identification is based on its physical properties, aphid transmission, morphology by electron microscope and serological properties. This study confirmed the presence of ZYMV for the first time on *C. pepo* in Saudi Arabia (Al-Shauwan 1990). ZYMV is reported from *C. pepo* and cucumber in Greece. This was the first record of this virus in Greece. This was detected in all eight countries in investigation and its incidence is thought to be very high (83%) (Kyriakopoulou and Varveri 1991). In Turkey, based on host range, serological and physical properties, the virus causing systemic mosaic of leaves and fruit blight on squash plants was identified as ZYMV. This was the first report of ZYMV in the Ankara province of Turkey. Later, infection of CMV on *C. pepo* was confirmed by Ertunc (1992) in Turkey. In Bulgaria, ZYMV was identified from squash, melon and cucumber for the first time by biotest on indicator plants and indirect ELISA (Dikova 1994). In Singapore, ZYMV was propagated in Pumpkin cv. first taste and mechanically transmitted to 12 sp. of 6 families. It induced milder symptoms than the Connecticut and Florida strains of ZYMV in infected leafs of marrow cv. zucchini elite. ZYMV-S was neither seed nor aphid transmissible. Immunoelectron microscopy revealed that ZYMV-S was distantly related to WMV-2, Moroccan WMV, and *Telfairia mosaic virus* but not to PRSV or ZYFV. Cytoplasmic pinwheels and scrolls were observed in the ultrathin section of infected leaves cells by light, confocal laser scanning and transmission electron microscopy. The molecular weights of the viral coat protein, cytoplasmic inclusion protein, RNA and dsRNA were estimated to be $3.2 \times 10^4$, $6.1 \times 10^4$, $3.23 \times 10^6$ and $6.53 \times 10^6$ Da, respectively (Wong et al. 1994). In France, it was reported by Desbiez et al. (1996) with the help of PCR analysis. In Jordan, ZYMV causes serious damage in *C. pepo* crop. It was found as a single or mixed infection in 82.6% of 910 samples. The second virus, WMV was detected in 13.5% tested samples as a single infection in 48.2% as a double infection with ZYMV. ELISA showed that some weeds may act as a primary source of inoculum to cucurbit viruses in Jordan (Mansoor et al. 1997). From Chile, ZYMV was first described in 1981 affecting squash and melon with severe yellowing symptoms. It was reported to be present in most countries where cucurbits were grown, and in Chile since 1995 through survey, infected samples were collected and ELISA assays were done and results revealed the presence of ZYMV on *C. pepo* (Prieto et al. 2001). From Serbia, Branka et al. (2002) reported that the most prevalent viruses infecting pumpkins were ZYMV (62%) and CMV (58%). In Egypt, double infection of CMV and ZYMV on *C. pepo* was reported by Fattouh (2003) through ELISA technique. Papayiannis et al. (2005) from Cyprus did molecular analysis and found the infection of ZYMV and several other viruses on *C. pepo* plants.

Yardumici and Ozgonen (2007) from Turkey confirmed the existence of ZYMV on *C. pepo*. In Hungary, ZYMV was isolated (as a new viral pathogen) in 1995. It generally developed into a devastating disease of several cucurbit crops. Its wide distribution suggested seed transmission. From a total of 2,879 seedlings grown from infected boll-less seeded oil pumpkin (*C. pepo*), only one showed symptoms and contained the ZYMV, indicating that the virus could be transmitted via the seed, but only at a very low rate (Tobias and Kovacs 2001). Later, ZYMV and CMV infection in pumpkin seedlings were identified by direct antibody coating (DAC) ELISA, test plants and RT-PCR. (Tobias et al. 2008). Recently from Brazil, Bonilha et al. (2009) reported the presence of ZYMV on *C. pepo* plants and found that this virus is severely damaging and reducing the quality of the crop.

### 32.2.5 Cucurbit Yellow Stunting Disorder Virus (CYSDV)

First detection of CYSDV was in the United Arab Emirates in 1982 (Hassan and Duffus 1991), where it remained in epidemic proportions. CYSDV has since spread throughout the Mediterranean region including Egypt, Israel, Jordan, Spain, Turkey, Lebanon, Portugal and Morocco, where it caused major economic damage to cucurbit crops. CYSDV has also been introduced in North America (Sinclair and Crosby 2002).

CYSDV produces initial symptoms of severe interveinal chlorosis and green spots on oldest leaves which appear between 14 and 22 days post inoculation; definite symptoms are visible after 30 days. Leaves may also develop prominent yellow sectors. Severe symptoms include complete yellowing of the leaf lamina (except for the veins) and rolling and brittleness of the leaves. Fruit quality is severely affected; yield, fruit size and sugar content are reduced, making fruits unacceptable for sale on the commercial market resulting in economic losses for melon growers. Since, Criniviruses produce symptoms mainly in older leaves, CYSDV symptoms may easily be confused with physiological disorders, nutritional deficiencies, inadequate water, insect damage, natural senescence or pesticide damage. Growers, diagnosticians and researchers may have a hard time visually recognizing such virus infections (Sinclair and Crosby 2002).

The three major cucurbit species under cultivation worldwide: *Cucumis sativus* (cucumber), *Citrullus lanatus* (watermelon) and *C. pepo* (squash) are affected by CYSDV. Cucurbits and lettuce are the only known hosts to date. CYSDV is a member of the newly assigned *Crinivirus* genus, in the *Closteroviridae* family (Sinclair and Crosby 2002).

CYSDV is a phloem-limited virus making diagnosis, isolation and purification difficult.

However, it has been purified with differential centrifugation and determined to have particle lengths ranging from 825 to 900 nm. The virus has a bipartite genome consisting of two ss plus sense RNA segments estimated at ~9 kb (RNA1) and ~8 kb (RNA2) encapsulated separately. Later, leaf dip preparations suggested somewhat shorter particle lengths from 750 to 800 nm (Liu et al. 2000; Sinclair and Crosby 2002). Papayiannis et al. (2005) did the survey and found severe infection of viruses on *C. pepo* and other cucurbits plants in a survey in Cyprus. For further confirmation he used molecular technique and phylogenetic analysis. He reported the presence of CYSDV on all the selected cucurbits plants including *C. pepo* plant in his study.

### 32.2.6 Cucurbit Aphid-Borne Yellows Virus (CABYV)

Yellowing diseases of cucurbit crops were noticed in Spain since last two decades. Due to the lack of appropriate diagnostic tools, these disorders were first attributed to nutritional deficiencies, but were later shown to be most often caused by viruses. During the 1980s, yellowing were found associated with the whitefly *Trialeurodes vaporariorum* vectoring the crinivirus *Beet pseudo-yellows virus* (BPYV) (Jordá-Gutiérrez et al. 1993). During the 1990s, *T. vaporariorum* was progressively displaced by *Bemisia tabaci* and consequently, CYSDV, a new crinivirus, spread over cucurbit crops (Célix et al. 1996). CABYV has been detected in protected and open field melon (*Cucumis melo*), cucumber (*Cucumis sativus*), squash (*C. pepo*) and watermelon (*Citrullus lanatus*) crops of southeastern Spain, with a very high incidence (Juarez et al. 2004).

CABYV is a member of the genus *Polerovirus* of the family *Luteoviridae*. Its viral particles are isometric, approximately 25 nm in diameter, and encapsidate the CABYV genome which consists of a ss positive sense RNA molecule of 5.7 kb which has neither a 5′ cap nor 3′poly(A) tail. The CABYV genome has been fully sequenced and a full-length infectious clone is available. It contains six major open reading frames (ORFs) (Guilley et al. 1994; Pfeffer et al. 2002; Taliansky et al. 2003).

CABYV is transmitted in a persistent and circulative manner by two aphid species, the black melon aphid *Aphis gossypii* and the green peach aphid *Myzus persicae* (Lecoq et al. 1992). CABYV cannot be mechanically transmitted, and no other transmission method has been described for this virus. The host range of CABYV includes the major cultivated cucurbit species (melon, cucumber, squash and watermelon), in which it remains confined to their phloem.

CABYV was first described by Lecoq and coworkers in 1992 in France where it affected open field cucurbit crops (Lecoq et al. 1992). Afterwards, CABYV was detected in many countries from time to time by different workers. CABYV can be readily diagnosed by ELISA. Desbiez et al. (1996) again established the presence of this virus through molecular detection from France. Papayiannis et al. (2005) also characterized CABYV from Cyprus through phylogenetic analysis.

CABYV was detected for the first time in cucurbit-growing areas of the Isparta region of Turkey with 94 of the 200 samples positive for CABYV. This survey detected the presence of CABYV at incidences between 30 and 100 % in the different cucurbit species tested (*C. maxima*, *C. melo* 'Hasanbey', *C. melo* var. Flexious, *C. sativus*, *C. pepo* and *C. lanatus* 'Paladin'). CABYV is transmitted only by aphids (*A. gossypii* and *M. persicae*) and in a persistent manner. The occurrence and wide distribution of CABYV in cucurbit plants was most likely related to the large abundance of aphids (Yardumici and Ozgonen 2007).

### 32.2.7 Squash Leaf Curl Begomovirus

It was first observed in 1977/1978 and is one of a group of viruses which have become very important since the spread of biotype B of *B. tabaci*. The original strain of the virus (SLCV-CA) was reported in California (USA) damaging crops of *C. pepo* (marrows). The virus is transmitted in a persistent manner by *B. tabaci*, especially biotype B. It is not transmitted by mechanical inoculation. At least two distinct strains of SLCV are known. The WCMoV strain, originally described as a distinct *watermelon curly mottle virus* in Arizona, has a wider host range than the type strain SLCV-CA in California. Narrow host-range and broad host-range strains have also been described on the basis of the molecular characterization of DNA components; it is not yet clear whether these coincide with the strains identified in the field. SLCV causes severe systemic stunting and leaf curl in cucurbits. The so-called silver leaf symptom is seen on cucurbits infested by *B. tabaci* biotype B, but is not associated with SLCV. It has been suggested that it could be due to infection by another virus, but it is now generally thought to be induced physiologically by the feeding of biotype B. SLCV has geminate particles, 22 × 38 nm in size. The virus is associated with maturing phloem sieve tube elements. SLCV is not mechanically transmissible. Recommended indicator plants are *Cucurbita maxima, C. moschata* or *C. pepo* (symptoms as above) or *Phaseolus vulgaris* (systemic green mosaic leaf symptoms and distortion). SLCV can be detected in *B. tabaci* by DNA spot hybridization assay. SLCV causes severe losses of squashes, melons and related cucurbits.

### 32.2.8 Bryonia Mottle Virus (BMV)

The host range of BMV is largely confined to cucurbits. Serologically, it is unrelated to PRSV-W, WMV and a few other potyviruses. It was found in Morocco, where it primarily affected *Bryonia dioica*. Symptoms consist of prominent foliar chlorotic and necrotic spotting, plant stunting and flower abortion. Apparently BMV is not seed transmitted. In a study involving many *Cucurbitaceae,* only *Momordica charantia* was not infected by this virus (Lockhart and Fisher 1979).

### 32.2.9 Clover Yellow Vein Virus (CYVV)

CYVV commonly occurs in legumes, in which it causes severe symptoms. In nature, it has been found to infect summer squash, causing numerous chlorotic leaf spots in yellow-fruited varieties. Fruits are not affected, but seed production may be reduced (Lisa and Dellavalle 1981). In a survey during 2000–2002, to determine the identity and prevalence of viruses affecting cucurbit crops in Cyprus, 2,993 samples of cucumber, zucchini, melon and watermelon were collected from the five major cucurbit-growing areas in Cyprus. Confirmation of viruses was made by molecular analysis and it resulted positive for CYVV in the selected plant (Papayiannis et al. 2005).

### 32.2.10 Beet Pseudo-Yellow Virus (BPYV)

In 2004, severe yellowing and chlorotic (and occasionally mosaic) symptoms were observed in field grown *C. pepo* and *C. maxima* in Costa Rica. Symptoms resembled those of the genus *Crinivirus* and large populations of the greenhouse whitefly, *Trialeurodes vaporariorum,* were observed in the fields and on symptomatic plants. Nucleotide sequence analyses of purified PCR products verified their identity as variants of BPYV, with 97 and 99 % sequence identity with reported CPm and HSP sequences, respectively (Hammond et al. 2005). Papayiannis et al. (2005) from Cyprus confirmed the existence of BPYV and other viruses on *C. pepo* plants with help of sequences and phylogenetic analysis. Recently, Ramirez et al. (2008) also reported the presence of BPYV on *C. pepo* in Costa Rica.

### 32.2.11 Bean Golden Mosaic Virus

Begomoviruses have emerged as constraints to the cultivation of a variety of crops in various parts of the world. Some of the diseases caused by Begomoviruses that are appearing, show that these viruses are still evolving and pose a serious threat to sustainable agriculture, particularly in the tropics and subtropics. Another concern is the emergence of diseases that are caused by a complex of *Begomovirus* and satellite DNA molecules (Varma and Malathi 2003; Stanley

2004). Some crops appear to be a paradise for *Begomoviruses*. So far, 45 recognized and 30 tentative species of *Begomoviruses* have been found to naturally infect tomato, pepper and cucurbits in the New and Old World. Some of the viruses have a large number of distinct strains (Jones 2003). *Begomoviruses* have bipartite genomes (A and B components), with some exceptions (e.g. *Tomato yellow leaf curl virus* (TYLCV), *Cotton leaf curl virus* (CLCuV), *Tomato leaf curl virus* (ToLCV)) for which no B component has been found (Fauquet et al. 2003). *Begomovirus* (type species *Bean golden mosaic virus*) is the only genus of the family to be either bipartite with virus genes resident on two different circular ssDNA molecules (DNA A, DNA B) each of about 2.6–2.8 kb, or monopartite with all genes resident on one (DNA A-like) ssDNA of about 2.8 kb. The twinned particles have diameter of 18–20 nm, 30 nm long, like most of the *Geminiviruses*. The *Begomoviruses* are all transmitted by the whitefly *B. tabaci* in a circulative manner and infect dicotyledonous plants. They have been considered as the most numerous and widespread group of whitefly-transmitted viruses causing severe epidemics in India. The fast spreading *Begomovirus* in India on cucurbits is *Tomato leaf curl New Delhi virus* (Sohrab et al. 2003; Tiwari et al. 2008, 2010a; Phaneendra et al. 2011), *Tomato leaf curl Palampur virus* (Tiwari et al. 2010b, 2012a), *Ageratum enation virus* (Raj et al. 2011; Tiwari et al. 2012b), *Pepper leaf curl Bangladesh virus* (Raj et al. 2010a), *Bitter gourd yellow mosaic virus* (Rajinimala et al. 2005) and *Squash leaf curl China virus* (Singh et al. 2008).

*Squash leaf curl geminivirus* was first observed in squash in California in 1977–1978 (Flock and Mayhew 1981). Virus symptoms including mosaics, yellowing, mottling and vein clearing were most commonly observed in pumpkin (*Cucurbita* sp.) and cucumber (*Cucumis sativa*). Zucchini (*C. pepo*), the only crop examined, stunting, twisting and plant death was observed in Vietnam (Revill et al. 2004). In Egypt, Farag et al. (2005) reported the detection of bipartite Geminiviruses on squash. Squash plants infected with whiteflies exhibited leaf curl and severe stunting symptoms, which were observed on newly grown leaves. Vein banding was also observed on leaves showing vein clearing. Furthermore, the enations often start from the lower surface of the symptomatic leaves.

Dafalla et al. (1998), from Sudan reported the association of Geminivirus with *Watermelon chlorotic stunt virus*. *C. pepo* plants were showing severe yellowing. All cultivars grown in affected regions were susceptible. Occurrence of the virus was associated with high populations of *B. tabaci*. Disease incidence reached 100 % in the late summer crop in Gezira and early winter crop in the Gash Delta. From Oman, *Squash yellow leaf curl virus* on *C. pepo* was reported through PCR analysis (Zouba et al. 1998). In November 2006, *Cucurbit leaf crumple virus* (CuLCrV), a new virus in Florida, was found in squash (*C. pepo*) fields in north central and northeast Florida. Leaves of yellow straight neck squash and zucchini were thickened and distorted, as well as curled and crumpled. The symptoms on infected yellow straight neck squash were slightly different from those on zucchini. Zucchini fruit did not show obvious symptoms, but the fruits from infected yellow straight neck squash were streaked with green, making them unmarketable. Feeding by whitefly nymphs' causes silvering of leaves of squash and blanching of yellow-fruited squash and yellow blotchiness of green-fruited squashes. The leaf silvering is distinct from cucurbit leaf crumple disease and should be not be confused with it. Samples collected from symptomatic plants failed to react in ELISA with antibodies to all cucurbit-infecting viruses known to be present in Florida. The samples also tested negative for the whitefly-transmitted potyvirus *Squash vein yellowing virus* but did test positive in PCR assays for *Begomovirus*. PCR followed by sequence analysis and whitefly transmission tests revealed that the plants were infected with *Cucurbit leaf crumple virus* (Webb et al. 2007).

Tahir et al. (2010) from Pakistan collected leaf samples of *C. pepo* plants showing yellow mosaic symptoms and completed full genome sequencing of *Begomovirus* and in phylogenetic analysis found the existence of *Squash leaf curl China virus* for the first time in Pakistan.

Recently, Tiwari et al. (2010b) reported infection of *Tomato leaf curl Palampur virus* on *C. pepo* in India. Pumpkin plants were found infected with mosaic, leaf curling, puckering and yellows. A ~800 bp product was amplified from symptomatic leaf samples in *C. pepo* through TLCV-CP primers, NCBI BLAST search analysis showed highest identities of 92 % with *Tomato leaf curl Palampur virus* isolate (AM884015) followed by 90 % with *Tomato leaf curl Palampur virus* isolate (EU547682 & EU547683). However, 88 % identities were recorded with *Squash leaf curl China virus* (AM260205) and *Squash leaf curl Philippines virus.* Phylogenetic analysis at nucleotide level, showed close relationship with *Tomato leaf curl Palampur virus* (AM8844015, EU547682 and EU547683) and making distance with *Tomato leaf curl virus* (AY691902), *Luffa yellow mosaic virus* (AF509739), *Squash leaf curl virus* (AF509739, AM260205, AM260206, EF197940, AB085793, EU487031, EU487033 and EU487041). Phylogenetic analysis and similarity at nucleotide level confirm the existence of *Tomato leaf curl Palampur virus* on *C. pepo*.

### 32.2.12 Trichosanthes Mottle Potyvirus (TMPV)

In Japan, two *potyviruses* were isolated from *T. rostrata* and *C. pepo*. Their host range and physical properties were investigated and were identified as TMPV and *Watermelon mosaic* I *potyvirus* (WMIV). Both viruses infected cucurbits and were transmitted by *A. gossypii* and *M. persicae* in a nonpersistent manner. TrMV was serologically related to *Papaya leaf distortion mosaic virus* but had no relationship with WMIV and other potyviruses in immune EM. It is suggested that TrMV is a new type of *potyvirus* (Yonaha et al. 1988).

### 32.2.13 Moroccan Watermelon Mosaic Potyvirus (MWMPV)

In Italy, unusual symptoms were observed in summer 1997 in field of *C. pepo* of several cultivars grown in central Italy. Symptoms included reduction in growth, severe mosaic, blistering and deformation of leaves and malformation on fruits. Plants gave negative results in ELISA for CMV, SqMV, PRSV, *Zucchini yellow fleck virus,* ZYMV and ZYMV. It was found positive with *Watermelon Moroccan mosaic virus*. MWMV is a tentative species in the genus *Potyvirus,* widely present in Africa and occasionally found in Spain (Roggero et al. 1998).

## 32.3 Management Approaches

Control of epidemic outbreak of virus diseases can be carried out theoretically either by controlling the vector or by eliminating the pathogen from the infected plants by meristem tip culture, antiviral or other chemicals. At present, insect vector control using pesticides is the tool of choice for limiting outbreaks of virus diseases. The vector control is difficult to achieve when wild reservoir plants are sources of contaminations. Viruses can be eliminated from their plant hosts, as they are not present in the shoot meristem (Tiwari et al. 2012b). Furthermore, they are sensitive to few antiviral chemicals. They can be eliminated by heat therapy followed by meristem tip culture. Now, the researchers are also applying cryotherapy with combination of shoot tip culture for removal of the virus. Several methods have been applied to clean plant material for virus; these include in vitro tissue culture such as shoot tip or micropropagation sometimes in combination with heat or antiviral treatment. Therefore, a real way to control virus infection is to prevent the outbreaks by producing clean material or by finding resistant varieties.

The virus diseases cannot be controlled by any chemical treatment in the field (Valkonen 1998). However, they can be managed based on strategies that prevent infection. Both conventional and nonconventional methods have been suggested and described over the years for management of viral diseases (Raj et al. 2010b).

This comprises controlling the vectors which act as carriers for transmitting the viruses from infected plant to the healthy ones. Killing the insect vectors by using insecticides is being used since 1930s as a means of direct attack on vectors (Du 1948). Bradbent et al. (1956) tried several insecticides, e.g. endrin, aalathion, DDT, etc. and showed successful prevention of PLRV spread from infected to healthy ones. However, excessive use of insecticides may also prove ecologically harmful in the long run.

The development of strategies for integrated disease management by spray of oils, viricides, pesticides and botanicals has been found to be significant in reducing the yield loss by many workers for various viral diseases (Verma et al. 1980; Kumar and Awasthi 2003; Kunkalikar et al. 2006). Three sprayings of malathion (50 % E.C.) insecticide (0.2 %) at a 21-day interval was successful in field conditions to minimize the whitefly population and to assess the improvement of their biomass and ultimately the fruit yield (Khan et al. 2006; Somvanshi et al. 2009). However, the regular use of insecticide should not be in practice because it adversely effects the environment, other way round, may diminish the quality of crops (Raj et al. 2010b).

Virus control strategies in cucurbits have been based on the use of cultural practices intent on preventing or delaying virus spread through vectors (Lecoq et al. 1998). Time of planting and other epidemiological factors may be important in determining virus severity (Berdiales et al. 1999). Genetic resistance is the most likely method for controlling the virus. Virus resistance may be related to the existence of mechanisms that inhibit vascular transport of the pathogen, changes in cellular membranes that impede the diffusion or transport of virus particles from cell to cell or an inhibition of virus particle replication in tissue of resistant hosts (Lopez-Sese and Gomez-Guillamon 2000).

Cucurbit viruses are one of the most complex pathosystems in the world, making breeding for virus resistance a challenge. Breeding virus resistant varieties is generally slow and inefficient due to several factors. First, environmental conditions may have a large effect on the expression of virus symptoms. Second, many viruses have multiple strains, some are able to overcome resistance genes. Locating molecular markers linked to virus resistance is expected to make breeding for virus resistance more efficient and will lead to faster development of resistant cultivars (Danin-Poleg et al. 2000). Effective breeding and marker utilization should be much easier than otherwise.

Weeds have been identified as key sources of viral infections for a long time. Eradication of perennial weeds from around greenhouses, gardens and fields to eliminate possible sources of virus therefore may prove helpful (Agrios 1978). Elimination of weed hosts harbouring cultivated fields has been found successful in reducing the incidences of virus in cucumber and celery (Rist and Lorbeer 1989). A number of weed plants were surveyed in different cucurbitacea crops and chilli-growing locations, the geminivirus infection was detected in them. These weed plants also showed typical symptoms of geminivirus infection. Various weeds, viz. *Croton bonplandianum, Acalypha indica, Malvasrtum coromandalianum, Eclipta alba, Ageratum conyzoides, Launaea procumbens, Jatropha gossypifolia, Luffa* spp., *Coccinia grandis Nicotiana plumbaginifolia, Sorghum vulgare, Parthenium hysterophorus, Physalis minima, Sida cordifolia, Solanum nigrum* and *Sonchus oleraceus* have been reported as wild hosts of the viruses (Khan 2006; Somvanshi et al. 2009; Raj et al. 2010b; Pandey et al. 2011; Khan et al. 2012). Therefore, such weed plants may be eliminated from and nearby cultivated fields for possible management of *Begomoviruses*. Furthermore, roguing, or immediate removal of infected individual plants, may assist in delaying virus spread once the infected material is immediately destroyed and not left to compost near adjacent, developing fields (Raj et al. 2010b).

Use of virus-free planting material and their transplantation in greenhouses that isolates crop from other plants which harbour or may harbour viral diseases, e.g. susceptible crops, should be practiced for better crop production yield (Agrios 1978). Using planting material from which all infected plants have been rogued, applying heat therapy (35–54 °C), use of meristem tip cultures, cold treatment and chemotherapy are other means suggested for obtaining virus free plants

(Raychaudhuri and Verma 1977). Nowadays, cryotherapy is another liable methods to produce virus free materials. For identifying the virus-free material, highly reliable and sensitive diagnostic techniques are available to accomplish this need (Raj et al. 2010b).

## 32.4 Conclusion

Reports confirmed the existence of several viruses on *C. pepo* from almost every part of the world. Many researchers reported the presence of mixed infection on single *C. pepo* plants. The most prevalent viruses on *C. pepo* in this literature survey were reported as WMV, CMV and PRSV. These three viruses have been reported from almost every part of the world on *C. pepo* plants. While, occurrence of other viruses are of less significance on *C. pepo*, some viruses like *Clover yellow vein, Squash leaf curl begomovirus, Trichosanthes mottle virus, Moroccan mosaic virus*, BMV etc. are reported only from limited countries. Their existence in other countries was not clearly known. *Begomovirus* infection was reported only by few researchers, and some of them are the first report.

As far the management of viral diseases on *C pepo* is concerned, the use of virus-free planting material seems to be the most effective control measure. The virus-free planting material could successfully be generated through meristem culture technique adopted in a number of plant species. The development of resistant genotypes through breeding methods is another approach. However, this approach may not be much effective in controlling the viral diseases due to continued introduction and spread of new viral strains from time to time. Whiteflies and other vectors play important role in virus transmission from infected crop to healthy one. The use of pesticides to check the populations of these vectors may also be an effective control measure. Besides running awareness programmes for the farmers, distribution of virus-free seeds should be ensured by the concerned authorities to control the spread of viral diseases.

## References


Agrios GN (1978) Plant Viruses. Plant Path 2:466–470

Al-Shauwan AM (1990) First report of *Zucchini yellow mosaic virus* on cucurbits in the central region of Saudi Arabia. J King Saud Uni 2(2):251–260

Andres TC (2000) An Overview of the Oil Pumpkin. Proceedings of the First International Pumpkin Conference Cucurbit Genetics Cooperative Report 23:87–88

Berdiales B, Bernal JJ, Saez E, Would B, Beitia F, Rodriguez-Cerezo E (1999) Occurrence of *Cucurbit yellow stunting disorder virus* (CYSDV) and *Beet pseudoyellows virus* (BPYV) in cucurbit crops in Spain and transmission by two biotypes of *Bemisia tabaci*. Eur J Plant Pathol 105:211–215

Bombardelli E, Morazzoni P (1997) Biologically active compounds in *C. Pepo*. Fitoterapia 68:291

Bonilha E, Goria R, Frobori J (2009) Yield of varieties of *Cucurbita pepo* preimmunized with mild strains of *Papaya ringspot virus*—type W and *Zucchini yellow moisak virus*. Sci Agric (Piracicaba, Braz) 66(3):419–424

Bradbent L, Burt PE, Heathcote CD (1956) The control of potato viruses by insecticides. Ann Appl Biol 44:256–273

Branka B, Dukia N, Katis NI, Papavassiliou C, Berenjii J, Vico I (2002) Identification of viruses infecting pumpkins (*Cucurbita pepo* L.) in Serbia. Proceedings for Natural Sciences. Matica Srpska 103:67–79

Célix A, López-Sesé A, Almarza N, Gómez-Guillamón ML, Rodríguez-Cerezo E, (1996) Characterization of cucurbit yellow stunting disorder virus, a *Bemisia tabaci*-transmitted closterovirus. Phytopath 86:1370–1376

Chakravarty HL (1982) Fascicles of flora of India, Botanical survey of India

Coutts BA, Jones RAC (2005) Incidence and distribution of viruses infecting cucurbit crops in Australia. Aust J Agric Res 56:847–858

Dafalla GA, Gronenborn B, Kheyr-Pour A, Lecoq H (1998) *Watermelon chlorotic stunt virus*: a new emerging epidemic in export melons in Sudan, p 37. In: 2nd Int. workshop *Bemisia* geminiviruses, San Juan, Puerto Rico

Danin-Poleg Y, Tzuri G, Reis N, Karchi Z, Katzir N (2000) Search for molecular markers associated with resistance to viruses in melon. Acta Hort 510:399–403

Delgadillo F, Garzon JA, Vega A (1988) Detection of cucurbit viruses in Mexico. Phytopath 78:626

Desbiez C, Wipf-Schibit C, Granier F, Robaglia C, Delaunay T, Lecoq H (1996) Biological and molecular variability of ZYMV on the island of Martinique. Pt Dis 80(2):203–207

Dikova B (1994) *Zucchini yellow mosaic virus* on Cucurbits in Bulgaria. In: international conference on Plant virology, 1994 in Bulgaria 32:101–104.

Du T (1948) The control of spotted wilt tomato. Farm S Africa 23:786–788

Ertunc F (1992) Pumpkin in Ankara in the diagnosis of infectious viral agent research on. A. Ü. Fak.Yay agriculture, No. 1252

Farag AG, Amer MA, Amin HA, Mayzad HM (2005) Detection of bipartite geminiviruses causing squash leaf curl disease in Egypt using polymerase chain reaction and nucleotide sequence. Egypt J Virol 2:239–354

Farhangi SH, Mosahebi MK, Okhowai SM (2004) Occurrence distribution and relative incidence of mosaic viruses infecting field—grown squash in Tehran Provinica, Iran. Agr Biol Sci 69(4):507–512

Fattouh FA (2003) Double infection of cucurbit hosts by *Zucchini yellow mosaic virus* and *Cucumber mosaic virus*. Pak J Pt Pathol 2(2):85–90

Fauquet CM, Bisaro DM, Briddon RW, Brown JK, Harrison BD, Rybicki EP, Stenger DC, Stanley J (2003) Revision of taxonomic criteria for species demarcation in the family Geminiviridae, and an updated list of *Begomovirus* species. Arch Virol 148:405–421

Fernandes FF, Valverde RA, Black LL (1991) Viruses infecting cucurbits in Louisiana. Pt Dis 75:431

Flock RA, Mayhew D (1981) Squash leaf curl: a new disease of cucurbits in California. Pt Dis 65:75–76

Freitas CMS, Rezende JAM (2008) Protection between strains of *Papaya ring spot virus* type W in Zucchini squash involves competition for viral replication sites. Sci Agri 65(2):183–189

French CJ, Elder M (1999) Virus particles in guttate and xylem of infected cucumber. Ann of Appl Biol 134:81–87

Guilley H, Wipf-Scheibel C, Richards K, Lecoq H, Jonard G (1994) Nucleotide sequence of *Cucurbit aphid-borne yellows luteovirus*. Virol 202:1012–1017

Hammond RW, Hernandez E, Mora F, Ramirez P (2005) First report of *Beet pseudo-yellows virus* on *Cucurbita moschata* and *C. pepo* in Costa Rica. Pt Dis 89:1130

Hassan AA, Duffus JE (1991) A review of a yellowing and stunting disorder of cucurbits in the United Arab Emirates. Emir J Agric Sci 2:1–16

Jones DR (2003) Plant viruses transmitted by whiteflies. Eur J Pt Pathol 109:195–219

Jordá-Gutiérrez C, Gómez-Guillamón ML, Juárez M, Alfaro-García A (1993) Clostero-like particles associated with yellows disease of melons in South-eastern Spain. Pt Pathol 42:722–727

Juarez M, Truniger V, Aranda MA (2004) First Report of Cucurbit aphid-borne yellows virus in Spain. Plant Dis 88:907

Khan MS, Raj SK, Bano T, Garg VK (2006) Incidence and management of mosaic and leaf curl diseases in cultivars of chilli (*Capsicum annuum*). J Food Agri Environ 4:171–174

Khan MS, Tiwari AK, Ji SH, Chun SC (2012) *Ageratum conyzoides* and its role in Begomoviral Epidemics, *Ageratum enation virus*: an emerging threat in India. Vegetos 24:20–28

Kumar P, Awasthi LP (2003) Management of infection and spread of bottle gourd mosaic virus disease in bottle gourd through botanicals. Ind Phyotopath 56:361

Kunkalikar S, Byadgi AS, Kulkarni VR, Reddy MK (2006) Management of *Papaya ringspot virus* disease. Ind Virol 17:39–43

Kyriakopoulou PE, Varveri C (1991) *Zucchini yellow mosaic virus* in Greece. Ann Inst Phytopathol Benaki 16:147–150

Lecoq H, Pitrat M (1984) Strains of *Zucchini yellow mosaic virus* in muskmelon (*Cucumis melo* L.). Phytopath Z 111:165–173

Lecoq H, Wisler G, Pitrat M (1998) Cucurbit viruses: the classics and the emerging. In: McCreight JD (ed) Cucurbitaceae' 98: evaluation and enhancement of cucurbit germplasm. ASHS Press, Alexandria, p 126–142

Lecoq H (1992) Viruses melon crops and full courgette (I and II). PHM-Horticultural Journal 323:23–28

Lisa V, Dellavalle G (1981) Characterization of two potyviruses in *Cucurbita pepo*, Phytopath. Z 100:279–286

Lisa V, Lecoq H (1984) *Zucchini yellow mosaic virus*. CMI/AAB, descriptions of plant viruses. No. 282

Lisa V, Boccardo G, D'Asostino G, Dellavalle G, D' Aquilio M (1981) Characterization of a potyvirus that causes *Zucchini yellow mosaic*. Phytopathol 71:667–672

Liu HY, Wisler GC, Duffus JE (2000) Particle lengths of whitefly-transmitted criniviruses. Plant Dis 84:803–805

Lockhart BEL, Fisher HU (1979) Host range and some properties of *Bryonia mottle virus*, a new member of the potyvirus group. Phytopathol Z 96:244–250

Lopez-Sese AI, Gomez-Guillamon ML (2000) Resistance to *Cucurbit Yellowing Stunting Disorder Virus* (CYSDV) in *Cucumis melo* L. Hort Science 35:110–113

Mansoor S, Khan SH, Saeed M (1997) Evidence for the association of a bipartite geminivirus with tomato leaf curl disease in Pakistan. Plant Dis 81:958

Pandey N, Tiwari AK, Shukla K and Rao GP (2011) Detection and identification of *Ageratum enation virus* infecting *Ageratum conzoides* in India. Acta Phytopathologica et Entomologica Hungrica 46(2):205–214

Papayiannis LC, Ioannou N, Boubourakas IN, Dovas CI, Katis NI, Falk BW (2005) Incidence of Viruses Infecting Cucurbits in Cyprus. J Phytopath 153:530–535

Paris HS, Maynard DN (2008) *Cucurbita* spp. Squash, pumpkin, gourds. In: Janick J, Paulls RE (eds) The encyclopedia of fruits & nutes. CAB Internationals, Wallingford, pp 292–299

Paris HS, Lebeda A, Kristkova E, Andres TC, Nee MH (2012) Parallel evolution under domestication and phenotypic differentiation of the cultivated sub species of *Cucurbita pepo*. Eco Bot 66(1):71–90

Pfeffer S, Dunoyer P, Heim F, Rechards KE, Jonard G, Ziegler-Graff V (2002) P0 of *Beet western yellows virus* is a suppressor of post transcriptional gene silencing. J Virol 76:6815–6824

Phaneendra C, Rao KRSS, Jain RK, Mandal B (2011) *Tomato leaf curl New Delhi virus* is associated with pumpkin leaf curl: a new disease in northern India. Ind J Virol 23(1):42–45

Prieto H, Bruna A, Hinrihsen P, Munoz C (2001) Isolation and molecular characterization of a Chilean isolate of *Zucchini yellow mosaic virus*. Plant Dis 85:644–648

Provvident R, Gonsalves D (1984) Occurrence of *Zucchini yellow mosaic virus* in cucurbits from Connecticut, New York, Florida and California. Plant Dis 68:443–446

Purcifull DE, Hibert E (1979) Serological distinction of *Watermelon mosaic virus* isolate. Phytopath 69:112–116

Purcifull DE, Hibert E, Edwardson J (1984) *Watermelon mosaic virus* 2, CMI/AAB Descriptions of Plant Viruses. No 293 (No. 63 revised)

Raj SK, Khan MS, Kumar S, Pratap D, Visnoi R (2008) *Cucumber mosaic virus* infecting vegetable and pulse crops in India. In: Rao GP, LavKumar P, Hluin RJ (eds) Characterization, diagnosis and management of plant viruses. Studium Press LLC, USA, pp 39–62

Raj SK, Snehi SK, Khan MS, Tiwari AK, Rao GP (2010a) First report of *Pepper leaf curl Bangladesh virus* strain associated with bitter gourd (*Momordica charantia* L.) yellow mosaic disease in India. Aus Plant Dis Notes 5:14–16

Raj SK, Snehi SK, Tiwari AK, Rao GP (2010b) Biological, molecular identification and management strategies of *Begomovirus* infecting Cucurbitaceous crops in India. In: Rao GP, Mandal B, Barnwal VK, Rishi N (eds) Plant viruses. Published from LLC Press, USA, pp 135–155

Raj SK, Snehi SK, Khan MS, Tiwari AK, Rao GP (2011) Molecular identification of *Ageratum enation virus* associated with mosaic disease of pointed gourd (*Trichosanthes dioica* Roxb.) in India. Phytoparasitica 39:497–502

Rajinimala N, Rabindran R, Ramiah M, Kamlakhan A (2005) Virus vector relationship of *Bitter gourd yellow mosaic virus* and whiteûy *Bemisia tabaci* germ. Acta Phytopathologica et Entomologica Hungarica 40:23–30

Ramirez P, Hernandez E, Mora F, Abraitis R, Hammond RW (2008) Limited geographic distribution of *Beet pseudo-yellows virus* in Costa Rican cucurbits. J Pt Pathol 90(2):331–335

Raychaudhuri SP, Verma JP (1977) Therapy by heat, radiation and meristem culture. In: Horsfall JG, Cowling EB (eds) Plant diseases—an advanced treatise. Academic Press, New York, pp 1015

Reddy KRC, Narriani TK (1963) Studies on mosaic disease of vegetable marrow (*Cucurbita pepo*). Ind Phytopath 8:99–104

Revill PA, Ha CV, Lines RE, Bell KE, Vu MT, Dale JL (2004) PCR and ELISA-based virus surveys of banana, papaya and cucurbit crops in Vietnam. Asia Pacific J Biotech 12(1 & 2):27–32

Rist DL, Lorbeer JW (1989) Occurrence and overwintering of *Cucumber mosaic virus* and *Broad bean wilt virus* in weeds growing near commercial lettuce fields in New York. Phytopath 79:65–69

Roggero P, Dellavalle G, Lisa V, Stravato VM (1998) First report of *Moroccan watermelon mosaic virus* in Zucchini. Pt Dis 82:351

Schrijnwerkers CCFM, Huijberts N, Bos L (1991) ZYMV, Two outbreaks in the Netherlands and seed transmissibility, Netherlands. J Pt Pathol 97(3):187–191

Sinclair JW, Crosby K (2002) A review of cucurbit yellow stunting disorder—a new virus affecting Melon in the lower Rio Grande Italy. Sub Trop Plant Sci 54:54–58

Singh R, Raj SK, Prasad V (2008) Molecular identification of a new strain of *Squash leaf curl china virus* infecting *Cucurbita maxima* in India. J Phytopath 156:222–228

Sohrab SS, Mandal B, Pant RP, Varma A (2003) First report of association of *Tomato leaf curl New Delhi* virus with yellow mosaic disease of *Luffa cylindrica* in India. Pt Dis 87:1148

Somvanshi P, Khan MS, Raj SK, Seth PK (2009) *Ageratum conizoides* and *Parthenium hystorophorous*: alternate hosts of *Begomovirus* and Phytoplasma. International day for Biological diversity. Invasive Alien Species, Souvenir, pp 44–45

Stanley J (2004) Sub viral DNAs associated with geminivirus disease complexes. Vet Microbiol 98:121–129

Tahir M, Haider MS, Briddon RW (2010) First report of *Squash leaf curl China virus* in Pakistan. Aus Plant Dis Notes 5:21–24

Taliansky M, Mayo MA, Barker H (2003) *Potato leafroll virus*: a classic pathogen shows some new tricks. Mol Plant Pathol 4(2):81–89

Tiwari AK, Snehi SK, Raj SK, Rao GP, Sharma PK (2008) *Begomovirus*: a major problem for cucurbitaceous crops in Eastern part of U.P. India. Ind J Virol 19(1):124

Tiwari AK, Sharma PK, Khan MS, Snehi SK, Raj SK, Rao GP (2010a) Molecular detection and identification of *Tomato leaf curl New Delhi virus* isolate causing yellow mosaic disease in bitter gourd (*Momordica charantia*), a medicinally important plant in India. Med Plants 2(2):117–123

Tiwari AK, Mall S, Khan MS, Snehi SK, Sharma PK, Rao GP, Raj SK (2010b) Detection and identification of *Tomato leaf curl Palampur virus* infecting *Cucurbita pepo* in India. Gunaxi J Agri Sci 40(12):1291–1295

Tiwari AK, Snehi SK, Singh R, Raj SK, Rao GP, Sharma PK (2012a) Molecular identification and genetic diversity among six begomovirus isolates affecting cultivation of cucurbitaceous crops in Uttar Pradesh. Arch Phytopath Pt Prot 45(1):62–72

Tiwari AK, Rao GP, Khan MS, Pandey N, Raj SK (2012b) Detection and elimination of begomovirus affecting *T dioica* plant in India. Arch Phytopath Pt Prot 45(9):1070–1075

Tóbiás I, Kovács G (2001) Seed transmission of *Zucchini yellow mosaic virus*—new viral pathogen—on cucurbitaceous plants. Növényvédelem (Plant Protection) 37:29–31 (in Hungarian)

Tobias I, Szabó B, Salánki K, Saria L, Kuhlmann H, Palkovics L (2008) Seed borne transmission of *Zucchini yellow mosaic virus* and *Cucumber mosaic virus* in Styrian Hulless group of *Cucurbita pepo*. Cucurbitaceae 2008, Proceedings of the IXth EUCARPIA meeting on genetics and breeding of Cucurbitaceae (Pitrat M, ed), INRA, Avignon (France), May 21–24th

Valkonen J (1998) Virus disease control in plants using natural and engineered resistance and some considerations regarding biosafety. Currents 17:51–55

Varma A, Malathi VG (2003) Emerging geminivirus problem: a serious threat to sustainable crop production. Ann App Biol 142:145–164

Verma HN, Awasthi LP, Kumar V, Chaudhary B, Rastogi P, Diwedi SD (1980) Control of plant virus diseases by extracts from higher plants. Ind Bot Soc 59:30

Webb SE, Akad F, Nyoike TW, Liburd OE, Polston JE (2007) Whitefly-transmitted *Cucurbit Leaf Crumple Virus* in Florida. http://edis.ifas.ufl.edu. Accessed 13 May 2013

Wong SM, Chng CG, Chng CY, Chong PL (1994) Characterization of an isolate of *Zucchini yellow mosaic virus* from cucumber in Singapore. J Phytopath 141:355–368

Xiao H, Fan H (1994) Studies on the cross-protection among strains of *Papaya ringspot virus*. Chin J Virol 10:164–171

Yardumici N, Ozgonen H (2007) First report of *Cucurbit aphid-borne yellows virus* in Turkey. Aust Plant Dis Notes 2:59

Yonaha T, Tamori M, Fujiwara Y, Ishimine M (1988) Trichosanthes mottle virus. Bull Coll Agric Univ Ryukus 35:1–15

Zouba AA, López MV, Anger H (1998) *Squash yellow leaf curl virus*: a new whitefly-transmitted poty-like virus. Pt Dis 82:475–478

# Plant Growth Promoting Rhizobacteria in Vegetable Disease Management

33

M. Loganathan, A. B. Rai, Arpita Singh and Sujoy Saha

**Abstract**

In recent years, microorganisms are being given emphasis, to employ them in cropping system in order to overcome the problems faced due to application of chemicals for nutrient and disease management. The microbes especially plant growth promoting rhizobacteria (PGPR) living in rhizosphere, are known to enhance the yield of the crops and protect them from diseases through several mechanisms, importantly by production of metabolites, volatile compounds and phytohormones and induced systemic resistance (ISR). ISR in plants is achieved through induction of defence-related proteins and chemicals by the PGPR upon pathogen attack. Though different genera of PGPR are known to be involved in ISR in wide range of crops against various diseases, the bacteria belonging to *Bacillus* and *Pseudomonas* are extensively studied. Application of PGPR either singly or in mixture through different formulation in vegetable ecosystem for sustainable production has been thoroughly discussed in this chapter.

**Keywords**

Plant growth-promoting rhizobacteria (PGPR) · Induced systemic resistance (ISR) · Biocontrol · Crop protection agents

## 33.1 Introduction

Vegetables are occupying a coveted position in Indian agriculture and the area under vegetables is accounted as 8.49 M ha area with annual production of 146.55 Mt (IHD 2011). Vegetable environment is severely threatened by abiotic and biotic factors. Among which crop loss due to biotic factors especially due diseases is huge and estimated as 50–80% from the heavily infected fields (Datar and Mayee 1981; Jones et al. 1991; Chellemi et al. 1997). Farmers are extremely concerned for their crop, and for a quick remedy they use excessive amount of chemicals to control the diseases which in turn leads to ill effects on environment and nontargeted organisms including animals and human beings (Agrios

A. B. Rai (✉) · M. Loganathan · A. Singh · S. Saha
Division of Crop Protection, Indian Institute of Vegetable Research, Varanasi 221 305, India
e-mail: abraiiivr@gmail.com

R. N. Kharwar et al. (eds.), *Microbial Diversity and Biotechnology in Food Security*,
DOI 10.1007/978-81-322-1801-2_33, © Springer India 2014

1988; Cook 1993; Heydari 2007; Heydari et al. 2007). To maintain an ecological balance, protection measures need to be resorted to which are based on less or zero use of chemicals. Alternatively, use of microbes is shown to be an attractive method because in this method, a beneficial interaction or mutual relationship between the host plant and microorganisms can be achieved (Bull et al. 2002; Katska 1994; Chisholm et al. 2006). PGPR are group of bacteria that generally colonize in the rhizospheric zone of the plant tissue, encourage plant growth in different ways and control various plant diseases (Weller 1988) and some of the best examples of these bacterial genera are *Bacillus, Pseudomonas* and *Bradyrhizobium*. PGPR facilitate to enhance plant growth through various activities like chelation of iron (siderophore), fixation of nitrogen, production of phytohormones and solubilization of insoluble phosphorus (P) (Zaidi et al. 2009) apart from controlling diseases through ISR (Kloepper and Beauchamp 1992; Liu et al. 1995; Chen et al. 2000; Saravanakumar et al. 2007; Sangeetha et al. 2010; Kumar et al. 2012).

## 33.2 Sustainable Production of Vegetables Through PGPR

Vegetables are short-duration crops and to obtain high yield, excessive chemical fertilizers are being used in the field. In the long term, due to this practice, the soil and water will be polluted. Rather, microbes should be used to enhance the crop productivity through fixation of nitrogen, production of siderophore and indole acetic acid (IAA) and solubilization of phosphate

### 33.2.1 Nitrogen Fixation

Fixation of nitrogen by bacteria in soil is widely distributed among the domain and is well documented. Normally nitrogen present in the soil cannot be utilized by the plant. So these nitrogen-fixing bacteria play an important role of converting free nitrogen to ammonia which can be utilized by the plants. The bacteria belonging to the genus *Rhizobium, Mesorhizobium, Bradyrhizobium, Azorhizobium, Allorhizobium* and *Sinorhizobium* are involved in this process. However, the use of inorganic fertilizers has been a major concern for decreasing the potentiality of nitrogen fixers in soil (Cummings et al. 2006).

Some of *Bacillus* species such as *B. megaterium, B. cereus, B. pumilu, B. circulans B.licheniformis, B. subtilis, B. brevis, B. firmus, B. sphaericu, B. fusiformis, B. marisflavi* and *B. alkalidiazotrophicus* were involved in nitrogen fixation (Xie et al. 1998; Park et al. 2005; Sorokin et al. 2008)

### 33.2.2 Siderophore Production

Iron is an essential nutrient for plant growth. Normally, iron present in the soil is in the form of ferric ion ($Fe^{3+}$), and its concentration is too low to be utilized by microorganisms. Under stress or competition, microbes, especially bacteria, produce siderophore (*sidero*=iron, *phores*=bearer) to chelate the iron with more affinity toward ferric ion ($Fe^{3+}$) than ferrous ion ($Fe^{2+}$) (Neiland and Nakamura 1997; Neiland 1995). Under the stress, $Fe^{3+}$ ion is transported inside the bacterial cell using specific receptor, and the solubilized form of $Fe^{+3}$ is delivered to the plant root surface where it is reduced to $Fe^{+2}$ and immediately absorbed (Von Wiren et al. 2000).

### 33.2.3 IAA Production

IAA is a natural auxin, responsible for plant growth promotion. Tryptophan is the precursor of IAA and greater accumulation of IAA was noticed in plants which were colonized by bacteria, and later an alternate tryptophan dependent pathway was also found in beneficial bacteria which are responsible for main production of IAA. The authentication for the production of IAA by bacteria was found out when isolates were grown in nutrient broth along with precursor tryptophan (Ahmad et al. 2005).

### 33.2.4 Phosphate Solubilization

Phosphorus is a major yield limiting factor in vegetable production and majority of P applied to the soil goes to immobile pool as it precipitated as orthophosphate or adsorbed by aluminium ion ($Al^{3+}$) or $Fe^{3+}$. Bacteria present in the soil convert the immobile P into available P by the process of solubilization and mineralization mainly by lowering the soil pH through production of acid phosphatases (Khan et al. 2009). The most powerful phosphate solubilizers are *Pseudomonas, Bacillus* and *Rhizobium* (Rodríguez and Fraga 1999).

## 33.3 Disease Suppression Through PGPR

Most of the vegetables like brinjal, tomato, cowpea, chilli, capsicum, etc. are generally infected by soil borne pathogens like *Rhizoctonia, Sclerotinia, Pythium, Fusarium,* etc. Application of PGPR leads to restriction of pathogen to attack the crop by creating unfavourable environment (Timmusk 2003). Like humans and animals, plants also respond to pathogen attack by local and systemic response. In local response, infected points as well as adjoining cells respond to the attack in the form of progressive cell death so as to restrict the disease spread (Kombrink and Schmelzer 2001). By the application of PGPR, the systemic response to the pathogen attacks received away from infection site is called induced systemic response (ISR). The most attractive ISR mechanism is induction of PR proteins (pathogenesis-related proteins) and chemicals in plants by PGPR upon infection of pathogen (von Loon and van Kammen 1970; Maurhofer et al. 1994). PGPR also involves strengthening the cell wall of plants apart from changing the physical or biochemical processes in the host in order to counter attack the invading pathogen (Ramamoorthy et al. 2001).

Application of PGPR leads to accumulation of callose or phenolics or formation of cell wall opposition at the site of pathogen attack which in turn results in development of resistance in plants against diseases. The mechanism of resistance has been demonstrated in peas against *Pythium ultimum* and *F. oxysporum* f. sp. *pisi* (Benhamou et al. 1996) and tomatoes against *F. oxysporum* f. sp. *radicis-lycopersici* (Benhamou et al. 1998). The cell wall of PGPR contains lipopolysaccharides (LPS) and the LPS plays a major role in the induction of ISR against diseases. LPS of *Pseudomonas fluorescens* induced ISR in radish against *F. oxysporum* f. sp. *raphani* (Leeman et al. 1995).

Phenylalanine ammonia-lyase (PAL) is an enzyme involved in synthesis of phenolic phytoalexins and PGPR-treated plants which showed post pathogen attack induction of PAL. Induction of PAL in cucumber plants treated with *Pseudomonas corrugata* has been reported against *Pythium aphanidermatum* (Chen et al. 2000). Simultaneous induction of PAL and salicylic acid was observed in bean roots treated with *P. aeruginosa* 7NSK2 (De Meyer et al. 1999). Similarly, activation of PAL was observed in tomato and pepper plants treated with *P. fluorescens* upon infection by *Pythium aphanidermatum, F. oxysporum* f. sp. *lycopersici* and *C. capsici* (Ramamoorthy and Samiyappan 2001; Ramamoorthy et al. 2002a; Ramamoorthy et al. 2002b).

Peroxidases (PO) and polyphenol oxidases (PPO) played a major role in biosynthesis of lignin and other oxidative phenols (Bruce and West 1989). PGPR-treated plants showed induction of PO and PPO upon challenge inoculation with pathogen. Root treatment of cucumber plants with *P. corrugata* induced PO and PPO when challenged with *P. aphanidermatum* (Chen et al. 2000). Ramamoorthy et al. (2002b) demonstrated that seed treatment with *P. fluorescens* isolate Pf1 inducted PO and PPO isoforms against *Pythium aphanidermatum* in tomato and hot pepper.

Chitinases and $\beta$-1, 3 glucanase are cell wall degrading enzymes and induction of these enzymes was directly correlated with disease resistance (Sahai and Manocha 1993; Jackson and Taylor 1996; Maurhofer et al. 1994). Similarly PGPR-treated plants expressed thaumatin-like proteins (TLP) when challenged with pathogens (Datta et al. 1999). Induction of chitinases, β-1,3 glucanase and TLP was observed in tomato plants treated with *P. fluorescens* against *F. oxysporum* f. sp. *lycopersici* (Ramamoorthy et al. 2002a).

PGPR are also known to attack the pathogens through production of certain metabolites and volatiles. Production of antimicrobial compounds (metabolites) has been reported in PGPR. *B. subtilis* produced Iturin A and Surfactin antimicrobial compounds against *Rhizoctonia solani* in tomatoes (Asaka and Shoda 1996). Monteiro et al. (2005) reported production of lipopetide by *Bacillus* spp against *Xanthomonas campestris* pv. *campestris* Leila in crucifers. Similarly, *B. subtilis* ME488 produced Bacilysin bacD against *F. oxysporum* f.sp. *cucumerinum* in cucumbers (Chung et al. 2008) and *B. subtilis* ZK8 induced jiean-peptide against *Rhizoctonia* rot in tomatoes (Zhang et al. 2010). PGPR are also involved in production of volatile compounds against the pathogens. Fiddaman and Rossall (1993) observed the production of antifungal volatile by *B. subtilis* against *Rhizoctonia solani* and *Pythium ultimum*.

Potentiality of PGPR has been well documented against insects (Zehnder et al. 1997); nematode (Sikora 1988) and diseases. However, control of fungal, bacterial and viral diseases of vegetables by PGPR has been listed in (Table 33.1.)

## 33.4 Formulations of PGPR

Suitability of PGPR for development of formulation depends upon certain ideal characters, they should have compatibility with environment and other beneficial microbes, high rhizosphere competence, ease to mass multiply and tolerance to adverse environment (Jeyarajan and Nakkeeran 2000). Different formulations of PGPR were reported *viz.*, talc preparation of *P. fluorescens* and *B. subtilis* (Vidhyasekaran and Muthamilan 1995; Amer and Utkhede 2000), lignite-based *P. fluorescens* (Vidhyasekaran and Muthamilan 1995), peat formulation of *P. chlororaphis* and *B. subtilis* (Nakkeeran et al. 2004), vermiculite-based *P. fluorescens, P. putida* and *B. subtilis* (Vidhyasekaran and Muthamilan 1995; Amer and Utkhede 2000) and kaolinite and farm yard manure-based *P. fluorescens* (Vidhyasekaran and Muthamilan 1995).

An ideal bio-formulation should retain its bacterial cells under viable condition for a long period, which is highly depended on the type of carrier and the bacterial strains. Talc-based *P. fluorescens* (P7NF, TL3) could retain 8.4 (Log cfu/g) for 12 months (Caesar and Burr 1991) while talc of *P. putida* could maintain only $10^3$ (cfu/g) for 45 days (Amer and Utkhede 2000). Similarly, wide variation was observed in vermiculite preparations because such preparation of *P. fluorescens* (Pf1) retained $10^6$(cfu/g) for 8 months (Vidhyasekaran and Muthamilan 1995) whereas *B. subtilis* maintained $10^6$ (cfu/g) only for 45 days (Amer and Utkhede 2000).

Though at many instances the bio-formulations were applied individually and found effective, the method failed to perform under adverse environmental conditions (Weller and Thomashow 1994). This problem can be solved by application of mixture of two or more bio-agents, because it ensures at least one strain fight against the disease under such environment. Different bio-formulations were applied against vegetable disease and they found to control diseases from 40–80 % and enhanced the crop yield by 5–70 % (Table 33.2).

PGPR bio-formulations are applied as seed treatment or foliar spray or soil application or in combination. Talc based formulations of *Bacillus amyloliquefaciens* (BA1) and *B. subtilis* (BS2) were tested against collar rot (*Sclerotium rolfsii*) of cowpea under field conditions (Loganathan et al. 2013). The bioformulations (minimum $10^8$ cfu/g) were applied as seed treatment (20 g/kg) and soil application (2.5 kg/ha formulation mixed with 50 kg well-decomposed farm yard manure). The isolate BS2 could reduce the collar rot (*Sclerotium rolfsii*) disease by 62.73 % as well as enhanced the yield of the crop to 76 % (Table 33.3).

### 33.4.1 Commercial Formulations

Though numerous reports are available for control of diseases by PGPR and only few are extended to commercial applications. Commercial products are registered items, which can be used against specific disease as per the recommendation. In India, only two PGPR microbes such as *P. fluorescens* 1.75 % WP (in-house isolated strain Accession No. MTCC 5176) and *P. fluorescens* 0.5 % WP (TNAU Strain Acces-

**Table 33.1** Spectrum of resistance shown by PGPR against various diseases of vegetables

| Bio-agent | Disease/Pathogen | Crop | Reference |
|---|---|---|---|
| *B. sublitis* | *Rhizoctonia solani* | Tomato | Asaka and Shoda (1996) |
| *B. pumilus* | *Colletotrichum orbiculare* | Cucumber | Wei et al. (1996) |
| *B. pumilus* | *F. oxysporum* | Tomato | Benhamou et al. (1998) |
| *B. pumilus*, *B. subtilis* | *Colletotrichum orbiculare* | Cucumber | Raupach and Kloepper (1998) |
| *B. subtilis* BS 107 | *Erwinia carotovora* subsp. *atroseptica* and *Erwinia carotovora* subsp. *carotovora* | Potato | Sharga and Lyon (1998) |
| *B. subtilis* strain LS213 | Bacterial spot and late blight of tomato Angular leaf spot of cucumber | Tomato and cucumber | Reddy et al. (1999) |
| *B. subtilis*, *B. cereus*, *P. putida* | *Pythium* sp | Cucumber | Uthede et al. (1999) |
| *P. putida* | *Fusarium oxysporum* f.sp. *cucurbitacearum* | Cucumber | Amer and Utkhede 2000 |
| *B. polymyxa* and *P. fluorescens* PRS9 | *F. oxysporum* f.sp. *lycopersici* | Tomato | Khan and Akram (2000) |
| *B. amyloliquefaciens* strain 1 N 937a | Tomato mottle virus | Tomato | Murphy et al. (2000) |
| *P. fluorescens* Pf1 | *F. oxysporum* f.sp. *lycopersici* | Tomato | Ramamoorthy et al. (2002a) |
| *P. fluorescens* Pf1 | *Pythium* sp | Tomato and hot pepper | Ramamoorthy et al. (2002b). |
| *B. pumilus* SE34 | *Phytophthora infestans* | Tomato | Yan et al. (2002) |
| *B. cereus* X16 | *Fusarium roseum* var. *sambucinum* | Potato | Sadfi et al. (2002) |
| *B. subtilis* and *P. chlororaphis* (PA23) | Damping off | Tomato | Kavitha et al. (2003) |
| *B. subtilis* BS 21; BS 22; BS 23 | *Colletotrichum lindemuthianum* | Cowpea | Adebanjo and Bankole (2004) |
| *P. putida* | *Fusarium oxysporum* f.sp. *melonis* | Muskmelon | Bora et al. (2004) |
| *B. subtilis* | *R. solani* | Tomato | Szczech and Shoda (2004) |
| *Bacillus* spp | *Xanthomonas campestris* pv. *campestris* Leila | Crucifers | Monteiro et al. (2005) |
| *B. subtilis* RB 14–CS | *R. solani* | Tomato | Mizumoto et al. (2007) |
| *B. subtilis* RB 14–CS | *R. solani* | Tomato | Mizumoto and Shoda (2007) |
| *B. subtilis* UM AF6614; UM AF6619; UM AF6639; UM AF8561 | *Podosphaera fusca* | Cucurbit | Romero et al. (2007) |
| *B. subtilis* EU07 | *F. oxysporum* f.sp. *radicislycopersici* | Tomato | Baysal et al. (2009) |
| *Bacillus* spp. | *F. oxysporum* f.sp. *lycopersici* and *Sclerotium rolfsii* | Tomato and cowpea | Loganathan et al. (2010) |
| *B. subtilis* ZK8 | *Rhizoctonia* rot | Tomato | Zhang et al. (2010) |

**Table 33.2** PGPR formulations on disease control and productivity in vegetables

| PGPR | Crop | Formulation | Target pathogen | Disease control (per centreduction over control) | Yield (per cent increase over control) | Reference |
|---|---|---|---|---|---|---|
| *B. japonicum* Tal 629 | Radish | Bacterial cell | – | – | 15 (dry matter) | Antoun et al. (1998) |
| *B. pumilus* strain SE34 or *B. amyloliquefaciens* strain IN937a or *B. subtilus* strain IN937 | Tomato | Bacterial cells | Cucumber mosaic virus disease | 40–60 | 5–55 | Zehnder et al. (2000) |
| Mixture of *Serratia plymuthica* strain C-1, *Chromobacterium* sp. strain C-61 and *Lysobacter enzymogenes* strain C-3+ Chitin | Pepper | Bioformulation+ chitin | *Phytophthora* blight | 80–90 | – | Kim et al. (2008) |
| *B. pumilus* (T4) + *B. subtilis* (GBO3) | Cowpea | Talc based | Bean common mosaic virus | 62.1 | – | Shankar et al. (2009) |
| *Bacillus* sp C2 + *Streptomyces* sp. C32 | Chilli | Bacterial suspension | – | – | 69.9 | Datta et al. (2011) |
| *P. fluorescens* | Tomato | Bacterial cells | *Ralstonia solanacearum* | 57.9 | >100 | Seleim et al. (2011) |

**Table 33.3** Effect of PGPR on collar disease and yield of cowpea under field conditions. (Data published in Microbial diversity and its applications (Eds. Barbuddhe SB, Ramesh R, Singh NP, 2013), New India Publishing Agency, New Delhi, 244 p)

| S. No | Treatments | *Incidence of collar rot (%) | Percent reduction over control | Yield (tonnes/ha) | Percent increase over control (%) |
|---|---|---|---|---|---|
| 1 | BA1 | $17.0^{c}$ | 47.20 | $4.0^{a}$ | 60.00 |
| 2 | BS2 | $12.00^{b}$ | 62.73 | $4.4^{a}$ | 76.00 |
| 3 | Carbendazim | $16.3b^{c}$ | 49.37 | $3.8^{a}$ | 52.00 |
| 4 | Carbendazim + mancozeb | $6.6^{a}$ | 79.50 | $4.5^{a}$ | 80.00 |
| 5 | Control | $32.2^{d}$ | – | $2.5^{b}$ | – |
| 6 | CD (5%) | 4.6 | – | 1.2 | – |

In a column a mean followed by common letters are not significantly different at 5% level by DMRT. Values were arc transformed before analysis

**Table 33.4** Some of the commercial biopesticides used against vegetable diseases. (http://www.omnilytics.com; http://naturalindustries.com; Backmann et al. 1994; Kilian et al. 2000)

| Name | Organism | Country | Application |
|---|---|---|---|
| Agriphage$^{TM}$ | Bacteriophages of *Xanthomonas* spp. and *Pseudomonas syringae* pv. tomato | USA | Bacterial spots in pepper and tomatoes, and bacterial speck in tomatoes |
| Actinovate® AG | *Streptomyces lydicus* WYEC 108 | USA | Broad spectrum *Pythium*, *Fusarium*, *Rhizoctonia*, powdery mildew, downy mildew, *Botrytis*, *Alternaria*, etc. |
| Actinovate® SP | *Streptomyces lydicus* WYEC 108 | USA | *Pythium*, *Fusarium*, *Rhizoctonia*, *Verticillium*, powdery mildew, downy mildew, *Botrytis*, *Alternaria*, etc. |
| Serenade® MAX™ | *B. subtilis* strain QST 713 | USA | *Sclerotinia*, rust, powdery mildew, bacterial spot and white mould, etc. |
| | *B. subtilis* | USA | Seed dressing |
| | *B. subtilis* FZB 24 | Germany | Seed dressing (potatoes) |

sion No. ITCC BE 0005) have been registered by the Central Insecticide Board and Registration committee, Government of India to use as biopesticide. The countries like the USA have registered many PGPR products for commercial use and PGPR bio-pesticides developed against vegetable diseases have been documented in Table 33.4.

## 33.5 Conclusion

PGPR induces spectrum of resistance against pest and diseases in wide range of crops apart from enhancing crop productivity. These microbes are good root colonizers and are easy to multiply and formulate and have proven their efficacy against diseases of fungal, bacterial and viral origin. Vegetables are short-duration crops and excessive use of chemicals to control diseases may lead to residual toxicity in the harvest produce. As agriculture is moving toward approaches for ecofriendly management of pests and diseases and so far only few attempts are made to manage vegetable diseases through PGPR. Hence, utilization of PGPR in vegetable cultivation system to manage various diseases will be a viable and potential option.

## References

Adebanjo A, Bankole SA (2004) Evaluation of some fungi and bacteria for biocontrol of anthracnose disease of cowpea. J Basic Microbiol 44(1):3–9

Agrios NA (1988) Plant pathology, 3rd edn. Academic Press, U S A pp 220–222

Ahmad F, Ahmad I, Khan MS (2005) Indole acetic acid production by the indigenous isolates of *Azotobacter* and fluorescent *Pseudmonas* in the presence and absence of tryptophan. Turk J Biol 29:29–34

Amer GA, Utkhede RS (2000) Development of formulations of biological agents for management of root rot of lettuce and cucumber. Can J Microbiol 46:809–816

Antoun H, Beauchamp CJ, Goussard N, Chabot R, Lalande R (1998) Potential of *Rhizobium* and *Bradyrhizobium* species as plant growth promoting rhizobacteria on non-legumes: Effect on radishes (*Raphanus sativus* L.). Plant Soil 204:57–67

Asaka O, Shoda M (1996) Biocontrol of *Rhizoctonia solani* Damping-Off of tomato with *Bacillus subtilis* RB14. Appl Environ Microbiol 62(11):4081–4085

Backmann PA, Brannen PM, Mahaffe WF (1994) Plant response and disease control following seed inoculation with *Bacillus subtilis*. In: Ryder MH et al (eds) Improving plant productivity with rhizosphere bacteria. CSIRO Division of soils, Glen Osmond

Baysal ÖM, Siragusa H, Ikten I, Polat E, Gümrükcü F, Carimi JA, da Silva T (2009) *Fusarium oxysporum* f.sp. *lycopersici* races and their genetic discrimination by molecular markers in West Mediterranean region of Turkey. Physiol Mol Plant Pathol 74:68–75

Benhamou N, Belanger RR, Paulitz TC (1996) Induction of differential host responses by *Pseudomonas yuorescens* in Ri T-DNAtransformed pea roots after challenge with *Fusarium oxysporum* f. sp. *pisi* and *Pythium ultimum*. Phytopathology 86:114–178

Benhamou N, Kloepper JW, Tuzun S (1998) Induction of resistance against *Fusarium* wilt of tomato by combination of chitosan with an endophytic bacterial strain ultrastructural and cytochemistry of the host response. Planta 204:153–168

Bora T, Ozaktan H, Gore E, Aslan E (2004) Biological control of *Fusarium oxysporum* f. sp. *melonis* by wettable powder formulations of the two strains of *Pseudomonas putida*. J Phytopathology 152:471–475

Bruce RJ, West CA (1989) Elicitation of lignin biosynthesis and isoperoxidase activity by pectic fragments in suspension cultures caster bean. Plant Physiol 91:889–897

Bull CT, Shetty KG, Subbarao KV (2002) Interactions between Myxobacteria, plant pathogenic fungi and biocontrol agents. Plant Dis 86:889–896

Caesar AJ, Burr TJ (1991) Effect of conditioning, betaine, and sucrose on survival of rhizobacteria in powder formulations. Appl Environ Microbiol 57:168–172

Chellemi DO, Olson SM, Mitchell DJ, Seeker J, McSorley R (1997) Adaptation of soil solarization to the integrated management of soil borne pests of tomato under humid conditions. Phytopathology 87:250–258

Chen C, Belanger RR, Benhamou N, Paulitz TC (2000) Defense enzymes induced in cucumber roots by treatment with plant-growth promoting rhizobacteria (PGPR). Physiol Mol Plant Pathol 56:13–23

Chisholm ST, Coaker G, Day B, Staskawicz BJ (2006) Host-microbe interaction: shaping the evolution of plant immune response. Cell 124:803–814

Chung S, Kong H, Buyer JS, Lakshman DK, Lydon J, Kim SD, Roberts D (2008) Isolation and partial characterization of *Bacillus subtilis* ME488 for suppression of soil borne pathogens of cucumber and pepper. Appl Microbiol Biotechnol 80(1):115–123

Cook RJ (1993) Making greater use of microbial inoculants in agriculture. Annu Rev Phytopathol 31:53–80

Cummings SP, Humphry DR, Santos SR, Andrews M, James EK (2006) The potential and pitfalls of exploiting nitrogen fixing bacteria in agricultural soils as a substitute for inorganic fertiliser. Environ Biotech 2:1–10

Datar VV, Mayee CD (1981) Assessment of losses in tomato yield due to early blight. Indian Phytopathology 34:191–195

Datta K, Velazhahan R, Oliva N, Ona I, Mew T, Khush GS, Muthukrishnan S, Datta SK (1999) Overexpression of cloned rice thaumatin-like protein (PR-5) gene in transgenic rice plants enhances environmental friendly resistance to Rhizoctonia solani causing sheath blight disease. Theor Appl Genet 98:1138–1145

Datta M, Palit R, Sengupta C, Pandit MK, Banerjee S (2011) Plant growth promoting rhizobacteria enhance growth and yield of chilli (*Capsicum annuum* L.) under field conditions. AJCS 5(5):531–536

De Meyer G, Capieau K, Audenaert K, Buchala A, Metraux JP, Höfte M (1999) Nanogram amounts of salicylic acid produced by the rhizobacterium Pseudomonas aeruginosa 7NSK2 activate the systemic acquired resistance pathway in bean. Mol Plant Microbe Interact 12:450–458

Fiddaman PJ, Rossall S (1993) The production of antifungal volatiles by *Bacillus subtilis*. J Appl Microbiol 74(2):119–126

Heydari A (2007) Biological control of Turfgrass fungal diseases In: Pessarakli M (ed) Turfgrass management and physiology. CRC Press, Florida

Heydari A, Misaghi IJ, Balestra GM (2007) Pre-emergence herbicides influence the efficacy of fungicides in controlling cotton seedling damping-off in the field. Int J Agric Res 2:1049–1053

IHD (2011) Indian Horticulture Database. National Horticulture Board, Ministry of Agriculture, Government of India, Aristo Printing Press, New Delhi-110020 p 273

Jackson AO, Taylor CB (1996) Plant microbe interaction: life and death at the interface. Plant cell 8:1651–1668

Jeyarajan R, Nakkeeran S (2000) Exploitation of microorganisms and viruses as biocontrol agents for crop disease management. In: Upadhyay et al (eds) Biocontrol potential and their exploitation in sustainable agriculture. Kluwer Academic/Plenum Publishers, U S A, pp 95–116

Jones JB, Jones JP, Stall RE, Zitter TA (1991) Compendium of tomato diseases. APS Press, St. Paul

Katska V (1994) Interrelationship between vesicular-arbuscular mycorrhiza and rhizoshpere microflora in apple replant disease. Biol Plant 36:99–104

Kavitha K, Nakkeeran S, Chandrasekar G, Fernando WGD, Mathiyazhagan S, Renukadevi P, Krishnamoorthy AS (2003) Role of antifungal antibiotics, siderophores and IAA production in biocontrol of *Pythium aphanidermatum* inciting damping off in tomato by *Pseudomonas chlororaphis* and *Bacillus subtilis*. In proceedings of the 6th International workshop on PGPR, Organised by IISR, Calicut 5-10 October 2003. pp 493–497

Khan AA, Jilani G, Akhtar MS, Naqvi SMS, Rasheed M (2009) Phosphate-solubilizing bacteria: occurrence, mechanisms and their role in crop production. J Agric Biol Sci 1(1):48–58

Khan MR, Akram M (2000) Effect of certain antagonistic fungi and rhizobacteria on wilt disease complex caused by *Meloidogyne incognita* and *Fusarium oxyspovium* f. sp. *lycopevsici* on tomato. Nematol Mediterr 28:139–144

Kilian M, Steiner U, Krebs B, Junge H, Schmiedeknecht G, Hain R (2000) FZB24 *Bacillus subtilis*—mode of action of a microbial agent enhancing plant vitality. Pflanzenschutz-Nachrichten, Bayer 1/00, 1:72–93

Kim YC, Jung H, Kim KY, Park SK (2008) An effective biocontrol bioformulation against Phytophthora blight of pepper using growth mixtures of combined chitinolytic bacteria under different field conditions. Eur J Plant Pathol 120:373–382

Kloepper JW, Beauchamp CJ (1992) A review of issues related to measuring colonization of plant roots by bacteria. Can J Microbiol 38:1219–1232

Kombrink E, Schmelzer E (2001) The hypersensitive response and its role in local and systemic disease resistance. Euro J Plant Pathol 107:69–78

Kumar P, Khare S, Dubey RC (2012) Diversity of *Bacilli* from disease suppressive soil and their role in plant growth promotion and yield enhancement. N Y Sci J 5(1):90–111

Leeman M, Van Pelt JA, Den Ouden FM, Heinsbroek M, Bakker PAHM, Schippers B (1995) Induction of systemic resistance against *Fusarium* wilt of radish by lipopolysaccharides of *Pseudomonas fluorescens*. Phytopath 85:1021–1027

Liu L, Kloepper JW, Tuzun S (1995) Induction of systemic resistance in cucumber against, *Fusarium* wilt by plant growth promoting rhizobacteria. Phytopathology 85:695–698

Loganathan M, Garg R, Saha S, Bag TK, Rai AB (2010) Selection of antagonistic rhizobacteria against soil borne pathogens. J Mycopathol Res 48(2):227–232

Loganathan M, Saha S, Bag TK, Venkataravanappa V, Garg R, Rai AB (2013) Efficient management of soil borne pathogens of vegetables through plant growth promoting rhizobacteria (PGPR). In: Barbuddhe SB, Ramesh R, Singh NP (eds) Microbial diversity and its applications. New India Publishing Agency, New Delhi, 244 p

Maurhofer M, Hase C, Meuwly P, Metraux JP, Defago G (1994) Induction of systemic resistance of tobacco to tobacco necrosis virus by the root-colonizing *Pseudomonas fluorescens* strain CHAO: influence of the gacA gene and of pyoverdine production. Phytopathology 84:139–146

Mizumoto S, Hirai M, Shoda M (2007) Enhanced iturin A production by *Bacillus subtilis* and its effect on suppression of the plant pathogen *Rhizoctonia solani*. Appl Microbiol Biotechnol 75:1267–1274

Mizumoto S, Shoda M (2007) Medium optimization of antifungal lipopeptide, iturin A, production by *Bacillus subtilis* in solid state fermentation by response surface methodology. Appl Microbiol Biotechnol 76:101–108

Monteiro L, Mariano RLR, Souto-Maior AM (2005) Antagonism of *Bacillus* spp. against *Xanthomonas campestris* pv. *campestris*. Braz Arch Boil Technol 48:23–29

Murphy JF, Zender GW, Schuster DJ, Sikora EJ, Polston JE, Kloepper JW (2000) Plant growth promoting rhizobacterial mediated protection in tomato against tomato mottle virus. Plant Dis 84:779–784

Nakkeeran S, Kavitha K, Mathiyazhagan S, Fernando WGD, Chandrasekar G, Renukadevi P (2004) Induced systemic resistance and plant growth promotion by *Pseudomonas chlororaphis* strain PA-23 and *Bacillus subtilis* strain CBE4 against rhizome rot of turmeric (*Curcuma longa* L.). Can J Plant Pathol 26:417–418

Neiland JB (1995) Siderophore: structure and function of microbial iron transport compounds. J Biol Chem 270:26723–26726

Neiland JB, Nakamura K (1997) Detection determination isolation characterization and regulation of microbial iron chelates. In: Winkelmann G (ed) CRC handbook of microbial iron chelates. CRC Press, Boca Raton, pp 1–14

Park M, Kim C, Yang J, Lee H, Shin W, Kim S, Sa T (2005) Isolation and characterization of diazotrophic growth promoting bacteria from rhizosphere of agricultural crops of Korea. Microbiol Res 160:127–133

Ramamoorthy V, Samiyappan R (2001) Induction of defense related genes in *Pseudomonas fluorescens* treated chilli plants in response to infection by *Colletotrichum capsici*. J Mycol Plant Pathol 31:146–155

Ramamoorthy V, Viswanathan R, Raguchander T, Prakasam V, Samiyappan R (2001) Induction of systemic resistance by plant growth promoting rhizobacteria in crop plants against pests and diseases. Crop Protec 20:1–11

Ramamoorthy V, Raguchander T, Samiyappan R (2002a) Induction of defense-related proteins in tomato roots treated with *Pseudomonas fluorescens* Pf1 and *Fusarium oxysporum* f. sp. *lycopersici*. Plant Soil 239:55–68

Ramamoorthy V, Raguchander T, Samiyappan R (2002b) Enhancing resistance of tomato and hot pepper to Pythium diseases by seed treatment with fluorescent pseudomonads. Eur J Plant Pathol 108:429–441

Raupach GS, Kloepper JW (1998) Mixtures of plant growth promoting rhizobacteria enhance biological control of multiple cucumber pathogens. Phytopath 88:1158–1164

Reddy MS, Rodriguez-Kabana R, Kenney DS, Ryu CM, Zhang S, Yan Z, Martinez-Ochoa N, Kloepper JW (1999) Growth promotion and induced systemic resistance (ISR) mediated by a biological preparation. Phytopathy 89:S65

Rodríguez H, Fraga R (1999) Phosphate solubilizing bacteria and their role in plant growth promotion. Biotechnol Advan 17:319–339

Romero D, de Vicente A, Rakotoaly RH, Dufour SE, Veening JW, Arrebola E, Cazorla FM, Kuipers OP, Paquot M, Perez- Garcia A (2007) The iturin and fengycin families of lipopeptides are key factors in antagonism of *Bacillus subtilis* toward Podosphaera fusca. Mol Plant Microbe Interact 20:430–440

Sadfi N, Cherif M, Hajlaou MR, Boudabbous, Belanger AR (2002) Isolation and partial purification of antifungal metabolites produced by *Bacillus cereus*. Ann Microbiol 52:323–337

Sahai AS, Manocha MS (1993) Chitinases of fungi and plants: their involvement in morphogenesis and host-parasite interaction. FEMS Microbiol Rev 11:317–338

Sangeetha G, Thangavelu R, Usha Rani S, Muthukumar A, Udayakumar R (2010) Induction of systemic resistance by mixtures of antagonist bacteria for the management of crown rot complex on banana. Acta Physiol Pt. doi:10.1007/s11738-010-0513-y

Saravanakumar D, Kumar CV, Kumar N, Samiyappan R (2007) PGPR-induced defense responses in the tea plant against blister blight disease. Crop Prot 26:556–565

Seleim MAA, Saead FA, Abd-Ed-Moneem KMH, Abo-Elyousr KAM (2011) Biological control of bacterial wilt of tomato by plant growth promoting rhizobacteria. Plant Pathol J 10(4):146–153

Shankar ACU, Nayaka SC, Niranjan-Raj S, Kumar HB, Reddy MS, Niranjanaa SR, Prakasha HS (2009) Rhizobacteria-mediated resistance against the blackeye cowpeamosaic strain of bean commonmosaic virus in cowpea (*Vigna unguiculata*). Pest Manag Sci 65:1059–1064

Sharga BM, Lyon GD (1998) *Bacillus subtilis* BS 107 as an antagonist of potato black leg and soft rot bacteria. Can J Microbiol 44:777–783

Sikora RA (1988) Interrelationship between plant health promoting rhizobacteria, plant parasitic nematodes and soil microorganisms. Med Fac Landbouww Rijksuniv Gent 53(2b):867–878

Sorokin ID, Kravchenko IK, Tatjana P, Tourova TV, Kolganova ES, Boulygina, Sorokin DY (2008) *Bacillusalkalidiazotrophicus* sp. nov., a diazotrophic, low salt-tolerant alkaliphile isolated from Mongolian soda soil. Int J Syst Evol Microbiol 58:2459–2464

Szczech M, Shoda M (2004) Biocontrol of *Rhizoctonia* Damping-off of tomato by *Bacillus subtilis* combined with *Burkholderia cepacia*. J Phytopathol 152(10):549–556

Timmusk S (2003) Mechanism of action of plant growth promoting bacterium *Paenibacillus polymyxa*. acta universitatis. Comprehensive summary of Uppsala dissertation from the faculty of science and technology. Uppsala ISBN 91-554-5802–5

Uthede RS, Koch CA, Menzies JG (1999) Rhizobacterial growth and yield promotion of cucumber plants inoculated with *Pythium aphanidermatum*. Can J Plant Pathol 21:265–271

van Loon LC, van Kammen A (1970) Polyacrylamide disc electrophoresis of the soluble leaf proteins from *Nicotiana tobaccum* var. "Samsum" and "Samsum NN"II. Changes in protein constitution after infection with tobacco mosaic virus. Virology 40:199–211

Vidhyasekaran P, Muthamilan M (1995) Development of formulations of *Pseudomonas fluorescens* for control of chickpea wilt. Pt Dis 79:782–786

Von Wiren N, Khodr H, Hider RC (2000) Hydroxylated phytosiderophore species possess an enhanced chelate stability and affinity for iron III. Plant Physiol 124:1149–1157

Wei G, Kloepper JW, Tuzun S (1996) Induction of systemic resistance to cucumber disease and increase plant growth by plant growth promoting rhizobacteria under field conditions. Phytopath 86:221–224

Weller DM (1988) Biological control of soil borne plant pathogens in the rhizosphere with bacteria. Ann Rev Phytopath 26:379–407

Weller DM, Thomashow LS (1994) Current challenges in introducing beneficial microorganisms into the rhizosphere. In: O'Gara F, Dowling DN, Boesten B (eds) Molecular ecology of rhizosphere microorganisms: biotechnology and the release of GMO's. VCH publishers, Weinheim, pp 1–18

Xie GH, Su BL, Cui ZJ (1998) Isolation and identification of N2-fixing strains of *Bacillus* in rice rhizosphere of the Yangtze River valley. Acta Microbiol Sin 38:480–483

Yan Z, Reddy MS, Ryu CM, McInroy JA, Wilson M, Kloepper JW (2002) Induced systemic protection against tomato late blight elicited by plant growth-promoting rhizobacteria. Phytopath 92:1329–1333

Zaidi A, Khan MS, Ahemad M, Oves M (2009) Plant growth promotion by phosphate solubilizing bacteria. Acta Microbiol Immunol Hung 56:263–284

Zehnder G, Kloepper J, Yao C, Wei G (1997) Induction of systemic resistance in cucumber against cucumber beetles (Coleoptera: Chrysomelidae) by plant growth promoting rhizobacteria. J Econ Entomol 90:391–396

Zehnder GW, Yao C, Murphy JF, Sikora ER, Kloepper JW (2000) Induction of resistance in tomato against cucumber mosaic cucumovirus by plant growth-promoting rhizobacteria. Biocontrol 45:127–137

Zhang X, Zhou J, Fu W, Li Z, Zhong J, Yang J, Xiao L, Tan H (2010) Response surface methodology used for statistical optimization of jiean-peptide production by Bacillus subtilis. Elect J Biotechnol 13(4):July

# Biological Control of Bacterial Wilt Disease-Causing Pathogens: A Sustainable Approach for Increasing Crop Production

34

R. Srinivasamurthy, J. Pratibha Singh and Ashwani K. Rai


**Abstract**

India is the second largest global producer of vegetables and accounts for about 15 % of the world's vegetables production. The production of vegetables is affected by infection of crops with several diseases throughout their life cycle. Among the diseases, bacterial wilt caused by *Ralstonia solanacearum* in crops such as tomato (*Lycopersicon esculentum*), potato (*Solanum tuberosum*), chili (*Capsicum annum*), tobacco (*Nicotiana tabacum*), eggplant (*Solanum melongena*), and pepper (*Capsicum annum*) is a major disease contributing to production loss of 10.80–92.62 % per unit area in India. The incidence of this disease is much severe during summer due to high temperature (28–36 °C) and high moisture (50–100 %), which favor the activity of the pathogen (*R. solanacearum*). Currently, adopted disease management practices like chemical application, use of resistance varieties, and manual removal of infected plants are of limited success to control the disease. The use of naturally occurring microorganisms in the rhizosphere of crop plants as a biocontrol agent offers an alternate source, and is gaining greater importance nowadays. Many effective plant growth-promoting rhizobacteria (PGPR) such as *Pseudomonas* spp., *Bacillus* spp., *Burkholderia* spp., *Serratia* spp., and *Streptomyces* spp. are abundant in rhizospheric soil. Moreover, rhizospheric soils are regarded as a source of



A. K. Rai (✉)
Department of Botany, Banaras Hindu University,
Varanasi 221005, India
e-mail: akrai.bhu@gmail.com

R. Srinivasamurthy
Institute Technology Management Unit,
Indian Agricultural Research Institute, Pusa campus,
New Delhi 110012, India
e-mail: srinrcpb@gmail.com

J. P. Singh
Rajiv Gandhi Cancer Institute & Research Center,
New Delhi 110085, India
e-mail: juliepratibhas@gmail.com



R. N. Kharwar et al. (eds.), *Microbial Diversity and Biotechnology in Food Security*,
DOI 10.1007/978-81-322-1801-2_34, © Springer India 2014

natural, effective, and valuable antagonists for the purpose of biological control. The use of PGPR to suppress pathogen (*R. solanacearum*)-causing bacterial wilt in crops has lately become successful, and thus is gaining greater importance. The PGPR control disease by producing siderophore, hydrogen cyanide (HCN), secondary metabolites/antibiotics such as pyoluteorin, phenazines, pyrrolnitrin, 2,4-diacetylphloroglucinol (2,4-DAPG), 1,2-benzene dicarboxylic 46 acid, bis(2-ethylexyl) ester, 2,6-di-T-butyl-4-methelyne-2,5-cyclohexadiene-1, and antifungal enzymes such as cellulase, chitinase, and protease. Production of antibiotics is one of the primary mechanisms involved in disease suppression. Among many antibiotics, 2,4-DAPG, a polyketide produced by bacteria showing broad-spectrum antiviral, antifungal, antibacterial, antitumor activities, and phototoxic properties, has received considerable attention. Thus, the 2,4DAPG-producing genotypes can be exploited to suppress bacterial wilt disease in crop plants.

**Keywords**

Bacterial wilt diseaseBiocontrol · Plant growth-promoting rhizobacteria (PGPR) · Crop protection agents

## 34.1 Introduction

The global food demand is increasing with progressive increase of population; however, the production of food is greatly affected due to damage caused by plant diseases. The use of chemicals to control the pathogens causing plant diseases is resulting in resistance development in pathogens as well as various adversatives to the environment. In order to overcome these problems, use of naturally occurring beneficial microorganisms present in rhizospheric soil as biological control agents is a more reliable and effective technique. Plant growth-promoting rhizobacteria (PGPR), which are eco-friendly in nature and effectively suppress the disease-causing plant pathogens, have proven to be beneficial. Biological control is thus considered as an alternative approach to control plant diseases and increase crop production in sustainable agricultural management system.

## 34.2 Bacterial Wilt

Bacterial wilt caused by *Pseudomonas solanacearum* was first reported by E. F Smith (1897; Rolfs 1898). The disease incidence was observed in many solanaceous species and several other plant families (Kelman 1954; Kucharek 1998). The occurrence of this bacterial pathogen has been reported across the world. Because of its extensive host range, it is known as a dangerous pathogen among the bacterial diseases (Buddenhagen et al. 1962; Hayward 1993). Among the vegetable crops, bacterial wilt caused by *Ralstonia solanacearum* in tomato is a serious disease and a major constraint in the production of tomatoes in tropical, subtropical, and warm temperate regions of the world (Buddenhagen et al. 1962; Hayward 1993). In India, bacterial wilt disease contributes to production loss of 10.80–92.62% (Mishra et al. 1995). The problem is severe, especially in summer season due to the presence of high temperature (28–36 °C) and high moisture (50–100%). Crop plants infected with wilt pathogens show stunted growth, poor fruit set, and complete wilting symptoms (Kucharek 1998). Bacterial wilt is initially characterized with wilting of upper leaves, followed by complete wilting of the plants. Brown discoloration in the vascular tissues of the lower stem of the wilted plants can also be observed. If the stem of a wilted plant is cut and immersed in clear water, it shows visible white or yellowish bacterial ooze.

The control of bacterial wilt pathogens is a much difficult task using chemical methods (Kucharek 1998). Management of bacterial wilt pathogen through integrated management approach like changing cultural practices, crop rotation, and use of resistant cultivars has provided some limited success, but failed to increase tomato production at commercial level (Kucharek 1998). The efficacy of current disease management methods employed for suppression of bacterial wilt pathogen is limited. No conventional practices are known to provide effective control of this soil-borne pathogen. Suppression of soil-borne wilt-plant pathogens using PGPR with increased crop production has been variously documented (Ciampi-Panno et al. 1989; Toyota and Kimura 2000). These can therefore be utilized as biocontrol agents. The PGPR have no side effects; hence, they are eco-friendly in nature (Bowen and Rovira 1999; Whipps 1997).

## 34.3 *Ralstonia solanacearum*

In the nineteenth century, for the first time, *R. solanacearum* was reported to cause bacterial wilt in potato, tobacco, tomato, and groundnut in Asia, southern USA, and South America. The pathogen was then described as *Bacillus solanacearum* by E. F. Smith (Rolfs 1898). Yabuuchi et al. (1992) described the pathogen as *Burkholderia solanacearum. R. solanacearum* is a Gram-negative rod with a polar tuft of flagella, and often produces nonfluorescent but diffusible brown pigments. Based on rRNA homology, *R. solanacearum* belongs to pseudomonad's group II and the *b*-subclass of proteobacteria. Different phenotypic and genotypic diversity in the same pathogen has been identified. The species are mainly divided into five races and five biovars depending on their host range and ability for oxidation/utilization of certain carbon sources (Hayward 1964). *R. solanacearum* mostly persists through soil, surface water, and infected crop residues (Granada and Sequeira 1983). The pathogen is also carried in the seeds of crop plants such as tomato and eggplant (Shakya 1993).

The most devastating and widespread bacterial wilt disease in many crop plants of tropical environment is due to the infection of *R. solanacearum* (He et al. 1983). This pathogen has worldwide distribution with host range of more than 50 different plant families (Hayward 1993). *R. solanacearum* gained its nuisance importance in the world due to its destructive nature, wide host range, and geographical distribution. It affects a wide range of economically important crops such as tomato, potato, eggplant, chili, and non-solanaceous crops such as banana and groundnut in India. The disease limits both the commercial and domestic level production.

## 34.4 Diversity of *R. solanacearum*

Buddenhagen et al. (1962) divided this pathogen into three races. Race 1 infects many solanaceous plants such as tomato, tobacco, pepper, and other plants including some weeds. It has a high temperature optimum (35 °C). Race 2 occurs mainly in tropical areas of South America and attacks bananas and *Heliconia* spp. (causing the so-called Moko disease), and in the Philippines (causing the so-called bugtok disease on plantains). In high tropics and subtropical and temperate areas, Race 3 is present and infects potato, tomato, occasionally capsicum and some solanaceous weeds like *Solanum nigrum* and *Solanum dulcamara* (Pradhanang et al. 2000; Strider et al. 1981; Janse et al. 2004). This race has a low temperature optimum (27 °C) and worldwide distribution, and appears to be mostly biovar 2A of RFLP group 26 (Cook and Sequeira 1994) and 27 (found in Chile and Colombia), or biovar 2T (sometimes also called 2N, found in tropical areas of South America). Race 4 infection is severe in *Zingiber* and *Morus*.

Another type of classification based on the pathogen's capacity to utilize or oxidize hexoses mannitol, dulcitol, and sorbitol classify *R. solanacearum* into four groups, while three groups are created on its utilization ability of disaccharides lactose, maltose, and cellobiose. Biovar I oxidizes hexose alcohols but not disaccharides, whereas biovar II oxidizes only disaccharides. Biovar III oxidizes both disaccharides and hexose alcohols, while biovar IV oxidizes only alcohols (Hayward 1964). Shrestha (1977) and Adhikari (1993) reported race 3 and the biovar II in potato from mid-to-high hill regions, and race 1 (biovar III) in eggplants, peppers, tomatoes, and marigolds from lowland areas of Nepal. Titatarn (1986) classified the bacterial wilt pathogen of potato as biovar III and IV from mid hills, and biovar II from high hills of Thailand.

Thus, there exist different biovars of *R. solanacearum,* which can adapt to a wide range of environments and may cause severe damage to crop production. There is an urgent need to understand the nature of interactions of pathogens to develop effective biocontrol agents to suppress its activity. The use of PGPR for suppressing bacterial wilt-causing pathogen is an important strategy of disease management and has a lot of scope to isolate, identify, and improve plant growth-promoting rhizobacterial strains specific to biovars of *R. solanacearum*.

## 34.5 Rhizospheric Soil

The entire soil mass along with roots is called rhizosphere (Hiltner 1904), and the bacteria present in the rhizosphere are called rhizobacteria, whose composition is influenced by root exudates (Kennedy 1999). Root exudates that contain different chemical compounds such as carbohydrates (sugars and oligosaccharides), vitamins, organic acids, flavonoids, enzymes, hormones, nucleotides, and volatile compounds (Prescott et al. 1999) influence the rhizobacterial community. The effect of rhizocomposition of the soil microbial population can be determined comparing the population density (colonies forming units) of the rhizospheric soil (*R*) and the bulk soil (*S*), for which the "*R*/*S* ratio" is employed (Atlas and Bartha 1997). The rhizospheric effect is more for bacteria, than that for fungi. Plant type and root exudates also affect the *R*/*S* ratio. The release of root exudates can be affected by various plant, soil, and environmental factors (Bowen and Rovira 1999; Whipps and Lynch 1986; Mc Cully 1999).

The colonization of individual microorganisms in the rhizosphere and its surviving capacity depends on the competence of the individual bacterial strains (Weller et al. 1988). Cellular activities like motility, chemotaxis, prototrophy, and the ability to sequester iron and oxygen (Lugtenberg et al. 2001) also contribute to the establishment of microorganisms in the rhizosphere. Other factors contributing colonization of bacteria are their ability to withstand physical stresses including heat, desiccation, and the presence of reactive oxygen molecules (Miller and Wood 1996; Sarniguet et al.1997; Schnider-Keel et al. 2001). The genes, which play a role in colonization, are *sss* (phenotypic variation; Sanchez-Contreras et al. 2002), *gacS, gacA, rpoS, algU* (global regulators; Sanchez-Contreras et al. 2002; Sarniguet et al. 1997; Schnider-Keel et al. 2001), *dsbA* (periplasmic disulfide-bond-forming protein), and *ptsP* (organic nitrogen utilization; Mavrodi et al. 2006). Moreover, plant genotype greatly influences the microbial community structure of the rhizosphere (Grayston et al. 1998; Smith and Goodman 1999; Smith et al. 1999; Weiland et al. 2001; Marschner et al. 2004). The presence of specific bacteria in the rhizosphere is decided by the type and nature of crop plants (Larkin et al. 1993; Smith and Goodman 1999, Raaijmakers and Weller 2001; Weiland et al. 2001; Marschner et al. 2004; Bergsma-Vlami et al. 2005). This proves the importance of the trait relationship between plant species and genotypes in the selection of specific groups of microbes in rhizospheric soil (Bergsma-Vlami et al. 2005; Landa et al. 2003; Mazzola et al. 2004; Okubara et al. 2004). Among the four *phlD*-positive bacterial genotypes (B, D, E, and L) present in the fields of wheat and barley affected by take-all decline (TAD) disease, genotype D was the most abundant in the soil in Washington State, USA (Raaijmakers and Weller 2001). However, depending upon the geographic location and the host plant, multiple genotypes of bacterial population have been reported in other crops like pea, flax, corn, and soybean (de Souza et al. 2003; Raaijmakers and Weller 2001; Landa et al. 2003). Plants also show a differential response to introduced and indigenous biocontrol agents (Landa et al. 2002; Maurhofer et al. 1995; Mazzola et al. 2004; Okubara et al. 2004). Hence, it may be inferred that microenvironment of the rhizospheric soil decides the diversity of microorganisms.

Microorganisms present in rhizospheric soil play an important role in promoting plant growth, directly as well as indirectly, by protecting them from disease-causing pathogens. There is a wide diversity of these beneficial bacteria associated with the rhizospheric soil. Proper agriculture practices should be followed to maintain a favorable microenvironment to build up beneficial plant growth-promoting microorganisms in the soil and sustain these communities for a long period. The sustainability of these colonized bacteria in the rhizosphere for a long duration is thus a key factor for a biocontrol agent to successfully protect the plant against soil-borne plant pathogens.

## 34.6 Plant Growth-Promoting Rhizobacteria (PGPR)

Naturally occurring bacteria in the rhizospheric soil beneficial to crop growth and development are often referred to as PGPR (Kloepper et al. 1989; Glick 1995). The major groups of PGPR in rhizospheric soil are *Azotobacter, Azospirillum, Pseudomonads, Acetobacter, Burkholderia, Enterobacter,* and *Bacillus* spp. (Brown 1974; Elmerich 1984; Kloepper et al. 1988, 1989; Bashan and Levanony 1990).

## 34.7 PGPR's Role in Relation to Plants

PGPR play a vital role in the promotion of plant growth, directly by fixing nitrogen and increasing phosphorus availability through solubilization of organic and inorganic phosphorus (Kim et al. 1998; El-Tarabily et al. 2008; Sabannavar and Lakshman 2009; Hariprasad et al. 2009). The phytohormones such as auxins, cytokinins, and gibberellins produced by the PGPR directly influence root and shoot growth (Asghar et al. 2002; Tanushree et al. 2007), and also indirectly by suppressing plant disease-causing organisms. A great diversity of biocontrol PGPR agents are reported in rhizospheric soil (Maria et al. 2005; Keel et al. 1996; Landa et al. 2002; Raaijmakers and Weller 2001; Bergsma-Vlami et al. 2005). Among the PGPRs, *Pseudomonas fluorescens* is widely studied as a biocontrol agent against seed and soil-borne plant pathogens. The control of bacterial wilt and bacterial blight of potato with inoculation of *P. fluorescens* has been recorded in field and laboratory trials (Ciampi-Panno et al. 1989). The strains, *P. fluorescens* effectively controls *Fusarium* wilt of radish (Leeman et al. 1995), bacterial wilt of tobacco (Liu et al. 1999) and cucumber (Liu et al. 1999), *Sclerospora graminicola* in pearl millet (Umesha et al. 1998), *Xanthomonas oryzae* pv. *oryzae* in rice (Vidhyasekaran et al. 2001), eucalyptus wilt (Ran et al. 2005), and bacterial wilt in chili (Umesha et al. 2005).

## 34.8 Mechanisms of Biological Control

Different mechanisms involved in biological control of plant pathogens by PGPR have been documented. They include a variety of cell wall-degrading enzymes, competition, plant ethylene levels, systemic acquired resistance, hydrogen cyanide (HCN), siderophore, and antibiotic production.

### 34.8.1 Antifungal Enzymes

Some PGPR strains produce antifungal enzymes like chitinase, b-1,3-glucanase, protease, and lipase that can lyse cell wall of fungi and prevent disease infection in plants (Chet and Inbar 1994). *Pseudomonas stutzeri* strain, which produces extracellular chitinase and laminarinase effectively lyse the cell walls of *Fusarium solani* mycelia and control root rot (Lim et al. 1991). Similarly, *Pseudomonas cepacia* enzymes damage fungal mycelia of *Rhizoctonia solani, Sclerotium rolfsii,* and *Pythium ultimum* (Fridlender et al. 1993). Furthermore, Chernin et al. (1995) showed chitinolytic activity in the PGPR strain *Enterobacter agglomerans* antagonistic to fungal pathogens *R. solani, Trichoderma harzianum,* and *Rhizobium meliloti* transformed with chitinase gene from the bacterium *Serratia marcescens*. These transformants displayed increased antifungal activity (Chet and Inbar 1994). Similarly, *P. fluorescens* transformed with chitinase gene was effective against the pathogen *R. solani* (Koby et al. 1994). *Cladosporium werneckii, P. cepacia,* and *P. solanacearum* hydrolyzes fusaric acid produced by *Fusarium,* which upon infection can damage the plants (Toyoda and Utsumi 1991).

### 34.8.2 Competition

The effective competitive nature of PGPR to utilize available nutrients efficiently and the ability to proliferate on the root surface plays an important role in disease suppression (Kloepper et al. 1988; O'Sullivan and O'Gara 1992). A limited surface area of leaf is invaded by phytopathogenic bacteria that cause disease in crop plants. This can be controlled by PGPR that can compete successfully with pathogens for these sites and thus often reduce disease incidence. The persistence and competition of a bacterium in the rhizosphere is influenced by a number of factors such as soil composition (Heijnen and van Elsas 1994; Bashan et al. 1995) and temperature (Sun et al. 1995; Chiarini et al. 1994). The rhizospheric soil contains a wide diversity of microorganisms, preferentiality those strains that are able to utilize an unusual carbon or nitrogen source such as an opine, a 1-aminocyclopropane carboxylate (ACC), or a xenobiotic compound (such as a herbicide or pesticide), and are able to proliferate and persist longer than other microorganisms in such rhizospheric soils (Jacobson et al. 1994). Effective suppression of *P. ultimum* by *Pseudomonas* spp. was dependent on the latter's ability to utilize seed exudates for the production of inhibitory compounds (Stephens et al. 1993). The saprophytic *Pseudomonas syringae* protected pears against gray mold and blue mold caused by *Botrytis cinerea* and *Penicillium expansum,* respectively, due to its high competitive nature (Janisiewicz and Marchi 1992).

### 34.8.3 Plant Ethylene Levels

In response to fungal infection, plants synthesize excess amount of ethylene, which leads to senescence, abscission of leaf or fruit, disease development, inhibition of growth and synthesis of antifungal enzyme (Abeles et al. 1992). Reports of this kind are available for various plants such as wheat plant infected by *Septoria nodorum* (Abeles et al. 1992), *Verticillium* wilt of tomato (Cronshaw and Pegg 1976), *B. cinerea* infection in roses, carnations, tomatoes, peppers, French beans, and cucumbers (Elad 1988). The PGPR *Pseudomonas putida* GR12–2producing enzyme ACC deaminase modulates the level of ethylene (Glick 1995; Jacobson et al. 1994) and promotes plant growth.

### 34.8.4 Systemic-Acquired Resistance

The infection of plant disease is also controlled by inoculating plants with PGPR strains, which induce plant defense system (systemic acquired resistance; van Peer et al. 1991; Tuzun and Kloepper 1994). *P. putida* and *S. marcescens*' inoculation protects cucumber plants from bacterial angular leaf spot disease caused by *P. syringae* pv. *Lachrymans* (Liu et al. 1995). The diacetyl phloroglucinol (DAPG)-induced systemic resistance (ISR) in *Arabidopsis thaliana* with root inoculation of *P. fluorescens* strain CHA0 protects the leaves from *Peronospora parasitica* infection (Iavicoli et al. 2003). DAPG significantly influences the net efflux (i.e., exudation) of amino acids from roots of plant species like alfalfa, maize, wheat, and medicago (Philips et al. 2004).

### 34.8.5 HCN

Some of the PGPR produce low-molecular-weight antifungal metabolites like HCN (Dowling and O' Gara 1994). *Pseudomonas* spp. produce antifungal metabolite substance HCN, which inhibits *Thielabiopsis basicola,* the causative agent of black root rot of tobacco (Voisard et al. 1989). An extensive role of HCN produced by PGPR in suppressing disease causing pathogen has been reported by various workers (Vansuyt et al. 2007; Chincholkar et al. 2007; Ramette et al. 2003).

### 34.8.6 Siderophore

Iron present in bound form in the soil is unavailable to the plants for direct assimilation. Some of the PGPR strains are able to produce a natural iron chelating agent siderophore, which has high affinity towards iron (Castignetti and Smarrelli 1986) making it readily available to the plants (Neilands and Leong 1986; Briat 1992). Thus, it affects the proliferation of phytopathogens and controls disease spread in plant (O' Sullivan and O' Gara 1992). The effect of siderophore on disease suppression depends upon the type of PGPR, the siderophore affinity to iron, specific crop plant, type of pathogen and soil composition. Bacteria that strongly control pathogen in vitro may not perform well under field conditions. Siderophore produced by *Pseudomonas putida* helps in suppression of *Fusarium oxysporum* and *Pythium* spp. infections in tomato and *Gaeumannomyces graminis* var. *tritici* infection in wheat plants (Vandenburgh and Gonzalez 1984; Buysens et al. 1994; Elsherif and Grossmann 1994).

### 34.8.7 Antibiotics

The primary mechanism involved in biocontrol of disease causing pathogens by PGPR is through the production of antibiotics. Bacteria synthesize variety of antibiotics, agrocin, 2,4-diacetylphloroglucinol (2,4-DAPG), herbicolin, oomycin, phenazines, pyoluteorin and pyrrolnitrin. They are highly specific in their action; a particular antibiotic acts on specific phytopathogens. The evidence of direct involvement of antibiotic production in PGPR-mediated disease-suppression has been reported by various workers (Carmi et al. 1994; Thomashow and Weller 1988; Haas et al. 1991; Keel et al. 1992; Pierson et al. 1994). The biocontrol agent *P. fluorescens* (BL915) produces the antibiotic pyrrolnitrin, which acts on *R. solani* and prevents damping-off of cotton plants. An increased production of pyoluteorin and 2,4-DAPG antibiotics by *P. fluorescens* protects cucumber plants against the disease caused by *P. ultimum* (Maurhofer et al. 1992; Schnider et al. 1994). The enhanced antibiotic production by PGPR strains through modification of global regulation of genes in *P. fluorescens* (CHAO) encoding the housekeeping sigma factor has improved the protection against *P. ultimum*-induced damping-off of cucumbers (Maurhofer et al. 1995; Schnider et al. 1995).

#### 34.8.7.1 Diacetylphloroglucinol (DAPG)

Antibiotic DAPG is a polyketide compound produced by bacteria. Among the antibiotics, it has received particular attention because of its broad-spectrum antiviral, antifungal, antibacterial, and antitumor activity and phytotoxic properties (Keel et al. 1992; Shanahan et al. 1992; Thomashow and Weller 1995; Bangera and Thomashow 1999; Isnansetyo et al. 2003; de Souza et al. 2003; Raaijmaker et al. 2002; Haas and Keel 2003). Production of 2,4-DAPG by *Pseudomonas* spp. is extensively studied (Shanahan et al. 1993, Bangera and Thomashow 1999; Schnider-Keel et al. 2000; Raaijmakers et al. 2002). This particular genotype of bacteria (*Pseudomonas* spp.) is associated with specific crop rhizosphere (Raaijmakers and Weller 2001; Landa et al. 2002; Okubara et al. 2004; Bergsma-Vlami et al. 2005). The DAPG-producing bacteria are highly rhizosphere competent (Mc Spadden et al. 2000; Raaijmakers and Weller 2001). This characteristic is an essential prerequisite of DAPG-producing organisms for successful biocontrol of plant diseases (Raaijmakers et al. 1995; Johnson 1994; Raaijmakers and Weller 1998; Lugtenberg et al. 2001). The competitive nature of DAPG-producing bacteria increases their ability to establish and maintain population densities sufficient to suppress disease in rhizosphere of pea and wheat (Landa et al. 2003; Raaijmakers and Weller 2001). DAPG-producing PGPR are found to be more competitive in nature compared to that of other bacteria (Mavrodi et al. 2006). Competence is important since direct correlation exists between the population size of the biocontrol strain and the level of biocontrol (Johnson 1994; Raaijmakers et al. 1995, 1998).

The role of DAPG produced by *Pseudomonas* strains has been reported in biological control of *Fusarium* crown and root rot, *Pythium* root rot (Rezzonico et al. 2007; Sharifi-Tehrani et al. 1998), black root rot of tobacco caused by *T. basicola* (Stutz et al. 1986; Keel et al. 1996; Ramette et al. 2003), and pea wilt caused by *F. oxysporum* (Landa et al. 2003). The dominating DAPG-producing *Pseudomonas* spp. in the rhizospheric soil of wheat effectively suppresses TAD disease of wheat caused by *G. graminis* var. *tritici* (Raaijmakers et al. 1997; Mc Spadden et al. 2000; Weller et al. 2002; de Souza et al. 2003).

#### 34.8.7.2 Diversity of 2,4-DAPG-Producing PGPR in Rhizospheric Soil

To study the diversity among the DAPG-producing rhizobacteria, various methods such as amplified ribosomal DNA restriction analysis (ARDRA; Keel et al. 1996; Mc Spadden et al. 2000; Picard et al. 2000), random amplified polymorphic DNA (RAPD; Raaijmakers and Weller 2001; Mavrodi et al. 2001), colony hybridization (Raaijmakers et al. 1997), direct characterization with whole-cell repetitive sequence-based PCR (rep-PCR; Mc Spadden et al. 2000; Landa et al. 2002), restriction fragment length polymorphism (RFLP; Mavrodi et al. 2001; Mc Spadden et al. 2001; Ramette et al. 2001; Wang et al. 2001), and BOX-PCR (McSpadden et al. 2001) analyses have been employed. Recently, the denaturing gradient gel electrophoresis (DGGE; Bergsma-Vlami et al. 2005) and allele-specific primer-based techniques (de la Fuente et al. 2006) have been developed to detect highly competitive and indigenous DAPG producers in natural environments. The diversity study of 200 *phlD*-positive strains collected across the world by BOX-PCR could distinguish 18 genotypes (A–Q and T; McSpadden et al. 2000; Landa et al. 2002) and were close to groups distinguished by RFLP (Mavrodi et al. 2001; McSpadden et al. 2001). Additional groups of DAPG-producing organisms (genotypes PspC, PspD, PspF, and PspZ) and (genotypes R and S) are given by Bergsma-Vlami et al. (2005). These techniques reveal the existence of diverse genotypic and phenotypic populations of DAPG-producing *P. fluorescens* in the environment (Thomashow and Weller 1995; Keel et al. 1996; McSpadden et al. 2000; Lee and Kim 2001; Weller et al. 2002; Isnansetyo et al. 2003). Identification of potential biocontrol agents capable of controlling disease-causing pathogens can be the core area of interest in maintaining sustainable agriculture production.

#### 34.8.7.3 Factors Influencing Production of DAPG

DAPG production can be modulated by a diverse array of abiotic and biotic factors, including carbon and nitrogen sources, metal ions and metabolites released by bacteria, fungi, and plants (Duffy and Défago 1999; Maurhofer et al. 2004; Notz et al. 2001). The impact of environmental factors on the production of DAPG has been studied both in vitro and in situ for a number of *Pseudomonas* strains (Duffy and Defago 1999; Notz et al. 2001; Shanahan et al. 1992). The mineral elements glucose, sucrose, fructose, and mannitol promote the production of DAPG in *P. fluorescens* (Shanahan et al. 1992). Zinc sulfate and ammonium molybdate supports 2,4-DAPG production; in contrast, organic phosphate and sorbose inhibit its synthesis (Shanahan et al. 1992). Fusaric acids produced by *F. oxysporum* inhibit the production of DAPG by *P. fluorescens* (Duffy and Defago 1999). The bacterial and plant metabolite salicylate inhibits DAPG production (Schnider-Keel et al. 2000). DAPG production is controlled by four global regulators in *Pseudomonas* spp. The two-component regulatory system composed of the sensor kinase GacS (formerly designated LemA) and the cognate response regulator GacA is required for the synthesis of 2,4-DAPG (Blumer et al. 1999; Corbel and Loper 1995; Laville et al. 1992; Whistler et al. 1998). Furthermore, its synthesis is influenced by the relative level of the housekeeping sigma factor RpoD and the stationary-phase and stress sigma factor RpoS. Mutational studies on *rpoS* gene of *P. fluorescens* indicated that overproduction of antibiotics 2,4-DAPG and pyoluteorin effectively suppressed certain root diseases (Maurhofer et al. 1992; Sarniguet et al. 1995; Whistler et al. 1998). The strain *P. fluorescens* Q2–87 contains gene *phlACBD* cluster involved in biosynthesis of 2, 4-DAPG and can gainfully be manipulated in PGPR to increase the production of 2,4-DAPG and its biocontrol efficiency.

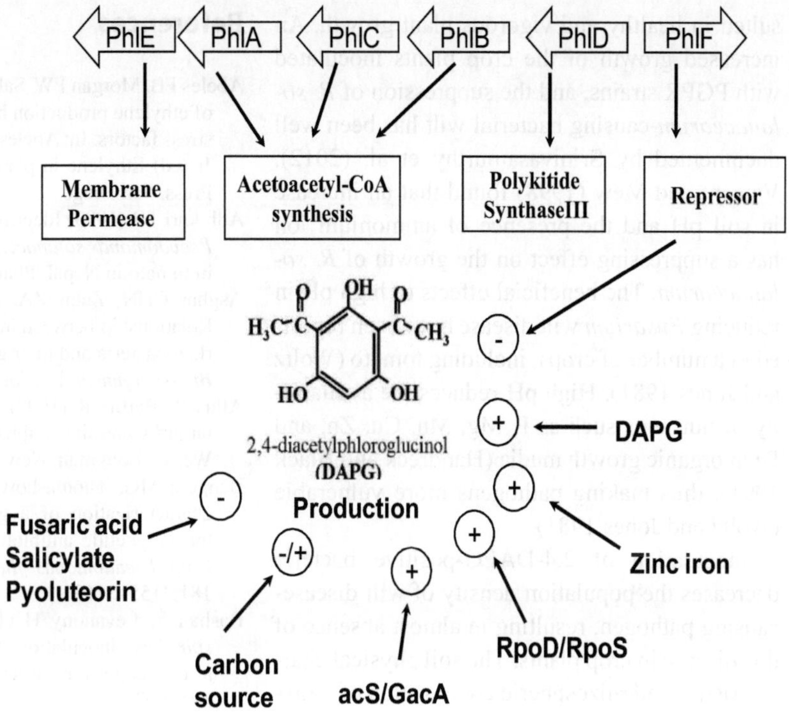


**Fig. 34.1** Factors of 2, 4-influencing biosynthesis DAPG

### 34.8.7.4 Biosynthesis and Regulation of 2,4-DAPG

Six genes (*PhlE, PhlA, PhlC, PhlB, PhlD,* and *PhlF*) are involved in the biosynthesis and regulation of 2, 4-DAPG (Fig. 34.1). Among them, four genes *PhlA, PhlC, PhlB,* and *PhlD* (identified in *P. fluorescens* strains Q2–87, F113, CHA0, and Pf-5) are transcribed as a single operon (Bangera and Thomashow 1999; Fenton et al. 1992; Bangera and Thomashow 1999; Schnider-Keel et al. 2000; Mavrodi et al. 2001). *phlD* acts as an initiation gene for the production of 2,4-DAPG; it encodes a polyketide synthase (PKS), which is involved in the production of monoacetylphloroglucinol (MAPG; Bangera and Thomashow, 1999). MAPG is converted to 2,4-DAPG by the action of *PhlA, PhlC,* and *PhlB* genes (Shanahan et al. 1993). The *phlE* gene encodes a putative transmembrane permease (Bangera and Thomashow 1999), which is involved in the exportation of 2,4-DAPG from the cell. The *phlF* gene encodes a pathway-specific transcriptional repressor of the 2,4-DAPG biosynthetic operon (Bangera and Thomashow 1999; Schnider-Keel et al. 2000).

## 34.9 Possible Action of 2,4-DAPG-Producing PGPR in Suppression of Wilt Disease Incidence in Crop Plants

Biocontrol potential of 2,4-DAPG in bacterial wilt disease suppression is reported by many early workers (Jian-Hua Guo 2004; Ran et al. 2005; Lemessa and Zeller 2007; Naser et al. 2008; Qing-Yun et al. 2009, Rashmi 2010). The evidence in favor of suppression of bacterial wilt disease of crop plants (caused by *R. solanacearum*) through antagonistic effects of inoculated 2,4-DAPG-positive bacteria came from high population density of 2,4-DAPG-positive bacteria in the crop rhizosphere. The suppression of pathogen in the rhizospheric soil has resulted in healthy and vigorous plant growth. An increased growth of the crop plants inoculated with PGPR strains, and the suppression of *R. solanacearum*-causing bacterial wilt has been well documented by Srinivasamurthy et al. (2012). Vincent and Mew (1998) found that an increase in soil pH and the presence of ammonium ion has a suppressing effect on the growth of *R. solanacearum*. The beneficial effects of high pH in reducing *Fusarium* wilt disease have been reported in a number of crops, including tomato (Woltz and Jones 1981). High pH reduces the availability of nutrients such as P, Mg, Mn, Cu, Zn, and Fe in organic growth media (Handreck and Black 1991), thus making pathogens more vulnerable (Woltz and Jones 1981).

Inoculation of 2,4-DAPG-positive bacteria decreases the population density of wilt disease-causing pathogen, resulting in almost absence of this disease in crop plants. The soil physical characteristics and rhizospheric composition of tomato crop plants support the activity of 2,4-DAPG-positive bacteria thereby increasing their population, which in turn suppresses *R. solanacearum.* Microbial competition of *P. fluorescens,* which plays a significant role in disease suppression, is decided by carbon source present in the media as well as on the role of root exudates (Lockwood 1988; Nelson 1990; Weller et al. 2002; Celia et al. 2004).

Rhizocompetence is a critical factor in the suppression of plant diseases (Dashti et al. 2000; Kamilov et al. 2005; Qing-Yun et al. 2009). Biocontrol efficiency of inoculated PGPR is influenced by the microenvironment of the rhizospheric soil (Weller et al. 2002; Celia et al. 2004). Rhizocompetence of *phlD*-positive bacteria and 2,4-DAPG are essential factors for the suppression of *R. solanacearum,* responsible for bacterial wilt disease. A direct relation exists between PGPR population density and prevention of wilt disease infection by *R. solanacearum* in tomatoes (Srinivasamurthy et al. 2012). Thus, the efficacy of biocontrol potential of *phlD*-positive bacteria can be exploited in future as a potential biocontrol measure in sustainable agriculture system to suppress bacterial wilt disease in a large number of crop plants.

## References


Abeles FB, Morgan PW, Saltveit ME Jr (1992) Regulation of ethylene production by internal, environmental and stress factors. In: Abeles FB, Morgan PW, Saltveit ME Jr (ed) Ethylene in plant biology, 2nd edn Academic Press, San Diego

Adhikari TB (1993) Identification of biovars and races of *Pseudomonas solanacearum* and sources of resistance in tomato in Nepal. Plant Dis 7:905–907

Asghar CHN, Zahir ZA, Arshad M, Khaliq A (2002) Relationship between in vitro production of auxins by rhizobacteria and their growth-promoting activities in *Brassica juncea* L. Biol Ferti Soils 35:231–237

Atlas R, Bartha R (1997) Microbial ecology, The Benjamin/ Cumming Publishing Company, Inc. Addison Wesley Longman, New York

Bangera MG, Thomashow S (1999) Identification and characterization of a gene cluster for synthesis of the polyketide antibiotic 2, 4-diacetylphloroglucinol from *Pseudomonas fluorescens* Q2-87. J Bacteriol 181:3155–3163

Bashan Y, Levanony H (1990) Current status of *Azospirillum* inoculation technology: *azospirillum* as a challenge for agriculture. Can J Microbiol 36: 591–608

Bashan Y, Puente ME, Rodriguez-Mendoza MN, Toledo G, Holguin G, Ferrera-Cerrato R, Pedrin S (1995) Survival of *Azospirillum brasilense* in the bulk soil and rhizosphere of 23 soil types. Appl Environ Microbiol 61:1938–1945

Bergsma-Vlami M, Prins ME, Raaijmakers JM (2005) Influence of plant species on population dynamics, genotypic diversity and antibiotic production in the rhizosphere of indigenous *Pseudomonas* spp. FEMS Microbiol Ecol 52:59–69

Blumer C, Heeb S, Pessi G, Haas D (1999) Global GacA-steered control of cyanide and exoprotease production in *Pseudomonas fluorescens* involves specific ribosome binding sites. Proc Natl Acad Sci U S A 96:14073–14078

Bowen GD, Rovira AD (1999) The rhizosphere and its management to improve plant growth. Adv Agronomy 66:1–102

Briat JF (1992) Iron assimilation and storage in prokaryotes. J Gen Microbiol 138:2475–2483

Brown ME (1974) Seed and root bacterization. Ann Rev Phytopath 12:181–197

Buddenhagen I, Sequeira L, Kelman A (1962) Designation of races of *Pseudomonas solanacearum* (Abstr.). Phytopath 52:726–728

Buysens S, Poppe J, Höfte M (1994) Role of siderophores in plant growth stimulation and antagonism by *Pseudomonas aeruginosa* 7NSK2. In: Ryder MH, Stephens PM, Bowen GD (eds) Improving plant productivity with rhizosphere bacteria. CSIRO, Adelaide

Carmi R, Carmeli S, Levy E, Gough FJ (1994) (+)-(*S*)-Dihydroaeruginoic acid, an inhibitor of *Septoria tritici* and other phytopathogenic fungi and bacteria, produced by *Pseudomonas fluorescens*. J Nat Prod 57:1200–1205

Castignetti D, Smarrelli J Jr (1986) Siderophores, the iron nutrition of plants, and nitrate reductase. FEBS Lett 209:147–151

Celia BM, Isabel T, José O, Julio CT, Manuel A (2004) Predictive factors for the suppression of *Fusarium* wilts of tomato in plant growth media. Phytopath 94:1094–1101

Chernin LS, Ismailov Z, Haran S, Chet I (1995) Chitinolytic *Enterobacter agglomerans* antagonistic to fungal plant pathogens. Appl and Environ Microbiol 61:1720–1726

Chet I, Inbar J (1994) Biological control of fungal pathogens. Appl Biochem Biotech 48:37–43

Chiarini L, Bevivino A, Tabacchioni S (1994) Factors affecting the competitive ability in rhizosphere colonization of plant-growth promoting strains of *Burkholderia cepacia*. In: Ryder MH, Stephens PM, Bowen GD (eds) Improving plant productivity with rhizosphere bacteria. CSIRO, Adelaide

Chincholkar SB, Chaudhari BL, Rane MR, Sarode PD (2007) Fungal phytopathogen suppression using siderophoregenic bio-inoculants. In: Chincholkar SB, Mukerji KG (eds) Biological control of plant diseases, current concepts. Haworth Press, U S A

Ciampi-Panno L, Fernandez C, Bustamante P, Andrade N, Ojeda SCA (1989) Biological control of bacterial wilt of potatoes caused by *Pseudomonas solanacearum*. American Potato J 66:315–332

Cook D, Sequeira L (1994) Strain differentiation of *Pseudomonas solanacearum* by molecular genetic methods, In: Bacterial wilt disease and its causative agent, *Pseudomonas solanacearum* (Ed. Hayward AC & Hartman GL), pp. 77–93, CAB International, Wallingford (GB)

Corbel N, Loper JE (1995) A global regulator of secondary metabolites in *Pseudomonas fluorescens* PF-5. J bacteriol 177:6230–6236

Cronshaw DK, Pegg GF (1976) Ethylene as a toxin synergist in *Verticillium* wilt of tomato. Physiol Plant Pathol 9:33–38

Dashti N, Prithiviraj B, Hynes RK, Smith DL (2000) Root and rhizosphere colonization of soybean glycine max (L) by plant growth promoting rhizobacteria at low root zone temperatures and under short season conditions. J Agron Crop Sci 185:15–22

de la Fuente L, Mavrodi DV, Landa BB, Thomashow LS, Weller DM (2006) *phlD*-based genetic diversity and detection of genotypes of 2,4-diacetylphloroglucinol-producing *Pseudomonas fluorescens*. FEMS Microbiol Ecol 56:64–78

de Souza JT, Weller DM, Raaijmakers JM (2003) Frequency, diversity and activity of 2,4-diacetylphloroglucinol-producing fluorescent *Pseudomonas* spp. in Dutch take-all decline soils. Phytopath 93:54–63

Dowling DN, O'Gara F (1994) Metabolites of *Pseudomonas* involved in the biocontrol of plant disease. Trends Biotech 12:133–141

Duffy BK, De'fago G (1999) Environmental factors modulating antibiotic and siderophore biosynthesis by *Pseudomonas fluorescens* biocontrol strains. Appl Environ Microbiol 65:2429–2438

Elad Y (1988) Involvement of ethylene in the disease caused by *Botrytis cinerea* on rose and carnation flowers and the possibility of control. Annual Appl Biol 113:589–598

Elmerich C (1984) Molecular biology and ecology of diazotrophs associated with non-leguminous plants. Biotech 2:967–978

Elsherif M, Grossmann F (1994) Comparative investigations on the antagonistic activity of fluorescent pseudomona's against *Graeumannomyces graminis* var. *tritici* in vitro and in vivo. Microbiol Res 149:371–377

El-Tarabily KA, Nassar AH, Sivasithamparam K (2008) Promotion of growth of bean (*Phaseolus vulgaris* L.) in a calcareous soil by a phosphate-solubilizing, rhizosphere competent isolate of *Micromonospora endolithica*. Appl Soil Ecol 39:161–171

Fenton AM, Stephens PM, Crowley J, O'Callaghan M, O'Gara F (1992) Exploitation of gene(s) involved in 2, 4-diacetylphloroglucinol biosynthesis to confer a new biocontrol capability to a *Pseudomonas* strain. Appl Environ Microbiol 58:3873–3878

Fridlender M, Inabar J, Chet I (1993) Biological control of soil borne plant pathogens by a β-13 glucanase producing *Pseudomonas cepacia*. Soil Biol Biochem 25:1211–1221

Glick BR (1995) The enhancement of plant growth by free-living bacteria. Can J Microbiol 41:109–117

Granada GA, Sequeira L (1983) Survival of *Pseudomonas solanacearum* in soil, rhizosphere and plant roots. Can J Microbiol 29:433–440

Grayston SJ, Wang S, Campbell CD, Edwards AC (1998) Selective influence of plant species on microbial diversity in the rhizosphere. Soil Biol Biochem 30:369–378

Haas D, Keel C (2003) Regulation of antibiotic production in root colonizing *Pseudomonas* spp. and relevance for biological control of plant disease. Annual Review of Phytopath 41:117–153

Haas D, Keel C, Laville J, Maurhofer M, Oberhansli T, Schnider U, Voisard C, Wuthrich B, Defago G (1991) Secondary metabolites of *Pseudomonas fluorescens* strain CHAO involved in the suppression of root diseases. In: Hennecke H, Verma DPS (eds) Advances of molecular genetics of plant-microbe interactions. Kluwer Academic Publishers, Dordrecht, pp. 450–456

Handreck K, Black N (1991) Growing media for ornamental plants and turf. New South Wales University Press, Kensington

Hariprasad P, Navya HM, Chandra NS, Niranjana SR (2009) Advantage of using PSIRB over PSRB and IRB to improve plant health of tomato. Biol Control 50:307–316

Hayward AC (1993) Bacterial Wilt. ACIAR proceedings, no. 45: Australian Centre for International Agricultural Research, Camera

Hayward AC (1964) Characteristics of *Pseudomonas solanacearum*. J Appl Bacteriol 27:365–277

He LY, Sequeira L, Kelman A (1983) Characteristic of *Pseudomonas solanacearum* from China. Plant Dis 67:1357–1361

Heijnen CE, van Elsas JD (1994) Metabolic activity of bacteria introduced into soil. In: Ryder MH, Stephens PM, Bowen GD (eds) Improving plant productivity with rhizosphere bacteria. CSIRO, Adelaide

Iavicoli A, Emmanuel B, Antony B, Jean-Pierre M (2003) Induced systemic resistance in *Arabidopsis thaliana* in response to root inoculation with *Pseudomonas fluorescens* CHA0. Mol Plant Microbe Interaction 16:851–858

Isnansetyo A, Cui L, Hiramatsu K, Kamei Y (2003) Antibacterial activity of 2,4-diacetylphloroglucinol produced by *Pseudomonas* sp. AMSN isolated from a marine alga, against vancomycin-resistant *Staphylococcus aureus*. Int J Antimicro Agents 22:545–547

Jacobson CB, Pasternak JJ, Glick BR (1994) Partial purification and characterization of ACC deaminase from the plant growth promoting rhizobacterium *Pseudomonas putida* GR12-2. Can J Microbiol 40:1019–1025

Janisiewicz WJ, Marchi A (1992) Control of storage rots on various pear cultivars with a saprophytic strain of *Pseudomonas syringae*. Plant Dis 76:555–560

Janse JD, van den Beld HE, Ephinstone J, Simpkins S, Tjou-Tam Sin NNA, van Vaerenbergh J (2004) Introduction to Europe of *Ralstonia solanacearum* biovar 2, race 3 in *Pelargonium zonale* cuttings. J Plant Path 87:147–155

Jian-Hua G, Hong-Ying Q, Ya-Hui G, Hong-Lian G, Long-Ying G, Li-Xin Z, Ping-Hua S (2004) Biocontrol of tomato wilt by plant growth-promoting rhizobacteria. Biol Control 29:66–72

Johnson KB (1994) Dose-response relationships and inundative biocontrol. Phytopath 84:780–784

Kamilov F, Validov S, Azarova T, Mulders I, Lugtenberg B (2005) Enrichment for enhanced competitive plant root tip colonizers selects for a new class of biocontrol bacteria. Environ Microbiol 7:1809–1817

Keel C, Schnider U, Maurhofer M, Voisard C, Laville J, Burger P, Wirthner P, Haas D, Défago G (1992) Suppression of root diseases of by *Pseudomonas fluorescens* CHA0: importance of the secondary metabolite 2,4-diacetylphloroglucinol. Mol Plant-Microbe Interact 5:4–13

Keel C, Weller DM, Natsch A, Défago G, Cook RJ, Thomashow LS (1996) Conservation of the 2, 4-diacetylphloroglucinol biosynthesis locus among fluorescent *Pseudomonas* strainsm from diverse geographic locations. App Environ Microbiol 62:552–563

Kelman A (1954) The relationships of pathogen city in *Pseudomonas solanacearum* to colony appearance on tetrazolium medium. Phytopath 64:693–695

Kennedy A (1999) The rhizosphere and spermosphere. In: Sylvia D, Fuhrmann J, Hartel P, Zuberer D (eds) Principles and applications of soil microbiology. Upper Saddle River, New Jersey p 389–407

Kim KY, Jordan D, McDonald GA (1998) *Enterobacter agglomerans*, Phosphate solubilizing bacteria and microbial activity in soil: effect of carbon source. Soil Sci Soc Am 30:995–1003

Kloepper JW, Lifshitz R, Zablotowicz RM (1989) Free living bacterial inocula for enhancing crop productivity. Trends Biotech 7:39–43

Kloepper JW, Lifshitz R, Schroth MN (1988) *Pseudomonas* inoculants to benefit plant protection, ISI Atlas of Science. Institute for Scientific Information, Philadelphia

Koby S, Schickler H, Chet I (1994) Oppenheim, AB, the chitinase encoding Tn7-based *chiA* gene endows *Pseudomonas fluorescens* with the capacity to control plant pathogens in soil. Gene 147:81–83

Kucharek T (1998) Bacterial wilt of row crops in Florida Circ-1207, University of Florida, IFAS, Cooperative Ext. Serv, Florida

Landa BB, Olga VM, Raaijmakers JM, Mc Spadden G, Thomashow LS, David MW (2002) Differential ability of genotypes of 2,4-Diacetylphloroglucinol-producing *Pseudomonas fluorescens* strains to colonize the roots of pea plants. Appl Envi Microbiol 68:3226–3237

Landa BB, Mavrodi DM, Thomashow LS, Weller DM (2003) Interactions between strains of 2, 4-diacetylphloroglucinol-producing *Pseudomonas fluorescens* in the rhizosphere of wheat. Phytopath 93:982–994

Larkin RP, Hopkins DL, Martin FN (1993) Effect of successive watermelon plantings on *Fusarium oxysporum* and other microorganisms in soils suppressive and conducive to *Fusarium* wilt of watermelon. Phytopathology 83:1097–1105

Laville J, Voisard C, Keel C, Maurhofer M, De´fago G, Haas D (1992) Global, stationary-phase control in *Pseudomonas fluorescens* mediating antibiotic synthesis and suppression of black root rot of tobacco. Proc Natl Acad Sci U S A 89:1562–1566

Lee ET, Kim SD (2001) An antifungal substance, 2,4-diacetylphloroglucinol, produced from antagonistic bacterium *Pseudomonas fluorescens* 2112 against *Phytophthora capsici*. Korean J Appl Microbiol Biotech 29:37–42

Leeman M, Van Pelt JA, Den Ouden FM, Heinsbroek M, Bakker PAHM, Schippers B (1995) Induction of systemic resistance against *Fusarium* wilt of radish by lipopolysaccharides of *Pseudomonas fluorescens*. Phytopath 85:1021–1027

Lemessa F, Zeller W (2007) Screening rhizobacteria for biological control of *R. solanacearum* in Ethopia. Biol control 42:336–344

Lim HS, Kim YS, Kim SD (1991) *Pseudomonas stutzeri* YPL-1 genetic transformation and anti-fungal mechanism against *Fusarium solani*, an agent of plant root rot. Appl Environ Microbiol 57:510–516

Liu L, Kloepper JW, Tuzun S (1995) Induction of systemic resistance in cucumber against bacterial angualar leaf spot by plant growth-promoting rhizobacteria. Phytopath 85:843–847

Liu QG, Li Z, Tang Z, Zeng XM (1999) Control of tobacco bacterial wilt with antagonistic bacteria and soil amendments. Chi J Biol Control 15:94–95

Lockwood JL (1988) Evolution of concepts associated with soil borne plant pathogens. Annu Rev Phytopath 26:3–121

Lugtenberg BJ, Dekkers L, Bloemberg GV (2001) Molecular determinants of rhizosphere colonization by *Pseudomonas*. Annu Rev Phytopath 39:461–490

Maria B-V, Prins ME, Raaijmakers JM (2005) Influence of plant species on population dynamics, genotypic diversity and antibiotic production in the rhizosphere by indigenous *Pseudomonas* spp. FEMS Microbiol Ecol 52:59–69

Marschner P, Crowley D, Yang CH (2004) Development of specific rhizosphere bacterial communities in relation to plant species, nutrition and soil type. Plant Soil 261:199–208

Maurhofer M, Keel C, Schnider U, Voisard C, Haas D, De′fago G (1992) Influence of enhanced antibiotic production in *Pseudomonas fluorescens* strain CHA0 on its disease suppressive capacity. Phytopath 82: 190–195

Maurhofer M, Keel C, Haas D, Defago G (1995) Influence of plant species on disease suppression by *Pseudomonas fluorescens* strain CHA0 with enhanced antibiotic production. Plant Pathol 44:40–50

Maurhofer M, Baehler E, Notz R, Martinez V, Keel C (2004) Cross talk between 2,4-diacetylophloroglucinol-producing biocontrol pseudomonads on wheat roots. Appl Environ Microbiol 70:1990–1998

Mavrodi OV, McSpadden Gardener BB, Mavrodi DV, Bonsall RF, Weller DM, Thomashow LS (2001) Genetic diversity of *phlD* from 2,4-diacetylphloroglucinol-producing fluorescent *Pseudomonas* spp. Phytopath 91:35–43

Mavrodi OV, Mavrodi DV, Weller DM, Thomashow LS (2006) Role of *ptsP*, *orfT*, and *sss* recombinase genes in root colonization by *Pseudomonas fluorescens* Q8r1-96. Appl Environ Microbiol 72:7111–7122

Mazzola M, Funnell DL, Raaijmakers JM (2004) Wheat cultivar-specific selection of 2, 4-diacetylphloroglucinol-producing fluorescent *Pseudomonas* species from resident soil populations. Microbial Ecol 48:338–348

Mc Cully M (1999) Roots in soil: unearthing the complexities of roots and their rhizospheres. Annu Rev Plant Physiol Plant Mol Biol 50:695–718

Mc Spadden-Gardener BB, Schroeder KL, Kalloger SE, Raaijmakers JM, Thomashow LS, Weller DM (2000) Genotypic and phenotypic diversity of *phlD*-containing *Pseudomonas* strains isolated from the rhizosphere of wheat. Appl Environ Microbiol 66:1939–1946

Mc Spadden-Gardener BB, Mavrodi DV, Thomashow LS, Weller DM (2001) A rapid polymerase chain reaction-based assay characterizing rhizosphere populations of 2,4-diacetylphloroglucinol-producing bacteria. Phytopath 91:44–54

Miller KJ, Wood JM (1996) Osmo adaptation by rhizosphere bacteria. Ann Rev Microbiol 50:101–136

Mishra A, Mishra SK, Karmakar SK, Sarangi CR, Sahu GS (1995) Assessment of yield loss due to wilting in some popular tomato cultivars. Environ Ecol 13:287–290

Naser A, Chemeda F, Yaynu H (2008) Evaluation of rhizosphere antagonist for their potential to bioprotect potato (*Solanaum tuberosum*) against bacteria wilt (*Ralstonia solanacearum*). Biol control 47:282–288

Neilands JB, Leong SA (1986) Siderophores in relation to plant growth and disease. Ann Rev Plant Physiol 37:187–208

Nelson EB (1990) Exudates molecules initiating fungal responses to seeds and roots. Pt Soil 129:61–73

Notz R, Maurhofer M, Schnider-Keel U, Duffy B, Haas D, Defago G (2001) Biotic factors affecting expression of the 2, 4-diacetylphloroglucinol biosynthesis gene *phlA* in *Pseudomonas fluorescens* biocontrol strain CHA0 in the rhizosphere. Phytopath 91:873–881

O'Sullivan DJ, O'Gara F (1992) Traits of fluorescent *Pseudomonas* spp. Involved in suppression of plant root pathogens. Microbiol Rev 56:662–676

Okubara PA, Kornoely JP, Landa BB (2004) Rhizosphere colonization of hexaploid wheat by *Pseudomonas fluorescens* Q8r1-96 and Q2-87 is cultivar-variable and associated with changes in gross root morphology. Biol Control 30:392–403

Philips DA, Fox TC, King MD, Bhuvaneswari TV, Teuber LR (2004) Microbial products trigger amino acid exudation from plant roots. Plant Physiol 136:2887–2894

Picard C, Di Cello F, Ventura M, Fani R, Guckert A (2000) Frequency and biodiversity of 2,4-diacetylphloroglucinol-producing bacteria isolated from the maize rhizosphere at different stages of plant growth. Appl Environ Microbiol 66:948–955

Pierson LS, Keppenne VD, Wood DW (1994) Phenazine antibiotic biosynthesis in *Pseudomonas aureofaciens* 30-84 is regulated by PhzR in response to cell density. J Bacteriol 176:3966–3974

Pradhanang PM, Elphinstone JG, Fox RTV (2000) Sensitive detection of *Ralstonia solanacearum* in soil: a comparison of different detection techniques. Plant Pathol 49:414–422

Prescott L, Harley J, Klein DA (1999) Microbiology. McGraw-Hill, Boston

Qing-Yun X, Yu C, Shi-Mo Li, Li-Feng C, Guo-Chun D, Da-Wei G, Jian-Hua G (2009) Evaluation of the strains of *Acinetobacter* and *Enterobacter* as potential biocontrol agents against *Ralstonia* wilt of tomato. Biol Control 48:252–258

Raaijmakers JM, Weller DM (1998) Natural plant protection by 2, 4-diacetylphloroglucinol-producing *Pseudomonas* spp. in take-all decline soils. Mol Plant-Microbe Interact 11:144–152

Raaijmakers JM, Weller DM (2001) Exploiting genotypic diversity of 2, 4-diacetylphloroglucinol-producing *Pseudomonas* spp.: characterization of superior root colonizing *P. fluorescens* strain Q8r1 M-96. Appl Environ Microbiol 63:881–887

Raaijmakers JM, Leeman M, van Oorschot M, van der Sluis I, Schippers B, Bakker PAHM (1995) Dose-response relationships in biological control of *Fusarium* wilt of radish by *Pseudomonas* spp. Phytopath 85:1075–1081

Raaijmakers JM, Weller DM, Thomashow LS (1997) Frequency of antibiotic producing *Pseudomonas* spp. in natural environments. Appl Environ Microbiol 63:881–887

Raaijmakers JM, Vlami M, Souza JT (2002) Antibiotic production by bacterial biocontrol agents. Antonie Van Leeuwenhoek 81:537–547

Ramette A, Frapolli M, Defago G, Monenne Y (2003) Phylogeny of HCN synthase-encoding *hcnBC* genes in biocontrol fluorescent pseudomonas and its relationship with host plant species and HCN synthesis ability. Mol Plant Microbe Interact 16:525–535

Ran LX, Liu CY, Wu GJ, van Loon LC, Bakker PAHM (2005) Suppression of bacterial wilt in *Eucalyptus urophylla* by fluorescent *Pseudomonas* spp in China. Biol Control 32:111–120

Ramette A, Moënne-Loccoz Y, Défago G (2001) Polymorphism of the polyketide synthase gene phlD in biocontrol fluorescent pseudomonads producing 2,4-diacetylphloroglucinol and comparison of phlD with plant polyketide synthases. Molecular Plant-Microbe Interactions 14:639–652

Rashmi S, Abdul K, Singh US, Sharma AK (2010) Evaluation of arbuscular mycorrhizal fungus, fluorescent Pseudomonas and *Trichoderma harzianum* formulation against *Fusarium oxysporum* f. sp. *Lycopersici* for the management of tomato wilt. Biol Control 53:24–31

Rezzonico F, Zala M, Keel C, Duffy B, Moenne-Loccoz Y, Defago G (2007) Is the ability of biocontrol fluorescent pseudomonads to produce the antifungal metabolite 2,4-diacetylphloroglucinol really synonymous with higher plant protection. New Phytol 173:861–872

Rolfs PH (1898) Disease of tomato, Fla. Agri. Exp. Sta. Bul. 47:128–136

Sabannavar SJ, Lakshman HC (1898) Effect of rock phosphate solubilization using mycorrhizal fungi and phosphobacteria on two high yielding varieties of *Sesamum indicum* L. World J Agri Sci 5:470–479

Sanchez-Contreras MM, Martín MV, Gara FO, Bonilla I, Rivilla R (2002) Phenotypic selection and phase variation occur during alfalfa root colonization by *Pseudomonas fluorescens* F113. J Bacteriol 184:587–1596

Sarniguet A, Kraus J, Henkels MD, Muehlchen AM, Loper JE (1995) The sigma factor s affects antibiotic production and biological control activity of *Pseudomonas fluorescens* Pf-5. Proc Natl Acad Sci U S A 92:12255–12259

Sarniguet A, Kraus J, Henkels MD, Muehlchen AM, Loper JE (1997) The sigma factor σs affects antibiotic production by Pseudomonas fluorescens Pf-5. Proc Natl Acad Sci USA 92:12255–12259

Schnider U, Blumer C, Troxler J, Défago G, Haas D (1994) Over production of the antibiotics 2,4-diacetylphloroglucinol and pyoluteorin in *Pseudomonas fluorescens* strain CHAO. In: Ryder MH, Stephens PM, Bowen GD (eds) Improving plant productivity with rhizosphere bacteria. CSIRO, Adelaide

Schnider U, Keel C, Blumer C, Troxler J, De´fago G, Haas D (1995) Amplification of the house-keeping sigma factor in *Pseudomonas fluorescens* CHA0 enhances antibiotic production and improves biocontrol abilities. J Bacteriol 177:5387–5392

Schnider-Keel U, Seematter A, Maurhofer M, Blumer C, Duffy B, Gigot-Bonnefoy C, Reimmann C, Notz R, Défago G, Haas D, Keel C (2000) Autoinduction of 2,4-diacetylphloroglucinol biosynthesis in the biocontrol agent *Pseudomonas fluorescens* CHA0 and repression by the bacterial metabolites salicylate and pyoluteorin. J Bacteriol 182:1215–1225

Schnider-Keel U, Lejbolle KB, Bachler E, Haas D, Keel C (2001) The sigma factor Algu (Algt) controls exopolysaccharide production and tolerance towards desiccation and osmotic stress in the biocontrol agent *Pseudomonas fluorescens* CHAO. Appl Environ Microbiol 67:5683–5693

Shakya DD (1993) Occurrence of *Pseudomonas solanacearum* in tomato seeds imported into Nepal. Bact wilt ACIAR 45:371–372

Shanahan P, O'Sullivan DJ, Simpson P, Glennon JD, O'Gara F (1992) Isolation of 2,4-diacetylphloroglucinol from a fluorescent pseudomonad and investigation of physiological parameters influencing its production. Appl Environ Microbiol 58:353–358

Shanahan P, Glennon JD, Crowley JJ, Donnelly DF, O'Gara F (1993) Liquid chromatographic assay for microbially derived phloroglucinol antibiotics for establishing the biosynthetic route to production, and the factors affecting their regulation. Anal Chim Acta 272:271–277

Sharifi-Tehrani A, Zala M, Natsch A, Moenne-Loccoz Y, Defago G (1998) Biocontrol of soil borne fungal plant diseases by 2,4-DAP producing fluorescent pseudomonads with different restriction profiles of amplified 16SDNA. Euro J Plant Pathol 104:631–643

Shrestha SK (1977) Preliminary study on brown rot of potato in Nepal. Nepal J Agri 12:11–21

Smith EF (1897) Description of *Bacillus phaseoli* n. sp. Bot. Ghz 24:192

Smith KP, Goodman RM (1999) Host variation for interactions with beneficial plant-associated microbes. Ann Rev Phytopath 37:473–491

Smith KP, Handelsman J, Goodman RM (1999) Genetic basis in plants for interactions with disease- suppressive bacteria. Proc Natl Acad Sci USA 96:4786–4790

Srinivasamurthy R, Rai AK, Jaat RS, Singh A, Rai R (2012) Isolation and screening of $phlD^+$ plant growth promoting rhizobacteria antagonistic to *Ralstonia solanacearum.* World J Microbiol Biotech 28: 1681–1690

Stephens PM, Crowley JJ, O'Connell C (1993) Selection of pseudomonad strains inhibiting *Pythium ultimum* on sugarbeet seeds in soil. Soil Biol Bioche 25: 1283–1288

Strider DL, Jones RK, Haygood RA (1981) Southern bacterial wilt of geranium caused by *Pseudomonas solanacearum*. Plant Dis 65:52–53

Stutz EW, Défago G, Kern H (1986) Naturally occurring fluorescent *pseudomonas* involved in suppression of black root rot of tobacco. Phytopath 76:181–185

Sun X, Griffith M, Pasternak JJ, Glick BR (1995) Low temperature growth, freezing survival and production of antifreeze protein by the plant growth promoting rhizobacterium *Pseudomonas putida* GR12-2. Can J Microbiol 41:776–784

Tanushree M, Goswami C, Talukdhar NC (2007) Characterization and screening of beneficial bacteria obtained on king's *B agar* from tea rhizosphere. Indian J Biotech 6:490–494

Thomashow LS, Weller DM (1995) Current concepts in the use of introduced bacteria for biological disease control. In: Stacey G, Keen N (ed.) Plant-microbe interactions, vol 1. Chapman and Hall, New York

Thomashow LS, Weller DM (1988) Role of a phenazine antibiotic from *Pseudomonas fluorescens* in biological control of *Gaeumannomyces graminis* var. *Tritici.* J Bacteriol 170:3499–3508

Titatarn V (1986) Bacterial wilt in Thailand. In: Persley GL (ed) Bacterial wilt disease in Asia and the South Pacific, Proceeding of an International Workshop held at PCARD, Los Banos, The Philippines

Toyoda H, Utsumi R (1991) Method for the prevention of *Fusarium* diseases and microorganisms used for the same, U.S. Patent # 4, 988, 586

Toyota K, Kimura M (2000) Suppression of *Ralstonia solanacearum* in soil following colonization by other strains of *R. Solanacearum.* Soil Sci Plant Nutr 46:449–459

Tuzun S, Kloepper J (1994) Induced systemic resistance by plant growth promoting rhizobacteria. In: Ryder MH, Stephens PM, Bowen GD (eds) Improving plant productivity with rhizosphere bacteria. CSIRO, Adelaide, pp 104–109

Umesha S, Shylaja MD, Sudheer AS, Krishnappa M, Shetty HS (1998) Biocontrol of downy mildew disease of pearl millet using *Pseudomonas fluorescens.* Crop Prot 17:387–392

Umesha S, Kavitha R, Shetty HS (2005) Transmission of seedborne infection of chilli by *Burkholderia solanacearum* and effect of biological seed treatment on disease incidence. Arch Phytopathol Plant Prot 38:281–293

Vandenburgh PA, Gonzalez CF (1984) Method for protecting the growth of plants employing mutant siderophore producing strains of *Pseudomonas putida*, US Patent #4, 479, 936

van Peer R, Niemann GJ, Schippers B (1991) Induced resistance and phytoalexin accumulation in biological control of *Fusarium* wilt of carnation by *Pseudomonas* spp strain WCS417r. Phytopath 81:728–734

Vansuyt G, Robin A, Briat JF, Curie C, Lemanceau P (2007) Iron acquisition from Fe pyoverdine by *Arabidopsis thaliana*. Mol Plant Microbe Interact 20:441–447

Vidhyasekaran P, Kamala N, Ramanathan A, Rajappan K, Paranidharan V, Velazhahan R (2001) Induction of systemic resistance by *Pseudomonas fluorescens* Pfl against *Xanthomonas oryzae pvoryzae* in rice leaves. Phytoparasitica 29:155–166

Vincent VM, Mew TW (1998) Effect of a soil amendment on the survival of *Ralstonia solanacearum* in different soils. Phytopathology 88:300–305

Voisard C, Keel C, Haas D, Defago G (1989) Cyanide production by *Pseudomonas fluorescens* helps suppress black root rot of tobacco under gnotobiotic conditions. EMBO J 8:351–358

Wang C, Ramette A, Punjasamarnwong P, Zala M, Natsch A, Moënne-Loccoz Y, Défago G (2001) Cosmopolitan distribution of *phlD*-containing dicotyledonous crop-associated biocontrol pseudomonads of worldwide origin. FEMS Microbiol Ecol 1267:1–12

Weiland G, Neumann R, Backhaus H (2001) Variation of microbial communities in soil, rhizosphere and rhizoplane in response to crop species, soil type and crop development. Appl Environ Microbiol 67:5849–5854

Weller DM (1988) Biological control of soil borne plant pathogens in the rhizosphere with bacteria. Ann Rev Phytopath 26:379–407

Weller DM, Raaijmakers JM, Mc S, Gardener BB, Thomashow LS (2002) Microbial populations responsible for specific soil suppressiveness to plant pathogens. Ann Rev Phytopath 40:309–348

Whipps JM (1997) Developments in the biocontrol of soil borne plant pathogens. Adv Bot Res 26:1–134

Whipps JM, Lynch JM (1986) The influence of the rhizosphere on crop productivity. Adv Microbial Ecol 9:187–244

Whistler CA, Corbell NA, Sarniguet A, Ream W, Loper JE (1998) The two-component regulators GacS and GacA influence accumulation of the stationary-phase sigma factors and the stress response in *Pseudomonas fluorescens* Pf-5. J Bacteriol 180:6635–6641

Woltz SS, Jones JP (1981) Nutritional requirements of Fusarium oxysporum: basis for a disease control system. In: Nelson PE, Toussoun TA, Cook RJ (eds) *Fusarium*: diseases, biology, and taxonomy. Pennsylvania State University Press, University Park p 340–349

Yabuuchi E, Kosako Y, Oyaizu H, Yano I, Hotta H, Hashimoto Y, Ezaki T, Arakawa M (1992) Proposal of Burkholderia gen. nov. and transfer of seven species of the genus Pseudomonas homology group II to the new genus, with the type species Burkholderia cepacia (Palleroni & Holmes 1981) comb. nov. Microbiol Immunol 36:1251–1275

# Part III

# Microbial Function and Biotechnology

# Bio-Fungicides: The Best Alternative for Sustainable Food Security and Ecosystem

35

C. Rettinassababady and C. Jeyalakshmi

**Abstract**

Plant diseases cause estimated yield reductions of almost 20 % in agricultural and horticultural crops worldwide. The discovery of synthetic fungicides has contributed greatly to the increase of food production by controlling diseases. However, the use of these synthetic fungicides during the last three decades has raised a number of ecological problems. According to CGIAR, sustainable agriculture is the successful management of resources to satisfy the changing human needs, while maintaining the quality of environment and conserving natural resources. The obvious choice for that will be the use of bio-fungicides to combat plant diseases, as they are relatively safe, biodegradable and eco-friendly. Bio-fungicides exert different mechanisms like antibiosis, competition, mycoparasitism and induce systemic resistance to encounter harmful plant pathogens. The simplified registration system of biopesticides in recent years allows for commercial pilot production of bio-fungicides in many countries. Accordingly, many commercial formulations are available in the market. Development of microbial consortia will improve disease control through synergy in signal cross talk between microbes. Genetically improved antagonistic microorganisms tend to increase their effectiveness as bio-fungicides with an enhanced antifungal activity. Application of genetically manipulated bio-fungicides in modern agriculture offers tremendous opportunities for expanding food production, reducing risks in food production, improving environmental protection, strengthening food marketing, enhancing the livelihood of farmers and ensuring food security in developing countries.

**Keywords**

Bio-fungicides · Plant diseases · Food production · Induced-systemic resistance (ISR) · Antibiosis · Mycoparasitism

C. Rettinassababady (✉) · C. Jeyalakshmi
Department of Plant Pathology, Pandit Jawaharlal Nehru College of Agriculture & Research Institute, Karaikal 609603
U.T. of Puducherry, India
e-mail: crsvaisu@yahoo.ac.in

## 35.1 Introduction

The green revolution since the early 1960s has resulted in a global food supply sufficient to provide adequate energy and protein for all. The

R. N. Kharwar et al. (eds.), *Microbial Diversity and Biotechnology in Food Security*, DOI 10.1007/978-81-322-1801-2_35, © Springer India 2014

introduction of high-yielding varieties together with increasing application of agrochemicals increased the productivity of land with a concomitant increase in the proportion lost to pests and diseases in India and other developing Asian countries. Plant diseases are estimated to cause yield reductions of almost 20% in the principal food and cash crops worldwide. Of the over 100,000 described species of fungi in the world, approximately 20,000 produce one or more diseases in various plants. Crop losses due to the diseases are very heavy and have caused many famines reported in the human history, viz. Irish famine (1845) due to late blight of potato and Bengal famine (1943) due to brown spot of rice. Over the past 150 years, much has been learnt about the control of plant diseases, and several complementary approaches such as cultural methods, biological methods, chemical methods and regulatory measures were developed for their control. Depending on the crop, the disease and the availability of control methods, a different set of approach is employed (Thind 2012). Despite the use of available means of plant protection, about one third of crop produces are lost due to pests and diseases.

Fungicides are considered to be the second line of defence in plant disease control programmes after disease resistance to increase the food production. The benefit of fungicides used in the US agriculture is reported to boost farm income by nearly US$ 13 billion annually (Gianessi and Reigner 2006). Agrochemicals' market in India is worth about 4,800 crore, out of which, fungicides account for nearly 12% of the sales (Thind 2005). It has been estimated that about 2.5 million t of pesticides are used on crops each year and the worldwide damage caused by pesticides reaches US$ 100 billion annually (Koul 2008). The use of synthetic fungicides in crop disease management has raised a number of ecological problems. The most serious problem is the development of fungicide-resistant strains leading to the failure of disease control (Thind 2008). The overuse of systemic fungicides for managing diverse fungal diseases in crops leads to residue build-up in harvested produce. In the annual EU report by European Food Safety Authority (EFSA), vegetables and fruits of 27 countries were surveyed for pesticide contamination, the results highlighted that dithiocarbamates are among the most common residual contaminants (EFSA 2009). Detection of residues of carbendazim and isoprothiolane above the maximum residue levels (MRLs) in basmati rice consignments exported from India to Germany in 2010 and that of tricyclazole in the consignments exported to the USA in 2011 has sent warning signals on fungicide residual toxicity among the agriculturists and traders in India (Thind 2012). In addition to residual toxicity, fungicides can adversely affect native soil microflora. For example, triadimefon and propiconazole, on repeated use, could alter the soil microbial ecology for a long period of time (Jui–Hung et al. 2009). Soil microbial populations including various fungi, denitrifying bacteria and aerobic diazotrophs are also greatly affected by the irrational application of mancozeb (Pozo et al. 1994). Benomyl, fenpropimorph, fenhexamid, mepronil, furalaxyl and thiram also reported to affect symbiotic relationship between the plants and mycorrhizae. Elemental sulphur, benomyl and formulations of difenoconazole are toxic to non-target parasitic and predatory arthropods (Thind 2012).

The environmental pollution caused by excessive use and misuse of agrochemicals, as well as fearmongering by some opponents of pesticides, has led to considerable changes in people's attitude towards the use of pesticides in agriculture. Today, there are strict regulations on chemical pesticide use, and there is political pressure to remove the most hazardous chemicals from the market (Pal and Gardener 2006). Some of the fungicides have either been banned or withdrawn (e.g. ethyl mercury chloride, phenyl mercury acetate, quintozene (PCNB), pentachlorophenol, ferbam and nickel chloride) or their use has been made restricted (e.g. methoxy ethyl mercury chloride, captafol) in India by the Central Insecticides Board (CIB) due to certain ill effects on environment and ecology (Anonymous 2006). According to CGIAR, sustainable agriculture is the successful management of resources to satisfy the changing human needs, while maintaining the quality of environment and conserving natural resources. Increasing consciousness about conservation of environment as well as health hazards associated with agrochemicals and consumers' preference towards safe and hazard-free food have forced the scientist to divert their attention towards exploring the potential of beneficial microbes as an alternative strategy to mitigate the problems emanating from the use of synthetic pesticides (Biswas 2011).

Bio-fungicides of microbial origin seem to be a potentially powerful alternative method to synthetic fungicides. The rich diversity, complexity of interactions and numerous metabolic pathways makes the microbes an amazing resource for biological activity (Mitchell et al. 2008). Over the past 30 years, several microorganisms have been described, characterized and tested for their use as biocontrol agents against plant pathogens. A wide range of biocontrol agents have been developed as commercial bio-fungicides and used as an alternative for managing seed-, soil-, and air-borne fungal diseases as well as post-harvest disease problems especially in crops which are stored under controlled temperature and high relative humidity (Janisiewicz and Korsten 2002; Fravel 2005; Nakkeeran et al. 2005; Kim and Hwang 2007; Kaewchai et al. 2009). Of which, 90% of bio-fungicides are being formulated using different strains of *Trichoderma* such as *Trichoderma harzianum, Trichoderma virens* and *Trichoderma viride* (Benítez et al. 2004). In addition to this, several species of plant growth-promoting rhizobacteria (PGPR) such as *Bacillus* spp. and *Pseudomonas* spp. are being commercialized as bio-fungicides. Bio-fungicides formulated with PGPR may promote plant growth directly in healthy plants or indirectly by controlling phytopathogens in different crops (Kloepper 1993).

In India, the CIB has greatly helped to spread and encourage the use of biopesticides in crop protection. A total of 65 antifungal products have been registered in India including four antifungal/antibacterial antibiotics and four biocontrol agents as on 20 January 2012 (CIB 2012). As the bio-fungicides of microbial origin are eco-friendly and environmentally safe, it is expected that they will find their greatest commercial application in protection of stored commodities and high-value row crops within organic food production systems. Further, higher commodity prices for organically produced products, changing consumer preferences towards the use of natural products and regulations restricting the use of fungicides improve the chances of successful commercialization of bio-fungicides in the markets. In India, more than 65% of the population is still living in villages, and more than 85% of the rural people are dependent on agriculture for their livelihood. More than 75% being small and marginal holders, most of their earnings are utilized to ensure food security. Therefore, it is essential to promote agricultural technologies that enable the rural poor to earn sustainable livelihood, while enhancing food production. Since the bio-fungicides seem to improve global food availability by reducing crop losses due to diseases in an economically viable and ecologically safe way, they are considered as the best alternative for sustainable food security and ecosystem.

## 35.2 Definitions

The terms "biological control" and its abbreviated synonym "biocontrol" have been used in different fields of biology, most notably entomology and plant pathology. In plant pathology, the term applies to the use of microbial antagonists to suppress diseases as well as to the use of host-specific pathogens to control weed populations. In general, an organism that suppresses pests or pathogens is referred to as a biological control agent (BCA). More broadly, the term biocontrol has been applied to the use of natural products extracted or fermented from various sources. Bio-fungicides are antagonistic fungi/bacteria and/or their derivatives which are used as an active ingredient to control fungal diseases. The concept of "bio-fungicide" widely refers to all types of bio-pharming products that reduce the fungal pathogen population and increase food production. Bio-fungicide formulations may be very simple mixtures of antagonistic fungi/bacteria with specific activities or complex mixtures with multiple effects on the host as well as the target pathogen (Pal and Gardener 2006).

## 35.3 Mechanism of Action of Bio-Fungicides

Pathogens are antagonized by the presence and activities of other organisms that they encounter. The mechanisms of biocontrol agents and reaction with the pathogen are many and complex. Understanding the mechanisms by which biocontrol agents suppress the plant pathogens is essential for the improvement and wider use of biological methods. Biocontrol may result from direct or indirect interactions between beneficial microorganisms and the pathogen. A direct interaction may involve physical contact and synthesis of hydrolytic enzymes, toxic compounds or antibiotics as well as competition. An indirect interaction may result from induced resistance in the host plant or the use of organic soil amendments to improve the activity of antagonists against the pathogens (Benítez et al. 2004; Pal and Gardener 2006; Viterbo et al. 2007). The major mechanisms of action of bio-fungicides are discussed as follows:

### 35.3.1 Antibiosis (Direct Inhibition by Secreting Secondary Metabolites, Mainly Antibiotics and Enzymes)

Antibiosis is an antagonism mediated by specific or non-specific metabolites of microbial origin, by lytic agents, enzymes, volatile compounds or other toxic substances, and it has a direct effect on other organisms (Benítez et al. 2004; Irtwange 2006; Viterbo et al. 2007; Haggag and Mohamed 2007). Antibiotics are generally considered to be organic compounds of low molecular weight produced by microbes, and at low concentrations they are deleterious to the growth or other metabolic activities of other microorganisms. Antibiotic production requires substantial substrates whereas the lifestyle of soil microbes is characterized by starvation with brief periods of activity in the saprophytic and/or parasitic condition. Therefore, the challenging strategy in biocontrol involves manipulation of such antagonists so that they will not only produce antibiotics for their survival, but also interfere with pathogenic activity in the infection court by reducing inoculum potential or by interfering with saprophytic phase of the pathogen. A mutant of *T. virens* which produces gliovirin showed better control of *Pythium ultimum* than the parental and gliovirin-negative strain (Howell and Stipnovic 1983). The various antibiotics produced by *Trichoderma* spp. include gliotoxin, gliovirin, viridin, viridiol, trichodermin, etc. (Harman 2006). Antibiotic production by certain strains of fluorescent pseudomonads is now recognized as an important mechanism in the biological control. A range of compounds like phenazine, pyoluteorin, tropolone, pyocyanin, 2,4-diacetyl phloroglucinal (DAPG) and pyrrolnitrin are known to be produced with broad spectrum activity against many plant pathogens. Bacteriocins are antibiotic-like compounds with bactericidal specificity, which restricts bacterial strains closely related to the bacteriocin producer (Handelsman and Stabb 1996; Benítez et al. 2004; Haggag and Mohamed 2007).

### 35.3.2 Induction of Systemic Resistance (ISR)

Induced resistance is a state of enhanced defensive capacity developed by a plant when appropriately stimulated. It is effective under field conditions and offers a natural mechanism for biological control of plant diseases (van Loon et al. 1998). PGPR, *Pseudomonas* spp., and *Bacillus* spp. play a major role in plant growth promotion and disease reduction by inducing systemic resistance (Zehnder et al. 2001; Kloepper et al. 2004; Mondal 2007). The ability of *Trichoderma* spp. to systemically activate plant-resistance mechanism against fungal pathogens has been demonstrated in Gramineae, Solanaceae and Cucurbitaceae against *Rhizoctonia solani, Botrytis cinerea, Alternaria* spp., *Colletotrichum* spp., *Magnoporthe grisea* and *Phytophthora* spp. (Woo et al. 2006). Induced resistance by biocontrol agents involves the same suite of genes and gene products which are involved in plant response known as systemic acquired resistance (SAR). This type of resistance shows host specificity, and this suggests that specific recognition between biocontrol agents and the plant is a prerequisite for the activation of signal cascade leading to induced-systemic resistance (Whipps 2001).

ISR defence responses in plant genes are induced by certain molecules called elicitors (Bostock 2005). Biocontrol agents or their products act as elicitors which results in the synthesis of phytoalexins, pathogenesis-related proteins (PR-proteins) and other defence-related compounds and in an increase in resistance against several plant pathogens. The lipopolysaccharides, siderophores and salicylic acid (SA) of PGPR and peptides or proteins of *Trichoderma* spp. act as elicitors to elicit defence in crop plants (Harmann et al. 2004; Bakker et al. 2007; Ongena et al. 2007). The defence responses of the plant may include the physical thickening of cell walls by lignification, deposition of callose, accumulation of antimicrobial low-molecular-weight substances (phytoalexins), defence enzymes such as phenyl-alanine ammonia-lyase (PAL) and chalcone synthase (CHS), and synthesis of various pathogenesis-related (PR) proteins such as chitinases, glucanases, peroxidases, etc. (van Loon et al. 1998; Khan et al. 2004; Viterbo et al. 2007).

### 35.3.3 Competition

Soil microbes compete for space and minerals and organic nutrients to proliferate and survive in their natural habitats. Competition has been suggested to play a key role in the biocontrol of species of *Fusarium* and *Pythium* by some strains of fluorescent pseudomonads. Competition for substrates is the most important factor for heterotrophic soil fungi (Whipps 2001; Benítez et al. 2004; Viterbo et al. 2007). Success in competition for substrate by any particular fungal species is determined by its competitive saprophytic ability (CSA) and inoculum potential of that species. Siderophore-mediated competition is an important mechanism of biocontrol agents. Siderophore (iron-chelating substances) are extracellular, low-molecular-weight iron transport agents, which selectively make complex with iron with very high affinity. It is a particular form of nutrient competition involving iron and this has been proposed as one of the mechanisms of biocontrol (Kloepper et al. 1980).

### 35.3.4 Mycoparasitism

It is the phenomenon of a fungus parasitizing another fungus. The parasitizing fungus is called as hyperparasite and the parasitized fungus as hypoparasite. In mycoparasitism, as a result of inter-fungus interaction several events take place which lead to predation, viz. coiling, penetration, branching and sporulation, resting body production, barrier formation and ultimately leading to lysis (Whipps 2001; Benítez et al. 2004; Viterbo et al. 2007).

Most of the effective biocontrol agents studied till now appear to antagonize pathogens using multiple mechanisms. These indirect and direct mechanisms may act coordinately and their importance in the biocontrol process depends on the strain of antagonist, the antagonized fungus, the crop plant and the environmental conditions, including nutrient availability, pH, temperature and iron concentration (Benítez et al. 2004; Viterbo et al. 2007). For instance, pseudomonads known to produce the antibiotic 2,4-diacetylphloroglucinol (DAPG) may also induce host defences. Additionally, DAPG producers can aggressively colonize roots, a trait that might further contribute to their ability to suppress pathogen activity in the rhizosphere of wheat through competition for organic nutrients (Iavicoli et al. 2003).

## 35.4 Characters of an Ideal Bio-Fungicide (Nakkeeran et al. 2005)

An ideal biocontrol agent should possess the following characters:

1. High rhizosphere competence
2. High competitive saprophytic ability (CSA)
3. Enhancer of plant growth
4. Amenable for mass multiplication
5. Broad spectrum of action
6. Excellent and reliable control
7. Safe to environment
8. Easily compatible with biofertilizers and rhizobacteria
9. Able to tolerate desiccation, heat, oxidizing agents and UV radiations

## 35.5 Commercialization of Bio-Fungicides

Recently, public health and safety concerns about the environmental impact of chemical pesticides have led to consideration of biological control as a natural approach to maintain crop health. Yet, there is a demand for biocontrol products, especially in the organic and agricultural niche markets, where there is no efficient chemical competitor. Growth in sales of biocontrol agents has significantly outpaced that of chemicals during the last decade (Reisch 2011). Currently, a major incentive favoring the development of biopesticides is the ease of federal registration in the USA. The Environmental Protection Agency (EPA) has established a Biopesticide Pollution and Prevention Division (BPPD) to manage accelerated registration of biopesticides. In the mid-1990s, the average duration for registration of a biopesticide was 12 months compared with 36 months for all new chemical pesticide registrations, and the cost of registration of a chemical was often more than eight times that of a biocontrol agent (Woodhead et al. 1990). In India, the CIB is responsible for the registration of all biopesticides under the Insecticide Act, 1968. It simplified the registration system to allow commercial pilot production in parallel with registration, which was particularly encouraging to small and medium enterprises. With the simplified registration system, four bio-fungicides, viz. *Bacillus subtilis, Gliocladium* spp., *Trichoderma* spp. and *Pseudomonas fluorescence* have been registered in India (CIB 2012).

The use of biocontrol agents to combat plant pathogens has been investigated for more than 70 years, and over 40 biocontrol products have been introduced into the market within the past 10 years (Kim and Hwang 2007). However, despite regulatory incentives, relatively few biological control agents have reached the market, often due to one or more of the following pitfalls: (1) poor choice of pathosystem for biological control, (2) availability of relatively few candidate microorganisms for testing, (3) microbes are selected based on the results of an assay that does not replicate under field conditions, (4) exclusion of amenability of microbes to commercial development as a selection criterion, (5) optimization of fermentation technology and mass production of biocontrol agents, (6) inconsistent performance and poor shelf life, (7) lack of patent protection, (8) awareness, training and education shortfalls, (9) lack of multidisciplinary approach and (10) technology constraints (Nakkeeran et al. 2005).

## 35.6 Formulations of Bio-Fungicides

An efficient formulation is essential to transfer the biocontrol agent from laboratory to the field. The important characteristics of successful commercial formulations are good market potential, simplicity in production and application, adequate product and stability, shelf life during transport and storage, efficacy over a long term, guaranteed propagule viability, economic, suitable and appropriate action, and compatibility with agronomic practices and equipment (Boyetchko et al. 1999). Bio-fungicide formulations are sold under various names in different parts of the world (Kaewchai et al. 2009). Details of commercial bio-fungicide available in the market are given in the web sites http://www.cibrc.nic.in; http://www.ippc.orst.edu/biocontrol/biopesticides/; and http://www.epa.gov/pesticides/biopesticides. Formulations of *T. viride* are marketed under trade names of Bioderma in India, Promote in USA; *T. harzianum* under the name of Binab T in Sweden, Bioderma-H in India, Fungi-Killer in Thailand, Promote, Root Shield and Plant Shield in USA, Trichodex in Belgium and *Trichoderma* spp. under the name of Biofungus in Belgium, Sentinel and Vinevax in USA, *G. virens* under the name of SoilGard in USA; *Gliocladium catenulatum* under the names of Prestop and Primastop in Finland and *Chaetomium cupreum* under the names of Ketocin, Ketomium and Novacide in Thailand. *B. subtilis* is sold under the names of Quantum, Kodiak, Epic, Campanion, Serenade, Rhapsody, Subtilex and System 3 in USA; Rhizo-Plus in Germany; *Bacillus pumilus* under the names of GB 34 and Sonata in USA; *Bacillus cereus* under the names of Pix plus plant regulator in Finland. *P. fluorescence* under the names of Bio-Save in Orlando; Blight Ban

**Table 35.1** Fungal formulations as bio-fungicides for packaging

| Sl. no. | Particulars | *Trichoderma* spp. | *P. fluorescens* |
|---|---|---|---|
| 1. | Packing material | White Polythene bags | White Polythene bags |
| 2. | Size of the carrier (talc) | 500 µg | 500 µg |
| 3. | Moisture content of the formulation | 20 % | 20 % |
| 4. | Storage period | 4 months | 3–4 months |
| 5. | Population in the fresh product | $28 \times 10^6$ cfu/g | $2.5 \times 10^8$ cfu/g |
| 6. | Population at the time of mixing with talc | $10^8$–$10^9$ conidia/ml | $9 \pm 2 \times 10^8$ cfu/ml |
| 7. | Population at the time of expiry | $20 \times 10^6$ cfu/g | $8$–$9 \times 10^7$ cfu/g |

A506 in the USA; Conquer and Victus in Australia; Cedomon in Sweden; *P. aureofaciens* under the names of BioJect Spot–less in San Diego, CA; *P. cepacia* as Deny in Shawnee, KS and Intercept in the USA; *P. chlororaphis* as AtEze in San Diego, CA and *P. syringae* under the names of Bio-save in Finland. There are many types of bio-fungicide formulations, viz. alginate prill formulation, fluid-bed granulation having dextrin as a binder, liquid formulation, water-dispersible granule formulation, wettable powder formulation, dusts, granular or powder products (Khetan 2001). The formulations comprise of active ingredients (microorganism or spores), adjuvants, dilution agents, bulking additives, membrane stabilizers, growth and contaminant suppressants, buffers, binders, dispersants, lubricants, activators, food sources and coatings. Bio-fungicide formulations can be applied to seeds, tubers, cuttings, seedlings, transplants, mature plants and soil (Nakkeeran et al. 2005). Even though liquid formulation is preferred with drip irrigation, granular formulations are more appropriate for combining with potting mix, while a wettable powder is more appropriate for root dips or foliar sprays (Spadaro and Gullino 2005).

## 35.7 Mixed-Strain Formulations of Bio-Fungicides

In general, biocontrol of crop diseases involves the application of single antagonistic organism, but in nature, the microbes do not reside alone. They live in close association with other species for nutrient production, waste removal, etc. It may be unrealistic to expect a single organism to perform better in terms of growth, spread and suppression of pathogens without the help of the members of the microbial community on which they depend (Raupach and Kleopper 1998). Combined applications of microbes are always found to be promising in the control of several pathogens. It may be due to the synergistic effects of biocontrol agents against pathogens. This approach will certainly lead to the selection of effective isolates, with sustainable performance under diverse environment. Development of cocktail formulation with compatible isolates will improve disease control through synergy in signal cross talk between the isolates that leads to increased production of antibiotics at the site of colonization and thereby suppresses the establishment of pathogenic microbes. Advantages of strain mixtures include a broad spectrum of action, enhanced efficacy, reliability and also allow the combination of various traits without genetic engineering (Janiseiwicz 1996). Hence, augmentation of bio-fungicides with compatible mixtures of biocontrol agents (microbial consortia) to infection court will mimic the natural environment and could broaden the spectrum of activity against different plant pathogens (Schisler et al. 1997).

## 35.8 Quality Control Specifications of Bio-Fungicides

As per the Gazette Notification No. G.S.R.224(E) dated 26.03.1999 of the CIB, Department of Agriculture and Cooperation, Ministry of Agriculture, Government of India, *B. subtilis, Gliocladium* spp., *Pseudomonas fluorescens* and *Trichoderma* spp. have been included under the Insecticide Act, 1968 as bio-fungicides for plant disease management. The guidelines requirement for provisional registration and regular registration of biopesticides are given under Sects. 9(3B) and 9(3) of the Insecticide Act, 1968 (www.cibrc.nic.in). The specifications given in Table 35.1 have to be followed strictly by the producer during commercial production of talc formulation of *Trichoderma* spp. and *P. fluorescens.*

## 35.9 Genetic Engineering of Bio-Fungicides

Genomic tinkering of naturally occurring biocontrol agents with genes that are beneficial to plants will lead to the accentuated expression of the genomic products which could alleviate the attack of diseases. Genetically improved antagonistic microorganisms tend to increase their effectiveness as bio-fungicides with an enhanced antifungal metabolite activity to control a broad spectrum of phytopathogens (Hornok 2000). Genetic improvement can be achieved by chemical and physical mutation, sexual hybrids, homokaryons and genetic manipulation by directed mutagenesis, protoplast fusion, genetic analysis of fusants, recombination, transformation or isolation of useful genes from biocontrol fungi without functional sexual stages (Mohamed et al. 2004; Palumbo et al. 2005). There are a number of reports of high-yielding *Trichoderma* mutants expressing altered morphological phenotype such as growth, sporulation, etc. and being able to over-secrete enzymes (Mukherjee et al. 1999). Ghosh et al. (1982) observed improved cellulose productivity of *Trichoderma* linked to increased endoplasmic reticulum content. Availability of a biocontrol-related promoter sequence combined with a reliable transformation system permits genetic improvement of fungal biocontrol agents through the use of novel transgenes to enhance disease control mechanisms (Mach et al. 1999).

Some genetically engineered biocontrol microorganisms with multi-copy chitinase genes have shown enhanced disease suppression activity (Simi 1994). An engineered strain of *Pseudomonas fluorescens* P5–1, which expressed a 6.5 kb DNA chitinase gene (chiB), showed improved suppression of cotton damping off and rice sheath blight, both caused by *R. solani* (Xu et al. 2005). At a molecular genetic level, attempts to increase the biocontrol ability of *Trichoderma* have been focused on increasing chitinase or proteinase activity either by increasing the number of copies of the appropriate genes or by fusing them with strong promoters (Brunner et al. 2005). Thus, the introduction of a single biocontrol agent with multiple modes of action will facilitate to achieve maximum benefit in terms of disease control. However, the release of genetically modified organisms (GMOs) is a policy decision. Hence, making the policymakers understand the safe usage of the bacteria and their benefits will go a long way for sustainable food security and the ecosystem.

## 35.10 Bio-safety of Bio-Fungicides

Application of biocontrol agents serves as an alternate to synthetic fungicides for the management of various diseases of plants. However, the safety of the microbes in terms of human welfare could not be neglected. Several biocontrol agents are also known to be opportunistic pathogens. Though *Pseudomonas aeruginosa* is a potential biocontrol agent of grey leaf spot of turf grasses, it is also a virulent opportunistic pathogen which infects wounds and severe burns. *B. cereus* being a potential antagonist for the management of damping off and root rot of soya bean, it is also a food contaminant and closely mimics *Bacillus anthracis,* the causal agent of anthrax diseases. The confusions involved in distinguishing between the related strains that turn as opportunistic pathogens for human beings has to be solved to convince the policymakers and environmental protection agencies to promote the acceptance, registration and transfer of technology and its adoption (Nakkeeran et al. 2011)

## 35.11 Bio-Fungicides for Food Security

The doubling of cereal and livestock production in the last half of the twentieth century should have resulted in a global food supply that was adequate for all and yet currently nearly a billion people are hungry, i.e. they do not have access to food (FAO 2008a, b). The predicted growth of world's population from 7 billion at present (1.2 billion in India alone) to 8.3 billion by 2030 presents a major global challenge to meet necessary food demand (Thind 2012). Low crop yields are common in many developing countries and improved productivity is vital to reduce rural poverty and increase the food security. While the causes of low productivity are complex, one major contributory factor is crop losses due to plant health problems. One major aspect of plant health concerns is crop losses due to pests which include insects, pathogens and weeds. Often accurate information on the extent of these losses is missing but estimates of 30–40% loss annually from "field to fork" are common (FAO 2008a).

In developing countries, losses are costly in terms of food security, foreign exchange requirements for food imports, income losses to farmers and others whose livelihoods depend on agriculture. Nevertheless, plant diseases have had enormous impact on livelihoods throughout human history. The Irish potato famine is an excellent example of the effect of plant disease on food security and changing demographics. The plant disease that could have severe impacts on global food production, livelihoods and changes in demographics in the twenty-first century is Ug99, a virulent strain of black stem rust (*Puccinia graminis tritici*) that has spread from Africa into the Middle East and has threatened the high-yielding wheat varieties developed under the Green Revolution in South Asia (FAO 2008b). Coffee wilt disease (*Fusarium xylarioides*) is yet another important disease, which affected producer livelihoods in central and eastern Africa and had cost an estimated US$ 1 billion and continues to spread to other parts of the world (Flood 2009).

The maximum yield reductions in crops due to fungal diseases compared to other pathogenic diseases need to be prevented, mitigated or controlled in order to maintain global food availability, one of the three pillars of food security (Gullino et al. 2000). Chemical fungicides have reduced crop losses in many situations, but even with a very substantial increase in fungicides use, the overall proportion of crop losses and the absolute value of these losses from the diseases appear to have increased over time (Oerke et al. 1994). There is consensus that indiscriminate, excessive and inefficient use of fungicides exerts too high a toll in terms of human health, environmental safety and ultimate diminishing returns to justify any short-term increases in farm income or food output. The usage of bio-fungicides in collaboration with traditional or conventional breeding methods, can raise crop productivity, increase resistance to diseases, develop tolerance to adverse weather conditions, improve the nutritional value of some foods and enhance the durability of products during harvesting or shipping. The application of genetically manipulated bio-fungicides in modern agriculture offers tremendous opportunities for expanding food production, reducing risks in food production, improving environmental protection, strengthening food marketing and ensuring food security in developing countries (Chet 1993; Cook 1993; Kaewchai et al. 2009; Flood 2010)

## 35.12 Conclusion

Recently, the use of bio-fungicides has been advanced as an alternative to synthetic fungicides to control plant diseases. They are easy to deliver, improve plant growth, induce resistance mechanism in the host with increased biomass production and yield. Genetically improved antagonistic microorganisms tend to increase their effectiveness as biocontrol agents. Recent surveys of both conventional and organic growers indicate an interest in biocontrol for suppressing plant diseases, suggesting that the market potential of bio-fungicides will increase further in coming years. The harmless cheaper bio-fungicides which are highly effective throughout the crop period will improve global food availability by reducing crop losses, will boost the farm income and enhance the livelihood of low-income farmers. So, the bio-fungicides can be concluded as the best alternative for sustainable food security and the ecosystem.

## References

Anonymous (2006) List of pesticides/pesticides formulations banned in India. Pesticide Res J 18:243

Bakker PAHM, Pieterse CMJ, Van Loon LC (2007) Induced systemic resistance by fluorescent *Pseudomonas* spp. Phytopathology 97:239–243

Benítez T, Rincón MA, Limón MC, Codón CA (2004) Biocontrol mechanisms of *Trichoderma* strains. Intl Microbiol 7:249–260

Biswas A (2011) Organic farming in relation to crop disease management. Pestology 11:44–49

Bostock RM (2005) Signal crosstalk and induced resistance: straddling the line between cost and benefit. Annu Rev Phytopathol 43:545–580

Boyetchko S, Pedersen E, Punja Z, Reddy M (1999) Formulations of biopesticides. In: Frinklin RH, Julius JM (eds) Methods in biotechnology Vol. 5: Biopesticides: use and delivery. Humana press, 487–508

Brunner K, Zeilinger S, Ciliento R, Woo SL, Lorito M, Kubicek CP, Mach RL (2005) Improvement of the fungal biocontrol agent *Trichoderma atroviride* to enhance both antagonism and induction of plant systemic disease resistance. Appl Environ Microbiol 71:3959–3965

Chet I (1993) Preface. In: Chet I (ed) Biotechnology in plant disease control. Wiley-Liss, New York, p XV

CIB (2012) Fungicides registered under section 9(3) of the Insecticides Act, 1968 as on 20/01/2012. http://cibrc.nic.in. Accessed Jan. 11, 2013

Cook RJ (1993) Making greater use of introduced microorganisms for biological control of plant pathogens. Annu Rev Phytopathol 3(1):53–80

EFSA (2009) European Food Safety Authority, Annual report on pesticide residues. EFSA scientific report vol. 305

FAO (2008a) The state of food insecurity in the world-high food prices and food security-threats and opportunities. http://www.fao.org/catalog/inter-e.html. Accessed Jan. 11, 2013

FAO (2008b) Report of Ug 99 in Iran. http://www.fao.org/newsroom/en/news/2008/1000805/index.html. Accessed Jan. 11, 2013

Flood J (2009) Coffee wilt disease. 200 pages. CABI. ISBN:978-1-84593-641-9

Flood J (2010) The importance of plant health to food security. Food Sec 2:215–231

Fravel RD (2005) Commercialization and implementation of biocontrol. Annu Rev Phytopathol 43:337–359

Ghosh A, Rabiai AS, Ghosh BK (1982) Increased endoplasmic reticulum content of a mutant of *Trichoderma reesei* (RUT-C30) in relation to cellulase synthesis. Enzyme Micro Technol 4:110–113

Gianessi L, Reigner N (2006) The importance of fungicides in U.S. crop production. Outlook Pest Manage 10:209–213

Gullino ML, Leroux P, Smith CM (2000) Uses and challenges of novel compounds for plant disease control. Crop Prot 19:1–11

Haggag WM, Mohamed HAA (2007) Biotechnological aspects of microorganisms used in plant biological control. Am-Eurasian J Sustain Agric 1:7–12

Handelsman J, Stabb VE (1996) Biocontrol of soil born plant pathogens. Plant Cell 8:1855–1869

Harman GE (2006) Overview of mechanisms and uses of *Trichoderma* spp. Phytopathology 96:190–194

Harmann GE, Howell CR, Viterbo A, Chet I, Lorto M (2004) *Trichoderma* species opportunistic, avirulent and plant symbionts. Nature Rev Microbiol 2:43–56

Hornok L (2000) Genetically modified microorganisms in biological control. Novenyvedelem 36:229–237

Howell CR, Stipnovic RD (1983) Gliovirin, a new antibiotic from *Gliocladium virens* and its role in the biological control of *Pythium ultimum*. Can J Microbiol 29:321–324

Iavicoli A, Boutet E, Buchala A, Métraux JP (2003) Induced systemic resistance in *Arabidopsis thaliana* in response to root inoculation with *Pseudomonas fluorescens* CHA0. Mol Plant-Microbe In 16:851–858

Irtwange VS (2006) Application of biological control agents in pre- and postharvest operations. Agricultural Eng Int: the CIGR E J. Invited Overview 3:1–12

Janisiewicz WJ (1996) Ecological diversity, niche overlap, and co-existance of antagonists used in developing mixtures for biocontrol of post harvest diseases of apples. Phytopathology 86(473–4):79

Janisiewicz WJ, Korsten L (2002) Biological control of postharvest diseases of fruits. Annu Rev Phytopath 40:411–441

Jui-Hung Y, Jin-Shu C, Pin-Jui H, Yei-Shung W (2009) Effects of fungicides triadimefon and propiconazole on soil bacterial communities. J Environ Sci Health 44:681–689

Kaewchai S, Soytong K, Hyde KD (2009) Mycofungicides and fungal biofertilizers. Fungal Divers 38:25–50

Khan J, Ooka JJ, Miller SA, Madden LV, Hoitink HAJ (2004) Systemic resistance induced by *Trichoderma hamatum* 382 in cucumber against *Phytophthora* crown rot and leaf blight. Plant Dis 88:280–286

Khetan SK (2001) Microbial pest control. Marcel Dekker Inc., New York p 300

Kim BS, Hwang BK (2007) Microbial fungicides in the control of plant diseases. J Phytopathol 155:641–653

Kloepper JW (1993) Plant growth promoting rhizobacteria as biological control agents. In: Metting FB Jr (ed) Soil microbial ecology applications in agricultural and environmental management. Marcel Dekker, New York, pp 255–274

Kloepper JW, Leong J, Teintze M, Schroth MN (1980) *Pseudomonas* siderophores: a mechanism explaining disease suppressive soils. Cur Microbiol 4:317–320

Kloepper JW, Ryu CM, Zhang SA (2004) Induced systemic resistance and promotion of plant growth by *Bacillus* spp. Phytopathology 94:1259–1266

Koul O (2008) Phytochemicals and insect control: an antifeedant approach. Crit Rev Plant Sci 27:1–24

Mach RL, Peterbauer CK, Payer K, Jaksits S, Woo SL, Zeilinger S, Kullnig CM, Lorito M, Kubicek CP (1999) Expression of two major chitinase genes of *Trichoderma atroviride* (*T. harzianum* P1) is triggered by different regulatory signals. Appl Environ Microbiol 65:1858–1863

Mitchell AM, Strobel GA, Hess WM, Vargas PN, Ezra D (2008) *Muscodor crispans*, a novel endophyte from *Ananas ananassoides* in the Bolivian Amazon. Fungal Divers 31:37–43

Mohamed HAA, Haggag Wafaa M, Abo-Aba SM (2004) Influence of salt stress on *Pseudomonas fluorescens* plasmids, some phenotypic traits and antibiosis against *Diplodia theobromae*. J Genet Eng Biotechnol 2:265–281

Mondal KK (2007) Bacterial antagonists for bacterial diseases in plants. In: Ahmad S, Narain U (eds) Ecofriendly management of plant diseases. Daya Publishing House, India, pp 156–177

Mukherjee PK, Sherkhane PD, Murthy NBK (1999) Induction of stable benomyl tolerant phenotypic mutants of *Trichoderma pseudokoningii* MTCC3011, and their evaluation for antagonistic and biocontrol potential. Indian J Exp Biol 37:710–712

Nakkeeran S, Fernando WGD, Siddiqui ZA (2005) Plant growth promoting rhizobacteria formulations and its scope in commercialization for the management of pests and diseases. In: Siddiqui ZA (ed) PGPR: biocontrol and biofertilization. Springer, Dordrecht, pp 257–296

Nakkeeran S, Renukadevi P, Muthukumar A, Chandirasekar G, Jonathan EI (2011) Exploring the biocontrol potential of PGPR for the management of plant diseases. In: Proceedings of the National Seminar on NSRAPDR, Annamalai University, India, pp 9–18, 25–26 Feb 2011

Oerke EC, Dehne HW, Schönbeck F, Weber A (1994) Crop production and crop protection–estimated crop losses in major food and cash crops. Elsevier, Netherlands (Reprinted in 1999)

Ongena M, Jourdan E, Adam A, Paquot M, Brans A, Joris B, Arpigny JL, Thonart P (2007) Surfactin and fengycin lipopeptides of *Bacillus subtilis* as elicitors of induced systemic resistance in plants. Environ Microbiol 9:1084–1090

Pal KK, Gardener BM (2006) Biological control of plant pathogens. Plant Health Instructor. doi:10.1094/PHI-A-2006-1117-02

Palumbo JD, Yuen GY, Jochum CC, Tatum K, Kobayashi DY (2005) Mutagenesis of beta- 1,3-glucanase genes in *Lysobacter enzymogenes* strain C3 results in reduced biological control activity toward *Bipolaris* leaf spot of tall fescue and *Pythium* damping-off of sugar beet. Phytopathology 95:701–707

Pozo C, Rodelas B, Salmeron V, Martinez-Toledo MV, Vela GR, Gonzalez-Lopez J (1994) Effects of fungicides maneb and mancozeb on soil microbial populations. Toxicol Environ Chem 43:123–132

Raupach GS, Kleopper JW (1998) Mixtures of plant growth promoting rhizobacteria enhance biological control of multiple cucumber pathogens. Phytopathology 88:1158–1164

Reisch MS (2011) Going mainstream biopesticide producers look beyond organic growers. Chem Eng News 89:16

Schisler DA, Sliminger PJ, Bothast RJ (1997) Effects of antagonistic cell concentration and two strain mixtures on biological control of *Fusarium* dry rot of Potato. Phytopathology 87:177–183

Simi K (1994) The chitinase encoding T7-based chiA gene endows Pseudomonas fluorescence with the capacity to control plant pathogens in soil. Gene 147:81–83

Spadaro D, Gullino LM (2005) Improving the efficacy of biocontrol agents against soil borne pathogens. Crop Prot 24:601–613

Thind TS (2005) Significant achievements and current status: Fungicide research. In: Chahal SS, Khetarpal RK, Thind TS (eds) One hundred years of plant pathology in India: an overview. Scientific Publishers, Jodhpur, pp 267–305

Thind TS (2008) Fungicide resistance: a perpetual challenge in disease control. Indian J Mycol Pl Pathol 38:407–417

Thind TS (2012) Fungicides in crop health security-the road ahead. Indian Phytopathol 65:109–115

van Loon LC, Bakker PAHM, Pieterse CMJ (1998) Systemic resistance induced by rhizosphere bacteria. Annu Rev Phytopathol 36:453–483

Viterbo A, Inbar J, Hadar Y, Chet I (2007) Plant disease biocontrol and induced resistance via fungal mycoparasites. In: Kubicek CP, Druzhinina IS (eds) Environmental and microbial relationships, the mycota, 2nd edn, vol IV. Springer-Verlag, Berlin, pp 127–146

Whipps JM (2001) Microbial interactions and biocontrol in the rhizosphere. J Exp Bot 52:487–511

Woo SL, Scala F, Ruocco M, Lorito M (2006) The molecular biology of the interactions between *Trichoderma* spp., phytopathogenic fungi and plants. Phytopathology 96:181–185

Woodhead SH, O'Leary DJ, Rabatin SC (1990) Discovery, development, and registration of a biocontrol agent from an industrial perspective. Can J Plant Pathol 12:328–331

Xu XJ, Zhang LQ, Zhu YY, Tang WH (2005) Improving biocontrol effect of *Pseudomonas fluorescens* P5 on plant diseases by genetic modification with chitinase gene. J Agric Biotechnol 12:460–463

Zehnder GW, Murphy JF, Sikora EJ, Kloepper JW (2001) Application of rhizobacteria for induced resistance. Eur J Plant Pathol 107:39–50

# PCR Amplification, Sequencing, and *In Silico* Characterization of Pectin Lyase Genes from *Aspergillus flavus* NIICC8142

36

Amit Kumar Dubey, Sangeeta Yadav,
Gautam Anand and Dinesh Yadav

**Abstract**

The genomic sequences of *Aspergillus flavus* NRRL 3357 revealed seven pectin lyase (*pnl*) genes, which were used to design gene-specific primers and these genes were amplified through polymerase chain reaction (PCR) from five strains of *A. flavus,* namely *A. flavus* MTCC 7589, *A. flavus* MTCC 10938, *A. flavus* MTCC 8836, *A. flavus* NIICC 8142 and *A. flavus* NIICC 8147. All the seven predicted *pnl* genes were amplified from the genomic DNA of *A. flavus* NIICC 8142, while six *pnl* genes were from *A. flavus* MTCC8837 and five *pnl* genes from the remaining strains, respectively. A total of five *pnl* gene sequences of *A. flavus* NIICC 8142 designated as *Afpnl*-1, *Afpnl*-2, *Afpnl*-3, *Afpnl*-4, and *Afpnl*-5 were submitted to GenBank and assigned accession numbers JQ735890 to JQ735894, respectively. These sequences were subjected to in silico characterization for homology search, multiple sequence alignment, phylogenetic tree construction, and motif analysis. The homology search revealed their identity to the predicted pectin lyase genes of *A. flavus* NRRL 3357, and multiple sequence alignment of these five genes showed various conserved residues. The phylogenetic tree revealed two distinct clusters and four subclusters with five pectin lyase genes of *A. flavus* occupying distinct position among the PCR-amplified pectin lyase genes from different fungi. The presence of unique pec_lyase C domain was observed among these sequences.

**Keywords**

Pectin lyase · Aspergillus flavus · Multiple sequence alignment · Motif · Phylogenetic tree

## 36.1 Introduction

Pectin lyase (PNL, E.C. 4.2.2.10) represents an important member of pectinases group of enzymes associated with degradation of pectin and has potential industrial applications in fruit juice

D. Yadav (✉) · A. K. Dubey · S. Yadav · G. Anand
Department of Biotechnology, D.D.U Gorakhpur University, Gorakhpur, Uttar Pradesh 273009, India
e-mail: dinesh_yad@rediffmail.com

R. N. Kharwar et al. (eds.), *Microbial Diversity and Biotechnology in Food Security*,
DOI 10.1007/978-81-322-1801-2_36, © Springer India 2014

clarifications and retting of fibers (Alkorta et al. 1998; Kashyap et al. 2001; Hoondal et al. 2002; Jayani et al. 2005; Pedrolli et al. 2009; Yadav et al. 2009a). These enzymes are unique owing to the fact that they degrade pectin polymers directly by β-elimination mechanism resulting in the formation of 4,5 unsaturated oligogalacturonides while other pectinases act sequentially to degrade pectin molecule totally. Pectin lyase is better suited for fruit juice clarification as they degrade pectin without disturbing the ester group responsible for the specific aroma of the juice and also do not form toxic methanol (Taragano and Pelosof 1999).

The production, purification, and biochemical characterization of pectin lyase from different fungal strains have been reported recently, and its application in fruit juice clarification and retting of natural fibers has been elucidated (Yadav et al. 2012; Yadav et al. 2009b, c; Pedrolli and Carmona 2009; Yadav et al. 2008). The fungi belonging to the genus *Aspergillus, Penicillium,* and *Fusarium* spp. are a major source of pectin lyase along with a few bacteria (Gummadi and Kumar 2005).

Pectinolytic genes from *Aspergillus niger* (Dean and Timberlake 1989; Harmsen et al. 1990), *Aspergillus oryzae* KBN 616 (Kitamoto et al. 2001) has been isolated and characterized. The pectin lyase gene family of *A. niger* and *Glomerella cingulata* (Templeton et al. 1994), bacterial sources like *Bacillus subtilis* (Sakamoto et al. 1996), *Erwinia caratovora* sub sp. *caratovora* (Chatterjee et al. 1991) has also been reported. Very recently, bioinformatics of pectinases and pectin lyase sequences has been reported (Yadav et al. 2009d; Dubey et al. 2010, 2012). This chapter reports polymerase chain reaction (PCR) amplification of putative pectin lyase genes from indigenous fungal strains of *Aspergillus flavus* NIICC8142 using predicted pectin lyase genes from genomic sequence of *A. flavus* NRRL3357, and sequencing of eluted amplicons and in silico characterization of PCR-amplified pectin lyase gene sequences for homology search, multiple sequence alignment, phylogenetic tree construction, and motif search.

## 36.2 Materials and Methods

### 36.2.1 Fungal Strain and Culture Conditions

The indigenously isolated five soil fungal strains, namely *A. flavus* MTCC 7589, *A. flavus* MTCC 10938, *A. flavus* MTCC 8836, *A. flavus* NIICC 8142, and *A. flavus* NIICC 8147, were deposited to Microbial Type Culture Collection and Gene Bank Centre, IMTECH, Chandigarh and National Institute of Interdisciplinary Culture Collection Trivendrum, India. The cultures were maintained on the Czapek yeast extract agar (CYA) medium and mycelia were produced in CY broth after incubating for 4–5 days at 28 °C. The mycelia were filtered using a cheesecloth, washed with chilled autoclaved water several times, dried, and stored at −20 °C till further use.

### 36.2.2 Genomic DNA Isolation

Genomic DNA was isolated from the mycelium of *A. flavus* NIICC8142 using the standard method with a slight modification in the extraction buffer (Dubey et al. 2012). Quality and quantity of genomic DNA were analyzed by agarose gel electrophoresis and UV-Vis spectrophotometer (Maniatis et al. 1982).

### 36.2.3 Primer Designing

A set of five primers were designed based on the available pectin lyase genes from the whole genome of *A. flavus* NRRL 3357 from NCBI (http://www.ncbi.nlm.nih.gov/) (Pruitt et al. 2007) subjecting to multiple sequence alignment using the tools DNA star (Burland 2000) and primer basic local alignment search tool (BLAST) (Altschul et al. 1990). These primers were synthesized by Merck Specialities Private Limited, Mumbai, India and diluted as per the instructions. The primer sequences along with the pectin lyase gene sequences from different source organisms used for primer designing are listed in Table 36.1.

**Table 36.1** List of primers designed for PCR amplification of pectin lyase genes

| Code | Primers (5′–3′) | Size (mer) | Tm | % GC | Source gene | Amplicon size (bp) |
|---|---|---|---|---|---|---|
| GCEL-PNL023-F | GCGGAGGCAAACCCTTCTCACT | 22 | 71.5 | 59 | AFLA_017180 | 1,697 |
| GCEL-PNL023-R | TGCGCATGCAAACCTGTCGGA | 21 | 76.1 | 57 | | |
| GCEL-PNL024-F | TGTTGGCCATACTCCCTTTC | 20 | 63.7 | 50 | AFLA_025400 | 1,863 |
| GCEL-PNL024-R | CAACCACCCATGTCACTGAG | 20 | 64.2 | 55 | | |
| GCEL-PNL025-F | AGCTAGGGACCTTCGCTCTC | 20 | 63.8 | 60 | AFLA_119860 | 1,717 |
| GCEL-PNL025-R | TTGTGTGACGCTGTGCATTA | 20 | 64.0 | 45 | | |
| GCEL-PNL026-F | CGTGTAGCCATGTTGGAGTG | 20 | 64.2 | 55 | AFLA_124660 | 1,371 |
| GCEL-PNL026-R | ATTGCTCAAGGCAGCAAGAT | 20 | 63.8 | 45 | | |
| GCEL-PNL027-F | TGGGACTCCCTAACAAGGCCGA | 22 | 72.5 | 59 | AFLA_116040 | 1,737 |
| GCEL-PNL027-R | ATCGTCGTCCCGTGCAGAGT | 20 | 69.4 | 60 | | |

### 36.2.4 PCR Amplification, Gel Elution, and Sequencing of PCR Amplicons

Standardization for PCR amplification of pectin lyase genes was carried out as per previous work (Dubey et al. 2012), and expected size bands were gel eluted by HiPurA$^{TM}$Agarose Gel DNA Purification Spin Kit (Hi Media, India) as per the instructions, quantified and sequenced using gene specific primers at National Institute of Plant Genome Research, JNU Campus, New Delhi, India.

### 36.2.5 Sequence Submission and *In Silico* Characterization

A total of five PCR-amplified partial pectin lyase gene sequences designated as *Afpnl*-1, *Afpnl*-2, *Afpnl*-3, *Afpnl*-4, and *Afpnl*-5 were submitted to the GenBank and assigned accession numbers JQ735890 to JQ735894, respectively. These sequences were subjected to homology search with NCBI database using BLAST. The program ClustalW (Thompson et al. 1994) was used for multiple sequence alignment. Mega 4 (Tamura et al. 2007) was used for dendrogram construction by neighbor-joining (NJ) method (Saitou and Nei 1987). These sequences were subjected to bioinformatics servers, namely FGENESH (Solovyev et al. 2006), for prediction of full-length genes with putative CDS and protein sequences. For domain search, the Pfam site (http://www.sanger.ac.uk/software/pfam/search.html) (Finn et al. 2010) was used. Domain analysis was done using multiple EM for motif elicitation (MEME) software (http://meme.sdsc.edu/meme/meme.html) (Bailey and Elkan 1994). The conserved protein motifs deduced by MEME were characterized for biological function analysis using protein BLAST, and the domains were studied with InterProScan version 24 (Quevillon et al. 2005).

## 36.3 Results and Discussion

### 36.3.1 PCR Amplification of Pectin Lyase Genes from Different Strains of *Aspergillus flavus*

Five pectin lyase genes with assigned accession numbers, namely AFLA_017180, AFLA_119860, AFLA_124660, AFLA_116040, AFLA_025400 predicted from the genome sequence of *A. flavus* NRRL 3357 were retrieved from NCBI databases and were used for designing primers (Table 36.1). PCR amplification of pectin lyase genes using these sets of primers were standardized with genomic DNA of *A. flavus* NIICC 8142 as template DNA (Fig. 36.1). The PCR amplification with primer sets GCEL-PNL023 at annealing temperature

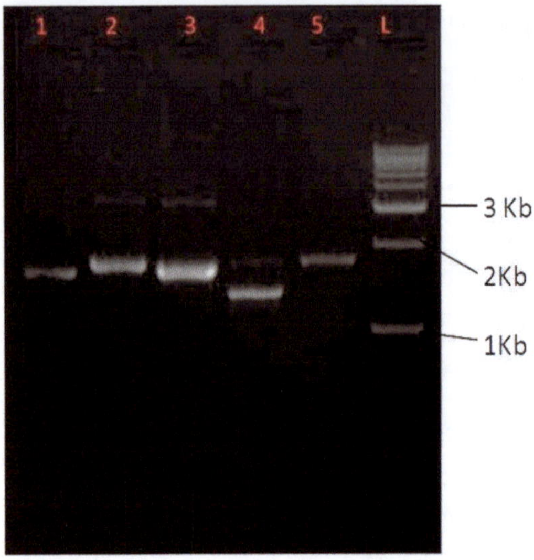


**Fig. 36.1** Pectin lyase gene from *A. flavus* NIICC 8142. *Lane 1—Afpnl-1* (1697 bp), *Lane 2—Afpnl-2* (1863 bp), *Lane 3—Afpnl-3* (1717 bp), *Lane 4—Afpnl-4* (1371 bp), *Lane 5—Afpnl-5* (1737 bp), *Lane L*—1Kb Ladder. (Merck Specialities, India)

57 °C and GCEL-PNL024, GCEL-PNL025, GCEL-PNL026, and GCEL-PNL027 at annealing temperature of 62 °C resulted in expected size bands. The other *A. flavus* strains, namely *A. flavus* MTCC 7589, *A. flavus* MTCC 10938, *A. flavus* MTCC 8836, *A. flavus* NIICC 8142, and *A. flavus* NIICC 8147 also gave the expected size bands with the same set of primers.

### 36.3.2 Gel Elution of Expected Size Amplicons and Sequencing

The expected size amplicons from different sets of primers representing multigene family of pectin lyase from *A. flavus* were gel eluted by HiPur A$^{TM}$Agarose Gel DNA Purification Spin Kit and analyzed on 1.5 % agarose gel, quantified spectrophotometrically and sequenced with respective primers used for amplification. These sequences were submitted to GenBank and were assigned accession number JQ35890 to JQ35894.

### 36.3.3 *In Silico* Analysis of Sequenced Pectin Lyase Genes

The sequences of five pectin lyase genes designated as *Afpnl*-1, *Afpnl*-2, *Afpnl*-3, *Afpnl*-4, and *Afpnl*-5 with assigned accession numbers JQ735890 to JQ735894, respectively, were subjected to BLAST for homology search. The BLAST analysis revealed its maximum identity to the predicted pectin lyase genes from the whole genome sequence of *A. flavus* NRRL 3357. The results of BLAST show best hit with maximum identity ranging from 99–100 % (Table 36.2).

The nucleotide sequences of five pectin lyase genes were then subjected to FGENESH to identify putative CDS and protein (Table 36.3). The amino acid residues ranged from 99 to 214 in the corresponding translated proteins of the sequenced pectin lyase genes of *A. flavus*. Except *Afpnl*-1 (JQ 735890), the remaining pectin lyase genes were predicted for putative exons in +ve (direct) strand of sequence. Sequencing of eluted amplicons with gene-specific primers has resulted in the partial sequences of pectin lyase genes ranging from 455 to 781 bp (Table 36.3). The cloning of these expected size amplicons in suitable vector and subsequent sequencing with the universal primer might result in the expected sequence length. The pectin lyase gene *Afpnl*-4 revealed two CDS regions while the remaining four genes have single CDS.

The multiple sequence alignment of five *pnl* gene sequences by ClustalW revealed a conserved nucleotides at variable positions, namely 137(A), 168(C), 179(C), 185(G), 197(G),218(G), 219(G), 244(C), 245(G), 268(C), 281(G), 316(C), 365(G), 376(), 377(), 387(C), 388(C), 407(G),409(T), 422(G), and 434(G) (Fig. 36.2).

The phylogenetic tree constructed by NJ method for five pectin lyase genes of *A. flavus* revealed two distinct clusters comprising three (JQ735892, JQ735894, JQ735890) and two genes (JQ735891, JQ735893), respectively (Fig. 36.3a). These sequences were also subjected to phylogenetic tree construction after incorporating sequences of 18 PCR-amplified pectin

**Table 36.2** BLAST analysis of PCR-amplified pectin lyase genes of *A. flavus*

| *Pnl* genes | Accession number | Maximum similarity | Max score | Total score | Query coverage (%) | E value | Max identity (%) |
|---|---|---|---|---|---|---|---|
| *Afpnl*-1 | JQ735890 | *Aspergillus flavus* NRRL3357 pectin lyase A XM_002375697.1 | 1,061 | 1,061 | 76 | 0 | 99 |
| *Afpnl-2* | JQ735891 | *Aspergillus flavus* NRRL3357 pectin lyase D XM_002374010.1 | 1,062 | 1,062 | 100 | 0 | 100 |
| *Afpnl-3* | JQ735892 | *Aspergillus oryzae* pel1 gene for pecyinlyase 1 AB029322.1 | 843 | 843 | 100 | 0 | 99 |
| *Afpnl-4* | JQ735893 | *Aspergillus flavus* NRRL3357 pectin lyase precursor XM_002381618.1 | 1,011 | 1,355 | 93 | 0 | 100 |
| *Afpnl-5* | JQ735894 | *Aspergillus flavus* NRRL3357 pectin lyase B XM_002384270.1 | 560 | 560 | 66 | 3e-156 | 100 |

**Table 36.3** The putative CDS and protein of five pectin lyase genes of *A. flavus* NIICC8142

| *Pnl* genes | Accession number | Size (bp) | Strand | CDS | Feature[a] | Amino acid |
|---|---|---|---|---|---|---|
| *Afpnl*-1 | JQ735890 | 760 | – | 170–646 (576 bp) | CDSl | 191 |
| *Afpnl*-2 | JQ735891 | 575 | + | 40–570 (528 bp) | CDSi | 176 |
| *Afpnl*-3 | JQ735892 | 472 | + | 48–467 (420 bp) | CDSf | 140 |
| *Afpnl*-4 | JQ735893 | 781 | + | 44–547 596–733 | CDSi | 214 |
| *Afpnl*-5 | JQ735894 | 455 | + | 153–451 | CDSf | 99 |

[a] *CDSf* first (starting with start codon), *CDSi* internal (internal exon), *CDSl* last coding segment (ending with stop codon)

lyase genes from different fungal sources as reported earlier (Dubey et al. 2012). The four pectin lyase genes sequences (JQ735891, JQ735892, JQ735893, JQ735894) of *A. flavus* occupied distinct position in the tree forming separate clusters comprising of two members each, respectively, while JQ735890 formed cluster with pectin lyase gene of *A. niger* (JF44761) (Fig. 36.3b).

The nucleotide sequences of these five pectin lyase genes were translated to respective protein sequences using translation tool and subjected to protein functional analysis using Pfam and INTERPROSCAN version 24. These protein sequences were analyzed for the presence of conserved motifs by means of MEME software. A total of three conserved motifs designated as I, II, and III occupied different positions in the five pectin lyase sequences (Fig. 36.4)

Motif I was observed in four of the pectin lyase genes, namely JQ35891-JQ35894 while motif II was observed in JQ35891 and JQ35893 and motif III was observed in JQ35890 and JQ35893. All the three motifs were found in the sequence JQ35893 (Fig. 36.4). The best possible matches of these three motifs are represented by respective motif logos (Fig. 36.5). The best possible amino acid sequences of these three motifs revealed their identity to pec_lyase_C superfamily on BLAST search indicating that they belong to a common group, i.e., trans-eliminases as

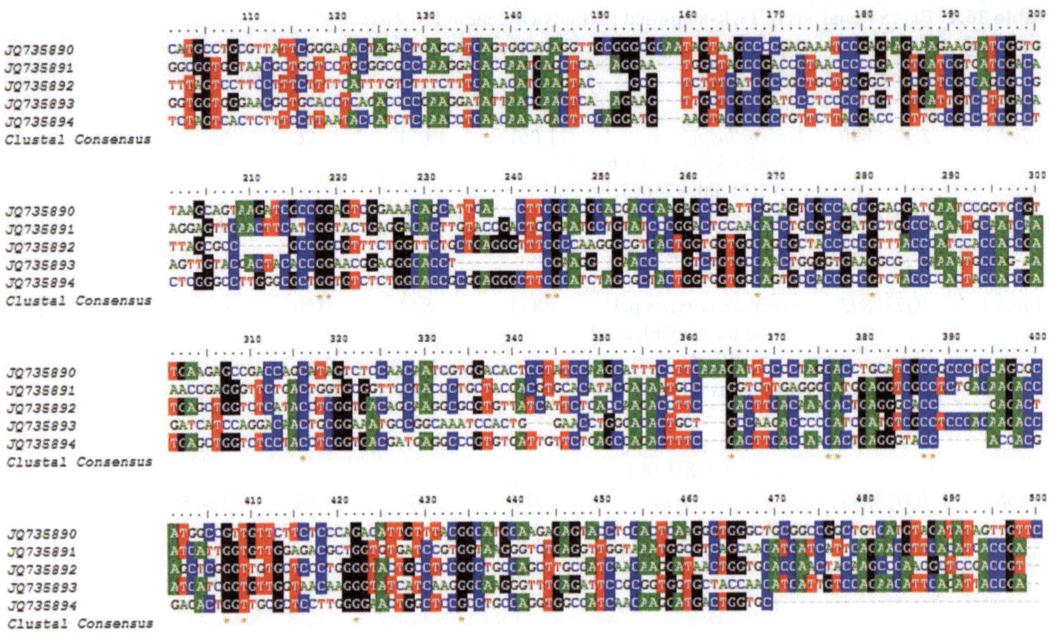


**Fig. 36.2** Multiple sequence alignment of five pectin lyase genes of *A. flavus* (* showed conserved nucleotide among these sequences)

reported earlier (Yadavet al. 2009; Dubey et al. 2010). Though these three motifs represents pec_lyase_C superfamily, their variable distribution among the sequenced PCR-amplified pectin lyase gene sequences needs to be further analyzed by cloning these genes in suitable vector and subjecting them to expression profiling.

## 36.4 Conclusions

The insights into the available genomic sequences of *A. flavus* NRRL 3357 revealed the presence of several pectin lyase genes. Based on the predicted pectin lyase gene sequences, specific primers were designed and an attempt was made to amplify the putative pectin lyase genes from different strains of *A. flavus.* A total of five genes from *A. flavus* NIICC 8142 were amplified, gel eluted, and sequenced and in silico characterized to reveal their identity. These putative genes need to be cloned in suitable vector, sequenced, and will be subjected to expression profiling. The *in silico* prediction of secondary, tertiary structure of these pectin lyase enzymes can further give an insight into the potential target for possible manipulation for enhancing the catalytic efficiency.

**Acknowledgments** The authors wish to acknowledge the Head of the Department of Biotechnology, D.D.U. Gorakhpur University, Gorakhpur for providing the infrastructural facilities. The financial support by the UGC, India in the form of UGC-Major Project (F. no. 37–133/2009-SR) to Dinesh Yadav and by the DST, India in the form of Fast Track Young Scientist Fellowship (FT/LS-125/2008) to Sangeeta Yadav is duly acknowledged.

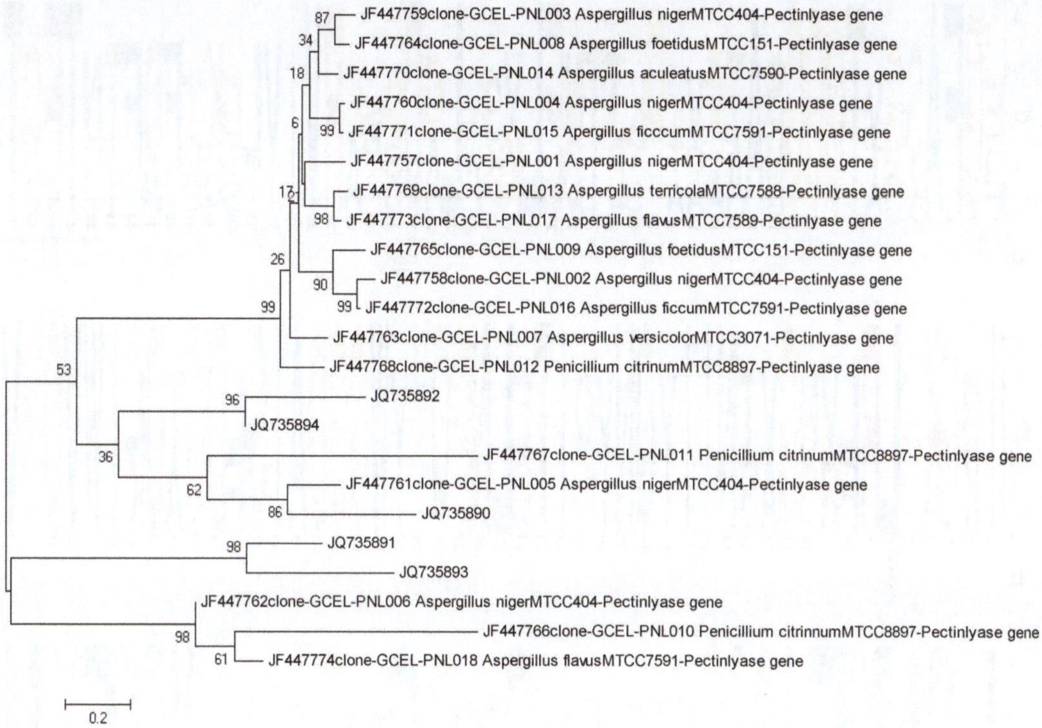


**Fig. 36.3** Phylogenetic tree constructed by NJ method using nucleotide sequences of pectin lyase genes of **a** *A. flavus* NIICC8142 and **b** *A. flavus* NIICC 8142 along with other fungal pectin lyase genes

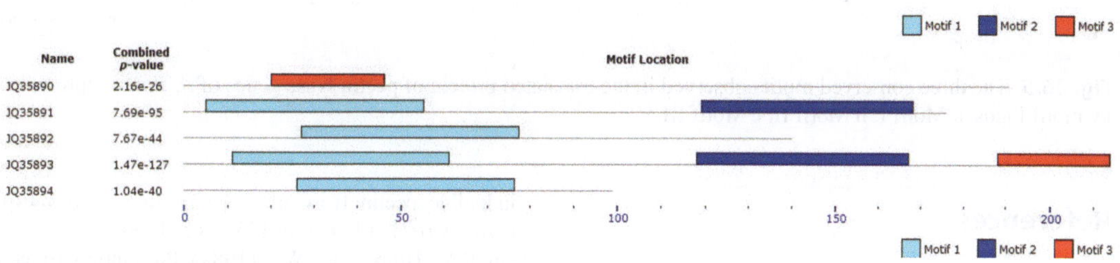


**Fig. 36.4** Distribution of three conserved motifs for PCR-amplified pectin lyase genes of *A. flavus*

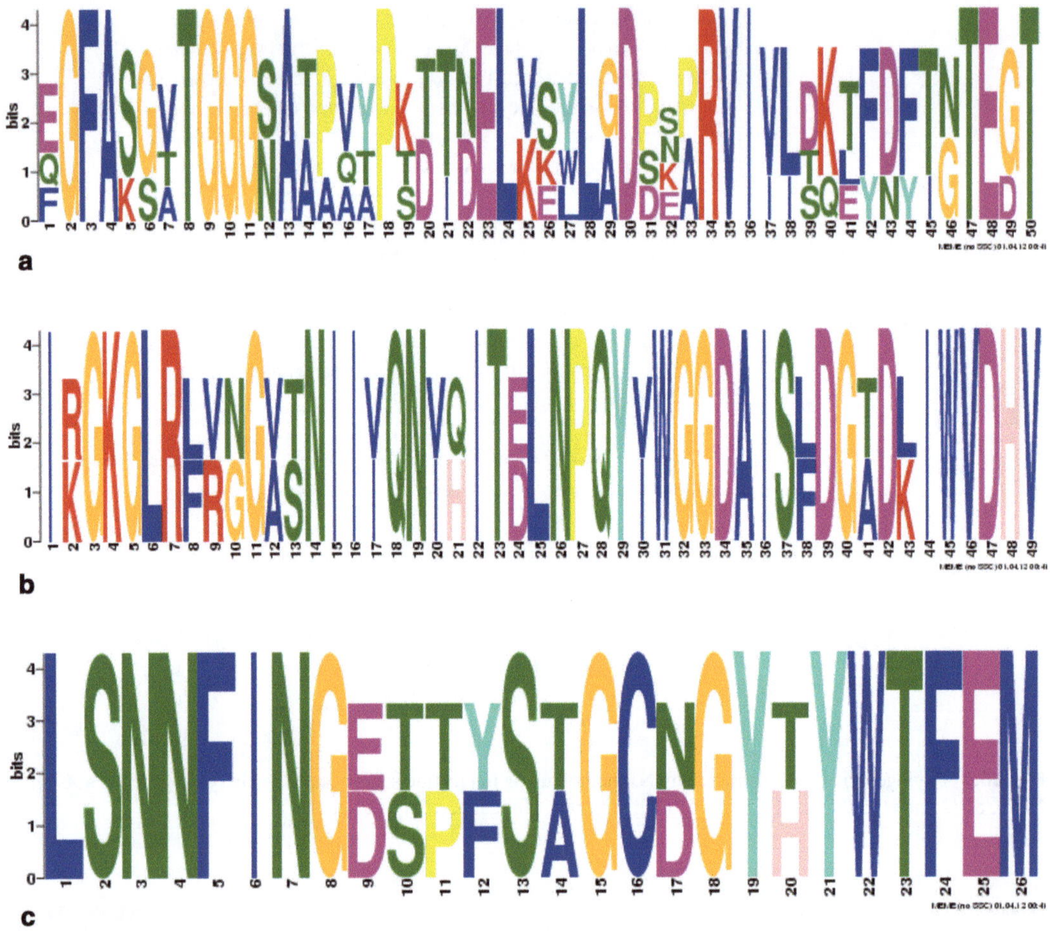

**Fig. 36.5** The three conserved motifs observed in the translated protein of pectin lyase genes of *A. flavus* represented by motif logos. **a** Motif I. **b** Motif II. **c** Motif III

## References


Alkorta I, Garbisu C, Llama MJ, Serra JL (1998) Industrial applications of pectic enzymes: a review. Pro Biochem 33:21–28

Altschul SF, Gish W, Miller W, Myers EW, Lipman DJ (1990) Basic local alignment search tool. J Mol Biol 215(3):403–410

Bailey TL, Elkan C (1994) Fitting a mixture model by expectation maximization to discover motifs in biopolymers. Proceedings of the Second International Conference on intelligent systems for molecular biology, AAAI Press, Menlo Park, 28–36

Burland TG (2000) DNASTAR's laser gene sequence analysis software. Methods Mol Biol 132:71–91

Chatterjee A, McEvoy JL, Chambost JP, Blasco F, Chatterjee AK (1991) Nucleotide sequence and molecular characterization of *pnlA*, the structural gene for damage inducible pectin lyase of *Erwinia carotovora* subsp. *carotovora*71. J Bacteriol 173:1765–1769

Dean RA, Timberlake WE (1989) Production of cell wall degrading enzymes by *Aspergillus nidulans*: a model system for fungal pathogenesis. Plant Cell 1:265–273

Dubey AK, Yadav S, Kumar M, Singh VK, Sarangi BK, Yadav D (2010) In silico characterization of pectatelyase protein sequences from different source organism. Enzyme Res (Article ID 950230)

Dubey AK, Yadav S, Anand G, Bisht NC, Yadav D (2012) In sights to sequences of PCR amplified pectin lyase genes from different fungal strains. J Bioinform 13(1):80–92

Finn RD, Mistry J, Tate J, Coggill P, Heger A, Pollington JE, Gavin OL, Gunesekaran P, Ceric G, Forslund K, Holm L, Sonnhammer EL, Eddy SR, Bateman A (2010) The Pfam protein families database. Nucleic Acids Res 38:211–222

Gummadi SN, Kumar DS (2005) Microbial pectic transeliminases. Biotechnol Lett 27(7):451–458

Harmsen JAM, Kusters-van Someren MA, Visser J (1990) Cloning and expression of a second *Aspergillus niger* pectin lyase gene (*pel*A): indications of a pectin lyase gene family in *Aspergillus niger*. Curr Genet 18(2):161–166

Hoondal GS, Tiwari R, Tewari R, Dahiya N, Beg Q (2002) Microbial alkaline pectinases and their industrial applications: a review. Appl Microbiol Biotechnol 259(4):409–418

Jayani RS, Saxena S, Gupta R (2005) Microbial pectinolytic enzymes: a review. Process Biochem 40(9):2931–2944

Kashyap DR, Vohra PK, Chopra S, Tewari R (2001) Applications of pectinases in the commercial sector: a review. Bioresour Technol 77(3):215–227

Kitamoto N, Yasuda-yashino S, Ohmiya K, Tsukagoshi NA (2001) Second pectin lyase gene (pel2) from *Aspergillus oryzae* KBN 616: its sequence analysis and over expression and characterization of the gene products. J Bioscience Bioeng 91:378–381

Maniatis T, Fritsch EF, Sambrook J (1982). Molecular cloning: a laboratory manual. Cold Spring Harbor Lab Press, Cold Spring Harbor

Pedrolli DB, Carmona EC (2009) Pectin lyase from *Aspergillus giganteus*: comparative study of productivity of submerged fermentation on citrus pectin and orange waste. Prikl Biokhim Mikrobiol 45(6):677–683

Pedrolli DM, Monteiro AC, Gomes E, Carmonia EC (2009) Pectin and pectinases: production, characterization and industrial application of microbial pectinolytic enzymes. Open Biotechnol J 3:9–18

Pruitt KD, Tatusova T, Maglott DR (2007) NCBI reference sequences (Ref Seq): a curated non-redundant sequence database of genomes, transcripts and proteins. Nucleic Acids Res 35D:61–65

Quevillon E, Silventoinen V, Pillai S, Harte N, Mulder N, Apweiler R, Lopez R (2005) Inter Pro Scan, protein domains identifier. Nucleic Acids Res 33(suppl 2):116

Saitou N, Nei M (1987) The neighbor-joining method: a new method for reconstructing phylogenetic trees. Mol Biol Evolut 4(4):406

Sakamoto T, Kawasaki H, Sakai T (1996) Molecular cloning and nucleotide sequence of the gene encoding phosphate inducible pectin lyase of *Bacillus subtilise*. FEBS Lett 398(2–3):269–273

Solovyev V, Kosarev P, Seledsov I, Vorobyev D (2006) Automatic annotation of eukaryotic genes, pseudogenes and promoters. Genome Biol 7:1–12

Tamura K, Dudley J, Nei M, Kumar S (2007) Molecular evolutionary genetics analysis (MEGA) software version 4.0. Mol Biol Evolut 24(8):1596

Taragano VM, Pelosof AMR (1999) Application of Doehlert design for water activity, pH and fermentation time, optimization for *Aspergillus niger* pectinolytic activity, production in solid state and submerged fermentation. Enzyme Microbiol Technol 25:411–419

Templeton MD, Sharrock KR, Bowen JK, Crowhust RN, Rikkerink EHA (1994) The pectin lyase-encoding gene(pnl) family from *Glomerella cingulata*: characterization of pnlA and its expression in yeast. Gene 142(1):141–146

Thompson JD, Higgins DG, Gibson TJ (1994) Improving the sensitivity of progressive multiple sequence alignment through sequence weighting positions-specific gap penalties and weight matrix choice. Nucleic Acids Res 22(22):4673

Yadav S, Yadav PK, Yadav D, Yadav KDS (2008) Purification and characterization of an alkaline pectin lyase from *Aspergillus flavus*. Process Biochem 43:547–552

Yadav S, Yadav PK, Yadav D, Yadav KDS (2009a) Pectin lyase: a review. Process Biochem 44(1):1–10

Yadav S, Yadav PK, Yadav D, Yadav KDS (2009b) Purification and characterization of pectin lyaseproducted by *Aspergillus terricola* suitable for retting of natural fibers. Appl Biochem Biotechnol 159:270–283

Yadav S, Yadav PK, Yadav D, Yadav KDS (2009c) Purification and characterization of pectin lyase secreted by *Penicillium citrinum*. Biochemistry (Moscow) 74(7):800–806

Yadav PK, Singh VK, Yadav S, Yadav KDS, Yadav D (2009d) In silico analysis of pectin lyase and pectinase sequences. Biochemistry (Moscow) 74(9):1049–1055

Yadav S, Dubey AK, Anand G, Yadav D (2012) Characterization of a neutral pectin lyase produced by Oidiodendron echinulatum MTCC 1356 in solid state fermentation. J Basic Microbiol 52:713–720

# 37 Antifungal Activity of *Agave* Species from Gujarat, India

Anjisha R. Maharshi and Vrinda S. Thaker

**Abstract**

In the present investigation, crude extracts of five different species of *Agave* (*Agave americana, Agave ferox, Agave montana, Agave scabra* and *Agave marginata*) have been examined against six plant pathogenic fungi, viz. *Macrophomina phaseolina, Alternaria porii, Aspergillus awamorii, Aspergillus niger, Fusarium udum* and *Fusarium solani* using media poisoning method. The percent inhibition of hyphal growth was measured after the seventh day of incubation. Vast variations were observed in their activities on different pathogens. Amongst the fungi studied, the highest inhibition (more than 50 %) of hyphal growth was observed for *M. phaseolina* by all the extracts. These data suggest that the different *Agave* species have potential as antifungal agents with a broad range of activity. The antifungal compound isolated from these plants can be used as possible ecofriendly plant-based fungicides to control plant diseases.

**Keywords**

*Agave* sp. · Antifungal activity · *Macrophomina phaseolina* · Inhibition

## 37.1 Introduction

Effective management of a plant disease is the key to save plants from microbes, as plants have both economic and aesthetic values. Nowadays, quick and effective management of plant diseases and microbial contamination in several agricultural commodities is generally achieved by the use of synthetic pesticides (Agrios 1997). Despite serious environmental implications associated with their use, chemical fungicides remain the first line of defense against effective and efficient management of pathogens. Many of these chemical fungicides cause tremendous environmental load, thereby adversely affecting the agro-ecosystem (Zadoks 1993; Guleria and Kumar 2009).

The recurrent and indiscriminate use of fungicides have posed a serious threat to human

V. S. Thaker (✉) · A. R. Maharshi
Centre for Advanced Studies in Plant Biotechnology and Genetic Engineering, Department of Biosciences, Saurashtra University, Rajkot 360005, Gujarat, India
e-mail: cpbge@yahoo.co.in

R. N. Kharwar et al. (eds.), *Microbial Diversity and Biotechnology in Food Security*,
DOI 10.1007/978-81-322-1801-2_37, © Springer India 2014

health and to the existing human eco-geographical conditions as some of them have already been proved to be mutagenic, carcinogenic, or teratogenic. Also, many pathogenic microorganisms and insect pests have developed resistance against these chemical pesticides (Williams and Heymann 1998; Witte 1998; Yadav 2010).

Due to increased awareness about the risks involved in the use of chemical pesticides, much attention is being focused on alternative methods of pathogen control. The spiralling cost of chemical fungicides, the threat to human health, particularly in developing countries, by the accumulation of obnoxious chemical residues due to continuous use of fungicides and the development of resistance races to these chemicals are now forcing scientists to look for methods, which are ecologically friendly, safe and specific for pathogens (Joseph et al. 2008; Mazid et al. 2011).

Keeping in view the drawbacks of chemicals, the use of plant extracts in the management of plant diseases is gaining importance. Various plant products like plant extracts, essential oils, gum, resins etc. were shown to exert biological activity in vitro and in vivo and are used as biofungicidal compounds (El-Mougy and Alhabeb 2009; Fawzi et al. 2009; Al-Askar and Rashad 2010).

Biological approaches for the control of pathogens on aerial surfaces have been worked out extensively over the past 30 years (Blakeman and Fokkema 1982; Elad 1993; Sahayaraj et al. 2011). During this period, most approaches employed for the biological control of diseases of aerial plant surfaces have concentrated on the use of a single, empirically selected biocontrol agent to antagonize a single pathogen (Wilson 1997).

Plants have been known for their medicinal and antimicrobial properties since ancient times. For this reason, attention has been diverted to alternative, safe and economic methods for the management of pathogenic microorganisms from plant products (Khallil 2001; Begum et al. 2010). The potential of plant extracts to control plant diseases has long been recognized (Ark and Thompson 1959). There are approximately 250,000 species of higher plants, of which only 5–15 % have been studied for their therapeutic value (Rojas et al. 2003). The use of plants for human disease control attracts more attention, compared with its use in that of plant and animal diseases (Newton et al. 2002; Cano and Volpato 2004). In crop protection studies, various natural plant products have been identified and employed to control postharvest diseases (Mekbib et al. 2007; Ilondu 2011).

*Agave* is a plant which produces medically useful substances such as vitamins and steroid precursors like saponins and fructans (Gentry 1982; Sanchez et al. 2005; Guleria and Kumar 2009). The antifungal activity of *Agave* is due to steroidal saponins (Yang et al. 2006). Saponins are secondary plant metabolites that occur in a wide range of plant species (Hostettmann and Marston 1995). The natural role of saponins in plants is to provide protection against attack by pathogens and pests (Morrissey and Osbourn 1999). The antifungal activity of steroidal saponins, particularly against agricultural pathogens, has been known for a long time (Wolters 1966; Imai et al. 1967; Dimoglo et al. 1985; Sahu et al. 2008; Zwane et al. 2011).

Other reported activities for this class of compounds include antibacterial (Chattopadhyay et al. 2001), anti-insect (Kozukue et al. 2004) and anti-yeast activities (Miyakoshi et al. 2000). The antifungal potency of these compounds is not associated with cytotoxicity to mammalian cells (Dini et al. 2001). Because of their anti-pest properties, saponins can be utilized for the development of insecticidal and fungicidal preparations (Sparg et al. 2004).

Natural plant products are usually inherently less toxic than conventional fungicides, often are effective in very small quantities, and generally decompose quickly, thereby resulting in lower exposures and largely avoiding pollution problems (Huang and Chou 2005; Gupta and Dikshit 2010). Considering these issues, the present study was undertaken to evaluate the effectiveness of the extract from five species of *Agave,* namely *Agave americana, Agave ferox, Agave montana, Agave scabra* and *Agave marginata* on the growth of six pathogenic fungi: *Macrophomina phaseolina,*

*Alternaria porii, Aspergillus awamorii, Aspergillus niger, Fusarium udum* and *Fusarium solani.*

## 37.2 Materials and Methods

### 37.2.1 Preparation of Fungal Inoculum

The fungi *M. phaseolina, A. porii, A. niger* and *F. solani* were isolated from their host. *A. awamorii* (MTCC-548) and *F. udum* (MTCC-2204) were obtained from Microbial Type Culture Collection (MTCC), Institute of Microbial Technology (IMTECH), Chandigarh, India. The lyophilized culture was revived on suitable media and then streaked that activated culture on potato-dextrose-agar (PDA) plate. After 7 days of growth, a disc of fungal culture was cut with 1 cm borer diameter and was placed in the centre of the other PDA plate. A disc from the peripheral margin of the 7-day-old growth of all fungi was used as inoculum.

### 37.2.3 Preparation of Media

The extracts of all the plants to be tested were mixed with PDA at 0.5/20 ml distilled water. The final volume was made up to 20 ml in glass tubes. The media in the tubes was autoclaved at 15 lb pressure and 121 °C temperature for 15 min. The sterilized medium (20 ml) was then poured into the radiation-sterile petri plates ($90 \times 15$ mm diameter) under aseptic conditions.

### 37.2.4 Inoculation of Fungi

The centre of each poisoned PDA plate (with plant extracts) was inoculated with one fungal disc (10 mm diameter) of 7-day-old cultured plate. Control plates lacked plant extracts. The plates were incubated at $28 \pm 2$ °C for 7 days. The antifungal activities of plant extracts were observed every 24 h and noted at the seventh day of incubation by measuring the diameter of test and control in millimetre. The toxicities of plant extracts were recorded in terms of percent mycelial inhibition (Al-Burtamani et al. 2005) against the test fungi. The percent inhibition of hyphal growth was calculated based on the following formula:

$$\text{Percent inhibition} = \frac{\text{Diameter of fungal colony in treatments}}{\text{Diameter of fungal colony in control}} \times 100\,\%$$

### 37.2.2 Preparation of Leaves Extract

The leaves of *A. americana, A. ferox, A. montana, A. scabra* and *A. marginata* were collected and washed with cleaned water. Fresh weights, 250 g, were measured and were kept to dry at 60 °C temperature. Dried leaves were grounded into fine powder and extracted with 80 % methanol for 2 days. The plant debris was removed by twice centrifuging at $5{,}000 \times g$ for 20 min. The supernatants were collected and were allowed to evaporate, then dissolved in final volume (i.e. 10 ml) of sterile distilled water. These stock solutions were used for an antifungal activity against all fungi by media poisoning method.

## 37.3 Results and Discussion

In the present study, the extracts of the five species of *Agave* showed antifungal activities against the six pathogenic fungi tested by media poisoning method, and they failed to inhibit *A. niger,* (Fig. 37.3) and the inhibitory levels were different in all cases. The radial growth assay was performed, and results were noted down. At the end of the seventh day of growth period, the highest percent inhibition was exhibited for *M. phaseolina* by all the extracts. The highest inhibition was observed in *A. marginata* (64.75 %), followed by *A. americana* (61.69 %), *A. ferox* (59.39 %), *A. montana* (58.24 %) and the lowest

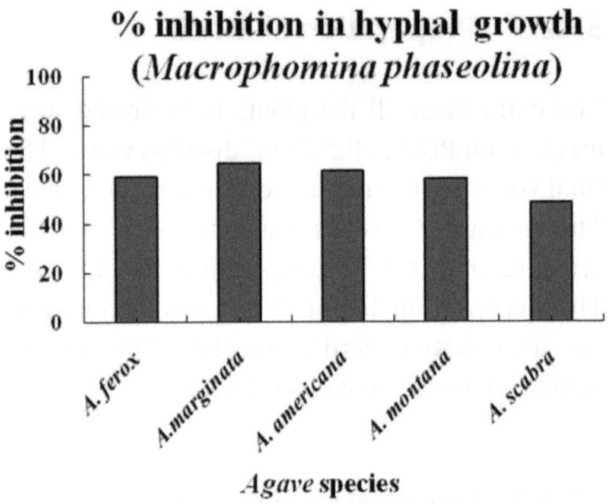


**Fig. 37.1** Percent inhibition in hyphal growth of *Macrophomina phaseolina* by different species of *Agave*

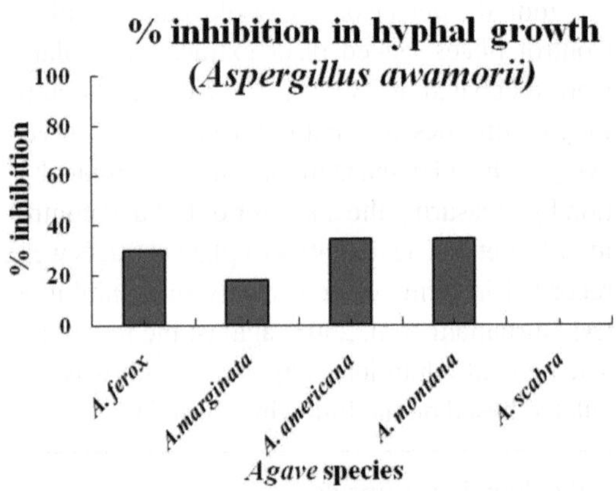


**Fig. 37.2** Percent inhibition in hyphal growth of *Aspergillus awamorii* by different species of *Agave*

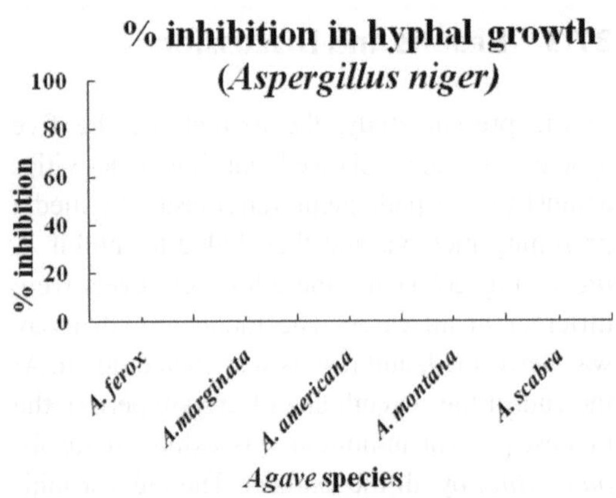


**Fig. 37.3** Percent inhibition in hyphal growth of *Aspergillus niger* by different species of *Agave*

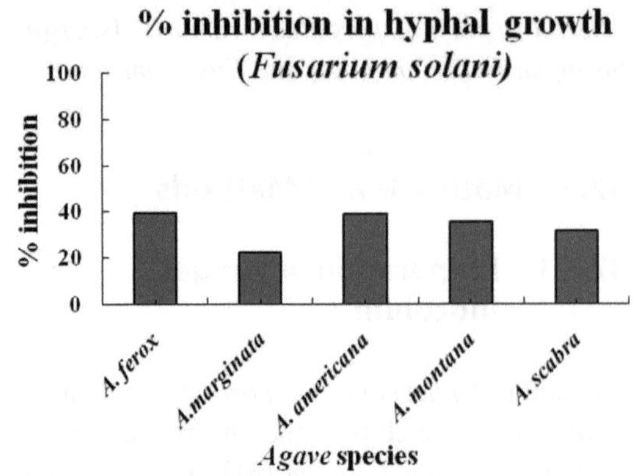


**Fig. 37.4** Percent inhibition in hyphal growth of *Fusarium solani* by different species of *Agave*

as 48.66% in *A. scabra.* So, four species of *Agave* showed more than 50% inhibition for *M. phaseolina* (Fig. 37.1).

*A. awamorii* showed higher growth inhibition by *A. americana* (34.44%) and *A. montana* (34.17%), followed by *A. ferox* (30.00%). There was no significant inhibition observed in *A. marginata* (18.06%), and *A. scabra* failed to show any inhibition (Fig. 37.2). For *A. niger,* no inhibition was observed in any extract (Fig. 37.3).

*F. solani* showed maximum inhibition by *A. ferox* (39.34%), followed by *A. americana* (38.86%), *A. montana* (35.55%), *A. scabra* (31.75%) and *A. marginata* (22.27%; Fig. 37.4). On the other hand, *F. udum* was the least affected by these extracts; *A. scabra* extract was found to be ineffective. Amongst these extracts, the highest activity was seen in *A. marginata* (27.61%), followed by *A. montana* (26.12%), *A. ferox* (23.88%), with the least in *A. americana* (8.00%; Fig. 37.5). *A. porii* was 29.05% inhibited by *A. ferox*; *A. americana* was only able to inhibit 10.14% of the hyphal growth, and negligible inhibition was shown by *A. marginata* (5.41%). *A. montana* and *A. scabra* act as growth-promoting agents (Fig. 37.6).

Bobbarala et al. (2009) reported in their study that significant reduction was seen in the growth of *M. phaseolina* with extracts of 49 plants. *M. phaseolina* appears to be a non-host-specific fungus. Physiological specialization of the fungus is not well demonstrated. High level of variation in morphology, physiology and

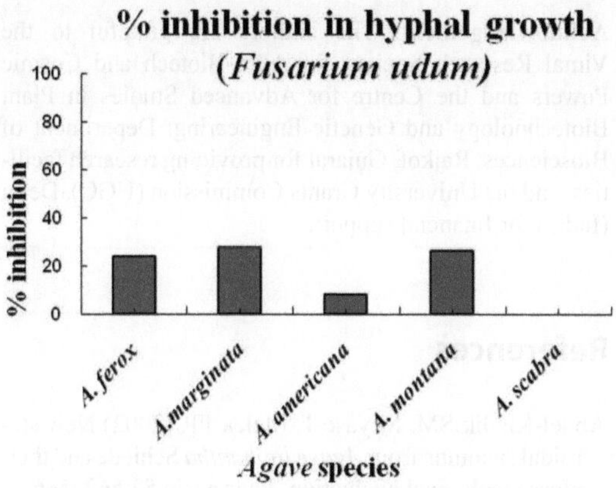


**Fig. 37.5** Percent inhibition in hyphal growth of *Fusarium udum* by different species of *Agave*

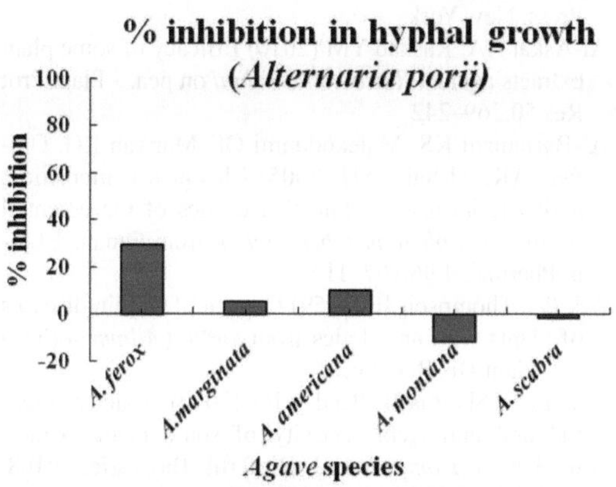


**Fig. 37.6** Percent inhibition in hyphal growth of *Alternaria porii* by different species of *Agave*

pathogenesis has been reported even when isolated from different parts of the same plant (Dhingra and Sinclair 1973; Khan 2007).

Totally contradictory to our finding, Guleria and Kumar (2009) have reported that leaf extracts of *A. americana* showed antifungal activity against *Alternaria brassicae,* a causal agent of Alternaria blight of *Brassica juncea*. The inhibitory effect of the plant extracts might be attributed to the presence of antifungal compounds. Currently, only limited information is available regarding the biological activity of compounds isolated from *Agave* sp. Many workers have reported the antifungal and antimicrobial activity of *A. americana* (Pandey et al. 1992; Jin et al. 2002; Guleria and Kumar 2009; Chetan et al. 2010; Khan et al. 2010), while very few studies have been carried out on other species of *Agave* (Abdel-Khalik et al. 2002; Verastegui et al. 2008; Santos et al. 2009; Hammuel et al. 2011).

*Agave lophanta* has steroidal saponins with activity against stomach ulcers and anti-inflammatory properties (Abdel-Khalik et al. 2002). Antifungal activity of steroidal saponins from *A. americana* has been reported by Yang et al. (2006) against *Candida albicans, C. glabrata* and *Aspergillus fumigatus*. The antimicrobial activity of *Agave lecheguilla* against pathogens such as *Clostridium perfringens, Salmonella enteritidis, Proteus vulgaris, Y. enterocolitica,* Actinomycetes and molds has been also reported (Verastegui et al. 1996). Verastegui et al. (2008) have also reported the antimicrobial activity of *A. scabra* and *Agave picta* extracts in *Escherichia coli, Listeria monocytogenes, Staphylococcus aureus* and *Vibrio cholerae*.

Saponins also exhibit anticholesterolemic, anticancer (Haridas et al. 2001; Afrose et al. 2009), adjuvant (Behboudi et al. 1999; Ragupathi et al. 2010) and haemolytic (Oh et al. 2000; Hassan et al. 2010) activities. Some saponins have negative effects and were detrimental to human health; steroidal glycoalkaloids can be toxic when they are ingested (Friedman 2002). Saponins produced complexes with sterols and cause sterol-dependent membrane permeabilization (Morrissey and Osbourn 1999). The antifungal activity of saponins was generally attributed to this membrane permeabilizing property. Some fungi were resistant to the toxic effects of saponins because they have little or no sterols in their membranes, while others produce enzymes which specifically detoxify the saponins of their host plant (Osbourn et al. 1995; Ito et al. 2004; Coleman et al. 2010).

Natural chemicals and their use for integrated plant protection is one of the main interest areas of research workers all over the world (Kiran et al. 2006; Joseph et al. 2008; Chandler et al. 2011). The results of present investigation are clear indication of the potential that plant extracts and their compounds can be used to control fungal pathogens. It is evident from the results that all the plant extracts significantly inhibited the radial growth of isolated fungus. The formulation studies of the plant extracts can be successfully devised as fungicides using a simple process with minimum instrumentation and few chemical agents.

An interesting observation was that some of the test chemicals showed stimulation rather than inhibition. Stimulation of radial mycelial growth of fungi with test chemicals may be due to utilization of the chemicals or/of their degradative products (fungal enzymes) by the fungi for their growth and development; for this, the mode of action should be worked out. The mode of action of saponins in imparting antifungal activity is not fully understood; according to one possible mechanism, their activity could be due to their ability to form complexes with sterol constituents of fungal cell membranes, leading to the loss of membrane structure (Keukens et al. 1995; Morrissey and Osbourn 1999; Simons et al. 2006; Shukla et al. 2011).

In view of the harmful effects associated with the use of synthetic chemical fungicides, the worldwide trend calls for environmentally safe methods of plant disease control such as biological control, use of induced resistance by biotic and abiotic means (Lyon et al. 1995; Guleria et al. 2005; Guleria and Kumar 2006a) and the use of biodegradable natural products, especially from medicinal plants (Prithiviraj and Singh 1995; Guleria and Kumar 2006b; Begum et al. 2010) for sustainable agriculture. The present work may add one more building block in this research field.

## 37.4 Conclusions

On the basis of results, it is concluded that the *Agave* species are important for the development of a bio-fungicide preparation for the management of fungal pathogens of crop plants. However, this activity could differ depending on the method used to determine it and on its applicability in different processes. The biological activity exhibited by the *Agave* species and their wide geographic distribution suggest that, in future, *Agaves* may have applications important to human health and medicine, in addition to their traditional uses.

**Acknowledgements** The authors are grateful to the Vimal Research Society for Agro-Biotech and Cosmic Powers and the Centre for Advanced Studies in Plant Biotechnology and Genetic Engineering, Department of Biosciences, Rajkot, Gujarat for providing research facilities, and the University Grants Commission (UGC), Delhi (India) for financial support.

## References

Abdel-Khalik SM, Miyase T, Melek FR (2002) New steroidal saponins from *Agave lophantha* Schiede and their pharmacological evaluation. Pharmazie 57:562–566

Afrose S, Hossain MS, Maki T, Tsujii H (2009) Karaya root saponin exerts a hypocholesterolemic response in rats fed a high-cholesterol diet. Nutr Res 29(5):350–354

Agrios GN (1997) Plant pathology, 4th edn. Academic Press, New York

Al-Askar AA, Rashad YM (2010) Efficacy of some plant extracts against *Rhizoctonia solani* on pea. J Plant Prot Res 50:269–242

Al-Burtamani KS, Majekodunmi OF, Marwah RG, Onifade AK, Al-Saidi SH (2005) Chemical composition, antibacterial and antifungal activities of the essential oil of *Haplophyllum tuberculatum* from Oman. J EthnoPharmacol 96:107–112

Ark PA, Thompson JP (1959) Control of certain diseases of plants with antibiotics from garlic (*Allium sativum* L.). Plant Dis Rep 43:276

Begum N, Sharma B, Pandey RS (2010) Toxicity potential and anti AchE activity of some plant extracts in *Musca Domestica*. J Biofertil Biopestici 2:108. doi:10.4172/2155-6202.1000108

Behboudi S, Morein B, Villacres-Eriksson M (1999) Quillaja saponin formulations that stimulate proinflammatory cytokins elicit a potent acquired cell-mediated immunity. Scand J Immunol 50:371–377

Blakeman JP, Fokkema NJ (1982) Potential for biocontrol of plant diseases on the phylloplane. Annu Rev Phytopathol 20:167–192

Bobbarala V, Chadaram RK, Vadlapudi V, Katikala PK (2009) Medicinal plants as alternative biocontrol agents in the control of seed borne pathogen *Macrophomina phaseolina*. J Pharm Res 2:1045–1048

Cano JH, Volpato G (2004) Herbal mixtures in the traditional medicine of eastern Cuba. J Ethnopharmacol 90:293–316

Chandler D, Bailey AS, Tatchell GM, Davidson G, Greaves J, Grant WP (2011) The development, regulation and use of biopesticides for integrated pest management. Phil Trans R Soc B 366:1987–1998

Chattopadhyay DK, Maiti AP, Kundu MS, Chakraborty R, Bhadra SC, Mandal AB (2001) Antimicrobial activity of *Alstonia macrophylla*: a folklore of bay islands. J Ethnopharmacol 77:49–55

Chetan AC, Patel RM, Dakhara SL, Jariwala JK (2010) In vitro cytotoxicity study of *Agave americana*, Strychnos nux-vomica and *Areca catechu* extract using MCF-7 cell line. J Adv Pharm Tech Res 1:245–252

Coleman JJ, Okoli I, Tegos GP, Holson EB, Wagner FF, Hamblin MR, Mylonakis E (2010) Characterization of plant-derived saponin natural products against *Candida albicans*. ACS Chem Biol 5:321–332

Dhingra CD, Sinclair JB (1973) Variation among the isolates of *Macrophomina phaseolina* (*Rhizoctonia bataticola*) from different regions. Phytopathol Z 76:200–204

Dimoglo AS, Choban IN, Bersuker IB, Kintya PK, Balashova NN (1985) Structure-activity correlations for the antioxidant and antifungal properties of steroid glycosides. Bioorg Khim 11:408–413

Dini I, Schettino O, Simioli T, Dini A (2001) Studies on the constituents of *Chenopodium quinoa* seeds: isolation and characterization of new triterpene saponins. J Agric Food Chem 49:741–746

Elad Y (1993) Microbial suppression of infection by foliar plant pathogens. IOBC Bull 16:3–7

El-Mougy NS, Alhabeb RS (2009) Inhibitory effects of powdered caraway and peppermint extracts on pea root rot under greenhouse conditions. J Plant Protec Res 49:93–96

Fawzi EM, Khalil AA, Afifi AF (2009) Antifungal effect of some plant extracts on *Alternaria alternata* and *Fusarium oxysporum*. Afr J Biotechnol 8:2590–2597

Friedman M (2002) Tomato glycoalkaloids: role in the plant and the diet. J Agric Food Chem 50:751–5780

Gentry HS (1982) *Agaves* of continental North America. The University of Arizona Press, Tucson

Guleria S, Kumar A (2006a) *Azadirachta indica* leaf extract induces resistance in sesame against *Alternaria* leaf spot disease. J Cell Mol Biol 5:81–86

Guleria S, Kumar A (2006b) Antifungal activity of some Himalayan medicinal plants using direct bioautography. J Cell Mol Biol 5:95–98

Guleria S, Kumar A (2009) Antifungal activity of *Agave americana* leaf extract against *Alternaria brassicae*, causal agent of *Alternaria* blight of Indian mustard (*Brassica juncea*). Arch Phytopathol Plant Protec 42:370–375

Guleria S, Sohal BS, Mann APS (2005) Salicylic acid treatment and/or Erysiphe polygoni inoculation on phenylalanine ammonia-lyase and peroxidase content and accumulation of phenolics in pea leaves. J Veg Sci 11:71–79

Gupta S, Dikshit AK (2010) Biopesticides: an ecofriendly approach for pest control. J Biopest 3:186–188

Hammuel C, Yebpella GG, Shallangwa GA, Magomya AM, Agbaji AS (2011) Phytochemical and antimicrobial screening of methanol and aqueous extracts of *Agave sisalana*. Acta Poloniae Pharmaceut Drug Res 68:535–539

Haridas V, Arntzen CJ, Gutterman JU (2001) Avicins, a family of triterpenoid saponins from *Acacia victoriae* (Bentham), inhibit activation of nuclear factor-kappa B by inhibiting both its nuclear localization and ability to bind DNA. Proc Nat Acad Sci U S A 98:11557–11562

Hassan SM, Haq AU, Byrd JA, Berhow MA, Cartwright AL, Bailey CA (2010) Haemolytic and antimicrobial activities of saponin-rich extracts from guar meal. Food Chem 119:600–605

Hostettmann KA, Marston A (1995) Saponins. Chemistry and pharmacology of natural products. Cambridge Univesity Press, Cambridge, pp. 239–284

Huang HC, Chou CH (2005) Impact of plant disease biocontrol and allelopathy on biodiversity and agricultural sustainability. Plant Pathol Bull 14:1–12

Ilondu EM (2011) Evaluation of some aqueous plant extracts used in the control of pawpaw fruit (*Carica papaya* L.) rot fungi. J Appl Biosci 37:2419–2424

Imai S, Fujioka S, Murata E, Goto M, Kawasaki T, Yamauchi T (1967) Bioassay of crude drugs and oriental crude drug preparations. XXII. Search for biologically active plant ingredients by means of antimicrobial tests. 4. Antifungal activity of dioscin and related compounds. Takeda Kenkyusho Nenpo 26:76–83

Ito S, Eto T, Tanaka S, Yamauchi N, Takahara H, Ikeda T (2004) Tomatidine and lycotetraose, hydrolysis products of alpha-tomatine by *Fusarium oxysporum* var. tomatinase, suppress induced defense responses in tomato cells. FEBS Lett 571:31–34

Jin JM, Liu XK, Yang CR (2002) New steroidal saponin from fermented leaves of *Agave americana*. Acta Bot Yunnanica 24:539–542

Joseph B, Dar MA, Kumar V (2008) Bioefficacy of plant extracts to control *Fusarium solani* f. Sp. *melongenae* incitant of brinjal wilt. Global J Biotechnol Biochem 3:56–59

Keukens EAJ, de Vrihe T, van den Boom C (1995) Molecular basis of glycoalkaloid induced membrane disruption. Biochem Biophys Acta 1240:216–228

Khallil A-RM (2001) Phytofungitoxic properties in the aqueous extracts of some plants. Pak J Biol Sci 4:392–394

Khan MTJ, Ahmad K, Alvi MN, Noor-ul-Amin, Mansoor B, Saeed MA, Khan FZ, Jamshaid M (2010) Antibacterial and irritant activities of organic solvent extracts of *Agave americana* Linn., *Albizzia lebbek* Benth. *Achyranthes aspera* Linn. and *Abutilon indicum* Linn—A preliminary investigation. Pak J Zool 42:93–97

Khan SN (2007) *Macrophomina phaseolina* as causal agent for charcoal rot of sunflower. Mycopath 5(2):111–118

Kiran K, Linguraju S, Adiver S (2006) Effect of plant extract on *Sclerotium rolfsii*, the incitant of stem rot of ground nut. J Mycol Pl Pathol 36:77–79

Kozukue N, Han J, Lee K, Friedman M (2004) Dehydrotomatineand α-tomatine content in tomato fruits and vegetative plant tissues. J Agric Food Chem 52:2079–2083

Lyon GD, Reglinski T, Newton AC (1995) Novel disease control compounds: the potential to 'immunize' plants against infection. Plant Pathol 44:407–427

Mazid S, Rajkhowa RC, Kalita JC (2011) A review on the use of biopesticides in insect pest management. Inter J of Sci Adv Technol 1:169–178

Mekbib SB, Regnier TJC, Korsten L (2007) Control of *Penicillium digitatum* on citrus fruit using two plant extracts and study of their mode of action. Phytoparasitica 35(3):264–276

Miyakoshi M, Tamura Y, Masuda H, Misutani K, Tanaka O, Ikeda T, Ohtani K, Kasai R, Kazuo Y (2000) Antiyeast steroidal saponins from *Yucca schidigera* (*Mohave Yucca*), a new anti-food-deteriorating agent. J Nat Prod 63:332–338

Morrissey JP, Osbourn AE (1999) Fungal resistance to plant antibiotics as a mechanism of pathogenesis. Microbiol Mol Boil Rev 63:708–724

Newton SM, Lau C, Gurcha SS, Besra GS, Wright CW (2002) The evaluation of forty-three plant species for in vitro antimycobacterial activities: Isolation of active constituents from *Psoralea corylifolia* and *Sanguinaria canadensis*. J Ethnopharmacol 79:57–67

Oh S, Kinjo J, Shii Y, Ikeda T, Nohara T, Ahn K, Kim J, Lee H (2000) Effects of triterpenoids from *Pueraria lobata* on immunohemolysis: β-D-glucuronic acid plays an active role in anticomplementary activity in vitro. Planta Med 66:506–510

Osbourn A, Bowyer P, Lunness P, Clarke B, Daniels M (1995) Fungal pathogens of oat roots and tomato leaves employ closely related enzymes to the detoxify different host plant saponins. Mol Plant Microbe Interact 8:971–978

Pandey JC, Kumar R, Gupta RC (1992) Possibility of biological control of rhizome rot of ginger by different antagonists. Progressive Hortic 24:227–232

Prithiviraj B, Singh UP (1995) Biological control of plant pathogens: a key to a serene agro-ecosystem. J Rec Adv Appl Sci 10:99–100

Ragupathi G, Damani P, Deng K, Adams MM, Hang J, George C, Livingston PO, Gin DY (2010) Preclinical evaluation of the synthetic adjuvant SQS-21 and its constituent isomeric saponins. Vaccine 28:4260–4267

Rojas R, Bustamante B, Bauer J, Fernandez I, Alban J, Lock O (2003) Antimicrobial activity of selected Peruvian medicinal plants. J Ethnopharmacol 88:199–204

Sahayaraj K, Namasivayam SKR, Rathi JM (2011) Compatibility of entomopathogenic fungi with extracts of plants and commercial botanicals. Afr J Biotechnol 10:933–938

Sahu NP, Banerjee S, Mondal NB, Mandal D (2008) Steroidal saponins. Prog Chem Organ Nat Prod 89:45–141

Sanchez E, Heredia N, Garcıa S (2005) Inhibition of growth and mycotoxin production of *Aspergillus flavus* and *Aspergillus parasiticus* by extracts of *Agave* species. Int J Food Microbiol 98:271–279

Santos JDG, Branco A, Silva AF, Pinheiro CSR, Neto AG, Uetanabaro APT, Queiroz SROD, Osuna JTA (2009) Antimicrobial activity of *Agave sisalana*. Afr J Biotechnol 8(22):6181–6184

Shukla S, Mehta P, Mehta A, Vyas SP, Bajpai VK (2011) Preliminary phytochemical and antifungal screening of various organic extracts of *Caesalpinia bonducella* seeds. Roman Biotechnol Lett 16:6384–6389

Simons V, Morrissey JP, Latijnhouwers M, Csukai M, Cleaver A, Yarrow C, Osbourn A (2006) Dual effects of plant steroidal alkaloids on *Saccharomyces cerevisiae*. Antimicrob Agents Chemother 50:2732–2740

Sparg SG, Light ME, Staden JV (2004) Biological activities and distribution of plant saponins. J Ethnopharmacol 94:219–243

Verastegui MA, Sanchez-Garcıa CA, Heredia NL, García-Alvarado JS (1996) Antimicrobial activity of three major plants of the Chihuahuan desert. J Ethnopharmacol 52:175–177

Verastegui A, Verde J, Garcıa S, Heredia N, Oranday A, Rivas C (2008) Species of *Agave* with antimicrobial activity against selected pathogenic bacteria and fungi. World J Microbiol Biotechnol 24:1249–1252

Williams RJ, Heymann DL (1998) Contamination of antibiotic resistance. Science 279:1153–1154

Wilson M (1997) Biocontrol of aerial plant diseases in agriculture and horticulture: Current approaches and future prospects. J Ind Microbiol Biotechnol 19:188–191

Witte W (1998) Medical consequences of antibiotic use in agriculture. Science 279:996–997

Wolters B (1966) Antimicrobial activity of plant steroids and triterpenes. Planta Med 14:392–401

Yadav SK (2010) Pesticide applications-Threat to ecosystems. J Hum Ecol 32(1):37–45

Yang C, Zhang Y, Jacob M, Khan S, Zhang YJ, Xing-Cong L (2006) Antifungal activity of C-27 steroidal saponins. Antimicrob Agents Chemother 50:1710–1714

Zadoks JC (1993) Antipodes on crop protection in sustainable agriculture. In: Corey S, Dall D, Milne W (eds) Pest control and sustainable agriculture. The Commonwealth Scientific and Industrial Research Organisation (CSIRO), Australia, pp. 3–12

Zwane PE, Masarirambi MT, Magagula NT, Dlamini AM, Bhebhe E (2011) Exploitation of *Agave americana* L. plant for food security in Swaziland. Am J Food Nutr 1:82–88

# Production of Extracellular Phytate Hydrolyzing Enzymes by Soil Fungi

38

Sapna, Jinender Jain and Bijender Singh

**Abstract**

A total of 299 fungal isolates were obtained from 33 soil samples collected from different regions of Haryana. All the isolates were screened for extracellular phytase production on phytase-screening medium. Among all the isolates, 97 isolates were found positive for phytase production. After repeated screening, ten potent phytase-producing fungi were selected. Selected fungi were screened for phytase production using quantitative screening. Phytase-producing fungal isolates belonged to the genera of *Aspergillus*, *Penicillium* and *Trichoderma*. *Aspergillus* sp. 50 was found to be the best phytase producer in liquid medium.

**Keywords**

Anti-nutritional factor · Environmental pollution · Monogastric animals · Phytic acid · Phytase

## 38.1 Introduction

Phosphorus is an essential element for the growth of all living organisms; therefore, feed must be supplemented with inorganic phosphorus in order to meet the requirements of animals (Vohra and Satyanarayana 2003; Vats and Banerjee 2004; Singh and Satyanarayana 2011, Singh et al. 2012). One-third of phosphorus is present in soil in digestible inorganic form and two-thirds of organic phosphorus is present in the form of phytin, which is a mixture of calcium–magnesium salts of inositol hexaphosphoric acid, known as phytic acid (Maga 1982; Singh and Satyanarayana 2011, Singh et al. 2012). Phytic acid is the main constituent of animal diet, but it cannot be digested by monogastric animals due to the lack of adequate levels of phytases in their digestive tract (Maenz and Classen 1998; Boling et al. 2000; Singh and Satyanarayana 2011). The unabsorbed phosphorus passes into the environment with faeces and causes environmental pollution due to eutrophication (Mullaney et al. 2000). Reduction of inorganic phosphate supplementation to animal feed reduces phosphorus in the manure by about 33 %, thus cutting the pollution burden by one-third. This type of pollution can be reduced to some ex-

B. Singh (✉) · Sapna · J. Jain
Department of Microbiology, Maharshi Dayanand University, Rohtak 124001, Haryana, India
e-mail: ohlanbs@gmail.com

R. N. Kharwar et al. (eds.), *Microbial Diversity and Biotechnology in Food Security*,
DOI 10.1007/978-81-322-1801-2_38, © Springer India 2014

tent with the help of phytate-degrading enzyme, i.e. phytase (Singh et al. 2011, 2012).

Phytases *(myo*-inositol hexaphosphate phosphohydrolase) hydrolyze phytic acid to *myo*-inositol and inorganic phosphates through a series of *myo*-inositol phosphate intermediates, and eliminate its anti-nutritional properties (Wodzinski and Ullah 1996; Singh et al. 2011). Phytic acid (*myo*-inositol 1,2,3,4,5,6 hexakisdihydrogen phosphate) is the major storage form of phosphorus abundantly present in plant-derived food materials (1–5 % by weight), such as edible legumes, cereals, oil seeds and nuts (Wodzinski and Ullah 1996; Vats and Banerjee 2004; Greiner and Konietzny 2006; Rao et al. 2009; Singh et al. 2011). Phytic acid form complexes with some metal ions ($Ca^{2+}$, $Mg^{2+}$, $Fe^{2+}$, $Zn^{2+}$) and proteins, resulting in their decreased dietary availability (Wodzinski and Ullah 1996; Vats and Banerjee 2004, Singh and Satyanarayana 2011; Singh et al. 2011). Phytic acid can be removed by some physical (autoclaving, cooking) or chemical (ion exchange, acid hydrolysis) methods, but these methods negatively affect the nutritional value of the food (Singh et al. 2011). Therefore, the reduction of phytic acid content in foods and feeds by enzymatic hydrolysis using phytase is desirable, which improves the nutritional value of the foods. This enzyme also has potential applications in other fields and is of immense commercial value primarily in feed and food industries (Singh et al. 2011).

A large number of fungi have been reported to produce phytases (Vohra and Satyanarayana 2003; Vats and Banerjee 2004; Singh and Satyanarayana 2011; Singh et al. 2011, 2012). Most of the scientific work has been done on microbial phytases, especially those originating from filamentous fungi such as *Aspergillus niger, Mucor piriformis* (Vats and Banerjee 2005), *Sporotrichum thermophile* BJTLR50 (Singh and Satyanarayana 2009), *Thermomucor indicae seudaticae* ATCC28404, *Myceliophthora thermophila* ATCC48102 (Mitchell et al. 1997), *Thermoascus aurantiacus* (Nampoothiri et al. 2004), *A. niger* NCIM 563 (Shah et al. 2009), *A. niger* st-6 (Tahir et al. 2010) and *A. tamari* (Shah and Trivedi 2012). In view of industrial importance, the ultimate objective of the present investigation is to isolate fungi for extracellular phytase production and establish conditions for its industrial application. So, one attempt was made to isolate and screen fungi for extracellular phytase production. This paper describes the isolation and screening of mesophilic moulds for extracellular phytase production.

## 38.2 Materials and Methods

### 38.2.1 Isolation and Maintenance of Mesophilic Fungi

Soil fungi were isolated from various soil samples collected from different parts of Haryana on phytase-screening medium (PSM) containing (g/L) D-glucose 15, calcium phytate 3, ammonium sulphate 5, magnesium sulphate 5, potassium chloride 5, ferrous sulphate 0.01, manganese sulphate 0.01, agar–agar 20 (pH 5.5). One gram of sample was suspended in 100 ml normal saline and incubated at 30 °C and 200 rpm for 1 h. The serially diluted soil samples were plated on PSM agar. The medium was supplemented with antibiotics like ampicillin/streptomycin (50 μg/ml) to suppress the bacterial growth. Any colony that developed in media and developed a clear zone in medium was taken as phytase positive. Phytase-producing fungus produced a hydrolysis zone by the activity of phytase. Stock cultures are maintained on potato dextrose agar (PDA) slants at 4 °C.

### 38.2.2 Qualitative Screening Using Double-Staining Method

The phytase-producing isolates were further screened in the medium containing sodium phytate instead of calcium phytate in order to avoid false selection in case of acid-producing microbes (Bae et al. 1999). These acids secreting into the medium caused solubilization of calcium phytate. After incubation at 30 °C for 48–72 h, the plates were flooded with 2 % (w/v) cobalt chloride and incubated at room temperature for 5 min. The cobalt chloride solution was replaced with a freshly prepared solution containing equal volumes of a 6.25 % (w/v) ammonium molybdate and a 0.42 % (w/v) ammonium vanadate solution. The plates were incubated for 5 min at room temperature and then the solution was decanted off, and the plates were observed for zone of hydrolysis (Bae et al. 1999).

### 38.2.3 Quantitative Screening: Screening for Phytase Production in Shake Flasks

The phytase-producing fungi selected by qualitative screening were screened for phytase production in liquid medium. A total of 50 ml of phytase production broth in 250 ml Erlenmeyer flasks containing (g/L) starch 10, glucose 30, magnesium sulphtate 0.5, potassium chloride 0.05, ferrous sulphate 0.1, ammonium sulphate 5 (pH 5.0) was autoclaved, inoculated with 1 ml spore suspension and incubated at 30 °C and 200 rpm for 3 days. The contents of the flasks were filtered through Whatman no. 1 filter paper, and cell-free culture filtrates were used in phytase assays as described earlier (Singh and Satyanarayana 2009).

### 38.2.4 Measurement of Phytase Activity

Phytase was assayed by determining the amount of phosphate liberated (Fiske and Subbarow 1925) using calcium phytate as the substrate (Singh and Satyanarayana 2009). The reaction mixture consisted of 0.5 ml acetate buffer (0.1 M, pH 4.0 containing 2.0 mM calcium phytate) and 0.5 ml of crude enzyme. After incubation for 10 min at 50 °C, the reaction was stopped by adding 1 ml of 10 % tricholoroacetic acid. The amount of free phosphate released was determined spectrophotometrically at 660 nm. One unit of phytase is defined as the amount of enzyme required to liberate 1 nmol of inorganic phosphate $ml^{-1}$ $s^{-1}$ under the assay conditions using $KH_2PO_4$ as the standard.

### 38.2.5 Selection of Potent Phytase-Producing Fungal Isolates

After repeated screening, the best phytase producing isolates were further screened in liquid medium. After incubation for 3 days, the cell-free culture filtrates were used in determination of phytase activities as described earlier.

## 38.3 Results and Discussion

Phytates have been considered a threat in human diet due to their anti-nutritional behaviour, known as strong chelators of divalent minerals, proteins, carbohydrates and enzymes (Wodzinski and Ullah 1996; Vohra and Satyanarayana 2003; Vats and Banerjee 2004; Singh and Satyanarayana 2011; Singh et al. 2011). The reduction of this phytates can be achieved through both enzymatic and nonenzymatic removal. Enzymatic degradation includes addition of either isolated form of wild-type or recombinant exogenous phytate-degrading enzyme carrying microorganisms in the food matrix. However, there are still limited sources of phytase suited for all food application. Thus, screening for ideal phytase with more improved properties and engineering phytases in order to optimize their catalytic and stability features are of research interest (Greiner and Konietzny 2006; Singh and Satyanarayana 2011).

In the present investigation, we have isolated 299 fungal cultures from 33 soil samples collected from different parts of Haryana. Extracellular phytase activity was displayed by a large number of isolates. During the initial screening of fungi for phytase production, 97 fungi showed zone of hydrolysis on PSM agar plates (Fig. 38.1). Among these isolates, 40 isolates of fungi showed high phytase production as compared to others on the basis of ratio of diameter of zone of hydrolysis and fungal colony (Table 38.1). Extracellular phytase production was shown clearly in 40 fungal isolates. Similar findings were observed earlier by Shieh and Ware (1968). They isolated more than 2,000 cultures from 68 soil samples in enrichment culture

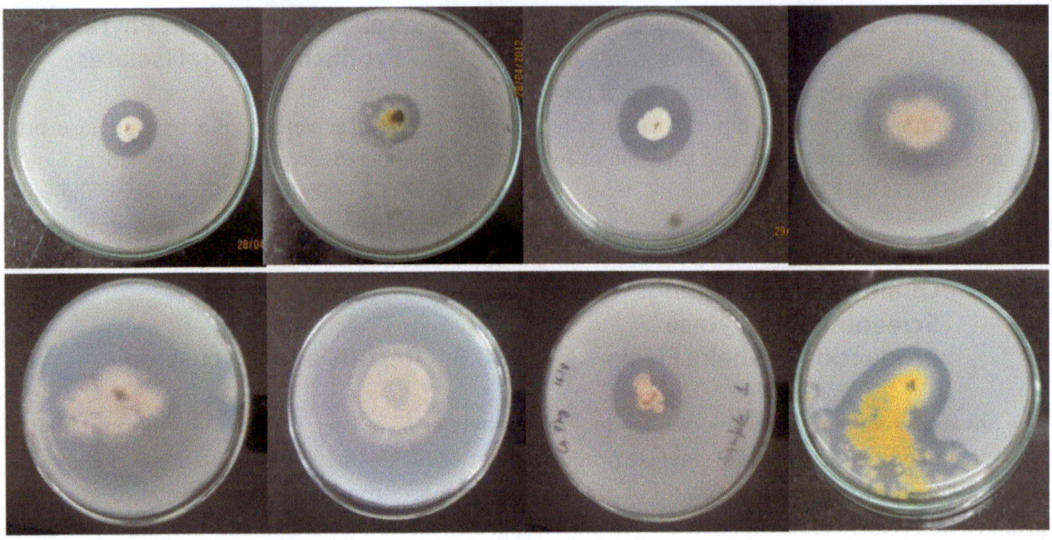

**Fig. 38.1** Calcium phytate agar plates showing zones of hydrolysis by different fungal isolates

medium and found extracellular phytase activity mainly among the various fungal isolates. However, the number of isolates-producing phytase was significantly lower than the total number. Howson and Davis (1983) examined 84 fungi from 25 species for the production of extracellular phytase. Only 58 fungal strains showed substantial phytase activity. Among the large number of fungal isolated from soils collected from Korea, only five colonies, fungal strain L002, L102, L116, L117 and L121, displayed the largest clearing zones in comparison to others (Lee et al. 2005). The screening of 203 fungal strains resulted in the selection of *Aspergillus* sp. 307, which synthesizes both acid phosphatase and phytase acting equally well at pH 5.0 and 2.5 (Gargova et al. 1997).

All the 40 fungal isolates were further screened on PSM agar containing sodium phytate instead of calcium phytate. It was observed that only 19 fungi were able to grow and show clear zone of hydrolysis after double-staining method (Bae et al. 1999; Table 38.2). This could be explained by the formation of various acids like acetic acid, mallic acid, etc., which solubilize calcium phytate and result in zone formation. These acids lower the pH of the medium and hence increase the solubility of calcium phytate (Bae et al. 1999). The number of phytase-producing fungi were reduced, when they were grown in PSM broth. Similar findings were made earlier (Shieh and Ware 1968, Howson and Davis 1983, Gargova et al. 1997, Lee et al. 2005, Kumar et al. 2011).

After final screening by double-staining method, 10 fungi were selected and phytase production was compared in liquid medium. Only four fungi were able to produce phytase extracellularly in which three isolates belong to genus *Aspergillus* and one to *Trichoderma* (Fig. 38.2). It has been evident from other report that *Aspergillus* spp. are the predominant phytase producers in majority of the studies carried out on fungi (Shieh and Ware 1968; Howson and Davies 1983, Lee et al. 2005).

## 38.4 Summary and Conclusions

In this investigation, we have isolated 299 fungal isolates from different soil samples collected from various regions of Haryana. Among all the isolates, 97 isolates were found positive for phytase production as revealed by zone of hydrolysis. These selected fungi were screened for phytase production using quantitative screening. Phytase-

**Table 38.1** Calcium phytate hydrolysis zone by various fungal isolates

| S. No. | Fungal isolates | Hydrolysis zone/colony diameter (cm) |
|---|---|---|
| 1 | Isolate-1 | 4/2 |
| 2 | Isolate-3 | 7/3.8 |
| 3 | Isolate-5 | 6.5/3 |
| 4 | Isolate-6 | 3.5/1.7 |
| 5 | Isolate-7A | 3/1.5 |
| 6 | Isolate-7B | 8/5.5 |
| 7 | Isolate-8A | 8.5/5.5 |
| 8 | Isolate-8B | 7.5/4.8 |
| 9 | Isolate-9 | No zone |
| 10 | Isolate-10 | 3.8/1.5 |
| 11 | Isolatee-12 | 3.5/1.7 |
| 12 | Isolate-13A | 4/1 |
| 13 | Isolate-13B | 3.6/1.5 |
| 14 | Isolate-14A | 4/2.5 |
| 15 | Isolate-14B | 6/3 |
| 16 | Isolate-14C | 3/1.4 |
| 17 | Isolate-14D | 3/1.6 |
| 18 | Isolate-15A | 2.1/1.8 |
| 19 | Isolate-15B | 2.4/1.4 |
| 20 | Isolate-16A | 3/2.4 |
| 21 | Isolate-16B | 6.7/3.0 |
| 22 | Isolate-16C | 4/1.8 |
| 23 | Isolate-16D | 2.7/1.7 |
| 24 | Isolate-16E | 2.2/1.2 |
| 25 | Isolate-18A | 3.2/1.8 |
| 26 | Isolate-18B | 3.5/2.1 |
| 27 | Isolate-18C | No zone |
| 28 | Isolate-19 | 3.5/1.9 |
| 29 | *Aspergillus* sp. 49 | 8/3.8 |
| 30 | *Aspergillus* sp. 50 | 7/2.5 |
| 31 | *Aspergillus* sp. 51 | 8/3.8 |
| 32 | *Trichoderma* sp. 199 | 8/4.5 |
| 33 | *Aspergillus* sp. 224 | C[a] |
| 34 | *Aspergillus* sp. 244 | 6.5/5.3 |
| 35 | *Aspergillus* sp. 251 | C[a] |
| 36 | *Fusarium* sp. 255 | 5.5/1.8 |
| 37 | *Trichoderma* sp. 256 | 6/3 |
| 38 | *Aspergillus* sp. 262 | 8.5/6.3 |
| 39 | *Penicillum* sp. 273 | 5.7/2.5 |
| 40 | *Aspergillus* sp. 284 | 8.5/4.6 |

[a] Complete hydrolysis of calcium phytate

producing fungal isolates belong to the genera of *Aspergillus, Penicillium, Trichoderma,* etc. Fungal isolate *Aspergillus* sp. 50 was found to be the best phytase producer in the liquid medium.

**Acknowledgement** The authors wish to thank the University Grants Commission (UGC), New Delhi for providing financial assistance during the course of this investigation.

**Table 38.2** Hydrolytic zone by selected fungal isolates on PSM agar containing sodium phytate using double-staining method

| Sl. no. | Fungal isolates | Hydrolytic zone/colony diameter (cm) |
|---|---|---|
| 1 | Isolate-1 | 2.8/1.1 |
| 2 | Isolate-7A | 2.8/1.3 |
| 3 | Isolate-7B | 2.0/1.8 |
| 4 | Isolate-8A | 2.3/1.6 |
| 5 | Isolate-8B | Full plate |
| 6 | Isolate-10 | No zone |
| 7 | Isolate-14A | 3.3/2.2 |
| 8 | Isolate-14B | 1.9/1.7 |
| 9 | Isolate-14C | 1.8/1.6 |
| 10 | Isolate-16A | 2.2/0.5 |
| 11 | Isolate-16B | 0.7/0.5 |
| 12 | Isolate-19 | 1.0/0.8 |
| 13 | *Aspergillus* sp. 49 | 2.9/2.7 |
| 14 | *Aspergillus* sp. 50 | 5.0/2.0 |
| 15 | *Aspergillus* sp. 51 | 3.3/1.8 |
| 16 | *Trichoderma* sp. 199 | 5.8/5.5 |
| 17 | *Aspergillus* sp. 251 | 2.4/2.2 |
| 18 | *Aspergillus* sp. 262 | No zone |
| 19 | *Penicillum* sp. 273 | 1.8/1.2 |

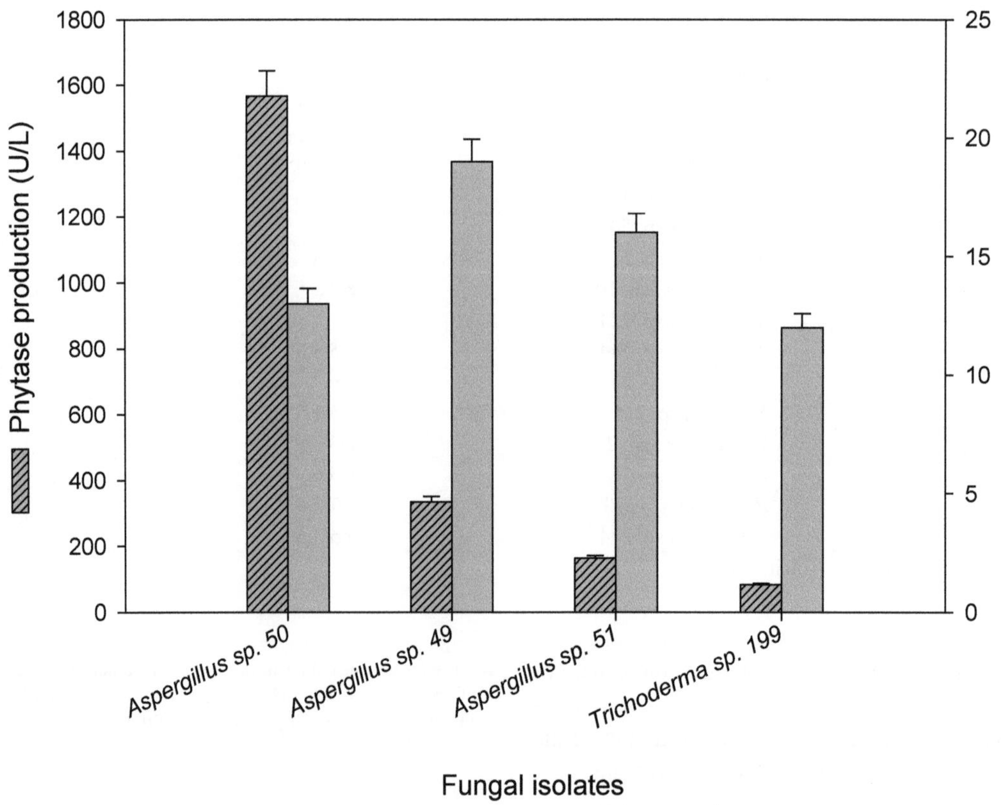


**Fig. 38.2** Phytase production by selected fungal isolates in liquid medium at 30 °C and 200 rpm after 3 days

## References


Bae HD, Yanke LJ, Cheng KJ, Selinger LB (1999) A novel staining method for detecting phytase activity. J Microbiol Methods 39(1):17–22

Boling SDD, Johnson MW, Wang ML, Parsons X, Koelkebeck CM, Zimmerman KW (2000) The effects of dietary available phosphorus levels and phytase performance of young and older laying hens. Poult Sci 79:224–230

Fiske CH, Subbarow YP (1925) The colorimetric determination of phosphorus. J Biol Chem 66:375–400

Gargova S, Roshkova Z, Vancheva G (1997) Screening of fungi for phytase production. Biotechnol Tech 11(4):221–224

Greiner R, Konietzny U (2006) Phytase for food application. Food Technol Biotechnol 44:125–140

Howson SJ, Davis RJ (1983) Production of phytate-hydrolysing enzyme by fungi. J Enzyme Microbiol Technol 5:377–382

Kumar J, Kumar A, Ali D, Singh S, Sharma PK, Sharma DC (2011) Isolation and screening of phytase producing fungi from soil and production of phytase in submerged and solid state fermentation. Adv Plant Sci 24(1):313–316

Lee DH, Choi SU, Hwang Y (2005) Culture conditions and characterizations of a new phytase-producing fungal isolate, *Aspergillus* sp. L117. Mycobiology 33(4):223–229

Maga JA (1982) Phytate: its chemistry, occurrence, food interactions, nutritional significance, and methods of analysis. Crit Rev Food Sci Nutr 16:1–48

Maenz DD, Classen HL (1998) Phytase activity in small intestine brush-border membrane of the chicken. Poult Sci 77:557–564

Mitchell DB, Vogel K, Weimann BJ, Pasamontes L, van Loon APGM (1997) The phytase subfamily of histidine acid phosphatase: isolation of genes for two novel phytases from fungi *Aspergillus terreus* and *Myceliophthora thermophila*. Microbiology 143:245–252

Mullaney EJ, Daly CB, Ullah AHJ (2000) Advances in phytase research. Adv Appl Microbiol 47:157–199

Nampoothiri KM, Tomes GJ, Roopesh K, Szakacs G, Nagy V, Soccol CR, Pandey A (2004) Thermostable phytase production by *Thermoascus aurantiacus* in submerged fermentation. Appl Biochem Biotechnol 118:205–214

Rao DE, Rao KV, Reddy TP, Reddy VD (2009) Molecular characterization, physicochemical properties, known and potential applications of phytases: an overview. Crit Rev Biotechnol 29(2):182–198

Shah P, Bhavsar K, Soni SK, Khire JM (2009) Strain improvement and up scaling of phytase production by *Aspergillus niger* NCIM 563 under submerged fermentation conditions. J Indust Microbiol Biotechnol 36:373–380

Shah KB, Trivedi R (2012) Purification and purification and characterization of an extracellular phytase from *Aspergillus tamari*. Int J Pharma Biosci 3(2):775–783

Shieh TR, Ware JH (1968) Survey of microorganism for the production of extracellular phytase. Appl Microbiol 16:1348–1351

Singh B, Satyanarayana T (2009) Characterization of a HAP-phytase from a thermophilic mould *Sporotrichum thermophile*. Bioresour Technol 100:2046–2051

Singh B, Satyanarayana T (2011) Phytases from themophilic molds: their production, characteristics and multifarious applications. Process Biochem 46(7):1391–1398

Singh B, Kaur P, Satyanarayana T (2006) Fungal phytases for improving the nutritional status of foods and combating environmental phosphorus pollution. In: Chauhan, AK, Verma A (eds) Microbes: health and environment. IK International Publishers, New Delhi pp. 289–326

Singh B, Kunze G, Satyanarayana T (2011) Developments in biochemical aspects and biotechnological applications of microbial phytases. Biotechnol Mol Biol Rev 6(3):69–87

Singh B, Sapna, Jain J, Satyanarayana T (2012) Fungal phytases for combating environmental phosphorus pollution and ameliorating the nutritional status of non-ruminants. In: Devi R, Kidwai MK, Rose, PK, Saran AK (eds) Energy-water-waste nexus for environment management. Narosa Publishing House, New Delhi, pp. 292–301

Tahir A, Mateen B, Saeed, Uslu H (2010) Studies on the production of commercially important phytase from *Aspergillus niger* st-6 isolated from decaying organic soil. Micol Aplic Int 22(2):51–57

Vats P, Banerjee UC (2004) Production studies and catalytic properties of phytases (*myo*-inositolhexakisphosphate phosphohydrolases): an overview. Enzyme Microb Technol 35:3–14

Vats P, Banerjee UC (2005) Biochemical characterization of extracellular phytase (*myo*-inositol hexakisphosphate phosphohydrolase) from a hyper-producing strain of *Aspergillus niger* van Teighem. J Indust Microbiol Biotechnol 32:141–147

Vohra A, Satyanarayana T (2003) Phytases: microbial sources, production, purification, and potential biotechnological applications. Crit Rev Biotechnol 23:29–60

Wodzinski RJ, Ullah AHJ (1996) Phytase. Adv Appl Microbiol 42:263–301

# Isolation, Characterization and Production of Bacterial Laccase from *Bacillus sp.* 39

Deepti Singh, Ekta Narang, Preeti Chutani, Amit Kumar, K. K. Sharma, Mahesh Dhar and Jugsharan S. Virdi

**Abstract**

Two species of *Bacillus* exhibiting laccase activity were screened from earthworm cast and soil samples. The M162 medium with 5 mM guaiacol was used for isolating bacterial strains capable of oxidizing guaiacol. One species was identified as *Bacillus pumilus* and the other as *Bacillus licheniformis* based on the result of biochemical tests and 16S rDNA analysis. *B. pumilus* and *B. licheniformis* could grow at temperature ranging from 30–55 °C and showed optimum growth at temperature 37 °C and pH 8.0 and 5.0. Laccase activity was maximum at 37 °C and pH 7.0. They were found positive for different hydrolytic enzymes. The dyes toluidine blue O and rose bengal were degraded within 24 h.

**Keywords**

*Bacillus pumilus* · *Bacillus licheniformis* · Bacterial laccase · Hydrolytic enzymes

## 39.1 Introduction

Laccases are defined in the Enzyme Commission (EC) nomenclature as oxidoreductases acting on diphenols and related substances using molecular oxygen as acceptor. Laccases (EC 1.10.3.2) couple the four electron reduction of molecular oxygen to water with the oxidation of a broad range of substrates including phenols, arylamines, anilines, and thiols (Thurston 1994). Furthermore, laccases are also capable of performing polymerization, depolymerization, methylation, and demethylation reactions (Solomon et al. 1996; D'Annibale et al. 2000; Ullah et al. 2000; Held et al. 2005). Because of their wide substrate activity, laccases have received broad interest for their biotechnological applications in paper pulping, dye decolourization, wood composite production, bioremediation, denim refining, textile cleaning, juice and wine clarification, biosensor and biofuel cell design (Couto and Herrera 2006; Desai and Nityanand 2011), synthesis of natural products like pigments and antioxidants

K. K. Sharma (✉) · D. Singh · E. Narang · P. Chutani · A. Kumar
Department of Microbiology, Maharshi Dayanand University, Rohtak, Haryana 24001, India
e-mail: kekul_sharma@yahoo.com

M. Dhar · J. S. Virdi
Department of Microbiology, University of Delhi, South Campus, Benito Juarez Road,
New Delhi 110021, India

R. N. Kharwar et al. (eds.), *Microbial Diversity and Biotechnology in Food Security*,
DOI 10.1007/978-81-322-1801-2_39, 

through dimerization of phenolic and nonphenolic acids, manufacture of new compounded material from lignin waste (Hüttermann et al. 2001), detoxification of environmental pollutants, and revalorization of wastes and wastewaters (Mayer and Staples 2002; Saparrat et al. 2002a; Jurado et al. 2009).

The first laccase was studied from *Rhus vernicifera* in 1883, a Japanese lacquer tree (Yoshida 1883). Laccases are widely distributed in plants, fungi, bacteria, and insects (Gregory and Bendall 1966; Givaudan et al. 1993; Alexandre and Zhulin 2000; Sato et al. 2001; Heinzkill and Messner 1997; Hattori et al. 2010), performing specific functions in each; in higher plants, these enzymes participate in the synthesis of lignin (Sato et al. 2001). In fungi, they play a role in lignin degradation, pigment formation, detoxification, and pathogenesis (Williamson 1994; Eggert et al. 1996; Nosanchuk et al. 1999). In bacteria, laccases are found to have roles in melanin production, spore coat resistance, involvement in morphogenesis, and detoxification of copper (Kuznetsov et al. 1984; Roberts et al. 2002; Sharma et al. 2007). In insects, laccases have been suggested to be active in cuticle sclerotization and catalyse cuticle tanning (Dittmer et al. 2004; Sharma and Kuhad 2008).

The bacterial laccases may have many advantageous properties compared to classical fungal laccases because of their highly efficient expression, much higher thermostability (example is CotA), which is higher than fungal laccases (Martins et al. 2002), and alkaline torelancy (such as Lbh1 from *Bacillus halodurans* C-125) that may be useful for paper pulp bleaching (Ruijssenaars and Hartmans 2004). Since spores allow microorganisms to survive under extreme conditions, spore coat enzymes might also withstand high temperatures or extreme pH values. Spore laccases which are active in the alkaline pH range could be used for industrial and biotechnological applications (Held et al., 2005).

Furthermore, existence of intron in fungal laccase genes, formation of disulfide bridges, and glycosylation of fungal laccase are also frequently obstructive. Despite such stability and the numerous advantages that prokaryotic enzyme production may offer over eukaryotic production, until now the majority of laccases characterized have been derived from fungi especially from white-rot basidiomycetes, and only a few bacterial laccases have been completely purified and characterized (Sharma et al. 2007; Singh et al. 2007; Koschorreck et al. 2008). Industrial or biotechnological use of bacterial laccase is still not viable because of its extremely low yield (Endo et al. 2003; Suzuki et al. 2003). Therefore, there is a need to find novel bacterial laccases with potential industrial relevance through the exploration of natural diversity and also to improve the yield which can have tremendous laccase applications.

Due to the diverse roles played by laccase in different organisms and its wide range of biotechnological applications, the present investigation was planned for the isolation and screening of laccase-producing bacteria from soil samples. In the present study, we have isolated two thermotolerant *Bacillus* strains having wide pH stability, capable of producing laccase, and having dye decolourization abilities. The isolates were also found to be positive for different hydrolytic enzymes like xylanase, cellulose, pectinase, and α-amylase.

## 39.2 Materials and Methods

### 39.2.1 Isolation and Screening of Bacteria

For isolation, earthworm cast and soil samples were collected from decomposing leaf litter from forest canopy, Rohtak, Haryana. The samples were diluted and plated on M 162 agar plates containing 5 mM guaiacol and incubated at 37 °C for 96 h and also on nutrient agar medium containing (g/L) 5 g peptone, 5 g sodium chloride, 1.5 g beef extract, 1.5 g yeast extract, and 2 g agar and were incubated at 37 °C for 24 h (Figs. 39.1 and 39.2). The bacterial colonies obtained were further screened for laccase production by plate assay using 25 mM guaiacol in 100 mM citrate-phosphate buffer (pH 5.2), used as substrate.

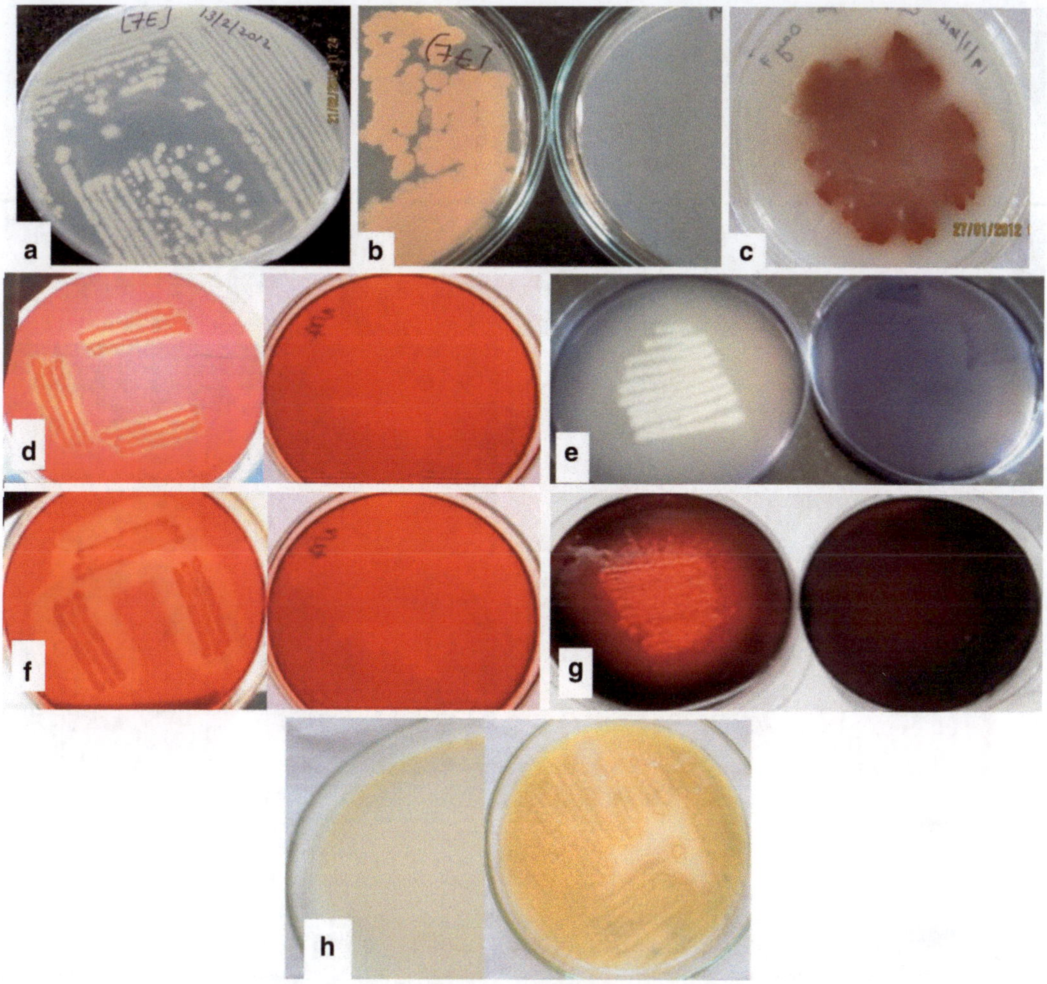

**Fig. 39.1** Growth of *Bacillus pumilus* DSKK1 on different media. **a** Morphological colony appearance on nutrient agar medium. **b** *Reddish brown* appearance of colonies on M162 medium indicates laccase production as compared with control. **c** Qualitative plate assay using guaiacol as a substrate gives intense *red* colouration which confirms the presence of intracellular laccase. **d** Qualitative plate assay for cellulose. **e** α-amylase. **f** Xylanase. **g** Pectinase. **h** Protease

### 39.2.2 16S rRNA Sequencing and Analysis

DNA was isolated using standard protocol (Sambrook et al. 1989). The polymerase chain reaction (PCR) products comprising the partial 16S rDNA sequence were generated and sequenced using 50 ng template, 20 pmol each of 16S rDNA primers (forward primer 5′AGAGTTTGATCCTGGCTCAG3′ and reverse primer 5′ACGGCTACCTTGTTACGAC3′), 10× PCR buffer (Tris/HCl, pH 8.4, 200 mM; KCl, 500 mM; $MgCl_2$, 15 mM), Taq DNA polymerase (1 unit) in 100 µl. The thermal program consisted of 30 cycles of 95 °C for 30 s, 60 °C for 40 s and 72 °C for 2 min). The last cycle was for 10 min at 72 °C. The amplified products were run on 1 % agarose electrophoresis gel (Fig. 39.3) and the desired band was purified using HiYield™ Gel/PCR DNA Mini Kit (Real Genomics, RBC). The purified products were sequenced from the University of Delhi, South Campus by an ABI PRISM 310NT Genomic Analyzer (PerkinElmer) using a BigDye Terminator Cycle-Sequencing Kit. The nucleotide sequence data obtained were then analyzed using the NCBI databases blastn. The sequences were submitted to GenBank (accession no. *Bacillus pumilus* strain

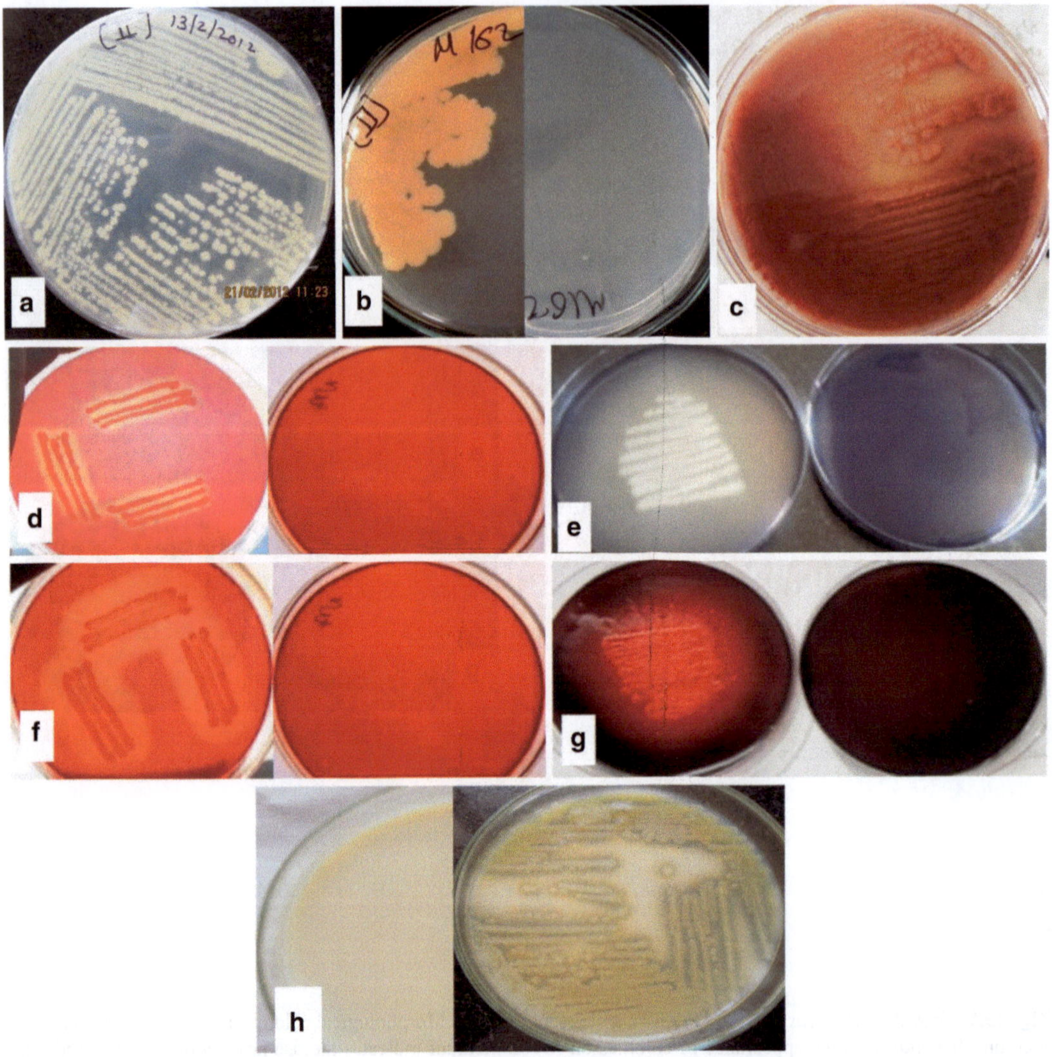

**Fig. 39.2** Growth of *Bacillus licheniformis* DSKK2. **a** Morphological colony appearance on nutrient agar medium. **b** *Reddish brown* appearance of colonies on M162 medium indicates laccase production as compared with control. **c** Qualitative plate assay using guaiacol as a substrate gives intense *red* colouration which confirms the presence of intracellular laccase. **d** Qualitative plate assay for cellulose. **e** α-amylase. **f** Xylanase. **g** Pectinase. **h** Protease

DSKK1: JQ639010 and *Bacillus licheniformis* strain DSKK2: JQ639011).

## 39.2.3 Phylogenetic Tree Constructions

Sequences were aligned using ClustalW with default settings. Phylogenetic relationships (Fig. 39.4) were inferred by two different methods using the MEGA analysis tool version 3.5: maximum parsimony with statistical support obtained by generating 500 bootstrap replicates or UPGMA clustering based on the Dayhoff PAM250 matrix with statistical support by generating 100 bootstrap replicates. Sequences have been deposited in the GenBank.

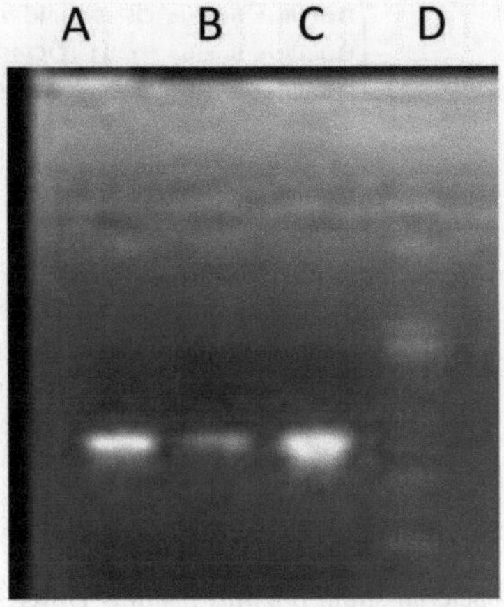


**Fig. 39.3** Agarose gel electrophoresis showing band of 16S rDNA sequence. **a** *Bacillus pumilus* DSKK1. **b** *Bacillus licheniformis* DSKK2. **c** *Bacillus firmus* DSKK3 and **d** 1 Kb marker

### 39.2.4 Optimization of Growth Conditions

#### 39.2.4.1 Effect of pH

The effect of pH range 3–10 on the growth of the bacteria was studied. The cultures were grown in 250 ml Erlenmeyer flask containing 50 ml nutrient broth media with different pH and incubated at 37 °C, 200 rpm for 24 h. The medium without inoculation was considered as control. The difference in growth at varying pH was measured using spectrophotometer at 600 nm.

#### 39.2.4.2 Effect of Temperature

1.1.1. The effect of varying temperatures 30, 37, 45, 50, and 55 °C on the growth of the bacteria was studied. The cultures were grown in 250 ml Erlenmeyer flask containing 50 ml nutrient broth media incubated at different temperatures at 200 rpm for 24 h. The medium without inoculation was considered as control. The difference in the growth at varying temperature was measured using spectrophotometer at 600 nm.

#### 39.2.4.3 Effect of rpm

1.1.2. The effect of varying rpm (150, 200, and 250 rpm) on the growth of the bacteria was studied. The cultures were grown in 250 ml Erlenmeyer flasks containing 50 ml nutrient broth media and incubated at 37 °C and varying rpm for 24 h. The medium without inoculation was considered as control. The difference in growth at varying rpm was measured using spectrophotometer at 600 nm.

#### 39.2.4.4 Effect of Amount of Inoculum

The effect of the amount of inocula at 0.2, 0.4, 0.6, 0.8, and 1 ml on the growth of the bacteria was studied at every 2 h interval. The cultures were grown in 250 ml Erlenmeyer flask containing 50 ml nutrient broth media incubated at 37 °C 200 rpm for 24 h. The medium without inoculation was considered as control. The difference in the growth with varying amount of inoculum was measured by using spectrophotometer at 600 nm.

### 39.2.5 Harvesting Intracellular Laccase

Bacterial isolates were grown in Luria-Bertani broth media and incubated at 37 °C for 72 h. Thereafter, the cultures were centrifuged at 10,000 rpm for 3 min at 4 °C. The pellet was dissolved in minimum amount of citrate-phosphate buffer, pH 7. The cells were then sonicated to harvest intracellular laccase. Laccase assay for extracellular and intracellular laccases was done with guaiacol as substrate.

### 39.2.6 Laccase Assay

#### 39.2.6.1 Qualitative Assay of Spore Laccase

The cultures were grown in three different 250 ml Erlenmeyer flasks containing 50 ml Luria-Bertani broth medium at 37 °C, 200 rpm for 72 h. Laccase is a secondary metabolite and is produced by the culture under stressed conditions. Heat shock treatment increases laccase production (Wang et al. 2012). Stress was given by transferring the flasks to higher temperature. One flask was transferred to 50 °C and one to 55 °C for 24 h. The culture was centrifuged at 10,000 × g for 3 min at 4 °C. The pellet was dissolved in minimal amount of citrate phosphate buffer, pH 7. A total of 200 µl of culture pellet

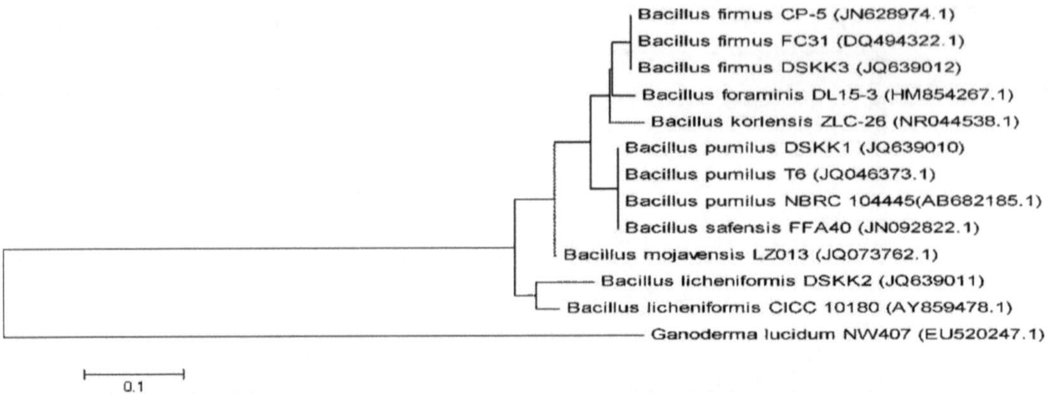


**Fig. 39.4** Phylogenetic tree using Mega 5 software of the bacterial strain *Bacillus pumilus* DSKK1 and *Bacillus licheniformis* DSKK2. The ITS sequence of *Ganoderma lucidium* was used as outgroup species

was added to 200 µl of guaiacol solution (25 mM guaiacol in 100 mM citrate-phosphate buffer, pH 7). After vortexing for few seconds at room temperature, incubation was done at 37 °C for 30 min. Appearance of reddish-brown colour showed the presence of laccase.

#### 39.2.6.2 Qualitative Assay of Spore Laccase

The cultures were grown in 250 ml Erlenmeyer flasks containing 50 ml Luria-Bertani broth medium at 37 °C, 200 rpm for 210 h. One flask was withdrawn every 24 h, and the culture was centrifuged at 10,000 × g for 5 min at 4 °C. The pellet was dissolved in minimal amount of citrate-phosphate buffer, pH 7.0. Thereafter, 250 µl of guaiacol solution was added to 250 µl of culture pellet (25 mM guaiacol in 100 mM citrate-phosphate buffer, pH 7.0). After vortexing for few seconds at room temperature, incubation was done at 37 °C for 30 min. The laccase activity was measured at 460 nm. One unit of laccase was defined as the change in absorbance of 0.01/ml/min at 460 nm.

### 39.2.7 Decolourization of Dyes by Bacterial Culture

Decolourization of dyes by the bacterial cultures for malachite green, congo red, xylidine ponceau, remazol brilliant blue R (50 mg/L), methylene blue, rose bengal, toluidine blue O, lissamine green B, bromophenol blue indicator, and crystal violet (25 mg/L) was studied on Luria-Bertani agar plates incubated at 37 °C for 48 h.

### 39.2.8 Production of Hydrolytic Enzymes

The qualitative plate assay for the production of hydrolytic enzymes by the cultures was studied on modified Horikoshi agar medium containing yeast extract 0.5 %, peptone 0.5 %, $MgSO_4.7H_20$ 0.01 %, $KH_2PO_4$ 0.1 %, and agar 2.0 % and birchwood xylan 0.25 %, carboxymethyl cellulose 0.5 %, starch 0.25 %, and citrus pectin 0.25 % for xylanase, cellulase, α-amylase, and pectinase, respectively. The plates were incubated at 37 °C for 24 h and the visible clear zone surrounding the culture showed the production of the hydrolytic enzyme.

## 39.3 Results and Discussion

Approximately 204 bacterial isolates were obtained from soil samples, of which 30 were found positive for laccase production. Out of 59 isolates obtained from earthworm cast, seven were found positive for laccase production as they gave an intense reddish-brown colony on M162 agar plates containing 5 mM guaiacol (Figs. 39.1b and 39.2b). Degryse et al. (1978) and Bains et al. (2003) used similar media for the screening

**Table 39.1** Characteristic features of *Bacillus* sp. isolated from soil samples

| Sl. no. | Characteristic feature | Isolate 1 | Isolate 2 |
|---|---|---|---|
| 1. | Gram stain | + | + |
| 2. | Morphology | Rod | Rod |
| 3. | Cellulose utilization | + | + |
| 4. | Xylan utilization | + | + |
| 5. | Pectin utilization | + | + |
| 6. | Starch utilization | + | + |
| 7. | Oxidase | + | + |
| 8. | Catalase | + | + |
| 9. | Indole production | − | − |
| 10. | Methyl red test | + | + |
| 11. | Voges–Proskauer | + | + |
| 12. | Citrate utilization | − | + |
| 13. | $H_2S$ production | − | − |
| 14. | Identified as | *Bacillus* | *Bacillus* |

of laccase-producing bacteria from industrial wastewater-drained soil. Laccase production by the isolates was confirmed by qualitative plate assay of 48-h-old culture with 25 mM guaiacol in 100 mM citrate-phosphate buffer, pH 5.2 (Figs. 39.1c and 39.2c).

Both the isolates were Gram-positive rods. They were endospore forming and catalase positive identified as *Bacillus*. Further, biochemical characterization showed them to be oxidase, methyl red, and Voges–Proskauer positive but indole and hydrogen sulfide negative. Citrate was utilized by the isolate from the soil sample but not by the isolate from the earthworm cast (Table 39.1).

Using 16S rDNA sequencing, the best laccase-producing isolate obtained from earthworm cast was identified as *B. pumilus* DSKK1 (accession no. JQ639010) and that from soil samples as *B. licheniformis* DSKK2 (accession no. JQ639011). Phylogenetic tree was constructed using MEGA analysis tool version 3.5. The phylogenetic tree showed close resemblance among the same species. *Bacillus safensis* FFA40 was in the same clad with *B. pumilus* species which reveals high evolutionary similarity. The ITS sequence of *Ganoderma lucidum* NW407 was used as outgroup species.

The growth conditions for the cultures were optimized for different parameters such as pH, temperature, rpm, and amount of inoculum. *B. pumilus* DSKK1 was able to grow at a pH range of 4–10, but the maximum growth was at pH 8 (Fig. 39.5) while *B. licheniformis* DSKK2 was able to grow at a pH range of 3–10, but the maximum growth was at pH 5. The cultures were found to be thermo tolerant as they showed growth at 50 °C but the optima for *B. pumilus* DSKK1 and for *B. licheniformis* DSKK2 was at 30 °C (Fig. 39.6). Previous workers have also reported thermostable cot A laccase from *B. pumilus* that showed biocatalytic activity (Reiss et al. 2011). The optimum agitation for growth was found to be 150 rpm. The effect of different amount of inocula was studied at every 2 h time interval for 24 h. The growth of bacteria was directly proportional to the amount of initial inocula. The growth was maximum in the case of the highest amount of primary inocula and vice versa. Previous studies show the inoculum size plays significant role in production of enzymes like xylanase (Mittal et al. 2013), protease (Daniel et al. 2013), and laccase (Yesilada et al. 1997; Patel et al. 2009).

Expression of laccase is induced in response of glucose starvation (Nurudeen and Ahearn 1979; Polacheck et al. 1982). So, the cultures were grown in Luria-Bertani broth medium as it lacks carbon source. With the increase in the incubation temperature, there was a gradual decrease in laccase activity. Activity was maximum at temperature 37 °C, and it decreased with increasing temperature from 50 to 55 °C (Fig. 39.7). Similarly, in earlier reports, the brown pigmentation of *Cryptococcus neoformans,* which is due to laccase, was enhanced by low glucose

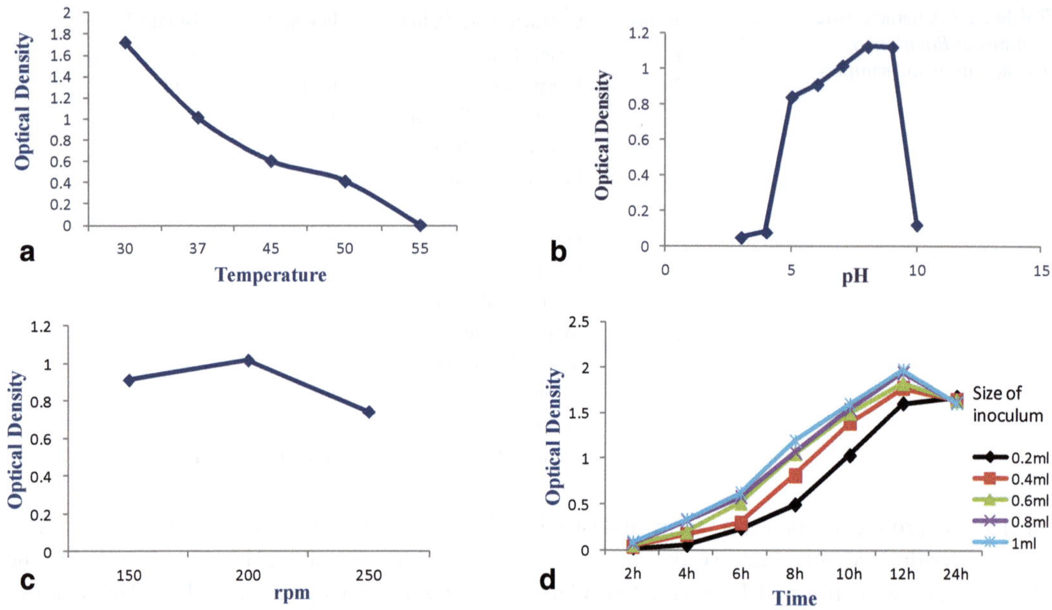


**Fig. 39.5** Effect of different physiological parameters on growth of *Bacillus pumilus* DSKK1. **a** Effect of temperature. **b** effect of pH. **c** effect of rpm and **d** effect of inocula

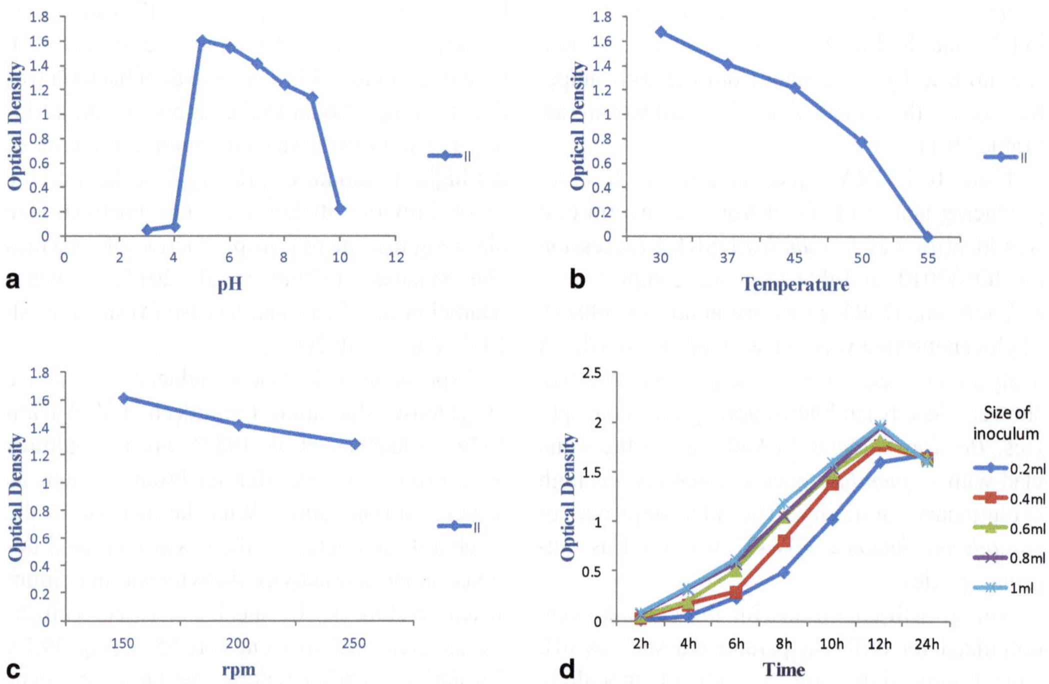


**Fig. 39.6** Effect of different physiological parameters on the growth of *Bacillus licheniformis* DSKK2. **a** Effect of pH. **b** Effect of temperature. **c** Effect of rpm and **d** effect of inocula

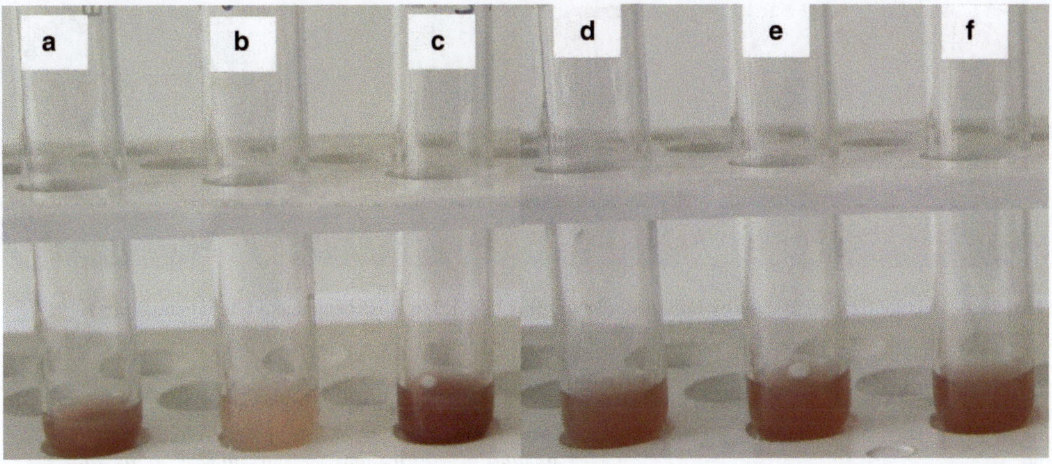


**Fig. 39.7** Laccase assay with 25 mM guaiacol in 100 mM citrate phosphate buffer as substrate, pH 7. **a** *Bacillus pumilus* DSKK1 at 50 °C. **b** *Bacillus pumilus* DSKK1 at 55 °C. **c** *Bacillus pumilus* DSKK1 at 37 °C. **d** *Bacillus licheniformis* DSKK2 at 55 °C. **e** *Bacillus licheniformis* DSKK2 at 50 °C and **f** *Bacillus licheniformis* DSKK2 at 37 °C

concentrations and low temperatures (Jacobson and Emery 1991). The production of laccase was associated with the formation of spores. The laccase activity was enhanced when the culture were grown for 72 h. Laccase assay was done with guaiacol as substrate, although we did not get any activity of extracellular laccase as well as intracellular laccase in nutrient broth. Further, laccase activity was detected at the sporulating stage. In *B. pumilus* DSKK1, the activity was found to be maximum at 5.4 units, after seventh day of incubation whereas in *B. licheniformis* DSKK2, the activity was maximum at 8.1 units, after 8 days of incubation.

Different hydrolytic enzymes like cellulose (Figs. 39.1d and 39.2d), xylanase (Figs. 39.1f and 39.2f), pectinase (Figs. 39.1g and 39.2g), and α-amylase (Figs. 39.1e and 39.2e) were also screened and were found positive on qualitative plate assays. Among the nine different dyes screened on plate assay, the clear zone surrounding the growth of bacteria showed positive dye decolourization. In rose bengal and toluidine blue O, good bacterial growth as well as dye decolourization was observed (Fig. 39.8). After 24 h, the dye was completely decolourized, whereas congo red was partially decolourized. In xylidine ponceau, bromophenol blue, and lissamine green bacterial growth occurred but no decolourization was observed. No growth was observed on malachite green and crystal violet which shows that the concentration of the dye used in the study was lethal to the bacteria (Table 39.2). Previous studies showed 92.79 % decolourization of reactive black 5, 86.40 % decolourization of reactive blue 19, and 91.99 % decolourization of indigo carmine by *B. licheniformis* LS04 at pH 9 (alkaline; Lu et al. 2011). Further, Wang et al. (2010) reported 90 % decolourization of remazol brilliant blue R and alizarin red by spore laccase of *Bacillus subtilis* WD23, whereas Cho et al. (2011) reported the decolourization of indigo carmine by spore laccase of *B. subtilis* within 2 h.

## 39.4 Conclusion

Spore laccase isolated from *B. licheniformis* DSKK2 was found to give more enzyme activity as compared to *B. pumilus* DSKK1. Moreover, the enzyme was found to be thermostable, pH stable, and efficient in dye decolourization. Due to the high stability of the enzyme, it can be applied under harsh conditions in textile industry, bioremediation as well as paper and pulp industry. Not much work has been done in bacterial laccase system, so the area remains unexplored and possesses further opportunities in the field of enzyme technology.

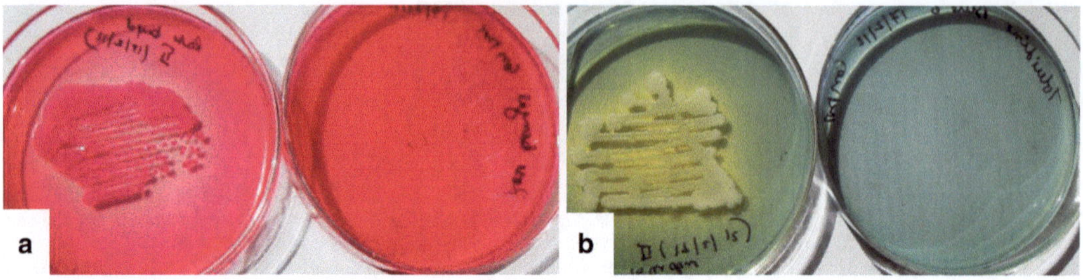

**Fig. 39.8** Decolourization of dyes by *Bacillus licheniformis* DSKK2. **a** Rose bengal and **b** toludiene blue O

**Table 39.2** Growth and dye decolourization of different dyes used in qualitative dye decolourization plate assay

| Sl. no. | Dye | Conc. (mg/L) | Growth | | Dye decolourization | |
|---|---|---|---|---|---|---|
| | | | *Bacillus pumilus* | *Bacillus licheniformis* | *Bacillus pumilus* | *Bacillus licheniformis* |
| 1. | Malachite green | 50 | + | – | – | – |
| 2. | Congo red | 50 | ++++ | ++++ | + | + |
| 3. | Xylidine ponceau | 50 | ++++ | ++++ | – | – |
| 4. | Methylene blue | 25 | + | – | – | – |
| 5. | Rose bengal | 25 | +++++ | +++++ | +++++ | +++++ |
| 6. | Toluidine blue O | 25 | +++++ | +++++ | +++++ | +++++ |
| 7. | Lissamine green B | 25 | +++ | +++ | – | – |
| 8. | Bromophenol blue indicator | 25 | +++ | +++ | – | – |
| 9. | Crystal violet | 25 | + | – | – | – |

**Acknowledgements** The financial support from University Grants Commission, Government of India is highly acknowledged.

## References


Alexandre G, Zhulin LB (2000) Laccases are widespread in bacteria. Trends Biotechnol 18:41–42

Bains J, Capalash N, Sharma P (2003) Laccase from a non-melanogenic, alkalotolerant γ-proteobacterium JB isolated from industrial wastewater drained soil. Biotechnol Lett 25:1155–1159

Cho E-A, Seo J, Lee D-W, Pan J-G (2011) Decolorization of indigo carmine by laccase displayed on *Bacillus subtilis* spores. Enzyme Micro Technol 49:100–104

Couto SR, Herrera JLT (2006) Industrial and biotechnological applications of laccases: a review. Biotechnol Adv 24:500–513

Daniel D, Rakhi BS, Subramaniyan S, Sandhia GS (2013). Optimisation of Cultural and Nutritional Parameters for the Production of Protease from Newly Isolated Bacterial Strain *Bacillus* SDR 10. In: Sabu A, Augustine A (ed). Prospects in Bioscience: Addressing the Issues, Springer, India, pp 63–77

D'Annibale A, Stazi SR, Vinciguerra V, Sermanni GG (2000) Oxirane-immobilized *Lentinula edodes* laccase: stability and phenolics removal efficiency in olive mill wastewater. J Biotechnol 77:265–273

Degryse E, Glansdorff N, Pierard A (1978) A comparative analysis of extreme thermophilic bacteria belonging to the genus *Thermus*. Arch Microbiol 117:189–196

Desai SS, Nityanand C (2011) Microbial laccases and their applications: a review. Asian J Biotechnol 3(2):98–124

Dittmer NT, Suderman RJ, Jiang H, Zhu YC, Gorman MJ, Kramer KJ, Kanost MR (2004) Characterization of cDNAs encoding putative laccase-like multicopper oxidases and developmental expression in the tobacco hornworm, *Manduca sexta*, and the malaria mosquito, *Anopheles gambiae*. Insect Biochem Mol Biol 34:29–41

Eggert C, Temp U, Dean JFD, Eriksson KEL (1996) A fungal metabolite mediates degradation of non-lignin structures and synthetic lignin. FEBS Lett 391:144–148

Endo K, Hayashi Y, Hibi T, Hosono K, Beppu T, Ueda K (2003) Enzymological characterization of EpoA, a laccase-like phenol oxidase produced by *Streptomyces griseus*. J Biochem 133:671–677

Givaudan A, Effosse A, Faure D, Potier P, Bouillant ML, Bally R (1993) Polyphenol oxidase in *Azospirillum lipoferum* isolated from rice rhizosphere: evidence for laccase activity in nonmotile strains of *Azospirillum lipoferum*. FEMS Microbiol Lett 108:205–210

Gregory RP, Bendall DS (1966) The purification and some properties of the polyphenol oxidase from tea (*Camella sinesis* L.). Biochem J 101:569–581

Hattori M, Tsuchihara K, Noda H, Konishi H, Tamura Y, Shinoda T, Nakamura M, Hasegawa T (2010) Molecular characterization and expression of laccase genes in the salivary glands of the green rice leafhopper, *Nephotettix cincticeps* (Hemiptera: Cicadellidae). Insect Biochem Mol Biol 40:331–338

Heinzkill M, Messner K (1997) The ligninolytic system of fungi. In: Anke T (ed) Fungal biotechnology. Chapman & Hall, Weinheim, pp 213–226

Held C, Kandelbauer A, Schroeder M, Cavaco-Paulo A, Guebitz GM (2005) Biotransformation of phenolics with laccase containing bacterial spores. Environ Chem Lett 3:74–77

Hüttermann A, Mai C, Kharazipour A (2001) Modification of lignin for the production of new compounded materials. Appl Microbiol Biotechnol 55:387–394

Jacobson ES, Emery HS (1991) Temperature regulation of the *cryptococcal phenoloxidase*. J Med Vet Mycol 29:121–124

Jurado M, Prieto A, Martínez-Alcalá MA et al (2009) Laccase detoxification of steam-exploded wheat straw for second generation bioethanol. Bioresource Technol 100:6378–6384

Koschorreck K, Richter SM, Ene AB, Roduner E, Schmid RD, Urlacher VB (2008) Cloning and characterization of a new laccase from *Bacillus licheniformis* catalyzing dimerization of phenolic acids. Appl Microbiol Biotechnol 79:217–224

Kuznetsov VD, Filippova SN, Rybakova AM (1984) Nature of the brown pigment and the composition of the phenol oxidases of *Streptomyces galbus*. Microbiol 53:193–197

Lu L, Zhao M, Wang T-N, Zhao L-Y, Du M-H, Li T-L, Li D-B (2011) Characterization and dye decolorization ability of an alkaline resistant and organic solvents tolerant laccase from *Bacillus licheniformis* LS04. Bioresour Technol doi:10.1016/j.biortech.2011.07.111

Martins LO, Soares CM, Pereira MM, Teixeira M, Costa T, Jones GH, Henriques AO (2002) Molecular and biochemical characterization of a highly stable bacterial laccase that occurs as a structural component of the *Bacillus subtilis* endospore coat. J Biol Chem 277:18849–18859

Mayer A, Staples R (2002) Laccase: new functions for an old enzyme. Phytochem 60:551–565

Mittal A, Nagar S, Kumar Gupta VK (2013). Production and purification of high levels of cellulase-free bacterial xylanase by *Bacillus* sp. SV-34S using agro-residue. Annals of Microbiology 63(3):1157–1167

Nosanchuk JD, Valadon P, Feldmesser M, Casadevall A (1999) Melanization of *Cryptococcus neoformans* in murine infection. Mol Cell Biol 19:745–750

Nurudeen TA, Ahearn DG (1979) Regulation of melanin production by *Cryptococcus neoformans*. J Clin Microbiol 10:724–729

Patel H, Gupte A, Gupte S (2009). Effect of different culture conditions and inducers on production of laccase by a basidiomycete fungal isolate *Pleurotus ostreatus* HP-1 under solid state fermentation. BioResources 4(1):268–284

Polacheck I, Hearing VJ, Kwon-Chung KJ (1982) Biochemical studies of phenoloxidase and utilization of catecholamines in *Cryptococcus neoformans*. J Bacteriol 150:1212–1220

Reiss R, Ihssen J, T-M L (2011) *Bacillus pumilus* laccase: a heat stable enzyme with a wide substrate spectrum. BMC Biotechnol 11:9

Roberts SA, Weichsel A, Grass G, Thakali K, Hazzard JT, Tollin G, Rensing C, Montfort WR (2002) Crystal structure and electron transfer kinetics of CueO, a multicopper oxidase required for copper homeostasis in *Escherichia coli*. Proc Natl Acad Sci U S A 99:2766–2771

Ruijssenaars HJ, Hartmans S (2004) A cloned *Bacillus halodurans* multicopper oxidase exhibiting alkaline laccase activity. Appl Microbiol Biotechnol 65:177–182

Sambrook J, Fritsch EF, Maniatis T (1989) Molecular cloning: a laboratory manual, 2nd edn Cold Spring Harbor Laboratory Press, Cold Spring Harbor

Saparrat MCN, Cabello MN, Arambarri AM (2002a) Extracellular laccase activity in *Tetraploa aristata*. Biotechnol Lett 24:1375–1377

Sato Y, Bao W, Sederoff R, Whetten R (2001) Molecular cloning and expression of eight laccase cDNAs in loblolly pine (*Pinus taeda*). J Plant Res 114:147–155

Sharma KK, Kuhad RC (2008) Laccase: enzyme revisited and function re defined. Ind J Microbiol 48:309–316

Sharma P, Goel R, Capalash N (2007) Bacterial laccases. World J Microbiol Biotechnol 23:823–832

Sharma R, Goel R, Capalash N (2007) Bacterial laccases. World J Microbiol Biotechnol 23:823–832

Singh G, Capalash N, Goel R, Sharma P (2007) A pH-stable laccase from alkalitolerant γ-proteobacterium JB: purification, characterization and indigo carmine degradation. Enzyme Microb Technol 41:794–799

Solomon EI, Sundaram UM, Machonkin TE (1996) Multicopper oxidases and oxygenases. Chem Rev 96:2563–2605

Suzuki K, Hirai H, Murata H, Nishida T (2003) Removal of estrogenic activities of 17 β-estradiol and ethynylestradiol by ligninolytic enzymes from white rot fungi. Water Res 37(8):1972–1975

Thurston CF (1994) The structure and function of fungal laccases. Microbiol 140:19–26

Ullah MA, Bedford CT, Evans CS (2000) Reactions of pentachlorophenol with laccase from *Coriolus versicolor*. Appl Microbiol Biotechnol 53:230–234

Wang C, Zhao M, Li D-B, Cui D-Z, Lu L, Wei X (2010) Isolation and characterization of a novel *Bacillus subtilis* WD23 exhibiting laccase activity from forest soil. Afr J Biotechnol 9(34):5496–5502

Wang F, Guo C, Wei T, Zhang T, Liu CZ (2012). Heat shock treatment improves *Trametes versicolor* laccase production. Appl Biochem Biotechnol 168(2):256–265

Williamson PR (1994) Biochemical and molecular characterization of the diphenol oxidase of *Cryptococcus neoformans*: identification as a laccase. J Bacteriol 176:656–664

Yesilada O, Sik S, Sam M (1997). Biodegradation of olive oil mill wastewater by *Coriolus versicolor* and Fu*nalia trogii*:effects of agitation, initial COD concentration, inoculum size and immobilization. World J Microbiol Biotechnol 14(1):37–42

Yoshida H (1883) Chemistry of Lacquer (Urushi) part 1. J Chem Soc 43:472–486

# Optimization of Protease Enzyme Production by the Halo-Tolerant *Vibrio alginolyticus* Isolated from Marine Sources

40

S. Malathi, D. Mohana Priya and P. Palani

**Abstract**

A renewed interest in the study of proteolytic enzymes is chiefly attributable to its surging industrial applications apart from its pivotal role in the cellular metabolic process. We report the screening, isolation and partial purification of a low-molecular-weight halo-tolerant alkaline protease from a bacterial isolate of marine environment. Twenty four bacterial isolates were collected from marine environment, of which *Vibrio alginolyticus* showed relatively higher proteolytic activity on agar plate and was subjected to further characterization. Enzyme production kinetics under submerged fermentation was optimized, that included incubation time, pH and heavy metal tolerance. The bacterium produced the highest protease activity (4650 U/mL) when incubated for 48 h at pH 8.0. The organism was found to be more tolerant to iron than other metals like $Cu^{2+}$, $Mn2^{+}$, $Zn^{2+}$, $Co^{2+}$ and $Hg^{2+}$. Natural substrates such as rice and wheat brans were evaluated for protease enzyme production under solid state fermentation. A mixture of rice and wheat bran at a ratio of 1:1 supported the higher enzyme production. Effect of NaCl, heavy metals and EDTA on the catalytic activity of the crude enzyme was assayed. The total activity of the crude enzyme was unaffected at elevated concentrations of NaCl and retained 40% of its activity even at 4 M concentration which indicated the halo-tolerant ability of the bacterium. An increase in the catalytic activity (about 30% or 1.2 fold) was recorded when incubated with 1 mM of $Co^{2+}$ while considerable loss in function was recorded with other heavy metals. Similar loss of function has been recorded with EDTA indicating that it is a metal ion-dependent enzyme. The partially purified enzyme was able to digest the meat and showed high capability of removing blood stains from fabric.

**Keywords**

*Vibrio alginolyticus* · Protease · Halo-tolerance · Marine bacterium

P. Palani (✉) · S. Malathi · D. Mohana Priya
Centre for Advanced Studies in Botany, University of Madras, Guindy Campus, Chennai 600 025, India
e-mail: palani7@unom.ac.in

R. N. Kharwar et al. (eds.), *Microbial Diversity and Biotechnology in Food Security*,
DOI 10.1007/978-81-322-1801-2_40, © Springer India 2014

## 40.1 Introduction

Proteases constitute one of the most important groups of industrial enzymes accounting for nearly 60 % of the total worldwide enzyme sales (Beg et al. 2003; Kalisz 1988; Ellaiah et al. 2003; Adinarayana and Ellaiah 2003). Of late, there has been growing interest in microbial proteases that are of commercial importance. Bacterial proteases are the most significant among the various proteases when compared to animal and fungal proteases (Ward 1985). Proteases produced by alkalophilic microorganisms are investigated not only in the scientific areas but also in the laundry, dishwashing, food and other such industries. In recent years, the use of thermostable alkaline enzymes has increased in a wide range of biotechnological applications such as silver recovery, feed and peptide synthesis. These enzymes are currently receiving increased attention in view of their inherent stability at high pH and temperature, and are active in the presence of surfactants, organic solvents and denaturing agents, which enable their use in processes that restrict conventional enzyme (Kumar et al. 1999). Marine microorganisms serves as a source of novel alkaline proteases with their inherent stability at a wide range of pH with the maximum activity between pH 9 and 11, cleavage of a wide spectrum of peptide bonds (Mao et al. 1992), temperature and salinity. Several proteolytic bacteria have been reported to be associated with both fresh water and marine fish processing wastes (Sudeepa et al. 2007). It has also been reported that aquatic microbes synthesize exoenzymes identified as inducible catabolic enzymes (Chrost 1990; Kim et al. 2002). In view of the above information, an attempt has been made in the present investigation to isolate bacteria from marine environment and screen them for the production of protease. A halo-tolerant bacterium isolated in the present study produced relatively higher protease enzyme than other bacteria. An attempt has also been made to characterize and study the efficacy of the protease produced by the above organism and was evaluated for industrial applications.

## 40.2 Materials and Methods

### 40.2.1 Microorganism and Growth Conditions

Water samples, soil sediments (collected at 30–40 cm depth), shells and fishes were collected from Marina beach, Bay of Bengal, Chennai, India. The water and soil sediment samples were serially diluted and spread on Nutrient Agar medium prepared in sea water and incubated at 37 °C for 48 h. The shells and fish samples were kept on the marine agar plate and incubated at 37 °C for 48 h. The bacterial colonies were isolated and maintained on marine Nutrient Agar slants for further study.

### 40.2.2 Screening of Microorganisms for Protease Production

Each of the bacterial isolates isolated from marine sources were streaked on to 1 % casein and gelatine agar plates and incubated at 37 °C for 24 h. A clear zone of hydrolysis gave an indication of protease production on the plates. The promising bacterial strain was chosen based on the size of zone of clearance.

### 40.2.3 Identification of Protease Producing Bacterium

#### 40.2.3.1 DNA Extraction, PCR Amplification and Sequencing

For genomic DNA isolation, the bacterium was grown on Nutrient Broth medium. The total genomic DNA was extracted (Ausubel et al. 1987) and then RNase enzyme was added to the sample to remove RNA, followed by incubation at 37 °C for 2 h. The intensity of DNA was examined by electrophoresis on 0.8 % agarose gel in tris-borate-ethelynediaminetetraacetic acid (TBE) buffer stained with ethidium bromide and visualized under ultraviolet (UV) light.

#### 40.2.3.2 Polymerase Chain Reaction

Amplification reactions were performed in a total volume of 20 µl consisting of the 1µl of genomic DNA, 1× polymerase chain reaction (PCR) buffer including 20 pmol of forward (0.5µl) and reverse primers (0.5µl), 10µl of 0.16mM dNTP mix and 8µl of milliQ water (MBI, Fermentas, Lithuania). The mixer was overlaid with two drops of mineral oil. PCR amplification was carried out in a thermocycler (Eppendorf Mastercycler 5330, Germany) for 30 cycles. The forward and reverse primers used were 27F bacterial 5′ AGA GTT TGA TCM TGG CTC 3′ and 1525-R bacterial 5′ AAG GAG GTG WTC CAR 3′. Four microliters of PCR amplified product was electrophoresed on 1% agarose gel in 1 × Tris-aceate-ethelynediaminetetraacetic acid (TAE) buffer (40 mM Tris, 1 mM EDTA pH 8.0) and visualized in an image analyzer (Chemilmager 5500, Alpha Innotech, CA, USA) after staining with ethidium bromide (0.5 µg $mL^{-1}$).

#### 40.2.3.3 DNA Sequencing and Phylogenetic Analysis

The PCR product was directly sequenced in an automatic DNA sequencer (Applied Biosystems, Inc., CA, USA). The sequence reactions were conducted using the Dye Terminator Cycle Sequencing Kit (Applied Biosystems, Inc., CA) following the manufacturer's protocol. Analyses of sequences were performed with the basic sequence alignment BLAST program run against the database NCBI. A tree showing the phylogenetic relatedness between the isolates was constructed from the distance matrix values by the neighbor-joining method using ClustalW.

### 40.2.4 Production of Protease

Production of protease was carried out in a Nutrient Broth medium and incubated at 37 °C for 48 h in a refrigerated shaker at 125 rpm. The pH of the medium was adjusted to 7.6 with 1N NaOH or 1N HCl. After the completion of fermentation, the broth was centrifuged at 10,000 rpm at 4 °C for 20 min and the clear supernatant was recovered and used as a crude enzyme source.

### 40.2.5 Determination of Total Protein and Protease Activity

The total protein content of the culture supernatant was estimated by the dye-binding method of Bradford (1976) using bovine serum albumin fraction (V) (Sigma Chemical Co., USA) as the standard.

The proteolytic enzyme activity was assayed using casein as the substrate. The reaction mixture contained a known amount of protein (50 µg) in 0.5 ml culture filtrate and 0.5 ml of 2% (w/v) casein. The mixture was incubated for 1 h at 37 °C and the reaction was terminated by the addition of 1 ml of cold trichloroacetic acid (10% w/v). The undigested casein was removed by centrifugation in a microfuge for 5 min at 5,000 rpm and the aliquots of the supernatant (0.5 ml) were taken in separate tubes and to this 2.5 ml of reagent (2.9% $Na_2Co_3$ and 0.3N NaOH) was added. Then, 0.75 ml of Folin–Ciocalteu's phenol reagent (1:2 diluted with glass distilled water) was incubated for 30 min at room temperature for colour development. The liberation of tyrosine equivalents was determined at 650 nm (Mc Donald and Chen 1965) in a spectrophotometer. A standard curve constructed with tyrosine served as a reference for protease-activity measurement. Heat-killed enzyme with the substrate served as control. One unit of protease activity was defined as the amount of enzyme liberating 1 mM tyrosine equivalent/ml under the assay condition.

### 40.2.6 Optimisation of Enzyme Production

#### 40.2.6.1 Effect of Incubation Time

The effect of incubation time on the production of protease enzyme was evaluated under submerged fermentation condition for 72 h. The samples were harvested at 12 h intervals (12, 24, 36, 48, 60, 72, 84, 96, 108 and 120) and the protease enzyme activity was measured.

#### 40.2.6.2 Effect of pH

The bacterium was grown in the basal medium adjusted to different pH from 6.0 to 10.0 for the time duration as optimized above. The samples were collected and the protease enzyme activity was measured.

#### 40.2.6.3 Effect of Heavy Metals

The effect of heavy metals like copper, cobalt, manganese, iron, zinc and mercury on the protease production of the organism was studied. The bacterium was inoculated in the medium containing respective heavy metal ions at a final concentration of 1 mM and the protease production of the organism under these heavy metal conditions was assayed after 48 h.

#### 40.2.6.4 Evaluation of Natural Substrates for Protease Production Under Solid State Fermentation (SSF)

The protease production of the test isolate was carried out in 500 ml conical flasks either individually with 20 g wheat or rice brans or in combination of 10 g of wheat bran and 10 g of rice bran. The agro waste was moistened with 10 ml of mineral salt medium containing $K_2HPO_4$, $MgSO_4$, $CaCl_2$, $FeSO_4$ and $ZnSO_4$. The flask was inoculated with 1 ml of 24-h-grown culture and kept for incubation at 37 °C for 48 h.

#### 40.2.6.5 Effect of NaCl on Protease Activity

The supernatant containing the enzyme was incubated with NaCl (1.0–4.0 M) at 37 °C for 30 min. After the incubation period the residual activity was assayed as described above. Casein was used as the substrate.

#### 40.2.6.6 Effect of EDTA and Metal Ions on Protease Activity

The effect of different heavy metals on the proteolytic activity of crude enzyme was investigated by incubating the enzyme in the presence of metal ions (1 mM) such as $Cu^{2+}$, $Mn^{2+}$, $Hg^{2+}$, $Fe^{2+}$, $Zn^{2+}$ and $Co^{2+}$. The protease inhibitor EDTA was also used at the same concentration. The enzyme supernatant containing the enzyme was incubated with metal ions for 30 min at 37 °C, and the residual activity was measured.

## 40.3 Partial Purification of Protease

The proteins in the culture supernatant was precipitated with 80 % saturation of ammonium sulphate and was allowed to stand overnight at 4 °C. The resulting precipitate was collected by centrifugation at 10,000 rpm for 10 min at 4 °C. The precipitate was dissolved in Tris-HCl buffer (50 mM; pH 7.2) and dialyzed at 4 °C against the Tris-HCl buffer (10mM; pH7.2). The dialysate was then lyophilized and used for further studies.

## 40.4 Applications of the Partially Purified Protease

### 40.4.1 Protease in Digestion of Natural Proteins

The lyophilized protease enzyme 4000U/mL was incubated with 2 ml of animal blood sample and coagulated egg white in Tris-HCl (10mM; pH7.2) at 37 °C. The blood clots and the coagulated egg white were examined for clot removal after 14 h of incubation.

### 40.4.2 Protease in Meat Tenderization

A small piece of meat (2 gm) was incubated with the lyophilized enzyme in Tris-HCl (10 mM; pH 7.2) at 37 °C for 14 h and the meat was then examined for tenderness. A piece of meat without enzyme treatment served as control.

### 40.4.3 Protease in Dehairing of Skin

A small piece of goat's skin (3.5 × 4 cm) with hair was incubated with the lyophilized enzyme 4000U/mL in Tris-HCl (10 mM; pH7.2) at 37 °C for 14 h, and the ability of the enzyme in removing the hairs was then visually monitored.

**Table 40.1** Morphological and growth characteristics of the bacteria isolated from marine source

| S. No. | Test isolates | Pigmentation | Growth | Motility |
|---|---|---|---|---|
| 1 | PC 1 | Yellow colony | Moderate growth | Non-motile |
| 2 | PC 2 | White colony | Slow growth | Non-motile |
| 3 | PC 3 | White colony | Fast growth | Non-motile |
| 4 | PC 4 | White colony | Fast growth | Non-motile |
| 5 | PC 5 | Pink colony | Fast growth | Non-motile |
| 6 | PC 6 | White colony | Fast growth | Non-motile |
| 7 | PC 7 | Yellow colony | Moderate growth | Non-motile |
| 8 | PC 8 | Orange colony | Fast growth | Non-motile |
| 9 | PC 9 | White colony | Moderate growth | Non-motile |
| 10 | PC 10 | White colony | Moderate growth | Non-motile |
| 11 | PC 11 | Pink colony | Slow growth | Non-motile |
| 12 | PC 12 | Orange colony | Fast growth | Non-motile |
| 13 | PC 13 | White colony | Slow growth | Non-motile |
| 14 | PC 14 | White colony | Fast growth | Non-motile |
| 15 | PC 15 | Yellow colony | Fast growth | Non-motile |
| 16 | PC 16 | Orange colony | Slow growth | Non-motile |
| 17 | PC 17 | White colony | Moderate growth | Non-motile |
| 18 | PC 18 | White colony | Moderate growth | Non-motile |
| 19 | PC 19 | White colony | Fast growth | Non-motile |
| 20 | PC 20 | White colony | Fast growth | Motile |
| 21 | PC 21 | White colony | Fast growth | Non-motile |
| 22 | PC 22 | White colony | Fast growth | Non-motile |
| 23 | PC 23 | White colony | Fast growth | Non-motile |
| 24 | PC 24 | White colony | Fast growth | Motile |

### 40.4.4 Protease in Removing Blood Stain

A clean piece of cotton cloth (4 × 4 cm) was soaked with goat's blood and dried at room temperature. Then, the cloth was soaked in 2% formaldehyde for 30 min and washed with water to remove excess formaldehyde. The cloth was cut into two equal halves, and the first half was incubated with partially purified enzyme 4000U/mL in Tris-HCl (10 mM; pH 7.2), and the second half was incubated with the buffer alone without enzyme at 45–50 °C. After 2 h, the cloth pieces were rinsed with water for 2 min and then dried. The removal of blood stain from the cloth pieces were examined visually.

## 40.5 Results and Discussion

### 40.5.1 Isolation of Microorganisms

There were about 24 isolates obtained from different marine sources. Their growth and morphological characteristics were studied and have been listed in Table 40.1. A clear zone of hydrolysis was observed around the bacterial strain PC 24 in agar plates containing casein and gelatine (Fig. 40.1). This zone of hydrolysis appeared after 24 h incubation indicating protease production on the plate and this isolate was used for further studies.

### 40.5.2 Identification of Protease Producing Bacterium

To solve the inherent doubts of classical identification of bacterial species, various molecular techniques are currently being used to characterize and analyze the taxonomic complexity of the bacteria. One of the frequently used techniques is PCR-based amplification of the 16S rRNA sequence with universal primers. The nucleotide sequence of the amplified product was obtained and contained 237 base pairs. The nucleotide base pairs were aligned with sequences of closely related bacteria using the software ClustalW and a phylogenetic tree was constructed using PHYLIP software. The above analyses have shown 99%

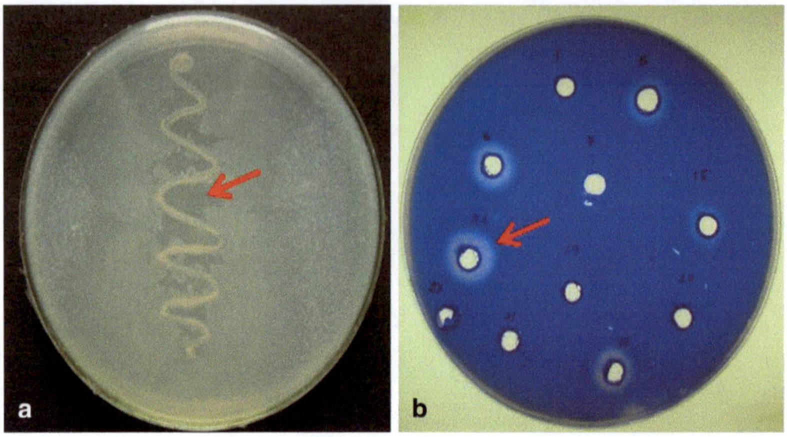


**Fig. 40.1** Plates showing hydrolytic zone of casein (**a**) and gelatin (**b**) by test *Vibrio alginolyticus*

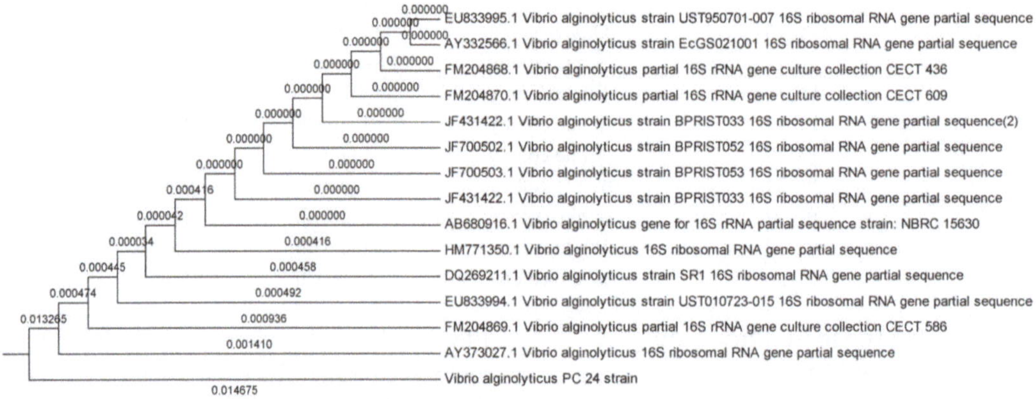


**Fig. 40.2** Phylogenetic tree showing taxonomic position of the isolate *Vibrio alginolyticus.* The evolutionary history was inferred using the UPGMA method (Sneath and Sokal 1973). The optimal tree with the sum of branch length = 0.03306149 is shown (next to the branches). The evolutionary distances were computed using the Maximum Composite Likelihood Method (Tamura et al. 2004) and are in the units of the number of base substitutions per site. The analysis involved 15 nucleotide sequences. All positions containing gaps and missing data were eliminated. There were a total of 1,203 positions in the final dataset. Evolutionary analyses were conducted in MEGA5. (Tamura et al. 2011)

homology with *Vibrio alginolyticus* (Fig. 40.2), and therefore the bacterium was identified as *V. alginolyticus*. The identified bacterial sequence was submitted to GenBank with the accession No. JQ 780446.

### 40.5.3 Effect of Incubation Time on Protein Content and Protease Activity

The total protein and the total enzyme activity were determined from the supernatants collected from 12 to 120 h. The supernatant collected after 48 h incubation showed maximum protease activity as well as protein content, and hence further assays were carried out with the incubation time of 48 h (Table 40.2).

### 40.5.4 Effect of pH on Enzyme Production

*V. alginolyticus* showed relatively higher activity at pH 8. Decrease or increase in pH from 8.0 resulted in decreased enzyme activity. As

**Table 40.2** Total protease activity and total protein of *Vibrio alginolyticus* from 12 to 120 h

| Time (in hours) | Total protease activity (U/ml) | Total protein (mg/ml) |
|---|---|---|
| 12 | 2650 | 0.016 |
| 24 | 2640 | 0.030 |
| 36 | 2750 | 0.020 |
| 48 | 4650 | 0.032 |
| 60 | 4200 | 0.014 |
| 72 | 3200 | 0.018 |
| 84 | 2200 | 0.020 |
| 96 | 3400 | 0.020 |
| 108 | 2500 | 0.024 |
| 120 | 2050 | 0.032 |

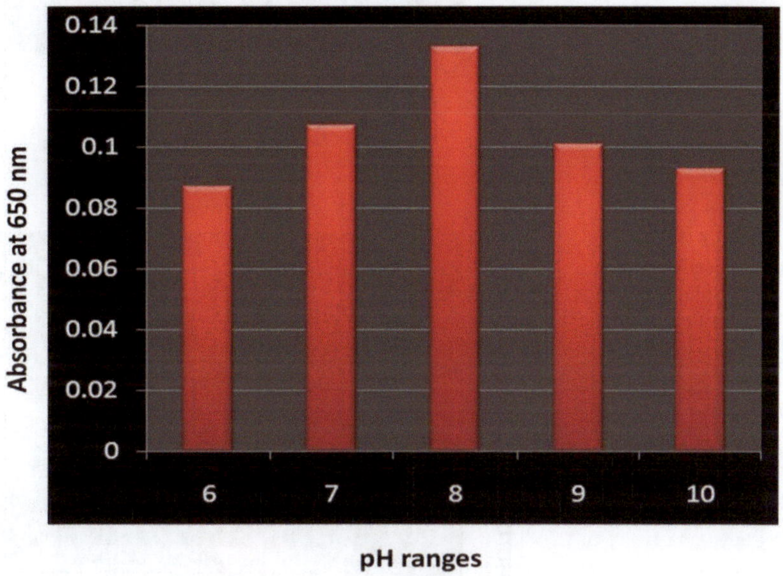


**Fig. 40.3** Effect of pH on the activity of protease

evident from Fig. 40.3, the enzyme remained active even at pH 10.0. This is comparatively higher with the protease from *Vibrio fluvialis* VM10 (Venugopal & Saramma 2006) with an optimum pH at 8, *V. fluvialis* TKU005 (Wang et al. 2007) at pH 7.5.

### 40.5.5 Effect of Heavy Metals on Enzyme Production

The bacterium was able to grow in the presence of iron, and the enzyme production was found to be unaffected. The organism was moderately tolerant to zinc and manganese while less tolerant to mercury and cobalt. In contrast, copper completely inhibited the growth of the organism (Fig. 40.4) and therefore no protease activity was measured.

### 40.5.6 Effect of Natural Substrates on Protease Production

The test bacterium produced relatively higher protease enzyme when incubated with a mixture containing equal proportion of wheat bran and rice bran. The protease activity, however, was relatively lower when the bacterium was grown on only one substrate. The organism showed the least enzyme activity on rice bran (Fig. 40.5).

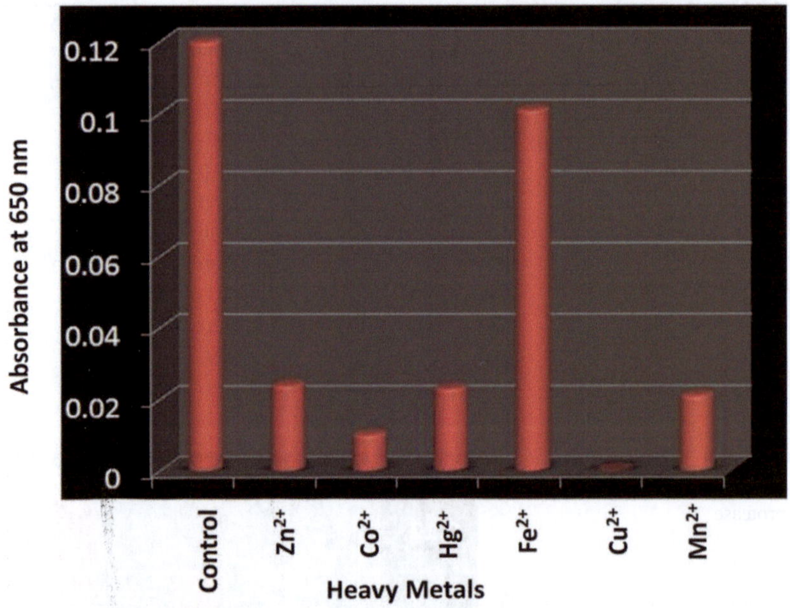


**Fig. 40.4** Heavy metal tolerances of the test isolate *Vibrio alginolyticus* against $Zn^{2+}$, $Co^{2+}$, $Hg^{2+}$, $Fe^{2+}$, $Cu^{2+}$ and $Mn^{2}$

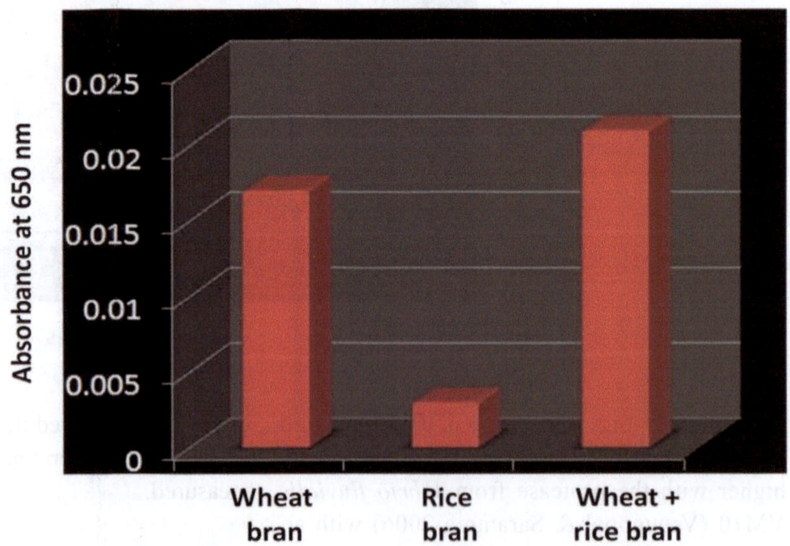


**Fig. 40.5** Production of extracellular protease on solid substrate fermentation

### 40.5.7 Effect of NaCl on Crude Enzyme Activity

The optimal activity of the enzyme remained unaffected in the presence of NaCl even at a concentration of 2 M (Fig. 40.6) and started decreasing when the concentration of NaCl was increased from 2 M and higher. However, the enzyme was not completely inhibited even at a concentration of 5 M. About 40 % of the activity was retained even in the presence of 4 M NaCl which unequivocally indicate that the bacterium is a halo-tolerant.

### 40.5.8 Effect of EDTA and Metal Ions on Enzyme Activity

Incubation of the enzyme with metal ions such as $Co^{2+}$, $Mn^{2+}$ and $Cu^{2+}$ enhanced the enzyme activity with the highest enhancement observed with $Co^{2+}$. Incubation of the enzyme with other metal ions such as $Fe^{2+}$, $Zn^{2+}$, $Hg^{2+}$ and EDTA has considerably reduced the activity. The inhibition by metals may be because the metals not only affect the active site of the protease, but also

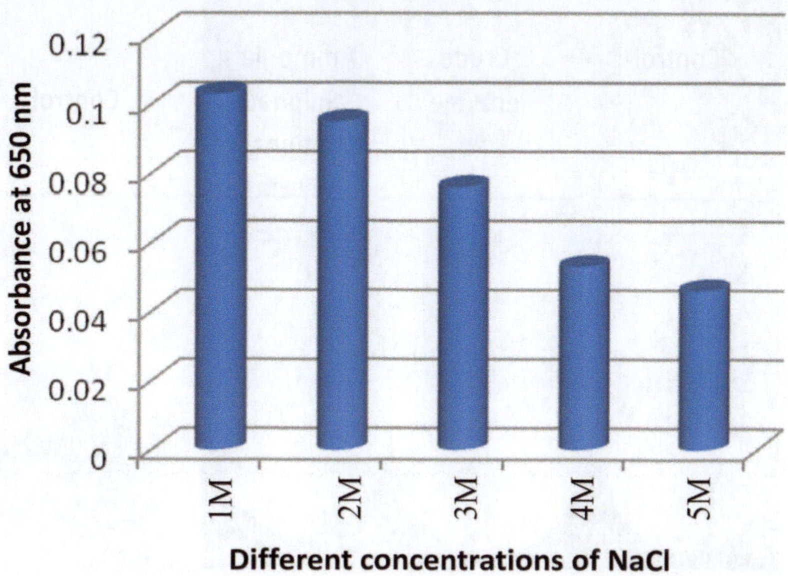


**Fig. 40.6** Effect of NaCl on the stability of protease

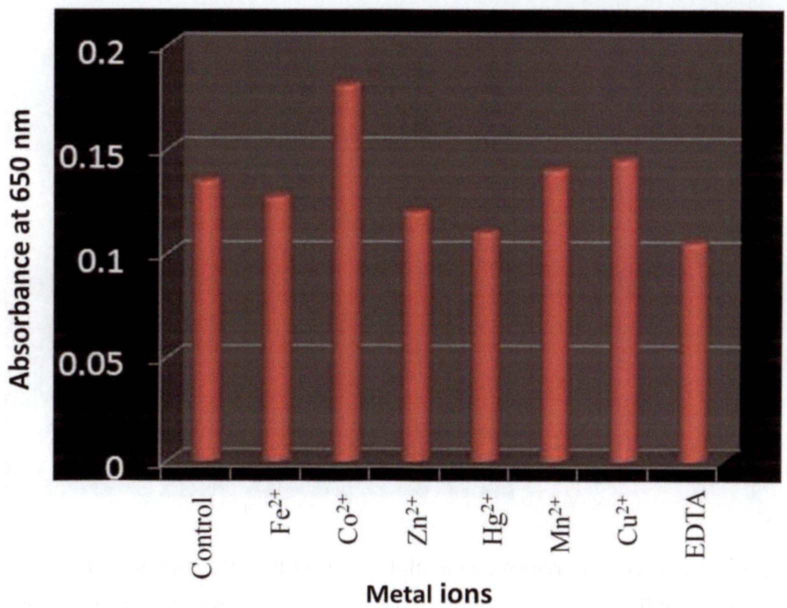


**Fig. 40.7** Effect of metal ions on the enzyme activity of the test isolate *Vibrio alginolyticus*

the non-catalytic protein-binding region, which is involved in efficient hydrolysis of the substrate (Subashini et al. 2012). EDTA was more effective in inhibiting the enzyme followed by $Hg^{2+}$ (Fig. 40.7). Activity of the crude enzyme was decreased in the presence of EDTA, indicating that it is a metal ion requring enzyme.

### 40.5.9 Digestion of Natural Proteins

The crude and ammonium sulphate fractionated proteins containing protease enzyme were incubated with coagulated egg white (Fig. 40.8a) and blood clot (Fig. 40.8b) for 14 h and examined for removal of coagulated egg protein and blood clot. As evident in Fig. 40.8, the crude enzyme as well as the ammonium sulphate fractionated

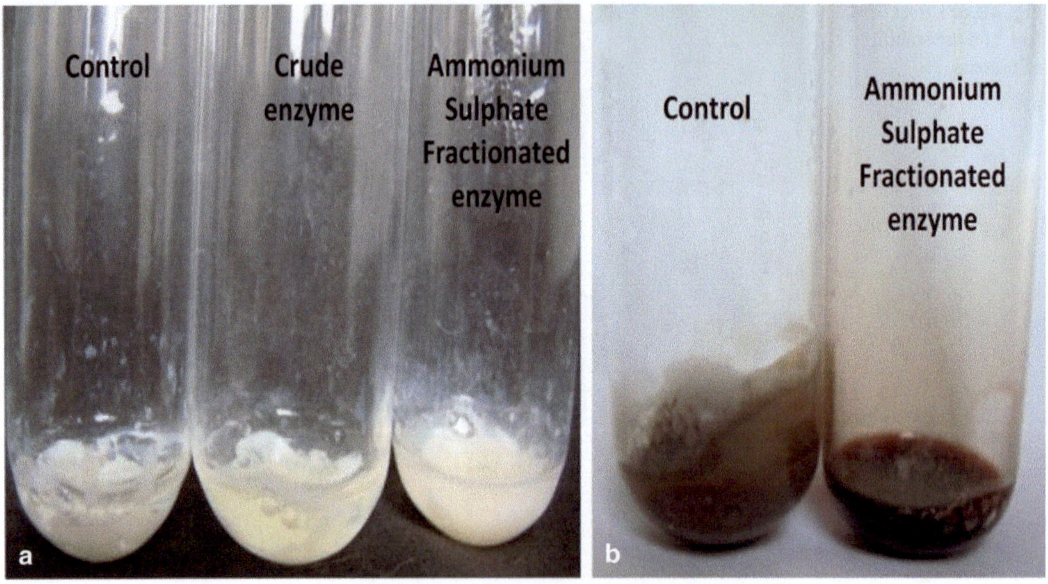


**Fig. 40.8** Effect of protease on the removal of clots of egg white albumin (**a**) and blood (**b**)

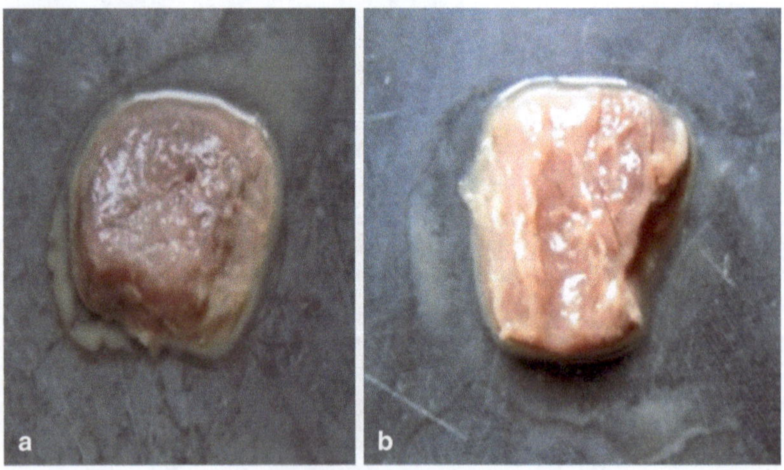


**Fig. 40.9** Analysis of tenderizing ability of protease enzyme. **a** Control. **b** Enzyme treated

enzyme appreciably removed the clots after 14 h of incubation.

### 40.5.10 Meat Tenderization and Dehairing of Skin

The ammonium sulphate fractionated protease enzyme was incubated with fresh meat for 14 h and the tenderizing capacity of the meat was examined. The meat incubated with the enzyme become soft (Fig. 40.9b). The partially purified enzyme was able to digest the meat when it was incubated with it. The capability of this enzyme to digest different natural substrates with base of fibrin, albumin and collagen suggests the usefulness of this enzyme for different applications such as extraction of collagen from skin for collagen replacement therapy, waste treatment and other related applications. As seen in Fig. 40.10b, the hair has been significantly removed from the skin treated with partially purified enzyme whereas it remained intact and fine in the skin treated with the buffer alone. the (Fig. 40.10a).

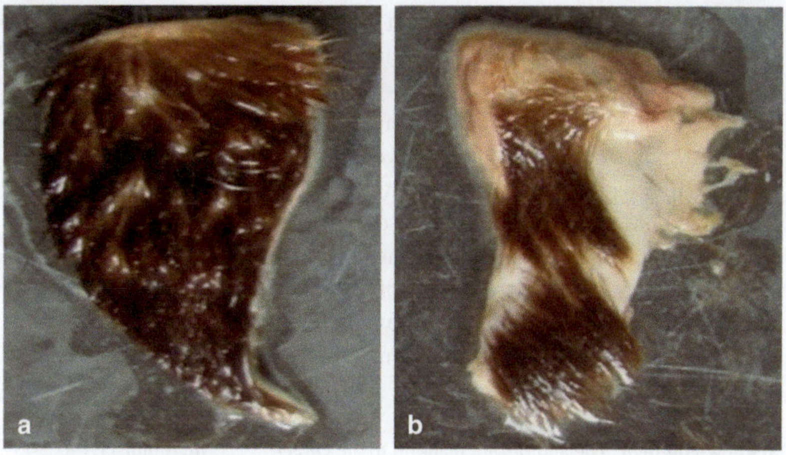


**Fig. 40.10** Analysis of dehairing potential of the enzyme. **a** Control. **b** Enzyme treated

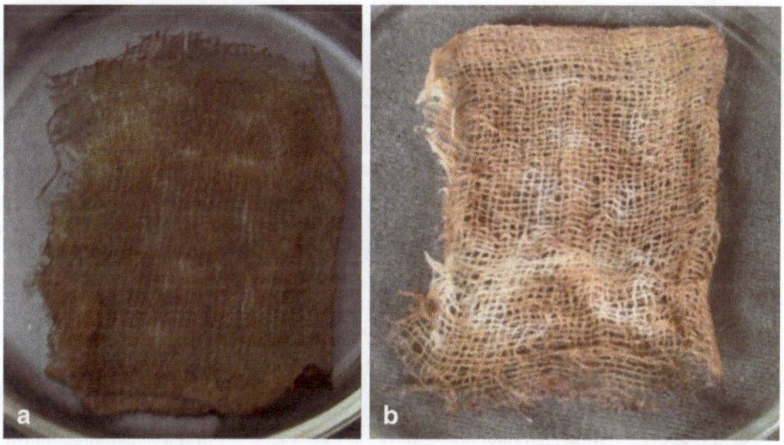


**Fig. 40.11** Removal of blood stain from fabric by protease enzyme. **a** Control. **b** Enzyme treated

### 40.5.11 Removal of Blood Stain

As is evident in Fig. 40.11a and 40.11b, the protease enzyme removed the blood stain from the fabric significantly, and the stain was intact in the fabric treated with the buffer alone. This enzyme showed high capability for removing proteins and stain from cloth. The properties of this enzyme such as metal tolerance, alkaline pH make it as a potential candidate as an alkaline protease in detergent powders or solutions. Same kind of results have already been obtained from *Bacillus cereus* isolated from marine samples which removed the animal blood stains from the fabric (Abou-Elela et al. 2011).

## 40.6 Conclusion

There is renewed interest in the study of proteolytic enzymes, mainly due to the recognition that these enzymes not only play an important role in the cellular metabolic processes but also considerable attention in the industrial applications. In this study, a low-molecular weight halo-tolerant alkaline protease-producing bacteria was isolated from marine environment. The organism showed maximum enzyme activity at pH 8, and it was resistant to the heavy metal $Fe^{2+}$. The metal ions like $Cu^{2+}$, $Co^{2+}$ and $Mn^{2+}$ enhanced the protease activity. The activity of the crude enzyme was decreased in the presence of EDTA, indicating that it is a metal ion-dependent enzyme. Even in the presence of 4 M NaCl, the crude retained 40% of its activity, indicating the halo-tolerant behaviour of the enzyme. The partially purified enzyme was able to digest the meat when it was incubated with it and showed high capability for removing blood stain from cloth which showed it could be used in detergent powder or solution. Thus, the remarkable activity and stability of alkaline protease from *V. alginolyticus* to higher pH, salt tolerance, heavy metal tolerance and digestion of natural proteins and capability of removing blood stain make this enzyme to be an alternative candidate in the industrial sector.

**Acknowledgment** The authors are thankful to the Director, Centre for Advanced Studies in Botany, University of Madras, Chennai for laboratory facilities.

## References

Abou-Elela M, Ibrahim Hah, Hassan SW, Abd-Elnaby H, El-Toukhy Nabil MK (2011) Alkaline protease production by alkaliphilic marine bacteria isolated from Marsa-Matrouh (Egypt) with special emphasis on *Bacillus cereus* purified protease. Afr J Biotechnol 10:4631–4642

Adinarayana K, Ellaiah P (2003) Production of alkaline protease by immobilized cells of alkalophilic *Bacillus* sp. J Sci Ind Res 62:589–592

Ausubel FM, Brent R, Kingston RE, Moore DD, Seidman JG, Smith JA, Struhl K (1987) Current protocol in molecular biology. Greene Publishing associates/ Wiley Interscience, New York

Beg KB, Sahai V, Gupta R (2003) Statistical media optimization and alkaline protease production from *Bacillus mojavensis* in a biore-actor. Process Biochem 39:203–209

Bradford MM (1976) A rapid and sensitive for the quantization of microgram quantities of protein utilizing the principle of protein dye binding analytical. Biochem 72:248–254

Chrost RJ (1990) Microbial ectoenzymes in aquatic environments. In: Overbeck J, Chrost RJ (eds) Aquatic microbial ecology: biochemical and molecular approaches. Springer, New York, pp 47–47

Ellaiah P, Adinarayana K, Rajyalaxmi P, Srinivasulu B (2003) Opti-mization of process parameters for alkaline protease production under solid state fermentation by alkalophilic *Bacillus* sp. Asian J Microbiol Biotechnol Environ Sci 5:49–54

Kalisz HM (1998) Microbial proteinases. Adv Biochem Eng Biotechnol 36:1–65

Kim YK, Bae JH, Oh BK, Lee WH, Choi JW (2002) Enhancement of proteolytic enzyme activity excreted from *Bacillus stearothermophilus* for a thermophilic aerobic digestion process. Biores Technol 82:157–164

Kumar CG, Tiwari MP, Jany KD (1999) Novel alkaline serine proteases from alkalophilic *Bacillus* sp. purification and some properties. Process Biochem 34:441–449

Mao W, Pan R, Freedman D (1992) High production of alkaline protease by *Bacillus licheniformis* in fed-batch fermentation using a synthetic medium. J Ind Microbiol 11:1–6

Mc Donald CE, Chen LL (1965) The Lowry modification of FOLIN reagent for determination of proteinase activity. Annal Biochem 10:175–177

Sneath PHA, Sokal RR (1973) Numerical taxonomy. Freeman, San Francisco

Subashini P, Annamalai N, Saravanakumar A, Balasubramanian T (2012) Thermostable alkaline protease from newly isolated Vibrio sp.: extraction, purification and characterisation. Biologia 67/4:629–635

Sudeepa ES, Rashmi HN, Tamilselvi A, Bhaskar N (2007) Proteolytic bacteria associated with fish processing wastes: isolation and characterization. J Food Sci Technol 44:281–284

Tamura K, Nei M, Kumar S (2004) Prospects for inferring very large phylogenies by using the neighbor-joining method. Proc Nat Acad Sci (USA) 101:11030–11035

Tamura K, Peterson D, Peterson N, Stecher G, Nei M, Kumar S (2011) MEGA5: Molecular evolutionary genetics analysis using maximum likelihood, evolutionary distance, and maximum parsimony methods. Mol Biol Evol 28(10):2731–2739

Venugopal M, Saramma AV (2006) Charaterisation of alkaline protease from *Vibrio fluvialis* strain VM10 isolated from a mangrove sediment sample and its application as a laundry detergent additive. Process Biochem 41:1239–1243

Wang SL, Chio YH, Yen YH, Wang CL (2007) Two novel surfactant stable alkaline proteases from *Vibrio fluvialis* TKU005 and their applications. Enzyme microb tech-nol 40:1213–1220

Ward OP (1985) Proteolytic enzymes. In: Blanch HW, Drew S, Wang DI (eds) Comprehensive biotechnol. Oxford, UK: Per-gamon Press, 3:789–818

# Antimicrobial Activity of Some Cyanobacteria

41

Rashmi Tyagi, B. D. Kaushik and Jitendra Kumar

**Abstract**

Many green and blue-green algae were examined for their antibacterial and antifungal properties using their extracts in three different solvents, ethanol, methanol-toluene (3:1) and acetone, on solidified agar nutrient medium by agar well diffusion method. Extracts of only five screened cyanobateria, *Westiellopsis prolifica* ARM 365, *Hapalosiphon hibernicus* ARM 178, *Nostoc muscorm* ARM 221, *Fischerella* sp. ARM 354 and *Scytonema* sp., were found to have antimicrobial activity against the bacteria *Pseuomonas striata, Bacillus subtilis, Escherichia coli, Bradyrhizobium* sp. and fungi *Aspergillus niger, Alternaria brassicae* and *Fusarium moniliforme*. Ethanol extracts of *Westiellopsis prolifica* ARM 365 and *H. hibernicus* ARM 178 showed the greatest antibacterial as well as antifungal activity against the tested bacteria and fungi as revealed by the largest inhibition zone formation on solidified agar nutrient medium. The test bacterium *P. striata* and the test fungus *A. niger* were found to be the most sensitive to ethanol extracts of *H. hibernicus* ARM 178 and *W. prolifica* ARM 365. These cyanobacterial extracts inhibited test bacteria more strongly than test fungi. The strongest antimicrobial activity of *H. hibernicus* ARM 178 and *W. prolifica* ARM 365 was observed on the tenth day of their growth and their ethanol extracts showed the maximum antimicrobial activity at 30 °C temperature. These antimicrobial compounds may have potential pharmaceutical, biocontrol or agricultural applications.

**Keywords**

Antimicrobial metabolites · Cyanobacteria · Bacteria · Fungi

## 41.1 Introduction

Due to the occurrence of cyanobacteria in diverse habitats, requirement of low-cost inorganic N and P sources and their intrinsic ability to take up $H_2O$ and $CO_2$ with the aid of

R. Tyagi (✉)
Department of Applied Sciences & Humanities, ITM University, Sector 23-A, Palam Vihar, Gurgaon, Haryana, India
e-mail: tyagi_rashmi63@rediffmail.com

B. D. Kaushik
Anand Engineering College Keetham, Agra 282007, U.P., India

J. Kumar
Department of Science, S. S. University, Varanasi, U.P., India

R. N. Kharwar et al. (eds.), *Microbial Diversity and Biotechnology in Food Security*, DOI 10.1007/978-81-322-1801-2_41, © Springer India 2014

solar energy, these organisms are a promising material for investigation for their antimicrobial activity. A number of cyanobacteria are known to produce intracellular and extracellular compounds which are accumulated and/or secreted as primary or secondary metabolites with diverse biological activities such as antialgal (John et al. 2003), antibacterial, antifungal (Ghasemi et al. 2007; Isnansetyo et al. 2003; Jaki et al. 1999; Kundim et al. 2003; Soltani et al. 2005) and antiviral (Moore et al. 1989). These metabolites are polysaccharides, nitrogenous compounds, bacteriocin-like proteins, bicyclic secondary amine anatoxin-a produced by *Anabaena flos-aquae* and may have vast potential in diverse fields, like antibiotics (Biswas 1957; Fogg 1952; Kulik 1995; Tuney et al. 2006), biocontrol activities in agriculture, control of mosquitoes, etc. These bioactive compounds play a major role in biological interactions in natural ecosystems.

Use of *Bacillus thuringiensis* and *Bacillus sphaericus* has been found to be successful in mosquito control. Blooms of certain cyanobacteria have been found to be associated with the absence of mosquito larvae from certain rice fields, and many cyanobacteria have been reported to be toxic to mosquito larvae. Cyanobacteria have better properties as biocontrol agents against mosquitoes as compared to toxic bacteria because of their natural presence in the habitats of mosquito larvae, ability to grow in these habitats and being suitable food for the mosquito larvae. Antifungal activity of some algal species has been reported against some dermatophyte strains and pathogenic yeasts (Calvo 1986; Tariq 1991). Analysis of bioactive metabolites have been studied mostly with marine algal species (Lima-Filho et al. 2002), and little is known about terrestrial algae and cyanobacteria (Burja et al. 2001). In aquaculture and agriculture fields, antimicrobial algae have been the best choice, and screening the bioactivity of algal crude extracts is mandatory in biomedical practice. The search for cyanobacteria with antimicrobial activity has gained importance in recent years, and in the present investigation, some strains of cyanobacteria have been screened and their pure cultures have been tested for their antimicrobial activity against some microorganisms.

## 41.2 Materials and Methods

Five axenic cyanobacterial strains *Westiellopsis prolifica* ARM 365, *Hapalosiphon hibernicus* ARM 178, *Nostoc muscorm* ARM 221, *Fischerella* sp. ARM 354 and *Scytonema* sp. screened for their antimicrobial activity and the sensitive test bacteria *Pseuomonas striata, Bacillus subtilis, Escherichia coli* and *Bradyrhizobium* sp. were obtained from CCUBGA, IARI, New Delhi. Cyanobacteria were grown in BG-11 medium (Stanier et al. 1971) at temperature 27 ± 1 °C and 14.4 ± 1 $Wm^{-2}$ light intensity for a 12/12 light–dark cycle. *P. striata* grew in Pikovaskaya's medium (Pikovaskaya 1948), *E. coli* in Luria broth (Sambrook et al. 1989) and *Bradyrhizobium* in Fred's medium (Fred et al. 1932) in dark at 32 ± 1 °C. *Aspergillus niger, Alternaria brassicae* and *Fusarium moniliforme* were isolated from soil samples and were kindly identified by Pathology Department, IARI, New Delhi and were grown in glucose-peptone broth at 30 ± 1 °C in dark.

For preparation of algal extracts, 0.5 g of dried algal powder of each of the five cyanobacteria was suspended in three different solvents, ethanol, methanol-toluene (3:1) and acetone. For testing antimicrobial activity of cyanobacteria, 100 µl of different sensitive indicator bacteria and fungi were inoculated on separate 1.5 % agar solidified nutrient medium in petri plates. Two wells of 6 mm diameter were made in each plate and were filled with 100 µl of cyanobacterial extracts. The plates were incubated for 2 days at 32 ± 1 °C for bacteria and 3 days at 30 ± 1°C for fungi. The diameter of inhibition zones was measured.

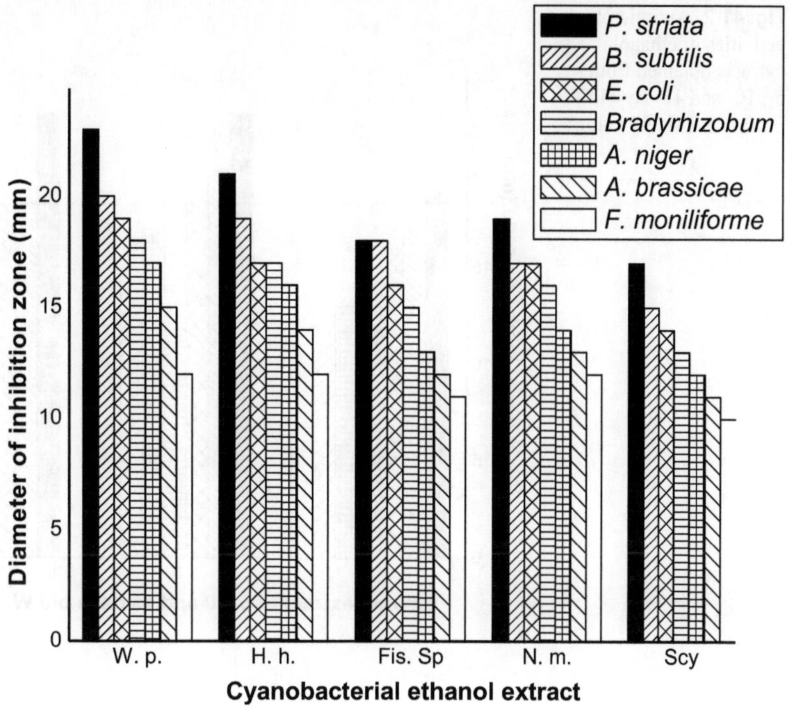


**Fig. 41.1** Antibacterial and antifungal activities of ethanol extracts obtained from 10-day-old cyanobacterial cultures (*W. p.-Westiellopsis prolifica* ARM 365, *H. h.-Hapalosiphon hibernicus* ARM 178, *N. m.-Nostoc muscorm* ARM 221, *Fis.* sp.-*Fischerella* sp. ARM 354 and *Scy-Scytonema* sp.)

## 41.3 Results and Discussion

Forty five cyanobacterial strains were screened for their antimicrobial activity using their extracts in three different solvents, ethanol, methanol-toluene (3:1) and acetone, on solidified agar nutrient medium by agar well diffusion method against various bacteria and fungi as sensitive test organisms. Out of these, metabolites of only five cyanobacteria, *W. prolifica* ARM 365, *H. hibernicus* ARM 178, *N. muscorm* ARM 221, *Fischerella* sp. ARM 354 and *Scytonema* sp., were found to have a weak-to-strong antimicrobial activity. Clear growth inhibition hyaline zones were formed around the wells made in petri plates inoculated with different sensitive strains *P. striata, B. subtilis, E. coli, Bradyrhizobium* sp., *A. niger, A. brassicae* and *F. moniliforme* due to the production of some extracellular substances by growing algae with antimicrobial activity. Ethanol extracts of *W. prolifica* ARM 365 and *H. hibernicus* ARM 178 showed the greatest antibacterial as well as antifungal activity against the tested sensitive bacteria and fungi as revealed by the largest inhibition zone formation on solidified agar nutrient medium. Plaza et al. (2010) identified different fatty acids and volatile compounds with antimicrobial activity, such as phytol, fucosterol, neophytadiene or palmitic, palmitoleic and oleic acids; and ethanol was found to be the most appropriate solvent for the extraction of these compounds. The test bacterium *P. striata* and the test fungus *A. niger* were found to be the most sensitive to ethanol extracts of *H. hibernicus* ARM 178 and *W. prolifica* ARM 365. These cyanobacterial extracts inhibited test bacteria more strongly than test fungi (Fig. 41.1). Das et al. (2005) examined acetone, ethanol and methanol extracts of some algae and showed from moderate-to-high activity against strains of virulent pathogens *Pseudomonas florescence, Aeromonas hydrophila, Vibrio anguillarum* and *E. coli.* Cyanobacteria *Anabaena oryzae, Tolypothrix ceytonica* and *Spirulina platensis* and green algae *Chlorella pyrenoidosa* and *Scenedesmus quadricauda* exhibited antibacterial and antifungal agent production on various organisms that incite diseases of humans and plants, such as *E. coli, B. subtilis,*

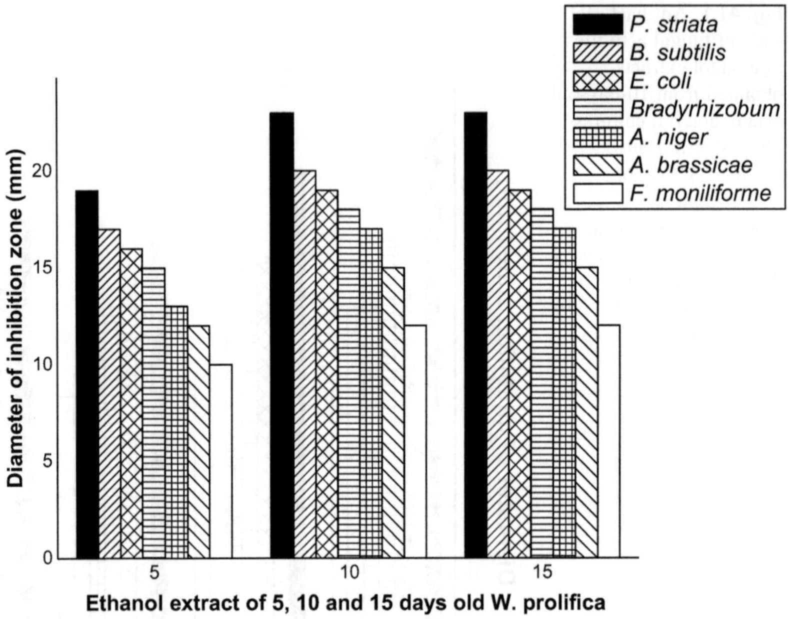


**Fig. 41.2** Antimicrobial activities of ethanol extracts obtained from 5-, 10- and 15-day-old *W. prolific* ARM 365 culture

*Staphyllococcus aureus, Pseudomonas aeruginosa, A. niger, Aspergillus flavus, Penicillium herquei, F. moniliforme, Helminthosporium* sp., *A. brassicae, Saccharomyces cerevisiae, Candida albicans* (Rania and Hala 2008). Harder in *Nostoc punctiforme* (Harder 1917) and Frankmolle et al. (Frankmolle et al. 1992) in *Anabaena laxa* have reported that these algae produce some metabolites which accumulate in the medium and inhibit the growth of other microorganisms. Cell extracts and active constituents of various algae may be potential bioactive compounds of interest in the pharmaceutical industry (Rodrigues et al. 2004).

In order to test the best growth phase for the strongest antibacterial and antifungal activity, ethanol extracts of *W. prolifica* ARM 365 and *H. hibernicus* ARM 178 obtained from different periods of growth were used against sensitive test bacteria and fungi in petri plates. The strongest antimicrobial activity of *H. hibernicus* ARM 178 (Fig. 41.2) and *W. prolifica* ARM 365 was observed on the tenth day of their growth (Fig. 41.3) for all the tested sensitive microorganisms.

To find the temperature stability of antimicrobial compounds of *W. prolifica* ARM 365 and *H. hibernicus* ARM 178, their ethanol extracts obtained from 10-day-old cultures were subjected to different temperature regimes. It was revealed that maximum antimicrobial activity was observed at 30 °C whereas decreased activity was found both at lower (15 °C) and higher (45 °C) temperatures (Figs. 41.4 and 41.5) as revealed by the diameter of clear hyaline areas. Factors like temperature, pH of the culture medium, incubation period, medium constituents and light intensity have been reported to influence biosynthesis of antimicrobial agent production by Noaman et al. (2004) where 35 °C temperature, pH 8 and 15-day-old cultures were found to be the best for antimicrobial agent production.

Thus, extracts of cyanobacteria, *W. prolific* ARM 365, *H. hibernicus* ARM 178, *N. muscorm* ARM 221, *Fischerella* sp. ARM 354 and *Scytonema* sp. were found to have antimicrobial activity against the test bacteria and fungi, and these antimicrobial compounds may have potential pharmaceutical, biocontrol or agricultural applications. Although the chemical nature of the antimicrobial metabolites produced by test cyanobacteria is unknown, they have potential as biological agents to control undesired microbial growth. Various strains of cyanobacteria are known to produce intracellular and extracellular metabolites with diverse biological activities such

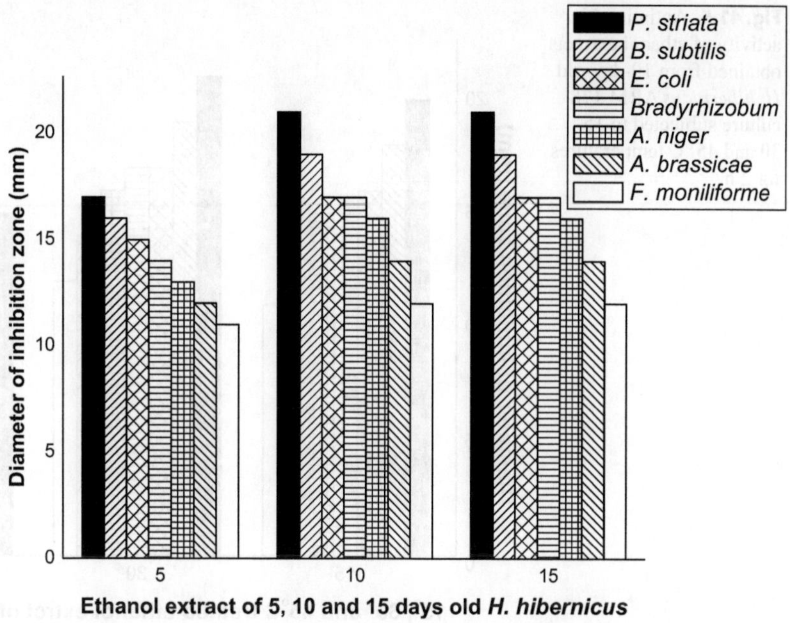


**Fig. 41.3** Antimicrobial activities of ethanol extracts obtained from 5-, 10- and 15-day-old *H. hibernicus* ARM 178 culture

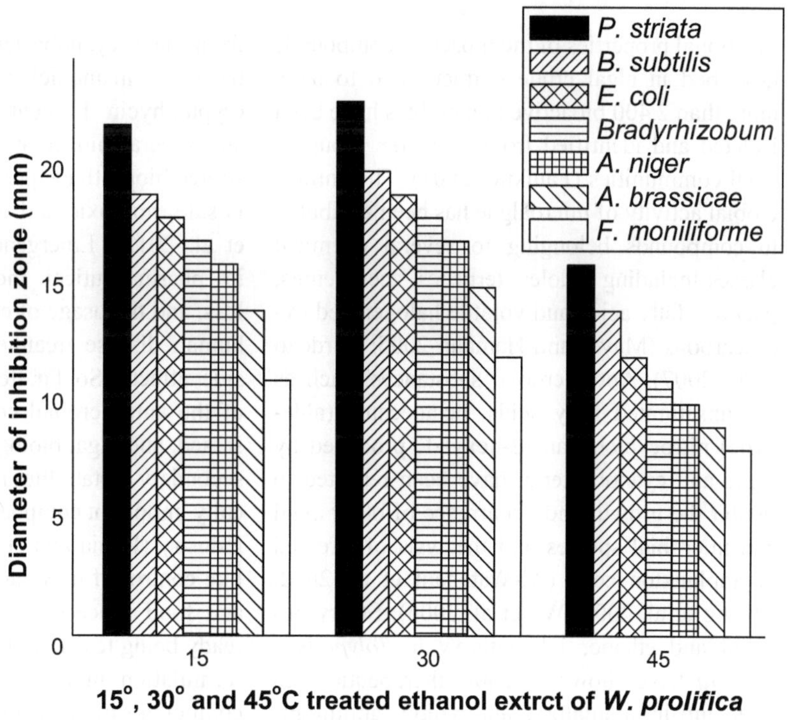


**Fig. 41.4** Antimicrobial activities of ethanol extracts obtained from 10–day-old *W. prolifica* ARM 365 culture subjected to 15, 30 and 45 °C temperatures for 2 h

as antialgal, antibacterial, antifungal and antiviral activities (Kalireioglu et al. 2006). Pratt et al. (1944) isolated the first antibacterial compound from a microalga, *Chlorella,* where a mixture of fatty acids, viz. chlorellin, was found to be responsible for the inhibitory activity against both Gram+ and Gram− bacteria. Emergence concerns have been raised to establish structural and

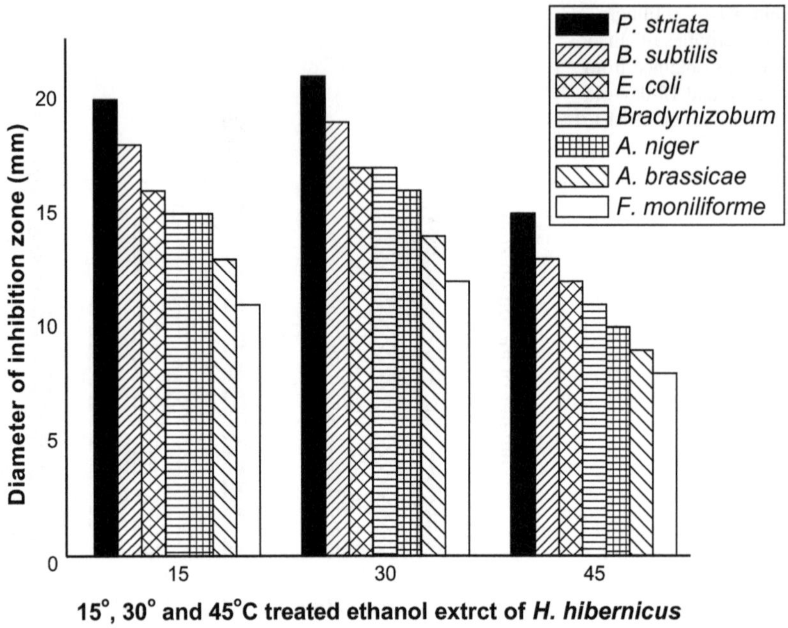


**Fig. 41.5** Antimicrobial activity of ethanol extracts obtained from 10-day-old *H. hibernicus* ARM 178 culture subjected to 15, 30 and 45 °C temperatures for 2 h

functional properties of the bioactive compounds described in algal crude extracts, and to date, more than 2,400 bioactive metabolites have been isolated and identified from a diverse group of algal communities (Faulkner 2001). The antimicrobial activity of microalgae has been attributed to compounds belonging to several chemical classes including indoles, terpenes, acetogenins, phenols, fatty acids and volatile halogenated hydrocarbons (Mayer and Hamann 2005; Cardozo et al. 2007). Antimicrobial substances, such as polyunsaturated fatty acids, octadecanal (aldehyde) nitrocyclohexane-2-hexyl-1, produced by algae and cyanobacteria, have been reported to inhibit the growth and occurrence of other algal and microbial species and the lysis of stressed phytoplankton cells (Al-Wathnani et al. 2012, Chiang et al. 2004, Wu et al. 2006). Stearyl alcohol and ethane, 1,1-diethoxy in *Tolypothrix distorta* have shown valuable therapeutic uses like, anti-inflammatory, antipyretic, antithrombotic and analgesic effects (Al-Wathnani et al. 2012). Some algal antimicrobial compounds have been shown to exhibit important biomedical characteristics such as butanal in controlling bovine soles symptoms in cattle (Gregory et al. 2006). Some properties of bioactive compounds from many cyanobacteria have been reported to be useful in medicine and agriculture, such as cryptophycin 1 agent with anticancer activity and several bioactive compounds with demonstrated biopesticide property in agriculture which is safe, nontoxic and efficient (Thummajitsakul et al. 2012). Emergence of microbial diseases in pharmaceutical industries implies serious loss, and the usage of commercial antibiotics for human disease treatment produces undesirable side effects. So far, few biomedical properties of the antimicrobial compounds have been assessed, and algal biologically active primary and secondary metabolites may be potential biomedically important compounds in the pharmaceutical industry (Rania and Hala 2008) and other areas like control of unwanted bacteria in aquaculture and food processing. Cell-free extracts are already being tested as additives for food and feed formulation, in attempts to replace antimicrobial compounds of synthetic origin currently in use—including subtherapeutical doses of antibiotics employed as prophylactic measure in animal breeding (Tramper et al. 2003).

The present study provides an understanding of the role of extracellular metabolites in natural selection of dominant algae. Fungal infection has been observed to be widely distributed in human pathology, and the potential application of the antifungal compounds produced by the cyanobacteria in medicines may have a very promising approach. Algae are valuable natural sources, effective against infectious agents, and so far no antifungal activity has been reported against many fungi such as *C. albicans* and *A. niger*. Nowadays, antibiotic resistance in bacteria is one of the emerging health-related problems owing to the widespread and almost unrestricted use of antibiotics. The study of resistance to antifungal agents has lagged behind the study of antibacterial resistance as fungi were not recognized as important pathogens until a few years ago. The associated increase in fungal infections has prompted the search for newer and safer agents to combat fungal infections, and extensive efforts for the identification of bioactive compounds derived from various algae have been made worldwide.

Though these findings would help in opening up new trends in biomedical and pharmaceutical industries, further study of the purification of the potent compounds to explain their usefulness in the pharmaceutical and biotechnological industry is required. We are attempting to characterize the antimicrobial compounds and to undertake small-scale studies to find out about the practical potential of these cyanobacteria as algicides, bacteriocides and fungicides for controlling undesired microorganisms.

**Acknowledgments** This study was performed by Rashmi Tyagi in the Division of Microbiology, Indian Agricultural Research Institute, New Delhi (CSIR-Senior Research Associate) with B. D. Kaushik (Head of the Department) and in the ITM University, Gurgaon.

## References

Al-Wathnani H, Ara I, Tahmaz RR, Al-Dayel TH, Bakir MA (2012) Bioactivity of natural compounds isolated from cyanobacteria and green algae against human pathogenic bacteria and yeast. J Medicinal Pt Res 6(18):3425–3433

Biswas BB (1957) Cyanobacterial studies on the central body of cyanophyceae. Cytologia 22:90–95

Burja AM, Banaigs B, Abou-Mansour E, Burgess JG, Wright PC (2001) Marine cyanobacteria-a prolific source of natural products. Tetrahedron 57:9347–9377

Calvo Ma A, Cabanes FJ, Abarca L (1986) Antifungal activity of some mediterranean algae. Mycopathologia 93(1):61–63

Cardozo KHM, Guaratini T, Barros MP, Falcão VR, Tonon AP, Lopes NP, Campos S, Torres MA, Souza AO, Colepicolo P, Pinto E (2007) Metabolites from algae with economical impact. Comparative Biochemistry Physiology C. Toxicology Pharmacology 146:60–78

Chiang IZ, Huang WY, Wu JT (2004) Allelochemicals of *Botryococcus braunii* (Chlorophyceae). J Phycol 40:474–480

Das B, Pradhan J, Pattnaik P, Samantaray B, Samal S (2005) Production of antibacterials from the fresh water alga *Euglena viridis* (Ehren). World J Microbiol Biotechnol 21:45–50

Faulkner DJ (2001) Marine natural products. Nat Prod Rep 18:1–4

Fogg GE (1952) The production of extracellular nitrogenous substances by a blue-green alga. Proc Roc Soc B 139:372–397

Frankmolle WP, Larsn LK, Caplan FR, Patterson GML, Knubel G, Levin IA, Moore RE (1992) Antifungal cyclic peptides from the terrestrial blue-green alga *Anabaena laxa* I isolation and biological properties. J Antibiot 45:1451–1457

Fred EB, Baldwin IL, Mc Coy E (1932) Root nodule bacteria and leguminous plants. In: Studies in Science No. 5, (Ed. Fred EB, Baldwin IL, Mc Coy E ), Univ Wis Madison, Wisconsin, pp 41

Ghasemi Y, Moradian A, Mohagheghzadeh A, Shokravi S, Morowvat MH (2007) Antifungal and antibacterial activity of the microalgae collected from paddy fields of Iran: characterization of antimicrobial activity of *Chroococcus disperses*. J Biol Sci 7:904–910

Gregory N, Craggs L, Hobson N, Krogh C (2006) Softening of cattle hoof soles and swelling of heel horn by environmental agents. Food Chem Toxicol 44:1223–1227

Harder R (1917) Ernahrungsphysiologische untersuchungen an cyanophycean hauptsachlich dem endophytischen Nostoc punctiforme. Z Biot 9:145–242

Isnansetyo A, Cui L, Hiramatsu K, Kamei Y (2003) Antibacterial activity of 2,4-diacetylphloroglucinol produced by *Pseudomonas* sp. AMSN isolated from a marine algae, against vancomycin-resistant *Staphylococcus aureus*. Int J Antimicrob Agents 22:545–547

Jaki B, Orjala J, Sticher O (1999) A novel extracellular diterpenoid with antibacterial activity from the cyanobacterium *Nostoc commune* EAWAG 122b. J Nat Prod 62:502–503

John DM, Whitton BA, Brook AJ (2003) The freshwater algal flora of the British isles, an identification guide to freshwater and terrestrial algae. Cambridge University Press, Cambridge, pp 117–122

Kalireioglu H, Beyatli Y, Aslim B, Yüksekdag Z, Atici T (2006) Screening for antimicrobial agent production of some microalgae in freshwater. Internet J Microbiol 2:235

Kulik MM (1995) The potential for using cyanobacteria (blue-green algae) and algae in the biological control of plant pathogenic bacteria and fungi. Eur J Plant Path 101:585–599

Kundim BA, Itou Y, Sakagami Y, Fudou R, Iizuka T, Yamanaka S, Ojika M (2003) New haliangicin isomers, potent antifungal metabolites produced by a marine myxobacterium. J Antibiot 56:630–638

Lima-Filho JVM, Carvalho AFFU, Freitas SM (2002) Antibacterial activity of extracts of six macroalgae from the Northeastern Brazillian coast. Braz J Microbiol 33:311–333

Mayer AMS, Hamann MT (2005) Marine pharmacology in 2001–2002: marine compounds with antihelmintic, antibacterial, anticoagulant, antidiabetic, antifungal, anti-inflammatory, antimalarial, antiplatelet, antiprotozoal, antituberculosis, and antiviral activities; affecting the cardiovascular, immune and nervous systems and other miscellaneous mechanisms of action. Comparative Biochemistry Phycology C 140:265–286

Moore RE, Cheuk C, Yang XG, Patterson GML (1989) Hapalindoles, antibacterial and antimycotic alkaloids from the cyanophyte *Hapalosiphon fontinalis*. J Org Chem 52:1036–1043

Noaman NH, Fattah A, Khaleafa M, Zaky SH (2004) Factors affecting ant *Synechococcus leopoliensis*. Microbiol Res 159:395–402

Pikovaskaya RI (1948) Mobilization of phosphorus in soil in connection with vital activity of some microbial species. Mikrobiologiya 17:362–370

Plaza M, Santoyo S, Jaime L, Garcia-Blairsy Reina G, Herrero M, Seriorans FJ, Ibanez E (2010) Screening for bioactive compounds from algae. J Pharm Biomed Anal 51(2):450–455

Pratt R, Daniels TC, Eiler JB, Gunnison JB, Kumler WD et al (1944) Chlorellin, an antibacterial substance from *Chlorella*. Science 99:351–352

Rania MA, Hala MT (2008) Antibacterial and antifungal activity of cyanobacteria and geen microalgae. Evaluation of medium components by Plackett-Burman design for antimicrobial activity of *Spirulina platensis*. Global J Biotech Biochem 3(1):22–31

Rodrigues E, Tilvi S, Naik CG (2004) Antimicrobial activities of marine organisms collected off the coast of East India. J Exp Biol Ecol 309:121–127

Sambrook J, Fritsch FE, Maniatis T (1989) Molecular cloning. In: A Laboratory Manual 2 Ed. (Ed. Sambrook J, Fritsch FE, Maniatis T) Cold Spring Harbour Lab Press

Soltani N, Khavari-Nejad RA, Tabatabaei Yazdi M, Shokravi SH, Fernandez-Valiente E (2005) Screening of soil cyanobacteria for antibacterial antifungal activity. Pharm Biol 43(5):455–459

Stanier RY, Kunisawa R, Mandel M, Cohen-Bazire G (1971) Purification and properties of unicellular blue-green algae (order Chroococcales). Bact Rev 35:171–205

Tariq V-N (1991) Antifungal activity in crude extracts of marine red algae. Mycol Res 95(12):1433–1435

Thummajitsakul S, Kun Silprasit K, Sittipraneed S (2012) Antibacterial activity of crude extracts of cyanobacteria *Phormidium* and *Microcoleus* species. Afr J Microbiol Res 6(10):2574–2579

Tramper J, Battershill C, Brandenburg W, Burgess G, Hill R, Luiten E, Müller W, Osinga R, Rorrer G, Tredici M, Uriz M, Wright P, Wijffels R (2003) What to do in marine biotechnology? Biomol Eng 20:467–471

Tuney I, Cadirci B, Unal D, Sukatar A (2006) Antimicrobial activities of the extracts of marine algae from the cost of Urla (Izmir, Turkey). Turk J Biol 30:171–175

Wu JT, Chiang YR, Huang WY, Jane WN (2006) Cytotoxic effects of free fatty acids on phytoplankton algae and cyanobacteria. Aquat Toxicol 80:338–345

# Reaction of Chickpea Varieties to *Macrophomina Phaseolina* and Their Effect on Peroxidase Activity

42

Preeti, Nilima Kumari and Vinay Sharma

**Abstract**

Fifty varieties of chickpea were screened against *M. phaseolina* and peroxidase activity of both control and inoculated plants were determined for its correlation with resistant response. Out of the 50 varieties, RSG-143 was found to be resistant. Two varieties RSG-896 and RSG-973 were found to be moderately resistant. The varieties Pusa-391, Pusa-362, Pusa-267, RSG-888 and RSG-963 were found to be susceptible and the varieties Pusa-256, Pusa-372 and Pusa-1003 were rated as highly susceptible to *M. phaseolina*. A positive correlation was found between peroxidase activity and resistance response of the varieties.

**Keywords**

Chickpea · *Macrophomina phaseolina* · Peroxidase

## 42.1 Introduction

The intimate interactions between cultivated crops and bacterial, viral and fungal pathogens often results in serious outbreak of diseases. 'Plant disease' is defined as 'physiological disorder' or structural abnormality that is harmful to the plant or any of its parts or products or that reduces its economic value (Guiñazú et al. 2012). Chickpea (*Cicer arietinum*) is an important leguminous crop. It not only serves as a good source of nutrition to the people but also improves the soil. Chickpea plays the pivotal role of supplying protein source in the vegetarian diet, it is also called as the 'poor man's meat' (Singh 1987; Smithson et al. 1985). *Macrophomina phaseolina* (Tassi) Goid is an important pest of chickpea causing significant loss in its yield. It is a soil borne, plant pathogenic fungus, ubiquitous in nature. It belongs to anamorphic ascomycetes. *M. phaseolina* induced disease, e.g. charcoal root rot (Smith et al. 1989), may occur in severe form when a proper combination of high temperature and low soil moisture exists, that is why this disease is quite prevalent in Rajasthan. Multicellular, typically dark, sclerotia of variable size enable the fungus to survive adverse environmental conditions and later serve as a source of inoculum for infection. Host plant resistance may offer a potential practical solution (Kesta and Atantee 1998) to resistant of varieties. Rapid screening methods to evaluate host resistance and pathogen variability are needed to hasten progress in developing resistant cultivars. *M. phaseolina* produces a number of phytotoxins,

V. Sharma (✉) · Preeti · N. Kumari
Department of Bioscience and Biotechnology, Banasthali University, Banasthali 304 022, Rajasthan, India
e-mail: vinaysharma30@yahoo.co.uk

R. N. Kharwar et al. (eds.), *Microbial Diversity and Biotechnology in Food Security*,
DOI 10.1007/978-81-322-1801-2_42, © Springer India 2014

namely asperlin, isoasperlin, phomalactone, phaseolinic acid, phomenon and phaseolinone (Dhar et al. 1982; Mahato et al. 1987; Bhattacharya et al. 1992). Phaseolinone appears to be the most important toxin that induces disease symptoms in plants. It is a non-specific exotoxin that is highly stable and non-biodegradable even at high temperatures. It affects seed germination, seedling growth and cause necrosis in tissue cultures, during callusing and regeneration. Screening of germplasms to get resistant or tolerant cultivars may be important because it can be used as a source of resistance in plant-breeding programs. Peroxidase may have an important role in the resistance mechanism of plant (Daly et al. 1971; Noel and McClure 1978; Siddiqui and Mahmood 1992; Mantoo and Siddiqui 1996). Higher peroxidase activity resulted in an increase in the phenolic contents of plant, which plays an important role in the resistance of cultivar (Mahmood and Saxena 1986). Resistance to some plant diseases is associated with increased peroxidase activity and expression of specific isoenzymes (Ye et al. 1990; Goy et al. 1992). In the present study, 50 varieties of chickpea were screened against *M. phaseolina*, and peroxidase activity of both control and inoculated plants were determined for its correlation with resistant response.

## 42.2 Material and Methods

### 42.2.1 Collection of Plant Material

Seeds of different varieties of chickpea were procured from Krishi Vigyan Kendra (KVK), Banasthali and Division of Genetics, Indian Agricultural Research Institute, New Delhi.

### 42.2.2 Activation of Fungal Strain

*M. phaseolina* was obtained from IMTECH Chandigarh (MTCC 2165). The lyophilized fungal strain was activated by inoculating in potato dextrose broth (PDB) under proper aseptic conditions. The flask was then incubated at a temperature of 25±2 °C at 120 rpm in the incubator shaker for 5 days. This activated fungal strain was then streaked on the PDA slants. These slants were then incubated for 7 days at 25±2 °C. Figure 42.1. The mycelium growth initiated after 2 days of incubation.

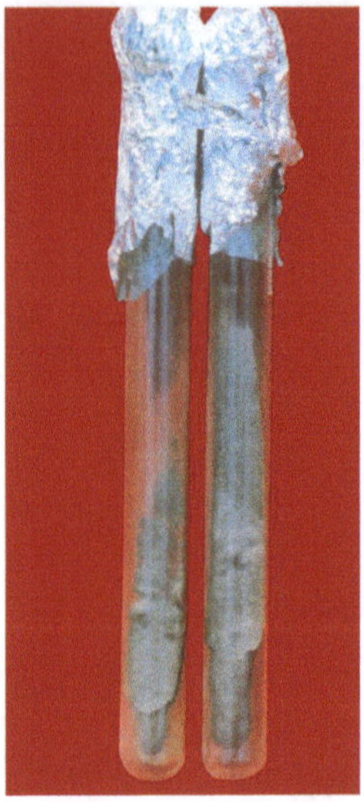

**Fig. 42.1** Slant of *Macrophomina phaseolina*

### 42.2.3 Preparation of Inoculum

The surface of the mycelium mat was scrapped gently 2–3 times with the sterilized inoculating loop and inoculated in the conical flask containing autoclaved distilled water. These flasks were then kept in the incubator shaker at 25±2 °C, 120 rpm. for 2 h.

### 42.2.4 Experiment Design

Ten seeds of each chickpea variety were sown in 15 cm diameter plastic pots, containing 1 kg sterilized soil and vermicompost (3:1) mixture. One-week-old seedlings were inoculated with 10 ml of mycelia suspension of *M. phaseolina*. The experiment was continued for 90 days after inoculation and data on dry shoot weight and root-rot index were recorded (Figs. 42.2–42.4). Peroxidase activity of control and inoculated plants was determined by the method of Putter (1974). Dry shoot weight and root-rot index was calculated according to Hussain (1986). Resistance-susceptibility rating

**Fig. 42.2** Symptom development on chickpea resistant line RSG-143, after 90 days. **a** Control **b** inoculated, after inoculation with *Macrophomina phaseolina*

**Fig. 42.3** Symptom development on chickpea moderately resistant line RSG-896, after 90 days. **a** Control **b** inoculated, after inoculation with *Macrophomina phaseolina*

**Fig. 42.4** Symptom development on chickpea highly susceptible line Pusa-1003, after 90 days. **a** Control **b** inoculated, after inoculation with *Macrophomina phaseolina*

**Fig. 42.5** Detection of hydrogen peroxide in chickpea resistant line RSG-143. **a** Control **b** inoculated by staining with3,3′-diaminobenzidine (DAB) after inoculation with *Macrophomina phaseolina*

**Fig. 42.6** Detection of hydrogen peroxide in chickpea moderately resistant line RSG-896. **a** Control **b** inoculated by staining with 3,3′-diaminobenzidine (DAB) after inoculation with *Macrophomina phaseolina*

were calculated on reduction in dry shoot weight and root-rot index according to the 0–5 scale proposed by Hussain (1986) with slight modification (using root-rot index in place of nematode reproduction), where 0 = no root-rot and 5 = severe root-rot. Suppression in dry shoot weight was also calculated on 0–5 scale as: 0 = no suppression in dry shoot weight = Immune (I); 5 = more than 25 % suppression in dry shoot weight = highly susceptible (HS). Detection of Hydrogen Peroxide by DAB (3,3′-diaminobenzidine) staining in chickpea leaves (Figs. 42.5–42.7).

## 42.3 Results and Discussion

Out of 50 chickpea varieties tested, only one variety, i.e. RSG-143 was found as resistant, two as moderately resistant (RSG-896 and RSG-973) and the varieties, viz. Pusa-391, Pusa-362, Pusa-267, RSG-888, RSG-963 were found to be susceptible and the remaining varieties Pusa-256, Pusa-372, Pusa-1003 were rated as highly susceptible to *M. phaseolina*. It is clear from the results presented in Table 42.1 that increase in peroxidase activity resulted in resistant response of the chickpea varieties, which is in agreement with the findings of Siddique and Mahmood (1993). Increase in peroxidase activity in response to *M. phaseolina* infection over control in resistant variety RSG-143 was 43.3 %. This increase was 38.38 and 38.54 % in moderately resistant varieties RSG-896 and RSG-973, respectively. In susceptible varieties 23, 19.95, 16.6, 20.11 and 22.81 % in Pusa-391, Pusa-362, Pusa-267, RSG-888 and RSG-963. However, in highly susceptible varieties the increase in peroxidase activity due to infection of *M. phaseolina* over control was 7.6, 5.1 and 12.5 % in Pusa-256, Pusa-372 and Pusa-1003. A positive correlation

**Fig. 42.7** Detection of hydrogen peroxide in chickpea resistant line Pusa-1003. **a** Control **b** inoculated by staining with 3, 3′-diaminobenzidine (DAB) after inoculation with *Macrophomina phaseolina*

was found between peroxidase activity and resistant response of the varieties. Peroxidase activity as a preliminary biochemical marker may predict resistance of uninfected muskmelon to *Pseudoperonospora cubensis* (Reuveni 1983) and resistance of maize to gray leaf spot (Garraway and Beltran 1997). Peroxidase activity could be related to a fast production of physical barriers involved in avoiding virus translocation through the plant restricting it to the localized region of infection. Lignin synthesis (Whitmore 1978; Siegel 1953), oxidative coupling reactions involving phenolics that are esterified to wall polysaccharides (Geissman and Neukon 1971; Hartley 1973; Fry 1982a), and the formation of isodityrosine bridges that are believed to cross-link extension molecules (Fry 1982b) are among the functions proposed for peroxidase that could be responsible for reinforcements of the cell wall as a barrier to the pathogen (Ride 1975). Similar results were obtained when peroxidase activities were calculated between groups of muskmelon genotypes varying in their resistance to *P. cubensis* (Reuveni 1983), between maize genotypes varying in their resistance to *Cercospora Zeae-maydis* (Garraway and Beltran 1997), between lettuce genotypes contrasting in their resistance to downy mildew (Reuveni 1983) and in Nicotiana hybrid resistance to several diseases has been linked to high levels of peroxidase (Goy et al. 1992). Some peroxidase isoforms have also been associated with disease resistance. In maize, resistance to northern leaf blight was associated to a specific isoform (Bar-Zur et al. 1998) and resistance to *Exesrohilium turcicum* was correlated to peroxidase-banding patterns (Shimoni et al. 1996). Collectively, these results indicate that increased peroxidase activity and expression of specific isoenzymes may be a constitutive mechanism used by plants against severe pathogen infection. These results are in agreement with the inoculation of tobacco leaves with tobacco mosaic virus (TMV) leading to an increase in peroxidase activity (Ye et al. 1990), inoculation of okra with yellow vein mosaic virus (Ahmed et al. 1992), and inoculation of maize with *E. turcicum* (Shimoni et al. 1996). These studies show that increased peroxidase activity occur in both resistant and susceptible genotypes. The role of oxidative enzymes such as peroxidase could be explained as an oxidation process of phenol compounds to oxidized products (quinones) which may limit the fungal growth. Vance et al. (1980) and Fry (1982b) stated that peroxidase is known to be involved in the oxidation of polymerization of hydroxycinnamyl alcohols to yield lignin and crosslinking isodityrosine bridges in cell wall. Peroxidase also produces free radicals and hydrogen peroxide which are toxic to many microorganisms (Pena and Kuc 1992). Another supportive suggestion was brought by Nawar and Kuti (2003) who stated that an increase in peroxidase activity is considered as a preliminary indicator for resistance of broad beans to chocolate spot disease (Tarred et al. 1993). These compounds act as barriers against pathogen invasion. Aly and Afify (1989) pointed out that the resistance induced in barley plants showed new

**Table 42.1** Reaction of 50 chickpea varieties to *Macrophomina phaseolina* and their effect on peroxidase activity

| | | | | | Peroxidase activity | | |
|---|---|---|---|---|---|---|---|
| Varieties | Treatments | Shoot dry weight (g) | % Reduction over control | Root-rot index | Activity/mg Protein/min | % increase over control | Reaction |
| RSG-143 | Control | 3.3 | – | – | 0.150 | – | Resistant |
| | Inoculated | 3.2 | 3.0 | 1 | 0.265 | 43.3 | |
| RSG-896 | Control | 5.2 | – | – | 0.130 | – | Moderately resistant |
| | Inoculated | 5.0 | 3.8 | 2 | 0.211 | 38.38 | |
| RSG-973 | Control | 7.9 | – | – | 0.118 | – | Moderately resistant |
| | Inoculated | 7.2 | 8.8 | 2 | 0.192 | 38.54 | |
| Pusa-391 | Control | 5.2 | – | – | 0.100 | – | Susceptible |
| | Inoculated | 4.0 | 23.0 | 4 | 0.130 | 23.0 | |
| RSG-888 | Control | 5.0 | – | – | 0.143 | – | Susceptible |
| | Inoculated | 3.8 | 24 | 4 | 0.179 | 20.11 | |
| RSG-963 | Control | 6.8 | – | – | 0.100 | – | Susceptible |
| | Inoculated | 5.4 | 20.52 | 4 | 0.129 | 22.81 | |
| Pusa-1003 | Control | 6.2 | – | – | 0.111 | – | Highly susceptible |
| | Inoculated | 4.0 | 35.4 | 5 | 0.127 | 12.5 | |

peroxidase isozymes in infected plants pretreated with ethephon. Bargabus et al. (2002) indicated that the increase in peroxidase-specific activity following benzoic acid derivatives (acibenzolar-S-methyl) treatment was due to the production of two isoforms not found in untreated plants. These findings indicate positive correlation between resistance response and peroxidase activity.

**Acknowledgments** The financial support provided by the Department of Science and Technology (DST), New Delhi for the project, Banasthali Centre for Education and Research in Basic Sciences under their CURIE (Consolidation of University Research for Innovation and Excellence in Women Universities) programme is gratefully acknowledged.

## References

Ahmed N, Thakur MR, Bajaj KL, Cheema SS (1992) Biochemical basis of resistance to yellow vein mosaic virus in okra. Pl Dis Res 9:20–25

Aly MM, Afify WM (1989) Induced resistance against plant disease using ethephon (2-chloroethyl phosphoric acid) treatment. I: Powdery mildew and net blotch of barley. Proc 7th Conference Microbial Cairo, Egypt, pp 299–314

Bargabus RL, Zidack NK, Sherwood JE, Jacobsen BJ (2002) Characterization of systemic resistance in sugar beat elicited by a non-pathogenic, phyllosphere-colonizing *Bacillus mycoides*, biological control agent. Physiol Mol Plant Path 61:289–298

Bar-zur A, Tadmor Y, Juvik JA, Shimoni M, Reuveni R (1998) Resistance to Northern leaf blight in maize (*Zea mays*) conditioned by the *HtN* gene and the association with isoperoxidases. Can J Plant Pathol 20:28–34

Bhattacharya D, Siddiqui KAI, Ali E (1992) Phytotoxic metabolites of *Macrophomina phaseolina*. Ind J Mycol Plant Pathol 22:54–57

Dhar TK, Siddiqui KAI, Ali E (1982) Structure of phaseolinone, a novel phytotoxin from *Macrophomina phaseolina*. Tetrahedron Lett 23:5459–5462

Daly J, Ludden MP, Seevers P (1971) Biochemical comparisons of resistance to wheat stem rust disease controlled by the Sr 6 and Sr 11 alleles. Physiol Plant Pathol 1:397–407

Fry SC (1982a) Isodytirosine, a new cross-linking amino acid from plant cell-wall glycoprotein. Biochem J 204:449–455

Fry SC (1982b) Phenolic components of the primary cell wall. Feruloylated disaccharides of D-galactose and L-arabinose from spinach polysaccharide. Biochem J 203:493–504

Garraway MO, Beltran JD (1997) Peroxidase activity in selected maize cultivars may predict resistance to gray leaf spot. Phytopathol 87:32 (Supplement)

Geissman T, Neukon H (1971) Vernetzung von Phenolcarbonsaureestern von Polysaccharidendurchnoxydative phenolicshe kupplung. Helv Chim Acta 54:112–113

Guiñazú LB, Andrés JA, Rovera M, Rosas SB (2012) Isolation and characterization of rhizobacteria antagonistic to *Macrophomina phaseolina* (Tassi) Goid., causal agent of Alfalfa damping-off. Environmental protection strategies for sustainable development strategies for sustainability 329–339

Goy P, Felix A, Metraux G, Meins JP, Jr F (1992) Resistance to disease in the hybrid *Nicotiana glutinosa x Nicotiana debneyi* is associated with high constitutive levels of Beta-1,3-glucanase, chitanase, peroxidase and polyphenoloxidase. Physiol Mol Plant Pathol 41:11–21

Hartley RD (1973) Carbohydrates esters of ferulic acid as components of cell-walls of *Lolium multiflorum*. Phytochemistry 12:661–665

Hussain SI (1986) Resistance-susceptibility rating for screening crop varieties against root-knot reniform and cyst nematode. Int Nematol Network News 3:15–16

Ketsa S, Atantee S (1998) Phenolics, lignin, peroxidase activity and increased firmness of damaged pericarp of mangosteen fruit after impact. Postharvest Biol Technol 14:117–124

Mahato SB, Siddiqui KAI, Bhattacharya G, Ghosal T, Miyahara K, Sholichin M, Kawasaki T (1987) Structure and stereochemistry of phaseolinic acid: a new acid from *Macrophomina phaseolina*. J Nat Prod 50:245–247

Mahmood I, Saxena SK (1986) Relative susceptibility of different cultivars of tomato to *Rotytenchulus reniformis* in reaction to changes in phenolices. Revue Nematol 9:89–91

Mantoo MA, Siddiqui MN (1996) Reaction of chickpea varieties to root-knot nematode and their effect on peroxidase activity and protein content. Indian J Plant Pathol 14(1 & 2):41–45

Nawar HF, Kuti JD (2003) Wyerone acid phytoalexin synthesis and peroxidase activity as markers for resistance of broad beans to chocolate spot disease. J Phytopathol 151:564–570

Noel GR, Mc Clure MA (1978) Peroxidase and 6 phosphogluconate dehydrogenase in resistance and susceptible cotton infected by *Meloidogyne incognita*. J Nematol 10:34–39

Pena M, Kuc JA (1992) Peroxidase-generated hydrogen peroxidase as a source of antifungal activity in vitro and on tobacco leaf disks. Phytopathol 82:696–699

Putter J (1974) Peroxidases. In: Bergmeyer HU (ed) Method of enzymatic analysis, vol 2. Academic Press, New York, p 685

Reuveni R (1983) Resistance reaction of *Cucumis melo* to inoculation with *Pseudoperonospora cubensis*. Ann Appl Biol 102:533–537

Ride JP (1975) Lignification in wounded wheat leaves in response to fungi and its possible role in resistant. Physiol Plant Pathol 5:124–134

Shimoni M, Reuveni R, Bar-Zur A (1996) Relation between peroxidase, Beta-1,3-glucanase the*se* gene and partial resistance of maize to *Exserohilium turcicum*. Canadian J Plant Pathol 18:403–408

Siegel SM (1953) On the biosynthesis of lignins. Physiol Plant 6:135–139

Siddiqui ZA, Mahmood I (1992) Response of chickpea cultivars to *Meloidogyne incognita* race 3 and their effect on peroxidase activity. Pak J Nematol 10:113–117

Siddiqui ZA, Mahmood I (1993) Response of some accessions of chickpea to *Macrophomina phaseolina* and their effect on peroxidase activity. J Indian Hot Soc 72:127–129

Singh KB (1987) Chickpea breeding. In: Saxena MC, Singh KB (eds) The Chickpea. CAB International, UK, pp 127–162

Smith RD, Hodges SC, Cordell CE (1989) Charcoal root rot and black root rot. In: Cordell CE, Anderson RL, Hoffard WH, Landis TD, Smith RS, Harvey T (eds) Forest nursery pest. Agriculture Handbook 680, US, pp 112–113

Smithson JB, Thompson JA, Summerfield RJ (1985) Chickpea (Cicer arietinum L). In: Summerfield RJ, Roberts EH (eds) Grain Legume Crops. Collins, London, pp 312–390

Tarred AM, El-Hyatemy YY, Omar SA (1993) Wyerone derivatives and activities of peroxidase and polyphenol oxidase in faba bean leaves as induced by chocolate spot disease. Plant Sci 89:161–165

Vance CP, Kirk TK, Sherwood RT (1980) Lignification as a mechanism of disease resistance. Annual Rev Phytopathol 18:259–288

Whitmore FW (1978) Lignin-protein complex catalyzed by peroxidase. Plant Science Lett 13:241–245

Ye XS, Pan SQ, Kuc J (1990) Activity, isozyme pattern, and cellular localization of peroxidase as related to systemic resistance of tobacco to blue mold (*Peronospora tabacina*) and to tobacco mosaic virus. Phytopathol 80:1295–1299

# Purification and Characterization of a Novel Thermostable β-Amylase from *Aspergillus foetidus* MTCC-508. β-Amylase from *Aspergillus foetidus* MTCC-508

43

Sarad Kumar Mishra, Kumar Shivam, Sanjeev Kumar Diwakar and Swati Shukla

**Abstract**

An extracellular β-amylase was produced from *Aspergillus foetidus* MTCC-508, and was purified 254.8-fold with 14.6 yields by precipitation with acetone and by column chromatographies with DEAE-Sephadex A-50 and Sephadex G-100. The purified enzyme was homogeneous on polyacrylamide gel electrophoresis. The band of enzyme was visible around 20 kDa on SDS-PAGE while around 80 kDa on Native-PAGE, showing its homotetrameric nature. The enzyme was optimally active at pH-6.0 and 50 °C temperature. It was fully stable at 50 °C for 2 h. The activity was strongly inhibited by $Hg^{2+}$, $Zn^{2+}$ and $Co^{2+}$, while $Mg^{2+}$ marginally enhanced the enzyme activity. The enzyme was able to hydrolyze the raw starches of potato, wheat, rice, maize, and *Trapa natans,* with the highest degree of saccharification of maize starch. The $K_m$ and $V_{max}$ values for this enzyme against boiled soluble starch were found 2.7 mg/mL and 2,100 U/mg of protein, respectively.

**Keywords**

β-amylase · *Aspergillus foetidus* · Enzyme activity · Enzyme purification · Native-PAGE

## 43.1 Introduction

Microorganisms are the most important sources for enzyme production. Selection of the right organism plays a key role in the high yield of desirable enzymes. The *Aspergillus species* produces a large variety of extracellular enzymes, of which amylases are of significant industrial importance (Pandey et al. 2000).

β-amylase (α-1, 4–glucan maltohydrolase, E.C. 3.2.1.2) is an exoamylase that catalyzes the hydrolysis of α-1, 4–glucan bonds in amylosaccharide chains from the nonreducing ends, producing β-maltose and β-limit dextrin as products (Thoma et al. 1971). It is used for the saccharification of liquefied starch to produce high maltose syrups and high conversion syrups used in pharmaceutical and clinical sectors and also increases

S. K. Mishra (✉) · K. Shivam · S. K. Diwakar · S. Shukla
Department of Biotechnology, D. D. U. Gorakhpur University, Gorakhpur 273009, U.P., India
e-mail: saradmishra5@rediffmail.com

R. N. Kharwar et al. (eds.), *Microbial Diversity and Biotechnology in Food Security*,
DOI 10.1007/978-81-322-1801-2_43, © Springer India 2014

fermentability of brewing wort (White and White 1997). In structural studies β-amylase is used for the structural analysis of starch and glycogen molecules (White and White 1997).

In fungi, detailed studies on amylases are in scarce and have largely been limited to only a few species of fungi (Abouzeid 1997). On the other hand, bacterial amylases have generally been produced from the strains belonging to genus *Bacillus* and *Clostridium*; several attempts have been made at their purification and characterization, from both mesophilic as well as thermophilic strains (Pandey et al. 2000). The thermostable β-amylase is needed especially in food and beverage industries. Therefore, in recent times most of the attention has been made to search β-amylases which can remain active at higher temperatures. Mostly thermostable β-amylases have been reported from few bacteria, i.e., *Bacillus* sp., *Clostridium thermosulfurogenes* (Kwan et al. 1994; Shen et al. 1988), etc. as well as plants, i.e., barley (Okada et al. 1995). Among fungi, extracellular thermostable β-amylase has been reported from *Thermoactinomyces* sp (Obi and Odibo 1984).

The amylase family of enzymes is of great biotechnological significance due to its wide area of potential application. Although a lot of research work has been done on α-amylase from various microbial sources, but less attention has been paid towards fungal β-amylase, which too can have enormous industrial applications. In addition to thermostability, such applications require highly purified amylases. Thus, it is significant to develop economical and convenient processes for their purification to obtain pure enzymes with maximum specific activity. The apparent paucity of information on thermostable β-amylase from mesophilic fungi prompted us to screen various *Aspergillus* sp., among which *Aspergillus foetidus* MTCC-508 was found to be an excellent source for thermostable β-amylase production. Moreover, *A. foetidus* can be considered an important source of enzyme as it does not produce ochratoxin A, as reported by Schuster et al. in 2002 *and* is a member of Aspergillus niger group which is considered as GRAS (generally regarded as safe) organism. To date, there is no report of thermostable β-amylase from *A. foetidus* in literature. This article describes the isolation, purification, and characterization of a thermostable β-amylase enzyme produced under submerged fermentation (SmF) conditions by *A. foetidus* MTCC-508.

## 43.2 Materials and Methods

### 43.2.1 Materials

*A. foetidus* MTCC-508 was procured from the Institute of Microbial Technology, Chandigarh, India. Fungal culture was maintained on potato-dextrose-agar medium. Sephadex G-100 and DEAE-Sephadex A-50 were obtained from GE Healthcare (UK). Soluble starch, $KH_2PO_4$, $K_2HPO_4$, $MgSO_4.7H_2O$, $(NH_4)_2SO_4$, $FeSO_4$, and all other chemicals used were of analytical grade purchased from Merck, India. All experiments were repeated thrice.

### 43.2.2 Organism Growth and β-Amylase Production

The liquid culture medium for β-amylase production was composed of (grams per liter) $KH_2PO_4$ 0.3; $K_2HPO_4$ 0.7; $MgSO_4.7H_2O$, 0.5; $(NH_4)_2SO_4$ 2.0; $FeSO_4$ 0.01; and 1 % (w/v) of soluble starch, pH 6.5. 250 ml Erlenmeyer glass flasks containing 50 ml of culture medium were autoclaved at 121 °C for 20 min. These flasks were inoculated with fungal spores ($1 \times 10^8$ $mL^{-1}$) and incubated at 30 °C for 24 h in an automatic incubator.

### 43.2.3 Enzyme Activity Determination

Amylase activity was routinely assayed by measuring the rate of generation of reducing sugars from starch. The enzyme assay for β-amylase activity used a reaction mixture (1 mL) containing boiled soluble starch (1 % (w/v)), and sodium acetate buffer (50 mM, pH 6.0). Reaction mixture was incubated at 30 °C for 30 min and the reducing sugar released by enzymic hydrolysis of soluble starch was determined by dinitrosalicylic acid method (Bernfeld 1955). One unit of β-amylase was defined as the amount of enzyme that produces 1 μmol of reducing sugar as maltose per min under the above assay conditions.

### 43.2.4 Protein Determination

Protein concentration was determined by the method of Lowry et al. (1951) using bovine serum albumin as the standard.

Except for the chromatographic experiments, the enzyme activity values presented are mean values of triplicate assays. Standard deviation values were always smaller than 10% of the mean value.

### 43.2.5 Purification of β-Amylase

The precooled acetone was slowly added to the crude extract with constant stirring to give a concentration of 75% (v/v) and this solution was left on ice for 10 min. The precipitate was collected by centrifugation at 10,000 rpm for 15 min at 4 °C, dissolved in acetate buffer (50 mM, pH 6.0) with 2 mM $CaCl_2$, dialyzed against the same buffer and then used as partially purified enzyme preparation.

The partially purified enzyme preparation was loaded on a DEAE-Sephadex A-50 column (45 × 1.5 cm), pre-equilibrated with 50 mM acetate buffer, pH 7.0. After complete loading of the enzyme, the unbound fraction was checked for amylase activity and then the column was washed with the equilibration buffer. The enzyme was eluted with linear gradient of NaCl (0–1.0 M), prepared with 210 mL of equilibration buffer and 210 mL of the same buffer containing 1.0 M NaCl. Samples of 3 mL were collected at a flow rate of ~30 mL/h. The single activity peak fractions were pooled, concentrated, and loaded on a Sephadex G-100 column (50 cm × 1.5 cm), which was pre-equilibrated with 50 mM acetate buffer (pH 6.0) with 2 mM $CaCl_2$. The enzyme was eluted with the same buffer at a flow rate of ~10 mL/h. The single activity peak fractions were collected, concentrated and dialyzed against the same buffer. The dialyzed enzyme solution was then used as almost purified enzyme preparation.

The purity of the enzyme was checked on 10% SDS-PAGE under denaturing conditions according to the method of Laemmli (1970). Gel was stained with Coommasie Brillant Blue R-250. The molecular weight markers used in SDS-PAGE were ovalbumin (43,000), carbonic anhydrase (29,000), soybean trypsin inhibitor (20,100), and lysozyme (14,300). Native-PAGE was performed using 10% polyacrylamide gel according to the method of Walker (2002). The procedure was same as SDS-PAGE except that SDS and β-mercaptoethanol were not added.

### 43.2.6 Effect of pH

The effect of pH on enzyme activity was assessed over a pH range from 3 to 9 by using either 50 mM potassium phosphate buffer or 50 mM citric acid as buffer (adjusted to the required pH by the other components) under otherwise standard enzyme assay conditions. In both buffers with same pH, the activity of enzyme was same.

### 43.2.7 Effect of Temperature and Thermal Stability

The effect of temperature on enzyme activity was assessed by incubating enzyme with the substrate (as previously described) at various temperatures ranging from 30 to 80 °C.

The effect of temperature on enzyme stability was assessed by incubating enzyme at 50, 60, and 70 °C for several time periods. The residual activity of the enzyme was estimated under standard assay conditions in each case and expressed as relative activity (%) referred to the activity observed before incubation.

### 43.2.8 Effect of Metallic Ions

A mixture consisting of 0.1 mL of the properly diluted enzyme solution and 0.1 mL of

**Table 43.1** Steps for purification of β-amylase from *Aspergillus foetidus* MTCC-508

| Purification step | Total activity (units) | Total protein (mg) | Specific activity (units/mg) | Purification fold | Recovery (%) |
|---|---|---|---|---|---|
| Crude extract | 6800 | 820 | 8.29 | 1 | 100 |
| Acetone precipitation | 3036 | 9.86 | 307.91 | 37.14 | 44.64 |
| DEAE-Sephadex A-50 | 1590 | 1.2 | 1325 | 159.8 | 23.4 |
| Sephadex G-100 | 993 | 0.47 | 2112.76 | 254.8 | 14.6 |

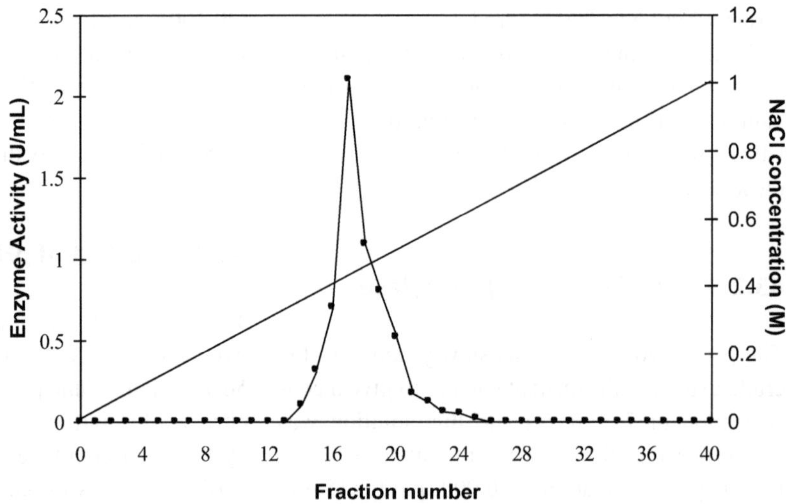


**Fig. 43.1** Elution profile of the β-amylase from *Aspergillus foetidus* MTCC-508 chromatographed on the DEAE-Sephadex A-50 column. β-amylase activity (•); linear gradient of NaCl (0–1.0M) (–)

metallic ions (final concentration 0.5 mM) was incubated for 30 min at room temperature and enzyme activity was checked under standard assay conditions.

### 43.2.9 Hydrolysis of Raw Starch

The ability of the enzyme to hydrolyze raw starch was studied using rice, wheat, *Trapa natans,* and maize starches. Commercial soluble starch was used as the standard.

### 43.2.10 Substrate Specificity

The $K_m$ and $V_{max}$ for β-amylase using soluble starch as a substrate were determined by the method described by Lineweaver and Burk (1934).

## 43.3 Results and Discussion

### 43.3.1 Enzyme Purification

The enzyme purification steps are summarized in Table 43.1. β-amylase was purified 254.8-fold with 14.6% yield from the crude enzyme extract. First, the proteins in the crude extract were precipitated with 75% acetone. The precipitation of proteins was also tried with 110% ethanol but in this case significant amount of enzyme got inactivated and also the percent recovery was comparatively low and hence precipitation with acetone was preferred. In this step, enzyme purification fold achieved was 37.14 and 44.64% enzyme could be recovered. Then the concentrated enzyme preparation was fractionated by anionic exchange chromatography using DEAE-Sephadex A-50 column, in which the enzyme was eluted by 0.43 M NaCl in the fractions from 14 to 24 as shown in Fig. 43.1.

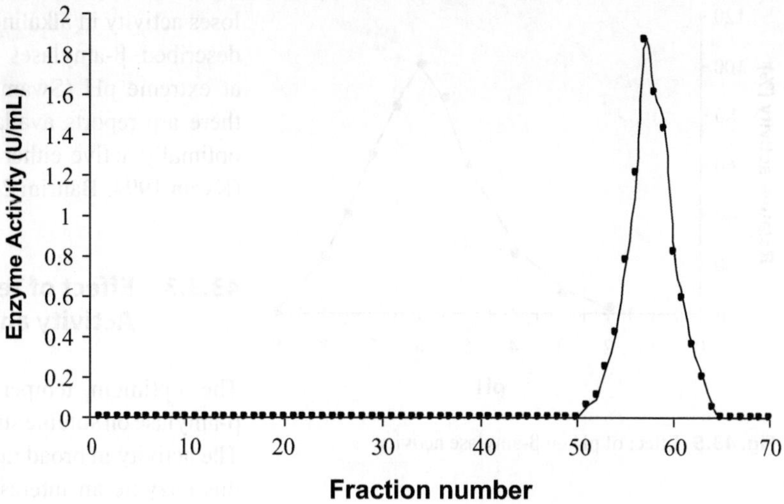


**Fig. 43.2** Elution profile of the β-amylase from *Aspergillus foetidus* MTCC-508 chromatographed on the Sephadex G-100 column

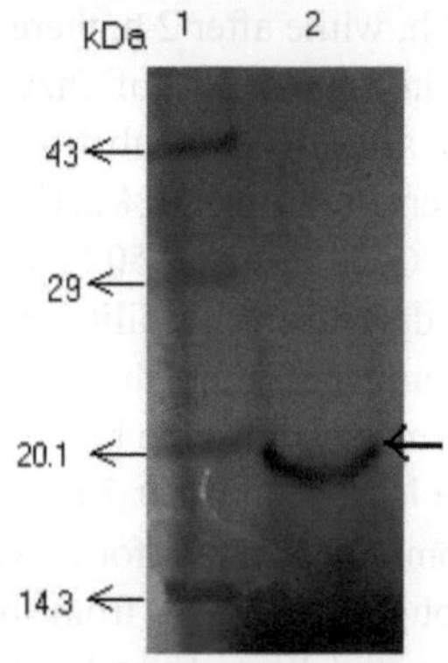


**Fig. 43.3** Ten percent SDS-PAGE of *Aspergillus foetidus* MTCC-508 β-amylase; *1*—molecular mass standards; *2*—purified β-amylase

After this step, the purification fold and enzyme recovery were 159.8 and 23.4%, respectively. To remove the remaining contaminating proteins, the partially purified enzyme was loaded on Sephadex G-100 column (Fig. 43.2). The enzyme was eluted in fraction numbers 51 to 64. All the fractions were pooled together, concentrated and loaded on 10% SDS-PAGE to check the purity. In 10% SDS-PAGE, only one protein band could be observed at around 20 kDa (Fig. 43.3), confirming the purity of β-amylase enzyme. The purification scheme employed here is relatively simple and used routinely in a general protein laboratory.

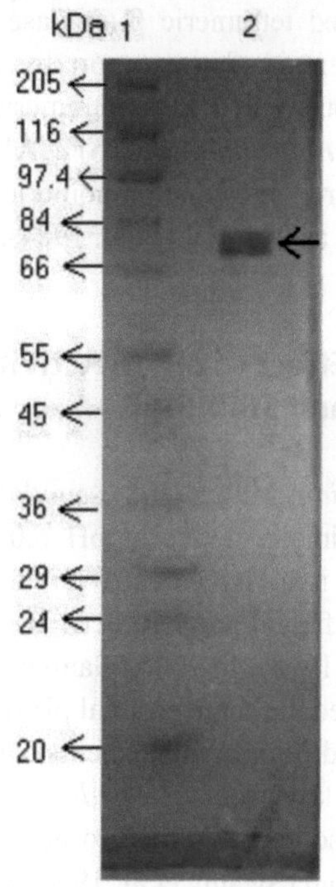


**Fig. 43.4** Ten percent Native-PAGE of *Aspergillus foetidus* MTCC-508 β-amylase; *1*—molecular mass standards; 2—purified β-amylase

When Native-PAGE of the purified enzyme was performed, the band came around 80 kDa (Fig. 43.4), suggesting that the enzyme is a homotetrameric protein. Our finding is in agreement with the earlier report by Shen et al. (1988),

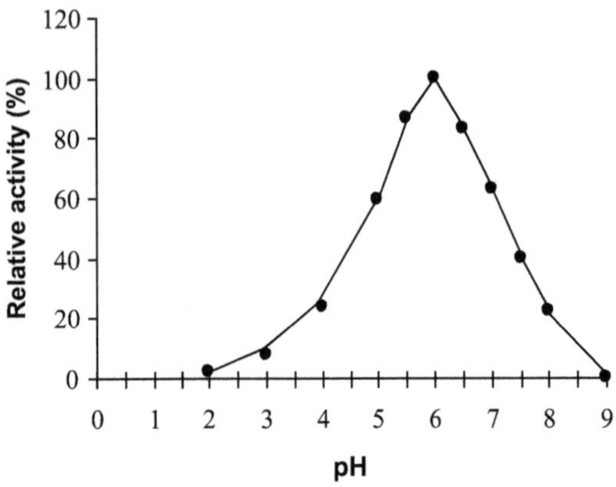


**Fig. 43.5** Effect of pH on β-amylase activity

who reported a thermostable β-amylase from *C. thermosulfurogenes* having homotetrameric structure. Colman and Matthews (1971) have also reported tetrameric β-amylase from sweet potato. There is also a report by Rama et al. (1998) about a 210 kDa tetrameric β-amylase from *C. thermosulfurogenes SV2*. Although there are several reports about monomeric and dimeric β-amylases as well (Obi et al. 1984).

### 43.3.2 Effect of pH on Activity and Stability

The *A. foetidus* MTCC-508 β-amylase was found to be maximally active at pH 6.0 (Fig. 43.5). This result is in agreement with the optimal pH for certain fungal amylase activities reported by Alli et al. (1998). In 2005, Ganiyu Oboh (2005) also reported the same optimal pH (6.0) for amylase isolated from fermented cassava wastewater. The β-amylase from *Clostridium thermocellum* SS8 has also been reported to be maximally active at pH 6.0 (Swamy et al. 1994). Optimum pH for an extracellular thermostable β-amylase from C. thermosulfurogenes has also been reported in the range of 5.5–6.0 by Hyun and Zeikus 1985. Figure 43.5 also indicates that the β-amylase is mostly active in acidic range of pH, showing >80% activity at pH 6.5 and 5.5 and almost 60% activity in the pH range of 5.0–7.0. However, the activity of the enzyme decreased sharply above pH 7.0, which indicates that the enzyme loses activity in alkaline region. Most commonly described β-amylases lack significant activity at extreme pH (Swamy et al. 1994). Although there are reports available of β-amylases being optimally active either at neutral or alkaline pH (Kwan 1994; Bahrim 2007).

### 43.3.3 Effect of Temperature on Activity and Stability

The optimum temperature for the action of β-amylase on soluble starch was 50 °C (Fig. 43.6). The activity in broad range of temperature makes this enzyme an interesting candidate for industrial applications.

When the thermal stability of the enzyme was checked, it was found to be completely stable at 50 °C up to 2 h, while after 2 h, there was a gradual decrease in the stability of enzyme showing 40% activity after 20 h incubation (Fig. 43.7). At 60 °C the enzyme was 80% active for 30 min while at 70 °C it lost its 50% activity after 30 min. The decrease in stability increases with increase in temperature as shown in Fig. 43.6, with the enzyme showing the highest stability at 50 °C and the least stability at 70 °C. Further, the enzyme is completely stable for longest duration at 50 °C although β-amylases from some thermophilic bacteria and fungi have been reported to be stable at higher temperatures (Shen 1988; Obi and Odibo 1984; Rama et al. 1998). Thermostability is an important attribute of this enzyme for its use in industries.

The temperature inactivation of the enzyme can be attributed to the formation of incorrect conformation due to processes such as hydrolysis of the peptide chain, destruction of amino acids and aggregation (Schokker and Van Boekel 1999).

### 43.3.4 Effect of Metallic Ions on Enzyme Activity

The effect of various metal ions (final concentration 0.5 mM) on the activity of β-amylase was investigated (Table 43.2). The enzyme was strongly inhibited showing 2% and 3% of its

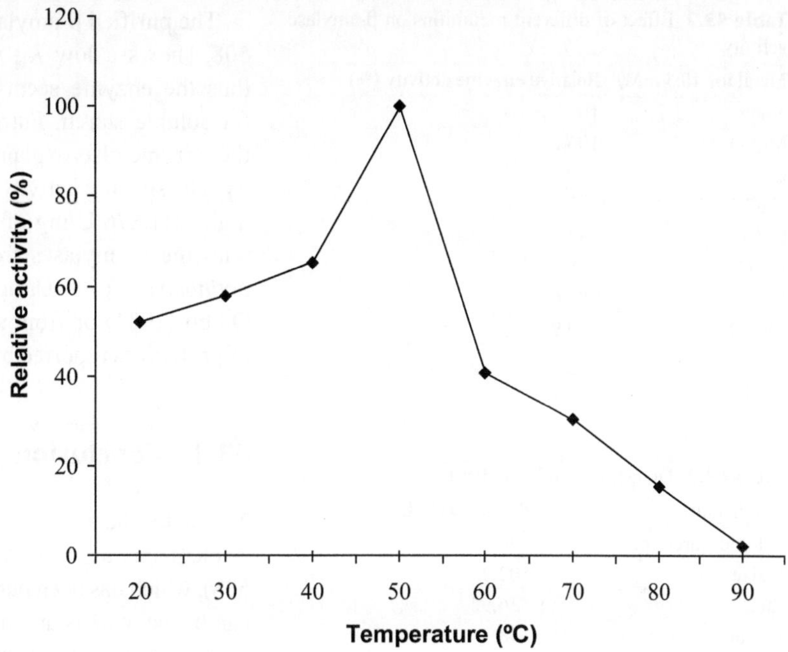


**Fig. 43.6** Effect of temperature on β-amylase activity

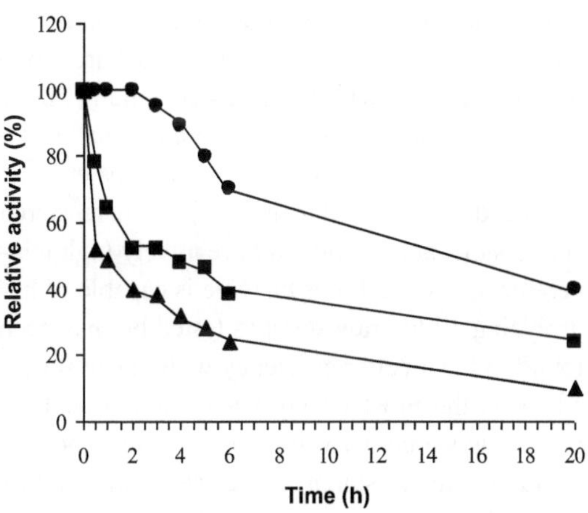


**Fig. 43.7** Stability of β-amylase from *Aspergillus foetidus* MTCC-508 as a function of temperature. Remaining β-amylase activity as percentage of initial activity at different temperatures (●) 50 °C, (■) 60 °C, (▲) 70 °C

original activity in presence of $Zn^{2+}$ and $Hg^{2+}$ respectively, while $Co^{2+}$ and $Fe^{3+}$ considerably inhibited the activity, whereas marginal inhibition was observed in the presence of $Cu^{2+}$ and $Mn^{2+}$. This result is in agreement with the report of Swamy et al. (1994), who reported that high molecular weight metal ions such as $Ag^{+}$, $Fe^{2+}$, $Zn^{2+}$, $Cu^{2+}$, and $Hg^{2+}$ inhibit the activity of β-amylase from *C. thermocellum* SS8. Chang et al. (1996) have also reported inhibition of amylase by $CuSO_4$ and $HgCl_2$. Further, the inhibition by $Hg^{+}$ may be attributed to their binding of the thiol group of the enzyme (Dey and Pridham 1977). $Mg^{2+}$ was found to be the only metal which enhanced the activity of enzyme up to 106.2 % of its original activity. Other metal ions, such as $Ca^{2+}$ and $Ni^{2+}$ did not show any considerable effect on enzyme activity.

### 43.3.5 Hydrolysis of Raw Starch

The ability of the β-amylase from *A. foetidus* MTCC-508 to digest raw starch was studied using starches from different sources like maize, wheat, rice, and *Trapa natans*. The result shown in Table 43.3, reveals that the relative activity of this enzyme is maximum with maize starch (102.3 %) as compared to other raw starches used (*Trapa natans*: 73 %, wheat: 48 %, rice: 26.92 %). Better efficiency of this enzyme with maize starch is remarkable as maize starch is the most wide spread raw material used to produce starch derivatives and hydrolysis products. This enzyme being efficient in converting maize starch into

**Table 43.2** Effect of different metal ions on β-amylase activity

| Metal ion (0.5 mM) | Relative enzyme activity (%) |
|---|---|
| None | 100 |
| $MgSO_4$ | 106.2 |
| $CaCl_2$ | 100 |
| $HgCl_2$ | 3 |
| $CoCl_2$ | 32 |
| $NiSO_4$ | 98.2 |
| $CuSO_4$ | 83.6 |
| $MnSO_4$ | 80.9 |
| $ZnCl_2$ | 2 |
| $FeCl_3$ | 70.1 |

**Table 43.3** Digestibility of raw starches

| Starch source | Relative activity (%) |
|---|---|
| Soluble starch (potato) | 100 |
| Maize | 102.3 |
| Rice | 26.92 |
| Wheat | 48 |
| *Trapa natans* | 73 |

maltose, it can be applied to produce high maltose syrups, one of the important maize sweeteners that can be adapted to various needs; can be used in food to provide clean, sweet taste, retain moisture, controlled crystallization and freezing points, inhibit spoilage and modified density etc. It is obvious from the result that when the starchy material is incubated with amylases, it gets hydrolyzed into smaller subunits, but the degree of hydrolysis greatly depends on the source of the starch which is used as a substrate. This agrees with earlier reports of Omemeu et al. (2005) that the susceptibility of raw starch granules to digestion by amylase is dependent on starch source.

### 43.3.6 Substrate Specificity

The effect of varying the substrate concentration on the rate of reaction was studied with soluble starch. Lineweaver-Burk plots of the data gave an apparent Michaelis constant ($K_m$) of 2.7 mg/mL and a maximum velocity ($V_{max}$) of 2,100 U/mg of protein (Fig. 43.8). No significant inhibition effect was observed at high substrate concentration.

The purified β-amylase of *A. foetidus* MTCC-508, shows a low $K_m$ value for soluble starch, thus the enzyme seems to have higher affinity for soluble starch. Further the low $K_m$ value of the enzyme also explains its high specific activity. The specific activity of the enzyme was fairly high (2112.76 U/mg of protein) when compared with the β-amylases from *Clostridium thermoactinomyces* (408 U/mg of protein) by Obi and Odibo (1984) or from sweet potato (0.06 U/mg of protein) as reported by Oboh and Ajele (1997).

## 43.4 Conclusion

This is for the first time we are reporting thermostable β-amylase from A. foetidus *isolate (*MTCC-508), which has been purified in simple steps. This can be claimed as an important source of this enzyme as A. foetidus does not produce Chratoxin A *and* is a member of Aspergillus niger group which is considered a GRAS organism. In industries, β-amylases active in acidic range of pH are mostly required. The activity of this enzyme in acidic range of pH and at higher temperature makes it a suitable candidate for industrial use. This enzyme can be exploited in the industries where starch conversions are performed at relatively high temperatures. Although this enzyme is capable of hydrolyzing all the raw starches tested but has been found to have better efficiency with maize starch, which is the most widely used raw material for producing starch derivatives and related products. Hence, the maize starch can be used in the production of high maltose syrups and high conversion syrups required in pharmaceutical and clinical sectors by using this enzyme. Further detail study is required regarding the conversing efficiency and hydrolyzing capability of this enzyme to maize starch to make it much more industrially suitable enzyme.

**Acknowledgments** The support from the Department of Biotechnology, DDU Gorakhpur University, Gorakhpur is duly acknowledged.

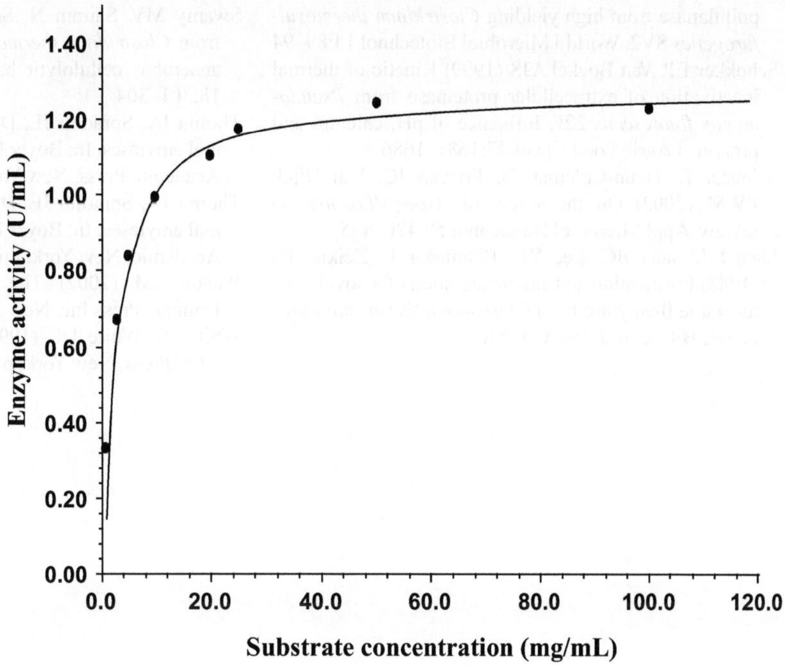


**Fig. 43.8** Michaelis Menten and Lineweaver—Burk plots for the β-amylase from *Aspergillus foetidus* MTCC-508, showing effect of substrate concentration on reaction velocity

## References


Abouzeid AM (1997) Production purification and characterization of an extracellular α-amylase enzyme isolated from *Aspergillus flavus*. Microbios 89:55–66

Alli AI, Ogbonna CIC, Rahman ATMF (1998) Hydrolysis of certain Nigerian cereal starch using crude fungal amylase. Nig J Biotechnol 9:24–36

Bahrim GE, Scantees M, Negoitan T (2007) Biotechnological conditions of amylase and complex production and utilization involving filamentous bacteria. The Annals of the University Dunarea de jos of Gulati-, Fascicle iv- Food Technol 76–81

Bernfeld P (1955) Amylase α/β. Methods Enzymol 1:149

Chang CT, Liou HY, Tang HL, Sung HY (1996) Activation, purification and properties of β-amylase from sweet potatoes (*Ipomea batatas*). Biotechnol App Bichem 24:113–118

Colman P, Matthews B (1971) Symmetry, molecular weight and crystallographic Data for sweet potato beta-amylase. J Mol Biol 60:163

Dey PM, Pridham JB (1977) Biochemistry of alpha galactosidase. Adv Enzymol 15:91–130

Hyun HH, Zeikus JG (1985) General biochemical characterization of thermostable extracellular β-amylase from *Clostridium thermosulfurogenes*. Appl Environ Microbiol 49:1162–1167

Kwan HS, So KH, Chan KY, Cheng SC (1994) Purification and properties of β-amylase from *Bacillus circulans* S31. World J Microbiol Biotechnol 10:597–598

Laemmli UK (1970) Cleavage of Structural Proteins during the Assembly of the Head of Bacteriophage T4, Nature 227:680–685

Lineweaver H, Burk D (1934) The determination of enzyme dissociation constants. J Am Chem Sci 56:658–666

Lowry OH, Rosenbrough NJ, Farr Al, Randall RJ (1951) Protein estimation with the Folin-phenol reagent. J Biol Chem 193:265–275

Obi SKC, Odibo FJC (1984) Partial purification and characterization of a thermostable actinomycete β-amylase. Appl Environ Microbiol 47:571–575

Oboh G (2005) Isolation and characterization of amylase from fermented cassava (*Manihot esculenta* Crantz) waste water. Afr J Biotechnol 4:1117–1123

Oboh G, Ajele JO (1997) Effect of some metallic chlorides on the activity of β-amylase from sweet potatoes. Nig J Biochem Mol Biol 12:73–75

Okada y, Yoshigi N, Sahara H, Koshino S (1995) Increase in thermostability of recombinant barley beta-amylase by random mutagenesis. Biosci Biotechnol Biochem 59:1152–1153

Omemu AM, Akpan I, Bankole MO, Teniola OD (2005) Hydrolysis of raw tuber starches by amylase of *Aspergillus niger* AM07 isolated from the soil. Afr J Biotechnol 4:19–25

Pandey A, Nigam P, Sccol CR, Sccol VT, Singh D, Mohan R (2000) Advances in microbial amylases (Review). Biotechnol Appl Biochem 31:135–152

Rama M, Swamy MV, Seenayya G (1998) Purification and characterization of thermostable β-amylase and

pullulanase from high yielding *Clostridium thermosulfurogenes* SV2. World J Microbiol Biotechnol 14:89–94
Schokker EP, Van Boekel AJS (1999) Kinetic of thermal inactivation of extracellular proteinase from *Pseudomonas fluorescens* 22F, Influence of pH, calcium and protein. J Agric Food Chem 47:1681–1686
Schuster E, Dunn-Coleman N, Frisvad JC, Van Dijck PWM (2002) On the safety of *Aspergillus niger*-a review. Appl Microbiol Biotechnol 59:426–435
Shen GJ, Saha BC, Lee YE, Bhatnagar L, Zeikus JG (1988) Purification and characterization of a novel thermostable β-amylase from *Clostridium thermosulphurogenes*, BiChemi. J 254:835–840
Swamy MV, Sairam N, Seenayya G (1994) β-amylase from *Clostridium thermocelluum* SS8 a thermophillic, anaerobic, cellulolytic bacterium. Lett Appl Microbiol 18:301–304
Thoma JA, Spradlin JE, Dygert S (1997) Plant and animal amylases. In: Boyer PD (ed) The enzymes, 3rd edn. Academic Press, New York, pp 115–189
Thoma JA, Spradlin JE, Dygert S (1971) Plant and animal amylases. In: Boyer PD (ed) The enzymes, 5th edn. Academic, New York, pp 115–189
Walker JM (2002) The protein protcols handbook. Humana Press Inc, New Jersey, pp 343–345
White JS, White DC (1997) Source book of enzymes. CRC Press, New York, p 572

# Effect of *Euphorbia pulcherrima* Leaf and Inflorescence Extracts on Spore Germination of *Alternaria solani*

**44**

Arti Goel and Kanika Sharma

**Abstract**

Early blight in solanaceous plants is a common and serious disease of potato and tomato caused by the pathogen *Alternaria solani*. Leaf blight caused by *A. solani* resulted in 50–80 % yield loss. Several fungicides have been used for a long time to prevent losses due to pre- and post-harvest diseases but have led to serious environmental threat to human life and microbial contamination. Therefore, the present study was done with the aim to analyse the effect of acetone extract of leaf and alcohol extract of inflorescence of *Euphorbia pulcherrima* on spore germination of *A. solani* by detached leaf assay method for the development of biologically safe fungicide. A significant reduction in infection as well as conidial germination of *A. solani* was observed on petiolated but detached leaves of *Lycopersicon esculentum*. Acetone extract of the leaf of *E. pulcherrima* was found to be 100 % inhibitory at 125 μg/ml concentrations whereas alcohol extract of inflorescence showed 100 % inhibition of spores at only 62.5 μg/ml concentration.

**Keywords**

Early blight · *Alternaria solani* · *Euphorbia pulcherrima* · Plant protection

## 44.1 Introduction

Diseases caused by fungi in plants are responsible for decrease in yield of crops. Economic losses due to fungal diseases in plants are of two types: pre-harvest and post-harvest. Pre-harvest diseases occur in the standing crop whereas post-harvest diseases occur during storage, so loss in crop yield is much more in these types of diseases. Among the various severe plant pathogenic fungi, *Alternaria* is the most common fungi which cause blight disease in solanaceous plants such as potato and tomato. Tomato is an important vegetable crop of the world and ranks next to potato in acreage and production (Saini et al. 2005). One of the most common diseases of tomatoes caused by the fungus *Alternaria solani* is known as the

A. Goel (✉)
Amity Institute of Microbial Biotechnology, Amity University, Noida, U.P., India
e-mail: agoel2@amity.edu

K. Sharma
Department of Botany, Mohanlal Sukhadia University, Udaipur, Rajasthan 313001

R. N. Kharwar et al. (eds.), *Microbial Diversity and Biotechnology in Food Security*,
DOI 10.1007/978-81-322-1801-2_44, © Springer India 2014

early blight of tomatoes. The early blight may affect foliage, stems and fruits (causing brown leaf spots and concentric black circles on the tomato fruit). This fungal disease can affect solanaceous crops (Karima and Sayeda 2007). Early blight is an important foliar pathogen of potatoes worldwide. Susceptibility to infection is primarily determined by the age of the host plant. *A. solani* causes collar rot in tomato seedlings, and early blight caused by *A. solani* is the most destructive disease of tomatoes in the tropical and subtropical regions. Each 1 % increase in intensity can reduce yield by 1.36 %, in mature tomatoes and potatoes (El-Mougy 2009; Babagoli and Behdad 2012). Complete crop failure due to early blight can also occur when the disease is the most severe (Pandey 2003). Yield losses of up to 79 % have been reported in the USA, of which 20–40 % is due to seedling losses (collar rot) in the field (Chaerani and Voorrips 2006).

*A. solani* is also one of the most important foliar pathogens of potato. In the USA, yield loss estimates attributed to foliar damage, which results in decreased tuber quality and yield reduction, can reach 20–30 %. In storage, *A. solani* can cause dry rot of tubers and may also reduce storage length, both of which diminish the quantity and quality of marketable tubers. Singh (1985, 1998) reported that production loss of tomatoes due to blight disease is as high as 78 %. Early blight is the most common disease of potatoes in South Africa and is caused by the fungus *A. solani*. Yield loss of up to 50 % in heavily infected fields has been recorded (OrzoLek et al. 2010).

Several fungicides have been used for a long time to prevent losses due to pre- and post-harvest diseases but have led to serious environmental threat to human life and microbial contamination (Agriose 1997). Pathogenic microorganisms and insect pests develop resistance due to repeated use of these chemical pesticides. Hence, there is a need to develop an alternative, safe and cheap method for the management of pathogenic microorganisms (Rice et al. 1998).

Hence, in the present study, the effect of acetone and alcohol extract of *Euphorbia pulcherrima* leaf and inflorescence on the germination of *A. solani* was studied by detached leaf assay method.

## 44.2 Materials and Methods

### 44.2.1 Inoculum Preparation

Inoculum of the pathogen was prepared by harvesting spores from 7 to 8-day-old fungal culture growing on potato-dextrose-agar (PDA) medium. Spores were suspended in 5 ml of distilled water and were gently shaken with a glass rod. The resultant spore suspension was filtered through double-layered cheesecloth to remove mycelial fragments and was subsequently centrifuged at 3,500 rpm for 10 min. The spore pellet so obtained was resuspended in sterile water and suitably diluted to obtain spore density of $1 \times 10^4$ spores/ml.

### 44.2.2 Effect of Plant Extract on Spore Germination

Leaves and inflorescence of *E. pulcherrima* were collected from the Botanical Garden of College of Science, Udaipur. Leaves and inflorescence were shade dried, mechanically powdered and were used for extract preparation. A total of 50 mg of extract was dissolved in 10 ml of dimethyl formamide (DMF) to obtain stock solution of 5 mg/ml concentration, and different concentrations such as 500, 250, 125, 62.5, 31.25, 15.62, 7.81 and 3.95 µg/ml were made from this stock solution by twofold serial dilution technique. Concentrations up to minimum inhibitory concentration (MIC) were further used for experimental purpose, i.e. up to 125 µg/ml concentration of acetone extract of leaves and up to 62.5 µg/ml concentration of alcohol extract of inflorescence were used.

Inhibition of spore germination was studied on petiolated but detached leaves of *Lycopersicon esculentum*. The leaves were washed in distilled water and the extracts were spread with a fine camel hairbrush on the adaxial surface of leaves. Three replicates were used for each concentration. A total of 1 ml of spore suspension was then spread on the surface of these leaves with the help of sterilized inoculation loop, untreated control was maintained to observe the normal germination of conidia, and Similarly negative control of DMF was also maintained to rule out the possibility of its inhibitory effect. The leaves were transferred to petri plates filled with water. The leaf lamina was supported by tilted glass slide and the petiole was kept dipped in water. The leaves were incubated at 20 °C for 24 h, and the experiment was repeated thrice. After incubation, the leaves were cut into small pieces (10 × 10 mm) and cleared by the method of Bruzzese and Hassan (1983). These pieces were placed in a hot solution of 1:1 of glacial acetic acid and 95 % ethanol and were boiled gently for few minutes until the pieces lost their chlorophyll and became opaque. These leaf bits were cleared for 30 min in strong solution of chloral hydrate and stained with cotton blue and observed under a light microscope. The number of germ tubes/conidium, the length of the germ tube and the per cent germination were recorded.

## 44.3 Results and Discussion

Results of inhibitory effect of acetone extract of *E. pulcherrima* leaf on the per cent conidial germination, the length of the germ tube and the number of germ tube/conidium are listed in Table 44.1. The percent germination decreased with increasing concentration of extract. Inhibition of conidial germination started at 3.95 µg/ml and 100 % inhibition was observed at MIC (125 µg/ml). The length and the number of germ tubes also decreased with increasing concentration of the extract. As compared to control germ tube length was decreased from 102.2 to 7.36 µm at 62.5 µg/ml (MIC). Similarly, the number of germ tubes also decreased from 2.0/conidium (control) to single-germ tube/conidium at sub-MIC, i.e. at 62.5 µg/ml. Mancozeb used as standard antifungal exhibited 75 % inhibition of conidia of *A. solani*.

Results of the effect of the alcohol extract of *E. pulcherrima* inflorescence on per cent conidial germination, length of the germ tubes and the number of germ tube/conidium are listed in Table 44.2. Results indicate that as compared to mancozeb (2,000 µg/ml), 15.62 µg/ml concentration of the extract is found to be more inhibitory against conidia of *A. solani*. A total of 43.34 % inhibition of *A. solani* conidia was observed at sub-MIC concentration, i.e. 31.25 µg/ml. A total of 100 % inhibition was observed at 62.5 µg/ml (MIC) concentration of the extract. Similarly, the germ tube length was also decreased from 99.16 µm (control) to 14.83 µm at sub-MIC, i.e. 31.25 µg/ml. Decrease in the number of germ tubes/conidium was also observed. In control, 1.9 germ tubes/conidia were observed, and at sub-MIC concentration of 31.25 µg/ml, a single germ tube/conidium was observed. At 62.5 µg/ml, germ tubes were not formed which indicates that spores failed to germinate due to the treatment of the extract.

Early blight in solanaceous plants is a common and serious disease of potatoes and tomatoes caused by the pathogen *A. solani*. Leaf blight caused by *A. solani* resulted in 50–80 % of yield loss (Mathur and Shekhawat 1986). A significant reduction in infection as well as conidial germination of *A. solani* was studied on petiolated but detached leaves of *L. esculentum* by treatment with the acetone extract of leaf and the alcohol extract of inflorescence of *E. pulcherrima*. Inhibitory effect of *E. pulcherrima* leaf and alcohol extracts on conidial germination increased with increasing concentrations of the extract.

Since ancient time, plants have been known for their medicinal and antimicrobial properties, and some plant extracts have been proven to be good hatching inhibitors and sporulation altering agents (Khalil 2001). Several plant products as therapeutic agents against various plant pathogens causing different diseases in cereal, oil seed and vegetable crops have been tried successfully (Ganesan 1994; Valarini et al. 1994). Hence, biorational alternative including plant extracts are the best alternative present today for solving vegetable disease problems (Tiwari et al. 2004).

Feng and Zheng (2007) investigated that *Cassia* oil completely inhibit the growth of *Alternaria alternata* at 300–500 ppm whereas spore germination and germ tube elongation of the pathogens was strongly inhibited in the presence of 500 ppm concentration of *Cassia* oil. Reduction in mycelial growth and spore germination of *Phytophthora infestans* and *A. solani* has been reported (Abd-El-Khair and Haggag 2007). Murugan et al. (2007) reported inhibition of spore formation and

**Table 44.1** Effect of *E. pulcherrima* leaf extract on conidial germination of *A. solani*

| Sl. no. | Extract concentration (μg/ml) | Per cent spore germination | | | Mean spore germination | Number of germ tubes/conidium | | | Mean no. of germ tubes/conidium | Length of germ tubes/conidium (μm) | | | Mean length of germ tubes/conidium (μm) |
|---|---|---|---|---|---|---|---|---|---|---|---|---|---|
| | | $R_1$ | $R_2$ | $R_3$ | | $R_1$ | $R_2$ | $R_3$ | | $R_1$ | $R_2$ | $R_3$ | |
| 1. | Control (extract free) | 100 | 100 | 100 | 100% | 2 | 2 | 2 | 2.0 | 102.4 | 101.9 | 102.3 | 102.2±0.264 |
| 2. | 3.95 | 88.0 | 87.0 | 88.0 | 87.6% | 2 | 1 | 2 | 1.66 | 99.9 | 99.8 | 100.0 | 99.9±0.1 |
| 3. | 7.81 | 79 | 81 | 80 | 80.0% | 2 | 1 | 1 | 1.33 | 44.3 | 45.0 | 42.7 | 44.0±1.17 |
| 4. | 15.62 | 49.9 | 51.1 | 50 | 50.3% | 1.0 | 1.0 | 1.0 | 1.0 | 25.6 | 26.4 | 25.2 | 25.73±0.611 |
| 5. | 31.25 | 40 | 39.9 | 41.1 | 40.3% | 1.0 | 1.0 | 1.0 | 1.0 | 12.8 | 12.4 | 12.5 | 12.56±0.208 |
| 6. | 62.5 | 28.2 | 29.0 | 32.8 | 30.0% | 1.0 | 1.0 | 1.0 | 1.0 | 7.4 | 7.3 | 7.4 | 7.36±0.057 |
| 7. | 125 | 0.0 | 0.0 | 0.0 | 0.00% | No germ tube | No germ tube | No germ tube | No germ tube | No germ gube | No germ tube | No germ tube | No germ tube |
| 8. | Mancozeb (2,000 μg/ml) | 75 | 75 | 75 | 75% | 1 | 2 | 1 | 1.33 | 15.5 | 14.9 | 15.3 | 15.23±0.305 |

**Table 44.2** Effect of *E. pulcherrima* inflorescence extract on conidial germination of *A. solani*

| Sl. no. | Extract concentration (μg/ml) | Per cent spore germination | | | Mean spore germination | Number of germ tubes/conidium | | | Mean no. of germ tubes/conidium | Length of germ tubes/conidium (μm) | | | Mean length of germ tubes/conidium (μm) |
|---|---|---|---|---|---|---|---|---|---|---|---|---|---|
| | | $R_1$ | $R_2$ | $R_3$ | | $R_1$ | $R_2$ | $R_3$ | | $R_1$ | $R_2$ | $R_3$ | |
| 1. | Control (extract free) | 99.0 | 100 | 99.0 | 99.3% | 1.9 | 2.0 | 1.9 | 1.9 | 98.7 | 99.9 | 98.9 | 99.16±0.642 |
| 2. | 3.95 | 85.0 | 84.9 | 85.1 | 85.0% | 1.6 | 1.7 | 1.6 | 1.6 | 52.4 | 51.6 | 52.5 | 52.16±0.493 |
| 3. | 7.81 | 74.0 | 73.5 | 73.4 | 73.63% | 1.3 | 1.2 | 1.1 | 1.2 | 44.3 | 45.0 | 43.6 | 44.3±0.7 |
| 4. | 15.62 | 62.5 | 62.4 | 62.6 | 62.5% | 1.0 | 1.0 | 1.0 | 1.0 | 26.8 | 26.2 | 26.4 | 26.46±0.305 |
| 5. | 31.25 | 58.0 | 57.9 | 57.4 | 57.76% | 1.0 | 1.0 | 1.0 | 1.0 | 15.4 | 14.8 | 14.3 | 14.83±0.550 |
| 6. | 62.5 | 0.0 | 0.0 | 0.0 | 0.0% | No germ tube | No germ tube | No germ tube | No germ tube | No germ tube | No germ tube | No germ tube | No germ tube |
| 7. | Mancozeb (2,000 μg/ml) | 75 | 75 | 75 | 75% | 1.34 | 1.33 | 1.33 | 1.33 | 15.5 | 14.9 | 15.3 | 15.23±0.305 |

thus 100 % inhibition of the growth of *Aspergillus flavus* and *A. parasiticus* by using the flower extracts of *Euphorbia milli* and *E. pulcherrima.* Deepak et al. (2007) reported that aqueous extracts of some plants have the ability to inhibit zoosporium formation and thus zoospore formation in *Sclerospora graminicola.* Zaker (2013) screened some medicinal plant extracts against *Alternaria sesami,* the causal agent of *Alternaria* leaf spot of sesame and reported that methanolic extracts of peppermint (15 and 10 %), lavandula (15 %) and eucalyptus (15 %) were more effective than methanol–water extracts and completely inhibited the growth of the pathogen. Among the tested extracts, methanolic extracts of peppermint (15 %) and eucalyptus (15 %) were the best in preventing spore germination of the pathogen ($P$ = 0.01).

## 44.4 Conclusion

It was concluded from the above-mentioned study that fungal diseases of tomatoes and potatoes are caused by several pathogenic fungi before and after storage conditions. Many studies have been carried out with respect to occurrence, causal organisms, severity, losses, pathogenicity and disease control with fungicides. The present study demonstrated that *E. pulcherrima* extracts can be used for bio-control of early blight disease. Thus, this method of control can contribute in minimizing the risks and hazards of toxic fungicides, especially on vegetables produced for fresh consumption. Further research into these extracts will identify the active compounds responsible for their fungicidal activity.

This chapter may help future researchers to devise a concrete strategy for evaluating different pathological aspects and manage the post-harvest fungal diseases of tomatoes. However, further study is needed to review all other recent reports about various pathological aspects of the fungal diseases of tomatoes and potatoes and the management strategies opted for pre-harvest diseases of tomatoes and potatoes.

## References


Abd-El-Khair H, Haggag WM (2007) Application of some Egyptian medicinal plant extracts against potato late and early blights. Res J Agric Biol Sci 3(3):166–175

Agriose GN (1997) Plant pathology: control of plant diseases, 4th edn. Academic Press, New York

Babagoli MA, Behdad E (2012) Effects of three essential oils on the growth of the fungus *Alternaria solani.* J Res Agric Sci 8:45–57

Bruzzese E, Hassan S (1983) A whole leaf clearing and staining technique for host specificity studies of rust fungi. Pt Pathol 32:335–338

Chaerani R, Voorrips R (2006) Tomato early blight (*Alternaria solani*): the pathogen, genetics, and breeding for resistance. J of Gen Plant Pathology 72:335–347

Deepak SA, Oros G, Sathyanarayana SG, Shetty SH, Shashikanth S (2007) Antisporulant activity of water extracts of plants against *Sclerospora graminicola* causing downy mildew disease of pearl millet. Am J Agric Biol Sci 2(1):36–42

El-Mougy NS (2009) Effect of some essential oils for limiting early blight (Alternaria solani) development in potato field. J Plant Protection Res 49:57–62

Feng W, Zheng X (2007) Essential oils to control *Alternaria alternata* in vitro and in vivo. Food Control 18(9):1126–1130

Ganesan T (1994) Antifungal properties of wild plants. Adv Pl Sci 7(1):185–187

Karima HEH, Sayeda FF (2007) Effect of metalaxyl and chlorpyrifos-methyl against early blight (*Alternaria solani*, Sor.) and whitefly (*Bemisia tabaci*, Genn.) in tomato and eggplant. J Appl Sci Res 3:723–732

Khalil ARM (2001) Phytofungitoxic properties in the aqueous extracts of some plants. Pakistan J Biol Sci 4(4):392–394

Mathur K, Shekhawat KS (1986) Chemical control of early blight in kharif sown tomato. J Mycol Pt Pathol 16(3):235–236

Murugan S, Anand R, Uma Devi P, Vidhya N, Rajesh KA (2007) Efficacy of *Euphorbia milli* and *Euphorbia pulcherrima* on aflatoxin producing fungi (*Aspergillus flavus* and *Aspergillus parasiticus*). Afr J Biotechnol 6(6):718–719

Olanya OM (2009) The effect of cropping systems and irrigation management on development of potato early blight. J Gen Plant Pathol 75:267–275

OrzoLek M, Greaser GL, Harper KJ (2010) Commercial Vegetable Production Guide. Penn state cooperative extension Agricultural Alternatives: The Pennsylvania state University

Pandey KK (2003) Resistance to early blight of tomato with respect to various parameters of disease epidemics. J Gen Pt Pathol 69:364–371

Rice MJ, Legg M, Powell KA (1998) Natural products in agriculture: a view from the industry. Pest Sci 51:227–234

Saini AK, Jalali I, Pal V (2005) Eco-friendly Management of *Fusarium* wilt root-knot nematode complex in tomato. J Mycol Pl Pathol 35(2):338–341
Singh RS (1985) Diseases of vegetable crops. Oxford and IBH Publishing Co. Pvt. Ltd., New Delhi
Singh BD (1998) Biotechnology: fuel biotechnology, 1st edn. Kalyani Publishers, New Delhi
Tiwari RKS, Chandravanshi SS, Ojha BM (2004) Efficacy of extracts of medicinal plant species on growth of *Sclerotium rolfsii* root rot in tomato. J Mycol Pl Pathol 34(2):461–464
Valarini PJ, Frighetto RTS, Melo IS, De Melo IS (1994) The effect of medicinal herb *Cymbopogon citratus* on control of phytopathogenic fungi in a bean crop. Revista de Agricultura Piracicaba 69(2):139–150
Zaker M (2013) Screening some medicinal plant extracts against *Alternaria sesami*, the causal agent of *Alternaria* leaf spot of sesame. J Ornament Horti Pl 3(1):1–8

# 45

# Fungal and Mycotoxin Contamination of Herbal Raw Materials and Prospects of Higher Plant Products as Plant-Based Preservatives During Post-Harvest Processing

Nawal Kishore Dubey, Prashant Kumar Mishra, Akash Kedia and Bhanu Prakash

**Abstract**

Herbal drugs have been used since ancient times for prevention and treatment of diseases as well as to promote health and healing. Generally, herbal drugs are considered to be free from side effects but the poor practices of their harvesting, collection, transportation and storage often lead to extensive fungal growth and accumulation of mycotoxins. Fungal and mycotoxin contaminations are the major cause of decline of market value of herbal drug raw materials. Such contamination degrades the quality of raw materials and the medicinal value of the formulated herbal drugs. Synthetic chemical preservatives have been prescribed to control different post-harvest fungal contaminations but due to their residual and mammalian toxicities, herbal pharmaceutical industries need some safer chemicals as preservatives during post-harvest processing of herbal raw materials. Currently, several plant-derived chemicals and their formulations are practically used on a large scale as antimicrobials and are recognized as safer alternatives of synthetic chemicals. Among the higher plant products, plant essential oils, being volatile in nature, may be recommended as botanical fumigants to minimize fungal growth and mycotoxin contamination of herbal drug raw materials. The present chapter deals with an account of fungal and mycotoxin contamination of herbal raw materials and the prospective of plant-derived chemicals as preservatives during post-harvest processing of herbal raw materials.

**Keywords**

Herbal drugs. Mycotoxins · Antimicrobial · Plant based preservatives · Botanical fumigants

N. K. Dubey (✉) · P. K. Mishra · A. Kedia · B. Prakash
Laboratory of Herbal Pesticides, Centre of Advanced Study in Botany, Banaras Hindu University, Varanasi 221005, India
e-mail: nkdubey2@rediffmail.com

R. N. Kharwar et al. (eds.), *Microbial Diversity and Biotechnology in Food Security,*
DOI 10.1007/978-81-322-1801-2_45, © Springer India 2014

## 45.1 Introduction

Herbal drugs have been popular in all cultures and are being used by about 80 % of the world's population especially in vast rural areas for curing their various health ailments and diseases (WHO 2001). Herbal drugs are prepared from different parts of medicinal plants such as roots, stems, barks, leaves, flowers, fruits, seeds or the whole plant, which are called herbal drug raw materials. Herbal medicine is gaining popularity day by day because of its efficacy, safety and lesser side effects. Furthermore, the herbal compounds such as phenolics, alkaloids, quinones, terpenoids, lectins and polypeptides have been found to be very effective alternatives to antibiotics and synthetic drugs (Citarasu 2010). They also offer therapy for cancer and some age-related disorders like memory loss, osteoporosis, immune disorders for which no effective synthetic drugs are currently available (Kamboj 2000).

India is rich in medicinal plant diversity distributed in different parts of the country, from alpine in the Himalayas to tropical hot–humid in the south, arid in the northwest and cold wet in the northeast, and is also rich in tribal and folk knowledge system (Dubey et al. 2004). Out of the 17,000 species of flowering plants, 7,500 plant species have been recognized for their medicinal value. With such wonderful biodiversity of medicinal plants, India has tremendous potential in the field of herbal medicine. About 300 plant species are used for herbal preparation by around 7,800 herbal drug manufacturing units in India which consume about 2,000 tons of herbs annually for the production of around 25,000 effective plant-based formulations (Bhandari et al. 2008; Aneesh et al. 2009).

Out of the ~US $ 900 billion share of global pharmaceutical market, the present global market of herbal medicine is around US $ 62 billion which is estimated to increase up to ~US $ 5 trillion in 2050 (Aneesh et al. 2009). The demand of medicinal plants among people is increasing continuously because of their increasing popularity as an easily available and safe remedy for various diseases in contrast to high prices and adverse side effects of synthetic medicines. Therefore, the countries rich in medicinal plant diversity will receive more revenue in near future following the widening of the market of herbal products in developed countries. India and China are the two largest producers of herbal raw materials; yet, in international markets, India accounts for only 1.61 %share, which is a very small portion of the world trade of medicinal plants, while China holds 30.65 % share of the total global herbal market.

The comparatively small share of India in global herbal market is due to the poorer quality of Indian herbal medicines than that of Chinese medicines and to some extent the poor foreign trade policies of India for export of herbal products. The USA is the largest market for Indian herbal products accounting for about 50 % of total exports of India, whereas the countries like Japan, Hong Kong, Korea, Singapore and even the USA are the major importers of Chinese herbal products which results in a smaller share of India in the global herbal market than that of China (Anon 2003). Microbial contamination of herbal raw materials is one of the major reasons for the very poor share of India in the world market because some foreign herbal pharmaceutical firms have decided to re-evaluate the Indian herbal raw materials to make healthy formulations (Dubey et al. 2008; Sahoo et al. 2010; Aneesh et al. 2009).

Medicinal plants after harvesting and grading need suitable post-harvest processing technologies to maintain their therapeutic value. In warm and humid areas like most parts of India, poor sanitary conditions during harvesting, collection, transportation and post-harvest processing may lead to microbial contamination of the harvested raw materials, and after long-term storage, these microbes especially molds secrete some undesirable toxic substances (mycotoxins)which make raw materials toxic to different human systems (Dubey et al. 2008). India can be a major herbal product exporter in the global herbal market if it adopts control measures to reduce fungal contamination of the raw materials. The present chapter deals with fungal and mycotoxin contamination of herbal raw materials and the prospects of higher plant products during post-harvest treatment.

## 45.2 Fungal Association and Mycotoxin Contamination of Herbal Raw Materials

In recent years, some reports are available on the adverse side effects of herbal drugs although they are generally considered as lesser or no side effects in contrast to synthetic drugs. Such adverse effects of herbal drugs are generally due to the consumption of herbal drug raw materials contaminated with molds and mycotoxins. The undesirable effects are not due to the consumption of herbal drugs but due to the consumption of certain toxic substances secreted by microbes associated with these drugs (Bugno et al. 2006; Mazzanti et al. 2008). A case of food poisoning and growth retardation of infants was found in some areas of South-eastern Africa when mycotoxin-contaminated herbal preparation was given to them for ritual protection (Delgado et al. 2011). The raw materials of herbal drugs have the highest chance of contamination with various fungi due to unscientific storage conditions especially in tropical and subtropical countries like India where high temperature and moisture content is conducive to fungal growth and mycotoxin production (WHO 1993; Roy 2003). Mycotoxin produced by storage fungi causes major health hazards to the consumers of herbal drugs. They affect liver, kidney, nervous system, muscular system, digestive system, genital system and respiratory organs of humans (Efuntoye 1996; Rizzo et al. 2004; Bugno et al 2006; WHO 2007; Dubey et al. 2008; Mishra et al. 2012b). The most common mycotoxins reported on stored herbal drug raw materials are aflatoxins, ochratoxin, citrinin, zearalenone and fumonisin produced by *Aspergillus flavus, Aspergillus ochraceous, Penicillium citrinum* and *Fusarium* spp., respectively (Dubey et al. 2004; WHO 2007; Dubey et al. 2008; Delgado et al. 2011). A large number of reports by researchers on the occurrence of mycoflora on herbal drug raw materials reveal the species of *Aspergillus* and *Penicillium* as the most dominant among the toxigenic molds associated with raw materials (Hitokoto et al. 1978; Roy and Chourasia 1990; Efuntoye 1996; Halt 1998; Abou-Arab et al. 1999; Rizzo et al. 2004; Bugno et al. 2006; Aquino et al. 2007; Singh et al. 2008; Kumar et al. 2009; Sahoo et al. 2010; Gautam and Bhadauria 2011; Mishra et al. 2012b). Aflatoxins are the most prevalent mycotoxin on different stored herbal raw materials (Roy et al. 1988; Roy and Chourasia 1990; Chourasia 1995; Aziz et al. 1998; Efuntoye 1999; Pathanadech et al. 2001; Kumar et al. 2009)and are produced by *A. flavus, A. parasiticus* and few strains of *A. Nomius* (Bhatnagar and García 2001; WHO-IARC 2002). Aflatoxins are highly toxic, mutagenic and carcinogenic in nature and are classified as group 1 human carcinogen by the International Agency for Research on Cancer (WHO-IARC 1993; Razzaghi-Abyaneh et al. 2008; Shukla et al. 2008; Prakash et al. 2010; Tian et al. 2011).

Fungal deterioration adversely affects the chemical composition of raw materials. During interaction with raw materials, the fungi absorb nutrients from raw materials by secreting hydrolytic enzymes, and these fungal enzymes change the complex active compounds of raw materials into simpler useless forms or less efficacious forms, thereby degrading their quality (Roy 2003; Kumar et al. 2009). Moreover, mycotoxin production makes the raw materials toxic due to which the finished herbal product also becomes toxic resulting in threat to human health. Thus, fungal proliferation on raw materials causes both quantitative and qualitative loss of herbal drug raw materials during storage.

## 45.3 Strategies to Minimize Fungal and Mycotoxin Contamination of Herbal Raw Materials

Control of molds and mycotoxin contamination during post-harvest storage of herbal raw materials can be achieved either by employing some physical and chemical methods or by using plant products as preservatives. Physical methods include heat therapy, low temperature storage, radiation treatment and modified atmospheric packaging. Chemical methods include application of some synthetic chemical fumigants like methyl bromide, ethylene dibromane, etc. Physical methods are quite expensive for small-scale storage of raw materials, and hence, after cost benefit analysis, herbal industries find appropriate reasons to reject these methods (Golob 1997; Chauhan and Ghaffar 2002). However, chemical methods have other drawbacks. The undesirable side effects of synthetic chemical preservatives on human health and environment in addition to development of resistance in microorganisms are well known, because of which these agents will be surely unsuitable for herbal raw materials treatment (Webley and Harris 1977; Hatton and Cubbedge 1979; Feng and Zheng 2007). Therefore, there is an urgent need to develop safer, more environment friendly and efficient alternatives to reduce post-harvest losses of herbal raw materials by fungal and mycotoxin contamination. Thus, plant products have got the top position as they fulfil the requirement of ideal preservatives for herbal raw materials.

## 45.4 Higher Plant Products for Post-Harvest Processing of Herbal Raw Materials

The main purpose of post-harvest processing is to increase the efficacy and to preserve the active ingredients of medicinal plants by reducing the microbial and toxic contaminants of raw materials. In addition, sometimes processing may improve the odor or flavor of the raw materials (Zhao et al. 2010; Chang et al. 2011; Zhan et al. 2011). Thus, the use of plant products as preservatives during processing chiefly during storage of herbal raw materials would be a better option in contrast to physical and chemical processing methods. Plants possess a variety of chemicals having the least role in their normal growth and metabolism called secondary metabolites, such as phenolics, flavonoids, quinones, tannins, terpenes, alkaloids, saponins and sterols. Such secondary plant chemicals may have evolved to protect the plants from the attack of microbial pathogens (Benner 1993). Hence, they may be exploited as better preservatives for herbal raw materials due to their reported antimicrobial action. These plant chemicals would be biodegradable, renewable in nature and safe to human health and environment if they are being treated as antimicrobials (Varma and Dubey 1999; Tripathi et al. 2004).

## 45.5 Historical Use of Plant Products as Antimicrobials

Since ancient times, aromatic plants mainly spices and herbs have been recognised as antimicrobials and have been used for the preservation of food and medicine (Conner 1993; Bauer et al. 2001). Later, it has been found that the volatile antimicrobial components of spices and herbs are mainly essential oils (EOs) which are a mixture of monoterpenes and sesquiterpenes, a class of organic compounds. Similarly, some non-volatile botanicals such as phenolics, flavonoids, quinones, alkaloids and saponins also show antimicrobial property and these have been reviewed in the past to have antimicrobial properties. Extracts of various plant parts have been used for pest control from ancient times(Benner 1993).The Bible and other early literature also revealed the description of plant diseases such as rusts, mildews, blights and blast and their control by using extracts and juices of locally available plants (Dubey et al. 2010). Application of leaf juice and seed oil of the Neem tree (*Azadirachta-indica*) as a natural pesticide is the best example of ancient use of botanical antimicrobials (Tripathi et al. 2002). Hence, by remembering their historical use as pesticides, such plant chemicals are virtually guaranteed to have antimicrobial activity, and undoubtedly, they can be used to control post-harvest pathogens of herbal raw materials during storage.

## 45.6 Current Status of Plant-Based Preservatives

Currently, different plant products and their formulations are widely used as botanical pesticides in eco-friendly management of plant pests on a large scale for the protection of crops as well as for storing food commodities and pharmaceuticals. These plant products have low mammalian toxicity and are cost effective. Some plant EO-based food preservatives such as 'DMC Base Natural' (50 % EO from rosemary, sage and citrus and 50 % glycerol) and carvone, a monoterpene of the EO of *Carum carvi* have already been commercially available and are used on a large scale. 'Protecta 1' and 'Protecta 2' are blended herb extracts produced by Bavaria Corp. Apopka, FL, USA and are classed as generally recognized as safe (GRAS) food additives in the USA (Burt 2004). Some plant products such as azadirachtin from *A. indica,* carvone from *C. carvi* and allyl isothiocynate from mustard and horseradish oil are being extensively used as food preservatives (Kumar et al. 2007). Some EO components and a large number of EOs have been reported by different workers to have in vivo antifungal efficacy, offering complete protection of stored grains, fruits, vegetables and herbal raw materials from post-harvest fungal deterioration.

Thymol is the main component of the EO of *Thymus* spp. and was found effective in controlling post-harvest gray mold rot caused by *B. cineria* and brown rot caused by *M. fructicola* (Chu et al. 1999). Thymol is considered as a food as well as food additive by the USA Food and Drug Administration (FDA) and was already registered as a pesticide in the USA since 1964 (Tripathi and Dubey 2004). Carvone is the main component of caraway (*C. carvi*) EO, which exhibited in vivo fungicidal activity protecting potato tubers from fungal rotting without altering their taste and quality. It has been sold as 'TALANT', the trade name in the Netherlands(Tripathi and Dubey 2004). EO components of *Lippiascaberrimawas* have been also found effective in vivo control measure during post-harvest processing of mango fruits (Combrinck et al. 2006).

Dubey et al. (1983) demonstrated the efficacy of the EOs of *Ocimumcanum* and *Citrus medica* as volatile fungi toxicants for the protection of some spices against post-harvest fungal deterioration. The EO of *Cymbopogon citratus* has shown in vivo fumigant activity for the control of storage fungi of some food items (Mishra et al. 1994). The EO of *Salvia officinalis* has shown practical potency in enhancing the storage life of some vegetables by protecting them from fungal rotting (Bang 1995). Varma and Dubey (2001) have determined the in vivo practical effectiveness of *Caesulia axillaris* and *Mentha arvensis* as fumigants for the protection of orange fruits against fungal deterioration. Tripathi and Kumar (2007) have found that the EO from the seed kernels of *Putranjiva roxburghii* offers complete protection of peanut seeds during 6 months of storage without causing any adverse effect on seed germination and seedling growth of peanuts. Tzortzakis (2007) has found a way of enhancing the shelf life of strawberry and tomato fruits by applying eucalyptus and cinnamon EOs. Lee et al. (2007) applied the EOs of two plants *Cuminum cyminum* and *Eucalyptus citriodora* for post-harvest preservation of artificially infested apples with *Botrytis cinerea.* The EO of mustard has shown in vivo fumigant activity for the control of fungal growth in stored shelled groundnuts (Dhingra et al. 2009). Plooy et al. (2009) applied EO-amended coatings on citrus fruits for their post-harvest preservation. During in vivo investigation, Singh et al. (2010a) have found the incidence of fungi and aflatoxin $B_1$ production decreased considerably on the root samples of *Asparagus racemosus* treated by EO of *Citrus reticulate* and *C. citratus.* The EO of *Cicutavirosa* has shown potent fumigant activity for the preservation of cherry tomatoes (Tian et al. 2011). Prakash et al. (2012a) have applied the fumigant antifungal efficacy of the EO of *Zanthoxylum alatum* in vivo study for the preservation of stored *Piper nigrum* fruits. The findings thus indicate the possibility of the exploitation of EO and EO components as effective post-harvest fungi toxicants for the control of storage fungi and mycotoxins and that they may be used on a large scale as plant-based safe preservatives for complete protection of herbal raw materials from fungal and mycotoxin contamination during post-harvest processing.

## 45.7 Future Prospects of Some Recently Explored Higher Plant Products for Preservation of Herbal Raw Materials

Plant chemicals, which are thought to play a major role in the defense system against plant pathogens during pre-harvest attack, may also be effective against post-harvest deteriorating microbes. If plant EOs and extracts having high concentration of antimicrobial constituents are being isolated and applied as antimicrobials, it would be a boon for agro-industries and pharma companies during post-harvest processing technologies and will replace synthetic chemical preservatives which are toxic to humans and the environment. In recent years, a large number of plant extracts, EOs and their components have been investigated for their antifungal properties against storage fungi isolated from food commodities (Sanchez et al. 2005; Joseph et al. 2005; Sharma and Tripathi 2006; Tzortzakis and Economakis 2007; Omidbeygi et al. 2007; Amiri et al. 2008; Zabka et al. 2009; Tatsadjieu et al. 2009; Corato et al. 2010; Combrinck et al. 2011). These plant products can also be applied for the preservation of herbal raw materials. However, complete protection of raw materials can be achieved only when the preservatives have the ability to check fungal growth as well as mycotoxin production. This will provide protection from quantitative as well as qualitative loss of herbal materials by deteriorating fungi.

Plant products selected as preservatives should be easily available and safe so that they can be applied on a large scale without any ill effect on the commodities during storage. For easy availability and the production of sufficient quantity of plant products, emphasis should be given to easily growing wild plants. Since hot and humid condition is more conductive to fungal and mycotoxin contamination on herbal raw materials, it would be beneficial to select naturally growing plants of hot and humid areas which contain antimicrobial components. Some recent literature show that several EOs isolated from higher plants easily grown in warmer areas have been found to possess antifungal as well as mycotoxin inhibitory potential (Kumar et al. 2007; Kumar et al. 2008; Shukla et al. 2009; Singh et al. 2010b; Kumar et al. 2010; Prakash et al. 2010; Prakash et al. 2011; Shukla et al. 2012; Prakash et al. 2012a; Prakash et al. 2012b). Since most of the deteriorating fungi are common for both food commodities and herbal raw samples, the plant products recommended for the preservation of food commodities may also be recommended for the preservation of herbal raw materials.

In addition, the plant products which are used to control fungi deteriorating herbal raw materials would be more beneficial for post-harvest preservation of herbal raw materials. By virtue of this, some plants which have been easily cultivated and naturally growing in tropical and subtropical parts of India have been selected by different workers for isolation of extracts and EOs and are applied as antifungal against fungi deteriorating herbal raw materials for their possible recommendation to preserve the raw materials by fumigation.

*Cinnamomum camphora* is a tree, native to China, Japan and Taiwan, which has been introduced to other countries such as India and Sri Lanka. The EO of the plant has been recorded for antifungal and antiaflatoxigenic efficacy against herbal raw material-deteriorating fungi and is recommended for protection of herbal raw materials without the possibility of change in the original quality of raw materials during storage as the plant is used in traditional medicines (Singh et al. 2008).The leaf extract of *Adenocalymma alliaceum* has shown antifungal and antiaflatoxigenic potency against fungi deterioration of some herbal raw materials, viz. *Acorus calamus, Boerhavia diffusa, Rauwolfia serpentina, Withania somnifera*. The plant is commonly known as garlic creeper, and has been traditionally used as an analgesic, antiarthritic, anti-inflammatory, antipyretic, antirheumatic, antitussive, depurative, purgative and vermifuge (Shukla et al. 2008). Due to easy cultivation, growth in hot and humid areas in India and traditional medicinal value, *A. alliaceum* leaf extract may be used to preserve herbal raw materials by spraying or dipping methods. *Cymbopogon flexuosus* (Cochin grass or Malabar grass) is a perennial grass naturally growing in India, Sri Lanka, Burma and Thailand. The EO of the grasses has been found to be inhibitory against fungi deteriorating herbal raw materials as well as aflatoxin produced by toxigenic *A. flavus* (Kumar et al. 2009).

*C. axillaris*, a common weed of paddy fields growing luxuriantly in central and north-central part of India, is selected for antifungal and antiaflatoxigenic activity. The plant has sufficient amount of EO which have the potential to control a large number of fungi isolated from herbal raw materials of *Andrographis paniculata, Terminalia bellirica* and *Tinospora cordifolia* (Mishra et al.

2012a). The EO of another plant, Jamrosa (*Cymbopogon khasans*), a hybrid grass widely grown in central and southern parts of India has shown very good antifungal activity against fungi isolated from five herbal raw materials, viz. *A. racemosus, Evolvulus alsinoides, Glycyrrhiza glabra, Holarrhena antidysenterica* and *R. serpentina* (Mishra et al. 2012b). The components of Jamrosa EO, Z-citral and linalyl acetate have been also found effective as antifungal and antiaflatoxigenic (Mishra et al. 2012b).Thus, because of significant antifungal and antimycotoxigenic efficacy and a broad spectrum of fungi toxicity against herbal raw material-deteriorating fungi, plant extracts, EOs and EO components after in vivo trial may be recommended as indigenous plant-based preservative for post-harvest preservation of herbal raw materials in order to enhance their shelf life.

## 45.8 Conclusion

Post-harvest fungal deterioration and mycotoxin accumulation lead to loss of quantity as well as quality of herbal raw materials. During infestation, fungal enzymes change the active principle of raw materials into inert or less efficacious compounds and mycotoxin secretion makes them toxic due to which they become unfit for human use for curing their diseases. During post-harvest processing of medicinal plants, fungal and mycotoxin contamination on herbal raw materials may occur at any stage during collection, transportation and storage; however, the highest chance of contamination occurs during storage. This problem is most common in tropical and subtropical countries due to hot and humid conditions which are more conducive for proliferation of different fungi. The fungal contamination of Indian herbal raw materials is the main reason of their declining market value. Due to this reason, some foreign pharmaceutical firms have rejected Indian herbal products as well as raw materials. Different plant products and their formulations are widely used as botanical fungicides for preservation of food items from food-spoiling microbes and can also be used to preserve stored herbal raw materials because most of the storage fungi are common in both cases. Some recently explored higher plant products having potent fungicidal and mycotoxin inhibitory capacity during in vitro conditions can also be recommended for post-harvest processing of herbal raw materials after large-scale in vivo trials and dose determination for complete protection from deteriorating fungi and mycotoxins as well as after their safety profile assessment.

## References


Abou-Arab AAK, Kawther MS, Tantawy MEE, Badeaa RI, Khayria N (1999) Quantity estimation of some contaminants in commonly used medicinal plants in the Egyptian market. Food Chem 67:357–363

Amiri A, Dugas R, Pichot AL, Bompeix G (2008) In vitro and in vitro activity of eugenol oil (*Eugenia caryophylata*) against four important postharvest apple pathogens. Int J Food Microbiol 126:13–19

Aneesh TP, Hisham M, Sekhar MS, Madhu M, Deepa TV (2009) International market scenario of traditional Indian herbal drugs—India declining. Int J Green Pharm 3:184–190

Anon (2003) A dream of globalizing the TCM market. Report of the Ministry of Commerce of the People's Republic of China

Aquino S, Gonçalez E, Reis TA, Araújo1 MM, Corrêa B, Villavicencio ALCH (2007) Evaluation of fungal burden of medicinal plants submitted to gamma radiation process after 30 days. International Nuclear Atlantic Conference—INAC 2007 Santos, SP, Brazil, September 30 to October 5. AssociaçãoBrasileira de Energia Nuclear-ABEN ISBN: 978-85-99141-02-1

Aziz NH, Youssef YA, El-FoulyMZ, Moussa LA (1998) Contamination of some common medicinal plant samples and spices by fungi and their mycotoxins. Bot Bull Acad Sinica 39:279–285

Bang U (1995) Essential oils as fungicides and sprout inhibitors in potatoes. In: Proceedings of the EAPR Pathology Section Meeting, Phytophthorainfestancs ISO, Dublin, 10–16 September

Bauer K, Garbe D, Surburg H (2001) Common fragrance and flavor materials: preparation, properties and uses. Wiley-VCH, Weinheim, pp 293

Benner JP (1993) Pesticidal compounds from higher plants. Pest Sci 39:95–102

Bhandari S, Dobhal U, Sajwan M, Bisht NS (2008) *Trichosanthestricuspidata*: a medicinally important plant. Trees Life 3:1–5

Bhatnagar D, Garcia S (2001) Aspergillus. In: Labbe RG, Garcia S (eds) Guide to foodborne pathogens. Wiley, New York, pp 35–49

Bugno A, Almodovar AAB, Pereira TC, Pinto TJA, Sabino M (2006) Occurrence of toxigenic fungi in herbal drugs. Braz J Microbiol 37:47–51

Burt S (2004) Essential oils: their antibacterial properties and potential applications in foods—a review. Int J Food Microbiol 94:223–253
Chang WT, Choi YH, VanderHeij- denR, Lee MS, Lin MK, Kong H, Kim K, Verpoorte R, Hanke- meierT, VanderGreef J, Wang M (2011) Traditional processing strongly affects metabolite composition by hydrolysis in *Rehmanniaglutinosa* roots. Chem Pharm Bull 59:546–552
Chauhan YS, Ghaffar MA (2002) Solar heating of seeds a low cost method to control bruchid (*Callosobruchus* spp.) attack during storage of pigeon pea. J Stored Prod Res 38:87–91
Chourasia HK (1995) Mycobiota and mycotoxins in herbal drugs of Indian pharmaceutical industries. Mycol Res 99:697–703
Chu CL, Liu WT, Zhou T, Tsao R (1999) Control of post harvest gray mold rot of modified atmosphere packaged sweet cherries by fumigation with thymol and acetic acid. Can J Plant Sci 79:685–689
Citarasu T (2010) Herbal biomedicines: a new opportunity for aquaculture industry. Aquacult Int 18:403–414
Combrinck S, Bosman AA, Botha DPlooyW, McCrindle RI, Retief E (2006) Effects of post-harvest drying on the essential oil and glandular trichomes of Lippiascaberrima Sond. J Essential Oil Res 18:80–84
Combrinck S, Regnier T, Kamatou GPP (2011) In vitro activity of eighteen essential oils and some major components against common postharvest fungal pathogens of fruit. Ind Crops Prod 33:344–349
Conner DE (1993) Naturally occurring compounds. In: Davidson PM, Branen AL (eds) Antimicrobials in Food. Marcel Dekker, New York, pp 441–468
Corato UD, Maccioni O, Trupo M, Sanzo GD (2010) Use of essential oil of *Laurusnobilis* obtained by means of a supercritical carbon dioxide technique against post harvest spoilage fungi. Crop Prot 29:142–147
Delgado S, Nunez F, Sanchez B, Bermudez E, Rodriguez JM (2011) Toxigenic microorganisms in medicinal plants used for ritual protection of infants. Food Res Int 44:304–309
Dhingra OD, Jham GN, Rodrigues FA, Silva GJ, Costa MLN (2009) Fumigation with essential oil of mustard retards fungal growth and accumulation of ergosterol and free fatty acid in stored shelled groundnuts. J Stored Prod Res 45:24–31
Dubey NK, Bhargava KS, Dixit SN (1983) Protection of some stored food commodities from fungi by essential oils of *Ocimumcanum*and *Citrus medica*. Int J Trop Plant Dis 1:177–179
Dubey NK, Kumar R, Tripathi P (2004) Global promotion of herbal medicine: India's opportunity. Curr Sci 86:37–41
Dubey NK, Kumar A, Singh P, Shukla R (2008) Microbial contamination of raw materials: a major reason for the decline of India's share in the global herbal market. Curr Sci 95:717–718
Dubey NK, Shukla R, Kumar A, Singh P, Prakash B (2010) Prospects of botanical pesticides in sustainable agriculture. Curr Sci 98(4):479–480
Efuntoye MO (1996) Fungi associated with herbal drug plants during storage. Mycopathologia 136:115–118
Efuntoye MO (1999) Mycotoxins of fungal strains from stored herbal plants and mycotoxin contents of Nigerian crude herbal drugs. Mycopathologia 147:43–48
Feng W, Zheng X (2007) Essential oils to control *Alternaria alternata* in vitro and in vivo. Food Control 18:1126–1130
Gautam AK, Bhadauria R (2011) Diversity of fungi and mycotoxins associated with stored Triphala churn and its ingredients. J BiolSci 11:226–235
Golob P (1997) Current status and future perspectives for inert dusts for control of stored product insects. J Stored Prod Res 33:69–79
Halt M (1998) Moulds and mycotoxins in herb tea and medicinal plants. Eur J Epidemiol 14:269–274
Hatton T, Cubbedge RH (1979) Phytotoxicity of methyl bromide as a fumigant for Florida citrus fruit. Proc Fla State Hortc Soc 92:167–169
Hitokoto H, Morazumi S, Wauke T, Sakai S, Kurata H (1978) Fungal contamination and mycotoxin detection of powdered herbal drugs. Appl Environ Microbiol 36(2):252–256
IARC Working Group (1993) International Agency for Research on Cancer (IARC). Some naturally occurring substances: food items and constituents. IARC monographs on the evaluation of carcinogenic risk to humans. World Health Organization, Lyon
IARC Working Group (2002) International Agency for Research on Cancer. Some traditional herbal medicines, some mycotoxins, Nnaphthalene and styrene IARC monographs on the evaluation of carcinogenic risks to humans World Health Organization, vol. 82, Lyon
Joseph GS, Jayaprakasha GK, Selvi AT, Jena BS, Sakariah KK (2005) Antiaflatoxigenic and antioxidant activities of *Garcinia extracts*. Int J Food Microbiol 101:153–160
Kamboj VP (2000) Herbal medicine. Curr Sci 78(1):35–39
Kumar R, Mishra AK, DubeyNK TripathiYB (2007) Evaluation of *Chenopodium ambrosioides* oil as a potential source of antifungal, antiaflatoxigenic and antioxidant activity. Int J Food Microbiol 115:159–164
Kumar A, Shukla R, Singh P, Prasad CS, Dubey NK (2008) Assessment of *Thymus vulgaris* L. essential oil as a safe botanical preservative against post harvest fungal infestation of food commodities. Innov Food Sci Emerging Technol 9:575–580
Kumar A, Shukla R, Singh P, Dubey NK (2009) Biodeterioration of some herbal raw materials by storage fungi and aflatoxin and assessment of *Cymbopogonflexuosus* essential oil and its components as antifungal. Int Biodeter Biodegrad 63:712–716
Kumar A, Shukla R, Singh P, Dubey NK (2010) Chemical composition, antifungal and antiaflatoxigenic activities of *Ocimum sanctum* L. essential oil and its safety assessment as plant based antimicrobial. Food Chem Toxicol 48:539–543
Lee SO, Choi GJ, Jang KS, Lim HK, Cho KY, Kim J (2007) Antifungal activity of five plant essential oils

as fumigant against postharvest and soilborne plant pathogenic fungi. Pt Pathol J 23:97–102
Mazzanti G, Battinelli L, Daniele C, Costantini S, Ciaralli L, Evandri MG (2008) Purity control of some Chinese crude herbal drugs marketed in Italy. Food Chem Toxicol 46:3043–3047
Mishra AK, Dubey NK (1994) Evaluation of some essential oils for their toxicity against fungi causing deterioration of stored food commodities. Appl Environ Microbiol 60:1101–1105
Mishra PK, Shukla R, Singh P, Prakash B, Dubey NK (2012a) Antifungal and antiaflatoxigenic efficacy of *Caesuliaaxillaris* Roxb. essential oil against fungi deteriorating some herbal raw materials, and its antioxidant activity. Ind Crops Prod 36:74–80
Mishra PK, Shukla R, Singh P, Prakash B, Kedia A, Dubey NK (2012b) Antifungal, anti-aflatoxigenic, and antioxidant efficacy of Jamrosa essential oil for preservation of herbal raw materials. Int Biodeter Biodegrad 74:11–16
Omidbeygi M, Barzegar M, Hamidi Z, Naghdibadi H (2007) Antifungal activity of thyme, summer savory and clove essential oils against *Aspergillus flavus* in liquid medium and tomato paste. Food Control 18:1518–1523
Pathanadech A, Petcharat V, Chuenchit S, Lim S, Chinaputi A (2001) Aflatoxin $B_1$-producing *Aspergillus* in sun-dried medicinal plant materials. Songklanakarin J Sci Technol 23:499–514
Plooy W, Regnier T, Combrinck S (2009) Essential oil amended coatings as alternatives to synthetic fungicides in citrus postharvest management. Postharvest BiolTechnol 53:117–122
Prakash B, Shukla R, Singh P, Kumar A, Mishra PK, Dubey NK (2010) Efficacy of chemically characterized Piper betle L. essential oil against fungal and aflatoxin contamination of some edible commodities and its antioxidant activity. Int J Food Microbiol 142:114–119
Prakash B, Shukla R, Singh P, Mishra PK, Dubey NK, Kharwar RN (2011) Efficacy of chemically characterized *Ocimumgratissimum* L. essential oil as an antioxidant and a safe plant based antimicrobial against fungal and aflatoxin $B_1$ contamination of spices. Food Res Int 44:385–390
Prakash B, Singh P, Kedia A, Dubey NK (2012b) Assessment of some essential oils as food preservatives based on antifungal, antiaflatoxin, antioxidant activities and in vivo efficacy in food system. Food Res Int 49:201–208
Prakash B, Singh P, Mishra PK, Dubey NK (2012a) Safety assessment of *Zanthoxylumalatum* Roxb. essential oil, its antifungal, antiaflatoxin, antioxidant activity and efficacy as antimicrobial in preservation of *Piper nigrum* L. Fruits. Int J Food Microbiol 153:183–191
Razzaghi-Abyaneh M, Shams-Ghahfarokhi M, Yoshinari T, Rezaee MB, Jaimand K, Nagasawa H, Sakuda S (2008) Inhibitory effects of Saturejahortensis L. essential oil on growth and aflatoxin production by *Aspergillus parasiticus*. Int J Food Microbiol 123:228–233
Rizzo I, Vedoya G, Maurutto S, Haidukowski M, Varsavsky E (2004) Assessment of toxigenic fungi on Argentinean medicinal herbs. Microbiol Res 159:113–120
Roy AK (2003) Mycological problems of crude herbal drugs—overview and challenges. Indian Phytopathol 56:1–13
Roy AK, Chourasia HK (1990) Mycoflora, mycotoxinproducibility and mycotoxins in traditional herbal drugs from India. J General Appl Microbiol 36:295–302
Roy AK, Sinha KK, Chourasia HK (1988) Aflatoxin contamination of some common drug plants. Appl Environmental Microbiol 54:842–843
Sa'nchez E, Heredia N, Garcı'a S (2005) Inhibition of growth and mycotoxin production of *Aspergillus flavus* and *Aspergillus parasiticus* by extracts of *Agave* species. Int J Food Microbiol 98:271–279
Sahoo N, Manchikanti P, Dey S (2010) Herbal drugs: standards and regulation. Fitoterapia 81:462–471
Sharma N, Tripathi A (2006) Fungitoxicity of the essential oil of *Citrus sinensis* on post-harvest pathogens. World J Microbiol Biotechnol 22:587–593
Shukla R, Kumar A, Prasad CS, Srivastava B, Dubey NK (2008) Antimycotic and antiaflatoxigenic potency of *AdenocalymmaalliaceumMiers.* on fungi causing biodeterioration of food commodities and raw herbal drugs. Int Biodeter Biodegrad 62:348–351
Shukla R, Kumar A, Singh P, Dubey NK (2009) Efficacy of*Lippia alba* (Mill.) N.E. Brown essential oil and its monoterpene aldehyde constituents against fungi isolated from some edible legume seeds and aflatoxin $B_1$ production. Int J Food Microbiol 135:165–170
Shukla R, Singh P, Prakash B, Dubey NK (2012) Antifungal, aflatoxin inhibition and antioxidant activity of *Callistemon lanceolatus*(Sm.) Sweet essential oil and its major component 1,8-cineole against fungal isolates from chickpea seeds. Food Control 25:27–33
Singh P, Srivastava B, Kumar A, Dubey NK (2008) Fungal contamination of raw materials of some herbal drugs and recommendation of *Cinnamomum camphora* oil as herbal fungitoxicant. Microb Ecol 56:555–560
Singh P, Shukla R, Kumar A, Prakash B, Singh S, Dubey NK (2010a) Effect of *Citrus reticulata* and *Cymbopogoncitratus* essential oils on *Aspergillus flavus* growth and aflatoxin production on *Asparagus racemosus*. Mycopathologia 170:195–202
Singh P, Shukla R, Prakash B, Kumar A, Singh S, Mishra PK, Dubey NK (2010b) Chemical profile, antifungal, antiaflatoxigenic and antioxidant activity of *Citrus maxima* burm. and*Citrus sinensis* (L.) osbeck essential oils and their cyclic monoterpene, DL-limonene. Food Chem Toxicol 48:1734–1740
Tatsadjieu NL, Dongmo PMJ, Ngassoum MB, Etoa FX, Mbofung CMF (2009) Investigations on the essential oil of *Lippiarugosa* from Cameroon for its potential use as antifungal agent against *Aspergillus flavus* Link ex. Fries. Food Control 20:161–166
Tian J, Ban X, Zeng H, He J, Huang B, Wang Y (2011) Chemical composition and antifungal activity of essential oil from *Cicutavirosa* L. var. *latisectaCelak.* Int J Food Microbiol 145:464–470

Tripathi P, Dubey NK (2004) Exploitation of natural products as an alternative strategy to control postharvest fungal rotting of fruit and vegetables. Postharvest BiolTechnol 32:235–245

Tripathi NN, Kumar N (2007) *Putranjivaroxburghii* oil-A potential herbal preservative for peanuts during storage. J Stored Prod Res 43:435–442

Tripathi P, Dubey NK, Pandey VB (2002) Kaempferol: the antifungal principle of *Acacia nilotica*Linn. Del J Indian Bot Soc 81:51–54

Tripathi P, Dubey NK, Banerji R, Chansouria JPN (2004) Evaluation of some essential oils as botanical fungitoxicants in management of post-harvest rotting of citrus fruits. World J Microbiol Biotechnol 20:317–321

Tzortzakis NG (2007) Maintaining postharvest quality of fresh produce with volatile compounds. Innov Food Sci Emerging Technol 8:111–116

Tzortzakis NG, Economakis CD (2007) Antifungal activity of lemongrass (*Cympopogoncitratus* L.) essential oil against key postharvest pathogens. Innov Food Sci Emerging Technol 8:253–258

Varma J, Dubey NK (1999) Prospectives of botanical and microbial products as pesticides of tomorrow. CurrSci 76:172–179

Varma J, Dubey NK (2001) Efficacy of essential oils of *Caesuliaaxillaris* and *Menthaarvensis*against some storage pests causing biodeterioration of food commodities. Int J Food Microbiol 68:207–210

Webley DJ, Harris AH (1977) A comparison of fumigants for in-bag fumigation. Trop Stored Prod Inf 33:9–18

WHO (1993) Regional office for the Western Pacific, research guidelines for evaluating the safety and efficacy of herbal medicines. World Health Organisation, Manila

World Health Organization (2001) General guidelines for methodologies on research and evaluation of traditional medicine. WHO, Geneva

WHO (2007) WHO guidelines for assessing quality of herbal medicines with reference to contaminants and residues, WHO Library Cataloguing-in-Publication Data, Spain pp 14–15

Zabka M, Pavela R, Slezakova L (2009) Antifungal effect of *Pimentadioica* essential oil against dangerous pathogenic and toxinogenic fungi. Ind Crops Prod 30:250–253

Zhan JY, Zheng KY, Zhu KY, Bi CW, Zhang WL, Du CY, Fu Q, Dong TT, Choi RC, Tsim KW, Lau DT (2011) Chemical and biological assessment of *Angelica sinensis* Radix after processing with wine: an orthogonal array design to reveal the optimized conditions. J Agric Food Chem 59:6091–6098

Zhao Z, Liang Z, Chan K, Lu G, Lee EL, Chen H, Li L (2010) A unique issue in the standardization of Chinese materia medica. Planta Med 76:1975–1986

# Isolation of Nonpathogenic Strain of Ballistosporous Yeast *Sporobolomyces salmonicolor* from House Mouse *Mus musculus* (Rodentia: Muridae)

46

Jyoti Rani and Karuna Singh

**Abstract**

This chapter deals with the isolation of nonpathogenic strain of ballistosporous yeast *Sporobolomyces salmonicolor* from the liver of a house mouse *Mus musculus* and its epidemiological significance. Micromorphology of the yeast showed characteristics pseudohyphae. Isolated *S. salmonicolor* was found to assimilate glucose, galactose, sucrose, maltose, and lactose sugars. The pathogenicity of the isolated strain of *S. salmonicolor* was also tested on Swiss mice. Experimental animals were divided into immunocompetant and immunocompromised groups. Infected animals of both groups exhibited neither behavioral nor physical or histopathological changes with respect to the control group. The present study suggests that this nonpathogenic strain of *S. salmonicolor* found in the liver of *M. musculus* was either present as a commensal or the animal served as the reservoir or passenger of this red yeast and may have epidemiological significance.

**Keywords**

*Sporobolomyces salmonicolor* · Non-pathogenic · Liver · *Mus musculus*

## 46.1 Introduction

*Sporobolomyces* is a unicellular free-living "red" yeast belonging to the class Urediniomycetes and family sporobolomycetaceae. The family Sporobolomycetaceae has two genera: *Sporobolomyees* and *Bullera,* which are distinguished from other yeasts by their reproductive method of ballistoconidia formation. *Sporobolomyees* produce visible carotenoid pigments, resulting in pink to orangeo color of colonies with asymmetric ballistospores. These conidia are produced on

K. Singh (✉) · J. Rani
Department of Zoology, Mahila Mahavidyalaya,
Banaras Hindu University, Varanasi 221005, India
e-mail: karuna@bhu.ac.in; karunasingh5@gmail.com

R. N. Kharwar et al. (eds.), *Microbial Diversity and Biotechnology in Food Security,*
DOI 10.1007/978-81-322-1801-2_46, © Springer India 2014

simple or branched denticles that arise from yeast and mycelial cells. At maturity, the ballistoconidia are discharged like small missiles through a water drop excretion method and form satellite colonies (Fell and Tallman 1984). At present, the core group of *Sporobolomyces*, i.e., the species belonging to the Sporidiobolales, encompasses 15 taxa (Sampaio 2001) comprising known species, which are commonly found in air, soil, plants, and other decaying organic material. Review of literature revealed that to date there have been six cases of documented *Sporobolomyces* infections: a nasal polyp (Dunnette et al. 1986), one case of dermatitis (Bergman and Kauffman 1984), one case of endophthaimitis (Rantala and Vaahtoranta-Lehtonen 1995), and two cases of disseminated infection (lymph node and bone marrow) in patients with AIDS (Plazas et al. 1994). Pathogenicity of *Sporobolomyces* spp. was also reported by many authors (Anaissie et al. 2003; Dismukes et al. 2003; Morris et al. 1991; Nakase and Suzuki 1987a, 1987b; Morrow 1994).

This chapter describes the isolation of non-pathogenic strain of ballistosporous yeast *Sporobolomyces salmonicolor* from the liver of a house mouse *Mus musculus* and its epidemiological significance.

## 46.2 Material and Methods

### 46.2.1 Isolation and Identification

*M. musculus* (8.75 g; Fig. 46.1) was collected from BHU Campus, Varanasi. The animal was killed. All the vital organs (brain, liver, kidney, and lung) were removed and homogenized. Plating was done on Sabouraud's dextrose agar (SDA) plates and the plates were kept in a BOD incubator at 28 °C for 3 days. For the identification, the colonies were subcultured on corn meal agar and potato glucose agar to determine ballistoconidia formation. Isolated yeast was tested for urease production and assimilation tests were carried out in liquid media and read visually according to the procedure of Wickerham (1951).

**Fig. 46.1** House mouse (*Mus musculus*)

### 46.2.2 Experimental Animal

For animal pathogenicity of isolated culture of *S. salmonicolor*, Swiss male mice weighing approximately 26 g were selected and fed on a pellet diet and water ad libitum. The mice were divided into two groups: control and experimental. Each group was divided further into two subgroups: immunocompetent and immunocompromised of three mice each. For immunosuppression, the animals were injected intraperitoneally (IP) with 125 mg $kg^{-1}$ body weight of aqueous cortisone acetate in three divided doses on alternate days (Naidu and Singh 1992).

### 46.2.3 Inoculum Preparation

Spore suspension was prepared from 7-day-old culture of the *S. salmonicolor* (Fig. 46.2) isolated from the liver of *M. musculus* by transferring two to three loopful of culture in 100 ml normal saline containing 0.05 mg $ml^{-1}$ chloramphenicol and filtered through a sterilized muslin cloth. The size of the test inoculum containing predominantly spores was adjusted approximately to $5.2 \times 10^6$ cfu $ml^{-1}$ with an hemocytometer.

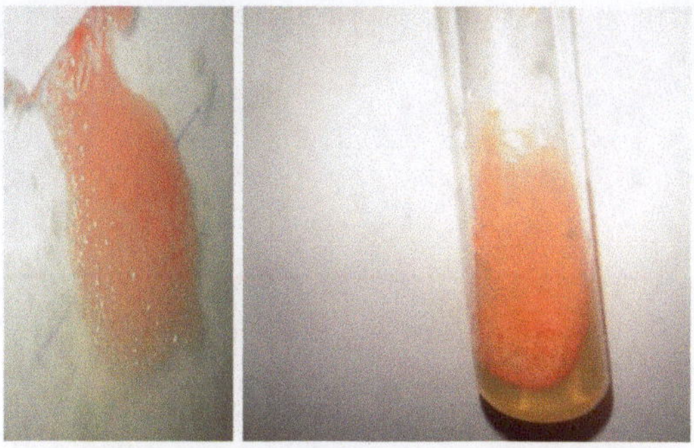

**Fig. 46.2** Culture of red yeast (*Sporobolomyces salmonicolor*)

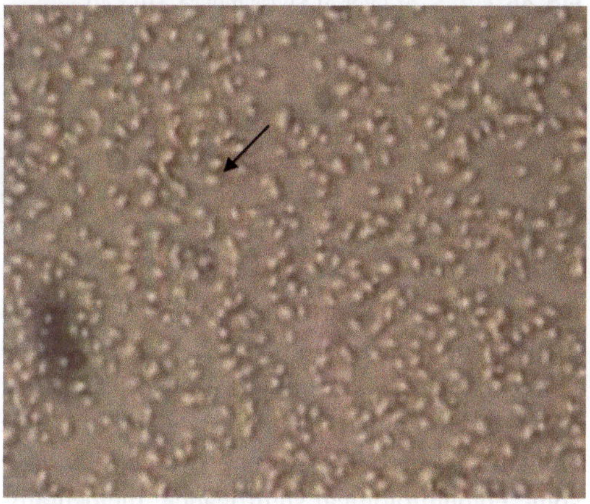

**Fig. 46.3** Micromorphology of *S. salmonicolor* showing pseudohyphae. Lacto × 40

### 46.2.4 Animal Pathogenicity

Twenty five hours after immunosuppression, the control mice were challenged with 0.5 ml of normal saline with 0.05 mg $ml^{-1}$ of chloramphenicol and the experimental group of animals were challenged IP with 0.5 ml of spore suspension. The animals were killed and autopsied after 5 weeks of inoculation. The aseptically removed liver, lung, brain, and kidney from each animal were homogenized and inoculated on SDA plates. Parts of the organs were also examined histopathologically using hematoxylin and eosin (HE) staining (Kwon-chung and Bennet 1992).

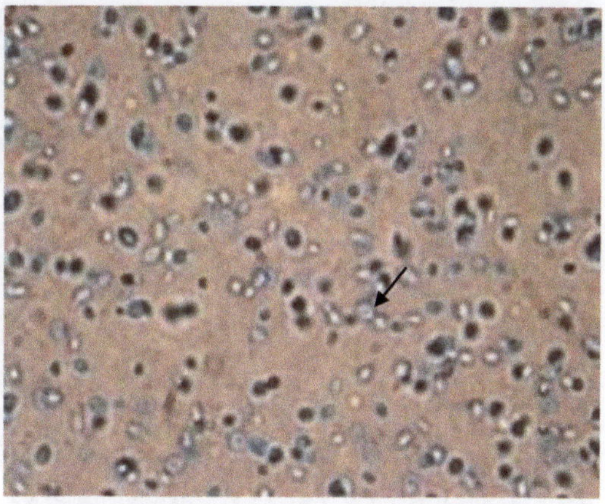

**Fig. 46.4** Micromorphology of *S. salmonicolor* showing ballistoconidia. Cotton blue × 100

## 46.3 Result and Discussion

A salmon-colored mucoid colony with a flat and fringe margin appeared on the SDA plates after 3 days of incubation. Ovoid to ellipsoid ballistoconidia, the formation of which was determined at 22 °C (Fig. 46.4), formation of pseudohyphae (Fig. 46.3) and the inability of the yeast to assimilate starch-like compounds suggested it to be a red yeast *S. salmonicolor*, which was isolated from the liver of *M. musculus*. Unlike the variety of *S. salmonicolor* isolated by Mishra and Randhawa (1976), this strain of *S. salmonicolor* was found to assimilate maltose and lactose (Table 46.1).

*Sporobolomyces* is often associated with other plant parasites and can be found in plant lesions as well as in trees and other wood substrates. As far as the isolation from veterinary samples is concerned, Satosh and Makimura (2008) isolated a new species *Sporobolomyces koalae* from

**Table 46.1** Showing result of carbon assimilation tests of isolated strain of *S. salmonicolor*

| S. No. | Carbon compounds | Assimilation test |
|---|---|---|
| 1 | Glucose | Positive |
| 2 | Lactose | Positive |
| 3 | Galactose | Positive |
| 4 | Maltose | Positive |
| 5 | Sucrose | Positive |
| 6 | Starch | Negative |

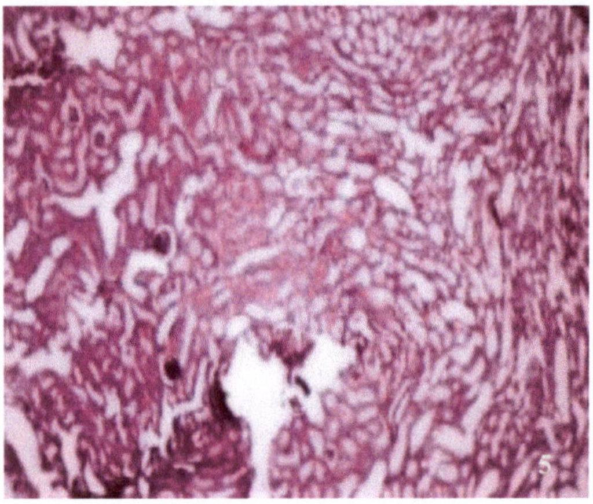

**Fig. 46.5** T. S. of lung of Swiss mouse. HE × 40

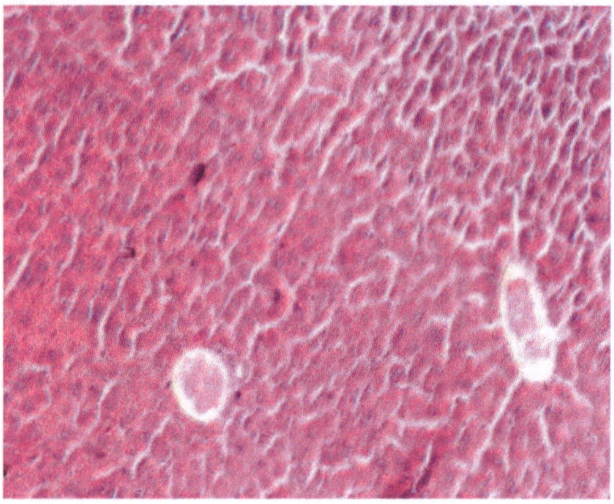

**Fig. 46.7** T. S. of liver of Swiss mouse. HE × 40

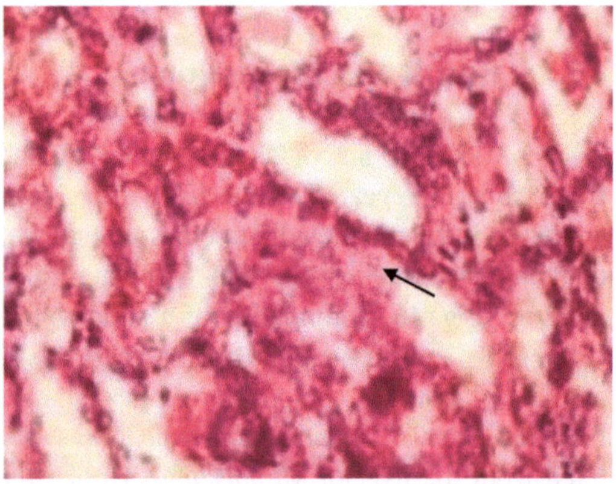

**Fig. 46.6** T. S. of lung of Swiss mouse showing alveolar macrophages. HE × 100

nasal smears of Queensland Koalas kept in the Japanese zoological park. However, the animal pathogenicity of *S. koalae* was not determined by the authors.

*Sporobolomyces* is also found as an environmental saprophyte and as a human commensal (Huxley and Hurd 1956). Three species of *Sporobolomyces,* viz. *roseus, holsaticus,* and *salmonicolor* have been reported to be pathogenic (Connell et al. 1953, 1954; Morrow and Fraser 2009). There are few cases of sporobolomycoses including reports of maduromycosis, dermatitis, fungemia, and lymphadentitis in both immunocompetent and AIDS patients (Plazas et al. 1994; Sharma et al. 2006).

*S. salmonicolor* was first time reported as a fish pathogen by Muench et al. in 1996.They isolated this red yeast from kidney swabs of heavily infected fries of Chinook salmon *Oncorhynchus tschawytscha.* In the present study, we also tested the pathogenicity of the isolated strain of *S. salmonicolor* on Swiss male mice. Though in the present study no behavioral, physical, and histopathological abnormalities were recorded in the experimental animals (Figs. 46.5, 46.6, 46.7), this study exhibits that wild and domestic animals can harbor such fungi which may have some epidemiological significance. Incidentally, to our knowledge, this is the first report of isolation of *S. salmonicolor* from the liver of house mouse *M. musculus.*

## 46.4 Conclusion

On the basis of the present study, we may conclude that the isolated strain of red yeast *S. salmonicolor,* which is nonpathogenic to both immunocompetent and immunocompromised hosts could be a commensal or a passenger. However, further work is required in this direction.

**Acknowledgments** Authors are thankful to the Principal, Mahila Mahavidyalaya, Banaras Hindu University, Varanasi, for providing facilities to conduct the experiments. One of the authors, Ms Jyoti Rani, is thankful to the University Grants Commission, New Delhi, for providing RG-SRF.

## References


Anaissie E, McGinnis M, Pfaller M (2003) Clinical mycology. Churchill Livingstone, New York

Bergman A, Kauffman C (1984) Dermatitis due to *Sporobolomyces* infection. Arch Dermatol 120:1059–1060

Connell G, Skinner C (1953) The external surface of the human body as a habitat for non fermenting non pigmented yeasts. J Bacteriol 66:627–633

Connell G, Skinner C, Hurd R (1954) Lipomycesstarkeyi on the skin surface of the human body. Mycologia 46:12–15

Dismukes W, Pappas P, Sobel J (2003) Clinical mycology. Oxford University Press, Oxford

Dunnette S, Hall M, Washington J (1986) Microbiologic analyses of nasal polyp tissue. J Allergy Clin Immunol 78:102–108

Fell J, Tallman A (1984) *Sporobolomyces* Kluyver et van Niel. In: Kreger-van Rij NJW (ed) The yeasts: a taxonomic study. 3rd ed. Amsterdam: Elsevier.911–914.

Huxley M, Hurd R (1956) Pink yeasts isolated from human skin surfaces. J Bact 71:492

Kwon-Chung K, Bennet J (1992) Medical mycology. Lea & Febiger, Philadelphia

Mishra V, Randhawa H (1976) *Sporobolomyces salmonicolor* var. fischerii, a new yeast. Arch Microbiol 108:141–143

Morris J, Beckius M, McAllister C (1991) *Sporobolomyces* infection in an AIDS patient. J Infect Dis 4:164–623

Morrow CA, Fraser JA (2009) Sexual reproduction and dimorphism in the pathogenic basidiomycetes. FEMS Yeast Res 9:161–177

Morrow J (1994) Prosthetic cranioplasty infection due to *Sporobolomyces*. J Tenn Med Assoc 87:466–467

Muench T, White M, Wu C (1996) Visceral mycosis in chinook salmon (*Oncorhynchus Tschawytscha*) due to *Sporobolomyces salmonicolor*. Vet Pathol 33:238–241

Naidu J, Singh SM (1992) Hyalohyphomycosis caused by *Paecilomycesvariotii*: a case report, animal pathogenicity and in vitro sensitivity. Antoine Van Leeuwenhoek 62:225–230

Nakase T, Suzuki M (1987a) *Sporobolomycesinositophilus*, a new species of ballistosporous yeast isolated from a dead leaf of *Sasa* sp. in Japan. Antonie Leeuwenhoek 53:245–251

Nakase T, Suzuki M (1987b) *Sporobolomycesnaganoensis*, a new species of ballistosporous yeast equipped with ubiquinone-9 isolated from a dead leaf of Sasa sp. in Japan. Trans Mycol Soc Jpn 28:1–8

Plazas J, Portilla J, Boix V, Perez-Mateo M (1994) *Sporobolomyces salmonicolor* lymphadenitis in an AIDS patient. Pathogen or passenger? AIDS 8:387

Rantala A, Vaahtoranta-Lehtonen H (1995) Hematogenous endophthalmitis in patients with postoperative septicemia. Clinic Infec Dis 20:472

Sampaio JP, Gadanho M, Santos S, Duarte FL, Pais C, Fonseca A, Fell JW (2001) Polyphasic taxonomy of the basidiomycetous yeast genus *Rhodosporidium*: *Rhodosporidium kratochvilovae* and related anamorphic species. Int J Syst Evol Micr 50:687–697

Satosh K, Makimura K (2008) *Sporobolomyces koalae* sp. Nov., a basidiomycetous yeast isolated from nasal smears of Queensland koalas kept in a Japanese zoological park. Int J Syst Evol Micr 58:2983–2986

Sharma V, Shankar J, Kotamarthi V (2006) Endogeneous endophthalmitis caused by *Sporobolomyces salmonicolor*. Eye 20:945–946

Wickerham LJ (1951) Taxonomy of yeasts. U S Dept Agric Tech Bull No 1029:11

# Part IV

# Microbes and Environment

# Fungal Toxins and Their Impact on Living Systems

# 47

Vivek Kumar Singh, Mukesh Meena, Andleeb Zehra, Arti Tiwari, Manish Kumar Dubey and R. S. Upadhyay

**Abstract**

Some of the most potent toxins are synthesized by fungi. Fungal toxins are the chemicals produced by fungi under certain conditions. They may be classified under different chemical classes. They are not essential for fungal growth or reproduction, but are toxic to plants, animals or humans. Fungal toxin contamination in certain agricultural commodities has been a serious concern for animal and human health. The major toxin-producing fungi are the species of *Aspergillus, Penicillium, Fusarium* and *Alternaria*. Aflatoxins, citrinin, fumonisins, fusaric acid, moniliformin, AAL-toxins and alternariol are some of the important fungal toxins responsible for causing economic losses to agriculture, spoilage of food that are often fatal to living systems. This review focuses on toxigenic fungi, toxins and their characteristics with biological effects.

**Keywords**

*Aspergillus* · *Penicillium* · *Fusarium* · *Alternaria* · Aflatoxins · Citrinin · Fumonisins · Fusaric acid · Moniliformin · AAL-toxins · Alternariol

## 47.1 Introduction

Several species of fungi produce toxins that have significant agricultural, epidemiological and economical impact. *Aspergillus*, *Fusarium, Penicillium* and *Alternaria* species are responsible for majority of agricultural mycotoxin contaminations. These fungi are common components of the microbial flora associated with many agronomic crops including maize, peanuts, tree nuts, grapes, coffee, cotton, wheat, barley and other cereal grains (Palencia et al. 2010). Depending on the host crop and the fungal species, mycotoxigenic fungi may cause plant diseases such as *Aspergillus* fruit rot of grapes, maize ear rots caused by *Aspergillus* and *Fusarium* species, and *Fusarium* head blight and seedling blight diseases on cereal crops (Palumbo et al. 2008). Thus, these interactions result in further crop losses resulting in economic loss. In contrast, other host–fungal interactions do not cause disease symptoms such as epiphytic growth of mycotoxigenic *Aspergilli* on tree nuts and asymptomatic endophytic growth of *Fusarium* in maize leaves (Palumbo et al. 2008). The major mycotoxins are aflatoxins, citrinin, fumonisins, fusaric acid, trichothecenes, deoxynivalenol, AAL-toxin, alternariol, ochratoxins,

R. S. Upadhyay (✉) · V. K. Singh · M. Meena · A. Zehra · A. Tiwari · M. K. Dubey
Laboratory of Mycopathology and Microbial Technology, Centre of Advanced Study in Botany, Banaras Hindu University, Varanasi 221005, India
e-mail: upadhyay_bhu@yahoo.co.uk

R. N. Kharwar et al. (eds.), *Microbial Diversity and Biotechnology in Food Security*,
DOI 10.1007/978-81-322-1801-2_47, © Springer India 2014

**Table 47.1** Some important physical and chemical properties of the aflatoxins (Reddy and Waliyar 2000)

| Aflatoxin | Molecular formula | Molecular weight | Melting point | UV absorption max (e), nm, methanol | |
|---|---|---|---|---|---|
| | | | | 265 | 360–362 |
| $B_1$ | $C_{17}H_{12}O_6$ | 312 | 268–269 | 12,400 | 21,800 |
| $B_2$ | $C_{17}H_{14}O_6$ | 314 | 286–289 | 12,100 | 24,000 |
| $G_1$ | $C_{17}H_{12}O_7$ | 328 | 244–246 | 9,600 | 17,700 |
| $G_2$ | $C_{17}H_{14}O_7$ | 330 | 237–240 | 8,200 | 17,100 |
| $M_1$ | $C_{17}H_{12}O_7$ | 328 | 299 | 14,150 | 21,250 (357) |
| $M_2$ | $C_{17}H_{14}O_7$ | 330 | 293 | 12,100 (264) | 22,900 (357) |

etc., each of which is produced by several fungal species. *Aspergillus flavus* and *Aspergillus parasiticus* are the major producers of aflatoxins; *Fusarium verticillioides* and *Fusarium proliferatum* produce fumonisins. The impact of these classes of mycotoxins on human and animal health has been extensively studied. Thus, these mycotoxins are of considerable food safety concern, which lead to regulatory action to limit contamination of agricultural commodities used for food and feed. Although many scientific reports have been published concerning the occurrence of mycotoxins in foods and feeds, their impact on human and animal health needs more documentation on the levels and effects of mycotoxin contamination in the environment. The aim of this chapter is to focus on the major fungal toxins which are contaminants of many agronomic crops worldwide and cause both economic losses and health effects.

## 47.2 Major Fungal Toxins

### 47.2.1 Aflatoxin

Aflatoxins (AF) are the best-known toxic secondary metabolites and are a major group of polyketide-derived mycotoxins produced by three species of *Aspergillus*: *A. flavus, A. parasiticus* and the rare *A. nomius* which have worldwide distribution. AF are able to contaminate a wide range of substrates including cereal grains, oilseeds, and tree nuts under favourable conditions of temperature and relative humidity (Bennett and Klich 2003). Under favourable environmental conditions, the propagules of these fungi grow on plants and other materials and produce carcinogenic AF. It has been estimated that the genes for AF biosynthesis have persisted for more than 100 million years (Cary and Ehrlich 2006). It is believed that AF are produced as a fungal defense response to stressful conditions, for protecting the fungus from UV damage, as virulence factors or as a part of defense mechanisms to protect the fungus from predators in the environment (Cary and Ehrlich 2006). Most of the people worldwide are exposed to dietary AF. Exposures are highest in tropical and subtropical regions of the world, where maize and peanuts are the dietary staple food. Poor harvesting practices, improper storage, and less than optimal conditions during transport and marketing can also contribute to proliferation of fungus and increase the risk of AF production (Bhat and Vasanthi 2003). Human exposure to AF can result directly from ingestion of contaminated foods, or indirectly from consumption of foods from animals previously exposed to AF in feeds.

#### 47.2.1.1 General Characteristics

Although 20 compounds, all designated as AF, have been isolated referring to four metabolites of this group of bis-furocoumarin metabolites produced by *Aspergillus flavus, A. parasiticus,* and *A. nomius*. These are designated as $B_1$, $B_2$, $G_1$ and $G_2$ (Table 47.1, Fig. 47.1), all of which occur naturally and are normally found in foods. *A. flavus* produces only B aflatoxins, while the other two species produce both B and G aflatoxins. The four compounds are distinguished on the colour of their fluorescence under long-wave ultraviolet illumination (B = blue; G = green), with the subscripts relating to their relative chromatographic mobility (Sargeant et al. 1963). Aflatoxin $M_1$ ($AFM_1$) and aflatoxin $M_2$ ($AFM_2$) are hydroxylated metabolites of $AFB_1$ and $AFB_2$. Aflatoxin

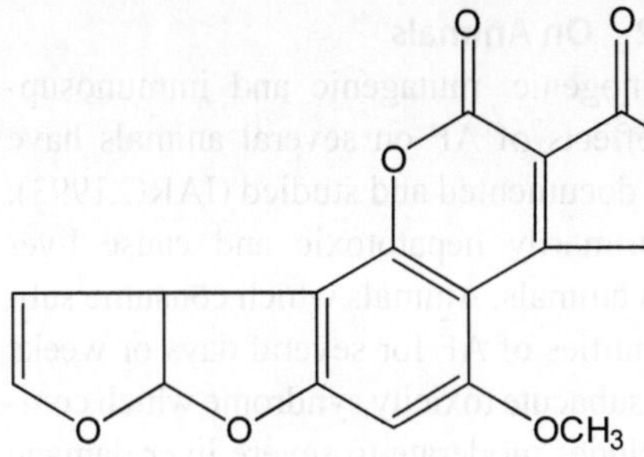

AFLATOXIN B1

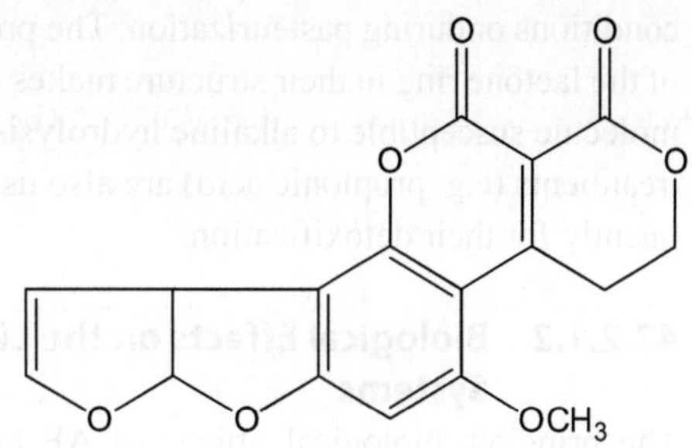

AFLATOXIN G1

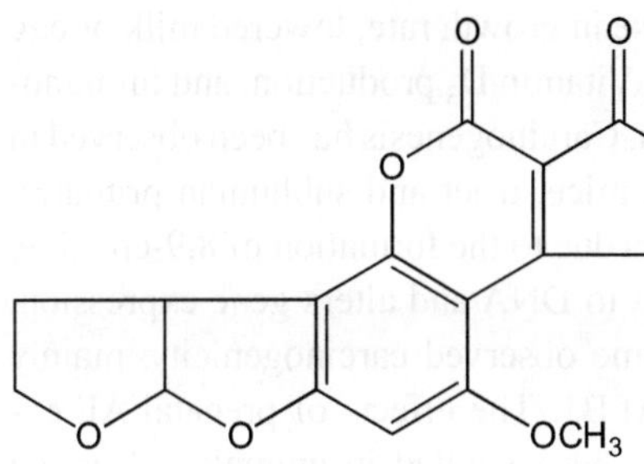

AFLATOXIN B2

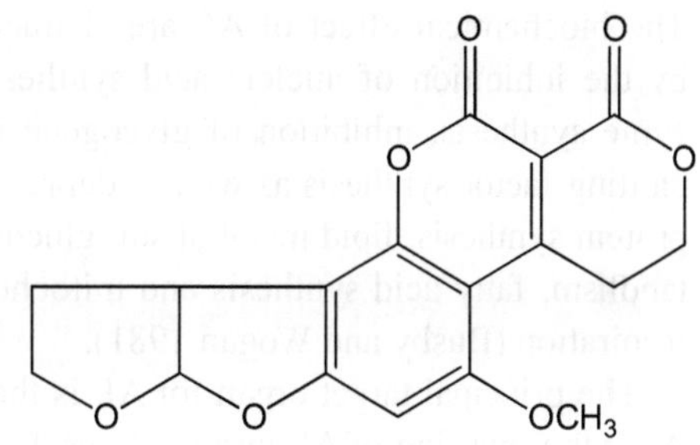

AFLATOXIN G2

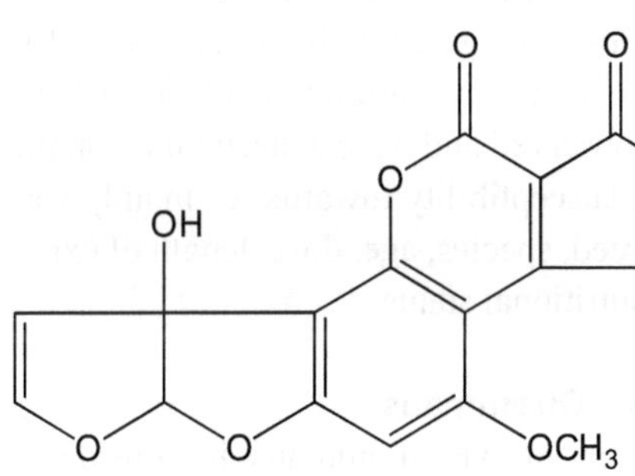

AFLATOXIN M1

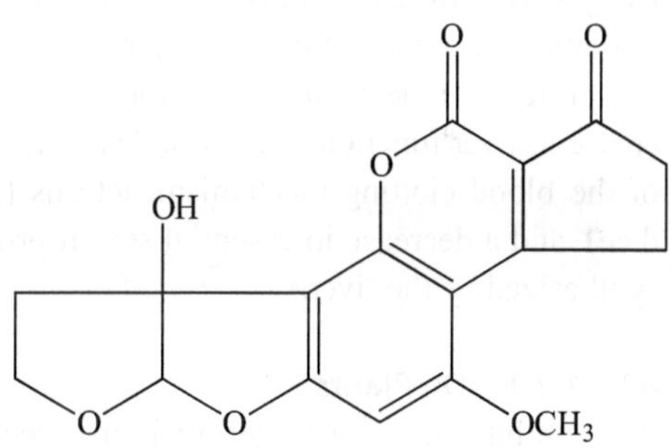

AFLATOXIN M2

**Fig. 47.1** Structures of aflatoxins *$B_1$, B2, $G_1$, $G_2$, $M_1$,* and *$M_2$* (Michelle McLean and Michael F. Dutton 1995)

$B_{2a}$ ($AFB_{2a}$) and aflatoxin $G_{2a}$ ($AFG_{2a}$) are 8,9-hydrated products of $AFB_1$ and $AFG_1$ (Dutton and Heathcote 1968). These compounds are relatively nontoxic when compared with $AFB_1$ and $AFG_1$. Aflatoxins $B_1$, $B_2$, $G_1$, and $G_2$ are classified as Group I human carcinogens, whereas $M_1$ is classified as Group 2B probable human carcinogen (Ioannou et al. 1999). These findings were confirmed in vitro (Terao and Ueno 1978) and in vivo (Cole and Cox 1981).

Chemically, AF are dihydrofuran or tetrahydrofuran moieties fused to a coumarin ring. AF are small molecular weight compounds, which are freely soluble in moderately polar solvents (e.g. chloroform and methanol), especially in dimethylsulphoxide, and also have some water solubility. AF are quite stable in many foods and are fairly resistant to degradation. These compounds are very stable at high temperatures, with little or no destruction occurring under ordinary cooking conditions or during pasteurization. The presence of the lactone ring in their structure makes the AF molecule susceptible to alkaline hydrolysis. Acid treatments (e.g. propionic acid) are also used frequently for their detoxification.

### 47.2.1.2 Biological Effects on the Living Systems

The principle biological effects of AF are carcinogenicity (WHO 1979), immunosuppression (Pier and McLoughlin 1985), mutagenicity (Ong 1975) and teratogenicity (Di Paolo et al. 1967). The biochemical effect of AF are characterised by the inhibition of nucleic acid synthesis, enzyme synthesis, inhibition of glycogenesis, and clotting factor synthesis as well as depression of protein synthesis, lipid metabolism, glucose metabolism, fatty acid synthesis and mitochondrial respiration (Busby and Wogan 1981).

The principal target organ for AF is the liver. After the invasion of AF into the liver, lipids infiltrate hepatocytes that lead to necrosis or liver cell death. The main reason is that AF react negatively with different cell proteins, which leads to inhibition of carbohydrate and lipid metabolism and protein synthesis. In correlation with the decrease in liver function, there is a dearrangement of the blood clotting mechanism, icterus (jaundice), and a decrease in essential serum proteins synthesized by the liver.

#### 47.2.1.2.1 On Plants

AF inhibits the seed germination, seedling growth, root elongation, chlorophyll and carotenoid synthesis as well as protein, nucleic acid and some enzyme synthesis in plants (Sinha et al. 1993). AF is reported to induce chromosomal aberrations in plants (Lin and Key 1968). At the cellular level, low concentrations of AF stimulate lipase activity in *Gossypium* and β-indole acetic acid activity in *Pisum,* whereas high concentration is inhibitory (Jones et al. 1967; Reiss 1971). Thus, AF restrict the plant growth by altering the physiological processes of plants. Plants contaminated with the above permissible level are not allowed to use in food and feed products.

#### 47.2.1.2.2 On Animals

The carcinogenic, mutagenic and immunosuppressive effects of AF on several animals have been well documented and studied (IARC 1993). AF are primarily hepatotoxic and cause liver damage in animals. Animals which consume sublethal quantities of AF for several days or weeks develop a subacute toxicity syndrome which commonly includes moderate to severe liver damage and teratogenic effects. Large doses of AF in diet have been shown to produce hepatic necrosis. Even with low levels of AF in the diet, there will be a decrease in growth rate, lowered milk or egg production, Vitamin $D_3$ production, and immunosuppression. Carcinogenesis has been observed in rats, ducks, mice, trout and subhuman primates, and it occurs due to the formation of 8,9-epoxide, which binds to DNA and alters gene expression. There is some observed carcinogenicity, mainly related to AFB1. The effects of prenatal AF exposure have been studied in animals. AF given to rat towards the end of the pregnancy, resulted in foetal growth retardation (Buttler and Wigglesworth 1966), but no effects were observed when the same dose was given early in pregnancy. Effects on the lungs, myocardium and kidneys have also been observed and AF can accumulate in the brain. The susceptibility towards AF mainly varies with breed, species, age, dose, length of exposure and nutritional status.

#### 47.2.1.2.3 On Humans

Consumption of AF in humans is considered not only as a risk factor for cancer but also has immunologic and nutritional effects associated modulation with the infectious diseases such as HIV (Williams et al. 2004, 2010). AF have been implicated in subacute and chronic effects in humans. These effects include primary liver cancer (Peers and Lindsel 1973; Shank et al. 1972), chronic hepatitis (Ngindu et al. 1982; Krishnamachari et al. 1975), jaundice, hepatomegaly and cirrhosis through repeated ingestion of low levels of AF. It is also considered that AF may play a role in a number of diseases, including Reye's syndrome (Dvorackova et al.1977), kwashiorkor (Hendricks et al. 1982), and hepatitis. AF can also affect the immune system (Pier 1991). In human, AF react not only with nucleic acids to generate the polynucleotide-base adducts responsible for the initiation of carcinogenesis, but also interacts with various liver and blood proteins, particularly with serum albumin to form a stable adduct. In addition, $AFM_1$, a metabolite of $AFB_1$ found in the milk of lactating mothers exposed to AF, is of concern due to its potential hepatotoxic and immunotoxic effects in infants and children. In general, at smaller concentrations the AF, and $AFB_1$ in particular, can affect male reproduction namely spermatogenesis (Egbunike et al. 1980), Leydig cell function (Egbunike 1982), and fertility (Ibeh et al. 1994).

##### 47.2.1.2.4 On Microorganisms

$AFB_1$ produces varied effects in microorganisms including reduction of the DNA-to-protein ratio, viability and oxygen uptake, inhibition of mRNA transcription, aberrant cell formation, and binding to cells with and without nucleic acids along with acid toxic effects. AF inhibit incorporation of labelled precursors into DNA, RNA and proteins and blocked induction as well as production of various enzymes at various levels in microorganisms.

### 47.2.2 Citrinin

Citrinin (CTN) is one of the potent and well-known quinone methide mycotoxins, which is possibly spread all over the world; *Penicillium citrinum* is a potent producer of CTN. Other CTN producing fungi include *Aspergillus niveus, A. niger, A. oryzae, Monascus ruber* and *P. camemberti*. CTN is generally formed after harvest under storage conditions and occurs mainly as a natural contaminant in stored grains such as wheat, rice, barley, maize, oat, peanut and rye and may occur as co-contaminant in cereals. It also occurs in other plant products such as beans, fruits and vegetable juices, herbs and spices, and also in spoiled dairy products. In addition, CTN is found as an undesirable contaminant in red mould rice (RMR) which is used as a food preservative and colourant in Asian foods (Fink-Gremmels et al. 1991). Thus, CTN is regarded as

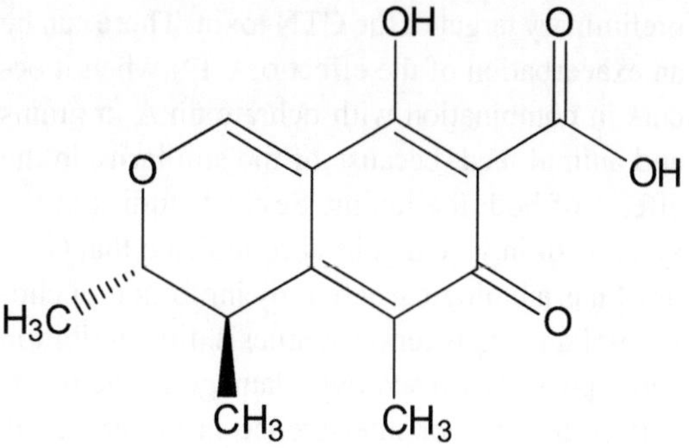


**Fig. 47.2** Chemical structure of Citrinin (Sabater-Vilar et al. 1999)

an important mycotoxin which may be ingested by humans and animals (Krogh et al. 1973).

#### 47.2.2.1 General Characteristics

CTN ($C_{13}H_{14}O_5$) is 4,6-dihydro-8-hydroxy-3,4,5-trimethyl-6-oxo-3H-2-benzopyran-7-carboxylic acid (Fig. 47.2). It is a simple, acidic, low molecular weight compound that crystallizes as lemon coloured needles melting at 175 °C. Alternatively, CTN is also known as antimycin. Its molecular weight is 250.25 g/mol and density 1.335 g/cm$^3$.

It is sparingly soluble in water, but totally soluble in dilute sodium hydroxide, sodium carbonate, sodium acetate, methanol, acetonitrile, ethanol and most other polar organic solvents (Deshpande 2002). Due to its conjugated double bonds, CTN absorbs the light in the visible wavelength range. Its colour varies from lemon yellow at pH 4.6 to cherry red at pH 9.9 and its absorption maxima are in the UV range 250–331 nm. Some photodecomposition occurs in fluorescent light both in solution and in the solid state. It can be degraded in acid or alkaline solution or by heat. CTN may be extracted with nonpolar solvents.

#### 47.2.2.2 Biological Effects on the Living Systems

CTN is known to cause a wide range of abnormalities in the living systems. However, susceptibility varies with sex, age and dosage. CTN acts as a nephrotoxin in all species in which it has been tested (Heperkan et al. 2006). Kidney is the preliminary target of the CTN toxin. There can be an exacerbation of the effect of CTN when it occurs in combination with ochratoxin A in grains and animal feed, because of the similarity in the effects of both the toxins. Several studies in the systems of in vivo and in vitro indicate that CTN itself has a biological action by inhibition of cholesterol and triglyceride synthesis, this inhibition being possibly caused by a damage to the transport system and/or interference in the energetic metabolism (Betina 1984).

#### 47.2.2.2.1 On Plants

CTN toxicity in plants is not much reported. CTN contaminates maize, wheat, rye, barley, oats, and rice (Nelson et al. 1985; Scott et al. 1972; Tanaka et al. 2007). In the endemic area in Bulgaria, CTN was more common and had higher concentration in maize and beans intended for human consumption as compared to the non-endemic area (Petkova-Bocharova et al. 1991). CTN causes pathological effect and ultrastructural changes in the young leaves of cultivars Nijisseiki and Chojuro of Japanese pear. It was reported that CTN caused necrosis and permeability changes in Nijisseiki pear but no effect was seen in Chojuro pear (Nishimura and Kohmoto 1983).

#### 47.2.2.2.2 On Animals

CTN has been found to be toxic to several animal species (Carlton 1980). It causes necrosis. In the chicken, hepatic and lymphoid necrosis were found in addition to the renal alterations (Mehdi et al. 1983). CTN exerts toxic effect on turkeys and ducklings by alteration in necrosis that is more severe in turkeys. Hepatic and lymphoid lesions occur in both of these (Mehdi et al. 1984). CTN is foetotoxic and embryocidal in mice and rats. Depending upon the concentration, effects include reduction in yolk sac diameter, crown-rump length, sornite number, protein and DNA contents (Yang et al. 1993). The toxic effects of CTN in chick embryos include growth retardation of foetuses, microphthalmia, cleft beak, deformities of the limb, etc. A strict additive effect was seen in combination with ochratoxin A (Vesela et al. 1983).

Significant effects of CTN toxin are also reported in rats and mice such as a decrease in DNA, RNA and protein content in kidney as well as in liver. Liver glycogen was significantly lowered. Surviving animals showed decreased body weight, food consumption, per cent liver to body weight and liver glycogen. (Hood et al. 1976; Phillips and Hayes 1978). Nephrotoxicity caused by CTN in rodents like rats and mice is due to slight stimulation of orotic acid incorporation into liver and kidney in the early stage (Sansing et al. 1976). CTN also produces renal tumours in a male rat by oral administration in diet.

#### 47.2.2.2.3 On Humans

The International Agency for Research on Cancer (IARC) classifies CTN in group 3 because of limited evidence on carcinogenicity for animals (Castegnaro and McGregor 1998). However, the presence of CTN and other toxic metabolites in food, regardless of the concentration, must be considered as a potential human health hazard. In humans, the renal system is primarily affected by CTN and the mitochondrial respiratory chain was identified as a possible sensitive target for this toxin. CTN also affects human neutrophils with regard to superoxide anion generation. It stimulates the superoxide anion production in resting neutrophils in whole blood but inhibits it in isolated cells. Exposure to CTN diminishes the number of T-helper type 1 (Th1) cells in the peripheral blood of children which results in a risk factor for the development of allergic diseases. It has been hypothesized that CTN being a mycotoxins is responsible for this effect (Wichmann et al. 2002).

Evidences support the role of CTN in combination with ochratoxin A in the etiology of Balkan endemic nephropathy (BEN) and associated urinary tract tumours (UTT). Both diseases occur in subjects born and/or living in certain rural areas where home-produced and home-stored stable foods were found to be more frequently contaminated by the ochratoxin A and CTN (Castegnaro et al. 1991).

#### 47.2.2.2.4 On Microorganisms

CTN is reported to be a broad spectrum antibiotic especially against Gram-positive bacteria (Raistrick and Smith 1941; Blanc et al. 1995). CTN is also found to be fungicidal against *Fusarium* spp. and *A. niveus* (Prabha et al. 2009).

In a study, CTN produced by *Penicillium citrinum* was found to have a newly identified function of inducing bacterial motility by transcriptional activation of some genes related to the expression of flagella. The swarming motility of *Paenibacillus polymyxa*, a Gram-positive low-G1C spore-forming soil bacterium belonging to the plant growth-promoting rhizobacteria was greatly induced by CTN in a dose-dependent manner. The development of lateral flagella by CTN toxin has also been reported in other bacterial cells with swarming motility (Merino et al. 2006).

### 47.2.3 Fusaric Acid

Fusaric acid (FA) is a host nonspecific toxin produced by a number of *Fusarium* species (Bacon et al. 1996; Rani et al. 2009). The most expansive producer of this toxin is *Fusarium oxysporum* and its special forms (f. sp.) *lycopersici*. *Fusarium* spp. are ubiquitous fungi found in soil worldwide, as both pathogenic and nonpathogenic strains. High production of FA has been correlated with the virulence of plant pathogenic strains of *Fusarium* spp. It is a natural contaminant that accumulates during infection in corn and cereal grains, which is extremely toxic to animals and human beings, by enhancing toxicity of other *Fusarium* metabolites (e.g. Trichothecenes). It is not only moderately toxic to animals but also has antibiotic, insecticidal, and pharmacological properties (Bacon et al. 1996). The species of *Fusarium* cause wilt diseases in pepper, corn, tomato, banana and in various other plants (Paterson and Rutherford 1991). FA causes necrotic spots on leaves, shrivelling and drying of leaves, as well as shrinking and wilting of stem and petioles of tomato plants. FA could elicit various plant defense responses at 100 nM concentration without any toxic effect. Increased FA concentration reduces root and root-hair growth inducing a rapid transient membrane hyperpolarization (Bouizgarne et al. 2006). They are toxic for eukaryotes and prokaryotes involved in fungal defense. They also act against *Pseudomonas* spp. (biocontrol strains) by repressing the production of antifungal metabolites (van Rij et al. 2005). FA showed higher nematicidal activity against *Bursaphelenchus* xylophilus. FA causes the rot of potato tubers (Venter and Steyn 1998). FA has a tumoricidal activity for head and neck squamous cell cancer (HNSCC). FA can be involved in fungal pathogenicity by decreasing cell viability. It could induce typical early defense response such as reactive oxygen species production (ROS) (Bouizgarne et al. 2006). FA can be detected using HPLC, TLC, Mass spectroscopy and NMR techniques.

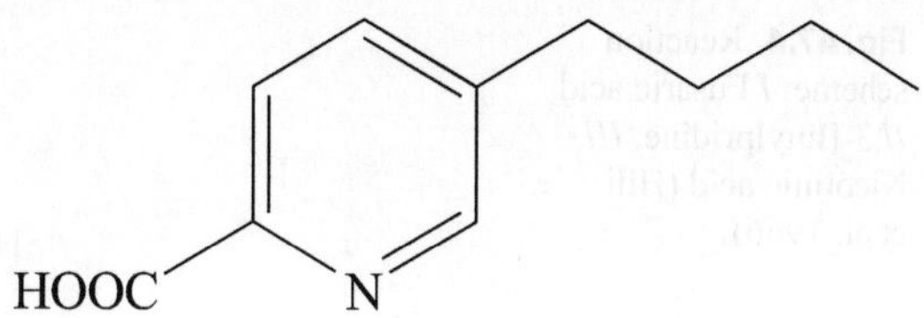


**Fig. 47.3** Structure of Fusaric acid (Yabuta et al. 1934)

#### 47.2.3.1 General Characteristics

FA is a mycotoxin with low to moderate toxicity. It might be synergistic with other mycotoxins. The chemical formula of FA is 5-butylpicolinic acid (Fig. 47.3). The biosynthesis of FA involves the condensation reaction involving a polyacetate unit and aspartic acid (Hill et al. 1966). Decarbonylation of FA gives $CO_2$ (C-7) and 3-butyl pyridine, which is oxidized with KMnO4 to nicotinic acid. The activity is found mainly in positions 2, 3, 5, 9 and 11 of FA. The low activity at C-2 and C-3 of pyridine ring, presence of pyridine ring, and activity at C-4 and C-7 are consistent with the participation of 4C Krebs's cycle acid or its equivalent (Hill et al. 1966) (Fig. 47.4).

#### 47.2.3.2 Biological Effects

##### 47.2.3.2.1 On Plants

FA causes diseases in wide varieties of plants. FA causes various symptoms due to *Fusarium* infection (leaf wilting and necroses), which confirm the role of FA in the disease progression. At

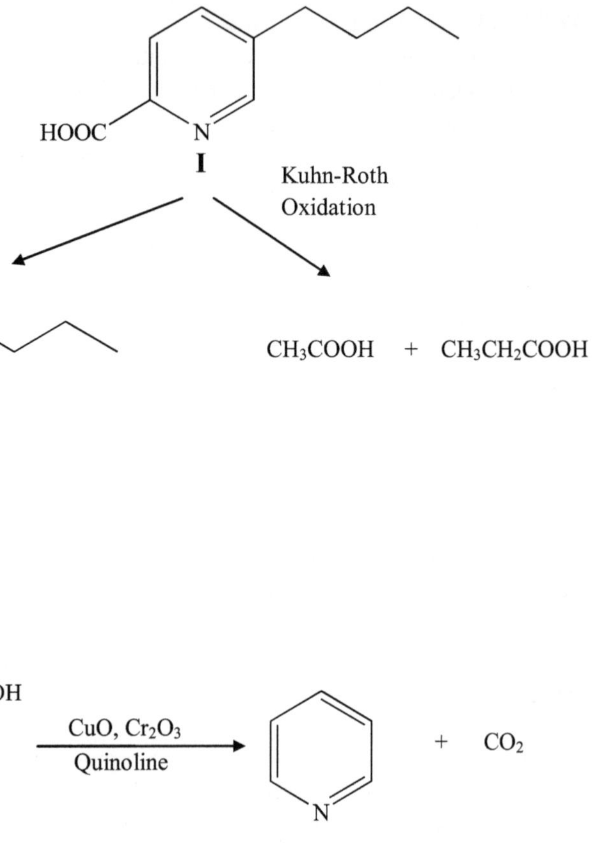


**Fig. 47.4** Reaction scheme: *I* Fusaric acid, *II* 3-Butylpridine, *III* Nicotinic acid (Hill et al. 1966)

the subcellular level, FA affects numerous biochemical processes related to membrane permeability changes, dysfunctions of mitochondrial activity, and inhibition of respiration (Kohler and Bentrup 1983). FA enhances the electrolyte leakage, disturbs the electrochemical gradients for $H^+$ and $K^+$ at the plasma membrane causes membrane depolarization that lowers the intracellular ATP content and inhibits some metal containing enzymes (e.g. cytochrome oxidase). It results eventually in respiratory disorders and cell death (Marre et al. 1993). The influence of toxic doses of FA on pro- and antioxidant systems was investigated with an example of leaves and the cultured cells of tomato (Kuzniak et al. 1999; Kuzniak 2001). At the same time, Bouizgarne et al. (2006) showed that low concentrations of FA may exert various protective responses in plant cells in the absence of toxic effects and may perform signalling function in host–pathogen interactions. Moderate FA doses (50–100 uM) induce apoptotic features, while high FA doses (>200 uM) stimulate necrosis. The phytotoxic pathogenicity factor FA represses the production of 2,4 diacetyl phloroglucinol (DAPG), a key factor in the antimicrobial activity of biocontrol strain *Pseudomonas fluorescence*.

#### 47.2.3.2.2 On Animals and Humans

FA is potentially toxic to animals. FA is mildly toxic to mice (Hidaka et al. 1969) and it has several important pharmacological properties (Hidaka et al. 1969; Malini 1966; Porter et al. 1990; Porter et al. 1995). It affects the nervous systems of horses and may cause cancer in rodents.

FA has been reported to have many primary and secondary effects in humans (Wang and Ng 1999). FA raises serum melatonin, 5-hydroxytryptamine, tyrosine and dopamine. These increments are thought to be secondary to FA inhibitory effects on tyrosine hydroxylase and dopamine beta-hydroxylase. FA has been shown to have an anti-hypertensive effect secondary to a reduction of peripheral vascular resistance through peripheral arteriolar dilation. FA reduces catecholamine synthesis by lymphocytes, which may affect their cytotoxic activity on solid tumours. FA may also be a risk factor for oesophageal cancer (Voss et al. 1999). FA has also been shown to induce DNA damage in vitro in cultured larynx cancer cells (Nadgornaya et al. 1981). FA is cytostatic to human fibroblasts and cytotoxic to colon and mammary adenocarcinomas as well as epidermoid carcinoma. Picolinic acid, a precursor carboxylic acid of FA, has been shown to be cytotoxic to many malignancies as well (Fernandez-Pol and Jonson 1977; Fernandez-Pol 1977). FA does not interfere with dopamine uptake. FA decreases contractile response elictor with norepinephrine, histamine, serotonin, acetylcholine and KCl. FA represses the PCN production under different environmental condition. It has been shown to produce both ethylene and ethylene-like symptoms (David et al. 1978).

### 47.2.4 Fumonisins

Fumonisins are mycotoxins produced by at least 11 species of the *Fusarium* including the maize pathogens *F. verticillioides* and *F. proliferatum* (Fotso et al. 2002; Rheeder et al. 2002; Leslie et al. 2004; Desjardins 2006). The consumption of food contaminated by fumonisins is implicated in the development of several human and animal diseases (Nelson et al. 1993). Fumonisins are the major contaminants of maize, ingestion of which leads to oesophageal cancer and neural tube defects in human populations (Marasas 1996; Hendricks 1999).

#### 47.2.4.1 General Characteristics

Fumonisins can be divided into structurally distinct groups, four of which have been designated A, B, C and P (Musser and Plattner 1997). Fumonisins consist of a linear 19 or 20-carbon, polyketide-derived backbone with one nitrogen, 3–4 hydroxyl, two methyl and two tricarballylic ester functions at the positions along the backbone. A, B, C, and P fumonisins differ in structure by differences in the nitrogen function and the length of the carbon backbone. For example, in B and C fumonisins the nitrogen function is a free amine, in A fumonisins it is an acetylated amine and in P fumonisins it is a 3-hydroxypyridinium (Musser and Plattner 1997; Sewram et al. 2005). In B fumonisins (FBs) the backbone is 20 carbon atoms long, whereas in C fumonisins (FCs) it is of 19 carbon atoms long (Fig. 47.5).

The B series of fumonisins are esters of 2-amino-12, 16-dimethyl-14, 15-dihydroxyecosane and propan-1, 2, 3-tricarboxylic acid. Fumonisin B1 has hydroxyl groups at C-3, C-5 and C-10. Fumonisins B2 and B3 are isomers with hydroxyl groups at C-3, C-5 and C-3, C-10. Fumonisin B4 has one less hydroxyl group than FB2 and FB3. Fumonisins A1 and A2 are N-acetyl derivates of FB1 and FB2. Fumonisin C1 differs from FB1 in a lack of methyl group at C-1, which is characteristic of the other fumonisins. A new fumonisin iso-FB1 differs from FB1 only in the placement of hydroxyl group at C-4 instead of C-5.

In comparison with other fumonisins, FB1 is the most polar compound. In polar solvent, FB1 exists as zwitterion because of carboxylic groups, which can have positive and negative charges, and also because of the existing free primer amine. Small changes of pH can produce alterations in the structure of tricarboxylic esters and it can cause interactions among active groups. However, the chemical structure of fumonisins has a high number of stereoisomers. Among them, FB1 is a predominant molecular form produced by *Fusarium moniliforme*, FB2 and FB3 appear to be active as FB1 although they occur in lower concentrations, FA1 and FA2 lack the toxicity and promotion activity of FB1 (Gelderblom et al. 1992).

#### 47.2.4.2 Biological Effects on the Living Systems

Fumonisin are potent toxins which cause wide range of abnormalities in the living systems. Fumonisin B1 is extremely toxic to horses, moderately toxic to swine, weakly toxic to cattle, and has been associated with oesophageal cancer in humans. Consequently, the Food and Drug Ad-

**Fig. 47.5** Chemical structure of *B* and *C* Fumonisins (Proctor et al 2008)

B Fumonisins

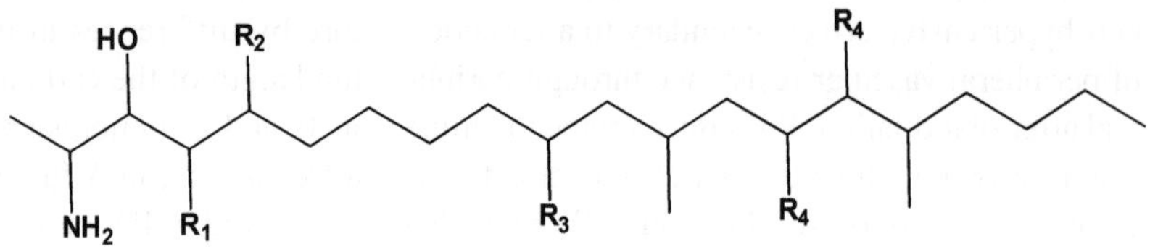

C Fumonisins

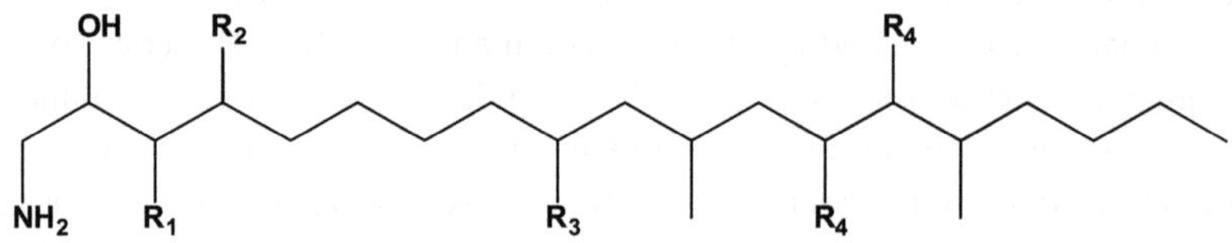

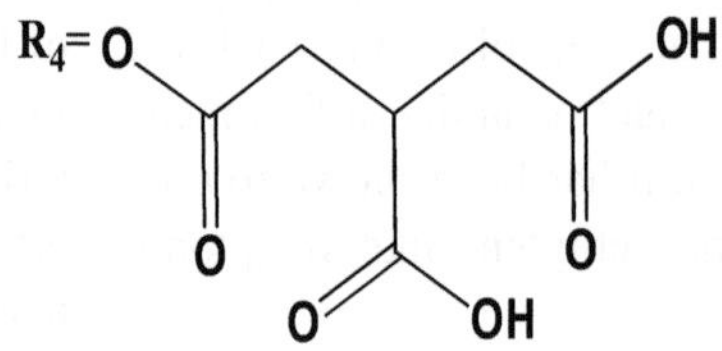

| | $R_1$ | $R_2$ | $R_3$ |
|---|---|---|---|
| $FB_1$ | H | OH | OH |
| $FB_2$ | H | OH | H |
| $FB_3$ | H | H | OH |
| $FC_1$ | H | OH | OH |
| iso$FC_1$ | OH | H | OH |
| $FC_1OH$ | OH | OH | OH |
| $FC_2$ | H | OH | H |
| $FC_3$ | H | H | OH |

ministration (FDA) recently established recommended action levels for fumonisins in human foods and animal feeds (Roohi et al. 2012). Food or feed containing levels of fumonisin higher than the action level could pose a health hazard.

#### 47.2.4.2.1 On Plants

Fumonisin are common contaminants of maize. FB1 has been reported to be phytotoxic to corn callus culture, tomato (*Lycopersicon esculentum* Mill.), jimson weed (*Datura stramonium* L.) leaves, tomato seedlings, corn seedlings, duckweed (*Lemna minor* L.) fronds and a variety of other weeds as well as crop plants (Abbas and Boyette 1992; Merrill 1991; Mirocha et al. 1992; Van Asch et al. 1992; Vesonder et al. 1992).

#### 47.2.4.2.2 On Animals and Humans

Fumonisins cause diseases in different animal species. In horses, it causes equine leukoencephalomalacia (ELEM); the primary target is the brain. Once clinical signs become obvious, chances of recovery are less. In pigs, it causes acute pulmonary edema; the lungs are the primary target. Clinical signs include laboured or difficult breathing, weakness, and greyish-blue or slate coloured skin or mucous membranes; other lung diseases also may cause these same symptoms. Pigs ingesting smaller amounts of fumonisins may perform poorly and develop icterus (pigmentation of tissues, membranes, and secretions with bile pigments). Skin, mucous membranes, secretions, and the whites of the eyes may be stained yellow. Ruminants do not appear to be adversely affected by fumonisins. Poultry are even more resistant to adverse health effects from fumonisins. Fumonisins can cause cancer and neural tube defects in experimental rodents. Fumonisins may also cause certain types of cancer in humans and is regarded as a potential carcinogen (Desjardins 2006).

### 47.2.5 Moniliformin

Moniliformin (MON) has been reported as a natural contaminant in maize and other cereals (rice, oats, rye, barley, wheat and triticale) in different parts of the world (Sanhueza and Degrossi 2004). MON contamination is higher in maize than in other substrates, particularly in visibly infected samples. MON is produced at least by 30 *Fusarium* spp. mainly produced by *F. proliferatum* and *F. subglutinans*.

#### 47.2.5.1 General Characteristics

In nature, MON occurs as a sodium or potassium salt of the semisquaric acid. The UV-absorbance of MON has a maximum absorption at 227 nm and a shoulder at 258 nm (Sydenham et al. 1996). MON or semisquaric acid are trivial names for 3-hydroxy-3 cyclobutene-1, 2-dione (Fig. 47.6). Due to low $pK_a$ value (>1.7) of the free acid, MON does not occur as such in nature but as a water soluble sodium or potassium salt.

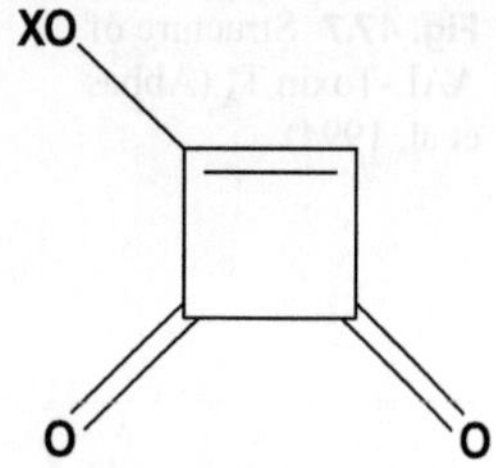


**Fig. 47.6** Chemical structure of Moniliformin. X=H (free acid), Na (sodium salt) or K (Potassium salt) (Jestoi 2008)

#### 47.2.5.2 Biological Effects on Plants, Animals, and Humans

MON can produce plant growth-regulating and phytotoxic effects on the plant systems. MON has also been proven to be toxic to several animal species (Peltonen et al. 2010). Muscular weakness, respiratory distress, cyanosis, coma, and death are symptoms described in animals. The pathology associated with MON toxicity in most species has primarily involved myocardial changes. Thiel et al. (1982) found that exceedingly low concentration of MON (<5 μM) selectively inhibited rat liver mitochondrial pyruvate and α-ketoglutarate oxidations to the level of 50% and suggested that these inhibitory effects could constitute the major molecule mechanism of toxicity. The high metabolic rate of cardiac tissue makes the heart a likely target for the toxic effects of the inhibited energy metabolism.

### 47.2.6 AAL-Toxin

Mycotoxins produced by the fungus *Alternaria alternate* f. sp. *lycopersici* collectively known as the AAL toxins, were described initially as host-specific determinants of the Alternaria stem canker disease of tomato (Gilchrist and Grogan 1976). AAL toxins consist of a family of structurally analogous metabolites, out of which AAL toxin $T_A$ is the major toxin (Bottini et al. 1981). AAL toxins are chemically related to the fumonisins, including fumonisin $B_1$ ($FB_1$) and share similar toxicological mechanisms of action. The toxicological properties and structural similarities of the AAL toxins and the fumonisins have raised concern about the potential effects of these mycotoxins on human and animal health (Moussatos et al. 1993b).

**Fig. 47.7** Structure of AAL-Toxin $T_A$ (Abbas et al. 1994)

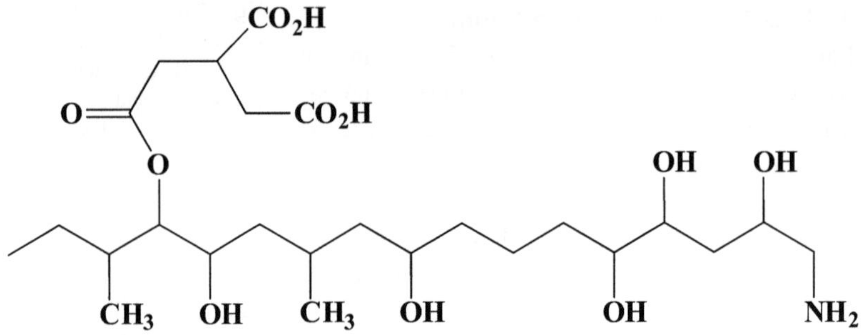


AAL-Toxin $T_A$

**Fig. 47.8** Structure of AAL toxins. Each major toxin fraction is composed of a pair of regioisomers, e.g. *TA* is a mixture of *TA₁* and *TA₂* (Ferenc Szurdoki et al. 1996)

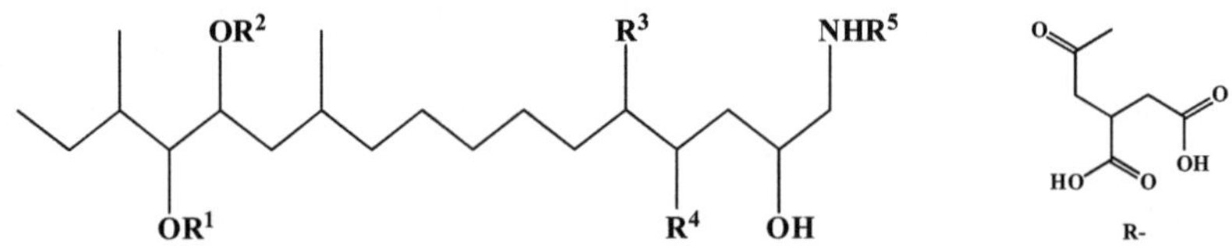


| Toxin | $R^1$ | $R^2$ | $R^3$ | $R^4$ | $R^5$ |
|---|---|---|---|---|---|
| $TA_1$ | H | R | OH | OH | H |
| $TA_2$ | R | H | OH | OH | H |
| $TB_1$ | H | R | H | OH | H |
| $TB_2$ | R | H | H | OH | H |
| $TC_1$ | H | R | H | H | H |
| $TC_2$ | R | H | H | H | H |
| $TD_1$ | H | R | H | OH | $COCH_3$ |
| $TE_1$ | H | R | H | H | $COCH_3$ |
| $TE_2$ | R | H | H | H | $COCH_3$ |

AAL toxin is toxic to a wide variety of weeds at a very low dose rates. AAL-toxin and its analogues kill plants by inhibiting a ceramide synthase-like enzymes causing rapid accumulation of free sphingoid bases that disrupt the membrane. These novel structures and sites of action are excellent and can lead to the discovery and development of environmentally safe and potential herbicides.

#### 47.2.6.1 General Characteristics

AAL-toxins are long-chain alkylamines with one tricarboxylic acid moiety attached (Fig. 47.7); five types of AAL-toxin have now been described, each with two isomers (Fig. 47.8). Structural characterization of AAL-toxins $T_A$ and $T_B$ indicated that each of the toxins existed as a pair of regioisomers of 1,2,3-propanetricarboxylic acid (tricarballylic acid) esterified to 1-amino-11,15-dimethylheptadeca-2,4,5,13,14-pentol ($T_A$) and 1-amino-11,15-dimethylheptadeca-2,4,13,14-tetrol ($T_B$) (Bottini and Gilchrist 1981; Bottini et al. 1981). The type $T_A$, the most active and produced in greatest quantities by *A. alternata,* has a relative molecular mass of 522. AAL-toxin is structurally related to the fumonisins (Fig. 47.7) which includes two tricarboxylic acid moieties. Sphingolipids are important constituents of cell membranes in both animals and plants (Duke and Dayan 2011), but their role in plants is not well studied. Both AAL-toxin and the fumonisins inhibit sphinganine (sphingosine) N-acyltransferase (ceramide synthase in animal), apparently as a result of structural similarities between the toxins and sphingolipids (Fig. 47.7), suggesting that these compounds are competitive inhibitors of the enzyme. Sphinganine and AAL-toxin have an 18-carbon and a 17-carbon alkylamine backbone, respectively, with sphinganine having an N-terminal methanolic substituent and AAL-toxin having multiple substituents at the C-terminal end. AAL-toxin apparently competes with sphinganine and other sphingolipids for the enzyme sphinganine (sphingosine) N-actyl transferase causing accumulation of sphinganine and depletion of complex sphingolipids. Therefore, the effect of AAL-toxin can be determined by measuring the build-up of free sphingoid bases in plant and animal systems (Wang et al. 1996).

#### 47.2.6.2 Biological Activity

AAL-toxin has been well documented to be phytotoxic to a variety of weed and crop species (Abbas et al. 1993a, b). AAL-toxin molecules posses high specific biological activity and disrupt cell homeostasis. It was initially thought to be a host-specific toxin for susceptible tomatoes (Kohmoto et al. 1982; Nishimura and Kohmoto 1983; Mirocha et al. 1992). AAL-toxin causes phototoxic damage on susceptible tomatoes including chlorosis, necrosis, stunting, leaf curl, wilt and mortality. The toxin is also phytotoxic to a variety of other plant species (Mašková et al. 2012). Subsequent studies have shown that AAL-toxin $T_A$ is highly phytotoxic to jimsonweed, black nightshade, prickly sida, duckweed, and other higher plant species (Abbas et al. 1993a, b). Not much is known about its toxicity to animals. It is responsible for the induction of cell death in rat liver, dog kidney, and African green monkey kidney cells (Mirocha et al. 1992).

AAL-toxins and the fumonisins share similar toxicological mechanisms of action as both inhibit ceramide synthase in animal cells (Merill et al. 1993), both inhibit cell proliferation in rat liver and dog kidney cells (Mirocha et al. 1992), and both induce cell death in tomato tissues and protoplasts (Moussatos et al. 1993b).

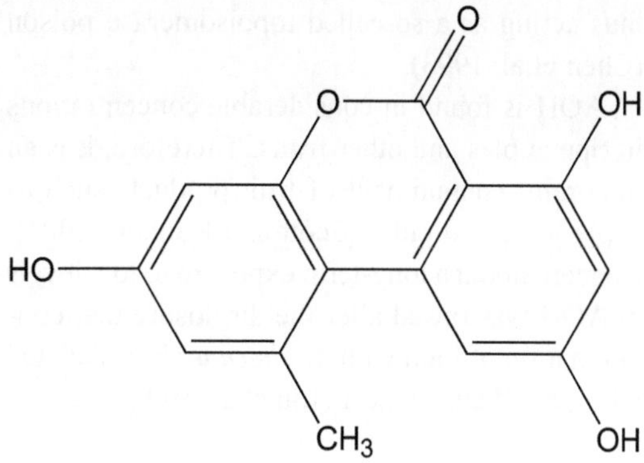


**Fig. 47.9** Structure of Alternariol (AOH) (Lehmann et al. 2006)

### 47.2.7 Alternariol

Alternariol (AOH) is a toxic metabolite of *Alternaria*. It is an important contaminant in cereals and fruits and also exhibits antifungal and phytotoxic activity. It is reported to inhibit cholinesterase enzymes. AOH [(3, 7, 9-trihydroxy-1-methyl-6*H*-dibenzo [*b, d*] pyran-6-one); molecular weight- 258; molecular formula $C_{14}H_{10}O_5$] which is one of the major secondary metabolites produced by various species of *Alternaria* (Fig. 47.9).

#### 47.2.7.1 Biological Activities

AOH possesses foetotoxic, teratogenic, genotoxic and mutagenic properties (Pero et al. 1973; Lehmann et al. 2006; Brugger et al. 2006) and also interferes with the activity of human topoisomerases. DNA topoisomerases are enzymes regulating DNA topology during transcription, replication, chromosome condensation, and the maintenance of genome stability (Wang et al. 1996). DNA-damaging activities of AOH (structurally derivative of dibenzopyrone), i.e. DNA single-stranded and double-stranded breaks, as well as induction of DNA repair synthesis and inhibition of DNA replication, were observed in *Escherichia coli*. Several studies have confirmed that AOH induced DNA strand breaks in cell-free systems (Xu et al. 1996) and in mammalian hepatocytes (Liu et al. 1992). AOH was found to stabilize the catalytically generated DNA topoisomerase intermediate of topoisomerase I and II, thus acting as a so-called topoisomerase poison (Chen et al. 1993).

AOH is found in considerable concentrations in ripe apples and other fruits. Therefore, it is an important contaminant of fruit products such as apple juice and cider (Scott and Kanhere 2001). Concern about a long-term exposure to low levels of AOH was raised after the disclosure that contamination of food with *A. alternata* is associated with oesophagal cancer (Liu et al. 1992).

## 47.3 Conclusion

In conclusion, It is apparent that fungal toxins (Aflatoxins, citrinin, fumonisins, fusaric acid, moniliformin, AAL-toxins, and Alternariol) cause contamination of agricultural commodities and pose a potential threat to the living systems resulting in economic losses due to their specific biological activities which disrupt cell homeostasis through fundamental mechanisms. Some of the fungal toxins have great potential to serve as future herbicides due to their novel structures and sites of action. With the informations about their charaterization and biological effects one can better understand the ecological roles of fungal toxins. More informations about biological activities and hazard characterization might lead to consideration of the need for regulations and development of strategies against mycotoxin contamination in the environment.

## References

Abbas HK, Boyette CD (1992) Phytotoxicity of fumonisin B1 on weed and crop species. Weed Technol 6:548–552

Abbas HK, Boyette CD, Vesonder RF (1993a) Biological control of weeds using AAL-toxin. United States Patent No. 5,256,628 October 26:1–10

Abbas HK, Vesonder RF, Boyette CD, Peterson SW (1993b) Phytotoxicity of AAL-toxin and other compounds produced by *Alternaria alternata* to jimsonweed (*Datura stramonium*). Can J Bot 71:155–160

Abbas HK, Tanaka T, Duke SO, Porter JK, Wray EM, Hodges L, Sessions AE, Wang E, Merrill AH, Riley RT (1994) Fumonisin and AAL-toxin-induced disruption of sphingolipid metabolism with accumulation of free sphingoid bases. Plant Physiol 106:1085–1093

Bacon CW, Porter JK, Norred WP, Leslie JF (1996) Production of fusaric acid by *Fusarium* species. Appl Environ Microbiol 62:4039–4043

Bennett JW, Klich MA (2003) Mycotoxins. Clin Microbiol Rev 16:497–516

Betina V (1984) Mycotoxins—production, isolation, separation and purification. Elsevier, Amsterdam, pp 217–236

Bhat RV, Vasanthi S (2003) Food safety and food security and food trade—mycotxin food safety risk in developing countries. International Food Policy Research Institute

Blanc PJ, Laussac JP, Le Bars J, Loret MO, Pareilleux A, Prome D, Santerre AL, Goma G (1995) Characterization of monascidin A from Monascus as citrinin. Int J Food Microbiol 27:201–213

Bottini AT, Gilchrist DG (1981) Phytotoxins I, a 1-aminodimethylheptadecapentol from *Alternaria alternate* f.sp. *lycopersici*. Tetrahedron Lett 22:2719–2722

Bottini AT, Bowen JR, Gilchrist DG (1981) Phytotoxins II, characterization of a phytotoxic fraction from *Alternaria alternate* f.sp. *lycopersici*. Tetrahedron Lett 22:2723–2726

Bouizgarne B, El-Maarouf-Bouteau H, Frankart C, Reboutier D, Madiona K, Pennarun AM, Monestiez M, Trouverie J, Amiar Z, Briand J, Brault M, Rona JP, Ouhdouch Y, El Hadrami I, Bouteau F (2006) Early physiological responses of *Arabidopsis thaliana* cells to fusaric acid: toxic and signaling effects. New Phytol 169:209–218

Brugger EM, Wagner J, Schumacher DM, Koch K (2006) Mutagenicity of the mycotoxin alternariol in cultured mammalian cells. Toxicol Lett 164:221–230

Busby F, Wogan GN (1981) Aflatoxins. In: Shank RC (ed) Mycotoxins and N-nitroso compounds: environmental risks, vol 2. Florida, CRC Press

Buttler WH, Wigglesworth JS (1966) The effects of aflatoxin $B_1$ on the pregnant rat. Br J Exp Pathol 47:242–247

Carlton W (1980) Penicillic acid, citrinin and xanthomegnin quinone metabolites a review. Conference on mycotoxins in animal feeds and grains related to animal health, FDA, Rockville, Maryland, US, 13 June 1980, pp 345

Cary JW, Ehrlich KC (2006) Aflatoxigenicity in *Aspergillus*: Molecular genetics, phylogenetic relationships and evolutionary implications. Mycopathologia 162:167–177

Castegnaro M, McGregor D (1998) Carcinogenic risk assessment of mycotoxins. Revue MPedecine VPet-Perinair 149:671–678

Castegnaro M, Maru V, Petkova-Bocharova T, Nikolov I, Bartsch H (1991) Concentrations of ochratoxin A in the urine of endemic nephropathy patients and controls in Bulgaria: Lack of detection of 4-hydroxyochratoxin A. IARC Sci Pub 115:165–169

Chen AY, Yu C, Gatto B, Liu LF (1993) DNA minor groove-binding ligands: a different class of mammalian DNA topoisomerase I inhibitors. Proc Natl Acad Sci 90:8131–8135

Cole RJ, Cox RH (1981) Handbook of toxic fungal metabolites. Academic Press, New York

David MW, Kays SJ, Etherton B (1978) The relationship between pathogenic fungal metabolities (fusaric and picolinic acid), endogenous ethylene evolution and the development of ethylene-like symptoms. Plant Soil 50:355–362

Deshpande SS (2002) Handbook of food toxicology. Marcel Dekker, Inc, New York, p 424

Desjardins AE (2006) *Fusarium* mycotoxins chemistry, genetics and boilogy. APS Press, St Paul

Di Paolo JA, Elis J, Ewin H (1967) Teratogenic response by hamsters, rats and mice to aflatoxin. Nature 215:638–639

Duke SO, Dayan FE (2011) Modes of action of microbially-produced phytotoxins. Toxins (Basel) 3(8):1038–1064

Dutton MF, Heathcote JG (1968) The structure, biochemical properties and origin of the aflatoxins $B_{2a}$ and $G_{2a}$. Chem Inds London 13:418

Dvorackova J, Kusak V, Vesely D, Vesela J, Nesnidal P (1977) Aflatoxin and encephalopathy with fatty degeneration of viscera (Reye). Ann Nutr Aliment 31:977–990

Egbunike GN (1982) Steroidogenic and spermatogenic potentials of the male rat after acute treatment with aflatoxin B1. Andrologia 14:440–446

Egbunike GN, Emerole GO, Aire TA, Ikegwuonu FI (1980) Sperm production rates, sperm physiology and fertility in rats chronically treated with sublethal doses of aflatoxin B1. Andrologia 12:467–475

Fernandez-Pol JA (1977) Iron: possible cause of the G1 arrest induced in NRK cells by picolonic acid. *Biochem. Biophys*. Res Comm 78:136–143

Fernandez-Pol JA, Jonson GS (1977) Selective toxicity induced by picolinic acid in simian virus 40-transformed cells in tissue culture. Cancer Res 37:4276–4279

Ferenc S, Eugene T, Barney W, Shirley JG, Bruce DH, David GG (1996) Synthesis of protein conjugates and development of immunoassays for AAL toxins. J Agric Food Chem 44:1796–1803

Fink-Gremmels J, Dresel J, Leistner L (1991) Use of Monascus extracts as an alternative to nitrate in meat products. Fleischwirtschaft 71:1184–1186

Fotso j, Leslie JF, Smith JS (2002) Production of beauvericin, moniliformin, fusaproliferin, and fumonisins $B_1$, $B_2$, and $B_3$ by fifteen ex-type strains of *Fusarium* species. Appl Environ Microbiol 68:5195–5197

Gelderblom WCA, Marasas WFO, Vleggaar R, Thiel PG, Cawood ME (1992) Fumonisins-isolation, chemical characterisation and biological effects. Mycopathologia 117:11–16

Gilchrist DG, Grogan RG (1976) Production and nature of a host specific toxin from *Alternaria alternate* f.sp. *lycopersici*. Phytopathology 66:165–171

Hendricks K (1999) Fumonisins and neural tube defects in south Texas. Epidemiology 10:198–200

Hendricks RG, Coulter JBS, Lamplugh SM (1982) Aflatoxins and kwashiorkor: a study in sudanese children. Br Med J 285:843–846

Heperkan D, Meric BE, Sismanoglu G, Dalkiliç G, Güler FK (2006) Mycobiota, mycotoxigenic fungi, and citrinin production in black olives. Adv Exp Med Biol 571:203–210

Hidaka H, Nagatsu T, Takeya K (1969) Fusaric acid, a hypotensive agent produced by fungi. J Antibiot 22:228–230

Hill RD, Unrau AM, Canvin DT (1966) The biosynthesis of fusaric acid from 14C-labeled acetate in *Gibberella fujikuroi*. Canad J Chem 44:2077–2082

Hood RD, Hayes AW, Scammell JG (1976) Effects of prenatal administration of citrinin and viriditoxin to mice. Food Cosmet Toxicol 14:175–178

IARC (1993) Chemicals, groups of chemicals, complex mixtures, physical and biological agents and exposure circumstances to be evaluated in future IARC monographs. Report of an ad-hoc Working Group (IARC intern. Rep. No. 93/005)

Ibeh IN, Uraih N, Ogonar JI (1994) Dietary exposure to aflatoxin in human male infertility in Benin city, Nigeria. Int J Fertil Menopausal Studies 39:208–214

Ioannou KL, Aletrali A, Christou E, Hadjioannou-ralli A, Koliou A, Akkelidou D (1999) Surveillance and control of aflatoxins B1, B2, G1, G2 and M1 in foodstuffs in the Republic of Cyprus: 1992-1996. J AOAC Int 82:883–892

Jestoi M (2008) Emerging fusarium-mycotoxins fusaproliferin, beauvericin, enniatins, and moniliformin—a review. Crit Rev Food Sci Nutr 48:21–49

Jones HC, Black HS, Altschul AM (1967) Comparision of the effects of gibberellic acid and aflatoxin in germinating seeds. Nature (London) 214:171–172

Kohler K, Bentrup FW (1983) The effect of fusaric acid upon electrical membrane properties and ATP Level in photoautotrophic cell suspension of *Chenopodium rubrum* L, Z. Pflanzenphysiol 109:355–361

Kohmoto K, Verma VS, Nishimura S, Tagami M, Scheffer RP (1982) New outbreak of Alternaria stem canker of tomato in Japan and production of host-selective toxins by the causal fungus. J Fac Agric Tottori Univ 17:1–8

Krishnamachari DAVR, Bhat V, Nayarjan V, Tilak TBG (1975) Investigation into an outbreak of hepatitis in part of Western India. Indian J Med Res 63:1036–1049

Krogh P, Hald B, Pedersen EJ (1973) Occurrence of ochratoxin A and citrinin in cereals associated with mycotoxic porcine nephropathy. Acta Pathol Microbiol Scand Sect B 81:689–695

Kuzniak E (2001) Effect of fusaric acid on reactive oxygen species and antioxidants in tomato cell culture. J Phytopathol 149:575–582

Kuzniak E, Patykowski J, Urbanek H (1999) Involvement of the antioxidative system in tomato response to fusaric acid treatment. J Phytopathol 147:385–390

Lehmann L, Wagner J, Metzler M (2006) Estrogenic and clastogenic potential of the mycotoxin alternariol in cultured mammalian cells. Food Chem Toxicol 44:398–408

Leslie JF, Zeller KA, Logrieco A, Mule` G, Moretti A, Ritieni A (2004) Species diversity of and toxin production by *Gibberella fujikuroi* species complex strains isolated from native prairie grasses in Kansas. Appl Environ Microbiol 70:2254–2262

Lin CY, Key JL (1968) Cell elongation in the soybean root: the influence of inhibitors of RNA and ptotein biosynthesis. Plant Cell Physiol 9:553–560

Liu GT, Qian YZ, Zhang P, Dong WH (1992) Etiological role of *Alternaria alternata* in human esophageal cancer. Chin Med J 105:394–400

Malini S (1966) Heavy metal chelates of fusaric acid: in vitro spectrophotometry. Phytopathol 57:221–231

Marasas WFO (1996) Fumonisins: history, world-wide occurrence and impact. In: Jackson LS, DeVaries W, Bullerman LB (eds) Fumonisins in food. Plenum Publishing Co, New York, pp 1–17

Marre MT, Vergani P, Albergoni FG (1993) Relationship between fusaric acid uptake and its binding to cell structures in leaves of *Egeria denca* and its toxic effects on membrane permeability and respiration. Physiol Mol Plant Pathol 42:141–157

Mašková Z, Tančinová D, Barboráková Z, Felšöciová S, Císarová M (2012) Comparison of occurrence and toxigenity of *Alternaria* spp. isolated from samples of conventional and new crossbread wheat of Slovak origin. J Microbiol Biotechnol Food Sci 1:552–562

Mehdi NA, Carlton WW, Tuite J (1983) Acute toxicity of citrinin in turkeys and ducklings. Avian Pathol 12:221–233

Mehdi NA, Carlton WW, Tuite J (1984) Mycotoxicoses produced in ducklings and turkeys by dietary and multiple doses of citrinin. Avian Pathol 13:37–50

Merino S, Shaw JG, Tomas JM (2006) Bacterial lateral flagella: an inducible flagella system. FEMS Microbiol Lett 263:127–135

Merrill AH Jr (1991) Cell regulation by sphingosine and more complex sphingolipids. J Bioenerg Biomembr 23:83–104

Merrill AH Jr WE, Gilchrist DG, Riley RT (1993) Fumonisins and other inhibitors of de novo sphingolipid biosynthesis. Adv Lipid Res 26:215–234

Michelle McLean, Michael FD (1995) Cellular interactions and metabolism of aflatoxin: an update. Pharmac Ther 65:163–192

Mirocha CJ, Gilchrist DG, Shier WT, Abbas HK, Wen Y, Vesonder RF (1992) AAL toxins, fumonisins (biology and chemistry) and host-specificity concepts. Mycopathologia 117:47–56

Musser SM, Plattner RD (1997) Fumonisin composition in cultures of *Fusarium moniliforme*, *Fusarium proliferatum* and *Fusarium nygami*. J Agric Food Chem 45:1169–1173

Moussatos VV, Lucas WJ, Gilchrist DG (1993b) AAL-toxin induced physiological changes in Lycopersicon esculentum Mill: differential sucrose transport in tomato lines isogenic for the Asc locus. Physiol Mol Plant Pathol 42:359–371

Nadgornaya NI, Barchtein US, Kurbastskaya ZA (1981) Disruptive effect of various mycotoxins on the mitotic regime of Her-2 cells. Mikrobiol Zh 43:114–118

Nelson TS, Kirby LK, Beasley JN, Johnson ZB, Ciegler A (1985) The effect of drying method and storage time on citrinin activity in corn. Poult Sci 64:464–468

Nelson EB, Burpee LL, Lawton MB (1993) Biological control of turf grass diseases. In: Leslie A (ed) Handbook of integrated pest management for turf and ornamentals. CRC Press, Boca Raton, pp 409–427

Ngindu A, Johnson BK, Kenya PR (1982) Outbreak of acute hepatitis caused by aflatoxin poisoning in Kenya. Lancet I:1346–1348

Nishimura S, Kohmoto K (1983) Host -specific toxins and chemical structures from *Alternaria* species. Ann Rev Phytopathol 21:87–116

Ong T (1975) Aflatoxin mutagenesis. Mutat Res 32:35–53

Palencia ER, Hinton DM, Bacon CW (2010) The black Aspergillus species of maize and peanuts and their potential for mycotoxin production. Toxins 2:399–416

Palumbo JD, O'Keeffe TL, Abbas HK (2008) Microbial interactions with mycotoxigenic fungi and mycotoxins. Toxin Rev 27:261–285

Paterson RR, Rutherford MA (1991) A simplified rapid technique for fusaric acid detection in *Fusarium* strains. Mycopathologia 113:171–173

Peers FG, Lindsel CA (1973) Dietary aflatoxins and liver cancer—a population based study in Kenya. Br J Cancer 27:473–484

Peltonen K, Jestoi M, Eriksen GS (2010) Health effects of moniliformin a poorly understood *Fusarium* mycotoxin. World Mycotoxin J 3(4):403–414

Pero RW, Posner H, Blois M, Harvan D, Spalding JW (1973) Toxicity of metabolites produced by the "*Alternaria*". Environ Health Perspect 4:87–94

Petkova-Bocharova T, Castegnaro M, Michelon J, Maru V (1991) Ochratoxin A and other mycotoxins in cereal from an area of Balkan endemic nephropathy and urinary tract tumours in Bulgaria. In: Castegnaro et al. (eds) Mycotoxins, endemic nephropathy and vurinary tract tumours. IARC, Lyon, pp 83–87

Phillips RD, Hayes AW (1978) Effect of the mycotoxin citrinin on composition of mouse liver and kidney. Toxicon 16:351–359

Pier AC (1991) The influence of mycotoxins on the immune system. In: Smith JE, Henderson RS (eds) Mycotoxins and animal foods. CRC Press, Boca Raton 489–497

Pier AC, McLoughlin ME (1985) Mycotoxic suppression of immunity. In: Lacey J (Ed) Trichothecenes and other mycotoxins. Wiley, New York

Porter JK, Bacon CW, Norred WP (1990) Effects of *Fusarium moniliforme* and corn associated with equine leukoencephalomalacia on rat neurotransmitters and metabolites. Proc Soc Exp Biol Med 194:265–269

Porter JK, Bacon CW, Wray EM, Hagler WM Jr (1995) Fusaric acid in *Fusarium moniliforme* cultures, corn,

and feeds toxic to livestock and the neurochemical effects in the brain and pineal gland of rats. Nat Toxins 3:91–100
Prabha D, Souza LD, Kamat T, Rodrigues C, Naik CG (2009) Batch culture fermentation of *Penicillium chrysogenum* and a report on the isolation of purification, identification and antibiotic activity of citrinin. Indian J Mar Sci 38:38–44
Proctor RH, Busman M, Seo JH, Lee YW, Plattner RD (2008) A fumonisin biosynthetic gene cluster in *Fusarium oxysporum* strain O-1890 and the genetic basis for B versus C fumonisin production. Fungal Genet Biol 45:1016–1026
Raistrick H, Smith G (1941) Antibacterial substances from mould. Citrinin, a metabolic product of *Penicillium citrinum*. Chem Ind 60:828–836
Rani TD, Savitha R, Lavanya L, Kamalalochani S, Bharathiraja B (2009) An overview of Fusaric acid production. Advanced Biotech Mini Rev 8:18–22
Reddy SV, Waliyar F (2000) Properties of aflatoxins and its producing fungi. International Crops Research Institute for the Semi-Arid Tropics (ICRISAT). URL: http://www.icrisat.org/aflatoxin/aflatoxin.asp
Reiss J (1971) Förderung der Aktivität von β-Indolylesigsäure durch Aflatoxin S1. Z Pflanzenphysiol 64:260–262
Rheeder JP, Marasas WFO, Vismer HF (2002) Production of fumonisin analogs by *Fusarium* species. Appl Environ Microbiol 68:2101–2105
Roohi S, Azizi IG, Hashemi M (2012) Fumonisin contamination based on flour quality used in bakeries and confectioneries in Qaemshahr (city of the Northern Iran). African J Microbiol Res 6(8):1815–1818
Sabater-Vilar M, Maas R, Fink-Gremmels J (1999) Mutagenicity of commercial Monascus fermentation products and the role of citrinin contamination. Mutat Res 444:7–16
Sanhueza CEP, Degrossi MC (2004) Moniliformin, a *Fusarium* mycotoxin. Rev Mex Micol 19:103–112
Sansing GA, Lillehoj EB, Detroy RW, Miller MA (1976) Synergistic toxic effects of citrinin, ochratoxin A, and penicillic acid in mice. Toxicon 14:213–219
Sargeant K, Carraghan RB, Allcroft R (1963) Toxic products in groundnuts: chemistry and origin. Chem And Ind 2:53–55
Scott PM, Kanhere SR (2001) Stability of *Alternaria* toxins in fruit juices and wine. Mycotoxin Res 17:9–14
Scott PM, Van Walbeek W, Kennedy B, Anyeti B (1972) Mycotoxins (Ochratoxin A, citrinin and sterigmatocystin) and toxigenic fungi in grains and other agricultural products. J Agric Food Chem 20:1103–1109
Sewram V, Mshicileli N, Shepard GS, Vismer HF, Rheeder JP, Lee YW, Leslie JF, Marasas WFO (2005) Production of fumonisin B and C analogues by several *Fusarium* species. J Agric Food Chem 53:4861–4866
Shank RC, Bhamarapravti N, Gordon JE, Wogan GN (1972) Dietary aflatoxins and human liver cancer. IV Incidence of primary liver cancer in two municipal populations in Thailand. Food Cosmet Toxicol 10:171–179
Sinha KK, Kumar N, Prasad G (1993) The use of mustard (Brassica juncea L.) and gram (Cicer arietinum L.) seedling germination inhibition assay for aflatoxin $B_1$. Mycopatghologia 121:175–178
Sydenham EW, Thiel PG, Vleggaar R (1996) Physicochemical data for some selected *Fusarium* toxins. J AOAC Int 79(6):1365–1379
Tanaka K, Sago Y, Zheng Y, Nakagawa H, Kushiro M (2007) Mycotoxins in rice. Int J Food Microbiol 119:59–66
Terao K, Ueno Y (1978) Morphological and functional damage to cells and tissues. In: Uraguchi K, Yamazaki M (eds) Toxicology, biochemistry and pathology of mycotoxins. Kodansha Press, Tokyo, pp 189–238
Thiel PG, Meyer CJ, Marasas WFO (1982) Natural occurrence of moniliformin together with deoxynivalenol and zearalenone in Transkeian corn. J Agric Food Chem 30:308–312
Van Asch MAJ, Rijkenberg FHJ, Coutinho TA (1992) Phytotoxicity of fumonisin B1, moniliformin and T-2 toxin in corn callus cultures. Phytopathology 82:1330–1332
Van Rij Tjeerd E, Girard G, Lugtenberg BJJ, Bloemberg GV (2005) Influence of fusaric acid on phenazine-1-carboxamide synthesis and gene expression of *Pseudomonas chlororaphis* strain PCL1391. Microbiology 151:2805–2814
Venter SL, Steyn PJ (1998) Correlation between fusaric acid production and virulence of isolates of *Fusarium oxysporum* that cause potato dry rot in South Africa. Potato Res 41:289–294
Vesela D, Vesely D, Jelinek R (1983) Toxic effects of achratoxin A and citrinin, alone and in combination, on chicken embryos. Appl Environ Microbiol 45:91–93
Vesonder RF, Labeda DP, Peterson RE (1992) Phytotoxic activity of selected water-soluble metabolites of *Fusarium* against *Lemna minor* L. (duckweed). Mycopathologia 118:185–189
Voss KA, Porter JC, Bacon CW (1999) Fusaric acid and modifications of the subchronic toxicity to rats of fumonisims in *F. moniliforme* culture material. Food Chem Toxicol 37:853–881
Wang H, Ng TB (1999) Pharmacological activity of fusaric acid (5-butylpicolinic acid). Life Sci 65:849–856
Wang H, Jones C, Ciacci-Zanella J, Holt T, Gilchrist DG, Dickman MB (1996) Fumonisins and *Alternaria alternata* lycopersici toxins: sphinganine analog mycotoxins induce apoptosis in monkey kidney cells. Proc Natl Acad Sci USA 93:3461–3465
Wichmann G, Herbarth O, Lehmann I (2002) The mycotoxins citrinin, gliotoxin, and patulin affect interferon-gamma rather than interleukin-4 production in human blood cells. Environ Toxicol 17(3):211–218
Williams JH, Phillips TD, Jolly PE, Stiles JK, Jolly CM, Aggarwal D (2004) Human aflatoxicosis in developing countries: a review of toxicology, exposure, potential health consequences, and interventions. Am J Clin Nutr 80:1106–1122

Williams JH, Grubb JA, Davis JW, Wang JS, Jolly PE (2010) HIV and hepatocellular and esophageal carcinomas related to consumption of mycotoxin-prone foods in sub-Saharan Africa. Am J Clin Nutr. doi:10.3945/ajcn.2009. 28761

World Health Organisation (1979) Environmental health criteria II: mycotoxins. World Health Organisation, Geneva

Xu DS, King TQ, Ma JQ (1996) The inhibitory effect of extracts from *Fructus lycii* and *Rhizoma polygonati* on in vitro DNA breakage by Alternariol. Biomed Environ Sci 9:67–70

Yabuta T, Kambe K, Hayashi T (1934) Biochemistry of the bakanae¨-fungus. I. Fusarinic acid a new product of the bakanae¨-fungus. J Agric Chem Soc Jpn 10:1059–1068

Yang YG, Mayura K, Spainhour CB Jr, Edwards JF, Phillips TD (1993) Evaluation of the developmental toxicity of citrinin using Hydra attenuata and post-implantation rat whole embryo culture. Toxicology 85(2):179–198

# Bacterial Degradation of Some Organophosphate Compounds

# 48

Deepak Kumar Malik, Divya Bhatia and Meenu Rathi

**Abstract**

Organophosphorus compounds cause high mammalian toxicity and therefore their detoxification from the environment is essential. Bioremediation can be an efficient and cheap option for decontamination of polluted ecosystems. Several bacterial species can degrade a wide range of organophosphorus compounds in liquid cultures and soil systems. Organophosphate hydrolase encoding opd gene has been isolated, sequenced, cloned in different organisms and altered for better activity and stability. Bacteria capable of complete mineralization have been constructed by transferring the complete degradation pathway for specific compounds in one bacterium.

**Keywords**

Chlorpyrifos · Malathion · Coumaphos · Fenamiphos

## 48.1 Introduction

Overall, organophosphorus compounds account for around 38% of total pesticides used worldwide (Singh 2009). Organophosphorus compound poisoning is a worldwide health problem with 200,000 deaths annually (Sogorb et al. 2004, update). Continuous and excessive use of organophosphorus compounds has led to the contamination of several ecosystems (Tse et al. 2004). As these pesticides cause extensive damage to nontarget organisms, studies regarding their degradation have received considerable attention (Goswami and Singh 2009). The use of microorganisms for bioremediation requires a complete understanding of physiological, microbiological, ecological, biochemical and molecular aspects (Iranzo et al. 2001) involved. Mixed bacterial cultures with pesticide degrading ability were isolated but individually were unable to utilize the chemical as an energy source (Roberts et al. 1993). The nonculturable bacteria in the laboratory with

D. K. Malik (✉) · D. Bhatia
Department of Biotechnology, University Institute of Engineering & Technology, Kurukshetra University, Kurukshetra, Haryana, India
e-mail: deepmolbio@rediffmail.com

M. Rathi
Department of Botany, University College, Kurukshetra University, Kurukshetra, Haryana, India

R. N. Kharwar et al. (eds.), *Microbial Diversity and Biotechnology in Food Security,*
DOI 10.1007/978-81-322-1801-2_48, © Springer India 2014

**Table 48.1** Bacteria isolated for the degradation of organophosphorus compounds (chlorpyrifos, parathion, glyphosate, coumaphos, monocrotophos and malathion)

| Compound | Bacteria | Reference |
|---|---|---|
| Chlorpyrifos | *Enterobacter* sp. | Singh et al. (2003) |
| | *Pseudomonas aeruginosa* | Fulekar and Geetha (2008) |
| | *Synechocystis* | Singh et al. (2011) |
| | *Bacillus licheniformis* | Zhu et al. (2010) |
| | *Klebsiella* | Ghanem et al. (2007) |
| Parathion | *Pseudomonas diminuta* | Serdar et al. (1982) |
| | *Agrobacterium radiobacter* | Horne et al. (2002b) |
| | *Serratia* sp. | Suresh et al. (2007) |
| | *Plesiomonas* sp. | Zhongli et al. (2001) |
| Glyphosate | *Pseudomonas* ssp. | Kertesz et al. (1994a) |
| | *Geobacillus caldoxylosilyticus* T20 | Obojska et al. (2002) |
| | *Acetobacter* sp. | Moneke et al. (2010) |
| | *Pseudomonas fluorescence* | Moneke et al. (2010) |
| | *Azotobacter* sp. | Moneke et al. (2010) |
| Monocrotophos | *Bacillus* spp. | Rangaswamy and Venkateswaralu (1992) |
| | *Pseudomonas mendocina* | Bhadbhade et al. (2002a) |
| | *Bacillus megaterium* | Bhadbhade et al. (2002b) |
| | *Pseudomonas aeruginosa* F10B | Singh and Singh (2003) |
| | *Pseudomonas aeruginosa* | Balamurugan et al. (2010) |
| Malathion | *Pseudomonas* sp. | Singh and Seth (1989) |
| | *Pseudomonas* sp. | Imran et al. (2004) |
| | *Pseudomonas* sp. | Sayed et al. (2010) |
| | *Acinetobacter johnsonii* MA19 | Xie et al. (2009) |
| | *Brevibacillus* sp. | Singh et al. (2012) |
| | *Bacillus cereus* | Singh et al. (2012) |
| | *Bacillus thuringiensis* MOS-5 | Zeinat et al. (2008) |

a simulated natural environment (Kaeberlein et al. 2002) may lead to isolation of several new chemical-degrading bacteria. Degradation of fenitrothion by *Burkholderia* sp. strain NF100 was reported (Hayatsu et al. 2000). Dimethoate degradation was reported to be carried out by a plasmid-based gene of *Pseudomonas aeruginosa* MCMB-427 (Deshpande et al. 2001). Utilization of ethoprophos as a sole source of carbon by *Pseudomonas putida* has been observed (Karpouzas et al. 2000). Isolation and metabolism of cadusafos by *Sphingomonas paucimobilis* and *Flavobacterium* sp. have been reported (Karpouzas et al. 2005). *Flavobacterium* sp. and *Pseudomonas diminuta* were isolated by diazinon and parathion enrichment but they can degrade a wide range of other organophosphorus compounds such as coumaphos, methyl parathion, chlorpyrifos and nerve agents (Singh et al. 1999). Most organophosphorus compounds are ester or thiol derivatives of phosphoric, phosphonic or phosphoramidic acid. It has been estimated that one enzyme can hydrolyze 300,000 molecules of acetylcholine every minute. This inhibition causes paralysis and finally death of insects and mammals. The principal reactions in organophosphate degradation are hydrolysis, oxidation, alkylation and dealkylation (Singh et al. 1999). A list of microorganisms capable of degrading these compounds is presented in Table 48.1. In this article, we review degradation of some organophosphorus compounds by bacteria.

## 48.2 Chlorpyrifos

Chlorpyrifos (O, O-diethyl O-(3, 5, 6-trichloro-2-pyridyl) phosphorothioate) is one of the most widely used insecticides effective against a broad spectrum of insect pests of economically important crops. A single application of chlorpyrifos poses risks to small mammals, birds, fish and aquatic invertebrate species (Anonymous 2002). The half-life of chlorpyrifos in soil varies from 10 to 120 days (Racke et al. 1988) with 3, 5, 6-trichloro-2-pyridinol (TCP) as the major degradation product. It was suggested that the accumulation of TCP, which has antimicrobial properties, prevents the proliferation of chlorpyrifos degrading microorganisms (Racke et al. 1990). A bacterial strain *Bacillus licheniformis* ZHU-1 capable of utilizing chlorpyrifos as the sole carbon sources and energy was isolated from the soil. The addition of ZHU-1 to soil treated with chlorpyrifos resulted in a higher degradation rate than noninoculated soils (Zhu et al. 2010). Three aerobic bacterial consortia developed from pesticide-contaminated soils of Punjab (India) were able to degrade chlorpyrifos after 21 days of incubation in basal medium by 54, 46 and 61 % and chlorpyrifos (50 mg/L) in soil after 30 days by 50, 56 and 64 %. *Pseudomonas aeruginosa, Bacillus cereus, Klebsiella* sp. and *Serratia marscecens* obtained from these consortia showed 84, 84, 81 and 80 % degradation of chlorpyrifos (50 mg/L) in liquid medium after 20 days and 92, 60, 56 and 37 % degradation of chlorpyrifos (50 mg/L) in soil after 30 days (Lakshmi et al. 2009). Savitha and Raman (2012) reported that three chlorpyrifos hyper-resistant bacteria, *Bacillus stearothermophilus, Bacillus circulans* and *Bacillus macerans,* were found resistant to 50, 55 and 60 mg/L of chlorpyrifos. GC-MS analysis showed that chlorpyrifos at 10, 25 and 50 mg/L degraded completely over a period of 1, 5 and 7 days, respectively. The intermediate 3, 5, 6 trichloro-2-pyridion, 2, 4-bis (1, 1 dimethyiethyl) phenol and 1, 2 benzenedicarboxylic acid persisted during bioremediation, but in the long run these get converted into $CO_2$, biomass and nutrients. *Pseudomonas aeruginosa* has been used in bioremediation of chlorpyrifos at concentrations up to 50 mg/L, but the organism is inhibited by higher concentrations (Fulekar and Geetha 2008). Hua et al. (2009) concluded that chlorpyrifos residues in the soil had a temporary or short-term inhibitory effect on soil microbial functional diversity. The repeated treatment with chlorpyrifos over many years in an Australian soil resulted in the development of some microorganisms with the capability to use the toxic compound as has been reported with organochlorine compounds (Singh et al. 2000). The aerobic bacteria tend to transform chlorpyrifos by hydrolysis to produce diethylthiophosphoric acid (DETP) and TCP. This transformation reaction removes chlorpyrifos and its mammalian toxicity but yields compounds that are not metabolized by the microorganisms (Horne et al. 2002b). Chlorpyrifos has been reported to be degraded co-metabolically in liquid media by *Flavobacterium* sp. and *P. diminuta,* isolated from a diazinon-treated field and by parathion enrichment (Serdar et al. 1982). An *Enterobacter* sp. degrades chlorpyrifos to DETP and TCP and utilizes DETP as a source of carbon and phosphorus (Singh et al. 2003). Ghanem et al. (2007) reported a chlorpyrifos degrading bacterial strain, *Klebsiella* sp., isolated from an activated sludge sample collected from wastewater treatment plant. Shelton (1988) isolated a consortium that could use DETP as a carbon source but was unable to degrade it when presented as a source of phosphorus or sulfur. Zhu et al. (2010) reported the isolation of *B. licheniformis* ZHU-1 is capable of utilizing chlorpyrifos as the sole carbon source and energy from the soil.

## 48.3 Parathion

Parathion (O, O-diethyl-O-p-nitrophenyl phosphorothioate) is one of the most toxic insecticides registered with the US Environmental Protection Agency (EPA). The microbial degradation of parathion produce hydrolytic product (p-nitrophenol). This pesticide is persistent and very toxic to humans and animals (Olvera-Velona et al. 2008), even at low concentrations. Several species of bacteria have been isolated that can hydrolyze parathion (Racke et al. 1996). Mineralization, where parathion was used as a source of carbon (Rani and Lalitha-kumari 1994) or phosphorus (Rosenberg and Alexander 1979), and co-metabolic hydrolysis (Horne et al. 2002b) have been reported. A *Pseudomonas* sp. and a *Xanthomonas* sp. were isolated, which can hydrolyze parathion and can further metabolize p-nitrophenol (Tchelet et al. 1993). A *Moraxella* sp. can use p-nitrophenol as the sole source of carbon and nitrogen (Spain and Gibson 1991). This bacterium degrades p-nitrophenol to p-benzoquinone using the enzyme p-nitrophenol monooxygenase. *Candida parapsilosis* has been reported to convert hydroquinone to cis, trans-4-hydroxymuconic semialdehyde. This is then metabolized to maleylacetate by semialdehyde dehydrogenase. A *Pseudomonas putida* strain was found to metabolize p-nitrophenol to hydroquinone and 1, 2, 4-benzenetriol, which was further cleaved by benzenetriol oxygenase to maleylacetate (Rani and Lalitha-kumari 1994). A soil bacterium *Serratia* capable of utilizing methyl parathion as sole carbon and energy source was isolated by selective enrichment on minimal medium containing methyl parathion. *Serratia* sp. strain DS001 utilized methyl parathion, p-nitrophenol, 4-nitrocatechol and 1, 2, 4-benzenetriol as the sole carbon and energy sources but could not grow using hydroquinone as a source of carbon. Further, p-nitrophenol and dimethylthiophosphoric acid were found to be the major degradation products of methyl parathion. The key enzymes involved in degradation of methyl parathion and conversion of p-nitrophenol to 4-nitrocatechol are, namely, parathion hydrolase and p-nitrophenol hydroxylase (Suresh et al. 2007). In the recent years, biocatalyst has become a cost-effective technology for solving ecological and energy issues (Chu et al. 2009). The microbial degradation of pesticides residues has become the focus of many studies because it is economical and effective. Therefore, organophosphorus degrading (opd) and methyl parathion-degrading (mpd) genes have been intensively researched (Fu et al. 2004; Zhang et al. 2006). The attempts had been made to enhance organophosphorus hydrolase (OPH) biodegradation efficiency by displaying the OPH onto the cell surface (Yang et al. 2008) or by secreting the OPH into the periplasmic space (Yang et al. 2009). Six carriers were compared in regards to the whole-cell activity of methyl parathion hydrolase (MPH). The surface display systems yielded one to three times of the whole-cell activity than the periplasmic secretion systems (Yang et al. 2011). A different pathway of degradation was reported in *Arthrobacter* sp. strain JS443 and *Arthrobacter protophormiae* RHJ100 where p-nitrophenol was mineralized via p-nitrocatechol. Nitrocatechol is converted into 1, 2, 4-benzenetriol by benzotriol dehydrogenase, which in turn is directly converted into maleylacetate by benzotriol dioxygenase (Chauhan et al. 2000). A consortium of two *Pseudomonas* sp. (strains S1 and S2) was isolated which can also metabolize p-nitrophenol via p-nitrocatechol (Qureshi and Purohit 2002) although in most of the studies on microbial degradation of parathion, the first reaction was hydrolysis of the phosphotriester bond. In one study, degradation of parathion by a mixed culture and a *Bacillus* sp. (Sharmila et al. 1989) was shown to occur by a reduction of the nitro group that was later hydrolyzed to p-aminophenol. Co-metabolic degradation of methyl parathion by *Plesimonas* sp. strain M6 was observed (Zhongli et al. 2001). They also isolated *Pseudomonas* sp. A3 which can utilize p-nitrophenol as a sole source of carbon and nitrogen. This isolate can also utilize a series of aromatic compounds as a sole source of carbon (Zhongli et al. 2002). Another strain of *Pseudomonas* sp. WBC was capable of completing the degradation of methyl parathion and could utilize it as a sole source of carbon and nitrogen (Yali et al. 2002).

## 48.4 Glyphosate

Glyphosate (N-(phosphonomethyl) glycine) is a broad-spectrum herbicide. The mode of action includes inhibition of the plant enzyme 5-enolpyruvylshikimate-3-phosphate synthase, which catalyzes synthesis of the aromatic amino acids (Lu et al. 2005). Microbial degradation of glyphosate produces the major metabolite aminomethylphosphonic acid (AMPA) and ultimately leads to the production of $CO_2$, phosphate and water (Araujo et al. 2003). Research on glyphosate degradation has mainly focused on wastewater treatment processes (Manassero et al. 2010; Leticia et al. 2008). Degradation of glyphosate in soil, under solar light photocatalyzed by $Fe_3O_4$/$SiO_2$/$TiO_2$, was also reported (Xuan et al. 2011). *Arthrobacter* sp. GLP-1 and *Pseudomonas* sp. PG2982 degrade glyphosate, resulting in the production of sarcosine (N-methylglycine) by C-P lyase activity (Dick and Quinn 1995). The second pathway involves the conversion of glyphosate to AMPA. This compound is then dephosphorylated by C-P lyase and further broken down to methylamine and formaldehyde (Lerbs et al. 1990). A thermophile, *Geobacillus caldoxylosilyticus* T20 was isolated that degrades glyphosate by the above pathway, utilizing the compound as a sole source of phosphorus (Obojska et al. 2002). Utilization of aminoalkylphosphonates as a source of nitrogen by different bacterial isolates has been reported (Ternana and McMullan 2000). A strain of *Kluyveromyces fragilis* has been shown to utilize AMPA as a source of nitrogen (Ternana and McMullan 2000). *Streptomyces morookaensis* DSM 40565 could degrade aminoalkylphosphonate as a sole source of nitrogen and phosphorus (Obojska and Lejczak 2003). Alkyl amines are intermediate degradation products for several xenobiotics such as carbofuran, atrazine and monocrotophos and have been reported for different microorganisms (Strong et al. 2002). Use of methylamine as a source of carbon is widespread in nature (Trabue et al. 2001). Three bacteria strains GDP1, GDP2 and GDA were isolated from agricultural soil heavily polluted with glyphosate, which are capable of degrading glyphosate pesticide. The bacteria strains GDP1, GDP2 and GDA were from *P. putida, P. aeruginosa* and *Acetobacter faecalis,* respectively. In *Ochrobactrum anthropi* GPK 3, the initial cleavage reaction is catalyzed by glyphosate oxidoreductase with the formation of AMPA and glyoxylate, whereas *Achromobacter* sp. MPS12 utilizes C-P lyase, forming sarcosine. *Ole K. Borggaard (*2011*)* concluded that the soil phosphorus status did not influence glyphosate and AMPA dissipation at that site. When the mixed bacteria inoculum size was increased, percentage of glyphosate degradation reached a maximum value of 99 % at an optimum pH 6–7 while for pH values higher than 9 or lower than 4, no degradation was observed (Nourouzi et al. 2012). Moneke et al. (2010) showed that *P. fluorescens* and *Acetobacter* sp. exhibited a high capacity to efficiently degrade glyphosate under the environmental conditions studied. Thus, the organisms can be exploited for biodegradation of glyphosate. Ersilia et al. (2010) reported that glyphosate biodegradation capacity in the experimental field indicates that the $CO_2$ fraction accumulated after 50 days is 28.02 %. Weather conditions, especially temperature variations between day and night, influence the activity of soil microorganisms and affect the biodegraded glyphosate percentage. The ability of manganese peroxidase (MnP), laccase, lignin peroxidase (LiP) and horseradish peroxidase (HRP) to degrade the widely used herbicide glyphosate and other pesticides was reported by Leticia et al. (2009).

## 48.5 Monocrotophos

Monocrotophos ((3-hydroxy-N-methyl-cis-crotonamide) dimethyl phosphate) is used to control aphids, leaf hoppers, mites and other foliage pests. Rangaswamy and Venkateswaralu (1992) isolated a monocrotophos degrading *Bacillus* sp. from previously treated soil. *P. aeruginosa* F10B and *Clavibacter michiganense* sp. can utilize monocrotophos as a phosphorus source but not as a carbon source (Singh and Singh 2003). A monocrotophos degrading *Paracoccus* sp. isolated from sludge collected from the waste water treatment pool of a pesticide factory was able to utilize microbial carbon pump (MCP) as sole carbon sources (Jia et al. 2007). Further studies demonstrated that *Pseudomonas mendocina* is the most efficient monocrotophos degrader among the isolated bacteria and its degrading capability is plasmid based (Bhadbhade et al. 2002a). The same group isolated another 17 bacterial isolates from previously exposed soils which can mineralize monocrotophos in liquid culture (Bhadbhade et al. 2002b). Tejomyee et al. (2009) isolated 25 fungal strains by an enrichment method using MCP as a carbon and phosphorus source. Balamurugan et al. (2010) evaluated *P. aeruginosa* and *Trichoderma viridae* for their potential to degrade monochrotophos and methyl parathion. In this investigation, *P. aeruginosa* was more efficient in degrading monochrotophos and methyl parathion. Dimethyl- and monomethyl phosphates were involved as intermediates in monocrotophos degradation in plants and animals (Muck 1994).

Another intermediate identified during monocrotophos degradation was methylamine, produced by an esterase enzyme. As with most of the other organophosphorus compounds, the first degradation step of monocrotophos should involve hydrolysis, which could produce N-methyl acetoacetamide and dimethyl phosphate (Beynon et al. 1973). Further, degradation of N-methyl acetoacetamide produced valeric acid in *A. atrocyaneus* and acetic acid in *B. megaterium* (Bhadbhade et al. 2002b).

## 48.6 Malathion

Malathion (S-(1, 2-dicarbethoxyethyl)-O, O-dimethyldithiophosphate) is used for the control of sucking and chewing insects (Imran et al. 2004). Malathion is one of the acetylcholinesterase (AChE) inhibitors and poses a hazard to humans. The loss of AChE results in acetylcholine accumulation, which interferes with muscular responses and produces serious symptoms in vital organs, eventually leading to death (Fahd and Ahmed 2009). Carboxyesterase activity, which degrades malathion to its monoacid and diacid derivatives, is the predominant metabolic mechanism (Singh and Seth 1989). Xie et al. (2009) reported the biodegradation of malathion by *Acinetobacter johnsonii* MA19 and optimization of co-metabolism substrates. Four compounds, sodium succinate, sodium acetate, glucose and fructose, were tested as the carbon source for the co-metabolism of the strain MA19. The results of sodium succinate concentrations on the malathion biodegradation indicated that more sodium succinate being supplied resulted in quick degradation of malathion and fast cells multiplication. Singh et al. (2012) isolated Brevibacillus sp. strain KB2 and B. cereus strain PU from the soil samples, collected from malathion contaminated field, and analyzed for carboxylesterase activity and maximum activity was observed for strains KB2 and PU, respectively. Cloning and sequencing of the putative malathion degrading carboxylesterase gene was done using primers-based PCR approach. Zeinat et al. (2008) isolated *B. thuringiensis* MOS-5 (Bt) for the degradation of malathion. It was able to utilize malathion as a sole carbon and energy source and to degrade it co-metabolically. Esterase activity involved in malathion degradation was also determined in culture filtrate of MOS-5. Results indicated that esterase activity was twofold more in the presence of yeast extract compared to glucose. These results indicate that Bt MOS-5 may be considered as a highly potential candidate in the biodegradation of organophosphorus in contaminated soil. Sayed et al. (2010) isolated five malathion-degrading bacterial strains from soil samples collected from different agricultural sites in Cairo, Egypt. Malathion was used as a sole source of carbon to enumerate malathion degraders. They were identified as *Pseudomonas* sp., *P. putida*, *Micrococcus lylae*, *Pseudomonas aureofaciens* and *Acetobacter liquefaciens*, respectively. LC/ESI-MS analysis confirmed the degradation of malathion to malathion monocarboxylic and dicarboxylic acids, which formed as a result of carboxylesterase activity. Bourquin (1977) isolated 11 bacterial isolates from salt-marsh environments which utilized malathion as a sole carbon source. The isolated bacteria possessed carboxyesterase activity, which metabolized malathion to its monoacid and diacid derivatives. Small amounts of other metabolites were also produced including desmethyl malathion, phosphorothionates and four carbon dicarboxylic acids which were probably formed as a result of phosphatase activity. Singh and Seth (1989) isolated a *Pseudomonas* M-3 strain which metabolized malathion to its monoacid derivative with the parallel formation of ethanol that was used by the strain as a sole carbon source. Guha et al. (1997) isolated a *Micrococcus* sp. from a malathion-enriched soil, which metabolized malathion. The malathion-degrading *Pseudomonas* strain was isolated from an agricultural soil (Imran et al. 2004).

## 48.7 Detoxifying Enzymes

Most studies of organophosphorus-degrading enzymes have focused on OPH and organophosphorus acid anhydrolase (OPAA). The effects of metal substitution on the catalytic activity of

OPH were studied by removing the native metal (Zn) from purified OPH (Benning et al. 2001). OPH has also been found to contain a two-oxygen-bridging mechanism in the active site suggesting subtle differences compared to OPAA (Vyas et al. 2010). It was suggested that divalent cations increased the activity of enzyme by assisting folding of expressed enzyme in the medium (Manavathi et al. 2005). Another enzyme OPAA from *Alteromonas undina* was isolated and purified (Cheng et al. 1993). It plays an important role in cellular dipeptide metabolism because all OPAAs were found to have activity against several dipeptides (DeFrank and White 2002). Three unique parathion hydrolases were isolated, purified and characterized from Gram-negative bacterial isolates (Mulbry 1992). A unique phosphotriesterase has been characterized from *Nocardioides simplex* NRRL B-24074 (Mulbry 2000). Another novel phosphotriesterase HocA (hydrolysis of caroxon) was isolated from *Pseudomonas monteilli* (Horne et al. 2002c). The first reported enzyme able to degrade the phosphonates, 2-phosphonoacetaldehyde hydrolase (phosphonatase), was isolated from *B. cereus* (La Nauze et al. 1970). Phosphonatase does not degrade phosphomonoesters and is not a metalloenzyme (Kononova and Nesmeyanova 2002). Several microbial isolates have been reported to have further novel enzyme/gene systems but most of these were not isolated or purified such as MPH (Zhongli et al. 2001) and chlorpyrifos-degrading enzyme (Singh et al. 2004).

## 48.8 Coumaphos

Coumaphos (O, O-diethyl-O-(3-chloro-4-methyl-2-oxo-2H-1-benzopyran-7-yl) phosphorothioate) is used for the control of cattle ticks. Three morphologically distinct bacteria were isolated, hydrolyzed coumaphos to DETP and chlorferon. Chlorferon was further metabolized to a-chloro-b-methyl-2, 3, 4-trihydroxy-trans-cinnamic acid (CMTC) (Shelton and Somich 1988). A similar breakdown pathway for the propenoic side chain of substituted cinnamic acid molecule, p-coumaric acid, has been observed in *Pseudomonas* sp. (Tse et al. 2004). Monooxygenase and dioxygenase catalyze the formation of the 2-, 3-, and 4- hydroxy derivatives as substituted acid and/or substituted catechol (Peng et al. 2003). Another microorganism, *Nocardiodes simplex* NRRL B-24074, was found to have a distinct enzymes system for coumaphos degradation (Mulbry 2000). Horne et al. (2002b) isolated an *Agrobacterium radiobacter* P230 capable of hydrolyzing coumaphos from an enrichment culture containing organophosphorus as the sole source of phosphorus. *P. monteilli* degrades coumaphos and diazinon but not parathion (Horne et al. 2002a). Coumaphos is degraded by the other microorganisms *Enterobacter* sp. B-14 (Singh et al. 2004), which were isolated for their ability to degrade other organophosphorus compounds.

## 48.9 Perspectives

Bioremediation with microorganisms is an attractive alternative to these conventional techniques for pollutant disposal. The use of whole living cells for bioremediation presents some difficulties such as delivery of fresh inocula and nutrient composition. To avoid these difficulties, the use of cell-free OPH was carried out successfully. Site-specific mutagenesis in organophosphorus-degrading enzyme OPH has been carried out successfully to increase the catalytic activity against poor substrates, and to decrease the stereoselectivity of the enzyme. Future areas of research include increasing enzyme activity against poor substrates and improving enzyme catalytic activities in mixtures of chemicals. DNA shuffling was successfully used to isolate an improved variant of opd cloned *Escherichia coli,* which can degrade methyl parathion 25 times faster than the wild type (Cho et al. 2002). Introducing cells containing C-P lyase activity in consortia or C-P lyase gene in degrading microorganisms might accelerate the overall degradation process. Screening of anaerobic microorganisms and extremophiles may be useful but this so far has received little attention for organophosphorus compound degradation. A microorganism engineered to complete mineralization of organophosphorus compounds would avoid the generation of toxic hydrolytic products. The introduction of all degradative genes into a single organism allows for future optimization of gene expression, and the potentials to utilize further directed evolution to optimize degradation rates and minimize the metabolic burden placed on the cell. The application of genetic engineering and biochemical techniques, to improve and evolve natural biodegradative capabilities, will ultimately create strains capable of degrading complex mixtures of compounds.

## References


Anonymous (2002) Chlorpyrifos Facts. USEPA 738-F-01–006.

Araujo ASF, Monteiro RTR, Abarkeli RB (2003) Effect of glyphosate on the microbial activity of two Brazilian soils. Chemos 52:799–804

Balamurugan K, Ramakrishnan M, Senthilkumarl R, Ignacimuthu S (2010) Biodegradation of methyl parathion and monocrotophos by *Pseudomonas aeruginosa* and *Trichoderma viridae*. Asian J Sci Technol 6:123–126

Benning MM, Sims H, Raushel FM, Holden HM (2001) High resolution X-ray structures of different metal-substituted forms of phosphotriesterase from *Pseudomonas diminuta*. Biochem 40:2712–2722

Beynon KI, Hutson DH, Wright AN (1973) The metabolism and degradation of vinyl phosphate insecticides. Residue Rev 47:55–142

Bhadbhade BJ, Dhakephalkar PK, Sarnik SS, Kanekar PP (2002a) Plasmid-associated biodegradation of an organophosphorus pesticide, monocrotophos, by soil bacteria. J Appl Microbiol 93:224–234

Bhadbhade BJ, Sarnik SS, Kanekar PP (2002b) Biomineralization of an organophosphorus pesticide, monocrotophos, by soil bacteria. J Appl Microbiol 93:224–234

Bourquin AW (1977) Degradation of malathion by salt-marsh microorganisms. Appl Environ Microbiol 33:356–362

Chauhan A, Chakraborti AK, Jain RK (2000) Plasmid encoded degradation of p-nitrophenol and 4-nitrocatechol by *Arthrobacter protophormiae*. Biochem Biophys Res Commun 270:733–740

Cheng T-C, Harvey SP, Stroup AN (1993) Purification and properties of a highly active organophosphorus acid anhydrolase from *Alteromonas undina*. Appl Environ Microbiol 59:3138–3140

Cho CM-H, Mulchandani A, Chen W (2002) Bacterial cell surface display of organophosphorus hydrolase for selective screening of improved hydrolysis of organophosphate nerve agents. Appl Environ Microbiol 68:2026–2030

Chu YF, Hsu CH, Soma PK, Lo YM (2009) Immobilization of bioluminescent *E. coli* cells using natural and artificial fibers treated with polyethyleneimine. Bioresour Technol 100:3167–3174

DeFrank JJ, White WE (2002) Phosphofluoridates: biological activity and biodegradation. In: Neilson AH (ed) The handbook of environmental chemistry. Springer-Verlag, Berlin

Deshpande NM, Dhakephalkar PK, Kanekar PP (2001) Plasmid-mediated dimethoate degradation in *Pseudomonas aeruginosa* MCMB-427. Lett Appl Microbiol 33:275–279

Dick RE, Quinn JP (1995) Glyphosate-degrading isolates from environmental samples: occurrence and pathway of degradation. Appl Microbiol Biotech 43:545–550

Ersilia A, Roxana M, Monica N, Renata N, Olimpia I (2010) Research on the weed control degree and glyphosate soil biodegradation an apple plantation (pioneer verity). Fascicula Biologie 17(1):5–8

Fahd A, Ahmed AM (2009) Phytoremediation and detoxification of two organophosphorous pesticides residues in Riyadh area. World Appl Sci J 6(7):987–998

Fu GP, Cui ZL, Huang TT, Li SP (2004) Expression, purification, and characterization of a novel methyl parathion hydrolase. Protein Expr Purif 36:170–176

Fulekar MH, Geetha M (2008) Bioremediation of chlorpyrifos by *Pseudomonas aeruginosa* using scale up technique. J Appl Biosci 12:657 -660

Ghanem I, Orfi M, Shamma M (2007) Biodegradation of chloropyriphos by *Klebsiella* sp. isolated from an activated sludge sample of waste treatment plant in Damascus. Folia Microbiol 52:423–427

Goswami S, Singh DK (2009) Biodegradation of α- and β-endosulfan in broth medium and soil microcosm by bacterial strain *Bordetella* sp. B9. Biodegrad 20:199–207

Guha A, Kumari B, Roy MK (1997) Possible involvement of plasmid in degradation of malathion and chlorpyrifos by *Micrococcus* sp. Folia Microbiol 42:574–576

Hayatsu M, Hirano M, Tokuda S (2000) Involvement of two plasmids in fenitrothion degradation by *Burkholderia* sp. strain NF1000. Appl Environ Microbiol 66:1737–1740

Horne I, Harcourt RL, Sutherland TD, Russell RJ, Oakeshott JG (2002a) Isolation of a *Pseudomonas monteilli* strain with a novel phosphotriesterase. FEMS Microbiol Lett 206:51–55

Horne I, Sutherland TD, Harcourt RL, Russell RJ, Oakeshott JG (2002b) Identification of an opd (organophosphate degradation) gene in an *Agrobacterium* isolate. Appl Environ Microbiol 68:3371–3376

Horne I, Sutherland TD, Oakeshott JG, Russell RJ (2002c) Cloning and expression of the phosphotriesterase gene hocA from *Pseudomonas monteilli* C11. Microbiol 148:2687–2695

Hua F, Yunlong Y, Xiaoqiang C, Xiaoe Y, Jingquan Y (2009) Degradation of chlorpyrifos in laboratory soil and its impact on soil microbial functional diversity. J Environ Sci 21(3):380–386

Imran H, Altaf KM, Kim JG (2004) Malathion degradation by *Pseudomonas* using activated sludge treatment system (biostimulator). Biotechnol 3:82–89

Iranzo M, Sain-Pardo I, Boluda R, Sanchez J, Mormeneo S (2001) The use of microorganisms in environmental remediation. Annals Microbiol 51:135–143

Jia KZ, Li XH, He J, Gu LF, Ma JP, Li SP (2007) Isolation of a monocrotophos degrading bacterial strain and characterization of enzymatic degradation. Huan Jing Ke Xue 28(4):908–912

Kaeberlein T, Lewis K, Epstein SS (2002) Isolating "uncultivable" microorganisms in pure culture in a simulated natural environment. Sci 296:1127–1129

Karpouzas DG, Morgan JAW, Walker A (2000) Isolation and characterization of ethoprophos-degrading bacteria. FEMS Microbiol Ecol 33:209–218

Karpouzas D, Fotopoulou A, Menkissoglu-Spiroudi U, Singh BK (2005) Non-specific biodegradation of the organophosphorus pesticides, cadusafos and ethoprophos by two bacterial isolates. FEMS Microbiol Ecol 53:369–378

Kertesz MA, Cook AM, Leisinger T (1994a) Microbial metabolism of sulfur and phosphorus-containing xenobiotics. FEMS Microbiol Rev 15:195–215

Kononova SV, Nesmeyanova MA (2002) Phosphonates and their degradation by microorganisms. Biochem (Moscow) 67:184–195

La Nauze JM, Rosenberg H, Shaw DC (1970) The enzymatic cleavage of the carbon-phosphorus bond: purification and properties of phosphonatase. Biochim Biophys Acta 121:332–350

Lakshmi CV, Kumar M, Khanna S (2009) Biodegradation of chlorpyrifos in soil by enriched cultures. Curr Microbiol 58:35–38

Lerbs W, Stock M, Parthier B (1990) Physiological aspects of glyphosate degradation in *Alcaligenes* sp. strain GL. Arch Microbiol 153:146–150

Leticia P, Pilar CMD, John S (2008) Degradation of Glyphosate and other pesticides by ligninolytic enzymes. Biodegrad 2:195–199

Leticia P, Marıa delPC, John S (2009) Degradation of glyphosate and other pesticides by ligninolytic enzymes. Biodegrad 20:751–759

Lu X, Zhao BZ, Zhang JB, Deng JC, Li P, Xin XL (2005) Property and environmental behavior of herbicide glyphosate. Chin J Soil Sci 5:785–790

Manassero A, Passalia C, Negro AC, Cassano AE, Zalazar CS (2010) Glyphosate degradation in water employing the H2O2/UVC process. Water Res 13:3875–3882

Manavathi B, Pakala SB, Gorla P, Merrick M, Siddavattam D (2005) Influence of zinc and cobalt on expression and activity of parathion hydrolase from *Flavobacterium* sp. ATCC27551. Pestic Biochem Physiol 83:37–45

Moneke AN, Okpala GN, Anyanwu CU (2010) Biodegradation of glyphosate herbicide in vitro using bacterial isolates from four rice fields. Afr J Biotechnol 9(26):4067–4074

Muck W (1994) Metabolism of monocrotophos in animals. Rev Environ Contam Toxicol 139:59–65

Mulbry WW' (1992) The aryldialkylphosphatase-encoding gene adpB from *Nocardia* sp. strain B-1: cloning, sequencing and expression in *Escherichia coli*. Gene 121:149–153

Mulbry WW (2000) Characterization of a novel organophosphorus hydrolase from *Nocardiodes simplex* NRRL B-24074. Microbiol Res 154:285–288

Nourouzi MM, Chuah TG, Choong TS, Rabiei F (2012) Modeling biodegradation and kinetics of glyphosate by artificial neural network. J Environ Sci Health B 3(5):455–465

Obojska A, Lejczak B (2003) Utilization of structurally diverse organophosphonates by *Streptomyces*. Appl Microbiol Biotechnol 62:557–563

Obojska A, Ternana NG, Lejczak B, Kafarski P, McMullan P (2002) Organophosphate utilization by the thermophile *Geobacillus caldoxylosilyticus* T20. Appl Environ Microbiol 68:2081–2084

Borggaard OK (2011) Does phosphate affects soil sorption and degradation of glyphosate—a review. Trends Soil Sci Plant Nutri 2(1):50–64

Olvera-Velona A, Benoit P, Barriuso E, Ortiz-Hernandez L (2008) Sorption and desorption of organophosphate pesticides, parathion and cadusafos, on tropical agricultural soils. Agron Sustain Dev 28:231–238

Peng X, Misawa N, Harayama S (2003) Isolation and characterization of thermophilic *Bacilli* degrading cinnamic, 4-coumaric and ferulic acids. Appl Environ Microbiol 69:1417–1427

Qureshi AA, Purohit HJ (2002) Isolation of bacterial consortia for degradation of p-nitrophenol from agricultural soil. Annals Appl Biol 140:159–162

Racke KD, Coats JR, Titus KR (1988) Degradation of chlorpyrifos and its hydrolysis products, 3,5,6-trichloro-2- pyridinol, in soil. J Environ Sci Health B 23:527–539

Racke KD, Laskowski DA, Schultz MR (1990) Resistance of chlorpyrifos to enhanced biodegradation in soil. J Agric Food Chem 38:1430–1436

Racke KD, Steele KP, Yoder RN, Dick WA, Avidov E (1996) Factors effecting the hydrolytic degradation of chlorpyrifos in soil. J Agric Food Chem 44:1582–1592

Rangaswamy V, Venkateswaralu K (1992) Degradation of selected insecticides by bacteria isolated from soil. Bull Environ Contam Toxicol 49:797–804

Rani NL, Lalitha-kumari D (1994) Degradation of methyl parathion by *Pseudomonas putida*. Can J Microbiol 4:1000–1004

Roberts SJ, Walker A, Parekh NR, Welsh SJ, Waddington MJ (1993) Studies on a mixed bacterial culture from soil which degrades the herbicide linuron. Pestic Sci 39:71–78

Rosenberg A, Alexander M (1979) Microbial cleavage of various organophosphorus insecticides. Appl Environ Microbiol 37:886–891

Savitha K, Saraswathi Raman DN (2012) Isolation, identification, resistance profile and growth kinetics of chlorpyrifos resistant bacteria from agricultural soil of Bangalore. Res Biotechnol 3(2):08–13

Sayed KG, Iman EE, Taha AK, Walaa El-S, Mervat EM (2010) Screening for and isolation and identification of malathion-degrading bacteria: cloning and sequencing a gene that potentially encodes the malathion-degrading enzyme, carboxylestrase in soil bacteria. Biodegrad 21:903–913

Serdar CM, Gibson DT, Munnecke DM, Lancaster JH (1982) Plasmid involvement in parathion hydrolysis by *Pseudomonas diminuta*. Appl Environ Microbiol 44:246–249

Sharmila M, Ramanand K, Sethunathan N (1989) Effect of yeast extract on the degradation of organophosphorus insecticides by soil enrichment and bacterial cultures. Can J Microbiol 35:1105–1110

Shelton DR (1988) Mineralization of diethylthiophosphoric acids by an enriched consortium from cattle dip. Appl Environ Microbiol 54:2572–2573

Shelton DR, Somich CJ (1988) Isolation and characterization of coumaphos-metabolising bacteria from cattle dip. Appl Environ Microbiol 54:2566–2571

Singh BK (2009) Organophosphorus-degrading bacteria:ecology and industrial applications. Nat Rev Microbiol 7:156–164

Singh AK, Seth PK (1989) Degradation of malathion by microorganisms isolated from industrial effluents. Bull Environ Contam Toxicol 43:28–35

Singh S, Singh DK (2003) Utilization of monocrotophos as phosphorus source by *Pseudomonas aeruginosa* F10B and *Clavibacter michiganense* subsp. *insidiosum* SBL 11. Can J Microbiol 49:101–109

Singh BK, Kuhad RC, Singh A, Lal R, Triapthi KK (1999) Biochemical and molecular basis of pesticide degradation by microorganisms. Crit Rev Biotechnol 19:197–225

Singh BK, Kuhad RC, Singh A, Tripathi KK, Ghosh PK (2000) Microbial degradation of the pesticide lindane. Adv Appl Microbiol 47:269–298

Singh BK, Walker A, Morgan JAW, Wright DJ (2003) Effect of soil pH on the biodegradation of chlorpyrifos and isolation of chlorpyrifos-degrading bacterium. Appl Environ Microbiol 69:5198–5206

Singh BK, Walker A, Morgan JAW, Wright DJ (2004) Biodegradation of chlorpyrifos by *Enterobacter* strain B-14 and its use in the bioremediation of contaminated soils. Appl Environ Microbiol 70:4855–4863

Singh DP, Khattar JI, Nadda J, Singh Y, Garg A, Kaur N, Gulati A (2011) Chlorpyrifos degradation by the cyanobacterium *Synechocystis* sp. strain PUPCCC 64. Environ Sci Pollut Res Int 18(8):1351–1359

Singh B, Kaur J, Singh K (2012) Biodegradation of malathion by *Brevibacillus* sp. strain KB2 and *Bacillus cereus* strain PU. World J Microbiol Biotechnol 28(3):1133–1141

Sogorb MA, Vilanova E, Carrera V (2004) Future application of phosphotriesterases in the prophylaxis and treatment of organophosphorus insecticide and nerve agent poisoning. Toxicol Lett 151:219–233

Spain JC, Gibson DT (1991) Pathway for biodegradation of p-nitrophenol in a *Moraxella* species. Appl Environ Microbiol 57:812–819

Strong LC, Rosendahl C, Johnson G, Sadowsky MJ, Wackett LP (2002) *Arthrobacter aurescens* TC1 metabolizes diverse s-triazine ring compounds. Appl Environ Microbiol 68:5973–5980

Suresh BP, Purushotham G, Aleem BP, Ravi Kumar K, Rajasekhar B, Mahesh A, Mike M, Dayananda S (2007) Biodegradation of methyl parathion and p-nitrophenol: evidence for the presence of a p-nitrophenol 2-hydroxylase in a gram-negative *Serratia* sp. strain DS001. Appl Microbiol Biotechnol 73:1452–1462

Tchelet R, Levanon D, Mingelrin D, Henis Y (1993) Parathion degradation by a *Pseudomonas* sp. and a *Xanthomonas* sp. and by their crude enzyme extracts as affected by some cations. Soil Biol Biochem 25:1665–1671

Tejomyee SB, Pravin RP (2009) Microbial degradation monocrotophos by *Aspergillus oryzae*. Int Biodeter Biodegr 63(4):503–508

Ternana NG, McMullan G (2000) The utilization of 4-aminobutylphosphonate as sole nitrogen source by a strain of *Kluyveromyces fragilis*. FEMS Microbiol Lett 184:237–240

Trabue SL, Ogram AV, Ou L-T (2001) Dynamics of carbofuran degrading microbial communities in soil during three successive annual applications of carbofuran. Soil Biol Biochem 33:75–81

Tse H, Comba M, Alaee M (2004) Methods for the determination of organophosphate insecticides in water, sediments and biota. Chemos 54:41–47

Vyas NK, Nickitenko A, Rastogi VK, Shah SS, Quiocho FA (2010) Structural insights into the dual activities of the nerve agent degrading organophosphate anhydrolase/prolidase. Biochem 49:547–559

Xie S, Liu J, Li L, Qiao C (2009) Biodegradation of malathion by Acinetobacter johnsonii MA19 and optimization of cometabolism substrates. J Environ Sci 21(1):76–82

Xuan X, Fangying J, Zihong F, Li H (2011) Degradation of glyphosate in soil photocatalyzed by $Fe_3O_4/SiO_2/TiO_2$ under solar light. Int J Environ Res Public Health 8:1258–1270

Yali C, Xianen Z, Hong L, Yinshan W, Xiangming X (2002) Study on *Pseudomonas* sp. WBC-3 capable of complete degradation of methyl parathion. Weishengwu Xuebao 42:490–497

Yang C, Cai N, Dong M, Jiang H, Li J, Qiao C, Mulchandani A, Chen W (2008) Surface display of MPH on *Pseudomonas putida* JS 4444 using ice nucleation protein and its application in detoxification of organophosphate. Biotechnol Bioeng 99:30–37

Yang C, Freudl R, Qiao C (2009) Export of methyl parathion hydrolase to the periplasm by the twin-arginine translocation pathway in *Escherichia coli*. J Agric Food Chem 57:8901–8905

Yang J, Liu R, Hong J, Yang Y, Qiao C (2011) Selection of a whole-cell biocatalyst for methyl parathion biodegradation. Appl Microbiol Biotechnol 10:3792–3801

Zeinat KM, Nashwa AH, Fetyan A, Mohamed AI, Sherif El-N (2008) Biodegradation and detoxification of malathion by of *Bacillus thuringiensis* MOS-5. Aust J Basic Appl Sc 2(3):724–732

Zhang RF, Cui ZL, Zhang XZ, Jiang JD, Gu JD, Li SP (2006) Cloning of the organophosphorus pesticide hydrolase gene clusters of seven degradative bacteria isolated from a methyl parathion contaminated site and evidence of their horizontal gene transfer. Biodegrad 17:465–472

Zhongli C, Shunpeng L, Guoping F (2001) Isolation of methyl parathion-degrading strain M6 and cloning of the methyl parathion hydrolase gene. Appl Environ Microbiol 67:4922–4925

Zhongli C, Ruifu Z, Jian H, Shunpeng L (2002) Isolation and characterization of a p-nitrophenol degradation *Pseudomonas* sp. strain p3 and construction of a genetically engineered bacterium. Weishengwu Xuebao 42:19–26

Zhu J, Zhao Y, Qiu J (2010) Isolation and application of a chlorpyrifos-degrading *Bacillus licheniformis* ZHU-1. Afr J Microbiol Res 4:2410–2413

# Biodiversity and Conservation of Forest Fungi of Central India

49

R. K. Verma

**Abstract**

Biodiversity of the forest fungi of central India (Chhattisgarh, Madhya Pradesh, and part of Maharashtra) has been recorded. A total of 838 fungi belonging to 321 genera (199 genera of ascomycetes, 107 basidiomycetes, 10 phycomycetes, and 3 myxomycetes) were recorded on different substrates from the forests. Among them, 529 species belong to ascomycete, 282 to basidiomycete, and 26 to lower fungi and myxomycete. The maximum number of fungi, 347 recorded on leaf, followed by 259 on stem and wood, 121 in soil/on ground, 33 in litter, 19 on roots, 24 on seed/pods/seedlings and 2 on insects, were reported on leaf causing common leaf diseases. Fungi recorded on stem, branches, twigs, culms, etc. are either causing cankers, twig blights, die back, etc. or decay and deterioration of these organs in standing trees as well as fallen parts on the forest floor. Top dying and root rot of teak caused by *Phomopsis tectonae* and *Helicobasidium compactum* are the diseases causing serious damage in plantations of Madhya Pradesh and Maharashtra, respectively. Some fungi, for example species of *Amanita, Astraeus, Boletus, Geastrum, Lepiota, Pisolithus, Ramaria, Russula, Scleroderma, Thelephora,* etc., produce ectomycorrhizae while species of *Coriolopsis, Daedalea, Daldinia, Earliella, Favolus, Flavodon, Funalia, Ganoderma, Hypoxylon, Hymenochaete, Inonotus, Irpex, Junghunia, Lenzites, Microporus, Navisporus, Phellinus, Polyporus, Polystictus, Poria, Pyrofomes, Rigidoporus, Skeletocutis, Stereum, Tremates, Xylaria*, etc. are the common wood-decay fungi. Common mushroom collected from the ground and decaying wood and litter are species of *Agaricus, Amanita, Agrocybe, Coprinus, Lepiota, Marasmius, Mycena, Pleurotus, Termitomyces, Tricholoma,* and *Volvariella*. Root rot of *Dalbergia sissoo* and *Acacia catechu* caused by *Ganoderma lucidum* is posing a serious threat in central India. *Spongipellis spumeus* causes root

R. K. Verma (✉)
Forest Pathology Division, Tropical Forest Research Institute, PO-RFRC, Jabalpur 482 021, MP, India
email: rkverma28@rediffmail.com; vermaramk@icfre.org

R. N. Kharwar et al. (eds.), *Microbial Diversity and Biotechnology in Food Security*,
DOI 10.1007/978-81-322-1801-2_49, © Springer India 2014

rot in the mature trees of *Albizia procera* and is also causing considerable damage. Some recently recorded new diseases include vascular wilt of aonla caused by *Fusarium solani* and root rot of teak seedlings (only occurs under water stress condition) caused by *Helicobasidium compactum* and *Tritirachium roseum.* Two new genera and 30 new species were described from central India during the last 5 years. Recently described new species include *Asterostomella shoreae, Cheilymenia jabalpurensis, Nitschkia tectonae,* and *Passalora emblicae.* In central India over 2,700 fungi were collected during the last 15 years, out of them 12.3 % fungi were collected only once and are considered as threatened and needs conservation. Only 1.8 % fungi were frequently collected (more than 11 times during this period) including *Phomopsis tectonae,* whose population is recorded increasing in teak growing areas. The data indicate that the population of major fungi is dwindling with time.

**Keywords**

Forest fungi · Fungal diversity · Conservation of fungi · Leaf spot · Root rot · Soil fungi · Twig spot · Wood rot

## 49.1 Introduction

Fungi occurring in forests of central India are not systematically explored; only some sporadic studies have been made from time to time. The population of soil fungi in Achanakmar and Lamni ranges was studied and 63 fungi were reported (Chakraborty et al. 1991). Shettyi (1957) also reported *Absidia butleri* and *Aspergillus versicolor* from the soil of Amarkantak. *Cercosporidium helicteri*, a new parasitic fungus, was reported from Amarkantak (Soni et al. 1984). The needle blight in different provenances of *Pinus roxburghii* due to the fungus *Pseudocercospora* (*Cercospora*) *pini-densiflorae* was studied (Jamaluddin et al. 1990). *Grevillea pteridifolia* planted on mined-over areas of Amarkantak suffer from charcoal root rot disease caused by *Macrophomina phaseolina,* leaf spot by *Phoma sorghina, Phoma glomerata, Cytospora* sp., and *Pestalotiopsis* sp. (Dadwal and Jamaluddin 1991). Leaf spot and blight diseases of three medicinal plants caused by *Alternaria alternata* were reported from Amarkantak and Lamni (Jamaluddin et al. 1993). Population of arbuscular mycorrhizal (AM) fungi and root colonization in bauxite mined-over areas of Amarkantak was studied (Jamaluddin and Chandra 1997). Upreti et al. (2007) and Nayaka et al. (2007) have made lichenological studies in Achanakmar–Amarkantak biosphere reserve and reported 37 lichens, belonging to different families. Harsh et al. (1989) reported two edible fungi from Madhya Pradesh. Market assessment and business potential of six edible fungi collected from Amarkantak biosphere reserve was also done (Harsh, Rai and Ayachi 1993).

A study on wide range of fungi occurring in forest of central India was undertaken to document the forest fungi. The fungi were identified, classified, and are reported in the present chapter. This information may be useful in further investigation on fungi and for better utilization of their potential, in managing forest ecosystem and economy of tribal people inhabiting these forests.

The study area falls in between 75–85° N and 18–26° E and includes three central Indian states, namely Chhattisgarh, Madhya Pradesh, and northeastern parts of Maharashtra. The forest area of Madhya Pradesh and Chhattisgarh is dominated by sal (*Shorea robusta*) forests followed by teak (*Tectona grandis*) while teak dominated in the forests of Maharashtra. Fungi were reported from different forest types of the above states in

**Table 49.1** Substrate-wise list of forest fungi of central India

| Substrate | Number of species | Number of genera |
|---|---|---|
| Bark | 1 | 1 |
| Entmogenous | 2 | 2 |
| Ground | 68 | 19 |
| Hyperparasites | 2 | 1 |
| Leaf | 347 | 106 |
| Litter | 33 | 26 |
| Pod | 3 | 1 |
| Root | 19 | 7 |
| Seed | 8 | 4 |
| Seedling | 13 | 5 |
| Soil | 53 | 20 |
| Stem/wood | 259 | 124 |
| Total | 838 | 321 |

different seasons from parts of trees, associated plants, woods, twigs, logs, forest litter, forest flora. Also, rhizosphere soils of tree species are included in the present chapter. The fungi were recorded from 161 species of trees, 8 species of bamboos and grasses, 26 shrubs, 15 climbers, 12 herbaceous plants, and 2 ferns. Maximum number of fungi (71) were recorded on sal followed by teak (33), *Terminalia tomentosa* (22) and *Diospyros melanoxylon* (18) (Verma et al. 2008).

## 49.2 Diversity of Forest Fungi

In the present study, a total of 838 fungi belonging to 321 genera (199 genera of Ascomycetes, 107 Basidiomycetes, 10 Phycomycetes, and 3 Myxomycetes) growing on different substrates in the forests have been collected. Among them, 529 fungi belong to Ascomycete, 282 to Basidiomycete, and 26 to lower fungi. Further, 347 fungi were recorded on leaf, 259 on stem and wood, and 121 in soil/on ground, 33 in litter, 19 on roots, 24 on seed/pods/seedlings, and 2 on insects (Table 49.1). The fungi causing diseases in seedlings not only affect the seedling growth in nursery but also get transferred to forest when such seedlings are planted in the field. For example, *G. pteridifolia* planted on mined-over areas of Amarkantak suffer from charcoal root rot disease by *M. phaseolina*, leaf spot by *P. sorghina*, *P. glomerata*, *Cytospora* sp., and *Pestalotiopsis* sp. These diseases were transmitted from the nursery (Dadwal and Jamaluddin 1991). A comprehensive list of identified forest fungi genera with number of species, their broad classification, and substrate is given in Table 49.2.

### 49.2.1 Soil Fungi

The forest trees of central India (most deciduous trees) prefer fungal soils. The trees standing for years support fungi to survive. It is also suspected that the decline in forest soil is related to the loss in fungal diversity. A study conducted on the forests of Achanakmar and Lamni ranges of Chhattisgarh reported 85 species of soil fungi, from both fenced and unfenced areas, and was correlated with various factors involved in the degradation of soil and their vegetation covers; and it was concluded that the fungi are responsible for improving the physicochemical properties of the soil, resulting in an overall improvement in soil fertility and productivity (Chakraborty et al. 1991).

### 49.2.2 Foliicolous Fungi

Among parasitic fungi, the leaf spot fungi are the most common, as the foliage are a major plant organ providing greater surface area for growth and development of fungi. Foliage fungi destroy the leaves in standing trees, for example, leaf spot caused by *Pseudocercospora* in *Anogeissus* spp. and *Rhytisma lagerstroemiae* tar spot in *Lagerstroemia* spp. are the most common in central India. In the present study, 347 fungi were collected on foliage of different plants (Table 49.1). *Cercosporidium helicteri* causing leaf spotting in *Helicteres isora* was reported from Amarkantak (Soni et al. 1984). Jamaluddin et al. (1986) had also reported 36 fungi as new host records from Madhya Pradesh including *Asteromella* sp. on *Casearia elliptica* Willd., *Casearia graveolens* Dalz., *Casearia glauca* (Roxb.) Kuntze, and *Garuga pinnata* Roxb; *Myrothecium roridum* Tode ex. Fr. on *Careya arborea* and *C. glauca*; *Phomopsis bauhiniae* on *Bauhinia retusa* Roxb., *Phomopsis natsume* on *Ziziphus xylopy-*

**Table 49.2** List of fungi recorded from forests of central India along with the main classification group and substrate

| S. No. | Fungi (genera) | Number of species | Broad group | Substrate |
|---|---|---|---|---|
| 1 | *Absidia* | 1 | Ascomycete | Litter |
| 2 | *Achlya* | 1 | Phycomycete | Litter |
| 3 | *Acarocybella* | 1 | Ascomycete | Leaf spot |
| 4 | *Acaulospora* | 2 | Phycomycete | Soil |
| 5 | *Achaetomium* | 1 | Ascomycete | Litter |
| 6 | *Acremonium* | 1 | Ascomycete | Hyperparasite |
| 7 | *Acrodictiella* | 1 | Ascomycete | Leaf spot |
| 8 | *Acrophialophora* | 1 | Ascomycete | Seed |
| 9 | *Acrostroma* | 2 | Ascomycete | Stem |
| 10 | *Aecidium* | 3 | Basidiomycete | Leaf |
| 11 | *Agaricus* | 5 | Basidiomycete | On ground |
| 12 | *Agrocybe* | 2 | Basidiomycete | Ground |
| 13 | *Alternaria* | 3 | Ascomycete | Leaf |
| 14 | *Amanita* | 7 | Basidiomycete | Ground |
| 15 | *Amerodiscosiella* | 1 | Ascomycete | Stem |
| 16 | *Anellophragmia* | 1 | Ascomycete | Stem |
| 17 | *Annellophora* | 1 | Ascomycete | Leaf |
| 18 | *Antennariella* | 1 | Ascomycete | Leaf |
| 19 | *Antrodia* | 1 | Basidiomycete | Stem |
| 20 | *Aplosporella* | 6 | Ascomycete | Stem, twigs |
| 21 | *Arthrinium* | 3 | Ascomycete | Stem/litter |
| 22 | *Ascochyta* | 3 | Ascomycete | Leaf |
| 23 | *Aspergillus* | 13 | Ascomycete | Soil/litter |
| 24 | *Asperisporium* | 1 | Ascomycete | Leaf |
| 25 | *Asterina* | 2 | Ascomycete | Leaf |
| 26 | *Asterinella* | 1 | Ascomycete | Leaf |
| 27 | *Astraeus* | 1 | Basidiomycete | Gorund |
| 28 | *Asteromella* | 3 | Ascomycete | Leaf/litter |
| 29 | *Asterostomella* | 1 | Ascomycete | Leaf/litter |
| 30 | *Auricularia* | 1 | Basidiomycete | Stem |
| 31 | *Aurificaria* | 1 | Basidiomycete | Stem |
| 32 | *Australohydnum* | 1 | Basidiomycete | Stem |
| 33 | *Bagnisiella* | 3 | Ascomycete | Stem |
| 34 | *Bahupaatra* | 1 | Ascomycete | Litter |
| 35 | *Balladynopsis* | 1 | Ascomycete | Stem |
| 36 | *Bartalinia* | 1 | Ascomycete | Leaf |
| 37 | *Basidiobotrys* | 1 | Ascomycete | Stem |
| 38 | *Beauveria* | 1 | Ascomycete | Entomo |
| 39 | *Beltrania* | 1 | Ascomycete | Litter |
| 40 | *Biscogniauxia* | 1 | Ascomycete | Branch, stem |
| 41 | *Boletus* | 1 | Basidiomycete | Ground |
| 42 | *Botryobasidium* | 1 | Basidiomycete | Stem |
| 43 | *Botryodiplodia* | 1 | Ascomycete | Leaf |
| 44 | *Botryosphaeria* | 2 | Ascomycete | Leaf |
| 45 | *Calvatia* | 1 | Basidiomycete | Ground |
| 46 | *Camarosporium* | 1 | Ascomycete | Leaf |
| 47 | *Camptomeris* | 3 | Ascomycete | Leaf |
| 48 | *Cantharellus* | 1 | Basidiomycete | Ground |

**Table 49.2** (continued)

| S. No. | Fungi (genera) | Number of species | Broad group | Substrate |
|---|---|---|---|---|
| 49 | *Capnodium* | 1 | Ascomycete | Stem |
| 50 | *Catenulaster* | 1 | Ascomycete | Leaf |
| 51 | *Catenulopsora* | 1 | Basidiomycete | Leaf |
| 52 | *Cercopsora* | 32 | Ascomycete | Leaf |
| 53 | *Cercosporella* | 1 | Ascomycete | Leaf |
| 54 | *Cerioporiopsis* | 1 | Basidiomycete | Stem, wood |
| 55 | *Cerotelium* | 1 | Ascomycete | Leaf |
| 56 | *Cephaliophora* | 1 | Ascomycete | Litter |
| 57 | *Ceuthospora* | 1 | Ascomycete | Litter |
| 58 | *Chaetomella* | 1 | Ascomycete | Seedling |
| 59 | *Chaetomium* | 5 | Ascomycete | Seed/litter |
| 60 | *Chaetopsina* | 1 | Ascomycete | Leaf |
| 61 | *Chaetosphaeronema* | 1 | Ascomycete | Litter |
| 62 | *Cheilymenia* | 1 | Ascomycete | Ground |
| 63 | *Ciliochorella* | 2 | Ascomycete | Leaf |
| 64 | *Circinella* | 1 | Phycomycete | Litter |
| 65 | *Cladobotrium* | 1 | Ascomycete | Stem, bark |
| 66 | *Cladosporium* | 10 | Ascomycete | Leaf |
| 67 | *Clitocybe* | 3 | Basidiomycete | Soil |
| 68 | *Coccomyces* | 1 | Ascomycete | Leaf |
| 69 | *Coleophoma* | 1 | Ascomycete | Litter |
| 70 | *Colletotrichum* | 12 | Ascomycete | Leaf, pod, stem |
| 71 | *Coltricia perennis* | 1 | Basidiomycete | Stem |
| 72 | *Coniella* | 5 | Ascomycete | Stem |
| 73 | *Coniothyrium* | 1 | Ascomycete | Leaf |
| 74 | *Coprinus* | 3 | Basidiomycete | Litter |
| 75 | *Coriolopsis* | 2 | Basidiomycete | Stem twig |
| 76 | *Corticium* | 1 | Basidiomycete | Stem |
| 77 | *Corynespora* | 10 | Ascomycete | Leaf |
| 78 | *Coryneum* | 1 | Ascomycete | Stem/Twig |
| 79 | *Cryptosphaeria* | 1 | Ascomycete | Stem |
| 80 | *Curvularia* | 9 | Ascomycete | Leaf/seed/litter |
| 81 | *Cylindrocarpon* | 1 | Ascomycete | Litter |
| 82 | *Cylindrocladium* | 1 | Ascomycete | Seedling, leaf |
| 83 | *Cylindrosporium* | 1 | Ascomycete | Leaf spot |
| 84 | *Cytospora* | 2 | Ascomycete | Stem, twigs |
| 85 | *Daedalea* | 2 | Basidiomycete | Stem |
| 86 | *Daldinia* | 2 | Ascomycete | Stem, wood |
| 87 | *Dasturella* | 2 | Basidiomycete | Leaf |
| 88 | *Datronia* | 1 | Basidiomycete | Stem |
| 89 | *Dendrographium* | 1 | Ascomycete | Twigs |
| 90 | *Dendrophoma* | 1 | Ascomycete | Leaf |
| 91 | *Denticularia* | 1 | Ascomycete | Leaf |
| 92 | *Diachanthodes* | 1 | Basidiomycete | Stem |
| 93 | *Dichomitis* | 1 | Basidiomycete | Stem, wood |
| 94 | *Diatrype* | 5 | Ascomycete | Stem, twig |
| 95 | *Diatrypella* | 2 | Ascomycete | Leaf, Stem |
| 96 | *Didymosphaeria* | 1 | Ascomycete | Stem |

**Table 49.2** (continued)

| S. No. | Fungi (genera) | Number of species | Broad group | Substrate |
|---|---|---|---|---|
| 97 | *Diplodia* | 1 | Ascomycete | Leaf |
| 98 | *Discosia* | 1 | Ascomycete | Leaf |
| 99 | *Dothidea* | 1 | Ascomycete | Stem |
| 100 | *Dothiorella* | 3 | Ascomycete | Pod, Stem, twig |
| 101 | *Drechslera* | 3 | Ascomycete | Leaf/litter |
| 102 | *Earliella* | 1 | Basidiomycete | Stem, twig |
| 103 | *Echinocatena* | 1 | Ascomycete | Litter |
| 104 | *Echinostelium* | 1 | Myxomycetes | Stem |
| 105 | *Endocalyx* | 2 | Ascomycete | Leaf |
| 106 | *Epicoccum* | 1 | Ascomycete | Twig |
| 107 | *Eriocercopsora* | 1 | Ascomycete | Leaf |
| 108 | *Erysiphe* | 2 | Ascomycete | Leaf |
| 109 | *Eupenicillium* | 1 | Ascomycete | Soil |
| 110 | *Eutypella* | 1 | Ascomycete | Stem |
| 111 | *Excipularia* | 1 | Ascomycete | Bark |
| 112 | *Exserohilum* | 1 | Ascomycete | Leaf |
| 113 | *Favolus* | 3 | Basidiomycete | Stem |
| 114 | *Flavodon* | 1 | Basidiomycete | Stem |
| 115 | *Fomes* | 5 | Basidiomycete | Stem |
| 116 | *Fumago* | 1 | Ascomycete | Stem |
| 117 | *Funalia* | 1 | Basidiomycete | Stem |
| 118 | *Fusarium* | 12 | Ascomycete | Basal canker/Foot/ Root/Culm rot |
| 119 | *Fusicoccum* | 1 | Ascomycete | Leaf |
| 120 | *Fusicladium* | 1 | Ascomycete | Leaf |
| 121 | *Ganoderma* | 3 | Basidiomycete | Stem, butt, Root |
| 122 | *Geastrum* | 2 | Basidiomycete | Ground |
| 123 | *Gigaspora* | 1 | Phycomycetes | Soil |
| 124 | *Gliocladium* | 1 | Ascomycete | Litter |
| 125 | *Glomerella* | 1 | Ascomycete | Stem |
| 126 | *Gloeophyllum* | 1 | Basidiomycete | Stem, wood |
| 127 | *Glomus* | 9 | Phycomycetes | Soil |
| 128 | *Goosiella* | 1 | Ascomycete | Leaf |
| 129 | *Graphiola* | 1 | Ascomycete | Leaf |
| 130 | *Haematostereum* | 1 | Basidiomycete | Stem |
| 131 | *Haplographium* | 1 | Ascomycete | Leaf |
| 132 | *Haplopilus* | 1 | Basidiomycete | Stem wood |
| 133 | *Harposporium* | 1 | Ascomycete | Litter |
| 134 | *Helicobasidium* | 1 | Basidiomycete | Root |
| 135 | *Helicomina* | 1 | Ascomycete | Leaf |
| 136 | *Helicosporium* | 1 | Ascomycete | Litter |
| 137 | *Helvella* | 1 | Ascomycete | Stem wood |
| 138 | *Helminthosporium* | 3 | Ascomycete | Stem & twigs, Leaf |
| 139 | *Hendersonula* | 1 | Ascomycete | Stem, twigs |
| 140 | *Heterobasidium* | 1 | Basidiomycete | Stump |
| 141 | *Hexagonia apiaria* | 2 | Basidiomycete | Branch |
| 142 | *Hjorststamia* | 1 | Basidiomycete | Stem wood |
| 143 | *Humicola* | 1 | Ascomycete | Litter |

**Table 49.2** (continued)

| S. No. | Fungi (genera) | Number of species | Broad group | Substrate |
|---|---|---|---|---|
| 144 | *Hymenochaete* | 3 | Basidiomycete | Twig, Branch |
| 145 | *Hyphodontia* | 1 | Basidiomycete | Stem, wood |
| 146 | *Hypoxylon* | 13 | Ascomycete | Stem |
| 147 | *Hysterium* | 4 | Ascomycete | Branch |
| 148 | *Irpex* | 2 | Basidiomycete | Stem |
| 149 | *Inonotus* | 2 | Basidiomycete | Stem wood |
| 150 | *Junghunia* | 3 | Basidiomycete | Stem |
| 151 | *Kamalomyces* | 1 | Ascomycete | Stem |
| 152 | *Kameshwaromyces* | 2 | Ascomycete | Leaf |
| 153 | *Kernkampella* | 1 | Basidiomycete | Leaf |
| 154 | *Lacterius* | 7 | Basidiomycete | Ground |
| 155 | *Laxitextum* | 1 | Basi | Stem/logs |
| 156 | *Lembosea* | 1 | Ascomycete | Leaf |
| 157 | *Lentinus* | 3 | Basidiomycete | Ground |
| 158 | *Lenzites* | 9 | Basidiomycete | Stem |
| 159 | *Lepiota* | 5 | Basidiomycete | Ground |
| 160 | *Leptodothiorella* | 1 | Ascomycete | Leaf |
| 161 | *Leptosphaeria* | 4 | Ascomycete | Stem twig |
| 162 | *Leptosphaerulina* | 1 | Ascomycete | Litter |
| 163 | *Leptoxyphium* | 1 | Ascomycete | Leaf |
| 164 | *Leucoagaricus* | 2 | Basidiomycete | Ground |
| 165 | *Leucocoprinus* | 2 | Basidiomycete | Stem |
| 166 | *Leucophellinus* | 1 | Basidiomycete | Ground |
| 167 | *Linospora* | 1 | Ascomycete | Stem |
| 168 | *Lophium* | 1 | Ascomycete | Leaf |
| 169 | *Lophodermium* | 2 | Ascomycete | Leaf |
| 170 | *Macrophoma* | 1 | Ascomycete | Stem canker |
| 171 | *Macrophomina* | 1 | Ascomycete | Root |
| 172 | *Marasmius* | 2 | Basidiomycete | Ground |
| 173 | *Maravalia* | 2 | Basidiomycete | Leaf |
| 174 | *Melanographium* | 1 | Ascomycete | Stem |
| 175 | *Meliola* | 2 | Ascomycete | Leaf |
| 176 | *Memnoniella* | 2 | Ascomycete | Twig |
| 177 | *Microdiplodia* | 2 | Ascomycete | Leaf |
| 178 | *Micropera* | 2 | Ascomycete | Stem |
| 179 | *Microporus* | 3 | Basidiomycete | Stem, Branch |
| 180 | *Microspora* | 1 | Ascomycete | Leaf |
| 181 | *Microstoma* | 1 | Basidiomycete | Leaf |
| 182 | *Microxyphiella* | 2 | Ascomycete | Stem |
| 183 | *Microxyphium* | 1 | Ascomycete | Stem |
| 184 | *Mitteriella* | 1 | Ascomycete | Leaf |
| 185 | *Monochaetia* | 1 | Ascomycete | Soil |
| 186 | *Monodictys* | 2 | Ascomycete | Twig/litter |
| 187 | *Moorella* | 1 | Ascomycete | Bark |
| 188 | *Mucor* | 1 | Phycomycete | Litter |
| 189 | *Mycena* | 2 | Basidiomycete | Litter |
| 190 | *Mycocarpon* | 1 | Ascomycete | Stem |
| 191 | *Mycocentrospora* | 1 | Ascomycete | Leaf |

**Table 49.2** (continued)

| S. No. | Fungi (genera) | Number of species | Broad group | Substrate |
|---|---|---|---|---|
| 192 | *Mycoleptodiscus* | 1 | Ascomycete | Stem |
| 193 | *Mycosphaerella* | 3 | Ascomycete | Leaf |
| 194 | *Myrothecium* | 1 | Ascomycete | Leaf |
| 195 | *Mystrosporiella* | 2 | Ascomycete | Leaf |
| 196 | *Nectria* | 2 | Ascomycete | Stem |
| 197 | *Nemania* | 1 | Ascomycete | Stem |
| 198 | *Nigrofomes* | 1 | Basidiomycete | Stem |
| 199 | *Nigroporus* | 2 | Basidiomycete | Stem |
| 200 | *Nigrospora* | 1 | Ascomycete | Twig |
| 201 | *Nitschkia* | 2 | Ascomycete | Stem |
| 202 | *Nomuraea* | 1 | Ascomycete | Entmogenous |
| 203 | *Obstipipilus* | 1 | Ascomycete | Stem |
| 204 | *Oedocephalum* | 1 | Ascomycete | Soil |
| 205 | *Oidium* | 6 | Ascomycete | Leaf |
| 206 | *Olivea* | 1 | Basidiomycete | Leaf |
| 207 | *Paathramaya* | 1 | Ascomycete | Twig |
| 208 | *Paecilomyces* | 1 | Ascomycete | Litter |
| 209 | *Passalora* | 2 | Ascomycete | Leaf |
| 210 | *Penicillium* | 6 | Ascomycete | Soil |
| 211 | *Perenniporia* | 1 | Basidiomycete | Stem, wood |
| 212 | *Periconia* | 2 | Ascomycete | Soil |
| 213 | *Pestalotia* | 9 | Ascomycete | Leaf |
| 214 | *Pestalotiopsis* | 11 | Ascomycete | Leaf |
| 215 | *Phaeoisariopsis* | 2 | Ascomycete | Leaf |
| 216 | *Phaeoseptoria* | 2 | Ascomycete | Leaf, Stem |
| 217 | *Phakopsora* | 3 | Basidiomycete | Leaf |
| 218 | *Phanerochaete* | 2 | Basidiomycete | Stem |
| 219 | *Phellinus* | 18 | Basidiomycete | Stem |
| 220 | *Phlebia* | 1 | Basidiomycete | Twig |
| 221 | *Phloeospora* | 1 | Ascomycete | Leaf |
| 222 | *Phlyctaeriella* | 1 | Ascomycete | Stem, logs |
| 223 | *Phoma* | 13 | Ascomycete | Leaf/Pod |
| 224 | *Phomopsis* | 19 | Ascomycete | Leaf |
| 225 | *Phyllachora* | 5 | Ascomycete | Leaf |
| 226 | *Phyllactinia* | 9 | Ascomycete | Leaf |
| 227 | *Phylloporia* | 1 | Basi | Stem Wood |
| 228 | *Phyllosticta* | 12 | Ascomycete | Leaf |
| 229 | *Phytophthora* | 2 | Phycomycetes | Leaf |
| 230 | *Pisolithus* | 1 | Basidiomycete | Ground |
| 231 | *Pithomyces* | 3 | Ascomycete | Leaf, seed |
| 232 | *Pleurotus* | 4 | Basidiomycete | Stem |
| 233 | *Podaxis* | 1 | Basidiomycete | Ground |
| 234 | *Podoscypha* | 1 | Basidiomycete | Litter |
| 235 | *Poitrasia* | 1 | Phycomycetes | Bark |
| 236 | *Polychaeton* | 1 | Ascomycete | Leaf |
| 237 | *Polyporus* | 9 | Basidiomycete | Twigs |
| 238 | *Polystictus* | 3 | Basidiomycete | Stem |
| 239 | *Poria* | 3 | Basidiomycete | Stump |

**Table 49.2** (continued)

| S. No. | Fungi (genera) | Number of species | Broad group | Substrate |
|---|---|---|---|---|
| 240 | *Porodaedalea* | 1 | Basidiomycete | Stem, wood |
| 241 | *Postia* | 1 | Basidiomycete | Stem wood |
| 242 | *Prathigada* | 1 | Ascomycete | Leaf |
| 243 | *Pseudocercospora* | 28 | Ascomycete | Leaf |
| 244 | *Pseudodiplodia* | 1 | Ascomycete | Leaf |
| 245 | *Pseudolachnea* | 1 | Ascomycete | Twig |
| 246 | *Pseudospiropes* | 2 | Ascomycete | Leaf |
| 247 | *Puccinia* | 6 | Basidiomycete | Leaf |
| 248 | *Pulveroboletus* | 1 | Basidiomycets | Ground |
| 249 | *Pycnoporus* | 3 | Basidiomycete | Stem |
| 250 | *Pycnothera* | 1 | Ascomycete | Leaf |
| 251 | *Pyrenochaeta* | 2 | Ascomycete | Litter |
| 252 | *Pyrofomes* | 2 | Basidiomycete | Stem |
| 253 | *Pythium* | 1 | Phycomycetes | Seedling |
| 254 | *Questieriella* | 1 | Ascomycete | Leaf litter |
| 255 | *Ramaria* | 1 | Basidiomycete | Ground wood stem |
| 256 | *Ravenelia* | 6 | Basidiomycete | Leaf |
| 257 | *Rehmiodothis* | 1 | Ascomycete | Stem |
| 258 | *Rhizoctonia* | 1 | Ascomycete | Collar rot |
| 259 | *Rhizopus* | 1 | Phycomycetes | Soil |
| 260 | *Rhizostilbella* | 1 | Ascomycete | Root |
| 261 | *Rhytisma* | 1 | Ascomycete | Leaf |
| 262 | *Ricinicum* | 1 | Basidiomycete | Bark |
| 263 | *Rigidoporopsis* | 1 | Basidiomycete | Stump |
| 264 | *Rigidoporus* | 2 | Basidiomycete | Stem |
| 265 | *Robillarda* | 2 | Ascomycete | Leaf |
| 266 | *Rosellinia* | 3 | Ascomycete | Stem |
| 267 | *Russula* | 18 | Basidiomycete | Ground |
| 268 | *Sarcinella* | 4 | Ascomycete | Leaf |
| 269 | *Sarocladium* | 1 | Ascomycete | Stem |
| 270 | *Schiffnerula* | 1 | Ascomycete | Stem |
| 271 | *Schizophyllum* | 1 | Basidiomycete | Stem |
| 272 | *Schizopora* | 2 | Basidiomycete | Branch |
| 273 | *Sclerotium* | 1 | Basidiomycete | Root |
| 274 | *Scleroderma* | 3 | Basidiomycete | Ground |
| 275 | *Scopella* | 1 | Basidiomycete | Leaf |
| 276 | *Scopulariopsis* | 2 | Ascomycete | Litter |
| 277 | *Scutellospora* | 2 | Phycomycete | Soil |
| 278 | *Scytalidium* | 2 | Ascomycete | Soil |
| 279 | *Seimatosporium* | 1 | Ascomycete | Leaf |
| 280 | *Septoria* | 2 | Ascomycete | Leaf |
| 281 | *Serpula* | 1 | Basidiomycete | Branch |
| 282 | *Sirosporium* | 2 | Ascomycete | Leaf |
| 283 | *Skeletocutis* | 1 | Basidiomycete | Stem |
| 284 | *Sphacelonema* | 1 | Ascomycete | Leaf |
| 285 | *Sphaeriopsis* | 1 | Ascomycete | Leaf |
| 286 | *Sphaerotheca* | 1 | Ascomycete | Leaf |
| 287 | *Spiropes* | 1 | Ascomycete | Stem |

**Table 49.2** (continued)

| S. No. | Fungi (genera) | Number of species | Broad group | Substrate |
|---|---|---|---|---|
| 288 | *Spongipellis* | 1 | Basidiomycete | Butt rot |
| 289 | *Stemonitis* | 1 | Myxomycete | Stem |
| 290 | *Stenella* | 7 | Ascomycete | Leaf |
| 291 | *Stereum* | 4 | Basidiomycete | Branch |
| 292 | *Stigmina* | 5 | Ascomycete | Leaf |
| 293 | *Syncephalastrum* | 1 | Phycomycetes | Litter |
| 294 | *Synnematium* | 1 | Ascomycete | Stem |
| 295 | *Teichospora* | 1 | Ascomycete | Stem |
| 296 | *Termitomyces* | 5 | Basidiomycete | Ground |
| 297 | *Thelephora* | 2 | Basidiomycete | Stem |
| 298 | *Torula* | 1 | Ascomycete | Twig |
| 299 | *Trametes* | 16 | Basidiomycete | Stem |
| 300 | *Tretospora* | 1 | Ascomycete | Leaf |
| 301 | *Trichaptum* | 3 | Basidiomycete | Stump |
| 302 | *Trichobotrys* | 1 | Ascomycete | Twig |
| 303 | *Trichoderma* | 6 | Ascomycete | Soil |
| 304 | *Tricholoma* | 4 | Basidiomycete | Stem |
| 305 | *Trichospermum* | 1 | Ascomycete | Stem |
| 306 | *Trichothecium* | 1 | Ascomycete | Leaf |
| 307 | *Trichothyrina* | 1 | Ascomycete | Stem |
| 308 | *Trichurus* | 1 | Ascomycete | Seed |
| 309 | *Trimmatostroma* | 1 | Ascomycete | Stem |
| 310 | *Tripospermum* | 7 | Ascomycete | Stem |
| 311 | *Tritirachium* | 1 | Basidiomycete | Root rot |
| 312 | *Tubercularia* | 1 | Ascomycete | Stem |
| 313 | *Uncinula* | 3 | Ascomycete | Leaf |
| 314 | *Uredo* | 3 | Basidiomycete | Leaf |
| 315 | *Uromyces* | 1 | Basidiomycete | Leaf |
| 316 | *Valsa* | 2 | Ascomycete | Stem, Twig |
| 317 | *Virgaria* | 1 | Ascomycete | Bark |
| 318 | *Verticillium* | 4 | Ascomycete | Seedling |
| 319 | *Volvariella* | 3 | Basidiomycete | Soil, wood |
| 320 | *Wiesneriomyces* | 1 | Ascomycete | Litter |
| 321 | *Xylaria* | 7 | Ascomycete | Stem, wood, ground |
| | **Total** | **838** | | |

*rus* Willd.; *Phomopsis* sp. on *C. elliptica, Phyllachora* sp. on *Bauhinia retusa,* and *Piggotia* sp. on *Gardinia latifolia* Ait. from Chilpi Ghati; *Pestalotiopsis foedans* and *Phoma herbarum* on *Pinus caribaea,* and *Phomopsis* sp. on *Grewia sclerophylla* from Amarkantak. Thirty-two different leaf spot fungi were recorded on different plants (Verma and Soni 2007). Jamaluddin et al. (1993) had also reported leaf spot and blight diseases of three medicinal plants, viz. *Hedychium spicatum, Indigofera tinctoria,* and *Acorus calamus* caused by *Alternaria alternata* from Amarkantak and Lamni. Four new hyphomycetes were reported causing leaf spots in the trees of Satpura forests (Sharma et al. 2006a). Leaf spot of *Manilkara hexandra* caused by newly described fungus, *Acrodictiella indica* has been reported from Machkot, Chhattisgarh (Verma et al. 2008).

### 49.2.3 Fungi Causing Root Diseases

*Macrophomina phaseolina* caused mortality in young plantation of *P. caribaea* in Chhattisgarh

(Jamaluddin et al. 1982, 1984a), *Acacia auriculiformis* and eucalyptus in Madhya Pradesh (Jamaluddin and Dadwal 2001; Soni et al. 1985). *Polyporus shoreae* caused root rot in dry and wet sal forests of Chhattisgarh and Madhya Pradesh (Jamaluddin 1991). Some recently recorded new diseases include vascular wilt of aonla caused by *Fusarium solani* (Soni and Verma 2010) and root rot of teak seedlings (underwater stress conditions) caused by *Helicobasidium compactum* and *Tritirachium roseum* (unpublished) (Fig. 49.1).

### 49.2.4 AM Fungi

Arbuscular Mycorrhizal (AM) fungi play an important role in the formation of plants, especially in disturbed and comparatively low fertile soils. In an earlier work, population of AM fungi and root colonization was found to vary in different plantations of *Grevillea pteridifolia* in bauxite mined-over areas of Amarkantak. *Grevillea pteridifolia* belonging to the family Proteaceae showed very little response to AM fungi and the population of AM spores was more in undisturbed sites as compared to mined-over areas (Jamaluddin and Chandra 1997). Development of arbuscular mycorrhizae and leaf blight disease in young plantation of 25 species of bamboos has been studied at Amravati (Verma and Soni 2008). Diversity of AM fungi in different forest types, teak, legumes, bamboo, and miscellaneous forests, along with the seasonal variations in spore population and root colonization, were reported from central India (Verma 2009).

### 49.2.5 Lichen Fungi

Upreti et al. (2007) and Nayaka et al. (2007) have made lichenological studies in Achanakmar–Amarkantak biosphere reserve and have reported 37 lichens, belonging to different families on different substrates. A new species of lichen, *Ramboldia amarkantakana* with whitish, granular thallus, is reported from Amarkantak region on *Shorea robusta* (Upreti et al. 2009). Upreti et al. (2005) has also enumerated lichens of Madhya Pradesh.

### 49.2.6 Macro Fungi

Fungi grown in the soil, wood, litter, etc., and produced from their fruiting bodies (sporophores) on the substrata, and are large enough to be viewed with the naked eyes are known as macro fungi. Common macro fungi recorded from central India are given below.

#### 49.2.6.1 Mushrooms

Common mushroom collected from the ground and decaying wood and litter were the species *Agaricus, Amanita, Agrocybe, Calvatia clytocy, Coprinus, Clarkeinda, Lepiota, Marasmius, Mycena, Pleurotus, Termitomyces, Tricholoma*, and *Volvariella.*

#### 49.2.6.2 Wood Decay Fungi

Fruiting bodies of wood decaying fungi appear on woods and woody debris on the forest floor. *Coriolopsis, Daedalea, Daldinia, Earliella, Favolus, Flavodon, Funalia, Ganoderma, Hypoxylon, Hymenochaete, Inonotus, Irpex, Junghunia, Lenzites, Microporus, Navisporus, Phellinus, Polyporus, Polystictus, Poria, Pycnoporus, Pyrofomes, Rigidoporus, Schizophyllum, Skeletocutis, Stereum, Trremates, Xylaria*, were the common wood decaying fungi. Three new records of wood decay fungi namely *Australohydnum dregeanum, Hjorstamia friesii*, and *Schizopora flavipora* were recently reported from India (Tiwari et al. 2010a; Tiwari et al. 2010b) studied the occurrences and distributions of wood decaying fungi in the forest wood depots of Chhattisgarh and reported 81 fungi occurring in different wood depots. Among them *Flavodon flavus, Daedalea flavida, Lenzites elegans, Pycnoporus sanguineus, Schizophyllum commune, Stereum hirusitum, Trametes cingulata* are the most frequently occurring species. Tiwari et al. (2008) reported 13 wood decaying fungi on sal (*Shorea robusta*) from Madhya Pradesh. Tiwari et al. (2009) also reported 20 wood decaying fungi on stored wood of *Pterocarpus marsupium* from Chhattisgarh.

**Fig. 49.1** *Helicobasidium compactum symptoms* produced on teak saplings under water stress condition

#### 49.2.6.3 Ectomycorrhizal Fungi

Ectomycorrhiza forming fungi occur only in the rainy season (June–August) and the common fungi are: Astraeus hygrometricus, Boletus fallax, Calvatia elata, Russula emetica, Scleroderma bovista, Scleroderma geaster, and Scleroderma verrucosum (Soni et al. 2011b; Pyasi et al. 2011, 2012). Species of Amanita, Boletus, Geastrum, Lepiota, Pisolithus, Ramaria, Thelephora, etc. also formed ectomycorrhizae in tropical trees. Ramaria flaccida was found to be associated with teak tree (Fig. 49.1).

#### 49.2.6.4 Edible and Medicinal Fungi

Fruiting bodies of edible mushrooms growing on woods and forest floor are frequently collected by local tribal folk. Wild animals and insects also eat these fungi. Tribal people of sal forest areas of central India collect mushrooms during rainy season (June–September) as a source of their livelihood and subsistence (Khera 1991). Harsh et al. (1989) reported two edible fungi, *Scleroderma texense* and *Termitomyces albuminosa*, as a minor forest product of Madhya Pradesh. Market assessment and business potential of six edible fungi, *Astraeus hygrometricus, Mycenastrum corium, Podabrella microcarpa, Russula* sp., and *Termitomyces heimii* collected from the Amarkantak biosphere reserve was also conducted (Harsh, Rai and Ayachi 1993). Fruiting bodies of *Ganoderma lucidum* and other wood decaying fungi are also used in medicine. In Madhya Pradesh and Chhattisgarh, wild mushrooms (species of *Clitocybe, Lentinus, Lepiota, Pleurotus, Scleroderma, Termitomyces, Volvariella*, etc.) are sold in the local markets and provide sustenance to the tribal people and forest dwellers (Harsh and Joshi 2008). The Baiga tribe of Madhya Pradesh and Abujhmad (Chhattisgarh) used *Ganoderma lucidum* growing on *Dalbergia sissoo* and *Terminalia bellerica* for the treatment of cataract, asthma, and gastric troubles (Harsh, Rai and Tiwari 1993).

### 49.2.7 Litter-Decomposing Fungi

Season-wise successional change in litter mycoflora has been studied in four main seasons of the year, namely spring (March–May), summer (June–August), autumn (September–November), and winter (December–February). Fungi like *Aspergillus flavus, Aspergillus niger,* and *Rhizopus stolonifer* were associated with sal litter decomposition throughout the year, while *Aspergillus fumigatus, Cladosporium cladosporioides, C. oxysporum, Curvularia indica,* and *Curvularia lunata* were recorded in three seasons . Fungi like *Alternaria citri, Gleocladium virens, Helicosporium phragmitis,* and *Pithomyces cortarum* were rarely recorded, only in one season. The former fungus was recorded during spring while the latter three fungi were recorded in winter. Some fungi occur only in the rainy season (June–August) like *Colletotrichum dematium, Corticium rolfsii, Periconia minutissima, Mycena roseus,* and *Scopulariopsis alba* (Soni et al. 2011a, b).

### 49.2.8 Fungi Causing Top Dying and Heart Rot in Standing Trees

Root rot of *Dalbergia sissoo* and *Acacia catechu* caused by *Ganoderma lucidum* is posing a serious threat in central India. *Spongipellis spumeus* causing root rot in the mature trees of *Albizia procera* and *Tectona grandis* is also causing a consider-

**Fig. 49.2** *Nitschkia tectonae* associated with dead teak stem

**Fig. 49.3** *Hypoxylon* associated with top dying tree of sal

**Fig. 49.4** *Ramaria flaccida* associated with teak tree

able damage. Top dying and root rot of teak was caused by *H. compactum* (Fig. 49.2). *Phomopsis tectonae* and *Phomopsis variosporum* caused top dying of teak (Fig. 49.3) in central India (Verma et al. 2008). Recently, a *Hypoxylon* sp. was also recorded to be associated with top dying of sal (Fig. 49.4). The heart rot fungi caused considerable wood damage in the standing trees due to decay in heart wood. Majority of heart rot fungi belonged to genera *Hymenochaete, Phellinus* and *Polyporus* (Harsh et al. 2000; Soni et al. 2010). *Hymenochaete rubiginosa* (Dicks.) Lév., *Trametes incerta* (Curr.) Cooke, and *Phellinus caryophylli* (Racib.) G. Cunn causes heart rot of sal and up to 10% loss in the wood volume (Bagchee 1959; Bakshi 1957). *Ganoderma lucidum* causing root rot of *Acacia catechu* and *Dalbergia sissoo* is a serious threat to plantations and natural trees of these species (Figs. 49.5, 49.6 and 49.7).

### 49.2.9 New Taxa Reported

Two new genera and 28 species of fungi were reported from the forests of Chhattisgarh and Madhya Pradesh (Sharma et al. 2005; Sharma et al. 2006b; Verma et al. 2008). The new genera reported are: *Acrodictiella* and *Kamalomyces,* while the species reported are *Acrodictiella indica, Acrostroma madhucae, Acrostroma sterculiae, Corynespora pogostemonis, Corynespora supkharii, Corynespora woodfordiae, Denticularia terminaliae, Diatrypella semecarpi,*

**Fig. 49.5** *Ganoderma lucidum* associated with root rot of *Acacia catechu* young tree (12 year)

**Fig. 49.6** *Ganoderma lucidum* associated with *Dalbergia sissoo* sporophores attached with living young tree

*Hypoxylon dendrocalmi, Hypoxylon spiralis, Hysterium adinae, jabalpurensis, Kamalomyces indicus, Kameshwaromyces butiicolous, Meliola ougeiniae, Mysterosporiella terminaliae, Phaeoseptoria shoreae, Phomopsis ougeiniae, Phyllachora ramicola, Pseudocercospora isorae, Pseudocercospora schleicherae-oleosae, Pseudospiropes shoreae, Rehmiodothis bambusae, Sirosporium aeglicola, S. xylopyrae, Stenella flacourticola, S. liliacearum,* and *S. satpurensis.* Dadwal et al. (2003) reported a new species, *Phomopsis acaciae* causing phyllode spot and top dying in *Acacia mangium.*

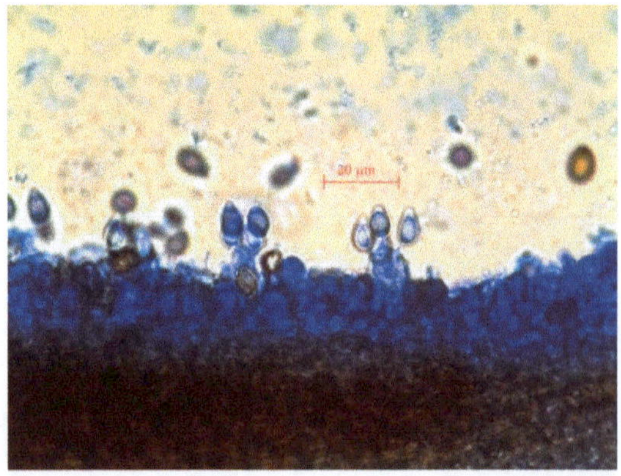

**Fig. 49.7** *Ganoderma lucidum* on *Dalbergia sissoo*, a cross section of sporophore showing basidiospores and basidia

*Nitschkia tectonae,* a new ascomycete on teak, was recently reported from central India (Verma 2010). *Cercosporidium helicteri* and *Sarcinella diospyri* causing leaf spot of *Helicteres isora* and *Diospyros melanoxylon,* and *Lophodermium shoreae* isolated from the leaf litter of *Shorea robusta* were reported from the forest of Amarkantak, Madhya Pradesh (Soni et al. 1984; Rajak and Soni 1981; Jamaluddin et al. 1984b). *Chaetoptela indica* on dead culms of *Dendrocalamus strictus*; *Goosiella helicospora* on living frond of unidentified fern; *Phaeoisariopsis lagerstroemiae* causing leaf spots in *Lagerstroemia parviflora* and *Kameshwaromyces globosus* causing leaf spots in *Elephantopus scaber* were reported from the forest of Pachmarhi, Madhya Pradesh (Rajak and Soni 1978; Morgan-Jones et al. 1986; Rajak et al. 1978; Kamal and Morgan-Jones 1986). *Lembosia shoreae* causing leaf spots in *Shorea robusta* was reported from sal forest of Jagdalpur, Chhattisgarh (Verma et al. 2001). A new and rare species of *Phlyctaeniella* occurring on logs of *Eucalyptus* sp. was recently reported from Chhattisgarh (Tiwari et al. 2012).

## 49.3 Conservation of Fungi

Mycology Herbarium, at the Tropical Forest Research Institute, Jabalpur, of over the last 15 years were scanned for determining the status of fungi that needs conservation. The forest areas from where the maximum number of fungi collected were recorded as the hot spots for collection of fungi. Fungi are rarely considered in the conservation initiatives of most countries. Mycologists have put limited efforts into conservation of fungi.

### 49.3.1 Status of Forest Fungi in Central India

In central India, forest fungi have been studied and collected for the last two decades, and at Division's Mycology Herbarium, 2,700 fungal species reported 12.3 % fungi which were collected only once during the last 15 years and this can be considered as threatened and so needs conservation. Further, 5.6 % fungi were collected two to five times and 1.6 % fungi six to ten times. Only 1.8 % fungi were frequently collected (more than 11 times during this period). The following four genera reported from central India with their frequency of occurrence are:

1. *Kameshwaromyces*—frequently collected
2. *Goosiella*—never collected since 1985
3. *Kamalomyces*—Collected only once
4. *Acrodicdiella*—Collected only once

The data indicates that the population of major fungi is dwindling with the time. However, some fungi showed spreading nature, for example, *Phomopsis tectonae*, causing top dying of teak is found throughout the teak growing areas in India. The fungus was reported for the first time on the leaf (Tiwari et al. 1981), and then spread to twigs, flowers, and to the main stem which later caused top dying in trees, mostly in young plantation (Figs. 49.8 and 49.9).

### 49.3.2 Hot Spots for Collection of Forest Fungi of Central India

The hot spots for the collection of fungi from Madhya Pradesh and Chhattisgarh are Amarkantak–Achmakmar, Pachmarhi and Bori, Chhindwada, Shivpur, Gwalior, Mada, Shidi, Sagar, Supkhar, Balaghat, Bastar, and Keshkalghat.

**Fig. 49.8** *Phomopsis tectonae*, top dying symptoms in young teak tree

### 49.3.3 Digitization of Records of Indian Fungi

At the Tropical Forest Research Institute, Jabalpur central Indian forest fungi is being documented with every details including the detailed taxonomic descriptions, photographs/drawings, references, hosts, etc. To date, over 400 fungi have been documented. Record of all Indian fungi with the details of host, synonymy, references, etc. is also being compiled. Over 15,000 fungi have been digitized covering 2,732 genera. The number of fungi recorded in India exceeds 27,000 species (Manoharachary et al. 2005).

### 49.3.4 Society for Conservation of Fungi

An *International Society for Fungal Conservation* was established in August 2010 at Edinburgh

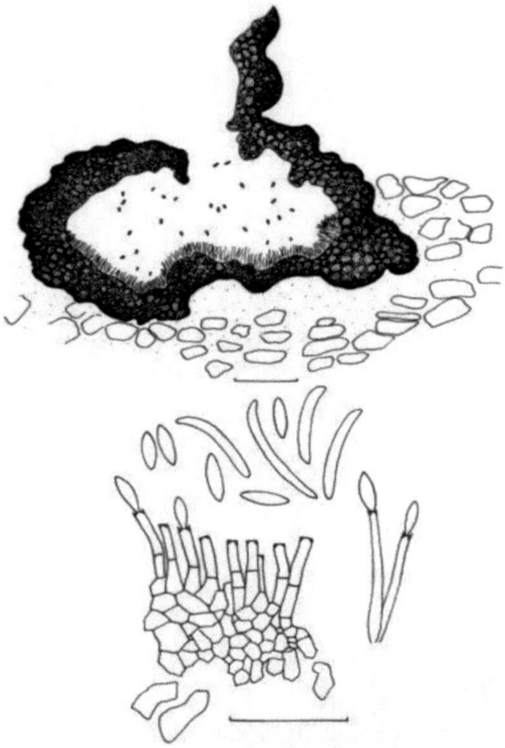

**Fig. 49.9** *Phomopsis tectonae*, vertical section of conidioma, conidiophores, and α and β conidia. Scales 20 μm

after thorough discussion in the second international congress on fungal conservation held at Whitby, the UK in October 2009, and now has members in over 50 countries. The objective of the society is to promote conservation of fungi globally. It publishes a journal named, "*Fungal Conservation*." It acts as a global federation for fungal conservation groups and to join the society one can contact Dr. D.W. Minter (D.MINTER@CABI.ORG).

## 49.4 Conclusion

Fungi play an important role in building good quality soil and maintaining its fertility. They are also a good source of food and income generation for tribal population; therefore, systematic study should be undertaken to document the whole forest fungi present in the central India.

**Acknowledgment** The authors are thankful to Dr. U. Prakasam, Director, Tropical Forest Research Institute, Jabalpur for providing the research facilities.

## References

Bagchee KD (1959) The fungal diseases of sal (*Shorea robusta*) V. The heart rot of sal caused by *Trametes incerta*. Indian Forest Record (NS). Mycology 2:61–69

Bakshi BK (1957) Heart rots in relation to management of sal. Indian Forester 83:651–661

Chakraborty L, Panwar SK, Shukla RV (1991) Effect of closer on soil properties and its fungal population in sal foest. J Trop Forestry 7:51–61

Dadwal VS, Jamaluddin (1991) Unrecorded diseases of *Grevillea pteridifolia*. J Trop Forestry 7:248–249

Dadwal VS, Verma RK, Jamaluddin (2003) New species of *Phomopsis* causing phyllode spot and top dying in *Acacia mangium*. J Mycol Pl Pathol 33:42–44

Harsh NSK, Joshi K (2008) Mushrooms: the vegetable of the future. India Sci Technol. www.nistads.res.in

Harsh NSK, Tiwari CK, Jamaluddin (1989) Prospects of wild edible fungi as minor forest product in Madhya Pradesh. Paper presented in the National Seminar on Minor Forest Produce and Tribal development held on 19–20 Oct. 1989 at IDF, Jabalpur

Harsh NSK, Rai BK, Ayachi SS (1993) Forest fungi and tribal economy—a case study in Baiga tribes of MP. J Trop Forestry 9:94–96

Harsh NSK, Rai BK, Tiwari DP (1993) Use of *Ganoderma lucidum* in folk medicine. Indian J Trop Biodiv 1(3-4):324–326

Harsh NSK, Soni KK, Tiwari CK, Verma RK, Jamaluddin (2000) Decline of sandal trees in Seoni district of Madhya Pradesh. J Trop Forestry 16(4):85–91

Jamaluddin (1991) Status of *Polyporus shoreae* causing root-rot in dry and wet sal forest. J Trop Forestry 7(4):342–344

Jamaluddin, Chandra KK (1997) Distribution of VAM fungi in bauxite mine over burden plantation of Amarkantak (M.P). Indian Forester 125(5):412–418

Jamaluddin, Dadwal VS (2001) Studies on charcoal root rot of *Acacia auriculiformis*. Indian J Trop Biodiv 9(1-4):61–65

Jamaluddin, Dadwal VS, Soni KK (1982) Studies on charcoal root rot of *Pinus caribaea*. Indian Forester 108(9):618–622

Jamaluddin, Dadwal VS, Soni KK (1984a) An observation on the incidence of charcoal root rot disease of *Pinus caribaea* plantation of Bastar (MP). Indian Forester 110(6):552–557

Jamaluddin, Dadwal VS, Soni KK (1984b) Two new Ascomycetes from India. Biol Bull India 6(3):323–326

Jamaluddin, Dadwal VS, Soni KK (1986) The status of diseases and their management in forests of Madhya

Pradesh. Research Report, Regional Forest Research Centre, Jabalpur
Jamaluddin, Dadwal VS, Soni KK (1990) Susceptibility of different provenances of *Pinus roxburgii* to Cercospora needleblight at Amarkantak (M.P). Indian Forester 116:58–61
Jamaluddin, Nath V, Namdeo RK (1993) Studied on diseases of some important medicinal plants. J Trop Forestry 9:270–279
Kamal VRK, Morgan-Jones G (1986) Notes on hyphomycetes L1. *Kameshwaromyces* a new foliicolous sooty mold like genus from Madhya Pradesh, India. Mycotaxon 25:247–250
Khera PD (1991) The Baigas and the sal forest. Indian J MFP 1:72–80
Manoharachary C, Sridhar K, Singh R, Adholeya A, Suryanarayanan TS, Rawat S, Johri BN (2005) Fungal biodiversity: distribution, conservation and prospecting of fungi from India. Curr Sci 89(1):58–71
Morgan-Jones G, Kamal, Verma RK (1986) *Goosiella*, a new pteridicolous helicosporous genus from Madhya Pradesh, India. Mycologia 78:496–500
Nayak S, Satya, Upreti DK (2007) Lichen diversity in Achanakmar wild life sanctuary, core zone area of proposed Achanakmar biosphere reserve, Chhattisgarh. J Econ Taxon Bot 31:133–142
Pyasi A, Soni KK, Verma RK (2011) Dominant occurrence of ectomycorrhizal colonizer *Astraeus hygrometricus* of sal (*Shorea robusta*) in forest of Jharsuguda Orissa. J Mycol Pl Pathol 41:222–225
Pyasi A, Soni KK, Verma RK (2012) A new record of *Boletus fallax* from India. J Mycol Pl Pathol 42(1):172–173
Rajak RC, Soni KK (1978) *Chaetopatella indica* sp. nov, a new generic record for India. Curr Sci 47(4):136
Rajak RC, Soni KK (1981) Follicolous ectoparasites from Jabalpur-1—some sarcinellae. Indian J Mycol Pl Pathol 2(1):89–91
Rajak RC, Soni KK, Pathak GP (1978) Two new species of hyphomycetes. Curr Sci 47:397–398
Sharma N, Soni KK, Jamaluddin, Verma RK (2005) A new species of *Corynespora* from central India. Indian Phytopath 58:503–504
Sharma N, Soni KK, Verma RK (2006a) Some new hyphomycetes from forests of Satpura. Indian J Trop Biodiv 14(1):34–40
Sharma N, Soni KK, Jamaluddin, Verma RK (2006b) A new species of *Mystrosporiella* causing leaf spots in *Terminalia bellerica*. Indian Phytopath 59(2):257–260
Shettyi PK (1957) Soil fungal flora of two forest compartments of Amarkantak, M.P. Bull Biol Soc Univ Saugar 9:40–47
Soni KK, Verma RK (2010) A new vascular wilt disease of aonla (*Emblica officinalis*) and its management. J Mycol Pl Pathol 40(2):187–191
Soni KK, Dadwal VS, Jamaluddin (1984) A new species of *Cercosporidium* from India. Curr Sci 53(16):877–878
Soni KK, Dadwal VS, Jamaluddin (1985) Charcoal root rot and stem rot of Eucalyptus. Euro J Forest Pathol 15:397–401
Soni KK, Tiwari CK, Verma RK (2010) Heart rot in Indian hard wood tree species. J Trop Forestry 26(2):15–21
Soni KK, Pyasi A, Verma RK (2011a) A new record of *Helicosporium phragmitis* from India. J Mycol Pl Pathol 41:330–331
Soni KK, Pyasi A, Verma RK (2011b) Litter decomposing fungi in sal (*Shorea robusta*) forests of central India. Nusantara Bioscience 3:136–144
Tiwari DP, Rajak RC, Nikhra KM (1981) A new species of *Phomopsis* causing leaf-spot disease on *Tectona grandis* L. Curr Sci 50(22):1002–1003
Tiwari CK, Verma RK, Ayachi A, Asaiya AJK (2008) Wood decaying fungi of sal from Madhya Pradesh, India. Sci-fronts 2:13–26
Tiwari CK, Parihar J, Verma RK (2009) Wood decaying fungi on stored wood of *Pterocarpus marsupium* from Chhattisgarh, India. Sci-fronts 3(3):109–121
Tiwari CK, Parihar J, Verma RK (2010a) Additions to wood decaying fungi of India. JoTT 2(6):970–973
Tiwari CK, Parihar J, Verma RK (2010b) Occurrence and distribution of wood decaying fungi in forest wood depots of Chhattisgarh. Indian Forester 136(4):476–486
Tiwari CK, Parihar J, Verma RK (2012) A new and rare species of *Phlyctaeniella* from Central India. Mycosphere 3(4):450–453. doi 10.5943/mycosphere/3/4/8
Upreti DK, Nayaka S, Satya (2005) Enumeration of lichens from Madhya Pradesh and Chhattisgarh, India. J Appl Biosci 31(1):55–63
Upreti DK, Satya, Joshi Y (2007) Lichenological studies in Achanakmar-Amarkantak Biosphere Reserve. Workshop on Research Needs for Amarkantak-Achamakmar biosphere reserve, 30th Aug. 2007, TFRI, Jabalpur
Upreti DK, Nayaka S, Joshi Y (2009) *Ramboldia amarkantakana* (Lecanoraceae, Ascomycota), a new lichen species from India. Mycotaxon 107:239–242
Verma RK (2009) Diversity of AM fungi in forests of central India. Indian J Trop Biodiv 17(1):37–46
Verma RK (2010) *Nitschkia tectonae*-a new ascomycete on teak from central India. Indian Phytopath 63:430–432
Verma RK, Soni KK (2007) Fungi of Achanakmar-Amarkantak biosphere reserve and Chilpi Ghati. Indian J Trop Biodiv 15:116–126
Verma RK, Soni KK (2008) Development of arbuscular mycorrhizae and leaf blight disease in young plantation of 25 species of bamboos. Indian Forester 134(9):1236–1244
Verma RK, Soni KK, Tiwari CK, Jamaluddin (2001) Two new ascomycetes from India. Kavaka 28-29:19–25
Verma RK, Sharma N, Soni KK, Jamaluddin (2008) Forest fungi of Central India. International Book Distributing Co., Lucknow

# Spatiotemporal Variations in Microbial Mediated Nitrogen (N) Release Under N-Fertilization Experiment from Banaras Hindu University, India

50

Punita Verma, R. Sagar, Kuldip and Dharmendra K. Singh

**Abstract**

Globally, atmospheric nitrogen depositions due to fossil fuel combustion, industrial, and agricultural activities have been identified as serious threats to soil, water, and vegetation. In soil, N-deposition affects the respiration, microbial activities, enzymes actions, litter decomposition, and N-mineralization. The process of N-mineralization involves ammonification and nitrification. Ammonification is mediated by *Clostridium* sp., *Micrococcus* sp., *Proteus* sp., etc. Nitrification is mediated by the activities of diverse group of microorganisms (*Nitrosomonas europaea*, *Nitrosococcus nitrosus*, *Nitrosospira briensis*, *Nitrosovibrio*, *Nitrocystis*, *Nitrobacter winogradski*, *Nitrospira gracilis*, *Nitrosococcus mobilis*, *Penicillium*, *Aspergillus*, *Streptomyces*, *Nocardia*, etc.). In the present study, spatiality, rates of ammonification, nitrification, and net N-mineralization were governed by the soil properties (pH, moisture, C, N, and litter quality) and temporally these processes are determined by the rainfall pattern. Further, the study suggested that the rates of ammonification, nitrification, and net N-mineralization were greater at moderate level of N application. This affinity can be speculated as: at low levels of N resource, soil-C and -N are not enough for the activities of nitrifiers to release them in available form. As N increases more, actively participating microorganisms are enabled to release the nutrients in available form through the process of ammonification, nitrification and thus net N-mineralization, at sufficiently high N level, nitrifier population as well as their activities could be limited and thus the process of N-mineralization is limited. On the other hand, excessive N-application may damage the natural flora and fauna of soil which depletes the soil fertility. It could be also speculated that the N-limited ecosystems keep the deposited N by using it for the growth and developments of plants and microbes, in addition to accumulation in biomass and soil organic matter. At a certain point, the deposited N commences to

R. Sagar (✉) · P. Verma · Kuldip · D. K. Singh
Department of Botany, Banaras Hindu University,
Varanasi 221005, India
e-mail: sagar@bhu.ac.in

R. N. Kharwar et al. (eds.), *Microbial Diversity and Biotechnology in Food Security*,
DOI 10.1007/978-81-322-1801-2_50, © Springer India 2014

go beyond the biotic and abiotic needs for N within the system and the ecosystem is predicted to fail its N-retention ability. As the capability to keep N exceeds, surplus N is offered to be vanished from the ecosystem through solution losses and gas flux. Thus, in this study moderate level of N accelerated the process of N-mineralization.

**Keywords**

Nitrogen deposition · Nitrifiers · N-mineralization

## 50.1 Introduction

Nitrogen (N) depositions in majority of the ecosystems on the globe have increased substantially over the last many decades and have had considerable impacts on the soil, water, air, and vegetation (Smith et al. 1999; Rao et al. 2009). Human-induced perturbation in the N cycle is adding substantial amount of reduced N to the biosphere due to immense agricultural activities (Vitousek et al. 1997; Bobbink et al. 2010; Kros et al. 2011), fossil fuel combustion, biomass burning, and changes in land use pattern (Waldrop et al. 2004, Verma et al 2013). Surplus N from anthropogenic forcing has started to overwhelm the normal N cycle causing reduction in soil fertility, eutrophication of water bodies, increase in concentrations of nitric and nitrous oxides in air, and changes in the vegetation due to reduction in biodiversity because every ecosystem has the limit to hold the amount of N, beyond which the ecosystem may be in an unbalanced state (Smith et al. 1999; Bobbink et al. 2010). Excess amount of N-fertilization in soil affects the soil respiration, soil microbial activities, soil pH, soil enzymes (Makoi and Ndakidemi 2008), soil nutrients, litter decomposition, and ultimately soil health (Wang et al. 2009) and changes in soil health alter the composition, productivity, and species diversity of terrestrial as well as aquatic ecosystems (Smith et al. 1999; Stevens 2009; Pan et al. 2010). Sala et al. (2000) have considered it as one of the leading threats to species diversity after habitat loss and global climate change. Therefore, uninterrupted human-induced N-deposition in soil is a solemn question for discussion.

Nitrogen (N) is an essential macronutrient used by plants from soil for their growth and survival. Its availability depends on the mineralization potential of the soil (Roy and Singh 1994; Singh and Kashyap 2007, Verma et al. 2013), which could be a measure of soil quality, because productivity of the soil depends heavily on the mineralization potential of organic matter (Gonzalez-Prieto et al. 1992). Soils with high N-mineralization potential tend to be inherently fertile (Fu et al. 1987) which is governed by large numbers of biotic and abiotic factors, viz. rainfall, soil moisture, temperature, soil texture, water holding capacity (Singh and Kashyap 2007), soil aeration, pH, nutrients (Fu et al. 1987), quality and quantity of organic matter (Verma et al. 2013), and soil flora and fauna (Mlambo et al. 2007; Vourlitis et al. 2007; Rao et al. 2009).

Understanding the impact of N-deposition on soil health, of many ecosystems, is becoming undoubtedly essential for tropical countries like India. The data on the impacts of N-fertilization as a source of N-deposition on the soil N-mineralization from Indian region are lacking (Verma et al 2013). Looking into the realized effects of increased N-fertilization on the soil, N-mineralization is a necessity in managing the soil productivity and fertility. Therefore, the objective of the present study was to examine the effects of N-treatment on the soil ammonification, nitrification, and N-mineralization and to detect the causal soil variables holding the N-mineralization processes in a dry tropical environment of India under the condition of N-perturbation.

## 50.2 Materials and Methods

### 50.2.1 Study Sites

The study was conducted at 15 locations within the campus of Banaras Hindu University (24°18′N and 83°03′E, and 129 m msl altitude), Varanasi, India, in January 2007. The study area experiences a seasonally dry tropical monsoon climate. The year is divisible into three seasons: a hot summer (April–June), a warm rainy season (July–September), and a cold winter (November–February). The months of March and October constitute transition periods, between winter and summer, and between rainy and winter seasons, respectively. The mean monthly minimum and maximum temperature varied from 7.3–25.4 °C to 25.6–35.6 °C, respectively, and the mean annual rainfall was 932 mm. The soil of the Banaras Hindu University campus is alluvial, well drained, pale brown, silty loam, and inceptisol (Buol et al. 2003). In general, the soil is moderately fertile being low in available nitrogen and medium in available phosphorus and potassium. The soil pH is neutral to alkaline (Sagar et al. 2008).

The campus of Banaras Hindu University covers more than 520 ha of land area having natural dense flora. The study area houses 329 vascular plant species distributed in 80 families (Singh 2011). The ground vegetation of the study area is locally dominated by *Alysicarpus monilifer, Cynodon dactylon, Cyperus compressus, Desmodium gangeticum, Dichanthium annulatum, Evolvulus nummularius, Imperata cylindrica, Malvastrum coromandelianum, Oplismenus burmannii, Sida acuta* (Sagar et al. 2008; Sagar and Verma 2010). During the summer season, the grassland vegetation becomes dormant and under shade and mesic conditions, it shows some green parts (Sagar and Verma 2010).

### 50.2.2 Study Design

A 10 × 10 m area was demarcated in the centre of each location. Within each 10 × 10 m area, nine 1 × 1 m experimental plots, arranged in three parallel rows (three 1 × 1 m plots in each row), were established. A 1.5 m distance between two 1 × 1 m plots was kept as the buffer zone to protect against boundary effects due to migration of N out of the sampling areas (Verma et al. 2013). Within each location three treatments of nitrogen, each replicated three times, were randomly established on the basis of lottery method: control (without N), low N (60 kg N $ha^{-1}$ $year^{-1}$), and high N (120 kg N $ha^{-1}$ $year^{-1}$). Thus, a total of 405, 1 × 1 m plots (15 locations × 3 treatments × 9 replicates) were used for the experiment.

### 50.2.3 N-Treatment

Since January 2007, commercial urea fertilizer as a source of nitrogen was applied to the plots in the evening, at 1 month interval in the form of split dose. In the evening time, temperature is low and at this low temperature, activation energy of the urease enzyme is low and that decreases the N loss by the volatilization process (Makoi and Ndakidemi 2008). Urea was used as a source of dry N due to its relatively high N content, easy handling, and affordable price (Jones et al. 2007).

### 50.2.4 Soil Sampling and Analyses

After 3 years of N-treatment, for each month of the year 2010, three soil samples (0–10 cm depth) were collected from each 1 × 1 m plot, using a corer of 100 $cm^3$ volume. The three soil cores were combined to form a composite soil sample for each plot. These composite samples were gently homogenized. Large roots, woods, litters, and all fine roots were removed from the composite soil samples carefully. One part of the soil sample was air dried, sieved through 2 mm mesh screen, and were analysed for soil pH, soil organic carbon, nitrogen, ammonium nitrogen ($NH_4^+$-N), and nitrate nitrogen ($NO_3^-$N).

Soil pH was determined by using a glass electrode (1:2; soil: water ratio). Soil organic carbon (soil-C) was analysed by using dichromate oxidation and titration with ferrous ammonium sulphate (Walkley 1947). Ammonium nitrogen

($NH_4^+$-N) was extracted by 2 M KCl and analysed by using the phenate method American Public Health Association (APHA 1985). Nitrate nitrogen ($NO^-_3N$) was analysed by the phenol disulphonic acid method after extraction by $CaSO_4$ (Jackson 1958). The in situ buried bag technique was adopted for N-mineralization. Before incubation the $NH_4^+$-N and $NO_3^-N$ were determined for 0 month sampling. Using a large sealed polythene bag, a portion of fresh soil sample (200 g) was incubated in the soil at a depth of 0–10 cm at the same microsite from which the samples had been collected for the analysis of $NH_4^+$-N and $NO_3^-N$. After 1 month of field incubation, the incubated bags were collected for the analyses of $NH_4^+$-N and $NO_3^-N$. Again, a portion of fresh soil sample (200 g) was incubated at the same microsite from which the samples had been collected for the analyses of $NH_4^+$-N and $NO_3^-N$ and after 1 month of incubation the incubated samples were recollected and analysed for the concentration of $NH_4^+$-N and $NO_3^-N$. These analyses were repeated for each month of the year 2010 after an interval of 30 days from 1st January 2010 to 27th December 2010. The increase in the concentrations of $NH_4^+$-N and $NO_3^-N$ after field incubation is referred to as ammonification and nitrification, respectively, and the increase in the amount of $NH_4^+$-N plus $NO_3^-N$ over the course of field incubation is defined as net N mineralization (Roy and Singh 1994) of each month.

### 50.2.5 Statistical Analyses

The N-treatment levels were ordinated by Principal Component Analysis (PCA) option in Biodiversity Pro. (McAleece et al. 1997), using average values of ammonification for the 12 months experienced by each site. Similarly, N-treatment levels for nitrification and net N-mineralization were ordinated. MANOVA (Multivariate Analysis of Variance) procedure of SPSS package (SPSS 1997) was used to see the effects of sites, months and N-treatment on the rates of ammonification, nitrification, and net N-mineralization. Tukey's honest significant difference (HSD) test was used to determine the significance of differences in mean values of these variables among different N-treatment levels. The ammonification, nitrification, and net N-mineralization were linearly regressed with studied soil variables using linear regression option in SPSS software (SPSS 1997).

## 50.3 Results and Discussion

The summary of selected soil variables of the studied plots is presented in Table 50.1. Soil pH ranged from 6.76 to 7.38 indicating neutral to slightly alkaline soil. The soil was characterized by low levels of soil-C (0.47–2.09 %), -N (0.03–0.22 %), and soil moisture (2.70–15.92 %). Table 50.1 depicted that soil pH decreased consistently due to increase in soil moisture and N-dose. Soil-C and -N increased as the soil moisture increased, while the C:N ratio exhibited inverse trend with soil moisture. Soil-C and -N were greater in 60 kg N treated plots compared to control and 120 kg N treated plots (Table 50.2).

The ammonification, nitrification, and net N-mineralization (μg $g^{-1}$ $month^{-1}$) across the plots varied from 4.96 to 10.94, 2.07 to 5.28, and 7.02 to 16.22, respectively (Table 50.1). Similar to soil-C and -N, the ammonification, nitrification and net N-mineralization yielded a humped-back trend in relation to N-application (Table 50.2). Across the months, the values of the ammonification, nitrification, and net N-mineralization ranged between 12.36–56.78, 4.65–35.85, and 17.01–92.63 (Table 50.3). The values of these variables were minimum in March and maximum in September.

Analysis of variance revealed that differences in the soil-C, -N, pH, ammonification, nitrification, and net N-mineralization were significantly different due to site, N-treatment, and month (Table 50.4). PCA ordination of the three N doses on the basis of ammonification, nitrification, and net N-mineralization of the component sites revealed three groups. In the ordination diagram, among these groups, the 60 kg N-treatment cluster in the case of ammonification, nitrification, and net N-mineralization occupied distant position from the control and 120 kg N doses, while

**Table 50.1** Mean values of soil variables and parameters of N-mineralization in different experimental plots (S1–S15) at the Banaras Hindu University under N-application study. The values of soil moisture, soil-C, and -N are in per cent. The values of Amm (Ammonification), Nitri (Nitrification), and Net N-min (Net N-mineralization) are in μg $g^{-1}$ $month^{-1}$. Values in parentheses are ±1SE

| Sites | Moisture | pH | Soil-C | Soil-N | C:N ratio | Amm | Nitri | Net N-min |
|---|---|---|---|---|---|---|---|---|
| S-1 | 2.70 | 7.38 | 0.47 | 0.03 | 15.73 | 4.96 | 2.07 | 7.03 |
| | (0.10) | (0.02) | (0.04) | (0.01) | (1.42) | (0.26) | (0.18) | (0.52) |
| S-2 | 3.50 | 7.31 | 0.78 | 0.05 | 15.62 | 5.12 | 2.54 | 7.66 |
| | (0.11) | (0.03) | (0.01) | (0.00) | (0.18) | (0.28) | (0.08) | (0.59) |
| S-3 | 4.60 | 7.28 | 0.89 | 0.06 | 14.84 | 6.13 | 2.72 | 8.85 |
| | (0.13) | (0.02) | (0.02) | (0.00) | (0.19) | (0.33) | (0.05) | (0.62) |
| S-4 | 5.80 | 7.22 | 0.97 | 0.07 | 13.89 | 6.83 | 3.05 | 9.88 |
| | (0.12) | (0.01) | (0.01) | (0.00) | (0.17) | (0.31) | (0.12) | (1.13) |
| S-5 | 6.40 | 7.18 | 1.11 | 0.08 | 13.81 | 7.01 | 3.18 | 10.19 |
| | (0.12) | (0.01) | (0.02) | (0.00) | (0.08) | (0.32) | (0.11) | (1.14) |
| S-6 | 7.90 | 7.14 | 1.23 | 0.09 | 13.64 | 7.18 | 3.25 | 10.43 |
| | (0.10) | (0.02) | (0.01) | (0.00) | (0.11) | (0.40) | (0.13) | (1.21) |
| S-7 | 8.70 | 7.02 | 1.41 | 0.11 | 12.83 | 7.61 | 3.58 | 11.19 |
| | (0.13) | (0.01) | (0.01) | (0.00) | (0.18) | (0.28) | (0.12) | (1.11) |
| S-8 | 9.40 | 6.96 | 1.39 | 0.11 | 12.64 | 7.89 | 4.03 | 11.92 |
| | (0.09) | (0.01) | (0.02) | (0.00) | (0.27) | (0.32) | (0.12) | (1.12) |
| S-9 | 10.60 | 6.92 | 1.45 | 0.12 | 12.09 | 8.05 | 4.05 | 12.10 |
| | (0.11) | (0.01) | (0.04) | (0.01) | (3.55) | (0.35) | (0.14) | (1.23) |
| S-10 | 11.22 | 6.88 | 1.68 | 0.14 | 12.00 | 8.15 | 4.12 | 12.27 |
| | (0.13) | (0.02) | (0.04) | (0.00) | (0.16) | (0.27) | (0.12) | (1.10) |
| S-11 | 12.54 | 6.85 | 1.82 | 0.16 | 11.38 | 8.55 | 4.41 | 12.96 |
| | (0.13) | (0.02) | (0.06) | (0.00) | (0.29) | (0.25) | (0.14) | (1.14) |
| S-12 | 13.36 | 6.82 | 2.09 | 0.19 | 11.00 | 9.03 | 4.41 | 13.44 |
| | (0.14) | (0.01) | (0.05) | (0.02) | (0.57) | (0.41) | (0.13) | (1.18) |
| S-13 | 14.66 | 6.78 | 2.00 | 0.19 | 10.50 | 9.45 | 4.85 | 14.30 |
| | (0.15) | (0.02) | (0.06) | (0.01) | (0.08) | (0.29) | (0.14) | (1.21) |
| S-14 | 15.14 | 6.76 | 2.02 | 0.21 | 9.60 | 10.20 | 5.12 | 15.32 |
| | (0.14) | (0.01) | (0.02) | (0.00) | (0.03) | (0.38) | (0.15) | (1.15) |
| S-15 | 15.92 | 6.76 | 1.91 | 0.22 | 8.67 | 10.94 | 5.28 | 16.22 |
| | (0.16) | (0.01) | (0.09) | (0.02) | (0.16) | (0.43) | (0.13) | (1.12) |

**Table 50.2** Mean values of the C:N ratio and parameters of N-mineralization in different levels of N-application at the Banaras Hindu University. The values of Ammonification, Nitrification, and Net N-mineralization are in μg $g^{-1}$ $month^{-1}$. Values in parentheses are ±1SE. Different superscript alphabets within a row (N-treatment levels) are significantly different at $P < 0.05$

| Parameters | Control | 60 kg N $ha^{-1}$ $year^{-1}$ | 120 kg N $ha^{-1}$ $year^{-1}$ |
|---|---|---|---|
| C:N ratio | $16.00^{a}$ | $9.88^{b}$ | $14.44^{a}$ |
| | (1.13) | (0.81) | (0.74) |
| Ammonification | $6.76^{a}$ | $9.41^{b}$ | $7.25^{a}$ |
| | (0.45) | (0.48) | (0.65) |
| Nitrification | $2.73^{a}$ | $5.38^{b}$ | $3.22^{a}$ |
| | (0.32) | (0.38) | (0.27) |
| Net N-mineralization | $9.49^{a}$ | $14.79^{b}$ | $10.47^{a}$ |
| | (0.46) | (0.52) | (0.41) |

**Table 50.3** Mean values of N-mineralization parameters in different months at the Banaras Hindu University under N-application study. The values of Ammonification, Nitrification, and net N-mineralization are in μg $g^{-1}$ $month^{-1}$. Values in parentheses are ± 1SE

| Months | Ammonification | Nitrification | Net N-mineralization |
|---|---|---|---|
| January | 12.55 | 4.74 | 17.29 |
| | (0.87) | (0.54) | (0.74) |
| February | 16.75 | 11.67 | 28.42 |
| | (0.98) | (0.78) | (0.92) |
| March | 12.36 | 4.65 | 17.01 |
| | (0.68) | (0.32) | (0.88) |
| April | 12.51 | 4.68 | 17.19 |
| | (0.56) | (0.31) | (0.76) |
| May | 15.47 | 5.76 | 21.23 |
| | (0.75) | (0.24) | (1.20) |
| June | 14.28 | 5.25 | 19.53 |
| | (0.65) | (0.28) | (0.94) |
| July | 53.04 | 29.07 | 82.11 |
| | (1.44) | (0.97) | (1.11) |
| August | 50.00 | 25.29 | 75.292 |
| | (1.14) | (0.99) | (1.16) |
| September | 56.78 | 35.85 | 92.63 |
| | (1.15) | (0.88) | (2.17) |
| October | 19.44 | 12.63 | 32.07 |
| | (0.59) | (0.43) | (0.87) |
| November | 13.14 | 4.86 | 18.00 |
| | (0.41) | (0.33) | (0.67) |
| December | 12.54 | 4.77 | 17.31 |
| | (0.39) | (0.30) | (0.59) |

**Table 50.4** Analysis of variance (ANOVA) showing effects of sites, months, and N-treatments and their interactions on the parameters of soil N-mineralization from the experimental plots of Banaras Hindu University. Single, double, and triple asterisks on different *F*-values suggest significant differences in the parameters of N-mineralization at ≤0.01, ≤0.001, and ≤0.0001 *P* levels, respectively and NS is used for insignificant differences in *F*-values at *P* ≤0.05

| Variables | *df* | Ammonification | Nitrification | Net N-mineralization |
|---|---|---|---|---|
| Site | 14 | 46.18*** | 23.02** | 128.37*** |
| Month | 11 | 19.21*** | 12.46** | 96.65*** |
| Nitrogen | 2 | 6.95** | 4.12* | 12.97*** |
| Site × Month | 154 | 2.65*** | 1.26* | 6.36*** |
| Site × Nitrogen | 28 | 1.58* | 1.51* | 3.24** |
| Month × Nitrogen | 22 | 2.14* | $0.78^{NS}$ | $1.48^{NS}$ |
| Site × Month × Nitrogen | 308 | $0.64^{NS}$ | 1.21* | $0.87^{NS}$ |
| Error | 1080 | | | |

the control and 120 kg N groups were closer to each other, suggesting differences in ammonification, nitrification, and net N-mineralization of control and 120 kg N treated plots with that of 60 kg N-treated plots (Fig. 50.1). The differences in ammonification, nitrification, and net N-mineralization between different pairs of N-treatment doses based on Tukey test are summarized in Table 50.2, which also suggested differences in the ammonification, nitrification,

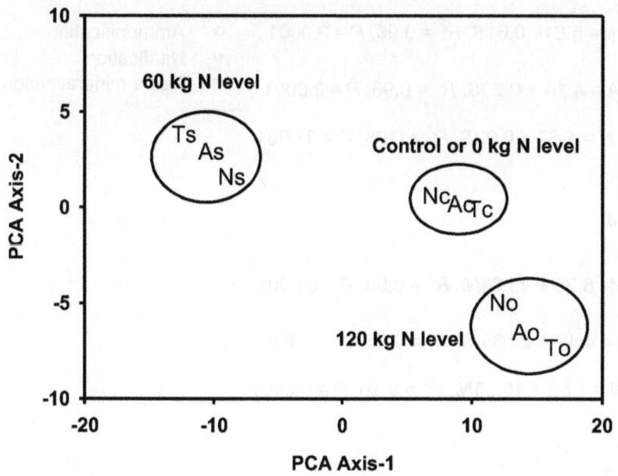


**Fig. 50.1** Principal Component Analysis (PCA) of the three N-treatment levels from different experimental plots at the Banaras Hindu University. In the diagram, *A*, *N*, *T*, *c*, *s*, and *o* stand for ammonification, nitrification, total N-mineralization, control, or 0 kg N $ha^{-1}$, 60 kg N $ha^{-1}$, and 120 kg N $ha^{-1}$

and net N-mineralization between control and 60 kg N-treated doses, and between 120 and 60 kg N-treated doses; nevertheless, the difference between 60 and 120 kg N-treated doses was insignificant (Table 50.2).

In the present study, increased soil moisture lowered the soil pH (approaches towards acidity). This situation could be due to the decomposition of leaf as well as woody litters at moisture-rich sites. It has been well established that moisture-rich soils enables the microorganisms for the process of decomposition than the moisture-poor soils and the carbonic acids released from litter decomposition may lower the soil pH (Rao et al. 2009). The plots treated with 120 kg N $ha^{-1}$ $year^{-1}$ experienced lower soil pH compared to control and 60 kg N-treatments. It is most likely due to easy breakdown of urea fertilizer and increased concentration of $H^+$ by decomposition of organic matters (Rao et al. 2009; Stevens 2009; Wang et al. 2009). Additionally, the available soil-N (Fang et al. 2006) and ammonium ions released during the process of ammonification by the activity of various species of *Clostridium*, *Micrococcus*, and *Proteus* may acidify the soil system (Matson et al. 1999).

The relationship between soil moisture and net N-mineralization had paradoxical results with few findings increased (Mazzarino et al. 1998; Singh and Kashyap 2007; Singh et al. (2009), Verma et al 2013) or decreased (Fisher et al. 1987) mineralization but few findings having no effect of soil moisture on N-mineralization (Whitford et al. 1995). In the present study, rates of ammonification, nitrification, and net N-mineralization were significantly greater in the moisture-rich plot than the moisture-poor plot which is in accordance with the studies of Singh and Kashyap (2007) and Singh et al. (2009). The monthly variation in the parameters of N-mineralization could be explained by the monthly variation in rainfall; evidently, there was a quadratic relationship between total N-mineralized and rainfall ($Y = 4.82 + 0.32X - 0.001X^2$, $R^2 = 0.99$, $P \leq 0.0001$). Principally, under the condition of high rainfall the absence of oxygen inhibits the activity of the *Nitrosomonas* which oxidizes $NH_4^+$ and finally the nitrogen mineralization is inhibited and further water drainage creates the aerobic condition and then microorganisms are enabled for the process of N-mineralization (Reddy et al. 1984). Thus, greater water availability inhibits the activity of microorganisms and reduces the N-mineralization.

As in the present study (Fig. 50.2), the trend of increasing N-mineralization with increasing C and N contents has been widely observed (Mlambo et al. 2007; Rao et al. 2009; Wang et al. 2009, Verma et al 2013). Greater amount of soil-C and -N improves soil structure, water retention capacity, and infiltration rates. The porous soils provide the aeration, soil moisture, and habitat for microbial community (Mlambo et al. 2007) which are largely responsible for the transformation of organic matters and facilitate N-mineralization potential of the soils (Mlambo et al. 2007). Further, the negative relationships between parameters of N-mineralization and quality of organic matter (C: N ratio), as shown in Fig. 50.2, suggested that organic matter rich in N content (low C: N ratio) has greater N-mineralization than the organic matter poor in N content as also reported in several studies (Vourlitis et al. 2007; Rao et al. 2009, Verma et al. 2013). Plants that exist in nutrient-rich habitats generally generate high quality tissues that are having high N content (Pan et al. 2010). The quantity of N in

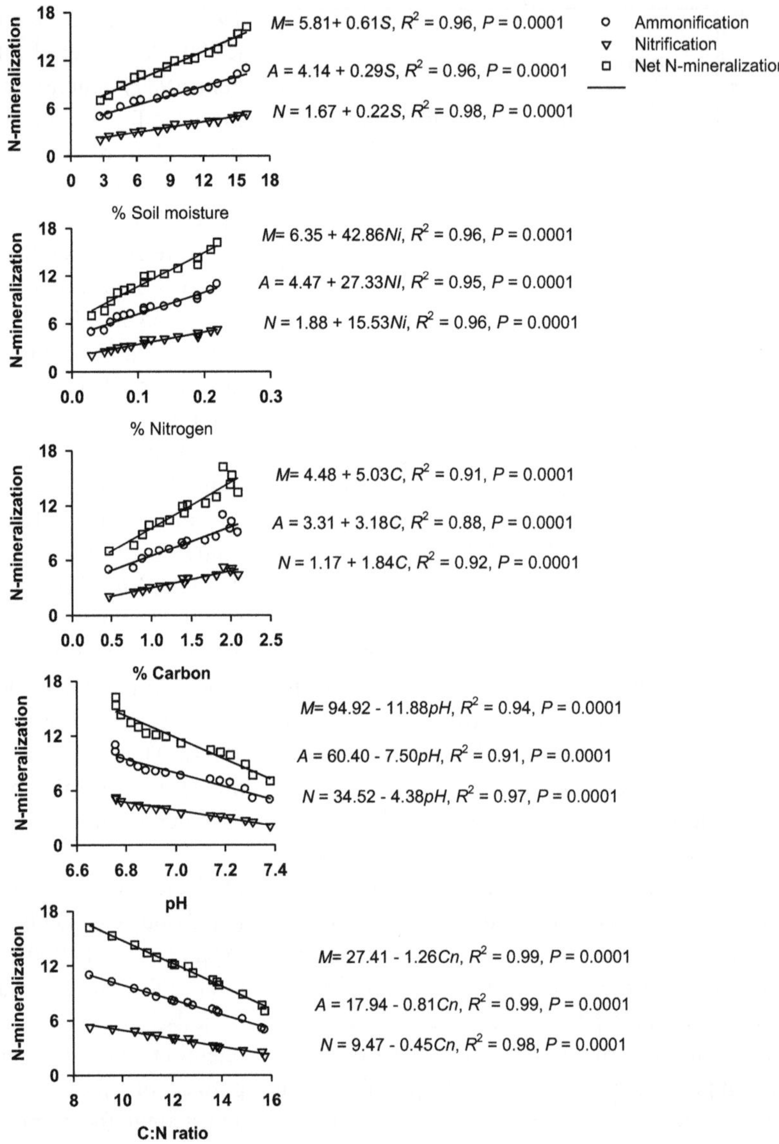


**Fig. 50.2** Linear relationships between N-mineralization parameters and selected soil variables from N-treatment experiments at Banaras Hindu University. The values of ammonification, nitrification, and net N-mineralization are in μg $g^{-1}$ $month^{-1}$

litter affects both decomposition rates and soil N-mineralization. High quality litter (low C:N ratio) is more quickly decomposed than low quality litter, resulting in more rapid nutrient cycling (Pan et al. 2010). Thus, in this study, the C:N ratio partially determined the rate of N-mineralization.

Soil-C, -N, and rates of ammonification, nitrification, and net N-mineralization yielded a humped-back pattern in relation to N-application. This conveys that these soil variables are low at low N level, increase to peak at moderate level, and decrease gradually at high N level. This affinity can be speculated as: at low level of N-application soil-C and -N are not enough for the activities of nitrifiers to release them in available form. As N increases, more microorganisms, viz. *Clostridium*, *Micrococcus*, *Proteus*, *Nitrosomonas europaea*, *Nitrosococcus nitrosus*, *Nitrosospira briensis*, *Nitrosovibrio*, *Nitrocystis*, *Nitrobacter winogradski*, *Nitrospira gracilis*, *Nitrosococcus mobilis*, *Penicillium*, *Aspergillus*, *Streptomyces*, and *Nocardia*, are enabled for the process of ammonification and nitrification and thus net N-mineralization (Bobbink et al. 2010) and sufficiently high N level, possibly, limits the nitrifier population as well as their activities and finally the processes of N-mineralization, because extreme N-application can damage the natural flora and fauna of soil which depletes soil fertility (Fu et al. 1987). The N saturation theory predicts that the N limited ecosystems keep the deposited N by using it for the growth and development of plants and microorganisms in addition to accumulation in biomass and soil organic matter. At a certain point, the deposited N commences to go beyond the biotic and abiotic needs for N within the system and the ecosystem is predicted to fail its N-retention ability. As the capability to keep N is exceeded, surplus N is offered to be lost from the ecosystem through solution losses and gas flux (Matson et al. 2002).

In conclusion, N fertilization changes the soil pH, soil nutrients, litter quality, and the process of N-mineralization. Further, N mineralization is spatially governed by the soil pH, moisture, nutrient availability, and litter quality and temporally it is governed by the variation in rainfall. Study suggested that moderate level of N fertilization is needed for the sustainability of microbial-mediated N release in the dry tropical environment.

**Acknowledgement** Funding support from the Department of Science and Technology, Government of India, is acknowledged.

## References

APHA (American Public Health Association) (1985) Standard methods for the examination of water and wastewater. American Public Health Association, Washington

Bobbink R, Hicks K, Galloway J, Spranger T, Alkemade R, Ashmore M, Bustamante M, Cinderby S et al (2010) Global assessment of nitrogen deposition effects on terrestrial plant diversity: a synthesis. Ecol Appl 20:30–59

Buol SW, Southard RJ, Graham RC, McDaniel PA (2003) Soil genesis and classification, 5th edition. Iowa State Press-Blackwell, Ames

Fang YT, Zhu WX, Mo JM, Zhou GY, Gundersen P (2006) Dynamics of soil inorganic nitrogen and their responses to nitrogen additions in three subtropical forests, south China. J Environ Sci 18:752–759

Fisher FM, Parker LW, Anderson JP, Whitford WG (1987) Nitrogen mineralization in a desert soil-interacting effects of soil moisture and nitrogen fertilizer. Soil Sci Soc Am J 51:1033–1041

Fu MH, Xu XC, Tabatabai MA (1987) Effect of pH on nitrogen mineralization in crop residue-treated soil. Biol Fertile Soils 5:115–119

Gonzalez-Prieto SJ, Villar MC, Carballas M, Carballas T (1992) Nitrogen mineralization and its controlling factors in various kinds of temperate humid-zone soils. Plant Soil 144:31–44

Jackson M (1958) Soil chemical analysis. Prentice Hall, Englewood Cliffs

Jones CA, Koenig RT, Ellsworth JW, Brown BD, Jackson GD (2007) Management of Urea Fertilizer to Minimize Volatilization. EB 173. Montana State University Extension and Washington State University Extension

Kros J, Frumau KFA, Hensen A, de Vries W (2011) Integrated analysis of the effects of agricultural management on nitrogen fluxes at landscape scale. Environ Pollut 159:3171–3182

Makoi JHJR, Ndakidemi PA (2008) Selected soil enzymes: Examples of their potential roles in the ecosystem. Afr J Biotechnol 7:181–191

McAleece N, Gage JD, Lambshead J, Patterson GLJ (1997) Biodiversity Professional. The Natural History Museum and The Scottish Association for Marine Science

Matson PA, McDowell WD, Townsend A, Vitousek P (1999) The globalization of nitrogen deposition: ecosystem consequences in tropical environments. Biogeochemistry 46:67–83

Matson PA, Lohse KA, Hall HJ (2002) The globalization of nitrogen deposition: consequences for terrestrial ecosystems. Ambio 31:113–119

Mazzarino MJ, Bertiller MB, Sain C, Satti P, Coronato F (1998) Soil nitrogen dynamics in northeastern Patagonia Steppe under different precipitation regimes. Plant Soil 202:125–131

Mlambo D, Mwenje E, Nyathi P (2007) Effects of tree cover and season on soil nitrogen dynamics and microbial biomass in an African savanna woodland dominated by *Colophospermum mopane*. J Trop Ecol 23:437–448

Pan JJ, Widner B, Ammerman D, Drenovsky RE (2010) Plant community and tissue chemistry responses to fertilizer and litter nutrient manipulations in a temperate grassland. Plant Ecol 206:139–150

Rao LE, Parker DR, Bytnerowicz A, Allen EB (2009) Nitrogen mineralization across an atmospheric nitrogen deposition gradient in Southern California deserts. J Arid Environ 73:920–930

Reddy KR, Patrick WH, Broadbent FE (1984) Nitrogen transformation and loss in flooded soils and sediments. Crit Rev Environ contr 13:273–309

Roy S, Singh JS (1994) Consequences of habitat heterogeneity for availability of nutrients in a Dry Tropical Forest. J Ecol 82:503–509

Sagar R, Verma P (2010) Effects of soil physical characteristics and biotic interferences on the herbaceous community composition and species diversity in the campus of Banaras Hindu University, India. Environmentalist 30:289–298

Sagar R, Singh A, Singh JS (2008) Differential effect of woody plant canopies on species composition and diversity of ground vegetation: a case study. Trop Ecol 49:189–197

Sala OE, Chapin FS III, Armesto JJ, Berlow E, Bloomfield J, Dirzo R, Huber-Sanwald E, Huenneke LF et al (2000) Global biodiversity scenario for the year 2100. Science 287:1770–1774

Singh A (2011) Natural floristic composition of Banaras Hindu University, India: an overview. Int J Peace Dev Stud 2:13–25

Singh Jay S, Kashyap AK (2007) Contrasting pattern of nitrifying bacteria and nitrification in seasonally dry tropical forest soils. Curr Sci 92:1739–1744

Singh Jay S, Singh DP, Kashyap AK (2009) A comparative account of the microbial biomass-N and N-mineralization of soils under natural forest, grassland and crop field from dry tropical region, India. Plant Soil Environ 55:223–230

Smith VH, Tilman D, Nekola JC (1999) Eutrophication: impacts of excess nutrient inputs on freshwater, marine and terrestrial ecosystems. Environ pollut 100:179–196

SPSS (1997) SPSS base 7.5 application guide. SPSS, Chicago

Stevens C (2009) The impact of atmospheric nitrogen deposition on grassland: species composition and biogeochemistry. Vdm Verlag

Verma P, Verma P, Sagar R (2013) Variations in N mineralization and herbaceous species diversity due to sites, seasons, and N treatments in a seasonally dry tropical environment of India. Forest Ecology an. Management 297:15–26

Vitousek P, Howarth RW, Likens GE, Matson PA, Schindler D, Schlessinger WH, Tilman D (1997) Human alteration of the global nitrogen cycle: cause and consequences. Issue Ecol 1:1–17

Vourlitis GL, Zorba G, Pasquini SC, Mustard R (2007) Chronic nitrogen deposition enhances nitrogen mineralization potential of semiarid shrubland soils. Soil Sci Soc Am J 71:836–842

Waldrop MP, Donald R, Zak RL, Sinsabaugh MG, Lauber C (2004) Nitrogen deposition modifies soil carbon storage through changes in microbial enzymatic activity. Ecol Appl 14:1172–1177

Walkley A (1947) A critical examination of a rapid method for determining organic carbon in soils effect of variations in digestion conditions and of inorganic soil constituents. Soil Sci 63:251–264

Wang H, Mo J, Xiankai L, Jinghua XUE, Jiong L, Yunting F (2009) Effect of elevated nitrogen deposition on soil microbial biomass carbon in major subtropical forest of southern China. Front For China 4:21–27

Whitford WG, Martinez-Turanzas G, Martinez-Meza E (1995) Persistence of decertified ecosystems: explanations and implications. Environ Monit Assess 37:319–332

# Influence of Crop Rotation and Intercropping on Microbial Populations in Cultivated Fields Under Different Organic Amendments

51

Haribashai Swer and M. S. Dkhar

**Abstract**

The present investigation was carried out with an aim to study the microbial populations (fungi and bacteria) of organically amended soils under three different crop cycles, i.e. maize (*Zea mays* L.), French bean (*Phaseolus vulgaris* L.) and soybean (*Glycine max*). From the selected crops, maize and French bean were grown in rotation with soybean as an intercropping crop. The different organic amendments incorporated into the experimental field include farmyard manure (FYM), plant compost (PC), vermicompost (VC) and integrated compost (INT) (combination of FYM, PC and VC in 1:1:1 ratio). A control (CTRL) plot without any fertilizer was also maintained. Treatment-wise, the fungal population was increased by FYM amendment; however, PC amendment resulted in higher bacterial population. Significant variations were observed in the microbial populations (fungi and bacteria) between the organic treatments according to Tukey's test (ANOVA) at $p \leq 0.05$. The dominant fungal species isolated include *Acremonium* spp., *Aspergillus* spp., *Cladosporium cladosporioides, Fusarium oxysporum, Gongronella butleri, Humicola* spp., *Mortierella gamsii, Phoma eupyrena, Paecilomyces carneus, Penicillium* spp., *Pythium irregulare, Rhizopus stolonifer* and *Trichoderma* spp., *Arthrobacter* sp., *Bacillus* spp. and *Pseudomonas aeruginosa* were the dominant bacterial species isolated. It can be suggested that diverse plant residues returned to the soil by rotation of crops and improved the organic matter resulting in higher microbial populations. The study also revealed that the types of crop grown and the degree of fertilizer decomposition have a significant impact on the microbial populations.

**Keywords**

Microbial · Compost · Crop rotation · Intercropping

H. Swer (✉) · M. S. Dkhar
Microbial Ecology Laboratory, Department of Botany, North-Eastern Hill University, Shillong, Meghalaya 793022, India
e-mail: haribashai@yahoo.com

R. N. Kharwar et al. (eds.), *Microbial Diversity and Biotechnology in Food Security*,
DOI 10.1007/978-81-322-1801-2_51, © Springer India 2014

## 51.1 Introduction

Crop rotation means changing the type of crop grown on a particular piece of land from year to year. However, the term includes both cyclical rotations, in which the same sequence of crops is repeated indefinitely on a field, and noncyclical rotations, in which the sequence of crops varies yearly (Mohler 2009). Intercropping is an all-encompassing term for the practice of growing two or more crops in close proximity: in the same row or bed, or in rows or strips that are close enough for biological interaction (Mohler and Stoner 2009). Both crop rotation and intercropping are essential features of all organic cropping systems because they provide the principal mechanism for building healthy soils, a major pathway to control pests, and have a positive impact on soil microbial composition and diversity (Acosta-Martínez and Scot 2010; Xuan 2012).

It is a well-known fact that all organisms in the biosphere depend on microbial activity (Pace 1997). Microorganisms constitute less than 0.5 % (w/v) of the soil mass, yet they have a major impact on soil properties and processes. Microbial communities, particularly bacteria and fungi, constitute an essential component of biological characteristics in soil. Fungi and bacteria also known as primary decomposers are the prime participants in soil processes and are the most commonly studied soil living communities and they are responsible for the breakdown of organic matter and release of nutrients particularly in the case of nitrogenous and phosphatic minerals (Tate 2000).

The size of the microbial populations in soil is of ecological importance because of the essential role that microorganisms play in the conservation and cycling of plant nutrients and exerts considerable influence upon the fertility of soil and consequently on the growth and development of plants. It is widely accepted that each type of vegetational community harbours a characteristic soil microfloral population.

Next to the cover crops, soil management practices also change the soil environment and affect the population dynamics of soil microbial communities. Management practices such as cultivation, irrigation, crop rotation (Larkin 2008; Gupta et al. 2010), tillage practices (Acosta-Martínez and Scot 2010), input of fertilizers and organic residue addition (Araújo et al. 2009; Elhottová et al. 2012; Lazcano et al. 2012) have a major impact on diversity of biological population in soil.

Continuous presence of legumes in the intercropping system could favour stability to microbial community relative to other systems. Intercropping has been widely accepted as a sustainable practice due to its yield advantage, high utilization efficiency of light and water, and pest and disease suppression (Mohler 2009; Hummel et al. 2009).

Organic amendments represent an important resource to maintain and restore soil fertility and are of great values nowadays, particularly in those countries where the organic matter content of the soil is low. However, not much has been reviewed on the microbial populations and diversity under organic farming systems in this part of the region. Therefore, a study to understand the microbial population in soils under crop rotation and intercropping under different organic treatments was undertaken.

## 51.2 Study Site

The field experiment was conducted at a lowland experimental block of Agronomy Division, Indian Council of Agricultural Research (ICAR) for North Eastern Hill (NEH), Umiam, Ri-Bhoi District, Meghalaya. The geographical location of the study site is 25°41′26.7″N latitude and 91°55′26.2″E longitude and at an elevation of 956 m (asl). The soil texture of the experimental site is silty loam (Clay—32.58 %; Sand—12.83 %; Silt—54.58 %).

### 51.2.1 Experimental Design and Treatments

For the experimental set up, three crops were selected viz. maize (*Zea mays* L.), French bean (*Phasoelus vulgaris* L.) and soybean (*Glycine*

**Table 51.1** Type of organic fertilizer and doses

| Sl. No. | Organic fertilizer | Source | Dose (tonnes/ha) | Dose (kg/plot) |
|---|---|---|---|---|
| 1. | Farmyard manure (FYM) | Dried cow dung | 5 t/ha | 48 |
| 2. | Plant compost (PC) | Weeds from the field | 5 t/ha | 48 |
| 3. | Vermicompost (VC) | Earthworm cast | 5 t/ha | 48 |
| 4. | Integrated compost (INT) | FYM:PC:VC (1:1:1) | 5 t/ha | 48 (16:16:16) |

*max*). Two of the selected crops were grown in rotation, i.e. maize (May–August) and French bean (August–September). Soybean (May–September) was grown as an intercropping crop in between the rows under rotation.

The experimental field was divided into five plots. The net plot size was 3 × 4 m. Three replicate plots were maintained each for the different organic amendment. A control plot (i.e. without organic fertilizer) was also maintained. According to the types of organic amendment, each experimental plot viz. farmyard manure, plant compost, vermicompost, integrated compost and control were designated as FYM, PC, VC, INT and CTRL, respectively. The fertilizers were applied during the pre-sowing period. Optimum fertilizer dosage was applied to the field as recommended by ICAR (Table 51.1).

### 51.2.2 Soil Sampling

Soil samplings were done from the upper 0–15 cm depth at monthly interval from pre-sowing to post-harvest period for a period of 2 years (2006 and 2007) for each crop cycle. Soil samples were randomly collected from each plot and pooled together to get a composite sample.

### 51.2.3 Microbial Population Count

The soil-plate method (Warcup 1950) using Rose Bengal Agar Medium (Martin 1950) and the serial dilution plate method (Johnson and Curl 1972) using Nutrient Agar Medium (Difco manual 1953) were followed for the isolation and estimation of microbial populations. The inoculated petri plates were incubated at 25 ° and 30 °C for a period of 5–7 days and 24–48 h for fungi and bacteria, respectively. Colonies formed were counted and expressed on dry weight basis.

Identification of fungi at the species level were carried out according to morphological characters found principally in publications by Gillman (1957), Barnett and Hunter (1972), Domsch et al. (1980), Subramanian (1983), Ellis (1993) and Watanabe (1994) whereas bacterial species were identified based on biochemical tests.

### 51.2.4 Physicochemical Analysis

Soil samples were sieved through a 2 mm sieve at field-moist conditions to determine soil moisture and pH. Air dried ground and sieved (0.25 mm) samples were used for the estimation of organic C, total N, available P and K content. Three replicate samples were used for each analysis. Moisture content was determined by weight loss after drying 10 g of soil at 105 °C for 24 h and expressed as percentage dry weight. Soil pH was measured in a 1:5 water suspension using a portable digital pH meter. Colorimetric method (Anderson and Ingram 1993), micro Kjeldahl distillation and titration method (Jackson 1973), Molybdenum blue method (Allen et al. 1974) and the ammonium acetate flame photometry method (Jackson 1973) were applied to estimate organic carbon ($C_{org}$), total nitrogen (N), available phosphorus (P) and exchangeable potassium (K), respectively.

### 51.2.5 Statistical Analysis

Analysis of Variance (ANOVA) for the microbial populations was performed using Statistica

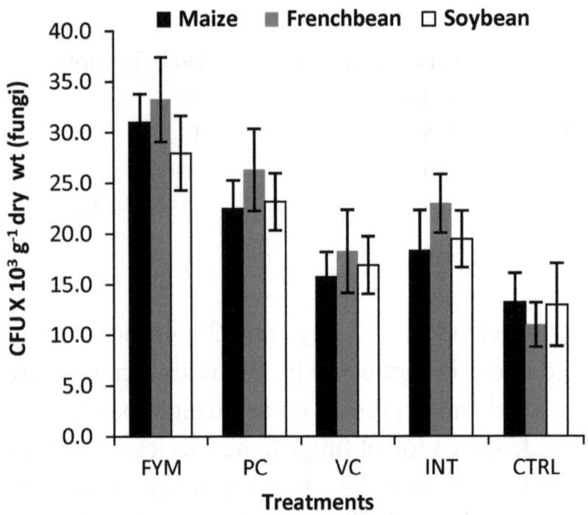


**Fig. 51.1** Colony-forming units of fungi (CF $\times 10^3$ $g^{-1}$ dry wt) in soil treated with different organic treatments

8.0 package. Tukey's test (post-hoc analysis) was done as a comparison between the treated means.

## 51.3 Results and Discussion

The colony-forming unit (CFU $\times 10^3$ $g^{-1}$ dry wt) of fungi was recorded to be maximum in FYM amended plot and minimum in the control plot in all the three crop cycles (maize–French bean in rotation and intercropping crop, soybean) under different organic amendments (Fig. 51.1). Improvement of soil organic matter and nutrient contents (Table 51.2) with FYM application resulted in an enhancement of soil quality and fertility which is reflected in increased fungal population. Studies of Larkin (2006) indicated that fungal population was significantly increased by dairy manure amendment. Chakrabarti et al. (2000) showed that soils treated with cow dung manure were found to be superior in accumulating higher total organic C as compared to soils treated with chemical fertilizers or no input.

The results of the bacterial CFU ($\times 10^5$ $g^{-1}$ dry wt) revealed that it was maximum in PC-treated plot and minimum in the control plot (Fig. 51.2). Significant variations in the fungal and bacterial CFU were observed in all the plots according to Tukey's test at $p \leq 0.05$ (Figs. 51.3 and 51.4). Bacteria are more competitive when labile (easy-to-metabolize) substrates are present which includes fresh, young plant residue and the compounds found near living roots. The addition of plant compost stimulated the bacterial activity as such; higher CFU was recorded with this treatment. Manici et al. (2004) showed that dry and fresh plant tissue-treated soil gave a significantly higher bacterial population than untreated soil. Kang et al. (2005) signified that the application of organic residues increased the number of bacteria due to better nutrient status and the effect was greater in the green manure treatment. Van Diepeningen et al. (2006) found that the bacterial species diversity and richness steadily increased, as years of organic management increased. Thus, application of FYM and PC to agricultural land has been viewed as an excellent way to recycle nutrients and organic matter that can support crop production and maintain or improve soil quality.

Comparing the three crop cycles, the CFU was comparatively higher during the French bean crop cycle followed by maize crop cycle and soybean crop cycle (Fig. 51.1). However, slightly higher bacterial CFU was observed during the soybean crop cycle (Fig. 51.2). It can be suggested that inclusion of legume crop (i.e. French bean) in the rotation increases the fungal population relative to monocropping. Crop rotation has been shown to have a direct impact on organic matter levels in soil (Bremer et al. 1994; Janzen et al. 1992; Larkin and Griffin 2007; Acosta-Martínez and Scot 2010) due to higher residue production (from maize plants) leading to greater soil organic carbon storage. This indicates the positive effect that crop rotation can play in enhancing fungal populations and improving soil fertility. Rotation of crop, thus, increased the fungal CFU and prevented a decrease in soil fertility as growing of the same crop repeatedly in the same plot depletes the soil of various nutrients. Havlin et al. (1990) and Larkin (2008) also showed that crop rotations have a positive effect on physical, chemical and biological properties of the soil. Further, they revealed that the effect was related to higher carbon inputs and diversity of plant residues returned to soils.

However, higher bacterial CFU during the soybean crop cycle could be due to higher available nutrients in the form of sugary exudates

**Table 51.2** Physicochemical characters of maize (M), French bean (FB) and soybean (S) field soils with standard error (SE). The range of the values are given in parentheses

| Soil properties | | Treatments | | | | |
|---|---|---|---|---|---|---|
| | | FYM | PC | VC | INT | CTRL |
| Soil temperature (°C) | M | 29.08±0.0 (26.17–31.67) | 28.79±0.0 (26.17–32.67) | 29.40±0.0 (26.67–33.33) | 29.48±0.0 (26.50–33.0) | 30.15±0.0 (26.67–34.33) |
| | FB | 29.47±0.00 (27.33–33.00) | 29.28±0.0 (26.00–32.67) | 29.56±0.0 (27.00–34.67) | 29.94±0.0 (26.33–34.33) | 30.78±0.0 (28.33–34.0) |
| | S | 29.67±0.00 (26.17–33.73) | 28.60±0.00 (26.00–32.33) | 30.35±0.00 (26.67–35.33) | 29.13±0.00 (26.33–33.0) | 30.08±0.00 (26.67–3.33) |
| Soil pH | M | 5.47±0.028 (4.70–5.90) | 5.35±0.031$^{a}$ (4.77–5.93) | 5.16±0.017 (4.50–5.70) | 4.86±0.028$^{b}$ (4.50–5.10) | 4.72±0.030$^{c}$ (4.30–5.13) |
| | FB | 5.49±0.028$^{a}$ (5.20–5.83) | 5.24±0.031$^{a}$ (4.77–5.53) | 5.14±0.017 (4.73–5.80) | 4.92±0.028 (4.37–5.63) | 4.63±0.030$^{c}$ (4.13–5.0) |
| | S | 5.32±0.02$^{a}$ (5.0–6.0) | 5.12±0.017$^{a}$ (4.3–5.8) | 5.0 8±0.017$^{a}$ (4.5–5.7) | 4.83±0.04$^{b}$ (4.2–5.3) | 4.62±0.013$^{d}$ (4.1–5.0) |
| Moisture content (%) | M | 24.60±0.244 (20.90–28.47) | 25.71±0.266 (22.20–31.33) | 23.78±0.162 (20.50–26.63) | 24.16±0.102 (21.17–26.17) | 23.77±0.091 (20.57–28.70) |
| | FB | 25.08±0.244$^{a}$ (23.67–26.67) | 25.77±0.266$^{a}$ (24.40–27.47) | 24.54±0.162 (22.63–26.70) | 23.43±0.102$^{b}$ (22.67–24.37) | 21.93±0.091$^{d}$ (20.50–23.27) |
| | S | 24.07±0.126 (20.10–28.10) | 25.02±0.304$^{a}$ (22.20–28.07) | 23.70±0.174 (20.50–27.47) | 23.71±0.16 (21.17–27.30) | 22.79±0.099$^{b}$ (19.87–25.67) |
| Organic carbon (%) | M | 1.342±0.015 (0.856–1.604) | 1.572±0.014$^{a}$ (1.144–1.752) | 1.291±0.018 (0.851–1.671) | 1.359±0.014 (1.113–1.640) | 1.172±0.014$^{b}$ (0.847–1.275) |
| | FB | 1.618±0.015$^{a}$ (1.252–1.860) | 1.696±0.014$^{a}$ (1.559–1.829) | 1.524±0.018$^{a}$ (1.288–1.779) | 1.468±0.014 (1.374–1.644) | 1.248±0.014$^{d}$ (1.081–1.477) |
| | S | 1.405±0.014$^{a}$ (0.856–1.568) | 1.583±0.011$^{a}$ (1.144–1.910) | 1.346±0.013$^{c}$ (0.851–1.568) | 1.337±0.015$^{c}$ (1.113–1.532) | 1.203±0.014$^{c}$ (0.847–1.401) |
| Total nitrogen (%) | M | 0.395±0.008$^{a}$ (0.280–0.658) | 0.323±0.008 (0.252–0.495) | 0.388±0.008 (0.253–0.460) | 0.270±0.007 (0.223–0.363) | 0.239±0.006$^{b}$ (0.130–0.368) |
| | FB | 0.478±0.008$^{a}$ (0.403–0.658) | 0.408±0.008 (0.295–0.568) | 0.454±0.008$^{a}$ (0.348–0.625) | 0.361±0.007 (0.200–0.540) | 0.264±0.006$^{c}$ (0.163–0.330) |
| | S | 0.384±0.009$^{a}$ (0.300–0.493) | 0.273±0.007$^{b}$ (0.220–0.325) | 0.338±0.006$^{b}$ (0.270–0.387) | 0.245±0.004$^{b}$ (0.200–0.313) | 0.202±0.006$^{e}$ (0.132–0.288) |
| Available phosphorus ($\mu g^{-1}$) | M | 29.83±0.22$^{a}$ (21.40–39.10) | 22.01±0.34$^{b}$ (16.90–28.50) | 20.37±0.23$^{b}$ (15.8–27.7) | 24.78±0.29$^{b}$ (18.30–32.50) | 16.75±0.28$^{c}$ (13.80–20.80) |
| | FB | 33.80±0.22$^{a}$ (21.40–47.70) | 26.97±0.34 (18.20–37.10) | 26.42±0.23 (17.0–37.1) | 29.72±0.29 (18.00–38.40) | 17.18±0.28$^{b}$ (14.90–21.60) |
| | S | 29.85±0.33$^{a}$ (21.40–47.60) | 23.59±0.22$^{b}$ (17.10–31.0) | 24.86±0.19$^{b}$ (16.70–39.90) | 24.17±0.37$^{b}$ (16.20–38.80) | 16.96±0.23$^{e}$ (13.80–23.80) |
| Exchangeable potassium (%) | M | 0.073±0.001$^{a}$ (0.040–0.104) | 0.039±0.0$^{b}$ (0.014–0.057) | 0.048±0.001$^{b}$ (0.025–0.059) | 0.035±0.0$^{b}$ (0.015–0.053) | 0.030±0.0$^{b}$ (0.014–0.048) |
| | FB | 0.101±0.001$^{a}$ (0.073–0.122) | 0.064±0.0 (0.022–0.098) | 0.072±0.001 (0.035–0.106) | 0.051±0.0$^{b}$ (0.018–0.092) | 0.030±0.0$^{b}$ (0.017–0.051) |
| | S | 0.061±0.001$^{a}$ (0.037–0.079) | 0.038±0.000$^{b}$ (0.014–0.067) | 0.042±0.001$^{b}$ (0.018–0.064) | 0.032±0.000$^{b}$ (0.014–0.052) | 0.025±0.001$^{c}$ (0.014–0.048) |

Mean ± SE in columns followed by the same letter do not differ significantly according to Tukey's test at $p \leq 0.05$
*FYM* farmyard manure, *VC* vermicompost, *INT* integrated compost, *PC* plant compost, *CTRL* control

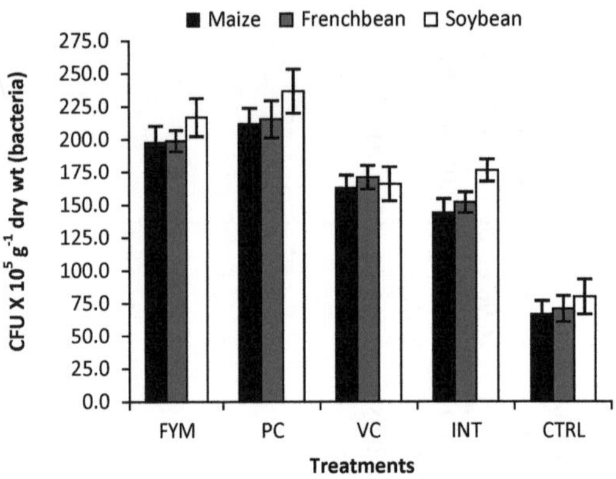


**Fig. 51.2** Colony-forming units of bacteria (CFU $\times 10^5$ $g^{-1}$ dry wt) in soil treated with different organic treatments

from this legume. Additionally, the presence of root nodules might have enriched the immediate soil environment through the rhizosphere effect thereby enriching the bacterial diversity. Song et al. (2007) also showed that intercropping of wheat and maize with faba bean or mixed cropping increased the diversity of the bacterial community significantly in the rhizosphere than in the respective sole crops.

Qualitatively, the composition of the fungal and bacterial species was almost similar in all the three crop cycles and in all the different organically treated plots. Most of the species isolated were common to all the soils except for few species which were confined to a particular crop cycle or to a particular organic treatment. Altogether, a total of 135 fungal species and 3 sterile mycelia were isolated from all the plots from both the soils under crop rotation and intercropping. Highest fungal species were isolated during the maize crop cycle (111 fungal species and 3 sterile mycelia) followed by soybean crop cycle (98 fungal species and 2 sterile mycelia) and minimum fungal species were isolated during the French bean crop cycle (82 fungal species and 2 sterile mycelia). The dominant fungal species isolated include *Acremonium cerealis, A. kiliense, Aspergillus flavus, A. fumigatus, A. niger, C. cladosporioides, F. oxysporum, G. butleri, Humicola fuscoatra, H. grisea, M. gamsii,*

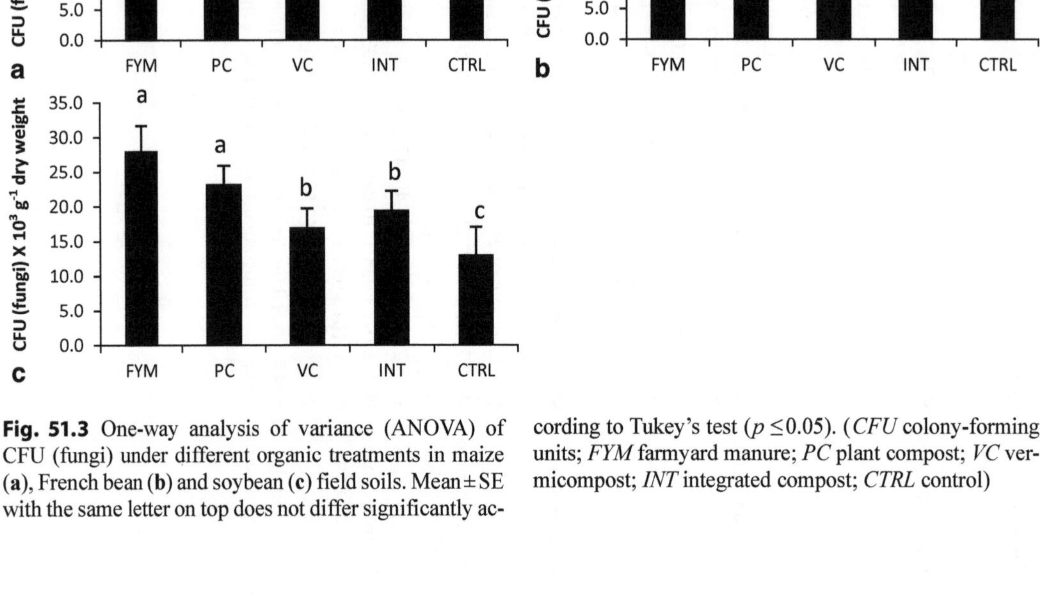


**Fig. 51.3** One-way analysis of variance (ANOVA) of CFU (fungi) under different organic treatments in maize (**a**), French bean (**b**) and soybean (**c**) field soils. Mean ± SE with the same letter on top does not differ significantly according to Tukey's test ($p \leq 0.05$). (*CFU* colony-forming units; *FYM* farmyard manure; *PC* plant compost; *VC* vermicompost; *INT* integrated compost; *CTRL* control)

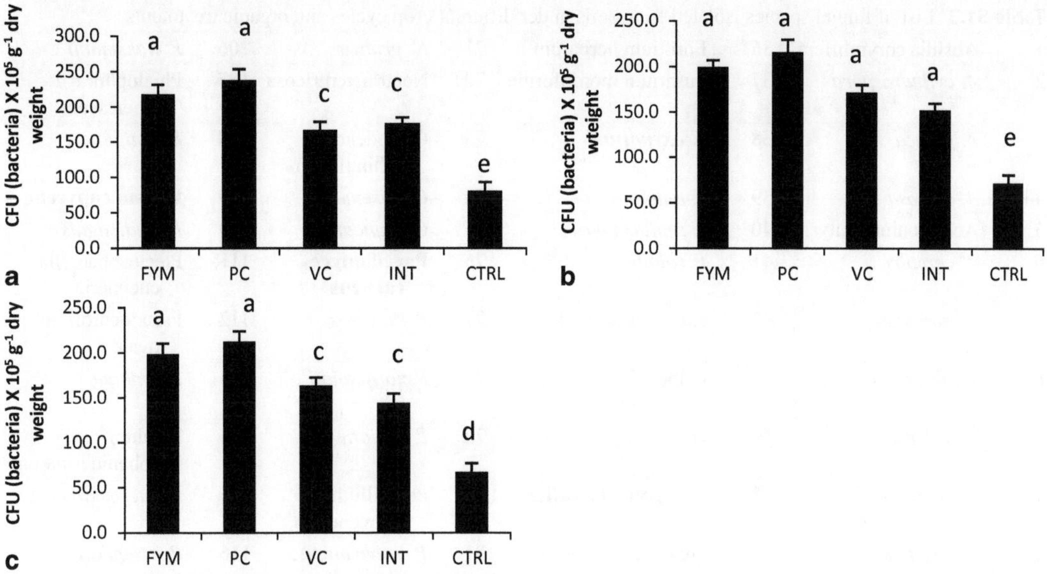


**Fig. 51.4** One-way analysis of variance (ANOVA) of CFU (bacteria) under different organic treatments in maize (**a**), French bean (**b**) and soybean (**c**) field soils. Mean ± SE with the same letter on top does not differ significantly according to Tukey's Test ($p \leq 0.05$). (*CFU* colony-forming units; *FYM* farmyard manure; *PC* plant compost; *VC* vermicompost; *INT* integrated compost; *CTRL* control)

*P. eupyrena, Paecilomyces carnues, Penicillium brevicompactum, P. daleae, P. janthinellum, P. rubrum, P. simplissicimum, P. verrucosum, P. irregulare, R. stolonifer, Trichoderma koningii* and *T. viride* (Table 51.2). A total of 16 bacterial species were isolated from all the soils under maize–French bean rotation and intercropping crop (soybean) (Table 51.3). Highest number of bacterial species were isolated during the soybean crop cycle (16) followed by French bean crop cycle (14) and maize crop cycle (13). The species commonly isolated from all the plots throughout the investigating period belonged to three genera viz. *Acetobacter* sp., *Arthrobacter* sp., *Bacillus* sp. and *Pseudomonas* sp. *Bacillus* was the dominant genus with four species. The dominant bacterial species isolated from all the three crop cycles include *Arthrobacter* sp., *Bacillus cereus, B. subtilis, Bacillus* sp. and *P. aeruginosa* (Table 51.4).

Similar composition of the fungal and bacterial species in both organically treated and control plot is due to similar microclimatic conditions or same type of crops grown. Hackl et al. (2000) indicated that the plant species growing on the soil also equally influence the population and species composition of the soil microbes. It can be proposed that incorporation of organic manures directly has an impact on the soil properties, the plant growth which in turn influences the microbial population and species. Entry and Emmingham (1996) reported that the rate of change in microbial population is attributed to the type of vegetation grown on a particular area and variation in physicochemical properties of the soil. Tangjang et al. (2009) noticed that plant residues, added organic matter, vegetation, plant species composition and soil mineral nutrients altered the microbial population as well as their species composition under traditional agroforestry system in Arunachal Pradesh. Lesser species composition of bacteria compared to fungal species could be due to the unculturable nature of these microbes (Wall and Virginia 2000). The selection of media and the method used were probably not conducive for the isolation of many bacterial species as some species required selective media for their growth.

**Table 51.3** List of fungal species isolated from soils under different crop cycles and organic treatments

| | | | | | | | |
|---|---|---|---|---|---|---|---|
| 1 | Absidia corymbifera | 36 | Eurotium herbarum | 71 | *N. grubya* | 106 | *P. waksmanii* |
| 2 | *A. cylindrospora* | 37 | Fusarium moniliforme | 72 | Nectria ventricosa | 107 | Phialophora cinerescens |
| 3 | *A. glauca* | **38** | ***F. oxyporum*** | 73 | Oidiodendron echinulatum | 108 | *P. festigiata* |
| 4 | *A. spinosa* | 39 | *F. redolens* | 74 | *O.griseum* | **109** | **Phoma eupyrena** |
| 5 | Acremonium butyri | 40 | *F. semitectum* | 75 | *O. truncatum* | 110 | *P. medicagnis* |
| 6 | *A. cerealis* | 41 | *F. solani* | **76** | **Paecilomyces carneus** | 111 | Plectosphaerella cucuneria |
| 7 | *A. furcatum* | 42 | *F. sporotrichioides* | 77 | *P. lilacinus* | 112 | Pseudoeurotium ovale |
| 8 | *A. fusidioides* | 43 | Gliocladium catenulatum | 78 | *P.marquandii* | 113 | *P. zonatum* |
| 9 | *A. kiliense* | 44 | *G. roseum* | 79 | *P. variotii* | 114 | Pythium aphanidermatum |
| 10 | *A. morurum* | **45** | **Gongronella butleri** | 80 | Penicillium atrovenetum | 115 | *P. intermedium* |
| 11 | *A. strictum* | 46 | Gymnoascus ressii | **81** | ***P. brevicompac-tum*** | **116** | *P. irregulare* |
| 12 | Allescheriella crocea | 47 | Helicosporium sp. | 82 | *P. canescens* | 117 | Ramichloridium schulzeri |
| 13 | Alternaria alternata | 48 | Helminthosporium sp. | 83 | *P. chrysogenum* | **118** | **Rhizopus stolonifer** |
| 14 | *A. citri* | **49** | **Humicola fuscoatra** | 84 | *P. citrinum* | 119 | Scopulariopsis brumptii |
| 15 | *A. longipes* | **50** | ***H. grisea*** | 85 | *P. coryliphylum* | 120 | *S. stercoraria* |
| 16 | Anthroderma cuniculi | 51 | Hyphomyces chrysospermus | **86** | ***P. daleae*** | 121 | Staphylotrichum coccosporum |
| 17 | *A. insingulare* | 52 | Mammaria echinobotryoides | 87 | *P. decumbens* | 122 | Talaromyces emersonii |
| 18 | Aspergillus clavatus | 53 | Mitteriella zizyphina | 88 | *P. digitatum* | 123 | *T. helicus* |
| **19** | *A. flavus* | 54 | Monilia sitophila | 89 | *P. fellutanum* | 124 | *T. stachyspermum* |
| 20 | *A. fumigatus* | 55 | Mortieralla alpina | 90 | *P. frequentans* | 125 | *T. wortmanii* |
| 21 | *A. japonicus* | 56 | *M. elongata* | 91 | *P. herquei* | 126 | Torula herbarum |
| **22** | *A. niger* | **57** | ***M. gamsii*** | 92 | *P. implicatum* | 127 | Trichoderma hamatum |
| 23 | *A. oryzae* | 58 | *M. hyalina* | 93 | *P. italicum* | 128 | *T. koningii* |
| 24 | *A. wentii* | 59 | *M. minutissima* | **94** | ***P. janthinellum*** | 129 | *T. polysporum* |
| 25 | *A. versicolor* | 60 | *M. nanna* | 95 | *P. jensenii* | **130** | ***T. viride*** |
| 26 | Beltrania sp. | 61 | *M. parvispora* | **96** | ***P. lanosum*** | 131 | Verticillium albo-atrum |
| 27 | Ceratocystis fimbriata | 62 | Mucor circinelloides f. circinelloides | 97 | *P. nigricans* | 132 | *V. chlamydosporium* |
| 28 | Chaetomium sp. | 63 | *M. circinelloides* f. griseo cyanus | 98 | *P. regulosum* | 133 | *V. dahliae* |
| 29 | Cladosporium cladosporioides | 64 | M.hiemalis f. hiemalis | 99 | *P. restrictum* | 134 | *V. nigrecens* |
| 30 | *C. herbarum* | 65 | *M. hiemalis f. silvaticus* | 100 | *P. roseo-purpu-reum* | 135 | Wardomyces humicola |
| 31 | *C. macrocarpus* | 66 | *M.mucedo* | **101** | ***P. rubrum*** | | Brown sterile mycelium |
| 32 | Cochliobolus sativus | 67 | *M. racemosus* | **102** | ***P. simplissicimum*** | | White sterile mycelium |

**Table 51.3** (continued)

| | | | | | | |
|---|---|---|---|---|---|---|
| 33 | Cunninghamella elegans | 68 | Myrothecium cinctrum | 103 | *P. stoliniferum* | Yellow sterile mycelium |
| 34 | Curvularia pallascens | 69 | *M. verrucaria* | 104 | *P. variabile* | |
| 35 | Cylindrocladium scoparium | 70 | Nannizia incurvata | **105** | ***P. verrucosum*** | |

Dominant species are highlighted in bold

**Table 51.4** List of bacterial species isolated from soils under different crop cycles and organic treatments

| | | | |
|---|---|---|---|
| 1 | *Acetobacter* sp. | 9 | *Escherichia coli* |
| **2** | ***Arthrobacter* sp.** | 10 | *Flavobacterium* sp. |
| 3 | *Azotobacter* sp. | 11 | *Micrococcus luteus* |
| **4** | ***B.acillus cereus*** | 12 | *Micrococcus* sp. |
| 5 | *B. mycoides* | **13** | ***Pseudomonas aeruginosa*** |
| **6** | ***B. subtilis*** | 14 | *Pseudomonas* sp. |
| **7** | ***Bacillus* sp.** | 15 | *Rhizobium* sp. |
| 8 | *Chromobacterium* sp. | 16 | *Xanthomonas* sp. |

Dominant species are highlighted in bold

It can be suggested that application of FYM and PC was proven to be the best nutrient input for the microbial populations (fungi and bacteria). Further, crop management practices such as organic amendments, crop rotation and intercropping are essential for restoration of microbial populations and diversity.

## References


Acosta-Martínez V, Scot ED (2010) Microbial community composition as affected by dryland cropping systems and tillage in a semiarid sandy soil. Diversity 2:910–931

Allen SE, Grinshaw HM, Parkinson JA, Quaramby C (1974) Chemical analysis of ecological materials. Blackwell Scientific Publications, Oxford

Anderson JM, Ingram JSI (1993) Tropical soil biology and fertility: a handbook of methods, 2nd eds. CAB Intenational, Oxford

Araújo ASF, Leite LFC, Santos VB, Carneiro RFB (2009) Soil Microbial Activity in Conventional and Organic Agricultural Systems. Sustainability 1:268–276

Barnett HL, Hunter BB (1972) Illustrated genera of imperfect fungi, 3rd edn. Burgess Publishing Company, Minneapolis

Bremer E, Janzen HH, Johnston AM (1994) Sensitivity of total, light fraction and mineralizable organic matter to management practices in a Lethbridge soil. Soil Sci Soc Am J 74:131–138

Chakrabarti K, Sarkar B, Chakraborty A, Banik P, Bagchi DK (2000) Organic recycling for soil quality conservation in sub-tropical plateau region. J Agron Crop Sci 184:137–142

Difco Manual (1953) Difco laboratories Defroit, Mich

Domsch KH, Gams W, Anderson TH (1980) Compendium of soil fungi. Academic Press, London

Entry JA, Emmingham WH (1996) Influence of vegetation on microbial degradation of atrazine and 2,4-dichlorophenoxyaceticacid acid in riparian soils. Can J Soil Sci 76:101–106

Elhottová D, Koubová A, Simek M (2012) Changes in soil microbial communities as affected by intensive cattle husbandry. Appl Soil Ecol 58:56–65

Ellis MB (1993) Demtiaceous hyphomycetes. CAB International, Wallingford

Gillman JC (1957) Manual of soil fungi, 2nd edn. Oxford and I.B.H Publishing company (Indian reprint)

Gupta VVSR, Hicks M, Kroker S, Davoren B, Roget D (2010) Crop rotation and fallowing can affect the functional resilience of microbial communities in a rainfed cropping system in southern Australia. 19th World Congress of Soil Science, Soil Solutions for a Changing World 1–6 August 2010, Brisbane, Australia.

Hackl E, Bachmann G, Boltenstern-Zechmeister S (2000) Soil microbial biomass and rhizosphere effects in natural forest stands. Phyton 40:83–90

Havlin H, Kissel DF, Maddux ID, Claassen MM, Long JH (1990) Crop rotation and tillage effects on soil organic carbon and nitrogen. Soil Sci Soc Am J 54:448–452

Hummel JD, Dosdall LM, Clayton GW, Turkington TK, Lupwayi NZ, Harker KN, O'Donovan JT (2009) Canola–wheat intercrops for improved agronomic performance and integrated pest management. Agron J 101:1190–1197

Janzen HH, Campbell CA, Brandt SA, Lafond GP, Townley-Smith L (1992) Light-fraction organic matter in soils from long-term crop rotations. Soil Sci Soc Am J 56:1799–1806

Johnson LF, Curl AE (1972) Method for the research on necology of soil borne plant pathogens. Burgess Publishing Company, Minnneapolis, p 249

Jackson ML (1973) Soil chemical analysis. Prentice Hall India (P) Limited, New Delhi

Kang GS, Beri V, Sidhu ES, Rupela OP (2005) A new index to assess soil quality and sustainability of wheat based cropping systems. Biol Fertil Soils 41:389–398

Larkin RP (2006) Biological amendments and crop rotations for managing soil microbial communities and soilborne diseases of potato. 18th World Congress of Soil Science, July 9th–15th 2006 Philadelphia, Pennsylvania, USA

Larkin RP (2008) Relative effects of biological amendments and crop rotations on soil microbial communities and soilborne diseases of potato. Soil Biol Biochem 40:1341–1351

Larkin RP, Griffin TS (2007) Control of soilborne potato using Brassica green manures. Crop Protec 26:1067–1077

Lazcano C, Brandón MG, Revilla P, Domínguez J (2012) Short-term effects of organic and inorganic fertilizers on soil microbial community structure and function. Biol Fertil Soils. doi 10.1007/s00374-012-0761-7

Manici LM, Caputo F, Babini V (2004) Effect of green manure of *Pythium* spp. population and microbial communities in intensive cropping systems. Plant Soil 263:133–142

Martin JP (1950) Use of acid, rose bengal and streptomycin in the plate method for estimating soil fungi. Soil Sci 69:215–232

Mohler CL (2009) Crop rotation on organic farms: a planning manual. Natural Resource, Agriculture and Engineering Service (NRAES). Cooperative Extension PO Box 4557, Ithaca, p 3

Mohler CL, Stoner KA (2009) Guidelines for intercropping. In: Mohler CL, Johnson SE (eds) Crop rotation on organic farms: a planning manual. Natural Resource, Agriculture and Engineering Service (NRAES). Cooperative Extension PO Box 4557, Ithaca, p 95

Pace NR (1997) A molecular view of microbial l diversity and the biosphere. Science 276:734–740

Song YN, Zhang FS, Marschner P, Fan FL, Gao HM, Bao XG, Sun JH, Li L (2007) Effect of intercropping on crop yield and chemical and microbiological properties in rhizosphere of wheat (*Triticum aestivum* L.), maize (*Zea mays* L.), and faba bean (*Vicia faba* L.). Biol Fertil Soils 43:565–574

Subramanian CV (1983) Hyphomycetes—Taxonomy and Biology. Academic Press, London

Tangjang S, Arunachalam K, Arunachalam A, Shukla AK (2009) Microbial population dynamics of soil under traditional agroforestry systems of northeast India. Res J Soil Biol 1:1–7

Tate RL III (2000) Soil microbiology. Wiley, USA

Van Diepeningen AD, De Vos OJ, Korthals GW, Van Burggen AHC (2006) Effects of organic versus conventional management on chemical and biological parameters in agricultural soils. Appl Soil Ecol 31:120–135

Wall DH, Virginia RA (2000) The world beneath our feet soil biodiversity and ecosystem functioning. In: Raven P, Williams TA (eds) Nature and human society: the quest for a sustainable world. National Academy Press, Washington DC, pp 225–241

Warcup JH (1950) The soil plate method for isolating of fungi from the soils. Nature 166:117–118

Watanabe T (1994) Pictorial atlas of soil and seed fungi: morphologies of cultured fungi and key to species. Lewis Publishers, USA

Xuan DT (2012) Microbial communities in paddy fields in the Mekong Delta of Vietnam functional and molecular diversity. PhD Thesis, Swedish University of Agricultural Sciences Uppsala

# Leaf Litter Breakdown by Two Earthworm species—*Eisenia foetida* (Exotic) and *Perionyx excavatus* (Indigenous) Under Laboratory Condition

52

Ruth Laldinthar and M. S. Dkhar

**Abstract**

A comparative study was performed to evaluate the breakdown of leaf litter of two broad-leaved tree species, i.e. *Polyalthia longifolia* and *Rhododendron arboreum* by an exotic earthworm species *Eisenia foetida* and an indigenous earthworm species *Perionyx excavatus* under laboratory condition. Methods of Haimi and Huhta (Biol Fertil Soil 10:178–183, 1990) were followed for the present study. The rate of decay, percent cellulose, hemicelluloses and lignin contents of the decomposing leaf litter of *P. longifolia* and *R. arboreum* were found to be higher in the sets treated with the two different earthworm species as compared to that of untreated sets where no earthworm was added. Though there was not much difference in the rate of leaf litter breakdown by two different earthworm species, the rate of litter breakdown by the exotic earthworm, *E. foetida*, was slightly higher as compared to the indigenous earthworm *P. excavatus*. Both the selected leaf litters treated with the different earthworm species exhibited significant positive correlations between percent biomass remaining, cellulose, hemicelluloses and lignin contents ($p \leq .001$). Significant variation was observed in hemicelluloses of leaf litters in sets treated with the two different earthworm species and in the untreated sets. It can be concluded that the indigenous earthworm species *P. excavatus* also has a high potential for leaf litter breakdown. It can act as a potential candidate in vermicompost technology—which will be of immense help in wastes management, resource recovery and environmental conservation.

**Keywords**

*Eisenia foetida* · *Perionyx excavatus* · Leaf litter breakdown

R. Laldinthar (✉) · M. S. Dkhar
Microbial Ecology Laboratory, Department of Botany, North-Eastern Hill University, Shillong, Meghalaya 793 022, India
e-mail: ruthteinbuon@aol.com

R. N. Kharwar et al. (eds.), *Microbial Diversity and Biotechnology in Food Security*, DOI 10.1007/978-81-322-1801-2_52, © Springer India 2014

## 52.1 Introduction

Degradation of organic matter using earthworm is one of the interesting and current area of research in the era of sustainable development. It is a known fact that the earthworms play an important role in the breakdown of organic matter in the soil. Involvement of earthworms for degradation of organic wastes and production of vermicompost is catching up with scientific investigation.

Understanding the influence of soil fauna, particularly the earthworms in plant litter breakdown are important in vermicompost technology. The thrust for searching localized species of earthworms is showing increasing trend because of the adaptability and suitability of the local earthworms for converting organic materials into valuable vermicompost. In India, more than 500 species have been distributed in ecologically different climatic conditions (Julka 1993). But only very few earthworm species (< 8 % of total species) were reported for their efficiency in converting organic substrates into vermicompost. The use of exotic earthworm in leaf litter breakdown has been well established but not much work has been done with respect to the leaf litter breakdown using indigenous earthworm species, particularly in the northeast region. Hence, much work is still needed to investigate the decomposition potential of some commonly distributed indigenous species of earthworms.

The information acquired will be useful in proper understanding of the role of earthworm species in leaf litter breakdown and for better utilization of the selected species in vermitechnology that can be subsequently used in Meghalaya where soil erosion is very high. Understanding of vermitechnology will help in wastes management, resource recovery and environmental conservation.

Keeping in view, the importance of earthworms in the breakdown of plant litter, the present study was carried out in laboratory conditions to study the effects of indigenous and exotic earthworm species on different leaf litter decomposition.

## 52.2 Materials and Methods

Exotic earthworm species, *Eisenia foetida*, was obtained from Rural Research Training Center, Umran, Meghalaya. Indigenous earthworm species collected from broad-leaved forest of Meghalaya were identified as *Drawida papillifer papillifer, Lumbricus terrestris, Perionyx excavatus, Pontoscolex corethrurus* of which the earthworm *P excavatus* found to be more frequently present was chosen for the study.

*Polyalthia longifolia* and *Rhododendron arboreum*, the commonly growing trees species in the selected sites, were chosen for the study.

For determining the role of earthworms in the breakdown of two types of leaf litters, i.e. *P. excavatus* and *R. arboreum* under laboratory condition, the method adopted by Haimi and Huhta (1990) was followed. All statistical analysis of data was done using Statistica 8.0 and Excel Stat. 2007 version.

The experiment was conducted in 72 plastic containers (29 × 23 cm diameter) for 6 months and the environment similar to the forest floor was created in these containers by including mineral soil and litter horizons. A total of 2 kg air-dried, sieved and sterilized mineral soil (about 3 cm layer) was spread on the bottom of each container covered with a nylon net (3 mm mesh) and watered with 450 ml of distilled water. Also 10 g of freshly fallen leaf litter (oven dried) was spread on the upper net. Ten adult specimens each of *E. foetida* and *P. excavatus* were introduced into half of the containers. Side by side untreated sets were also maintained where no earthworm was added. The vessels were irrigated with 450 ml after every 30-day intervals. The leaf litter was collected at every 30-day interval, washed with tap water and oven dried in a hot air oven at 60 °C for 48 h. The change in chemical composition of the decomposing leaf litter, viz. cellulose, hemicelluloses and lignin content,

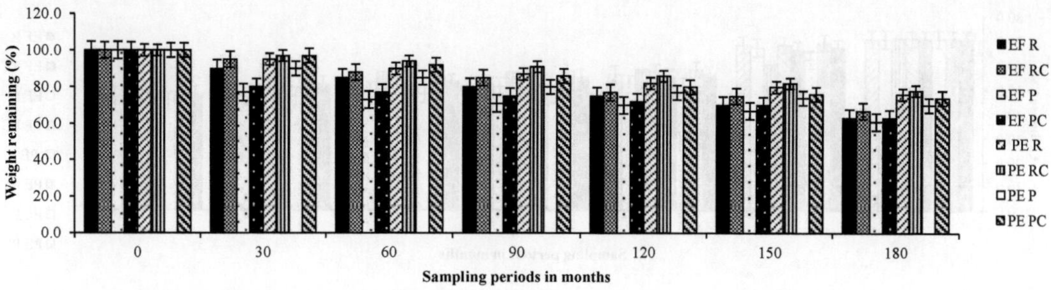


**Fig. 52.1** Biomass remaining of decomposing leaf litters of *R. arboreum* and *P. longifolia* treated with *E. foetida* and *P. excavatus* and untreated (without earthworms) sets under the laboratory condition

was estimated by the method of Peach and Tracy (1955).

## 52.3 Results

The rate of decay, percent cellulose, hemicelluloses and lignin contents of the decomposing leaf litter of *P. longifolia* and *R. arboreum* were found to be higher in the sets treated with the two different earthworm species as compared to that of untreated sets where no earthworm was added. The leaf litter treated with the exotic earthworm species, *E. foetida*, exhibited higher rate of leaf litter breakdown as compared to the leaf litter treated with the indigenous earthworm species, *P. excavatus*. There was not much difference in the rate of leaf litter breakdown by two different earthworm species. The rate of the selected leaf litter breakdown by the exotic earthworm, *E. foetida*, was slightly higher as compared to the indigenous earthworm, *P. excavatus*. The rate of leaf litter breakdown was higher in *P. longifolia* than *R. arboreum*. The biomass of remaining leaf litter was slightly less in the treated sets as compared to that of the untreated sets.

### 52.3.1 One Percent Weight Remaining of Decomposing Leaf Litter

Biomass remaining was slightly less in the treated sets as compared to the untreated sets. It was also lesser in the sets treated with *E. foetida* as compared to *P. excavatus*. Biomass remaining of decomposing leaf litters of *R. arboreum* and *P. longifolia* was maximum at the initial stage and decreased towards the end of decomposition.

In the sets treated with *E. foetida,* biomass remaining of decomposing leaf litter of *R. arboreum* was 63.00 % in the treated sets and 67.00 % in the untreated sets, whereas in *P. longifolia,* biomass remaining at the end of decomposition was 61.00 % in the treated sets and 63.00 % in the untreated sets (Fig. 52.1).

In the sets treated with *P. excavatus,* biomass remaining of decomposing leaf litter of *R. arboreum* was 76.00 % in the treated sets and 78.00 % in the untreated sets, whereas in *P. longifolia,* biomass remaining at the end of decomposition was 70.00 % in the treated sets and 74.00 % in the untreated sets.

Biomass remaining of *R. arboreum* in the sets treated with *E. foetida* ranged between 63.00–100.00 % and 67.00–100.00 % in the untreated sets. In *P. longifolia*, it ranged between 61.00 and 100.00 % in the treated sets and 63.00 and 100.00 % in the untreated sets. Biomass remaining of *R. arboreum* in the sets treated with *P. excavatus* ranged between 76.00 and 100.00 % and 78.00 and 100.00 % in the untreated sets. In *P. longifolia*, it ranged between 70.00 and 100.00 % in the treated sets and 74.00 and 100.00 % in the untreated sets.

### 52.3.2 Cellulose

Percent cellulose remaining was maximum in the initial stage and decreased towards the end of de-

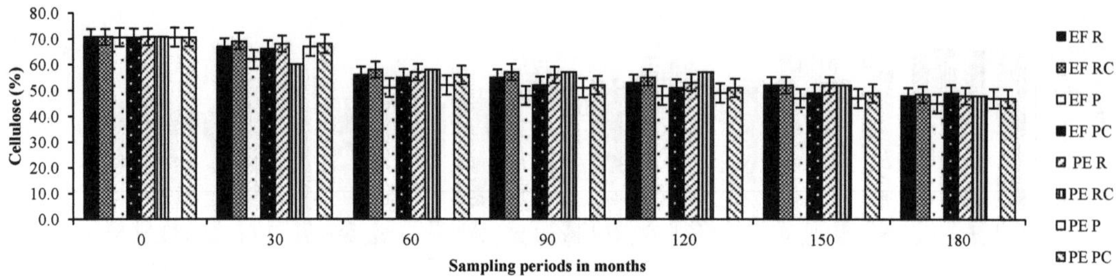


**Fig. 52.2** Percent cellulose remaining of decomposing leaf litters of *R. arboreum* and *P. longifolia* treated with *E. foetida* and *P. excavatus* and untreated (without earthworms) sets under the laboratory condition

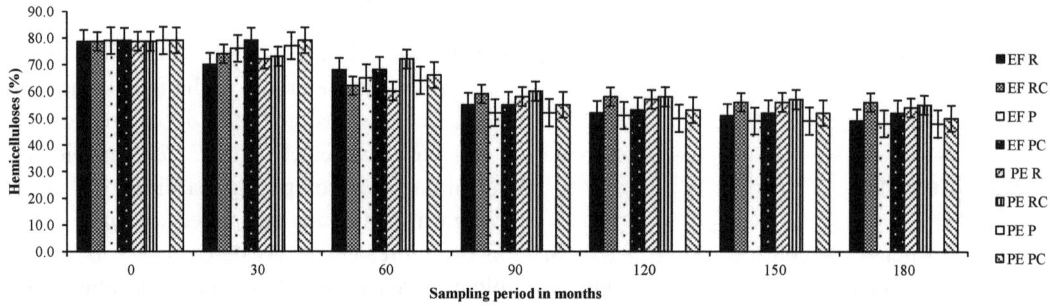


**Fig. 52.3** Percent hemicelluloses remaining of decomposing leaf litters of *R. arboreum* and *P. longifolia* treated with *E. foetida* and *P. excavatus* and untreated (without earthworms) sets under the laboratory condition

composition. In the sets treated with *E. foetida,* percent cellulose remaining of *R. arboreum* at the end of decomposition was 48.00% in the treated sets and 48.50% in the untreated sets, whereas in *P. longifolia,* percent cellulose at the end of decomposition was 45.00% in the treated sets and 49.00% in the untreated sets.

In the sets treated with *P. excavatus,* percent cellulose remaining of *R. arboreum* at the end of decomposition was 48.00% in the treated sets and 48.50% in the untreated sets, whereas, in *P. longifolia,* percent cellulose at the end of decomposition was 47.00% in the treated and 48.50% in the untreated sets (Fig. 52.2).

Percent cellulose remaining of *R. arboreum* in the sets treated with *E. foetida* ranged between 48.00 and 70.70% in the treated sets and 48.50 and 70.70% in the untreated sets. In *P. longifolia,* it ranged between 45.00 and 70.60% in the treated sets and 49.00 and 70.60% in the untreated sets. Percent cellulose remaining of *R. arboreum* in the sets treated with *P. excavatus* ranged between 48.00 and 70.70% in the treated sets and 50.00 and 70.70% in the untreated sets. In *P. longifolia,* it ranged between 47.00 and 70.60% in the treated sets and 48.50 and 70.60% in the untreated sets.

### 52.3.3 Hemicelluloses

In the sets treated with *E. foetida,* percent hemicelluloses remaining of decomposing leaf litter of *R. arboreum* at the end of decomposition was 49.00% in the treated sets and 56.00% in the untreated sets, whereas in *P. longifolia,* percent hemicelluloses at the end of decomposition was 48.00% in the treated sets and 52.00% in the untreated sets.

In the sets treated with *P. excavatus,* percent hemicelluloses remaining of *R. arboreum* at the end of decomposition was 54.00% in the treated sets and 55.00% in the untreated sets, whereas in *P. longifolia,* percent hemicelluloses at the end of decomposition was 48.00% in the treated sets and 50.00% in the untreated sets (Fig. 52.3).

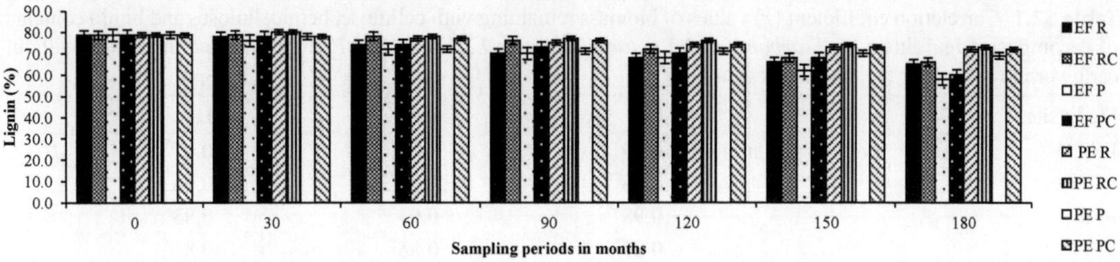


**Fig. 52.4** Percent lignin remaining of decomposing leaf litters of *R. arboreum* and *P. longifolia* treated with *E. foetida* and *P. excavatus* and untreated (without earthworms) sets under the laboratory condition. *EFR E. foetida* with *R. arboreum*; *EFRC R. arboreum* untreated; *EFP E. foetida* with *P. longifolia*; *EFPC P. longifolia* untreated; *PER P. excavatus* with *R. arboreum*; *PERC R. arboreum* untreated; *PEP P. excavatus* with *P. longifolia; PEPC P. longifolia* untreated

Percent hemicelluloses remaining of *R. arboreum* in the sets treated with *E. foetida* ranged between 49.00 and 78.70 % and 56.00 and 78.70 % in the untreated sets. In *P. longifolia,* it ranged between 48.00 and 79.00 % in the treated sets and 52.00 and 79.00 % in the untreated sets. Percent hemicelluloses remaining of *R. arboreum* in the sets treated with *P. excavatus* ranged between 54.00 and 78.70 % and 55.00 and 78.70 % in the untreated sets. In *P. longifolia,* it ranged between 48.00 and 79.00 % in the treated sets and 50.00 and 79.00 % in the untreated sets.

### 52.3.4 Lignin

Lignin also followed the same trend as that of cellulose and hemicelluloses. It was maximum at the initial stage and decreased towards the end of the decomposition. In the sets treated with *E. foetida,* percent lignin remaining of *R. arboreum* at the end of decomposition was 65.00 % in the treated sets and 66.00 % in the untreated sets, whereas, in *P. longifolia,* percent lignin at the end of decomposition was 58.00 % in the treated sets and 60.00 % in the untreated sets.

In the sets treated with *P. excavatus,* percent lignin remaining of *R. arboreum* at the end of decomposition was 72.00 % in the treated sets and 73.00 % in the untreated sets, whereas in *P. longifolia,* percent lignin at the end of decomposition was 69.00 % in the treated sets and 72.00 % in the untreated sets (Fig. 52.4).

Percent lignin remaining of *R. arboreum* in the sets treated with *E. foetida* ranged between 65.00–78.70 % and 66.00–78.70 % in the untreated sets. In *P. longifolia,* it ranged between 58.00 and 78.67 % in the treated sets and 60.00 and 78.67 % in the untreated sets. In the sets treated with *P. excavatus* containing *R. arboreum,* it ranged between 72.00–78.70 % and 73.00–78.70 % in the untreated sets. In *P. longifolia,* it ranged between 69.00–78.67 % in the treated sets and 72.00–78.67 % in the untreated sets.

## 52.4 Statistical Analysis

Table 52.1 depicts the correlation coefficient ($r$) values of percent biomass remaining with cellulose, hemicelluloses and lignin of decomposing leaf litters with the different components in both the treated and untreated sets.

In the sets treated with *E. foetida* containing decomposing leaf litter of *R. arboreum,* biomass remaining was positively correlated with cellulose ($r=0.99$; $p<0.001$), hemicelluloses ($r=0.99$; $p<0.001$) and lignin ($r=0.97$; $p<0.001$. In the untreated sets, biomass remaining was positively correlated with cellulose ($r=0.96$; $p<0.001$), hemicelluloses ($r=0.96$, $p<0.001$) and lignin ($r=0.96$; $p<0.001$). In the sets treated with *E. foetida* containing decomposing leaf litter of *P. longifolia,* percent weight remaining was positively correlated with cellulose ($r=98$; $p<0.001$), hemicelluloses ($r=0.93$; $p<0.001$) and lignin

**Table 52.1** Correlation coefficient ($r$) values of biomass remaining with cellulose, hemicelluloses and lignin contents of decomposing leaf litters of *R. arboreum* and *P. longifolia* with *E. foetida* and *P. excavatus* and untreated (without earthworms) sets under the laboratory condition

| Study sites | | C | H | L |
|---|---|---|---|---|
| EFR | Biomass remaining | 0.99*** | 0.99*** | 0.97*** |
| EFRC | | 0.96*** | 0.96*** | 0.96*** |
| EFP | | 0.98*** | 0.93** | 0.83* |
| EFPC | | 0.89* | 0.85* | 0.81* |
| PER | | 0.97*** | 1.00*** | 0.98*** |
| PERC | | 0.94** | 1.00*** | 0.99*** |
| PEP | | 0.97** | 0.99*** | 0.94** |
| PEPC | | 0.99*** | 0.99*** | 0.99*** |

*EFR Eisenia foetida* with *R. arboreum*, *EFRC R. arboreum* untreated, *EFP E. foetida* with *P. longifolia*, *EFPC P. longifolia* untreated, *PER P. excavatus* with *R. arboreum*, *PERC P. longifolia* untreated, *PEP P. excavatus* with *P. longifolia*, *PEPC P. longifolia* untreated, *WTR* weight remaining, *C* cellulose, *H* hemicelluloses, *L* lignin
*$p<0.05$, **$p<0.01$, ***$p<0.001$, respectively. Insignificant values are marked with '–'

($r=0.83$; $p<0.01$). In the untreated sets, biomass remaining was positively correlated with cellulose ($r=0.89$; $p<0.01$), hemicelluloses ($r=0.85$, $p<0.01$) and lignin ($r=0.81$; $p<0.01$). In the sets treated with *P. excavatus* containing decomposing leaf litter of *R. arboreum,* biomass remaining was positively correlated with cellulose ($r=97$; $p<0.001$), hemicelluloses ($r=0.1.0$; $p<0.001$) and lignin ($r=0.98$; $p<0.001$). In the untreated sets, biomass remaining was positively correlated with cellulose ($r=0.94$; $p<0.01$), hemicelluloses ($r=0.1$; $p<0.001$) and lignin ($r=0.99$; $p<0.001$). In the sets treated with *P. excavatus* containing decomposing leaf litter of *P. longifola,* biomass remaining was positively correlated with cellulose ($r=97$; $p<0.001$), hemicelluloses ($r=0.99$; $p<0.001$) and lignin ($r=0.94$; $p<0.001$). In the untreated sets, biomass remaining was positively correlated with cellulose ($r=0.99$; $p<0.01$), hemicelluloses ($r=0.99$; $p<0.001$) and lignin ($r=0.99$; $p<0.001$).

The one-way analysis of variance (ANOVA) of biomass remaining, cellulose, hemicelluloses and lignin of decomposing leaf litters showed no significant variation at $p\leq 0.05$. The ANOVA between hemicelluloses of decomposing leaf litters with the different components in both the treated and untreated sets showed significant variations at $p\leq 0.05$ (Table 52.2).

## 52.5 Discussion

From the present study, it was observed that there was not much difference in the rate of leaf litter breakdown by the two different earthworm species selected. The rate of the selected leaf litter breakdown by the exotic earthworm *E. foetida* was slightly higher as compared to the indigenous earthworm *P. excavatus*. It was observed that the percent of leaf litter breakdown was higher in the presence of earthworm as compared to the untreated sets (without earthworm) due to ingestion of leaf litter by earthworm and microbial activity. Edwards and Bohlen (1996) also reported that earthworms can consume very large amount of leaf litter and the amount they ingest seem to depend on the total amount of suitable organic matter available than on any other factors. It appears that this earthworm species consumed more leaf litter and due to their utilization by the earthworm—the crucial drivers as they are involved in the stimulation of microbial populations through ingestion and fragmentation (Domínguez et al. 2010). The earthworm also modified the physical and chemical structure of leaf litter as compared in a study made earlier by Raphael and Velmourougane (2011) who reported that earthworms act as mechanical blenders and by comminuting the

**Table 52.2** One way analysis of variance (ANOVA at litters with the different components in both the treated and *R. arboreum* and *P. longifolia* with *E. foetida* and *P. excavatus* and untreated (without earthworms) sets under the laboratory condition

| | Sources of variation | *F* value | *P* level |
|---|---|---|---|
| Biomass remaining (%) | EFR × EFRC × EFP × EFPC × PER × PERC × PEP × PEPC | – | – |
| | EFRC × EFPC × PERC × PEPC | – | – |
| | EFR × EFP × PER × PEP | – | – |
| | EFR × EFRC | – | – |
| | EFP × EFPC | – | – |
| | PER × PERC | – | – |
| | PEP × PEPC | – | – |
| Cellulose (%) | EFR × EFRC × EFP × EFPC × PER × PERC × PEP × PEPC | – | – |
| | EFRC × EFPC × PERC × PEPC | – | – |
| | EFR × EFP × PER × PEP | – | – |
| | EFR × EFRC | – | – |
| | EFP × EFPC | – | – |
| | PER × PERC | – | – |
| | PEP × PEPC | – | – |
| Hemicelluloses (%) | EFR × EFRC × EFP × EFPC × PER × PERC × PEP × PEPC | 3.57 | 0.003606 |
| | EFRC × EFPC × PERC × PEPC | 4.075 | 0.017931 |
| | EFR × EFP × PER × PEP | – | – |
| | EFR × EFRC | – | – |
| | EFP × EFPC | – | – |
| | PER × PERC | – | – |
| | PEP × PEPC | – | – |
| Lignin (%) | EFR × EFRC × EFP × EFPC × PER × PERC × PEP × PEPC | – | – |
| | EFRC × EFPC × PERC × PEPC | – | – |
| | EFR × EFP × PER × PEP | – | – |
| | EFR × EFRC | – | – |
| | EFP × EFPC | – | – |
| | PER × PERC | – | – |
| | PEP × PEPC | – | – |

Insignificant values are marked with '–'

*EFR Eisenia foetida* with *R. arboreum*, *EFRC R. arboreum* untreated, *EFP E. foetida* with *P. longifolia*, *EFPC P. longifolia* untreated, *PER P. excavatus* with *R. arboreum*, *PERC ERClongifolia* untreated, *PEP P. excavatus* with *P. longifolia*, *PEPC P. longifolia* untreated

organic matter they modify its physical and chemical status increasing the surface area exposed to microorganisms, thus making it much more favourable for microbial activity and further decomposition. The higher rate of leaf litter breakdown was seen in *P. longifolia* than *R. arboreum*. This could be due to the composition of the plant material. Slower rate of decomposition of *Rhododendron* litters may be due to a high fibre and polyphenolic content which reduced the palatability for earthworms (Slapokas and Granhall 1991; Dkhar and Dkhar 2000). It was also reported that *R. arboreum* litters contain lots of fibres that slow decomposition and are especially known to carry lingering toxin traces. Since the leaves contain toxins, the extra time needed also ensures these trace chemicals thoroughly degrade before the compost is used.

## 52.6 Conclusion

The exotic earthworm species, *E. foetida*, exhibited a slightly higher rate of leaf litter breakdown as compared to the indigenous earthworm species *P. excavatus*. *P. excavatus,* an indigenous earthworm species, also shows a high potential in leaf litter breakdown. It can be suggested that this indigenous earthworm species could also be useful in vermicompost technology which will be of immense help in waste management, resource recovery and environmental conservation. The role of *P. excavatus* in leaf litter breakdown and better utilization of the selected species in vermitechnology will subsequently be used in Meghalaya where soil erosion is very high. An understanding of vermitechnology will help in waste management, resource recovery and environmental conservation. The study also suggests that inoculation with earthworms accelerate the decomposition process and the large-scale adoption of vermicomposting may help to sustain soil quality and better agricultural productivity.

## References

Dkhar DN, Dkhar MS (2000) Microfungi in different regions of the gut of the earthworm species *Drawida papillifer papillifer* Steph. Soil Biol and Ecol 20 (1, 2):60–66

Domínguez J, Aira M, Gómez-Brandón M (2010) Vermicomposting: earthworms enhance the work of microbes. In: Insam H, Franke-Whittle I, Goberna M (eds) Microbes at work: from wastes to resources. Springer, pp 93–114

Edwards CA, Bohlen PJ (1996) Biology and Ecology of earthworms, 3rd edn. Chapman and Hall, London

Haimi J, Huhta V (1990) Effects of earthworms on decomposition processes in raw humus forest soil: a microcosm study. Biol Fertil Soil 10:178–183

Julka JM (1993) Earthworm resources of India and their utilization in vermiculture. In: Julka M (ed) Earthworm resources and vermiculture. Zoological Survey of India, Calcutta, pp 51–56

Peach K, Tracey MV (eds) (1955) Modern methods of plant analysis, vol 1. Springer, Berlin, pp 542

Raphael K, Velmourougane K (2011) Chemical and microbiological changes during vermicomposting of coffee pulp using exoticn (*Eudrilus eugeniae*) and native earthworm (*Perionyx ceylanesis*) species. Biodegradation 22:497–507

Slapokas T, Granhall U (1991) Decomposition of willow-leaf litter in a short rotation forest in relation to fungal colonization and palatability for earthworms. Biol Fertil Soil 10:241–248

# Chilli Anthracnose: A Review of Causal Organism, Resistance Source and Mapping of Gene

53

R. Garg, M. Loganathan, S. Saha and B. K. Roy


**Abstract**

Anthracnose disease is one of the major economic constraints to chilli production in tropical and subtropical regions of the world and it is gaining much attention towards causes of damage in the field. Growing understanding has been based on conventional methods of characterisation of *Colletotrichum* species and its interaction with the host but it was not clear enough to recognise the differentiation among species, host–pathogen relationship and genetics of resistance in chilli. In this chapter, emphasis has been made on the evaluation of the isolates of *Colletotrichum capsici* causing chilli anthracnose for their morphological and cultural characteristics, pathogenic variability on chilli fruits and genetic diversity with the help of random amplified polymorphism (RAPD-PCR) analysis and designated into different major clusters. Simultaneously, screening of Capsicum genotypes against anthracnose for testing the resistance has been highlighted under *in vitro* condition. Further, on the basis of inheritance and the segregation ratio of resistance to susceptibility, gene controlling resistance at different fruit maturity stages has been discussed. More importantly, by QTL mapping, distribution of resistance gene/s located on chromosomes by using simple sequence repeats (SSR) primers, linkage groups are indicated. A number of complementary resistant component (host–parasitic interaction) controlled by one or multiple genes with small quantification effects have been emphasized. This information will be valuable to overcome the use of agrochemicals, impact of environmental factors and in the management of this serious threat to chilli through the development of resistant varieties as a donor candidate in commercial and resistance-breeding program.



M. Loganathan (✉) · S. Saha
Indian Institute of Vegetable Research, Varanasi,
Uttar Pradesh, India
e-mail: logumuruga@gmail.com

R. Garg · B. K. Roy
Department of Botany, Banaras Hindu University,
Varanasi, Uttar Pradesh 221005, India



R. N. Kharwar et al. (eds.), *Microbial Diversity and Biotechnology in Food Security*,
DOI 10.1007/978-81-322-1801-2_53, © Springer India 2014

Keywords

Chilli anthracnose · *Colletotrichum capsici* · RAPD-PCR

## 53.1 Introduction

It is an established fact that chilli (*Capsicum* spp .) is one of the important economic crops and reported to be cultivated worldwide. The production of the crop is challenged by several biotic factors (Isaac 1992) such as *Phytophthora* root rot (Leonian 1922), *Rhizoctonia* root rot (Muhyi and Bosland1992), *Fusarium* wilt (Rivelli 1989), *Verticillium* wilt (Sanogo 2003), bacterial wilt (Yabuuchi et al. 1995), *Chilli veinal mottle virus* (Ong et al. 1979), *Cucumber mosaic virus* (Ong et al. 1979), *Pepper veinal mottle virus* (Brunt et al. 1978) and anthracnose (Paul and Behl 1990; Ramachandran and Rathnamma 2006).

Among the diseases, anthracnose is the most important constraint to chilli (*Capsicum annum*) production in most of the major chilli growing region of the world and often results in high yield losses (Voorrips 2004; Manandhar et al. 1995). Several species of *Colletotrichum* have been reported as causal agents of chilli fruit rot worldwide. The disease causes severe problem on mature fruits and hence it is also called ripe fruit rot of chilli (Agrios 2005) (Fig. 53.1). As far as its distribution and diversity is concerned, several species of *Colletorichum* have been reported from different parts of the world. Simmonds (1965) reported *Colletotrichum acutatum, Colletotrichum atramentarium, Colletotrichum dematium, Colletotrichum gloeosporioides* var. *minor* and *C. gloeosporioides* var. *gloeosporioides* from Australia. Similarly Voorrips (2004) has reported *C. acutatum, Colletotrichum capsici* and *C. gloeosporioides* from Indonesia. Park and Kim (1992) reported *C. acutatum*, *C. gloeosporioides*, *Colletotrichum Coccodes* and *C. dematium* from Korea. Long back, Dastur (1920) had reported two different species *Gloeosporium piperatum* and *C. nigrum* from Myanmar (Burma). Pearson et al. (1984) reported *C. capsici* and *C. gloeosporioides* from Papua New Guinea. Johnston and Jones (1997) reported *C. coccodes* from New Zealand. Manandhar et al. (1995) reported *C. acutatum, C. capsici* and *C. gloeosporioides* from Taiwan. Than et al. (2008a) reported *C. acutatum, C. capsici and C. gloeosporioides* from Thailand. Adikaram et al. (1983) reported *C. acutatum* and *Glomerella cingulata* from the UK and Roberts et al. (2001) reported *C. acutatum* from the USA and Don et al. (2007) reported *C. acutatum, C. capsici, C. gloeosporioides* and *C. nigrum* from Vietnam. Among all the species, *C. capsici* is common and dominant in most part of the world. Population of this species has been reported from India (Thind and Jhooty 1990; Hedge and Srikant 2002; Paul and Behl 1990). In addition to *C. capsici,* other species viz. *C. dematium, C. gloeosporioides, C. graminicola, C. acutatum, C. piperatum and C. atramentaum* were also reported from India (Selvakumar 2007; Thind and Jhooty 1990; Kaur and Singh 1990; Ramachandran and Rathnamma 2006). Species of *Colletotrichum* are the most effective plant pathogenic fungi, attacking an extremely wide range of plants growing in temperate and tropical environments. These pathogen cause damage to most parts of the plants including roots, stem, leaves flowers and fruits, but are often highly specific to individual tissues.

In several cases the pre- and post-harvest infection together account for more than 50 % losses of crops. In India, the yield loss ranged from 10–54 % due to this disease (Lakshmesha et al. 2005, Ramchanderan and Rathnamma 2006). The average annual loss due to anthracnose disease has been estimated up to 29.5 % with economic loss equivalent to US$ 491.67 million (Ramachandaran et al. 2007) and at present the value must have gone up. The post-harvest loss estimated was 3.5 to 75 % from various states of India (Bagri et al. 2004, Lakshmesha et al. 2005). The severity of this disease has been reported from worldwide, including Asian countries (Widido 2007; Zhang et al. 2007), where the disease incidence was globally found, for example, the crop loss was around 10 % in Korea (Byung-

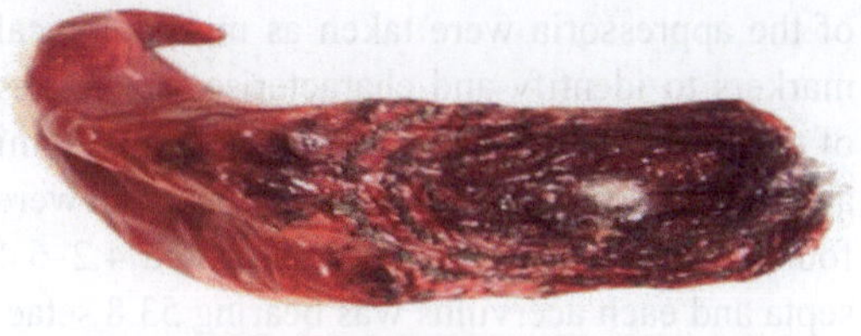

**Fig. 53.1** Characteristic symptom of *Colletotrichum capsici* on a red ripe fruit of chilli

sookim 2007), 20–80% in Vietnam (Leh Don et al. 2007) and 100% in Brazil (Lopes and Villa 2003). The severity of the disease varies depending upon cultivars grown and weather condition prevailing in a particular region.

Since long back, Pring et al. (2002) has described pathogenicity, host range and infection process of three isolates of *C. capsici,* from cowpea (*Vigna unguiculata*), bean (*Phaseolus vulgaris*) and betel vine (*Piper betle*). In this background, it was observed that the initial infections occurred after production of appressoria and deposition on plant surface and complex differentiation at the time of conidial germination. Appresorium is an important means for infection as it is the deciding factor of basic morphological characteristics of the genus *Colletotrichum*. They are globose or subglobose and with or without lobes. Mechanism of infection has been described as first hyphae penetration in cuticles and initial growth beneath the cuticle that extend to the anti- and periclinal-walls of epidermal cells causing extensive wall degradation. Subsequently, growing hyphae destroy the walls of underlying cortical cells that associated with extensive death of adjacent cells and led to the production of water-soaked lesions. When tissues are extensively rotted, hyphae enter sclerenchymatous fibres by direct growth through their walls and freeze-substituted preparations revealed vesicles and plasmatubules in infection hyphae. It has been suggested that successful pathogenesis is superior in the sense of suppression of host defence responses through rapid killing of infected tissues. In comparison, this infection strategy with those of other *Colletotrichum* species indicates towards the knowledge of a pathogen's infection process which may be used to predict about narrow or wide host range. Therefore, species identification has been must for disease control and its epidemiology. In this context, Cai et al. (2009) has used various parameters for morphology, pathogenicity, physiology, phylogenetics and production of secondary metabolite for the identification of *Colletotrichum* complexes. Besides the above, molecular approaches have been used and a backbone phylogenetic tree using Internal transcribed spacers (ITS) sequence data from 42 ex-type specimens was also generated, for identification of *Colletotrichum* species to place them in species complexes as multigene phylogenetic data and for better understanding of the relationships within *Colletotrichum species*. Multigene phylogeny, comparison with type specimens and a well-defined phylogenetic lineage in conjunction with recognisable polyphasic characters, such as morphology, physiology, pathogenicity, cultural characteristics and secondary metabolites have been proposed as an ideal approach for *Colletotrichum* systematics. Further, with the advancement of research molecular-based identification was undertaken, where molecular-based designed primer set-based sequences of the ribosomal internal transcribed spacer (ITS1 and ITS2) regions were used in a conventional polymerase chain reaction (PCR) assay, the primer set (CcapF/CcapR) was amplified into a single product of 394 bp of DNA extracted from 20 Mexican isolates of *C. capsici* (Torres Calzada 2011). The occurrence of different virulent strains of *C. capsici* has been well documented. However, information on distribution of races or pathotypes in chilli-growing areas and their accurate method of identification and characterisation made easier for effective disease management by the development of host resistance in breeding programme. Besides the above, frequent application of fungicides, which is expensive and undesirable for ecosystem, is also leading to development of its resistance.

Therefore, in the present scenario, understanding of the interaction between host and anthracnose pathogen complexes of species involved in chilli is needed for accurate identification and their effective disease control. In this review, we will emphasize on different aspects of alternative control measures for resistance development: (a) accurate identification of pathogen, (b) host and anthracnose pathogen interaction through pattern of inheritance and resistance development, (c) genetic basis of resistance development involving superior genotypes and their characterisation by implicating the molecular technologies.

### 53.1.1 Morphological and Cultural Characteristics of *C. capsici*

Correct and accurate identification of pathogen has always been a basis for developing strategies for management. Considering the diversity in morphology, efforts have been made in developing subgeneric groups of species based upon similar colony character. For identification and characterisation of *Colletotrichum,* many morphological and molecular techniques were used in different laboratories of the world. For culturing and sporulating *Colletotrichum* at 30 °C, potato dextrose agar (PDA) was found to be the best medium (Yoon-joe and Park 2001; Jayalakshmi and Seetharaman 1999). Richard's agar medium was found to be best for growth and sporulation of the pathogen (Wasantha and Rawal 2008), whereas Czapeck's Dox Agar was used for purification of culture by Selvakumar (2007). The suitable temperature for growth of the fungus ranged from 28 to 30 °C (Wasantha and Rawal 2008; Sinha 2004) but the development of disease has been found to be completely arrested at 0–5 °C and slowed down at 10–15 °C (Datar 1996), whereas the thermal death point of conidia has been reported to be 47 °C (Jayalakshmi and Seetharaman 1999). In specific case, the optimal increase in colony diameter for *C. capsici* was reported between 28 and 32 °C, while for *C. gloeosporioides* it was 28 °C (Hartman and Wang 1992). Besides, above pH 5.0 was found best for the growth of *C. gloeosporioides* (Wasantha and Rawal 2008), while sporulation was found to be better at pH 6.0, whereas according to Sinha et al. (2004), pH 7.0 was found to be most suitable for the maximum radial growth of *C. capsici.* Diversity in shape and size of conidia, conidiophores and setae in culture and size and shape of the appressoria were taken as morphological markers to identify and characterise the species of the *Colletotrichum*. According to Jayalakshmi and Seetharaman (1999), *C. capsici* conidia were found to be 22.9 × 3.94 um in size and 4.2–6.4 septa and each acervulus was bearing 53.8 setae. Long ago, Butler and Bisby (1960) reported that the conidia of size 18–23 × 3.5 formed from the pale buff of salmon masses were falcate, fusiform and gradually tapered towards each end. Few years back, Selvakumar (2007) reported the size of the conidia ranging 25–26 × 3.2–3.72 μm at the tip of unbranched conidiophores and each conidium was found to be hyaline, single celled, fusoid and aseptate. Gehlot and Purohit (2001) studied the ultra structure of conidium in *C. capsici* and found that the conidiogenous cell of the fungus resembles phialids and the formation of the primary conidium was found to be holoblastic. At the same time during the secondary conidium development, the outer layer of phialids wall ruptured and completed the formation of the septum at the base and Woronin bodies near the septum.

Long ago, morphological characteristics of the colony of *Colletotrichum* was studied by Butler and Bisby (1960) who reported that the colonies formed were dense, white to dark grey, reverse dark brown, had no sclerotia, abundant setae and appressoria, sepia brown clavate to ovate margin, and entire. Selvakumar (2007) reported that the colonies of *C. capsici* were circular, smooth, white having thick texture and the colour varied from grey, greenish to white and the growth rate varied from media to media composition. Sharma et al. (2005) had observed the cultural and morphological traits and reported that the isolates produce cottony, fluffy or suppressed colonies without any significant differences in shape and size of conidia and on morphological basis isolates were categorised into different groups by several workers. Based on cultural variability, categorisation of isolates was also done by Garg (2011).

Khirbhat et al. (2004) characterised in detail the nine isolates of *C. capsici* and categorised into four different groups on the basis of pathogencity: GP1 (CC9 characterised by significantly

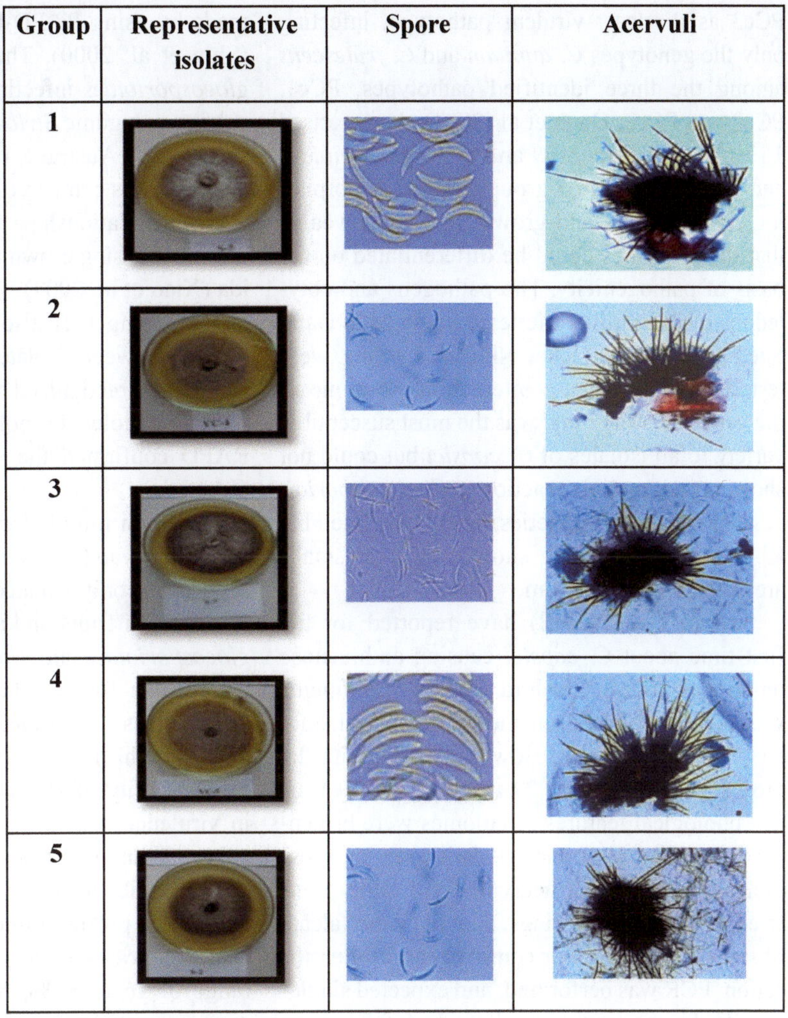


**Fig. 53.2** Morphological grouping of different isolates of *C. capsici*

higher growth and capability of producing susceptible reaction in all capsicum cultivars), GP 11 (CC8, capable of producing susceptible reaction in Sadabahar, Hisar vijay and Kiran but resistant to Hisar sakti) and GP 111 and IV (Cc-1, Cc-2, Cc-3, Cc4, Cc-5 and Cc-7 incapable of producing susceptible reaction in all capsicum cultivars). The categorization was also done by Sharma et al. (2005) for 37 isolates of *C. capsici* into five groups, Cc-I, Cc-II, Cc-III, Cc-IV and Cc-V. Similarly, categorisation of *C. capsici* isolates was also done by Garg (2011) based on radial growth (2 groups) and conidial size (4 groups) (Fig. 53.2). While testing the pathogenicity, the use of differential hosts proved to be a viable option for the evaluation of pathogenic variability. It was felt that combined application of molecular diagnostic tools along with the use of morphological variabilities among the different isolates could be an appropriate and reliable approach for studying pathological variability in *Colletotrichum* species.

The differential reaction against six isolates of *C. capsici* was studied by Deshpande and Ram (2007) and concluded that none of the released/notified varieties were resistant to any of the six used isolates. Whereas, Khirbhat et al. (2004) reported CcP1 isolate to be the most virulent among 15 pathotypes of the pathogen that were characterised by differential reaction on a different set of capsicum cultivars in this field. Montri et al. (2009) also reported PCc1 as another most virulent pathotype, infecting all genotypes of *C. annuum, C. chinense* and *C. frutescens,* whereas

PCc3 as the least virulent pathotype, infecting only the genotypes *C. annuum* and *C. frutescens* among the three identified pathotypes, PCc1, PCc2 and PCc3. Oanh et al (2004) characterised 15 isolates of *C. capsici* and *C. gloeosporioides* and categorised into 5 groups based on morphological characters and growth rate and revealed that these isolates could be differentiated on the basis of pathogenicity. The pathogens and cotyledons of five chilli varieties, namely Mun Dam, She Fha, Mae Ping, Khee Nhu and Louang, were tested for their disease interactions. Among all the varieties 'Mae Ping' was the most susceptible variety to all isolates of *C. capsici* but could not show any susceptible reaction to *C. gloeosporioides*. The other four varieties, Mun Dam, She Fha, Khee Nhu and Louang showed tiny (or small) area of necrotic symptom.

Sawant et al. (2012) have reported for the first time about *C. capsici* causing anthracnose on grapes in Maharashtra, India. The fungus was identified based on morphological characters. The isolates were slowly growing with the growth rate of 3.86+0.29 mm/day at 30+1 °C. Morphological features of colonies were brownish white and later turned dark grey or moss green in colour. The acervuli were either scattered or in concentric rings. Conidia were falcate of size 21, 7/5.1 m. For confirmation of identification, PCR was performed, and expected single-specific fragment of approximately 460 base pair was amplified from all tested *C. capsici* isolates, whereas the primer pair did not amplify DNA from isolates belonging to *C. gleosporiodes*. Symptoms were confirmed after observation up to 7 days and again re-isolated. Although *C. capsici* has been reported on many hosts from India and other countries of the world, this is the first report of *C. capsici* affecting grapes.

Apart from pathogenicity, molecular techniques have also been used to differentiate among isolates of *Colletotrichum*. Backman et al. (1999) used the RAPD technique to determine *C. graminicola* isolates infecting annual bluegrass and creeping bent grass. A comparative study of anthracnose pathogens, *C. gloeosporioides, C. coccodes, C. dematium, Glomerella cingulata* and *C. acutatum,* infecting *Capsicum* was made by using RAPD-PCR in Korea and China (Shin et al. 2000). The genetic diversity in *C. gloeosporioides* infecting species with the tropical forage legume *Stylosanthes* by the researchers from Brazil, Australia, China and India (Weeds et al. 2003) was compared using molecular markers. A genetic relationships among *C. gloeosporioides* isolates, causing crown rot of straw-berry in Florida (Xiao et al. 2004) was establish using RAPD and grouping was also made by Sharma et al. (2005) between isolates and five morphological groups recognized within *C. capsici.* Along with this, molecular polymorphism generated by RAPD confirmed the variation in virulence of *C. capsici*.

Based on morphological and cultural characters Madhavan (2010) found that *C. capsici* is the most commonly isolated fungal species from infected chilli fruits and apart from *C. capsici, C. gloeosporioides* and *Alternaria alternata* isolates where also found. The virulence between the isolates was determined by inoculating them on detached chilli fruits (cv. K2) *in vitro*. Molecular variability of isolates of *C. capsici* differing in virulence was analysed by means of RAPD, using 22 random primers. After analysis of a genetic coefficient matrix derived from scores of the RAPD profile, minimum and maximum per cent similarities among the *C. capsici* were in the range of 16 to 81%, respectively. The isolates were saperated using the unweighted pair-group method with arithmetic average (UPGMA) for cluster analysis, which clearly separated the isolates into three clusters (I, II and III) and confirmed the genetic diversity among the isolates of *C. capsici* from chilli. Cluster III consisted of only one isolate (CBE1), cluster II with two isolates (TEN 3 and TEN 4) and all the remaining isolates included in cluster I. According to his study no relationship was observed between clusters in the dendrograms and virulence of *C. capsici* isolates. Ratanacherdchai et al. (2007) had applied RAPD analysis on 18 isolates including 2 species, *C. gloeosporioides* and *C. Capsici* and isolates from three varieties of chilli, i.e. chilli pepper (*C. annuum*), long cayenne pepper (*C. annuum var acuminatum*) and bird's eye chilli (*C. frutescens*). UPGMA was used to construct the dendogram of RAPD to show the relationship among the species. A clear difference was observed between the species viz. *C. gloeosporioides* and *C. capsici* in the RAPD analysis but isolates of *C. capsici* were more closely related among each other than *C. gloeosporioides* isolates.

Torres Calzada et al. (2008) reported that *Colletotrichum* species isolated from papaya fruits showing 450 bp PCR product but DNAs from isolates with a typical lesions on fruit failed to produce an amplified product. For further characterisation, the rDNA 5.8S-ITS region was amplified by PCR and processed for sequencing and restriction fragment length polymorphism (RFLP) analysis indicated the existence of two *Colletotrichum* species *C. gloeosporioides* and *C. capsici* on papaya fruits causing anthracnose lesions, whereas, during PCR-RFLP analyses, using the restriction endonuclease MspI, reproduced restriction patterns specific for *C. capsici* or *C. gloeosporioides*. In this case, the generation of RFLP patterns by MspI (or AluI or RsaI) was found to be rapid, accurate and unequivocal for the detection and differentiation of *C. gloeosporioides* and *C. capsici*. Further, Moriwaki et al. (2002) sequenced the internal transcribed spacers (ITS) of the ribosomal RNA gene (rDNA) of 236 isolates covering 26 *Colletotrichum* species. The isolates were grouped into 20 ribosomal groups (RGs) based on the sequences of ITS1, correlated the species based on morphology. *C. gloeosporioides* were separated into three RGs that were morphologically different, which indicated that possibly *C. destructivum, C. linicola* and *C. higginsianum* were nonspecific. *C. dematium* sensu lato including *C. capsici* and other species producing falcate conidia except for graminicolous ones were separated into three RGs but were difficult to distinguish morphologically. In the phylogenetic study using ITS2 and the 285 rDNA domain 2 region, topologies compiled by neighbour-joining and maximum-parsimony methods showed almost the same, reflecting the conidial morphology. The phylogenetic group 1 (PG1) produced conidia with acute ends for *C. acutatum, C. destructivum* and *C. graminicola*; PG2 produced those with obtuse ends for *C. gloeosporioides* and *C. orbiculare. Colletotrichum theae-sinensis,* which produced the smallest conidia, was grouped as PG3, far from other species, indicated non-belonging to *Colletotrichum.* Thirty four isolates of *Colletotrichum* spp. from anthracnose on Bell pepper, Long cayenne pepper and Bird's eye chilli, which included two species, *C. gloeosporioides* and *C. capsici* were collected by Ratanacherdchai et al. (2010) and were divided into low, medium and high virulence groups based on their pathogenic potential as tested *in vitro* on fruit. *C. capsici* isolated from three tested hosts expressed the highest virulence. Cross-inoculation of three high virulent isolates of *C. capsici* in accordance with the three chilli varieties showed that all isolates could produce anthracnose symptom in the same lesions. All tested isolates developed lesions after co-inoculation of all hosts. Based on a molecular study, inter simple sequence repeat (ISSR) analysis indicated two distinct groups of *C. gloeosporioides* and *C. capsici*. Furthermore, genetic diversity was correlated with geographic distribution, but lacking clear relationship between genetic diversity and pathogenic variability. Surprisingly, *C. gloeosporioides* and *C. capsici* appearing in the same geographic area cause lower disease incidence. Isolates of *C. capsici* were also categorised by Garg (2011) based on RAPD analysis using a similarity co-efficient. *C. capsici* isolates were divided into two major clusters, viz. cluster I (S-5, VC-1, Ccf and VC-3) and cluster II (Vc-4, Ccc-2, S-7 and S6) (Figs. 53.3a and b).

### 53.1.2 Identification of Resistant Source Against Anthracnose

According to the earlier reports, anthracnose disease can be managed through appropriate use of fungicides, or by the long-lasting resistant cultivars (Whitelaw-Weckert et al. 2007). Recent findings suggest that there are centres having unique diversity in terms of existence of naturally occurring interspecific landraces, which are resistant to diseases (Rai 2010; Kumar et al. 2011). Hence, search for long-lasting resistant cultivars through screening of existing cultivars/wild lines

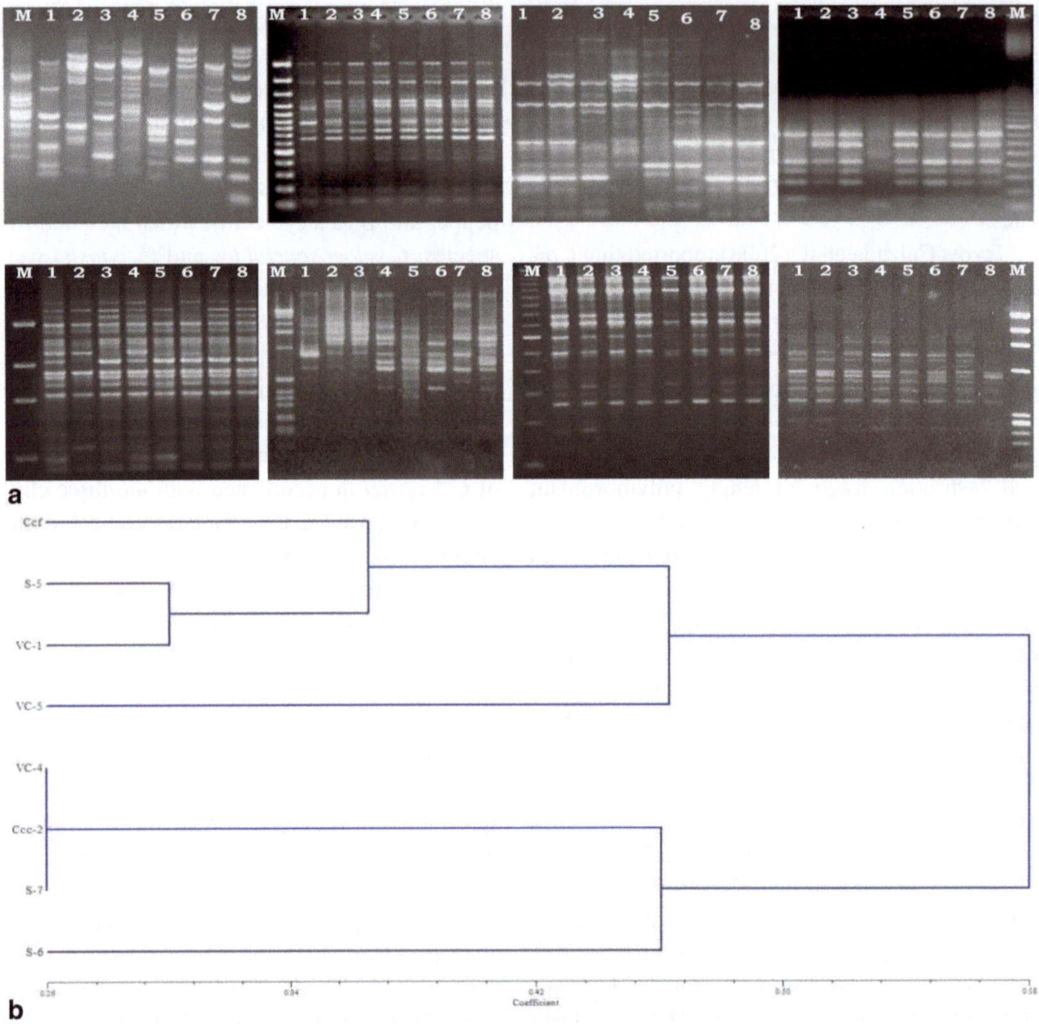


**Fig. 53.3** **a** RAPD for genetic diversity study of *C. capsici* isolates. **b** Genetic relationship among different isolates of *C. capsici* using UPGMA cluster analysis of the distance matrix

can be the best viable option. Nanda (2007) has identified the contrasting lines of hot pepper (*Capsicum* spp.) as parents of a population and after mapping and tagging of genes conferred resistance to anthracnose (*Colletotrichum* spp.) in Southern India. Among 74 lines/germplasm used for field screening, seven lines PBC 80, PBC 81 (*C. baccatum*), PBC 142, Kunchangi Local, Puri Local, male-sterile line CCA 7244 and its maintainer 9907–9611 (*C. annuum*) were found to be resistant, whereas Pusa Jwala, Utkal awa and CA14 (all *C. annuum*) were found to be susceptible. Similarly, Ekbote et al. (2002) conducted field experiment for resistance to fruit rot caused by *C. capsici* in 51 (*C. annuum*) cultivars. Of the cultivars tested none were tolerant, one was resistant, three were moderately resistant, five were moderately susceptible, seven were susceptible and nine were highly susceptible to the disease. On the basis of the infection percentage, Roy et al. (1998) evaluated 24 chilli genotypes for fruit rot incidence (*C. gloeosporiodes*) among which six genotypes DC1, DC2, DC3, DC4, DC14 and DC24 were moderately resistant. Basak (1997) screened ten cultivars of chilli against three major fruit pathogens *C. capsici, C.*

*gloeosporiodes and Fusarium semitectum*. None of the cultivars were found to be immune but the cultivars C-011 and C-045 were susceptible to *G. cingulata* and *C. capsici*, while C-123, Chitagong local and Bogra were highly susceptible to both of them. Pcrane and Jai (1986) found in Pant C1 and B7–9 self-crossed generation as resistant and 7262, Deglur, B79A, LIC 24, 764, Guntur and 574 Thirumalapuram as moderately resistant under natural infection in the field. In a similar way, Jayalakshmi and Seetharaman (1998) reported on the reaction of 40 cultivars, out of which only one (CA-874) was reported to be highly resistant, 7 resistant, 19 moderately susceptible followed by 13 as susceptible against *C. capsici* in pot culture. The screening method for resistance to pepper fruit anthracnose was also demonstrated by Yoon-joe and Park (2001). Wound inoculation through microinjector was more repeatable and distinguishable to the study for symptom development, at 30 °C for post-inoculation as optional condition, whereas Singh (1993) reported at 28 °C and also 6 days after inoculation (DAI) a suitable *in vitro* observation. Gniffke et al. (2007) evaluated a green and red ripe fruit in the field and *in vitro* condition using two isolates of *C. acutatum,* apart from discrepancies between the result of the field and *in vitro* results. Among the 40 accessions, PBC 932 was the only accession where amplified fragment length polymorphism (AFLP)-derived molecular markers were found to be valid in tracing their resistance. Deshpande and Ram (2007) showed similar results of screening in the field and laboratory conditions and reported that none of the released/notified varieties were resistant to any of the isolates, except two accessions of *C. baccatum* IHR 951 and 1263 which were found to be resistant against all six isolates used for screening. Singh (1993) has screened 19 varieties of *in vitro* conditions, among them BGI and Lorai were graded resistant. In addition to the above, Singh (1979) has also demonstrated that Celctio2, Seeswal and Patna chilli were found to be most resistant to 11 varieties tested and none proved to be immune. Nanda (2007) observed on the basis of *in vitro* screening of isolates of both *C. capsici* (Cc) and *C. gleosporioides* (Cg) that PBC 80 was immune to two isolates tested (Cc and Cg) and resistant to other three isolates. Further, in another step of investigation, tested five lines (PBC 142, Kunchangi Local, Puri Local, male-sterile line CCA 7244 and its maintainer 9907–9611) showed their range from moderate resistance to complete susceptibility, where three lines (Utkal awa, Pusa Jwala and CA 14) were recognised to be completely susceptible to all five isolates. With the advancement of the research, combined efforts for the evaluation of isolates have been made by taking the parameters such as morphological and cultural characteristics, pathogenic variability on chilli fruits along with genetic characterisation using DNA (RAPD-PCR). Molecular polymorphism generated by RAPD confirmed the variation of different isolates and grouped into two clusters. Garg (2013) in *in vitro* studies found that among 41 genotypes tested, 33 were *C. annuum* genotypes and the majority of them were categorised into moderately susceptible (11) followed by moderately resistant (7) and highly susceptible (7) and less number of highly resistant (6) and symptom-less (2) categories, whereas in the non-*annuum* group and its crosses out of 8 tested, except one moderately resistant, all were highly resistant or symptom-less. In total, 24.2 % of *C. annuum* genotypes (33) and 87.5 % of non-*annuum* genotypes (8) were found to be highly resistant including few symptom-less categories (Figs. 53.4a and b). However, morphological (Fig. 53.2), pathological and RAPD grouping (Fig. 53.3a) of isolates suggested non-correlation among test isolates. On the basis of the above view, search for a resistance gene associated with anthracnose of chilli and a number of resistant component (host–parasite interaction) controlled by multiple genes with contributory effects have been undertaken.

### 53.1.3 Genetics of Inheritance of the Resistance Gene

In the present scenario, the resistant sources obtained from the screening of existing germplasm/wild lines can be used for the resistance breeding programme and it could be an important

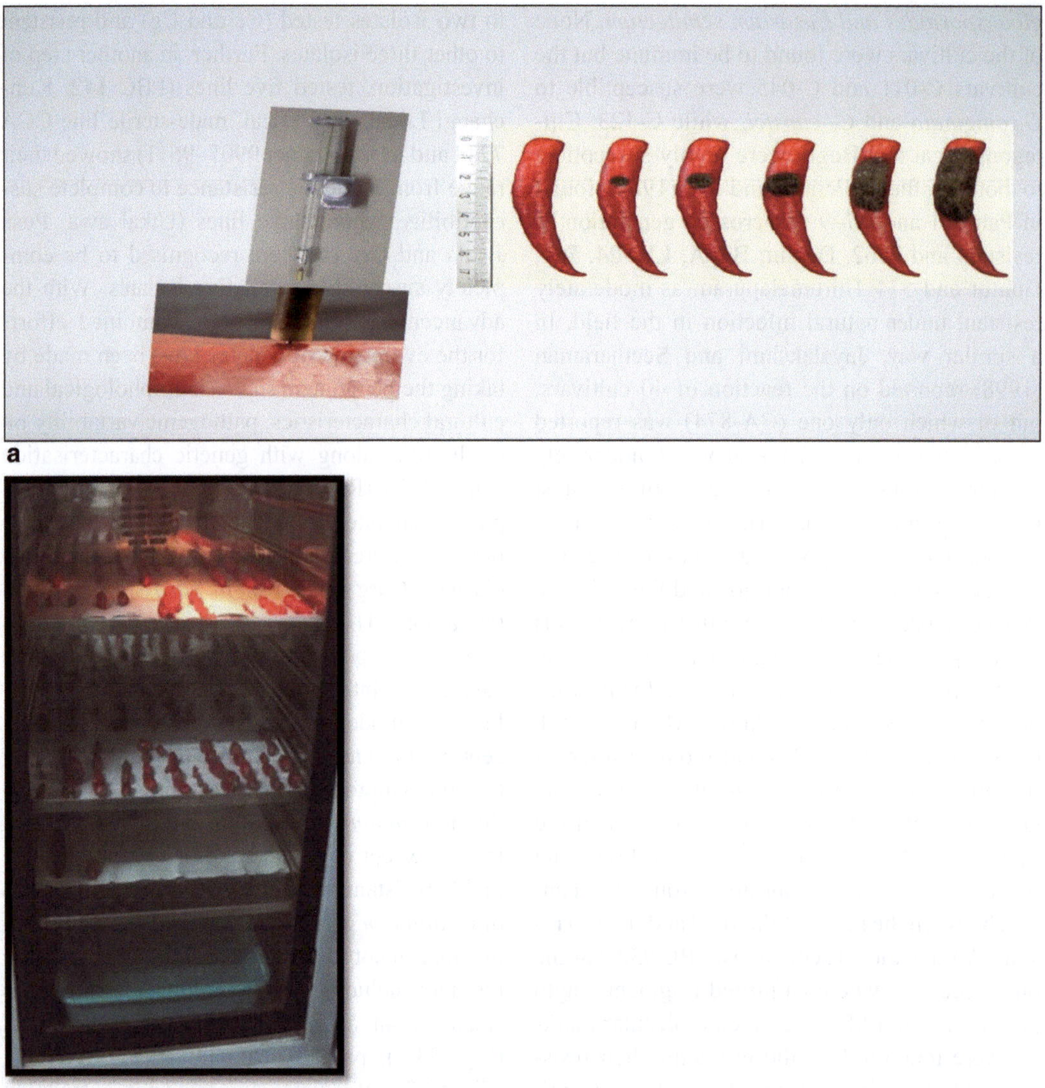

a

b

**Fig. 53.4** Screening of chilli genotypes for resistant source against anthracnose. **a** Pin-prick method and scoring parameter. **b** Inoculated fruits *in vitro* condition

approach to limit the use of pesticides, as both economically and environmentally viable options. Apart from developing resistant varieties, the study of tagging and mapping of the genes, responsible for resistance and pattern of inheritance, to observe the regulation of dominant or recessive gene for the traits of interest is becoming important. Polygenic or monogenic control over the trait in crosses developed is also important to tag the gene through the genetic markers linked to the trait of interest.

Genetic markers are well defined as a specific gene that produces a recognisable trait which can be used in the family or population studies. Presently, there are different genetic markers available for evaluating genetic variations of different traits: morphological, biochemical and DNA markers (Demissie and Bjornstrand 1996; Dulloo et al. 1997; Staub et al. 1997; Gwanme et al. 2000). Long ago, by using such markers, genetic variation studies were conducted on different crops, including cereals, pulses, horticultural crops and fruits (Cipriani et al. 1996; Margal'e et al. 1995; Sharman et al. 1997; Staub et al. 1997; Gwanme et al. 2000). DNA markers are widely accepted as the potentially valuable tools for crop improvement in rice (Mackill et al. 1999), wheat (Eagles et al. 2001; Koebner and Summers 2003), maize (Stuber et al. 1999; Tuberosa et al. 2003), barley (Thomas 2003; Williams 2003), tuber crops (Barone 2004; Fregene et al. 2001; Gebhardt and Valkonen 2001), pulses (Kelly et al. 2003; Muehlbauer et al. 1994; Svetleva et al. 2003;), oilseeds (Snowdon and Friedt 2004), horticultural crop species (Baird et al.1996; Mehlenbacher 1995) and pasture species (Jahufer et al. 2002). It was also suggested that enhancement in global food production can be done by improving the conventional plant breeding programmes using DNA markers (Kasha 1999; Ortiz 1998).

With reference to the above, management of a serious threat to chilli through the development of resistant variety can be implicated with the study of genetics of this disease, in relation to several species of pathogen to host.

The continuous effort on the construction and application of linkage maps has facilitated greatly in the tagging of many agronomically important genes in a variety of crop plants, including capsicum. The two new genes, *co4* and *Co5,* responsible for resistance were recently identified in chilli from PBC80 which were located at different loci from the previously identified resistant genes (*co1* and *co2*) of PBC932 (Mahasuk et al. 2009). Gene Anr-1 identified from *C. annum* cv. chungryong was resistant against *C. dematium* (Park et al. 1992). Genes, Anr-2, Anr-3 and Anr-4, resistance to *C. gloeosporoides* were reported in *C. annuum* lines BGH3077, BGH28850 and BGH5085, respectively (Fernandes and Ribero 1998). In continuation to the above, line 83–168 with gene Anr-5 resistant to *C. capsici* has also been reported (Lin et al. 2002).

Mahasuk et al. (2009) had analysed the anthracnose resistant gene and its position on chromosome by tagging and mapping the gene in an intraspecific cross between PBC80 and PBC1422 which was used to develop resistance and inheritable transfer of resistance to *C. acutatum*. It was evidenced with frequency distribution of the disease scores in the $F_2$ and $BC_1$ populations, which indicated that a single recessive gene is responsible for the resistance at a mature green fruit stage and a single dominant gene for the resistance at a ripe fruit stage. Linkage analysis between the two genes identified in both fruit maturity stages showed an independent behaviour. Further, Kim et al. (2008a) showed inheritance of anthracnose resistance to *C. acutatum* in an intraspecific population derived from a cross between *C. baccatum* Golden-aji and PI594137 in the detached mature green fruits. The segregation ratio of resistance and susceptibility in the $F_2$ and $BC_S$ population was significantly fitted with 3:1 Mendelian ratio and indicated that the resistance of PI594137 to *C. acutatum* was controlled by a single dominant gene. In the next phase of his research, Kim et al. (2008b) also studied the inheritance of the resistance to *C. capsici* in the cross between Yejoo (susceptible) and Deepong chou (resistant) followed by *C. acutatum* in a cross between HN1 (susceptible) and AR (resistant), and the developed generation $F_2$ and BC indicated the resistance controlled by a single recessive gene where the segregation ratio of resistance to susceptibility fitted with the 1:3 and 1:1 Mendelian ratio, respectively. In addition, it was also indicated that the gene responsible for resistance against *C. capsici* in both the resistant cultivars was the same, and an additional gene responsible for resistance against *C. acutatum* was also reported in AR parent. The inheritance of resistance to anthracnose at fruiting and seedling stages in a cross between Bangchang (*C. annuum* cv.) and PBC932 (*C. chinense*) was demonstrated by Mahasuk et al. (2009b). The $F_2$ population was used in the study and the stage of mature green and ripe red fruit, as well as at both seedling and fruiting stages in the second $F_2$ population were also considered. Against the above, the single gene model has been suggested for each trait, based on the distribution of the disease scores on green and red fruit and seedlings in both $F_2$ population and three different recessive genes were indicated for the development of individual resistance from the crosses. The resistance genes at green and red fruit was found to be linked (recombination frequency 0.25), whereas at seedling stage, the resistance gene was not linked to the fruit resistances.

Further, Yoon et al. (2004) had contributed heritability of anthracnose resistance in backcross population, which did not fit to Mendelian manners in $BC_1F_1$ and $BC_2F_2$ but the variation in progenies was similar to the normal distribution. According to the equation $h^2 = \Delta G/i$, broad sense heritability of anthracnose resistance using disease incidence (%) and incidence index was found to be about 0.61 and 0.56, respectively. Finally, the resistance of *C. baccatum* var. PBC 81 has been indicated as quantitatively controlled with more than three genes. Lin et al. (2006) studied the inheritance of resistance twice in a cross between resistant cultivar 'IR' (BC F4) and susceptible cultivar 'Susan's Joy' at the green and red fruit stages; out of ten, seven $F_1$ plants were evaluated as resistant, and three displayed a susceptible reaction in green fruit stage and vice versa for red fruit stage. The ratio of resistant to susceptible plants in the $F_2$ generation approximated a ratio of 7R:9S in green and 9R:7S model in red fruit. While in the BCsF generation, the ratio (1R:3S) suggested the regulation of resistance by the two complementary dominant genes. However, backcross to the resistant parent (BCrF) failed to fit and predicted a ratio of 1R:0S, and almost all the BCsF progenies were susceptible by producing 4R:96S plant population, suggesting that the resistance might be controlled by duplicate recessive genes. However, the BCrF generation showed 84R:15S population that could not fit with the expected 1R:1S ratio. Thus, the gene pair commencing resistance at the immature green fruit stage appears to be distinct, while independent pair of genes was responsible for resistance controlling in mature red fruit. Many similar trials for finding resistance controlling genes were made. Lin et al. (2006) studied the inheritance of anthracnose resistance against *C. acutatum* in a cross between the resistant line '0038–9155' (P1), a $BC_3F_4$ selection derived from an interspecific cross between PBC 932 (*C. chinense*) (p1) and the cultivar 'IR' (*C. annuum* PBC 535) and the susceptible line '9955–15' (*C. annuum*) (P2). During the green fruit stage, seven out of ten $F_1$ plants were evaluated as resistant. The finding of segregation ratio in the $F_2$ generation was fitted with a 9R:7S model and $BC_{1P2}F_1$ generation showing 1R:3S model of segregation indicated an involvement of two complementary dominant genes, whereas at the red fruit stage, seven out of ten $F_1$ plants were proven susceptible. The ratio of resistant to susceptible in the $F_2$ generation approximated as 7R:9S ratio and almost all the $BC_{1P2}F_1$ progeny were susceptible (4R:96S plants), suggesting that the resistance might be controlled by a duplicate recessive gene. However, the $BC_{1P1}F_1$ generation (84R:15S) did not fit the expected 1R:1S ratio. Reports on the finding of expression of different models of ratio have been referred to as one of the species of *Nicotiana* which is incompatible with *Peronospora tobacina,* causing blue model disease. There was an involvement of a single dominant gene in this species and its accession was recognised as NIRPT by the linkage map (Zhang et al. 2007). Plants with altered resistance at different fruit maturity were found far in excess of expectations based on linked inheritance. Thus, the genes conferring resistance at the immature green fruit stage have been reported to be distinct and independent of those controlling resistance genes in mature red fruit. Recently, Garg (2011) studied pattern of inheritance performed by developing three crosses between KA-2 × Taiwan-2PT 12–3 × Bhut Jolokia, PT-12–3 × Punjab lal from resistant and susceptible parents (selected from screening of various germplasm). They were used for mapping the population to study the inheritance in natural as well as *in vitro*. The segregation of resistance to susceptibility ratio obtained fitted with 3:1 and indicated that resistance is regulated by a single dominant gene. Whereas, at green fruit stage the segregation ratio was just reverse expressing 1:3 model and indicated a single recessive gene conferring the resistance to anthracnose. The evidence suggested that different genes are responsible for resistance at different stages of fruit maturity.

On the basis of segregation ratio obtained, inheritance is not dependent on the host species used for crosses, neither the pathogen used for inoculation according to the performance of interaction between *C. annum* × *C. annuum* and *C. annuum* ×interderivatives of *C. fruitscence* × *C. chinense* inoculated by *C. capsici* (Garg 2011). The inheritance of resistance to anthracnose against *C. capsici* was studied in interspecific crosses of capsicum derived from a cross between a Thai elite cultivar *C. annuum* L. 'Bangchang' and a resistant line *C. chinense* Jacq. 'PBC932' by Pakdeevaraporn et al. (2005). The resistance was assessed by measuring lesion diameter per fruit area (LFA) on detached green chilli fruits. After crossing, segregation of resistance (nil LFA) and susceptibility in the $F_2$ were fitted with 1:3 Mendelian ratio, indicating that the resistance was governed by a single recessive gene. The segregation of the trait in the test crosses in both BC1s also confirmed the 1:3 gene segregating model in $F_2$. Lin et al. (2002) studied the inheritance of the resistance to anthracnose, caused by *C. capsici,* in *C. annuum* L. populations established from a cross between accession '83–168' and cv. 'KKU-Cluster' and their progenies in $F_1$s, $F_2$s and BC sub(1)s on green fruits. The segregation ratio of resistance to susceptibility appeared to be 3:1 in the $F_2$s and 1:1 in the BC sub (1) (F sub(1) × KKU-Cluster). Crosses indicated that one dominant gene is responsible for the resistance to *C. capsici* in the breeding line '83–168'. Park et al. (1990b) reported that the lines developed from *C. annuum* were resistance to *C. capsici* in $F_2$ and BC. These lines also segregated in a Mendelian fashion. Hence, the resistance was likely to be controlled by a single dominant gene. Polygenic resistance against *C. capsici* has also been reported by Ahmed et al. (1991) in a cross between Kolascai E-14 (susceptible) and perennial resistant genotype. The mean reaction of the $F_1$ generation was intermediate due to additive gene action governing resistance. Finding a continuous variation for resistance in $F_2$, $BC_1$ and $BC_2$ supported polygenic control of resistance. Park et al. (1990) studied the inheritance of resistance to *C. dematium* in cross PI244670 (susceptible) and chungryong (resistant) on the basis of lesion diameter and it was concluded that in both $F_1$ and $F_2$ mean deviated towards resistance was partially dominant, whereas Park et al. (1990) in their further study on *C. gloeosporioides* showed that resistance was partially overdominant.

Besides the above, a decade ago Fernandes and Ribeiro (1998) have already indicated that resistance to *C. gloeosporioides* has been controlled by a single dominant gene in one of the three crosses used, whereas in the case of the other two crosses a pair of dominant gene controlled the resistance. Kim et al. (2008) reported the inheritance of resistance to *C. acutatum* in segregating populations derived from the two crosses HN 11 AR and Daepoong-cho AR on detached mature green fruits. The segregation ratio of resistance and susceptibility to *C. acutatum* in the $F_2$ and BCR populations derived from the two crosses fit significantly to a ratio of 1:3 Mendelian model indicating a role of single recessive gene. Cheema et al. (1984) further added it to be inherited recessively with epistatic effect.

In the study of Garg (2011), inheritance of the three crosses (KA-2 × Taiwan-2, PT-12–3 × Bhut Jolokia, PT-12–3 × Punjab Lal) was developed from the resistant and susceptible parents and was used for developing mapping population to study inheritance in natural as well as *in vitro*. The segregation ratio of resistance to susceptibility obtained was 3:1 and hence resistance is likely to be controlled by a single dominant gene. Interestingly, at the green fruit stage the segregation ratio of resistance to susceptibility was found to be 1:3 and hence resistance is likely to be controlled by a single recessive gene. Thus, it can be concluded that different genes are responsible for resistance at different stages of fruit maturity. It was also concluded from the result of segregation ratio that the inheritance does not depend on the species of the host used for the development of cross, neither on the specific species of the pathogen used for inoculation, that is why in both the types of interactions, i.e. *C. annuum* × *C. annuum* (KA-2 × Taiwan-2, PT-12–3 × Punjab Lal) and *C. an-*

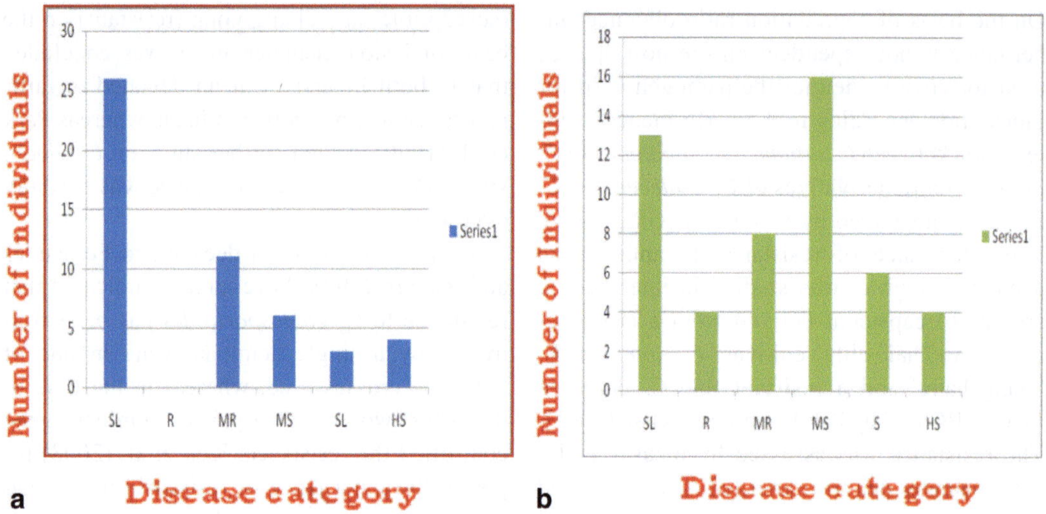


**Fig. 53.5** Frequency distribution of resistance in red (**a**) and green (**b**) fruit stage

*nuum* × interderivative of *C. fruitescence* × *chinense* (PT 12–3 × Bhut jolokia), were inoculated by *C. capsici.* (Figs. 53.5a and b).

### 53.1.4 Linkage Analysis and QTL Mapping

Genetic linkage map is a physical representation of the key principle of Mendelian genetics, which is used for a genome mapping representing relativity of the positions of specific DNA markers. It also draws the inheritance of genetic trait based on segregation among progeny, on which all computational programmes rely for genetic map construction. In order to identify a marker, it must be a polymorphic means of any variations in the parental DNA of a particular gene leading to polymorphism. Since closely linked markers experience fewer recombination events, linked markers are inherited together and marker's recombination frequency has been indicated as a function of this concept. Recombination frequencies between two markers decrease as the markers are closely located on the chromosome. Basically, two markers are inherited together depending on the distance between each other in the genome. In the genetic linkage maps, tightly linked markers are very beneficial for marker-assisted selection (MAS) breeding programmes. Researchers employ molecular markers to improve the efficiency for selecting the genes of interest. This application has been well established in most of the higher plants, for example, rice, tomato, wheat and barley (Motto and Marsan 2002). Molecular markers technologies are employed to improve the Plant breeding programmers ability to identify breeding lines, hybrids and cultivars, to assess genetic diversity in a germplasm and phylogenetic relationship (Gupta et al. 1999; Jain et al. 2002). Molecular markers have been categorised into two broad classes, hybridization-based DNA markers and PCR-based DNA markers. The primary purpose of all markers is to create a detailed genetic linkage maps that can be employed for gene research. PCR-based markers include a wide range of techniques that differ mainly in the sequences and position of the primers utilized. Besides the above, a few common types of techniques have been implicated as RAPDs), simple sequence repeats (SSRs) and AFLPs among which microsatellite or SSR is being most useful as they refer to tandem repeated sequences of 1–6 nucleotides that repeat up to a few dozen times per sites (Litt and Luty 1989). According to earlier reports, repetitive sequences are easily mutated during replication, due to polymerase slippage, that resulted into highly polymorphic regions (Guyomarc'h et al. 2002). Generally, a repeat is gained or lost which produces variable lengths at that locus. SSRs appear to be randomly dispersed throughout the genome and are generally flanked by conserved regions (Winter et al. 2002). The other characteristics of SSRs, leading to their extensive development as molecular markers, display hypervariability, co-dominant inheritance, multiallelism, reproducibility and good genome coverage, making them a preferred marker system (Powell et al. 1996). SSRs have been useful in a variety of applications such as linkage map construction, gene tagging and studies of genetic diversity and evolution (Powell et al. 1996; Varshney et al. 2005). Genetic linkage information has also proven to be a very powerful tool for accelerating pepper breeding through MAS. Several genetic linkage maps have been constructed in pepper using predominantly RFLP and AFLP markers, as well as RAPD, isozyme and morphological markers (Tanksley et al. 1988; Prince et al. 1993; Lefebvre et al. 1995; Lefebvre et al. 1997; Livingstone et al. 1999; Kang et al. 2001). In an illustrative research an integrated map based on six population and consisting of 2,262 markers covering 1,832 cM has recently been assembled (Paran et al. 2004). Molecular techniques and marker technology have been directed towards studying genes with agriculturally or economically important traits. Genetic linkage maps have been the cornerstone for this goal over the past decade. They provide information for analysis of QTLs, understanding genetic variation in germplasm collections, gene tagging, gene cloning and MAS (Motto and Marsan 2002). Besides the above, the known function of gene markers have also contributed to constitute a functional genetic map. On the above view, molecular techniques were used to locate the genes of different traits on chromosomes especially resistant and susceptibility development in relation to host–pathogen (anthracnose–chilli) relationship.

Kim et al. (2002) identified a Thaumatin-like gene in a non-climacteric pepper fruit used as a molecular marker in probing disease resistance. They reported a gene designated Pep TLP (for pepper thaumatin-like protein) which was isolated and characterised by using mRNA differential display. Pep TLP gene expression is developmentally regulated during ripening. The accumulation of Pep TLP mRNA and Pep TLP protein in the incompatible interaction (interaction between anthracnose fungus and ripe fruit) was found to be higher than that in a compatible one. Kim et al. (2008) developed mapping population consisting of 192 $F_2$ plants which were developed from crosses between *C. baccatum* pendulum (Cbp) and *C. baccatum* 'Golden Aji', susceptible to anthracnose; a linkage map with 14 linkage groups was constructed. Using this molecular linkage map, number, location and the effect of QTL for artificial inoculation and field inoculation were compared and several significant QTL detected were found to be stable under different inoculation conditions. Studies were also done by Voorips (2004) to identify the QTLs for disease resistance by inheritance of the resistance in $F_2$ population and QTL mapping of Anthracnose (*Colletotrichum* spp.) resistance in a cross between *C. annum* and *C. chinense*. Three resistance-related traits were scored the infection frequency, the true lesion diameter and overall lesion diameter. One main QTL was identified with a large effect on true lesion diameter after inoculation with *C. capsici*; no significant QTL was identified for overall lesion diameter or infection frequency. Lia et al. (2002) analysed QTL for anthracnose resistance in an $F_2$ population derived from an interspecific cross between *C. annum* and cv. Jatilaba and *C. chinensis*. For AFLP and microsatellites, a total of 238 markers were used to construct pepper genetic map with 982 cM total length. Five QTLs for *C. gloeosporioides* were identified by Join Map 3.0/Map QTL 4.0 software, using LOD score $>3.0$ and $\theta < 30$ cM as the threshold for QTL detection (both QTLs one for lesion diameter and other for disease incidence) which were found to be linked tightly with the same marker in the linked group N, and three $QTL_S$ associated with disease severity in linked group G, K and L, respectively. Based on the lesion diameter QTL for *C. capsici*, *C. gloeosporioides* resistance was the same. Lee (2010) worked on an introgression

BC1F2 population which was developed from interspecific crosses between *C. annuum* SP26 (susceptible) and *C. baccatum* PBC 81 for QTL mapping and analysis of anthracnose resistance. Both green and red fruits were inoculated with *C. accutatum* KSCa-1 and *C. capsici* ThSCc-1 and the disease reaction was evaluated by the disease incidence, true lesion diameter and overall lesion diameter. On the whole, distribution of anthracnose resistance was skewed towards anthracnose resistant parent and the presence of one or two major QTLs was indicated. The introgression map consisting of 13 linkage groups with a total of 218 markers (197 AFLP and 21 SSRs), covering a total length of 325 cM, was constructed. The study revealed four QTLs for resistance to KSCa-1 and three QTLs for ThSCc-1 isolate, respectively in composite interval analysis. Interestingly, the major QTLs (CaR12.2 and CcR9) for resistance to *C. acutatum* and *C. capsici* were positioned differently but there were close links between the minor QTL CcR1 2.2 for *C. capsici* and major QTL CaR1 2.2 as well as minor QTL CaR9 for *C. acutatum* along with major QTL CcR9. These results may be helpful for marker-assisted selection and pyramiding of two different anthracnose resistant genes in commercial pepper breeding.

During the past decades, MAS by using molecular markers has been a helpful tool in plant breeding, as it determines the position of gene in chromosomes. Molecular markers simplify the screening; if molecular marker takes its position close to the genes of interest, it indicates the presence of those resistant genes which are unable to accomplish the disease screening. According to Voorips (2004), the resistance associated with anthracnose of chilli follows quantitative resistance and a continuous range of variation within a range in resistance, that is, from extremely susceptible to fairly resistant. General resistance consisting of a number of complementary resistant components (host–parasite interaction) is controlled by multiple genes, each having small quantification effects (Thursten 1971). Recently, Garg (2011) have used a cross PT-12–3 × Punjab lal for QTL mapping using polymorphic markers. During linkage map and QTL analysis, out of 200 markers 11 polymorphic markers were selected which were mapped on linkage groups LG1 and LG2. Six and three SSR markers were mapped on linkage 1 and linkage 2, respectively. CIM analysis revealed eight significant $QTL_S$ for resistance against *C. capsici* on LG1 and detected two different genomic region of above linkage group. Trait infection percentage belonging to two $QTL_S$, QCcR-ifp-iivr-1.1 and QCcR-ifp-iivr 1.2, contributed phenotypically 68 and 7.2% for anthracnose resistance in red fruit stage, while lesion area QCcG-la.iivr1.1 and QCcG-la.iivr1.2, contributed 71 and 18% phenotypic variations of resistance. Simultaneously, infection percentage in the green fruit stage was also detected in the same genomic region with flanking markers CAMS020 and HPMSE016 of LG1. A disease was categorised on the basis of identification of two QTLS, individually. According to QTL mapping, QCcR-def-iivr-1.2 detected on LG1 contributed 15–30% of resistance against anthracnose disease in red chilli. Overall observation suggested that one or more QTL (QCGR-ifp-iivr 1.1:QCcR-la-iivr 1.1,QCcG-ifp-iivr) were co–localised at the same genomic region (CAMS020-HPMSE016) and indicated a common QTL and supported the earlier findings (Lee et al. 2010). Further, they have also added by taking a cross PT-12–3 × Punjab lal which was used for QTL mapping. Out of 11 polymarkers only 9 markers were mapped on linkage groups LG1 and LG2. Nine QTLS were significant for resistance in *C. capsici,* located at LG1 and LG2 irrespective of the traits infection percentage in red fruit and green fruit stages lesion area and for disease categorisation during both *in vitro* and field evaluation (Figs. 53.6, 53.7, 53.8). In the preceding discussions, we have described the aspects of conventional methods of identification and characterisation of *Colletotrichum* species based on morphological traits such as the size and shape of conidia, existence of setae and cultural expression like colony colour, growth rate and texture. As was felt, these criteria are not clear enough to differentiate among species and genetics of host–pathogen relationship. We have tried to cover the contributions made on molecular basis of understanding since a decade. In this molecular era,

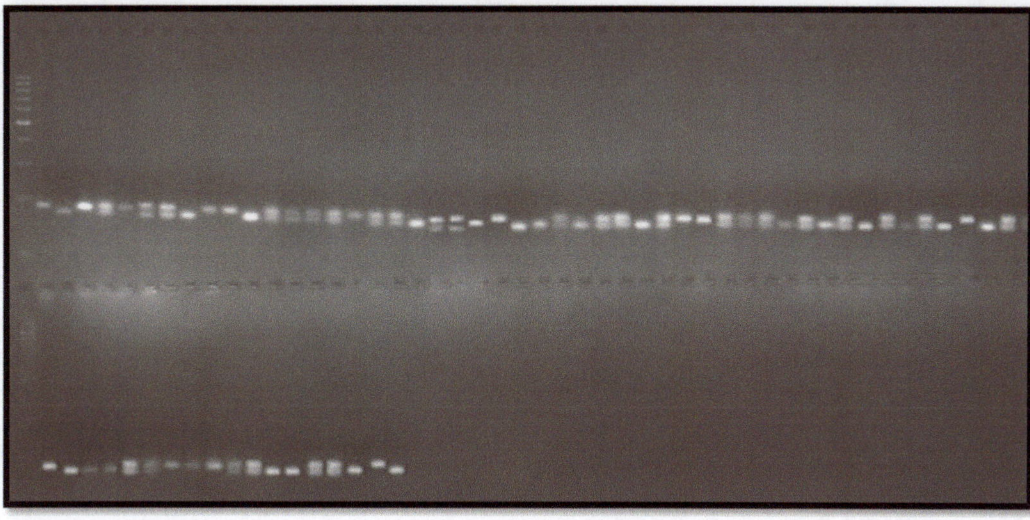

**Fig. 53.6** Segregation pattern of polymorphic markers in $F_2$ population. Where P1 parent 1, P2 parent 2 and H heterozygous marker, HPMS 016

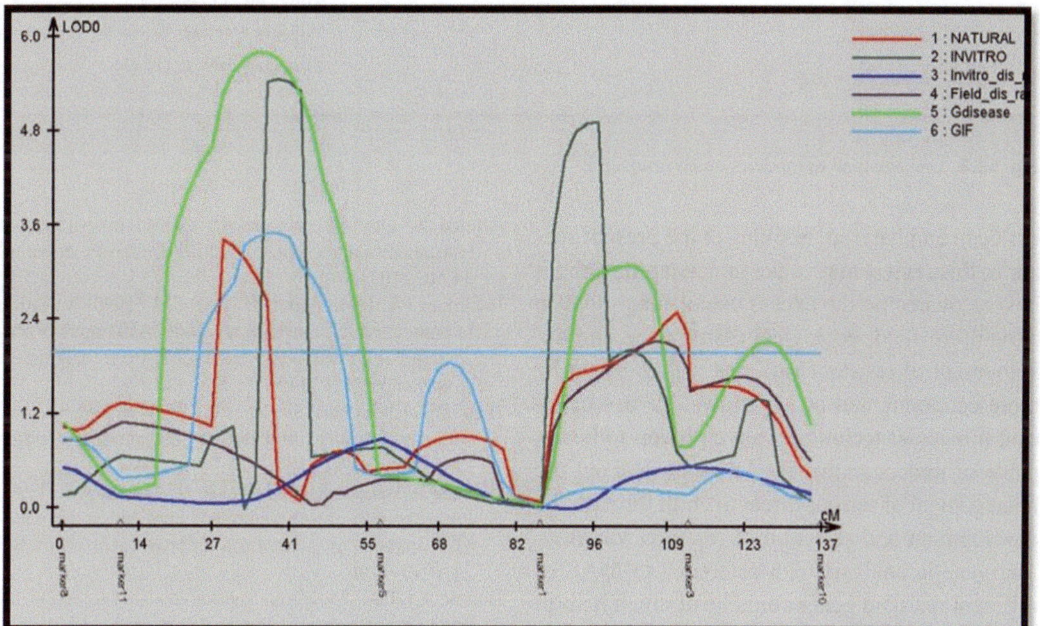


**Fig. 53.7** QTL of traits on the linkage group

MAS using molecular markers for the determination of the position of gene in chromosome has been undertaken. Molecular markers authenticated the screening on the basis of position of markers close to gene(s) of interest especially resistant gene without having to accomplish the disease screening. A picture of resistance-associated anthracnose of chilli followed by quantitative resistance and continuous range of variation derived from extremely susceptible to fairly resistant has been drawn. The resistance covered by a number of complementary resistant component (host–parasite interaction) controlled by one or multiple genes with small quantification effects

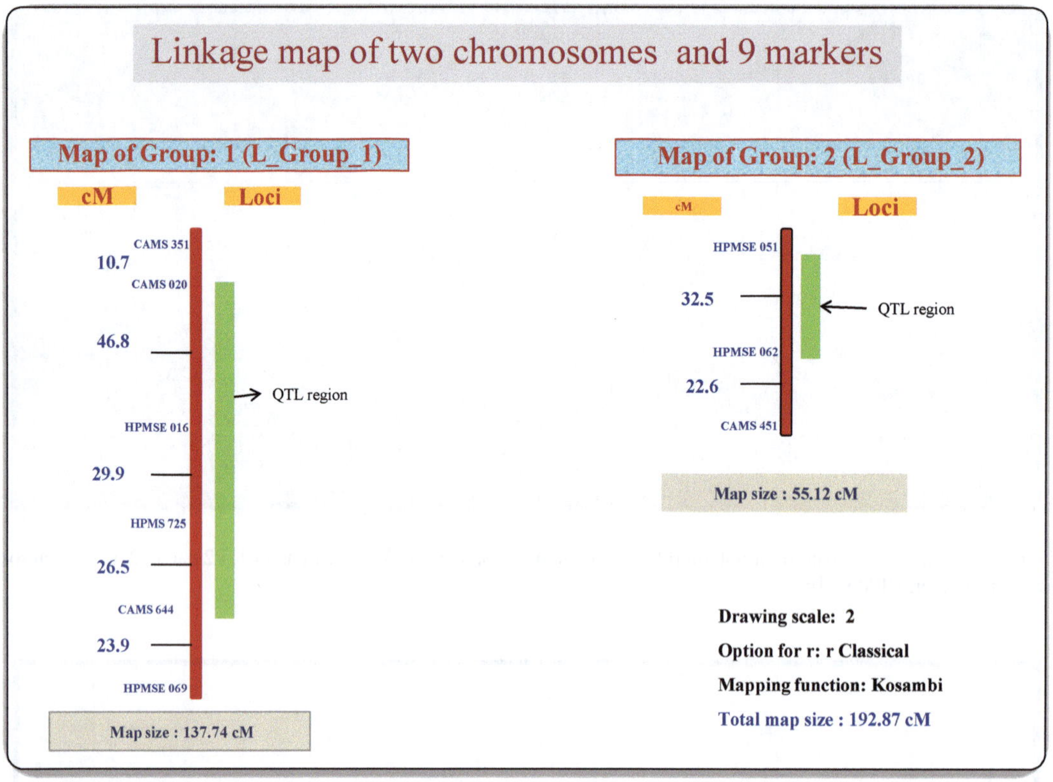


**Fig. 53.8** Linkage map of markers on chromosome

has been emphasized. Looking at the present scenario, this review may make understanding about how to overcome the excess use of fungicides in agriculture field, ecological disturbance and development of resistant varieties, which might be more economic and eco-friendly. The use of recent molecular techniques have proven to be capable of understanding the know-how about the management of serious threat to chilli through the development and selection of resistant varieties. The genetic analysis with reference to MAS of different resistant genes could be of much helpful in a breeding programme.

## References


Adikaram NKB, Brown A, Swinburne TR (1983) Observations on infection of *Capsicum annuum* fruit by *Glomerella cingulata* and *Colletotrichum capsici*. Trans Brit Mycol Soc 80:395–401

Agrios GN (2005) Plant pathology 5th edn. Academic Press, San Diego, p 922

Ahmed N, Dey SK, Hundal JS (1991) Inheritance of resistance to anthracnose in chilli. Indian Phytopathol 44:402–403

Backman PA, Landschoot PJ, Huff DR (1999) Variation in pathogenicity, morphology and RAPD marker profiles in *Colletotrichum graminicola* from turfgrasses. Crop Sci 39:1129–1135

Bagri RK, Choudhary SL, Rai PK (2004) Management of fruit rot of chilli with different plant products. Indian Phytopathol 57(1):107–109

Baird WV, Ballard RE, Rajapakse S, Abbott, AG (1996) Progress in Prunus mapping and application of molecular markers to germplasm improvement. Hort Sci 31:1099–1106

Barone A (2004) Molecular marker-assisted selection for potato breeding. Am J Potato Res 81:111–117

Basak AB (1997) Reaction of some chili germplasm to major fruit rotting fungal Pathogen. Chittagong—Univ-Stud Sci 21(1):123–125

Brunt AA, Kenten RH, Phillips S (1978) Symptomatologically distinct strains of pepper veinal mottle virus from four West Africa solanaceous crops. Ann Appl Biol 88:115–119

Butler EJ, Bisby GR (1960) The fungi of India. Indian Council of Agricultural Research, New Delhi

Byung SK (2007) Country report of anthracnose research in Korea first international symposium on chili anthracnose. Hoam Faculty House, Seoul National University, Seoul, 17–19 Sept 2007

Cai L, Hyde KD, Taylor PWJ, Weir BS, Waller J, Abang MM, Zhang JZ, Yang YL, Phoulivong S, Liu ZY, Prihastuti H, Shivas RG, McKenzie EHC, Johnston PR (2009) A polyphasic approach for studying *Colletotrichum*. Fungal Divers 39:183–204

Cheema DS, Singh DP, Rawal RD, Deshpande AA (1984) Inheritance of resistance to anthracnose disease in chillies. *Capsicum* Eggplant Newsl 3:44

Cipriani G, Di Bella R, Testolin R (1996) Screening RAPD primers for molecular taxonomy and cultivars fingerprinting in genus *Actinidia*. Euphytica 90:169–174

Dastur JF (1920) *Glomerella cingulata* (Stoneman) Spald and its conidial form, *Gloesporium piperatum and and Colletotrichum nigrum* and Hals on chillies and *Carica papaya*. Ann Appl Biol 6(4):245–268

Datar VV (1996) Pathogenicity and effect of temperature on six fungi causing fruit rot of chili. Indian J Mycol Plant Pathol 25(3):195–197

Demissie A, Bjornstrand A (1996) Phenotypic diversity of Ethiopian barely in relation to geographical regions, altitudinal range and agroecological zones: as an aid to germplasm collection and conservation strategy. Hereditas 124:17–29

Deshpande A, Ram DR (2007) Resistant sources of chili (*Capsicum annuum* L.) Anthracnose Fruit Rot Disease (*Colletotrichum capsici* (Syd.) against different isolates collected from commercial chili growing areas of India. First International Symposium on Chili Anthracnose. Hoam Faculty House, Seoul National University, Seoul, 17–19 Sept 2007

Don LD, Van TT, Phuong VY TT, Kieu PTM (2007) *Colletotrichum* spp attacking on chilli pepper growing in Vietnam. Country Report. In: Oh DG, Kim KT (eds) Abstracts of the first international symposium on chilli anthracnose. Held at Seoul National University, Korea, p 42 (17–19 Sept 2007)

Dulloo ME, Guarino L, Ford-Lioyed BV (1997) A bibliography and a review of genetic diversity studies of African germplasm using protein and DNA Markers. Genetic Reso Crop Evol 44:447–470

Eagles H, Bariana H, Ogbonnaya F, Rebetzke G, Hollamby G, Henry R, Henschke P, Carter M (2001) Implementation of markers in Australian wheat breeding. Aust J Agric Res 52:1349–1356

Ekbote SD, Jagadeesha RC, Patil MS (2002) Reaction of chili germplasm to fruit rot disease. Karnataka J Agric Sci 15(4):717–718

Fernandes R, Ribeiro de LD (1998) Mode of inheritance of resistance in *Capsicum annuum* accessions to *Colletotrichum gloeosporioides*. In: Proc. 10th Eucarpia Meeting on Genetics and Breeding of Capsicum and Eggplant, p 711

Fregene M, Okogbenin E, Mba C, Angel F, Suarez MC, Janneth G, Chavarriaga P, Roca W, Bonierbale M, Tohme J (2001) Genome mapping in cassava improvement: challenges, achievements and opportunities. Euphytica 120:159–165

Garg R (2011) Genetics of Host-Pathogen interaction: resistance to anthracnose in chilli (Capsicum annuum L.). Ph. D. Thesis, Banaras Hindu University, Varanasi, India, p 210

Garg R, Kumar S, Kumar R, Loganathan M, Saha S, Kumar S, Rai AB, Roy BK (2013) Novel source of resistance and differential reactions on chilli fruit infected by *Colletotrichum capsici*. Aust Pt Pathol 42:227–233

Gebhardt C, Valkonen, JPT (2001) Organization of genes controlling disease resistance in the potato genome. Ann Rev Phytopathol 39:79–102

Gehlot P, Purohit DK (2001) Ultra structure of conidium ontogency in *C. capsici.* Indian Phytopathology 54(2):215–218

Gniffke P A, Lin SW, Wang TC (2007) Evaluation of diverse chili pepper sources for resistance to anthracnose. First International Symposium on Chili Anthracnose held at Seoul National Univ. Seoul, 17–19 Sept 2007, p 42

Gupta PK, Varshney RK, Sharma PC, Ramesh B (1999) Molecular markers and their applications in wheat breeding. Plant Breed 118:369–390

Guyomarc'h H, Sourdille P, Charmet G, Edwards KJ, Bernard M (2002) Characterization of polymorphic microsatellite markers from *Aegilops tauschii* and transferability to the D-genome of bread wheat. Theor Appl Genet 104:1164–1172

Gwanme C, Labuschangne MJ, Botha AM (2000) Analysis of genetic variation in *Cucurbita moschata* by random amplified polymorphic DNA (RAPD) Markers. Euphytica 113:19–24

Hartman GL, Wang TC (1992) Characteristics of two *Colletotrichum* species and evaluation of resistance to anthracnose in pepper. Proc 3rd Intl Conf Plant Protection in the Tropics, vol 6. Malaysian Plant Protection Society, Kuala Lumpur, pp 202–205

Hedge GM, Kulkarni S (2002) Vulnerable infection stage of chili fruit by *C. capsici* (Sydow.) Butler and Bisby. Karnatak J Agr Sci 14(1):162–163

Isaac S (1992) Fungal Plant Interaction. Chapman and Hall Press, London, p 115

Jahufer M, Cooper M, Ayres J, Bray R (2002) Identification of research to improve the efficiency of breeding strategies for white clover in Australia: a review. Aust J Agric Res 53:239–257

Jain SM, Brar DS, Ahloowalia BS (2002) Molecular techniques in crop improvement. Kluwer Academic Publishers, Boston, p 616

Jayalakshmi C, Seetharaman R (1998) Evaluation chilli genotypes against fruit rot disease incited by *C.Capsici*. South-Indian-Hortic 46(1–2):104–105

Jayalakshmi C, Seetharaman K (1999) Qualitative losses of chili fruits due to infection by *Collectorichum capsici (syd)* Butler and Bisby. *Capsicum* Eggplant Newsl 18:80–82

Johnston PR, Jones D (1997) Relationships among *Colletotrichum* isolates from fruit-rots assessed using rDNA sequences. Mycologia 89(3):420–430

Kang BC, Nahm SH, Huh JH, Yoo HS, Yu JW, Lee MH, Kim BD (2001) An interspecific *Capsicum annuum* x *C. chinense* $F_2$ linkage map in pepper using RFLP and AFLP markers. Theor Appl Genet 102:531–539

Kasha KJ (1999) Biotechnology and world food supply. Genome 42:642–645

Kaur S, Singh J (1990) *C.acutatum*, a threat to chilli crop in Punjab. Indian Phytopathol 43:108–110

Kelly JD, P Gepts, PN Miklas, DP Coyne (2003) Tagging and mapping of genes and QTL and molecular marker-assisted selection for traits of economic importance in bean and cowpea. Field Crops Res 82:135–154

Khirbhat SK, Vajnana T, Mehra R (2004) Cultural and pathogenic variation among the nine isolates of *Colletotrichum capsici* causing fruit rot of *Capsicum*. *Capsicum* Eggplant Newsl 24:131–134

Kim SH, Yoon JB, Park HG (2008) Inheritance of anthracnose resistance in a new genetic resource, *Capsicum baccatum* PI594137. J Crop Sci Biotech 11:13–16

Kim S H, Yoon JB, Do JW, Park HG (2008a) Resistance to anthracnose caused by *Colletotrichum acutatum* in chili pepper (*Capsicum annuum* L.). J Crop Sci Biotech 10(4):277–280

Kim SH, Yoon JB, Do JW, Park HG (2008b) A major recessive gene associated with anthracnose resistance to *Colletotrichum capsici* in chilli pepper (*Capsicum annuum* L.). Breed Sci 58:137–141

Koebner RMD, Summers RW (2003) 21st century wheat breeding: plot selection or plate detection? Trends Biotechnol 21:59–63

Kumar S, Kumar R, Kumar S, Singh M, Rai AB, Rai M (2011) Incidences of leaf curl disease on *Capsicum* germplasm under field conditions. Indian J Agric Sci 81:187–189

Lakshmesha K, Lakshmidevi K, Aradhya N, Mallikarjuna S (2005) Changes in pectinase and cellulase activity of *Colletotrichum capsici* mutants and their effect on Anthracnose disease on capsicum fruit. Arch Phytopathol Plant Prot 38:267–279

Lee J, Jee-Hwa H, Jae WD, Jae BY (2010) Identification of QTLs for resistance to anthracnose to two *Colletotrichum* species in pepper. J Crop Sci Biotech 13(4):227–233

Lefebvre V, Palloix A, Caranta C, Pochard E (1995) Construction of an intra-specific integrated linkage map of pepper using molecular markers and doubled haploid progenies. Genome 38:112–121

Lefebvre V, Caranta C, Pflieger S, Moury B, Daubèze AM, Blattes A, Ferriere C, Phaly T, Nemouchi G, Ruffinatto A, Palloix A (1997) Updated intra-specific maps of pepper. Capsicum Eggplant Newsl 16:35–41

Leonian LH (1922) Stem and fruit blight of chillies caused by *Phytophthora capsici* sp. nov. Phytopathol 12:401–408

Lia S, Wattimena GA, Guhrja E, Yusuf M, Aswidinoor dan Piet A (2002) Mapping QTLs for anthracnose resistance in anthracnose spp. J Bioteknol Pertan 7(2):43–54

Lin Q, Kanchana UC, Jaunet T, Mongkolporn O (2002) Genetic analysis of resistance to pepper anthracnose caused by *Colletotrichum capsici*. Thai J Agric Sci 35:259–264

Lin SW, Gniffke PA, Wang TC (2006) Inheritance of resistance to anthracnose in chili pepper. 27th International Horticultural Congress & Exhibition, Seoul, 13–19 Aug 2006, p 14

Litt M, Luty JM (1989) A hypervariable microsatellite revealed by *in vitro* amplification of a dinucleotide repeat within the cardiac muscle actin gene. Am J Hum Genet 44:397–401

Livingstone KD, Lackney VK, Blauth JR, Van Wijk R, Jahn MK (1999) Genome mapping in *Capsicum* and the evolution of genome structure in the *Solanaceae*. Genetics 152:1183–1202

Lopes, Vila (2003) First International Symposium on Chilli Anthracnose held at Seoul National University, Seoul, 17–19 Sept 2007

Mackill DJ, Nguyen HT, Zhan J (1999) Use of molecular markers in plant improvement programs for rainfed lowland rice. Field Crops Res 64:177–185

Madhavan S, Vaikuntavasan P, Rethinasamy V (2010) RAPD and virulence analyses of *Colletotrichum capsici* isolates from chilli (*Capsicum annuum*). J Plant Dis Prot 117:253–257

Mahasuk P, Taylor PWJ, Mongkolporn O (2009a) Identification of two new genes conferring resistance to *Colletotrichum acutatum* in *Capsicum baccatum*. Phytopathol 99(9):1100–1104

Mahasuk P, Khumpeng S, Wasee PW J Taylor, Mongkolporn O (2009b) Inheritance of resistance to anthracnose (*Colletotrichum capsici*) at seedling and fruiting stages in chilli pepper (*Capsicum* spp.) Plant Breed 128(6):701–706

Manandhar JB, Hartman GL, Wang TC (1995) Anthracnose development on pepper fruits inoculated with *Colletotrichum gloeosporioides*. Plant Dis 79:380–383

Margale E, Herve Y, Hu J, Quiros CF (1995) Determination of genetic variability by RAPD markers in cauliflower, cabbage, and Kale local cultivars from France. Genetic Res Crop Evol 42:281–289

Mehlenbacher SA (1995) Classical and molecular approaches to breeding fruit and nut crops for disease resistance. Hort Sci 30:466–477

Montri P, Taylor PWJ, Mongkolporn O (2009) Pathotypes of *Colletotrichum capsici*, the causal agent of chilli anthracnose, in Thailand plant disease. Plant Dis 93(1):17–20

Moriwaki J, Tsukiboshi T, Sato T (2002) Grouping of *Colletotrichum* species in Japan based on rDNA sequences. J Gen Plant Pathol 68(4):307–320

Motto M, Marsan PA (2002) Construction and use of genetic maps in cereals. In: Jain MS (ed) Molecular techniques in crop improvement. Kluwer Academic Publishers, Netherlands, pp 347–370

Muehlbauer F, Kaiser W, Simon C (1994) Potential for wild species in cool season food legume breeding. Euphytica 73:109–114

Muhyi R, Bosland PW (1992) Evaluation of *Capsicum* germplasm for sources of resistance to *Rhizoctonia solani*. Hort Sci 30:341–342

Nanda C, Mohan Rao A, Ramesh S, Pratibha VH, Shivakumara AP (2007) Identification of parents suitable for mapping and tagging genes conferring resistance to anthracnose in hot pepper. First International Symposium on Chili Anthracnose, Hoam Faculty House, Seoul National University, Seoul, 17–19 Sept 2007, p 26

Oanh L T K, Korpraditskul V, Rattanakreetakul C (2004) A pathogenicity of anthracnose fungus, *Colletotrichum capsici* on various Thai chilli varieties. Kasetsart J (Nat Sci) 38(6):103–108

Ong CA, Varghese G, Poh TW (1979) Aetiological investigations on a veinal mottle virus of chilli *(Capsicum annuum* L.) newly recorded from Peninsular Malaysia. Malay Agric Res Dev Inst (MARDI) Res Bull 7:78–88

Ortiz R (1998) Critical role of plant biotechnology for the genetic improvement of food crops: perspectives for the next millennium. Electron J Biotechnol 1(3, Issue of August 15), pp 1–8

Pakdeevaraporn P, Wasee S, Taylor PWJ, Mongkolporn O (2005) Inheritance of resistance to anthracnose caused by *Colletotrichum capsici* in *Capsicum*. Plant Breed 124:206–208

Paran I, Van der Voort JR, Lefebvre V, Jahn M, Landr, L, van Schriek, M, Tanyolac B, Caranta C, Ben-Chaim A, Living stone K, Palloix A, Peleman J (2004) An integrated genetic linkage map of pepper (*Capsicum* spp.). Mol Breed 13:251–261

Park KS, Kim CH (1992) Identification, distribution, and etiological characteristics of anthracnose fungi of red pepper in Korea. Korean J Plant Pathol 8:61–69

Park HK, Kim BS, Lee WS (1990a) Inheritance of resistance to anthracnose (*Colletotrichum* spp.) in pepper (*Capsicum annuum* L.) I. Genetic analysis of anthracnose resistance by diallel crosses. J Kor Soc Hort Sci 31:91–105

Park HK, Kim BS, Lee WS (1990b) Inheritance of resistance to anthracnose (*Colletotrichum* spp.) in pepper (*Capsicum annuum* L.) II. Genetic analysis of resistance to *Colletotrichum dematium*. J Kor Soc Hort Sci 31:207–212

Paul YS, Behl MK (1990) Some studies on bell pepper anthracnose caused by *Colletotrichum capsici* and its control. Seed Res 1:656–659

Pcrane RR, Jai MB (1986) Reaction of chilli Cultivars to fruit rot and die-back of chili incited by *C.capsici*. Curr-Res-Report 2:1, 52–53

Pearson MN, Bull PB, Speke H (1984) Anthracnose of *Capsicum* in Papua, New Guinea; varietal reaction and associated fungi. Trop Pest Manag 30:230–233

Powell W, Machery GC, Provan J (1996a) Polymorphism revealed by simple sequence repeats. Trends Genet 1:76–83

Powell W, Morgante M, Andre C, Hanafey M, Vogel J, Tingey S, Rafaski A (1996b) The comparison of RFLP, RAPD, AFLP and SSR (microsatellite) markers for germplasm analysis. Mol Breed 2:225–238

Prince JP, Pochard E, Tanksley SD (1993) Construction of molecular linkage map of pepper and a comparison of synteny with tomato. Genome 36:404–417

Pring RJ, Nash C, Zakaria M, Bailey JA (2002) Infection process and host range of *Colletotrichum capsici*. Physiol Mol Plant Pathol 46(2):137–152

Rai VP (2010) Genetic and molecular analysis of pepper leaf curl resistance in chilli (Capsicum annuum L.). Ph. D. Thesis, Banaras Hindu University, Varanasi

Ramachandran N, Rathnamma K (2006) *Colletotrichum acutatum*—a new addition to the species of chilli anthracnose pathogen in India. Paper presented at the Annual meeting & symposium of Indian Phytopathological society, Central Planation Crops Research Institute, Kasaragod, 27–28 Nov 2006

Ramachandran N, Madhavi Reddy K, Rathnamma K (2007) Current status of chilli anthracnose in India. Paper presented at the First International Symposium on Chilli Anthracnose held at Seoul National University, Seoul, 17–19 Sept 2007

Ratanacherdchai K, Wang HK, Lin FC, Soytong K (2007) RAPD analysis of *Colletotrichum* species causing chilli anthracnose disease in Thailand. J Agric Technol 3(2):211–219

Ratanacherdchai K, Wang HK, Lin FC, Kasem S (2010) ISSR for comparison of cross-inoculation potential of *Colletotrichum capsici* causing chilli anthracnose. Afri J Microbiol Res 4(1):076–083

Rivelli VC (1989) A wilt of pepper incited by *Fusarium oxysporum* f. sp. *capsici* f. sp. nov. M.S. Thesis, Louisiana State University, Baton Rouge

Roberts PD, Pernezny K, Kucharek TA (2001). Anthracnose caused by *Colletotrichum* sp. on pepper [Online]. J Univ Florida/Inst Food Agric Sci. http://edis.ifas. ufl. edu/PP104. Accessed 25 Dec 2007

Roy A, Bordoloi DK, Paul SR (1998) Reaction of chili (*C.annum* L.) genotypes to fruit rot under field condition. PKV-Res-J 22:1, 155

Sanogo S (2003) Chile pepper and the threat of wilt diseases. Online. *Plant Health Progress*. doi:10.1094/PHP-2003-0430-01-RV

Sawant IS, Narker SP, Shetty DS, Upadhyay A, Sawant SD (2012) First report of *Colletotrichum capsici* causing anthracnose on grapes in Maharastra. New Disease report 25:2. http://dx.doi.org/10.5197/j2044-0588.2012.025.002

Selvakumar R (2007) Variability among *Colletotrichum capsici* causing Chilli Anthracnose in North Eastern India In first international symposium on chilli Anthracnose held at Seoul National University, Seoul, 17–19 Sept 2007

Sharma PN, Kaur M, Sharma OP, Sharma P, Pathania A (2005) Morphological, pathological and molecular variability in *Colletotrichum capsici*, the cause of fruit rot of chillies in the subtropical region of north-western India. J Phytopathology 153(4):232–237

Sharman PC, Winter P, Bunger T, Huttel B, Kahl G (1997) Expanding the repertoire of molecular markers for resistance breeding in chickpea. In: Udupa SM, Weigand F (eds) DNA Markers and breeding for resistance to ascochyta blight in chickpea. Proceedings of the symposium on " application of dna fingerprinting for crop improvement of: marker assisted selection of

chickpea for sustainable agriculture in the dry areas." ICARDA, Aleppo, pp 175–198 (11–12 April 1994)
Shin HJ, Xu T, Zhang CL, Chen Z J (2000) The comparative study of capsicum anthracnose pathogens from Korea with that of China. J Zhejiang Univ (Agric Life Sci) 26:629–634
Singh AP, Kaur S, Singh J (1993) Determination of infection in fruit rot (*C.capsici*) of chilli (*C.annum*). Indian J Agric Sci 63(5):310–312
Singh A, Thakur DP (1979) Reaction of chili (*C. fruitescens*) varieties to *C. capsici* (Syd.) Butler and Bisby. Curr Sci 48(11):512–513
Sinha AK (2004) Factors influencing growth sporulation and spore germination of *C Capsici*. Adv Plant Sci 17(1):71–73
Simmonds JH (1965) A study of the species of *Colletotrichum* causing ripe fruit rots in Queensland. Qld J Agric Anim Sci 22:437–459
Snowdon R, Friedt W (2004) Molecular markers in *Brassica* oilseeds breeding: current status and future possibilities. Plant Breed 123:1–8
Staub JE, Box J, Meglic V, Horejsi TF, Mc Creight JD (1997) Comparison of isozymes and random amplified polymorphic DNA data for determining interspesfic variation in cucumis. Gene Reso Crop Evol 44:557–564
Stuber CW, Polacco M, Senior ML (1999) Synergy of empirical breeding, marker-assisted selection, and genomics to increase crop yield potential. Crop Sci 39: 1571–1583
Svetleva D, Velcheva M, Bhowmik G (2003) Biotechnology as a useful tool in common bean (*Phaseolus vulgaris* L.) improvement: a review. Euphytica 131:189–200
Tanksley SD, Bernatzky R, Lapitan N, Prince JP (1988) Conservation of gene repertoire but not gene order in pepper and tomato. Proc Natl Acad Sci USA 85:6419–6423
Than PP, Shivas RG, Jeewon R, Pongsupasamit S, Marney TS, Taylor PWJ, Hyde KD (2008a) Epitypification and phylogeny of *Colletotrichum acutatum* JH Simmonds. Fungal Divers 28:97–108
Thind TS, Jhooty JS (1990) Studies on variability in two *Colletotrichum* spp. causing anthracnose nose and fruit rot of chilli in Punjab. Indian Phytopathol 43:53–58
Thomas W (2003) Prospects for molecular breeding of barley. Ann Appl Biol 142:1–12
Thurston HD (1971) Relationship of general resistance: late blight of potato. Phytopathol 61:620–626
Torres-Calzada C, Tapia-Tussell R, Quijano-Ramayo A, Martin-Mex R, Rojas-Herrera R, Higuera-Ciapara I, Perez-Brito D (2011) A species-specific polymerase chain reaction assay for rapid and sensitive detection of *Colletotrichum capsici*. Mol Biotechnol 49(1):48–55
Tuberosa R, Salvi S, Sanguineti MC, Maccaferri M S, Giuliani Landi P (2003) Searching for quantitative trait loci controlling root traits in maize: a critical appraisal. Plant Soil 255:35–54
Tussell RT, Ramayo AQ, Velazquez AC, Lappe P, Saavedra AL, Brito DP (2008) PCR-Based detection and characterization of the fungal pathogens *Colletotrichum gloeosporioides* and *Colletotrichum capsici* causing anthracnose in papaya (*Carica papaya* L.) in the Yucatan Peninsula. Mol Biotechnol 40:293–298
Varshney RK, Graner A, Sorrells ME (2005) Genic microsatellite markers in plants: features and applications. Trends Biotechnol 23:48–55
Voorrips RE (2004) QTLs mapping of Anthracnose (*Collectorichum* sp) resistance in a cross between *Capsicum annum* and *C. Chinense*. Theor Appl Genet 109(6):1275–1282
Wasantha KL, Rawal RD (2008) Influence of carbon, nitrogen, temperature and pH on the growth and sporulation of some Indian isolates of *Colletotrichum gloeosporioides* causing anthracnose disease of papaya (*Carrica papaya* l). Trop Agric Res Ext 11:7–12
Weeds P L, Chakraborty S, Fernandes CD, Charchar MJ dÎ„A, Ramesh CR, Kexian Y, Kelemu S (2003) Genetic diversity in *Colletotrichum gloeosporioides* from *Stylosanthes* spp. at centers of origin and utilization. Phytopathology 93:176–185
Whitelaw-Weckert MA, Curtin SJ, Huang R, Steel CC, Blanchard CL, Roffey PE (2007) Phylogenetic relationships and pathogenicity of *Colletotrichum acutatum* isolates from grape in subtropical Australia. Plant Pathol 56:448–463
Widodo WD (2007) Status of Chili Anthracnose in Indonesia, First International symposium on chilli Anthracnose. Seoul, 17–19 Sept 2007
Williams KJ (2003) The molecular genetics of disease resistance in barley. Aust J Agric Res 54:1065–1079
Winter P, Huttel B, Weising K, Kahl G (2002) Microsatellites and molecular breeding: exploitation of microsatellite variability for the analysis of a monotonous genome. In: Jain MS (ed) Molecular techniques in crop. Improvement. Kluwer Academic Publishers, Norwell, pp 85–138
Xiao CL, MacKenzie, S J, Legard DE (2004) Genetic and pathogenic analyses of *Colletotrichum gloeosporioides* from strawberry and non cultivated hosts. Phytopathology 94:446–453
Yabuuchi EY, Kosako I, Yano H Hotta, Y Nishiuchi (1995) Transfer of two Burkholderia and an *Alcaligenes* species to Ralstonia gen. nov: proposal of Ralstonia pickettii (Ralston, Palleroni and Douderoff 1973) comb. nov., *Ralstonia solanacearum* (Smith 1896) comb. nov. and *Ralstonia eutropha* (Davis 1969) comb. nov. Microbiol Immunol 39:897–904
Yoon JB, Park HG (2001) Screening method for resistance to pepper fruits anthracnose: pathogen sporulation, inoculation methods related to inoculum concentrations, post-inoculation environments. J Korean Soc Hortic Sci 42:389–393
Yoon JB, Yang DC, Lee WP, Ahn SY, Park HG (2004) Genetic resources resistant to anthracnose in the genus *Capsicum*. J Korean Soc Hortic Sci 45:318–323
Zhang D, Chunhui Zhu, Yong Liu (2007) Chilli Anthracnose Research in China: an overview. First International Symposium on Chilli Anthracnose. Seoul, 17–19 Sept 2007

MIX
Papier aus verantwortungsvollen Quellen
Paper from responsible sources
FSC® C105338

If you have any concerns about our products,
you can contact us on
**ProductSafety@springernature.com**

In case Publisher is established outside the EU,
the EU authorized representative is:
**Springer Nature Customer Service Center GmbH**
**Europaplatz 3, 69115 Heidelberg, Germany**

Printed by Libri Plureos GmbH
in Hamburg, Germany